#9627 £18.50

THE BOOKS OF ASSUMPTION
OF THE
THIRDS OF BENEFICES

Scottish Ecclesiastical Rentals
at the Reformation

RECORDS OF SOCIAL AND ECONOMIC HISTORY
NEW SERIES 21

THE BOOKS OF ASSUMPTION OF THE THIRDS OF BENEFICES

Scottish Ecclesiastical Rentals at the Reformation

EDITED BY

JAMES KIRK

Published for THE BRITISH ACADEMY
by OXFORD UNIVERSITY PRESS

Oxford University Press, Walton Street, Oxford OX2 6DP

Oxford New York
Athens Auckland Bangkok Bombay
Calcutta Cape Town Dar es Salaam Delhi
Florence Hong Kong Istanbul Karachi
Kuala Lumpur Madras Madrid Melbourne
Mexico City Nairobi Paris Singapore
Taipei Tokyo Toronto

and associated companies in
Berlin Ibadan

© *The British Academy, 1995*

All rights reserved. No part of this publication may be reproduced, stored in a retrieval system, or transmitted, in any form or by any means, without the prior permission in writing of the British Academy

British Library Cataloguing in Publication Data
Data available

ISBN 0-19-726125-6

Typeset by the editor
Printed in Great Britain
on acid-free paper by
The Cromwell Press Limited
Melksham, Wiltshire

Foreword

This volume takes the New Series of the *Records of Social and Economic History* north of the Border for the first time; although a transatlantic bridgehead was established by the travel diaries of Jabez Maud Fisher, *An American Quaker in the British Isles* (Volume XVI) and another in France through the *Charters and Custumals of the Abbey of Holy Trinity, Caen* (Volumes V and 22).

The *Books of Assumption* contain a unique survey of the income of church properties in Scotland (Argyll and the Isles apart) in the 1560s, driven by the characteristic imperative to document, and thus secure, revenue in a time of turmoil. The historian, following in the footsteps of the tax-gatherers, is offered a window through which to explore the wider world of the resources which sustained the church. The *Books of Assumption* document in detail the distribution of the Scottish Church's wealth drawn from a thousand parishes and a population of 800,000 living mainly in small rural communities. Apart from levies for papal taxation these revenues supported the superstructure of the cathedrals and abbeys, the wealth of the richest monastic establishments, as in England, being greater than that of the dioceses. (Arbroath Abbey, for example, enjoyed annual revenues of £13,000 while those of the archbishopric of St Andrews exceeded £6,000.)

Apart from revealing the distribution of annual income within the church – itself of importance because the church was the largest organisation and wealthiest single landowner in Scotland – the entries provide much evidence about land which was leased or 'feued', together with the nobles and lairds who acquired the 'feu' charters. The whole pyramid rested upon the yields of a myriad levies. Every conceivable surplus of man and nature was mulcted, from land, river, sea and air (hawking was dunned). There were levies on fish and fowl, mills, forests, hay, straw, every variety of crop and animal, peat, lime, salt-works and mines of all kinds. Nothing escaped: there were even 'corpse presents'.

In the background is visible the medieval heritage of land use, tenurial cutoms, the distribution of crops and surviving labour services. Dr Kirk has provided a huge, fascinating and varied database of importance for historians of all interests.

April 1995

Peter Mathias
Chairman, Records of Social and Economic History Committee

To the Memory of my mentor in Scottish History
GORDON DONALDSON
who died on 16 March 1993

Contents

Foreword	v
Acknowledgments	ix
Abbreviations	x
Introduction	
1 Antecedents to the 'assumption of thirds'	xi
2 The levy of a third on revenues	xiv
3 The survey of benefices	xvi
4 Submission of rentals	xx
5 Rentals entered and omitted	xxvi
6 The record's significance	xxix
7 The medieval heritage	xxx
8 Parochial resources	xl
9 Episcopal wealth	xlvi
10 Monastic patrimony	li
11 The social setting	lxv
12 The manuscripts and editorial method	lxxxiv
13 Note on vulgar fractions	lxxxvii
14 Note on indices	lxxxviii
The Books of Assumption	
Fife	1
Edinburgh	91
Haddington (1)	145
Linlithgow	151
Haddington (2)	161
Berwick	183
Roxburgh (1)	207
Peebles	247
Roxburgh (2)	257
Selkirk	263

Annandale	265
Dumfries	269
Perth	281
Forfar (1)	351
Kincardine	403
Forfar (2)	409
Aberdeen	413
Moray (1)	453
Banff	455
Moray (2)	461
Lanark	495
Renfrew	527
Stirling (1)	537
Dunbarton	539
Stirling (2)	543
Kyle	557
Carrick	567
Cunninghame	573
Wigtown	585
Kirkcudbright	609
Inverness	625
Orkney	655

Appendix 1
    Bishopric of the Isles   673

Appendix 2
    Bishop of Orkney to King James VI   677
    Sir Gideon Murray of Elibank to King James VI   681

Glossary   685

Bibliography   699

Indices
    Index of Persons   703
    Index of Places   755

Map
    Diocesan boundaries at the Reformation   *at end of volume*

Acknowledgments

I would record my gratitude to the British Academy, first, for the award of a Major Research Grant over two years to prepare a critical edition of the text offered below, and, secondly, for undertaking to publish this material in its series on the Records of Social and Economic History. The encouragement I received from Professor G. W. S. Barrow, Professor J. K. Cameron and Professor T. C. Smout is deeply appreciated. Invaluable help was forthcoming, too, from Professor D. C. Coleman, Professor A. A. M. Duncan, Professor P. Mathias, and Dr Athol Murray.

Dr John Durkan, with characteristic generosity, resolved many ambiguities in my text; and my former teacher, Professor Gordon Donaldson, H. M. Historiographer in Scotland, has my admiration and warm thanks for assuming the arduous task of reading the text in its entirety. I would place on record also the help received from my research assistant, Mrs Helen Cummings, whose efforts extended to typesetting a significant portion of the text; and I should be guilty of ingratitude were I not to extend my appreciation to Mr James Rivington, the Academy's Publications Officer, for his helpful advice on presentation.

Grateful acknowledgment is made to the Keeper of the Records of Scotland; to the Librarians of Edinburgh University Library and the National Library of Scotland for permission to publish manuscript material in their possession; and to Sheriff Peter McNeill for permission to utilise the substance of a map I prepared for a forthcoming revised edition of his *Historical Atlas of Scotland*.

JAMES KIRK

June 1992

Abbreviations

APS	*Acts of the Parliaments of Scotland*
BUK	*The Booke of the Universall Kirk: Acts and Proceedings of the General Asssemblies of the Kirk of Scotland*
EUL	Edinburgh University Library
NLS	National Library of Scotland, Edinburgh
RPC	*Register of the Privy Council of Scotland*
RMS	*Registrum Magni Sigilli Regum Scotorum: Register of the Great Seal of Scotland*
RSS	*Registrum Secreti Sigilli Regum Scotorum: Register of the Privy Seal of Scotland*
SHR	*Scottish Historical Review*
SRO	Scottish Record Office, Edinburgh

Introduction

1 Antecedents to the 'assumption of thirds'

'The Books of Assumption of the Thirds of Benefices' - to assign the record its full and exact title - owe their origin to the crown's interest in ecclesiastical finances at the Reformation. More specifically, the 'Books of Assumption' derive their existence from the crown's determination to acquire for taxation a systematic knowledge of the income and certain categories of expenditure of the church's livings across the kingdom by conducting a survey of ecclesiastical rentals similar to that achieved for England and Wales in the *Valor Ecclesiasticus* of 1535.[1]

Among the issues raised by the Reformation was the fate of the church's extensive property. The late medieval church, after all, was the largest owner of property in the kingdom, with an annual income at least ten times that of the crown, and for generations before the protestant revolution in 1560, crown, nobility and gentry had fully exploited for personal profit the church's wealth at every level. Any statesmanlike settlement which might ensue, in the aftermath of the protestant victory, had therefore to reconcile the genuine claims of the reformed church, whose *Book of Discipline* in 1560 asserted a right to inherit the entire 'spirituality',[2] consisting principally of teinds (or tithes), with the material interests of sovereign, magnates, lairds and others who, as prominent beneficiaries of the existing system, stood to lose financially in any comprehensive redistribution of ecclesiastical resources which reformers favoured. Nor was it evident how the reformers' demands for the extrication of the teinds as 'spirituality' from the 'temporality', or lands and their rents, could readily be accommodated, except by conceding their drastic programme for subverting the ancient financial structure of benefices in its entirety. Such a prospect, at any rate, of dismantling so complex an organisation which had evolved over the centuries seemed remote so long as the dominant interests in society stood to gain financially by exploiting for their own ends the old system of benefices.

The 'Reformation parliament', which met in Edinburgh in July and August 1560, was careful to repudiate Rome, by legislating against papal jurisdiction and

---

[1] *Valor Ecclesiasticus*, eds. J. Caley and J. Hunter, 6 vols. (London, 1810-1834).
[2] *The First Book of Discipline*, ed. J. K. Cameron (Edinburgh, 1972), pp. 156-164.

the celebration of mass, and in favour of a protestant confession of faith, but it conspicuously avoided any pronouncement on either polity or endowment.³ It offered no solution for the reallocation of ecclesiastical finances which - were that to occur - had clearly to await further determination. Not only did the medieval structure of benefices survive the Reformation intact but existing holders retained a legal title to their livings. In any event, there were plainly limitations on what could be achieved by a provisional government of protestant lords who, in successful rebellion, had 'suspended' the queen mother, Mary of Guise, from the regency in October 1559 and who, with her death in June 1560, had transferred effective power to a 'great council of the realm' under the presidency of the duke of Chatelherault, heir presumptive to the throne. In turn, Mary, Queen of Scots, then queen of France, declined to ratify the work of the Reformation parliament, even though she had duly authorised its meeting. Only with the death of her husband Francis II in December 1560 and Mary's hitherto unexpected return to her native land in August 1561 did there emerge the prospect of a stable government which alone could hope to resolve the many pressing issues and competing claims arising from the Reformation.

Amid the uncertainty and dislocation resulting from the revolution against Rome and France, existing beneficed clergy frequently found difficulty in securing payment of their revenues, as parishioners and tenants withheld their dues and rents, and as lords too readily intromitted with the fruits of benefices. The old remedy of raising an action for the recovery of rent or teind in the court of the official was effectively ended; the threat of excommunication or 'cursing' no longer carried weight; and resort to a civil action rarely produced a speedy solution. Besides, the activities of protestant magnates, where they gained control, had resulted in the effective confiscation, for their own use, of much ecclesiastical property. The archbishop of St Andrews lamented how petitions to parliament in 1560 had failed to have the livings of clergy 'restorit'; and the archbishop of Glasgow was informed by his chamberlain how he, with the archbishop of St Andrews, the bishops of Dunkeld and Dunblane, and the 'abbot', or commendator, of Dunfermline, had sought in vain to secure restitution of certain property and how the authorities in 1560 refused to answer churchmen unless they first accepted the reformed confession of faith. Clergy who did not conform to the new regime, it was remarked in 1560, would 'get nathing this yeir to cum mair than thai gat the yeir bygane'.⁴ Not only so, leases and feus (the latter conferring proprietorship) granted by a number of prominent prelates since 6 March 1558/9 were held to be invalid; and official recognition in 1560 that, for the present, every man should retain his own teinds looked like a prelude to

---

³ *Acts of the Parliament of Scotland*, eds. T. Thomson and C. Innes, 12 vols. (Edinburgh, 1814-1875), ii, pp. 525-534.
⁴ R. Keith, *History of the Affairs of Church and State in Scotland*, eds. J. P. Lawson and C. J. Lyon, 3 vols. (Edinburgh, 1844-1850), iii, pp. 5, 8-9.

INTRODUCTION   xiii

reforming the teind system as favoured in the *Book of Discipline*.⁵ All this was followed in January 1560/1 by a convention of protestant lords who accepted the proposals of the *Book of Discipline* on the understanding that beneficed men who conformed to the new regime should retain their livings for life, if they promised to support ministers' stipends from their revenues.⁶

Some of these prospects were arrested, however, with Mary's return to rule in person. The former ascendency of the insurgent lords of the Congregation came to an end as more moderate counsels began to prevail; and beneficed clergy from the old regime expected to find in Mary, who was allowed her private mass in Holyroodhouse, a protector of their interests. Within months, 'the Bischoppis', it was noted, 'began to grypp agane to that which most injustlie thei called thair awin'; and the earl of Arran was discharged from intromitting with the revenues of St Andrews and Dunfermline.⁷ Yet, by her actions, Mary showed herself to be no protagonist of a counter-Reformation; one of her first measures, on returning, had been to reassure protestants by issuing a proclamation prohibiting, on pain of death, any attempt to alter the state of religion and form of public worship which she had found at the time of her arrival;⁸ and more was to follow as Mary and her councillors explored arrangements for dividing the wealth of the pre-Reformation church by assigning shares to the crown and reformed church as well as to former priests and members of the religious orders. If theory suggested that the continued existence of the old financial system held open the possibility of a papalist reaction, practice showed that Mary, by conferring on the reformed church a measure of financial support and official recognition, was prepared to act as if she had a vested interest in the Reformation.

Partly as a consequence of pressure from the barons who had played so prominent a part in the Reformation parliament and who petitioned the council for financial provision for the reformed ministry,⁹ Mary and her privy council agreed with some beneficed clergy that a 'general convention' of the entire ecclesiastical estate of the old church should be held on 15 December 1561. 'The bishops sought to be restored', it was observed (presumably in the sense of regaining effective control of revenues expropriated by powerful laymen), and they were prepared to offer, it was claimed, a 'large contribution to be put in possession'; they also expected the matter to be considered at the forthcoming convention.¹⁰

---

⁵ Ibid., i, p. 325; *First Book of Discipline*, pp. 108-113, 156-164. Various enactments, including the act of Oblivion, recognised the date 6 March 1558/9; but by 1564 the date was held to be 8 March 1558/9. *APS*, ii, pp. 536-537; *Register of the Privy Council of Scotland*, eds. J. H. Burton *et al.*, 1st ser., 14 vols. (Edinburgh, 1877-1898), i, pp. 162-163; cf *APS*, ii, p. 545.
⁶ *The Works of John Knox*, ed. D. Laing, 6 vols. (Edinburgh, 1846-1864), ii, pp. 257-258; *John Knox's History of the Reformation in Scotland*, ed. W. C. Dickinson, 2 vols. (Edinburgh, 1949), i, p. 345.
⁷ Knox, *Works*, ii, p. 298; Knox, *History*, ii, p. 28; cf *Calendar of State Papers relating to Scotland and Mary, Queen of Scots, 1547-1603*, eds. J. Bain *et al.*, 13 vols. (Edinburgh, 1898-1969), i, no 1071.
⁸ Knox, *Works*, ii, pp. 272-273; Knox, *History*, ii, pp. 9-10; *RPC*, i, pp. 209, 267-268, 356.
⁹ Knox, *Works*, ii, p. 298; Knox, *History*, ii, p. 28.
¹⁰ *CSP Scot.*, i, no 1049 (p. 575).

The crown, for its part, on seeking the convention's advice, ordered that no one should be dispossessed of holdings on kirklands feued (that is, disponed on a perpetual heritable tenure for an annual, fixed money-rent) since 6 March 1558/9 (the date considered to mark the start of the troubles and the point at which the crown's supervision of feuing was thought officially to have begun), a respite which was to extend at least until Whitsun 1563.[11] Many churchmen, after all, anticipating imminent revolution, had resorted to feuing further lands (replacing lease-holding by a form of heritable tenure) in return for such capital sums as they were able to secure; and the crown was ready to intervene in an effort to protect existing tenants.

As discussions proceeded among the council, the nobility and clerical estate at the convention on 22 December, the archbishop of St Andrews, with the bishops of Dunkeld, Moray and Ross, offered the crown a fourth part of their incomes 'to be employit as hir Grace thocht expedient'. That concession, however, was deemed insufficient, and owing to the council's uncertainty of the amount required to sustain the reformed ministry and support the needs of the royal household, it was decided by queen and council that necessity might dictate a third, or even more, of the fruits from all benefices as an annual contribution till 'ane generale ordoure be takin thairin'.[12]

2 The levy of a third on revenues

The decision to fix the levy on the fruits of benefices at exactly one third of their annual value was announced on 15 February 1561/2 when an act of council, regulating the 'assumption', that is, the uplifting or collection of revenues, placed the share needed to meet the requirements of the crown and reformed ministry at one third precisely, starting with the crop for 1561. The 'old possessors' were confirmed in the remaining two-thirds of the fruits of their benefices for their lifetimes; and, reaffirming the penalties imposed on 12 February, the council announced that holders who failed to produce rentals were to lose their entire revenues, as were holders whose rentals contained fraudulent omissions. The annual rents and other duties in towns belonging to chaplainries, prebends and friaries, along with the rents derived from lands owned by the friars, were to be transferred to hospitals, schools and 'utheris godlie usis', and the buildings of friaries still standing were to be maintained by town councils for the benefit of schools and colleges.[13]

The ingenuity of the settlement, so devised, went some way towards reconciling divergent financial interests. John Knox, it is true, denounced the

---

[11] *RPC*, i, pp. 162-163, 192-194.
[12] Ibid., i, pp. 192-193.
[13] Ibid., i, pp. 201-203.

compromise, for he saw 'twa partis freely gevin to the Devill, and the thrid maun be devided betwix God and the Devill'.[14] Yet the scheme did represent an attempt to harmonise the claims of competing parties; and if it were not exactly an equitable settlement, it was at least a statesmanlike solution (for which it might be hard to find a parallel in western Europe) which divided up the church's wealth into shares for Roman bishops, priests, monks, friars and nuns (whose religious duties were deemed to have ceased), for protestant ministers, exhorters and readers (active in the reformed church), and for the sovereign (to help meet the expenses of government). In such a division of the spoils, at any rate, none of the contenders need feel unduly aggrieved for being wholly ignored or excluded.

Nor was so substantial an impost altogether without precedent. The crown, after all, was accustomed to drawing money from the church by direct taxation; and in the sixteenth century, the church's proportion in taxation, assessed on the basis of Bagimond's roll of 1275, was increased from two-fifths to a half. Most notably, James V had persuaded Pope Clement VII in 1531 to grant him a tenth of all ecclesiastical revenues for three years (from benefices worth more than £20), ostensibly for the defence of the realm, and a permanent tax on prelates of some £10,000 annually, on the pretext of establishing a College of Justice. This latter sum amounted approximately to a sixth of valued revenue, and was soon adjusted to a composition of £72,000, plus an annual liability of £1,400 (paid by the prelates from the fruits of benefices in their patronage).[15]

Renewed pressures in the late 1540s, and the imposition of further taxes, led the prelates to grant a contribution of £30,000 in tax 'for the resistance to England';[16] in 1549, the governor, council and nobility assigned the clergy's share of taxation at £16,000, deducting payments already made;[17] and in 1550 a tax of £4,000 was imposed on the clergy 'for resisting of our auld ynemyis of Ingland'.[18] Then, in 1556 and again in 1557, the pope permitted Mary of Guise a subsidy of a twentieth of ecclesiastical revenues.[19] More generally, the idea of conducting a national assessment for a perpetual tax was seriously proposed in 1556 when a parliamentary committee recommended a survey of all lands and inhabitants within the realm to be conducted by sheriffs and nominated commissioners in the shires and by rural deans, vicars, curates and parish clerks in every diocese, who were to return to the treasurer the names and property of all freeholders, feuars, tenants, parishioners, craftsmen and cottars throughout the country.[20] The proposal proved abortive but the idea served as a precedent for the ecclesiastical survey six years later. In all, there was little originality, though much common

---

[14] Knox, *Works*, ii, p. 310; Knox, *History*, ii, p. 29.
[15] R. K. Hannay, *The College of Justice* (Edinburgh, 1933), pp. 52ff.
[16] *Acts of the Lords of Council in Public Affairs, 1501-1554*, ed. R. K. Hannay (Edinburgh, 1932), p. 583.
[17] *APS*, ii, p. 600.
[18] *RPC*, i, p. 83.
[19] *Papal Negotiations with Mary Queen of Scots during her reign in Scotland, 1561-1567*, ed. J. H. Pollen (Edinburgh, 1901), p. 3.
[20] *APS*, ii, pp. 604-605.

sense, in the scheme of 1562, and possibly the main novelty was that the terms of the compromise were honoured in practice.

3 The survey of benefices

With the intention, therefore, of levying a tax on the revenues of benefices to assist the needs of the crown and reformed church, which still looked for endowment, Mary and her privy council initiated a scheme to survey all Scottish benefices. Its purpose was to yield an assessment of the wealth of the pre-Reformation church, whose financial structure, of course, remained intact regardless of the Reformation, and whose benefice-holders, far from being dispossessed, actually retained a legal entitlement to revenues from their livings, irrespective of any attachment or antipathy they might exhibit toward serving the reformed church, whose work, in turn, remained hampered by a lack of finance through failure to secure its claim to the teinds from the old church.

Accordingly, the queen and council in December 1561 ordered beneficed men to produce for the crown's inspection full rentals of their revenues. Holders of benefices lying to the south of the Mounth - the mountainous range dissecting the country from Loch Linnhe in the west toward Stonehaven in the north-east - were to submit their rentals on 24 January 1561/2; and those north of the Mounth on 1 February. The sheriffs' officers were charged with responsibility for notifying holders of these arrangements. Thereafter, the onus was placed on beneficed men, their chamberlains and factors to ensure that accurate rentals were returned.[21] The announcement of these procedures on 22 December 1561, in the middle of winter, gave little enough opportunity for sheriffs' officers to intimate to all relevant parties throughout the country what the crown expected to be done; far less did it allow sufficient time for the preparation of up-to-date accounts for transmission to officials in Edinburgh at a time of year when prolonged travel was notably arduous and even hazardous.

Given the obstacles, the compilation of such a formidable record derived from so many different sources and within so restricted a timescale was a herculean task which might almost be reckoned to test the administrative capacity of the severely limited bureaucracy in Edinburgh. Certainly, within a decentralised framework in which the crown lacked resources for a multitude of officials at the centre, there was no possibility of imitating the example of England where a strong centralised machinery enabled the chancellor to authorise inquisitors to visit ecclesiastical properties and so ascertain the value of all benefices; and the commissioners, so appointed for each diocese and shire before whom all clergy were obliged to appear with full information, were likely in assessments to give benefit of doubt to the treasury. In Scotland, reliance instead

---

[21] *RPC*, i, pp. 193-194.

was firmly placed on beneficed men returning 'just and trew' rentals of their properties; and though penalties were imposed, by 12 February 1561/2, on those fraudulently concealing revenues or failing to render rentals,[22] the temptation for the tax-payers undergoing assessment, in the absence of official investigations in the localities, was plainly to undervalue or minimize their net incomes. Yet, even here, there were limitations on how far attempted evasion or concealment could be carried, for any rental which officials in the Exchequer considered suspicious or improbable was liable to be queried and checked with earlier rentals of the benefice.[23]

In any event, a simplified procedure to that employed in England was no doubt thought to be preferable, for Scotland had just over a thousand medieval parishes, compared with England's 10,000 or so parishes. The two archdioceses of St Andrews and Glasgow accounted for almost half the parishes in the country; and although Scotland's thirteen dioceses contrasted favourably with England's 21 (before Henry VIII's reforms), England's smallest dioceses had each around 200 parishes (apart from Rochester with 92) and so were comparable with Scotland's largest. Besides, two or three English sees - Lincoln, Norwich and possibly London - had as many parishes as the whole of Scotland.

Initially, for conducting the survey, all that was established at the centre in Edinburgh was a small committee of officials, comprising James McGill, Clerk Register, John Bellenden, Justice Clerk, Robert Richardson, the Treasurer, William Maitland of Lethington, the Secretary, and John Wishart of Pittarow, whom the queen appointed on 24 January to receive and scrutinize the rentals submitted, and to obtain details from superintendents, elders and deacons of all ministers in the reformed church eligible for support from the levy on benefices.[24] Soon afterwards, on 1 March, Wishart, the new Comptroller (appointed in February 1561/2), was chosen Collector general of the 'thirds', accountable to the Exchequer; and a new department, the Collectory, responsible for the revenues collected, acquired as staff a clerk to the Collectory, a keeper of the register of stipends (of ministers and their assistants, the exhorters and readers in the reformed church, sustained from the thirds) and officers to liaise with the Exchequer. In the localities, too, provision was made for deputy collectors, who accounted to the Collector general, for each of the twelve regions into which the country was divided for gathering the funds; each collector, in turn, had his staff; the services of messengers and macers, who executed charges against those in arrears with their contributions to the Collectory, had also to be met.[25]

The administrative areas into which the country was divided by the crown's officials in Edinburgh for the purpose of recording the fruits of benefices departed from the old diocesan structure, so familiar to the clerics, or their procurators, who rendered their returns. The clerk with responsibility for entering

---

[22] Ibid., i, pp. 199-200.
[23] See below, pp. xix-xx, lvii-lviii.
[24] *RPC*, i, pp. 196-197.
[25] *Accounts of the Collectors of Thirds of Benefices, 1561-1572*, ed. G. Donaldson (Edinburgh, 1949), pp. x, xv-xvi, xl.

the rentals in the 'Books of Assumption' arranged the material on a topographical basis which adhered neither to ecclesiastical units of administration nor wholly to the sheriffdoms with which the crown's servants might be expected to have more affinity. The units selected comprised: Fife, Edinburgh, Haddington, Linlithgow, Berwick, Roxburgh, Peebles, Selkirk, Annandale, Dumfries, Perth, Forfar, Kincardine, Aberdeen, Moray, Banff, Lanark, Renfrew, Stirling, Dunbarton, Ayr (with its three divisions of Kyle, Carrick and Cunninghame), Wigtown, Kirkcudbright, Inverness (including Ross and Cromarty, Sutherland and Caithness), and Orkney (with Shetland).[26] Such an arrangement broadly coincided with the twelve districts allocated to the sub-collectors of the thirds of benefices: Orkney and Shetland; Inverness; Moray; Aberdeen and Banff; Forfar and Kincardine; Fife, Fothrik and Kinross; Perth and Strathearn; Lothian; Roxburgh, Selkirk and Peebles; Ayrshire; Stirling, Dunbarton, Renfrew and Lanark; Dumfries, Annandale, Kirkcudbright and Wigtown.[27]

In registering the rentals, the clerk, James Nicolson, had the task of copying the material submitted in a style which recorded usually the diocese or sheriffdom in which the benefice lay, the revenues accruing (which if large or complex took the form of a detailed 'charge' and 'discharge'), the calculation of the third to be rendered in tax, and usually the signature of the benefice-holder or responsible agent. In some entries, the holder of a benefice might sign his rental but assign the task of delivering the evidence to another who was prepared to travel to Edinburgh. These procurators might be fellow clerics or notaries public.[28] In other instances, responsibility for furnishing the accounts for Dunfermline abbey, Jedburgh abbey and Restenneth priory, Lesmahagow priory, Aberdeen and Brechin bishoprics lay with the chamberlains.[29] At Haddington priory, a notary public extracted the evidence from the chamberlain's accounts in the registers of the nunnery on behalf of the prioress, 'who cannot write'.[30] Again, the prioress of St Bathans signed her rental 'with my hand on the pen led be the notar underwrittin at my command becaus I could nocht writ my selff'.[31] The services of a notary were also required by Alexander Hume of Heuch, parson of Polwarth, who signed his rental 'with my hand led at the pen be Robert Lauder, notar'.[32] A servant to the provost of Trinity College in Edinburgh took responsibility for signing that rental; and the parson of Govan, 'becaus he may not travell', entrusted delivery of his rental to his 'tacksman'(who also happened to be chamberlain to the absentee archbishop of Glasgow).[33] In Dunkeld, the factor to

---

[26] A heading for the 'Isles' (or Hebrides) was entered then abandoned, and the folio utilised for other purposes. See below, p. 281, n. 2.
[27] *Thirds of Benefices*, p. xl.
[28] See below, pp. 78, 79, 84-87, 105, 299, 301, 311, 318-21, 344, 368, 380, 406-407, 432, 436, 444, 451, 457, 480-482, 501, 504-506, 512, 517, 525, 533, 534, 542, 553, 554, 561, 636, 639, 641-645.
[29] See below, pp. 23, 40, 45; cf pp. 73; 217-218; 243; 416; 385.
[30] See below, p. 164.
[31] See below, p. 193.
[32] See below, p. 195.
[33] See below, pp. 118, 533.

the chancellor signed and produced the accounts; the factor for a prebendary of Dunkeld did likewise; and the factor, who 'cannot write', to the treasurer of Ross employed the services of a notary.[34] Rentals for Sweetheart abbey and Parton parsonage were delivered by Peter Thomson, Islay herald.[35] In Brechin, where John Hepburn was treasurer, it was evidently his kinsman, James Hepburn, who signed the rental for the treasurership, as well as for a chaplainry in the cathedral.[36] At Eccles priory, James Hamilton of Kingscavil, sheriff of Linlithgow, tacksman, and a kinsman of the prioress, signed the rental.[37]

Frequently, where the whole fruits of a benefice were leased, certain burdens of taxation might accrue to the tacksman, or lease-holder;[38] and where tacksmen had responsibility for paying the thirds, it was often the tacksman, rather than the holder of the benefice, who prepared and presented the rental. Accordingly, the rental for Glenluce abbey was presented in name of the earl of Cassillis, who had leased the abbacy; the earl of Crawford signed the rentals for two parsonages whose fruits he had leased; the farmer, or lease-holder, of the subdeanery of Brechin assumed responsibility for signing and delivering that rental; the tacksman of a prebend of Dunbar collegiate church affixed his signature to the rental; so, too, did the lease-holder of Tough parsonage.[39] At Ardclach, the farmer of the fruits of the vicarage signed the rental 'with my hand at the pen led be the notar underwrittin at my request becaus I cane nocht write', in June 1562, four months after the vicar had produced a rental.[40] Even the returns for a chaplainry of Luss, leased to a Colquhoun, were presented by the tacksman in name of the chaplain.[41] The rental for Inverkip vicarage, however, was jointly signed by vicar and tacksman.[42]

Attached to the copies of numerous rentals prepared by the clerk to the Collectory is assorted information in the form of royal warrants, proclamations, precepts, instructions from the crown to the lords of Exchequer, and from the Clerk Register and Comptroller to the clerk of the Collectory, the texts of certain leases collated by the clerk, and various memoranda, letters, petitions, and judgments.[43] Many entries, too, contain specific directions from the Clerk Register to the clerk of the Collectory requiring him to receive rentals which had been accepted.[44] Omissions and shortcomings detected in rentals were often directed to the clerk's attention for remedying. A conspicuous feature in

---

[34] See below, pp. 308, 366, 637.
[35] See below, pp. 612, 616.
[36] See below, pp. 377, 379; cf Watt, *Fasti*, p. 51.
[37] See below, p. 184; cf *Thirds of Benefices*, pp. 163, 285.
[38] See below, pp. 374-375, 401-402, 491-492.
[39] See below, pp. 600; 381; 394; 171; 425.
[40] See below, pp. 479, 488.
[41] See below, p. 542.
[42] See below, p. 535.
[43] See below, for example, pp. 5, 7-8, 21, 44-45, 54, 57, 59-60, 129, 140, 149-50, 201-202, 218, 221-222, 260-261, 281-282, 286, 371-372, 374-375, 401-402, 411, 446-447, 459, 465, 467, 473, 491-492, 524, 537, 543-545, 574-575, 609-610, 621, 651.
[44] See below, for example, pp. 132, 133, 135, 137, 140, 157, 248, 257, 264, 300, 328, 338, 339, 341, 343, 391, 395, 398, 399, 402, 411, 444, 449, 450, 535, 556, 610.

numerous accounts, especially from prelacies, was the failure to declare the value of receipts derived from 'grassums' (down payments for leases or feus), 'entry silver' (money paid on gaining possession), 'kanes' (customary payments by tenants, usually in kind), 'customs' (dues rendered by use and wont), fishings, capons, poultry, marts, muttons, kids, carriage and other labour services and dues; and it fell to the clerk of the Collectory to identify and remedy these deficiencies in the rentals.[45] More serious instances of inadequate returns, requiring revision, were investigated by the lords of Exchequer.[46]

A few of the smaller original rentals have survived, tucked within the folios of the 'Books of Assumption', where they were evidently left after copying.[47] Some of the copies, too, preserve the endorsement on the original rentals submitted.[48]

4 Submission of rentals

From so inauspicious a start, what may surprise is not the council's lament on 12 February 1561/2 that 'ane verray small nowmer' of rentals had been lodged on time[49] - that surely was to be expected given the exigencies of bad weather, difficulties in travel, and the short period available to produce rentals - but the significant returns which were forthcoming as requested. Many of those rentals which bear a date belong to the early months of 1562. Other undated rentals, for the most part, seem to have followed soon after; and a residue of decidedly late entries or, sometimes, rentals updating or supplementing or even superseding earlier ones were added, usually with a date supplied.

Late rentals may be readily identified. Attached to the entry for St Salvator's altar in St Giles' collegiate church in Edinburgh is a precept, dated 1 July 1563, from the Comptroller to the clerk of the Collectory, instructing him to receive the rental; and the accounts for the chaplainry of St Roch's altar in the same church were presented on 31 August 1564.[50] The entry for Stobo vicarage contains an addition to the rental, following the appointment of a new vicar in 1563.[51] The value of Hassendean vicarage was presented on 28 January 1563/4; and the proceeds of two altars in Sanquhar parish church were recorded at Edinburgh on 18 December 1564.[52]

---

[45] See below, for example, pp. 5, 9, 17, 22, 34, 57, 165, 185, 347, 352, 364, 468, 473, 497, 539, 595, 612, 629.
[46] See below, pp. 8, 459, 573-575.
[47] See below, for example, p. 117, n. 88; p. 183, n. 1.
[48] See below, for example, pp. 7, 8, 149, 272, 550, 610.
[49] *RPC*, i, pp. 199-200.
[50] See below, pp. 129, 138.
[51] See below, p. 252.
[52] See below, pp. 264, 273.

INTRODUCTION   xxi

The entry for Balmerino abbey, whose rental seems to have been composed by 1562,[53] contained the text of a lease of the fruits of the abbey in November 1564, which was collated with the original by the clerk to the Collectory in January 1565/6, and later instructions from the Regent Moray were added in a letter of 20 September 1568.[54] One rental for the abbey of Coupar Angus was certainly in the possession of officials at the Exchequer by 13 January 1564/5.[55] For Cambuskenneth abbey, four rentals were engrossed. Attached to one was a letter from the queen to the Comptroller, dated 1567, to another a letter dated 20 April 1565, and to a third a letter of 1567; but the earliest rental for that house is the fourth in sequence, signed not by Adam Erskine, appointed commendator in 1562, but by Andrew Hagy, oeconomus and procurator during the vacancy until discharged by Adam Erskine in 1562.[56]

In one rental for the Charterhouse in Perth, the prior declared himself ready, on 26 May 1565 (or possibly 1563), to alter and renew his 'assumption' as the Comptroller or his chamberlain might require; and a second rental for the same house bears the date 30 January 1569/70.[57] One rental for Melrose abbey seems to have been supplied by 1566.[58] Attached to another rental for Melrose abbey is a letter from King James to the Comptroller, dated 31 January 1577/8.[59] The rental for Brechin vicarage, submitted by the new vicar (appointed in April 1565[60]), contained the text of a lease granted by his predecessor on 12 March 1562/3, and was duly authorised by the Comptroller for registering by the clerk of the Collectory on 19 January 1565/6.[61] The valuation of Manuel priory and the proceeds of two chaplainries in Glasgow and Lanark were also presented in Edinburgh in January 1565/6.[62]

The figures for Auchtermuchty vicarage were rendered on 16 July 1566.[63] A 'new' rental for the bishopric of Galloway, the third in sequence entered for the bishopric, was received on 8 September 1566.[64] Production of the rental for the parsonage of Rothes took place at Edinburgh on 27 September 1566.[65] By 5 December 1566, the prebendary of St Catherine's altar in St Magnus' cathedral in

---

[53] The rental fails to record the yearly pension gifted to David Watt by October 1562, cf *Registrum Secreti Sigilli Regum Scotorum*, eds. M. Livingstone *et al.*, 8 vols. (Edinburgh, 1908-1982), v, no 1115. The calculation of the third conforms to the figures in the Collectors' early accounts. By 1568, the figure for silver had changed and the amounts of victual were by then omitted. See *Thirds of Benefices*, pp. 12, 30, 33, 36, 40, 52, 59, 66, 71.
[54] See below, pp. 57, 58, 61.
[55] See below, pp. 409-411.
[56] See below, p. 547; cf *Registrum Monasterii S. Marie de Cambuskenneth, A.D. 1147-1535*, ed. W. Fraser (Edinburgh, 1872), pp. cvii, cviii.
[57] See below, pp. 283, 286.
[58] See below, p. 209.
[59] See below, p. 261.
[60] *RSS*, v, no 2006.
[61] See below, pp. 401-402.
[62] See below, pp. 523-524, 549-550.
[63] See below, p. 76.
[64] See below, p. 606.
[65] See below, p. 491.

Kirkwall had submitted his rental.[66] The value of one prebend of Glasgow cathedral was declared on 5 March 1567/8.[67] Two rentals for the bishopric of Brechin were prepared at a point after Alexander Campbell's appointment as bishop in 1566; and a third, dated 28 January 1573/4, disclosed that the bishop was absent 'in Genewy at the scullis'.[68] The rental for Soulseat abbey was registered when Harry Smith was collector for Galloway from 1568;[69] and the assumption of the third for Sweetheart abbey, belonging to one of the rentals produced for that house, was received in Edinburgh on 13 July 1570; another rental is no earlier than 1565.[70]

An entry of the assumption of the third for Kirkbean vicarage is endorsed with the date 1567 (though an earlier rental for that vicarage had been rendered in February 1561/2); and the record for Collessie vicarage contains a marginal note, later inserted by the clerk of the Collectory, dated 13 May 1573.[71] The clerk to the Collectory was also ordered, on 9 July 1573, to register the rental submitted for the parsonage of Auldhame.[72] The figures for Duddingston vicarage were received, at the Comptroller's command, on 26 July 1573.[73] The 'assumption' of the third for Kirknewton parsonage was also presented in 1573.[74] One later rental for North Berwick priory was rendered on 18 December 1573; and one for Haddington priory on 28 December 1573.[75] The accounts for Deer abbey, presented on 21 December 1573, were considered 'unsufficient' when scrutinised by the lords of Exchequer, on 22 December.[76]

Particulars of Carriden vicarage were supplied on 25 December 1573.[77] One rental for the provostry of Methven collegiate church was furnished for 1573, though an earlier rental, submitted by the preceding provost, claimed to be 'trew and justlie gevin in but ony fraud'.[78] The vicar of Pencaitland rendered a brief statement of his earnings on 2 January 1573/4; the parson of Tulliallan did likewise on 12 January 1573/4.[79] An added incentive for the submission of fresh rentals in 1573 was assuredly the imposition by King James' government of a religious test on holders of benefices.[80] At that point, too, action was taken to sequester the fruits of benefices whose holders had failed to make true returns.[81]

---

[66] See below, p. 663.
[67] See below, p. 525.
[68] See below, pp. 383, 385, 388.
[69] See below, p. 600; *Thirds of Benefices*, p. xl.
[70] See below, pp. 611, 622.
[71] See below, pp. 609-610, 614; 81.
[72] See below, p. 179.
[73] See below, p. 99.
[74] See below, p. 104.
[75] See below, pp. 145, 180.
[76] See below, p. 459.
[77] See below, p. 157.
[78] See below, pp. 298; 296.
[79] See below, pp. 165, 297.
[80] *APS*, iii, p. 72.
[81] *RPC*, ii, pp. 263-264.

INTRODUCTION   xxiii

The rental for the prebends of Guthrie collegiate church, prepared at Haddington on 13 January 1573/4, was belatedly registered at Edinburgh, three years later, on 17 January 1576/7.[82] The accounts for prebends of Dunglass collegiate church were lodged on 20 January 1573/4.[83] One rental for Kelso abbey was supplied on 1 March 1573/4.[84] A marginal note inserted in the registered entry for Arbroath vicarage pensionary indicated that the vicar had augmented his rental for teind fish on 13 March 1573/4.[85] At that date, too, the vicar of Monzievaird delivered his rental to Holyroodhouse.[86]

A record of the fruits of Whitekirk vicarage was forthcoming on 23 March 1573/4.[87] The parson of Linton (in Roxburgh) declared the value of his benefice on 24 June 1577.[88] The rental for Moonzie vicarage pensionary was prepared on 25 February 1573/4.[89] Details of a pension from Aboyne vicarage were presented on 20 March 1573/4.[90] The submission for certain prebends of Crail collegiate church was made on 1 April 1574.[91] A letter attached to the entry for Inchcolm abbey is dated 22 May 1574.[92] Similarly, the vicar of Newburn furnished the rental of his benefice on 5 August 1574.[93]

The holder of two chaplainries in Dalmeny and Cramond churches, who sought to have his proceeds 'rentalled' in October 1573 (thereby avoiding the threat of forfeiture), complained that the clerk had failed so to do, and it was only in November 1576 that his accounts were officially registered.[94] The vicar of Rothiemurchus and Kincardine produced his rental on 12 March 1574/5.[95] The vicar of Dumfries is said to have produced his accounts at Edinburgh on 3 May 1575.[96] An assumption of the third for St Bathans priory bears the date 1575;[97] and a statement by the earl of Huntly on a lease of the fruits of Kingussie vicarage is dated 9 June 1575, though that rental superseded an earlier one submitted.[98] The rental for Symington vicarage in Kyle was delivered in May 1580.[99] A note

---

[82] See below, p. 390.
[83] See below, p. 166.
[84] See below, p. 222.
[85] See below, pp. 395-396.
[86] See below, p. 300.
[87] See below, p. 169.
[88] See below, p. 215.
[89] See below, p. 83.
[90] See below, p. 451.
[91] See below, p. 65.
[92] See below, pp. 63-64.
[93] See below, p. 69.
[94] See below, p. 149.
[95] See below, p. 465.
[96] See below, p. 611.
[97] See below, p. 195.
[98] See below, pp. 491-492. The vicar in the rental of earlier provenance, Mr Hercules Carnegie, sixth son of Robert Carnegie of Kinnaird, was dead by 1566. Cf *RSS*, v, no 2828; *Registrum Magni Sigilli Regum Scotorum*, eds. T. Thomson et al., 8 vols. (Edinburgh, 1882-1914), iv no 1597.
[99] See below, p. 569.

affixed to the rental for Whithorn priory has the date 9 March 1584/5.[100] The royal warrant for one of the rentals of Dunfermline abbey bears the date 20 January 1586/7, though an extract, attached to an earlier rental, of a decision regarding the fruits is dated 18 January 1571/2.[101] The returns from certain prebends of Corstorphine collegiate church were received at the command of the king's warrant on 11 February 1586/7.[102] The assumption of the third for St Andrews priory is dated 10 February 1586/7.[103] An entry for Colvend vicarage was registered in 1596.[104] Besides, so late as 1600, rentals for Coldingham priory and Jedburgh abbey were the subject of scrutiny by parliamentary commissioners,[105] though rentals of earlier provenance for both houses are also entered.[106]

The inclusion of rentals presented at dates later than those initially prescribed is not an indication, however, that these were necessarily the first rentals received for particular benefices. Sometimes they updated and displaced earlier rentals submitted; and the scribe's practice was often to insert a more recent rental for a benefice immediately preceding an earlier rental which the new one superseded.[107] In any event, instances of late submissions, enumerated at length above, ought not to obscure the many rentals which were promptly entered.

Rentals, for example, of the archbishopric of St Andrews, the bishopric of Dunkeld, with its appropriated churches, the abbeys of Culross and Paisley, the provostries of Corstorphine and St Mary of the Fields' collegiate churches, the parishes of Ardrossan,[108] Botary, Clerkington, Douglas, Forteviot, Glamis, Inchcailloch, Linton (now West Linton), Logie-Montrose, Nevay, and Stevenston, and Trinity chaplainry were produced in January 1561/2.[109] Arbroath and Kilwinning abbeys, and Pluscarden priory, also appear to have submitted rentals at that point;[110] accounts for Dunfermline abbey, which refer to 'this instant lxj yeris crop', were presumably rendered promptly;[111] and returns for the archdeaconry of Glasgow, comprising the parsonage of Peebles with Manor, were dated 26 March 1560.[112] One rental for Cambuskenneth abbey bore the signature of Andrew Hagy, whose office of oeconomus ended on Adam Erskine's

---

[100] See below, p. 595.
[101] See below, pp. 54; 44.
[102] See below, p. 97.
[103] See below, p. 8.
[104] See below, p. 272.
[105] See below, pp. 197-202, 218-222.
[106] See below, pp. 188, 194, 202-205 (where two of Coldingham's rentals were prepared between 1567 and 1571, and a third post-1568); 216-217 (Jedburgh).
[107] See below, for example, pp. 609-614 (Kirkbean); 491-492 (Kingussie).
[108] A copyist's evident slip has rendered the date January 1551/2, instead of January 1561/2. See below, p. 581.
[109] See below, pp. 1; 341, 343, 345; 289; 527; 100; 105; 581; 478; 100; 506; 317; 392; 540; 108; 373; 367; 540; 324.
[110] See below, pp. 358; 575; 471.
[111] See below, p. 40.
[112] See below, p. 253.

INTRODUCTION						xxv

appointment as commendator in 1562.[113] At Coupar Angus, an early rental for the abbey accounted for the crop of 1561.[114] Although one set of returns for Kilwinning abbey had been prepared in 1561, a later 'assumption' for the abbey was approved by the lords of Exchequer on 26 September 1566.[115] Again, the parson of Moy in Moray diocese produced his rental for the crop of 1560; and accounts for the parsonage of Dallas were prepared in August 1561, duly attested by a notary public.[116] The valuation of the fruits for the parsonage of Newdosk was similarly based on the crops of 1559, 1560 and 1561.[117] Besides, the rentals, for the vicarages of Perth and Monimail, which allude to the 'tumult' of the Reformation during the preceding year, were evidently produced in the course of 1561.[118]

By February 1561/2, returns, with dates recorded, had been made for the priory of Haddington, and the parishes of Arbuthnott and Glenbervie (annexed to the deanery of Aberdeen), Ardclach, Ayr, Dolphinton, Dun, Durris, Fetteresso, Idvie, and Kirkbean, as well as a prebend of Lincluden collegiate church,[119] several chaplainries in Elgin cathedral and the chaplainry of Tullypowrie and Fyndynate.[120] The rental for Glenluce abbey seems to have been produced when a new lease of the abbacy was arranged in Edinburgh in February 1561/2.[121] Returns for a prebend of the New College in Glasgow were rendered in '1561' (presumably early in 1561/2); and the valuation of the chaplainry of St Roch in Glasgow was prepared in '1561'.[122] A rental for Irvine vicarage followed on 3 March 1561/2; one for Cargill was lodged before 1 May 1562; the vicar of Straiton submitted his accounts on 16 June 1562; and the crown had knowledge of Inchaffray abbey's rental which was in the Comptroller's possession by October 1562.[123] Two early valuations for the parsonage of Fern, seemingly composed by 1562, were followed by the parson's complaint (whose text the clerk of the Collectory duly registered in 1570) to the general assembly of the reformed church about the levying of his third by the collector for Angus in 1567.[124] Again, the rental for Arbirlot vicarage extended over the years 1561 and 1562.[125]

Most of those rentals, bearing no date, were evidently registered within an acceptable time and without undue delay, if holders were to escape the penalty of losing the entire fruits of their benefices. Besides, the surviving records of the collectors of the 'thirds' (who could only assess the charge to be levied by first

---

[113] See above, p. xxi; below, p. 547.
[114] See below, p. 368.
[115] See below, pp. 575; 573.
[116] See below, pp. 470; 492-493.
[117] See below, p. 382.
[118] See below, p. 74.
[119] A copyist's evident slip has rendered the date February 1551/2, instead of February 1561/2.
[120] See below, pp. 161; 406, 422-424; 488; 561; 504; 393; 405, 408; 404; 376; 614; 615; 476; 324.
[121] See below, p. 600.
[122] See below, pp. 519; 514.
[123] See below, pp. 582; 326; 569; 347.
[124] See below, pp. 371-372, 398, 641.
[125] See below, p. 397.

inspecting the rentals produced) confirm the range of benefices whose revenues, from rentals submitted, were known to the Collectory in 1562 when preparing its accounts for the crop of 1561.

Numerous parochial benefices, not accorded separate entries in the register, were subsumed in the entries of the appropriating institutions to which they were attached, and are therefore to be located in the rental of the appropriating body. Accordingly, Aberdalgie is included in the rental for the bishopric of Dunkeld, Airlie is entered with Coupar Angus abbey, Old Cambus appears with Coldingham priory, Arngask is engrossed in the rental for Cambuskenneth abbey, Auchinleck appears with Paisley abbey, Auchtertool with Inchcolm abbey, Barnwell with Failford, Barro with Holyroodhouse abbey, Blairgowrie with Scone abbey, Bourtie with St Andrews priory, and so forth.

Occassionally in producing rentals for presentation in Edinburgh, reliance was placed on utilising pre-Reformation rentals without noticeable revision. The rental furnished for Morebattle parsonage was that compiled by the last archdeacon of Teviotdale in 1541, but bearing the signature of his successor (who ceased to be archdeacon by 1565).[126] Again, the accounts for Holywood abbey, prepared in 1544, were simply used by the commendator who succeeded in 1550.[127] The rental submitted by the Black friars of Elgin was prepared in 1555.[128] Similarly, the returns for Lesmahagow priory were those supplied in a rental of February 1556/7.[129] A few beneficed men chose to end their returns with the year 1558, before the onset of the troubles. The prebendary of Craigie submitted a valuation of his fruits 'befoir the yeir of God' 1558.[130] Two entries for the vicarage of Inchture valued the benefice for the period before June 1556 and 1558 respectively; and the vicar, who also held the parsonage of Crimond and the vicarage of Aboyne, likewise valued the fruits of these benefices for the period before June 1558.[131] Finally, the 'draucht and copie' of the rental of the Trinitarian house in Aberdeen extended over eighteen years from 1557.[132]

5 Rentals entered and omitted

Even so, rentals are missing for certain benefices which presumably were registered, for they occur in the early accounts of the collectors of thirds who used the rentals to assess the 'third'. Ignoring the northern isles, no entries for the following parishes on the Scottish mainland survive in the 'Books of Assumption'

---

[126] See below, pp. 214-215.
[127] See below, pp. 274-278.
[128] See below, p. 453.
[129] See below, pp. 243-245.
[130] See below, p. 312.
[131] See below, pp. 317, 339; pp. 431, 440.
[132] See below, pp. 420-422.

INTRODUCTION   xxvii

(despite their existence in the early accounts of the collectors of thirds): Abernethy, Altyre, Alvie, Bona in the diocese of Moray, Annan in Annandale, Campsie in the archdiocese of Glasgow, Colstone and Cruden in Aberdeen, Dunlappie in Angus, Dunlichity in Moray, Dupplin in Dunblane, Durness in the diocese of Caithness, Farnua in Moray, Flisk in Fife, Forbes and Forvie in Aberdeen, Garvald in Nithsdale, Kirkton in Teviotdale, Libberton in Glasgow, Lochbroom in Ross, Lochindilloch in Nithsdale, Lochmaben in Annandale, Logiebride in Dunkeld, Longcastle in Galloway, Old Roxburgh in Teviotdale, Peterculter in Aberdeen, Petty in Moray, Quothquan in Lanark, Rannoch in Dunkeld, Ruthwell in Annandale, Southwick in Nithsdale, Stoneykirk in Galloway, Tinwald in Nithsdale, Trinity Gask in Dunblane, and Wamphray in Annandale.

At the same time, collation of the benefices recorded in the 'Books of Assumption' with those in the early accounts of the collectors (charged with gathering the tax of a third of the annual revenues of benefices) discloses a significant number of entries for parochial benefices in the 'Books of Assumption' which are not to be found in the early accounts of the collectors of thirds. These parishes, on the Scottish mainland, may be identified as: Abercrombie, Aberdalgie, Abertarff, Airth, Airlie, Aldbar, Alford, Alves, Ardersier, Arngask, Athelstaneford, Auchinleck, Auldearn, Auldhame, Ayton, Beith, Bendochy, Bethelnie, Birse, Blairgowrie, Bourtie, Bowden, Bower, Calder-Clere, Carluke, Carnock, Carriden, Castleton, Cavers, Ceres, Channelkirk, Clackmannan, Clerkington, Closeburn, Cockpen, Conveth (in Moray), Conveth (in the Mearns), Crailing, Crathie, Crombie, Crossmichael, Dailly, Dalarossie, Dalcross, Dalgarno, Dalgetty, Dalry (in Galloway), Daviot (in Moray), Dornoch, Dowally, Drumoak, Drymen, Dull, Dumfries, Dungree, Dunipace, Dunino, Dunninald, Dyce, Earlston, Eastwood, Ecclesiamagirdle, Eckford, Ednam, Edrom, Errol, Ettrick, Fala, Farnell, Fetterangus and Longley, Fintray, Fintry, Fishwick, Fogo, Forres, Fossoway, Fyvie, Galston, Galtway, Garvald (in Lothian), Girthon, Glassford, Glen Tanar, Glencorse, Glenisla, Gordon, Greenlaw, Heriot, Hirsel, Hobkirk, Horndean, Hownam, Hume, Inchaiden, Inverallan, Inveresk, Invergowrie, Kailzie, Keith-Marischal, Kennethmont, Kettins, Kilbryde (in Dunblane), Kilchrist, Kilgour, Kilmahog, Kilmalemnock, Kilmarnock, Kilmaronock, Kincardine (in Menteith), Kinclaven, Kinfauns, Kinnellar, Kinnernie, Kinnettes, Kinross, Kintore, Kirkcolm, Kirkdale, Kirkoswald, Lagganallachie, Lamberton, Largs, Lempitlaw, Lennel, Lesmahagow, Lessudden, Lethendy, Lhanbryde, Liff, Lintrathen, Linton (in Teviotdale), Little Dunkeld, Lochwinnoch, Logie-Dundee, Logie-Wester, Longnewton, Loudon, Lumphanan, Markinch, Mauchline, Maxton, Maxwell, Meathie, Mertoun, Migvie, Minnigaff, Monkegy, Moonzie, Moulin, Mount Lothian and St Catherine's in the Hopes, Mow, Neilston, Nenthorn, Newburn, Nigg (in the Mearns), Nigg (in Ross), Old Cambus, Ordiquhill, Orwell, Pencaitland, Perceton, Port of Menteith, Premnay, Prestwick, Rafford, Rait, Rathmuriel, Ratho, Rescobie, Riccarton, Roberton, Rosyth, Rothiemurcus and Kincardine, Roxburgh, Ruthven (in Angus), St Quivox, Seton, Sibbaldbie, Simprin, Skene, Staplegorton, Stichill, Strachan, Strafontain, Strathardle, Strathblane, Strageath, Swinton, Tarbat, Tibbermore, Tillicoultry, Torthorwald,

Trailflat, Tranent, Tullibole, Urquhart (in Moray), Watten, Wauchope, Whitekirk, Woolmet, and Westerkirk (a benefice which, the 'Books of Assumption' disclose, had been 'out of use of payment ony kynd of teyndis sen Fluddoun' in 1513[133]).

Discounting rentals with late dates (such as Markinch, Migvie, Strafontain), no simple explanation may be offered for the discrepancies, if that is what they are, between the benefices entered in the 'Books of Assumption' and those recorded in the accounts of the collectors of thirds between 1561 and 1572. For a start, there is no direct correlation between those benefices in the 'Books of Assumption' omitted from the collectors' accounts and those 'new enterit' or 'new rentallit' benefices, belatedly recorded in the collectors' accounts after a reorganisation of the Collectory in 1573. The vicarage of Athelstaneford, for example, was incorporated in the collectors' accounts as late as 1573, but was registered, at a much earlier point, in the 'Books of Assumption' in a rental for Haddington priory, to which it was appropriated, dated February 1561/2.[134] Again, the entry for the vicarage of Airth in the 'Books of Assumption' was engrossed in the rental for Holyroodhouse and signed by Robert Stewart (who ceased to be commendator in 1568), but the benefice was not included in the collectors' accounts until 1576.[135] Although the vicarage of Galston occurs in the collectors' accounts only in 1573, the rental for the vicarage, annexed to the Trinitarian house at Fail, in the 'Books of Assumption' was plainly lodged with officials at the Exchequer by February 1562/3.[136] Dalcross in Moray is first registered in the collectors' accounts for 1578, but the benefice was incorporated in the rental for Pluscarden priory in the 'Books of Assumption' for the crop of 1561.[137] Kinclaven, which first appears in the collectors' accounts in 1573, was already entered in the 'Books of Assumption' in the early 1560s when the benefice-holder, William Adamson (who died by 1565) submitted his rental for assessment.[138] Similarly, Kinnettles vicarage, first recorded in the collectors' accounts in 1576, was registered in the 'Books of Assumption', during Wishart's tenure as Comptroller between 1562 and 1565.[139] Yet again, Kirkliston vicarage, which appears in the 'Books of Assumption' in the rental for St Andrews archbishopric in January 1561/2, occurs in the collectors' accounts only in 1573.[140]

Besides, another category of benefices comprises those which failed to figure in either the 'Books of Assumption'[141] or the collectors' accounts before

---

[133] See below, p. 209.

[134] G. Donaldson, 'The "New Enterit Benefices", 1573-1586', *Scottish Historical Review*, xxxii (1953), pp. 93-98, at 93; see below, p. 162.

[135] See below, pp. 91, 93; *SHR*, xxxii, p. 93; I. B. Cowan and D. E. Easson, *Medieval Religious Houses Scotland* (London, 1976), p. 90.

[136] *SHR*, xxxii, p. 95; see below, pp. 563-565.

[137] *SHR*, xxxii, p. 94; see below, pp. 471-472.

[138] *SHR*, xxxii, p. 96; see below, pp. 306-307; *RSS*, v, no 2113.

[139] *SHR*, xxxii, p. 96; see below, p. 391.

[140] See below, p. 3; *SHR*, xxxii, p. 96.

[141] Other mainland benefices missing from the 'Books of Assumption', not discussed above, include Aldcathie, Balfron, Covington, Dalton (Magna and Parva), Dunnottar, Kintail, Kirkmichael (in Ross) and Lairg.

1573.[142] Such a list, for the mainland, includes the parishes of Applecross, Aldcathie, Ardeonaig, Balfron, Bass, Brachlie, Carruthers, Castlemilk, Clyne, Corrie, Crawford-John, Cults, Daviot in Aberdeen, Dornock and Ecclefechan in Annandale, Drumdelgie and Essil in Moray, Ettleton in Teviotdale, Ewes in Eskdale, Fetternear and Forglen in Aberdeen, Gairloch in Ross, Gretna in Annandale, Hownam in Teviotdale, Hutton in the Merse, Inveravon in Moray, Keig in Aberdeen, Kemback and Kilmany in Fife, Kindrocht in Mar, Kinkell in Dunblane, Kirkmabreck in Galloway, Kirkpatrick-Fleming in Annandale, Kirkpatrick-Irongray in Nithsdale, Leochel in Mar, Lethnot in Brechin, Lochalsh and Lochcarron in Ross, Luce in Annandale, Madderty in Dunkeld, Maryculter in Aberdeen, Middlebie in Annandale, Minto in Teviotdale, Navar[143] in Brechin, Penersax in Annandale, Pitcairn in Dunkeld, Rankleburn in Teviotdale, Rerrick in Galloway, Rogart in Caithness, Rosyth[144] in Fife, the provostry of St Giles' in Edinburgh, Soutra in Lothian, Strathfillan in Dunkeld, Tullich in Aberdeen, Tullichettle in Dunblane, Tundergarth in Annandale, Urquhart (or Glenurquhart) in Moray, and Wemyss in Fife. Rentals for some of these benefices, admittedly, are obscured where benefices formed cathedral prebends whose accounts may not disclose the appropriated parishes, but others, sometimes those for independent, unappropriated parishes, were negligently or fraudulently withheld. Thus, Kirkpatrick-Irongray first appears among the 'new enterit benefices' in 1576.[145]

6 The record's significance

Arranged topographically for the whole kingdom (except for Argyll and the Hebrides, and for a few benefices whose value was carelessly or dishonestly not returned), the series of rentals recorded in the 'Books of Assumption' affords an almost comprehensive insight into the church's wealth - its income and certain categories of expenditure - at the Reformation. Besides illuminating the evolution of the church's financial structure, as it took shape during the middle ages, and the integration of diverse elements in a highly complex organisation, the 'Books of Assumption' offer valuable evidence for examining the society of which the church formed so conspicuous a part in a land which, on the eve of the Reformation, could claim a complement of some 3,000 clergy in a population of no more than 800,000 souls who lived mainly in small rural communities and who largely derived their livelihood from agriculture.

---

[142] Further benefices on the mainland, omitted from the accounts of the collectors of thirds before 1573, include Applegarth, Bassendean, Beath, Cambusmichael, and St Vigeans in Angus.
[143] The chancellory of Brechin, to which Navar was annexed, is recorded in the collectors' early accounts but it does not figure in the 'Books of Assumption'. See *Thirds of Benefices*, p. 11.
[144] The valuation of Rosyth glebe is recorded in the rental for Inchcolm abbey in the 'Books of Assumption'. See below, p. 62.
[145] *SHR*, xxxii, p. 96.

Thus, beyond providing the basis for calculating the distribution of ecclesiastical wealth, the record reveals the personnel who held the livings; many of the laymen who held the church's property by lease or feu, the proprietors who benefited from the changed conditions in land tenure through feuing, that decisive movement of social change in rural society in the sixteenth century, which accelerated the secularisation of so much ecclesiastical property; and, to the extent that it records early payments made to ministers and readers, the source also charts the early work of the reformed ministry after the Reformation. Apart from the record's evident value to the ecclesiastical, and even to the political, historian, the rich variety of rentals offers rewarding study for early-modern social and economic history by revealing patterns of rural settlement (in the enumeration of the 'ferme touns' or agricultural townships on ecclesiastical estates), and as a proportion of the rents and teinds is expressed in kind (in quantities of wheat, barley, meal, malt, oats, rye, butter, cheese, marts, sheep, kids, swine, poultry, geese, salmon, and, for the northern isles, oil), the work sheds further light on the balance between arable and pastoral farming, the predominance of particular crops (and sometimes their prices) and, therefore, regional variations in farming practice. Similarly, the proportion of rent paid in silver also signifies the growing, though still limited, level to which the economy was monetized. The size and staff of religious houses, as well as the monastic economy, are placed on record; and the nexus linking the finances of ecclesiastical institutions and broader economic trends in society is often discernible. Insights, too, are afforded on the distribution of alms for the poor by religious houses and of the contributions made by hospitals and collegiate churches; and the work of schoolmasters, attached to ecclesiastical institutions, is incidentally glimpsed. In a host of ways, the material presented refines and amplifies an appreciation of the particular characteristics and intersecting forces which helped shape Scotland's history not only in the sixteenth century but in the preceding medieval period as a whole.

7 The medieval heritage

A starting point in an assessment of the church's financial structure may be located in the parishes which were the foundations of the entire ecclesiastical system in so far as the 'superstructure' - monasteries, cathedrals, collegiate churches and even university colleges - was sustained from resources originating in the parishes. The appropriation of parochial revenues, during the middle ages, as endowment for the support of monastic institutions, which readily acquired the characteristics of property-owning corporations, could only be damaging to the well-being of the parishes, which, as a consequence, suffered financial starvation. Besides, as the monastery or appropriating body which acquired the parsonage teinds took the place of the parson ('rector' in Latin) but could not fulfil a parson's parochial duties, a vicar, maintained from residual teinds, had to be appointed to serve the cure. Where the vicarage settlement was intended to be enduring, the vicar, so appointed, was recognised as vicar perpetual; he held his benefice with the same security as a parson held his. If a vicar were sustained from the lesser teinds composed of farm produce (other than the parson's corn teinds), from various offerings and from any other revenues of the benefice, he was designated a vicar portionary (or 'vicar portioner'); and if supported by an annual sum of money, or victual, assigned as his stipend from the revenues, he was known as a vicar pensionary (or 'vicar pensioner'). At Kilspindie, in Gowrie, the vicar portionary's income was derived from such 'small teindis as lynt, hemp, leix, sybous, scheis, eggis and utheris small teindis and offerandis wount to be payit of the said vicarage with stirk and stang penny'; but the non-resident vicar did not serve and merely delegated his duties to a curate to whom he paid £13.[146] Vicars, however, might also occur in parishes where a church had not been annexed to a higher institution but where a non-resident parson wished to be relieved of the burden of parochial work by means of a vicarage settlement.[147] Absenteeism, of another variety, arose when a benefice was conferred on a student, perhaps to sustain him at the schools: the parsonage of Dunbar, which yielded 200 merks, was held by John Hamilton, himself a graduate, who was studying abroad as a 'student in Pareis'.[148] In general, the facility enabling a parson or vicar in the middle ages to provide a substitute, who might even be a poorly-paid assistant curate or chaplain of lowly status, was readily exploited.

The process of appropriation could be carried a stage further with the annexation of the vicarage to another ecclesiastical institution which then made provision for service in the parish by appointing a vicar pensionary and assigning him a pension from the vicarage revenues. At Dysart in Fife, whose parsonage and vicarage revenues of 300 merks helped finance the collegiate church of St Mary on the Rock in St Andrews, a vicar pensionary undertook the parochial duties for a salary of 41 merks and a manse at the church.[149] The same collegiate church also appropriated the parsonage and vicarage of Benholm, yielding 260 merks at the Reformation, and paid 20 merks to the vicar pensionary who served the cure.[150] Sometimes a cathedral engrossed both parsonage and vicarage revenues: Aberdeen, for example, laid claim, among other benefices, to the entire fruits of Aberdour and Belhelvie, valued at 220 merks and 260 merks respectively, and paid two vicars pensionary 20 merks each for their work in the parishes.[151] Responsibility for diverting all the resources of a parish might also rest with a university college. In St Andrews, St Mary's College acquired the entire fruits of five benefices;[152] and in Aberdeen, several livings were annexed to King's College, which was variously reported to have 46 or 48 masters and

---

[146] See below, p. 307.
[147] Cf *RSS*, ii, no 1480.
[148] See below, p. 169; cf p. 614.
[149] See below, pp. 70-71.
[150] See below, p. 403.
[151] See below, pp. 424, 442-443.
[152] See below, pp. 64-65.

students 'in divers faculteis and science'.[153] All this meant that the care of individual parishes was delegated either to a vicar pensionary or to a simple curate.[154] Again, when the small priory of Strathfillan acquired the parsonage and vicarage of Killin in Perthshire, a curate was hired to serve the needs of the parish.[155]

In unappropriated parishes, too, similar practices were apt to occur. Neither the parson nor his substitute, the vicar, wished to be burdened with the parochial duties at Slamannan (a free benefice valued at £53 and held by the one man at the Reformation), and so a vicar pensionary had been assigned the work for £8 with manse and glebe (worth an extra £2).[156] At Glass, an independent parish in Moray, the parsonage and vicarage which drew £70 were held by one individual who paid his vicar pensionary 12 merks to serve the parish; and the parson-vicar of Glass was also vicar of neighbouring Kirkmichael, where his reluctance to serve that parish, too, meant employing a curate for the modest outlay of £10.[157] Similarly, in the unattached parish of Baldernock, the incumbent, who held both parsonage and vicarage, delegated his duties to a curate for £16; and he concurrently possessed the vicarages of Lanark and Wiston where his curates were each paid £10 for their services.[158] In mountainous Atholl, too, where the patronage of several unappropriated churches - Blair, Struan, Kilmoveonaig and Lude - lay with the earl of Atholl, the parsons and vicars all devolved their work either on a vicar pensionary or, more usually, on curates.[159] As a whole, the system, by diverting parochial resources from their original purpose of endowing the cure of souls in the parishes, was liable to serious abuse.

As ecclesiastical resources became concentrated at the higher levels, the neglect and poverty of the parishes became manifest. A vicar's stipend, in these circumstances, could hardly be other than meagre. Although efforts had been made to raise the minimum stipend for vicars from 10 merks, free of burdens (originally set in the thirteenth century), to 20 merks in 1549 and to 24 merks (for all but five sees) in 1559,[160] rentals in the 'Books of Assumption' reveal many examples of vicars' stipends which failed even to reach the minimum decreed by the provincial councils of the pre-Reformation church. The vicar of Bonhill in Dunbarton, who had a miserable ten merks (£6 13s 4d), a room and acre of land, appropriately described himself 'the viccar of puire Bullill'.[161] The vicar of nearby Cardross fared only marginally better with an income of £10.[162] In

---

[153] See below, pp. 440-442; 405.
[154] For parishes entrusted to simple curates, see below, pp. 65, 78, 80, 86, 106, 214, 226, 307, 319, 321, 324, 326, 334, 340, 375, 378, 395, 403, 429, 432, 435, 438, 441, 442, 485, 508, 514, 541, 626, 632, 633, 642.
[155] *RSS*, vi, no 578; see below, p. 321.
[156] See below, p. 552.
[157] See below, p. 485.
[158] See below, pp. 507-508.
[159] See below, pp. 318-319.
[160] *Statutes of the Scottish Church, 1225-1559*, ed. D. Patrick (Edinburgh, 1907), pp. 11-12, 112, 169.
[161] See below, p. 541.
[162] See below, p. 542.

prosperous Lothian and the Merse (though vulnerable to English invasion), the vicar of Binnie could expect a pitiful £5; his colleague in Abercorn, £9; at Whitekirk the vicarage was leased for an income of £10; Ratho also yielded £10; Musselburgh, £13; and the vicarages of Kilspindie and Currie were each valued at 20 merks.[163] Matters were no better in Dunblane where the parsonage belonged to the bishop, and the vicarage, worth 320 merks, formed the dean's prebend, which left the vicar pensionary who had the cure of souls with a derisory £8.[164] These exemplify the worst instances, but a system which encouraged glaring inequalities merely contributed to the poverty of parishes.[165]

Though a majority of appropriations favoured the religious houses, bishops and cathedral establishments were prominent among the beneficiaries. In the course of the middle ages, some parishes had been assigned to the mensa of a bishop (who was expected to exercise charity and provide hospitality to poor people at his table); others were allocated to maintain the many prebends of cathedrals, collegiate kirks and the Chapel Royal for the performance of divine service. Cathedral establishments were financed, in the main, through the concentration of parochial benefices held by the canons either individually or in common. Part of the revenues might form the prebends of individual canons or they might form a fund from 'common kirks' belonging corporately to the chapter, which served as the cathedral's electoral and governing body. Sometimes, too, a prebend might be founded on a pension arising from rents derived from land. At Dunkeld, the chapter's gross income of over £230 was acquired from annexed benefices, pensions and the duties accruing from lands which had been feued;[166] and the revenues of Caputh, a prebend of the cathedral, were derived from the lands of certain 'touns' or agricultural townships, feued to tenants and others for £48, from which £5 was deducted for the services of a 'stallar' or vicar choral in the cathedral.[167]

The cathedral dignitaries, whose offices figure in the pages of the rentals, comprised the dean (or 'provost' in Orkney[168] from 1544), the administrative head of the cathedral and its clergy, president of the chapter and vicar general during episcopal vacancies; the chanter (or 'precentor'), in charge of music and of the song school attached to the cathedral and president of the chapter were the dean's office vacant; the chancellor, who was secretary, keeper of the seal, librarian and superintendent of the grammar school; the treasurer, who had custody of the relics, ornaments and vestments; and, where appropriate, the three minor dignitaries of subdean, subchanter and sacrist who deputed for the dean, chanter and treasurer. The archdeacon, who was also a canon, was not so much a

---

[163] See below, pp. 153; 170; 123; 120; 110.
[164] See below, p. 300.
[165] For examples of vicars' incomes less than 24 merks, see below, pp. 68, 69, 76, 79, 83, 84, 90, 99, 110, 118, 120, 123, 153, 170, 174, 192, 193, 195, 213, 250, 265, 266, 300, 310, 312, 318, 326, 329, 335, 338, 339, 344, 390, 395, 397, 399, 407, 420, 424, 425, 428, 429, 432-433, 457, 465, 485, 489, 493, 513, 518, 519, 532, 541, 542, 552, 553, 605, 606, 611, 618, 636, 639, 640, 644.
[166] See below, pp. 300-301.
[167] See below, p. 318.
[168] See below, p. 659.

cathedral dignitary, for his work was diocesan and lay in supervising the parish clergy within his jurisdiction; and, at a lower level, similar duties were undertaken by 'deans of Christianity' (or rural deans), Christianity denoting the district over which the dean had jurisdiction. The only dean of Christianity, however, who figures in the rentals is the dean of the Christianity of Angus,[169] though fleeting allusions to the 'dean' testify to the activities of the dean of the Christianity of Fife and of Gowrie, and to his counterpart within the dioceses of Glasgow and Galloway.[170]

The emoluments enjoyed by the dignitaries could be considerable. In Glasgow, the dean of the cathedral could expect an income of about £350 in cash plus assorted victual (composed of meal, oats and capons) which might fetch upwards of £50 at market.[171] The dean of Aberdeen could claim a competence of over £330 in silver and a substantial return in kind whose estimated value might reach more than £200.[172] His counterpart in Dunkeld declared an income of £257 in cash and some victual (worth perhaps £18), together with a pension of £133.[173] Elsewhere, the cash element in a dean's income was less significant: in Dunblane, the dean's income, in all, was £63; and the deans of Caithness and Brechin had each about £35 in silver supplemented by a modest amount of victual.[174] A chanter could gain just over £100 at Dornoch, £173 in Dunkeld and £66 with 8 chalders' victual (worth at least £100) in Fortrose.[175] A chancellor's gross earnings might range from just over £270 in Ross, and £260 in Aberdeen, to around £70, plus victual, in Caithness and Dunkeld.[176] The treasurers of Glasgow and Aberdeen each received £200; their counterparts at Fortrose and Brechin had £132 and £76 respectively; and, though accustomed to a return of £66 or thereabouts, the treasurer of Dunkeld, under the 'new order', had to make do with a mere £24.[177] Deductions, of course, were forthcoming from the gross incomes of the principales personae. These might take the form of payments to sustain chaplains of the choir and vicars pensionary in the parishes, annexed to form their prebends, or the customary 'burdens' on benefices associated with visitations but such items of expenditure were generally none too taxing.[178]

The archdeacon, though a diocesan official, had his cathedral prebend whose annual gross value might amount to £600 for the archdeacon of St Andrews; over £220 for the archdeacon of Teviotdale but, surprisingly, only £200 for his colleague in Glasgow; £240 for the archdeacon of Shetland; £213 for the archdeacon of Dunblane; £176 for the archdeacon of Brechin; and about £150 for

---

[169] See below, p. 375.
[170] See below, pp. 5, 81, 321, 620; cf pp. 66-67.
[171] See below, p. 504.
[172] See below, p. 424.
[173] See below, p. 301.
[174] See below, pp. 315; 630; 364-365.
[175] See below, pp. 630; 307-308; 642.
[176] See below, pp. 634-635; 433 (but cf p. 428); 630; 308.
[177] See below, pp. 510; 444-445; 637; 377; 311.
[178] See below, for example, pp. 307-308, 635.

the archdeacon of Moray.[179] Each diocese possessed a single archdeacon, except the archdioceses of St Andrews and Glasgow, which, on account of their size, each had two (St Andrews and Lothian; and Glasgow and Teviotdale respectively), and Orkney which also had two, on account of the diocese's two groups of islands, separated by fifty miles of sea.

Among the minor dignitaries, the subdean of Ross had £200; in Brechin he had £46; in Dunkeld, £33; and in Dunblane a mere 10 merks (£6 13s 4d).[180] The subchanter of Glasgow gained £133 in gross income; his counterpart in Aberdeen received £50.[181] All had similar burdens on their earnings. The 'common kirks' which helped support the cathedral establishment of canons were capable of yielding substantial sums (until diverted for the queen's use). In Aberdeen, they furnished over £800; in Glasgow, the figure was £640; and in Dunkeld, £230.[182] Again, there were expenses to be met: for Dunkeld, these included payments for riding the teinds (at harvest for assessment and collection), for carriage and porterage, as the crops were removed and transported; fees for the collector who gathered the revenues accruing, for the clerk and notary of the chapter, for the beadle and other servants, and for the boatmen on the River Isla.[183]

With the increase in canonries over the years, Aberdeen gained some 30 prebends funded from 46 churches; Glasgow had about 32 and Moray 25. Despite the manses and crofts assigned to them in the vicinity of a cathedral enabling them to take part in services, canons readily devolved their duties on deputies known as 'stallars' or vicars choral or chaplains of the choir (who might be priests or deacons), who might hold their property in common, often also held chaplainries in the cathedral and came to have separate and necessary functions in the services. In Glasgow cathedral, ten vicars choral held the benefice of Dalziel in common; in Dunblane, nine chaplains of the choir held revenue in common from the fruits of several vicarages and 'annuals' (or annual payments from land or property).[184] Similarly, the parsonage and vicarage revenues of Abernyte were assigned at £20 a head to four chaplains of the choir in Dunkeld cathedral, with 12 merks left aside for a vicar pensionary in the parish; a canon who held the prebend of Craigie earmarked 12 merks from his income of 45 merks for a stallar in the cathedral; and, again, the treasurer sustained a stallar whose stipend was set at £10.[185] From the teind sheaves, yielding £100 merks, of the united parsonage of Skirdustan and Botriphnie, erected as a prebend of Elgin cathedral, 10 merks

---

[179] See below, pp. 67; 214-215, 251, 253; 665; 300; 380; 487, 494.
[180] See below, pp. 651; 394; 341; 314.
[181] See below, pp. 272-273; 426, cf p. 434.
[182] See below, pp. 437-438; 499; 300.
[183] See below, p. 301.
[184] See below, pp. 519; 343.
[185] See below, pp. 310; 312; 311.

were allocated to a chaplain, as his 'stall fee' in the cathedral.[186] Other canons and cathedral dignitaries did likewise.[187]

Canons were accustomed also to appoint procurators to take their place at chapter meetings. Nor had they any responsibilities in the appropriated parishes, where the work was undertaken by their parochial vicars. The benefice of Muckersie, a prebend of Dunkeld cathedral, leased for £100, enabled the canon to set aside £10 for a vicar pensionary to undertake the cure of souls in the parish, and to maintain a 'stallar' in the cathedral for a further six merks.[188] Not surprisingly, as canonries were without cure of souls and were considered compatible with the holding of other offices, they proved attractive to non-resident careerists and to unqualified men.[189] In most cathedrals, the chapter consisted of secular canons but in two sees, St Andrews and Galloway, the chapter was composed of a priory of canons regular, Augustinian canons in St Andrews, and Premonstratensian canons at Whithorn.

The fashion in the fifteenth and earlier sixteenth centuries for erecting collegiate churches, where priests with choristers said masses for the souls of the founder and his family, attracted significant endowment, again from the parishes. This late development gave rise to new churches such as Dunglass, Guthrie, Roslin, Trinity College and the kirk of St Mary of the Fields, both in Edinburgh, and to existing parish churches converted to collegiate status, like Bothwell, Haddington, Seton, St Nicholas' in Aberdeen, St Giles' in Edinburgh and the church of the Holy Rude in Stirling. In all, some 40 collegiate kirks were so established,[190] where a high standard in services for worship was observed. Their constitutions resembled those of cathedrals. Each college formed a corporation, with a provost (or a dean, as at Restalrig,[191] or with the vicar as president, at the Holy Rood in Stirling), a number of prebendaries, choristers and sometimes bedesmen or almsmen. Again, the members of the collegiate body were sustained by the appropriation of parish revenues or by revenues derived from land. The prebends were held either individually or communally.

At Lincluden in Nithsdale, the provostry commanded a significant income of over £420 in silver plus victual.[192] Other well-endowed collegiate kirks included Dunbar, Methven, Restalrig, Trinity College in Edinburgh and St Mary on the Rock in St Andrews.[193] The provost of Trinity College (whose revenues as an institution amounted to £362) had an income of £160 (though, as part was

---

[186] See below, p. 320.
[187] See below, for example, pp. 307-308, 311, 312, 318, 320, 322, 325, 326, 341, 375, 378, 391, 426, 429, 430, 431, 432, 433, 439, 440, 470, 474, 483, 484, 498, 503, 519, 580, 627, 633, 635, 637, 638, 641, 642, 644.
[188] See below, p. 326.
[189] See below, for example, pp. 159, 236, 366.
[190] See below, pp. 71-72, 77, 80, 82, 86, 97, 105, 106-107, 109, 111-113, 115-116, 118, 119-120, 121, 122, 123-125, 130, 134, 139, 143-144, 166, 171, 172, 173, 175, 176, 186, 273, 296-299, 322, 338, 340, 390-391, 394, 511, 515, 516, 519, 521, 523-524, 586, 613, 615, 620-621, 640.
[191] See below, pp. 116, 141.
[192] See below, p. 613; cf p. 620.
[193] See below, pp. 171-172, 175; 296-299; 116, 141-144; 111, 118-119, 175; 55, 56, 71, 77, 80, 86, 87.

INTRODUCTION   xxxvii

derived from the parish of Lempitlaw in the Borders where collection proved difficult, he could not be assured of gaining his full earnings); the dean of Restalrig received £93 plus victual; the small foundation at Bothans in East Lothian provided its provost with £100, though the provostry of the larger establishment at Crail yielded just the same sum in income.[194] The provost of Dunglass in East Lothian received £82, which might be reckoned a reasonable competence for a professional man.[195]

Some collegiate kirks, however, had rather less to offer as financial reward. In Edinburgh, the provostry of St Mary of the Fields' yielded some £48, largely from offerings, small teinds and from the manse which was leased, an income which was liable to fall 'be ressoun of the dimolitioun of the papistrie'.[196] The provost of Guthrie in Angus could expect around £47 but his counterpart at Dirlton in East Lothian had to content himself with just £20.[197] The value of individual prebends might range from £133 at Dunbar to £6 at Lincluden and at Bothans.[198] Some prebendaries financed priests, vicars choral and choristers from their earnings.[199] The song schools and almshouses attached to collegiate kirks had also to be maintained. In Edinburgh, the prebend of St Vincent in St Mary of the Fields' collegiate kirk was assigned to a teacher in the song school; and Lincluden made provision for twenty-four bedesmen and a chaplain, as well as the provost and eight prebendaries.[200] From the detailed rentals[201] for some of the prebends, a fuller picture emerges of the nature of the revenues accruing and of their varied sources. As well as the fruits from appropriated benefices, collegiate kirks derived income from rents for lands and property, feu duties, mill dues, grassums and offerings. For most of the collegiate churches, income took the form of cash, not kind, as teinds and property were leased for money payments.

In constitution and in financial arrangements, the Chapel Royal also bore the characteristics of a collegiate church, with endowment for a dean, subdean, sacrist, canons and choristers. Among the prebends of the Chapel Royal were Crieff primo and secundo. Coylton sustained two canons of the Chapel Royal; Kirkinner and Kirkcowan were held by the subdean and sacristan; and Kells belonged to a stallar in the Chapel.[202]

Another development of the fifteenth and sixteenth centuries was the foundation of university colleges whose maintenance was met at the expense of the parishes. In St Andrews, St Mary's College derived at least £575 plus victual from the revenues of appropriated churches and altarages; and in Aberdeen, the yield for King's College was upwards of £300, plus victual, from annexed

---

[194] See below, pp. 111, 118; 116; 172; 82.
[195] See below, p. 186.
[196] See below, p. 105; cf pp. 106-107, 119-120.
[197] See below, pp. 390, 394; 173.
[198] See below, pp. 171; 615; 122.
[199] See below, pp. 109, 108, 322, 340.
[200] See below, pp. 124; 613; 620-621.
[201] See below, for example, pp. 296-299.
[202] See below, pp. 323, 560, 602, 603.

parishes.²⁰³ In all, by the later middle ages, some 880 parish churches (approximately 86 per cent) had their parsonage revenues annexed to another ecclesiastical institution, and over half of the appropriated parishes (at least 56 per cent) had their vicarage revenues similarly diverted.²⁰⁴

Associated with activities in cathedrals, collegiate churches, university colleges, parochial, and other, churches, chapels, hospitals and fraternities was the humble chaplain, who might be hired or fired and on whose shoulders so much of the church's work was placed. The proliferation of perpetual chaplainries and altarages in great cathedrals,²⁰⁵ affluent collegiate kirks²⁰⁶ and well-to-do burgh churches,²⁰⁷ endowed by patrons for the multiplication of masses, swelled the ranks of chaplains, many of whom gained a measure of security in tenure. Service at an altar was not normally maintained from a church's endowment but from revenues accumulating from annual rents assigned at the foundation of a chaplainry or altarage from lands and other property.²⁰⁸ Their value might vary in scale from six merks at St Mungo's altar in Currie to £60 for a chaplainry at Crieff.²⁰⁹ In Edinburgh, the Barras' chaplainry, worth 20 merks, was financed by customs from the tron.²¹⁰

Most chaplainries provided only a slender income; and chaplains were sometimes apt to acquire several chaplainries to increase their earnings.²¹¹ The evidence for the numerous chaplainries in Glasgow cathedral is particularly striking.²¹² There, revenues for two altars (together with an altar in Culross abbey kirk) were forthcoming from rents derived from lands so far away as Perthshire which had been annexed and mortified to the altars.²¹³ In the cathedrals of Elgin, Dunkeld and Dunblane, a proportion of a chaplain's income was sometimes still assigned in kind.²¹⁴ University colleges were also repositories of endowments for chaplainries and altarages.²¹⁵ Again, a chaplainry might be attached to a hospital: in Stirling, the chaplainry of St Thomas was associated with the almshouse.²¹⁶

Although the services of chaplains in cathedrals, collegiate kirks and burgh churches could hardly be other than conspicuous, numerous rural

---

²⁰³ See below, pp. 64-65, 364; 405, 425, 441-442.
²⁰⁴ I. B. Cowan, *The Parishes of Medieval Scotland* (Edinburgh, 1967), p. v.
²⁰⁵ See below, pp. 314, 320, 323, 324, 327, 328, 335, 379, 400, 476, 477, 478, 481, 490, 500, 501, 502, 515, 522, 523, 535, 615, 663, 664.
²⁰⁶ See below, pp. 95, 96, 117, 123-126, 129-141, 143-144, 161, 175-176, 254, 545-546, 554-556, 647.
²⁰⁷ See below, pp. 78, 81, 84, 85, 128-129, 154, 306, 320, 377, 380, 381, 395, 399, 401, 521, 535, 540, 551, 554-557, 560, 571, 636.
²⁰⁸ See below, pp. 125-126, 132-133, 135-137, 154-155, 545-546, 560, 571.
²⁰⁹ See below, pp. 139; 328.
²¹⁰ See below, p. 96.
²¹¹ See below, pp. 95; 117-118, 123; 128-129; 134; 149-150; 254; 273; 314, 328; 334, 501, 502; 377, 395, 408; 476; 477-478; 500, 502; 521; 551.
²¹² See below, pp. 501, 502, 515, 522, 523, 525, 615.
²¹³ See below, p. 522.
²¹⁴ See below, pp. 490; 323, 336; 327.
²¹⁵ See below, pp. 65, 96.
²¹⁶ See below, p. 551.

parishes,[217] non-parochial churches and chapels in towns and in the countryside[218] were attended by one or several chaplains. In the Mearns, Garvock kirk possessed St Andrew's chaplainry, worth 40 merks a year; at Newburgh in Fife, the chaplainry of St Catherine had more the complexion of a benefice, for besides £16 in revenue it possessed a manse and yard as well; Dalmeny had chaplains serving the altars of Our Lady and St Cuthbert; at Linton in East Lothian, a chaplainry's income was reduced 'be ressoun of the burnyng of Hadingtoun', and halved when payment was made 'to the substitute' who performed the duties in place of the holder.[219] Elsewhere a chaplainry might belong to 'ane auld blind man that seis nocht perfytlie'; or, again, five merks from the revenues might be 'evir payit to the pure folk and nocht to the cheplan'.[220]

The country's parishes, which contributed so much to sustaining the whole financial edifice, were distributed among two archdioceses (St Andrews and Glasgow) and eleven dioceses of unequal size and irregular boundaries. St Andrews, the most populous and wealthiest, with over 270 parishes at the Reformation, covered much of the east coast from the Tweed to the Dee, embracing most of Berwickshire, Lothian and Fife, extending along the River Forth past Stirling, and up the River Tay to Scone and beyond into Angus and the Mearns. Intersecting it, in the north, was the small diocese of Brechin with over two-dozen parishes; and entangled with it lay detached parishes belonging to the dioceses of Dunkeld and Dunblane.[221]

The heartland of Dunkeld diocese, with almost 70 parishes, was highland, stretching from Atholl to Breadalbane. To its immediate south lay Dunblane, a small diocese with just over three-dozen parishes. In the north east, Aberdeen, with over 100 parishes, embraced Buchan, Garioch, Strathbogie and mountainous Mar. Neighbouring Moray, with some 75 parishes, covered the coastal plain south to Strathspey and Badenoch as well as territory to the west of Loch Ness. Further west lay Ross, with its cathedral seat at Fortrose on the Black Isle administering three-dozen parishes, whose bounds extended northward to the southern shore of the Dornoch firth and across country to the west coast from Coigach and the Summer Isles south to Kintail. In the far north, the twenty or so parishes in the diocese of Caithness included highland Sutherland as well as lowland Caithness.

Orkney diocese, whose parishes were often united to form 30 or so cures, contained both Orkney and Shetland, each with its own archdeaconry; and the far-flung diocese of the Isles, straddling the Hebrides from the Butt of Lewis to

---

[217] See below, pp. 65, 82, 139, 149-150, 154-155, 174-175, 273, 314, 325, 328, 329, 337, 377-378, 475, 525, 526, 542.
[218] See below, pp. 65, 85, 96, 97, 108, 120, 157, 173, 187, 253-254, 324, 345, 408, 433, 477, 481, 482, 484, 501, 503, 514, 520, 521, 523, 524, 540, 546, 550, 551, 553, 639, 644, 646, 649.
[219] See below, pp. 82; 378; 149-150, 154-155; 171.
[220] See below, pp. 81, 138.
[221] Aberlady, Inchcolm, Cramond and Abercorn, for example, were among Dunkeld's detached parishes; and Abernethy and Culross were detached parishes belonging to Dunblane. See the ecclesiastical map at the end of the volume.

Kilmory on Arran, had about 47 parishes, which yielded the least in revenue.[222] On the mainland, adjacent Argyll, with around 44 parishes, almost matched the Isles in number; and its territory spanned from Glenelg and Knoydart to Kintyre and from Glengarry to Bute. The large archdiocese of Glasgow, with more than 230 parishes, covered territory from Loch Lomond south eastwards to Sprouston on the Tweed near Kelso. In the extreme south west lay the small diocese of Galloway, with less than fifty parishes, reaching from Loch Ryan through Glentrool and Dalry to Sweetheart or New Abbey.

8 Parochial resources

Throughout the later middle ages, one of the main characteristics of the whole ecclesiastical structure had been the impoverishment of the parishes through the constant drain of revenues from the parochial ministry and church to support other institutions. The responsibility, of course, for paying the teinds, which were so readily diverted, rested on the parishioners. In a land with just over 1,000 parishes and a population of no more than 800,000, a parish on average would contain fewer than 800 inhabitants, or 150 households whose heads were accountable for the various dues and exactions demanded by the pre-Reformation church. In the small coastal parish of Duffus, near Elgin, so few as twenty tenants contributed the parsonage teinds;[223] and at Airth, in Stirlingshire, parishioners making the customary offerings at the obligatory Easter communion could number six dozen or so.[224] The enormous variety in the size of parishes from the inordinately large highland parishes to the compact parishes so characteristic of the east-coast lowlands meant that some rural parishes might have fewer than two or three hundred inhabitants, and merely a few dozen households. Again, most towns and burghs were each contained within a single parish. Urban parishes like Edinburgh with over 10,000 inhabitants, or New Aberdeen, Dundee and Perth, each with perhaps 5,000 parishioners, could prove crowded, even congested, but the population of most other urban parishes ranged from 3,000 to 300 people.

The church was legally entitled to the teinds of the produce of the country's arable land, pasture, rivers and sea, and it exacted offerings, fines and fees from all who required its offices. The predial teinds derived from all that nature renewed from year to year included corn and other crops, hay, lint, young animals, wool, milk, butter, cheese, fowls, eggs, peat, wood, fishing and hawking, and were divided into the greater or garbal teinds, or corn teinds, and the lesser teinds of other farm produce, which often were commuted into a money payment. In the coastal parish of Anstruther, the vicarage teinds were derived from 'salt,

---

[222] See below, Appendix I, pp. 673-675.
[223] See below, p. 468.
[224] See below, pp. 157-158.

wool, flax, hemp, with fresh and dried fish and herring', and were rendered as a money payment.[225]

Besides predial teinds, there also existed personal teinds, the fruits of labour, or wages, for which offerings may have been substituted. Despite the lack of evidence for their exaction in Scotland, the presumption is that, in some form, they were exacted: the rental of Collessie vicarage, at any rate, reveals that with the Reformation 'Easter dues, personal teinds, corpse offerings, kirk cows and uppermost cloths' were no longer paid,[226] from which claim it is not unreasonable to deduce that until the Reformation personal teinds were still a reality. A third variety of teinds consisted of second teinds, the tenths of rents, and of fines, escheats and feudal casualties, which may have proved difficult to exact. Nonetheless, 'second teinds' formed a recognisable element in the income of the bishop of Aberdeen and of the commendator of Scone abbey.[227]

The various obligatory offerings, oblations or obventions, which were also reckoned as spirituality, could form a significant element in a priest's income. For the vicar of Aberdour in Fife, £20 in dues accruing from 'pasche reckonings, kirk cow and uppermost cloth' formed the major portion of his income; the remainder, derived from the lesser teinds of 'hemp, lint, stirks, grass and geese', amounted merely to £10, for the abbot of the appropriating house, Inchcolm, had laid claim to the wool, lamb and fish teinds of the parish.[228] Offerings were made at weddings, churching of women, baptism, and Easter communion. A surviving record of the Easter offerings or 'pasche reckonings' for Airth, extracted from the vicar's 'pasche book' and presented in his rental as evidence of his income, must be exceedingly rare if not unique; in all 75 individuals, male and female, in the parish presented the vicar with a variety of offerings - lambs, cattle, lint and hemp.[229]

Besides these 'offerings', originally voluntary but through time compulsory, a vicar's income was augmented by 'pasche fines' levied on non-communicants at Easter, and by burdensome mortuary or burial dues, imposed at a parishioner's death, in theory for unpaid teinds. These took the form of 'the corspe present', usually a cow or the uppermost cloth on the bed or best garment of the deceased.[230] Only with the Reformation were these clerical exactions discharged, as the *Book of Discipline* had recommended, by an act of the privy council (whose text has not survived).[231] The consequent loss of income for beneficed men, subjected to the added impost of the 'thirds', could prove acute:

---

[225] See below, p. 22.
[226] See below, p. 81.
[227] See below, pp. 416; 331.
[228] See below, p. 84; cf p. 64.
[229] See below, pp. 157-158; cf. p. 479.
[230] See below, for example, pp. 76, 78, 79, 81, 82, 84, 88, 90, 99, 105, 106, 111, 115, 144, 152, 153, 154, 172, 191, 193, 227, 248, 251, 265, 266, 273, 311, 316, 330, 331, 339, 341, 373, 375, 380, 381, 391, 398, 399, 403, 407, 424, 425, 428, 432, 435, 443, 450, 478, 479, 481, 483, 489, 503, 515, 518, 520, 532, 541, 550, 560, 580-582, 587, 588, 597, 601, 604, 605, 612, 614, 616, 617, 620, 638, 640, 668.
[231] *First Book of Discipline*, pp. 157-158; cf below, pp. 248, 373.

the vicar of Bathgate, whose income consisted of 'wool, lamb, hay, corpse presents, uppermost cloths and Pasche pennies', leased for a cash payment of 50 merks a year, discovered that, with the abolition of these exactions at the Reformation, his 'principal profit and commodity' had disappeared, with no payments forthcoming for the last three years.[232] At Abercorn, the vicar found his income, with the loss of this source, cut by a quarter to a meagre £9 6s 8d; in Peebles the vicar's proceeds dropped by a third; and in Annandale, the parson of Mousewald's earnings fell from a modest 40 merks to 30 merks (or £20).[233]

Yet the suppression of these dues could hardly have come as an unforeseen shock. Nor could they have taken men wholly by surprise. After all, as early as 1536, James V had proposed the abolition of mortuaries and had advocated a radical reform of the teind system in order that 'every man should have his own teinds'.[234] In all of this, the reformers in their *Book of Discipline* were simply borrowing and giving effect to the ideas of James V, who had been no friend to protestant reform.

Unlike these assorted dues levied ostensibly for parochial services rendered, parishioners were also subjected to papal taxation, 'Peter's pence', nominally an imposition of one penny from each house but probably commuted for much less. Nor was this all. Parishioners were obliged not only to contribute to maintaining the fabric of the church building, but to bear the greater share of the burden, as the parson's responsibility extended only to repairing the chancel (or choir). The parson of Kirknewton, accordingly, assigned 40 shillings 'for pointing and gleissing of ane queir and kirk yeirlie'.[235] Where parsonage revenues were diverted to a cathedral, a parson's first allegiance was sometimes to the cathedral. At Moy, a prebend of Elgin cathedral, the parson, himself a canon, assigned 10 merks at the Reformation 'for thekin of the queir of Murray and uther reformatioun within the samyn'; and the parson of Rhynie, another prebend of Elgin, also gave £8 for thatching the cathedral's chancel.[236]

There also existed a series of 'burdens' on benefices which directly affected parsons and vicars (though sometimes vicarages were declared free[237]). These burdens, which ceased at the Reformation, formed the 'internal taxation' of the church, and consisted of 'synodals', yearly payments to the bishop from each church; 'visitations' and 'procurations', sums paid as commutation for the duty of entertaining the bishop, archdeacon and dean at their visitations; and the 'subsidium charitativum', sometimes exacted from the clergy by a bishop on his promotion to a see. The amounts paid in procurations and synodals ranged from five merks (£3 6s 8d) for Muckersie, Luss and Dolphinton, six merks for Kinclaven, £2 for Benvie and £4 to the archbishop of Glasgow for the vicarage of

---

[232] See below, p. 152.
[233] See below, pp. 154, 251, 265.
[234] *Concilia Scotiae: Ecclesiae Scoticanae Statuta tam provincialia quam synodalia quae supersunt, MCCXXV-MDLIX*, ed. J. Robertson (Edinburgh, 1866), pp. cxxxvi-cxxxvii.
[235] See below, p. 104.
[236] See below, pp. 470, 486.
[237] See below, pp. 81, 429, where the vicar of Collessie claimed his vicarage was a non-taxable benefice, and the parson of Tyrie pronounced his living was 'nevir taxit be kynge nor bischope'.

Dreghorn. The recipients - bishops, deans and archdeacons - could therefore derive a significant annual income from this source until its abolition at the Reformation. The charitable subsidy, by contrast, could be levied once only in each episcopate. Sometimes, the burden of payment fell not on the benefice-holder but on the lease-holder: the parson of Menmuir, on leasing his benefice in 1550 for nineteen years, took the precaution of requiring the lease-holders to pay any charitable subsidy exacted by the bishop of Dunkeld as well as synodals, procurations and expenses incurred by visitations.[238]

That apart, a vicar was usually responsible for maintaining his manse which he was expected to make available to the bishop and archdeacon on their visitations. Again, arising from the clergy's bargain with James V for the extinction of the 'great tax' of 1536, payments for maintaining the senators of the College of Justice - the lords of Session - were met from an annual levy on the fruits of benefices in the patronage of prelates.[239] Generally, prelates recorded their contributions to the College of Justice in their rentals without specifying the parochial benefices from which these funds were drawn;[240] but the vicar of Carstairs, whose living lay in the patronage of the abbot of Coupar Angus, did place on record his contribution of 22 shillings annually towards meeting the judges' salaries.[241] Here was a further drain on certain parochial resources.

The economic problems of the parochial clergy were apt to be associated with clerical poverty. Serious underpayment led priests to augment their income by a variety of means. One way of making ends meet was to accumulate several livings. Alexander Pedder in Ross acquired the vicarage of Avoch on the Black Isle and that of Urray near Contin, whose combined income yielded £26; Hercules Carnegie held the vicarages of Ruthven and Kingussie, together worth about £42; George Dunbar, as parson of Kilmuir-Easter and vicar of Rosemarkie, was able to draw £86; and Robert Auchmuty was not only vicar of Dun and of Arbroath but also vicar of Stirling and chaplain at Bannockburn, with an income of £100, or so.[242] Careerists were able to exploit pluralism for financial advantage. Arthur Telfer collected the parsonages of Crimond and Inchture and the vicarage of Aboyne; George Hay, a connoisseur of benefices, held the parsonages of Eddleston, Kingussie, Rathven, Dundurcas and Kiltarlity; James Strachan was parson of Fettercairn, Belhelvie, Botary and Elchies; Henry Abercrombie, prior of Scone abbey, was parson of Kildonan and vicar of Logierait; John Stevenson, chanter of Glasgow, included among his string of livings the parsonage and vicarage of Muckersie and the vicarage of Mochrum; and John Thornton, elder, a notorious benefice-monger, picked up the chantory of Moray, the subdeanery of Ross, and the parsonages of Ancrum and Forteviot, while John Thornton, younger, acquired the parsonage of Advie and Cromdale

---

[238] See below, pp. 3, 4, 6, 66, 81, 307, 321, 326, 334, 374-375, 382, 394, 416, 504, 541, 581, 620.
[239] Hannay, *College of Justice*, p. 58; *Acts of the Lords of Council*, p. 498.
[240] See below, pp. 6, 39, 58, 92, 103, 149, 163, 182, 200, 217, 218, 221, 228, 291, 304, 332, 356, 360, 370, 409, 463, 466, 529, 538, 545, 558, 563, 574, 578, 591, 595, 626, 629, 634, 650, 671.
[241] See below, p. 509.
[242] See below, pp. 640; 428, 492; 643; 381, 396, 550.

(forming part of the chantory at Elgin) and the vicarage of Aberchirder.[243] In some cases, it proved possible for vicarages to remain in the possession of a family and even to pass from father to son (if legitimised) or from uncle to nephew. John Thornton, elder and younger, and James Thornton in the north certainly showed what could be achieved in keeping the fruits of benefices firmly within their family.

The temptation for priests to supplement their income by dilapidating their benefices for short-term advantage was conspicuous. Not only was it customary for the teinds to be leased to a farmer,[244] thus freeing the benefice-holder from the burden of collection, but the houses and lands - the manses and glebes - assigned to parsons and vicars were also liable to be set in tack for a rent in silver.[245] The need to secure a living wage, when faced with the effects of taxation and inflation, or again a wish to make provision for a kinsman by conferring on him a lease or pension were contributory factors in the resort by clergy to dilapidating their benefices. Unease, as the storm clouds gathered, at the approaching Reformation, of which few clerics could have been wholly ignorant, provided an added incentive for beneficed men to make what profit they could while they still could.

Where livings were leased for silver, the holders were dependent on money payments which might satisfy immediate needs but which, for the duration of the lease, would not increase with any rise in the cost of living. But where they retained their glebes for their own use or received rents in kind, beneficed men were better placed in the longer term to take advantage from any extra demand for agricultural produce and to be cushioned from the impact of rising prices. The size of parsons' glebes was usually more extensive than that of vicars' glebes and therefore capable of producing a greater yield or, alternatively, a higher rent. Similarly, the corn teinds, nominally in the possession of parsons but normally leased to laymen, were generally of greater value, so long as grain prices held, than the vicars' lesser teinds derived from produce other than grain. The real beneficiaries, however, were not the clergy but those laymen who acquired the leases. In areas where farmers, on balance, concentrated on arable rather than pastoral farming, the vicar (unlike the parson) might be offered a lower rent for his less attractive vicarage teinds from a prospective lease-holder than his counterpart in an upland pastoral area might be able to command. This, at any rate, might be one of several factors at work (such as a parish's size and its population) in accounting for the number of exceptionally low vicarage rentals in one of the most prosperous areas of the country, Lothian and the Merse, renowned for its fertile cornfields.

Another strategy which some priests adopted in a bid to improve their basic earnings was to engage in secular occupations, a practice most recently

---

[243] See below, pp. 317, 431, 440; 248, 456, 487; 407, 442, 489; 332, 335, 631; 277-278, 326, 340, 499, 597; 159, 479; 480-481.
[244] See below, for example, pp. 3, 16, 22, 26-27, 62, 63, 84, 106, 203, 204, 213, 260, 305, 307, 362, 368, 376, 464, 577-578.
[245] See below, for example, pp. 86, 88, 99, 104, 105, 119, 174, 188, 191, 272, 305, 338, 376, 396, 397, 505, 513, 571, 614.

denounced by the provincial councils of 1549 and 1559[246] but still adopted, with some purpose, by clerics like William McDowall, who was not only the queen's master of works but also chaplain of St Giles' in Edinburgh and master of the hospital of St Paul's Work in Leith.[247]

Again, economic difficulties were heightened with the Reformation, when parsons and vicars were faced with conscientious objection by parishioners who refused to pay any further dues for services which they did not receive and which, indeed, they had been prohibited from receiving by act of parliament in 1560. The frequent complaint in the rentals of beneficed men was that they had been unable to secure their regular income, usually for the three years preceding the submission of their rentals in 1562.[248] The action of parishioners and others in withholding revenues seems to have been widespread, though it could hardly be construed merely as illegal 'civil disobedience', for in 1560 parliament itself was reported to have authorised all possessors of teinds to retain payment in their hands until the privy council issued further instructions.[249] What is also clear (from the resumption of the council's register after a loss of the record from 1554 to 1561) is the council's decision in 1562 that no one should obey any title to benefices preceding 1 March or make payments, pending a settlement of the 'thirds'.[250] Similarly, the abolition at the Reformation of clerical exactions meant a further loss in earnings for existing clergy.

The persistent refusal by critical - or hypocritical - laymen to render payments not only gained a measure of official countenance at the Reformation but was closely linked, in many instances, to the inroads which protestantism had clearly made in local communities. The vicar of Perth and Monimail explained how he had failed to obtain his rents 'for the past term, and that with a quarter part of the year already gone, by reason of the tumult' occasioned by the Reformation; in Fife, the vicar of Kilconquhar contrasted the revenues he received 'befoir the rysing of the Congregatioun' with the non-payment of dues by parishioners thereafter; the vicar of Kinghorn Easter remarked how 'nothing had been received for the past term by reason of the tumult'; and the vicar of Dysart attributed his failure to secure revenues to 'sic inobedience as hes bene thir divers yeris'.[251] In Lothian, the vicar of Borthwick encountered similar difficulties for which he saw no remedy unless 'the parochineris wer examinit thairon in thair conscience'; the vicar of Linton, in East Lothian, reflected on his finances 'befoir the tyme of trubill'; and the story was the same in Dundee, where the vicar complained that 'befoir the rysing of the Congregatioun', he had obtained his dues but since had

---

[246] *Statutes of the Scottish Church*, pp. 15-16, 36, 65, 71, 92, 166.
[247] See below, p. 95, and n. 18.
[248] See below, pp. 56, 67, 70, 74, 77, 79, 80, 87, 89, 106, 110, 111, 112, 118, 120, 152, 188, 266, 293, 302, 309, 313, 315, 325, 344, 364, 380, 391, 392, 394, 396, 404, 407, 426, 435, 437, 443, 444, 445, 483, 486, 488, 499, 503, 515, 531, 532, 533, 534, 540, 561, 578, 588, 597, 615, 623, 636, 639, 640, 641, 643, 646, 648, 667.
[249] Keith, *History*, i, p. 325.
[250] *RPC*, i, p. 205.
[251] See below, pp. 74, 315; 77, 80; 70.

been 'payit not ane penny thir last iij yeiris bipast'.[252] Reports from elsewhere - from Ardclach to Aberfoyle, from Cushnie to Currie, from Dundurcas to Dalkeith, from Fettercairn to Forgandenny, from Glasgow to Glencairn, from Kilmorack to Kilmarnock, from Lagganallachy to Lilliesleaf - all confirmed the resolution of parishioners to withhold revenues and the resignation of clergy who awaited a 'general ordour', 'a new ordour, reformatioun'. Few had doubts about what was involved: it amounted to 'the dimolitioun of the papistrie'.

This remarkable phenomenon at the Reformation of a systematic boycott of payments to beneficed men was, of course, quite distinct from periodic non-payment which was apt to arise, perhaps as a consequence of a particular dispute, and which could often lead to prolonged litigation through the failure of a particular possessor to obtain revenues to which he was entitled from fruits of his benefice. At Canisbay, the failure of the earl of Caithness' tenants to pay their teinds arose simply 'throw default of justice'.[253] In the parish of Abertarff, Clanranald and other clans proved a stumbling-block to the collection of vicarage revenues; and in Orkney, the parish of Westray was so 'wastit be the Lewis men' that no payments were possible.[254] At the other extremity of the kingdom, it was disclosed that no teinds had been paid at Westerkirk and at New Kirk of Ettrick, in vulnerable Border territory, since the battle of Flodden in 1513, and that Lempitlaw had received no dues for seven years 'becaus the samin lyis on the Bordouris'.[255]

All in all, the valuation of parochial benefices reveals a wide fluctuation in finances from well-off to impoverished parishes. The variations in resources were not merely between poorer highland parishes and richer parishes in the eastern lowlands. The uneven distribution of parochial finances was all too apparent within a single geographical area in any diocese.

9 Episcopal wealth

On the eve of the Reformation, the economic and social circumstances of the episcopate differed little from those of other great landowners. The men chosen to serve as bishops in the sixteenth century were drawn largely from noble and other landed families and so shared a common social status which set them apart from lesser men. Episcopal property conferred lordship and jurisdiction, and offered its holders wealth, influence and social authority. The temporality of the archbishop of St Andrews covered ten lordships; the bishop of Aberdeen's patrimony extended over ten baronies, and included extensive fishing rights on

---

[252] See below, pp. 111, 170, 394.
[253] See below, p. 638.
[254] See below, pp. 488, 666-667.
[255] See below, pp. 118, 209.

the Rivers Dee and Don; the bishop of Moray drew rents from nine baronies and an income from fishing on the Spey; and the bishop of Dunkeld lived in a manner which enabled him to employ a master of works, a forester, two guardians of woods, a constable, chamberlain, graniter (responsible for collecting grain paid as teind or as rent from land), and assorted servants whose ranks included a porter, wright, mason, slater, gardener and fisherman.[256] Similar glimpses, in the rentals, of the chamberlain, graniter, secretary, master of works and gardener of the archbishop of St Andrews reveal some of the figures at work in the archbishop's household.[257]

Apart from income derived from teinds and from the lands and their rents, which constituted the main sources of episcopal wealth, bishops could also expect dues such as the levying of 'procurations' at the triennial visitation of their bishoprics, of 'synodals' paid by each church in a see,[258] and from 'quots' paid for registering testaments.[259] These dues could represent a significant element in a bishop's income: in St Andrews the archbishop drew £494 from procurations and synodals.[260]

The social standing and ecclesiastical office also made bishops, in the eyes of the sovereign, appropriate candidates for exercising political power. Apart from organising their dioceses and supervising their clergy, bishops were expected to sit in parliament, were eligible for election to the privy council, acted as royal servants, judges, politicians and sometimes as statesmen, too. This meant that they were often absent from their dioceses and therefore were apt to be absentee landlords who were obliged to delegate authority for managing their properties to trusted officials: to chamberlains, graniters, and also to bailies of episcopal baronies and regalities.

Plainly, to a careerist intent on gaining promotion in the church, the road to wealth and power lay in the acquisition of a bishopric. The plumb post of St Andrews, the primatial seat, claimed gross revenues in 1562 of £2,900 in silver and 140 chalders of assorted victual, worth at least a further £3,200, making the income of St Andrews comfortably in excess of £6,000 per annum.[261] The archbishop of Glasgow, from the evidence of the rentals, might count on an income yielding £1,000 in cash and the equivalent of £2,000 in kind,[262] a figure rivalled in the gross receipts for the bishopric of Moray of over £2,000 in money, and victual worth another £2,400 or so in cash.[263] Dunkeld yielded £1,500 plus victual whose value may be reckoned at £3,000.[264] Clearly, a bishop who received a fair proportion of his revenues in kind was better placed to cope with

---

[256] See below, pp. 1-2; 413-417; 465-466; 304.
[257] See below, p. 5.
[258] See above, pp. xlii-xliii.
[259] See below, p. 416.
[260] See below, p. 3; cf p. 5.
[261] See below, pp. 2-5.
[262] See below, p. 496.
[263] See below, p. 467.
[264] See below, p. 343; cf pp. 304, 347.

the impact of inflation than one wholly dependent on fixed money payments whose value steadily depreciated.

Further down the scale came Aberdeen with £1,600 in silver and rents in kind worth perhaps another £2,200; Ross whose patrimony may be valued approximately at £2,600; and Brechin with just over £2,000 or thereabouts, based on the cash equivalent of the victuals rendered in the accounts.[265] Despite fluctuating prices for victual, butter, oil and other commodities, Orkney was probably capable, in favourable circumstances, also of yielding an income of £2,000, as was the small bishopric of Dunblane, whose revenues in silver amounted only to £312.[266] Among the poorest sees were Caithness, valued wholly in money, at £1,300, and Galloway (with the annexed abbacy of Tongland) whose income of £1,200 in cash was supplemented £500 or so from victual and salmon.[267] In each case, commuted money payments had left the bishop poorly placed to weather the storm of rising prices. Though excluded from the survey, the bishopric of the Isles, to which was annexed the abbacy of Iona, remained the most impoverished bishopric,[268] and Argyll, also excluded, may be ranked among the poorer sees.

Princes of the church required princely patrimony not to fulfil their religious duties as shepherds of the flock but to maintain their superior standing in society. The temporality - the lands and their rents - which bishops enjoyed was a source not only of income but of provisions, especially as a proportion of rents continued to be paid in kind and not in cash. This afforded bishops ample food and fuel, and they might readily staff their households and estates with servants to whom they could easily provide free quarters. This could make for sizeable establishments. Apart from his palace beside the cathedral, the archbishop of Glasgow had a nearby residence, his manor house at Lochwood, six miles to the north east of Glasgow, not to mention outposts scattered across his archdiocese; and, beyond his cathedral city, the bishop of Dunkeld had residences at Cluny, Perth, Tullilum, Cramond and Edinburgh. Episcopal palaces and other residences, designed to assist bishops in travelling through their dioceses and even, on occasion, to attend court in Edinburgh, reinforced episcopal standing in localities, impressed lesser men by their size and grandeur and were sometimes coveted by those who hoped to acquire them by lease, particularly where bishops no longer wished to retain all their residences for regular use.

As landlords, however, bishops possessed no heritable rights to the property they held; besides, their academic training and clerical attire served to differentiate them from their lay counterparts. They tended to live as rentiers from revenues derived from property which they, or their predecessors, had leased or feued to laymen. Among the beneficiaries of episcopal property were nobles and lairds who had secured a lease of the teinds or had acquired a feu charter to some of the lands, an even more advantageous move, as it conveyed a heritable

---

[265] See below, pp. 417; 627; 383-385.
[266] See below, pp. 661-662; 295-296.
[267] See below, pp. 629; 592.
[268] See below, Appendix 1, pp. 673-675.

grant, in perpetuity, for a fixed annual payment. Most bishops found the strong arm of a powerful lay bailie to be an indispensable instrument in a period when episcopal authority was apt to be weakened by political circumstance and economic adversity, arising from high taxation and inflation. In Galloway, the bishop might count on support from the earl of Cassillis and lord Maxwell as his bailies; the bishop of Caithness had the services of the earl of Sutherland as heritable bailie for an annual fee of £100; and the earl of Argyll, who had a financial interest in the bishopric of Brechin, was also the bishop's bailie.[269] Bailies, however, were unlikely to act from altruistic motives, and, in any event, bailiaries tended to become hereditary.

The financial demands of the crown and nobility often induced bishops to grant lavish pensions which sometimes amounted to a significant part of their revenues. The bishop of Orkney awarded pensions totalling £1,020 from his see: £400 to John Stewart, prior of Coldingham, £200 to lord Ruthven's sons, £400 to the sons of Bellenden of Auchnoule and £20 to the benefice-monger, John Thornton.[270] From St Andrews, the archbishop's kinsman, Gavin Hamilton, commendator of Kilwinning (and coajutor of St Andrews), was the recipient of a pension of £400, and smaller sums assigned to others included £53 to lord Seton.[271] In Dunkeld, the earl of Atholl was the beneficiary of an annual pension of 100 merks; one of the earl of Huntly's sons had a yearly pension of 500 merks from the bishop of Aberdeen; the master of Maxwell was favoured with a pension of £400 from the bishopric of Galloway; and the bishop of Galloway, Alexander Gordon, once claimant to the see of Caithness, enjoyed a pension of 500 merks from the bishop of Caithness.[272] All this contributed to the well-established process of 'milking' the bishoprics and to the general secularisation of episcopal resources.

Not only had the crown gained access to episcopal wealth by nominating bishops, reserving pensions and by direct taxation, but the nobility was also firmly entrenched in episcopal property, profiting from feus and leases granted by bishops anxious to realise what ready cash they could, and from acquiring the office of bailie or the acquisition of a pension. For their part, the bishops by dispensing patronage and by making provision for kinsmen all too readily succumbed to lay demands by granting away their most valuable patrimony. Plainly, a division of the spoils was already far advanced before the Reformation took effect in 1560.

In Dunkeld bishopric, income forthcoming from the spirituality of appropriated churches was largely in cash: much of the teinds had been leased to laymen, and a fixed annual sum of money had replaced the need for the bishop's graniter to collect a precise tenth of the crop. Among the beneficiaries were the earl of Morton, lord Oliphant, lord Ruthven, Gavin Hamilton, abbot of Kilwinning, and William Chisholm, bishop of Dunblane. Possibly arising from

---

[269] See below, pp. 591; 629; 389.
[270] See below, p. 659.
[271] See below, p. 4.
[272] See below, pp. 304; 416; 591; 629.

the unsettled conditions prevailing at the Reformation, Archibald Douglas had succeeded in forcibly seizing the teinds of Cramond and victual belonging to Aberlady, a detached parish of the diocese. In all, only the crops of the kirks of Dowally, Alyth, Little Dunkeld and Caputh remained in the parishioners' hands. Similarly, most of the temporality of the bishopric had been feued for cash returns to local lairds and lesser folk, though the earls of Argyll, Atholl and Erroll, the abbot of Coupar Angus and the friars of Tullilum were among the recipients of feu charters.[273] Prevailing practice in the bishopric of Brechin, however, was for the teinds to be paid in kind by the tenants; most of the temporal lands seem to have been leased to local men, though the earl of Argyll and Bellenden of Auchnoule had acquired feus of certain properties.[274] In Moray, too, the teinds were rendered in the accounts in kind, though in Galloway, they had been leased for money payments, and the lands of most baronies had been feued to lairds for cash returns.[275] In northern Caithness, where the earl of Caithness had acquired a feu of the barony of Mey, returns were entirely monetized, to the detriment of episcopal patrimony and to the bishop's successors. There the bishop had failed to compensate for continuing inflation.[276]

In many sees, where episcopal property had largely fallen to nobles and lairds who had gained feu charters of lands, leaving bishops with the superiority and annual feu duties, control of the property effectively had been surrendered to lay proprietors, and the claims of the old tenants were liable to be disregarded. Only in the bishopric of Dunblane is there significant evidence in the rentals of episcopal property remaining with the existing tenants, who seem to have enjoyed favourable conditions with moderate rents and some security of tenure.[277] Elsewhere, as bishops resorted to feuing, tenants who were unable to afford the terms demanded found themselves at the mercy of feuars keen to recoup their outlay.

Among the numerous inroads on episcopal finance were the obligations of prelates to contribute annually to the upkeep of the Court of Session, the central civil court in Edinburgh, which James V had endowed in the form of a College of Justice, with papal approval from ecclesiastical resources. This perpetual subsidy meant that each year St Andrews handed over £70, Dunkeld £42, Aberdeen £28, Moray £28, Galloway £22 8s, Ross £16 16s, and Caithness £14.[278] An added burden was the expense associated with maintaining episcopal establishments. The bishop of Ross, in particular, had difficulty defending his castle at the Chanonry of Ross which 'lyis in ane far heland cuntrie' and apt to be occupied by 'brokin men' when the bishop was absent at the Court of Session or otherwise employed in royal service in the south. He was therefore obliged to maintain 'a guid cumpany of men' at his residence, whose provisions, consisting of 12 chalders of victual, 20 marts, 80 carcases of mutton, and 20 dozen poultry, and

---

[273] See below, pp. 302-304; cf pp. 341-343, 345-346.
[274] See below, pp. 383-389.
[275] See below, pp. 466-467; 589, 607.
[276] See below, pp 627-629.
[277] See below, pp. 294-6, 348-349.
[278] See below, pp. 4, 304, 416, 466, 591, 626, 629.

other necessities cost the bishop £300 a year.[279] In Dunkeld, the bishop incurred the extra cost of repairing the stone bridge across the River Tay, for which he assigned 300 merks; and the archbishop of St Andrews omitted as a charge in his accounts all capons, poultry and gifts of augmentation (in rent) which customarily went towards the upkeep of his castle, on the grounds that income of that sort was not subject to be sold or leased.[280] The cost of legal services had also to be met: the archbishop of St Andrews paid £50 in stipends to the advocates and procurators fiscal of his see, and the bishop of Galloway met the fees of three procurators.[281]

How far these bishops extended hospitality or dispensed charity is not revealed in the rentals. The bishop of Dunkeld, however, is on record contributing 30 merks to 'certane puir childer callit blew freris', presumably inmates of the hospital of St George in Dunkeld.[282] A commitment to maintaining educational provision in cathedral cities is evident, too, in the bishop of Dunkeld's support, at a cost of 100 merks, of masters for the grammar and song schools, and in the bishop of Moray's modest contribution towards the salary of the master of the grammar school in Elgin and his support for 'thre childir sangstarris in the queir' of his cathedral.[283]

10 Monastic patrimony

On the eve of the Reformation, the religious orders in Scotland were represented by four houses of Benedictine monks (Coldingham, Dunfermline, Iona and Pluscarden), two houses of Clunaic monks (Crossraguel and Paisley), five houses of Tironensian monks (Arbroath, Kelso, Kilwinning, Lesmahagow and Lindores), 11 houses of Cistercian monks (Balmerino, Beauly (in origin Valliscaulian), Coupar Angus, Culross, Deer, Dundrennan, Glenluce, Kinloss, Melrose, Newbattle, and Sweetheart or New Abbey), one house of Valliscaulian monks at Ardchattan, a solitary Carthusian priory at Perth, 17 houses of Augustinian canons (Blantyre, Cambuskenneth, Canonbie, Holyroodhouse, Inchaffray, Inchcolm, Inchmahome, Jedburgh, Loch Leven or Portmoak, Monymusk, Oronsay, Pittenweem (with May), Restenneth, St Andrews, St Mary's Isle or Trail, Scone and Strathfillan), six houses of Premonstratensian canons (Dryburgh, Fearn, Holywood, Soulseat, Tongland and Whithorn), four Trinitarian houses (Aberdeen, Fail, Peebles and Scotlandwell), and a preceptory at Torphichen, belonging to the Knights of St John of Jerusalem. Besides, there were seven Cistercian nunneries

---

[279] See below, p. 627.
[280] See below, p. 3.
[281] See below, pp. 4, 591.
[282] See below, p. 304.
[283] See below, pp. 304, 466.

(Coldstream, Eccles, Elcho, Haddington, Manuel, North Berwick and St Bathans or Abbey St Bathans), an Augustinian nunnery at Iona, a house of Dominican nuns - Sciennes priory - near Edinburgh, and Franciscan nunneries at Aberdour and Dundee. At the Reformation, the friaries, whose income from endowments remained slender, comprised 13 Dominican houses (Aberdeen, Ayr, Dundee, Edinburgh, Elgin, Glasgow, Inverness, Montrose, Perth, St Andrews, St Monance, Stirling and Wigtown), perhaps five Franciscan houses of Friars Minor (Dumfries, Dundee, Haddington, Kirkcudbright and Lanark), possibly eight Observant friaries (Aberdeen, Ayr, Edinburgh, Elgin, Glasgow, Perth, St Andrews and Stirling), and ten Carmelite houses (Aberdeen, Banff, Edinburgh, Inverbervie, Irvine, Kingussie, Linlithgow, Luffness, Queensferry and Tullilum).

Threatened with confiscation of their revenues if beneficed men failed to comply with the crown's plans for valuation, all the houses of monks, canons regular and Trinitarians conscientiously submitted rentals for the survey recorded in the 'Books of Assumption' (except for impoverished Strathfillan whose community had withered away). No accounts, of course, could be expected from Iona,[284] Ardchattan or Oronsay (if indeed it were still in business), for Argyll and the Isles lay outside the scope of the assessment. A few houses of nuns, however, omitted to produce valuations either through negligence or through fraudulence. There is no mention, for example, of any returns from the small Cistercian priory of Elcho in Perthshire (unlike the other Cistercian nunneries which did produce accounts[285]). Rentals for the two Franciscan houses of nuns - at Aberdour[286] and Dundee - are also missing. Nor was there any return for Torphichen, whose accounts presumably remained with Sir James Sandilands, the preceptor and sole surviving officer of the order.[287] Of the three dozen friaries, rentals were forthcoming from merely a few: the Dominican houses at St Andrews (with whose accounts were engrossed those of St Monance), Edinburgh, Elgin, Glasgow and Stirling;[288] the Franciscan friary at Lanark;[289] and the Carmelite house in Linlithgow.[290]

By the sixteenth century, monastic houses had acquired many of the characteristics of property-owning corporations. After all, during the middle ages, monasteries not only had been endowed with extensive lands, donated by generations of pious benefactors, but had gained the teinds appropriated from so many parish churches. At their head in each case was an abbot or prior, essentially a great landlord, who might be an efficient administrator of monastic estates and finances and who was usually a political figure with a part to play in public affairs. Sometimes he was already a monk, or, again, a secular cleric, or even a minor and so could be no more than titular head of the house. Great

---

[284] See below, Appendix 1, pp. 673-675.
[285] See below, pp. 145, 161, 166, 176, 180,183, 186, 187, 192, 194, 549.
[286] Cf below, p. 63.
[287] See below, pp. 11, 110; cf *The Knights of St John of Jerusalem in Scotland*, edd. I. B. Cowan, P. H. R. Mackay and A. Macquarrie (Edinburgh, 1983), pp. 1-40.
[288] See below, pp. 89, 126, 453-454, 522, 554.
[289] See below, p. 522.
[290] See below, p. 155.

abbeys were expensive institutions to sustain, and lordly abbots were apt to spend their wealth in ways similar to lordly bishops. Certainly, in neither case were prelates disposed to see their revenues devoted to the good of people in the parishes from whom their wealth ultimately derived. Nor was there difficulty for a noble family to obtain the prize of an abbacy for a kinsman through the system of commendatorship, whereby the commendator or titular head of the house assumed the duties normally exercised by a regular abbot.[291] The crown, in effect, had asserted a right to nominate, and those magnates who could most readily influence the government were best placed to succeed in promoting their choice of candidate as abbot or commendator.

Among the abbots, canonically appointed, were Quentin Kennedy, son of the 2nd earl of Cassillis, at Crossraguel, and Donald Campbell, son of the 2nd earl of Argyll, at Coupar Angus.[292] A few like Adam Forman at the Charterhouse in Perth,[293] John Philp at Lindores[294] or John Brown at Sweetheart[295] were raised to the office of prior or abbot from the ranks of monks. Others (including Quentin Kennedy, cleric of Glasgow diocese, and Donald Campbell, cleric of Argyll diocese) were recruited from the secular clergy: Malcolm Fleming, dean of Dunblane, was prior of Whithorn; Walter Reid, cleric of St Andrews diocese, was abbot of Kinloss; and Thomas Hay, canon of Moray and of Ross, was abbot of Glenluce.[296] Besides, it was not uncommon for a bishop also to obtain control of a religious house by securing appointment as commendator: Patrick Hepburn, bishop of Moray, was commendator of Scone abbey; and Alexander Gordon, bishop elect of Galloway, to which the abbacy of Tongland was united, was also commendator of the abbey of Inchaffray.[297] At Fearn, the commendator, Nicholas Ross, was also provost of Tain collegiate kirk.[298] On the eve of the Reformation, Arbroath, Coldingham, Culross, Deer, Dryburgh, Dundrennan, Dunfermline, Holyroodhouse, Holywood, Inchaffray, Inchcolm, Inchmahome, Kelso, Kilwinning, Melrose, Newbattle, Paisley, St Andrews, Scone and Soulseat were among the houses controlled by commendators.[299]

The crown itself had benefited substantially by acquiring commendatorships for James V's illegitimate sons: James, the elder, received Kelso and Melrose abbeys,[300] James, the younger, gained St Andrews priory,[301] Robert obtained Holyroodhouse,[302] and John became commendator of

---

[291] See M. Dilworth, 'The Commendator System in Scotland', *Innes Review*, xxxvii (1986), pp. 51-72.
[292] See below, pp. 568; 352, 353.
[293] See below, p. 285.
[294] See below, p. 29.
[295] See below, p. 612.
[296] Cf below, pp. 568; 352, 353, 356; 595; 464, 650; 488, 600.
[297] See below, pp. 331; 348.
[298] See below, pp. 634, 651, 652.
[299] See below, pp. 358; 188, 194, 198; 289; 459; 197; 606; 40; 93; 271, 278, 624; 348; 63; 548; 223, 228; 575, 578; 207, 257, 260; 103; 527; 9, 20; 331; 599.
[300] See below, pp. 208, 243.
[301] See below, p. 9.
[302] See below, p. 93.

Coldingham.[303] The family with the next best record was probably the Hamiltons who stood next to the royal house. Their fortunes had been founded on the position of their head, James, 2nd earl of Arran and duke of Chatelherault, governor for part of Mary, Queen of Scots' minority: not only did one of Chatelherault's half-brothers became abbot of Paisley and archbishop of St Andrews, and another bishop elect of Argyll and subdean of Glasgow, but one of his sons obtained Arbroath, reputed the wealthiest abbey in Scotland, as well as Inchaffray, and another son gained Paisley in succession to his uncle the archbishop.[304] If no other family equalled that record, the Erskines became entrenched at Dryburgh, Cambuskenneth and Inchmahome,[305] the Kennedies at Crossraguel,[306] the Humes at Jedburgh,[307] and the Colvilles at Culross.[308] Something like hereditary succession operated where abbots or commendators resigned their office in favour of a kinsman, enabling the post to pass from one member of a family to another. At Crossraguel, Quentin Kennedy was the nephew of the preceding abbot; Robert Pitcairn, commendator of Dunfermline, succeeded his uncle, whose office had been inherited from his uncle in turn; similarly, Andrew Hume, commendator of Jedburgh at the Reformation, was nephew of the preceding abbot; and Robert Keith succeeded his uncle to the commendatorship of Deer. The richer the benefice the stronger was the desire to keep it in the family.

Whether an abbot or a commendator controlled a religious house made little appreciable difference to the disposition of monastic property in favour of family and friends. The earl of Cassillis, head of the Kennedy family, was the recipient of a lease of Crossraguel abbey from its abbot, Quentin Kennedy, at a rent of 700 merks, as well as a lease of Glenluce abbey, granted by one abbot and sanctioned by another abbot, for 1,000 merks.[309] Inchaffray abbey was leased in liferent by its bishop-commendator to lord Drummond and his wife for 1,000 merks a year;[310] and John Johnston, commendator of Soulseat abbey, had assigned a lease of his abbacy to kinsman, James Johnston of Wamfray, for 380 merks.[311] At Balmerino, in August 1559 - perhaps in anticipation of the deluge - Robert Foster, the abbot, leased the entire revenues and buildings of the abbacy for five years to James Balfour, parson of Flisk (later commendator of Pittenweem), and his father Andrew Balfour of Mountquhanie for 900 merks; in 1564, for the same sum, the new commendator, John Hay, leased the fruits and property of the abbacy for a further five years to John Kinnear of that Ilk (who later acquired the commendatorship).[312] Coupar Angus, with its Campbell

---

[303] See below, p. 198, n. 61.
[304] See below, p. 527; cf below, pp. 498; 358; 527, n. 2.
[305] See below, pp. 260, 537-538, 543, 546, 548, and n. 26.
[306] See below, pp. 567-568.
[307] See below, pp. 216-221.
[308] See below, pp. 260, 294.
[309] See below, pp. 568, 600.
[310] See below, pp. 347-348.
[311] See below, p. 599.
[312] See below, pp. 59-61.

commendator, fell victim to clan Campbell whose head, the earl of Argyll, acquired a financial interest in the abbacy.[313] The Humes were among the beneficiaries of the Cistercian nunnery of North Berwick, where Margaret Hume was prioress, through the acquisition of leases, feus and pensions from the property: there Adam Hume, parson of Polwarth, enjoyed a handsome annual pension of £200 from the fruits of the appropriated church of North Berwick.[314] Haddington priory, perhaps the wealthiest nunnery, whose prioress was Elizabeth Hepburn, had passed into the control of the Hepburn family whose head, the earl of Bothwell, was bailie, for an annual fee of £100, and Patrick Hepburn of Beanston was his deputy at £40 a year.[315] Even so, competing claims to the property were advanced by William Maitland of Lethington after the Reformation.[316]

Other magnates, too, might expect to exercise influence over monastic property in their capacity as bailies, especially so as ecclesiastical authority waned and the power of an influential lay bailie was seen to be essential in defending the privileges of religious communities. Accordingly, Dunfermline abbey had the duke of Chatelherault as bailie of Musselburgh; the earl of Argyll acted as bailie of Culross abbey; lord Maxwell was heritable bailie of Holywood abbey (with a Maxwell kinsman as bailie depute), and he served also as hereditary bailie of Sweetheart abbey; lord Forbes was bailie of Lindores abbey; at Inchmahome, where the Erskine commendators held sway, James Erskine acted as bailie; Lindsay of Dunrod occupied the office of bailie at Blantyre priory; Newbattle abbey, with Mark Ker as commendator, had the services of Ker of Cessford as bailie principal, though Kelso, too, had claims on the laird as its bailie; and at Fearn abbey, where Nicholas Ross was commendator, Ross of Balnagown held office as bailie.[317]

The diversion of monastic patrimony into the hands of laymen was advanced, too, through grants of property and gifts of annual pensions on a generous scale. The earl Marischal was in effective possession of the property of Deer abbey whose commendator was his son, Robert Keith;[318] Blantyre priory's patrimony was leased to Hamilton of Bothwellhaugh;[319] and generally monastic lands were leased or feued to laymen.[320] Besides, as a rule, lairds and others had gained leases of the teinds of appropriated churches.[321] Yet another inroad into monastic resources was the acquisition of pensions by laymen. Lord Seton and his children reaped the fruits of what looked like a lavish pension of 1,600 merks

---

[313] See below, p. 410.
[314] See below, pp. 147-149.
[315] See below, pp. 161, 163.
[316] See below, p. 179.
[317] See below, pp. 39; 292; 271; 277, 612; 32; 548; 505; 103; 225; 633-634.
[318] See below, p. 457-459.
[319] See below, p. 505.
[320] See below, for example, pp. 10, 18-19, 29-32, 35, 145-147, 184, 352-355, 457-458.
[321] See below, for example, pp. 12-18, 26, 36-38, 48, 91, 145-148, 166-169, 198-200, 208, 216, 458, 573.

from Melrose abbey;[322] Paisley abbey, controlled by the Hamiltons, awarded a lucrative pension to the duke of Chatelherault, the head of their house;[323] Kilwinning, also in Hamilton hands, granted £100 in pension to Gavin Hamilton and over £200 to the bishop of Ross;[324] another Hamilton (perhaps to be identified with the captain of Dumbarton castle) was the beneficiary, at the queen's behest, of 300 merks in pension from Balmerino;[325] Holywood abbey, under Maxwell influence, assigned a pension of 200 merks to a son of lord Maxwell;[326] the earl of Glencairn's brother claimed pensions from both Kelso and Melrose abbeys;[327] and among pensions later bestowed by Kelso were £200 to Kennedy of Bargany, £200 each to two Humes, and £100 to Ker of Ancrum.[328] Prominent families, entrenched in monastic property at almost every level, were only too ready to facilitate the process of secularisation in every possible way.

Certainly, the wealth of the religious houses was conspicuous by any standard. From the evidence of the rentals, the richest religious house, if based on gross income, was Arbroath abbey, with revenues (in cash and kind) worth £13,000 a year; next came St Andrews priory with more than £10,600; and Dunfermline abbey's fruits readily surpassed £7,500; no others could rival these.[329] Paisley abbey yielded more than £5,600; and a further group of four - Holyroodhouse, Coupar Angus, Scone and Melrose - all comfortably exceeded £5,000 in annual receipts,[330] followed by Lindores with returns of £4,700 and Kelso with more than £4,100.[331] Kinloss had drawings in excess of £3,500, and Cambuskenneth could expect about £2,900.[332] The next tier, headed by Kilwinning abbey with an income of £2,800, included Whithorn priory whose revenues exceeded £2,600, the abbeys of Jedburgh and Deer each with £2,400, and Dryburgh with £2,100.[333] At a lower level, came houses with more moderate resources: the nunneries of Haddington and North Berwick each drew revenues of approximating £1,800; Balmerino abbey's income topped £1,700; the Charterhouse in Perth might expect £1,400; Newbattle's revenues also reached £1,400; Culross could reckon on £1,300; Inchcolm had possibly a competence of just over £1,000; as had Pittenweem and Holywood.[334] Among the decidedly less affluent houses, if judged by gross income, were Sweetheart, Glenluce, Inchaffray

---

[322] See below, p. 210.
[323] See below, p. 529.
[324] See below, p. 577.
[325] See below, p. 58.
[326] See below, pp. 277, 624.
[327] See below, pp. 224, 228-229, 257; cf p. 210.
[328] See below, pp. 228-229.
[329] See below, pp. 351-352, 358-364; 8-21; 23-29, 37-45. Calculations of the value of victual are based on the prices in *Thirds of Benefices*. (The rental for St Andrews priory is of a later date than that for Arbroath abbey.)
[330] See below, pp. 527-531; 91-93; 352-364, 368-371, 409-412; 331-334; 207-211, 257-261.
[331] See below, pp. 29-37; 222-243.
[332] See below, pp. 461-464; 537-538, 543-548.
[333] See below, pp. 573-579; 592-596; 457-459; 189-191, 197.
[334] See below, pp. 161-165, 176-179, 180-182; 145-149, 166-169; 57-61; 283-289; 100-103; 289-294; 62-64; 22; 269-272, 274-278.

and the nunnery at Eccles all valued above £650; the Trinitarian house at Fail was worth £600; Beauly yielded receipts of £500; and Coldstream nunnery also grossed £500.[335] The gross income, of course, constituted only part of the value of abbeys, whose buildings, contents and plenishings, treasures, stores, stables, barns, byres, equipment, and surplus cash in hand were not subject to assessment. Again, it was simply the income derived from such assets as coal-mines, salt-works, fishings and forests, fulling and corn mills, and brewhouses, which figured in the balance sheets, and not their capital value.

The valuation of monastic income presented in the rentals is, if anything, an underestimation. For a start, as the onus was placed on the tax-payers undergoing assessment to present a record of their income, beneficed men had an interest in minimising their gains and so were liable to err by producing conservative figures when submitting what, in effect, were income-tax returns. In the rental prepared in February 1561/2, the prioress of Haddington managed to make her expenses almost equal her income, and so claimed that the 'money and victuallis will skantlie pay the deductiounes and necessar ordinar chargis of the place sum yeiris'.[336] In instances, where submissions appeared to be blatantly fraudulent, officials in the Exchequer were empowered to scrutinize accounts and to compare the sums with earlier rentals. In the accounts for Lindores abbey, the omission from the rental of 'grassoumes, entrie silver, yairdis, fischeingis, caponis, pultrie, caynis, custoumes, mairtis, cariagis and uther dewteis' led the Collectory to issue instructions to its clerk to 'inserche thame out diligentlie'.[337] Similar omissions were identified in the accounts for St Andrews priory and Pittenweem, the nunneries of Haddington and Eccles, Dryburgh and Jedburgh abbeys, Pluscarden priory, Cambuskenneth abbey, Inchmahome priory, Kilwinning abbey, Whithorn priory, and Sweetheart abbey.[338]

At Dunfermline, where the earls of Arran and Argyll with the master of Lindsay had intromitted with monastic revenues for 1561, the commendator's chamberlain reported that he had made no return in his accounts for the coal of Wallford, partly because of the expenses, in excess of £500, incurred in mining the coal, and partly because the earl of Arran had 'wastit the said coill in sict soirt that na profeit can be had thairof without greit expensis in tyme cuming, as the haill cuntrie knawis', but the auditors of the Exchequer insisted on the inclusion of coal with other missing items from the rental.[339] Revenues derived from fishing rights could also prove a significant, if elusive, element of income. The rentals for Kinloss and Beauly were considered 'suspicious anent the fischeingis';[340] and, in examining the accounts for Pluscarden priory, Exchequer officials resolved 'to tak ordour withe the salmond fyscheingis' which had fallen into the hands of the

---

[335] See below, pp. 622-624; 611-612, 622; 600; 347-348; 183-185; 558-559, 562-563; 464-465, 649-651; 186-187.
[336] See below, pp. 164.
[337] See below, p. 34.
[338] See below, pp. 9, 22, 164, 185, 191, 218, 473, 547, 548, 578-579, 595-596, 612.
[339] See below, pp. 40-41.
[340] See below, pp. 465, 651.

sheriff of Moray.³⁴¹ Other shortcomings detected in the figures for Coldstream priory were identified for investigation;³⁴² the commendator of Jedburgh was charged to pay £560 and 39 chalders of victual when deficiencies were detected in his rental;³⁴³ and the lords of Exchequer refused to accept the accounts for Deer abbey as an accurate record of its revenue.³⁴⁴

With the exceptions noted, the generality of monastic rentals seem to have satisfied officials in the Exchequer; and where suspicions were aroused resort was had to earlier rentals in an effort to gain verification and so establish a basis of veracity for the figures rendered. Thus, in so far as the government's officials were vigilant and diligent in their duties, the incoming valuations may be considered as moderately accurate. At the same time, in compiling assessments some genuine mistakes and omissions were made. Excuses were offered to the auditors of Exchequer for imprecision in the figures supplied for Lesmahagow.³⁴⁵ In the rental for Arbroath abbey, it was noted that in the discharge of revenues and victual no account had been taken of the stipends paid by the commendator to ministers in the reformed church whose parishes were annexed to the abbey; nor were the revenues of certain appropriated benefices assigned to pensionaries properly rendered in the returns.³⁴⁶ Scrutiny of the accounts for Coupar Angus revealed that some horsecorn, geese, capons and poultry not offered for sale had been reserved for the household and for hospitality; and, when the 'auld rentall' was duly examined and the accounts checked, the Clerk Register and Comptroller, with evident approval, instructed their clerk to 'resave this as the just and actentik rentall of Coupar and keip it sua that the chalmarlane thairof be chargit thairwith'.³⁴⁷

At Kilwinning, the commendator in submitting his returns declared himself willing to produce, if the Exchequer auditors so wished, 'the yeirlie comptis that I haif tain of my chalmerlandis in divers yeiris bygaine'; and subsequent submissions in 1564 and 1566 passed the scrutiny of the Comptroller and lords of Exchequer.³⁴⁸ The commendator of Newbattle was ready to testify that his accounts formed 'the just rentale that I get payt, bayth of auld and augmentatioun, and siclyk the just sowme debursit yeirlie of debt furth of the samin';³⁴⁹ and the prioress of Haddington nunnery, 'who cannot write', enlisted the services of a notary to affirm that her submission was 'an extract from the registers of the said monastery and the annual accounts of the same, properly charged and discharged as they are at present, according to the account and reckoning of the chamberlain', though scrutiny showed she had failed to include income arising from 'canes, gressumes, entre silver, custoumes, caponis, pultrie,

---

³⁴¹ See below, pp. 471-472.
³⁴² See below, pp. 187-188.
³⁴³ See below, p. 218.
³⁴⁴ See below, p. 459.
³⁴⁵ See below, pp. 231-232.
³⁴⁶ See below, p. 363.
³⁴⁷ See below, pp. 371, 411.
³⁴⁸ See below, pp. 578; 573, 579.
³⁴⁹ See below, p. 103.

etc.'[350] In Perth, the prior of the Charterhouse, regardless of the current assumption of the third, undertook 'to alter and renew the sam[e] sa oft as my lord Comptrolare or Alexander Quhytlaw, his chalmerlane, sall desyre';[351] and at Kelso a moralising note was struck in an aphorism, thoughtfully incorporated, to the effect that 'if through honest labour something is done, the labour departs and the honesty remains; but if through dishonourable pleasure something is done, the sweetness departs and the dishonour remains'.[352]

Deficiencies of a different variety are detectable, too, in the shortcomings of the scribes who drafted the accounts and, especially perhaps, in the copyists who in transcribing the submitted rentals (sometimes at considerably later dates) were apt to misread a set of figures, mistake a 'x' for a 'v', insert a wrong figure, omit an entry from a rental, or supply erroneous totals to a series of additions and subtractions. Errors of that sort, which abound, and to which editorial comment is directed in the text, do not invalidate the general disposition of revenues in money and kind which forms the substance of the rentals and whose valuation serves as a basis for comparison.

It is also evident from the accounting which took the form of a 'charge' and 'discharge' that the total wealth of a monastic house was not disclosed: the valuation excluded capital assets and debts owing to and owed by the convent. Nor is it always easy to discover, with precision, the net income from the gross or, for that matter, the balance between income and expenditure. After all, valuation for levying the tax of a third of the revenues proceeded on the basis of gross income, with certain permitted allowances deducted. These deductions comprised the fees of recognised officials (such as bailies, chamberlains, graniters, clerks of regality, advocates, and where relevant the stipends of ministers and their assistant exhorters and readers in the reformed church), pensions awarded to individuals, and contributions to the College of Justice. Besides these disbursements, such expenses as charity, hospitality, and, not least, the maintenance of the monks and nuns were also advanced for inclusion among the allowances for deduction before tax. Other expenses were disallowed; and general household expenditure including repair of the buildings was not admissible for deduction.

Besides serving as a basis for assessing the income of the religious houses, the usually well-documented rentals impart a wealth of evidence on aspects of the monastic economy at the Reformation. Not only did they derive their income from the teinds of appropriated parishes and from the produce and rents obtained from the lands they possessed, religious houses also had rights over coal-mining, salt-panning, fishing, corn and waulk (or fulling) mills, woods and peat from which they derived a revenue usually in rent, were in receipt of various casualties, customary dues and labour services, such as arriage and carriage, often commuted into money payments. Through extensive leasing and feuing of their property, however, the monks relied primarily on an income from rents derived from leases,

---

[350] See below, pp. 164-165.
[351] See below, p. 286.
[352] See below, p. 239.

feus and augmentations. An income in cash was forthcoming from 'mails' or money rents, from feu duties and grassums, augmentations and casualties, and from those teinds leased for money; but a significant proportion of monastic revenues was composed of payments in kind: the 'fermes' or grain rents, the 'kanes' or customary dues often rendered in poultry, provisions or fish, and the revenues derived from those teinds leased for a payment in kind - in wheat, bere (or barley), oats, meal or malt. Teinds and rents in kind might be sold for cash, or the rents themselves might be leased for cash. Some produce (grain, livestock, fish and dairy products), not offered for sale, was retained for monastic consumption.

St Andrews priory, for example, derived its main income from fermes or grain rent, often leased for money payments amounting to some £640, from teinds of appropriated churches leased for £1,568 and for 439 chalders of assorted victual, from payments (derived from property, houses, booths, mills and customs, in such towns as Haddington, Linlithgow, Perth, Crail, Dundee and Aberdeen) amounting merely to £23, and from kane, or customary dues from tenants, valued at £6.[353] The patrimony of Dunfermline abbey was composed of 'penny maill' and annual payments accruing from property, customs from Kirkcaldy, burgh mails from Dunfermline, Musselburgh, and Kirkcaldy parish, customary payments commuted to silver, rents from lime kilns, rents rendered in kind (in wheat, bere, oats, oatmeal, salt, lime, cheese and butter), and, the most valuable single source of income, teinds, often leased either for victual or for a money income of over £900.[354] At Lindores in Fife, the abbey drew rents in silver of over £2,200 from six baronies and from teinds leased for money; and it could count on as much again from the sale of assorted victual. It also possessed a 'greit ludgeing in Perthe', another 'ludgeing in the Watergait', a third 'ludgeing in the Foirgait', and a fourth at Falkland, all of which proved superfluous to the convent's needs and so were set in feu.[355]

For estate management, monasteries required trusted officials responsible for collecting revenues, keeping accounts, administering baronial and regalian jurisdiction: officers in the form of chamberlains, graniters, cellarers, bailies principal and depute and all their subordinates. Besides, a community of religious required a community of servants to cater for their manifold needs. At Arbroath, apart from the bailie of regality, chamberlain, graniter and master of works, the more significant officers and servitors listed comprised the subprior, subchanter, sacrist, third prior, the cellarer who purchased victual for the kitchen, an almoner, janitor, baker, brewer, master builder, and launderer.[356] The twenty-six inmates of the convent in Dunfermline had the attention of a porter, plumber, glasswright, forester, a slater and his servants, a barber, miller, smith, wright, mason, carpenter, beadle, keepers of the seal and of the vessel, as well as chamberlain, bailie, advocate, officers, scribes of regality court and of bailiary and other

---

[353] See below, pp. 8-21.
[354] See below, pp. 23-29, 37-54.
[355] See below, pp. 29-37.
[356] See below, pp. 360, 362.

unspecified servants.[357] The principal servants at Pluscarden to attend to five monks included a master cook and a master baker as well as a porter and gardener;[358] and a brewster or brewer, cook, porter and barber were employed for six monks at Balmerino abbey.[359] The needs of 18 Cistercian nuns at Haddington priory (whose number fell to ten by 1573) were met by a master cook and two further cooks, two porters, a pantryman, maltman, brewer, common launderers, a gardener, smith, carter and boys, ploughmen, stable boys and other servitors.[360]

Plainly, the revenues of the religious houses went far beyond the needs of religious communities professedly devoted to rules of self-denial and austerity. The comforts enjoyed by monastic communities - and not least by the nuns of Haddington - were certainly hard to reconcile with the demands of their rule. That the rank and file in the monasteries had lost sight of their ideals of simple living and of a community of goods is suggested by the assignment to each monk of an individual 'portion' or salary to meet their living expenses. On the whole, monks enjoyed comfortable security, with no shortage of food and a seeming superfluity of servants. At Balmerino, a Cictercian house, the portion allocated to each of the six monks (who had the services of a brewer, cook, porter and barber) consisted of 17 bolls of bere, 5 bolls of wheat, 12 bolls of meal, with £10 in money for the 'kitchen', 8 merks as 'habit silver', plus £15 6s 8d to each monk annually.[361] It looked almost as if the days of strict observance and self-denial were at an end. The 18 nuns in Haddington were each assigned 4 bolls of wheat, 7 bolls of bere and 3 bolls of meal a year, each received 8 pence a day for fish and meat, and each nun gained a clothing allowance of 6 merks.[362] Coldstream in the Borders was a smaller Cistercian nunnery, with 'nyne agit wemen sisteris in the place', who, even with the Reformation, continued to reside in the conventual buildings and were sustained from its revenues.[363]

At Augustinian Scone, claims for the living expenses of 18 members, presumably mainly canons regular, totalled £352, with 'the priour haiffand dowble'; that apart, quantities of victual, poultry and 12 barrels of salmon were 'nevir sauld for money bot spendit in sustentatioun of the place'.[364] The allowance at Jedburgh, another house of Augustinian canons (with its attendant cook, housekeeper, gardener, wright and lorimer), meant that the '4 brether, ane of thame ane auld man that hes double portion', could each expect to receive six bolls of wheat, 16 bolls of bere, and £20 in silver, 'and the auld man double'.[365] Kilwinning provided annually more than 9 bolls of meal, 8 bolls of bere and 8 bolls of wheat for each of its monks, whose number is variously recorded as either 6 or 7 in 1561; a further 40 stones of cheese were earmarked for the convent's use;

---

[357] See below, pp. 39-40.
[358] See below, p. 472.
[359] See below, p. 58.
[360] See below, pp. 163, 182.
[361] See below, p. 58.
[362] See below, p. 163.
[363] See below, p. 186.
[364] See below, pp. 332-333.
[365] See below, p. 216.

and for habit and kitchen silver, each monk received just over £16 annually.[366] In Galloway, where the abbacy of Tongland was annexed to the bishopric, seven 'brethren', evidently Premonstratensian canons, were each allocated some £9 in kitchen silver and 4 merks as habit silver; and four novices received 'tua haill portiounis of silver', worth more than £23.[367]

In his rental, composed in January 1561/2, the commendator of Culross abbey explained how he had paid his monks' portions, each amounting to £20 in cash, to five monks who had recanted their papistry but had discharged the portions of the remaining four monks 'quha wald nocht recant' (and who took the commendator to court to secure their entitlement). All nine were also in receipt of wheat, bere, oats and cheese.[368] The 15 canons of Cambuskenneth also had resorted to law to ensure they each obtained payment of 80 merks from the commendator.[369] The sum of £275 was disbursed at Dryburgh to the canons regular, pensionaries and factors;[370] 9 canons at Inchmahome gained 18 chalders of meal and 180 merks;[371] and a total of £240 was deducted at Newbattle to support 'sex decraipit and recantit monkis'.[372]

At the Benedictine abbey in Dunfermline, £394 in silver, 9 chalders of wheat, 32 chalders of bere, 38 chalders of oats, with 26 chalders of horsecorn, formed the allocation for the annual sustenance of 26 monks.[373] The monks of Arbroath abbey, whose numbers are not disclosed in 1561, continued to receive their habit silver and 'pittances'; and some 11 chalders of wheat, 42 chalders of bere, 8 chalders of meal, and 24 barrels of salmon were allotted for the convent's upkeep.[374] Again, at Coupar Angus, 19 'brethren' in 1561 gained their customary portions of wheat, meal and bere, for 'brede and drink', and shared £312 in silver.[375] Fourteen monks in Kinloss abbey each gained 50 shillings a year in habit silver, 8 pence a day for meat, two pence a day for fish, plus victual for bread and drink; and collectively a further £12 for heating, lighting, butter, spice and 'Lentroun meitt'; but at Beauly priory, a dependency of Kinloss, 8 monks there had an allowance merely of 40 shillings a year as habit silver, and 3 pence a day for meat, and 2 pence for fish.[376] Five monks of Pluscarden each gained £16 as their kitchen and habit silver, plus victual.[377] The kitchen and habit silver of an undisclosed number of monks at Paisley totalled £473, and 14 chalders of victual

---

[366] See below, p. 577.
[367] See below, p. 591.
[368] See below, pp. 291-293.
[369] See below, p. 538; cf p. 545 (where the sum is 60 merks); cf p. 547 (where, in the earliest rental, the canons are numbered 17, and assigned £293 in silver, with each allotted 17 chalders of bere, 5 chalders of meal, and 4 chalders and eleven bolls of wheat).
[370] See below, p. 190.
[371] See below, p. 548.
[372] See below, p. 103.
[373] See below, pp. 39-40.
[374] See below, pp. 360-362.
[375] See below, p. 370.
[376] See below, p. 463.
[377] See below, p. 472.

were disbursed for the convent's upkeep.[378] By 1565, 11 'religious monkis and thre portionares seculares' at Melrose abbey still received 20 merks a year, one chalder of bere, four bolls of wheat, two chalders of meal, and three stones of butter;[379] and a later rental for the same house valued the portions for eight surviving monks at £207 a year.[380] So late as 1573, the convent at North Berwick, consisting of eleven women - evidently Cistercian nuns - whose names are recorded, and two men, also named, allowed £20 in cash for each inmate, presumably as an annual retirement pension.[381]

Apart from household expenses of this sort (which met the needs of feeding and clothing the religious, paid the fees of their officers, and qualified as tax-free income), there were other approved deductions before tax in the form of pensions granted by the house, and any hospitality and charity which the religious might bestow. The readiness with which pensions might be granted to family, friends and protectors represented a significant drain on resources. In contrast, the obligation to extend hospitality to guests and travellers, and to provide for the poor made noticeably less heavy demands on monastic revenues.

Though monastic expenditure on charity might be hard to quantify - not least how far food from the monks' table found its way to the poor - the rentals nonetheless offer some guidance. The almoner of the wealthy abbey of Abroath, it was revealed, distributed 5 chalders of victual yearly to the poor.[382] Less wealthy Pluscarden donated 2 chalders of victual 'to the puir folkis and utheris puir tennentis and passingerris' or travellers.[383] At Haddington nunnery, the prioress complained in February 1561/2 that no payment was forthcoming for two years from the customs of the burgh, assigned in alms by the crown to the priory; but, in her income-tax returns, she also claimed as non-taxable expenses the distribution of 2 chalders of meal and £40 in silver, 'pairt les pairt mair', to 'auld failyeit men and wedowis' of the four parishes appropriated to the house, and to 'indigent persounes cumand daylie to the place'. An additional £66 13s 4d was entered as alms for 'pure folkis'; and 10 chalders of oats was disbursed for horsecorn and 'in ressaving of strangeris'. But none of that, the prioress remarked, took account of the cost of hospitality extended to guests, in 'ressaving of our freindis and strangeris'. She also recorded the burden of supporting six 'auld servitouris and gentilmen to thair claithes and hous' at £20 a head.[384] For Arbroath, it was remarked that there was scarcely sufficient victual and money to sustain the commendator and his servants, far less the expense incurred when 'grite lordis and utheris strangeris' arrived or when the superintendent of Angus paid a visit.[385]

---

[378] See below, pp. 528-529.
[379] See below, pp. 210-11.
[380] See below, p. 260.
[381] See below, p. 148.
[382] See below, p. 362.
[383] See below, p. 472.
[384] See below, pp. 163-164.
[385] See below, p. 363.

Elsewhere, Cambuskenneth abbey gave 3½ chalders of meal 'to the puris';[386] Dunfermline abbey merely assigned 8 bolls of wheat 'to the puir in ordinar baikin breid' and 2 bolls to the lepers of St Mary Magdalene's hospital near Musselburgh;[387] two poor men of advanced years gained support from Inchcolm abbey;[388] Holyroodhouse contributed 36 shillings to the bedesmen of St Leonard's hospital in Edinburgh, a dependency of the abbey.[389] Beyond that there is little trace of monastic expenses for charity.

As prelates, abbots like bishops were obliged to contribute annually to the upkeep of the lords of Session, the senators of the College of Justice. The contribution levied was broadly in proportion to the wealth and standing of the house: Arbroath abbey paid £84; Dunfermline abbey, £70; Holyroodhouse and Kelso abbeys, £56 each; Balmerino abbey, £48; Whithorn priory, £42; Cambuskenneth abbey, £36; Coupar Angus and Scone abbeys, £35 each; Kilwinning and Newbattle abbeys, and Coldingham priory, £28 each; Jedburgh abbey, Haddington and North Berwick priories, £21 each; Kinloss abbey, £18 4s; Culross abbey, £14; Restenneth priory, £11 4s; the Trinitarian house at Fail, £7; Fearn abbey, £5 12s; and Beauly priory, £4 4s.[390]

If endowment and income for most monasteries and nunneries proved excessive for houses professing discipline and self-denial, the situation in the friaries was noticeably different. The more modest revenues accruing to the friaries (which eschewed endowment and were located usually in the burghs) consisted mainly of annuals, forthcoming from tenements lying in the burghs. Few friaries could expect an annual income of more than £100, and most appreciably less.[391] Again, the houses of the Trinitarians (who were not mendicant friars) at Aberdeen, Fail, Peebles and Scotlandwell were sustained from annuals, and, as they were permitted endowment, from feu duties and the profits of teinds from a few appropriated parishes.[392] At Fail, the most prosperous house of the order, 'four ald men of the convent', later identified as bedesmen, each received 11 bolls of meal and 13 bolls of malt, and 8 merks each for habit and kitchen silver. The impression is not one of a wasteful household and lavish consumption; indeed, if correctly recorded, £22 were assigned to 'tua puir men in the place to sustain them', a sum rather more than the salary allotted by the head of the house to the protestant minister of the annexed parish of Inverchaolain in Argyll who gained £10.[393]

---

[386] See below, p. 547.
[387] See below, p. 39.
[388] See below, p. 62.
[389] See below, p. 92.
[390] See below, pp. 360; 39; 92; 228; 58; 595; 538; 545; 356, 370; 332; 574, 578; 103, 200; 217; 163, 182; 149; 463; 291; 218; 558; 563; 634; 650. Coupar Angus' contribution in 1565 is recorded as £36, see below, p. 409.
[391] See below, pp. 89, 126-128, 155-156, 453-454, 522, 554-555.
[392] See below, pp. 558-559, 562-563; 249; 56.
[393] See below, pp. 558, 563.

11 The social setting

The retention of ancient terms for the measurement of land, the principal source of wealth, is apparent in repeated reference in the rentals to 'ploughgates'[394] (land cultivable by a plough and team of eight oxen in a year, notionally 104 Scots acres) or the synonymous 'plough'[395] (with its fractions) or 'ploughlands'[396] (also denoting, as a unit of assessment, land capable of being tilled by one plough-team of eight oxen in the year); 'husbandlands'[397] (in theory, land cultivable by two oxen, or a quarter of a ploughgate, or 26 Scots acres); 'oxgates' or 'oxgangs'[398] (an eighth of a ploughgate or notionally what a single ox might plough, or 13 acres); 'acres', each denoting a group of perhaps four 'rigs', long ridged strips of cultivated land; land might be measured, therefore, as 'ane aker and ane rig';[399] and, north of the Forth and Clyde, in Gaelic-speaking areas 'davachs',[400] with their subdivisions, denoting units of agricultural capacity, broadly equivalent to ploughgates in the south. A reminder, too, of early methods of assessment for taxation was the continued use of such terms as 'pennylands',[401] 'shilling lands',[402] and 'merklands' of old extent,[403] depending on the original unit of valued rental.

With so extensive property in its preserve, the church could not be other than prominent as a landlord. The produce and the profits won from the land were of necessity the work of many hands employed in the innumerable 'ferme touns', small townships - the farmsteads and cottages, which formed the basic social and economic units - scattered across the countryside, in whose names are often clues to their location: Bogtoun, Brigtoun, Cheppeltoun, Fischartoun, Freirtoun, Furdtoun, Hilltoun, Kirktoun, Ladytoun, Langtoun, Leidtoun, Meiklefield toun, Milntoun, Munktoun, Muretoun (Over and Nether), Myddiltoun, Newtoun, Orcharttoun, Overtoun, Smythtoun, Tempiltoun.[404] At the bottom of the social scale, among the employed, were the stable boys, carters,

---

[394] See below, for example, pp. 29, 30, 35, 305.
[395] See below, for example, pp. 28, 29, 30, 33-35, 168, 305, 385.
[396] See below, for example, pp. 101, 223, 230, 231, 233, 234, 237, 239, 240.
[397] See below, for example, pp. 147, 162, 165, 167, 173, 177, 178, 181, 186, 187, 192, 202, 203, 212, 215, 223, 231, 233, 234, 236, 237, 239, 240, 242, 366.
[398] See below, for example, pp. 162, 178, 461, 511.
[399] See below, pp. 13-15, 18, 21, 24, 26, 28-30, 33, 38, 46, 48, 50, 51, 99, 120, 147, 155-156, 162, 168-169, 180-181, 192, 249, 259, 285, 289, 292, 293, 305, 353, 383-389, 456, 537, 544-545, 555; 180.
[400] See below, pp. 651, 652.
[401] See below, pp. 496, 628, 633.
[402] See below, pp. 184, 271, 275, 276, 605, 622, 623, 567, 588, 598.
[403] See below, pp. 214, 215, 227, 270, 271, 284, 349, 513, 564, 567, 569, 598, 609, 611, 612, 617, 622, 623.
[404] See below, for example, pp. 497-498; 274; 354, 357, 383, 405, 463; 491; 10, 284-285, 287; 383, 387, 388; 497; 10, 284, 357, 366, 367, 376, 412, 470, 492, 497, 628, 630; 539; 241; 105; 270; 355, 366, 633, 652; 562; 354; 366, 633, 652; 270; 355; 619, 622; 104; 275, 376; 367. In Berwickshire, the toun of Eyemouth contained 16 husbandlands, Ayton 14½, West Reston 23, Swinwood 16, and Renton, 14 husbandlands. See below, pp. 202-203. At Newbattle abbey, the named occupants, or main tenants, of Newbattle toun, responsible for paying their rent to the abbey, numbered 59; tenants in other touns are also named. See below, pp. 100-102.

fishermen, ploughmen and other hired agricultural labourers, who, with domestic servants, figure fleetingly in abbatial and episcopal rentals.[405] The property of prelates also required the services of assorted craftsmen who settled within the community.[406] Among the tenants on church lands, the cottars,[407] who derived their name from the unit of land they occupied, the 'cotland',[408] had each the occupancy of a small holding of land for cultivation, with dwelling attached. One cottar of the Mains of Ruthven in Forfar occupied 14 acres of land.[409] Sometimes, too, an area of settlement might give rise to a 'cottoun',[410] composed entirely of lands originally set to cottars; and there is mention of cottars' teinds, cottars' mills, and of the cottages of an individual toun or of a monastery.[411] Other smaller holdings might be described as 'crofts';[412] a croft might be conveniently measured in furlongs;[413] and reference is also made to a 'toft', a piece of land attached to a house: in Perth, one toft extended to two acres, with pasturage.[414] The holdings of husbandmen, the more substantial tenants, could be sublet, so that a husbandman might become the immediate landlord of a cottar. Husbandmen might hold the lands of a toun as co-tenants, with a share of the scattered rigs of arable land, and rights to pasture for grazing, or sometimes they might hold the land as single tenants. Even quite small areas of land were farmed by a mixture of tenants and subtenants, occupying various levels in the rural hierarchy. The kirklands of Ruthven vicarage in Forfar contained the Mains or home farm of 4 ploughgates, in the hands of the laird and his son; another portion of the same size was cultivated jointly by two occupants; a third section of two ploughgates was farmed by the laird's brother and four other occupants; two further pieces of land each had one occupant; and the final portion was held by three farmers; all were mundanely described as 'labowraris of the ground', in the sense that they themselves worked the ground directly.[415] Again, most of the kirklands and teinds belonging to the parsonage of Moy, in Moray diocese, had fallen to the possession of a local laird, Dunbar of Durris, as feuar, and his 'subtennentis, lauboraris of the ground'.[416] At Inchcolm abbey, some lands in Fife

---

[405] See below, for example, pp. 39, 58, 163, 216, 218, 304, 360, 472, 591.
[406] See below, pp. 5, 6, 12, 39, 40, 163, 216, 304, 360, 362, 472, 591.
[407] See below, pp. 15, 150, 151, 305.
[408] See below, p. 33.
[409] See below, p. 305.
[410] See below, pp. 422, 424.
[411] See below, pp. 230, 233, 239, 296, 298, 412, 450; 236; 202, 203, 222, 634.
[412] See below, pp. 11, 13, 14, 18-20, 25, 30, 32, 33, 35-37, 47, 62, 83, 89, 94, 102, 127, 147, 155, 162, 167, 168, 174, 177, 178, 180, 181, 184, 198, 207, 212, 217, 222, 223, 226, 227, 230, 233, 234, 237, 239, 240, 254, 257-259, 270, 271, 275, 276, 284, 285, 288, 296, 298, 307-309, 322, 336, 342, 345, 395, 396, 405, 406, 412, 413, 421, 438, 449, 463, 478, 479, 490, 491, 498, 516, 531, 545, 546, 550, 555, 556, 586, 587, 589, 590, 592, 593, 595, 596, 598, 599, 619, 620, 623, 628, 633, 634.
[413] See below, p. 174.
[414] See below, p. 74; see also pp. 30, 32-35, 37.
[415] See below, p. 305.
[416] See below, p. 470.

were occupied by 'four men', others by a single tenant.[417] A hill croft in the Grange belonging to the priory of St Mary's Isle was occupied by a 'subtenant' for £2 in rent. Various ferme touns and crofts of parishes appropriated to Whithorn priory were occupied, it seems, by individual tenants.[418]

The conditions on which farmers held land were generally specified in a 'tack' or lease, either of long or of short duration, or in a feu charter; and the rent, in kind or cash, and feu duties paid annually for the lands they occupied are illustrated abundantly throughout the volume. Initially at least, before the drive toward consolidating their properties took full effect, feuars sometimes might hold their lands as scattered possessions, intermingled with portions occupied by tenants who continued to lease their holdings. At any rate, the feuar of five merklands - whether as a block or assorted rigs is unclear - of the homeland of Soulseat abbey was surrounded by tenant occupiers who merely rented their land from the abbey.[419] Certainly, lands feued by St Andrews priory in portions consisting of 'the 4 pairt and 8 pairt', 'the ane halff with ane aucht pairt', 'fyve sex pairtis', 'ane quarter except the xxx pairt thairof', do not suggest a tidy allocation of consolidated blocks of land.[420] There is mention, too, of the 'steilbo guidis' of one peasant who evidently held his land by steelbow tenancy, a customary form of tenure in which the peasant farmer gained stock and seed along with the farm whose value he returned on the expiry of the lease.[421] Of the numerous customary labour services owed by peasants to their landlord, often commuted for payment in money or kind, there is ample evidence: arriage and carriage;[422] bonage silver (in lieu of harvest work);[423] castle wards;[424] 'dawarks'[425] (a day's labour); fed oxen and boars;[426] (or, in commutation) fed oxen silver;[427] rin mart silver (in place of a beast at Martinmas);[428] multures, or dues paid on cereal ground at the mill to which tenants were thirled, and dry multures paid by those relieved of thirlage to a mill;[429] ploughing and harrowing dargs[430] (calculated on a day's labour); and the ward for feeding a horse.[431]

Groups of ferme touns might compose a sizeable estate, co-extensive sometimes with a parish or, more often, corresponding to an episcopal or abbatial barony, and, in turn, the scattered baronies held by prelates might be recognised as lordships. Such a pattern of landholding merely underlined the main division

---

[417] See below, p. 64.
[418] See below, pp. 592-595.
[419] See below, p. 598.
[420] See below, p. 18.
[421] See below, p. 496.
[422] See below, pp. 104, 165, 209, 244, 301.
[423] See below, pp. 633, 634.
[424] See below, pp. 202-203.
[425] See below, pp. 352, 364, 547, 549.
[426] See below, pp. 305, 663.
[427] See below, pp. 23, 25, 42, 45-47, 51, 53, 331, 332.
[428] See below, pp. 461, 462; cf. p. 47.
[429] See below, pp. 26, 30, 32, 37, 43, 44, 48, 276, 462, 465, 471, 473, 613, 619.
[430] See below, p. 549.
[431] See below, p. 634.

between the proprietors, on the one hand, and the peasants, on the other. The gulf between proprietorship and tenancy was not unbridgeable, and the resort by clerics to extensive feuing enabled richer peasants to convert their leases of the lands they occupied into heritable feus. The feuing of so substantial a proportion of ecclesiastical estates led to decisive social change, as the movement accelerated in the sixteenth century, and resulted in the effective secularisation of much of the church's patrimony.

Alongside those magnates, lairds and speculators in land who purchased feus, former tenants on kirklands who had the wherewithal to buy feu charters joined the ranks of the gentry, even if occasionally just as a 'guidman'[432] or as a bonnet laird, and so established themselves as proprietors. Some feuars were plainly outsiders but others were residents, the former sitting tenants. In Dunkeld, feus of episcopal lands were assigned to lesser men - and to one woman - who have the appearance of tenants in the locality, as well as to great lords - Argyll and Atholl - local lairds and to the Carmelite friars of Tullilum.[433] Prominent among the feuars of episcopal (with abbatial) lands in Galloway were the earl of Cassillis, numerous local lairds and also the sheriff there; smaller men were far from evident.[434] Among the lands attached to Caputh, a prebend of Dunkeld cathedral, one ferme toun was feued 'to the tenants thereof', and another toun to lord Gray.[435] At Sweetheart abbey, lands, depicted as 'fourscoir markland', in the barony of Lochindilloch were feued to 'the inhabitaris of the maist pairt for dowble maile', of which five merklands' worth were feued to lord Maxwell as his bailie fee.[436] At another level, the kirkland of Carriden vicarage, amounting to three and a half acres, was feued to the old possessors, the laird of Bonhard ('Ballinhard') and the guidman of the Grange.[437] The dispersal of so much land into the hands of others through lease and feu meant that, for the most part, kirkmen derived their income as rentiers and indulged in little direct cultivation of the property they held. At North Berwick, the prioress acknowledged that merely half a ploughgate - some 50 acres or so - was 'manurit with our ain guidis'.[438] Increasingly, as churchmen capitalised on their property by granting feus, a wedge of new proprietors, with security of possession, came into being where hitherto there had been tenants only.

Valuable though the benefits were which feuing conferred, many nobles, lairds and small tenant farmers also continued to be recipients of leases for lands which they held either as middlemen or as occupiers and cultivators. The fruits of the provostry of Bothans fell to lord Yester, Edzell parsonage was let to the earl of Crawford, the earl Marischal had a nineteen years' lease of Fetteresso parsonage and vicarage, the earl of Atholl gained the benefice of Fortingall, and the parsonages and vicarages of Inch and Leswalt in Galloway, and also Straiton

---

[432] See below, pp. 157, 184, 189, 200, 226, 326, 499.
[433] See below, pp. 303-304.
[434] See below, p. 607.
[435] See below, p. 318.
[436] See below, p. 612.
[437] See below, p. 157.
[438] See below, p. 168.

vicarage in Carrick, were among the churchlands set to the earl of Cassillis.[439] The laird of Buccleuch's wife gained a lease of Hawick parsonage; and lady Carlyle obtained the lease of Cummertrees parsonage in Annandale.[440] Lairds were conspicuous beneficiaries of leases of the kirklands of Largo, Kilconquhar, North Berwick, Logie, and Maybole, benefices appropriated by North Berwick priory;[441] a pattern which recurred in countless parochial benefices across the land.[442] Even townsfolk might acquire such a lease: the parson of Govan leased the property of his benefice to his kinsman, a burgess in Glasgow; Inchcailloch parsonage and vicarage were set to a Dumbarton burgess; half of Dreghorn vicarage was leased to a burgess of Irvine; a lease of a share in Lyne parsonage and vicarage was purchased by a burgess of Peebles; a burgess of Perth possessed part of the fruits of Forteviot parsonage; and another Perth burgess had an interest in Kinnoull parsonage and vicarage.[443]

By contrast, the kirklands of Inchture, Longforgan and Rossie were leased to resident parishioners; the lands of Peebles vicarage were also set to 'the parochineris'; the parson of Linton in Lothian agreed 'to lait and suffer the tennentis to intromet' with his lands for £100 in rent; and a lease of Lasswade vicarage was held by parishioners.[444] The lands of Dalkeith provostry were also in the hands of tenant occupiers; and the same was true of a prebend of Corstorphine.[445] In Dunkeld diocese, the parsonage and vicarage of Kinclaven (the chanter's prebend) were leased mainly to 'the tenants', evidently kindly tenants with a claim to customary inheritance through kinship.[446] Aucherhouse parsonage (a common church of Dunkeld) was also set to the 'tennentis and occupyaris thairof'; so, too, was Cluny parsonage (belonging of the dean's prebend).[447] Again, the kirks of Alyth, Caputh, Dowally and Little Dunkeld (annexed to the bishopric) were all 'in the parishioners' hands'.[448] In general, bishops and cathedral dignitaries, with a reputation as lenient or inactive landlords, were disposed to respect kindly tenure; and in Dunblane, the 'temporal' lands of the bishopric were leased to 'the tennentis and possesouris of the samin ... as the use and custume is and hes bene in Stratherne'.[449]

The produce of land and water, which, except in years of scarcity, sustained the population, features prominently in the assorted rentals. As a sizeable proportion of the income derived from kirklands consisted of rents and teinds paid in kind, an impression may be formed of the distribution of crops and

---

[439] See below, pp. 172; 379; 404; 300; 607; 569.
[440] See below, pp. 263, 265.
[441] See below, pp. 145-148.
[442] See below, for example, pp. 17-18, 32, 48, 72, 82, 102, 118, 157, 171, 195, 212, 250, 251, 263, 312, 340, 379, 382, 394, 396, 401, 403, 424, 425, 430, 435, 470, 480, 506, 510, 513, 541, 544, 546-548, 559, 562, 570, 571, 580, 587, 600-605, 607, 612, 616.
[443] See below, pp. 533; 540; 581; 252; 317; 323.
[444] See below, pp. 18; 251; 108; 115.
[445] See below, pp. 115, 130.
[446] See below, p. 306.
[447] See below, p. 301.
[448] See below, p. 303.
[449] See below, p. 349.

of regional variations in husbandry dictated by geography, climate and soil. Settlements of ferme touns on the most fertile land for cultivation enabled the cropping of oats and bere, and where possible, in more favoured areas, of wheat, rye, peas and beans. The run-rig husbandry, which produced the crops, is commemorated in such place-names as Todrig, Quitrig, Bastanerig (Bastleridge), Benerig, Over Brumrig (Broomrig), Spittelrig, Cokrig (Cock Rig), Sande Riggis, Crammond Riggis, Mekill Rig, Carsrig and Fogo Rig (Fogorig).[450] Literally further afield beyond the arable, rougher pasture lent itself to the rearing of cattle and sheep, with some goats and swine. The 'commonty'[451] is also a reminder of the common pasture land on which tenants had grazing rights; while place-name evidence[452] serves as a guide to the use of upland shielings or summer pasture settlements on the hills: at Cockburnspath in coastal Berwickshire beyond the Lammermuir Hills, the content of a steading included two 'hoprigs' (or sloping hollows between rigs), and two 'scheillis' or shielings.[453] As well as dairy produce in the form of butter, cheese and eggs, rents might be paid in fowl and fish, and in peats, coal, lime and salt.

Upland areas from Galloway to Ross favoured the production of livestock, though any low-lying, ill-drained land, unsuitable for tilling, beside rivers provided grazing and meadow hay, conserved for winter fodder: thus, in Ayrshire, the meadow known as Ellingtoun Bog produced its crop of hay, from which teinds were extracted.[454] Elsewhere even moorland might be used as meadow;[455] and straw, collected in rent, might fetch 12 shillings for a dozen bundles.[456] At Melrose abbey, a 'small quantitie' of hay was reserved unsold for 'the use of the place'.[457] At the same time, in hilly areas, wherever possible, grain crops were grown; there is repeated mention of moorland oats;[458] and in the absence of specialisation, reliance was placed on bere (a four-rowed barley) and oats, the basic drink and food crops. Higher-yielding white oats were grown in more favoured locations such as the carse of Perth, Fife or East Lothian;[459] hardier black oats were more common, especially in exposed areas and on poorer ground.[460] Even so, no oats, as such, appear in rentals for Galloway and Dumfriesshire (unlike Kirkcudbright), though modest quantities of bere and meal (which might be derived from either oats or bere) figured as income. Again, in the far north, payment of oats and meal in rent was confined to the more

---

[450] See below, for example, pp. 186, 189, 198, 199, 204, 276; 94, 98, 103, 150, 180, 511; 226.
[451] See below, p. 202.
[452] See below, pp. 166, 167, 184, 190, 191, 198, 200, 204, 223, 224, 226, 233, 237, 239, 242, 249, 258, 309, 353, 409, 508; cf. pp. 33, 37.
[453] See below, p. 192.
[454] See below, p. 569.
[455] See below, p. 192. For meadow, in general, see below, pp. 14, 30, 32, 36, 92, 147, 167, 169, 208, 209, 222, 235, 238, 241, 242, 249, 291, 370, 535.
[456] See below, p. 316.
[457] See below, p. 210. For other references to hay, see below, pp. 106, 152, 188, 204, 210, 235, 241, 242, 245, 503, 518.
[458] See below, for example, pp. 39, 42.
[459] See below, pp. 39, 40-43, 54, 285, 288-289.
[460] See below, pp. 43, 54, 57, 285, 288-289.

productive soil of Easter Ross.[461] Similarly, oats for horsecorn occurred as rent in Forfar,[462] Perth,[463] Fife,[464] Lothian,[465] Renfrew[466] and Lanark,[467] with fleeting mention in Moray, Cunninghame and Roxburgh,[468] but, apparently, not elsewhere. The brewing of barley is testified not only in payments of malt (in Orkney,[469] Ross,[470] Moray,[471] Aberdeen,[472] Perth,[473] Fife,[474] Stirling,[475] Lothian,[476] Roxburgh,[477] Lanark,[478] Kyle,[479] Dumfries[480] and Wigtown[481]) but in mention of alehouses,[482] brewhouses,[483] brewers[484] and maltmen,[485] of 'brewlands',[486] and of such place-names as 'Ailhoushill' and 'Ailhous croft'.[487]

Rents derived from the most valuable crop of all, wheat, were limited (on the evidence of the rentals) to the fertile soils of the eastern coastal plain from Berwickshire, and parts of Roxburgh and Peebles, Lanark and Cunninghame, to Lothian, Stirling, Perth and Fife, onwards to Forfar (though seemingly not Kincardine), Aberdeen, Banff and Moray.[488] Rye figured only in accounts for Berwick, Peebles, Edinburgh and Moray.[489] An impression of the proportion of crops cultivated in particular districts is apparent, too, in the ratios of cereal

---

[461] See below, pp. 625-627, 633-634, 647, 649-650 (oats); 627, 649 (meal).
[462] See below, pp. 356-358, 365, 368, 369, 371, 384-386.
[463] See below, pp. 294-296, 303, 308, 316, 327, 329, 333, 341-342, 345-349.
[464] See below, pp. 29, 37, 39, 40-42, 51.
[465] See below, pp. 98, 99, 104, 164.
[466] See below, pp. 528, 530, 531.
[467] See below, pp. 495-497.
[468] See below, pp. 467, 468; 576-577; 225.
[469] See below, pp. 663-665.
[470] See below, p. 649.
[471] See below, pp. 471-473, 475.
[472] See below, pp. 422, 424, 439.
[473] See below, p. 304.
[474] See below, pp. 32, 34.
[475] See below, pp. 549, 555.
[476] See below, pp. 125, 168-169, 180, 182.
[477] See below, p. 216.
[478] See below, pp. 495, 496, 522.
[479] See below, pp. 558, 563. (Wine fleetingly occurs in the rental for Kilwinning abbey, whose monks received four hogsheads of wine from Irvine. See below, pp. 576, 578.)
[480] See below, pp. 277, 278.
[481] See below, pp. 590-592, 606.
[482] See below, p. 634.
[483] See below, pp. 30-37, 342, 346, 384, 386, 389, 511.
[484] See below, pp. 58, 360.
[485] See below, pp. 155, 156, 163, 180.
[486] See below, pp. 24, 29, 47, 51, 102, 316, 366-367.
[487] See below, pp. 376, 405.
[488] See below, pp. 2-5, 6, 9, 12-20, 22, 25-29, 32-34, 36-43, 47-53, 58-59, 61, 63-65, 69-70, 89, 91-94, 103, 107, 109, 116, 141-143, 146-148, 151, 162-165, 167-169, 177-179, 181-182, 186-196, 199-201, 204-205, 208-210, 216-221, 224-226, 228, 230, 235-236, 238-239, 241-242, 245, 249, 281, 285, 288-289, 292-297, 309, 332-333, 343, 348-349, 351, 355, 360-361, 363, 365-367, 369-371, 376, 383-384, 387-388, 393, 409-411, 415-418, 458-459, 466-468, 471-473, 479-480, 547-548, 550, 576, 578.
[489] See below, pp. 116, 141, 142, 143, 193, 196, 199, 204, 205, 249, 466.

rendered in certain rentals. In Fife, the leased kirklands of Abdie parish yielded in rent almost 19 chalders of bere, over 17 chalders of meal but less than 5 chalders of wheat;[490] at Balmerino abbey, the rent in victual comprised some 21 chalders of bere, 15 chalders of meal, and four chalders of wheat.[491] The farmers in over fifty settlements within Riccarton parish in Kyle paid their rents wholly in meal and bere, totalling more than 16 chalders of meal, and two chalders of bere.[492]

The appearance in rentals of legumes - peas and beans - is a reminder, too, of an element of diversity in cropping, and presumably in crop rotation, adopted in some areas where the emphasis on arable was most pronounced: Berwick,[493] Haddington,[494] Linlithgow,[495] Roxburgh,[496] and Fife.[497] The production of lime, not just for use in mortar, carried the prospect of increased yields of cereal and improved pasture in accessible areas, notably in coastal Fife, also at Melrose and in Galloway (where a boat belonging to a lime kiln appears on record).[498] Smaller quantities of flax, or lint, and hemp, which were also grown, were apt to figure among vicarage teinds.[499] Orchards, often associated with monastic houses and coveted by laymen, were also capable of yielding produce. The mains farm of Strathisla, with 'tour fortalice and orchard' was leased from Kinloss abbey;[500] Coupar Angus abbey had an orchard at Carsegrange;[501] Dryburgh had its 'New Orchard';[502] the Sciennes nunnery rented its orchard;[503] and to Kelso abbey belonged the 'King's orchard' in Maxwell parish;[504] settlements, too, might derive their name from their connection with an orchard: Orchardton connected with Sweetheart abbey and with the priory of St Mary's Isle;[505] 'Orcheart of Cleyne', attached to Scone abbey; and 'Yettoun Orchart', associated with St Mary's Isle.[506]

It is a reasonable inference from the evidence of the rentals, too, that the ready availability of peat[507] and heather, and of access to wood in certain areas,[508] together with the exploitation of surface seams of coal around the estuary of the

---

[490] See below, pp. 32-33.
[491] See below, p. 59.
[492] See below, pp. 531-532.
[493] See below, pp. 193, 196 (beans); 192-196, 199-201, 204-205 (peas).
[494] See below, pp. 168-169 (peas and beans).
[495] See below, p. 151 (peas).
[496] See below, p. 209 (peas).
[497] See below, pp. 2, 5, 9, 12, 13, 16, 17, 19, 22, 146, 167 (peas); 9, 13, 16, 17, 22, 146, 167 (beans).
[498] See below, pp. 23, 24, 26, 27, 42, 44, 45, 48, 49, 52, 53; 209; 586, 589.
[499] See below, pp. 22, 79, 84, 157, 188, 204, 203, 305, 307, 503, 550, 639, 640.
[500] See below, p. 461.
[501] See below, pp. 369; cf p. 354.
[502] See below, p. 189.
[503] See below, p. 94.
[504] See below, p. 225.
[505] See below, pp. 622; 619.
[506] See below, pp. 331; 619.
[507] See below, pp. 209-211, 260, 336, 585, 586, 589, 591, 592, 656-658, 661, 669.
[508] See below, pp. 30, 31, 35, 36, 314, 464, 617.

Forth and in Lanark,[509] helped keep the population warm. Around the Firth of Forth, with the mining of coal, was the allied industry of salt panning: salt featured as rent accruing to Pittenweem priory; teinds were levied on salt at Anstruther; Dunfermline abbey leased 23 salt pans; Culross abbey had 11 pans to let; Melrose had four at Preston in East Lothian, which were 'out of use of payment'; and in Edinburgh the salt tron for weighing was situated 'on the north syd of the gait' or port, near the Nether Bow and the adjacent burgh of Canongate.[510]

The rearing of livestock involved the management not only of herds of cattle and flocks of sheep - the 'marts' and 'muttons' of the rentals - but also large quantities of geese, capons and hens, and some goats and pigs. Payments due from tenants on kirklands were sometimes forthcoming in marts and wedders in Orkney, Ross, Moray, Aberdeen, Perth, Fife and Lanark.[511] Most rural families possessed at least a cow, a heifer and a calf: in the parish of Airth, 75 parishioners produced some 60 lambs and 127 cows, with lint and hemp, in Easter offerings alone; and elsewhere there is incidental mention of teind stirks (and milk) among vicarage payments.[512] There is also mention of 'fed oxen',[513] as distinct from 'fed-oxen silver'. Horses are seldom listed: apart from mention of horsecorn,[514] horse grass,[515] horse fodder,[516] and stables,[517] incidental reference is made to horses for carrying fuel,[518] to the pasturing of a horse,[519] the feeding of a horse,[520] the chamberlain's horse at Pluscarden priory,[521] the horse belonging to the bishop of Galloway's chamberlain,[522] and the minister's horse at Kirknewton.[523]

The appearance of lambs and 'muttons', and also wool, in rentals where churchmen continued to be paid in kind, and not in cash, was pronounced in parts of Ross, Aberdeen, Kincardine, Forfar, Fife, Lothian, Berwick, Dumfries and Kirkcudbright.[524] In addition, in the northern isles, 'flesh' was a recurring item paid as rent.[525] 'Hogs', 'fed boars', 'pigs' and 'swine' appear on record, notably in

---

[509] See below, pp. 40, 41, 53, 103, 233, 236, 239, 509, 549.
[510] See below, pp. 22, 24, 26, 48, 30, 35, 44, 46, 52, 91, 93, 140, 209-211, 260, 289, 626.
[511] See below, pp. 658, 661; 625-627, 633-634, 649-651; 467; 413-415, 417; 293, 305; 34; 495-496.
[512] See below, pp. 157-158, 576-577; 407.
[513] See below, p. 663.
[514] See above, pp. lviii, lxii-lxiii, lxxi.
[515] See below, p. 84.
[516] See below, p. 2.
[517] See below, p. 163.
[518] See below, p. 362.
[519] See below, p. 74.
[520] See below, p. 634.
[521] See below, p. 473.
[522] See below, p. 591.
[523] See below, p. 104.
[524] See below, pp. 64, 79, 84, 106, 110, 122, 127, 128, 152, 157, 158, 168, 188, 203, 213, 272, 374, 407, 431, 609, 639, 640 (lambs); 4, 625-627, 647, 649 ('muttons'); 22, 64, 79, 84, 85, 106, 110, 152, 168, 188, 203, 272, 374, 407, 431, 434, 609, 610, 614, 639 (wool). See also below, p. 580.
[525] See below, pp. 655-658; 662, 664, 668-669, 671.

Kyle, Orkney, Aberdeen and Perth.[526] Kids, on occasion, figured as payment in kind, particularly in Ross.[527] Geese,[528] capons,[529] hens,[530] 'fowls',[531] moorfowl[532] and poultry[533] generally were plentiful in most regions and continued to be used in paying rent. Dairy produce - eggs, [534] butter[535] and cheese[536] - also formed an element in rents and teind; and in the northern isles oil[537] and wax[538] were distinctive items in rentals. There, too, rabbits ('cunings') also featured as payment in kind.[539]

A ready availability of fish from the rivers, lochs and coastal waters - of salmon[540] and grilse,[541] trout,[542] white fish,[543] and herring,[544] salted and dried fish,[545] 'scray' or saithe[546] and also eel[547] - was yet another ingredient in ecclesiastical rentals, and a source of sustenance for churchmen, and others too, not least on Fridays, 'halydayis'[548] and during Lent. 'Hand fishing',[549] in one instance, was distinguished presumably from net fishing. Water also formed a highway of communication; and fishing boats and 'cobles', ferry boats and cargo boats (which were subject to teinding) also feature in the rentals.[550]

Obstacles in winning a living from the soil are apparent, too, in observations on adverse weather, floods, soil erosion, poor pollination and a

---

[526] See below, pp. 92, 305, 336, 385, 385, 414-415, 417, 562-564, 656, 658, 661-663, 669.
[527] See below, pp. 625-627.
[528] See below, pp. 82, 165, 168, 203, 378, 385-386, 414, 415, 417, 418, 628-629.
[529] See below, pp. 3-5, 9, 17, 34, 44, 53, 58, 59, 91-93, 185, 191, 202, 203, 204, 211, 352, 363-364, 371, 385-386, 413-415, 417, 418, 539, 579, 582-583, 626-627, 647, 649-651.
[530] See below, pp. 92-93, 583.
[531] See below, p. 211.
[532] See below, p. 414.
[533] See below, pp. 3, 4, 34, 41, 44, 53, 57, 58, 59, 165, 185, 191, 202-205, 352, 363-364, 371, 385-386, 413-415, 417-418, 579, 583, 662.
[534] See below, pp. 307, 626.
[535] See below, pp. 23, 44, 46, 53, 209, 211, 252, 260, 494, 639, 656-658, 661-667, 669-671.
[536] See below, pp. 23, 44, 46, 53, 79, 84, 168, 188, 203, 209, 252, 292, 293, 307, 407, 413, 431, 464, 494, 531, 562, 564, 577, 589, 592, 614, 639.
[537] See below, pp. 656, 658, 660-663, 669, 670.
[538] See below, pp. 657, 658, 661-663, 669. For wax elsewhere, see below, pp. 283, 287, 382, 479, 549, 599.
[539] See below, pp. 656, 658, 668.
[540] See below, pp. 58, 91, 93-94, 145, 167, 169, 210, 222, 277, 333, 351, 359, 362-364, 385-386, 393, 415-417, 423-424, 432, 465-467, 471, 473, 496-497, 530, 549-550, 562-564, 586-587, 589-592, 606, 626-628, 649-651.
[541] See below, pp. 562-563.
[542] See below, pp. 145, 167.
[543] See below, pp. 13, 359.
[544] See below, pp. 22, 512.
[545] See below, p. 22; cf. p. 626.
[546] See below, pp. 656, 658, 661, 663, 668.
[547] See below, p. 145.
[548] See below, pp. 416, 417.
[549] See below, p. 471.
[550] See below, pp. 56, 168, 203, 222, 230, 233, 236, 239, 301, 419, 422, 466, 467, 471, 473, 503, 555, 586, 589, 663, 639, 666.

dearth of salmon in certain rivers. At Holywood abbey, some land was 'worne away with the water'.[551] In Moray, difficulties also arose from 'inundatioun and sandinge of the landis be watteris';[552] and in Perthshire flood-water from the River Tay carried off crops of wheat, bere and white oats.[553] High winds in Orkney had blown sand over lands in Deerness, Burray, and Ronaldsay.[554] The complaint recorded in the rental for Holyroodhouse concerned the 'sterilitie of the fluris';[555] and at Paisley abbey it was claimed that owing to the 'sterility of the river' few salmon were caught 'these three and four years'; the burns of Paisley and Blackston were also said to be 'barren', and merely 32 salmon had been caught.[556] In other cases, human agencies were to blame: at Kelso abbey, the laird of Cessford, the abbey's bailie, had withheld payment for certain lands 'sen the deceis of King James the fyft'.[557]

For those in gainful employment, fear of 'broken men', masterless men, was a recurring theme in certain areas. The bishop of Ross complained that his palace at Fortrose, 'quhilk lyis in ane far heland cuntrie', was attacked and held for nine months or so by 'brokin men' who oppressed tenants on surrounding land.[558] Nor was lowanders' distrust of highland clans altogether unfounded: Orcadians complained of the depredations of Lewis men and of how, on Westray, 'the rowmes ar devastat be the hiland men';[559] the parson of Croy and Moy, 'lyand amang the Clan Chattan', was bereft of revenues;[560] and the vicar of Abertarff, at the head of Loch Ness, lamented he had gained not 'ane grot' in income on account of Clanranald 'and utheris sik clanis'.[561]

The disabled and deserving poor, who might expect to receive support from a wealthy church, also fleetingly cross the pages of the record. Whatever assistance monasteries may have given the poor,[562] other ecclesiastical institutions - hospitals, almshouses and leper houses - had a specific role in caring for the infirm and sick. Over fifty hospitals can be documented at the Reformation.[563] The rental submitted for the hospital of St Paul's Work in Leith records an income of £20 assigned to William McDowall, a chaplain, who also held offices in St Giles' collegiate church and at the abbey of Holyroodhouse.[564] The more detailed rental for the hospital of St Anthony for the poor and infirm, adjacent to Leith, whose income totalled £70, derived modest revenues from 'procurations or

---

[551] See below, p. 276.
[552] See below, p. 467.
[553] See below, p. 289.
[554] See below, p. 656, 658, 661, 671.
[555] See below, pp. 92-93.
[556] See below, p. 530.
[557] See below, p. 225.
[558] See below, p. 627.
[559] See below, p. 666-667.
[560] See below, p. 639.
[561] See below, p. 488.
[562] See above, p. lxiii.
[563] I. B. Cowan and D. E. Easson, *Medieval Religious Houses Scotland* (London, 1976), pp. 163-168.
[564] See below, p. 95, and n. 18.

pardons'; mendicants, authorised to beg, whose names are given, were assigned particular areas: Cunninghame, Carrick, Kyle and Galloway; Annandale, Nithsdale, Wauchopedale, Eskdale, Ewesdale, Clydesdale and Lennox; Merse and Teviotdale; east Lothian; west Lothian; Fife, Stirlingshire and half of Strathearn; Aberdeen, Moray, Ross, Caithness, Orkney and Shetland. Most income, however, accrued from annual payments from land or property; but there is no indication in the accounts of monies disbursed in caring for the poor of the hospital.[565] In the north, the rental for Rathven parsonage, 'with the annexis thairof', reveals payments to six bedesmen (of Rathven hospital) who were assigned 42 merks and, for their 'habits', £7 4s.[566] A subsequent rental for the kirks of Rathven, Dundurcas and Kiltarlity (whose revenues with those of the hospital of Rathven during the middle ages had been assigned as a prebend of Aberdeen cathedral) also reveals the continuing allocation of pensions to bedesmen of the hospital after the Reformation.[567] There is mention, too, of 'the hospitall of Glesgow' which received a meagre pension from Renfrew parsonage;[568] of the 'lipper men' at the hospital of St Mary Magdalene near Musselburgh;[569] of the poor children or 'blue friars' of the hospital of St George at Dunkeld;[570] the bedesmen of St Leonard's hospital in Edinburgh;[571] the hospital of St Nicholas in St Andrews.[572] At Brechin, the preceptory or maison dieu at Brechin had been leased to Carnegie of Kinnaird for £40, and a pension of £20 paid to the parson of Cookston;[573] at Stirling, the maison dieu, belonging to the Black friars, was occupied by the earl of Montrose for 50 shillings in rent.[574]

The role of collegiate churches in assisting the poor is sometimes demonstrable. Corstorphine collegiate kirk had a hospital attached;[575] Lincluden collegiate kirk had facilities for a provost, 8 priests and 24 bedesmen;[576] and in Edinburgh, the chaplains of St Salvator's altar and of St John the Evangelist in St Giles' collegiate kirk each contributed five merks yearly to the poor.[577] More generally, certain ecclesiastical revenues from property in Edinburgh were assigned to the 'bedmen yeirlie', apparently bedesmen attached to Dunbar collegiate kirk.[578]

Again, the lands of Massindew, belonging to the Black friars of Elgin, recall the hospital of Maison Dieu in Elgin;[579] and near Aberdeen 'the kirk of

---

[565] See below, p. 113.
[566] See below, p. 456.
[567] See below, p. 248.
[568] See below, p. 532.
[569] See below, p. 39.
[570] See below, p. 304.
[571] See below, p. 92.
[572] See below, pp. 13, 20, 21.
[573] See below, p. 390.
[574] See below, p. 554.
[575] See below, p. 100.
[576] See below, pp. 613, 620-621.
[577] See below, pp. 96, 138.
[578] See below, p. 122.
[579] See below, p. 453.

Spittall callit the hospitall' is a reminder of the hospital of St Peter there.[580] In the record, the place-name of Spittal is associated both with Jedburgh abbey[581] and with the parish of that name in Caithness, to which was attached the hospital of St Magnus.[582] After the Reformation, Queen Mary's privy council decreed in February 1561/2 that revenues from chaplainries in burghs should be applied to hospitals, schools and 'other godly uses', and, in January 1566/7, the council determined that the annuals of altarages, chaplainries and obits (on the decease of possessors) should be granted to the burghs for the support of the ministry, the poor and hospitals.[583] In a rental for Restalrig collegiate church, it was noted that certain revenues from chaplainries were 'now pertenand to the ministrie and to the puir'; and a later rental for 1574 disclosed that a prebend of Crail collegiate church had been allocated to the minister of Crail.[584]

Signs of a solicitude for certain disadvantaged folk are discernible in the depiction of a chaplain as 'ane auld blind man that seis nocht perfytlie';[585] in payments to 'ane auld man' in Orkney[586] in pensions to 'tua puir men' considered to be over 76 years of age;[587] in the claims of two old men for pensions for former services at Kelso abbey;[588] in the double portion allotted to 'ane auld man' at Jedburgh abbey;[589] and in the pension of £40 awarded by Mary of Guise to a Scottish soldier 'lamit in France'.[590]

Scotland's close ties with the continent, especially with France, may be traced in the visits to France of the parson of Ashkirk, John Mun,[591] and of the parson of Dunbar, John Hamilton, depicted as a 'student in Pareis'.[592] The young bishop of Brechin, Alexander Campbell, it was noted in 1574, was 'at the scullis' in Geneva.[593] Much earlier the Piedmontese scholar, Giovanni Ferrerio, who taught the monks of Kinloss, gained a pension for life of £40;[594] and a Frenchman was in receipt of £48 in pension from Coldingham priory.[595]

Of schools and schoolmasters, there is only incidental mention: the school at Arbroath, and grammar schoolmaster there;[596] the schoolmaster of Crail;[597] the

---

[580] See below, p. 434.
[581] See below, pp. 216, 218.
[582] See below, p. 631.
[583] *RPC*, i, pp. 202, 497-498.
[584] See below, pp. 144; 66.
[585] See below, p. 81.
[586] See below, p. 659.
[587] See below, p. 62.
[588] See below, p. 229.
[589] See below, p. 216.
[590] See below, p. 363.
[591] See below, p. 212.
[592] See below, p. 169.
[593] See below, p. 388.
[594] See below, p. 463.
[595] See below, pp. 200-201.
[596] See below, pp. 79, 83.
[597] See below, p. 66.

Aberdeen grammar schoolmaster;[598] the masters of the grammar and song schools at Dunkeld;[599] the grammar schoolmaster of Elgin;[600] the master of Stirling grammar school;[601] and the teacher of the song school in St Giles' collegiate church in Edinburgh.[602] There is mention, too, of David Vocat, an earlier grammarian in Edinburgh, who founded the manse of the vicarage at Livingston (annexed to the provostry of St Mary in the Fields' collegiate kirk in Edinburgh).[603] The appearance of the rental of a solitary parish clerkship - Fordyce - is a reminder of the one office in the late medieval church where appointments lay entirely at the disposal of parishioners.[604]

More promisingly, insofar as the record serves as a guide in documenting the early work of ministers, exhorters and readers in the reformed church, the 'Books of Assumption' supplement, and amplfy, information contained in the accounts of the collectors of thirds of benefices, and the register of stipends (which, though in existence in 1561, survives only from 1567).[605] Independently of the impost of the thirds in 1562 to support the royal household and reformed ministry, benefice-holders were already accustomed, in numerous cases, to contributing voluntarily a portion of their income towards reformed service in the parishes. It was a small price to pay, some may have reflected, for continued possession of their livings after the Reformation. Accordingly, those possessors who already supported ministers, exhorters and readers from their revenues expected these subventions to be taken into account when assessed for the thirds. Deductions in favour of the reformed ministry were usually counted (and accepted) as non-taxable allowances in monastic rentals and in those of other higher institutions, though frequently in parochial rentals allowances of that sort were ignored in calculating thirds. Nonetheless, rentals showing deductions for supporting the ministry are a valuable guide to the pioneering work of the reformed church. The pronouncement, favouring the *Book of Discipline*, by the privy council and convention of lords in January 1560/1, that beneficed men (who had 'adjoyned' or identified themselves with the protestant revolution) should retain their livings for life, provided they gave financial support to the reformed ministry,[606] did not become mandatory but it helped to foster a climate of co-operation between many benefice-holders of the old regime and ministers of the protestant church who lacked adequate endowment.

Many heads of religious houses showed a readiness to support reformed service in parishes appropriated to their houses, either by assigning, or by

---

[598] See below, pp. 425, 442.
[599] See below, pp. 304.
[600] See below, p. 466.
[601] See below, p. 545.
[602] See below, p. 124.
[603] See below, p. 106.
[604] See below, pp. 456-457.
[605] *Thirds of Benefices, passim; Register of Ministers, Exhorters and Readers*, ed. A. Macdonald (Edinburgh, 1830); *The Miscellany of the Wodrow Society*, ed. D. Laing (Edinburgh, 1844), pp. 329-394.
[606] Knox, *Works*, ii, pp. 129-130; *The First Book of Discipline*, pp. 210-211.

contributing towards, stipends for ministers, exhorters and readers in attached parishes. In his rental submitted in January 1561/2, the commendator of Culross abbey, who opted for a policy of awarding pensions to monks conforming to protestantism but not to those refusing to recant,[607] voluntarily paid the stipends of ministers in two appropriated parishes, in accordance with the 'Buk of Reformatioun' (or *Book of Discipline*): the minister at Culross received a generous £200, and the minister of Crombie gained £100, by no means an unsatisfactory salary.[608] The minister of Dundee also obtained £100 from Lindores abbey,[609] the appropriating house, whose abbot had conformed to the new regime. The commendator of Arbroath abbey, in the accounts for 1561, also indicated his willingness to support 'the ministeris in every kirk that is sustenit thair' in appropriated parishes.[610] In the north east, Kinloss abbey, with appropriated revenues from two parishes, contributed £26 13s 4d to maintaining 'the redare of the commoun prayaris' at Ellon;[611] Balmerino abbey in Fife, which drew its revenues from the parishes of Balmerino, Logie-Murdoch and Barry, provided £10 for the reader at Balmerino; £100 for the minister of Balmerino and Logie-Murdoch; and another £100 for the minister at Barry;[612] and the small Borders nunnery of Coldstream, with its parishes of Bassendean, Hirsel and Lennel (or Coldstream), made provision for 'a ministeris fie'.[613]

The same practice was observed at Scone abbey, which offered £23 6s 8d, plus 5 chalders of victual, to the minister of Scone, Cambusmichael and Kinfauns; £23 to the reader at Kinfauns; £58 to the minister of Logierait; £60 to the minister of Blairgowrie; £60 6s 8d for reformed service at Redgorton; 4 chalders of victual for service at Kilspindie and Rait; and £20, plus 4 chalders of victual, for service at Logie-Dundee, Liff and Invergowrie.[614] Blantyre priory, in origin a cell of Jedburgh abbey, devoted 40 merks to the minister of its solitary attached parish of Blantyre.[615] At Dundrennan abbey, the commendator displayed a willingness to support both a minister and a reader at each of its two annexed parishes: the minister of Dundrennan was allocated £20 6s 8d, and his assistant, the reader at Drundrennan, gained only marginally less with a stipend of £20; at Kirkmabreck, the minister obtained £30 and the reader there, £20.[616] The abbot of Sweetheart complained not of supporting protestant ministers but of 'evill payment' from the annexed parishes of Buittle and Crossmichael, which made it hard for him to 'sustein tua qualifiet ministeris'.[617] The small Trinitarian house of Scotlandwell in

---

[607] See above, p. lxii.
[608] See below, p. 292.
[609] See below, p. 32.
[610] See below, pp. 361, 363.
[611] See below, p. 463.
[612] See below, pp. 58-59.
[613] See below, p. 186.
[614] See below, pp. 333-334.
[615] See below, p. 505.
[616] See below, p. 606.
[617] See below, pp. 611-612.

Kinross contributed 20 merks to the reader's stipend at Carnock.[618] The pattern was repeated at the Trinitarian house at Fail, where Rankin Davidson, 'minister' at Galston, obtained £40; the minister of Symington, 50 merks; the minister of Barnwell, 42 merks; the minister of Torthorwald, 24 merks; and the minister of Inverchaolain in Argyll, £10.[619] A following rental for Fail, dated February 1562/3, disclosed that £130 was disbursed in ministers' stipends in 1561: Rankin Davidson, as 'exhorter' at Galston, received a stipend of £50; John Millar, exhorter at Barnwell, gained £40; Thomas Carrington, reader at Symington, had £20; and John Wallace, reader at Torthorwald, also received £20.[620] At Holyroodhouse, too (whose commendator had attached himself to the protestant cause), stipends were forthcoming for the minister of St Cuthbert's (£80), beside Edinburgh; the minister of Holyroodhouse, or Canongate (£80); the minister of Falkirk (£46 13s 4d); the 'minister of Galloway' (£50); and the minister of Barro (£3 6s 8d).[621] Holywood abbey in Galloway, which drew revenues from five annexed parishes, sustained a minister at Holywood.[622]

Where religious houses had leased the fruits of appropriated parishes, the tacksman frequently assumed responsibility for paying stipends. At Monymusk, the commendator entered into an arrangement with the lease-holder of the priory at the Reformation whereby the tacksman became liable for paying the stipend of 'ane sufficient precheour' at Monymusk, and the salaries of ministers at the other churches annexed to the priory.[623]

Bishops and cathedral dignitaries of the old regime, some of whom were won over to protestantism and to service in the reformed church, also adopted a strategy of assigning a small portion of their incomes to assist early ministers in the reformed church who served in appropriated parishes. At Ashkirk, a prebend of Glasgow cathedral, the archbishop's factor deducted from an income of £120 (derived from a lease of the benefice) 20 merks as payment to the minister of the parish.[624] At Govan, another prebend of the cathedral, the minister collected 40 merks from the benefice.[625] In Ross, the bishop (who did not conform to the Reformation) gave £50 to 'the preacher' at the kirks of Nigg and Tarbat, two parishes assigned to the bishop's prebend.[626] The chanter at Dunkeld cathedral granted £10 to the minister of Kinclaven, whose benefice formed the chanter's prebend.[627] At Methlick, where the parsonage and vicarage were annexed to Aberdeen cathedral, the prebendary paid the parish minister 20 merks.[628] Provision was also made at Auchterless, the prebend of Aberdeen's chanter, for

---

[618] See below, p. 56.
[619] See below, p. 558.
[620] See below, p. 563.
[621] See below, p. 92.
[622] See below, pp. 277, 624.
[623] See below, pp. 446-447.
[624] See below, p. 212.
[625] See below, p. 533.
[626] See below, p. 627.
[627] See below, p. 306.
[628] See below, p. 435.

'ane resonable fie to ane minister'.[629] The subchanter of Aberdeen, who held the vicarage of Kinairney in Mar, gave £13 6s 8d 'to ane chaplane and minister to serve the cuir'.[630] The archdeacon of Brechin set aside £26 for 'the ministeris' (presumably to officiate at Strachan, annexed to the archdeaconry).[631] Cookston, a prebend of Brechin leased to Carnegie of Kinnaird, set aside £10 (which was half the income from the lease) for a minister of the parish.[632] A deduction of 50 merks from the chancellory of Ross, leased for 460 merks, was allocated to 'the vicaris and chaplandis, ministaris of the samin', at the appropriated churches of Suddie and Kinnettes;[633] and 100 merks was set aside from the revenues of the treasurership of Ross, leased for 300 merks, 'for the uphold of the kirkis and to the ministaris' of the annexed parishes.[634] At Kiltearn, also in Ross, the parson of the leased benefice undertook to pay the minister who served the parish.[635] As the vicarage revenues of Moy (itself a prebend of Elgin cathedral) 'may nocht sustene ane minister of the tendis', the parson contributed a scanty 8 merks towards the minister's 'fie' from the crop of 1560.[636] The leased vicarage of Linlithgow (whose parsonage was annexed to St Andrews priory) yielded 10 merks for the vicar and 20 merks 'to be payit yeirlie to the man that sayis the commoun prayeris';[637] £10 was allocated 'to a curet or a minister' from the parsonage and vicarage revenues of Melville in Lothian.[638] In the Merse, the reader at Stow, a mensal church belonging to St Andrews cathedral, obtained 20 merks from the fruits of the vicarage pensionary.[639]

Again, in other cases, where benefices were leased, churchmen were apt to transfer responsibility for paying a minister to the tacksman of the living. At Walston in Lanark where the vicarage was leased to Michael and John Leishman, 'and utheris thair coligis', for 50 merks, the tacksmen paid 20 merks to Laurence Leishman, the parish minister.[640] In Galloway (whose bishop opted for service in the reformed church), the tacksman of Senwick vicarage, a parish appropriated to Tongland abbey (itself annexed to the bishopric), reported that an income of 40 merks was allocated so that 'ane reader within the said kirk may be sustenit to reid the commoun prayeris'.[641] At Straiton in Carrick, the vicar reported in 1562 that from the income of his benefice leased to the earl of Cassillis, he devoted 20 merks to the minister serving the parish.[642]

---

[629] See below, p. 430.
[630] See below, p. 434.
[631] See below, p. 380.
[632] See below, p. 382.
[633] See below, pp. 634-635.
[634] See below, p. 637.
[635] See below, p. 635.
[636] See below, pp. 470-471.
[637] See below, p. 153.
[638] See below, p. 118.
[639] See below, p. 120.
[640] See below, p. 513.
[641] See below, p. 604.
[642] See below, p. 569.

In Lanark, the burgh took the initiative in diverting £2 in annuals from the chaplainry of Our Lady altar in St Nicholas' kirk 'quhilk was wont to be givin to ane preist to say mes twys in the week' and applied the meagre proceeds 'to help to say the comoun prayeris within the kirk'.[643]

In Orkney diocese (whose bishop joined the ranks of the reformers), where most churches were either mensal or prebendal, the rental of the bishopric identified those parish churches in both Orkney and Shetland 'that hes neid of ministeris to serve the people, minister the sacramentis, to instruct and teach thame in the knawledge of the word of God',[644] a statement which should not be misconstrued as evidence that the parishes lacked reformed service - for Orkney's record here was exemplary - but interpreted as an indication of the churches, in insular communities, selected for protestant worship conducted by a minister and not, as was often the case, by an assistant exhorter or reader.

Collegiate churches, sustained from parochial resources, and unappropriated parishes also gave some financial succour to early recruits as ministers, irrespective of the levy of the thirds. A prebend of Lincluden collegiate kirk yielded £20 to the prebendary and 22 merks 'to ane redar and ane sangstar'; another prebend helped sustain a reader at Parton with a salary of £10.[645] The provostry of Crail collegiate kirk sustained a reader with a salary of £20.[646] At Kirkinner and Kirkcowan in Wigtown, prebends of the subdean and sacristan of the Chapel Royal in Stirling, the holders each contributed 50 merks as payment to 'the minister and precher'.[647] The parson of Strathbrock awarded the minister serving the parish £20;[648] his counterpart at Hawick affirmed that from an income of £163 6s 8d 'I sustein a minister'.[649] The parson of Newdosk, an unappropriated living, financed 'the minister's fie' from the crop of 1561;[650] at Torrie in Fife, the parson took 20 merks and the minister collected 18 merks;[651] and the vicar of Stirling kept £20 and gave the reader £16.[652] In his rental for Dolphinton, in February 1561/2, the parson recorded payment to the minister for the past year of £13 6s 8d;[653] at Luncarty, an unannexed benefice in Gowrie, the parson supported 'the curat and ridar';[654] and at Benvie, also unappropriated, the parson gave 20 merks 'to be payit to ane minister'.[655] At Peebles, too, the vicar considered 20 merks an appropriate sum for 'the minister'.[656] Nearby, at Lyne the parson

---

[643] See below, p. 524.
[644] See below, pp. 659-660.
[645] See below, pp. 272-273.
[646] See below, p. 82.
[647] See below, p. 602.
[648] See below, p. 153.
[649] See below, p. 263.
[650] See below, p. 382.
[651] See below, p. 73.
[652] See below, p. 550.
[653] See below, p. 504.
[654] See below, p. 324.
[655] See below, p. 334.
[656] See below, p. 251.

contributed £10 to the reader of the parish.[657] The minister of Wigtown gained 40 merks from the parson, who (as his rental submitted in February 1561/2 disclosed) also assigned 30 merks to the vicar pensionary who served as reader there.[658] At Morham in Lothian, the parson paid £16 to 'the redar and minister of the sacramentis'.[659]

Beneficed clergy of the old regime who conformed to the Reformation to serve the parishes whose benefices they held gained the income from their livings as stipends: in Moray, the vicar of Birnie, who opted for reformed service, styling himself 'vicar and minister of the said kirk and reader and exhorter', was financed from the fruits of his vicarage;[660] the vicar of Logie-Montrose, who also became minister of that parish, was supported by the parson and, as his rental reveals in January 1561/2, 'by order' of Erskine of Dun, a leading reformer (and hereditary possessor of the parish's teinds);[661] Andrew Hay, parson of Renfrew, who served the charge as minister also paid for a reader in the parish;[662] and Hay's brother (who, too, became a minister), as parson of Eddleston near Peebles, recognised that the first charge on his revenues was the maintenance of a reader in the parish 'according to the ordour and Buik of Discipline' (and that a second charge was payment 'to ane preist of the chore of Glasgow' and several vicars choral).[663] James Walker, parson of Stevenston and vicar of Inchcailloch (who entered the reformed ministry) indicated in his rental, submitted in January 1561/2, that he already sustained an 'under reider' from his revenues;[664] at Ratho in Lothian, the vicar pensionary, with an income of £10, served as reader of the parish.[665]

On the other hand, the vicar of Dingwall, who served as minister there, complained that he gained merely 5 merks, with some fishing, as income and so appealed for assistance to the queen that he might 'have ane lyf lyk ane minister'.[666] At Gamrie, in Aberdeen diocese, the vicar perpetual reported that his vicar pensionary, who joined the reformers to serve the parish as 'minister', lacked an adequate income, exacerbated with the extinction of clerical exactions.[667]

Finally, the unbeneficed curate of Conveth in the Mearns (a parish appropriated to St Mary's College in St Andrews University) entered reformed service as reader in the parish for a salary of £12;[668] and, at Tyninghame in Lothian, a benefice also annexed to the college, the curate acted as reader with a stipend of £12 13s 4d.[669] In all, this was a remarkable undertaking by beneficed men in providing a measure of financial support for the early protestant ministry,

---

[657] See below, p. 252.
[658] See below, p. 597.
[659] See below, pp. 170-171.
[660] See below, p. 487.
[661] See below, pp. 373-374.
[662] See below, p. 532.
[663] See below, p. 248.
[664] See below, p. 540.
[665] See below, p. 123, cf p. 131.
[666] See below, p. 639.
[667] See below, p. 443.
[668] See below, p. 64.
[669] See below, p. 65.

before the 'assumption of thirds' of 1562 (which gave rise to the 'Books of Assumption') took full effect. 'Si per laborem honeste quippiam feceris labor abit honestum manet quod si per voluptatem turpe quippiam feceris quod suave est abit quod turpe est manet.'[670]

12 The manuscripts and editorial method

The present work reproduces the substance of 929 folios contained in four separate manuscript volumes of the 'Books of Assumption', located in three archival repositories. Two volumes, E48/1/1 and E48/1/2, are preserved in the Scottish Record Office, H.M. General Register House, Edinburgh. The first contains an inscription on the cover to the effect that the earl of Dalhousie presented the volume to the Register House on 18 November 1864. Measuring 29.7 cms by 20 cms, and approximately 5 cms in depth (excluding the leather covers), the volume extends to 417 folios, and possesses an index prepared in the nineteenth century. The other volume, of earlier provenance, formed part of the Cunninghame of Caprington papers deposited in the Scottish Record Office in 1935 by Lt. Col. W. W. Cunninghame of Caprington Castle. In size, 28.7 cms by 19.7 cms, and in depth approximately 3.5 cms, the volume runs to 190 folios, containing the original record for Fife, Lothian, the Borders and Dumfriesshire.[671] This material is complemented by another volume in the National Library of Scotland, Adv. MS 31.3.12, which preserves the record for the east and north-east: Perth, Angus, Kincardine, Aberdeen, Banff and Moray.[672] In dimension, 29.7 cms by 20 cms, and approximately 5 cms in depth, the volume contains 174 folios. It is stamped 'Ex Libris Bibliothecae Facultatis Juridicae Edinburgi, 1808', but it was already in the possession of the Advocates' Library by 1807 when it was noted in that year's catalogue. (Three other references have been assigned to this manuscript at various times by the Avocates' Library: Jac.V.6.20, Historical Catalogue 79, and Wodrow Catalogue 300. Two further manuscripts in the National Library, Adv. MS 31.3.13 and 31.3.16, contain abridgments of part of the valuations, written in an eighteenth-century hand, but have no relevance in preparing the present edition.) SRO, E48/1/2 and NLS, Adv. MS 31.3.12 are the two earliest surviving manuscript texts; and SRO, E48/1/1, which engrosses so much of the material in these two volumes is a copy, prepared in 1605. The fourth manuscript volume, MS Dc.4.32, in Edinburgh University Library, is a copy prepared in 1624 of the record for the south west and north, from Wigtown to Inverness, Ross, Caithness and the northern isles.[673] The volume measures

---

[670] See below, p. 239.
[671] See above, pp. xvii-xviii; below, p. 1, and nn. 1-2.
[672] See above, p. xviii; below, p. 281, and nn. 1-2.
[673] See above, p. xviii; below, p. 495, and n. 1.

29.4 cms by 18.7 cms, and approximately 3 cms in depth. It consists of 196 folios, of which the first 148 folios contain rentals belonging to the 'Books of Assumption', and the remainder of the volume comprises extraneous material.

The text offered below, in calendared form, retains the order of the original manuscripts and as much of the 'flavour' as possible. Because of wide variations in the entries in the manuscripts, it has not been possible to adhere to a set form in calendaring the text, but each entry contains some or all of the following elements: the title of the benefice (rendered in bold, and usually found as a marginal rubric in the manuscripts); the source or sources of the manuscripts used in preparing the entry for the calendar (with the principal text placed in the superior position); foliation; description of the entry (rental, account etc.); the valuation of the benefice; deductions; the signature of the benefice-holder or person presenting; the calculation of the third; related royal warrants or similar documents, letters or precepts. In long entries these elements are set out in separate paragraphs, and large subdivisions of the valuation and deductions also have been assigned separate paragraphs. In short entries, some elements have been run together.

In each entry, the title of the benefice is placed in bold capitals, and spelling is modernised except for place names whose original orthography is retained. If the rubric is long and contains information about the benefice, all or part is treated as a description. Where no rubric exists in the original text, a title has been supplied. For entries which occur in more than one manuscript text, there are normally two foliations cited (SRO, Vol. a, denoting MS E48/1/1; SRO, Vol. b, denoting MS E48/1/2; NLS, denoting Adv. MS 31.3.12; EUL, denoting MS Dc.4.32 (for which there is no overlap with the other texts). The source of text and foliation is shown on the same line as the title, within round brackets, on the right-hand side of the page. Where an entry extends to more than one folio, the covering foliation is supplied. Reference is also made where an entry is defective or missing in one or other manuscript, or appears on a separate sheet inserted elsewhere in the volume, and where text has been supplied from the other manuscript.

The description of an entry is normally quoted as it appears in the main manuscript, but dates are modernised, as explained below. The valuation of the benefice is normally calendared, but following the order and wording of the original. Where, for example, the manuscript text reads 'Of the lands', which may imply that the income is not rent but an annual, the word 'Of' has been retained in the calendar. Ambiguous or unusual phrases have been retained as quotations. In the manuscript texts, individual items often appear on a separate line but in calendaring they have been run on as a single paragraph, using a full stop to separate items. If an item is subdivided, or if two or more consecutive items are related, the different parts have been separated by semi-colons. If a long valuation is shown in the manuscript as different branches of revenue, these have been treated as separate paragraphs in the calendar, with any general descriptive rubric set at the beginning of the paragraph. Marginal notes relating to, or qualifying, particular items appear in the calendar after the items and are introduced by the words '*In margin*' within square brackets.

Deductions have been treated in the same way as valuations. The phraseology associated with signatures in the text is rendered in the calendar in the form which occurs in the original. Designations (except in quotations) have been modernised. The calculation of the third is normally provided in calendared form. If no calculation appears in the manuscript, none has been supplied in the calendar. The text of related royal warrants, letters, precepts etc., is usually transcribed in full, without calendaring. In rendering names, personal surnames and place names are normally calendared in the form in which they are found in the main manuscript. Variant manuscript readings unless significant are not recorded. Christian names are modernised unless unusual. Colloquial forms are retained in modern spelling. Latin names have been translated; and contracted forms of names have been extended.

Figures are given in Arabic numerals. The following special applications may be noted. For dates, the day and year are rendered in figures, and the month in words. Before 1600 (the year when 1 January was substituted), the new year in Scotland officially began on 25 March. For clarity, dates between 1 January and 24 March are given in double form, e.g. 1561/2, both in the Introduction and in the text. In rendering money, the symbols '£' 's' 'd' are used to denote pounds, shillings and pence. All references are to pounds Scots. The use of 'merk' (denoting two thirds of a pound Scots) in the original has been retained in the calendar. Land measurements in acres denote Scots acres. Fractions, which can present special difficulty, have usually been rendered as vulgar fractions in the calendar, though for clarification some fractions are also quoted in the form in which they occur in the manuscript, e.g. '3 part ob.', etc. Such measurements of victual as chalders, bolls, firlots and pecks have been abbreviated to 'c.', 'b.', 'f.', and 'p.' The distinctive measurements employed in the northern isles of settins and meils are abbreviated as 'sts.' and 'm.'; marks and lispunds are rendered in full.[674] In citing numbers of livestock, fish etc., to avoid possible confusion over the use of the long hundred, amounts may be given, where appropriate, in a mixture of numerals and words, using 'hundred' ('c'), 'score' ('xx'), and 'dozen' where these forms occur in the manuscript; but 'xx' on its own is given as '20' and 'xij' as 12.

Some passages of text have been quoted, using single quotation marks, where they are of particular interest or where difficult to calender because of latent ambiguities. In quotations, the original orthography has been retained, except for the standardisation of 'i' and 'j', 'u', 'v' and 'w', 'y' and 'z', where appropriate. Quotations do not normally include dates or figures (except in the case of some fractions). Contractions and abbreviations, including thorn, yoch, superscript 't' and 'r', have been extended throughout, except for 'etc.' and 'viz'. Punctuation has been modernised. Editorial corrections and insertions are placed within square brackets. Where any gaps in the text can be filled editorially, square brackets are also employed. Throughout, every effort has been made to ensure that the calendar is a full one, and that nothing of significance is omitted. Indeed, many sections consist of a transcription of the original text.

---

[674] See below, p. 655, and n. 2.

The Latin text of some entries, usually those of religious houses, has been translated in calendaring, but a preliminary Latin phrase has been retained to indicate that the original entry was in Latin. A few entries in the original text are in a mixture of Latin and Scots. A glossary is provided for the many words and phrases in Scots which occur throughout the text.[675]

Editorial comment has been reserved for drawing attention to significant variant readings of manuscript texts, scribal errors (including arithmetical calculations), the identification of persons and places, and the elucidation of passages of text. Mistakes by scribes in calculating revenue have often been compounded by copyists who transcribed the documents. Where sixteenth-century calculations are at variance with the editor's, attention is drawn to the discrepancies in footnote references, but it is not always clear whether slips occur in the total figure or in component entries. No assurance is offered that editorial calculations are without blemish.

13 Note on vulgar fractions

It is worth placing on record some of the principles in the sixteenth century governing the use of fractions in calculation of the thirds. Where the original figure is a whole number (in either pence or chalders, bolls, firlots or pecks), the only possible fractions are '3' or '3 part' (one third) and '2' or '2 part'. Complications, however, arise in expressing a third of an original fraction. In money, these occur as fractions of the halfpenny and farthing, '3 part ob.' (one sixth of a penny) and '3 part f.' (one twelfth of a penny). They may occur on their own or in combination: for example, 'ob 3 d.' (five sixths of a penny), or even 'ob. 3 ob.' (two thirds of a penny). Although this appears a complex way of working, the method may result from breaking down the final remainder to produce a smaller figure which divides exactly by three: for example, $2 = 1\frac{1}{2} + \frac{1}{2}$, or $2\frac{1}{2} = 1\frac{1}{2} + 1$. With victual, there may be fractions of half, '3 part half' (one sixth), or fractions of third, '3 part 3' or '3 of 3 part' (one ninth), or fractions of quarter, '3 part fourth part' (one twelfth). These, too, may appear in combination: 'half part 3 part' (five sixths), '3 part peck 3 thereof' (four ninths). Here also there may be combinations which add up to what might seem simpler fractions: for example, '2 part half peck' (one third), 'half peck 3 thereof' (two thirds), or 'half peck 3 half peck' (also two thirds).

---

[675] See below, p. 685.

14 Note on indices

The principles adopted in arranging the Indices of Persons and Places broadly conform to those observed in the *Register of the Great Seal of Scotland* and in most volumes of *The Register of the Privy Seal of Scotland*. Entries for families with territorial titles are grouped before the remaining alphabetical entries for each surname. Personal names and place names are rendered, wherever possible, in standardised modern form. The Ordnance Survey spellings of place names are preferred to other forms. Variant sixteenth-century spellings of personal names and place names occurring in the text are placed in round brackets. Wherever possible Ordnance Survey national grid references are cited for the location of place names. Where a place name has not been identified, the name of the benefice where the place is cited (though not necessarily geographically adjacent) is entered in italicised form. Where the identity of a place name may be in doubt, the probable grid reference is supplied within square brackets and is preceded by an asterisk. Where an adjacent place name (of a castle, cottage, farm, house, mains, moor, etc.) is located in the Ordnance Survey's *Gazetteer of Great Britain*, the grid reference is provided within square brackets.

No separate index of offices is offered, for reference to all ecclesiastical offices may be traced appropriately under the entries for individual ecclesiastical institutions in the Index of Places. As with the *Privy Seal Register*, the contents of the Introduction are excluded from the indices.

Fife

ST ANDREWS, ARCHBISHOPRIC OF, (SRO, Vol. a,[1] fos 3r-7r)
(SRO, Vol. b,[2] fos 2r-6r)

'Rentale archiepiscopatus Sancti Andree respective'

Rental of the archbishopric of St Andrews respectively

The archbishopric of St Andrews at the present time, including all its fermes, sums of money, both for augmentations and grassums by reason of certain lands placed in feuferme and perpetual lease, both of the temporality and of the teinds of kirks set for victuals and sums of money annually, according to the rental and annual accounts of the chamberlains and the graniters of the same see, briefly written down, 6 January 1561/2.

---

[1] This denotes Scottish Record Office, E48/1/1. The description of the volume reads: 'Ecclesiastical Records. Book of Assumption to Benefices. For Counties of Fife, Edinburgh, Linlithgow, Haddington, Berwick, Roxburgh, Peebles, Dumfries, Perth, Forfar, Kincardine, Aberdeen, Moray. 1561 (Contemporary copy 1605).' This is followed, in a modern hand, by a list of sheriffdoms and similar administrative units: 'Fife, Edinburgh, Haddington, Berwick, Roxburgh, Peebles, Dumfries, Perth, Forfar, Kincardine, Aberdeen, Murray', with a note, 'See the end for the contents of the volume in the Advocates' Library.' Fo 1r reads 'Fyiff, Edinburgh, Hadingtoun, Beruick, Roxburgh, Peiblis and Dumfreis, 27 April 1605, Thomas Hope [with signature] Thomas Hope.' Fos 1v-2v are blank.

[2] This denotes SRO, E48/1/2 which previously formed part of the Cunninghame of Caprington Collection. The description of the volume reads: 'Partial Copy of Book of Assumption of an earlier date than 1/1 [*i.e.* SRO, E48/1/1]. (Another part of the same volume is in the National Library.)' This is followed by 'Assumptions of Thirds of Benefices in the sheriffdoms of Fife, Edinburgh, Haddington, Berwick, Roxburgh, Peebles, Selkirk and Dumfries. This volume is complementary to National Library of Scotland, MSS 31.3.12.* The two volumes (they are not two parts of the same volume) together cover the same parts of the country as the principal Register House volume. Several folios are awanting. The date of the volume is 1600 or later. *Containing rentals for Perth, Forfar, Kincardine, Aberdeen, Banff and Moray.' Fo 1r consists of various jottings unrelated to the text; fo 1v reads 'Thir ar the names of the schirrefdomes contenit within this buik 17 December 1573 deliverit [*damaged*] ordainis lettres of pensioun of Selkirk to his servand Mawnis per ... [*damaged*] of the parrys kirk. Memorandum, Sanct Magdalenis chaiplenrie thairof is nocht gyffin up be sir Thomas Kynneir, chaiplan thairof. Confermo^r [*sic*]. Fyfe, Edinburgh, Hadingtoun, Berwyck, Roxburgh, Peblis, Selkirk, Annandardaill, and Dumfreis, and na mae.' [Thomas Kinnear was chaplain of the Magdalene altar in the parish church of St Andrews in 1566, *RSS*, v, no 1974.]

The lordship of Monymaill in money with augmentations of the feufermes of the lands of Petcuntlie and Murehall extending in all, according to the rental and annual accounts, in money, £99 13s 4d [£99 14s 4d, *fo 5v*]. Total wheat of this lordship, 2 c. 12 b. Total bere of the same, 1 c. 8 b. Total oatmeal of the same lordship, 12 b.

The lordship of Byrhillis and Polduff in money according to the rental and annual accounts extending in all to £75 17s 8d. Total wheat of this lordship, 9 c. 5 b. 1½ f. Total bere of the same, 13 c. 5 b. 1½ f. Total oats of the same, 21 [c.] 5 b. 1½ f. No meal in this lordship.

Kyngorne Eister in all according to the rental and annual accounts in money, £10.

The lordship of Dersy in money and fermes of the lands of Mydil and Craigfudie set ('assedat') for the total of the free rents, thus augmented, and according to the annual accounts extending to £284 16d. Total wheat of the same lordship except the Fudeis, 8 c. Total bere of the same, 8 c. Total oats of the same, 2 c.

The lordship of Bischopschyre according to the rental and annual accounts in all extending to £119 4s 8d.

The lordship of Myckartschyre according to the rental and annual accounts in all extending to £104 2s 8d.

The lordship of Schottiscraig according to the rental and annual accounts, £148 7s 1d and in white peas of this lordship, 4 b.

The mair of Fyfe for kanes and annualrents of this, extending according to the rental and annual accounts to the sum of £57 18s.

Burgh fermes of the city of St Andrews, £8.

The great customs of the city of St Andrews according to the rental and annual accounts, £50.

The lordship of Angus in all, according to the rental, both for fermes and sums of money with augmentations, grassums, annualrents and kanes of the same lordship, similarly reckoned to extend, less than the old rentals, to £361 5s 5d.

The lordship of Kega and Monymusk in all, similarly reckoned according to the rental and annual accounts, both for ferme and for sums of money, augmentations and grassums extending to £154 5s 8d. Oats of the same lordship for horse fodder ('pro pabulis equorum'), 4 c.

The lordship of Stow according to the rental and annual accounts extending in all to £218 5s 4d.

The lordship of Tinnyghame in all with annualrents of the same according to the rental and annual accounts £84 4s 4d.

Total ferme wheat ('frumenti firmalis'), 20 c. 1 b. 1 f. 1½ p. Total ferme bere, 22 c. 13 b. 1 f. 1½ p. Total ferme oats, 27 c. 5 b. 1 f. 1½ p. Total ferme meal, 12 b. Total white peas meal ('pisorum albarum farinalis'), 4 b. Total money of the temporality of St Andrews, £2,080 3s 10d.[3]

---

[3] There seems to be a discrepancy here; the correct calculation appears to be £1,775 5s 6d.

Not charged with capons and poultry and gifts of augmentations because they were not accustomed to be sold or set for money but reckoned towards the upkeep of the castle of St Andrews annually.

Rental of the property of the spirituality of the archbishopric of St Andrews
Vicarage of Kirklistoun according to the rental and assedation, £13 6s 8d. Kirk of Cranstoun according to the rental and assedation, £26 13s 4d. The lordship of Litillprestoun in the parish of Cranstoun according to the rental, 40s. Kirk of Stow leased according to the rental for £133 6s 8d. Teinds of all and sundry of the lands of the lordship heritably belonging to John, lord Borthuik,[4] within the parish of Stow, according to the rental, 40s. Kirk of Monymaill in teinds according to the rental and annual accounts with defalcations and other non payments ('non solutis') in annual teinds, 4 c. 11 b. wheat; and in bere of the said kirk, 10 c. 8 b. 2 f.; and in oats of the same kirk 25 c. 5 b. Kirk of Kirklistoun in teinds leased for victuals as well as leased for money as follows according to the rental and annual accounts: Teinds of Eist Feild leased for £14. Teinds of Brigis remitted to Robert Hammiltoun of Brigis for annual service by reason of our letter written to him about the above matter. Teinds of Aldlistoun, Cliftoun and Cliftounhall leased for £40. Teinds of Elistoun, £16. Teinds of Cotlaw likewise remitted for the annual service of James Dundas of Newlistoun by reason of our letter to him about the above matter. Teinds of Kinpunt leased to the laird of Calder for £40. Teinds of Over Newlistoun leased to the son of the former laird of Dundas ('assedat quondam domino de Dundas filio') £4. Teinds of Wincheburcht and Humbie leased to lord Setoun[5] for £53 6s.

Total wheat of the same kirks except the assedation above, 2 c. 12 b. Total bere of the same kirks besides the assedation above, 4 c. 4 b. 1 f. Total oats besides the assedation above, 9 c. 11 b. 2 f.

Kirk of Prestoun Kirk alias Hauche, together with the teind sheaves of the lands of Eister Hallis, Trapren, Half Houstoun and Gurlay Bankis annexed to the archbishopric according to the rental and annual accounts, 3 c. wheat; and in bere, 4 c.; and in oats, 5 c. 8 b. [*In margin*, 'Nota, the thrid of thame ar deducit becaus thai ar nocht payit'.]

And also procurations and synodals and kanes of the archdeaconry of the archbishopric of the same in the year, calculated together according to the rental and annual accounts, £494.

Total of all teind victuals of the kirks of Monymaill and Kirklistoun and of Eister Halis, Trapren, Half Houstoun and Gurlawbankis yearly besides the victual leased

---

[4] John, 6th lord Borthwick, *The Scots Peerage*, ed. J. B. Paul, 8 vols. (Edinburgh, 1904-1914), ii, pp. 109-110.
[5] George, 5th lord Seton, *Scots Peerage*, viii, pp. 585-588.

above, calculated together, namely, in wheat, 10 c. 7 b.; and in bere, 18 c. 12 b. 3 f.; and in oats, 40 c. 8 b. 2 f.

Total of the money of the property of the spirituality of the archbishopric of St Andrews for kirks and for teinds set in assedation and procurations and synodals and kanes and fermes of the archdeaconry as above, calculated together in the year, £824 13s 4d.[6]

Sum of the totals, taking the temporality and the spirituality of the whole archbishopric of St Andrews in money, both in fermes of lands and lordships and of kirks and teinds leased for money, and feufermes, grassums, rentals, augmentations, annualrents, kanes, great and small customs and fermes from the burghs, and also with capons, poultry, oats, marts, muttons, customs in feuferme of the lands above set for money and procurations and synodals of all the lordships of the said archbishopric calculated together in the year, £2,904 17s 2d.[7]

Of which total sum in fermes there is to be deducted as follows

For the pension of Gavin,[8] commendator of Kylwynning and coadjutor in the archbishopric of St Andrews according to his provision and assignation, particularly concerning divers fermes of lands and teinds of kirks extending to the annual payment of £400. For payment of a pension and annual fee [to] lord Setoun, subscribed and sealed with our hand, £53 6s. For payment of a pension to Alexander Somervell by reason of our letters subscribed and sealed to him about the matter, £66 13s 6d. For the stipend of the advocates and procurators fiscal of the archbishopric according to use and custom, £50. For the annual stipend of the chamberlain according to use and custom, £66 13s 4d. For the payment to the graniter for his annual stipend, £50. For the stipend of the bailie of the lordship of Rescoby, £10. For the pension of Mr Andrew Davidsoun according to our letters to him about the above matter, £40. For annual contribution to the Lords of Council, £70. For the annual fees of the ordinary officials and mairs of the archbishopric from the old allowances according to the accounts of the chamberlain of the same extending to the sum of £56 3s 6d. Total deductions, £862 6s 10d.[9] So rests free in money, £2,042 0s 4d.[10]

Total wheat in ferme and teind except for the assedation above, according to the rental and accounts, 30 c. 8 b. 1 f. 1 p.[11] Bere, in ferme and teind besides the assedation above, 41 c. 10 b.[12]

---

[6] The correct calculation appears to be £838 12s 8d.
[7] This is the sum of the totals in the manuscript.
[8] Gavin Hamilton was appointed commendator of Kilwinning in 1550, I. B. Cowan and D. E. Easson, *Medieval Religious Houses, Scotland* (London, 1976), p. 69; D. E. R. Watt, *Fasti Ecclesiae Scoticanae Medii Aevi ad annum 1638* (Edinburgh, 1969), p. 299.
[9] The correct calculation appears to be £862 16s 4d.
[10] The correct calculation appears to be £2,042 0s 10d.
[11] ¼ p. has been omitted from this calculation.
[12] The correct calculation appears to be 41 c. 10 b. 1¼ p.

Oats, in ferme and teind besides the assedation above, 67 c. 13 b.[13] Of which sum is allocated according to the accounts and also according to use and custom, to the chamberlain and graniter, 2 c., and also the secretary, 1 c., and the master of works of the same, 1 c. 8 b., and the gardener of St Andrews, 1 c. Total deductions, 6 c. oats.[14]

So rests free and clear, 61 c. 13 b. 3 f. 1 p. Total meal according to the rental and annual accounts, 12 b. Total white peas, 4 b.

Assignation of the third of the archbishopric of Sanctandrois, £803 12s 4⅔d. [*In margin*, 'To be deducit for the thrid of the procurationis, synodallis and denis cryis quhilkis ar nocht payit, the sowme of' £164 13s 4d.]

Third of the money, £968 5s 8⅔d. Take the lordships of Monymeill with the augmentations of the feus of Petcuntly and Murehall for £99 14s 4d [£99 13s 4d, *fo 3r*]; Byirhillis and Polduff for £75 17s 8d; Dersye, Midlfudie and Craigfudie for £284 16d; Scottiscraig for £148 7s 1d; Bischopschyre for £119 4s 8d; Mikartschyre for £104 2s 8d; 'tak the mair of Fyffe for caynes and annuellis', £57 18s. 'Gif in' 20s '2 part ob'.[15] 'Eque.'[16]

Third of the wheat, 10 c. 2 b. 3 f. 1 '3 of a 3 part peck' [*i.e.* ⅓ p.].[17] Take the wheat of Byrehillis and Polduff giving by year, 9 c. 5 b. 1⅓ f.; wheat of Monemeill, 2 c. 12 b. 'Giff in out of Byirhillis', 1 c. 14 b. 2 f. 'and the 3 of a 3 part pect'.

Third of the bere, 13 c. 14 b. '3 part pect etc. [*blank*] 3 part pect.'[18] Take Monymeill, 1 c. 8 b.; Byrehillis, 13 c. 5 b. 1⅓ f. 'Gif in' 15 b. 1 f. Third of the oats, 22 c. 9 b. 3 f. '3 pect of a 3 part pect'.[19] The Byrehillis, 21 c. 5 b. 1 f. '1 pect 3 part pect'; Darsie, Midlfyd and Craigfudie, 2 c. 'Giff in' 16 b. 1 f. 2 p. '2 part of the 3 part pect.'

Third of the meal, 4 b. Out of Monemeill giving 8 b. meal.

Third of the white peas, 1 b. 1 f. 1⅓ p. Out of Scottiscraig paying 4 b. 'Gif in the rest'.

'Omittit capones, customes and all small dewteis that the Collector[20] for the haill' [*sic*].

---

[13] The correct calculation appears to be 67 c. 13 b. 3 f. 1⅓ p.
[14] The correct calculation appears to be 5 c. 8 b.
[15] These figures would give a total of £890 5s 9d, not £968 5s 8⅔d, without taking into account the fraction '2 pt ob'. 'Obol' denotes ½.
[16] 'Eque' frequently appears at the end of calculations but as it does not necessarily follow that the calculations are correct, 'eque' has been omitted from the transcription.
[17] '10 c. 2 b. 3 f. 1⅓ p.', *Accounts of the Collectors of Thirds of Benefices, 1561-1572*, ed. G. Donaldson (Edinburgh, 1949), p. 29.
[18] '13 c. 14 b. 3⅓ p.', *Thirds of Benefices*, p. 33.
[19] '22 c. 9 b. 3 f. 3⅓ p.', *Thirds of Benefices*, p. 40.
[20] John Wishart of Pittarow was appointed Collector general on 1 March 1562, *RSS*, v, no 998. William Murray of Tullibardine succeeded in 1565, *Thirds of Benefices*, p. xxvii.

'Memorandum, this assumptioun and divisioun is to be reformed as followis conforme to the quenis majesteis lettres and delyverance of the Lordis of Checker gevin thairupoun, off the quhilkis the tenouris ar heireftir insert[21] and according thairto the divisioun and assumptioun as [sic] in this maner to be reformed, that is to say of the silver thire now extend to' £803 12s 4¾d, 'the procuratiounes being deducit. Thair is to be yit deducit the thrid of the contributioun [to the] College of Justice extending to' £23 6s 8d. 'Sua restis of the said thrid of silver' £780 5s 8d,[22] 'quhairunto thair is to be augmentit for the victuallis of Hailles, Trapern, Half Houstoun and Gorlabank, converted in money at command of our soverane ladys lettre and ordinance of the Checker, the sowme of' £40. 'Sua the haill thrid of the money of the said bischoprick as it sould be chargit this lxvj yeir extendis to' £820 5s 8d.[23]

'Item, the quheit of the said arbischoprick [sic] quhilk extendit of befoir to' 10 c. 2 b. 3 f. 'a peck 3 of a 3 part peck, deduceand thairof the quheit of Eister Hailles, Traperne, Half Houstoun and Gourlabank converted in silver, as said is, now extends allanerlie to be chargit bot to' 7 c. 2 b. 3 f. 1 p. '3 thairof'.

'Item, the beir of the said arbischoprick [sic] quhilk extendit of befoir to' 13 c. 14 b. '3 part peck and 3 thairof, deduceand thairof the beir of the landis foirsaidis quhilk is' 4 c. 'will extend allanerlie to be chargit this lxvj yeir and in tyme cuming yeirlie bot to' 9 c. 14 b. '3 part peck 3 thairof'.

'Item, the aites thairof quhilk extendit of befoir to' 22 c. 9 b. 3 f. '3 partis thairof a 3 part peck, the aites of the landis foirsaidis extending to' 5 c. 8 b. 'now being deducit will extending [sic] now and yeirlie in tyme cuming to be chargit bot to' 17 c. 1 b. 3 f. '3 p. 3 of a 3 part peck'.

'Followis the tennoris of our soverane lordis [and ladyis] writting and the ordinances of the Checker befoir mentionat.

> My Lordis Auditouris of our soverane lord and ladyis Checker, unto your lordschippis humlie menis and schawis, I Johnne, archibischope of Sanctandrois, abbat of Paislay,[24] that quhair we have payit the haill thrid of the said archibischoprik and abacie according to the greit rentalles gevin in be us thairupoun and albeit thair is tane payment of the utermaist penny of the samin yeirlie be my lordis Comptrolleris bygane and present and the Collectouris of the contributiounes payit out of the samin to the College of Justice extending for the said bischoprick yeirlie to the sowme of' £70, 'and for the abacie yeirlie' £56 'money of this realme, beseikand your lordschippis heirfoir that ye will giff command to my lord

---

[21] The warrant follows on fo 7v (see below, p. 7).
[22] The correct figure appears to be £780 5s 8⅔d.
[23] The correct calculation appears to be £820 5s 8⅔d.
[24] John Hamilton, archbishop of St Andrews, 1546 - 1571, had been admitted commendator of Paisley in 1525 and in 1553 had resigned the abbacy in favour of Claud Hamilton, reserving to himself the fruits for life and also the administration. Watt, *Fasti*, pp. 298-299; Cowan and Easson, *Medieval Religious Houses*, p. 65.

Comptroller[25] and his Collectores to defese and allow to us yeirlie the thridis of the saidis tua contributiounis and sowmis foirsaidis and the samin to be [samin tak, *SRO, Vol. b*] allowit to thame in thair comptis lyk as your lordschippis hes done to the thridis of all uther greit prelaceis of this realme laitlie, of Brichen, Melros, Abirbrothe and utheris, and your lordschippis answer humblie we beseik.

Sic habetur a tergo [*i.e.* Thus it is endorsed]. Apud Edinbrucht', 3 February 1566/7. 'Fiat ut petitur [*i.e.* May it be done as asked].'

Huntlie; Galloway; Argyll; Athoill; Jo[hannes] E[piscopus] Rossensis.'[26]

ROYAL WARRANT[27]

(SRO, Vol. a, fo 7v)
(SRO, Vol. b, fo 6v)

'Regina.

Comptroller,[28] forsamekill as we understand that at the first ingiveing of the rentallis [the, *SRO, Vol. b*] teyndis of the landis of Trapren, Hailles, Half Houstoun and Gourlawbank pertenyng her[itably] to our traist cousing and counsalour, James, erle Bothwell,[29] wes gevin in be the maist reverend in God, Johnne, arbischope [*sic*] of Sanctandrois, in the rentall of his said bischoprick for victuallis be negligence of his servandis ingifares thairoff, albeit the samin teyndis wes then set in tak and assedatioun to our cousingnes dame Agnes Sinclar, countes of Bothwell, mother to our said traist cousing, for yeirlie payment of fourtie pundis monie, lyk as the samin wes possessit be the hous of Hailes of auld for yeirlie payment of the said sowme and willing that scho be nocht prejudgit of hir rycht of the said teyndis nor na new novatioun nor extortioun usit upoun hir thairfoir bot that scho may peciablie bruik and joyse the samin for yeirlie payment of the dewtie foirsaid contenit in hir said tak nochtwithstanding the ingeving of the teyndis for victuall and that the said maist reverend father be nocht hurt for ingeving heir[of] be negligence, as said is, or utherwayis compellit in tymes cuming to mak payment for the said victuallis, sua thairby he may tyne actioun or actes to our said cousing for his releiff, it is our will and we charge you that ye

---

[25] Sir William Murray of Tullibardine, Comptroller, 1565 - 1582; James Cockburn of Skirling also served as Comptroller, 1566 - 1567, *Handbook of British Chronology*, eds. E. B. Fryde, E. E. Greenway, S. Porter and I. Roy (London, 1986), p. 191.

[26] George Gordon, 5th earl of Huntly; Alexander Gordon, bishop of Galloway; Archibald Campbell, 5th earl of Argyll; John Stewart, 4th earl of Atholl; John Leslie, bishop of Ross. *Scots Peerage*, i, pp. 340-343, 444-445; iv, pp. 539-541. Watt, *Fasti*, pp. 132, 270. Following this entry, SRO, Vol. b, fo 6r has the beginning of the rental for the archdeaconry of St Andrews, but this is deleted and contains the marginal note, 'This is writin heireftir'. The rental, however, is not included in Vol. b, but can be found in Vol. a, fo 78r (see below, p. 66).

[27] Cf above, p. 6, n. 21.

[28] See above, n. 25.

[29] James Hepburn, 4th earl of Bothwell, *Scots Peerage*, ii, pp. 161-167.

caus the clerk, keipar of the rentallis, quhome we also charge deleit the said victuall furthe of the rentall of the said bischoprick els gevin in and ingros and insert in the said place thairof the said' £40 'monie and caus sua charge the samin in the Checker comptis baithe for all yeris bygane restand awand and siclyk yeirlie in tymecuming quhill forder ordour be tane sua that the said bischope have na occasioun heireftir to trubil our said cousingnes for payment of the saidis victuallis nor he [sic] compellit to satisfie and fulfill the said rentale gevin in be negligence, as said is, and faill nocht to caus excluid and sua deleit the saidis victuallis furth of the said rentale that it sall appeir heirefter that thair wes the said sowme of fourtie pundis gevin in the place thairof for sua ressoun requirit and we mynd nocht to hurt auld possessouris, chargeing also the Auditouris of our Checker to defeis and allow the saidis victuallis to you in your comptis and charge and accept the said' £40 'thairfoir conforme to the saidis takis. Thir presentis beand anes producit befoir thame for thair and your warrand as ye and thay will answer to us upoun your offices in that pairt, subscryvit with our hand at Edinbrucht the thrid day of Februar and of our regne the xxv yeir, 1566. Sic habetur a tergo [i.e. Thus it is endorsed]. Apud Edinbrucht', 14 February 1566/7.

'The quhilk daye the Lordis Auditouris of the Checker ordanes the Comptroller and his clerk to reforme the rentall within writtin for this present yeir and all yeiris to cum, according to the tennour of this our soverane ladeis rychtin.

Sic subscribitur: Huntlie; Jo[hannes] Episcopus Rossensis; Bellenden.'[30]

ST ANDREWS, PRIORY OF,

(SRO, Vol. a, fos 10v-25r[31])
(SRO, Vol. b, fos 7r-19v)

'The assumptioun of the thrid of the priorie of Sanctandrois reformed at command of the kingis majesties warrand of the dait at Halyruidhous', 10 February 1586/7, 'quhairof the tennour efter followis.'[32]

Third of the money, £745 19s 4½d. 'Tack the kirk of Migvie and Tarlane', £60; the kirk of Dow in Atholl, £93 6s 8d; the kirk of Foullis in Gowrie, £40; the kirk of Lythgow £246 13s 4d; the teinds of Wester Binning £10; the small teinds of Lythgow, £12; the teinds of Prestoun, £10; the teinds of Hadingtoun, £266 13s 4d; the teinds of Clerkingtoun, £6 13s 4d; 12s 8d out of the annuals of the toun of Lythgow.

---

[30] For Huntly and Ross, see above, p. 7, n. 26; Sir John Bellenden of Auchnoule was then Justice Clerk, *RSS*, v, no 589.
[31] SRO, Vol. a, fos 8r-10r, 11v, 22v are blank.
[32] This sentence does not appear in SRO, Vol. b. The warrant follows on p. 9 below.

Third of the wheat, 12 c. 11 b. 1 f. ⅓ p. Take out of the kirk of Leucheris, 8 c. 6 b.; out of the kirk of Forgound in Fyffe, 10 b.; the rest out of Sanctandrois kirk,[33] 6 c. 10 b. 2 f.[34]

Third of the bere, 44 c. 3 b. 1 f. 1⅔ p. Take the bere of Sanctandrois kirk, 27 c. 8 b. 1 f.; the bere of the kirks [recte, kirk] of Leuchers, 23 c. 14 b. 2 f. 'Gif in' 7 c. 3 b. 1 f. 2⅓ p.

Third of the meal, 38 c. 1 b. ⅓ p. Take the kirk of Lauth[r]ask, 20 c. 1 b.; the kirk of Kilgour, 8 c. 12 b.; the kirk of Skoniycht [i.e. Scoonie], 1 c. 7 b. 'Giff in' 1 c. 2 b. 3 f. ⅓ p.[35]

Third of the oats, 50 c. 8 b. 3 p. 'half pect 3 part pect' [i.e. five sixths p.]. Take Sanctandrois kirk paying 58 c. 1 f. 2 p. 'Gif in', 7 c. 7 b. 2 f. 1 p. '3 part half pect' [i.e. ⅓ p.].[36]

Third of the peas and beans, 1 c. 2 b. 1 f. 1⅓ p. Take them out of Sanctandrois kirk giving 3 c. 3 b.

'Omittit cainis, capones and all utheris dewteis.'

'Rex.

Collectour generall we greit you weill, forsamekill as the thrid of the pryorie of Sanctandrois, nochtwithstanding of the assumptioun of the thrid hes bene bruikit and possessit with the tua pairt thairof be the prioris of Sanctandrois for the tyme be vertew of the giftis and dispositiounes of the said thrid maid and gevin to thame be our selff and our darrest mother thairupoun, in respect quhairof the auld assumptioun of the said thrid hes nevir taken effect bot hes remanit with the possessouris of the said benefice and at thair dispositioun quhilk hes gevin thame occasioun to set ane greit pairt of the said thrid pairt in tak and assedatioun and to dispone thairupoun at thair plesour, as namlie the kirk of Markinsche assumit of auld for' 25 c. 2 b. meal 'wes disponit in pensioun be umquhill James, erle of [Moray],[37] commendatour of the said pryorie, to George Douglas, brother germane to the laird of Lochlevin,[38] and thairefter ratefeit and approvit be the last prior[39] with our darrest mother and our confirmatioun following thairupoun, be vertew thairof he hes bene in peceable and continuall possessioun of the same in all tymes bygane sen the dait of the provisioun but interruptioun, stope, trubill or impediment quhatsumevir, oure will is thairfoir and we command you to alter and change that pairt of

---

[33] This kirk is named 'Ecclesia Trinitatis' in SRO, Vol. a, fo 14r, that is Holy Trinity parish kirk, St Andrews.
[34] The total adds up to 15 c. 10 b. 2 f.
[35] These figures do not tally.
[36] These figures do not tally.
[37] James Stewart, earl of Moray, was assassinated at Linlithgow in January 1570, *Scots Peerage*, vi, p. 316.
[38] I.e. Sir William Douglas of Lochleven, *Scots Peerage*, vi, pp. 371-372.
[39] Robert Stewart, bishop of Caithness and earl of Lennox, was commendator from 1570 till his death on 29 March 1586. *RSS*, vi, no 930; *Scots Peerage*, v, p. 355.

the auld assumptioun of the said priorie pu[r]tenyng to the kirk of Markinsche with ony uther, the lyk rent ferme or dewtie to be tane out of the tua pairt of the said priorie quhairevir the samin may be had maist commodious and that ye deleit the said kirk of Markinsche furth of the assumptioun of the said priorie, keiping thir presentis for your warrand.

Subscryvit with our hand at Halyruidhous', 10 February 1586/7. 'Sic subscribitur: James R.; Linclowden, Collectour.'[40]

Rental of the priory of St Andrews

Rental of the fermes ('Rentale firmarum') in silver of all the lands in the lordship of the priory of St Andrews.

Northbank feued for £6 14s. Claremounthe feued for £11 3s 11½d. Dunnorke feued for £9 7s 5½d. Balgoiff feued for £14. Stralrynnis leased for £13 6s 8d. Paremounthe of Stralrynnes feued for 50s. Cokstoun leased for £9 6s 8d. Tenenbray [*in margin*, 'Dambray'] leased for £9 6s 8d. Ballonen leased for £5 5s. Tenenheid [*in margin*, 'Damheid'] minus one quarter and one eighth parts leased for £5. One quarter and one eighth parts of the same feued for £4 10s. Trumcarro leased for £5 6s 8d. Cassindonat feued for £7 13s 4d. Unthank [*i.e.* Winthank] feued for £14 6s 8d. For one pound of bere ('pro una libra bere'), 2s 4d. Craigtoun leased for £7. Lumbo leased for 40s. Balrymounthe Wester leased for £10 13s 4d. Langraw leased for £8. Prior Lathyne feued for £13 6s 8d. Grange [New Grange, *fo 20v*], minus a fourth part and except for a third part of the same fourth part, ('dempta quarta parte excepta tertia parte eiusdem quarta partis') leased for £23 11s 9d. A fourth part of the same minus a third part of the same fourth part ('quarta pars eiusdem dempta tertia parte eiusdem quarta partis') feued for £10 6s. Balrymounthe Eister feued for £10 6s 9d. Stravathy feued for £22 8s. Freirtoun feued for £33 13s 4d. Kirktoun feued for £15. Potlathy leased for £26 13s 4d. Innerbrig leased for 20s. New Mylne near Dersy leased for £8. Walkmylne feued for £3 16s.

Kirkland of Cupar ('Terra ecclesiastica Cupri') feued for £19. Kathlok leased for £6 13s 4d. Chapelkettill leased for £4. Kirkland of Markinche for 40s. Kilmukles feued for £10. Yaird Mylne leased for £4. New Mylne feued for £6. Law Mylne feued for £4. Kirkland of Dersy feued for 24s. Mills of the monastery ('molendina monasterii') feued for £9 10s. Priorwell of Balmeryinoch entered for 10s. Land of St Andrews in Lundie entered for 20s. Acre near the city ('Acra prope civitatem') feued for £149 5s 5d. Clerkingtoun feued for £20. Morestoun feued for 43s 4d. Trem feued for 26s 8d. Perkly feued for £8. The granary ('granaria') in Leith leased for £13 10s. Pilmure, Inclistuir [*i.e.* Inchture] and Inschyray feued for £30. Elenenhill leased for £8. Chape[l]toun in Mernis feued for £5 15s 6d. Kirkland of Bourty feued for £6 13s 4d. Littil Lone feued

---

[40] Robert Douglas, provost of Lincluden, was Collector general, *RPC*, iv, p. 167.

for 15s. Prior's croft in Craill under the cliff ('sub rupe') feued for 7s. The croft in the same place called Balcomy's croft feued for 7s. Another croft in the same place leased for 16s. Out of the lordship of Ochiltre, 8s 8d. Haddow in Mernia feued for £5 6s 8d. The croft of the monastery feued for £6 13s 4d.

Total of the fermes of the priory of Sanctandrois in money £639 18s 6d.[41]

Rental of the kane ('Rentale canas') of the priory of Sanctandrois
    Kane of Potmulye, 13s 4d. Kane of Banofeild, 6s 8d. Kane of Feddinche, 3s 4d. Kane of Kinnynmonth, 26s 8d. Kane of Ballinbrach, 10s. Kane of Seraffeat, 7s. Kane of Rossyclero, 13s 4d. Kane of Petpunty, 40s. Total, £6 0s 4d.

Rental of the annualrents
    Annualrents of Hadingtoun: Out of the croft near the kirk called Byris Orchard, 2s. Of the tenement of John Smyth, 2s. Out of the third portion of land in the High Street ('Ex tribus particalis terre in vico regio'), 2s. Out of the tenement of William Cokburne, 20d. Out of the land of St Andrews in Poldroch, 12d. Total of these annualrents, 8s 8d.
    Annualrents within the burgh of Linlythgow: Out of the tenement, formerly of Henry Levingstoun, afterwards of William Oustane, 9d. Out of the land nearest which is the end of the said tenement presently of Castren[42] Rusell, one pound of oats. Out of the bakehouse of the laird of Hacket, formerly of Thomas Spens now of Robert Carmichaell, 2s '1 particularie'.[43] Out of the nearest lodging house, formerly of John Muir, now of John Sinclar, 2s. Out of the tenement of Thomas Farar, now of Robert Jamesoun, 2s. Out of the tenement, formerly of Andrew Roth, now of George Sinclar, 4s. Out of the tenement, formerly Harkeris, now of John Hammiltoun, 6d. Out of the tenement of John Gibsoun, 3d. Out of the tenement of John Nasmyth, 3d. Out of the tenement of Thomas Johnnestoun, 3d. Out of the tenement of Robert Hammiltoun, 3d. Of the land of Gilbert Anderstoun, 3s 4d. Out of the tenement of John Hammiltoun, 3s. Total of these annualrents, 19s 7d.
    Annualrents outside the burgh of Lynlythgow: Out of the monastery ('monasterio') of Torphechin, 16s. Out of the crofts, formerly of Andrew Roth, now of William Hammiltoun, 2s. Out of our croft in the hands of the heir of Robert Wodderspune, 2s. Out of Culdrachy, 6s 8d. Out of Ogilface, 40s. Out of the mill of Manwell, 26s 8d. Total, £4 13s 4d.
    Annualrents of Perthe: Out of the great royal customs, 40s. Out of the king's fermes of the same town paid annually out of the mill, 13s 4d. Out of the booth in the corner ('Ex botha in angula'), 6s 8d. Out of the land formerly of John

---

[41] The correct calculation appears to be £640 5s 2d.
[42] Probably 'Christian'.
[43] One meaning of the Scots word 'particularie' is 'separately'. If '1 particularie' is taken to mean 1s, and then added to the 2s, the total of 19s 7d is correct.

Pyper, called Myllikynnis land 2s. Out of the land, formerly of Thomas Lamb, now Pincland [Pancland, *SRO, Vol. b*], 3s 4d. Total, £3 5s 4d.

Annualrents of Craill: Out of the king's fermes of the mill of the same, 53s 4d. Out of the tenement of James Moreis, 2s 6d. Total, 55s 10d.

Annualrents of Dundie: Out of the great royal customs of the said town, 13s 4d. Out of the land formerly of John Cowpar, 2s 6d. Out of the land formerly of Walter Blak, 2s. Out of the land formerly of David Rollok in Murraygait, 3s 4d. Out of the land formerly of James Newman, 6d. Out of the land formerly of James Craill alias James Keir, 20d. Total, 23s 4d.

Annualrents of Abirdeane: Out of the tenement [of] Craib in Abirdeane, '1 libera piperis. Summa particule, 1 libra piperis.'[44]

Annualrent pertaining to the master of works of the place, £5 5s 8d. Annualrent pertaining to the sacristan of the place specified, 58s 10d. Annualrent out of the mill of Leuchris, 40s. Total of these annualrents, £10 4s 6d.

Total of all annualrents, £23 10s 8d.[45]
Total of all the foresaid silver, £669 9s [*plus* 5d, *SRO, Vol. b*].

Ferme rentals of victuals paid yearly

S[t]rakynnes leased for 2 c. wheat; 4 c. bere; 4 c. oatmeal. Ballone leased for 1 c. 9 b. 1 p. wheat; 3 c. 2 b. 1 f. 1 p. bere; 3 c. 2 b. 1 f. 1 p. oatmeal. Drumcarro leased for 1 c. wheat; 1 c. bere. Lumbo leased for 8 b. bere; 8 b. meal. Craigtoun leased for 8 b. wheat; 12 b. bere; 1 c. 2 b. meal. Freirtoun feued for 2 b. peas. Kirktoun feued for 2 b. peas. Kirksyd within the parish of Aglisgrig [*i.e.* Ecclesgreig] feued for 7 b. bere; 6 b. meal. Total wheat, 5 c. 1 b. 1 p. Total bere, 9 c. 13 b. 1 f. 1 p. Total meal, 9 c. 2 b. 1 f. 1 p.[46] Total peas, 4 b.

Rental of the teind sheaves of the priory of St Andrews, both within and outside Fyff, leased for money and victual ('tam infra partes de Fyff quam extra tam pro pecuniis quam pro victualibus assedatur').

Ecclesia Trinitatis [Sanctandrois kirk, *fos 10v, 18v, 23v*]: Magask Superiour leased for 6 b. bere; 1 c. 8 b. oats. Magask Inferiour leased for 2 b. wheat; 4 b. bere; 14 b. [oats]. Clatto leased for 2 b. wheat; 2 b. bere; 1 c. oats. Wilkestoun leased for 1 b. wheat; 4 b. bere; 1 c. oats. Northbank leased for 1 b. wheat; 4 b. bere; 14 b. oats. Strakynnis added together extending to 7 b. 2 b. [*recte*, 2 f.] wheat; 1 c. 4 b. 1 f. [bere]; 3 c. 1 f. 2 p. oats; 3 b. peas. Paremounth leased for 1 b. wheat; 4 b. bere; 8 b. oats [Monthis pairt, 2 b. oats, *fo 24v*]. Ballone leased for 9 b. wheat; 1 c. 7 b. bere; 2 c. oats. Balgoiff leased for 4 b. wheat; 6 b. bere; 1 c. 8 b. oats. Sastoun [Goukstoun, *fo 20v*; Cokstoun, *fos 12r, 24r, 24v*] leased for 1 b. wheat; 6 b. bere; 1 c. oats. Denebray leased for 1 b. wheat; 6 b. bere; 1 c.

---

[44] I.e. 1 pound of pepper or spice.
[45] The correct calculation appears to be £23 10s 7d.
[46] Including oatmeal.

oats. Denneheid leased for 4 b. bere; 1 c. 4 b. oats. Dunnork leased for 5 b. bere; 2 c. oats. Drumcarro leased for 6 b. wheat; 10 b. bere; 1 c. 10 b. oats. Ladeddy leased for 9 b. bere; 1 c. 14 b. oats. Balduny leased for 4 b. wheat; 6 b. bere; 2 c. 2 b. [oats]. Kynnynmounth leased for 4 b. wheat; 12 b. bere; 3 c. 2 b. [oats]. Cassindonat leased for 1 c. oats. Lumbo leased for 3 b. bere; 8 b. meal. Craigtoun leased for 2 b. wheat; 6 b. bere; 1 c. 2 b. [oats]. Fadonch [*i.e.* Feddinch] leased for 3 b. bere; 1 c. 2 b. oats. Balrymounth Wester leased for 4 b. wheat; 5 b. bere; 1 c. 2 b. [oats]. Hospital of St Nicholas leased for 6 b. bere; 4 b. oats. Laderne leased for 8 b. bere; 3 c. 3 b. oats. Lathonis leased for 2 b. bere; 12 b. oats. Lathockker leased for 2 b. bere; 2 c. 12 b. oats. Hasildene leased for 1 b. bere; 4 b. oats. Ballmoungy leased for 3 b. [*plus* 1 f., *fo 24r*] bere; 14 b. oats. Gilmerstoun leased for 4 b. bere; 1 c. oats. Langraw leased for 2 b. wheat; 4 b. bere; 1 c. 4 b. [oats]. Stravethy leased for 4 b. wheat; 1 c. bere; 3 c. 8 b. [oats]. Byrehillis leased for $17\frac{1}{4}$ b. wheat; 2 c. $13\frac{1}{4}$ b. bere; 4 c. $5\frac{1}{4}$ b. oats. Polduff leased for 8 b. 2 f. $2\frac{2}{3}$ p. wheat; 1 c. $6\frac{2}{3}$ b. bere; 2 c. $2\frac{2}{3}$ b. oats. Bonytoun leased for 4 b. bere; £3 6s 8d money. Newgrange leased for 9 b. wheat; 1 c. 2 b. bere; 4 c. oats. Kynawdy leased for £15 money. Prior Lathynen leased for 40s money. Elenenhill leased for £5 money. White fish teinds of St Andrews and Byirhillis. Teinds of the gardens and crofts leased for 12 b. bere. Teinds of the acres near the city added together, 1 c. wheat; 9 c. bere; 2 c. oats; 3 c. beans and peas. Inschemurtho part [*recte*, park],[47] 1 b. bere; 4 b. oats.[48] Total wheat, 6 c. 10 b. 2 f. Total bere, 27 c. 8 b. 1 f.[49] Total meal, 8 b. Total oats, 58 c. 1 f. 2 p.[50] Total beans and peas, 3 c. 3 b. Total money, £26 6s 8d.[51]

Kirk of Leuchris: Brodland leased for 10 b. wheat; 4 c. bere; 4 c. oats. Bawenzie leased for 9 b. bere; 1 c. oats; £3 money. Ardat leased for 2 c. wheat; 2 c. bere; 5 c. oats. Myltoun leased for 10 b. bere; 10 b. meal; £5 money. Sagy leased for 6 b. wheat; 1 c. bere; 3 c. oats. D[r]one leased for 1 c. 2 b. wheat; 1 c. 12 b. bere; 2 c. oats; 8 b. oats. Pursk leased for 18 b. wheat; [8 b. bere, *deleted*] 3 c. bere; 3 c. oats. Petcullo leased for 5 b. wheat; 8 b. bere; 1 c. 2 b. oats. Ballmullo leased for 10 b. wheat; 14 b. bere; 1 c. 8 b. oats. Cragy leased for 12 b. wheat; 1 c. 8 b. bere; 3 c. oats. Cowbaikye leased for 13 b. 2 f. bere; 1 c. 12 b. oats. Kattely leased for 5 b. wheat; 8 b. bere; 1 c. 8 b. oats. Straburne leased for 3 b. bere; 1 c. oats. Fordell leased for 12 b. bere; 2 c. oats. Scottiscraig leased for 14 b. wheat; 1 c. 8 b. bere; 3 c. oats. Schanwell leased for 8 b. bere; 8 b. meal.

---

[47] The lordship of Byrehill contained the episcopal residence of Inchmurdo, *An Historical Atlas of Scotland c.400 - c.1600*, eds. P. McNeill and R. Nicholson, (St Andrews, 1975), p. 41.
[48] SRO, Vol. b has a folio, numbered 113, folded and bound at right angles to the other folios. It is inserted between fos 9v and 10r, and consists of a list of leases identical to that found below under 'Kirk of Leuchris', followed by the words 'Extractit furth of the registeris and Buikis of Assumptioun of benefices be me, Johne Gilmour, wryter to our soverane lordis signet, nottar publict, clerk deputt and keipar of the saidis registris, under my signet and subscriptioun manuall, Gilmour.'
[49] The correct calculation appears to be 27 c. 9 b. 1 f.
[50] 10 b. has been omitted from this calculation.
[51] The correct calculation appears to be £25 6s 8d.

Gorpot leased for 5 b. bere; 4 b. meal. Muirtoun leased for 4 b. bere; 6 b. meal. Wester Fotteris leased for 5 b. bere; 5 b. meal. Rynd leased for 12 b. 3 f. bere; 12 b. 3 f. meal; 11 b. oats. Brigend and Quhite croft leased for 3 b. 1 f. bere [4 b. 1 f., *fo 23v*]; 3 b. 1 f. meal; 3 b. oats. Kynstar leased for 5 b. bere; 6 b. oats. Ferytoun leased for 1 c. 4 b. bere; 1 c. 4 b. oats. Kynschawdy leased for 5 b. [*recte*, £5] money. Munsymylne leased for 4 b. wheat; 6 b. bere; 8 b. oats. Auldmore leased for £8 money. Nathe[r]mure leased for £6 13s 4d money. Brakmounth, Luklaw and Southeild [Southfeildis, *SRO, Vol. b*] leased for £14 money. Eister Fotteris leased for £6 money. Medowis leased for 50s money. Chapelyaird leased for 12s money. Total wheat, 8 c. 6 b. Total bere, 23 c. 14 b. 2 f.[52] Total meal, 3 c. 1 b. Total oats, 36 c. 6 b. Total money, £50 15s 4d.

Kirk of Forgound: Innerdonet [*i.e.* Inverdovat] leased for 4 b. wheat; 1 c. 6 b. bere; 4 c. oats. Newtoun Eister leased for 1 b. bere; 6 b. oats. Newtoun Wester leased for 4 b. 2 f. bere; 1 c. 2 b. oats. Midle Newtoun leased for 2 b. wheat; 6 b. 2 f. bere; 1 c. 8 b. oats. Sandfurd Nauchlame leased for 2 b. wheat; 6 b. bere; 1 c. 5 b. oats. Sandfurd Hay leased for 2 b. wheat; 6 b. bere; 1 c. 5 b. oats. Sandfurd Narne leased for 4 b. bere; 13 b. oats. Flanschill leased for 4 b. bere; 13 b. oats. Little Freirtoun leased for 3 b. bere; 10 b. oats. Nauchtane with pendicles ('cum sequelis') leased for £20 money. Ballodmounth, Cauldyhame, Schiris, Landris leased to David Balfour for £13 6s 8d. Wodhavin leased for 53s 4d. Total wheat, 10 b. Total bere, 3 c. 9 b. Total oats, 11 c. 14 b. Total money, £36.

Kirk of Dersy: Toun ('villa') of Dersy leased for 8 b. wheat; 14 b. bere; 3 c. oats. Craigfudy leased for 18 b. wheat; 1 c. 5 b. bere; 2 c. 12 b. oats. Mydlefudy leased for 1 c. wheat; 1 c. 1 b. bere; 3 c. oats. Fingask leased for 6 b. wheat; 8 b. bere; 1 c. 10 b. oats. West Fudy leased for 6 b. wheat; 8 b. bere; 3 c. 6 b. oats. Pottormie leased for 2 b. wheat; 3 b. bere; 1 c. oats. Newmylne leased for 2 b. wheat; 4 b. bere; 12 b. oats. Mydern leased for 3 b. bere. Total wheat, 3 c. 10 b. Total bere, 4 c. 14 b. Total oats, 15 c. 8 b.

Kirk of Cupar: Acres of Cowpar leased for 3 b. wheat; 3 c. 1 b. bere; 3 c. 5 b. oats. Pettincreiff leased for 4 b. wheat; 1 c. bere; 12 b. meal. Petblad leased for 8 b. bere; 1 c. 9 b. 2 f. meal. Ballgarvy leased for 4 b. wheat; 1 c. meal. Kylmarone leased for 8 b. bere; 12 b. meal. Tor leased for 4 b. oats. Kingask leased for 2 b. wheat; 4 b. bere; 10 b. meal. Foxtoun leased for 3 b. bere; 10 b. meal. Thomastoun leased for 10 b. wheat; 1 c. 8 b. bere; 3 c. 2 b. oats. Russellis Mylne leased for 8 b. meal. Carslogy leased for 8 b. meal. Total wheat, 1 c. 7 b. Total bere, 7 c. Total meal, 6 c. 5 b.[53] Total oats, 6 c. 12 b.[54]

Kirk of Lawthrisk: Lawthrisk Wester leased for 12 b. bere; 1 c. 6 b. meal. Lawthrisk Eister leased for 1 c. bere; 2 c. 9 b. meal. Riggis with Lythqu lands leased for 5 b. bere; 7 b. meal. Orky with mill leased for 4 b. bere; 1 c. 8 b. oats. Ramorgny, Burneturk and Ballingall leased for 1 c. bere; 4 c. [meal]. Douyne

---

[52] The correct calculation appears to be 22 c. 14 b. 2 f.
[53] 2 f. has been omitted from this calculation.
[54] The correct calculation appears to be 6 c. 11 b.

leased for 6 b. bere; 1 c. 11 b. meal. Ramboy [i.e. Rameldry] leased for 8 b. bere; 1 c. 8 b. meal. Kingis Kettill leased for 1 c. 8 b. bere; 3 c. meal. Cletty leased for 4 b. bere; 1 c. meal. Forthir Ramsay leased for 2 b. wheat; 12 b. bere; 2 c. 1 b. [meal]. Tounfeild leased for 11 b. meal. Hoilkettill leased for 10 b. bere; 1 c. 12 b. meal. Total wheat, 2 b. Total bere, 7 c. 5 b. Total meal, 20 c. 1 b. Total oats, 1 c. 8 b.

Kirk of Kilgour [in margin, 'non seminantur', i.e. not sown]: Toun of the same leased for 5 b. bere; 8 b. meal. Falkland with cottars leased for 2 c. 12 b. bere; 2 c. 12 b. [meal]. Newtoun Falkland leased for 1 c. 12 b. bere; 2 c. 4 b. [meal]. Frouchie leased for 2 c. bere; 3 c. 4 b. meal. Condeland, Pourndrem [i.e. Purin, Drums], Ballo, Glaslie and Pettillo leased for £33 6s 8d. Total bere, 6 c. 13 b. Total meal, 8 c. 12 b. Total money, £33 6s 8d.

Kirk of Skuny: Toun of the same leased for 10 b. wheat; 14 b. bere; 1 c. meal. Monthflowrie, Burnemill, Acre of Levyn, Ballinbrocht leased for 6 b. wheat; 1 c. 4 b. bere; 1 c. 6 b. meal. Swynishauch leased for 5 b. bere; 11 b meal. Dury leased for 12 b. wheat; 1 c. bere; 2 c. 8 b. meal. Adorny leased for 8 b. wheat; 1 c. 2 b. bere; 1 c. 10 b. meal. Ballgrummo leased for 8 b. bere; 1 c. 2 b. meal. Lathamen [i.e. Letham] leased for 14 b. bere; 2 c. 2 b. meal. Total wheat, 2 c. 4 b. Total bere, 5 c. 15 b. Total meal, 10 c. 7 b.

Kirk of Kennowy: Toun of the same leased for 8 b. wheat; 12 b. bere; 1 c. 12 b. meal. Drummard leased for 4 b. wheat; 12 b. bere; 2 c. 4 b. meal. Ochtermerny leased for 10 b. bere; 2 c. 4 b. meal. Lalathyne with Awdy leased for 6 b. wheat; 12 b. meal. Brountoun and Dallginche leased for 6 b. wheat; 1 c. bere; 3 c. meal. Newtounes Eister and Wester leased for 1 b. wheat; 9 b. bere; 1 c. 6 b. meal. Trottoun leased for 4 b. wheat; 1 c. bere; 1 c. 14 b. meal. Ballinkirk leased for 6 b. meal; 2 b. bere. Ballbroky Myll leased for 3 b. bere; 4 b. meal. Ballbroky leased for 5 b. bere; 1 c. 2 b. meal. Total wheat, 1 c. 7 b. Total bere, 5 c. 11 b. Total meal, 15 c.

Kirk of Markinche: Dunyface leased for 8 b. bere; 1 c. 6 b. meal. Ballcurvie Minor leased for 4 b. bere; 9 b. meal. Ballcurvie Major leased for 14 b. bere; 2 c. meal. Balfour leased for 4 b. wheat; 8 b. bere; 1 c. 4 b. meal. Awchmawtie leased for 6 b. bere; 1 c. 8 b. meal. Camerone leased for 4 b. bere; 17 b. meal. Ballqhany[55] leased for 1 b. wheat; 4 b. bere; 1 c. 2 b. meal. Ballbirny leased for 7 b. bere; 2 c. 4 b. meal. Bandonen [i.e. Bandon] leased for 3 b. bere; 1 c. 8 b. meal. Wester Markinche leased for 8 b. bere; 1 c. 4 b. meal. Kirkmarkinche, Nather Markinche, Newtoun Markinsche and Bythlie with mill leased for 12 b. bere; 2 c. 4 b. meal. Ballgony, Spitill, Myltoun, Schettoun with mill and Cauldhame leased for 2 c. 8 b. bere; 6 c. 8 b. meal oats ('farine avenarum'). [In margin, Balcurvehauch] Bawenevy Hauch leased for 4 b. bere; 6 b. meal. Caldycoites leased for 4 b. bere; 14 b. meal. Tullibrek leased for 2 b. bere; 8 b. meal. Maw and Littleron [i.e. Little Lun] leased for 7 b. bere; 1 c. 2 b.

---

[55] Cf Mountquhanie; *RMS*, iv, no 191 equates Balquhany with Mountquhanie.

meal. Coull leased for 2 b. bere; 10 b. meal. Total wheat, 5 b. Total bere, 8 c. 9 b. Total meal, 25 c. 2 b.[56]

Kirk of Eglisgrig [*i.e.* Ecclesgreig]: Manyis Morphy, Stane Morphy, Cronystoun, Morphy Fraser and Cantiland leased for £26 13s 4d; 5 b. [wheat]. Wardraptoun leased for 2 b. wheat; 8 b. 2 f. bere; 14 b. [meal]. Littil Quhi[t]soun leased for 1 b. wheat; 3 b. bere; 5 b. meal. Halquhitsoun leased for 2 b. wheat; 7 b. bere; 12 b. meal. Hilend for 6 b. bere. Cragy for 4 b. wheat; 1 c. 4 b. bere; 2 c. 4 b. meal. Kirkhauch for 5 b. bere. Scottistoun for 1 b. 1 f. wheat; 6 b. bere; 12 b. 3 f. meal. Laurenstoun with pendicles, 8 b. wheat; 2 c. bere; 3 c. 6 b. [meal]. Marboy, 1 b. bere; 4 b. meal. Eister Mathouris, 10 b. wheat; 2 c. 10 b. bere; 2 c. 8 b. meal. Wester Mathouris, 3 b. wheat; 11 b. 2 f. bere; 11 b. 2 f. meal. Nather Quhitsoun, 4 b. wheat; 11 b. bere; 15 b. meal. Total wheat, 2 c. 8 b. 1 f. Total bere, 9 c. 9 b. Total meal, 12 c. 12 b. 1 f. Total money, £26 13s 4d.

Kirk of Fordoun in Marnis [*i.e.* Mearns], £266 13s 4d; 2 c. meal. Kirk of Bourthy, £40. Kirk of Migvy and Terland, £40 [£60, *fos 10v, 18v, 20r*]. Kirk of Dull in Athoill, £93 6s 8d. Kirk of Langforgound, 3 c. 14 b. wheat; 7 c. 2 b. bere; 17 c. 14 b. oats; £100 money. Kirk of Rossy in Gowry, 1 c. 12 b. wheat; 4 c. 12 b. bere; 1 c. oatmeal; 3 c. 12 b. oats; £32 money. Kirk of Inchstoure, £147 [£178, *fo 20r*]. Kirk of Foullis in Gowry, £40. Kirk of Portmoak, £51. Kirk of Abircrummy, £40. Kirk of Lynlythqu, £246 13s 4d. Teinds of Wester Bynnyng, £10. Small teinds ('Minute decime') of Lynlythgow, £12. Teind sheaves of Prestoun, £10. Kirk of Hadyngtoun, £266 13s 4d. Teind sheaves of Clerkingtoun, £6 13s 4d.

Total wheat out of the teinds of the kirks, 33 c. 3 f. Total wheat in ferme and in teinds, 38 c. 1 b. 3 f. 1 p. Total bere in teinds, 122 c. 9 b. 3 f. Total bere in ferme and in teinds, 132 c. 7 b. 1 p. Total meal in teinds, 105 c. 3 f. Total meal in ferme and in teinds, 113 c. 3 b. 1 p. Total oats in teinds, 151 c. 10 b. 1 f. 2 p. Total beans and peas in teinds, 3 c. 3 b. Total beans and peas in ferme and in teinds, 3 c. 7 b. Total of all the foresaid victuals, 439 c. 13 b. 1 f. 1 p. Total of the silver out of the teinds of kirks within and outside Fyffe, £1,568 8s 8d. Total of the silver of fermes of lands, kanes, annualrents, as well as teinds of kirks, £2,237 18s 1d.[57]

Assignation of the third of the priory of Sanctandrois:

Third of the money, £745 19s 4¼d. Take the kirk of Migvy and Terland for £60; the kirk of Dull in Athoill, £93 6s 8d; the kirk of Foullis in Gowrie, £40; the kirk of Lynlythgow, £246 13s 4d; the teinds of Wester Bynnyng, £10; the small teinds of Lynlythgow, £12; the teinds of Prestoun, £10; the teinds of Hadingtoun, £266 13[s 4d]; the teinds of Clerkingtoun, £6 13s 4d; 12s 8d 'out of the annuellis of the toun of Lynlythgow'.

---

[56] The correct calculation appears to be 26 c. 2 b., including oatmeal.
[57] These calculations are difficult to verify because of discrepancies in the sub-totals.

Third of the wheat, 12 c. 11 b. 1 f. ⅓ p. Out of the kirk of Lucheris, 8 c. 6 b.; out of the kirk of Forgound in Fyiff, 10 b.; take the rest out of the kirk of Sanctandrois giving 6 c. 10 b. 2 f.

Third of the bere, 44 c. 2 b. 1 f. 1⅔ p. Take the bere of Sanctandrois kirk for 27 c. 8 b. 1 f.; the bere of the kirks of Lucheris for 23 c. 14 b. 2 p. 'Giff in' 7 c. 3 b. 1 f. 2 p. '2 part halff peck' [*i.e.* ⅓ p.].[58]

Third of the meal, 38 c. 1 b. ⅓ p.[59] The kirk of Lauthreisk, 20 c. 1 b.; the kirk of Markinche, 25 c. 2 b. 'Gif in' 6 c. 15 b. 3 f. 3⅔ p.[60]

Third of the oats, 50 c. 8 b. 3 f. ⅔ p. Take these oats of Sanctandrois kirk paying yearly 58 c. 1 f. 2 p. 'Giff in' 7 c. 7 b. 2 f. 1⅓ p.

Third of the peas and beans, 1 c. 2 b. 1 f. 1⅓ p. Take them out of Sanctandrois kirk giving 3 c. 3 b.

'Omittit canis, caponis and all utheris dewteis.'

'The set of landis in few and teyndis in assedatioun pertenyng to the abay of Sanctandrois.'

The kirks and teinds thereof set in assedation

The kirk and teinds of Sanctandrois set in assedation to [*blank*] for £5. The kirk and teinds of Luchris are set in assedation to [*blank*] for £4. The kirks of Forgoun and Dersy are set to £186. The kirk of Cowper is set in assedatioun to [*blank*] for £108. The kirk of Lathrisk is set in assedation to [*blank*] for £160. The kirks of Kilgour and Scowny are set in assedation to [*blank*] for £160.[61] The kirks of Markinche and Kennochie are set to [*blank*] for £160. The kirk of Eglisgrig is set in assedation to the commendator of Deir[62] for £66 13s 4d. The kirk of Hadingtoun 'set to the ladie Lochlavin eldare'[63] for £266 13s 4d. The kirk of Fordorno is set in assedation to the laird of Thornntoun for £266 13s 4d; 2 c. meal. The kirk of Lynlythgow and teind of the kirk of Bynnie and Prestoun are set to Mr William Creichtoun and Mr Robert Wynrame for £266 13s 4d. The kirk of Pertmork is set in assedation to the laird of Lundie for £51. The kirk of Dull is set in assedation to the laird of Grantillie, Alexander Stewart his brother and James Makgregour for £93 6s 8d. The kirk of Abircrummy set in assedation to the laird of Sanct Monanis for £40. The kirks of Mygvie and Tarlane are set in assedation to Robert Urwing [*i.e.* Irving] for £60. The kirk of Burthy is set in

---

[58] These figures differ slightly from those on fo 10v (see above, p. 9).
[59] The correct calculation appears to be 37 c. 11 b. 2 f. 2⅔ p.
[60] The correct calculation appears to be 7 c. 1 b. 3 f. 3⅔ p.
[61] Kirk of Kilgour, £33 6s 8d, fo 16v (p. 15 above) where no money total is given for Scoonie.
[62] Robert Keith, son of William, 4th earl Marischal, was appointed commendator of Deer in 1552, and in 1587 the abbacy was erected into a temporal lordship in his favour, as lord Altrie, Cowan and Easson, *Medieval Religious Houses*, p. 74.
[63] Agnes Leslie, daughter of George, 4th earl of Rothes, married Sir William Douglas of Lochleven c. 1565; their eldest son, Robert, contracted to marry Jean Lyon, 2nd daughter of John, 8th lord Glamis, in 1583. *Scots Peerage*, vi, pp. 371-375.

assedation to Alexander King for £40. The kirk of Foullis set in assedation to my lord Gray[64] for £40. 'Item, to the parochineris of Inchstuir set to the parochineris for' £178 [£147, *fo 18r*]. The kirk of Langforgound set in assedation to the parishioners for £125 13s 4d; 3 c. 7 b. 1 p. wheat; 6 c. 8 b. bere; 16 c. 10 b. 3 f. ¼ p. ('fourth part peck') oats. The kirk of Rossy is set to the parishioners for £47 6s 8d; 1 c. 12 b. wheat; 4 c. 12 b. bere; 1 c. meal; 3 c. 8 b. oats. The small teinds of Lythgow set to Mr Robert Wynrame for £12. The teind of Clerkingtoun to the laird for £6 13s 4d. The teinds of Sanct Laurence House to James Wilkie for £10. 'The sowme of the haill kirkis preceding is' £3,244 13s 4d.[65] Total wheat, 5 c. 3 b. 1 p. Total bere, 11 c. 4 b. Total meal, 3 c. Total oats, 20 c. 2 b. 3 f.[66]

The lands set in feu and assedation

Northbank set in feu to George Lermonth for £6 14s. Clarmonthe set to Sir Patrick Lermonth in feu for £11 3s 11d. Innerork in feu to the said Patrick for £9 7s 5d. Balgoiff to James Colvill in feu for £14. Poffill of Strakynnes in feu to James Month for 50s. Deneheid 'in few for the 4 pairt and 8 pairt to Johnne Forret for' £4 10s 'and the ane halff with ane aucht pairt togither with the De[n]bray and Goukstoun for fyve sex pairtis set in few to [*blank*]' for £33 8s. 'Ane uther sex pairt of Gokstoun set in few to Johnne Dewar for' 46s 8d. Cassindonald set in feu to James Mar for £7 13s 4d. Drumcarro set in feu to Thomas Barklay for £8; 1 c. wheat; 1 c. bere. Unthank [*i.e.* Winthank] in feu to John Forret for £14 9s. Priorlethame set in feu to David Orame for £13 8s 8d. New Grange [Grange, *fo 12r*] 'for ane quarter except the xxx pairt thairof' is set in feu to John Forret for £10 6s. The rest of the Grange in feu to Mr Nicol Elphingstoun for £34 16s. Balrymonth Eister to William Myrtoun in feu for £10 6s 9d. Strafuthy in feu to Andrew Wod for £22 8s. Freirtoun to Mr Robert Creichtoun in feu for £33 13s 4d. Kathlok with Freirsflat to John Kynner for £10. Kirktoun of Forgoun set in feu to David Balfour for £15. Newmyll of Dersy to Walter Melvill for £11 6s 8d. Walkmyll of Dersy to the said Walter for £3 16s. Kirkland of Cowper to John Spens for £19. The Heling Hill to Henry Lawmonthe for £8 6s 8d. Chapell Kettill to John Arnocht for £5 12d. Kylmukles to the laird of Dury for £10. Newmylne of Sanctandrois to Mr Allan Lamonthe for £6. The acres of Sanctandrois to the tenants for £149 11s 4d. Abbey Mylnes to Henry Carnes for £9 10s. Clerkingtoun to the laird thereof for £20. Morestoun to John Cranstoun for 43s 4d. Drem to my lord Lindesay[67] for 26s 8d. Parklye to the laird of Calder for £8. Pulmior, Inchisture and Inchetheray[68] to William Hammiltoun for £30. Chapeltoun to Robert Keyth for £5 15s 4d. Land of Borthy to Alexander King for £6 13s 4d. Littil Lune to David Guthrie for 15s. The Prior's croft in Craill to the laird of Ardree for 7s. Balcomy croft in Craill to the laird of Ardree for 7s. Haddo to the laird of Throntoun for £5 6s 8d. The Abay

---

[64] Patrick, 5th lord Gray, *Scots Peerage*, iv, pp. 283-284.
[65] The correct calculation appears to be £2,353 13s 4d.
[66] The 'fourth part peck' has been omitted from this calculation.
[67] Patrick, 6th lord Lindsay of the Byres, *Scots Peerage*, v, pp. 399-400.
[68] I.e. Inchyra; cf Inschyray, fo 12v (see above, p. 10).

croft to David Orme for £6 13s 4d. The lands of Strakynnes 'except Month pairt' set in feu to [blank] for £77 13s 3d. The lands of Craigtoun and Lumbo set in feu to [blank] for £28 20d. Ballone set in feu to James Reid for £47 2s 5d. Petlethie in feu to Robert Bruce for £27. The Kirktoun of Forgoun pays of peas, 2 b. The Kirksyd of Eglisgreig in feu to Arthur Stratoun for 7 b. bere; 6 b. meal. 'Sowme of the silver for landis set in few', £769 13s 4d.[69] Total wheat, 1 c. Total bere, 1 c. 7 b. Total peas, 2 b. Total meal, 6 b.

Langrow in assedation to James Sandelandis for £8. Innerbrig in assedation to David Gulane for 20s. Kirkland of Markinche to David Bennet for 40s. Yairdmyll to Thomas Diksoune and David Walker for £4. Lawmyll to Robert Murray and James Yeman for £4. Prior Wall to the abbot of Balmerynoche[70] for 10s. Sanctandrois land at Lundy to the laird thereof for 20s. The Freris croft in Craill occupied by sir David Broman for 16d. 'Sanctandrois land in the eist end of Ochiltrie occupyit be the laird of Keir', 8s 8d. 'Sowme of the landis set in assedatioun', £21.

'The caynes pertenyng to the abbay of Sanctandrois': The kane of Pitmyllie by the laird thereof, 13s 4d. The kane of Bannafeild by John Methven, 6s 8d. The kane of Fedinche by Patrick Lermonth, 3s 4d. The kane of Kynnynmonthe by the laird of Craighall, 26s 8d. The kane of Balbreiche by my lord of Rothes,[71] 10s. The kane of Rossythletay by the laird of Moncur, 13s 4d. The kane of Pitpuntie by the laird [of] Stricmartine [i.e. Strathmartine], 40s. Total of these kanes, £5 13s 4d.

'The annuellis of Hadingtoun', [blank]. 'The annuellis of Lynlythgow', [blank]. 'The annuellis of Perthe', [blank]. 'The annuellis of Dundie', [blank]. 'The annuellis of Abirdeane', [blank].

'The sowme of the haill money preceding is', £4,040 19s 11d.[72] Total wheat, 6 c. 3 b. 1 p. Total bere, 12 c. 11 b. Total meal, 3 c.[73] Total oats, 20 c. 2 b. 3 f. $\frac{1}{4}$ p. Total peas, 2 b.

'Thrid thairof according to this rentale in money', £1,346 19s 4d.[74] Third of the wheat, 2 c. 1 b. $\frac{1}{3}$ b. [recte, $\frac{1}{3}$ p.] Third of the bere, 4 c. 3$\frac{1}{3}$ b.[75] Third of the meal, 1 c. Third of the oats, 6 c. 11 b. 2 f. 1 p.[76] Third of the peas, $\frac{2}{3}$ b.[77]

---

[69] The correct calculation appears to be £753 17s 10d.
[70] John Hay, parson of Monymusk, received a gift of the abbacy of Balmerino in 1561, Cowan and Easson, *Medieval Religious Houses*, p. 73.
[71] Andrew Leslie, 5th earl of Rothes, *Scots Peerage*, vii, pp. 292-294.
[72] MS totals would give £4,041 exactly. Editorial calculations yield £3,134 4s 6d.
[73] The correct calculation appears to be 3 c. 6 b.
[74] The correct calculation appears to be £1,346 19s 11$\frac{1}{3}$ d.
[75] The correct calculation appears to be 4 c. 3$\frac{2}{3}$ b.
[76] $\frac{5}{12}$ p. have been omitted from this calculation.
[77] 'Third of the peas, $\frac{2}{3}$ b.' is missing from the text in SRO, Vol. b.

'The speciall assumptioun of the victuall of the priorie of Sanctandrois gevin [up] be Walter, prior of Blantyre,[78] [in] namis of the priorie to Lodavick, duke of Lennox,[79] commendatour thairof, according to ane charge gevin and direct be the Lordis of Checker to the prelate for inpreving of the speciall assumptiounes of the thrid of the benifices quhair thay have bene particularlie assumit of befoir.'

Third of the wheat, 12 c. 11 b. 1 f. ⅓ p.

Take Lucheris kirk: Cragy, 4 b.; Ardat, 2 c.; Segy, 6 b.; Dron, 1 c. 2 b.; Pursk, 1 c. 2 b.; Pitcullocht, 5 b.; Ballmullo, 10 b.; Scottiscraig, 14 b.; Brodland, 10 b.; Kettety, 5 b. Total, 7 c. 10 b.

Forgund kirk in Fyffe, namely, Innerdovet, 4 b.; Middle Newtoun, 2 b.; Sandfurd Nachtan, 2 b.; Sandfurd Hay, 2 b. Total, 10 b.

Sanctandrois kirk [Ecclesia Trinitatis, *fo 14r*], namely: Byrehillis, 7⅓ b.; Clatto, 2 b.; Wester Balrymonth, 4 b.; Strakinnes, 12 b.; Ballonie, 9 b.; Balgoiff, 4 b.; Nather Magus, 2 b.; Wilkestoun, 1 b.; Northbank, 1 b.; Jokstoun [Cokstoun, *fos 12r, 24r, 24v*], 1 b.; Denebray, 1 b.; Drumcarro, 6 b.; Balduny, 4 b.; Kininmonthe, 4 b.; Craigtoun, 2 b.; Langraw, 2 b.; Newgrange, 9 b. Total, 4 c. 7⅓ b.

'Summa of thir thrie kirkis', 12 c. 11⅓ b. 'Gif in' 1 p.

Third of the bere, 44 c. 2 b. 1 f. 1⅔ p.[80]

Take Lucheris kirk, namely: Brodland, 4 c.; Ardat, 2 c.; Myltoun, 10 b.; Segy, 1 c.; Dron, 1 c. 12 b.; Pursk, 3 c.; Pitcullo, 8 b.; Ballmullo, 14 b.; Cowbaky, 13 b. 2 f.; Kittuthy, 8 b.; Straburne, 3 b.; Fyrdaill, 12 b.; Scottiscraig, 1 c. 8 b.; Schanwell, 8 b. Garpet, 5 b.; Muirtoun, 4 b.; Wester Fetteris, 5 b.; Rynd, 12 b. 3 f.; Brigend and Quhyt croft, 4 b. 1 f. [3 b. 1 f., *fo 15r*]; Ferrytoun, 1 c. 4 b.; Munsymyll, 6 b.; out of Cragy, 3 b. 3 f. Total, 21 c. 12 b. 1 f.[81]

Sanctandrois kirk: Over Magus, 6 b.; Nather Magus, 4 b.; Clatto, 2 b.; Wilkestoun, 4 b.; Strathkynnes, 1 c. 8 b.; [B]alonie, 1 c. 7 b.; Mounthis pairt of Strakinnes, 4 b.; Balgoiff, 8 b.; Deneheid, 4 b.; Cokstoun, 6 b.; Denebray, 3 b.; Dunnork, 5 b.; Drumcarro, 10 b.; Ladeddy, 9 b.; Balduny, 6 b.; Kynninmonthe, 12 b.; Lumbo, 3 b.; Craigtoun, 6 b.; Faddinche, 3 b.; Balrymonthe Wester, 5 b.; Hospitall of Sanct Nicolas, 6 b.; Laderny, 8 b.; Lathonis, 2 b.; Lathokker, 2 b.; Hasildene, 1 b.; Ballmongy, 3 b. 1 f.; Gilmerstoun, 4 b.; Langraw, 4 b.; out of the Byrehillis, 1 c. 2 b. 1⅓ p.; Polduff, 1 c. 6⅔ b.; Bonytoun, 4 b.; Newgrange, 1 c. 2 b.; the teinds of the crofts and yairds, 12 b.; Inchemurtho Park, 1 b.; 'out of the toun akeris', 6 c. 9 b. 1⅔ p. Total, 22 c. 6 b. 1⅔ p.

'Summa of thir tua kirkis', 44 c. 2 b. 1 f. 1⅔ p.

---

[78] Walter Stewart, son of Sir John Stewart of Minto, gained confirmation of his rights as commendator in 1569, *RSS*, vi, no 732.

[79] Ludovic Stewart, 2nd duke of Lennox, received a gift of the priory of St Andrews in 1586, *Scots Peerage*, v, p. 357.

[80] This calculation is correct.

[81] The correct calculation appears to be 21 c. 13 b. 1 f.

Third of the meal, 38 c. 1 b. ⅓ p.

Take Lauthrisk, namely, Lauthrisk Wester, 1 c. 6 b.; Eister Lauthrisk, 2 c. 9 b.; Riggis with Lythgw lands, 7 b.; Ramorgny, Burneturk and Half Ballingall, 4 c.; Dounie, 1 c. 11 b.; Ramelry, 1 c. 8 b.; Kingis Kettill, 3 c.; Cletty, 1 c.; Forther Ramsay, 2 c. 1 b.; Dounfeild, 11 b.; Halkettil, 1 c. 12 b. Total, 20 c. 1 b.

Kilgour kirk, namely, the toun of Kilgour, 8 b.; Falkland with the cottars, 2 c. 12 b.; Newtoun of Falkland, 2 c. 4 b.; Freuchy, 3 c. 4 b. Total, 8 c. 12 b.

Scony kirk, namely, the toun of Skony, 1 c.; Mounthflowry, Burnmylne, acres of Levin, 1 c. 6 b. Swyinshauche 11 b.; Dury, 2 c. 8 b.; Aderny, 1 c. 10 b.; Balgrummo, 1 c. 2 b.; out of Letham, 15 b. ⅓ p. Total, 9 c. 4 b. ⅓ p.[82]

'Summa of thir thrie kirkis', 38 c. 1 b. ⅓ p.

Third of the oats, 50 c. 8 b. 3⅔ p.

Take Sanctandrois kirk, namely, Over Magus, 1 c. 8 b.; Nather Magus, 14 b.; Clatto, 1 c.; Wilkestoun, 14 b.; Strakynnes, 3 c. 8 b.; 'out of Monthis pairt', 2 b. [Paremounth, 8 b., *fo 14r*]; Ballone, 2 c.; out of Deneheid, 11 b.; out of Denebray, 3 b.; Cokstoun, 1 c.; Dunmork, 2 c.; Drumcarro, 1 c. 10 b.; Ladeddey, 1 c. 14 b.; Balduny, 2 c. 2 b.; Kynnymonthe, 3 c. 2 b.; Cassindonat, 1 c.; Craigtoun, 1 c. 2 b.; Northbank, 8 b.; Lathonis, 12 b.; Feddinch, 1 c. 2 b.;[83] Balrymonthe Wester, 1 c. 2 b.; Hospital of St Nicholas, 4 b.; Laderny, 3 c. 4 b.; Lathocker and Hessilden, 3 c.; Ballmongy, 14 b.; Gilmerstoun, 1 c.; Langraw, 1 c. 4 b.; Byrehillis, 4 c. 5⅓ b.; Polduff, 2 c. 2⅔ b.; Newgrange, 4 c.; Inchemurtho Park, 4 b.; 'the toun akeris', 2 c. 3⅔ p. Total 50 c. 8 b. 3⅔ p.

'Sederunt: Cancellarius; Thesaurarius; Comptroller; Collector; Culluthie; Clerk of Registre.'[84]

25 February 1588/9. 'Producit be the pryour of Blantyre and considerit be the Lordis of the Checker and ordanit to be insert and registrat in the Buikis of Assumptioun and the thrid of the priorie of Sanctandrois to be chargit in all tyme cuming according to this particular assumptioun.'

---

[82] These calculations are correct.
[83] The text from here to the end of this rental does not appear in SRO, Vol. b.
[84] Chancellor, Sir John Maitland of Thirlestane, keeper of the Great Seal; Treasurer, Thomas Lyon of Baldukie, master of Glamis (later Sir Thomas Lyon of Auldbar); Comptroller, David Seton of Parbroath; Collector general, Robert Douglas of Lincluden; David Carnegie of Colluthie; Clerk Register, Alexander Hay of Easter Kennet. *Handbook of British Chronology*, pp. 183, 188, 191, 197; *RPC*, iv, p. 381; p. 365.

PITTENWEEM, PRIORY OF,

(SRO, Vol. a, fos 29r-v[85])
(SRO, Vol. b, fos 20r-v)

'Priorie Pittinweme'

'Sequitur rentale omnibus firmarum'

Follows the rental of all fermes, kanes and annualrents and also of the teinds of all the kirks and lands of the priory of Peitinweyme, in money and victuals, both within and outside Fife.

Rental of all fermes, kanes and annualrents of the said priory. Total: Money, £233 0s 6d. Wheat, 2 c. 12 b. Bere, 2 c. 12 b. Meal, 4 c. 12 b. 2 f. 2 p. Oats, 5 c. Salt, 24 c.

Rental of the teind sheaves of the said priory in money and victuals
    The parish kirk of Anstruder, 1 c. 9 b. wheat; 4 c. 6 b. bere [14 c. 6 b., *fo 29v*]; 2 c. 2 b. 2 f. oats; 1 c. 11 b. beans and peas. The kirk of Ryndis Eister and Wester, £60.

Rental of the teinds of the vicarage of Anstruder coming from salt, teind wool, flax, hemp, with fresh and dried fish and herring etc., and other emoluments of the same vicarage in a single common year, extending in just estimation to £119 12s or thereabouts.

Total: Money, £412 12s 8d.[86] Wheat, 4 c. 5 b. Bere, 7 c. 2 b. Meal, 4 c. 12 b. 2 f. 2 p. Oats, 7 c. 2 b. 2 f. Beans and peas, 1 c. 11 b. Salt, 24 c.

Assignation of the priory of Pettinweyme, third thereof:
    Third of the money, £137 10s 10d. 'Tak it of the haill fermes, cane and annuellis extending in the yeir to' £233 6d.
    Third of the wheat, 1 c. 7 b. Take the kirk of Anstruder for 1 c. 9 b. 'Giff in' 2 b.
    Third of the bere, 2 c. 6 b., out of the kirk of Anstruder paying 14 c. 6 b. [4 c. 6 b., *fo 29r*]
    Third of the peas and beans, 9 b. out of the same kirk paying 1 c. 11 b.
    Third of the meal, 1 c. 9 b. 2 f. $\frac{2}{3}$ p., out of the ferme victual giving 4 c. 12 b. 2 f. 2 p.[87] Third of the oats, 2 c. 6 b. $2\frac{2}{3}$ p., out of the same fermes giving 5 c.
    Third of the salt, 8 c., out of the same fermes giving 24 c.
    'Omittit caynis, custoumes and all uther dewteis.'

---

[85] SRO, Vol. a, fos 25v-28v are blank.
[86] The correct calculation appears to be £412 12s 6d.
[87] This figure is given above for meal.

DUNFERMLINE, ABBEY OF,[88]   (SRO, Vol. a, fos 30r-35v)
                                                      (SRO, Vol. b, fos 21r-26v)

'The haill rentall of the patrimonie of the abay of Dumfermeling in penny maill, annuellis, customes, burrow mailles, feid oxin silver, lymekill mailles, canis, fermis, teyndis [of] kirkis and teyndis of townes set in assedatioun for money as followis particularlie in everie schyre be the selff, gevin in and be Allane Coutes, chalmerlane, subscryvit, etc.'

Dumfermelyneschyre: Maistertoun in penny mail 'in anno', £26 13s 4d. Eist Barnis in penny mail, £21 7s. Patortheis [*i.e.* Pitcorthies] in penny mail, £17 6s 8d. Sanct Margarettis Stane 'in anno', £6 13s 4d. The Walkerland in Burne Moutht, 53s 4d. Stenstanes [in] the Burne Mowthe, 40s. Selletoun, £17 6s 8d. Legattis Brig, £3 6s [£3 6s 8d, *fo 54r*]. Wondmylhill, £4 16s. Kinkas [*recte*, Knokas, *i.e.* Knockhouse], £13 8s 8d. Primrois, £8 2s. Pitbawillie, £3 7s 8d. Wester Bawdrick, 20s. Eister Bawdrick, 40s. Midle Bawdrick, £4 6s 8d. Wester Baith, £8 2s. Blaxlaw, £3 7s 8d. Nather Baith, £6 2s. Bayth under the Hill, £7 2s. Luscerewert, £7 8s 8d [£7 8s, *fo 54r*]. Northfaid, £6 15s 4d. Toucht, £4 2s. Wyndeage, 40s 6d. Cocklaw, 40s 6d. Over Lasoidey, 40s 6d. Eister Lasoidey, £4 2s. Nather Lasoiday, £6 18s [£6 18d, *fo 54v*]. Bawmill, £4 12d. Blairathy, £4 12d. Craigluscour, £8 15s 4d. Lathamounth, £4 12d. Craigduky Wester,[89] £3 7s 8d. Eister Craigduky, £3 6s 8d [£3 7s 8d, *fo 54v*]. Couch [*recte*, Touch] Myln, 53s 4d. Wester Luscour, £8 2s. Northtlethenes, £4 0s 8d [£4 12d, *fo 54v*].[90] 'The thrie mylnes of the toun of Dunfermeling', £13 6s 8d. Lathangy, £6 13s 4d. The Oucht, £8. Southfaid,[91] £6 15s 4d. 'The burrow mailles of Dunfermelyng toun', £8 13s 4d. 'The custoumes of the burcht of Dunfermelyng', £4. 'Summa of the penny maill, burrow mailles and custoumes of Dunfermeling', £286 17s 4d.[92]

The annuals pertaining to Dumfermling: Loggy, 26s 8d. Crummy, £10. The glebe of Stirling, £10. The annual of Scone, £3 6s 8d. The annual of Pittenweim, £3. The annual of Gartinfallo [*recte*, Gartinkeir, *fo 54v; SRO, Vol. b, fo 21v*], £4. Total, £31 13s 4d.

'Item, of feid oxin silver within the lordschip of Dumfermeling', £13. The bow places, wedder gangis within the parish of Dumfermeling 'be yeir', £96. 'Item, the bow places foirsaidis gevis be yeir' 35 stones [of butter]; 'and of cheis', 130 stones, 'videlicet': Rescoby, 30 stones cheese; 7 stones butter. The Gask, 7 stones butter; 30 stones cheese. Southtlothenis, 7 stones butter; 30 stones cheese. Tonigask, 7 stones butter; 25 stones cheese. Blairnbothie, 7 stones butter; 25 stones cheese. Total butter, 35 stones. Total cheese, 130 stones.[93]

---

[88] Cf SRO, Vol. a, fos 44r, 54r (see below, pp. 37, 45); see also *Registrum de Dunfermelyn*, ed. C. Innes (Edinburgh, 1842), pp. 425-462.
[89] 'Craigenky', *Registrum de Dunfermelyn*, p. 426.
[90] Included at this point on fo 54v, 'The Knok, £3 6s 8d' (see below, p. 45).
[91] 'South ford', *Registrum de Dunfermelyn*, p. 426.
[92] This calculation is correct, taking the figures on fo 54r-54v (see below, p. 45).
[93] The correct total should be 140 stones.

Dolour 'in penny maill be yeir': The Horhart 'in anno', 26s 8d. Dolour Brig [recte, Beig, fo 55r], £4 13s 4d. The Mylne of Dolour, 26s 8d. Schyirdaill, £13 6s 8d. The Mylne of Dolour Brig [recte, Beig, fo 55r], £3 6s 8d. The Bank 'in anno', 40s [recte, £40, fo 55r]. Total of the penny mail of Dolour with the Bank, £64.[94]

Newburneschyre in penny mail: Newburne in penny mail 'in anno', £6 13s 4d. The Mylne, £4. Manturpe, £6 13s 4d. Lathallane, £8. Hallhill, £10 13s 4d. Balbardy, £5 6s 8d. Melgum and Lawgrenis, 40s. Drummeldre, £10. Balcristie, £10. Cott, £10. The acres, 11s 8d. Total of the penny mail of Newburneschyre, £73 18s 4d.

Kynglassyschyre in penny mail: Pityowquhair 'in anno', £17 6s 8d. Stentoun 'in anno', £18 13s 4d. Inche Durne 'in anno', £18. Kinglassie toun, £8. The Smyddy land, 20s. The Broister land, 13s 4d. Cassebarrian, £13 6s 8d. Fynglassy, £16. Pitlochie, £6 13s 4d. The Walk Mylne, 40s. Gaitmilk, £15 4s. Finmont, £13. The annual of Cluny, 20s. Total of penny mail and annual of Kinglassie parish, £132 17s 4d.[95]

Kyngorne Wester in penny mail and annual: The Wingraig[96] [recte, Over Grange, fo 35r] 'in anno', £40. Weltoun 'in anno', £20. Eister Quarter, £20. The Mylne, £6 13s 4d. Moy Hous, 53s 4d. The Lymkill mail, £13 6s 8d. The annual of Orrok, £3 6s 8d. The annual of Sellebair, 33s 4d. The annual of Abirdour, 53s 4d. Total of penny mail and annual, £110 6s 8d.

Kirkcaldie: Abbottis Hall and Mylntoun, £34 13s 4d. The West Mylne, £4. Dunekeir, £8. Bogy, £40. The Smiddie Landis,[97] 53s 4d. The annual of Balwery, £4. The annual of Ratht, £5. 'The burrow maill of the burcht', 33s 4d. The mail of Salt Panes, £16 [recte, £46, fo 56r]. The Eist Mylne, 53s 4d. The annual of Sandyruidis ['certane ruidis', fo 56r], 13s 4d. The customs of Kirkcaldie, £5. 'Summa of penny maill, annuell, custoumes and burrow mailles of the parochin of Kirkcaldie', £154 6s [8d].[98]

Kynnenndir: Over Kinnedder, £8 8s 8d. Bandrum, £6 15s 4d. Drumkappie, £3 7s 8d. The Mylne, £3 6s 8d. Nather Kinnedder, £7 8s 8d. The Brewland, 13s 4d. Killarne, £13 15s 4d. Total of penny mail in Kinnedderschyre, £43 15s [8d].

The penny mail and annual within Mussilburche: Pinkie [Inveresk, fo 56v], 'in anno', £71 16d. Monktounhall, £43 6s 8d. Littill Monktoun, £18. Natoun, £29 3s [in margin, 'augmentit be Archibald [Pre]stoun', £7 16s [4]d, 'and comptit in this' [£29] 3s]. Smetoun, £30 13s 4d. Caldcoittis, £6. Wonet Bank, £6 13s 4d. Carbarry, £40. Cestertoun [i.e. Costerton], £10. Pikin and Cars,[99] £23. 'The fiching', £8. Womet, £44 16s. Sanehill [recte, Stanehill, i.e. Stoneyhill], £6

---

[94] This calculation is correct, taking £40 for the Bank.
[95] The correct calculation appears to be £130 17s 4d.
[96] SRO, Vol. b, fo 22v reads 'Uver Grang'.
[97] 'The syndry landis', *Registrum de Dunfermelyn*, p. 429.
[98] This calculation is correct, taking £46 for the Salt Pans.
[99] 'Pynkin and Kers', fo 56v; 'Pynkin and Cars', SRO, Vol. b, fo 46v; 'Wowmet ... Wowmetbank ... Pinkie et Cars', *RMS*, v, no 1305.

13s 4d. 'The mylnes', £66 13s 4d. 'The burrow maill of Mussilburcht toun', 53s 4d [*in margin*, 'payit be the balleis']. The annual of Hill, 20s [*in margin*, 'be the laird Hill']. The annual of Edmestoun, 13s 4d [*in margin*, 'be the laird thairof']. The annual of Edinbrucht, £5 [*in margin*, 'be the thesaurer thairof']. The annual of Hadingtoun, 40s [*in margin*, 'be the balleis']. The annual of Quhytsyd, 20s [*in margin*, 'be the laird Fawsyd, now be Mr David and his spous']. The annual of the Pannes, 20s [*in margin*, 'be the laird of Craigmiller']. The annual of Newbottell, 20s [*in margin*, 'the abbat of Newbottle'].[100] The Terreris crofts, 26s 8d [*in margin*, 'be Bessie Froig']. Total of penny mail and annual within Mussilburcht, £419 13s 8d.[101]

'The haill money within the paroche of Dunfermelyng of penny maill, annuell, custoume, burrow maill, feid oxin silver, mart silver with the wedder gang extendis to the sowme of' £427 10s 8d.[102] The penny mail of Dolour with the Bank of the same extending 'be yeir' to the sum of £64. The penny mail of Newburneschyre 'be yeir' extends to the sum of £73 18s 4d. The penny mail of Kynglassie yearly extends to the sum of £132 17s 4d. The penny mail of Wester Kinghorne 'be yeir' and annual extends to the sum of, £110 6s 8d. The penny mail, annual, burgh mail and custom of Kirkaldie extends to the sum of £154 6s 8d. The penny mail and annual of Kynnedderschyre extends 'be yeir' to the sum of £43 15s 8d. The penny mail of Mussilburcht with annuals [and] burgh mails, extends to the sum of £365.[103] The penny mail of Southferrie with annuals extends 'be yeir' to the sum of £23 9s 8d.[104] The penny mail of Stermontht extends 'be yeir' to the sum of £47 12s.[105] 'Kirkis and teyndis of townes set for money extendis be yeir to the sowme of', £908 13s 4d.

'The ferme quheit within the paroche of Dunfermelyng, Mussilburghe and Newburneschyre as followis particularlie etc. Item in the first the Gullattis within the paroche of Dunfermelyng of ferme quheit extending to' 5 c.

Newbyrneschyre: The Coit, 1 c. Drumneldrie, 1 c. Few, [*blank*]. Newburnetoun, 10 b. Manturpe, 8 b.

Mussilburcht: Pynkie [Inveresk, *fo 57v*] in ferme wheat, 2 c. 8 b. Monktounhall, 1 c. 4 b. Smetoun, 1 c. 4 b. Carbarry, 12 b. Naitoun, 8 b.

Total of ferme wheat within the parishes of Dunfermlyne, Newburne and Mussilburcht, 15 c. 6 b.[106]

---

[100] Mark Ker was appointed commendator of Newbattle in 1547; and his son, also Mark, was presented to the abbacy in 1567. Cowan and Easson, *Medieval Religious Houses*, p. 77.
[101] This calculation is correct, but see next paragraph where the penny mail of Musselburgh is given as £365.
[102] This calculation is correct for MS totals.
[103] Cf above, n. 101.
[104] The penny mail of Southferrie is not included in the above entries.
[105] The penny mail of Stermontht is not included in the above entries.
[106] The correct calculation appears to be 14 c. 6 b.

Ferme bere within the parishes of Dumfermelyng, Kingorne, Newburne and Dolour. Within the parish of Dunfermelyng: The Gullatis, 9 c. Girsmuir Land, 6 b. Lymekillis, 2 c. 8 b. Lymkilhill, 1 c. 5 b. The New Landis, 1 c. 2 b. Dolourschyre, 1 c. The Nedder Graig [recte, Grange] of Kingorne, 6 c. The dry multure of Wester Kingorne, 8 b. Manturpe, 8 b. Total of the ferme bere within the parishes of Dumfermeline, Dolour, Kingorne and Newburne, [blank].

The ferme oats within the parish of Dumfermeling. The two Gullates, 15 c. Total of the ferme oats, 15 c.

Ferme oatmeal within the parishes of Dolour, Kinglassie and Kirkcaldie: In Dolourschyre, 5 c. In Kirkaldie parish, 2 c. 8 b. In Kinglassie parish, 6 c. 3 b. 1 f. 3 p. Total of the oatmeal within the parishes of Dolour, Kirkcaldie and Kynglassie, [blank].

Ferme salt within the parish of Kirkcaldie, 23 pans, 'ilk ane of thame payit of ferme salt be yeir' 8 b. Total, 11 c. 8 b.

'Ferme kane lyme within the parochin of Kingorne Wester', 20 c.

'Kirkis [and, *fos 32v, 58r*] teyndis of townes set in assedatioun for money as followis particularlie'

The kirk of Stirling set to the laird [of] Carden, £80. The kirk of Hailles set to Mr Henry Foulis, 100 merks. The kirks of Uruquhill and Kynros to the lady Lochlavin,[107] £120. The kirk of Sanct Johnnestoun to the laird of Cragy, £1[00]. The kirk of Innerkeddy to John Suetoun [Swyntoun, *fo 58r*], £53 6s 8d. The kirks of Molloun and Straithardil, £120. The kirk of Wester Kingorne to Mr John Wemis, £12. The teind of Mon[k]tounhall set for £20. The teind of Innernisk set for £53 6s 8d. The teind of Smetoun set for £20. The teind of Naitoun set for £13 6s 8d. The teind of Carberie set for £33 6s 8d. The teind of Cosland set for £66 13s 4d. The teind of Pynkie [Pynkin, *fo 58v*] set for £10. The teind of Edmestoun set for £26 13s 4d. The teind of Wonet set for £33 6s 8d. The teind of Hallhill, £40. The teind of the burgh acres of Kirkcaldie set to David Kyngorne for £13 6s 8d. The teind of the acres of the burgh of Dunfermelyng set for £26 13s 4d. 'Summa of kirkis [and] teyndis of tounes set for money extendis to' £908 13s 4d.

Teind wheat of Mussilburcht: Littill Monktoun, 6 b. Hill, 2 b. Caldcoittis, 3 b. Stanehill, 4 b. Quhythill, 4 b. Schirefhall, 2 b. 'Summa of teynd quheit within the parochin of Mussilburcht by that is set for money', 1 c. 5 b.

Teind bere of Mussilburcht 'that is nocht set for money': Littill Monktoun, 11 b. Hill, 2 b. Caldcoittis, 3 b. Stanehill, 15 b. Quheit Hill, 14 b. Schirefhall, 4 b. Quhytsyd, 2 f. The Sande, 1 c. 'Summa of the teynd beir within Mussilburchtschyre nocht set for money', 4 c. 1 b. 2 f.

Teind oats within Mussilburchtschyre: Littill Monktoun, 1 c. 12 b. Hill, 6 b. Caldcoittis, 12 b. Stanehill, 15 b. Quheithill, 14 b. Schirefhall, 8 b. Quhytsyd, 4 b. Total of the teind oats of Mussilburcht, 5 c. 7 b.

---

[107] Margaret, 2nd daughter of John, 5th lord Erskine, was the widow of Sir Robert Douglas of Lochleven, who was killed at Pinkie in 1547; she died on 5 May 1572. *Scots Peerage*, vi, p. 369.

Teind wheat within the parish of Dumfermeling 'as followis particularlie': Eister Gullet of teind wheat, 11 b. Wester Gellet, 11 b. Eist Barnes, 5 b. North Ferrie, 2 b. Maistertoun, 8 b. Pitlever, 2 f. Pittravie, 3 b. Pitfirrane, 2 b. 2 f. Prymros, 5 b. Randellis Craigis, 3 f. Sanct Margaretis Stane, 1 b. 2 f. Selletoun, 1 b. Total of teind wheat within the parish of Dumfermelyng, 3 c. 3 b. 1 f.

Teind bere within the parish of Dumfermelyng 'as followis particularlie': Wester Gellet, 1 c. 2 b. Eister Gellet, 1 c. 2 b. Bur[n]e Mowtht, 2 f.[108] Fowillfuird, 1 b. Bayithe Maistertoun, 1 b. Blaklaw, 2 b. Bayith Stewart, 4 b. Bayth Creir [Baith Keir, *fo 59v*], 1 b. 2 f. Bayth Trumbill, 1 b. Bayth Stevinstoun, 1 b. Bayth Bell, 2 f. Blairbothy Nather [Northir, *fo 59v*], 2 f. Blairathy, 2 f. Bayth Danyell, 2 f. Bayth Bonalla, 6 b. Bayth Persoun, 3 b. 3 f. Coklaw, 1 b. 2 f. Cawill, 2 b. Crownarland, 3 f. Craigluscour, 4 b. Clunie, 2 b. Craigduky Yester, 1 b. 2 f. [1 b., *fo 59v*]. Craigduky Wester, 1 b. 1 f. Dunduff, 2 b. 2 f. Drumthuthill, 2 f. Eist Barnes, 10 b. Galrik, 2 b. Gask, 1 b. Halbaudrik [Hoill Baudrik, *fo 59v*], 1 b. 2 f. Knokas, 6 b. Kirk [*recte*, Knok, *fo 34v*], 1 f. Legattis Brig, 2 b. Lymekillis, 10 b. Lathallmonth, 1 b. Wester Luscour, 8 b. Luscour Ewart, 2 b. The Lasoiddeis, 8 b. The Foudeis [*i.e.* Northfod and Southfod], 10 b. North Ferry, 4 b. Mylhillis, 3 b. Maistertoun, 1 c. 8 b. Morlabank, 2 f. Middylbawdrik, 3 b. Meldrumes Mylne, 1 f. Northtlethenis, 2 f. Oucht, 2 p. Pitlever, 2 b. Pittoravy, 6 b. 2 f. Pitcorthis, 8 b. [10 b., *fo 60r*]. Pitfirrane, 5 b. Prymros, 5 b. [6 b., *fo 60r*]. Pitdunes, 10 b. Pitbawlie, 1 b. 3 f. Pitconquhy, 3 b. Pittincreif, 12 b. Randell Craigis, 4 b. Rascobie, 1 b. Stelletoun Hakheid [Selletoun Halkheid, *fo 60r*], 3 f. Sanct Margaretis Stane, 4 b. Stelletoun Sanctandrois [Selletoun Sandis *fo 60r*],[109] 1 b. Toucht, 2 b. 2 f. Tunegask, 2 f. Wyndeage, 1 b. 2 f. Wendmethill, 2 f. Walkarland, 1 b. Total of the teind bere within the parish of Dumfermeling, 15 c. 6 b. 3 f. 2 p.[110]

Teind oats within Dumfermelyng parish: Eister Gellet, 3 c. 4 b. Wester Gellet, 4 c. 2 b. Burne Mouth, 1 b. 2 f. Actcorward [Accorne Waird, *fo 60v*], 1 b. Bawmell, 8 b. Fowilfuird, 6 b. Bayth Maistertoun, 8 b. 2 f. Blaklaw, 10 b. Brounhill [Brumhill, *fo 60r*] 3 b. Bayth Stewart, 10 b. Bayth Ker, 10 b. Bayth Trumbill, 10 b. Bayth Stevinstoun, 6 b. Bayth Bell, 4 b. Blairboth Norther, 4 b. Blairrathie, 4 b. 2 f. Baith Danyell, 5 b. Blairinbother Souther, 2 b. Bayth Bownalla, 1 c. 14 b. Bayth Persoun, 1 c. 2 b. Coklaw, 10 b. Carrill, 10 b. Crownarland, 1 b. Craigluscour, 1 c. 4 b. Clunie, 10 b. Craigduky Eister, 8 b. Cragduky Wester, 8 b. Dunduff, 1 c. Druthuill [Drumthuthill, *fo 34r*; Drumthuill, *fo 60v*], 8 b. Eist Barnis, 1 c. 8 b. Galrik, 11 b. Girsmuirland, 1 b. Gask, 6 b. Hoill Bawidrik, 12 b. Knokas, 1 c. 12 b. Knok, 3 b. Kelty Wod [Kelte Wod, SRO, Vol. b, *fo 25v*], 1 b. Legattis Brig, 6 b. Lymekillis, 11 b. Lawthallmonth, 5 b. Wester Luscour, 1 c. 12 b. Luscour Evert, 1 c. 4 b. Lasoiddeis Eister and Wester, 2 c. 8 b. Fouddeis South and North, 4 c. 8 b. North Ferry, 8 b. Mylhillis,

---

[108] Included at this point on fo 59v is 'Bawmyll, 1 b. 2 f.' (see below, p. 49).
[109] SRO, Vol. b, fo 50r reads 'Selletoun [Sanctandrois, *deleted*] Sandis'; *RMS*, vi, no 75 reads 'Sillietoun-Sanderis'.
[110] The correct calculation appears to be 15 c. 2 b. 3 f. 2 p.

7 b. Maistertoun, 4 c. 8 b. Morlay Bank, 6 b. Mydlbawid[r]ik, 14 b. Meldrum Mylne, 1 b. 'North pairt quheit feild', 1 b. Newlandis, 1 c. Norlethenis, 6 b. Oucht, 2 b. Pitliver, 5 b. Pitrovey, 2 c. 2 b. [1 c. 2 b., *fo 61r*]. Pitcorthis Eister and Wester, 2 c. Pitfirrane, 1 c. Prymros, 2 c. Pitdunis Eister and Wester, 2 c. Pitbaulie, 7 b. Pitconochy, 1 c. Pittincreiff, 14 b. Randellis Craigis, 12 b. Rescoby, 8 b.[111] Sanct Margaretis Stane, 10 b. Selletoun Sanctandrois,[112] 11 b. South Lethenes, 6 b. Toucht, 10 b. Toucht Mylne, 2 b. Twnegask, 4 b. 2 f. Wyndeage, 1 b. 2 f. 'Wyndeage pairt Kellok',[113] 8 b. Windmilhill, 6 b. Woidaker, 3 f. Walkarland, 3 b. 2 f. Total of teind oats within the parish of Dunfermelyng, 'quhyt and blak', 62 c. 14 b. 2 f.[114]

Teind wheat of Kingorne Wester: Overgrange, 6 b. Weltoun, 6 b. Eist Quarter, 1 b. Orrok, 5 b. Sellebalbie, 1 b. 2 f. Total of teind wheat in Kingorne, 1 c. 3 b. 2 f.

Teind bere of Kingorne: Over Grange, 1 c. 7 b. Weltoun, 12 b. Eist Quarter, 4 b. Orrok, 8 b. Sellebalbie, 3 b. Total of the teind bere in Kingorne, 3 c. 2 b.

Teind oats in Kingorne: Over Grange, 3 c. 15 b. Weltoun, 2 c. 6 b. Eist Quarter, 1 c. 8 b. Orrok, Stanhous, Monyhous [Moyhous, *fo 61v*] and Drumharne [Dunhair, *fo 61v*; *i.e.* Dunearn], 2 c. 4 b. Sellebalbe, 1 c. 4 b. The Terrouris acres, 2 b. Total of the teind oats within Kingorne, 11 c. 7 b.

Teind wheat of Kirkaldie: The Abbottis Hall, 6 b. Balwery, 6 b. Bogy, 2 b. Raucht, 4 b. Total of the teind wheat in Kirkaldie, 1 c. 2 b.

Teind bere of Kirkcaldie: Abbottis Hall and Myltoun, 15 b. The West Mylne, 1 b. 2 f. Balwery, 8 b. Bogy, 12 b. Raucht, Turbane and Pittinmork, 20 b. The Murtoun of Kirkaldie, 1 b. Total of the teind bere in Kirkaldie, 3 c. 9 b. 2 f.

Teind oats of Kirkcaldie: Abbottis Hall and Myltoun, 3 c. 14 b. Balwery, 2 c. Bogy, 3 c. Racht, Pittinmark and Turbane, 4 c. 14 b. The Muirtoun, 8 b. Total of the teind oats in Kirkaldie, 14 c. 4 b. [*In margin*, 'Be ressoun of' 1 c. 2 b. 'deducit for the Racht.']

Teind wheat in Newburneschyre: The Cot, 14 b. Ballcrist, 4 b. Drummeldrie, 1 c. Newburnetoun, 5 b. Mantirope, 4 b. Total of teind wheat in Newburne, 2 c. 11 b.

Teind bere of Newburne: The Cot, 1 c. 8 b. Ballcreiff [*recte*, Balchrystie], 1 c. 8 b. Bawbaird, 1 b. 1 f. Lathalland, 4 b. Drummeldrie, 2 c. 2 b. Johnnestounes Mylne, 1 b. Newburnetoun, 12 b. Manturpe, 8 b. The Terrouris acres, 1 b. Total of teind bere of Newburne, 6 c. 13 b. 1 f.

---

[111] Included at this point on fo 61r is 'Selletoun Halkheid, 6 b.' (see below, p. 50).
[112] Cf above, p. 27, n. 109.
[113] 'Sillietoun-Sanderis', *Registrum de Dunfermelyn*, p. 439.
[114] The correct calculation appears to be 63 c. 14 b. 3 f.

Teind oats of Newburne: Cott, 2 c. 8 b. Ballcriste, 1 c. 1 b. Lathallane, 1 c. 8 b. Bawbard, 9 b. Drummeldrie, 2 c. 2 b. Lawgrenis, 6 b. Newburnetoun, 2 c. 3 b. Manturpe, 1 c. 4 b. The acres ['Terrouris' acres, *preceding paragraph*], 4 b. Total of teind oats in Newburne, 11 c. 13 b.

Teind wheat of the parish of Carnbie 'as it wes reddin this last yeir be honest men and suorne extenden to' 2 c. 2 b. Of teind bere, 7 c. 1 b. Of teind oats, 29 c. 4 b.

Teind bere of Kinglassie: Pityoquhair, 3 c. 2 b. Ky[nnyn]month, 1 c. 12 b. Gaitmilk, 4 c. Pitlowchy, 2 c. 2 b. [1 c. 2 b., *fo 63r*]. Fynglassy, 2 c. Inchedury, 2 c. 8 b. Stentoun, 3 c. Fynnocht, 8 b. Mildames, 9 b. Kinglassy toun, 1 c. 8 b. Brewland, 1 b. Sandlandis [Smiddeland, *fo 63r*; Sandelandis, *SRO, Vol. b, fo 26v*], 2 f. Craigsyd, 2 b. Walkmylne, 2 b. Feistertoun, 2 b. Cluniemylne, 1 b. Murtoun, 2 b. Cassebarrian, 3 b. Total of [teind] bere in Kinglassy, 21 c. 4 b. 2 f.[115]

Teind oats within Kinglassy: Fynmonth, 3 c. Feistertoun, 1 c. 3 b. 2 f. Clunie Mylne, 8 b. 2 f. Halltoun, 2 b. Muirtoun, 1 c. 6 b. Cassebarrian, 2 c. Total of the teind oats of Kinglassie, 8 c. 4 b.

'Summa totalis of teynd aittes within the parochinis of Dunfermelyng, Kingorne, Kirkcaldie, Kinglassie, Newbirne, Carnebie, Mussilbrucht, quheit and hors corne, extendis to' 143 c. 11 b. 2 f.[116]

'Summa totalis of aittes, ferme and teynd', 158 c. 11 b. 2 f., 'quhairof of quhyt aites to the thrid.'

[*In margin*, 'Summa of aittes', 114 c. 1 b. 2 f. 'by Carneby.'[117]]

LINDORES, ABBEY OF,[118]

(SRO, Vol. a, fos 36r-40v)
(SRO, Vol. b, fos 27r-31v)

'The rentale of the place of Lindores of money, teyndis and fermes pertenying to the samin.'

'Inprimis, the baronie of the Grange lyand within the schirefdome of Fyffe.

In the first, the Grange, viij plewis sett in few for mailles yeirlie', £68. The Beriehoill, 4 ploughgates set in feu, pays yearly £34. Ormestounes, 2 ploughgates set in feu for mail yearly, £15 12s 8d. Haltounhill, 2 ploughgates

---

[115] The correct calculation appears to be 21 c. 14 b. 2 f.
[116] The correct calculation, from the MS sub-totals, appears to be 143 c. 5 b. 2 f.
[117] This rental ends abruptly here; the next folio contains the rental of Lindores; and another rental of Dunfermline begins on fo 44r (see below, p. 37).
[118] Cf SRO, Vol. a, fo 41r (see below, p. 34); cf A. Laing, *Lindores Abbey and its Burgh of Newburgh* (Edinburgh, 1876), pp. 418-422. John Philp was abbot at the Reformation.

set in liferent, pays yearly, £11 4s. Lumquhat, 2 ploughgates set in feu for mail yearly, £8 7s 4d, 'the remanent of thir tua plewis is payit to the queinis gracis chapell in Falkland and to the lord Angus college in Abernethie'.[119] The toft of the Wodheid set in feu with the Southwod pays yearly £8 10s. The Eist Wod with the teinds of the same set in feu for mail yearly, £10 13s 4d. The brewhouse of the Grange set in feu for mail yearly, 20s. The burgh mails of the New Burcht set in feu for mail yearly, £4 16s 8d. The 'tennentis' [*recte*, tenements] of the New Burcht set in feu for mail yearly, £3 5s. The tenement in Sanctandrois set in feu for mail yearly, 26s 8d. The Derache land of Creiche[120] 'set for xix yeiris takis' for mail yearly, 40s. The toft of Cullessie set in feu for for mail yearly, 46s. The toft of Kynloche set for 19 years' tack for mail yearly, 10s. The brewhouse of Ochtermouthie [*i.e.* Auchtermuchty] set in feu for mail yearly, 24s. The toft of Auld Lindores set for 19 years' tack for mail yearly, 6s 8d. Clwny Estir set in feu for mail yearly, 24 merks 6s 8d. 'The luiging in Falkland' set in feu for mail yearly, 40s. The Craig End and Keiggis Holl set in feu for mail yearly, £4. Marie croft set for 19 years' tack for mail yearly, 40s. The Craig Mylne 'with the seggis by the dry multure' set in in feu for mail yearly, 50 merks 9s 4d. 'The Clayis with ix akeris of the Medowis' set in feu for mail yearly, 26 merks. The teind sheaves of the same set for 19 years' tack for mail yearly, 10 merks. 'The Reidinche with the yairdis and fructes thairof with salt girsis of the said inche' set in feu for mail yearly, £80. 'Threttie ane aker and ane halff set in few of our Hauch, Brodlandis and Medowis, for mail yeirlie ilk aker ij merkis, summa', £42. The Almery Cruik set in feu for mail yearly, 10 merks. The acres under the wood set in feu for mail yearly, 9 merks. 'The Kow Inche with the Reid Inche callit the Hillok Park and Eist Yaird with the fischeingis in Tay' set in feu for mail yearly, £32. [*In margin*, 'Particule', £226 9s 4d. 'Summa of the haill barony of Grange', £421 18s 4d.[121]]

The barony of Angus: Balmallo[122] and Newtyle set in feu for mail yearly, £17 8s. 'Oure pairt of the Hiltoun and Myltoun of Craigy' set in feu for mail yearly, £25 6s 8d. The Clay Pottis and Ferietoun set in feu for mail yearly, £11 13s 4d. Ennervarit[123] [Inneraritie, *fo 41v*] set in feu to the lord Craufuird[124] for 5 merks 'quhilk he hes in his ballie fie yeirlie'. The vicarage of Dundie should pay ten merks yearly 'bot sen the burning of the said toun we have gottin na payment thairoff.' The annuals of the town of Dundie should be yearly £6, 'bot

---

[119] I.e. Abernethy collegiate kirk, endowed by the earls of Angus, Cowan and Easson, *Medieval Religious Houses*, p. 215.

[120] The term 'Derache land' is understood to denote the land pertaining to the office of *toschederach*, a coroner, sergeand, or mair. W. C. Dickinson, 'The *Toschederach*', *Juridical Review*, liii (1941), pp. 85-111.

[121] The correct calculation appears to be £405 18s 4d.

[122] 'Balmaw in the parish of Newtyle', Laing, *Lindores Abbey*, p. 460; 'Balmaw', *Chartulary of Lindores Abbey 1195 - 1479*, ed. J. Dowden (Edinburgh, 1903), p. 293. It is possibly now Belmont.

[123] 'Inderraritie', Laing, *Lindores Abbey*, p. 412.

[124] David Lindsay, 10th earl of Crawford, *Scots Peerage*, iii, pp. 29-30.

sen the burnyng of the samin toun we have gottin na payment thairoff.' [*In margin*, 'Summa the baronie of Angus by the vicarege', £57 14s 8d.]

The barony of the Mairnes: The Hall of Witstones Hill End and Fischer Hill set in feu for mail yearly, £26 16s 8d. Little Wistones set in feu for mail yearly, £7 18s 10d. Nedder Witstones, Pittangus and Pittenhous[125] set in feu for mail yearly, £20 13s 4d. Scottistoun and Mertory[126] set in feu for mail yearly, £22 17s 6d. The Miltoun of Witstones set in feu for mail yearly, £5 8s 4d. The brewhouse of Witstones set in feu for mail yearly, £3 12d. Ardoiche set in feu for mail yearly, £6. The annuals of Barvy yearly, 8s. [*In margin*, 'Summa of the haill baronie of Mernis', £93 3s 8d.]

The barony of Feddellis 'with uther certane landis within Stratherne': Wester Feddellis set in feu for mail yearly, £26 6s 8d. Eister Feddellis set in feu for mail yearly, £26 13s 4d. The Mylne of Feddellis set in feu for mail yearly, £8. Bene set in feu for mail yearly, £12. Eglismagall [*i.e.* Ecclesiamagirdle], 'the toun thairof with the mylne and brewhous thairof' set in feu for mail yearly, 100 merks 8s 8d [100 merks 6s 8d, *fo 40r*]. 'The greit ludgeing in Perthe' set in feu for mail yearly, £10. 'The ludgeing in the Watergait' set in feu for mail yearly, £5. 'The ludgeing in the Foirgait' set in feu for mail yearly, 5 merks. 'The annuellis of Perthe sould be yeirlie sex pundis bot we get na payment thairoff.' [*In margin*, 'Summa of the barony of Feddellis and Stratherne', £159 10s.[127]]

The barony of Wranghame beyond the Montht[128] within the sheriffdom of Abirdeane: The Craigtoun set in feu for mail yearly, 20 merks. The Kirkhill set in feu for mail yearly, £6 6s 8d. The toun of Chrystis Kirk with Hedelik set in feu for mail yearly, £16 4s 8d. The Mylne of Leslie pays of annual yearly, 2 merks. Largy set in feu for mail yearly, 5 merks. The Newtoun and Wranghame 'with the water mylne [walkar mylne, *fo 42r*] and waird thairof' set in feu for mail yearly, 100 merks. The Kirktoun of Colsamond set in feu for mail yearly, £8 4s 6d. Powquhyit set in feu for mail yearly, £24. Leddingame and Williamestoun set in feu for mail yearly, £42. The mill and brewhouse of Williamestoun set in feu for mail yearly, 10 merks. Melingsyde set in feu for mail yearly, £28 9s 8d. Flynderis set in feu for mail yearly, £24 6s. Logydornocht with the brewhouse of the same set in feu for mail yearly, £20 18s. The Kirktoun of Inche with the mill and brewhouse of the same set in feu for mail yearly, £13 10s. The Kirktoun of Premeth set in feu for mail yearly, £4 13s 4d. Tullimorgoun set in feu for mail yearly, £16. [*In margin*, 'Summa this barony', £302 17s 10d.[129]]

The barony of Fyntray: Logy Fyntray with the Frostar Sait, set in feu for mail yearly, £8. The Haltoun of Fintray and Wester Fyntray with the place and wood of the same set in feu for mail yearly, £87 2s 8d. The Myll of Fyntray set in feu for mail yearly, £7 6s 8d. Balbuchein [*i.e.* Balbithan], Haddirlik and

---

[125] 'Pittareis, Pittargus, Pittamous', Laing, *Lindores Abbey*, p. 500.
[126] 'Marcarey', Laing, *Lindores Abbey*, p. 411.
[127] The correct calculation appears to be £158 8s 8d.
[128] I.e. the Mounth, the mountain chain stretching easterly from Loch Linnhe toward Stonehaven.
[129] The total given on fo 40r is £301 6s 2d; editorial calculations yield £295 19s 6d (see below, p. 34).

Craigforthy set in feu for mail yearly, £18. Baldiforrow set in feu mail, £5 7s 9d. Monkegy and the Bowndis set in feu for mail yearly, £20 13s 4d. Kymmocht [Kymmok, *fo 42v*] set in feu for mail yearly, £21 9s 2d. Tullie Carie [Tyllikere, *fo 42v*] set in feu for mail yearly, £5 6s 8d. Wester Displair set in feu for mail yearly, £24. Midle Disblair set in feu for mail yearly, £24 5s 8d. Eister Disblair with Cavillis Myln set in feu for mail yearly, £16 12s. The two brewhouses, meadow and crofts set in feu for mail yearly, £8 3s 4d 'quhilk the chalmerlane hes in fie'. The annuals of Ballaggarthy, Kelley and Ennerawrie pay yearly 20 merks. [*In margin*, 'Summa of this baronie', £258 13s 11d.[130]]

'The rentall of the kirkis and teyndis of the samin pertenyng to the place of Lindores set for money'

The kirk of Kynnathmonth beyond the Month pays yearly £28, 'thairfoir to my lord Forbes[131] for his ballie fie yeirlie' 20 merks, 'and sua we get bot' £14 of the said kirk. Christis Kirk pays yearly £32 'quhilk is bot ane pendicle of Kynnathmonth.' The kirk of the Inche pays yearly £77 6s 8d. The kirk of Leslie pays yearly £45 6s 8d. The kirk of Premeth pays yearly £50. The kirk of Culsalmonth pays yearly £91 6s 8d. The kirk of Logydornoch pays yearly £161 13s 4d. Ennerawrie pays yearly £62 13s 4d. The kirk of Monkegy pays yearly £57 'quhilk is bot ane pendicle of Ennerawrie.' The kirk of Fentrey pays yearly £60. For the teinds of Baddyforrow yearly, 4 merks. For the sheaves of Wester Disblair, 12 b. malt; 4 b. meal. The kirk of Dundy, 300 merks 'in all tyme bypast and this last crope and all tymes to cum I traist it salbe ane hundreth merkis mair, heirof the minister getis yeirlie' £100. The kirk of Eglismagall, 'viz., the toun of the samin payis yeirlie', 20 merks. The teind sheaves and vicarage of the toun of Claivag yearly, £18. The teind sheaves of the Eister Feddellis and Bene yearly, £18. The teind sheaves of Forret yearly, 20 merks.

The kirks of Fyffe set for victuals

The kirk of Ebdy: 'The Wodruif with the Hill and Thraiplandis payis of ferme yeirlie', 2 c. 8 b. bere. The Brodland, ferme yearly, 1 c. 7 b. bere. The Hauch, ferme yearly, 1 c. 15 b. bere. The east part of Dunmure pays yearly of teind, 8 b. wheat; 15 b. bere; 1 f. 20 b. [*sic*] oatmeal. The west part of the same, 8 b. wheat; 1 c. [bere]; 24 b. meal. Carpowy, 3 b. wheat; 3 b. bere. Kynnard, 10 b. wheat; 20 b. bere; 2 c. meal; and the said toun of Kynnard pays yearly of dry multure, 8 b. bere; 8 b. meal. The Wodmylne, 3 b. 2 f. wheat; 13 b. bere; 24 b. meal. The Freland, 2 f. wheat; 3 b. bere; 4 b. meal. Inchery, 1 b. wheat; 5 b. bere; 10 b. meal. Lundoris, 1 c. wheat; 24 b. bere; 2 c. meal. The Denemylne, 1 b. wheat; 2 b. bere; 1 f. 2 b. [*recte*, 12 b.] meal. The Parkhill, 3 b. wheat; 5 b. bere; 8 b. meal. The toft of Andrew Downy with the South Wod, 8 b. meal; 3 b. wheat; 3 b. bere. The tofts of Henry Phillip and Ninian Blyth, 2 f. 2 p. bere; 1 b. meal. The toft of the Craigmylne with the Segis, 1 b. bere; 1 b. 2 f. meal. The

---

[130] The total given on fo 40r is £245 8s 3d (see below, p. 34).
[131] William, 7th lord Forbes, *Scots Peerage*, iv, pp. 55-56.

Mary croft, 1 b. bere; 1 b. meal. The Cartward, 4 b. bere. The Craigend, 1 b. bere; 2 f. meal. The teind of the New Burcht, 2 c. 8 b. bere; 2 c. 8 b. meal. The teind of the barony of the Grange, 'viz., viij plewis of the Grange, four of the Berieholl, twa of Ormestoun, and tua of the Haltoun Hill, payis of teynd, viij ch., thairof 1 ch. quheit; ij ch. xij b. beir; iiij ch. meill'.[132] [Total:] Bere, 18 c. 15 b. 2 p. Wheat, 4 c. 9 b. Meal, 17 c. 12 b. 2 f.[133]

The parish of Cullesse: Halhill pays yearly, 1 b. wheat; 7 b. bere; 10 b. meal. The Mylhill, Scheillis and Bowis [brewhous, *fo 43v*], 22 b. bere; 22 b. meal. The croft of sir John Young, 1 b. meal. The Newtoun of Cullesse, 6 b. bere; 8 b. meal. Pitlair, 1 b. wheat; 7 b. bere; 14 b. meal. Daftmyll, 2 b. 2 f. bere; 4 b. meal. Maristoun, 1 b. bere; 11 b. meal. Lawfeild and Menysgrene, 8 b. oats. Bellow Mylne, 4 b. bere; 7 b. meal. Drumtennent, 6 b. bere; 18 b. meal. The east part of Kinlocht, 4 b. [wheat]; 13 b. 3 f. 2 p. bere; 1 c. 7 b. 1 f. meal. The west part of Kinlocht, 4 b. 2 f. wheat; 13 b. bere; 1 c. 10 b. meal. The toft of Kinloich, 2 b. bere; 2 b. oats. Rosse Eister, 4 b. wheat; 14 b. bere; 14 b. meal. Rosse Wester, 6 b. wheat; 14 b. bere; 20 b. meal. Weddersbie, 8 b. wheat; 1 c. 12 b. bere; 2 c. 4 b. meal. The pendicles of the same, 2 b. 1 f. bere; 2 b. 1 f. 2 p. meal. Lumquhat, 7 b. bere; 14 b. meal. Kylquhis Eister, 1 b. bere; 10 b. oats. Total wheat, 1 c. 11 b. 2 f.[134] Total bere, 9 c. 7 b. 2 f. 2 p.[135] Total meal, 14 c. 11 b. 2 f. 2 p. meal.[136] Total oats, 1 c. 15 b. oats.[137]

The parish of Auchtirmothie: The north part of the toun of Auchtermouthie, 10 b. wheat; 22 b. 1 f. 2 p. bere; 28 b. meal. The south part of the same pays yearly 10 b. wheat; 21 b. 2 p. bere; 31 [b.] 2 p. meal. The Bondhalff[138] pays yearly 10 b. 3 f. 2 p. wheat; 1 c. 8 b. 2 f. bere; 2 c. 4 b. 2 f. meal. Cotlandis with the Marislandis[139] pays yearly 9 b. bere; 8 b. 2 f. meal. Gerusland[140] pays yearly 3 b. wheat; 6 b. 2 f. bere; 7 b. 2 f. meal. The Myris Over and Nather pays yearly 6 b. wheat; 10 b. bere; 18 b. meal. Burngrenis pays yearly 8 b. oats. Dempstertoun pays yearly 8 b. wheat; 1 c. bere; 2 c. 8 b. meal. Redy and Langis Waird, 3 b. wheat; 5 b. bere; 1 c. meal. Kilquhyis Wester pays yearly 4 b. bere; 8 b. meal. Total wheat, 3 c. 2 b. 3 f. 2 p. Total bere, 6 c. 2 b. 2 f.[141] Total meal, 12 c. 1 b. 2 f. 2 p. Total oats, 8 b.

The parish of Creiche: The toun of Creich pays yearly 2 b. wheat; 4 b. bere; 12 b. meal. Perbroithe pays yearly 4 b. wheat; 12 b. bere; 1 c. meal. Luthre pays yearly 10 b. wheat; 2 c. 4 b. bere; 3 c. 4 b. meal. Kynsleif Eister pays yearly

---

[132] This would total 7 c. 12 b, not 8 c.
[133] The correct calculation appears to be 18 c. 14 b. 2 f. 2 p. bere; 17 c. 8 b. 3 f. meal.
[134] The correct calculation appears to be 1 c. 12 b. 2 f.
[135] The correct calculation appears to be 9 c. 6 b. 2 f. 2 p.
[136] The correct calculation appears to be 14 c. 6 b. 2 f. 2 p.
[137] The correct calculation appears to be 1 c. 4 b.
[138] 'Bond-half of Auchtermuchty called Jervese-lands', Laing, *Lindores Abbey*, p. 454.
[139] 'Marislands (the Virgin Mary's Lands) is now corrupted to maislands', Laing, *Lindores Abbey*, p. 454.
[140] See above, p. 33, n. 138.
[141] The correct calculation appears to be 7 c. 6 b. 2 f.

3 b. wheat; 6 b. bere; 10 b. meal. Kynsleif Wester pays yearly 3 b. wheat; 6 b. bere; 10 b. meal. Balmadysyd pays yearly 12 b. wheat; 18 b. bere; 20 b. meal. Total wheat, 2 c. 2 b. Total bere, 5 c. 2 b. Total meal, 5 c. 8 b.[142]

Total: Wheat, 11 [c.] 12 b. 3 f. 3 p.[143] Bere and malt, 40 c. 7 b. 1 f.[144] Meal, 49 c. 5 b. 3 f.[145] Oats, 2 c. 7 b.

Assignation of the third of the abbacy of Lundores:
    Third of the money, £746 18s 1½d.[146] The barony of Wranghame, £301 6s 2d [£302 17s 10d, *fo 37v*]; the barony of Fyntray, £245 8s 3d [£258 13s 11d, *fo 37v*]; the lands of Balmallo and Newtyle, £17 8s; Hyltoun and Myltoun of Cragy, £25 6s 8d; Claypottis and Freirtoun [Ferietoun, *fo 36v*; Ferritoun, *fo 41v*], £11 13s 4d; the barony of Mernes 'by the annuellis of Bervy', £92 15s 8d; Egleismagreltoun, mill and brewhouse, 100 merks 6s 8d [100 merks 8s 8d, *fo 37r*]. 'Giff in' £14 1s 11d '2 part ob.'[147]
    Third of the wheat, 3 c. 12 b. 1 f. 1 p.[148] Lundores, 1 c.; the east part of Dunmure, 8 b.; the west part, 8 b.; Kynnardie, 10 b.; the toft of Andrew Dow[n]y with the South Wod, 3 b.; the teinds of the Grange, 1 c. 'Giff in' 2 f. 3 p.
    Third of the bere, 13 c. 7 b. 3 f. Take the Wodrufhill and Treplandis for 2 c. 8 b.; the Brodland, 1 c. 7 b.; the Hauch, 1 c. 15 b.; the west part of Dunmure, 1 c.; Kynnaird, 20 b.; Newburcht, 2 c. 8 b.; the teind of the Grange, 2 c. 12 b.; the east part of Dunmure, 15 b. 'Giff in' 13 b. 1 f. 'The beir that is tane mair is in respect of the malt.'
    Third of the meal, 16 c. 7 b. 1 f. 'Tak the haill meill of the paroche of Ebdy givand be yeir', 17 c. 12 b. 2 f. 'Giff in' 1 c. 5 b. 1 f.
    Third of the oats, 14 b.[149] Take it out of Cullessy parish giving 1 c. 15 b.

'Omittit grassoumes, entrie silver, yairdis, fischeingis, caponis, pultrie, caynis, custoumes, mairtes, cariagis and uther dewteis. Inserche thame out diligentlie'.

LINDORES, ABBEY OF,[150]   (SRO, Vol. a, fos 41r-43v)
                                                    (SRO, Vol. b, fos 32r-34v)
The rental of Lundores in money, teinds and fermes
    'The baronie of Grange lyand within the schirefdome of Fyffe: Inprimis, the toun of the Grange, aucht pleuchis set in few, tua thairof for yeirlie maill ilk

---

[142] The correct calculation appears to be 4 c. 12 b. bere; 7 c. 8 b. meal.
[143] The correct calculation appears to be 11 c. 9 b. 1 f. 2 p.
[144] The correct calculation appears to be 39 c. 11 b. 1 f.
[145] The correct calculation appears to be 50 c. 1 b. 3 f.
[146] This is difficult to verify; no total is given.
[147] These figures do not tally.
[148] This is not one third of 11 c. 9 b. 1 f. 2 p. on fo 40r.
[149] One third of 2 c. 7 b. is 13 b.
[150] Cf SRO, Vol. a, fo 36r (see above, p. 29); Laing, *Lindores Abbey*, pp. 422-427.

pleuch' £8 10s, total £17; 'and sex pleuchis payis bot ilk pleuch' £5 14s, total £34 4s, 'during the tennentis lyftymes'. The Beriehoill, 4 ploughgates set in feu, three thereof for yearly mail, each ploughgate £8 10s, total £25 10s; one ploughgate pays only £5 14s, 'during the tennentis lyftymes'. Ormestoun, 2 ploughgates set in feu for yearly mail, £15 12s 8d. Haltoun Hill, 2 ploughgates set in liferent for yearly mail, £11 8s. Lumquhat, 2 ploughgates set in feu for yearly mail, £5 7s 8d, 'the remanent of the tua pleuchis is payit to the quenis gracis chapell in Falkland and to my lord Angus college in Abirnethy.' The toft of the Wodheid set in feu with the Southwod for yearly mail, 8 merks. The Eist Wod with the teinds thereof set in feu for yearly mail, £10 13s 4d. The brewhouse of the Grange set in feu for yearly mail, 17s. The burgh mails of New Burcht yearly, £4 12s. 'The tennentis of New Burcht yeirlie', £3 5s. The annuals in Sanctandrois yearly, 26s 8d. 'The Deratland of Creycht payit nathing thir xij yeiris.' The toft of Cusselie [*recte*, Cullessie] yearly, 46s. The toft in Kinloche yearly, 10s. The brewhouse of Ouchtermuchtie yearly, 24s. The toft of Auld Lundores yearly, 8s. 'The ludgeing in Falkland payit nathing thir xxij yeiris.' Cluny Eister yearly, £16 6s 8d. Craigend and Keggieshoill yearly, 42s. Marie croft 'and seggis' yearly, £4. Craigmyln yearly, £32. 'The Clayis with nyne akeris of the West Medow yeirlie', 6s 8d; the teind thereof, £6 13s 4d. 'Threttie ane akeris and ane halff in the Hauch, Medowis and Brodlandis respective', each acre 2 merks, total £42. The Almerie Cruik yearly, £6 13s 4d. The acres under the wood yearly, £6. 'The Reid Inchis salt gres, Kow Inchis, Park, Eistyaird, fructyairdis, fischeingis in Tay', £100. The teind of Auld Lundoris yearly, £30. The teinds of Forret yearly, 20 merks. The teinds of Den Mylne yearly, £4.

The mails of the lands in Angus: Balmallo and Newtyle yearly, £17 8s. 'Our pairt of the Hiltone and Mylntoun of Cragy', £25 6s 8d. Clay Pottis and Ferritoun, £11 13s 4d. Inneraritie, £3 6s 8d, 'payit nathing thir xxij yeiris.' The vicarage of Dundie 'payit nathing sen the burnyng of the said toun.' The annuals of Dundie 'payit x merkis bot nathing sen the burnyng of the samin toun.'

The mails of the lands in the Mernis: The Hall of Wistownis Hilend and Fischer Hill yearly, £26 16s 8d. Littill Wistownis yearly, £7 18s 10d. Nedder Wistownis, Pittangus and Pittennus yearly, £20 13s 4d. Scottistoun and Mertory yearly, £22 17s 6d. Myltoun and [of, *fo 37r*] Wistounes yearly, £5 8s 4d. Brewhouse and [of, *fo 37r*] Westownes yearly, £3 0s 2d. Ardoche yearly, £6. Annuals of Bervy yearly, 8s.

The mails and annuals of the lands lying within the sheriffdom of Stratherne:[151] Wester Feddellis yearly, £26 6s 8d. Eister Feddellis yearly, £26 14s 8d. Mylne of Feddellis yearly, £8. Benie yearly, £12. Eglismagrill [*i.e.* Ecclesiamagirdle], 'the toun thairof with the mylne and brewhous', £67 2s. 'The greit ludgeing in Perthe yeirlie', £10. 'The ludgeing in the Water Gait yeirlie', £5. 'The Foir [Foirgait, *fo 37r*] ludgeing yeirlie', £3 6s 8d 'thairof to Dunkell be the tennentis', 40s. The annuals of Perthe, 5 merks 5s.

---

[151] The earldom, lordship and stewartry of Strathearn lay in the sheriffdom of Perth, *RMS*, iv, no 303.

The barony of Wranghame lying within the sheriffdom of Abirdeane: The Craigtoun yearly, £30 6s 8d. Kirkhill yearly, £6 6s 8d. The toun of Chrystis Kirk with Hedelyk, £6 4s 8d. The Mylne of Leslie, 26s 8d, 'na payment'. Largy yearly, £3 6s 8d. Newtoun and Wranghame yearly, 'with the walkar mylne [water mylne, *fo 37r*] and waird thairof', £66 13s 4d. Kirktoun and [*recte*, of] Kilsalmound, £8 4s 6d. Powquhyt yearly, £24. Ledinghame and Williamstoun yearly, £48 13s 4d, 'to my lord Rothes fie.' The mill and brewhouse of Williamstoun yearly, £6 13s 4d. Malingsyd yearly, £28 9s 8d. Flendaris yearly, £24 6s. Logydornocht with the brewhouse of the same, £20 18s. The Kirktoun of Inche with the mill and brewhouse of the same, £13 10s. The Kirktoun of Premeth, £4 13s 4d. Tullymorgoun yearly, £16.

The barony of Fintray: Logyfintray with the Frostar Sait yearly, £8. Haltoun of Fintray and Wester Fintray with the place and wood thereof yearly, £87 2s 8d. Mylntoun of Fintray yearly, £7 6s 8d. Balbuthine, Haddowerk and Craigforthie, £18. Badyforrow yearly, £5 7s 9d. Monkegy and West Boundis, £19 13s 4d. Kymmok yearly, £21 9s 2d. Tyllikere yearly, £5 6s 8d. Wester Disblaire yearly, £24. Midle Disblair yearly, £24 5s 8d. Eister Disblair yearly, with Cavillis Mylne, £16 12s. 'The tua brewhoussis of Fintrayis with medow and croftis respective', £8 13s 4d, 'assignit to the the chalmerlane'. The annuals of Bathalgarthy, Kelley and Innerowry, £13 6s 8d.

The rental of the kirks lying within the sheriffdom of Abirdeane and teind sheaves thereof set for money

The kirk of Kynnathmound, £14, 'the rest thairof to my lord Forbes[152] ballie fie extending in the first rentall to xx merkis'. Chrystis Kirk yearly, £32. The kirk of Insche, £76 6s 8d. The kirk of Lesly, £45 6s 8d. The kirk of Premeth yearly, £50. The kirk of Culsalmound, £91 6s 8d. The kirk of Logydornocht, £161 13s 4d. The kirk of Innerowry, £62 13s 4d. The kirk of Monkegy yearly, £57. The kirk of Fintray yearly, £60. The teinds of Badyforrow, 53s 4d. The teinds of Wester Disblair yearly, £8.

The kirks within the sheriffdom of Forfar
The kirk of Dundy yearly, £200.

The kirks within the sheriffdom of Perthe
The kirk of Eglismagreill [*i.e.* Ecclesiamagirdle], £13 6s 8d. The teinds and vicarage of Clavege, £18. The teinds of Eist Feddellis and Bene, £18.

The kirks within the sheriffdom of Fyff set for victuals
Ebdy kirk, the teind sheaves thereof: the Wodrwff Hill and Thraiplandis, 2 c. 4 b. bere. The Brodland, 1 c. 7 b. bere. The Hauche, 1 c. 12 b. bere. The east part of Dunmure, 8 b. wheat; 15 b. 1 f. bere; 24 b. meal. The west part of Dunmure, 8 b. wheat; 1 c. bere; 24 b. meal. Carpowye, 3 b. wheat; 3 b. bere.

---

[152] William, 7th lord Forbes, *Scots Peerage*, iv, pp. 55-56.

Kynnaird, 10 b. wheat; 20 b. bere; 2 c. meal. The dry multure of Kynnaird, 8 b. bere; 8 b. meal. Wodmylne and Freland, 3 b. wheat; 13 b. bere; 24 b. meal. [*In margin*, 'be ane boll' [wheat]; [3] b. bere; 4 b. meal.[153]] Inchyray, 1 b. wheat; 5 b. bere; 10 b. meal. Parkhill, 2 b. wheat; 5 b. bere; 6 b. meal. The toft of Andrew Downy with the South Wod, 3 b. wheat; 3 b. bere; 8 b. meal. The toft of James Philip, 2 f. 2 p. bere; 1 b. meal. The toft of Craigmylne 'and seggis', 1 b. bere; 1 b. 2 f. meal. Mary croft, 1 b. bere; 1 b. meal. Cantward, 4 b. bere. Craigend, 1 b. bere. Newburcht, 2 c. 8 b. bere; 2 c. 8 b. meal. Beriehoill and the Grange, 1 c. wheat; 2 c. 10 b. bere; 4 c. meal. [*In margin*, 'los be' 2 f. meal; 'los be' 2 b. bere].

Cullesse kirk, teind sheaves thereof: Halhill, 2 b. wheat; 4 b. bere; 6 b. meal. [*In margin*, 'los in the auld', 1 b. wheat; 7 b. bere; 10 b. meal.] Mylhill, Scheillis and 'brewhous' [Bowis, *fo 39r*], 22 b. bere; 22 b. meal. Mylne croft of Cullesse, 1 b. meal. Newtoun of Cullesse, 6 b. bere; 8 b. meal. Pitlare, 1 b. wheat; 7 b. bere; 12 b. meal. [*In margin*, 'los be' 2 b. meal.] Daftmylne, 2 b. 2 f. bere; 4 b. meal. Maristoun, 1 b. bere; 11 b. 'aites for hors corne'. Laufeild and Menisgrein, 8 b. oats for horse corn. Balbow Mylne, 4 b. bere; 7 b. meal. Drumtennent, 6 b. bere; 18 b. meal. Easter part of Kinlocht, 4 b. wheat; 13 b. bere; 1 c. 6 b. meal. [*In margin*, 'los be' 3 f. 2 p. bere; 1 b. 1 f. meal.] Wester part of Kynlocht, 3 b. wheat; 12 b. bere; 1 c. 6 b. meal. [*In margin*, 'los be' 2 f. wheat; 1 b. bere; 4 b. meal.] The toft of Kynlocht, 2 b. bere; 2 b. oats for horse corn. Rosse Eister, 4 b. wheat; 14 b. bere; 14 b. meal, 'assignit to James Calwie and sua payis nathing to the parte'. Rosse Wester, 6 b. wheat; 14 b. bere; 20 b. meal. Weddersbie, 'James Sandelandis pairt thairof', 2 b. wheat; 8 b. bere; 12 b. meal. [*In margin*, 'in the auld', 8 b. wheat; 2 c. 12 b. bere; 2 c. 4 b. meal.] 'The uther half', 4 b. wheat; 12 b. 2 f. bere; 1 c. 4 b. meal. Pendicles of the same, 2 b. 1 f. bere; 2 b. 1 f. meal.

DUNFERMLINE, ABBEY OF,[154]   (SRO, Vol. a, fos 44r-52r[155])
(SRO, Vol. b, fos 35r-43r)

[*In margin*, 'Abacie Dunfermelyng']

'This is the haill rentall pertenyng to the abay of Dumfermeling in penny maill, annuellis, fermes, teyndis, kirkis and tounes set in assedatioun for money.'

The parish and barony of Dumfermeling in penny mail and annuals, with the burgh mail of the burgh of Dumfermelyng with the customs of the same, £294 18s 5d. Dolourschyre in penny mail and annuals, £28 13s. Mussilburcht in penny mail and annuals, £364. Newburne in penny mail and annuals, £73 3s 4d [£73 18s 4d, *fos 31r, 32r, 50r, 55v, 57r*]. Kynglassie in penny mail and annuals, £132

---

[153] These marginal notes refer to the differences between this rental and that on fo 39r (see above, p. 32), but the calculations are not always correct.
[154] Cf SRO, Vol. a, fos 30r, 54r (see above, p. 23; below, p. 45).
[155] SRO, Vol. a, fos 52v-53v are blank.

17s 4d. Kyngorne Wester in penny mail and annuals, £110 6s 8d. Kirkcaldie in penny mail, annuals and customs, £159 11s 3d. Kynneddir in penny mail, £42 15s 8d. Southe and Northe the Quene Ferreis in penny mail, £20 19s 5d. Starmonthe in penny mail, £47 12s 4d. 'The haill annuellis aucht to the said abbay extendis to' £31 6s 8d. The bow places paying 'be yeir' in money, £117 13s 4d. Total of penny mail and annuals, £1,424 17s 5d [£1,497 10s 4d, *fo 49v*].[156]

The kirks set in assedation for silver
 The kirk of Stirveling, £80. The kirk of Wester Haillis, £66 13s 4d. The kirks of Kinros and Urwale, £120. The kirk of Sanct Johnnestoun, £100. The kirk of Innerkeithing, £53 6s 8d. The kirk of Wester Kingorne £12. The kirks of Mollane and Strathardill, £120. The kirk of Carnebie, £266 13s 4d. Total of the kirks set for silver, £818 13s 4d.

'The teyndis of townes set for sylver': The teind of Smetoun, £20. The teind of Mounktounhall, £20. The teind of Carbery, £33 6s 8d. The teind of Cousland, £66 13s 4d. The teind of Natoun, £13 6s 8d. The teind of Inneresk [Pinkie, *fo 32r*; Pynkie, *fo 32v*], £53 6s 8d. The teind of Pinkin, £10. The teind of Hallhill, £40. The teinds of the acres of Kirkcaldie, £13 6s 8d. 'Summa of thir teyndis', £270.

Total in penny mail, annuals, kirks and teinds set for silver, £2,513 10s 9d.

'The haill victuall alsweill ferme as teynd in quheit, beir, meill and aites as followis'

Wheat: The parish of Dumfermelyng in ferme and teind, 8 c. 7 b. 3 f. [*In margin*, 'heirof in ferme' 5 c.] The parish of Mussilburcht in ferme and teind, 9 c. 2 f. [*In margin*, 'heirof in ferme' 6 c. 3 b.] The parish of Newbirne in ferme and teind, 5 c. 15 b. [*In margin*, 'heirof in ferme' 4 c. 2 b.] The parish of Kirkcaldie in teind, 1 c. 2 b. The parish of Kingorne Wester, 1 c. 2 b. The parish of Carnebie, 3 c. Total 'of the haill quheit in ferme and teynd', 28 c. 11 b. 1 f.
 Bere: The parish of Dumfermeling in ferme and teind, 45 c. 3 b. 3 p. [*In margin*, 'heirof in ferme' 22 c. 8 b.] The parish of Mussilburcht in teind, 5 c. 12 b. The parish of Newbirne, 5 c. 8 b. 1 f. [*In margin*, 'heirof in ferme' 8 b.] The parish of Kinglassie in teind, 21 c. 8 b. 2 f.[157] The parish of Kirkcaldie in teind, 6 c. 5 b. 2 f. Kingorne Wester in teind, 3 c. 2 b. Carnebie, 9 c. Total 'of the haill beir, ferme and teynd', 96 c. 7 b. 1 f. 2 p.[158]
 Meal: 'the haill ferme meill of Kinglassie, Dolourschyre and the mylnis extendis to' 15 c.

---

[156] The same places are listed on fo 32r (see above, p. 25), though some figures differ.
[157] 21 c. 4 b. 2 f., fo 35v (see above, p. 29); the correct calculation appears to be 21 c. 14 b. 2 f.
[158] The correct calculation appears to be 96 c. 7 b. 1 f. 3 p.

Oats: In Dumfermeling parish in ferme and teind 'of quhite aites', 32 c. 4 b. [*In margin*, 'heirof in ferme' 15 c.]; 'off mureland aites for horscorne', 62 c. 1 f. The parish of Kingorne Wester, 7 c. 15 b. white oats; 3 c. 8 b. horsecorn. The parish of Kirkcaldie, 8 c. white oats; 8 c. 10 b. horsecorn. The parish of Newbirne, 6 c. 12 b. 2 f. white oats; 4 c. 14 b. horsecorn. The parish of Kinglassie, 8 c. 4 b. horsecorn The parish of Mussilburcht, 9 c. 13 b. white oats. Total white oats, 64 c. 12 b. 2 f. Total horsecorn, 87 c. 4 b. 1 f.

'Off this foirsaid rentale thair is to be deducit in ordinar chargis and feis as efter followis':

Money: 'In silver to the sustentatioun of the convent beand in number xxvj, quhilk thai have had of lang tyme assignit to thame of the parochin of Dunfermelyng, Dolour and Kynnedderis with certane uther small annuellis, the sowme of' £394. 'To my lord duik[159] in bailyie fie for Mussilburcht', £20. 'In contributioun to the Lordis of the Sessioun', £70. 'To sir Adame Kingorne in pentioun quhairupoun he hes the commoun sele', £40. 'To Mr James Mowbray in siclyk pentioun', £30. 'To the porter of the yet of Dunfermeling under the commoun sele', £4. 'To the plumber and glasynwrycht under the commoun sele', £13 6s 8d. 'To the foster of the wod under the commoun sele', £4. 'To the baillie of the regalitie of Dunfermeling', £20. 'To the haill officeris feis of Mussilburcht, Dunfermeling, Dolour, Kingorne, Kirkcaldie, Kinglassie and Newbirne', £20. 'To the sklater and his servandis', £42 13s 4d. 'To the procuratour of the actiounes of the place', £20. 'To the Ruid Kirk of Stirveling of annuellis', 40s. 'To the chalmerlane', £40. 'To the barbour', £4. Total silver 'that is to be deducit of this rentale extendis to' £714;[160] so rests free, £1,789 10s 9d.

Wheat: 'To the sustentatioun of the convent', 9 c. 'To the puir in ordinar baikin breid', 8 b. 'To the lipper men at the Magdalenis[161] besyd Mussilburcht', 2 b. To Richard Prestoun, 'quhairupoun he hes my lordis hand writt', 4 b. Total wheat, 9 c. 14 b.; so rests free, 18 c. 13 b. 1 f.

Bere: 'To the conventis sustentatioun', 32 c. 'To the porter of Dunfermeling', 8 b. 'To the plumber and glasinwreycht', 2 b. 'To the carpentar', 8 b. 'To the maissoun', 8 b. To Richard Prestoun, 'quhairupoun he hes my lordis hand writ', 15 b. Total, 34 c. 9 b.; so rests free, 61 c. 14 b. 1 f.[162]

Oatmeal: 'To the plumbar and glasynwrycht', 24 b. 'To the keipar of the taipes[163] under the commoun sele', 1 c. 'To the miller of the Abay Myln', 1 c. 'To the forrester of the wod', 12 b. 'To the smyth of the abay', 8 b. 'To the lauender of the abay', 12 b. 'To the wrycht', 12 b. 'To the maissoun', 12 b. 'To the barbour', 24 b. 'To the keipar of the veschell', 4 b. 'To the beddall', 8 b. 'Assignit to the convent to be gevin to thair servandis feis', 5 c. 12 b.

---

[159] James Hamilton, 2nd earl of Arran and duke of Chatelherault, Scots Peerage, iv, pp. 366-368.
[160] The correct calculation appears to be £724.
[161] I.e. the hospital of St Mary Magdalene, Cowan and Easson, *Medieval Religious Houses*, p. 186.
[162] 2 p. have been omitted from this calculation.
[163] 'Tapis', *Registrum de Dunfermelyn*, p. 455.

Oats: 'To the convent in assignatioun of quheit aites', 38 c.; 26 c. horsecorn. 'To the porter of the yet of Dunfermelyng, of quhyte aites',[164] 3 c. 'To the scrybe of the court of regalitie', 8 b. white oats. 'To the officer of Dunfermelyng', 8 b. horsecorn. 'To the sklater', 1 c. white oats. To Mr John Spens of Condy, 24 b. white oats. 'To the chalmerlane', 1 c. horsecorn. To John Young, 'scribe of the ballerie of Mussilburcht', 8 b. [white oats]. To John Wallace, 4 b. white oats. To Richard Prestoun 'quhairupoun he hes my lordis hand writ', 14 b. [white oats]. Total white oats, 45 c. 10 b.; so rests free, 19 c. 2 b. 2 f. Total horsecorn, 27 c. 8 b.; so rests free, 59 c. 12 b. 1 f.

'Item, that samekill ather of silver or victuallis as is intromettit with and tane up of this instant lxj yeris crope be my lordis Arrane,[165] Argill,[166] the maister of Lindesay[167] or ony uther in thair names, be deducit to us and allowit of the quenis gracis pairt. Item, we have nocht chargit the ferme and teynd of the Nedder Grange of Wester Kingorne quhilk extendis to' 6 c. 8 b. bere 'becaus Petir Dury hes my lordis discharge thairof for all the dayis of his lyff under the subscriptiounes and seillis of my lord of Dunfermelyng[168] and his successouris.

We mak na rentall of the coill of Wellifuird [i.e. Wallyford] becaus it is bot ane casualitie and in the yeiris of God' 1556, 1557 and 1558 'large expensis wes maid in wynnyng thairof abone the sowme of' £500, and in 1559 'the erle of Arran[169] and his factouris intromettit thairwith and sensyne hes wastit the said coill in sict soirt that na profeit can be had thairof without greit expensis in tyme cuming, as the haill cuntrie knawis'.

Signature: 'Allane Coutis, with my hand, in the name of my lord of Dunfermeling.'

Assignation of the third of the abbacy of Dumfermeling:
Third of the money, £837 16s 11d. Take Mussilburcht for £364;[170] Kinglassie, £132 17s 4d; Kingorne Wester, £110 6s 8d; the bow places, £117 13s 4d; the teinds of Cousland, £66 13s 4d; the teind of Carbarie, £33 6s 8d; the teind

---

[164] 'Quheit aites' are the same as 'quhyte aites'.
[165] James Hamilton, 3rd earl of Arran, *Scots Peerage*, iv, pp. 368-369.
[166] Archibald Campbell, 5th earl of Argyll, *Scots Peerage*, i. pp. 340-343.
[167] Patrick, master of Lindsay, son of John, 5th lord Lindsay of the Byres, *Scots Peerage*, v, pp. 397-398.
[168] At the Reformation the commendator, George Dury, left for France in January 1561 and Robert Pitcairn exercised authority as commendator-designate. Cowan and Easson, *Medieval Religious Houses*, p. 58.
[169] See above, n. 165.
[170] SRO, Vol. a, fo 117v, at the end of the rental for Newbattle, has the following entry for Dunfermline abbey: [*In margin*, 'Of Dunfermelyng':] 'Mussilburcht', £364; 'Quheit', 6 c. 3 b.; 'Beir', 3 c. 5 b. 2½ p.; 'Quhyt aites', 5 c. 10 b. 2 f. 'Item, inhibit the haill coill quhill thai cum and condiscend upoun the thrid. Item, that thai answer nane bot the quenis grace of aragies, caragies, grassoumes, caynes, entrie silver and utheris dewteis.'

of Mounktounhall, £20; the teind of Natoun, £13 6s 8d. Total, £858 4s. 'Giff in' £20 7s 1d.

Third of the wheat, 9 c. 9 b. 1½ p. 'The ferme quheit of Mussilburcht', 10 c. 2 b. 1 f. 1½ p.[171]

Third of the bere, 34 c. 5 b. 2⅓ p.[172] Parish of Kinglassie, 21 c. 8 b. 2 f.; parish of Kirkcaldie, 6 c. 5 b. 2 f.; Wester Kingorne, 3 c. 2 b.; teinds of Mussilburcht, 3 c. 5 b. 2⅓ p.

Third of the meal, 5 c. 'Of the fermes of Kinglassie, Dolourschyre and the mylnes', 5 c.

Third of the white oats, 21 c. 9 b. 2 f. Wester [Kin]gorne, 7 c. 15 b.; take out of Kirkcaldie 2 c. 10 b., giving 8 c.; the teinds of Mussilburchtschyre for 10 c. 15 b. [9 c. 13 b., *fo 46r*], 'by Edmestoun, Stayhill [*recte*, Stanyhill, *i.e.* Stoneyhill] and Quhythill'.[173]

Third of the horsecorn,[174] 29 c. 1 b. 6⅔ p. Wester Kingorne, 3 c. 8 b.; parish of Kirkcaldie, 8 c. 10 b.; parish of Newbyrne, 4 c. 14 b.; parish of Kinglassie, 8 c. 4 b.; 'the merkland [muirland, *SRO, Vol. b, fo 39v*] aites of Dunfermeling', 3 c. 13 b. 1 f. 2 p.

'Bring in the hail coill as superplus with caynis, capounis,[175] pultrie, carragis, grassoumes, entrie silver and all uther dewteis omittit. Item, the New Mylne of Mussilburcht omittit. Item, out of Inneresk', 116 'thravis stray and x c.[176] laid of coill of Carbery'. [*In margin*, 'Haill to the quein.']

'Nota, tane doun of the quheit aites be the Auditouris of the Checker', 18 b., 'and sua the haill charge of the quhyt aites extendis to' 20 c. 7 b. 2 f., namely, Wester Kingorne, 7 c. 15 b.; Mussilburcht, 10 c. 15 b.; Newbyr[n]e, 2 c. 11 b. 2 f. 'Deduce heirof' 18 b. 'gevin doun to the laird of Raithe.'

'The alteratioun of the assignatioun for Newbyrne':

Penny mail and annuals, £73 3s 4d. The parish of Newbyrne ferme and teind: Wheat, 5 c. 15 b. Bere, 5 c. 8 b. 1 f. White oats, 6 c. 12 b. 2 f. Horsecorn, 4 c. 14 b.

'Giff for this silver: the bow places quhilk wes tane of befoir for' £117 13s 4d, 'and now out of the bow places tak' £43 10s 'allanerlie. Gyff in the haill quheit extending to' 5 c. 15 b. 'Gyff for the beir of Newbirne the beir of

---

[171] Musselburgh gave 6 c. 3 b. ferme wheat, fo 45r (see above, p. 38).

[172] Total bere, 96 c. 7 b. 1 f. 2 p., fo 45v (see above, p. 38).

[173] This would give a total of 21 c. 8 b.

[174] Horsecorn seems to be identified with black oats.

[175] SRO, Vol. a, fo 48v and SRO, Vol. b, fo 39v have inserted above 'capounis': 'xx' and 'xiiij'. If this were intended to mean 18 score and 14, the total of 374 capons on fo 51r would be correct; cf below, p. 44, n. 196.

[176] Presumably 10 chalders. The Culross chalder was the measure of coal for payment of custom and bullion. R. E. Zupko, 'The weights and measures of Scotland before the Union', *SHR*, lvi (1977), pp. 124-145, at p. 140.

Kirkcaldy quhilk wes tane of befoir for' 6 c. 5 b. 2 f. 'and tak now out of it allanerlie' 14 b. 1 f.

White oats: 'Giff for it the' 2 c. 11 b. 2 f. 'quhilk wes takin out of Kirkcaldie and sua restis to giff out of Newbyrne', 4 c. 1 b. white oats.

Horsecorn: 'Giff the muirland aites of Dumfermeling extending to' 3 c. 13 b. 1 f. 2 p., 'and giff out of Kirkcaldie quhilk gave of befoir in our thrid', 8 c. 10 b., for the rest of Newbyrne, 2 c. 2 f. 2 p., 'sua Kirkcaldie sall giff onlie' 1 c. 9 b. 1 f. 2 p.

'The assignatioun of the abay of Dunfermeling conforme to the last rentall thairof gevin in sen the setting of the fermes and few to the thesaurer.'

'The haill penny mailles and annuellis of the haill patrimonie of the abay of Dunfermelyng extendis to' £1,497 10s 4d [£1,424 17s 5d, *fo 44v*].

'The teyndis thairof set for silver, bayth kirkis and tounes comptand thairin': the teinds of Edmestoun for 40 merks; the teinds of Wolmet for 50 merks; and the teinds of Carneby to £253 6s 8d, 'be reasoun of the dounryding to' 20 merks 'les nor it wes of befoir and also comptand thairin the teyndis of the akeris of Dunfermelyng to' 40 merks extending in the whole to £1,162. 'Summa of the haill money', £2,549 10s 4d.[177]

'Item, to be eikit to this silver the pryces of' 11 c. 11 b. 3 f. ferme wheat 'set to the thesaurer in few for' 20 merks the chalder, extending to £156 9s 2d. 15 c. 9 b. ferme bere set to the same for £10 the chalder, 'inde moneta' [*i.e.* thereof in money] £155 12s 6d. 5 c. 4 b. 2 p. ferme meal set to the same at 10 merks the chalder, 'inde moneta' £35 12½d.[178] 15 c. oats set to the same at 10 merks the chalder, 'inde moneta' £100.[179]

'Summa of the thesaureris fermes', £447 2s 8d.[180]

'Summa of the haill silver with the thesaureris fermes extending to' £3,106 13s 4d;[181] third thereof, £1,035 11s 'ob'.[182] Take the penny mails and annuals within Mussilburchtschyre, Edinbrucht, Newbottell and Hadingtoun for £419 13s 8d; take the teind of Mounktounhall for £20; take the teind of Carbarie for £33 6s [£33 6s 8d, *fos 33r, 45r, 48r*]; take the teind of Cousland for £66 13s 4d; take the teind of Pinkie for £10; take the mail of the New Mylne of Mussilburcht for £20; take the mail of Dolourschyre for £64; 'tak the penny maill, annuellis, lymekill maill and fed oxin silver in Wester Kingorne, forby the fre beir set to the thesaurer in few and with the penny maill of the kirk for' £122 6s; the penny mail and annual of Newbyrneschyre, £73 18s 4d; the kirk of Sanct Johnnestoun, £100;

---

[177] The correct calculation appears to be £2,659 10s 4d.
[178] 'Obol' denotes ½.
[179] The arithmetic for this paragraph is correct, despite the complicated calculations.
[180] The correct calculation appears to be £447 2s 8½d.
[181] £2,549 10s 4d plus £447 13s 8d equals £2,997 4s.
[182] Instead of 'obol', the correct figure is 1¼d.

the kirk of Halis, £66 13s 4d. Total, £996 12s.[183] Take the rest extending to £38 19s 'fra the abbot[184] and his chalmerlandis.'[185]

'The haill ferme quheit and teynd within the patrimonie of the said abbay exceptand the ferme quheit set to the thesaurer in few extending to' 16 c. 15 b.;[186] third thereof, 5 c. 10¼ b. Take the teind of the Littill Mounktoun for 6 b.; take the teind of the Hill for 2 b; take the teind of Caldcoites for 3 b.; take the teind of Stanyhill for 4 b.; take the teind of Schirefhall for 2 b.; take the ferme wheat of Carbarry for 12 b.; take the ferme wheat of Natoun for 8 b.; take the ferme wheat of Innerresk and Mounktounhall 'by the thesaureris few' for 2 c. 18 b. 1 f. [2 c. 13 b. 1 f., *fo 65v*[187]]; take out of Wester Kingorne 8 b. 1¼ p.

'The haill beir within the lordschip of Dumfermeling, by the ferme set to the thesaurer, extending to' 55 c. 15 b. 1 f. 3 p.;[188] third thereof, 21 c. 15 b. 3 f. 1 p.[189] Take the teind of Littill Monktoun for 11 b.;[190] take the teind of the Hill for 2 b.; take the teind of Caldcoites for 3 b.; take the teind of Stanehill for 15 b. [*In margin*, 'Teynd beir of Quhythill assumit in place of the saidis for' 15 b. 'conforme to a speciall warrand direct to that effect.']; take the teind of Quhytsyd for 2 f.; take the teind of Schirefhall for 4 b.; take the whole teind bere of Kingorne for 3 c. 2 b.; take out of Newbyrne 5 c. 8 b. 1 f.; take the rest out of Kinglassie which is 10 c. 2 b. 1 p. 'Dischargit' 6 c. 8 b. 'to be tane of the fermes of Nather Kingorne and restis be tane out of Kinglassie bot' 3 c. 10 b. 1 p. 'as insert beiris'. [*In margin*, 'Nota, of this beir of Kinglassie thair is dry multures quhairof the writting heireftir.']

'The haill aittes within the patrimonie of the abbay of Dunfermeling, by the frie aites set to the thesaurer in few is' 114 c. 1 b. 2 f.;[191] third thereof, 38 c. 2 f., 'quhairof ar quhyt aites' 22 c., 'and thay ar to be takin within Mussilburchtschyre'. Of the Hill, 6 b.; take the teind of Cauldcoit for 12 b.; take the teind of Stanehall [*recte*, Stanehill, *i.e.* Stoneyhill] for 15 b.; take the teind of Quhytsyd for 4 b.; take the teind of Schirefhall for 8 b.; take Wester Kinghorne in white oats for 7 c. 15 b.; take out of Newbyrne 6 c. 12 b. 2 f. white oats; take out of Kirkcaldie, 4 c. 7 b. 2 f.

Third of black oats, 16 c. 2 f.[192] Take the black oats of Wester Kingorne for 3 c. 8 b.; take the black oats of Newbirneschyre for 4 c. 14 b.; 'tak out of the

---

[183] The correct calculation appears to be £996 10s 8d.

[184] See above, p. 40, n. 168.

[185] These figures do not tally.

[186] The total of ferme wheat is given as 15 c. 6 b. on fo 32v (see above, p. 25).

[187] The figure from fo 65v (see below, p. 53) would give correct total.

[188] The correct calculation appears to be 65 c. 15 b. 1 f. 3 p.

[189] The correct calculation appears to be 21 c. 15 b. 1 f. 2¼ p.

[190] SRO, Vol. b, fo 41v has the following deleted entry, preceding the entry for Little Monkton: 'Tak the teind of the Sandis of Mussilburcht for 1 c.'

[191] The total of oats, white and black, given on fo 46r (see above, p. 39), is 152 c. 3 f., minus 15 c. set to the treasurer on fo 49v (see above, p. 42), should leave 137 c. 3 f.

[192] The total oats of 38 c. 2 f. minus 22 c. white oats, equals 16 c. 2 f. black oats, though the calculation for horsecorn amounts to 16 c. 1 b. 1 f. 2 p. (see above, p. 42; see also below, p. 43, n. 193, for variant calculation).

aites of Kinglassie' 3 c. 15 b. 2 f.; take the oats of Turbane and Pittormok for 4 c.[193]

'The haill aitmeill within the patrimonie of the abbay of Dumfermeling, by the fre meill set to the thesaurer in few is' 7 c. 15 b.;[194] third thereof, 2 c. 15 b. 1 f. 3 p.[195] Take the meal out of Kinglassie, 'by the thesaureris few', 2 c. 15 b. 1 f. [*plus* 3 p., *fo 66r*]

'Caponis in the haill', 374; third thereof, $124\frac{1}{3}$. Take out of Mussilburchtschyre, 36; out of Wester Kingorne, 108.[196]

Poultry, 747; third thereof, 245.[197] Take Dolourschyre for 100; take Newbyrneschyre for 138; take Wester Kingorne, namely, Stanehous for 8 poultry, the rest extending to 9 to be taken out of Kinglassie.

'Cheis in the haill', 150 stones; third thereof, 50 stones. Take it out of the bow places.

'Butter in the haill', 35 stones; third thereof, $11\frac{2}{3}$ stones. Take it out of the bow places.

'Lyme in Kingorne', 20 c.; third thereof, 6 c. $10\frac{2}{3}$ b. Take the same out of Kingorne.

Salt, 11 c. 8 b.; third thereof, 3 c. $13\frac{1}{3}$ b. Take the same out of Kirkcaldie.

'At Leyth', 18 January 1571/2

'The commissiouneris of the kirk for certane ressonable causis and consideratiounes moving thame ordanes the Collectour[198] and clerk over the rentallis to exchange of thair assumptioun of the beir of the abay of Dunfermelyng' 6 c. 8 b. 'beir quhilk wes takin out of befoir of the parochin of Kinglassie to be now takin out of the ferme beir of the Nedder Grange of Kingorne givand be yeir' 6 c. 'and of the dry multures of Wester Kingorne giveand be yeir' 8 b., 'and sua Kynglassie quhilk payit in the kirkis assumptioun of befoir' 10 c. 2 b. 1 p. 'sall now pay bot' 3 c. 10 b. 1 p. 'and the remanent of the beir quhilk wes ass[ignit] out of Kinglassie of befoir extending to the saidis' 6 c. 8 b. 'to be now tane out of Wester Kingorne, and dry multures befoir specifeit, makand in the haill' 10 c. 2 b. 1 p. 'quhilk wes tane halelie of Kinglassie of before, provyding alwayis giff the kirk findis thame selffis hurt be this exchange throw evill payment or utherwayis that thei and thair Collectour sall have recours agane to the auld assumptioun or ellis the commendatour of Dunfermeling[199] sall

---

[193] These figures total 16 c. 5 b. 2 f.

[194] 7 c. 15 b. is the figure given throughout for the white oats of Kinghorn Wester.

[195] The correct calculation appears to be 2 c. 10 b. 1 f. $1\frac{1}{2}$ p.

[196] Capons do not appear in the rental except as part of the list on fo 48v. 18 score and 14 would give 374; cf above, p. 41, n. 175.

[197] One third of 747 is 249.

[198] This presumably refers to Robert Winram, Collector for Fife. No Collector general existed from 1568 to 1572. *Thirds of Benefices*, pp. xv-xvi, xl.

[199] See below, p. 45, n. 200.

assigne thame better payment at thair optioun quhill thai find thame selffis in als guid cais as thei wer be the auld assumptioun.

Sic subscribitur: R. Dumfermling, M[r] Johnne Wynrame, Johnne Erskin.'[200]

DUNFERMLINE, ABBEY OF,[201] (SRO, Vol. a, fos 54r-66v)
(SRO, Vol. b, fos 44r-56v[202])

'The rentale of the haill partrimonie [sic] of the abacie of Dunfermeling in penny maill, annuellis, customes, burrow maile, fed oxin silver, lymekill maill, cayne fermes, teyndis [of] kirkis and teyndis of tounis set in assedatioun for money as followis particularlie in everie schyre be the selff gevin in and subscryvit be Allane Coutes, chalmerlane thairoff.'

Dumfarmlingschyre

Maistertoun in penny mail 'in anno', £26 13s 4d. Eist Barnes in penny mail, £21 7s. Pitcorthies in penny mail, £17 6s 8d. Sanct Margarettis Stane 'in anno', £6 13s 4d. The Walkarland in Burnemouth, 53s 4d. Stenstanis in Burnemouth, 40s. Selletoun, £17 6s 8d. Legattis Brig, £3 6s 8d. Wyndmilhill, £4 16s. Knokas, £13 8s 8d. Primros, £8 2s. Pitbawillie, £3 7s 8d. Wester Bawdrik, 20s. Eister Bawdrik, 40s. Midle Bawidrik, £4 6s 8d. Wester Baith, £8 2s. Blaklaw, £3 7s 8d. Nather Bayth, £6 2s. Bayth under the Hill, £7 2s. Luscerewert, £7 8s. Northfad, £6 15s 4d. Touch, £4 2s. Windeage, 40s 6d. Coklaw, 40s 6d. Uver Lasoiddy, 40s 6d. Eister Lasoiddy, £4 2s. Nather Lasoiddy, £6 18d. Bawmill, £4 12d. Blairathie, £4 12d. Craigluscer, £8 15s 4d. Lathamontht, £4 12d. Wester Craigduky, £3 7s 8d. Eister Craigduky, £3 7s 8d. Wester Luscour, £8 2s. Touch Myln, 53s 4d. Northlethunes, £4 12d. The Knok, £3 6s 8d. 'The thrie mylnes of the toun of Dunfermeling', £13 6s 8d. Lathangy, £6 13s 4d. The Oucht, £8. Southfad, £6 15s 4d. 'The burrow mailles of Dunfermeling toun', £8 13s 4d. 'The custumes of the burcht of Dunfermeling', £4. Total of the penny mail, burgh mails and customs of Dunfermelyng, £286 17s 4d.

---

[200] Robert Pitcairn, commendator of Dunfermline; John Winram, superintendent of Fife and Strathearn; John Erskine, superintendent of Angus. Winram and Erskine were two of the church's commissioners who met at Leith in January 1572 with representatives from the government, whose number included Pitcairn, to secure a settlement of ecclesiastical finance. *The Booke of the Universall Kirk of Scotland: Acts and Proceedings of the General Assemblies of the Kirk of Scotland*, ed. T. Thomson, 3 vols. and appendix vol. (Edinburgh, 1839 - 45), i, pp. 207-208.
[201] Cf SRO, Vol. a, fos 30r, 44r (see above, pp. 23, 37).
[202] SRO, Vol. b, fo 43v is blank.

The annuals pertaining to Dumfarmling
   Logy, 26s 8d. Crummye, £10. The glebe of Striveling, £10. The annual of Scone, £3 6s 8d. The annual of Pettinweme, £3. The annual of Gartinkeir, £4. Total of annuals, £31 13s 4d.

'Item, of fed oxin silver within the lordschip of Dumfarmling', £13. 'Item, the bow places, wodder gangis within the parochin of Dumfarmling', £96. 'Item, the foirsaidis bow places gevis be yeir' 35 stones butter; 130 stones cheese, namely, Rescobe, 30 stones cheese; 7 stones butter. The Gask, 7 stones butter; 30 stones cheese. Southlethinis, 7 stones butter; 30 stones cheese. Tunegask, 7 stones butter; 25 stones cheese. Blairnbothie, 7 stones butter; 25 stones cheese. Total: Butter, 35 stones. Cheese, 130 stones.[203]

Doloure in penny mail 'be yeir': The Horhart 'in anno', 26s 8d. Dolour Beig, £4 13s 4d. The Mylne of Dolour, 26s 8d. Schyirdaill, £13 6s 8d. The Mylne of Dolour Beig, £3 6s 8d. The Bank 'in anno', £40. Total of the penny mail of Dolour with the Bank, £64.
   Newbirneschyre in penny mail: Newbyrne toun 'in anno', £6 13s 4d. The Mylne of Newbyrne, £4. Manturpe, £6 13s 4d. Lathallane, £8. Hallhill, £10 13s 4d. Balbard, £5 6s 8d. Melgum and Lawgrenis, 40s. Drummeldrie, £10. Balcristie, £10. Cot, £10. The acres, 11s 8d. Total of penny mail of Newbyrneschyre, £73 18s 4d.
   Kynglassieschyre in penny mail: Pityowquhair 'in anno', £17 6s 8d. Stentounes 'in anno', £18 13s 4d. Inchedarnie 'in anno', £18. Kinglassie toun, £8. The Smyddie Land, 20s. The Browistane Land, 13s 4d. Cassebarrian, £13 6s 8d. Fynglassie, £16. Pitlowchy, £6 13s 4d. The Walk Mylne, 40s. Gaitmilk, £15 4s. Fynnocht, £13. The annual of Clunie, 20s. Total of the penny mail and annual in Kinglassie parish, £132 17s 4d.[204]
   Kyngorne Wester in penny mail and annual: The Over Grange 'in anno', £40. Weltoun 'in anno', £20. Eister Quarter, £20. The Mylne, £6 13s 4d. Moy Hous, 53s 4d. The Leyme Kill mail, £13 6s 8d. The annual of Orrok, £3 6s 8d. The annual of Sellebalbie, 33s 4d. The annual of Abirdour, 53s 4d. Total of penny mail and annual, £110 6s 8d.
   Kirkcaldie: Abbottis Hall and Myltoun, £34 13s 4d. The West Mylne, £4. Dunekeir, £8. Bogy, £40. The Smiddie Landis 53s 4d. The annual of Balwery, £4. The annual of Raithe, £5. 'The burrow maill of the burcht', 33s 4d. The mail of the Salt Pannes, £46. The Eist Mylne, 53s 4d. 'The annuellis of certane ruidis' [Sandyruidis, *fo 31v*], 13s 4d. The customs of Kirkcaldie, £5. Total of penny mail, annuals, customs and burgh mails of the parish of Kirkcaldy, £154 6s 8d.
   Kynnedderis: Over Kinneddir, £8 8s 8d. Bandrum, £6 15s 4d. Drumkeppe, £3 7s 8d. The Mylne, £3 6s 8d. Nather Kinneddir, £7 8s 8d. The

---

[203] The correct calculation is 140 stones.
[204] The correct calculation appears to be £130 17s 4d.

Brewland, 13s 4d. Killarne, £13 15s 4d. Total of penny mail in Kinneddirschyre, £43 15s 8d [*recte*, £53 15s 8d].

The penny mail and annual within Mussilburghschyre: Inneresk 'in anno', £71 16d. Monktounhall, £43 6s 8d. Littill Monktoun, £18. Natoun, £29 3s. [*In margin*, 'augmentit be Archibald Prestoun', £7 16s 4d, 'and comptid in this' £29 3s.] Smetoun, £30 13s 4d. Caldcottis, £6. Wonet Bank, £6 13s 4d. Carbarry, £40. Costertoun, £10. Pynkin and Kers, £23. 'The fischeing', £8. Womet, £44 16s. Stanihill, £6 13s 4d. 'The mylnis', £66 13s 4d. 'The burrow maillis of Mussilburcht toun', 53s 4d. [*In margin*, 'payit be the balleis'.] The annual of Hill, 20s. [*In margin*, 'payit be the laird Hill'.] The annual of Adnistoun [Edmestoun, *fo 32r; SRO, Vol. b, fo 46v*], 13s 4d. [*In margin*, 'payit be the laird thairof'.] The annual of Edinbrucht, £5. [*In margin*, 'payit be the thesaurer thairof'.] The annual of Hadingtoun, 40s. [*In margin*, 'be the balleis'.] The annual of Quhitsyd, 20s. [*In margin*, 'be the laird Fausyd, now be Mr David and his spous'.] The annual of the Panes, 20s. [*In margin*, 'be the laird Craigmiller'.] The annual of Newbotle, 20s. [*In margin*, 'the abbot of Newbotle'.] The Terrouris croft, 26s 8d. [*In margin*, 'be Bessy Frog'.] Total of the penny mail and annuals within Mussilburcht, £419 13s 8d.[205]

'The haill money within the parochin of Dunf[erme]lyng of penny maill, annuell, custum, burrow maill, fed oxin silver, mart silver with the [wedder]gang extendis to the sowme of' £427 10s 8d. The penny mail of Dolour with the Bank of the same extends 'be yeir' to the sum of £64. The penny mail of Newbirnschyre 'be yeir' extends to the sum of £73 18s 4d. The penny mail of Kynglassye and annual 'be yeir' extends to the sum of £132 17s 4d. The penny mail of Wester Kingorne and annual 'be yeir' extends to the sum of £110 6s 8d. The penny mail, annual, burgh mail and custom of Kirkaldie extends to the sum of £154 6s 8d. The penny mail and annual of Kenneddirschyre extends 'be yeir' to the sum of £43 15s 8d. The penny mail of Mussilburcht with annuals and burgh mails extends to the sum of £365.[206] The penny mail of Southfery with annuals extends 'be yeir' to the sum of £23 9s 8d. The penny mail of Stermonth extends 'be yeir' to the sum of £47 12s. Kirks and teinds set for money extends 'be yeir' to the sum of £908 13s 4d.

The ferme wheat within the parishes of Dunfermeling, Mussilburcht and Newbyrne 'as followis particularlie. Item in the first, of the Gullattis within the [paroche of] Dunfarmling of ferme quheit extendis till' 5 c.

Newbyrne: The Cot, 1 c. Drummeldrie, 1 c. Newburne toun, 10 b. Manturpe, 8 b.

Mussilburgh: Inneresk in ferme wheat, 2 c. 8 b. Mounktounhall, 1 c. 4 b. Smetoun, 1 c. 4 b. Carbarry, 12 b. Nattoun, 8 b.

---

[205] This calculation is correct; cf above, p. 25, n. 101.

[206] See variant figures for Musselburgh in preceding paragraph, and see above, p. 25, n. 101.

Total of the ferme wheat within the parishes of Dumfermeling, Newburne and Mussilburcht, 15 c. 6 b.[207]

The ferme bere within the parishes of Dumfermeling, Kingorne, Newbyrne and Dolour. Within the parish of Dunfermeling: The Gulatis, 9 c. Girs Muirland, 6 b. Lymekillis, 2 c. 8 b. Lymekilhill, 1 c. 5 b. The New Landis, 1 c. 2 b. Dolourschyre, 1 c. The Neddir Grange of Kingorne, 6 c. The dry multure of Wester Kingorne, 8 b. Manturpe, 8 b. Total of ferme bere within the parishes of Dunfarmline, Dolour, Kingorne and Newbirne, 22 c. 5 b.

The ferme oats within the parish of Dumfarmling. The two Gullattis, 15 c. Total of ferme oats, 15 c.

Ferme oatmeal within the parishes of Dolour, Kinglassie and Kirkcaldie. Dolourschyre, 5 c. Kirkaldie parish, 2 c. 8 b. Kinglassie parish, 6 c. 3 b. 1 f. 3 p. Total oatmeal within the parishes of Dolour, Kinglassie and Kircaldie, [blank].[208]

Ferme salt within the parish of Kirkcaldie, 23 pans, 'ilk pan payis for ferme salt be yeir' 8 b. Total, 11 c. 8 b.

'Ferme kayne lyme within the parochin of Kingorne Wester', 20 c.

'Kirkis and teyndis of townes set in assedatioun for money as followis particularlie'.
The kirk of Striveling set to the laird of Carden, £80. The kirk of Halis set to Mr Henry Foullis for 100 merks. The kirks of Overquhill and Kynros to the lady Lochlavin, £120. The kirk of Sanct Johnnesoun to the laird of Cragy, £100. The kirk of Innerkeithing to John Swyntoun [Suetoun, *fo 33r*], £53 6s 8d. The kirks of Mollane and Strathardill, £120. The kirk of Wester Kingorne to Mr John Wemis, £12. The teind of Mounktounhall set for £20. The teind of Inneresk set for £53 6s 8d. The teind of Smetoun set for £20. The teind of Natoun set for £13 6s 8d. The teind of Carbarry set for £33 6s 8d. The teind of Cowisland set for £66 13s 4d. The teind of Pynkin set for £10. The teind of Edmestoun set for £26 13s 4d. The teind of Wonet set for £33 6s 8d. The teind of Hallhill, £40. The teinds of the burgh acres of Kirkaldie set to David Kingorny for £13 6s 8d. The teinds of the acres of the burgh of Dunfarmeling set for £26 13s 4d. Total of kirks and teinds of towns set for money, £908 13s 4d.

Teind wheat of Mussilburgh: Littill Monktoun, 6 b. Hill, 2 b. Caldcottis, 3 b. Stanihill, 4 b. Quhythill, 4 b. Schirefhall, 2 b. Total of teind wheat within the parish of Mussilburghschyr 'by that is sett for money', 1 c. 5 b.

---

[207] The correct calculation appears to be 14 c. 6 b.
[208] The correct calculation appears to be 13 c. 11 b. 1 f. 3 p.

Teind bere of Mussilburgh 'that is nocht set for money': Littill Monktoun, 11 b. Hill, 2 b. Caldcottis, 3 b. Stanyhill, 15 b. Quhythill, 14 b. Quhytsyd, 2 f. Schirefhall, 4 b. The Sandis, 1 c. Total of teind bere within Mussilburghschyr not set for money, 4 c. 1 b. 2 f.

Teind oats within Mussilburghschyr: Littill Monktoun, 1 c. 12 b. Hill, 6 b. Caldcottis, 12 b. Stanyhill, 15 b. Quhithill, 14 b. Quhitsyd, 4 b. Schirefhall, 8 b. Total of teind oats of Mussilburghschyre, 5 c. 7 b.

Teind wheat within the parish of Dumfarmling 'as followis particularlie': Eister Gullet of teind wheat, 11 b. Waster Gullet, 11 b. Eist Barnis, 5 b. North Ferry, 2 b. Maistertoun, 8 b. Pitlever, 2 f. Pittrave, 3 b. Pitferren, 2 b. 2 f. Prymros, 5 b. Pandellis [*recte*, Randellis] Craigis, 3 f. Sanct Margaretis Stane, 1 b. 2 f. Selletoun, 1 b. Total of teind wheat within the said parish of Dumfarmling, 3 c. 3 b. 1 f.

Teind bere within the parish of Dumfarmling 'as followis particularlie': Wester Gullet, 1 c. 2 b. Eister Gullet, 1 c. 2 b. Burn Mouth, 2 f. Bawmyll, 1 b. 2 f. Fowilfurd, 1 b. Baith Maistertoun, 1 b. Blaklaw, 2 b. Baith Stewart, 3 b. [4 b., *fo 34r*] Baith Keir [Bayth Creir, *fo 34r*], 1 b. 2 f. Baith Trumbill, 1 b. Baith Stenestoun, 1 b. Baith Bell, 2 f. Blairbothy Northir, 2 f. Blairrathy, 2 f. Baith Danyell, 2 f. Baith Bonolla, 6 b. Baith Persone, 3 b. 3 f. Coklaw, 1 b. 2 f. Cawill, 2 b. Crawnarland, 3 f. Cragieluscur, 4 b. Clunie, 2 b. Craigduky Eister, 1 b. [1 b. 2 f., *fo 34r*] Craigduky Wester, 1 b. 1 f. Dunduff, 2 b. 2 f. Drumthuthill, 2 f. Eist Bairnis, 10 b. Galrik, 2 b. Gask, 1 b. Hoill Baudrik, 1 b. 2 f. Knokis, 6 b. Knok, 1 f. Legattis Brig, 2 b. Lymekillis, 10 b. Lathamonth, 1 b. Luscer Wester, 8 b. Luscer Ewat, 2 b. 2 f. The Lasoiddeis, 8 b. The Fouddis [*i.e.* Northfod and Southfod], 10 b. Northe Ferrie, 4 b. Milhillis, 3 b. Maistertoun, 1 c. 8 b. Mirlabank, 2 f. Midle Bawdrik, 3 b. Meldrumis Mylne, 1 f. Northlethinis, 2 f. Oucht, 2 p. Pitlever, 2 b. Pittravy, 6 b. 2 f. Pitcortheis, 10 b. [8 b., *fo 34r*]. Pitfirrane, 5 b. Prymrois, 6 b. [5 b., *fo 34v*]. Pitduneis, 10 b. Pitbawlie, 1 b. 3 f. Pitconchy, 3 b. Pittincreiff, 12 b. Pandell [*recte*, Randell] Craigis, 4 b. Raslok [Rascobie, *fo 34v*; *i.e.* Roscobie], 1 b. Selletoun Hallheid, 3 f. Sanct Margaretis Stane, 4 b. Selletoun Sandis, 1 b. Touch, 2 b. 2 f. Tunegask, 2 f. Wendeage, 1 b. 2 f. Wondmilhill, 2 f. Walkerland, 1 b. Total of teind bere within the parish of Dunfermeling, 15 c. 6 b. 3 f. 2 p.

Teind oats within Dumfarmling parish: Eister Gullet, 3 c. 4 b. Wester Gullet, 4 c. 2 b. Burne Mouth, 1 b. 2 f. Accorne Waird, 1 b. Bawmeill, 8 b. Fowlisfuird, 6 b. Baithmaisterstoun, 8 b. 2 f. Blaklaw, 10 b. Brumhill [Brounhill, *fo 34v*], 3 b. Baith Stewart, 10 b. Baith Keir, 10 b. Baith Trumbill, 10 b. Baith Steinstoun, 6 b. Baith Bell, 4 b. Blairbothy Norther, 4 b. Blairathy, 4 b. 2 f. Baith Danyell, 5 b. Blairbothy Souther, 2 b. Baith Bonella, 1 c. 14 b. Baith Persoun, 1 c. 2 b. Coklaw, 10 b. Cawill, 10 b. Crownarland, 1 b. Craigluscour, 1 c. 4 b. Clunie, 10 b. Craigduky Eister, 8 b. Cragduky Wester, 8 b. Dunduff, 1 c. Drumthuill [Drumthuthill, *fo 34r*], 8 b. Eist Barnes, 1 c. 8 b. Galrik, 11 b. Girsmuirland, 1 b. Gask, 6 b. Hoill Baudrik, 12 b. Knokas, 1 c. 12 b. Knok, 3 b. Kelly [Kelty, *fo 34v*] Wod, 1 b. Legattis Brig, 6 b. Lymekillis,

11 b. Lathawmonth, 5 b. Wester Luscour, 1 c. 12 b. Luscour Evert, 1 c. 4 b. Lasoiddeis, Eister and Wester, 2 c. 8 b. Foirrdis, South and North, 4 c. 8 b. Northferry, 8 b. Milhillis, 7 b. Maistertoun, 4 c. 8 b. Morlay Bank, 6 b. Midle Bawdrik, 14 b. Meldrumis Mylne, 1 b. 'North pairt quheit feild', 1 b. Newlandis, 1 c. Northlethemes, 6 b. Oucht, 2 b. Pitlover, 5 b. Pittrayvie, 1 c. 2 b. [2 c. 2 b., *fo 35r*]. Pitcorthris Eister and Wester, 2 c. Pitfirrane, 1 c. Prymros, 2 c. Pitdunes, Eister and Wester, 2 c. Pitbawkie, 7 b. Pitconnochy, 1 c. Pittincreiff, 14 b. Randellis Craigis, 12 b. Rescobe, 8 b. Selletoun Halkheid, 6 b. Sanct Margaretis Stane, 10 b. Selletoun Sandis, 11 b. South Lethemes, 6 b. Touch, 10 b. Touchmylne, 2 b. Tunegask, 4 b. 2 f. Wendeage, 1 b. 2 f. 'Wendeage pairt Kellok', 8 b. Wyndmilhill, 6 b. Wodaker, 3 f. Walkerland, 3 b. 2 f. Total of teind oats within the parish of Dumfermeling, 'quhyt and blak', 62 c. 14 b. 2 f.[209]

Teind wheat of Kingorne Wester: Overgrange, 6 b. Weltoun, 6 b. Eist Quarter, 1 b. Orrok, 5 b. Seillebalbie, 1 b. 2 f. Total of teind wheat in Kingorne, 1 c. 3 b. 2 f.

Teind bere of Kingorne: Over Grange, 1 c. 7 b. Weltoun, 12 b. Eister Quarter, 4 b. Orrok, 8 b. Seillebalbie, 3 b. Total of teind bere in Kingorne, 3 c. 2 b.

Teind oats of Kingorne: Over Grange, 3 c. 15 b. Weltoun, 2 c. 6 b. Eist Quarter, 1 c. 8 b. Orrok, Standhous [mylne, *deleted*] Moyhous [Monyhous, *fo 35r*] and Dunhair [*i.e.* Dunearn], 2 c. 4 b. Seillybalbie, 1 c. 4 b. The Terrouris acres, 2 b. Total of teind oats within Kingorne, 11 c. 7 b.

Teind wheat of Kirkaldie: The Abbottis Hall, 6 b. Balwery, 6 b. Bogy, 2 b. Raith, 4 b. Total of teind wheat in Kirkaldie, 1 c. 2 b.

Teind bere in Kirkaldie: Abbottis Hall and Mylntoun, 15 b. The West Mylne, 1 b. 2 f. Balwery, 8 b. Bogy, 12 b. Raith Turbane and Pittonmark, 20 b. The Muirtoun of Kirkaldie, 1 b. Total of teind bere in Kirkaldie, 3 c. 9 b. 2 f.

Teind oats of Kirkaldie: Abbottis Hall and Mylntoun, 3 c. 14 b. Balwery, 2 c. Bogy, 3 c. Rathe, Pittonmark and Turbane, 4 c. 14 b. The Muirtoun, 8 b. Total of teind oats in Kirkaldie, 14 c. 4 b. [*In margin*, 'Be ressoun of' 1 c. 2 b 'deducit for the Rathe.']

Teind wheat of Newbyrneschyre: The Cott, 14 b. Ballchrist, 4 b. Drummeldrie, 1 c. Newbirntoun, 5 b. Manturbe, 4 b. Total of teind wheat in Newbirne, 2 c. 11 b.

Teind bere of Newbirne: The Cott, 1 c. 8 b. Ballcrist, 1 c. 8 b. Ballbaird, 1 b. 1 f. Lathalland, 4 b. Drummeldrie, 2 c. 2 b. Johnnestounes Mylne, 1 b. Newbirnetoun, 12 b. Manturpe, 8 b. The Terrouris acres, 1 b. Total of teind bere of Newbyrne, 6 c. 13 b. 1 f.

---

[209] The correct calculation appears to be 63 c. 4 b. 3 f.

Teind oats of Newbyrne: Cott, 2 c. 8 b. Ballcriste, 1 c. 1 b. Ballbard, 9 b. Lathalland, 1 c. 8 b. Drummeldrie, 2 c. 2 b. Lawgrenis, 6 b. Newbirtoun [*recte*, Newbirnetoun], 2 c. 3 b. Manturpe, 1 c. 4 b. The acres, 4 b. Total of teind oats in Newbyrne, 11 c. 13 b.

Teind wheat of the parish of Cambie 'as it wes riddin this last yeir be honest men and sworn extending in quheit till' 2 c. 2 b. 'Of teynd beir', 7 c. 1 b. 'Of teynd aites', 29 c. 4 b.

Teind bere of Kinglassie: Pithowcher, 3 c. 2 b. Kynnynmont, 1 c. 12 b. Gaitmylk, 4 c. Pitlochie, 1 c. 2 b. [2 c. 2 b., *fo 35r*]. Kinglassie [*recte*, Fynglassie, *fo 35v*], 2 c. Inchdarnye, 2 c. 8 b. Stentounes, 3 c. Fynmont, 8 b. Mildamis, 9 b. Kinglassie toun, 1 c. 8 b. Brewland, 1 b. Smiddeland, 2 f. Craigsyid, 2 b. Walkmylne, 2 b. Foistertoun, 2 b. Clunie Mylne, 1 b. Muirtoun, 2 b. Cassebarean, 3 b. Total bere in Kinglassie, 21 c. 4 b. 2 f.[210]
Teind oats within Kinglassie: Fynmont, 3 c. Fostertoun, 1 c. 3 b. 2 f. Clunie Mylne, 8 b. 2 f. Halltoun, 2 b. Muirtoun, 1 c. 6 b. Cassebarrian, 2 c. Total of the teind oats of Kinglassie, 8 c. 4 b.

'Summa totalis of teynd aites within the parochinis of Dunfermlyng, Kingorne, Kirkcaldie, Kinglassie, Newbyr[n]e, Cambie and Mussilburcht, quheit and hors corne, extendis to' 703 c. 11 b. 2 f. [7 score and 3 c. 11 b. 2 f., *fo 35v*]

'Summa totalis of aites, ferme and teynd', 7 score and 18 c. 11 b. 2 f.

'The haill money within the parochin of Dumfermeling in penny maill, annuellis, customes, burrow maill, fed oxin silver, cayne mart silver, extendis be yeir to the sowme of' £427 10s 8d. The penny mail of Dolour with the Bank mail extends 'be yeir' to the sum of £64. The penny mail of Newbyrne 'be yeir', 73s [£73 18s 4d, *fos 31r, 32r, 50r, 55v, 57r*; SRO, Vol. b, *fo 53r*]. The penny mail and annual of Kinglassie 'be yeir' extends to the sum of £130 17s 4d. The penny mail and annual of Kingorne, £110 6s 8d. The penny mail, custom, burgh mail and annual of Kirkaldie extends 'be yeir' to £154 6s 8d. The penny mail of Kynnedderschyre 'be yeir', £43 15s 8d. The penny mail of Mussilburcht, burgh mails and annuals extends 'be yeir' to the sum of £419 13s 8d. The penny mail and annual of Southferry, £23 9s 8d. The penny mail of Stormoth 'be yeir', £47 12s. The kirks and teinds of towns set for money extends 'be yeir' to the sum of £908 13s 4d. 'Summa of the haill money of the patrimonie of the abacie of Dunfermelyng', £2,404 4s;[211] third thereof, £801 8s.

Signature: 'Allane Coutis, chalmerlane of Dumfarmling, with my hand.'

---

[210] The correct calculation appears to be 20 c. 14 b. 2 f.
[211] This total is correct taking £73 18s 4d for Newburn.

Of ferme wheat within the parish of Dunfarmling extends 'be yeir' to 5 c. Of ferme wheat within the parish of Newbyrne extends 'be yeir' to 4 c. 2 b. Of ferme wheat within the parish of Mussilburcht 'be yeir' extends to 6 c. 6 b. Total of ferme wheat within the parishes of Dunfarmling, Newbyrne and Mussilburcht extends to 15 c. 8 b.

Of teind wheat within Dunfarmling parish, 3 c. 3 b. 1 f. Of teind wheat within Kingorne, 1 c. 3 b. 2 f. Of teind wheat within Kirkcaldie, 1 c. 2 b. Of teind wheat within Newbyrne, 2 c. 11 b. Of teind wheat within Mussilburcht, 1 c. 5 b. Of teind wheat in Carnbie, 2 c. 4 b. Total of teind wheat within the parishes of Dunfermling, Kingorne, Kirkcaldie, Newbyrne and Mussilburcht extends 'be yeir' to 11 c. 12 b. 3 f.

'Summa of the haill quheit pertenyng to the patrimonie of Dumfarmling extendis be yeir in ferme and teynd till' 27 c. 4 b. 3 f.; third thereof, 9 c. 1 b. 2 f. $1\frac{1}{3}$ p.

Of ferme bere within the parish of Dunfarmling extends 'be yeir' to 14 c. 5 b. Of ferme bere in Dolour, 1 c. Of ferme bere in Kingorne, 6 c. 8 b. Of ferme bere in Newbyrne, 8 b. Total of ferme bere within the parishes of Dunfermling, Dolour, Kingorne and Newbyrne, 22 c. 5 b.

Of teind bere within the parish of Dunfarmling, 15 c. 6 b. 3 f. 2 p. Of teind bere within the parish of Kingorne, 3 c. 2 b. Of teind bere within Kinglassie, 21 c. 4 b. 2 f. Of teind bere within the parish of Newbirne, 6 c. 14 b. 1 f. Of teind bere within the parish of Mussilburcht, 4 c. 1 b. 2 f. Of teind bere within the parish of Carnbie, 8 c. 1 b. Of teind bere within the parish of Kirkcaldie, 3 c. 9 b. 2 f. Total of teind bere within the parishes of Dumfermeling, Kingorne, Kirkcaldie, Kinglassie, Newbyrne and Mussilburcht, 61 c. 6 b. 2 f. 2 p.[212]

'Summa totalis of beir, ferme and teynd', 83 c. 11 b. 2 f. 2 p.; third thereof, 27 c. 14 b.[213]

'Les nor the first', 6 c. 6 b. 2 f. 2 p.[214]

'Summa totalis of aites ferme and teynd', 7 score and 18 c. 5 b. 2 f.; 'quhairof of quhyt aites to the thrid', 28 c.; 'and of blak aites to the thrid', 24 c. 14 b. 2 f.[215]
    'Item, of kaynis lyme', 20 c.; 'quhairof to the thrid', 6 c. 10 b. 2 f. 2 p.[216]
    'Item, of ferme salt', 11 c. 8 b.; 'quhairof to the thrid', 3 c. 13 b. 1 f. 1 p.[217]

---

[212] The correct calculation appears to be 62 c. 7 b. 2 f. 2 p.
[213] 2 f. $\frac{2}{3}$ p. have been omitted from this calculation.
[214] The correct calculation appears to be 6 c. 6 b. 2 f. $2\frac{2}{3}$ p.
[215] One third of 158 c. 5 b. 2 f. is 52 c. 12 b. 2 f.; the thirds of white and black oats added together give 52 c. 14 b. 2 f.
[216] $\frac{2}{3}$ p. has been omitted from this calculation.
[217] $\frac{1}{3}$ p. has been omitted from this calculation.

'Item, the coill pot [*i.e.* coal pit] of Wallyfurd to put ane man thairintill for uptaking of the thrid. The New Mylne of Mussilburcht siclyk.'

Third of oatmeal, 4 c. 9 b. 2 f. Third of butter, 11 stones 10 lb.[218] Third of cheese, 50 stones.[219]

'Off capones within Mussilburcht', 228; of capons within Dunfermling parish, 22; of capons within Kingorne, 108; of capons in Kirkcaldie, 24; of capons within the parish of Kinglassie, 12. Total of capons, 374; third thereof, 124.

'Item, of pultrie within Dunfermeling', 348; of poultry within Dolour and Kynnedderis, 100; poultry within Kingorne, 8; poultry within Kinglassie, 172; of poultry in Newburne, 138. Total of poultry, 746; third thereof, 255.[220]

'The alteratioun of the assumptioun of Dunfermlyng for Newbirnschyre according to the kingis warrand direct to that effect of the dait at Halyruidhous the [*blank*] day of [*blank*].'

Third of the money, £1,035 11s 'ob. etc.'[221] Take the penny mails and annuals within Mussilburgheschyre, Edinbrucht, Newbotle and Hadingtoun, £419 13s 8d; the teind of Mounktounhall, £20; the teind of Carbarrie, £33 6s 8d; the teind of Cousland, £66 13s 4d; the teind of Pinkie, £10; the mail of the New Mylne of Mussilburcht, £20; the mail of Dolourschyre, £64; 'the penny maill, annuellis, lymkilmaill, fed oxin silver in Wester Kingorne, by the beir set to the thesaurer in few and with the penny maill of the kirk', £122 6s 8d; the kirk of Sanct Johnnestoun, £100; the kirk of Halis, £66 13s 4d; and the rest extending to £112 17s 4d 'to be payit be the abbot and his chalmerlandis out of the best and reddiest payment of the tua pairt, viz.', £73 18s 4d 'quhilk wes assumit of auld out of the kirk of Newbirne and sensyne is set in tak and assedatioun to Andro Wod of Largo, Comptroller[222] to our so[verane] lord, with the victuallis thairof, be the abbot, quhilk he is obleist to warrand, in respect quhairof the auld assumptioun is ordanit to be alterit conforme to ane warrand subscrivit be the king and the Collectour heirupoun.'

Wheat, third thereof, 5 c. 10$\frac{1}{3}$ b. Take the teind of Littill Monktoun, 2 b.; the teind of Quhytehill, 4 b.; the teind of the Hill, 2 b.; the teind of Caldcottis, 3 b.; the teinds of Staniehillis, 4 b.; the teind of Schirefhall, 2 b.; the ferme wheat of Carbery, 12 b.; the ferme wheat of Natoun, 8 b.; the ferme wheat of Inneresk and Mounktoun [Mounktounhall, *fo 50v*] 'by the thesaureris few', 2 c. 13 b. 1 f.; out of Wester Kingorne, 8 b. 1$\frac{1}{3}$ p.

Bere, third thereof, 21 c. 15 b. 3 f. 1 p. Take the teind of Quhythill, 15 b.; the teind of Littill Monktoun, 11 b.; the teind of the Hill, 2 b.; the teind of Cauldcoites, 3 b.; the teind of Staniehill, 15 b.; the teind of Quhytsyd, 2 f.; the teind of Schirefhall, 4 b.; the teind bere of Kingorne, 3 c. 2 f.; out of

---

[218] One third of 35 stones of butter is 11$\frac{2}{3}$ stones.
[219] One third of 140 stones of cheese is 46$\frac{2}{3}$ stones.
[220] One third of 746 is 248$\frac{2}{3}$.
[221] The significance of 'obol' here is unclear.
[222] Andrew Wood of Largo, Comptroller, 1585 - 1587, *Handbook of British Chronology*, p. 191.

Kinglasschyre, 10 c. 3 b. 1 p.; 'and the rest for Newbyrneschyre fra the abbot for caussis abonspecifeit', 5 c. 8 b. 1 f.

Meal ['aitmeill', *fo 51r*], third thereof, 2 c. 15 b. 1 f. 3 p. Take the meal of Kinglassy 'by the thesaureris few', 2 c. 15 b. 1 f. 3 p.

Oats, third thereof, 38 c. 2 f., 'quhairof ar quhyt aites', 22 c. Take the teind of Quhythill, 15 b.; take out of Mussilburchtschyre: of the Hill, 6 b.; the teind of Cauldcot, 12 b.; the teind of Staniehill, 15 b.; the teind of Quhytsyd, 4 b.; the teind of Schirefhall, 8 b.; Wester Kingorne, 7 c.; out of Kirkcaldie, 4 c. 7 b. 2 f.; 'and the rest for Newburneschyre fra the abbot', 6 c. 7 b. 2 f.

Third of the black oats, 16 c. 2 f. Take the black oats of Wester Kingorne, 3 c. 8 b.; out of Kinglassie, 3 c. 15 b. 2 f.; the oats of Turbane, Pitcormokt, 4 c.; 'and fra the abbot in place of Newbirne', 4 c. 9 b.

'Custumes etc. ut antea [*i.e.* as before].'

'Followis the tennour of the warrand.

Lordis Auditouris of Checker, we greit you weill, forsamekill as in the first assumptioun of the thrid of the abay of Dunfermeling the kirk of Newbirnschyre assumit for' 5 c. 8 b. 1 f. bere, 6 c. 12 b. 2 f. white oats and 4 c. 14 b. black oats 'as the said auld assumptioun in the selff at mair lenth purportis, and now the said kirk with the haill fructes, pertinentis, with the personage and vicarege thairof, is set in tak and assedatioun be Patrik, maister of Gray,[223] commendatour of Dunfermeling, to Andro Wod of Largo, Comptroller, be ressoun quhairof the auld assumptioun of the thrid is vitiat be the said commendatouris proper deid to our greit prejudice and hurt albeit the haill prelaceis of our realme ar bund and obleis to warrand the thridis to us fra thair awin deidis, thairfoir it is our will and we command you to ass[ume] alsmeikill beir and aites out of the best and reddiest payment of the tua pairt of the abay of Dunfermling as the said kirk of Newbirneschyre wes assumit for of auld, extending to the quantitie abonewrittin, to the effect that the haill thrid of the said abay may remane [with] us to our use in the same state and integritie as the samin wes in the xlj [*recte*, lxj] yeir of God unvitiat or diminischit in ony maner of way, quhilk landis and teyndis to be assumit out of the tua pairt in place of the said kirk of Newbirneschyre salbe compitt and estemit in the thrid in all tyme cuming. Subscrivit with our hand at Halyruidhous', 20 January 1586/7.

'Sic subscribitur: James R.; Thirlstane; Thomas, maister of Glammis; Linclouden, Collectour.'[224]

---

[223] Patrick, eldest son of 5th lord Gray and later 6th lord Gray, became commendator of Dunfermline in 1583. *Scots Peerage*, iv, pp. 284-285; Cowan and Easson, *Medieval Religious Houses*, p. 58.

[224] See above, p. 21, n. 84.

ST ANDREWS, ST MARY ON THE ROCK COLLEGIATE KIRK, PROVOSTRY OF,

(SRO, Vol. a, fo 67r-v)
(SRO, Vol. b, fo 57r-v)

'This is the rentall of the provestrie of Kirkhill [*i.e.* Kirkheuch] besyd Sanctandrois, gevin in be the proveist[225] of the samin.'

'In Hiltattis towne for the teyndis', 8 b. bere; 24 b. meal. 'In Hiltattis [Haltatis, *SRO, Vol. b*] toun for the teynd thairof',[226] 3 b. bere; 12 b. meal. Tokles, 2 b. bere; 7 b. meal. Balderane [*i.e.* Bandirran], 8 b. bere; 22 b. meal. Tattis Mylne [*i.e.* Teassesmill], 3 b. meal. Carskerdo, 8 b. bere; 28 b. meal. Craigrothe, 2 b. bere; 5 b. meal. Kingarro, 12 b. meal; 6 b. bere. [*In margin*, 12 b. meal; 6 b. bere.] Baltulye, 10 b. meal; 4 b. bere. [*In margin*, 10 b. meal; 4 b. bere.] Seres, 12 b. meal; 8 b. bere. [*In margin*, 12 b. meal; 8 b. bere.] Eister Petscote, 8 b. bere; 20 b. meal. Total: Bere, 3 c. 9 b. Meal, 9 c. 11 b.

In silver for the teinds of Auchterstruder and Cassindulie, £30. Craighall, £20. Mylis Tarvet, 40 merks. Calludye, £20. Eister Petscottie, £10. Total money, £106 13s 4d.

'In few mailles pertenying to the said provestrie', £40, 'for the few of Kynnaldie and Gilmestoun. Of the saidis tounes', 22 b. 'of cayne aites with fyve dussoun of foullis. Pertenying to the said provestrie of annuellis within the toun of Sanctandrois', £3 8s 'of evill payment. The vicarege of Seres pertenying to the said provestrie in tyme of papistrie', 50 merks 'with the vicar pensiounes [pensionaris, *SRO, Vol. b*] dewtie and thir fyve yeiris bypast hes nocht bein payit.'

Signature: 'James Lermonth.'

Assignation of the third of the provostry of Kirkhill:
Third of the bere, 1 c. 3 b. Take Hiltattis for 8 b.; Seres, 8 b.; 'off Hiltattis [*recte*, Hall Teasses] toun', 3 b.
Third of the meal, 3 c. 3 b. 2 f. 2⅔ p. Take Hiltattis toun, 36 b.;[227] Seres, 12 b.; Craigrothe, 5 b. 'Giff in' 1 b. 1 f. 1⅓ p.
Third of the money, £61.[228] Take the feu mails of Kynnaldie and Gilmestoun for £40; the teinds of Auchterstruder and Cassindellie, £30. 'Giff in' £9.
Third of the oats, 7 b. 1 f. 1⅓ p.
Third of the kane fowl, 20 fowl 'to be liftit of Kynnaldie and Gilmestoun.'

---

[225] James Lermonth, provost of St Mary on the Rock collegiate kirk, 1540 - 1578, Watt, *Fasti*, pp. 372-373.
[226] SRO, Vol. a repeats 'Hiltattis' but SRO, Vol. b renders the second entry 'Haltatis'. They are now probably to be identified with Hill Teasses and Hall Teasses respectively.
[227] This is the meal of Hill Teasses and that of Hall Teasses added together.
[228] This calculation is correct.

SCOTLANDWELL, PRIORY OF,

(SRO, Vol. a, fo 68r)
(SRO, Vol. b, fo 58r)

Rental of Scotlandwell

The kirk of Auchtirmonsy pays £66 13s 4d. The lands of Fynave pay 40 b. meal; 20 b. bere with £11 6s 8d. The toun of Scotlandwell pays £24. The kirk of Carnok pays 52 b. meal; 20 b. bere.

'To be takin of the sowmes abonewrittin this that efter followis.'
To the reader at the kirk of Carnok, 20 merks. To friar John Dalzell in Scotlandwell, 20 b. meal; 4 b. bere; 8 merks silver. To sir George Hog there, 12 b. meal; 4 b. bere; 8 merks silver. To William Foullar 'for keiping of the place and yairdis', 6 b. meal; £4.

Signature: 'George Arnot.'

Assignation of the third of Scotlandwell:
Third of the money, £34. Take it out of the Kirkhill and Auchtirmousall 'and gleib thairof callit Lawteis Hoill' paying £66 13s 4d.
Third of the meal, 30 b. 2 f. $2\frac{2}{3}$ p.   Third of the bere, 14 b. 1 f. $1\frac{1}{3}$ p.[229]
'Tak bayth thir of the landis of Fynnave gevand be yeir' 40 b. meal; 20 b. bere.

PORTMOAK, PRIORY OF,

(SRO, Vol. a, fo 68v)
(SRO, Vol. b, fo 58v)

'Priorie of Portmork'[230]

'Magister Joannes Wyrame habet prioratum insule Sancti Servani infra lacum de Levin'

Rental of the priory on St Serf's Island in Loch Leven pertaining to Mr John Wy[n]rame.

'Rentale dicti prioratus sequitur [*i.e.* Rental of the said priory follows.]'

The fermes of the toun of Kir[k]nes paid annually by the laird of Lochlevin, feuar of the same, £60. Feuferme of Arthmur, £5. Annualrent of Awgmowty, 40s. Annualrent of Wester Markinche, 20s. Annualrent of Balchrystie, 6s 8d. Ferme of the boat on the loch ('firma cumbe super lacum'), 10 merks. Teind sheaves of Arthmur, £10. The vicarage of Portmook, which formerly paid freely 80 merks, now, however, has paid next to nothing, but if those things were paid as are set

---

[229] The correct calculation appears to be 13 b. 1 f. $1\frac{1}{3}$ p.
[230] Portmoak (Loch Leven) was a dependent house of St Andrews priory, Cowan and Easson, *Medieval Religious Houses*, p. 93.

down in the Book of Reformation,[231] it would still be worth ('vicaria de Portmook olim solvebat libere lxxx merkis nunc autem quasi nihill solvit sed tamen debite solveretur illa que exprimuntur in libro reformationis ad huc valeret') 40 merks. Total in silver, £111 13s 4d.

The teinds of Kirknes, 1 c. 12 b. bere; 4 c. 8 b. black oats. Total bere and oats, 6 c. 4 b.

Signature: 'Mr Johnne Wynrame.'[232]

Assignation of the third of the priory of Portmork:

Third of the money, £37 4s 5½d. Take it out of the feu mails of Kirknes giving £60. Third of the bere, 9 b. 1 f. 1½ p. Third of the oats, 1 c. 8 b. Take them out of the teinds of Kirknes.

BALMERINO, ABBEY OF,   (SRO, Vol. a, fos 69v-72v[233])
                                                          (SRO, Vol. b, fos 59v-61v[234])

'James Nicolsoun,[235] it is our will that ye onnawyse pas ony lettres in the four formes at the instance of the laird of Kynneir or ony utheris upoun ony transsumpt furth of the register of the greit seill upoun the informatioun of the xix yeris tak quhilk he alledgis him to have of the frutis and dewteis of the abacie of Balmerynocht, bot giff ony sict lettres beis socht at your handis that the samin be onlie lettres to summound all parteis havand entres to heir and sie sict lettres past and in speciall the commendatar of Balmerynocht[236] quhilk hes speciall entres or sict lettres pas, and failyie nocht heirin as ye will answer to us thairupoun nochtwithstanding ony command gevin or to be gevin be us, our Lordis of the Sessioun, to you in the contrair quhill ye speik our selff anent the geving thairof. Be thir presentis subscrivit with our hand at Edinbrucht, the twenty day of September 1568. Sic subscribitur, James, Regent.'[237]

'Rentale monasterij de Balmerynoch'

'Item, the baronie of Balmerynocht payis yeirlie in teynd and ferme of victuall', 30 c. 7 b. 3 f.; the said barony in money, £109 7s 9d; the said barony pays in poultry, 382; the said barony pays of fodder yearly, 49 'turssis'. 'The fischeingis

---

[231] I.e. the *First Book of Discipline* (1560).
[232] John Winram was the last pre-Reformation prior of Portmoak, *RMS*, v, no 1.
[233] SRO, Vol. a, fo 69r is blank.
[234] SRO, Vol. b, fo 59r is blank.
[235] James Nicolson was clerk of the Collectory, *Thirds of Benefices*, p. 62.
[236] See above, p. 19, n. 70.
[237] James Stewart, earl of Moray, was Regent, 1567 - 1570, *Scots Peerage*, vi, pp. 313-316.

of Kilburnis set to the laird of Nauchtane payand in anno at Witsonday',[238] £23 4s 4d. The fishings of Bromepark and Quhytquarrellhoip set to Robert Forester paying 2 dozen salmon; 2 dozen 'grississ'. The Barden fishing 'assignit to the conventis expenssis'. The lands of Littill Kinneir set to Andrew Kinloch paying £10. The lands of the tour set to the laird of Kynneir paying £3. The lands of Galstoun [*i.e.* Gilston] set to Mr George Strang paying £3 6s 8d. The fishings beside Sanct Johnnestoun 'callit the Stok and Garthe set to James Campbell of the Laweris payand in anno at the Assumptioun of Our Lady',[239] £50. Poldrait set to Thomas Flemyng paying 41 merks. The fishings of the Cruik set to the laird of Balumbie, £7. The barony of Pitgorno pays yearly in money, £100 13s 4d; the same barony pays in capons and poultry yearly, 5 score and 16 poultry [*i.e.* 116]. The teind of the kirk of Logymurtho pays yearly of victual, 10 c. The barony of Barrie with the kirks thereof extends in money yearly to the sum of £266 14s 8d; of victual in the said barony yearly, 3 c.; of poultry in the said barony yearly, 265. The fishings of Barry 'set to my lord dukis[240] sone quha payis yeirlie for the samin', £48. The lands of Johnnestoun and Gadden [*i.e.* Gadwen] with the Lochmylne set to the laird of Creycht, Robert Young and Robert Lumisden paying therefore yearly, £40 13s 4d, 'payand thairfoir of capones yeirlie', £40 [*sic*]. The annuals of Dundie, Perthe, Sanctandrois and Carraill, £14 15s 2d. Total money, £704 2s 11d. Total victual, 43 c. 7 b. 3 f. Total poultry, 763.

'Memorandum, to be deducit yeirlie heirof ane pensioun assignit be the quenis majestie furth of the samin to Capitane Hepburne quhilk extendis to the sowme of' 300 merks 'conforme to hir hienes gift maid to him thairupoun.' To the Lords of Session for their contribution yearly, £48. 'Sex mounkis portiounes conforme to thair assignatioun maid and subscrivit to thame under thair commoun seill be the last abbot[241] of the said place, extending yeirlie to evrie ane of thame', 17 b. bere; 5 b. wheat; 12 b. meal; £10 money 'for thair kiching'; with 8 merks 'for thair hebbet silver, makand in the haill in victuall', 12 c. 12 b.; £15 6s 8d money yearly 'to ilk ane of thame, summa of money to thame yeirlie', £92. 'To thair brewster yeirlie', 14 b. meal. 'To thair cuik yeirlie', 12 b. meal. 'To thair porter and barbour yeirlie', 12 b. meal. To sir Alexander Ker[242] in yearly pension of 40 merks, 'conforme to his assignatioun maid to him thairof be the last abbot[243] and convent of the said place as his gift seillit and subscrivit in the selff at mair lenth beiris.' To sir Henry Olyphant, 'conforme to his assignatioun maid to him as is abonewrittin be the last abbot and convent', 14 b. victual. 'To the reidair yeirlie of Balmerynocht', £10. 'To ane minister at the kirkis of Balmerynocht and Logie',

---

[238] I.e. seven weeks after Easter.

[239] I.e. 15 August.

[240] See above, p. 39, n. 159.

[241] Robert Foster was abbot of Balmerino, 1511 - 1561, J. Campbell, *Balmerino and its Abbey* (Edinburgh, 1899), p. 257.

[242] Alexander Ker, notary at Balmerino, *RMS*, v, no 841.

[243] Robert Foster, the last pre-Reformation abbot of Balmerino, died before 5 February 1561, Cowan and Easson, *Medieval Religious Houses*, p. 73.

£100. 'To ane minister at the kirk of Barry in lyk maner', £100, 'as presentlie is appoyntit to thame be the superintendentis.'

'Memorandum, the haill victuall of Balmerynocht in quheit, beir, meill and aites extendis as efter followis.'
     Wheat, 4 c. Bere, 21 c. 12 b. 3 f. 2 p. Meal, 15 c. 12 b. 2 f. Oats, 30 b. 2 f. 'Summa of the haill victuall', 43 c. 7 b. 3 f.[244] 'Off this victuall abonewrittin the laird of Forret hes' 10 b. meal and 6 b. bere, 'for the quhilk he payis yeirlie' 10 merks silver, 'conforme to ane assedatioun obtenit be him of the last abbot of the said abacie for the space of nyntein yeiris.'

Signature: 'M[r] J. Hay.'[245]

Assignation of Balmerinoch: 'The haill in money', £704 2s 11d; third thereof, £234 14s 3⅔d. Wheat, 4 c.; third thereof, 1 c. 5 b. 1 f. 1⅓ p. Bere, 21 c. 12 b. 3 f. 2 p.; third thereof, 7 c. 4 b. 1 f. ⅔ p. Meal, 15 c. 12 b. 2 f.; third thereof, 5 c. 4 b. 2⅔ p. Oats, 30 b. 2 f.; third thereof, 10 b. 2⅔ p.

'Omittit caynes, customes, capones, pultreis and all small dewteis, togither with the grassumes.'

[*In margin*, 'Ressavit at our soverane ladeis command and Lordis Auditouris of hir Checker.'[246]]

> 'Be it kend till all men be thir presentes, me, Mr Johnne Hay, commendatar of Balmerynocht, now being astricted to attend and await upoun the quenis grace daylie service in court, in consideratioun quhairof may nocht gui[d]lie be absent thairfra to remane at the said place of Balmerynocht for introme tting and uptaking of the profeitis and dewteis pertenyng thairto be my selff, and being addebtit and obleist to my weilbelovit cousing Johnne Kynneir of that Ilk, quha hes bestowit large sowmes of money besyd his travellis and labouris for outsetting of my effaires to the greit weill of me, utilitie and profeit of the said place, thairfoir and for divers utheris ressonable caussis and consideratiounes moving me thairto to have set and for maill lattin and be the tenour heirof settis and for maill lattis to my weilbelovit cousing Johnne Kynneir of that Ilk, his aires and assignaris, all and haill the fructes, rentis, proventis, teyndis, fischeingis and utheris dewteis quhatsumevir pertenyng or that richteouslie aucht and sould pertein to my said abacie, for all the dayis space and termes of fyve yeris nixt and immediatlie following thair entreis in and to the samin, quhilk entres wes and began at Martingmes[247]

---

[244] The correct calculation appears to be 43 c. 7 b. 3 f. 2 p.
[245] I.e. John Hay, see above, p. 19, n. 70.
[246] This refers to the following letter.
[247] I.e. 11 November.

lastbypast in this instant yeir of God' 1564, 'quhilk entres wes the ische
and outrynnyng of the tak and assedatioun of Mr James Balfour, persoun
of Flisk, and Andro Balfour of Montquhanny his father, thair aires and
assignaris, had of the fructes and dewteis of the said abacie, obtenit be
thame of umquhill Robert, last abbet of the samin, of the dait at the said
abay' 4 August 1559 'and thairefter to induir and to be peciablie bruikit,
joysit, intromettit with be thame ay and[248] quhill the said fyve yeiris be
full and togither compleitlie outrun with all and sindrie commoditeis,
fredomes, eismentis and richteus pertinentis quhatsumevir pertenyng or
that richteuslie aucht and sould pertein to my said abacie frelie, quyetlie,
weile and in peax but ony revocatioun, obstacle, impediment or agane
calling quhatsumevir and als makis and constituites the said Johnne
Kynneir of that Ilk, his aires and assignaris, halderis and keiparis of my
said place, yairdis and houssis of Balmerynocht withe full powar to him
and thame to intromet, occupy, use and joise the samin at thair plesure be
thame selffis induring the said space, reservand nevirtheles to me and my
successouris ane sufficient hall, chalmer, kicheing, stable, yaird with
utheris commoditeis at my and my successouris cuming and remanyng
thairinto at all tymes as I or thay sall think gud, payand thairfoir yeirlie the
said Johnne Kynneir of that Ilk, his aires and assignaris foirsaidis, to me
and my successouris, assignaris and factouris, according to the payment
contenit in the foirsaid tak and assedatioun obtenit be the said Mr James
Balfour and his foirsaidis, of the fructes of the said abacie quhilk is nyne
hundreth merkis usuall money of Scotland at tua termes in the yeir, viz.,
Witsonday and Mertinmes in winter, be equall portiounes allanerlie
begy[n]and the first termes payment thairof at the feist of Witsonday
nixtocum, and sua furth yeirlie and termlie at the saidis termes induring
the space foirsaid of fyve yeiris, and als payand yeirlie to six of the brether
now duelland presentlie in the said place in yeirlie portioun to thair
sustentatioun, conforme to the contract and appoyntment maid betuix me
and thame registrat in the register buik of the said abacie, and siclyk to
uphauld the houssis, yairdis and dykis thairof in als guid state as the samin
ar in presentlie, provyding alwayis that incaice I happin to withdraw my
selff of court for sict occasiounes as may occur and of mynd to remane at
my said place or quhair it sall pleis me best and than may the better attend
to have the oversicht and halding of my rent and dewteis pertenyng to my
said abbay, in that caice this present tak and assedatioun to be of nane
availl, force nor effect with all that followit or may follow thairupoun, and
I to be fre and have full powar to intromet and dispone upoun the fructes
and dewteis of my said abacie, siclyk and in the samin maner as giff this
present tak and assedatioun had nevir bene maid, I advertesand the said
laird of Kynneir, his aires and assignaris, fourty dayis befoir ane terme of

---

[248] In SRO, Vol. b the rental finishes here at the end of fo 61v, and the next entry, on fo 62r, corresponds to SRO, Vol. a, fo 104r (see below, p. 91).

Witsonday and I foirfurthe and my successouris bindis and oblissis me and thame to warrand, acquyet and defend this my present tak and assedatioun to the said Johnne Kynneir, his aires and assignaris, induring the yeiris foirsaidis in all and be all thingis as is abone expremit aganes all deidlie as law will but fraud or gyll, in witnes of the quhilk thing to this my present tak and asedatioun subscrivit with my hand, my signet is affixit at Edinbrucht', 20 November 1564 'befoir thir witnessis, Johnne Kynneir, George Reid, Robert Ramsay, David Wat and Thomas Rolland, with utheris divers.

Collatum cum suo originali, tenet Nicolsoun.'[249]

At Edinburgh, January 1565/6

'Memorandum, to deduce of the charge of the abay of Balmerinocht'

Wheat, 1 c. 5 b. 1 f. $\frac{1}{3}$ p. Bere, 7 c. 4 b. 1 f. $\frac{2}{3}$ p. Meal, 5 c. 4 b. $2\frac{2}{3}$ p. Oats, 10 b. $2\frac{2}{3}$ p.[250]

'For the quhilk victuall thair is to be eiked in the silver' £31 19s 3d 'and this alteratioun in respect of the abbay of Balmerynocht wes set for' 900 merks 'and sustenand the convent be the commendatouris predicessouris and now is set agane and confirmit be our soveranes lord and lady conforme to the sowme abonewrittin.'

Calculation of the third of Balmerynocht 'as it wes gevin up in charge be Pittarro, Comptroller.'[251]

Third of the money, £234 14s $3\frac{2}{3}$d. Third of the wheat, 1 c. 5 b. 1 f. $1\frac{1}{3}$ p. Third of the bere, 7 c. 4 b. 1 f. $\frac{2}{3}$ p. Third of the meal, 5 c. 4 b. $2\frac{2}{3}$ p. Third of the oats, 10 b. $2\frac{2}{3}$ p.

'Summa of the haill abbay and convent and all of the thrid thairof is' 400 merks.

Signature: 'M[r] J. Hay. James Nicolsoun, ressave this rentale. Clericus Registri.'

---

[249] This appears to be an instruction to the clerk to keep this copy which has been collated with the original.
[250] These deductions equal the thirds which follow.
[251] Sir John Wishart of Pittarow, Comptroller, 1562 - 1565, *Handbook of British Chronology*, p. 191.

INCHCOLM, ABBEY OF,[252] (SRO, Vol. a, fos 73r-74v)

Rental of Sanct Colmis Inche

The teind sheaves of Leslie set for £100. The teind sheaves of Rassyth set to the laird of that Ilk[253] for £34. The teind sheaves of Logy and Orquhat set to the laird of Dowhill[254] for £6. The teind sheaves of the kirks of Abirdour, Dalgathe and the chapel of Bayth set to John Stewart for £80. The lands of Bouipre and Inchebardy set in feu to John Burne for £14 yearly. The Quhytill yearly, £12. The Newtoun, £6. Croftgary and Brego, £10. Inchekery, £7 5s 4d. The Mylne of Abirdour, £6 6s 8d. The Cutilhill, £4. The kirk glebe of Rossyth, £8. The kirk glebe of Auchtertule, £3. The Cakinche, 20s. Pascar Mylne, £4. Bacclero, £5. Prinles, £12. The Colferie, 33s. Duddingstoun, 43s 4d. The Mylne of Crawmond, 53s 4d. The Hiltoun of the Coilheuche, £6. Eister Bothylokis, £4. The Schelis, £4. The Nethertoun, £3 6s 8d. The Mylne of Lossedy, 40s. The Murntoun, 26s 8d. The Kirktoun, 53s 4d. Craigbayth, £6. Eister and Westir Bochlaweis [*i.e.* Bucklyvie], £25. Clairbastoun, 40s. The annuals of Abirdour, £6. The Grange, £18. Donibirsall, £15. Killore, £13.

'Memorandum, thir landis ar all set in few and confeirmit.'

'Item, the kirk of Auchtertule and Dolour assignit in pensioun to Johnne Steill[255] befoir my interes, quhat thay extendis to I cannot tell.'

Calculation of third:
    'This rentall is just extending in the haill to' £427 8s 4d; 'of the quhilk the thrid extendis to' £142 9s 4d '1 fardein, halff ane fardein' [*i.e.* ⅓d; *recte*, ⅓d].

'Off the quhilk thrid your lordschippis man dedus to me the pensioun gevin to Johnne Abircrumbie, brother germane to my predicessour,[256] quha wes provydit of the samin afoir my entres and assignit to the mailles of Bayth for the payment of fourtie merkis yerlie, and is yit in possessioun of the samin, ane aigit and decraipit man of' four score 'yeiris, attour the pensioun gevin to Williame Cambroun of' £13 'yerlie, quha wes servand to my said predicessour on his enteres to the benifice, and is and hes bene afoir and efter my enteres in peicable possessioun of the samin, ane auld decraipit man, baith lyand in ane hous, sua this beand deducit the thrid extendis to' £102 16s, 'and as for the tua puir men your lordschippis may gar visie thame giff ony of thame be les nor' 76 'yeiris of aige,

---

[252] Cf *Charters of the Abbey of Inchcolm*, eds. D. E. Easson and A. Macdonald (Edinburgh, 1938), pp. 217-229.
[253] Robert Stewart of Rosyth.
[254] James Lindsay of Dowhill was bailie of the abbey in 1570, *Charters of Inchcolm*, p. 215.
[255] See below, p. 63, n. 257.
[256] Richard Abercrombie was abbot of Inchcolm, 1532 - 1549, *Charters of Inchcolm*, pp. xxxvi, 241-242.

and as for the uther tua kirkis afoir namit of Auchtertwill and Dolour, Johnne Steill, sone to George Steill, wes provydit in Rowme[257] of the samin or I wes provydit of benifice.'

Signature: 'James Steuarte.'[258]

'The teyndis of the parochinis of Abirdour and Dalgatie'

[Aberdour:] Cowcarny, 16 b. meal; 10 b. bere; 'and furletis of quheit'. Otterstoun, 6 b. meal; 3 b. bere; 3 f. wheat. Coustoun, 12 b. meal; 9 b. bere; 2 b. wheat. Bowpre, 10 b. meal; 3 b. bere; 1 b. wheat. Quhythill and Hesyd,[259] 10 b. meal; 3 b. bere; 1 b. wheat. The Newtoun 'led estimat to' 10 b. meal; 3 b. bere; 1 b. wheat. Crogary and Brego, 6 b. meal; 4½ b. bere; 3 b. corn. Balmullis, Balmyll Mylne, Tempillall, Motcay, 19 b. meal; 6½ b. bere. Culilow, 12 b. meal; 3 b. bere. Bakran, 11 b. meal; 4 b. bere; 2 b. wheat. Humbie, Cannyhillis [Calyhillis, *fo 74v*] and Deachy, 60 b. corn; 14 b. meal; 6 b. bere. Maines of Abirdour 'led estimat to' 3 c. bere, meal and wheat. 'The myln land, sisteris land [*i.e.* land of the nuns at Aberdour] and burrowland estimat to' 1 c. victual, bere, meal and wheat.

Dalgaty: Lowchat, 17 b. corn; 6 b. bere; 12 b. wheat. Dalgaty, 20 b. corn; 3 b. bere; 2 b. wheat. Dunnybirsall pays of ferme 8 c. meal, bere and wheat. Barnehill and Grange pay of ferme, 16 c. meal, bere and wheat. The barony of Fordell, namely, Lethame, Cluikhill 'with the rest of the tounes payis of teynd', 44 b. meal; 18 b. bere; 6 b. wheat; 8 b. corn. Cuthilhill, Downingis 'and the mylne land and the akeris and burrow ruidis on the west syd of the burne of Abirdour estimat to' 24 b. victual, meal, bere and wheat.

'Summa by the mans of teyndis': 2 c. wheat; 7 c. bere; 13 c. meal; 6 c. 12 b. oats [*i.e.* corn].[260]

'Schir, p[l]esyth your grace, wit anentis your mandat to inquyre the estimatioun or quantitie of the teyndis of Dynybirsall, Grange and Barnhill communibus annis thai wilbe worthe' 3 c. 12 b. wheat, bere and meal; 'to wit' Dynbirsall, 20 b.; Barnehill and Grange, 40 b.; 'this is the estimatioun that I may have be inquisitioun and as to the inquisitioun of the kirk or chapell quhat thei micht give, thair is bot tua kirkis your grace desyres the

---

[257] In 1530 James V had petitioned the pope in favour of John Steel, aged six (son of the king's familiar, George Steel) and in 1532 had requested the reservation of a yearly pension of 110 merks for Steel from the churches of Auchtertool and Dollar. *Charters of Inchcolm*, p. 203; cf *RSS*, ii, no 4131; vi, no 1973.

[258] James Stewart of Doune, son of the laird of Beath and titular abbot since 1544, assumed the commendatorship on the death of abbot Richard Abercrombie in 1549, *Charters of Inchcolm*, pp. 242-243.

[259] 'Sesyd' (i.e. Seaside), *Charters of Inchcolm*, p. 219.

[260] These totals are difficult to verify because some of the figures represent a combination of assorted victuals, and the respective amounts of wheat, bere, oats and meal are not specified.

rentall of, to wit, Abirdour and Dalgaty quhairof he is persoun of bayth[261] and I gave your grace the rentall of the haill, and as [to] the vicarege of Dalgady thair is ane vicar levand and as to Abirdour, my lord hes woll and lamb[262] quhilk wilbe worth' £20 yearly, 'and the redar hes the rest quhilk will nocht pas' £10 yearly 'and as to chapellis I knaw nane in the tua parochines except ane that gives na profeit to him.[263] This efter my maist humble commendatioun of service unto your guid grace, commitis your grace to the richtfull protectioun of God, at Abirdour', 22 May 1574.

Manis of Abirdour, 52 b. oats; 1 c. bere; 14 b. meal; 5 b. wheat. Barnhill and Grange 'extendit to' 4 c. namely, $\frac{1}{2}$ c. wheat; 1 c. bere; $2\frac{1}{2}$ c. meal. Dynibirssall, 2 c. 4 b. wheat; 8 b. bere; 20 b. meal. The kirk of Bayth teinds: the Eister Toun, William Colyeir and Robert Fillane, pay £3; the Hiltoun, James Bawe[r]age, £3; the Cheles, 'four men occupeis', £3; the Nathertoun of Bathedloskis,[264] £3; the Holl Myln pays 8s; Knoksudrattoun [i.e. Southerton] and Leuchquhattis Bayth pay £5; Craigbayth pays £3; the Kirktoun and West Mylne occupied by Robert Muirtoun, pays £3 'be estimatioun'; Kelleuch extends to 10s money. Dachy, Humbie and Calyhillis [Cannyhillis, fo 74r] in my lord Regent's hand, 'thir payis of teynd in the yeir', 1 c. bere; 5 b. wheat; 14 b. meal; 3 c. oats.

ST ANDREWS, ST MARY'S COLLEGE IN, (SRO, Vol. a, fos 75r-76r)

'The New College'

'The rentale of the New College, als caled the Pedagogie, foundit within the Universitie of Sanctandrois, yeirlie in fermes, teyndis, alteragis and victuallis, gevin in be Mr Johnne Douglas, principall maister of the samin, as followis.'

'The fermes payit in money
    The fundatioun of ane bursare student in the said college maid be Maister Henry Quhyt of the landis of Carnurrie occupyit be Elizabeth Creichtoun, lady of Ardros', £13 6s 8d.

Kirks
    The parsonage of Inchebryok, £206 13s 4d. The parsonage and vicarage of Conveth, £240 6s 8d, 'off this is gevin to him that wes curet and now redar', £12. The kirk of Tennedais, 'occupyit be Mr Henry Lummisdale, usufructuer

---

[261] The parsonages of Aberdour and Dalgety were appropriated to Inchcolm abbey, I. B. Cowan, *The Parishes of Medieval Scotland* (Edinburgh, 1967), pp. 2, 43.

[262] Cf 'My lord Sanctcolmisinche hes lambwoll and teynd fische of the haill parochin', fo 100r (p. 85, below). James Stewart of Doune was commendator of Inchcolm, see above, p. 63, n. 258.

[263] Beath was a chapel of Dalgety, Cowan, *Parishes*, p. 15.

[264] 'Bothedlach', *Charters of Inchcolm*, p. 249; cf 'Bothedillach' (in the parish of Ballingry), W. Ross, *Aberdour and Inchcolme* (Edinburgh, 1885), p. 119.

thairof, and payis to the college yeirlie quhilk hes the title thairof', £66 13s 4d. The kirkland of Tynninghame set in feu to Patrick Hepburne, £12 13s 4d, 'the quhilk sowme sir Thomas Manderstoun quha wes curate thairof and now redar ressavis for his fie.'

'Fundatioun of alteragis to the said college:
 Sanct Johnne the Evangeliste altare within the abbay kirk of Sanctandrois haldin violentlie fra the college be the archedene of Sanctandrois, scilicet [*i.e.* namely] M[r] Williame Murray',[265] £16. Sanct Johnne the Baptiste altar within the said abbey, held by the said archdeacon and Mr Thomas Methven,[266] £13 6s 8d. 'Sanct Anthoneis altare within the paroche kirk of Sanctandrois, of the quhilk alterage the tennentis [*sic*] ar reuenous, decayit and fallin doun', 53s 4d. 'Sanct Johnne the Evangeliste altare foundit within the said college, of the quhilk the tennentis [*sic*] ar decayit and fallin doun', £3 6s 8d. Total money, £575.[267]

Victual paid in teinds: The parsonage and vicarage of the kirk of Tunnynghame 3 c. 8 b. wheat; 6 c. bere; 5 c. oats [5 c. 10 b., *fo 76r*]. Total, 14 c. 8 b.

Signature: 'Maister Johnne Douglas, principall maister of the college.'

Assignation of the third of the New College of Sanctandrois: Money, £575; third thereof, £191 13s 4d. Wheat, 3 c. 8 b.; third thereof, 1 c. 2 b. 2 f. 2⅔ p. Bere, 6 c.; third thereof, 2 c. Oats, 5 c. 10 b. [5 c., *fo 75v*]; third thereof, 1 c. 2 f. 2⅔ p.[268]

CRAIL COLLEGIATE KIRK, PREBENDS OF,[269] (SRO, Vol. a, fos 76v-77r)

'The rentall of the prebendareis of the college kirk of Craill gevin in at Hadingtoun the first day of Apryll 1574.'

'Inprimis, Johnne Broun, prebendar of Our Lady altare in the new ile, hes of propertie pertenyng to his prebendarie the sowme of' 20 merks 'quhairof he hes ane annuell of the landis of Cambo of ten bollis beir with the cheritie.' Mr John Buthile, prebendary of Our Lady altar within the said aisle, has of property pertaining to his prebend, 20 merks. John Mortounes prebend of Sanct Michaellis altar has of property pertaining to his prebend, 20 merks. George King, another prebendary of the same altar, has of property pertaining to his prebend, 20 merks.

---

[265] He is not listed as archdeacon of St Andrews in Watt, *Fasti*; a person of the same name was treasurer of Dunblane, 1534 - 1563 x 1567, Watt, *Fasti*, pp. 86-87.
[266] Thomas Methven was prebendary of Kingask and Kinglassie, pertaining to the collegiate kirk of St Mary on the Rock, St Andrews, otherwise known as Kirkheuch; see below, p. 77.
[267] This figure is correct when the £12 paid to the reader is deducted.
[268] The correct calculation appears to be 1 c. 14 b.
[269] Cf SRO, Vol. a, fo 98v (see below, p. 82).

John Davidsone, prebendary of Sanct James altar, has of property pertaining to his prebend, 20 merks. Thomas Kynneir, prebendary of Sanct Nicolace altar, has of property pertaining to his prebend, 20 merks. John Hereis, 'maister of the gramer schole and prebendar of the prebendreis of Sanct Johnne the Baptiste and the Halyruid service, hes of propertie pertenyng to ilk ane of the saidis prebendareis, the sowme of' 20 merks. William Bousie, prebendary of Sanct Johnne the [Baptiste, *deleted*] Evangelist prebend, has of property pertaining to the same, 20 merks. Mr David Myretoun, prebendary of Sanct Cathereinis altar, has of property pertaining to his prebend, 20 merks. William Corstorphin, prebendary of 'Our Ladie servant at the hie altar', has of property pertaining to his prebend, 20 merks.

'Item, the saidis prebendareis hes the annuellis rentis underwrittin pertenyng to thame as propper commonis to the saidis prebendareis to be equallie distributit amangis thame': An annualrent of the lands of Cambo extending to the sum of 53 merks. Of the lands of Balcomy, an annualrent of 20 merks. An annualrent of the lands of Ardros, 16 merks. Of the lands of Donyface, an annualrent of 16 merks. Of the lands of Pitmillie, an annualrent of 51 merks. 'Extending to ilk prebendarie of the said college to the sowme of' 29 merks 5s, 'by the prebendarie of Sanct Nicolas altar possessit be the minister of Darreill [*recte*, Crail] and the prebendaries of Sanct Johnne the Baptist and the Haliruidis altar quhilk extendis to the sowme of' 32 merks 'quhilk ane of the saidis prebendareis pertenyng to the said minister and scuilmaister.'

ST ANDREWS, ARCHDEACONRY OF,   (SRO, Vol. a, fo 78r-v[270])

Rental of the archdeaconry of Sanctandrois

'Item, the haill archdeanrie set to Mr Richard Schoriswod quha hes had the samin assedatioun be the space of xxij yeiris bygane for yeirlie payment of the sowme of' £600. 'Item, tua paroche kirkis within Angus and Mernis, viz., Rescoby and Kynneiff [*blank*]. Item, of temporall landis within Fyffe, Scatirrun [*i.e.* Strathtyrum] and Wilkiestoun [*blank*].'

'Heirof to be deducit': In pension to sir William Austeand, £17 6s 8d. To sir Alexander Galloway in pension, £17 6s 8d. 'To be deducit in annualrent of Sanctandrois', 32s. 'To be deducit in procurationis' £35 7s 4d. 'To be deducit of the vicarege of Rescoby', £53 6s 8d. 'To be deducit of the vicarege of Kynneiff', £72 13s 4d. 'The jurisdictioun of the denereis within the parochinis of Rescoby and Kynneiff extending yeirlie to the valour of' £30.

---

[270] SRO, Vol. a, fo 77v is blank except for a deleted entry: 'The assignatioun of the thrid of the archidenrie of Sanctandrois. In the haill', £600; third thereof, £200.

'The quhilkis annuellis, vicaregis, procurationis and jurisdictioun of denry hes nocht bene payit in the said parochinis of Rescobie and Kynneiff within Angus and Mernis be the space of vij yeris bygane extending yeirlie to the sowme of' £227 12s 8d, 'quhilk sould be deducit to the archidein of this rentale or ellis giff him ane sufficient ordour to get payit thairof and sua the said taxman payit to the archideane allanerlie the sowme of' £360 0s 4d.[271]

Signature: 'Robert Pitcarne.'[272]

Assignation of the third of the archdeaconry of Sanctandrois: In the whole, £600; third thereof, £200.

KINGLASSIE, VICARAGE OF, (SRO, Vol. a, fo 79r)

Vicarage of Kynglassie pertaining to sir Adame Kingorne.

Calculation of third: In the whole, £25 16s; third thereof, £8 12s 8d.[273]

KIRKCALDY, VICARAGE OF, (SRO, Vol. a, fo 79r)

Vicarage of Kirkaldie pertaining to Mr James Multray.

Calculation of third: In the whole, £120; third thereof, £40.

LAUDER, PARSONAGE OF,[274] (SRO, Vol. a, fo 79r)

[*In margin*, 'Note, this lyis in Beruick.']

Calculation of third: In the whole, £133 6s 8d; third thereof, £44 8s 10⅔d.[275]

---

[271] £227 12s 8d plus £360 0s 4d equals £587 13s, not £600.
[272] Robert Pitcairn was archdeacon of St Andrews, 1539 - 1584. His predecessor, George Dury, was also a claimant, Watt, *Fasti*, pp. 308-309.
[273] The correct calculation appears to be £8 12s.
[274] Cf SRO, Vol. a, fo 196v (see below, p. 194). This benefice in the sheriffdom of Berwick was in St Andrews diocese, which may explain its inclusion in rentals for Fife.
[275] This fraction is rendered 'ob 3 ob'.

ABDIE, VICARAGE PORTIONARY OF, (SRO, Vol. a, fo 79r)

Vicarage portionary of Ebdy pertaining to William Symmer.

Calculation of third: In the whole, 53s 4d; third thereof, 17s 9⅓d.

NEWBATTLE, (SRO, Vol. a, fo 79r)
CHAPLAINRY OF ST CATHERINE IN,[276]

Chaplainry of Sanct Katherynne in Newbotle

Calculation of third: In the whole, £16; third thereof, £5 6s 8d.

DUNFERMLINE, VICARAGE OF,[277] (SRO, Vol. a, fo 80r[278])

Vicarage of Dunfermeling [*In margin*, 'alitter rentaleit heireftire.']

Calculation of third: In the whole, 20 merks; third thereof, £4 13s 4d.[279]

DUNBOG, VICARAGE OF, (SRO, Vol. a, fo 80r)

Vicarage of Dumbug

Calculation of third: In the whole, £13 6s 8d; third thereof, £4 8s 10⅔d.

INVERKEITHING, VICARAGE OF, (SRO, Vol. a, fo 80r)

Vicarage of Innerkeithing

Calculation of third: In the whole, £100; third thereof, £33 6s 8d.

---

[276] This benefice, in the sheriffdom of Edinburgh, was in St Andrews diocese, which may explain its inclusion in rentals for Fife.
[277] Cf SRO, Vol. a, fo 84r (see below, p. 72).
[278] SRO, Vol. a, fo 79v is blank.
[279] The correct calculation appears to be £4 8s 10⅔d.

CARNBEE, VICARAGE OF, (SRO, Vol. a, fo 80r)

Vicarage of Carnbie

Calculation of third: In the whole, £42; third thereof, £14.

NEWBURN, VICARAGE OF, (SRO, Vol. a, fo 80r)

Rental of the vicarage of Newbirne given in by David Baxter, vicar thereof, 5 August 1574.

'The haill vicerege of Newbirne extending to' 20 merks; third thereof, £4 8s $10\frac{2}{3}$d.

Signature: 'David Baxter, redar at Newbyrn.'

ABERCROMBIE, VICARAGE OF, (SRO, Vol. a, fo 80v)

'The just rentale of the vicarege of Abircrummie in Fife within the diocie of Sanctandrois, gevin up be Mr Thomas Young, vicar thairof.'

'The haill fructes, rentis and emolumentis of the said vicarege of Abercrommy extendis to four akeris of land and' £3 silver.

Signature: 'Mr Thomas Yong, vicar of Abircrommy, with my hand.'

STRATHMIGLO, BENEFICE OF, (SRO, Vol. a, fo 81r-v)

Strameglo

Rental of Strathmiglo parish

[*In margin*, 'Nota, this is a kirk of the bischoprik of Dunkeld and nocht tane up in the thrid.']

Eister Casche pays 48 b. meal. Wester Casche pays 8 b. meal. Corstoune pays 12 b. meal. Drumdreel pays 12 b. meal. Ovir Urquhard pays 20 b. meal. Nather Urquhard pays 20 b. 2 f. meal. Loppe Urquhart and Lawsonestoun, 10 b. meal. Gospert pays 8 b. meal. Bonnat pays 14 b. meal. Belcanquell pays 28 b. meal. Pitlochy pays 30 b. meal. Bannate Myll pays 6 b. meal. Craigfud pays 24 b. meal. Ladyn Urquhart pays 7 b. [meal]. Pitgornow pays 40 b. meal. Kyncragy pays 18 b. meal; 8 b. bere; 2 b. wheat. Freirmyll pays 7 b. meal; 2 b. bere; 2 b. wheat. Eister Pitlour pays 10 b. meal; 2 b. bere; 2 b. wheat. Stedinurland pays

7 b. meal. Stramiglo pays 16 b. meal; 16 b. bere. Northmure pays of silver, 40s. Total meal, 17 score and 5 b. 2 f., 'that is', 21 c. 9 b. 2 f. Total bere, 28 b. Total wheat, 6 b. Total of the bere, wheat and meal, 23 c. 11 b. 2 f. and 40s of silver.

Signature: 'Mr Williame Scott.'

Assignation of the third of Strathmiglo: Meal, 21 c. 9 b. 2 f.; third thereof, 7 c. 3 b. 2⅔ p. Bere, 28 b.; third thereof, 9 b. 1 f. 1⅓ p. Wheat, 6 b.; third thereof, 2 b. Money, 40s; third thereof, 13s 4d.

DYSART, PARSONAGE AND VICARAGE OF, (SRO, Vol. a, fo 83r-v[280])

Rental of the parsonage and vicarage of Dysert pertaining to Robert Danielstoun, parson thereof 'and now being in his awin handis.'

'The personage and vicarege thairof wes in all tymes set be my predecessouris for the sowme of thrie hundreth merkis money allanerlie. Item, becaus the samin wes ovre deir and the fermararis could nocht pay the samin it is cassin in the persounes handis and now the rentale thairof extendis as followis': The 'lard [of] Stathvey', 14 b. meal; 2 b. bere. Strathoire, 14 b. meal; 4 b. bere. 'Gibbis wyff thair', 9 b. meal; 3 b. bere. Babey, 12 b. meal; 2 b. bere. Touch, 12 b. meal. Orres Myllis, 2 b. meal. 'The burrow rudis about the toun being gadderit in totalibus annis extendis to' 6 c. victual, 'deduceand the expenssis maid thairof furth of the samin in collecting and gaddering thairof', £20. 'Item, because of sic inobedience as hes bene thir divers yeris bigane quhairin I have nocht bene of the vicarege answerit I have no just rentale that can be gevin thairof, bot quhen the vicarege wes dewlie answerit in all thingis it extendit to' 40 merks 'or thairby.'

Signature: 'R. Danyelston, mea manu.'

Calculation of third of the parsonage of Dysart: In the whole, 3 c. 15 b.; third thereof, 1 c. 5 b. 'Victuall undevidit in the haill', 6 c.; third thereof, 2 c.[281]

Calculation of third of the vicarage of Dysart: In the whole, £26 13s 4d; third thereof, £8 17s 9⅓d.

---

[280] SRO, Vol. a, fo 82r-v is blank.
[281] The bere has been omitted from this total.

DYSART, VICARAGE PENSIONARY OF,[282] (SRO, Vol. a, fo 83v)

Rental of the vicarage pensionary of Dysart pertaining to sir George Strauthauchin, vicar thereof, extends yearly to 41 merks, 'to be payit at four termes equallie with ane mans at the kirk'.

Signature: 'G. Strauchin. Clericus Registri.[283] John Wyischert, Comptroller. James, ressave cautioun for the thrid to caus him be obeyit of lettres.'

ST ANDREWS, (SRO, Vol. a, fo 83v)
CHAPLAINRY OF THE TRINITY IN,[284]

Chaplainry of Trinitie in Sanctandrois

Calculation of third: In the whole, £10; third thereof, £3 6s 8d.

'Seik Crawfurd Lyndesay in Lanark.'[285]

SCOONIE, VICARAGE OF, (SRO, Vol. a, fo 83v)

Vicarage of Scunye

Calculation of third: In the whole, £53 6s 8d; third thereof, £17 15s $6\frac{2}{3}$d.

'Seik the dewrie [*recte*, deanery] of Glasgow in Lanrak' [*sic*].[286]

ST ANDREWS, (SRO, Vol. a, fo 83v)
DURA AND RUMGALLY, PREBEND OF ST MARY ON THE ROCK COLLEGIATE KIRK

Prebend of Dury and Rumgallie

---

[282] Cf SRO, Vol. a, fo 190r (see below, p. 188).

[283] James McGill of Nether Rankeilour was Clerk Register at the Reformation, *Handbook of British Chronology*, p. 197.

[284] This probably denotes the chaplainry of the lesser altar of the Blessed Virgin in Holy Trinity parish church, founded by Duncan Yellowley, vicar of Crawford Lindsay, in 1478. W. E. K. Rankin, *The Parish Church of the Holy Trinity, St Andrews* (Edinburgh, 1955), p. 59; *Liber Cartarum Sancte Crucis. Munimenta Ecclesie Sancte Crucis de Edwinesburg*, ed. C. Innes (Edinburgh, 1840), p. 148.

[285] The rental for the vicarage of Crawford Lindsay appears among entries for Lanark, in EUL, fo 17v (see below, p. 512).

[286] James Balfour, vicar of Scoonie, was dean of Glasgow from 1561, Watt, *Fasti*, p. 156; cf EUL, fo 12r (see below, p. 504).

Calculation of third: In the whole, £20; third thereof, £6 13s 4d.

CLEISH, PARSONAGE OF, DUNFERMLINE, VICARAGE OF,[287]
(SRO, Vol. a, fo 84r)

Parsonage of Cleische and vicarage of Dunfermelyng pertaining to William Lummisden, sacristan of Dunfermeling.

Parsonage of Cleische
    'Inprimis, the personage of Cleische set in assedatioun to the laird of Cleische, payand thairfoir yeirlie', £53 6s 8d.

Calculation of third: In the whole, £53 6s 8d; third thereof, £17 15s 6⅔d.

Vicarage of Dunfermelyng
    'The vicarage of Dunfermeling nocht set in assedatioun extending be estimatioun to' £40 'yeirlie and skant sameikill.'

Calculation of third: In the whole, £40; third thereof, £13 6s 8d.

EASSIE, PARSONAGE AND VICARAGE OF,[288]
(SRO, Vol. a, fo 84v)

Parsonage and vicarage of Essy

[*In margin*, 'Nota this lyis in Angus.']

Calculation of third: In the whole, £145 15s 9d; third thereof, £48 11s 11d.

AUCHTERDERRAN, PARSONAGE OF,
(SRO, Vol. a, fo 85r)

Parsonage of Ochtirdery

'Item, the samin set in assedatioun to the laird of Balmowto be my lord of Dunfermeling of auld for the sowme of' £120.

---

[287] Cf SRO, Vol. a, fo 80r (see above, p. 68).

[288] Cf NLS, fo 112r (see below, p. 411). This benefice, in Forfar, was in St Andrews diocese, which may explain its inclusion in rentals for Fife.

Signature: 'Alane Couttis, in the name of my lord of Dunfermeling.'[289]

Assignation of the third: In the whole, £120; third thereof, £40.

TORRIE, PARSONAGE AND VICARAGE OF, (SRO, Vol. a, fo 86r[290])

The rental thereof is 50 merks set in assedation, and of that Mr David Gairlie has 12 merks in pension and the minister has 18 merks 'and the persoun bot' 20 merks.

Signature: 'Maister Edward Bruce, factar.'[291]

Assignation of the third: In the whole, £33 6s 8d; third thereof, £11 2s 2⅔d.

INVERKEITHING, ST JOHN'S ALTAR IN, (SRO, Vol. a, fo 86v)

'Sanct Johnnes altare in Innerkeithing'

Calculation of third: In the whole, 17 merks 11s; third thereof, 'xv' [recte, £3 19s 2⅔d].

TARVIT, PARSONAGE AND VICARAGE OF,[292] (SRO, Vol. a, fo 86v)

The parsonage and vicarage of Tervat kirk

'In the haill' 60 'of merkis' and 15 merks, pertaining to sir John Eclynsoun [Achesoun, fo 102v]; third thereof, 25 merks, 'ressavit at the Comptrolleris[293] hand.'

---

[289] Allan Coutts, chamberlain of Dunfermline abbey; see above, fo 30r (p. 23). At the Reformation, George Dury, commendator, left for France on 29 January 1561, and Robert Pitcairn exercised authority as commendator designate, Cowan and Easson, *Medieval Religious Houses*, p. 58.

[290] SRO, Vol. a, fo 85v is blank.

[291] Edward Bruce held the parsonage, *Thirds of Benefices*, p. 247; *RSS*, vi, no 1025.

[292] Cf SRO, Vol. a, fo 102v (see below, p. 88).

[293] Bartholomew de Villemore, Comptroller, 1555 - 1562, with Thomas Grahame of Boquhaple, 1561 - 1562. They were succeeded by Sir John Wishart of Pittarow, 1562 - 1565. *Handbook of British Chronology*, p. 191.

PERTH, (SRO, Vol. a, fo 87r)
VICARAGE PENSIONARY OF ST JOHN'S KIRK IN,[294]

'Rentale pro Magistro Georgio Cok, vicario ecclesiarum infra scriptarum.'

Rental on behalf of Mr George Cok,[295] vicar of the kirk within written.

Vicarage pensionary of the kirk of St John in the burgh of Perth, £20 money Scots, payable annually to him by [blank] Ros of Cragy[296] of the fermes of the said kirk, of which foresaid pension he received nothing at all of the said ferme for the past term, and that with a quarter part of the year already gone, by reason of the tumult of the said year.

'In the haill', £20.

MONIMAIL, VICARAGE PERPETUAL OF, (SRO, Vol. a, fo 87r[297])

'Vicaria ecclesie de Monymeill'

Vicarage perpetual of the parish kirk of Monymele extending yearly to the sum of 40 merks as it was last set, of which he has received nothing for the term gone by and earlier, by reason of the foresaid tumult.

Signature: 'Ita est M[r] G. Cok, vicar of super [sic], manu propria.'

Calculation of third: In the whole, £26 13s 4d; third thereof, £8 17s $9\frac{1}{3}$d.

ANSTRUTHER, VICARAGE PENSIONARY OF, (SRO, Vol. a, fo 88r)

Vicarage of Anstruther

The vicarege pensionary of Anstruther in the whole, £20, 'with the tua akeris toft and the pasturage of a hors, xij kyis gers.'

Calculation of third: In the whole, £20; third thereof, £6 13s 4d, 'withe the thrid of tua akeris toft and the thrid of the pasturing of ane hors and thrid of xij kyis gers.'

---

[294] This vicarage pensionary of the church of St John the Baptist in Perth, was appropriated to Dunfermline abbey, which may explain its inclusion among the rentals for Fife.
[295] Mr George Cook was appointed vicar pensionary of Perth in June 1535 and retained the office in 1566, *Registrum de Dunfermelyn*, no 525; *RMS*, iv, no 1753.
[296] I.e. John Ross of Craigie, cf *RSS*, v, nos 3-4.
[297] SRO, Vol. a, fo 87v is blank.

FIFE

KILRENNY, VICARAGE OF, (SRO, Vol. a, fo 88r)

'The haill vicarege of Kylrynnie', 40 merks.

Signature: 'Joannes Formane, manu propria.'

Calculation of third: In the whole, £26 13s 4d; third thereof £8 17s 7¼d.[298]

[*In margin*, 'Memorandum, this vicarege is worth' £100 'and leid.']

LATHRISK, VICARAGE OF, (SRO, Vol. a, fo 89r[299])

Vicarage of Lauthreisk

'The rentale of Mr David Methven, vicaregis of Lauthrisk yeirlie and Forgound.'

'Item, the said vicarege of Lauthreisk', £40.

Calculation of third: In the whole, £40; third thereof, £13 6s 8d.

'Item, ressavit ane uther rentale at my lord Comptrolleris[300] command.'

'Rentale vicarie de Lauthrisk in Fyffe'

[*In margin*, 'Aliter gevin in vicarage Lauthrisk.']

Vicarage of Lauthreisk pertaining to Mr David Methven extending in the year, deducting what should be deducted by order of the lords, to £22; third thereof, £7 6s 8d.

FORGAN, VICARAGE OF, (SRO, Vol. a, fo 89r)

'Vicaria de Forgound'

Vicarage of Forgone pertaining to the said Mr David Methven extending in the year, deducting what should be deducted by order of the lords, 30 merks.

[*In margin*, 'Ressavit at the Comptrolleris[301] command.']

---

[298] The correct calculation appears to be £8 17s 9¼d.
[299] SRO, Vol. a, fo 88v is blank.
[300] See above, p. 73, n. 293.
[301] See above, p. 73, n. 293.

The said vicarage of Forgound, 53 merks.

Calculation of third: In the whole, £35 6s 8d; third thereof, £11 15s 6⅔d.

INVERKEITHING, VICARAGE OF, (SRO, Vol. a, fo 89v)

Rental of the vicarage of Innerkeithing given in by William Boswell, vicar thereof, extending to £10.

AUCHTERMUCHTY, VICARAGE OF, (SRO, Vol. a, fos 89v-90r)

'The rentale of the vicarege of Auchtermuchtie according as it now gevis sen the dountaking of the corpis presentis, umestclaithes, Pasche fynes and offerandis,[302] will extend yeirlie to ten pundis allanerly, and giff onie man can find superplus or mair thairof I am content that the samin be halelie tane up be the Comptroller and his chalmerland.'

Signature: 'schir Wilyame Scott. Tullybardin, Comptroller.[303] Ressavit at command of my lord Comptroller', 16 July 1566.

'The haill vicarege of Auchtirmuchtie', 45 merks.

Calculation of third: In the whole, £30; third thereof, £10.

BALLINGRY, VICARAGE PENSIONARY OF,[304] (SRO, Vol. a, fo 90r)

Rental of the vicarage pensionary of Ballingery pertaining to Mr Alexander Wardlaw.

'Inprimis, ane aker of land in my gleib extending to' 13s 4d, 'and in money' £8.

---

[302] Such clerical exactions as the Easter offering at Communion services, the 'uppermost cloth' as a mortuary due and the 'corpse present', or funeral fine for unpaid teinds, were among the grievances presented to the provincial council of the pre-Reformation church in 1559, and were condemned as oppressive by the protestant reformers in their *Book of Discipline* in 1560. The lords who gave qualified approval to the *Book of Discipline* in January 1561 declared that exactions of this sort 'be cleane discharged, and no more taken in times comming'. *Statutes of the Scottish Church, 1225-1559*, ed. D. Patrick (Edinburgh, 1907), p. 158; *The First Book of Discipline*, ed. J. K. Cameron (Edinburgh, 1972), pp. 156-158.

[303] Sir William Murray of Tullibardine, Comptroller, 1565 - 1582; James Cockburn of Skirling also served as Comptroller, 1566 - 1567, *Handbook of British Chronology*, p. 191.

[304] Cf SRO, Vol. a, fo 101r (see below, p. 86).

'Summa', £8 13s 4d. 'Ressavit at command of the Clerk Registre.'

'Ane uther rentall gevin up of the same be sir James Stanis.'

Calculation of third: In the whole, £8 13s 4d; third thereof, 53s 4d.[305]

ST ANDREWS,   (SRO, Vol. a, fo 91r[306])
KINGLASSIE AND KINGASK, PREBEND OF ST MARY ON THE ROCK COLLEGIATE KIRK

Prebend of Kinglassie and Kyngask

'The haill prebendarie of Kynglassie in money', 37 merks; 'and Kyngask, the haill victuall thairof', 28 b. meal, 'pertenyng to Mr Thomas Methven.'

Calculation of third: In the whole, £24 13s 4d; third thereof, £8 4s 5½d. In victual, 28 b. meal; third thereof, 9 b. 1 f. 1⅓ p.

KILCONQUHAR, VICARAGE OF,   (SRO, Vol. a, fo 91v)

'The vicarege of Kynnunquhar payit befoir the rysing of the Congregatioun, to Mr Johnne Hammiltoun' £60. 'Of pensioun to the priorassie of North Berwik',[307] £20. Total, £80, 'and hes payit nathing thir thrie last yeris bygane, and the rest in the parochineris handis.'

Signature: 'Joannes Hammiltoun of Samuelstoun.'

Calculation of third: In the whole, £80; third thereof, £26 13s 4d.

ROSSIE, VICARAGE OF,[308]   (SRO, Vol. a, fo 92r)

Vicarage of Rossy [*In margin*, 'Angus.']

Rental of Mr David Henrysoun

---

[305] The correct calculation appears to be £2 17s 9¼d.
[306] SRO, Vol. a, fo 90v is blank.
[307] Margaret Hume was appointed prioress in 1544 and demitted office before 30 June 1566, Cowan and Easson, *Medieval Religious Houses*, p. 148.
[308] This benefice, in the sheriffdom of Forfar, was appropriated to St Andrews priory, which may explain its inclusion in rentals for Fife.

'The vicarege of Rossy worth' 15 merks 'be yeir and set to Mr George Scot and sir James Scot for the sowme foirsaid.'

Calculation of third: In the whole, £10; third thereof, £3 6s 8d.

CREICH, VICARAGE OF, (SRO, Vol. a, fo 93r[309])

The vicarage of Creycht pertaining to Mr John Seytoun, 'titular thairof.'

'The yeirlie availl of the said vicarege extendit to' £36 'and wes set in assedatioun for the samin sowme, bot now be ressoun that be the last act of Parliament Pasche fynis, corspresendis, upmest claithes and uther small offerandis and dewteis wer dischargit, the fructes thairof ar of les availl.'

Signature: 'David Sibbald, brother germane to the laird of Rankelour, with my hand, delyverit be me in name of Mr Thomas [recte, John] Seytoun, vicar thairof.'

Calculation of third: In the whole, £36; third thereof, £12.

'And the haill vicarege in respect of the Pasche fynis, corpis presentis, humest claithes, offeringis, is bot now gevand be the yeir the sowme of' £20 'allanerlie and thairfoir I, the vicar thairof under subscryvand, is content that giff onie mair be gottin of the said vicarege nor the said' £20 'that the Collectour tak the haill superplus by the thrid, quhilk thrid now extendis bot to' 10 merks.

Signature: 'Mr Johnne Seytoun, vicar of Creycht, with my hand.'

LARGO, VICARAGE PERPETUAL OF, (SRO, Vol. a, fo 94r[310])

'Rentale vicarie de Largo, Sanctiandree diocesis'

Vicarage perpetual of the kirk of Largo pertaining ('spectante') to Mr Alexander Wod, extending commonly in the year as also it was last set for the sum of 100 merks Scots money, thereof deducting for the stipend of the curate of this kirk, 20 merks, and so the sum distinct and free for the vicarage extends to 80 merks.

Assignation of the vicarage of Largo: In the whole, £53 6s 8d; third thereof, £17 15s 6⅔d.

'Ane uther rentall of the vicarege of Largo pertenyng to Mr Alexander Wod.'

---

[309] SRO, Vol. a, fo 92v is blank.
[310] SRO, Vol. a, fo 93r is blank.

'The cros [*recte*, corpse] presentis, umaist claithes, Pasche fynes and offerand silver taken of, of the quhilk now na payment is maid, the haill rest of the profeitis and emolumentis thairof consistes allanerlie in teynd woll, teynd lamb, teynd cheis, teynd fische, teynd lint and hemp with the hallobrokes will extend be yeir be guid estimatioun allanerlie to' £30.

Calculation of third: In the whole, £66 13s 4d; third thereof, £22 4s 5⅓d.

DAIRSIE, VICARAGE OF, (SRO, Vol. a, fo 94v)

'Rentale vicarie de Dersy, Mr Robert Wynrame possessor.'

The vicarage of Dersy formerly paid £10 but for the last three years nothing at all has been paid but if free payment were made of that which is expressed in the Book of Reformation[311] it would be worth 10 merks.

Calculation of third: in the whole, £6 13s 4d; third thereof, 44s 5⅓d.

LOGIE-MURDOCH, VICARAGE OF, (SRO, Vol. a, fo 95r)

The vicarage of Logymurtho pertaining to Mr Thomas Forret, 'titular thairof.'

'The yeirlie availl of the said vicarege extendit to' 24 merks 'and wes of na mair availl, bot now be ressoun that be the last act of Parliament Pasche fynis, corspresendis, upmest claithes and utheris small offerandis and dewteis wer dischargit the fructes thairof ar of les avail.'

Signature: 'Johnne Forret of that Ilk with my hand.'

Calculation of third: In the whole, £16; third thereof, £5 6s 8d.

STRATHMARTINE, VICARAGE OF,[312] (SRO, Vol. a, fo 95r)

Rental of the vicarage of Straythmertyne [*In margin*, 'Nota, this lyis in Forfar.']

The vicarage of Strathmerting within the diocese of Sanctandrois and the sheriffdom of Fyffe 'for the teiching of the schole of Abirbrothok callit the prepositive', 30 merks.

---

[311] I.e. the *First Book of Discipline* (1560).
[312] This benefice, in the sheriffdom of Forfar, was in St Andrews diocese, which may explain its inclusion in rentals for Fife.

ST ANDREWS, (SRO, Vol. a, fo 95v)
KINKELL, PREBEND OF ST MARY ON THE ROCK COLLEGIATE KIRK

'The prebendarie of Kinkell within the college kirk of the Lady Heuch besyd Sanctandrois at the presentatioun of our soverane ladie, to the quhilk James Henrysoun is lauchtfullie provydit thairto under his [*recte*, hir] graces privie seall etc., and possessour sen Pinkie feild. The fructes of the samin is the teynd schaves of Kinkell togither with the vicarege thairof and uther pertinentis thairto limitat be ressoun of our soveranes predicessouris fundatioun quhen the samin is led is worthe be yeir' £40.

Signature: 'This is subscryvit be the Clerk of Register.'

Calculation of third: 'The 3 heirof' £13 6s 8d, 'the thrid is diminischit at command of the Lordis of the Checker and extendis bot to the thrid of the baire [*i.e.* bere] thairof, [*blank*] tak quhilk is' £8 17s 9½d.[313]

KINGHORN EASTER, VICARAGE OF, (SRO, Vol. a, fos 95v-96r)

'Ane uther rentale of the vicarege of Kingorne'[314]

Rental of the vicarage of Kingorne Eister

Calculation of third: In the whole, £40; third thereof, 20 merks.

[*In margin*, 'Vicarege Kingorne.']

'Rentale vicarii de Kingorne Ester Sancti[andree] diocesis'

Rental of the vicarage of Kingorne Ester in the diocese of St Andrews

The vicarage perpetual of the parish kirk of Kingorne Ester pertaining to sir John Wolsoun extending in three years to 100 merks, thereof deducting for the stipend of a curate for a single year, 20 merks.

And the foresaid sum free for the vicarage in the year, 80 merks, of which nothing has been received for the past term by reason of the tumult of the said year.

Signature: 'Domine Joannis Wolsoun,[315] vicarius of [*sic*] supra, manu propria.'

---

[313] £40 minus £13 6s 8d is £26 13s 4d, divided by 3 gives £8 17s 9½d.
[314] The only other rental for Kinghorn Easter is that following this entry.
[315] I.e. John Wilson.

Calculation of third: In the whole, £66 13s 4d; third thereof, £22 4s 5⅓d.

CUPAR, (SRO, Vol. a, fo 96r)
CHAPLAINRY OF OUR LADY IN ST CATHERINE'S KIRK IN,

'Chapellanerie of Our Lady in Sanct Catherenis kirk in Cowper'

Calculation of third: In the whole, £11 6s 8d; third thereof, £3 11s 2⅓d.[316]

COLLESSIE, VICARAGE OF,[317] (SRO, Vol. a, fo 97r[318])

'Vicaria de Colassie diocesis Sancti[andree]'

Vicarage of Colassie in the diocese of St Andrews set in assedation for usual payment, 48 merks.

Deducting procurations, synodals and visitations of the dean for it is a non-taxable benefice, but now Easter dues, personal teinds, corpse offerings, kirk cows and uppermost cloths are not paid any longer.

Signature: 'Magister Thomas Scot.'

[*In margin*, 'Nota, to be chargit for viij merkis in tyme cuming', at Halyruidhous, 13 May 1573, 'Per dominos auditores scaccarij,[319] sic subscribitur Nicolsoun'.]

Calculation of third: In the whole, £32 6s 8d;[320] third thereof, £10 13s 4d.

KIRKCALDY, (SRO, Vol. a, fo 97r)
CHAPLAINRY OF THE HOLY BLOOD ALTAR IN,

Rental of the chaplainry of the Halybluid altar within the parish [of] Kirkcaldie yearly 'to be tane up of Puderneis Eister and Wester' pertaining to John Ballincanquall[321] 'ane auld blind man that seis nocht perfytlie.'

'Item, payis yeirlie' 13 merks 'togither with ane tenement of land occupyit be him selff lyand within the burcht of Kirkcaldie betuix etc.'

---

[316] The correct calculation appears to be £3 15s 6⅔d.
[317] Cf SRO, Vol. a, fo 103r (see below, p. 89).
[318] SRO, Vol. a, fo 96v is blank.
[319] I.e. 'By order of the Lords Auditors of the Exchequer.'
[320] 48 merks equals £32.
[321] John Balcanquell, *RMS*, iv, no 2499.

Signature: 'James resaiff this rentall, J. Clericus Registri.'

GARVOCK, VICARAGE OF,   (SRO, Vol. a, fo 97v)

Rental of the vicarage of Garvok within the diocese of Sanctandrois, pertaining to Mr John Wardlaw, which is set in assedation to James Keyth of Drumtiktie [*i.e.* Drumtochty] and sir James Symmer for £40.

Calculation of third: In the whole, £40; third thereof, £13 6s 8d.

GARVOCK, CHAPLAINRY OF ST ANDREW IN,   (SRO, Vol. a, fo 97v)

'Sanct Androis chapell'

Rental of Sanct Androis chaplainry extends to 40 merks.

Signature: 'M[r] J. Wardlaw.'

Calculation of third: In the whole, £26 13s 4d; third thereof, £8 17s $6\frac{2}{3}$d.[322]

METHIL, PARSONAGE AND VICARAGE OF,   (SRO, Vol. a, fo 98r)

Rental of the parsonage and vicarage of Mathill, £40 set in assedation in the hands of George Swine 'and of this deducit for the kow and upmest claithe and Pasche pennis', £10.

Calculation of third: In the whole, £40; third thereof, £13 6s 8d.

CRAIL COLLEGIATE KIRK, PROVOSTRY OF,[323]   (SRO, Vol. a, fo 98v)

Rental of the provostry of Craill 'quhilk consistes allanerlie into [*recte,* of] the vicarege, the quhilk is set in takis to sindrie men', pertaining to Patrick Myrtoun, for £100, 'and of this payit to the redar' £20; third thereof, £33 6s 8d.

LESLIE, VICARAGE OF,   (SRO, Vol. a, fo 98v)

Vicarage of Leslie pertaining to Andrew Angus extending to £13 6s 8d.

---

[322] The correct calculation appears to be £8 17s $9\frac{1}{3}$d.
[323] Cf SRO, Vol. a, fo 76v (see above, p. 65).

MOONZIE, VICARAGE PENSIONARY OF, (SRO, Vol. a, fo 98v)

Rental of the vicarage pensionary of Auchtermonsie given in by Mr Robert Patersoun, vicar pensionary of the same, 25 February 1573/4, 12 merks silver, 'with a littill croft'; third thereof, 4 merks.

Signature: 'Mr Robert, with my hand.'

KENNOWAY, VICARAGE OF, (SRO, Vol. a, fo 99r)

Vicarage of Kennochquhy set before for 45 merks, 'and now in respect of the thingis defalkit extendis to the sowme of [*blank*]'.

Signature: 'Ita est Joannes Row, manu sua.'

Calculation of third: In the whole, £30; third thereof, £10.

LEUCHARS, VICARAGE OF, (SRO, Vol. a, fo 99r)

Vicarage of Lucheris pertaining to Mr Robert Carnegy extending to £46 13s 4d; third thereof, £15 11s 1½d.

INVERKEILOR, VICARAGE OF,[324] (SRO, Vol. a, fo 99v)

'Vicaria de Innerkelour, monasterij de Abirbrothock'

The vicarage of Innerkelour of the monastery of Abirbrothock, of which Mr Alexander Forrest, provost,[325] is vicar perpetual by reason of his provision [and] after his lifetime, by reason of its union (with the consent of the lord abbot and convent of the monastery of Abirbrothok) to the said provostry,[326] apart from the next first fruits of the same, and such that the master of the grammar school can, because of his yearly stipend from the said provost, teach the boys and youth, of

---

[324] This benefice, in the sheriffdom of Forfar, was in St Andrews diocese, which may explain its inclusion in rentals for Fife.
[325] Alexander Forrest, provost of Fowlis Easter, 1549 - 1552; provost of St Mary of the Fields, 1552 - 1561, Watt, *Fasti*, pp. 358-359.
[326] This union has not been traced. The family of Beaton held the office of abbot from 1517 till 1551, when John Hamilton, son of 2nd earl of Arran, became commendator. Cowan and Easson, *Medieval Religious Houses*, p. 67.

which the fruits in wool and lamb and other small teinds extended of old to £40.[327]

Calculation of third: In the whole, £40; third thereof, £13 6s 8d.

KIRKFORTHAR, VICARAGE PENSIONARY OF,   (SRO, Vol. a, fo 99v)

'The rentale of the cuyr of the vicar pensionar of Kirkforthour pertenyng to sir David Donnald induring his lyftyme.'

'Inprimis thrie akeris of land with thrie kyis gers and thair followares with ane hors gers estimat to the sowme of' £4. 'Item, communibus annis', 15 teind lambs, price 30s. 'Item, communibus annis', 2 stones of teind wool, price £3. 'Item, communibus annis', 3 stones of teind cheese, price 30s. Total, £10.

Signature: 'James, ressave this rentall. J. Clerk Registre.'

KIRKCALDY,   (SRO, Vol. a, fo 99v)
CHAPLAINRY OF THE HOLY ROOD ALTAR IN,[328]

'This is the rentall of ane cheplanrie situat at the Ruid altar within the paroche kirk of Kirkcaldie pertenyng to sir Edward Leyne. In primis, iij akeris and ane halff of land now occupyt be [blank] Houchoun, burges of the burcht of Kirkcaldie, extending to the sowme of' £10.

Signature: 'Ita est Edward Leyne.'

ABERDOUR, VICARAGE OF,   (SRO, Vol. a, fo 100r)

Rental of the vicarage of Abirdour

'The said vicarege of auld extendit to' £35 'in Peasche rokningis [i.e. reckonings], kirk cow and upmest claith commounlie extendit to' £20 yearly, 'quhilkis now ar nocht payit and by the premissis onlie restis teynd hemp, lint, stirk, gryis and guis quhilk will extend to' 15 merks 'or thairby commoun yeris for the same wes nevir

---

[327] The Latin text reads thus: 'Vicaria de Innerkelour monasterii de Abirbrothock cuiusquidem Magister Alexander Forrest prepositus vicarius perpetuus extat ratione provisionis post eius vitam ratione unionis cum consensu domini abbatis et conventus monasterii de Aberbrothok ad dictam preposituram praeter propinquos eiusdem fructus priores et quatenus magister scole grammaticalis possit ex annuo stipendio dicti prepositi pueros et iuvenes erudire cuiusquidem fructus in lana et agnis ac ceteris aliis minutis decimis ab antiquo extendentes', £40.
[328] Cf SRO, Vol. a, fo 100v (see below, p. 85).

set in tak nor assedatioun. My lord Sanctcolmisinche[329] hes lambwoll and teynd fische of the haill parochin.'[330] Total, 15 merks.

Signature: 'Walter Robertsoun, vicar of Abirdour, with my hand. Magister Henry Kinros, presentar of the samin, with my hand.'

Calculation of third: In the whole, £35;[331] third thereof, £11 13s 4d.

ST JAMES' ALTAR, CHAPLAINRY OF,[332]   (SRO, Vol. a, fo 100r)

'The rentall of Sanct James altar and patrimonie thairof in the haill annuellis as the chartour of fundatioun beiris extendis to' £8 13s 4d.

Signature: 'Henry Davidsoun, chaiplane thairof.'

'Thrid thairof', 57s 9½d.

AUCHTERMUCHTY, VICARAGE OF,   (SRO, Vol. a, fo 100v)

Rental of the vicarage of Auchtirmuktie

'The yeirlie availl of the said vicarage extendit of auld to' 45 merks, 'and now be ressoun of defalcatioun of sindrie dewteis it is of small or nane availl.'

Signature: 'Maister Williame Scott.'

KIRKCALDY, HOLY ROOD ALTAR IN,[333]   (SRO, Vol. a, fo 100v)

Rental of the Rude altar in the parish kirk of Kirkcaldy, pertaining to Henry Young extending to £10.

---

[329] See above, p. 64, n. 262.
[330] Cf Abbey of Inchcolm, SRO, Vol. a, fo 73v (see above, p. 64).
[331] In valuing the benefice, no allowance was made for the loss of income from the abolition of clerical exactions at the Reformation.
[332] Cf 'chaplainry of St James, in the sheriffdom of Fyff and regality of Dunfermling', *RSS*, vi, no 1509.
[333] Cf SRO, Vol. a, fo 99v (see above, p. 84).

ST ANDREWS, LAMBELETHAM, (SRO, Vol. a, fo 100v)
PREBEND OF ST MARY ON THE ROCK COLLEGIATE KIRK

Rental of the prebend of Lamelethame pertaining to Mr Adam Foulis and set in assedation to James Benstoun for £20 yearly, and extends in the whole to £28.

KIRKFORTHAR, (SRO, Vol. a, fo 101r)
PARSONAGE AND VICARAGE OF,

Rental of the parsonage and vicarage of Kirkforther is £40, 'and thairof payit to the curat for his fe' 20 merks 'by the gleib, and sua restis to the persoun' 40 merks.

Signature: 'Mr Henry Lummisden.'

Calculation of third: In the whole, £40; third thereof, £13 6s 8d.

BALLINGRY, PARSONAGE AND VICARAGE OF,[334] (SRO, Vol. a, fo 101r)

Parsonage and vicarage of Ballingery

Calculation of third: In the whole, £100; third thereof, £33 6s 8d.

[*In margin*, 'Memorandum, this is a gleib.']

'Set be Mr Alexander Wardlaw to Johnne Wardlaw in Leyth.'

ST ANDREWS, VICARAGE OF, (SRO, Vol. a, fo 101r)

Vicarage of Sanctandrois

Calculation of third: In the whole, £66 13s 4d; third thereof, £22 4s 5½d.

DULL, VICARAGE OF,[335] (SRO, Vol. a, fo 101v)

'Magister David Guthrie habet vicariam de Dow in Athoill.'

---

[334] Cf SRO, Vol. a, fo 90r (see above, p. 76).

[335] This benefice, in Atholl in the sheriffdom of Perth, was appropriated to St Andrews priory, which may explain its inclusion in the rentals for Fife.

This vicarage formerly paid 135 merks but now for the past term nothing at all was paid, but if that which is owed was paid as is set down in the Book of Reformation[336] it would be worth £40.

'Summa totalis', £40.

Signature: 'M[r] Johnne Wynrame. Producit be me, M[r] Johnne Wod.'

Assignation of the third: In the whole, £40; third thereof, £13 6s 8d.

MONEYDIE, PARSONAGE OR VICARAGE, (SRO, Vol. a, fo 101v)

[*In margin*, 'Perthe.']

Calculation of third: In the whole, £100; third thereof, £33 6s 8d.

'Nota, it is uncertan quhither this rentall be personage or vicarege'.[337]

ST ANDREWS, CAIRNS, (SRO, Vol. a, fo 101v)
PREBEND OF ST MARY ON THE ROCK COLLEGIATE KIRK

'Prebendarie of Kirkheuchis callit Kernis'

Calculation of third: In the whole, £17 6s 8d; third thereof, £5 15s 6⅔d.

CUPAR, VICARAGE OF, (SRO, Vol. a, fo 102r)

'The rentale of the vicarege of Cowper presentlie, the thingis being deducit furth of the samin quhilk exceptit and dischargit be the reformatioun maid thairupoun. The small teyndis thairof will extend in the haill to' £30, 'the quhilk I wilbe hertlie contentit to find cautioun to pay the thrid thairof to the quenis grace siclyk as utheris dois, and this my rentall subscryvit with my hand.'

Signature: 'Williame Inglis, with my hand.'

Calculation of third: In the whole, £30; third thereof, £10.

---

[336] I.e. the *First Book of Discipline* (1560).
[337] Cf NLS, fo 52v (see below, p. 336). This benefice, in the sheriffdom of Perth, was a prebend of Dunkeld cathedral.

DOLLAR, VICARAGE PENSIONARY OF,[338] (SRO, Vol. a, fo 102r)

[*In margin*, 'Perthe.']

Rental of the vicarage pensionary of Dolour pertaining to Robert Burne, lying within the sheriffdom of Clakmannan, extends to in the whole 12 merks, 'with the mans and gleib thairof worthe' 2 merks.

Signature: 'Robert Broun [*recte*, Burne], with my hand. Ressavit at command of the Clerk Registre.'

KINROSS AND ORWELL, (SRO, Vol. a, fo 102r)
VICARAGE PENSIONARY OF,

'I, Mr Walter Balfour, makis manifest that my vicar pensioun of Kynros and Uruell, now vacand in my handis be the deceis of umquhill sir Johnne Mows is worth be yeir xl merkis money and servis thairfoir twa kirkis. Sic subscribitur, Mr Walter Balfour.'

TARVIT, (SRO, Vol. a, fo 102v)
PARSONAGE AND VICARAGE OF,[339]

Parsonage and vicarage of Tervat pertaining to sir John Achesoun [Eclynsoun, *fo 86v*].

'The said personage and vicarege wount to gif of auld' £50, 'and wes set thairfoir in assedatioun, bot now becaus of the deductioun of the corspresentis, umest claythis, Pasche fynes and offerandis, it is les worth be four pundis yeirlie and sua is agreit on betuix the said sir Johnne and the Comptroller.'[340]

Calculation of third: In the whole, £46; third thereof, £15 6s 8d.

---

[338] Cf NLS, fo 54r (see below, p. 339, and n. 214). This benefice, in the sheriffdom of Clackmannan and diocese of Dunkeld, was appropriated to Inchcolm abbey, and is entered among rentals for both Fife and Perth. The Books of Assumption contain no separate list of entries for the sheriffdom of Clackmannan, whose benefices are to be found among the rentals for Fife, Perth and Stirling.

[339] Cf SRO, Vol. a, fo 86v (see above, p. 73).

[340] See above, p. 73, n. 293.

ST ANDREWS, BLACK FRIARS OF,[341] (SRO, Vol. a, fos 102v-103r)

'The rentale of the Blak Freiris of Sanctandrois gevin up be Freir Johnne Greirsoun.'

'The place of Sanct Monanis: Inprimis, the hous and yaird and ane aker of arabill land on the north syd of the yaird set in few to James Sandelandis, laird of Sanct Monanis, for' 4 merks yearly, 'nocht ane penny payit thir thrie yeiris bigane', 12 merks. 'Item, of the toun of Sanct Monanis land and houssis yeirlie', £20, 'of that nocht gottin ane penny bot' £3 'thir fyve yeiris bygane, summa', £97. 'Item, of Ardros, annuell yeirlie' 7 merks, 'of that nocht gottin ane penny v yeiris bygane, summa', 35 merks.

'The rentale of the Blak Freiris of Sanctandrois gevin up be the young laird of Largo.'

The Kingis Bernis, 20 merks. Sanct Monanis, £20 4 merks. Henry Kempis place in Couper, 4 merks. Petmolye, 40s. Sanct Nicolace, £20. 'Of Medy', 30s. Rathibeath, £5. Feyldie and Cloichrie Steyne, £20. Kelour, 'the laird of Lochlevinis place be West Mephen', 18 merks. Alison Lyndesayis croft beside Couper, 4 b. wheat.

'The names of the tennentis assumed for payment of the annuellis of the landis of Feldye and Clochrie Stane, viz', John Fethringhame, William Fothringhame, Ninian Forthringhame.'

COLLESSIE, VICARAGE OF,[342] (SRO, Vol. a, fo 103r)

Rental of the vicarage of Culessie lying within the sheriffdom of Fyffe, pertaining to Mr Thomas Scot, vicar thereof.

'The vicarege of Cullessie olim assedatur[343] for' 48 merks, 'and now presentlie deductioun beand had conforme to the Buik of Reformatioun,[344] is worthe be yeir' 24 merks.

Signature: 'Mr Thomas Scot. Ressavit at command of my lord Comptroller.'[345]

---

[341] I.e. Dominicans.
[342] Cf SRO, Vol. a, fo 97r (see above, p. 81).
[343] I.e. formerly leased; this is an example of the mixed use of Scots and Latin terms, found throughout the text.
[344] I.e. the *First Book of Discipline* (1560).
[345] See above, p. 73, n. 293.

DUNINO, VICARAGE OF, (SRO, Vol. a, fo 103v)

Rental of the vicarage of Dunnennio lying within the sheriffdom of Fyffe, pertaining to Robert Smyth, 'wes wont to give yeirlie' 22 merks 'and now be ressoun of deductioun of the cropes [*recte*, corpse] presentis and umest claithes is straitlie worthe' £8.

Calculation of third: In the whole, £8; third thereof, 53s 4d.

Signature: 'Robert Smyth, vicar of Dunnennio. Ressavit at command of the Clerk of Registre.'

PENSION[346] (SRO, Vol. a, fo 103v)

'The rentale of ane pensioun of my lord Sanctandrois,[347] payit be him out of the landis of Scotiscraig within the schirefdome of Fyffe, and be Henry Durie, laird of that Ilk, to Mr Andro Davidsoun, pensionar of the samin, extending yeirlie to the sowme of' £40.

Signature: 'Ita est Magister Andreas Davidsoun, manu propria. Ressavit at command of the Clerk Registre. J. Clericus Registri.'

Calculation of third: In the whole, £40; third thereof, £13 6s 8d.

NOTE

Rentals for the following benefices are located elsewhere in the text:

ABDIE, VICARAGE PORTIONARY OF (NLS, fo 90v), p. 378.
DYSART, VICARAGE PENSIONARY OF (SRO, Vol. a, fo 190r), p. 188.
NEWBURGH, CHAPLAINRY OF ST CATHERINE IN (NLS, fo 90v), p. 378.

---

[346] No information on the location of this pension is forthcoming, except that Scotscraig lay in that part of Leuchars parish which after 1602 formed the parish of Ferryport-on-Craig.
[347] See above, p. 6, n. 24.

Edinburgh

HOLYROODHOUSE, ABBEY OF,   (SRO, Vol. a, fos 104r-107v)
                                                     (SRO, Vol. b, fos 62r-65v)
Halyruidhous

'Rentale monasterij Sancte Crucis'

'Item, the baronie of Quhitkirk set to Oliver Sinclar in few for' £92 money; 10 c. wheat; 10 c. bere of feuferme and 12 score and 12 capons kane, and the teind of the same set to him for 2 c. wheat; 3 c. bere and £58 13s 4d.

Kirks
      Barro, the teinds thereof set to Robert Carncerce [*i.e.* Cairncross] for £13 6s 8d. Boltoun, the same set for £26 13s 4d. Montlothiane and the Hopes set for £25. Tranent pays 13 b. wheat; 13 b. bere; 3 c. oats, £246 8s money 'for certan teyndis' and £4 13s 4d of feu mail for Littill Fallsyd and 33s 4d of annual of Prestoun and 12 'laid salt for the myldis' [myles, *SRO, Vol. b*]. Libertoun pays 9 c. 8 b. wheat; 12 c. 8 b. bere; 25 c. 8 b. meal;[1] £45 16s 8d [*in margin*, 'Nota, iij c. in pley with Nudry']. Sanct Cuthbertis kirk pays 5 c. 5 b. wheat; 13 c. 8 b. bere; 6 c. 8 b. oats; £252 3s 4d. Corstorphin set for £40 money 'allanerlie'. Carridein and Kynnell set for £106 13s 4d. Kirklevingstoun set for £66 13s 4d. Falkirk pays £239 in assedation. Kingorne Eister set for £100 13s 4d. Arthe set for £60. Dunrode set for £32. Craufurd Lindesay set for £86 13s 4d, 'bayth personage and vicarege'.[2] Dalgarno set, both parsonage and vicarage for £46 13s 4d, therefore for the parsonage, £20, and for the vicarage, 40 merks. Twynem set for £33 6s 8d. Keltoun set for £32. Ur set for £80. Kirkcormo set for £21 13s 4d; 'Pensioun of Tungland', 2 c. meal; £10 money. Kirkcudbrycht set for £54 6s 8d. Balmagy set for £40. Teind fish of Leyth set for £24. Melginche set for £33 6s 8d. Barony of Dunroid set in feu for £88 6s 8d. Barony of Kers with Ogilface and the Falkirktoun pays yearly £357 13s 2d; 5 score and 19 capons; 24 salmon; 12 b. bere; 2 c. oatmeal. Barony of Brochtoun pays yearly £545 0s 2d; 6 score

---

[1] This figure for meal is counted as oats in the thirds, fo 107v (see below, p. 93).
[2] The parish of Crawford Lindsay, or Crawford Douglas, is now known as Crawford; cf EUL, fo 17v (see below, p. 512).

and 12 capons; 24 hens and 3 swine. 'Item, of annuellis within Edinbrucht, Leyth, Cannogait, with the foir entres', £62 0s 6d.

'Expenssis and ordinar allowance under the commoun seill'
'Item, to the bedmen of Sanct Leonard', 36s. To the convent of Newbotle, £46 0s 8d for the annual of Pendrech. To Mr John Spens, £20. To Mr David Borthweik, £20. 'Item, [in, *SRO, Vol. b*] the quenis gracis handis for Abbatis Medo and Salisbery', £32; 'and for the foir enteres', £14 6s 8d; 'and for the tua yairdis new tane in be hir gracis mother', £10. To Mr Nicol Elphingstoun, £30. To Mr Alexander Chalmer, £50. To John Stewart, £24. To Stephen Litster, £24. To John Wolsoun, £24. To Alexander Harcas, £24. To David Guidsoun, £24. To the clerk of the regality, £8 17s 9d. To the cloister servant, £12. To the laird [of] Arthe, bailie of Tros[3] [Thros, *SRO, Vol. b*], £10. To John Logane, bailie of the regality, £20. To Mr David McGill, 20s. 'To my lord Flemyng[4] and laird of Bardoy for thair few annuell of Treis',[5] £46 13s 4d. To Alexander Forrester, £24. To the Lords of Session, £56 for their contribution. To Mr John Stewart, £10. To Patrick Ballenden, £3 11s 1d 'deducit of George West and Thomas Pratt few during thair tak of the samin'. To George Levingstoun, 17s 8d 'siclyk deducit during Williame Levingstoun lyftyme. Deducit of my lord dukes[6] few during my lordis lyftyme', £10. 'To the convent, of Crumzeatis [*i.e.* Crumzeane's] land, 40s; 'and of brint annuellis in Leith onbiggit', £6 8s 8d. To Mr James Makgill, £20. 'To my lord[is] moder',[7] £33 6s 8d. To the laird of Sanct Monanis, £50. To Robert Ormestoun, £9 6s 8d. To Mr George Strang, 'vicar portioner' of Craufurd [*i.e.* Crawford Lindsay], £32. To the bishop of Galloway,[8] £10. To Harry Sinclair, William Heslope, James Abircrumbie, John Gady, Alexander Ur, Andrew Blakhall, Peter Blakwod, Thomas Maxwell, Patrick Menteyth, Harry Sibbald, [*blank*]. [*In margin*, 'channones that gets na thing']. To the minister of Sanct Cuthbertis kirk, £80. To the minister of the abbey, £80. To the minister of Falkirk, £46 13s 4d. To the minister of Galloway [*sic*], £50. To the minister of Barro, £3 6s 8d. 'Summa', [*blank*].

'Ordinar expenssis and allowance of quheit'
To John Stewart, 9 b. 2 f. 3 p. To Stephen Litster, 9 b. 2 f. 3 p. To Alexander Harcas, 9 b. 2 f. 3 p. To David Guidsoun, 9 b. 2 f. 3 p. To John Wolsoun, 9 b. 2 f. 3 p. To Mr James McGill, 8 b. To Mr David McGill, 8 b. To Mr Nicol Elphingstoun, 8 b. To Robert Ormestoun, 12 b. 'Deducit for sterilitie of the fluris', 2 b. To the laird [of] Braterstoun [*i.e.* Brotherstone], 4 b. 'To Francie, closter servand', 5 b. To Alexander Forester, 9 b. 2 f. 3 p.

---

[3] Presumably Throsk.
[4] John, 5th lord Fleming, *Scots Peerage*, viii, pp. 543-544.
[5] Cf 'terras de Torbanes alias Treis', in the barony of Bathgate, *RMS*, vii, no 497; now Trees.
[6] James Hamilton, 2nd earl of Arran and duke of Chatelherault, *Scots Peerage*, iv, pp. 366-368.
[7] I.e. Euphemia Elphinstone, mother of Robert Stewart, commendator of Holyroodhouse, an illegitimate son of James V. *Scots Peerage*, i, p. 24.
[8] Alexander Gordon was bishop elect of Galloway at the Reformation, Watt, *Fasti*, p. 132.

'Ordinar allowance of beir'

To John Stewart, 20 b. 3 f. 1 p. To Stephen Litster, 20 b. 3 f. 1 p. To David Guidsoun, 20 b. 3 f. 1 p. To John Wolsoun, 20 b. 3 f. 1 p. To Alexander Harcas, 20 b. 3 f. 1 p. To Mr Nicol Elphingstoun, 8 b. To the Clerk of Register,[9] 8 b. To the clerk of regality, 9 b. To Mr David McGill, 8 b. To the laird [of] Bruterstoun [*i.e.* Brotherstone], 4 b. To Robert Ormestoun, 20 b. 'Deducit of the fluris for sterilitie', 2 b. To the cloister servant, 11 b. To Alexander Forrester, 20 b. 3 f. 1 p.

'Ordinar allowance of aites'

To the Clerk Register, 8 b. To the laird [of] Nudry, 3 c. 'in pley'. To the laird [of] Craigmiller, 5 c. To Mr John Spens, 12 b. To the officer of regality, 8 b. To the laird [of] Bruterstoun, 1 c. To Mr David McGill, 1 c. To the laird [of] Blaverne [*i.e.* Blanerne], 20 b. To the laird [of] Innerleith, 4 c. 'To the chalmerlane', 1 c. 'To the ministeris of Sanct Cuthbertis, Libertoun, and the abay', 2 c. 4 b.

Signature: 'Robert Stewart.'[10]

Assignation of the third of Halyruidhous:

Halyruidhous, third thereof, £975 9s 6d. Take the mails of the barony of Brochtoun for £545 0s 2d; the barony of Kers with Ogilface and Falkirktoun for £357 13s 2d; the barony of Dunrod for £88 6s 8d. 'Gif in' £15 10s 6d.

Third of wheat, 9 c. 3 b. 1 f. $1\frac{1}{2}$ p. Take Libertoun kirk for 9 c. 8 b., 'gif in the rest'.

Third of bere, 13 c. 8 b. 1 f. $1\frac{1}{2}$ p. Take Sanct Cudbertis kirk giving 13 c. 8 b., the rest out of Libertoun.

Third of oats, 11 c. 10 b. 2 f. $2\frac{2}{3}$ p. Take it out of Libertoun kirk paying 25 c. 8 b.[11]

Third of the meal, 1 c. 5 b. 1 f. $1\frac{1}{2}$ p.[12] Out of Tungland paying 2 c. meal.

Third of capons, 8 score and 7.[13] Out of Brochtoun, 6 score and 12, and the rest out of Ogilface giving 5 score and 19.

Third of hens, 8. Out of Brochtoun giving 24, 'and a suyne out of the same.'

Third of salt, 4 laids. 'Out of Prestoun for the myldis gevand' 12 laids.

Third of salmon, 8. Out of the barony of Kers.

---

[9] James McGill of Nether Rankeilour was Clerk Register at the Reformation, *Handbook of British Chronology*, p. 197.
[10] Robert Stewart, though a minor, was appointed commendator in 1539 and retained possession till 1568, Cowan and Easson, *Medieval Religious Houses*, p. 90.
[11] This figure for oats is given for meal on fo 104r (see above, p. 91).
[12] This is one third of 2 c. meal from Tongland, added to 2 c. oatmeal from Kers.
[13] This should read '$7\frac{2}{3}$'.

SCIENNES, PRIORY OF, (SRO, Vol. a, fo 108r-v)
(SRO, Vol. b, fo 66r-v)
'The rentall of the prioress[14] of the Sennis'

'In the first, Sanct Laurence Hous with the Spittelrig and the pertinentis set for' £44. 'Item, xv or thairby akeris of land quhairupoun our place is biggit, haldin of ane chapellane in Glasgw in few, payand thairfoir yeirlie to the said cheplane', 10 merks 'and set be us in few for doubling of the few maill, viz.,' £13 6s 8d. 'Item, ane croft within the wallis of Edinbrucht besyd the Grayfreris Port set for' 8 b. wheat and 6 b. bere. Of the laird of Glenberveis lands, £20 annual yearly. Of the laird of Calderis lands, £13 6s 8d. 'Item, of our soverane ladeis thesaurerie be fundatioun and gift', £24. 'Siclyk fra the Comptroller[15] yeirlie', 1 barrel of salmon. 'Item, of Cranstoun of that Ilk callat Kaa Cranstoun', £6 13s 4d annual. Of the lands of Ardros yearly, £20 annual. Of Alexander Murray of the Orchardis lands, £5 6s 8d. Of Adam Herkes in Abirladeis lands, £6. Of the lands of Gilmertoun 'sumtyme pertenyng to Issobell Mauchane', £13 6s 8d annual. 'Item, of the landis within Edinbrucht of umquhill Thomas Carkettill', £20 annual. 'Item, of ane land lyand at the fute of Nudreis Wynd occupyit be Johnne Hothe and utheris', £10 annual. 'Item, of landis pertenyng to the sisteris of the Naiperis[16] in sindrie places', £10 annual. 'Item, of ane hous besyd the Trinitie College', £4 annual. 'Item, of ane hous in Merleyonis Weynd presentlie occupeit be Mr Archibald Grahame', £4 13s 4d. 'Item, of ane hous at the kirk end sumtyme pertenyng to Mr Henry Spittell', 10 merkis annual; 'thairof defalcit for the birning', 40s.

Calculation of third: 'In the haill', £219 6s 8d '3';[17] third thereof, £73 2s $2\frac{2}{3}$d. Wheat, 8 b.; third thereof, 2 b. 2 f. $2\frac{2}{3}$ p. Bere, 6 b.; third thereof, 2 b. Salmon, 1 barrel; third thereof, '3 pairt barrell.'

---

[14] Christine Bellenden was prioress if this Dominican nunnery at the Reformation, *Liber Conventus S. Katherine Senensis Prope Edinburgum*, ed. J. Maidment (Edinburgh, 1841), p. 52; *RMS*, iv, no 1980.

[15] Bartholomew de Villemore, Comptroller, 1555 - 1562, with Thomas Grahame of Boquhaple, 1561 - 1562. They were succeeded by Sir John Wishart of Pittarow, 1562 - 1565. *Handbook of British Chronology*, p. 191.

[16] This presumably denotes property granted by the priory to the Napiers of Merchiston. Cf *Liber Conventus S. Katherine Senensis*, pp. lx-lxiii.

[17] This '3' presumably means divided by 3.

HOLYROODHOUSE, (SRO, Vol. a, fo 109r)
'VICARAGE OF THE PALACE OF',[18] (SRO, Vol. b, fo 67r)

Vicarage of the palace of Halyruidhous pertaining to sir William McDowgall [McDowell, *SRO, Vol. b*].

Calculation of third: In the whole, £13 6s 8d; third thereof, £4 8s 10⅔d.

EDINBURGH, (SRO, Vol. a, fo 109r)
ST GILES' COLLEGIATE KIRK, ALTARS IN,[19] (SRO, Vol. b, fo 67r)

'Tua altares in Sanct Jellis kirk pertenyng to the said sir Williame.'[20]

Calculation of third: In the whole, £31 6s 8d; third thereof, £10 8s 10⅔d.

LEITH, HOSPITAL OF ST PAUL'S WORK IN, (SRO, Vol. a, fo 109r)
(SRO, Vol. b, fo 67r)

'The hospitall of Sanct Paules Wark pertenyng to the said sir Williame.'[21]

Calculation of third: In the whole, £20; third thereof, £6 13s 4d.

KEITH, PARSONAGE AND VICARAGE OF,[22] (SRO, Vol. a, fo 109r)
(SRO, Vol. b, fo 67r)

Parsonage and vicarage of Keyth

Calculation of third: In the whole, £50; third thereof, £16 13s 4d.

---

[18] The text clearly reads 'Vicarege of the palace of Halyruidhous', though no such vicarage existed. William McDowall, chaplain, was master of works at Holyrood, Inchkeith and Edinburgh Castle, *Accounts of the Masters of Works*, eds. H. M. Paton, J. Imrie and J. G. Dunbar, 2 vols. (Edinburgh, 1957-82), i, p. xxvii.
[19] Cf *Registrum Cartarum Ecclesie Sancti Egidii de Edinburgh*, ed. D. Laing (Edinburgh, 1859). For a list of the altars, see G. Hay, 'The late medieval development of the High Kirk of St Giles, Edinburgh', *Proceedings of the Society of Antiquaries of Scotland*, cvii (1975-76), pp. 242-260, at p. 255.
[20] I.e. William McDowall.
[21] I.e. William McDowall.
[22] I.e. Keith-Marischal.

ST ANDREWS, (SRO, Vol. a, fo 109r)
ST SALVATOR'S COLLEGE, (SRO, Vol. b, fo 67r)
CRANSTOUN RIDDEL, CHAPLAINRY OF,

'The chapellanrie of Cranstoun Riddell consisting of annuellis in Edinbrucht.'

Calculation of third: In the whole, £7 8s; third thereof, 49s 4d.

EDINBURGH, (SRO, Vol. a, fo 109v)
ST GILES' COLLEGIATE KIRK, (SRO, Vol. b, fo 67v)
CHAPLAINRY OF ST SALVATOR'S ALTAR IN,[23]

'The rental of the chapellanrie of Sanct Salvatouris altar within the paroche kirk of Sanct Jeillis in Edinbrucht, extending to the sowmes efter following to be tane up of the landis efter specifeit, that is to say, of all and haill ane foir tenement of land lyand within the burcht of Edinbrucht on the south syd of the Quenis Streit at the heid of Mr James McGillis Clois pertenyng now to James Rynd in heritage, extending to the sowme of' 12 merks. 'And of all and haill sex tenementis of landis lyand in the said burcht on the south syd in the foirsaid clois, to be upliftit be the handis of the said cheplane the sowme of' 4 merks, 'with ane merk to be upliftit be the handis of the said cheplane of the landis pertenyng to umquhill Johnne Spens, lyand in the Cowgait of the said burcht on the north syd of the Hie Streit', 1 merk.

'Heirof fyve merkis to be deducit and to be gevin to the pure yeirlie, sua the haill alterage, deduceand the saidis v merkis, extendis to' 12 merks; 'and thrid thairof extendis to' 4 merks.

'Presentit be James Cor, cheplane thairof.'

Signature: 'James Cor, with my hand.'

EDINBURGH, (SRO, Vol. a, fo 109v)
THE BARRAS, CHAPLAINRY OF,[24] (SRO, Vol. b, fo 67v)

'The rentall of the cheplanrie of the Barras under the castell wall of Edinbrucht pertenyng to James Marjoribankis.'

'Item, twentie merkis yeirlie to be upliftit of the customes of the tronn of Edinbrucht fra the custumares thairof as ane yeirlie pensioun ordinar assignid and

---

[23] Cf SRO, Vol. a, fos 140v, 145r, 149r (see below, pp. 129, 135, 141.
[24] 'Capelle Beate Marie sub castro ... viz. *lie* Barres', *RMS*, iv, no 2813; now King's Stables Rd.

grantit to the sustentatioun of the cheplane thairof, conforme to the fundatioun and checker rollis porportand the same.'

Signature: 'James Marjoribankis, with my hand.'

POLTON, ST LEONARD'S CHAPEL OF, (SRO, Vol. a, fo 110r)
(SRO, Vol. b, fo 68r)

Rental of Sanct Leonardis chapel of Poltoun pertaining to the parson of Fulden, 'the haill dewtie thairof', £16.

Signature: 'Alexander Ramsay, persoun of Fulden.'

HALF HATTON AND HALF DALMAHOY, (SRO, Vol. a, fo 110r)
PREBEND OF CORSTORPHINE (SRO, Vol. b, fo 68r)
COLLEGIATE KIRK[25]

Prebend of Half Hatoun and Dalmahoy

'The rentall of ane prebendarie of Half Hatoun and Dalmahoy gevin in be Thomas Inglis, portiouner of Aldlistoun, as tutour and administratour to James Inglis his sone, prebendar thairof. Ressavit at command of the kingis warrand of the dait at Halyruidhous' 11 February 1586/7, 'the tennour quhairof followis.'

'The haill prebendarie of Half Hatoun and Dalmahoy is set in tak and assedatioun to Thomas Douglas in Dalmahoy for' £20.

Signature: 11 February 1586/7 'Thomas Inglis, with my hand.'

> 'Rex.
> Collectour generall[26] of the thridis, we greit you weill. Forsamekill as ane of the prebendareis of Corstorphing, callit Half Hatoun and Dalmahoy, presentlie bruikit and possessit be James Inglis, sone to Thomas Inglis, portiouner of Aldlistoun, as ane lait patronage and hes bene thir' 24 'yeiris bypast annuallie set in tak to Thomas Douglas in Dalmahoy for the yeirlie payment of twenty pundis lyk as the thrid thairof extending to' 10 'merkis hes bene sindrie yeiris assignit to the minister of Ratho, nevertheless as we ar informit the said prebendarie being disponit to George Douglas, sone to Williame Douglas of Quhittinghame, last possessour thairof, wes chargit yeirlie in the comptis with the sowme of'

---

[25] Cf SRO, Vol. a, fos 120v, 125v, 133v (see below, pp. 107, 112, 121).
[26] Robert Douglas, provost of Lincluden, was Collector general, *RPC*, iv, p. 167.

£40 'without ony respect had to the said tak and assedatioun and assignatioun of the thrid thairof quhilk provis manifestly the said prebendarie to extend to' £20 'allanerlie, as said is, lyk as the said George maid na payment of the thrid thairof to us or to the minister during his provisioun bot the same wes defesit and allowit to him yeirlie as gevin fre be us to ane student and bursar in respect quhairof na regaird wes takin of the just valour and estimatioun thairof to the greit prejudice of his successour quha is content to pay the just thrid thairof to us according to the new rentall of the same, thairfoir it is our will and we command you to ressave the just rentale of the said prebendarie fra the said James extending to' £20 'and according thairto that ye charge your selff on the comptis of the Collectour as of the yeris of God' 1585 and 1586 'and siclyk yeirlie in tyme cuming with the said sowme of twenty pundis allanerlie according to the tak and assedatioun thairof. Subscrivit with our hand at Halyruidhous', 11 February 1586/7. 'Sic subscribitur: James R.'

CALDER, PARSONAGE OF, (SRO, Vol. a, fos 111r-112r)
(SRO, Vol. b, fos 69r-70r)
Rental of the parsonage of Calder

Harperrig, 40s. Schelhill, 30s. Carnis Eister, 5 merks. Carnis Wester, 4 merks. Colzeame Eister, 20s. Colzeame Wester, 22s. Corswodburne, 32s. Corswod Midle, 39s. Corswodhill, 24s. Cobinschaw Ovir, 46s 8d. Cobinschaw Nethir, 20s. Camiltie, 40s. Hardwod Mekill, 12 merks. Hardwod Littill, 40s. Baddis, £3. The Dikheid, 30s. Ruschaw Eister, 2 merks. Ruschaw Wester, 2 merks. Longfuird, 4 merks. Murisdykis, 33s 4d. Mulroun, Handerswod Eister and Wester, £3 2s. Burnegranis and Breichmylne, £6 4s 8d. Sclintheuchis, 44s. Hirdinescheillis, £4 16s. Yallowstruther, 2 merks. Brethertoun, £3. Pallbethe, £4 5s. Galbesyd, 30s. Cleucheid, 27s. The Grange, 4 merks. The Commoun of Caldermure, £8 14s. Cokrig Wester, 2 merks. Cokrig Eister, 40s. Muristonis, £10. Braidschaw and Annatis Croce, £3 9s. Torphynis, £7. 'Summa lateris', £105 14s.[27]
Addeweill, £4. Lynhous, £5. Williamstoun Over, 10 merks. Longholm Myln, 2 merks. Adyistoun Over, 20 merks. Adyistoun Nather, 8 merks. Hobbetstoun, £10. Holden Over, 8 merks. Holden Nather, 6 merks. The Craig Over, 6 merks. The Craig Nather, 6 merks. Lethame, 16 merks. Pansterstoun [*i.e.* Pumpherston], 9 merks. 'Summa particule', £79 13s 4d.

'Summa of this haill rentale salvo justo calculo [*i.e.* saving a just calculation] extendis to' £185 6s 4d.[28] 'And by this sowme abonewritin thair is tua tounis quhilk payis of hors corne' 36 b. oats 'togither with half ane chalder beir; als thair

---

[27] The correct calculation appears to be £106 2s.
[28] The correct calculation appears to be £185 7s 4d.

is the vicarege, quhairof the sowme is uncertan becaus it is quhyles mair and quhylis les according to the number and plentie of store, sum yeir' £20, 'sum yeir mair; and last of all thair is a gleib with a mans of 6 acres of land 'or thairby.'

Signature: 'Johnne Spottiswode.'

Assignation of the third: In the whole, £185 5s 4d [£185 6s 4d, *preceding paragraph*]; third thereof, £61 15s 1½d. Bere, 1 c. [½ c., *preceding paragraph*]; third thereof, 5 b. 1⅓ p. Horsecorn, 26 b. [36 b., *preceding paragraph*]; third thereof, 8 b. 2 f. 2 p. '3 part half pect' [*i.e.* ⅙ p.; *recte*, ⅔ p.]

CALDER, VICARAGE OF, (SRO, Vol. a, fo 112r)
(SRO, Vol. b, fo 70r)

Vicarage of Calder

'In the haill, by vj akeris of gleib but a mans extendis to' £20; third thereof, £6 13s 4d.

DUDDINGSTON, VICARAGE OF, (SRO, Vol. a, fo 112v)
(SRO, Vol. b, fo 70v)

Vicarage of Duddingston pertaining to sir William Blakwod.

Calculation of third: In the whole, £20; third thereof, £6 13s 4d.

'Aliter gevin up and ressavit at command of the Comptroller.'[29]

In the whole, £13 6s 8d; third thereof, £4 8s 10⅔d.

'Apud Edinbrucht', 26 July 1573.

SETON COLLEGIATE KIRK, (SRO, Vol. a, fo 112v)
PROVOSTRY OF, (SRO, Vol. b, fo 70v)

Rental of the provostry of Seytoun pertaining to Mr Thomas Raithe.[30]

'The said provestrie wes set be umquhill Mr Williame Cranstoun, proveist thairof, for' £40 yearly 'and now is nocht sa mekill worth be ressoun of the doun taking of the corpis presentis, umaist claythis, Peasche fynis and offerandis quhairin it consistit maist be ressoun of the inhabitantes of the tounes of Seytoun and

---

[29] I.e. Sir William Murray of Tullibardine, *Handbook of British Chronology*, p. 191.
[30] Thomas Raith was provost, 1567 - 1582, Watt, *Fasti*, p. 373.

Wynstoun quhilkis now payis nathing. Sic subscribitur, Mr Thomas Rathe with with [sic] hand.'

'James Nicolsoun,[31] I pray you ressave this rentale. Youris at his uter powar, the Justice Clerk.'[32]

CORSTORPHINE COLLEGIATE KIRK, PROVOSTRY OF, CLERKINGTON, PARSONAGE OF,

(SRO, Vol. a, fo 113r)
(SRO, Vol. b, fo 71r)

'The provestrie of Corstorphin pertenyng sumtyme to M[r] James Scot, now to Robert Douglas.'[33]

'Personage of Clerkingtoun or Corstorphing'

'The rentale of the benifices pertenyng to M[r] James Scot, proveist of Corstorphing, maid the' 24 January 1561/2.

'Imprimis, the personage of Clerkingtoun pertenyng to the said provestrie quhilk sould giff yeirlie' £40, 'the laird of Corstorphin hes intromettit and tane up the samin thir thrie yeiris bigane and nevir pait ane penny. Item, the teynd schaves of the toun of Ratho sould giff yeirlie' 124 merks 'in the handis of Mr Robert Wynrame and the laird of Haltoun and his tennentis quhilkis ar awin thir foure yeiris bygane and als of this thair is gevin furth to the hospitall and prebendaries of Corstorphin yeirlie' 3 c. meal 'to be distributit amangis thame in money as may be cost and sauld.'

Calculation of third: In the whole, £122 13s 4d; third thereof, £40 17s 9½d.

NEWBATTLE, ABBEY OF,

(SRO, Vol. a, fos 113v-117v)
(SRO, Vol. b, fos 71v-75v)

Rental of Newbotle 'that it payis in the yeir.'

Newbotle toun: George Megot, £2 16s. James Gifhert, £2 19s. John Prymrois, 8s. William Allane, 16s 4d. John Tait, 32s 8d. Robert Thurbrand, £2 8s. George Richardsoun, 27s. James Wrycht, 10s. David Richardsoun, 20s 1d. John Huntar, 13s 4d. Janet Quhytlaw, 14s. John Bannantyne, £3 3s 6d. James Ker, 17s. John Trent, £3 11s 4d. David Skowgall, 30s 4d. John West, 17s 6d. Robert Craufuird,

---

[31] James Nicolson, clerk of the Collectory, *Thirds of Benefices*, p. 62.
[32] Sir John Bellenden of Auchnoule was Justice Clerk, *RSS*, v, nos 589, 1284.
[33] James Scott was provost, 1532 - 1565; Robert Douglas was provost, 1568 - 1585. Watt, *Fasti*, p. 348.

32s 3d. William Portius, £2 17s 8d. Patrick Johnnestoun, 6s 8d. John Moffet, 6s 8d. John Thomsoun, 9s. Robert Diksoun, 13s 4d. John Liddell, 19s 4d. James Knowis, 4s 8d. 'Wedow Lyntoun', 28s 4d. James Arrois, £3 8s 8d. Robert Moffet, £4 6s 8d. John Reid, £2 14s. 'Wedow Neilsoun', 19s 6d. Bartholomew Diksoun, £2 5s 6d. 'Lateris' , £47 17s.[34]

Robert Arrois, 9s 8d. James Yut, 19s 4d. John Jerdein, £2 13s 4d. Thomas Baldreny, 16s 8d. James Henrysoun, 6s 8d. Ninian Listoun, 6s 8d. James Youngar, 6s 8d. Oswald Portius, 19s 4d. Patrick Nicolsoun, 26s 8d. George Canny, 37s. William Cannes, 6s. George Cossour, 26s 8d. John Stevin, 13s 4d. John Henry, 13s 4d. Thomas Steill, 13s 4d. 'Wedow Falconer', 13s 4d. Thomas Carnis, 13s 4d. John Torrance, 13s 4d. David Kirkcaldie, 6s 8d. Robert Clerk, 6s 8d. George Jamesoun, 13s 4d. Thomas Turing, 13s 4d. 'Wedow Smyth', 3s. William Kay, 25s. Thomas Stevinsoun, 2s. William Diksoun, 22s 8d. Alexander Hunter, 22s 6d. Alexander Smyth, 18s. 'Wedow Brounleis', 13s 4d. 'Lateris', £23 2s 2d.[35]

'Summa of Newbotle toun in the yeir', £70 19s 2d.[36]

'The tua corne mylnes and walkmylne in the yeir', £35 6s 8d. The Eist Houssis and West Houssis in the year, £100. The Newtoun in the year, £20. Total in the year, £155 6s 8d.

Coittes: Mr John Henrysoun, £15 6s 8d. George Harrot, £6. William Harrot, £6. Thomas Wolsoun, £3 16s 8d. Total in the year, £31 3s 4d.

Maistertoun, £5. Newbyre, £6. South Sydie, Robert Reid, £6 13s 4d. Stephen Hunter, £12 13s 4d. Total, £30 6s 8d.

Morphetland in the year: Maldislie, £17 6s 8d. Huntlawcoit to Kysidhill with Mylne of Gledhoussis, £21 6s 8d. Cokside Nether, £4 12s. Coksydholl, £4. Morphet toun with the mains, £13 6s 8d. Gledhous and Coitlaw, £13 6s 8d. Herringden, James Adamsoun, £11 6s 8d. Lethenehoippes, £132. William Bischope, £12. 'Romanogra* and plewlandis with the myln in the yeir', £27 3s. Coitcoit, £3 6s 8d. Eister Denshoussis, £6 13s 4d. Wester Denshoussis, £6 13s 4d. Spirelandis, 26s 8d. Kirklandhill, £2. Curry, £3 13s 4d. Clerkingtoun Mylne, £3 6s 8d. Todhillis, £2 16s 8d. Gilmertoun Grange, £5 6s 8d. Dalhousie [*in margin*, 'kirkis'], £2 6s 8d. Laswaid [*in margin*, 'kirk'], £2 6s 8d. Badnormy, 13s 4d. Carlinglippes, £2. Kirkland of Bathcat, £21. Cranstoun, 20s. William Kernis, 14s. William Ogill, 5s. Kynpont, £3 6s 8d. Total, £325 4s.[37]

Prestoun toun in the year: Alexander Achesoun, £9. Janet Herot and her son, £18 12s. Adam Weddell, £4 2s 4d. Alexander Hammiltoun, £7 8s 8d.

---

[34] The correct calculation appears to be £47 16s 4d.
[35] The correct calculation appears to be £23 1s 2d.
[36] The sub-totals in the text yield this figure.
[37] This calculation is correct.

'Wedo Cuby', £7 12s. Barbara Douglas, 32s. John Robesoun, 16s. Alexander Reid, 16s. John Reid, 16s. James Myller, 16s. George Mot, 16s. John Woddell [Weddell, *SRO, Vol. b*], 35s 4d. Richard Henrysoun, 32s. Total, £54 14s 4d.[38]

'Prestoun be the se in the yeir', £10. 'The akeris of Prestoun with the teyndis callit the akeris of Mussilburcht pertenyng in few to James Adamsoun', £13 17s 4d. Cowthrople set in feu to James Adamsoun in the year with the teind, £21. 'Item, the rest of the akeris of Prestoun, with the mane place', £30 6s 8d. 'The annuell of ane hous in Edinbrucht pertenyng to James Adamsoun', £4. Total, £79 4s.

The barony of Barnefurd in the year: John Kirkpatrik, £4 17s. James Adamsoun, 30s. Thomas Henrysoun, £3. Andrew Murray, £16. John Fourhous, £11 4s. James Wilkie, £2 14s 8d. John Carkettle, £5 6s 8d. Richard Egger, £13 6s 8d. Total, £57 19s.

The Manes of Barnfuird: James Hammiltoun, £66 13s 4d. 'Summa', [*blank*].

Crawfuirdmure in the year: Glencaple, £13 6s 8d. The Schort Cleuch, £13 6s 8d. Glenumquhair, £13 6s 8d. Glengeith, £13 6s 8d. Pothowan, £16 5s. Fingling, £8. Over Glengouer, £8 6s 8d. Nather Glengouer, £6 13s 4d. Smythwode 'and towne', £18 13s 4d. Total in the year, £111 5s.

The barony of Monkland[39] in the year: Auchingray, £4 6s 8d. Caldercruikis, £2 6s 8d. Arnbukill, £5. Brownsyd, £5. Pedderburne, £6 6s. Fosken, £10 19s 4d. Gartlie, Rouchsollo with the mill, £6 13s 4d. Ardry with the mill, £11. Lizard, 34s 4d. Coittes and Gartturk, £3 6s 8d. Nather Coittes, £3 15s. Coittes and Overhous, 36s 4d. Dundyvane, 30s. Souterhous, 37s. Haggis, Hagmylne, Rydane, Drumgray 'with brewlandis', £26 13s 4d. Kirkwod and Drumpender, £10 5s. Potdothwan, £2 2s. Dene Bank, 20s. Garthery [Gartchery, *SRO, Vol. b*; Gartscharie, *EUL, fo 5r*], Baiklandis and Crumlat, £8 6s 8d. Kilgarth, 36s 8d. Carling croft, 8s 8d. Argownzie, 30s 8d. John Craig, 14s. William Bell, 18s 8d. William Salmand, 7s 5d. Gartwery, Gartlusken, £7 6s 8d. Kynnaird, £10 13s 4d. Gartmyllane, £3 13s 4d. Inchnok and Gayne, £6. Midrois, Myvat, Blairlynis and Gartyngailboik, £38 13s 4d. Glenhuif, £2 6s 8d. Glencors, £4. Rouchsollis, £5 0s 8d. Kypbyre, £10. Brydenhill, £5. Kyppis, £5 6s 8d. Hollynghirst, 28s. 'Summa in the yeir', £219 3s 1d.

'The teyndis of the kirkis in the yeir set in assedatioun to the persounes underwrittin': Bathcat kirk, William Home of Plandergaist, £80. Hereot kirk, John Borthwik in the Raschaw, £53 6s 8d. The parsonage of the kirk of Newbotle set in assedation to Walter Ker of Cosfuird [*i.e.* Cessford], £80. The kirk of Cowpen [Cokpen, *SRO, Vol. b*] 'offerit be the parochineris', 120 b. 'meill aites.' The vicarage of Newbotle set in assedation to George Adamsoun for 50 merks 'and now worthe' 30 merks.

---

[38] The correct calculation appears to be £55 14s 4d.
[39] Many of these place-names also appear in the rental for the subdeanery of Glasgow, parish of Monkland, EUL, fo 4r (see below, p. 497).

'Summa totalis', £1,413 1s 2d.[40]

'Debursit of this rentale yeirlie to thir persounes underwrittin as efter followis: Item, of sex decraipit and recantit monkis', £240. 'Item, to the laird [of] Cosfurd, ballie principall of the baronie of Newbotle, and his depuiteis yeirlie', £33 6s 8d. 'Item, to the ballie of Craufurdmure, his officer and servandis yeirlie', £20 13s 4d. 'Item, to Johnne Crawfurd, ballie of the Monkland, his depuiteis and officiares', £42. 'Item, to Bartilmo Layne, pensionar, in the yeir', £10. 'Item, to my Lord [Lordis, *SRO, Vol. b*] of the College of Justice in the yeir for the contributioun', £28.

'This is the just rentale that I get payt, bayth of auld and augmentatioun, and siclyk the just sowme debursit yeirlie of debt furth of the samin, subscrivit with my hand. Mark of Newbotill.'[41]

Assignation of the third:
  Third of the money, £444 13s 4d.[42] Take the barony of Bar[n]fuird for £124 12s 4d; the Lethinhopes for £132; the Williamhope[43] for £12; the Eist Hous and West Hous for £100; 'the corne mylne and walkmylne', £35 6s 8d; the Newtoun, £20; the Coites, £31 3s 4d. 'Gif in' £10 9s 1d.[44]
  Third of the victual, 40 b. Out of the kirk of Cokpen giving 120 b.; 'the denerie of Restalrig', £31 2s 2⅔d;[45] 'out of the handis of the erle of Mortoun[46] for the teyndis of Dalkeyth, Sande Riggis and tua coldennis quha gevis thairfoir', £93 6s 8d; 'the victuallis contenit in the roll man be chargit in generall becaus thai ar nocht speciallie condiscendit on'.

*Mussilburcht, £364.

'Of Dunfermelyng': Wheat, 6 c. 3 b. Bere, 3 c. 5 b. 2⅓ p. 'Quhyt aites', 5 c. 10 b. 2 f.

'Item, inhibit the haill coil quhill thai cum and condiscend upoun the thrid.

---

[40] The addition of the sub-totals in the text is £1,448 11s 11d.
[41] Mark Ker was appointed commendator of Newbattle in 1547; and his son, also Mark, was presented to the abbacy in 1567. Cowan and Easson, *Medieval Religious Houses*, p. 77.
[42] One third of £1,413 1s 2d is £471 0s 4⅔d.
[43] This is rendered 'Williame Bischope' on fo 114v, and in SRO, Vol. b, fo 75v.
[44] These figures do not tally.
[45] Cf SRO, Vol. a, fos 129v, 149v (see below, pp. 116, 141, and nn. 87, 149), rental for Restalrig: 'the teyndis of Dalkeyth, Sanderigis and tua coldenis set to the erle of Mortoun in assedatioun for' £93 6s 8d.
[46] I.e. James Douglas, 4th earl of Morton, *Scots Peerage*, vi, pp. 362-363.

Item, that thai answer nane bot the quenis grace of aragies, caragies, grassoumes, caynes, entrie silver and uther dewteis.'[47]

KIRKNEWTON, PARSONAGE OF,   (SRO, Vol. a, fo 118r-v)
                                                              (SRO, Vol. b, fo 76r-v)

Rental of the parsonage of Kirknewtoun pertaining to Mr James Broun.

'Item, Eister Newtoun payis be yeir' 40 merks. Kirknewtoun pays 'be yeir', 40 merks. Overtoun pays 'be yeir', 12 merks. Leidoun pays 'be yeir', 12 merks. Lethinseid pays 'be yeir', 14 merks. 'And the meill thairof payis' 30s. The Hous of the Muir pays 30s. Ormestoun Hill pays 30s. The Humbie with the Newlandis pays 12 merks, 'albeit sumtyme it payt mair bot I gat nevir mair of it.' The glebe pays 'be yeir', 16 merks for teind and meal. 'Item, the tounes abonewrittin payis certane hors corne existand [*recte*, extendand] to' 7 b. 'skarslie will sustein the ministeris hors.' Total, 152 merks 10s.

'Item, wes gersum out of this yeirlie to the bischope',[48] 2 merks 6s 8d. 'Item, to the prebendareis of Sanct Jeillis kirk', 40s. 'Item, to the vicar pensionarie', £10, 'and ane pairt of the glieb [*i.e.* glebe] land. Item, for pointing and gleissing of ane queir and kirk yeirlie', 40s.[49]

Signature: 'Jacobus Broun, manu propia.'

Calculation of third: In the whole, 176 merks 16s 8d; third thereof, 58 merks 14s 5⅓d, 'convertit in pundis extendis to' £39 6s 9⅓d.[50]

'Assumptioun of the personage of Kirknewtoun gevin in be the persoun him selff, 1573.'

Janet Purves, 33s 4d. James Biggar, 33s 4d. Edward Youngar, 20 merks. John Cunyghame, 16s 8d. Thomas Scot, 16s 8d. Alexander Borthuik, £5. Janet Flent, 16s 8d. Giles Allane, 16s 4d. Philip Thomsoun, 33s 4d.
      Leidtoun, 12 merks: James Cudbertsoun, 3 merks; Andrew Johnnestoun, 3 merks; John Mure, 3 merks; Thomas Cunyghame, 3 merks.

---

[47] This intrusive section from * belongs to the rental of Dunfermline where its exact location in the text is indeterminate; fo 44r reads 'Mussilburcht in penny maill and annuals, £364'; fo 45r reads 'Mussilburcht ... 6 c. 3 b.' (see above, pp. 37, 38).

[48] I.e. John Hamilton, archbishop of St Andrews, 1546 - 1571, Watt, *Fasti*, pp. 298-299.

[49] These items added to 152 merks 10s total 175 merks 16s 8d.

[50] 58 merks 14s 5⅓d is one third of 176 merks 16s 8d; if converted into pounds, the figure would be £39 7s 9⅓d.

Overtoun pays £4:[51] Donald Aikman, 'wodouie',[52] 33s 4d. Robert Smyth, 16s 8d. Margaret Pharar, 'wedow', 16s 8d. James Greg, 20s. John Purves, 6s 8d. 'The sowme of the haill', £39 6s 8d.

Signature: 'I assignne this for the thrid of my personage of Kirknewtoun, sic subscribitur, Mr James Broun, persoun of Kirknewtoun, with my hand.'

ST CUTHBERT'S KIRK, VICARAGE OF,   (SRO, Vol. a, fo 119r)
        (SRO, Vol. b, fo 77r)

Rental of Sanct Cuthbertis Kirk

'Item, the haill dewteis thairof extendis to' 100 merks 'in tymes bygane quhen all dewteis wes payit thairof and now croce presend and upmest claithe and Pasche offering with the kirkland and sua defalkand the croce presend, upmest claithe and Pasche offering, restis de claro' 50 merks.

Signature: 'M[r] Archibald Hammiltoun, vicar, Sanct Cuthbertis Kirk.'

Calculation of third: In the whole, £33 6s 8d; third thereof, £11 2s 2⅔d.

'The vicarege of Sanct Cuthbert kirk gevin in be M[r] Archibald Hammiltoun.'

EDINBURGH,   (SRO, Vol. a, fo 119v)
ST MARY OF THE FIELDS' COLLEGIATE KIRK,   (SRO, Vol. b, fo 77v)
PROVOSTRY OF,[53]

Kirkafeild

'Rentale prepositure collegiate ecclesie Beate Marie de Campis'

Rental of the provostry of the collegiate kirk of Blessed Mary of the Fields near Edinburgh, in St Andrews diocese, entirely at the provision and ordinary collation of the lord archbishop of St Andrews[54] by reason of the foundation and erection of the same, of which Mr Alexander Forrest is lawfully provost[55] and possessor from 1553. At Lynlythgow, 23 January 1561/2.

---

[51] The figures given immediately below total £4 13s 4d.
[52] This probably should read 'wedow', which in Scots also denoted 'widower'.
[53] Cf SRO, Vol. a, fos 120r, 132r (see below, pp. 106, 119).
[54] See above, p. 104, n. 48.
[55] Alexander Forrest was provost of St Mary of the Fields, 1552 - 1561, Watt, *Fasti*, pp. 358-359; cf below, p. 120, n. 99.

'Prepositura que habet vicariam de Levingstoun'

The provostry which has the vicarage of Levingstoun annexed and united to it for the upkeep of the said provost alone, by reason of the foundation of the same, extending annually in wool, lambs and Pasch fines and certain other small teinds along with fermes of kirk lands of the said vicarage from of old, and in all calculated together in the year according to the assedation and leasing apart from the manse of the same founded by Mr David Vocat, extending to £48.

'Memorandum quod prebendarij mihi fundati'

Memorandum, that the prebendaries [sic] founded to me according to the foundation of the patrons of the vicars of the same, according to their foundations, are ten,[56] some of which have the sum of 20 merks, others £10, others £20, along with a manse and yards appropriated to each.

CURRIE, VICARAGE OF,[57]
(SRO, Vol. a, fo 120r)
(SRO, Vol. b, fo 78r)

Vicarage of Curry [*In margin*, 'The archedeane of Lowthian.']

'The vicarege of Curry pertenyng to sir Mare [*recte*, Mark] Jamesoun as vicar pensionar of the samin in small teyndis with teynd hay of Curry that Kinspindie hes nocht set in assedatioun laitlie, payand the curat his fe for' £20 'and I gat nevir penny thir thrie yeiris of baithe the saidis vicaregis.'

[*In margin*, 'Aliter gevin up efterward.'[58]]

Signature: 'sir Mare [*recte*, Mark] Jamesoun, with my hand.'

Calculation of third: In the whole, £20; third thereof, £6 13s 4d.

EDINBURGH,
ST MARY OF THE FIELDS' COLLEGIATE KIRK,
PROVOSTRY OF,[59]
(SRO, Vol. a, fo 120r)
(SRO, Vol. b, fo 78r)

'The rentall of the provestrie of Kirkafeild gevin in be Williame Pennycuik, persoun thairof.'[60] [*In margin*, 'Aliter gevin up Kirkafeild']

---

[56] Cf Cowan and Easson, *Medieval Religious Houses*, p. 220.
[57] Cf SRO, Vol. a, fo 124r (see below, p. 110).
[58] I.e. on fo 124r (see below, p. 110).
[59] Cf SRO, Vol. a, fos 119v, 132r (see above, p. 105; below, p. 119).
[60] See below, p. 120, n. 99.

'In the haill yeirlie extendis to' £16 'computand the vicarege of the kirk of Levingstoun annexit thairto.'

Signature: 'Williame Pennycuik. Ressavit at command of the Clerk Registre.'

Calculation of third: In the whole, £16; third thereof, £5 6s 8d.

LIBERTON, VICARAGE OF, (SRO, Vol. a, fo 120v)
(SRO, Vol. b, fo 78v)

'The benifice and cheplanreis pertenyng to Mr Alexander Chalmer.'

The vicarage of Libertoun set for £28 13s 4d.

Calculation of third: In the whole, £28 13s 4d; third thereof, £9 11s 1⅓d.

CORSTORPHINE COLLEGIATE KIRK, (SRO, Vol. a, fo 120v)
PREBENDS OF,[61] (SRO, Vol. b, fo 78v)

'The thrid of the prebendareis of Corstorphin: The thrid of the prebendarie of Corstorphin callit Half Haltoun [*i.e.* Hatton] and Half Dalmahoy pertenyng sumtyme to sir Johnne Greinlaw', £13 6s 8d. 'The thrid of the prebendarie of Corstorphin callit Half Gogar and Half Aldstoun [*i.e.* Addiston] to Mr Alexander Coill', £8 17s 9d. 'The thrid of the prebendarie of the college kirk of Corstorphin callit Half Mortoun [*recte*, Norton] and Half Byres [*recte*, Ratho Byres] pertenyng to Mr George Lauder extending to' £10 4s 6d. 'The thrid of the prebendarie of Corstorphin callit Half Bruntoun [*recte*, Bonnington], Half Plat pertenyng to Mr Thomas Marjoribankis', £8 17s 9d. 'The thrid of the prebendarie of Corstorphin callit Half Mortoun, Half Byres pertenyng to Mr James Gray', 2⅔ b. wheat, 4 b. bere, 10⅔ b. oats. 'The prebendarie of Half Dalmahoy and Half Hatoun pertenyng to Mr James Wilkie, the thrid extending to' £13 6s 8d. 'The prebendarie of Half Gogar and Half Adestoun pertenyng to Mr Niniane Borthuik', £11 2s 2⅓d. 'The thrid of the prebendarie pertenyng to Robert Douglas of Half Bruntoun [Half Lunetoun, *SRO, Vol. b*] and Half Plat extending to' £8 17s 9⅓d.[62]

---

[61] The prebends of Corstorphine were Half Hatton and Half Dalmahoy; Half Gogar and Half Addiston; Half Norton and Half Ratho Byres; Half Bonnington and Half Platt; cf SRO, Vol. a, fos 110r, 122v, 125v, 126r-v, 128v, 129r, 133v, 141v (see above, p. 97; below, pp. 109, 112, 115, 116, 121, 130).

[62] The next entry in SRO, Vol. b, fo 79r, corresponds to SRO, Vol. a, fo 123r (see below, p. 109).

LINTON, PARSONAGE OF,[63]   (SRO, Vol. a, fo 121r)

'I, Mr Walter Balfour, persoun of Lintoun, makis knawin that the personage of Lintoun wes nevir estimat in the auld rentale of Kelso[64] above the sowme of' £36 13s 4d 'and in my assedatioun it is set for the sowme of' £100 'bot I esteim it na better nor' £80 'quhilk I manifestit to lait and suffer the tennentis to intromet thairwith for the sowme of' £80 'and this I did ane yeir bygane and promesit the samin to the parochineris and this I mak manifest be this my writting writtin and subscrivit with my hand 1561, 27 Januar [*i.e.* 1561/2]. Sic subscribitur, Maister Walter Balfour, with my hand.'

Calculation of third: In the whole, £80; third thereof, £26 13s 4d.

NEWHAVEN, CHAPLAINRY OF,[65]   (SRO, Vol. a, fo 121v)

'Chapell of Newhevin'

'The rentale of the cheplanrie of the chapell of the Newhevin extendis to' £16 yearly.

Signature: 'Maister Johnne Balfour, with my hand.'

Calculation of third: In the whole, £16; third thereof, £5, 6s 8d.

MIDDLETON,   (SRO, Vol. a, fo 122r)
PREBEND OF CRICHTON COLLEGIATE KIRK

'The rentale of the prebendarie of Midltoun with the pertinentis extendis in the yeir to' £40, 'of the quhilk thair is gevin furth to the tua preistis within the college kirk of Creichtoun' 8 merks yearly.

Signature: 'Thomas Nicolsoun, prebendar of Midltoun.'

Calculation of third: In the whole, £40; third thereof, £13 6s 8d.

---

[63] Cf SRO, Vol. a, fo 253v (see below, p. 253). Linton (now West Linton) lay in the sheriffdom of Peebles; it is unclear why it is placed among rentals for Edinburgh.
[64] Linton was appropriated to Kelso abbey, Cowan, *Parishes*, p. 133; cf SRO, Vol. a, fo 224r (see below, p. 224).
[65] I.e. the chapel of Our Lady and St James at Newhaven, *RMS*, vii, no 1015; cf F. H. Groome, *Ordnance Gazetteer of Scotland* (Edinburgh, 1901), p. 1237.

VOGRIE, (SRO, Vol. a, fo 122v)
PREBEND OF CRICHTON COLLEGIATE KIRK

The prebend of Vogarie pertaining to Mr Thomas Bannatyne extends to £20.

Calculation of third: In the whole, £20; third thereof, £6 13s 4d.

HALF NORTON AND HALF RATHO BYRES, (SRO, Vol. a, fos 122v-123r)
PREBEND OF CORSTORPHINE (SRO, Vol. b, fo 79r)
COLLEGIATE KIRK[66]

'Ane uther rentall of the prebendarie of Corstorphin callit Half Mortoun [*i.e.* Norton] and Half Rathobyers ressavit at command of the Clerk Registre, pertenyng to Mr James Gray.' The teind sheaves thereof extending in wheat to 8 b.; in bere, 12 b.; in oats, 2 c.

'Aliter. Ane prebendarie of Corstorphin callit Half Mortoun [*i.e.* Norton] and Half Rathobyers pertenyng to Mr James Gray set in assedatioun to the tennentis for' 85 merks yearly.

Signature: 'James Nicolsoun, ressave this rentale of silver and the uther of vict[ual] and charge bot this for the thrie scoir tua yeris [*i.e.* 1562] and in tyme cuming. Johnne Wischart, Comptroller.'[67]

'The teynd schaves of the Half of Mortoun [*i.e.* Norton] and siclyk of the Half of Rathabyeris pertenyng to Maister James Gray, extending to' 36 b. oats; 14 b. bere; 10 b. wheat; 'and payit out of this to the stallar' 24 merks, 'and to the boyis that sang in the queir', 12 b. meal.

Signature: 'Mr James Gray, with my hand.'

Calculation of third: 'In the haill of aites', 36 b.; third thereof, 12 b. Bere, 14 b.; third thereof, 4 b. 2 f. $2\frac{2}{3}$ p. Wheat, 10 b.; third thereof, 3 b. 1 f. $1\frac{1}{3}$ p. [*In margin*, 'The haill to the quein.']

---

[66] Cf SRO, Vol. a, fos 120v, 126v, 141v (see above, p. 107; below, pp. 113, 130).
[67] The corresponding text in SRO, Vol. b resumes at the start of the next paragraph. See above, p. 107, n. 62.

TEMPLE, VICARAGE OF, (SRO, Vol. a, fo 123v)
(SRO, Vol. b, fo 79v)

Rental of the vicarage of the Tempill 'apertenyng to the lord Sanct Johnnes'[68] extends to £30.

Calculation of third: In the whole, £30; third thereof, £10.

'Aliter gevin up: In the haill', £10; third thereof, £3 6s 8d.

DALKEITH, VICARAGE OF, (SRO, Vol. a, fo 124r)
(SRO, Vol. b, fo 80r)

'Rentale pro Magistro Da[vido] [An[dreo], *SRO, Vol. b*] Davidsoun.'[69]

The vicarage of the kirk of Dalkeith in St Andrews diocese in each year, 50 merks according to common estimation, of which nothing has been received annually these last two years.

Calculation of third: In the whole, £33 6s 8d; third thereof, £11 2s 2⅔d.

KILSPINDIE, VICARAGE PORTIONARY OF,[70] (SRO, Vol. a, fo 124r)
(SRO, Vol. b, fo 80r)

[*In margin*, 'This lyis and is writin in Perth']

'The rentale of the vicarege portionarie of Kilspindie, wa[n]tand lamb and woll, extendis yeirlie to' 20 merks 'pertenyng to Mark Jamesoun'.

CURRIE, VICARAGE PORTIONARY OF,[71] (SRO, Vol. a, fo 124r)
(SRO, Vol. b, fo 80r)

'The vicarege portionarie of Curry pertenyng to the said Marc [*i.e.* Jamieson] wantand lamb and woll extendis to' 20 merks.

'Thir ar ressavit at command of the Clerk Register.'

---

[68] I.e. James Sandilands, preceptor of Torphichen, otherwise lord St John, Cowan and Easson, *Medieval Religious Houses*, p. 161.
[69] Andrew Davidson, *RSS*, v, no 3042.
[70] This benefice, in the sheriffdom of Perth, is included in rentals for Edinburgh because the benefice-holder was Mark Jamieson, who also held Currie.
[71] Cf SRO, Vol. a, fo 120r (see above, p. 106).

EDINBURGH, TRINITY COLLEGIATE KIRK[72]

(SRO, Vol. a, fo 124r)
(SRO, Vol. b, fo 80r)

Rental of the Trinitie College beside Edinbrucht pertaining to sir Patrick Spreull.

[*In margin*, 'Nota, this is for the haill college, [bayth, *SRO, Vol. b*] provest and prebendaries.']

Calculation of third: In the whole, £362 6s 8d; third thereof, £120 15s 6d.[73]

BORTHWICK, VICARAGE PORTIONARY OF,

(SRO, Vol. a, fo 124v)
(SRO, Vol. b, fo 80v)

Vic[arage of] Borthuik

Rental of the vicarage portionary of Borthuik

'Inprimis the teynd schaves of the landis of Over and Nather Halkarstonis and Torcraik with the teind vicarege of the samin quhen the kirk rychtis, offerandis and uther thingis usit and wont wer payit of vicaregis within the said paroche of Borthuik and throucht all Scotland, the haill extendis be yeir supposit I imbursit nocht ane greit [*recte*, groat] thairof thir four yeiris bypast to' £20 'bot now sen kirk richtis, offerandis and the best pairt of the vicarege teyndis ar dischargit I can nocht wit quhat the valour thairof may extend to without the parochineris wer examinit thairon in thair conscience.'

Signature: 'Ita est Nicolaus Hay, vicarius de Borthuik, subscripsit.'

Calculation of third: In the whole, £20; third thereof, £6 13s 4d.

EDINBURGH, RAVELSTON, PREBEND OF ST GILES' COLLEGIATE KIRK

(SRO, Vol. a, fo 125r)
(SRO, Vol. b, fo 81r)

'The rentale of ane prebendarie of Sanct Jeillis kirk'

'Inprimis, Railstoun set in few to Williame Symsoun and to Johnne Heriot, quha payis for the samin yeirlie the sowme of' £42 13s 4d. 'Item, pertenyng to the prebendarie, ane pairt and portioun of the teynd schaves of Damberny [*i.e.* Dunbarny[74]], Potye and Moncreif, the quhilkis teynd schaves ar set in tak and

---

[72] Cf SRO, Vol. a, fo 131r (see below, p. 118).
[73] The correct calculation appears to be £120 15s 6⅔d.
[74] 'Haill in annuellis and commoun of Dumbarny', fo 126r (see below, p. 113).

assedatioun to the maister of Oliphant[75] for' £180 'quhilk pairt and portioun extendis yeirlie to' £10, 'of the quhilk I have gottin na payment thir thrie yeiris and ane half bypast.'

Signature: 'Maister Niniane Hammiltoun.'

Calculation of third: In the whole, £42 13s 4d; third thereof, £14 4s 5$\frac{1}{3}$d, 'by commoditeis.'

HALF DALMAHOY AND HALF HATTON, PREBEND OF CORSTORPHINE COLLEGIATE KIRK[76]

(SRO, Vol. a, fo 125v)
(SRO, Vol. b, fo 81v)

[*In margin*, 'Prebend callit Dalmahoy and Half Haltoun']

'The prebendarie of the college kirk of Corstorphin callit Dalmahoy and Half [Ha]toun extendis to' £40.

Calculation of third: In the whole, £40; third thereof, £13 6s 8d.

CRANSTON, VICARAGE OF,

(SRO, Vol. a, fo 125v)
(SRO, Vol. b, fo 81v)

The vicarage of Cranstoun, £33 6s 8d.

Calculation of third: In the whole, £33 6s 8d; third thereof, £11 2s 2$\frac{2}{3}$d.

HALF GOGAR AND HALF ADDISTON, PREBEND OF CORSTORPHINE COLLEGIATE KIRK[77]

(SRO, Vol. a, fo 126r)
(SRO, Vol. b, fo 82r)

'The rentale of the prebendarie of Corstorphin callit Half Gogar and Alderstoun pertenyng to Mr Alexander Coill.'

'Item, the profeitis of the said prebendarie set to James Forrester of Corstorphin for' £26 13s 4d.

Calculation of third: In the whole, £26 13s 4d; third thereof, £8 17s 9$\frac{1}{3}$d.

---

[75] Laurence, master of Oliphant, son of Laurence, 3rd lord Oliphant, *Scots Peerage*, vi, pp. 544-548.

[76] Cf SRO, Vol. a, fos 110r, 120v, 133v (see above, pp. 97, 107; below, p. 121).

[77] Cf SRO, Vol. a, fos 120v, 128v (see above, p. 107; below, p. 115).

[*In margin,* 'The prebendarie in Corstorphin callit Half Gogar and Alderstoun pertenyng to Ninian Borthwik.']

Calculation of third: In the whole, £33 6s 8d; third thereof, £11 2s 2⅔d.

EDINBURGH, (SRO, Vol. a, fo 126r)
ST GILES' COLLEGIATE KIRK, (SRO, Vol. b, fo 82r)
PREBEND OF ST GILES IN,

'The prebendarie of Sanct Jeillis kirk pertenyng to the said Niniane [*i.e.* Borthwick]. Haill in annuellis and commoun of Dumbarny[78] to' £30; third thereof, £10.

HALF RATHO BYRES AND HALF NORTON, (SRO, Vol. a, fo 126v)
PREBEND OF CORSTORPHINE (SRO, Vol. b, fo 82v)
COLLEGIATE KIRK[79]

'The rentale of the prebendarie of the college kirk of Corstorphin callit Byres and Half Mortoun [*recte*, Nortoun] set for' 46 merks money.

Calculation of third: In the whole, £30 13s 4d; third thereof, £10 4s 5⅓d.

LEITH, HOSPITAL OF ST ANTHONY (SRO, Vol. a, fos 127r-128r)
(SRO, Vol. b, fos 83r-84r)

'This is the rentale of the Hospitalitie of Sanct Anthonis besyd Leith.'

'The procurationis or pardones heirof: Item, inprimis, Alexander Curry, lemitur in Cunyghame, Carrick, Kyle and Galloway, for the boundis of the samin payand yeirlie', 20 merks. 'Item, Alexander Pettie and Symon Henrysoun for the boundis of Annandaill, Nythdisdaill, Wachopdaill, Esdaill, Ewisdaill, and the north syd of Cliedisdaill with the Lennox', 20 merks. 'Item, sir Neill Wody, haveand the Mers and Teviotdaill with the rest of his boundis payand' 20 merks. 'Item, the eist end of Lothiane set to Johnne Forsyth for' 10 merks. 'Item, set to Johnne Johnnestoun, the west pairt of Lowthiane with the rest of his boundis for' 10 merks. 'Item, David Quhytheid and Johnne Forrester payand for the boundis of Fyffe, Strivelingschyre and the halff of Stratherne' 20 merks. 'Johnne Barclay for the half of Stratherne payis' 22 merks. 'David Gordoun payand for the boundis of the

---

[78] Cf fo 125r (see above, p. 111), 'ane pairt and portioun of the teynd schaves' of Dunbarny.
[79] Cf SRO, Vol. a, fos 120v, 122v, 141v (see above, pp. 107, 109; below, p. 130).

north countrye, the haill diocie of Abirdene, Murray, Ros, Caithnes, Scheitland, Orknay', 30 merks. 'The bred and offerand stok', 40 merks.

'The annuellis pertenyng to the said hospitalitie: In the first, Williame Symsounes land callit the Stobis, of annuell yeirlie', 6 merks. 'Johnne Balfour quhair he duellis, payand yeirlie of annuell', 5 merks. 'For his over place lyand nixt Sanct A[n]thonis yaird,' 20 merks. 'Mair for his kill', 3d. John Mathesounes land, 6s 8d. 'For his obite yeirlie' 10s. John Homes land, 3 merks. Richard Youngis land, 2 merks. John Dikesounes land, 4 merks 8d. William Someris land, 20 merks. John Crokatis land, £3 4s. Walter Keris land 'that wes Jeffrayis land', 20s. David Quhyt, 'his land on hill', £3 12s. John Gallawayis land, 40s. Andrew Blairis land, £4. Andrew Grayis land, 7 merks. Culros land, 6s 8d. David Currellis land, 6s 8d. William Bartanes land 'at the altar stane', 10s. Thomas Martonis land, 3d. John Keris land, 4d. Robert Bartanes land 'that is Comptrollar',[80] 13s 4d. Alexander Lyallis land 'that is on the sandis', 40d. John Boymanis land 'on the sandis', 40d. Robin Aldokes land, 10s. Alexander Lyellis land 'nixt the linkis', 10s. George Knychtsones land, 12d. 'Johnne of Stanis land at the alter stane besyd Johnne Loganes land', 4s. 'Ane peice west land besyd Sandie Lambis land foiranent his awin hous', 4s. 'Carkettillis land on the schoir', 8 merks 2s. John Cantes land 'on the schore nixt Carkettillis land', 46s. John Orknayis land, 36s. Patrick Grayis land, 2 merks. John Wauchis land in Edinbrucht, 5 merks. Mavie Brounes land 'behind the tolbuith', 5 merks. 'Humes the cuitleris land under the wall', 4s. Cowperis land in the Cannogait, 9s. Prestones land in Mussilburcht, 15s. 'The abbat Abirbrothok[81] land at the brig end of Leith', 13s. John Curreis land, 'our procuratour in Air', 6s 8d. 'Item, the pairt of the kirk of Haillis pertenyng to the said hospitalitie of Sanct A[n]thonis set to thir persounes, that ar to say, Mathow Spreull, sir Williame Younger, sir Thomas Michelsoun, extending yeirlie payment as it is set for' 100 merks, with £10 'for the vicaris pensioun.'

Signature: 'Matthew Forrester.'

Calculation of third: In the whole, £70 11s 10d;[82] third thereof, £23 10s 7½d.

---

[80] Robert Barton of Over Barnton held the office of Comptroller, 1516 - 1525; and 1529 - 1530, *Handbook of British Chronology*, p. 191.
[81] John Hamilton was commendator of Arbroath at the Reformation, *RSS*, v, no 837.
[82] This appears to be a miscalculation.

LASSWADE, VICARAGE OF,

(SRO, Vol. a, fo 128v)
(SRO, Vol. b, fo 84v)

'The vicarege of Laswaid. It wes set in auld to Johnne Ballantyne and uther parochineris for' £40 'quhen all dewteis wes payit as corspresent and umest claithe.'

Signature: 'Ita subscribitur, M[r] Johnne Manderstoun, vicar of Lassuaid.'

Calculation of third: In the whole, £40; third thereof, £13 6s 8d.

HALF GOGAR AND HALF ADDISTON, PREBEND OF CORSTORPHINE COLLEGIATE KIRK[83]

(SRO, Vol. a, fo 128v)
(SRO, Vol. b, fo 84v)

'The rentale of the prebendarie of Corstorphin callit Half Gogar and Half Alderstoun pertenyng to Mr Alexander Coill.

Item, the said haill prebendarie is set be me, the said Alexander, to James Forrester of Corstorphine for payment of the sowme of' 40 merks 'as the assedatioun heir present to schaw proportis', 40 merks.

Calculation of third: In the whole, £26 13s 4d; third thereof, £8 17s 9⅓d.

DALKEITH COLLEGIATE KIRK, PROVOSTRY OF,

(SRO, Vol. a, fo 129r)
(SRO, Vol. b, fo 85r)

Rental of the provostry of Dalkeyth pertaining to Mr Robert Douglas, provost thereof.[84]

'The said provestrie is worthe be yeirlie maill becaus the samin is all temporall landis and tane up fra the occupyeris thairof extending yeirlie lyk as in all tymes hes bene yeirlie upliftit and tane up thairof', 40 merks.

Calculation of third: 'In monie', £26 13s 4d; third thereof, £8 17s 9⅓d.

---

[83] Cf SRO, Vol. a, fos 120v, 126r (see above, pp. 107, 112).
[84] Robert Douglas is not listed among the provosts of Dalkeith, but no names are entered for 1559 - 1619. Watt, *Fasti*, p. 351.

HALF BONNINGTON AND HALF PLATT, PREBEND OF CORSTORPHINE COLLEGIATE KIRK[85]

(SRO, Vol. a, fo 129r)
(SRO, Vol. b, fo 85r)

'The rentale of the prebendarie of Corstorphin callit Half Bonitoun [Bonyntoun, *SRO, Vol. b*], Half Platt.'

'The said prebendarie is in all tymes bygane payit be the persounes that hald the samin in possessioun and tak to the sowme of' 40 merks 'lik as I and my predicessoris hes set the samin of befoir.'

Signature: 'Robert Douglas, proveist and prebendar foirsaid, withe my hand.'

Calculation of third: In the whole, £26 13s 4d; third thereof, £8 17s 9⅓d.

RESTALRIG COLLEGIATE KIRK, DEANERY OF,

(SRO, Vol. a, fo 129v)
(SRO, Vol. b, fo 85v)

'The denerie of Restalrig'

'The fructes of the deanrie of Restalrig as it payis presentlie and commounlie thir divers yeiris bygane immediatlie preceding quhen my maister the deane[86] gat full payment thairof, extendis to' 9 b. rye; 2 c. 4 b. 2 f. wheat; 6 c. 6 b. bere; 20 c. 14 b. oats. 'Item, certin teyndis led be my said maister the dene on his awin expenssis commounlie extending to' 3 b. 3 f. 2 p. rye; 6 b. wheat; 22 b. bere; 2 c. 8 b. oats. 'Item, the teyndis of Dalkeyth, Sanderigis and the tua coldenis set to the erle of Mortoun in assedatioun for' £93 6s 8d.[87]

Calculation of third: Money, £93 6s 8d; third thereof, £31 2s 2⅔d. Rye, 12 b. 3 f. 2 p.; third thereof, 4 b. 1 f. ⅔ p. Wheat, 2 c. 10 b. 2 f.; third thereof, 14 b. 2⅔ p. Bere, 7 c. 12 b.; third thereof, 2 c. 9 b. 1 f. 1⅓ p. Oats, 23 c. 6 b.; third thereof, 7 c. 12 b. 2 f. 2⅔ p.

PENTLAND, PARSONAGE AND VICARAGE OF,

(SRO, Vol. a, fo 130r)
(SRO, Vol. b, fo 86r)

'The personage and vicarege of Pentland annexit to the college of Roslein', 160 merks.

---

[85] Cf SRO, Vol. a, fos 120v, 133r (see above, p. 107; below, p. 121).
[86] John Sinclair was dean of Restalrig, 1542 - 1566, Watt, *Fasti*, p. 370.
[87] Cf SRO, Vol. a, fo 117v (see above, p. 103), rental for Newbattle: 'Out of the handis of the erle of Mortoun for the teyndis of Dalkeyth, Sande Riggis and tua coldennis quha gevis thairfoir' £93 6s 8d; cf pp. 103, 141, nn. 45, 149.

Signature: 'Joannes Robesoun, prepositus de Roslin, manu propria.'

Calculation of third: In the whole, £106 13s 4d; third thereof, £35 11s 1⅓d.

LOTHIAN, ARCHDEACONRY OF, (SRO, Vol. a, fo 130r)
CURRIE, PARSONAGE OF, (SRO, Vol. b, fo 86r)

'The archidenrie of Lowthiane, personage of Curry', £240.

Calculation of third: In the whole, £240; third thereof, £80.

GOGAR, PARSONAGE AND VICARAGE OF, (SRO, Vol. a, fo 130v)
(SRO, Vol. b, fo 86v)
Rental of the parsonage and vicarage of Gogar

'The yeirlie availl of the saidis benifices extendis in the haill profeitis to' 100 merks, 'and is set in assedatioun for the samin sowme.'

Signature: 'Joannes Lermonth.'

Calculation of third: £66 13s 4d; third thereof, £22 4s 5⅓d.

EDINBURGH, (SRO, Vol. a, fo 130v)
ST GILES' COLLEGIATE KIRK, (SRO, Vol. b, fos 86v, 89b[88])
CHAPLAINRY OF NOMINE JESU[89] AND THE HOLY BLOOD[90] IN,

'The chapellanrie of Nomine Jesu and Halywod [Halyblude, *SRO, Vol. b*] pertenyng to sir George Littiljhonne.'

Calculation of third: In the whole, £28 12s; third thereof, £9 10s 8d.

'Ane uther rentale of the samin cheplanrie gevin in be Mr Johnne Craig.'

---

[88] SRO, Vol. b, preceding fo 89r, contains a scrap of paper. One side [fo 89a] reads 'The personage and vicarage of Haldain [*i.e.* Auldhame] sett in assedatioun to the lady Reidhaw [i.e. Reidhall] for xx merkis.' The other side [fo 89b] reads 'Rentall of the chappellenry of Sanct Blais alter callit Nomine Jesu. In the first, Frances Lyntoun for his hous, xvj merkis. George Litiljhone, x merkis. Alexander Bruce, v merkis. [Signature:] J. Craig.'
[89] Cf 'dominum Georgium Littiliohne capellanie altaris Sanct Blasii fundate sub invocatione dulcissimi nominis Jhesu situate infra ecclesiam', *Registrum Cartarum Ecclesie Sancti Egidii*, p. 257.
[90] Cf SRO, Vol. a, fo 141r-v (see below, pp. 130, 131).

'Item, of Frances Lantonis hous', 16 merks. 'Item, of George Littiljhonnes hous', 10 merks. 'Item, of Alexander Bruces hous', 5 merks. 'In the haill', [*blank*].

MELVILLE, PARSONAGE AND VICARAGE OF,   (SRO, Vol. a, fo 131r)
   (SRO, Vol. b, fo 87r)

'The personage and vicarege of Melvile, set in assedatioun be Mr Johnne Hay[91] to the laird of Lugtoun and Johnne Young in Laswaid, extendis yeirlie to' £50, and £10 'to a curet or a minister.'

Calculation of third: In the whole, £60; third thereof, £20.

AULDHAME,[92]   (SRO, Vol. a, fo 131r)
PARSONAGE AND VICARAGE OF,[93]   (SRO, Vol. b, fo 87r)

'The rentale of the personage and vicarage of Haldan, set in assedatioun to the lady Reidhall[94] for' £6 13s 4d.

EDINBURGH,   (SRO, Vol. a, fo 131r)
TRINITY COLLEGIATE KIRK, PROVOSTRY OF,   (SRO, Vol. b, fo 87r)

Rental of the provostry of the Trinitie College

'The said provestrie wes in tymes bygane worth' 240 merks, 'of the quhilk thair is ane paroche kirk callit Lemplaw annexit to the said provestrie of the valour of' 40 merks 'quhilkis hes nocht bene payit thir' 7 'yeiris bygane becaus the samin lyis on the Bordouris.'

Signature: 'Williame Smyth, servand to the said proveist.'[95]

Calculation of third: In the whole, £160; third thereof, £53 6s 8d.

---

[91] I.e. John Hay, parson of Monymusk, who received a gift of Balmerino abbey from the crown in September 1561, *RSS*, v, no 845; vi, no 2333; Cowan and Easson, *Medieval Religious Houses*, p. 73.
[92] The benefice is entered as 'Haldan', which is to be identified with the parish of Auldhame or Haldame (see Groome, *Gazetteer*, p. 39). It also occurs among rentals for the constabulary of Haddington, in SRO, Vol. a, fo 184v (see below, p. 179).
[93] Cf SRO, Vol. a, fo 184v (see below, p. 179).
[94] Jean Stewart, lady Reidhall, fo 184v (see below, p. 179); the barony of Reidhall lay in the sheriffdom of Edinburgh, *RSS*, iv, no 685.
[95] George Clapperton was provost 1540 - 1566, and was succeeded by Laurence Clapperton, 1566 - 1571 x 1572. Watt, *Fasti*, p. 359.

'Nota, Mr Robert Pont,[96] as his entres to the said provestrie declarit, the rentale thairof to extend to' 205 merks, 'the gleib set in few for' 40 merks, 'quhairof foure akeris man be gevin to the redar.'

RESTALRIG, PARSONAGE OF,   (SRO, Vol. a, fo 131v)
(SRO, Vol. b, fo 87v)

Rental of the parsonage of Restalrig

'The quhilk personage is set in assedatioun for the sowme of' 300 merks 'yeirlie be my [*recte*, me] and my predicessores this lang tyme bypast. The gleib and mans of the said personage wes gevin in few be my predicessoris for yeirlie payment of' £47 6s 8d 'deducit of the personage foirsaid.' 80 merks yearly given in pension to Mr Andrew Logane.

Signature: 'Mr Johnne Logane, persoun of Restalrig.'

Calculation of third: In the whole, £247 6s 8d;[97] third thereof, £82 8s 10⅔d.

CRICHTON COLLEGIATE KIRK,   (SRO, Vol. a, fo 131v)
PROVOSTRY OF,   (SRO, Vol. b, fo 87v)

Provostry of Creichtoun

Calculation of third: In the whole, £133 6s 8d; third thereof, £44 8s 10⅔d.

PENICUIK, PARSONAGE OF,   (SRO, Vol. a, fo 132r)
LIVINGSTON, VICARAGE OF,   (SRO, Vol. b, fo 88r)
EDINBURGH, ST MARY OF THE FIELDS, PROVOSTRY OF,[98]

'This is the rentale of the personage of Penny Cuik, provestrie of the Kirk of Feild and vicarege of the kirk of Levingstoun.'

The said parsonage extends to the sum of 138 merks 4s.

Calculation of third: In the whole, £92 4s; third thereof, £30 14s 8d.

Provostry of Kirk of Feild with the vicarage of Levingstoun.

---

[96] Robert Pont was presented to Trinity College in 1572, Watt, *Fasti*, p. 359.
[97] In this case, the pension of 80 merks has been deducted from the rental.
[98] Cf SRO, Vol. a, fos 119v, 120r (see above, pp. 105, 106).

The said provostry and vicarage 'payit of befoir yeirlie' £46 'to me and my predicessouris and now payis nocht samekill be ressoun of the dimolitioun of the papistrie.'

Provostry of Kirk a Feild

'In the haill', £48 'secundum Forrest,
Bot secundum Pennycuik',[99] £46.

'Tertia maioris summe [*i.e.* Third of the greater sum]', £16.

[*In margin*, 'Aliter gevin up befoir.']

MUSSELBURGH, VICARAGE OF,   (SRO, Vol. a, fo 132v)
(SRO, Vol. b, fo 88v)

'The vicarege of Mussilburcht will nocht now be worthe' 20 merks 'or thairby.'

Calculation of third: In the whole, £13 6s 8d; third thereof, £4 8s 10⅔d.

EDINBURGH, ALTARAGE IN,[100]   (SRO, Vol. a, fo 132v)
(SRO, Vol. b, fo 88v)

Altarage in Edinbrucht

'Ane altarage of' 10 merks 'within Edinburcht, pertenyng to Frances Wilsoun.'

Calculation of third: In the whole, £6 13s 4d; third thereof, 44s 5⅓d.

STOW, VICARAGE PENSIONARY OF,   (SRO, Vol. a, fo 133r)
(SRO, Vol. b, fo 89r)

'The vicarege pensionar of Stow pertenyng to sir Thomas Godrell', £10, with 4 acres of land 'quhilk payis' £12, 'and thairof is delyverit to the redar' 20 merks. 'Sua restis' £8 13s 4d.

Calculation of third: In the whole, £22; third thereof, £7 6s 8d.

---

[99] Alexander Forrest, provost of St Mary of the Fields, 1552 - 1561; William Penicuik, provost, c. 1550 - 1566. Watt, *Fasti*, p. 358.

[100] The location of this altar has not been traced. A priest of the same name appears as prebendary of Dunglass collegiate kirk, near Cockburnspath; cf SRO, Vol. a, fo 169r (see below, p. 166).

HALF BONNINGTON AND HALF PLATT, PREBEND OF CORSTORPHINE COLLEGIATE KIRK[101]

(SRO, Vol. a, fo 133r)
(SRO, Vol. b, fo 89r)

'The prebendarie of Corstorphin pertenyng to Mr Thomas Marjoribankis extendis to' £26 13s 4d.

Calculation of third: In the whole, £26 13s 4d; third thereof, £8 17s 9$\frac{1}{3}$d.

HALF HATTON AND HALF DALMAHOY, PREBEND OF CORSTORPHINE COLLEGIATE KIRK[102]

(SRO, Vol. a, fo 133v)
(SRO, Vol. b, fo 89v)

'The rentale of the prebendarie of Corstorphin wes yeirlie the sowme of fiftynyne merkis usuall money upliftit of the half teyndis of Haltoun and Dalmahoy and payit to me, M[r] James Wilkie. The thrid of the prebendarie of Corstorphin callit Half Haltoun and Dalmahoy.'

Calculation of third: In the whole, £39 6s 8d; third thereof, £13 2s 2$\frac{2}{3}$d.

CARRINGTON, PARSONAGE OF,

(SRO, Vol. a, fo 133v)
(SRO, Vol. b, fo 89v)

Parsonage of Caringtoun

The kirk of Caringtoun pertaining to Mr Robert Hammiltoun.

Calculation of third: In the whole, £66 13s 4d; third thereof, £22 4s 5$\frac{1}{3}$d.

KITTYMUIR, PREBEND OF BOTHWELL COLLEGIATE KIRK

(SRO, Vol. a, fo 133v)
(SRO, Vol. b, fo 89v)

'The prebendarie of Bothwell callit Kettymure' [*In margin*, 'Kettemure, Lanark'.[103]]

Calculation of third: In the whole, 40 merks; third thereof 13 merks 2$\frac{1}{3}$d.[104]

---

[101] Cf SRO, Vol. a, fos 120v, 129r (see above, pp. 107, 116).
[102] Cf SRO, Vol. a, fos 110r, 120v, 125v (see above, pp. 97, 107, 112).
[103] This benefice, in the sheriffdom of Lanark, was evidently submitted with rentals for Edinburgh.
[104] The correct calculation appears to be 13 merks 4s 5$\frac{1}{3}$d.

BOTHANS COLLEGIATE KIRK, PREBENDS OF,

(SRO, Vol. a, fo 133v)
(SRO, Vol. b, fo 89v)

'Tua prebendareis in the college kirk of Bothanes gevin in be the laird of Tallow.[105]

Item, in the first in the haill', 23 merks 10s; third thereof, [blank].

'The uther, in the haill', 10 merks; third thereof, [blank].[106]

DUNBAR, COMMONS OF,

(SRO, Vol. a, fo 134r-v)

'The Commonis of Dunbar'

'Item in the first, the teynd schaves and lambis of the Loche Heid and the teynd schaves of the kirkland of the same. The teynd schaves of the Newtoun Eist with the teynd schaves of the kirkland of the same and lambis. The teynd schaves of the feyid landis of the West Barnes usit and wont withe the teynd lambis of the same and the teynd schaves of the kirkland of the same. The sowme of' £20 'usuall money for the teynd schaves of the Grange quhilk Thomas Wod laboures and payis yeirlie. The sowme of' £20 'usuall money of the Eister Hartsyde payit for the teynd of the same yeirlie be the tennentis of the same.'

'This is the rentale pertenyng to sir Adame Syme as followis.'

'Inprimis, in Leyth: furth of umquhill James Humes land yeirlie', 11 merks. 'Item, furthe of Johnne Blakburne land, younger, in Leyth yeirlie', 20s. 'Item, furth of Johnne Stanyis land', 20s. 'Item, furth of Thomas Syme land', 13s 4d. 'Item, furthe of Johnne Blakburnes land, aldar, yeirlie', 2s.

'In Edinbrucht: furth of James Thomsounes land yeirlie', 40s. 'Item, furth of Jone Neilsoune land yeirlie', 40s. 'Item, furth of Williame Akins land', £3 15s. 'Item, furthe of James Elwart land yeirlie', 15s. 'Item, furth of Marteyne Nicolsoun, als Cuikland, yeirlie', 13s 4d. 'Item, furth of Williame Ureis land, goldsmith, yeirlie', 18s. 'Item, furth of David Blakstokis land yeirlie', 8s 4d. 'Item, furth of Johnne Murrayis land yeirlie', 18s. 'Item, furthe of Mr Williame Creychtounes land at Bellis Wynd heid and now pertenyng to the bedmen yeirlie', £3. 'Item, furth of umquhill James Lawsoun land yeirlie', 40s. 'Item, furth of John Slowane land yeirlie', 20s. 'Item, furth of my lord Dalkeyth[107] land yeirlie', 26s 8d.

Total, £28 16s 4d.[108]

---

[105] I.e. William Hay of Tallo.

[106] The next entry in SRO, Vol. b, fo 90r corresponds to SRO, Vol. a, fo 135r (see below, p. 123 and n. 109).

[107] James Douglas, as 4th earl of Morton, was lord of Dalkeith. *RMS*, iv, no 2842.

[108] This calculation is correct.

Signature: 'sir Adame Syme, with my hand. Ressavit at command of the Clerk Registre.'

EDINBURGH, ST MARY OF THE FIELDS' COLLEGIATE KIRK, PREBEND OF, ST GILES' COLLEGIATE KIRK, ST COLM'S ALTAR IN,

(SRO, Vol. a, fos 134v-135r)
(SRO, Vol. b, fo 90r)

'The rentale of sir Johnne Richie, prebendar of Kirk Feild and Sanct Colmes altar within Sanct Geillis kirk.'

'Item inprimis, the prebendarie of the Kirk of Feild payis yeirlie furth of Robert Haddowis hous be west the croce', £10. 'Item, furth of Williame Symsounes hous in the Cowgait', £10. 'Item, the altar of Sanct Colme within Sanct Geillis kirk payis yeirlie furth of Mr Thomas Marjoribankis', [*blank*].[109] 'Item, furth of Johnne Symes houssis in the Cowgait', 13s 4d. 'Item, James Marjoribankis hous in the wynd be west Baithes Wynd payis yeirlie', 40s.

Signature: 'sir Johnne Richie, with my hand.'

RATHO, VICARAGE OF,[110]

(SRO, Vol. a, fo 135r)
(SRO, Vol. b, fo 90r)

Rental of the vicarage of the parish of Rotho pertaining to James Byschope, vicar thereof.

'The sowme of the haill be just compt and calculatioun extendis to the sowme of' £10. 'The quhilk foirsaid vicarege pertenis to me, James Bischope, as vicar pensionar and in witnes of the foirsaidis hes subscrivit this present with my hand.'

Signature: 'James Bischop, with my hand, readar and vicar of Ratho.[111] Ressavit at command of the Clerk Registre.'

---

[109] SRO, Vol. b recommences at the next sentence.
[110] Cf SRO, Vol. a, fo 142r (see below, p. 131).
[111] 'Minister and vicar', fo 142r (see below, p. 131).

EDINBURGH, ST GILES' COLLEGIATE KIRK, CHAPLAINRY OF ST DUTHAC'S ALTAR IN,

(SRO, Vol. a, fo 135r-v)
(SRO, Vol. b, fo 90r-v)

'The rentale of the cheplanrie of Sanct Duthois altar, sumtyme cituat within the paroche kirk of Sanct Jeill of Edinbrucht pertenyng to Maister Thomas Wastoun, quhairof the patronage apertenis to Martein Creychtoun of Cranstoun Ryddell.'

'Item in the first, to be uptane yeirlie of the tenement of umquhill James Lawsoun in the Over Bow', 14s. 'Item, of Archibald Watsounes tenement yeirlie', 12s. 'Item, of Bissattis land lyand in James Aikmanis Clois yeirlie', 20s. 'Item, of Richard Blaklaikis land yeirlie', 20s, 'defalcatioun being maid of the feird penny to the byggin conforme to the actis of Parliament', [*blank*]. 'Item, of the bischope of Glasgowis[112] ludgeing yeirlie', 5 merks; 'and defalcatioun being maid yeirlie', 3 merks 8s 10d. 'Item, of Williame Fethis land now pertenyng to Adame Symer yeirlie, defalcatioun being maid', 12s. 'Item, of Henry Cantes land now pertenyng to Thomas Hammiltoun yeirlie', 20s, 'defalcatioun being maid of the ferd penny for the byggin conforme to the actis of Parliament. Summa of this rentale', £8 0s 10d.[113]

Signature: 'M[r] Thomas Westoun.'

Calculation of third: In the whole, £8 0s 10d; third thereof, 53s $7\frac{1}{3}$d.

EDINBURGH, ST MARY OF THE FIELDS' COLLEGIATE KIRK, PREBEND OF ST VINCENT,

(SRO, Vol. a, fos 135v-136r)
(SRO, Vol. b, fos 90v-91r)

'The rentale of Sanct Vincentis prebendarie in the college Kirk of Feild within Edinbrucht pertenyng to sir Thomas Bathcat fes for teiching of ane sangschole.'

'Item inprimis, Arthure Hammiltounes hous lyand in Nudreis Wynd, in the yeir', £8 6s 8d. Arthur Grayis house in the same wynd in the year, 8 merks. Allan Kanocht,[114] his house in Peiblis Wynd in the year, 5 merks. In the Bakraw, Stephen Wolsoun in the year, 20s. James Russell in the year, 20s. John Thomsoun in the year, 20s. John Fendar in the year, 30s.

Signature: 'Sir Thomas Bathcat, with my hand. Ressavit at command of the Clerk Registre.'

---

[112] James Betoun, archbishop of Glasgow, 1550 - 1570, retired to France in July 1560, Watt, *Fasti*, pp. 149-150.

[113] These figures do not tally; the correct calculation appears to be £8 4s 8d, without taking into account the defalcations.

[114] Allan Canoch, 'carnifex' [public executioner], *RMS*, iv, no 2436.

EDINBURGH,
ST MARY OF THE FIELDS'
COLLEGIATE KIRK, PREBEND OF OUR LADY,

(SRO, Vol. a, fo 136r-v)
(SRO, Vol. b, fo 91r-v)

'This is the just rentale of Our Lady prebendarie in the college Kirk of Feild pertenyng to sir Nicoll Huchesoun for the tyme.'

'Item inprimis, out of the malt mylne pertenyng to the burcht of Edinbrucht, callit ane of the commoun mylnes in use of payment, payit be the thesaurer of the said burcht yeirlie', 10 merks. 'Item, in the Kirk of Feild Wynd furth of the houssis occupyit and set be Johnne Mure and his wyff on the west syd, be yeir', 33s 4d. 'Item, furth of Adame Muris land lyand in the same wynd yeirlie', 10s. 'Item, furth of Mathow Lilyes land in the same wynd yeirlie', 20s, 'quhairof is to be defalkit yeirlie' 5s 'for the birnyng. Item, furth of Thomas Creychtounes wyffis land yeirlie the sowme of', 15s. 'Item, furth of ane land occupyit be Johnne Scot in the eist syd of the said Kirk a Feild Wynd be yeir', £4, 'of the quhilk thair is payit out of annuell in the yeir to Johnne Stoddert, burges of Edinbrucht', 24s 4d.

Signature: 'sir Nicoll Huchesoun, with my hand.'

Calculation of third: In the whole, £13 2s 4d;[115] third thereof, £4 7s 5⅓d.

EDINBURGH,
ST GILES' COLLEGIATE KIRK,
CHAPLAINRY OF ST LAURENCE'S ALTAR IN,[116]

(SRO, Vol. a, fos 136v-137r)
(SRO, Vol. b, fos 91v-92r)

'The just rentale of ane cheplanrie of Sanct Laurence altar within Sanct Geillis kirk of Edinbrucht pertenyng to sir Williame Bannatyne.'

'In the first, ane annualrent of' 40s 'to be yeirlie upliftit of ane tenement of land of umquhill James Aikman lyand in the northe syd of the Queinis Streit of Edinbrucht. Item, ane annualrent of' 20s 'to be yeirlie upliftit of ane tenement of land of umquhill James Cant lyand on the southe syd of the Queinis Streit of Edinbrucht. Item, ane annualrent of' 20s 'to be upliftit of ane tenement of land pertenyng to Helene Littil and Maister Johnne Skrymgeour, hir spous. Item, ane annualrent of' 20s 'to be yeirlie upliftit of ane tenement of land, pertenyng to James Johnnestoun and utheris, lyand on the north syd of the Queinis Streit of Edinbrucht. Item, ane annualrent of' 30s 'to be yeirlie upliftit of ane tenement of land pertenyng to Johnne Carkettill lyand on the northe syd of the [Queinis] Streit of Edinbrucht. Item, ane annualrent of' 30s 'to be yeirlie upliftit of ane tenement

---

[115] In this calculation, the 24s 4d paid to John Stoddart has been duly subtracted, but the deduction of 5s 'for the birynyng' has not.
[116] Cf SRO, Vol. a, fo 142v (see below, p. 132).

of land pertenyng to Johnne Cowtis and James Henrysoun of Fordaill. Item, ane annualrent of' 20s 'to be yeirlie upliftit of ane tenement of land pertenyng to James Harlaw and his mother liand in the Cannogait. Item, ane annualrent of' 6s 'to be yeirlie upliftit of ane tenement of land pertenyng to Mr Johnne Westoun lyand in the Freir Wynd heid. Summa,' £9 6s.

Signature: 'Ita est Williame Bannatyne. Ressavit at command of the Clerk Registre.'

Calculation of third: In the whole, £9 6s; third thereof, £3 2s.

CRAMOND, VICARAGE OF, (SRO, Vol. a, fo 137v)
(SRO, Vol. b, fo 92v)

Rental of the vicarage of Crawmond within the sheriffdom of Edinbrucht given in by Mr Thomas Scot, vicar of the same.

'The vicarege of Crawmond olim assedat[117] for' £40, 'and now presentlie deductioun had is worthe be yeir' £20.

Signature: 'Magister Thomas Scot. Ressavit at command of my lord Comptroller.'[118]

Calculation of third: In the whole, £20; third thereof, £6 13s 4d.

EDINBURGH, BLACK FRIARS OF,[119] (SRO, Vol. a, fos 137v-140r)
(SRO, Vol. b, fos 92v-95r)

'The rentale of the Freiris Predicatouris of Edinbrucht, of all annuellis and fermes pertenyng to thame and thair place yeirlie and first in the towne of Edinbrucht.'

'Inprimis, of Muncurris alias in Johnne Thornetoun land', 20s. Of John Wrichtis land, 20s. Of Andrew Symsones land, 7s. Of Mr Alexander Currour land, 6s 8d. Of Andrew Harvy, now James Harlawis land, 13s 4d. Of Laurie Haliburtoun, now Markie Broun and William Lawrie land, 13s 4d. Of William Davidsoun, now Hendersones land, 10s. Of Fairnlie land, 26s 8d. Of Thomas Castelhillis, now James Johnnestounes land, 13s 4d. Of James Townis land, 16s 8d. Of Gilbert Knokis land, £24. Of Edward Littillis land, 5 merks. Of William Foularis

---

[117] I.e. formerly set in assedation; this is an example of the mixed use of Scots and Latin terms, found throughout the text.
[118] See above, p. 94, n. 15.
[119] I.e. Dominicans; cf W. Moir Bryce, *The Black Friars of Edinburgh* (Edinburgh, 1911), p. 71.

land, 2 merks. Of Thomas Diksoun land, 40s. Of John Mar land, 40s. Of Donald Kylis land, 40s. Of Robin Skarthmure land, 13s 4d. Of Alexander Park land, 40s. Of Thomas Russellis land, 20s. Of Sanders Adamsones land, 40s. Of Helen Ros land, 20s. Of Henry Ramsyis land, 10 merks. Of Roddis land, now James Bellentyne, 20s. Of Mr Thomas Marjoribankis land, 20s 'and a pund of pepper'. Of Michael Tullos, now Alison Cokburne land, 20s. Of Peter Marche, now David Kinloches land, 16s 8d. 'Of the payntit chalmer', 20s. Of George Todrikis land, 20s 4d. Of Thomas Dynsyre, now William Ker land 16s 8d. Of Patrick Tennentis land, 5s. Of Pyncartonis alias Noryis land, 6s 8d. 'Of Robert Lyddellis, now Andro Mowbrayis aires', 6s 8d. Of William Lokhartis land, 26s 8d. Of Patrick Flemyngis land, 20s. Of John Watsounes land, 10s. Of Gallowayis land, 7s. Of William Adamsones land, 56s 8d. Of Lambis land in Leyth Wynd, 42s. Of Nicol Porteris land, 20s. Of John Joyis alias Thomas Maltmakeris land, 10s. Total, £72 11s 4d.

'The annuellis and mailles of the gleib of the saidis freiris west the Cowgait and under the wall': Of John Henrysounes land, 4s. Of George Cowparis land, 12s. Of sir Symon Blyth land, 9s 6d. Of John Mudeis land, 4s. Of Will Cadeis land, 2d. Of Patokis land, 10s. Of Wyntreppis, now Stevinsoun land, 40s. Of Richie Grayis land, 20s. Of David Chepmanes land 'be defalcatioun', 17s 6d. 'Of the land quhair Walter Bynnyng duellis', 5s. Of Cavouris land,[120] 6s 8d. Of Pyottis land, 33s 4d. 'Of the land that Patersoun duellis in', 14s. Of James Bassondyne land, 20s. 'Of our awin land quhair the sklateris wyff duellis with ane pairt of the querrell yaird', £8. Of Willie Andersounes land, 12 merks. Of Archibald Leith land, [blank]. Of John Spottiswod land, 13s 4d. Of Rolland Gardneris land, 14 merks. Of Bellis land, 20s. Of sir Alexander Jardein land, [blank]. Of Dundas land, 8s 4d. Of Melros land, 35s. Of Adam Purves land, 8 merks. Of Andrew Craiges land 'be defalcatioun', 15s. 'Our eist yaird', 7 merks. 'Our west yaird and Bowquannis yaird', 5 merks. 'Of the kirkyaird', 40s. Of John Hopperis land, 13s 4d. Of the Magdaline chapel, 10s. Of the laird of Innerleithis land, 36s 8d. Of Watsones land at the Muiswall, 8s. Of the laird of Fentoun land, 20s. Of Mawsy Lyne land 9s 6d. Of Cowchrenis land, 12s. Of John Blakstokis land, 16s. Of Adam Spens land, 2s. 'Of our croft', 18 b. bere. 'Of Adam Gibbis wyff hous maill', 10 merks. Of John Wauchis land, £7.

'The annuellis pertenyng to the Freiris Predicatouris on the south syd of the gait of Edinbrucht': Of Lawsounes land, 13s 4d. Of Fallsydis land in the Over Bow, 5s. Of Tuedeis land, 20s. Of Clement Littillis land, 10s. Of the laird of Corstorphines land, 10s. Of Willie Ra land, 10s. Of John Bestis land, 10s. 'Of Robein Cokstoun buith now Robene Dennun land', 6s 8d. Of William Andersoun land, 13s 4d. Of John Dee land now Sande Guithreis, 13s 4d. Of Gilbert Hayis land, 20s. Of Mr Thomas Marjoribankis land in Bellis Wynd, 13s 4d. Of William Lawsones land, 13s 4d. 'Of the Magdalene land', 18s. Of William

---

[120] 'Canours land', *Black Friars of Edinburgh*, p. 72

Lauderis 'foirland', 30s. Of William Nesbit land, 15s. Of Tavernouris land,[121] 13s 4d. Of Butlaris land, 8s. 'Of Walter Bertremis over and nather landis', 40 merks. Of James Ryndis, 5s. Of David Melros land, 20s. Of Neilsounes land, 2 merks. Of Auldoucht land now Marion Scottis, 13s 4d. Of the college of Creichtounes land, 20s. Of Mr John Prestounes land, £5. Of John Vernouris land, 20s. Of Robin Dawgleissis land, 40s. Of George Porokis[122] lands, 32s 8d. 'Of the baikhous fra George Gibsoun', 12s. Of Lowchis land, 5s. Of Lambis land, now George Gibsoun, 2s. 'Of our awin land quhair Thomas Jaksoun duellis', 4 merks. Of John Tonnohillis[123] land, 5 merks. Of William Lindesay land, 26s 8d. Of Andrew Murrayis land, 10s 6d. Of Walter Wichtis land, 8 merks. 'Of Robesoun the cuitleris [cultellaris, *SRO, Vol. b*][124] land', 40s. Of Andrew Mowbray land, 8 merks. Of James Bassynden land, 10s. Of Purves land, 13s 4d. Of Sandie Youngis land, 20s. Of Mr Thomas Marjoribankis, now Helen Reid land, 2 merks. Of Methestoun, now John Sprot land, 6s 8d. Of Fawsydis now Willie Mailles land, 26s 8d. Of William Elphingstounes land, 35s. Of Nicol Borthwikis land, 6s 8d. Of Muriel Kincaidis land, 5s. Of Loppie Stane,[125] 40s. Of Willie Hillis land, 10 merks. 'Of the custoumes of the toun of Edinbrucht', 10 merks. Of John Johnnestoun, 4 merks. 'Summa of thir annuellis abonewrittin', [*blank*].

'The annuellis in Leyth and to landwart pertenyng to the saidis Freiris Predicatouris of Edinbrucht': Of Findguidis land, now John Carkettillis, 13s 4d. Of Halkerstones land, now Anne Hammiltones, 26s 8d. Of William Clappertonis land, 20s. Of Todrikis land, 10s. Of William Fawsydis land, 10s. Of Michael Gilbertis land, 12 merks. 'Of the akeris in the New Hevin', £10. Of Gosfuird, £16. Of Hartisheid and Clyntes, 20 merks. Of Ratho Byeris, £12. Of the Burnefute, £4. Of Littill Barnebowgall, 20s. Of the laird of Bowardis land in Duddingstoun, 10s. Of Mr John Hayes land in Peblis, 20s. Of West Gordoun in the Mers, 4 b. bere.

'This is the rentale of the Freiris Predicatouris of Edinbrucht subscryvet be me, Bernard Stewart, priour of the saidis fewares' [*recte*, freiris, *SRO, Vol. b*].

Signature: 'Ita est Bernardus Stewart, manu sua.'

ST CUTHBERT'S KIRK, TRINITY ALTAR IN,   (SRO, Vol. a, fo 140v)
   (SRO, Vol. b, fo 95v)

'The rentale of ane altare in Sanct Cuthbertis kirk callit the Trinitie altar pertenyng to sir Symon Blyth.'

---

[121] 'Cavernours land, £1 13s 4d', *Black Friars of Edinburgh*, p. 73.
[122] 'Pecoks', *Black Friars of Edinburgh*, p. 74.
[123] 'Connohills', *Black Friars of Edinburgh*, p. 74.
[124] 'Tinclers', *Black Friars of Edinburgh*, p. 74.
[125] 'Lappie Stane', *Black Friars of Edinburgh*, p. 74.

'Item, ane land of umquhill James Fingland lyand in Gilbert Lauderis Clois foment the Skynkand [*recte*, Stynkand, *SRO, Vol. b*] Styll', £4. 'The land of umquhill Johnne Broun liand in Johnne Fischeris Clois', 15s. 'Ane land of umquhill Johnne de Peiblis, Clement Lytill in the Luckin Buthes', 43s 6d. 'Ane land of umquhill Johnne Hayne, sir Walter Kers in the Cannogait', 6s 8d. 'Ane land of umquhill Johnne Haynenis landis at the West Port cheik', 10s. 'Ane land of umquhill Johnne Wischart yaird within the same porte at the cheik thairof', 5s. 'Ane land of umquhill sir Robert Hyntrodis, now Jame[s] Hendersones in Forresteris Wynd fute', 12s. Total, £8 12s 2d.

Signature: 'Ita est Symon Blyth.'

EDINBURGH, ST GILES' COLLEGIATE KIRK, ST SALVATOR'S ALTAR IN[126]

(SRO, Vol. a, fo 140v)
(SRO, Vol. b, fo 95v)

'Ane uther rentale of the said Symon Blyth in Sanct Jeillis kirk callit Salvatouris altare.'

'Item, ane land of umquhill Henry Rynd of James Ryndis foirbuthe at thair or Gillaspeis Clois heid', £7 10s.

Signature: 'Ita est Symoun Blyth.'[127]

PRECEPT

(SRO, Vol. a, fo 141r)
(SRO, Vol. b, fo 96r)

'Ane precept of the Comptrolleris.'

'James Nicolsoun, ye sall ressave this rentale and tak cautioun and tak cautioun [*sic*] for the queinis graces thrid and answer him of lettres for his tua pairt conforme to the commoun ordour and keip this for your warrand. Subscrivit with my hand at Edinbrucht the first day of July, 1563.'

Signature: 'Johnne Wyscharte, Comptrollare.'

---

[126] Cf SRO, Vol. a, fos 109v, 145r, 149r (see above, p. 96; below, 135, 141).

[127] SRO, Vol. b, preceding fo 96r, contains a scrap of paper, consisting of one of the original rentals submitted, with the following entries, which do not appear in Vol. a: 'To sir Johnne Richie suld be payit bot nocht gevin in rentall. Johnne Robertsoun land in the Stra Mercat payis in commoun yeirlie', 2 merks. 'Item, my lord duk luging payis' 40 merks, 'for the quhilk Maister Alexander Forres hes the vicarage of Clatt'; and on the other side, there are some scribbles followed by 'l s. land of Curling and Myln of Kilmawris to Margarat Wallace lyferentar for all the dayis of his [*sic*] lyfe. To my [*blank*].'

'Memorandum, thair is ane annuell dedicat to ane daill of this last cheplanrie betuix Thomas Codynnaris and the Cowgait in Gillespikis Clois extending yeirlie to' 4 merks.

'Johnne Hoppes, the smythes landis, umquhill callit Spens land at Bellis Wynd fute', 1 merk.

'Nota this, elemosinarie.'

EDINBURGH, (SRO, Vol. a, fo 141r)
ST GILES' COLLEGIATE KIRK, (SRO, Vol. b, fo 96r)
CHAPLAINRY OF THE HOLY BLOOD ALTAR IN,[128]

'The rentale of the cheplanrie of the Halybluid altare within Sanct Jeillis kirk in Edinbrucht pertenyng to sir Johnne Lokart, cheplane thairof.'

'Item, the said cheplanrie foundit be umquhill Michaell Makquhen of' 18 merks 'be yeir, to be tane up, viz.,' 16 merks 'of ane tenement of land pertenyng to umquhill Williame Adamsoun and' 2 merks 'in ane buithe in ane tenement of land pertenyng to umquhill Mr Frances Bothwell, occupiet be Laurence Symsoun, quhilk makis' 18 merks 'be yeir.'

Signature: 'Johannes Lokhart, manu propria.'

HALF NORTON AND HALF RATHO BYRES, (SRO, Vol. a, fo 141v)
PREBEND OF CORSTORPHINE (SRO, Vol. b, fo 96v)
COLLEGIATE KIRK[129]

'The rentale of Florence Douglas provisioun within the queir of Corstorphin as ane of the barnes fundit to sing thairintill extendis yeirlie in all profeitis to' 12 b. meal 'payit to him yeirlie be the tennentis and occupyeris of the halff landis of Nortoun and Rothobiares lyand within the parochin of Ratho and schirefdome of Edinbrucht, and be Mr James Gray, prebendar of the prebendarie callit Half Nortoun and Rathobyares lyand within the college kirk of Corstorphin, of the quhilkis' 12 b. meal 'the said Florence hes bene in possesioun be uptaking thairof furth of the foirsaidis half landis be the space of' 6 'yeris lastbypast.'

Signature: 'Florence Douglas, with my hand. Ressavit at command of the Clerk Registre. J. Clericus Registri.'

---

[128] Cf SRO, Vol. a, fos 130v, 141v (see above, p. 117; below, p. 131).
[129] Cf SRO, Vol. a, fos 120v, 122v, 126v (see above, pp. 107, 109, 113).

Calculation of third: In the whole, 12 b. meal; third thereof, 4 b.

EDINBURGH,   (SRO, Vol. a, fos 141v-142r)
ST GILES' COLLEGIATE KIRK,   (SRO, Vol. b, fos 96v-97r)
CHAPLAINRY AT THE HOLY BLOOD ALTAR IN,[130]

'This is the rentale of ane cheplanrie cituat [at] the Halybluid altar within Sanct Jeillis kirk of Edinbrucht pertenyng to me, sir Williame Johnnestoun, extending yeirlie to the sowme of' £9.

'The names of thame that aucht [*i.e.* owed] the annuellis to the said cheplanrie efter followis: Williame Flemyng and Elizabeth Park, his spous, for the buithe the quhilk Alexander Park occupyeis now', 9 merks. 'Ane land under the castell wall, the quhilk Johnne Loche, burges under the wall, bruikis now', 2 merks. 'Of Sanct James land at the Over Bow, the quhilk Margaret Cant, spous sumtyme to James Fairlie, burges of Edinbrucht, now bruikis', 13s 4d. 'Ane land at the Over Bow quhilk Margaret Diksoun, spous sumtyme to James Hammiltoun, now occupyis and bruikis', 6s 8d. 'Ane land in Sanct Marie Wynd quhilk James Curle bruikis', 6s.[131] 'Ane land besyd the Blak Freir Wynd quhilk Jonet Langland occupyis', 6s 8d.

Signature: 'Ita est dominus Gulielmus Johnnestoun. Ressavit at command of the Clerk Registre. J. Clericus Registri.'

Calculation of third: In the whole, £9; third thereof, [*blank*].

RATHO, VICARAGE OF,[132]   (SRO, Vol. a, fo 142r)
                                                       (SRO, Vol. b, fo 97r)

Rental of the vicarage of the parish kirk of Ratho

'The sowme of the haill be just compt and calculatioun extendis to the sowme of' £10, 'the quhilk foirsaid vicarege pertenyis to me, James Bischope, as vicar pensionar and in witnes of thir foirsaidis hes subscrivit this present with my hand.'

Signature: 'James Bischope, with my hand, minister and vicar of Ratho.'[133]

---

[130] Cf SRO, Vol. a, fos 130v, 141r (see above, pp. 117, 130).
[131] If this were 6s 8d, the total of £9 would be correct.
[132] Cf SRO, Vol. a, fo 135r (see above, p. 123).
[133] 'Readar and vicar', fo 135r (see above, p. 123).

EDINBURGH, (SRO, Vol. a, fo 142v)
ST GILES' COLLEGIATE KIRK, (SRO, Vol. b, fo 97v)
CHAPLAINRY OF ST LAURENCE'S ALTAR IN,[134]

'Ane rentale of the yeirlie annuellis of ane cheplanrie foundit at Sanct Laurence altar be umquhill Walter Bartane with[in] the kirk of Edinbrucht pertenyng to me, sir James Huntar.'

'Item, in the first, James Watsounes land of Sauchtonehall of tua tenementis lyand in Toddis Clois fornent the Croce on the south syd thairof within Edinbrucht awand yeirlie' 5 merks. 'The landis ba[i]th befour and bak of umquhill Walter Blaiklokis aires lyand within Edinbrucht beneth Nidreis Wynd', £3. 'The land of Williame Brokes, smyth, lyand in the feit of Nidreis Wynd on the south syd thairof yeirlie', 16s. 'The landis of umquhill Johnne McGaw in the Cannogait, now occupyit be Robert Aytoun, yeirlie', 30s. 'Williame Duffis landis in Leyth yeirlie', 20s. 'The landis of umquhill Archibald Edmestoun and now pertenyng to Williame Dawsoun, his guidsone, in Leyth yeirlie', 50s.

Signature: 'Ita est Jacobus Huntar. James, ressave this rentale. J. Clericus Registri.'

EDINBURGH, (SRO, Vol. a, fo 143r-v)
ST GILES' COLLEGIATE KIRK, (SRO, Vol. b, fo 98r-v)
CHAPLAINRIES OF THE ALTARS OF ST DENIS AND ST FRANCIS IN,

'The rentallis of the chaiplanreis foundit sumtyme in the kirk of Sanct Jeill at the altares namit Sanct Diones and Sanct Frances be umquhill Maister Richard Robesoun, chapellane, and Thomas Cant, burges of Edinbrucht.'

'Inprimis, infeft be the said umquhill Mr Richard Robesoun', 20 merks 'now onlie' £10 'be ressoun of the birnyng, to be takin and upliftit of his awin tenement of land than pertenyng to him in heritage lyand in the towne of Edinbrucht on the south syd of the Queinis Streit of the samin, foiranent the kirk of Sanct Jeill, betuix the landis of umquhill Robert Vaus on the eist pairt and the kirk yaird on the west pairt, extending fra the eist end of the kirk to the Cowgait Strand on the south pairt, now heritour of the saidis landis Nicoll Scot, Margaret Fentoun and Robert Gray, hir spous, for his enteres Alexander Duncane, Johnne Davidsoun and the relict of umquhill Donnald Maxwell, the said Donaldis aires. Item,' 40s 'feft be umquhill Thomas Cant, burges of Edinbrucht, of his awin land in the Meill Mercat on the southe syd of the Queinis Streit betuix the landis of umquhill Johnne Yule on the eist pairt and the landis of umquhill Alexander Tuedye on the west pairt, extending fra the Cowgait Strand on the south pairt to the Meill Mercat

---

[134] Cf SRO, Vol. a, fo 136v (see above, p. 125).

on the north pairt. Item, ane annualrent of' 33s 4d 'feft be the said Thomas Cant of his' 5 merks 'annuellis quhilk the said Thomas had of the landis of umquhill James Cliddisdale now pertenyng to [blank] Wilsoun occupyit be [blank] Marjoribankis, the relict of umquhill Patrik Wilsoun, lyand on the south syd of the Queinis Streit betuix the landis of umquhill Hew Bar on the north pairt, the landis of umquhill James Portius on the south pairt and the trans on the eist pairt. Item, ane annualrent of' 25s 'feft be the said Thomas Cant of ane uther land of the said James Cliddisdaill, pertenyng to Andro Fischear, lyand in Nyddreis Wynd on the west syd of the trans betuix the landis of umquhill Adame Halkerstoun on the north pairt and the landis of umquhill sir Alexander Barcare on the south pairt. Item, ane annuelrent of' 26s 8d 'feft [be] the said Thomas Cant quhilk he had of the landis of umquhill Thomas Purves, now pertenyng to Alexander Moresoun, lyand in Bellis Wynd on the west syd of the trans betuix the land of umquhill [blank] Mekill on the south pairt and the landis of Adame Bell on the north pairt. Item, ane annuelrent of' 42s 8d 'feft be the said Thomas Cant quhilk he had of the land of umquhill Johnne Joly, now pertenyng to [blank], lyand in the Cannogait besyd Edinbrucht on the north syd of the Queinis Streit betuix the landis of umquhill Walter Bartrahame on the eist pairt and the landis of umquhill Martine Gordoun on the west pairt and the Cowgait Strandis on the north pairt.' Total, £18 7s 8d.

Signature: 'James Nicolsoun ressave this rentale. J. Clericus Registri.'

EDINBURGH, ANNUALRENTS IN,[135]   (SRO, Vol. a, fo 144r)
(SRO, Vol. b, fo 99r)

'Ane annuelrent of' 7 merks 'feft be umquhill Thomas Cant to Sanct Frances altaire to be takin and upliftit of the landis of the said Thomas as conqueis be him fra umquhill Thomas Morie lyand on the west syd of Leyth Wynd, the trans of the vennale on the eist syd and the trans of the tenement of umquhill Frances Prest on the west syd and the landis of umquhill James Morie on the southsyd and the landis of Williame Glasfuird on the north.'

'Item, ane annualrent of' 13s 4d 'to be tekin and upliftit of the landis of umquhill Johnne Fouleit and Jonet Law, his spous, lyand in the towne of Leyth on the south syd of the water within the baronie of Lestalrig[136] betuix the tenement of Walter Logane of the Flures on the [north] pairt and the landis of umquhill Andro Mowbray on the south pairt.'

---

[135] The two annualrents which follow are not included in the total of £18 7s 8d in the preceding entry, though they may still belong to that rental.
[136] This was a common representation of Restalrig.

EDINBURGH, (SRO, Vol. a, fo 144r)
ST MARY OF THE FIELDS' (SRO, Vol. b, fo 99r)
COLLEGIATE KIRK, PREBEND OF,

'The rentale of George Forret, prebendar in the Kirk Feild of Edinbrucht.'

'Item, ane annual rent of' 20 merks 'yeirlie to be upliftit of ane tenement of land abak and foir pertenyng to umquhill sir Robert Lillie, prebendar of the said kirk lyand in the Cowgait of [Edinbrucht] upoun the north syd of the Queinis Streit of the samin betuix the land of Sanct Catherines altare upoun the eist, the landis of umquhill Mr David Bothwell upoun the west, the land of umquhill Anthonie Brissat on the northe and the Queinis Streit upoun the south.'

Signature: 'James ressave this rentale. J. Clericus Registri.'

EDINBURGH, (SRO, Vol. a, fo 144r-v)
ST GILES' COLLEGIATE KIRK, (SRO, Vol. b, fo 99r-v)
ALTARS OF ST NINIAN AND ST PATRICK IN,

'Thir ar the rentales of the altares pertenyng to sir Johnne Broun lyand within the burcht.'

'Item, ane foundit at Sanct Ninianes altar of' 20 merks 'yeirlie to be upliftit of all and haill ane tenement of land bak and foir lyand at the Nather Bow pertenyng to umquhill Andro Mowbray on the south syd of the Queinis Streit.'

'Item, ane uther foundit at Sanct Patrikis altar of' 20 merks 'to be upliftit of all and haill ane tenement of land pertenyng to Hew Broun, lyand at the heid of Robert Bruces Clois on the north syd of the Queinis Streit, of the quhilk' 20 merks 'to be upliftit of this land' £10 yearly and 5 merks 'of the land of Andro Cor lyand in Todis Clois on the north syd of the gait.'

Signature: 'Mr Johnne Broun, with my hand. James Nicolsoun, ressave this rentale. J. Clericus Registri.'

EDINBURGH, (SRO, Vol. a, fos 144v-145r)
ST GILES' COLLEGIATE KIRK, (SRO, Vol. b, fos 99v-100r)
ALTARS OF ST NINIAN AND ST ELOY IN,

'Sir Johnne Scottis rentales of Sanct Niniane altar and Sanct Eloyis altar in Sanct Geillis kirk in Edinbrucht.'

'Item, of the ludgeing and tenement of umquhill Williame Ferrie in Peiblis Wynd on the southe syd of the Hie Gait and on the eist syd of the samin wynd Edward Tomsoun and his spous, heritouris of the samin, occupyit be the subtennentis yeirlie', 32s. 'Item, of umquhill Margaret Bannatynes land lyand in the said wynd betuix the land of umquhill Johnne Pendrech on the south pairt and the landis of umquhill Williame Ferrie on the north pairt Johnne Hammiltoun, baxter, and his spous, heritouris, occupiet be [blank] yeirlie', 15s. 'Item, of umquhill Johnne Pendreich landis lyand in the said wynd betuix tua double landis of umquhill Williame Ferrie on the south and north pairtis, Mr Robert Glen and his spous, heritouris, occupeit be [blank] yeirlie, 6s. Item, of the said umquhill Williame Ferreis land and tenement in the said wynd betuix the landis of umquhill Williame Ridwall on the south pairt and the said umquhill Johnne Pendreichis landis on the north pairt occupyit be Johnne Young and his subtennentis yeirlie', 15s. 'Item, of Mr Johnne Prestoun and his spous of ane ludgeing in Foster Wynd yeirlie', 26s 8d. 'Item, of the landis of umquhill Wiliame Rydwall lyand at the fute of the said Peiblis Wynd on the north syd of the Cowgait and the landis of umquhill Williame Ferrie on the said north pairt, Michaell Nasmyth, heritour, occupyit be [blank] yeirlie', 5s 4d. 'Item, of the landis of umquhill Andro Mowbray lyand in Soltrayis Wynd on the west pairt thairof betuix the landis of umquhill Andro Mowbrayis proper landis that he hes in few [of] the Blak Freiris occupyit be [blank] yeirlie', 18s. 'Item, of the said umquhill Andro Mowbrayis land in the said wynd nixt adjacent to the said Andro Mowbrayis dwelling occupyit be [blank] and his spous yeirlie', 18s. 'Item of the said umquhill Andro Mowbrayis hailyaird[137] lyand contigue to the saidis thrie landis on the south pairt thairof occupyit be [blank] Smyth, yeirlie', 10s 'and now pertenyng to Robert Mowbray in heritage. Item, of the said umquhill Andro Mowbrayis tenement and land lyand at the fute of the said yard on the north pairt of the Cougaite occupyit be [blank] Hoge and his spous, [blank] Payet and his spous, Robert Moubray heritour yeirlie', 30s. Total, £10 3s 8d.[138]

Signature: 'James Nicolsone, resaiff this rentall. J. Clericus Registri.'

EDINBURGH, (SRO, Vol. a, fos 145r-146v)
ST GILES' COLLEGIATE KIRK, (SRO, Vol. b, fos 100r-101v)
CHAPLAINRY OF ST SALVATOR IN THE HOLY BLOOD AISLE[139]

'The rentall of ane chaplanrie fundit in the Halye Bluid Ile in Sanct Gelis kirk of Edinbrucht, callit Iconie Salvatouris, pertenyng to schir Williame Makkye.'

---

[137] Both texts read 'hailyaird' but 'kailyaird' may have been intended.
[138] This total would be £8 10s without the sum payable by Robert Glen.
[139] Cf SRO, Vol. a, fos 109v, 140v, 149r (see above, pp. 96, 129; below, p. 141).

'Item in the first, ane annuellrent of fyve markis to be takin up of the landis of umquhile Georg Halkarstoun in Nedryis Wynd on the west syd of the trans thairof betuix the landis of Johne Fescher on the southe and the landis of the said umquhile Georg on the northe and west pairtis quhair of the fourth pennye is defalkit be re[asoun] of the burnyng. Item, ane annuelrent of' 20s 'to be takin and upliftit of the land of umquhile Georg Ramsay lyand on the southt syd of the Queinis Streit betuix the landis of umquhile Thomas Craufurd on the eist pairt and the landis of Hectour Thomson on the west pairt. Item, ane annuelrent of' 10s 'yeirlie takin and upleftit of the landis of umquhile Andro Mathesone on the southt syd of the Queinis Streit betuix the landis of umquhile Elizat Baudoun on the eist pairt and the landis of Luk Saidlar and Mertane Henter [Mertyne Hunter, *SRO, Vol. b*] on the west pairt. Item, ane annualrent of' 3s 'of the landis of umquhill [*blank*] Ednem lyand on the southe syd of the Queinis Streit betuix the landis of Williame Craik and the landis of Robert Carmechaell on the west syd [*in margin*, brint]. Item, ane annualrent of' 6s 8d 'of the landis of James Falcoun of the southe syd of the Queinis Streit betuix the landis of Johnne Watsoun on the eist pairt and the said James Falcoun on the west pairt. Item, ane annualrent of' 20s 'of the landis of umquhill [Johne, *SRO, Vol. b*] Lychtoun betuix the landis of the said umquhill Johnne on the south, the landis of Williame Lauder on the [north] pairtis. Item, ane annuelrent of' 26s 8d 'to be takin up of umquhill Thomas McLilland lyand betuix the landis of the said umquhill Thomas on the eist and west pairtis and the landis of Johnne Cranstoun on the south and the landis of Thomas Cranstoun on the north. Item, ane annuelrent of' 2 merks 'of the landis of umquhill George Tailyefeir, lyand in the landis of umquhill George Robesoun, Adame Carkettill and Richard Gammell on the south pairt and the clois of the said tenement on the eist pairt. Item, ane annuelrent of' 9s 'to be tane and upliftit of the landis of umquhill Johnne Sellar betuix the wall of the toun on the west and the landis of David Kincaid on the eist. Item, ane annuelrent of' 13s 8d 'of umquhill David Kincaid lyand betuix the landis of umquhill Johnne Malesoun on the eist and the land of Johnne Sellar on the west pairtis. Item, ane annuelrent of' 3s 'of the landis of umquhill Adame Robesoun lyand betuix the landis of umquhill Thomas Brethocht on the eist and the landis of the said Johnne Malisoun on the west pairtis. Item, ane annuelrent of' 10s 'to be tane of the landis of umquhill sir James Guild, cheplane, lyand betuix the landis of umquhill Johnne Crounzeong on the eist and the landis of Thomais Barthren on the west pairtis [*in margin*, brint]. Item, ane annuelrent of' 40s 'to be tane and upliftit of the landis of James Stury lyand [betuix] the castell wall on the north syd of the Queinis Streit, Johnne Crunzeome land on the west and the land of David Falcoun on the eist. Item, ane annuelrent of' 40s 'to be upliftit of David Falcones land lyand within the said burcht under the said castell betuix the vennall that passis to the castell on the eist pairt and the landis of James Sturry on the west pairt. Item, ane annuelrent of' 13s 4d 'to be upliftit of the landis of sir Johnne Dunkany lyand in the Cowgait on the south syd of the Queinis Streit betuix the landis of Thomas Laverok on the eist syd and Williame Harlaw on the west syd. Item, ane annuelrent of' 20s 'to be tane and upliftit of the landis of [Johne, *SRO, Vol. b*] Tevyotdalye lyand in the burcht of the Cannogait foirnent the croce of the samin on the south syd of the Queinis Streit betuix the landis of Jelis Harlie on the eist syd and the landis of Adam Carnis on the west [*in margin*, brint]. Item, ane annuelrent of' 6s 6d 'to be tane and upliftit of the landis of umquhill Johnne Barcare lyand under the castell wall on the north syd of the Queinis Streit betuix the landis of Tho[m]as Kincaid on the eist pairt and the landis of James Layng on the west pairt. Considder the brunt and defalk the fourth thairof quhat sall be my pairt to pay for the thrid.' Total, £11 15s 2d.[140]

Signature: 'James Nicolsoun, ressave this rentale. J. Clericus Registri.'

EDINBURGH, ST GILES' COLLEGIATE KIRK, ALTAR OF ST NICHOLAS IN,

(SRO, Vol. a, fo 147r-v)
(SRO, Vol. b, fo 102r-v)

'Thir ar the annuellis underwrittin that pertenis to Sanct Nicolas altare.'

'Item in the first, lyand at the West Port on the south syd in Marjorie Akinheidis land, the spous of Andro Lyndesay, quhar James Aikman, litster, duellis, in the yeir', 6s 8d. 'In David Falcounis land in the eist syd of the land abonewrittin', 18s. 'In Thomas Watsones land in the Meill Mercat', 16s. 'Umquhill Symon Prestounes land quhair the lady Stanehous duellis, now Bessie Otterburnis for the tyme', 13s 4d. Alexander Masones land, 10s. Robert Grayis land 'lyand in the Buythraw', 8s. William Melvinis land 'benethe the Stinkand Style quhair Adame Allane haldis buyth', £3. James Rigis land 'quhair Williame Patersoun haldis buyth', 13s 4d. Alexander Scharpes land 'quhair Martein the cuik hald[is] buyth', 6s 8d. John Newlandis land 'at Halkerstounes Wynd heid', 6s 8d. James Lowreis land 'beneth the Blakfreir Wynd heid', 40s. George Maynis land and Patrick Flemyngis land, 5s. 'Quhair Thomas Mow haldis buyth in the Fische Mercat', 26s 8d. Nicol Scottis land in the Cannogait 'quhair Williame Lausoun, bonnet maker, duellis', 6s 8d. 'Of the commoun guid of the toun, the 3 pairt of', 14 merks, 'quhilk cumis to iiij pairt in yeir' £3 2s 2d. 'Summa of the haill of this abonewrittin extendis to' £14 19s 6d.[141]

Signature: 'James, ressave this rentall. J. Clericus Registri.'

'The haill extendis to' £12 5s 2d, 'rest deducit for the burnyng.'

---

[140] The arithmetic does not seem to tally.
[141] The correct calculation appears to be £14 19s 2d.

EDINBURGH, (SRO, Vol. a, fo 147v)
ST GILES' COLLEGIATE KIRK, (SRO, Vol. b, fo 102v)
CHAPLAINRY OF ST JOHN THE EVANGELIST IN,

'Followis the rentale of the annuelrentis pertenyng to the cheplanrie of Sanct Johnne the Evangelist situat within Sanct Jeillis kirk quhilkis war tane up of before be sir David Scot, chaiplane thairof, of landis and tenementis underwrittin lyand within the burcht of Edinbrucht.'

'In the first, ane annuelrent of' 4 merks yearly 'tane up of all and haill the tenement of land and yaird of umquhill Nicoll Broun, pertenyng to Johnne Betty.' Another annualrent of 3 merks money 'to be upliftit of umquhill Williame Rais landis.' Another annualrent of 7 merks money 'to be upliftit of the haill cave roum or volt, bak and foir of umquhill Stevin Law, now pertenyng to Johnne Stodert.' Another annualrent of 2 merks 'to be upliftit yeirlie of the tenement of land bak and foir under and abone with the pertinentis of umquhill George and James Halkarstounes, now in the [handis] of the aires of Robert Chepman.' Another annualrent of 6 merks 'to be upliftit yeirlie of the bakland under and abone of umquhill Williame Carnecors and now in the handis of Robert Carnecors, his sone.' Another annualrent of 4 merks 'to be upliftit yeirlie of the over pairt of the tenement of umquhill Adame Fawsyd, now being in the handis of Johnne Robesoun, cuitlair.' Another annualrent of 5 merks money 'to be upliftit yeirlie of umquhill Mr Johnne Chepmanes awin land, now being in the handis of the aires of umquhill Robert Chepman.'

Signature: 'sir David Scot, with my hand.'

'Of the quhilk annuellis abonewrittin thair wes' 5 merks 'payit out of Mr Johnne Chepmanes land evir payit to the pure folk and nocht to the cheplan.'

EDINBURGH, (SRO, Vol. a, fo 148r)
ST GILES' COLLEGIATE KIRK, (SRO, Vol. b, fo 103r)
CHAPLAINRY OF ST ROCH'S ALTAR IN,

Rental of sir William Murray

'The rentale of sir Williame Murrayis chaiplanrie of Rocheis aultar situat within the college kirk of Sanct Geill in Edinbrucht, maid and gevin in be him and subscrivit with his hand at the said burcht the last day of August the yeir of God' 1564, 'as his fundatioun beiris.'

'Inprimis, ane annuelrent of' £4 yearly 'to be upliftit at tua usuall termes in the yeir etc. of umquhill Johnne Bateis land, burges of Edinbrucht, lyand within umquhill Johnne Barcaris Clois on the north syd of the Kingis Streit.' An annualrent of £4 yearly 'to be upliftit at tua usuall termes in the yeir etc., of umquhill Thomas Turnouris land, burges of Edinbrucht, lyand within the said burcht on the south syd of the Kingis Streit in the Freir Wynd on the west syd of the samin.' An annualrent 'of umquhill Williame Malcomes land, burges of Edinburcht, of' 13s 4d 'quhilk wes brint and thairfoir rebaittes' 3s 4d 'lyand within the said burcht betuix the landis of umquhill Johnne Coipland on the south and west land of umquhill Johnne Anderson on the northe.' An annualrent of 7 merks 'of umquhill Alexander Levingtounes land, burges of Edinbrucht, lyand within the burcht quhilk wes brint and thairfoir rebates of the annuell yeirlie', 23s 4d 'lyand on the south syd of the streit betuix the land of umquhill Edward Lamb on the eist syd and the eist kirk gavill of Sanct Jeill on the west', £3 10s. An annualrent of 13s 4d 'of the said umquhill Johnne Bateis land lyand within this burcht within the clois of [blank]', 13s 4d. An annualrent of 5 merks 'to be upliftit of umquhill Johnne Couplandis land lyand in the said burcht on the north syd of the Kingis Streit quhilk wes brint and thairfoir rebaites' 17s, 'restis frie' 50s. 'Summa of the haill', £15 3s 4d.[142]

Signature: 'sir Williame Murray, with my hand.'

EDINBURGH, (SRO, Vol. a, fo 148v)
ST MARY OF THE FIELDS' (SRO, Vol. b, fo 103v)
COLLEGIATE KIRK, PREBEND OF ST MATTHEW OF,

'The rentale of Sanct Mathow prebendrie of the Kirk of Feild pertenyng to Mr Alexander Chalmer. Item, of few maill and annuellis', £20.

CURRIE, ST MUNGO'S ALTAR IN, (SRO, Vol. a, fo 148v)
(SRO, Vol. a, fo 103v)

'The rentale lyand to Sanct Mungois altare in Currie pertenyng to me, sir Thomas Knowis. Thair is ane tenement of land lyand on the south syd of the gait of the Meill Marcat of Edinbrucht occupyit be Mr Petir Spens of ane pairt of it and ane uther parte of it occupyit be Johnne Gibsoun, bowar, payand to me yeirlie' 6 merks of annual.

Signature: 'Thomas Knowis, cheplane, with my hand.'

---

[142] The arithmetic does not seem to tally.

EDINBURGH, (SRO, Vol. a, fo 148v)
ST GILES' COLLEGIATE KIRK, (SRO, Vol. b, fo 103v)
ALTARAGE OF ST MICHAEL IN,

'This is the rentale of Sanct Michaellis alterage, sumtyme situat within Sanct Jeillis kirk pertenand to me, extending to the soum yeirlie', £10.

Signature: 'sir Thomas Gray, with my hand. James Nicolsoun, ressave this rentale. Clericus Registri.'

'Item, Williame Tamsones land at the Over Bow yeirlie', 40s. Euphemia Stirlingis land at the Nather Bow, 16s. James Baronis land 'foirnent the salt trone on the north syd of the gait', 6 merks. Andrew Murray land of Blakbarrony 'foirnent the Mercat Croce', 30s. John Youngis land in the Flesche Mercat Clois, 14s 4d. John Watsoun land at the Nather Bow 'on the north syd of the gait', 6s 8d.[143]

EDINBURGH, (SRO, Vol. a, fo 149r)
ST GILES' COLLEGIATE KIRK, (SRO, Vol. b, fo 104r)
CHAPLAINRY OF ST JOHN'S AISLE IN,

'My lord Clerk of Register, I, your lordschippis rectour,[144] Niniane Brydin, cheplane of Sanct Johnnes Ile cituat within the kirk of Sanct Jeill, foundit be umquhill Walter Chepman, lauchfullie provydit thairto and be vertew of my pensioun wes in possessioun of uptakin of the annuellis underwrittin.

Item, ten merkis of Walter Chepman grete tenement in the Kowgait. Item, tua merk of Flemyngis land nixt adjacent it in the Kowgait. Item, four merkis in Sklateris Clois of ane tenement of David Corsbeis. Item, thrie merkis of Willie Littillis land upoun the Castelhill. Item,' 13s 4d 'of Franche Antanes land in the heid of Merleyones Wynd. Item, thrie merk of Walter Chepmanes land in the Freir Wynd.

Heirfoir I beseik your lordschip for the rewaird of God to giff command to James Nicolsoun to giff me lettres in foure formes, as use is, to caus me be answerit of the saidis annuellis to my leving, I satisfeand my lord Comptrollar[145] for the queinis grace thrid thairof, and I sall pray for your lordschip quhill I leve.'

Signature: 'Clericus Registri. James Nicolsoun, ressave this rentale.'

---

[143] These sums, presumably belonging to the altarage of St Michael, total £9 7s, not £10 as stated above.
[144] This should read 'oratour'.
[145] See above, p. 94, n. 15.

EDINBURGH, (SRO, Vol. a, fo 149r)
ST GILES' COLLEGIATE KIRK, (SRO, Vol. b, fo 104r)
ALTARAGE OF ST SALVATOR IN,[146]

'Followis the rentale of the annuellis dotit to [the] alterage of Sanct Salvatoure foundit within the college kirk of Sanct Jeill pertenyng to Andro Naper.'

'Inprimis, ane annuellrent of' £10 yearly 'to be upliftit and tane at tua termes in the yeir, Wit Sonday[147] and Martinmes[148] in winter be equall portiounes of all and haill the said [foir, *SRO,Vol. b*] land of umquhill Alexander Rynd, viz., fra the mydis of the stair to the grund thairof lyand in the burcht of Edinbrucht upoun the south part of the Queinis Streit thairof betuix the landis of umquhill Thomas Butlaire upoun the eist, the saidis landis of Williame Borthuik in the west.' Another annualrent of 13s 4d money 'foirsaid yeirlie to be upliftit and tane at the termes foirsaidis of all and haill the landis of umquhill Johnne Spens with thair pertinentis lyand in the Cowgait upoun the north syd of the streit thairof betuix the landis of umquhill Adame Carkettill upoun the eist, the landis of umquhill Andro Bell upoun the west, and commoun streit of the Cowgait upoun the southe and the landis of Williame Tod upoun the north.' Another annualrent of 4 merks money 'foirsaid yeirlie to be upliftit at termes foirsaid of all and haill the landis of David Gillespie under and above with the pertinentis lyand within the land of the said umquhill Alexander Fynd [*recte*, Rynd, *SRO, Vol. b*] foirsaid betuix the land of Hesper [Jesper, *SRO, Vol. b*; *i.e.* Jasper] Maine upoun the north and the said Cowgait upoun the south partis. Summa totalis', 20 merks.

Signature: 'Andro Naper, with my hand.'

RESTALRIG COLLEGIATE KIRK, (SRO, Vol. a, fos 149v-151v)
DEANERY OF, (SRO, Vol. b, fos 104v-106v)

'The rentale of the deanrie of Restalrig with the parochinis of Leswaid and Glencors as it presentlie payis to me, Mr Johnne Sinclair, dean thairof.'

'Item inprimis, the teyndis of Dalkeyth, Sanderigis and tua coldenis set to the erle of Mortoun in assedatioun for' £93 6s 8d.[149] 'Johnne Bannatyne for the teyndis of Pettindreich occupyit be him', 1 c. oats; 8 b. bere; 8 b. rye. William Sinclar in Over Lassuaid, 6 b. oats; 2 b. bere; 6 f. wheat [*in margin*, 'Tak']. Francis Tennent in Over Lessuaid, 10 b. oats; 2 b. bere; 2 b. wheat [*in margin*, 'Tak']. Loynheid and Dryden, 36 b. oats; 9 b. bere; 3 b. wheat [*in margin*, 'Tak']. Laird of Rosling

---

[146] Cf SRO, Vol. a, fos 109v, 140v, 145r (see above, pp. 96, 129, 135).
[147] I.e. 15 May.
[148] I.e. 11 November.
[149] Cf above, pp. 103, 116, nn. 45, 87.

for the maines and toun thairof and pendicles', 27½ b. oats; 18½ b. bere; 6 b. 1 p. wheat; 8½ b. rye [*in margin*, 'Tak']. Carkettill [*i.e.* Kirkettle], 20 b. oats; 7 b. bere; 3 b. wheat [*in margin*, 'Tak']. Forthe [*i.e.* Firth, *SRO, Vol. b*], 11 b. oats; 3 b. bere; 2 b. wheat [*in margin*, 'Tak']. Mylne of Carkettill, 2 b. oats; 1 b. bere [*in margin*, 'Tak']. Laird of Over Auchindonye [*i.e.* Auchendinny], 1 c. oats; 3 b. 3 f. bere. William Forrester in Nather Auchindonye, 6 b. oats; 2 b. bere; 1 b. wheat. Laird of Hauthornden, 12 b. oats; 2 b. bere; 2 b. wheat. Laird of Poltoun for his teinds thereof, 8 b. oats. Roslyng Wodheid, ½ b. oats; 1 f. bere. Walkmylne of Rosling, ½ b. oats; 1 f. bere. Lands of Ley, 4 b. oats; ½ b. bere; ½ b. wheat. Newbigging, 3 b. oats; 1 b. bere. A[n]crumlaw, 3 b. oats; 1 b. bere. Fairhillis, 6 f. oats; ½ b. bere. Heycorne, 6 f. oats. Wowterschill, 4 b. oats; 1 b. bere. Bateis land, 2 b. oats; 1 b. bere. Gortoun, 20 b. oats; 7 b. bere; 2 b. wheat [*in margin*, 'Tak']. Gorlaw, 20 b. oats; 7 b. bere; 3 b. wheat [*in margin*, 'Tak']. Gorsnewt, 15 b. oats; 4 b. bere; 3 b. rye [*in margin*, 'Tak']. Cuikkin, 6 b. oats; 6 f. bere. Lennestown Law, 8 b. oats; 2 b. bere. Towrmonth, 9 b. oats; 2 b. bere. Grenelaw, 12 b. oats; 3 b. bere. Dawmoirmyln, 2 b. oats; ½ b. bere. Eistraw, Myltoun and Wodhouslie, 18 b. oats; 8½ b. bere; 3½ b. 3 p. wheat. Kirkland of Glencors, 6 b. oats; 2 b. bere. Glencors, 10 b. oats; 4 b. bere. Kyrnslaw, 3 b. oats; ½ b. bere. West Marche, 4 b. oats; 1 b. bere [*in margin*, 'Tak']. Castellaw, 40 b. oats; 12 b. bere; 10 b. wheat [*in margin*, 'Tak']. Fowlfurde, 2 c. oats; 12 b. bere; 4 b. wheat [*in margin*, 'Tak']. Scahyntie, Corshous, 13 b. oats; 4 b. bere; 2 b. wheat [*in margin*, 'Tak'].

Total money, £93 6s 8d. Total oats, 25 c. 8 b. 2 f.[150] Total bere, 8 c. 6 b. 2 f.[151] Total wheat, 2 c. 13 b. 2 f.[152] Total rye, 1 c. 3 b. 2 f.

'Assignatioun of the deanrie of Restalrig':

Third of money, £31 2s 2⅔d. 'Tak it out of the teyndis of Dalkeyth, Sanderigis and tua coldeinis payand be yeir' £93 6s 8d.

Third of oats, 8 c. 2 b. 3 f. 1⅓ p.[153] Take the teinds of Pettindreich occupied by John Bannatyne, for 1 c.; the teinds of Over Auchyndunye occupied by the laird thereof and his mother, for 1 c.; Nather Auchindunnye, occupied by William Foster [Forrester, *fo 149v*], 6 b.; Hawthornedeyn, occupied by the laird thereof, for 12 b.; Greynlaw for 12 b.; Cwickin for 6 b.; Wodhouslie, Mylntoun and Eistraw for 1 c. [18 b., *fo 150r*]; Ley and Bateislandis for 6 b.; Towrmonth and Lennestounlaw for 1 c. 1 b.; Glencors for 10 b.; Kirkland of Glencors for 6 b.; Wowterischill for 4 b.; Krinslaw for 3 b. Total, 8 c. 3 b.[154] 'Gif in' 1 b. 2⅔ p.[155]

Third of bere, 2 c. 12 b. 3 f. 1⅓ p. Take the teinds of Pettindreich for 8 b.; the teinds of Ovir Auchinebuye [*recte*, Auchindunye] for 3 b. 3 f.; Nather Abchindunye [*recte*, Auchindunye] for 2 b.; Hawthornedyne for 2 b.; Greynlaw

---

[150] This calculation is correct.
[151] The correct calculation appears to be 8 c. 6 b. 3 f.
[152] The correct calculation appears to be 2 c. 13 b. 3 f.
[153] The correct calculation appears to be 8 c. 8 b. 2⅔ p.
[154] The correct calculation appears to be 8 c. 2 b.
[155] 8 c. 3 b. minus 1 b. 2⅔ p. would give 8 c. 1 b. 3 f. 1⅓ p.

for 3 b.; Cwickin for 2 b. [6 f., *fo 150r*]; Wodhouslie, Mylntoun, Preistlaw [*recte*, and Eistraw] for 8 b. 2 f.; Lay and Bateislandis for 1 b. 2 f.; T[ow]rmonth and Leniestounlaw for 4 b.; Glencors for 4 b.; Kirkland of Glencors for 2 b.; Wowterischill, 1 b.; Newbigging for 1 b.; Ancrumlaw [*i.e.* Ankrielaw] for 1 b.; Fairhillis for 2 f.; Dawmoirmylne for 2 f.; Kyrnslaw for 2 f. Total, 2 c. 13 b. 1 f. 'Gif in' 1 f. 2⅔ p.

Third of wheat, 15 b. 2⅔ p. Take Nather Auchindony for 1 b.; Hawthorneden for 3 b. [2 b., *fo 149v*]; Wodhouslie, Myltoun and Eistraw for 3 b. 2 f. 3 p.; Lay and Bateislandis for 2 f.; Glencors for 2 b.; Gourtoun and Gowirlaw for 5 b. 1 p. [2 b. and 3 b., *fo 149v*]; Over Laswaid occupied by Frances Tennent for 2 b. Total of wheat, 1 c. 1 f. 'Giff in out of Frances Tennentis quheit', 1 b. 2 p.[156]

Third of rye, 6 b. 2 f. Take the teinds of Leswaid and Pettindreich for 8 b. 'Giff in' 1 b. 2 f.

RESTALRIG COLLEGIATE KIRK,   (SRO, Vol. a, fo 151v)
PREBENDS AND CHAPLAINRIES OF,   (SRO, Vol. b, fo 106v)

'The rentale of the prebendareis of Restalrig and cheplanes thairof.'

'Item inprimis, the sex prebendareis of Buit foundit upoun the personage and vicarege of Rothsay in Buit and upoun the personage and vicarege of Ellem kirk, Williame Barbour haveand thrie of thame, ane provydit of auld and tua provydit of new be the first Regentis grace,[157] quha is with God, the tua part allowit to him in his stipend be the kingis thrid; Lanceolat Gibsoun, prebendar of Buite provydit of auld; Cuthbert Suyntoun, prebendar of Buit, provydit of auld; Johnne Nasmyth, prebendar of Buit, provydit of auld; all being in lyff; Johnne Hoggarth, prebendar of the Kingis Wark of L[e]yth', £20 'out of the said wark and is departit in Ingland and is, as I am informit, is [*sic*] with God. Maister Hew Congiltoun, chaipland of Sentredwallis Yle,[158] foundit upoun Johnne Dalmahoyis hous in Leyth', 12 merks and 7 merks in the Cannogait 'with the offerantis that he gat with the relikis, quha is with God. Mr Williame Henrysoun, chaipland for Mr Thomas Diksoun foundit be the said Mr Thomas in Connis Clois', 10 merks; 12 merks 'out of the landis of the laird of Stenehous in the west land'; 5 merks 'out of Bruchtoun. Cuthbert Suyntoun haveand in custodie the cheplanrie of Ros foundit be ane bischope of Ros callit Fresall,'[159] 20 merks 'lyand in the Cannogait, now pertenand to the ministrie and to the puir, baith it and Congiltounis chaiplanry be vertew of the act

---

[156] There are discrepancies in the arithmetic.
[157] James Stewart, earl of Moray, Regent, 1567 - 1570, *Scots Peerage*, vi, pp. 315-316. This entry was thus composed at a point after Moray's murder in January 1570.
[158] Sentredwall is otherwise known as St Triduana.
[159] John Fraser or Frissell was bishop of Ross, 1497 - 1507, Watt, *Fasti*, p. 269.

of Parliament[160] and thir ar the prebendareis and cheplanreis that lay to the college kirk of Restalrig with thair chalmeris and yairdis and sum wantit chalmeris but sa mony pertenit to the kingis fundatioun.'

Signature: 'Williame Barbour, with my hand.'

TRANENT, VICARAGE OF, (SRO, Vol. a, fo 152r)
(SRO, Vol. b, fo 107r)

'The rentale of the vicarege of Tranent extendis sen the corpes presentis, small offerantis and Pasche rakiningis wer deducit be yeir to the sowme of' £38 'or thairby.'

Signature: 'Mr Johnne Ra, with my hand.'

NOTE

Rentals for the following benefices are located elsewhere in the text:

CRAMOND KIRK, CHAPLAINRY OF ST THOMAS' ALTAR IN (SRO, Vol. a, fo 157r), p. 149.
NEWBATTLE, CHAPLAINRY OF ST CATHERINE IN (SRO, Vol. a, fo 79r), p. 68.

---

[160] Cf *Acts of the Parliament of Scotland*, eds. T. Thomson and C. Innes, 12 vols. (Edinburgh, 1814-1875), iii, p. 169. See also, *RPC*, i, pp. 202, 497-498.

Haddington (1)

NORTH BERWICK, PRIORY OF,[1]   (SRO, Vol. a, fos 152v-156v)
(SRO, Vol. b, fos 107v-111v)

'The rentale of the abbay of Northberuik gevin in be Alexander Home at Hadingtoun', 18 December 1573.

The kirk of Largo

'Inprimis, set to Androw Wod of Largo, the teynd schaves of Largo for yeirlie payment of' £26 13s 4d, 'with the fischeing of Levyn for yeirlie payment of sevin barrellis salmond or truittis or ellis' 40s 'for ilk barrell, extending to' £14. Set to James Wod, his son, the teind sheaves of Balcarmo and Straarlie [*i.e.* Strathairly] for yearly payment of £22. Set to the laird of Lundie, the teind sheaves of Lundy 'with thair pertinentis' for yearly payment of £26 13s 4d. Set to Mr George Lundy, his brother, the teind sheaves of Over Prateris [*i.e.* Pratis], Nether Prateris and Monthrys for yearly payment of £30. Set to my lord Lindsay,[2] the teind sheaves of Pitcurvie [*i.e.* Pitcruvie] for yearly payment of £10.[3] Set to John Traill, the teind sheaves of Boseymylne [Bussie Mylne, *fo 156r*] for yearly payment of £4.[4] Set to Janet Lundy of Baldastard, the teind sheaves thereof for yearly payment of 53s 4d. Set to the laird of Lundy, the teind sheaves of Gilstoun for yearly payment of £6 13s 4d.

The temporal lands within the said parish of Largo

'Item, set in few ferme to the lady Durie, the landis of Monthris for yeirlie payment of' £24. Set in feuferme to Richard Carmichaell, the lands of Atherny for yearly payment of £11.

'Summa of the haill money of the teyndis and temporall landis within the said parochin of Largo', £177 13s 4d.

---

[1] Cf SRO, Vol. a fo 170r (see below, p. 166). An earlier rental, c. 1550, of this Cistercian nunnery is printed in *Carte Monialium de Northberwic*, ed. C. Innes (Edinburgh, 1847), pp. xxii-xxvi.
[2] Patrick, 6th lord Lindsay of the Byres, *Scots Peerage*, v, pp. 399-400.
[3] 'Set to Patrick Lindesay, "now maister", the teind sheaves of Petcrowye for the sum of £10', fo 170r (see below, p. 167).
[4] 'Set "to the guidwife of Bowsie" the teind sheaves thereof for £4', fo 170r (see below, p. 167).

'Off the rentis of the kirk foirsaid gevin furth and disponit in pensioun to [*blank*] Lyndesayis, dochteris to Thomas Lindesay, Snawdoun herauld', £40. 'Mair in pensioun to [*blank*] Wod, dochter to the laird of Largo', £20.

The kirk of Kilconquhair
'Item, set to the laird of Sanctmonanis, the teynd schaves thairof for yeirlie payment of' £6 13s 4d. Set to the laird of Sandfuird, the teind sheaves of Hilhous and Balbuthy [Bawbuthe, *fo 170v*] for yearly payment of £7. Set to the laird of Kincraig [lady Kincraig, *fo 170v*], the teind sheaves of Murecambuies for yearly payment of £6 13s 4d. Set to Alexander Home of Heuch, now in the hands of Patrick Home of Polwarth, the teind sheaves of Brintschellis and Banneill for yearly payment of £6 13s 4d. Set to Mr Alexander Wod of the Grange, the teind sheaves of Lathalland for yearly payment of £5 6s 8d. Set to John Betoun of Pitlochy, the teind sheaves of Balcarres for yearly payment of £13 6s 8d. Set to the said John Betoun, the teind peas and beans of Kilconquhair for yearly payment of 20s. Set to Arthur Forbes of Reres, the teind sheaves of Reres[5] for yearly payment of £16. Set to Elizabeth Creichtoun, lady Ardros, the teind sheaves of Kilbrachmont and Carmure [Carnie, *fo 170v*] for yearly payment of £20. Set to Mr Alexander Wod, the teind sheaves of Nather Reres for yearly payment of £6 13s 4d. Set to the said Mr Alexander, the teind sheaves of Kilconquhair for yearly payment of £20. 'Item, set to him', the teind sheaves of Newtoun of Reres for yearly payment of £10. Set to the laird of Kincraig [lady Kincraig, *fo 170v*], the teind fish of the Erlisferry for yearly payment of 14s. Set to Mr William Cok, the teind sheaves of Fawfeildis and Frostleyis for yearly payment of £6 13s 4d.

The temporal lands within the said parish
Set in feuferme to Mr Alexander Wod, the lands of the Grange paying yearly £6 13s 4d.
'Summa of the haill money of the teyndis and temporall landis within the parochin foirsaid', £133 7s 4d.

The victual of the said kirk of Kilconquhair
'Inprimis, set to the ladie of Ardros, the teynd schaves of the manes thairof for yeirlie payment of the victuall following, viz.', 4 b. wheat; 18 b. bere; 18 b. oats. Set to the laird of Sandefuird, the teind sheaves thereof for yearly payment of 8 b. bere; 8 b. oats. Set to the laird of Kincraig, the teind sheaves of that and half of the lands of Kincraig for payment yearly of 4 b. 2 f.[6] bere; 3 b. 2 f. oats. Set to the laird of Lundie, the teind sheaves of the other half of Kincraig for yearly payment of 3 b. 2 f. bere; 3 b. 2 f. oats. Total victual within the said parish: Bere, 2 c. 1 b. Oats, 2 c. 1 b. Wheat, 4 b. 'In the haill', 4 c. 6 b.

---

[5] The teind sheaves of Over Reres were set to John Duddingston for £16, fo 170v (see below, p. 167).
[6] This should read 3 b. 2 f. to give the correct total.

The kirk of Northberuick

Set to Robert [Robene, *fo 171r*] Ker of Ancrum, the teind sheaves of the Leuchin, Sydserff and Fentoun Tour for yearly payment of £10. Set to Walter Malvile, the teind sheaves of the lands of Gleghorne for yearly payment of £3 6s 8d. Set to Alexander Home of Northberuik Maines, the teind sheaves thereof for yearly payment of £6 13s 4d. Set to the said Alexander, 'the teynd fische of the Heavin of Northberuik' for yearly payment of £13 6s 8d. Set to Andrew Home the teind sheaves of the crofts of Northberuik for yearly payment of 33s 4d. To the laird of Eistcraig, 'the teynd schaves of the ane half thairof' for yearly payment of £6 13s 4d. Set to the laird of Smetoun, the teind sheaves of the other half of Eist Craig, paying yearly £6 13s 4d. 'My lord Regentis[7] grace paying yeirlie for the teynd schaves of Tantalloun', £50.[8]

The temporal lands within the parish foresaid

Set in feuferme to Robert Home of the Heuche, the land of the Heuch with the Law and Northmedow thereof, for yearly payment of £57. Set in feuferme to Mr Alexander Wod of the Grange, the lands of Northberuik 'callit Adames croft, with the landis occupyit be Alexander Maxwell on the north syd of the maynes, paying yeirlie' £6 13s 4d. Set in feuferme to the said Robert Home, Kynkaith Mylnes with the crofts thereof, paying yearly £5 6s 8d. Set to the said Robert in feuferme 2 husbandlands in Benestoun, paying yearly 26s 8d. Set in feuferme to Alexander Home, son to Patrick Home of Polwarth, the Manes of Northberuik and ferme acres thereof for yearly payment of £35. Set in feuferme to the said Alexander, the lands of the Grange of Breich, paying yearly £4. Set to Isobel Home in Kingstoun, the Mylne of Lyntoun, paying yearly £3 6s 8d.

'Summa of the haill money of the teyndis of the kirk foirsaid and few mailles of the temporall landis abonespecifeit', £211.

The victual of the said kirk of Northberuick

'Inprimis, set to the laird of Bas, the teynd schaves of Balgonye for yeirlie payment of the victuall following, viz.', 1 c. wheat; 1 c. bere; 2 c. 8 b. oats. Set to Robert Home of Heuch, the teind sheaves thereof for yearly payment of 1 c. wheat; 1 c. bere; 2 c. oats. 'Summa of the teynd victuall within the parochin foirsaid', 2 c. wheat; 2 c. bere; 4 c. 8 b. oats. 'Extending in the haill to' 8 c. 8 b.

'Off the fruites and rentis of the kirk foirsaid Maister Adame Home haveing ane yeirlie pensioun of' £200 'and for payment thairof specialie assignit to him the few mailles and teynd schaves of the landis of the Heuch and als the few mailles of the Manes of Northberuik and dewtie of the teynd schaves thairof.'

---

[7] James, 4th earl of Morton, Regent, 1572 - 1578, *Scots Peerage*, vi, pp. 362-363.
[8] The correct calculation appears to be £98 6s 8d.

The kirk of Logy[9]
Set to Patrick Home of Polwarth for yearly payment of £58.

'Summa of the haill money of the teyndis of the saidis kirkis of Largo, Kilconquhair, Northberuik and Logy intromettit with be the priores[10] and hir chalmerlane and of the haill few mailles of the temporall landis of the said abbay', £580 0s 8d.

'Summa of the haill victuallis of the teyndis of the kirkis foirsaidis': Wheat, 2 c. 4 b. Bere, 4 c. 1 b. Oats, 6 c. 9 b.

The kirk of Mayboill
Set to Thomas Kennadie of Barganie for yearly payment of £30 'togither with xx oxin and tuelf ky yeirlie, quhilk haill rent of the kirk foirsaid, bayth tua pairt and thrid pairt, is gevin up in pensioun to Williame Cunyghame, broder sone to my lord erle of Glencarne,[11] and in consideratioun of the priores and hir chalmerlane hes had na intromissioun thairwith sen the geving thairof in pensioun wes nevir burdenit with the thrid of the samin.'

'The assumit thrid of the said abbay':
'Item, be the laird of Lundie for the teynd schaves of Lundy with the pertinentis yeirlie', £26 13s 4d. Mr George Lundie, his brother, for the teind sheaves of Over Prateris, Nather Prateris and Monthryis, £30. The laird of Sandfuird for the teind sheaves of Hilhous and Balbuthy, £7. Arthur Forbes of Reres for the teind sheaves of Reres, £16. Patrick Home of Polwarth for the kirk of Logy, £58. 'Be my lord Regentis grace for the teynd schaves of Tantalloun', £50. John Traill for the teind sheaves of Bussie Myln [Boseymylne, *fo 152v*], £4.

The third of the victual of the said abbay:
Wheat, 12 b. Bere, 1 c. 5 b. 2 p.[12] Oats, 2 c. $\frac{2}{3}$ b.,[13] 'quhilk is payit yeirlie to the priores and hir chalmerlane.'

'Item, to the convent of the said abbay and utheris provydit in pensioun equallie with thame beand in number of the haill' 13 'persounes, ilk ane presentlie as followis': To Margaret Sinclar, £20. To Margaret Craufuird, £20. To Helen Schaw, £20. To Janet Creichtoun, £20. To Alison Puntoun, £20. To Helen Derling, £20. To Elizabeth Rentoun, £20. To Jean Home, £20. To Margaret Donnaldsoun, £20. To Marion Wod, £20. To Catherine Drummound, £20. To

---

[9] I.e. Logie-Atheron in the sheriffdom of Stirling, diocese of Dunblane; also known as Logie-Wallach. *RMS*, vi, no 46; Cowan, *Parishes*, p. 136.
[10] Margaret Hume, prioress, 1568 - 1587, Cowan and Easson, *Medieval Religious Houses*, p. 148.
[11] Alexander Cunningham, 4th earl of Glencairn, *Scots Peerage*, iv, pp. 239-240.
[12] The correct calculation appears to be 1 c. 5 b. 2 f. $2\frac{2}{3}$ p.
[13] The correct calculation appears to be 2 c. 3 b.

Alexander Storie, £20. To Alexander Patersoun, £20. To Mr Adam Home, parson of Polwarth, a yearly pension of £200.

'In the haill kirk of Mayboill and rentis thairof in pensioun to Williame Cunyghame. Item, to the Lordis of Sessioun for contributioun yeirlie', £21.

DALMENY KIRK,[14] CHAPLAINRY OF (SRO, Vol. a, fo 157r-v)
ST CUTHBERT'S ALTAR IN, (SRO, Vol. b, fo 112r-v)
CRAMOND KIRK,[15] CHAPLAINRY OF ST THOMAS' ALTAR IN,

'The warrand and supplicatioun for ressaving of the rentales of Sanct Cuthbertis altar in Dummanie and rentall of Sanct Thomas altar in Crawmond pertenyng to sir Patrik Murray' [Mowbray, *below*].

'My lord Collectour,[16] unto your lordschip humilie meinis and schawis your lordschippis servitour Patrik Mowbray in Dummanie, that quhair I have tua cheplanreis, ane callit the cheplanrie of Sanct Cuthbertis altare within the kirk of Dummanie, ane uther callit the cheplanrie of Sanct Thomais altar within the kirk of Crawmond, baith inwithe the valour of' 23 merks 'and hes wantit' 11 merks 'thairof thir' 20 'yeiris bygane, unpayit be ressoun the samin lyis in the southe pairtis besyd Peiblis quhair sict men as I that duellis in the inland can get na payment, albeit I have na mair to live on, and in the moneth of October' 1573 'I menit me to your lordschip for feir of tynsall thairof in respect of the ordinance maid anent the upgeving of rentales of benifices within this realme, desyering my saidis cheplanreis to be rentallit conforme to the rentall than presentit to your lordschip, at the quhilk tyme your lordschip gave command to Alexander Hay,[17] scribe, to rentale my saidis cheplanreis, quha promesit to do the samin, howbeit as I am informit he hes nocht done the samin as yit, heirfoir I beseik your lordschip to caus the said Alexander Hay or James Nicolsoun,[18] clerk, to rentale my saidis cheplanreis of the dait and tyme abonewrittin, conforme to your lordschippis command than gevin to him, and my rentallis thairof than producit be your lordschip, conforme to justice and your lordschippis answer, humblie I beseik.

Sic habet a tergo [*i.e.* Thus it is endorsed]. Apud Edinbrucht', 10 November 1576.

---

[14] Dalmeny lay in the sheriffdom of Linlithgow; the link with Haddington is unclear.
[15] Cramond lay in the sheriffdom of Edinburgh; the link with Haddington is unclear.
[16] Robert, lord Boyd, Collector general, *RSS*, vii, no 487; cf below, p. 150, n. 19.
[17] Alexander Hay, clerk of the Privy Council (till June 1572), was appointed director of Chancery and keeper of the quarter seal for life in September 1567, *RSS*, vi, no 20; *RPC*, ii, p. 139.
[18] James Nicolson, clerk of the Collectory, *Thirds of Benefices*, p. 62.

'Ordanes James Nicolsoun, clerk over the rentales, to ressave this complenaris rentale and tak cautioun for payment of the yeiris bigane. Sic subscribitur: R. Dumfermling, R. Boyd.'[19]

Rental of Sanct Cuthbertis altar in Dummanie now pertaining to sir Patrick Mowbray [Murray, *fo 157r*].

'Item, the landis of Hopcalzocht pertenyng to the laird of Drummelzer, heritour of the samin, ar addebtit in nyne merkis. Item, the landis of Skirling pertenyng to the laird of Skirling is addebtit in tua merkis. Item the landis of Dummanie occupyit be the said Patrik is yeirlie worthe' 40s 'as the samin wes set of auld.'

Rental of Sanct Thomas altar in Crawmond pertaining to the said sir Patrick.

'Item, Johnne Thomsounes landis in Crammond Riggis is addebtit in' 2 merks. 'The landis of Skinne[ri]s landis that now pertenis to Johnne Wardlaw in Leyth, heritour thairof, is addebtit in' 2 merks. 'Johnne Loganes land thair is addebtit in' 2 merks. 'The landis of Nather Crammond, occupyit be the said sir Patrik and his cotte[ri]s, is yeirlie worth and payit of auld' 3 merks.

Signature: 'Patrik Mowbray, with my hand.'

Calculation of third:
    'Sanct Cuthbertis altar in the haill', £9 6s 8d; third thereof, £3 2s $2\frac{2}{3}$d.
    'Sanct Thomas altar in the haill', £6; third thereof, 40s.[20]

---

[19] Robert Pitcairn, commendator of Dunfermline, Cowan and Easson, *Medieval Religious Houses*, p. 58; Robert, 5th lord Boyd, *Scots Peerage*, v, pp. 155-161.
[20] Rentals for Haddington resume at SRO, Vol. a, fo 165r; SRO, Vol. b, fo 118r (see below, p. 161).

Linlithgow

ECCLESMACHAN, PARSONAGE OF, (SRO, Vol. a, fo 158r-v)
 (SRO, Vol. b, fo 113r-v)
Rental of the parsonage of Eglismachane

'Item inprimis, the haill toun of Eglismachane payis' 40 merks. Eister Bengour 'with the cottares', 24 merks. Bengour Law, 16 merks. The Mydblak Craigis, 9 merks. The Wester Blak Craigis, 17 merks. Wester Bengour, 'the haill toun, coittares and all', 22 merks 6s. The Quhytlaw, 15 merks. Total, £95 12s 8d.[1]

'Item, the sowme abonewrittin the persounes usis to leid the teyndis of the ane half of the toun callit Walderstoun extending to' 2 c. 'of aites togither with the teynd aites of the gleib of the same benifice, the teynd beir of baithe the ane and the uther' 11 b.; 5 b. wheat; 4 b. peas 'or thairby, and this is the haill baithe of silver and victuall pertenyng to the said benifice of Eglismachane. Item, the gleib payis' 18 b. bere and meal. 'Item, by the money and victuallis abonewrittin the vicarege of the said benifice extendis yeirlie communibus annis to' 10 merks 'or thairby.'

Signature: 'Johnne Mowbray, persoun of Eglismachane.'

Calculation of third: Money, £102 6s; third thereof, £34 2s. Oats, 2 c.; third thereof, 10 b. 2 f. $2\frac{2}{3}$ p.,[2] 'by the teynd aites of the gleib.' Bere, 11 b.; third thereof, 3 b. 2 f. $2\frac{2}{3}$ p.[3] Wheat, 5 b.; third thereof, 1 b. 2 f. $2\frac{2}{3}$ p.[4] Peas, 4 b.; third thereof, 1 b. 1 f. $1\frac{1}{3}$ p.

---

[1] In the text this is rendered 7 score and 3 merks 6s.
[2] In the text this is rendered 'x b. ij f. ij pect, 3 pect quarter 3 pairt' [i.e. $\frac{1}{3}$ p. + $\frac{1}{4}$ p. = $\frac{2}{3}$ p.].
[3] In the text this is rendered 'iij b. ij f. ij pect, 3 pect 3 pairt 3 pect' [i.e. $\frac{4}{6}$ p.; *recte*, $\frac{2}{3}$ p.].
[4] In the text this is rendered 'j b. ij f. ij pect, 3 pect 3 of a 3 pect' [i.e. $\frac{4}{6}$ p.; *recte*, $\frac{2}{3}$ p.].

DALMENY, VICARAGE OF, (SRO, Vol. a, fo 158v)
(SRO, Vol. b, fo 113v)

Rental of the vicarage of Dunnany pertaining to William McDowyll.

Calculation of third: In the whole, £20; third thereof, £6 13s 4d.

STRATHBROCK, VICARAGE OF, (SRO, Vol. a, fo 158v)
(SRO, Vol. b, fo 113v)

Rental of the vicarage of Strabrok pertaining to sir Patrick Ogstoun.

Calculation of third: In the whole, £20; third thereof, £6 13s 4d.

TORPHICHEN, VICARAGE OF, (SRO, Vol. a, fo 158v)
(SRO, Vol. b, fo 113v)

Rental of the vicarage of Torphichane pertaining to sir Thomas Dikesoun.

Calculation of third: In the whole, £26 13s 4d; third thereof, £8 17s 9⅓d.

BATHGATE, VICARAGE OF, (SRO, Vol. a, fo 159r)
(SRO, Vol. b, fo 114r)

Rental of the vicarage of Baithcat pertaining to Mr John Layng.

'Consistes in woll, lamb, hay, corspresentis, umest claithes and Pasche penneis set of auld be my predicessouris for payment yeirlie of' 50 merks 'of the quhilk aucht to be deducit nonpayment of corpes presentis, umest claithes, and Pasche penneis quhilk wes the principall profeit and commoditie of the said vicarege and now hes payit na thing thir thrie yeiris bygane.'

Signature: ' Presentit be me, Mr Andro Layng.'

Calculation of third: In the whole, £33 6s 8d; third thereof, £11 2s 2⅔d.

'Aliter gevin up and ressavit at command of the Comptroller':[5] In the whole, £26 13s 4d; third thereof, £8 17s 9⅓d.

---

[5] Bartholomew de Villemore, Comptroller, 1555 - 1562, with Thomas Grahame of Boquhaple, 1561 - 1562. They were succeeded by Sir John Wishart of Pittarow, 1562 - 1565. *Handbook of British Chronology*, p. 191.

BINNIE, VICARAGE OF, (SRO, Vol. a, fo 159r)
(SRO, Vol. b, fo 114r)

Rental of the vicarage of Bynnie

Calculation of third: In the whole, £5 6s 8d; third thereof, 35s 6⅔d.

STRATHBROCK, PARSONAGE OF, (SRO, Vol. a, fo 159v)
(SRO, Vol. b, fo 114v)

Rental of the parsonage of Strabrok pertaining to Robert Pitcarrie [*recte*, Pitcarne].

'Item, the haill personage set in assedatioun to my lord Merschell[6] for the sowme of' £200. 'Item, heirof to be deducit to Mr Gilbert Kaith, quhilk he hes had in pensioun thir' 20 'yeiris and mair bipast and hes his provisioun thairupoun the sowme of' £30. 'Item, to the minister', £20. 'Sua restis frie' £150.

Signature: 'Robert Pitcarne.'

Calculation of third: In the whole, £200; third thereof, £66 13s 4d.

LINLITHGOW, VICARAGE OF, (SRO, Vol. a, fo 159v)
(SRO, Vol. b, fo 114v)

Rental of the vicarage of Lynlithgw

'Set be Patrik Frenche, vicar thairof, to James Hereot of Trabroun for yeirlie payment of' 10 merks, with 20 merks 'to be payit yeirlie to the man that sayis the commoun prayeris.'

Calculation of third: In the whole, 30 merks; third thereof, 10 merks.

ABERCORN, VICARAGE OF, (SRO, Vol. a, fo 160r)
(SRO, Vol. b, fo 115r)

'The rental of the vicarege of Abircorne quhairof I, Johnne Lynlythgu, is provydit as use wes for the tyme.'

'The said vicarege, the cors presentis, umest claythis, Pasche fynes and offerandis thairof being payit to me extendit yeirlie to the sowme of' 20 merks 'allanerlie and

---

[6] William Keith, 4th earl Marischal, *Scots Peerage*, vi, pp. 46-48.

corpes presentis, umest clayth and Pasche fynes now being deducit, will extend to' 14 merks.

Calculation of third: In the whole, £9 6s 8d; third thereof, £3 2s 2⅔d.

KINNEIL, VICARAGE OF,   (SRO, Vol. a, fo 160r)
                                                                       (SRO, Vol. b, fo 115r)

'The rentale of the vicarege of Kynneill gevin up be me, Johnne Johnnestoun, vicar of the same, extending in all thingis to' 45 merks; third thereof, £10.

Signature: 'Jhonne [*sic*] Johnnstoun, vicar of Kynneill. Ressavit at command of the Clerk Registre.'[7]

LINLITHGOW KIRK, CHAPLAINRY OF   (SRO, Vol. a, fo 160r-v)
ST BRIDE'S ALTAR IN,   (SRO, Vol. b, fo 115r-v)

'The rentale of Sanct Brydis altar cituat within the paroche kirk of Lynlythgu pertenyng to sir George Ros', £10; 'the thrid thairof extending to' £3 6s 8d.

Calculation of third: *see below*, Chaplainry of Allhallow altar.

LINLITHGOW KIRK, CHAPLAINRY OF   (SRO, Vol. a, fo 160r-v)
ALLHALLOW ALTAR IN,   (SRO, Vol. b, fo 115r-v)

'The rentales of Alhallow altar cituat within the said paroche kirk extendis to' 10 merks.

Calculation of third:
      Chaplainry of 'Brydis altare', in the whole, £10; third thereof, £3 6s 8d.
      Chaplainry of 'Alhallow altar thair', in the whole, £6 13s 4d; third thereof, 44s 5⅓d.

DALMENY KIRK, ALTARAGE OF   (SRO, Vol. a, fos 160v-161r)
OUR LADY IN,   (SRO, Vol. b, fos 115v-116r)

'The rentale of the alterage of Dummany p[ertaining to] sir James Wychtmuir.'[8]

---

[7] James McGill of Nether Rankeilour was Clerk Register at the Reformation, *Handbook of British Chronology*, p. 197.

[8] The name 'Wychtmuir' or 'Wychtirmuir' has not been traced; 'Wightman' may be intended.

'This is the tennour of the fundatioun of Our Lady altare of Dumany possessit be sir James Wychtirmuir, the mailles and annuellis thairof, lyand within the burcht and fredome of the South Queinis Ferry, now possessit and occupyit be thir persounes underwrittin. Item inprimis, be Catherine Logie, ane hous and ane yaird lyand on the west syd of Robert Gibsounes land with ane croft callit the Schotsoun croft, extending yeirlie to' 28s 8d of annual. 'Item, be Marioun Scharpe, ane hous and ane yaird lyand on the eist syd of Johnne Lytillis land extending of annuell yeirlie to' 11s. 'Item, be Johnne Inche of Archie Dawsones land, ane hous and ane yaird lyand on the eist syd of Robe Dawleinis land extending of annuell yeirlie to' 2s. 'Item, be Robe Dawlein, ane hous and ane yaird lyand on the west syd of Archie Dawsones land extending of annuell yeirlie to' 5s. 'Item, be Williame Lawrie, ane hous and ane yaird lyand on the eist syd of the commoun vennall of the kirk yaird extending of annuell yeirlie to' 15s. 'Item, be Petir Logy, ane hous and ane yaird lyand on the eist syd of Robe Allanes land extending of annuell yeirlie to' 40d. 'Item, be Ellene Calein, ane hous and ane yaird lyand on the eist syd of Petir Logeis land extending of annuell yeirlie to' 20d. 'Item, be George Hill, ane hous and ane yaird lyand on the west syd of Robert Allanes land extending of annuell yeirlie to' 6s. 'Item, be Alexander Bell, now possessed be Johnne Gairtner, ane hous and ane yaird lyand on the eist syd of the commoun vennall extending of annuell yeirlie to' 12d. 'Item, of Effie Logie, ane hous and ane yaird lyand on the west syd of umquhill Johnne Dawsonis land extending yeirlie to' 26s 8d of mail. 'Item, be Johnne Michelsoun in the same tenement, ane hous extending to' 8s yearly. 'Item, be Johnne Weir in the same tenement, ane hous extending yeirlie to' 10s.

'Ressavit at command of the Clerk Registre.'

Calculation of third: In the whole, £5 18s 4d; third thereof, 39s 5⅓d.

LINLITHGOW, CARMELITE FRIARY OF, (SRO, Vol. a, fos 161v-162r)
(SRO, Vol. b, fos 116v-117r)

'The rentale of the Freiris Carmeleittis besyd the burcht of Lynlythgw.'

'Item inprimis, Robert Young, maltman, four akeris of land', 4 b. bere; 2 merks money. James Hammiltoun of Sandhill, 3 acres, 3 b. oatmeal. John Adansoun [Adamsoun, *SRO, Vol. b*] and Elizabeth Lytstar, 3 acres, 3 b. bere; 20s money. Edward Wolsoun, 4 acres, 15 f. meal; 1 b. bere. Patrick Coill and [*blank*] Johnnestoun for the Lawerk Mure,[9] 6 merks money, and for the teind, 2 b. meal.

---

[9] This place has not been traced; cf 'the piece of land called Laverokmure, lying contiguous near and about the Place of the Carmelite Friars near Linlithqw ... ', *Protocol Books of Dominus Thomas Johnsoun, 1528 - 1578*, eds. J. Beveridge and J. Russell (Edinburgh, 1920), no 509.

John Gibbesoun, 2 acres, 2 b. meal; 1 merk money. David Gray, 1 acre, 1 b. bere 'and half merk money.' Janet Barry, 1 acre, 1 b. bere 'and half merk money.'

Cudbertstoun: 'Item, Johnne Wallane[10] for the wester syd of the Kirk Hill', 7 f. bere. James Dalzell 'for the eister syd of the Kirkhill and for ane uther aker', 10 f. meal. Mr John Kello, 1 acre, 1 b. meal. William Caling, 1 acre, 1 b. meal. James Caling, 1 acre, 1 b. meal. Michael Gibbesoun, 1 acre, 20s money. John Forrest, 2 acres, 20s money. David Pollok, 1 acre, 10s money. Richard Balderstoun, 1 acre, 5 f. meal. Robert Gairdner, 'tua yairdis quhilkis we lauborit our selffis and now set to him for debt awand for furnesing of our place, thir yairdis beand set as the rest of our akeris for' 6 f. bere. 'Item, our hous of Sanct Michaellis Wynd quhilk is almois fallin to the erde', 6 merks money.

'The annuelrentis of the toun of Lynlythgw': Thomas Watt, 10s. Robert Gairdner 'for the hous sumtyme pertenyng to Robert Gray', 13s 4d. 'Item, sir Thomas Kent land', 30d. 'Item, of ane hous pertenyng to James Wydderspyne with the taill riggis', 24s. 'Item, of the waist land and yaird betuix Mr Williame Creychtounes hous and sir James Hammiltounes hous', 18s 'payit to us yeirlie be the said Mr Williame Creychtoun. Item, of the waist land and yairdis quhilk my lord Sanctandrois[11] occupyis', 5s. 'Item, Andro Leverance hous callit Davidsounes land quhair he duellis', 10s. Beggeis Cornewaill house, 30d. Thomas Glennis house, 3s. Thomas Toddog[12] house, 30d. [Thomas Glenis, *deleted*] Grotpokis[13] house 'at the West Port', 2s. John Froster land, 3s 4d. 'Johnne Thomsoun elder land quhair Archibald Wolsoun duellis', 16s. Robert Younger land, maltman, 6s. Michael Gibbesones house 'quhair he duellis and the bakhous of the said Michaellis land, ilk ane payis' 8s; 'summa', 16s. Ebby Miller house, 8s. Archie Gleghorne house, 3s. Thomas Gudlad house 'callit the bakhous land', 2s. James Dennystoun lands 'quhair Robert Hammiltoun duellis', 6s. 'Item, the land quhair Camsy duelt outwith the Eist Port', 12d. Libra Bynnie house, 7s. Robert Reddeine house, 6s.

Signature: 'Freir James Hoppar, priour of the Carmeleitis besyd Lynlythgw, with my hand.'

---

[10] SRO, Vol. b also renders this 'Wallane'.
[11] John Hamilton, archbishop of St Andrews, 1546 - 1571, Watt, *Fasti*, pp. 298-299.
[12] Thomas Toddoch, burgess of Linlithgow, *Johnsoun Protocol Books*, no 549.
[13] This name has not been traced; Gilbert Grot appears in *Johnsoun Protocol Books*, no 195; see also *Protocol Book of Mr Gilbert Grote, 1552 - 1573*, ed. W. Angus (Edinburgh, 1914).

CHAPLAINRY[14]

(SRO, Vol. a, fo 162v)
(SRO, Vol. b, fo 117v)

'Thomas Kent alias sir Thomas, copy of his rentale'

'Inprimis, out of Patrik Flemyngis land yeirlie', 40s and 40d. Out of William Lawsones land in the Cowgait in Edinbrucht yearly, 38s. John Sprottis land in Sanct Mary Wynd yearly, 14s. John Dagleis there yearly, 16d.

In Lynlythgow: Out of Robert Youngis land yearly, 4s. Out of John Adamsounes land there yearly, 4s. Out of John Ramsayis land yearly, 13s 4d. Out of James Hammiltounes land yearly, 10s. Out of Robert Gardneris land yearly, 5s. Total, £6 17s.[15]

Signature: 'James Nicolsoun,[16] ressave this rentale. J. Clericus Registri.'

CARRIDEN, VICARAGE OF,

(SRO, Vol. a, fo 162v)
(SRO, Vol. b, fo 117v)

'The rentale of the vicarege of Carriddin gevin up be Mr Alexander Hammiltoun, vicar thairof una voce', 25 December 1573.

'Inprimis, the kirkland thairof set of auld in few ferme be umquhill sir Archibald Wetherspune, my predicessour, extending to thrie aker of land and ane half or thairby payand yeirlie', 44s 6d. 'Item, the minute landis thairof set to the laird of Ballinhard and the guidman of the Grange, auld possessouris thairof, for' 8 merks. 'The quhilk to be in veritie I testifie be this my present rentale.'

Signature: 'Mr Alexander Hammiltoun, vicar of Carriddin, with my hand.'[17]

AIRTH, VICARAGE OF,

(SRO, Vol. a, fos 163r-164r)

'The Pasche reknyngis pertenyng to the vicar of Airthe, in the quhilk ar contenit the names of the pairteis with thair profeit efter following, viz.':

'Item, Thomas Yair, ane lamb, pryce' 28d; 'ij ky, ane forrow kow and ane nuckald, pryce' 7d; and for his lint and hemp, 5d. Alexander Malcolme, 2 cows, 7d; and for his lint and hemp, 5d. Robert Smyth, 'ane forrow kow', price 3d; lint and

---

[14] This chaplainry has not been traced. 'Dominus Thomas Kent, chaplain' occurs in *Johnsoun Protocol Books*, nos 18, 39, 217, 408 and *passim*; cf *RMS*, v, no 572.
[15] The correct calculation appears to be £6 13s.
[16] James Nicolson, clerk of the Collectory, *Thirds of Benefices*, p. 62.
[17] The next entry in SRO, Vol. b, fo 118r is the rental for Haddington priory, which corresponds to the text in SRO, Vol. a, fo 168r (see below, p. 164).

hemp 5d; 3 lambs, 3d. James Malcolme, 'half lamb', 14d; 2 cows, 7d; lint and hemp, 5d. John Esplinis, 4 lambs, 4d; 2 cows, 7d; lint and hemp, 5d. Robert Hall, 3 lambs, 3d; 2 cows, 8d; teind lint, 5d. John Bathok, 1 lamb, 1d; 2 cows, 7d; lint, 5d. William Hall, 4 lambs, 3 cows, 11d; lint, 5d. Alexander Allane, 1 cow, 3d; lint, 4d. Robert Gilmour 'nichell lynt', 5d. John Allane, 'nichell teynd lynt', 5d. James Brony, 1 cow, 3d; lint, 5d. Adam Mortoun, 1 cow, 3d; lint, 5d. James Smyth, 1 'nuckald kow', 4d; lint, 5d. James Lamb, younger, 'nichell lynt', 5d. Duncan Baird, 1 cow, 3d; lint, 5d. Robert McKie, 1 cow, 4d; lint, 5d. Robert Bisset, 1 cow, 3d; lint, 5d. Thomas Smyth, 4 lambs, 4d; 1 cow, 3d; lint, 5d. William Wallace, 1 cow, 3d; lint, 5d. William Baird, 1 cow, 3d; lint, 5d. Richard Rannell, 2 cows, 7d; lint, 5d. Thomas Symsoun 'nichell'. Thomas Young, 1 cow, 4d; lint, 5d. Alexander Pawtoun, 1 cow, 3d; lint, 5d. John Bowar 'nichell'. Lawrie Rannald, 1 lamb, 1d; 1 cow, 3d; lint, 5d. William Ra, 1 cow, 3d; lint, 5d. Willie Henry, 1 cow, 3d; lint, 5d. Thomas Adame younger, 'teynd lamb', 28d; 4 cows, 14d; lint, 5d. John Logane, 4 lambs, 4d; 3 cows, 11d; lint, 5d. John Williame, 4 lambs, 4d; 3 cows, 11d; lint, 5d. Thomas Adame, elder, 4 cows, 14d; lint, 5d. Robert Hog, 2 cows, 7d; lint, 5d. John Dalrumple, 2 cows, 7d; lint, 5d. Alexander Hoge, 4 lambs, 4d; 3 cows, 11d; lint, 5d. Andrew Hog, 4 cows, 14d; lint, 5d. John Henry, 'nichell'. Adam Henry, 1 cow, 4d; lint, 5d. Thomas Dyk, 2 cows, 7d; lint, 5d. John Hegeyn, 1 cow, 4d; lint, 5d. [*blank*] Forsyth, 1 cow, 4d. John Small, 3 lambs, 3d; 1 cow, 3d; lint, 5d. Alexander McKie, 1 cow, 3d; lint, 5d. James Auldcorne, 1 cow, 3d; lint, 5d. David Hardie, lint, 5d. Willie Logane, 2 cows, 7d; 4 lambs, 4d; lint, 5d. John Gairdner, 1 cow, 3d; lint, 5d. John Murray, 3 lambs, 3d; 5 cows, 18d; lint, 5d. Patie Wat, 1 cow, 3d; lint, 5d. Alexander Drylaw, 'nichell'. Marion Maily, 1 lamb, 1d; 1 cow, 3d; lint, 5d. Alexander Logane, 4 cows, 14d; lint, 5d. Willie Broun, 1 cow, 3d; lint, 4d. Helen Broun, 4 lambs, 4d; lint, 5d. Charles Allane, 1 cow, 4d; lint, 4d. John McKnellane, 1 cow, 4d; lint, 4d. Andrew Kwnisoun, lint, 4d. John Cowye, 1 cow, 4d; lint, 8d. Archibald Bowyie, 3 lambs, 3d; 2 cows, 7d; lint, 5d. Patrick Hegeyn, 3 cows, 11d; lint, 5d. David Adame, 1 cow, 4d; lint, 5d. John Thomsoun, 1 cow, 3d; lint, 5d. Marion Cowy, 2 cows, 7d; lint, 5d. James Benny, 1 lamb, 1d; 2 cows, 7d; lint, 7d. William Gardinar, 2 lambs, 2d; 1 cow, 3d; lint, 5d. John Mathow, 1 cow, 3d; lint, 5d. Andrew Gardinar, 1 lamb, 28d; 3 cows, 11d; lint, 5d. Allan Russell, 1 cow, 4d; lint, 4d. John Hog, 4 cows, 16d; lint, 5d. The laird of Lugtoun, 3 'forrow ky', 9d; 'nukald ky', 16d. Patrick Hoge, 3 'forrow ky', 9d; 4 'nuckald ky', 16d. William Callender, 3 'forrow ky', 9d; 4 'nuckald ky', 16d. James Duncane, 1 cow, 3d. William Henry, 4 lambs, 4d; 2 cows, 7d.

'This is the verie trew extract of the Peasche buik[18] of the vicarege of Airthe.'

Signature: 'Johnne Bruce.'

---

[18] I.e. the register recording payments of the Easter offerings.

ANCRUM, PARSONAGE OF,[19] (SRO, Vol. a, fo 164v)
ROSS, SUBDEANERY OF,[20]
ABERCHIRDER, VICARAGE OF,[21]
FORTEVIOT, PARSONAGE OF,[22]
PENSIONS FROM,

'The pensiounes provydit be Mr Johnne Thorntoun in favouris of Mr Henry Thornetoun of the benifices underwrittin.'[23]

'In the first, ane pensioun of' £100 'of the fruites of the personage of Ancrum.'

'Item, ane uther pensioun of' £100 'in favouris of the said Mr Henry of the subdenrie of Ros.'[24]

'Item, ane pensioun of' 40 merks 'of the fruites of the vicarege of Abirkerdour.'

'To Gilbert Thorntoun. Item, ane uther pensioun of ane hundreth pundis provydit be Mr Johnne Thornetoun in favouris of Gilbert Thornetoun furth of the fruites of the personage of Forteviot.'

NOTE

Rental for the following benefice is located elsewhere in the text:

DALMENY KIRK, CHAPLAINRY OF ST CUTHBERT'S ALTAR IN (SRO, Vol. a, fo 157r), p. 149.

---

[19] Cf SRO, Vol. a, fos 212v, 239v; NLS, fo 82r (see below, pp. 213, 236, 365). The benefice of Ancrum, in the sheriffdom of Roxburgh, was a prebend of Glasgow Cathedral.
[20] Cf EUL, fo 126v (see below, p. 651); cf above, n. 19.
[21] Cf NLS, fo 167v (see below, p. 480); cf above, n. 19.
[22] Cf NLS, fo 37v (see below, p. 317); cf above, n. 19.
[23] The pensions from the parsonage of Ancrum and subdeanery of Ross had been granted at Rome in 1549, *RSS*, vi, no 775.
[24] James Thornton succeeded John Thornton as subdean of Ross in 1565. Watt, *Fasti*, p. 283. See below, p. 652.

Haddington (2)

MARKLE, CHAPLAINRY OF, (SRO, Vol. a, fo 165r)

'The rentale of the chepellanrie of Markille callit the provestrie[1] thairof pertenyng to Robert Kempt extendis yeirlie in all profeitis to the sowme of' 24 merks.

Signature: 'Robert Kempt, with my hand.'

Calculation of third: In the whole, £16; third thereof, £5 6s 8d.

KEITH-HUMBIE, VICARAGE OF, (SRO, Vol. a, fo 165r)

Vicarage of Keith Humbie pertaining to sir John Grenelaw extends to £20; third thereof, £6 13s 4d.

HADDINGTON, PRIORY OF,[2] (SRO, Vol. a, fos 165v-169r)
(SRO, Vol. b, fos 118r-119r)

'The rentale of the abbay of Hadingtoun pertenyng to ane venerable lady, Elizabeth Hepburne, prioress thairoff, contenyng all and sindrie the fermes, mailles, victuallis, alsweill of the temporalitie as the spiritualitie and teyndis set for money and victuallis gevin in befour our soverane ladyis commissiouneris in Edinbrucht, the [*blank*] day of Februar' 1561/2.

'The rentale of the temporalitie of the said abbay within Loudiane.'

[*In margin*, 'Victuallis, this awayis, utherwayis rentallit and ressavit at the queinis command as the rentall efter following beris.']

---

[1] See Cowan and Easson, *Medieval Religious Houses*, pp. 223-224; Watt, *Fasti*, p. 366.
[2] Cf SRO, Vol. a, fos 182r, 184v (see below, pp. 176, 180). The priory was a Cistercian nunnery.

The Manes [*i.e.* Abbey Mains] and lands of the said abbey set for yearly payment of victuals, namely, 4 c. wheat; 4 c. bere; 8 c. oats. The Mylne [*i.e.* Abbey Mill] thereof set for money, £13 9s 8d. The lands of Correllpittis and Walkmylne set for money, £8. The Nonigait [*i.e.* Nungate] acres set for money, £16 4s 8d. Gynne Mylne with the acres thereof set for money, £25 4s 1d. The mails and annuals of the burgh of Hadingtoun and Nonigait 'that payis now presentlie to us' £4. 'Memorandum, we had the kingis and queinis customes of the burcht of Hadingtoun in almus yeirlie' 40s 'and now thay refuis to pay us the samin thir tua yeiris bygane in falt of letteris that we can nocht obtene of the lordis.' The lands of Bagbe set for money, 71 merks. The Mounkland [Nunland, *fo 168v*; Mureland, *fo 182r*] set for money, 8 merks. The lands of Quhytcastell alias Munnraw [*i.e.* Nunraw], together with the 7 husbandlands of Garwat [*i.e.* Garvald] toun and mill thereof set for money, £20. The kirkland of Barro set for money, 20s. The husbandland of Steintoun set for money, 29s 8d. The husbandland of Beinstoun set for money, 40s. The lands of Slayid set for victuals, 2 c. bere; 4 c. oats [*in margin*, 'Victuallis']. The lands of Newtoun [Little Newtoun, *fo 182v*], Carfra, Newlandis and Snawdoun with their teinds 'siclyk set for victuallis', 5 c. bere; 10 c. oats. The lands of West Hope and Eist Hope with the teinds set for victuals, 8 b. bere; 2 c. oats. The lands of Ryndislaw with the teinds set for victuals, 11 b. 2 f. bere; 12 b. oats. The lands of Garwat [*i.e.* Garvald] Grange with one husbandland in Garwat toun set for money, £8. The four husbandlands in Garvok [*i.e.* Garvald] toun 'with the hauche at the kirk of Garvok' set for money, £5. 'The oxgait of land of Bagbie set for money', 20s.

The rental of the temporal lands beside the kirk of Craill in Fyffe
The lands of Petcorthy set for money, 16 merks 8s. The lands of Meikill Troistrie set for money, 8 merks. The lands of Furdefeild set for money, £11 16s 8d. The lands of Sauchope 'payis be yeir' 55 merks 6s 8d. The lands of Patefeild 'payis be yeir' 4 merks. 'The teynd fisches of our pairt of the kirk of Craill payis' £12. The Abbay croft of Craill 'waist and payis conforme to the rentale' 6s 8d. The lands of Newtoun of Randelstoun 'payis be yeir' £6.

'The rentale of the spiritualitie and kirkis thairto pertenyng bayth in teyndis and money yeirlie.'
Sanct Martines kirk[3] 'commonlie ryddin', the teinds thereof extending to 4 b. wheat; 1 c. bere; 1 c. 8 b. oats. The lands of Byeris 'ryddin and set for' 1 c. 4 b. wheat; 1 c. 4 b. bere; 2 c. oats. The lands of Barnes and half Harperdene 'rydin commounlie to' 8 b. wheat; 1 c. bere; 2 c. 8 b. wheat [*recte*, oats]. The lands of Bagbe set for 3 b. wheat; 4 b. bere; 8 b. oats. The kirk of Garvoll 'by the Stedingis teynd fra ar set for money', 20 merks [*in margin*, 'Kirk Garvok']. The kirk of Elstanfuird set for money, £50. The kirk of Craill 'commounlie riddin yeirlie to' £100; 8 b. wheat; 25 c. bere; 11 c. oats; 11 c. meal.

---

[3] I.e. the chapel of St Martin in the Nungate, Cowan, *Parishes*, p. 79.

'Summa totalis of money', £308 17s 5d 'ob.'[4]   Wheat, 7 c. 11 b.[5]   Bere, 40 c. 1 b. 2 f.[6]   Oats, 12 c. 4 b.[7]   Meal, 11 c.

'Of the quhilkis sowmes of money and victuallis thair is to be deducit as herefter followis: The conventis sustentatioun of the number of' 18 'persounes, ilk ane haveand in the yeir', 4 b. wheat 'extending in the haill to' 4 c. 8 b. 'And in beir ilk ane haveand in the yeir', 7 b. extending to 7 c. 14 b. 'And in meill ilk ane haveand in the yeir', 3 b. extending to 3 c. 6 b. 'And in money for flesche and fische ilk ane haveand in the yeir ilk day' 8d extending to £219 12s.[8] 'And in money ilk ane to thair claithes in the yeir haveand' 6 merks extending to £72. 'Summa particule, in money' £291 12s.

'The pensiounes be ressoun of our lettres and commoun seillis yeirlie payit to my lord Bothwell,[9] our ballie for his yeirlie fe' £100. To Patrick Hepburne of Beynstoun 'his ballie depuite', £40. To the laird of Ungiltoun, £50. To the lord Sempill[10] 'be ressoun of his laubouris to get us agane our place on Nountlaw[11] and landis to be nocht oppressit, conforme to the queinis grace dowares writting', £60. 'To vj auld servitouris and gentilmen to thair claithes and hous in the yeir, ilk ane' £20, 'extending to the sowme of' £120. Total of yearly pensions, £370.

'Pensiounes rasit upoun the said place yeirlie payit in lykwayis: To the Lordis of the Sessioun for thair yeirlie contributioun', £21. 'To the vicares pensiouneris of our kirkis by Craill yeirlie', £80. 'To our advocattis in Edinbrucht in procuratioun for the defence and ingetting of our dewteis yeirlie at the leist', £20. 'Summa particule', £121.

'The yeirlie feallis of the sober servitouris of the place conforme to use and wont: To the maister cuik and utheris tua cuikis', £10. 'To tua porteris', 10 merks. 'To the paintar', £4. 'To the pantreman', 40s. 'To the maltman', 10 merks. 'To the brouster', £4. 'To the brouster', 6 merks. 'To the commoun levanderis of the place', 40s. 'To the gairdner', 26s 8d. 'To the smyth of the place feall', 40s. 'To the cartar and the boyis', 40s. 'To the plewmen and utheris servitouris and boyis of the stabillis', £6 13s 4d. 'Summa of feallis of sober servandis', £51 6s 8d.[12]

'To auld failyeit men and wedowis of our four parochinis and kirkis, viz., of Nonngait, Elstanefuird, Garvak and Abbay and puir pilgrames and indigent

---

[4] The correct calculation appears to be £409 12s 9d.
[5] The correct calculation appears to be 6 c. 11 b.
[6] The correct calculation appears to be 40 c. 11 b. 2 f.
[7] The correct calculation appears to be 42 c. 4 b., as on fo 168v (see below, p. 165).
[8] These figures would be correct for a leap year, i.e. 1560.
[9] James Hepburn, 4th earl of Bothwell, *Scots Peerage*, ii, pp. 161-167.
[10] Robert, 3rd lord Sempill, *Scots Peerage*, vii, pp. 538-542.
[11] This should probably read 'Nunraw'.
[12] This is correct if the payment to the brewer was counted twice.

persounes cumand daylie to the place according to the chalmerlandis compt at the leist in meit', 2 c. 'meill togither with silver extending to' £40, 'distribuite pairt les pairt mair. And to the pure folkis almous daylie, conforme to the diet comptis hard be auditouris, at the leist yeirlie' £66 13s 4d. 'The horss corne and in ressaving of strangeris commounlie at the leist in the yeir', 10 c. oats.

'Memorandum, thair is na thing deducit for our awin sustentatioun and sober servitouris, ressaving of our freindis and strangeris in the yeir by uther greit ex[tra]ordinarie thingis that is sustenit be the grace of God in quheit, beir, meill and money, etc.'

'Memorandum, we have gottin litle thir thrie yeris of our kirk of Craill resting in the parochineris handis in fault of lettres to be obtenit at the Lordis of Counsall.'

'Summa totalis of money that is necessarlie deducit', £873 17s.[13] 'Summa of quheit deducit', 4 c. 8 b. 'Summa of beir deducit', 7 c. 14 b. 'Summa of meill deducit', 5 c. 6 b. 'Summa of aites deducit', 10 c.

'And we have bot of fre money bayth of our temporall mailles and kirkis set for money', £308 17s 5d.[14] 'And of quheit fre, the quheit foirsaid being deducit', 3 c. 3 b. 'And of beir fre', 22 c. 2 b.[15] 'And of meill fre', 5 c. 10 b. 'And of aites fre', 31 c. 4 b.[16] 'The quhilk money and victuallis will skantlie pay the deductiounes and necessar ordinar chargis of the place sum yeiris.'

'Sic subscribitur'

'Hoc est extractum de registris dicti monasterij'

This is an extract from the registers of the said monastery and the annual accounts of the same, properly charged and discharged as they are at present, according to the account and reckoning of the said chamberlain, drawn up by order of the said prioress by me, notary public, testifying under my subscription manual and seal.

To these, Maister William Waterstoun, notary public of the said lady prioress[17] who cannot write.

Assignation of the third of the 'abbay' of Hadingtoun:
    'The abacie of Hadingtoun, thrid thereof', £103 4s 3d '3 ob.' [*i.e.* $\frac{1}{6}$d.]

---

[13] The correct calculation appears to be £940 12s.
[14] This is the total given on fo 166v (see above, p. 163). SRO, Vol. b recommences at the next sentence (see above, p. 157, n. 17).
[15] The correct calculation appears to be 32 c. 3 b. 2 f.
[16] 42 c. 4 b. (see above, p. 163, n. 7) minus 10 c. is 32 c. 4 b.
[17] I.e. Elizabeth Hepburn, fo 165v (see above, p. 161).

'The haill money', £309 12s 9d 'ob' [*i.e.* ½d]; third thereof, £103 4s 3d '3 ob' [*i.e.* ¾d]. Wheat, 7 c. 11 b.;[18] third thereof, 2 c. 9 b. Bere, 40 c. 1 b. 2 f.;[19] third thereof, 13 c. 5 b. 3 f. 1⅓ p.[20] Oats, 42 c. 4 b.; third thereof, 14 c. 1 b. 1 f. 1⅓ p. Meal, 11 c.; third thereof, 3 c. 10 b. 2 f. 2⅔ p.[21]

'Omittit grassumes, caragies, caponis and pultrie.'

[*In margin*, 'Aliter assignatioun of Hadingtoun':]

Third of money, £103 14s 4d '3 part ob': Take the lands of Bagbie for £47 6s 8d; Gummeris Mylne for £25 4s 1d; the Nunland [Mounkland, *fo 165v*; Mureland, *fo 182r*] for £5 6s 8d; the lands of Garvet Grange with one husbandland in Garvet toun for £8; the four husbandlands in the toun of Garvet 'with the hauche at the kirk of Garvet' for £5; for the rest extending to £12 6s 10d '3 pairt ob.'[22] 'Tak sa mony akeris of the Nungait and particular occupyers as sall serve for the said sowme of rest and mak evin.'

Third of wheat, 2 c. 9 b. 'Tak this quheit furthe of the Maynes of the Abbay gevand' 4 c.

Third of bere, 13 c. 5 b. 3 f. 1⅓ p. Take the bere of the lands of Slaid for 2 c.; the bere of Newtoun, Carfra, Newlandis and Snawdoun for 5 c.; the West Hope and Eist Hope for 8 b.; Ryndislaw, 1 b. 2 f. [11b. 2 f., *fo 166r*]; 'the remanent beir extending to' 5 c. 12 b. 1 f. 1⅓ p.[23] 'furth of the Manes of Hadingtoun gevand' 4 c.

'Tak the meill out of the kirkis of Carraill', 3 c. 10⅔ b.

Third of oats, 14 c. 1 b. 1 f. 1⅓ p. Take the oats of the Slaid for 4 c.; the Newtoun, Carfra, Newlandis and Snawdoun for 10 c.; the rest extending to 1 b. 1 f. 1⅓ p. out of the Manes [*i.e.* Abbey Mains, *fo 118r*] giving yearly 8 c.

'Remember of the canes, gressumes, entre silver, custoumes, caponis, pultrie, etc.'

PENCAITLAND, VICARAGE OF,   (SRO, Vol. a, fo 169r)
(SRO, Vol. b, fo 119r)

'The rentale of the vicarege of Pencaitland in all profeitis, tuentie pundis, gevin in be Johnne Chatto, vicar thairof, at Hadingtoun', 2 January 1573/4.

Signature: 'Johnne Chatto.'

---

[18] The correct calculation appears to be 6 c. 11 b.
[19] See above, p. 163, n. 6.
[20] This calculation is correct.
[21] In the text this is rendered 'half of 3 pairt half p.', i.e. one twelfth p., but ⅔ p. is arithmetically correct. Presumably 'of' is a scribal slip for 'p.'; the phrase would then signify ⅔ p.
[22] This calculation is correct.
[23] This calculation is correct, counting 1 b. 2 f. for Ryndislaw.

DUNGLASS COLLEGIATE KIRK, PREBENDS OF,[24]

(SRO, Vol. a, fo 169r-v)
(SRO, Vol. b, fo 119r-v)

'The rentale of the prebendareis of the college kirk of Dunglas, gevin in at Hadingtoun', 20 January 1573/4.

'Inprimis, Frances Wolsoun,[25] the prebendarie callit Spittell,[26] £22. Sir Nicol Michelsoun, 'ane prebendarie haveing of the rentis of Kello and Nather Manes of Chyrnsyd', 17 merks 6s 8d. John Frost, 'ane prebendarie haveing of the teyndis of Strafontanes, personage and vicarege landis of Goddiscroft, Heland and Channonbank' [*i.e.* Shannabank], 16 merks 6s 8d; 8 b. oats; 2 b. bere. Patrick Herot, 'ane prebendarie haveing the dewtie of the landis of Barnesyd', 16 merks. Of the lands of Aldincraw [*recte*, Auchincraw, *fo 178v*] 'possessit be lard Paxtoun', £5. Of the lands of Foulches pertaining to Alexander Home of Manderstoun, 30s. Sir Hew Hudson, 'the prebendarie of Halieweill', £20. John Moncuir, 'ane prebendar furth of Chrinesyd', 18 merks. Of Upsatlingtoun Scheillis, 13s 4d. 'Item, the prebendarie quhilk pertenit to umquhill sir Archibald Ewein, furthe of Aldhamestok', 13 merks; and of Aldcambuies, 7 merks. 'This prebendarie disponit to Petur Hewat, conforme to his provisioun and donatioun gevin to him thairof.'

NORTH BERWICK, PRIORY OF,[27]

(SRO, Vol. a, fos 170r-172v)
(SRO, Vol. b, fos 120r-121v)

'The priores[28] and convent hes fyve kirkis, viz., Mayboill, Logy, Largow, Kilconquhair and Northberwik.'

The said kirk of Mayboill: 'Item, set to Thomas Kennady, laird of Bargany, the said kirk of Mayboill for' 32 'ky and oxin' and £22 of money and £8 for the mail of the kirkland and augmentations of the said feu 'extending in toto[29] the sowme of' £30.

The kirk of Logy: 'Item, set in assedatioun the kirk of Logy to Patrik Home of Polwarth, Patrik his sone and appeirand, and Patrik his oo for' £58.

---

[24] Cf SRO, Vol. a, fo 178v (see below, p. 173).
[25] A priest of the same name also held an altarage in Edinburgh; cf SRO, Vol. a, fo 132v (see above, p. 120).
[26] 'Reidspittall', *RSS*, v, no 3188.
[27] Cf SRO, Vol. a, fo 152v (see above, p. 145 and n. 1). The priory was a Cistercian nunnery.
[28] Margaret Hume, prioress, 1568 - 1587, Cowan and Easson, *Medieval Religious Houses*, p. 148.
[29] SRO, Vol. b text reads 'extending in toto to'.

The kirk of Largow: Set of the said kirk of Largow to Andrew Wod of Largow the teind sheaves of Largow, the teind sheaves of Balcarmo, Straayrlie for £48 13s 4d, 'and for the fischeing of the Leving', 7 'barrellis of salmond fische and troutis, or' 40s 'for the barrell, of the quhilk sowme James Wod, his sone, payis' £16 13s 4d, and the laird, £32. Set to the laird of Lundie for the teind sheaves of Lundie for £26 13s 4d. Set to Mr George Lundie, the teind sheaves of Over Pratas, Nather Pratas and Mountreis, Altane and the kane of Lundie for £30. Set to Patrick Lindesay, 'now maister',[30] the teind sheaves of Petcrowye for the sum of £10.[31] Set 'to the guidwife of Bowsie' the teind sheaves thereof for £4.[32] Set to Janet Lundie of Bandastet [*recte*, Baldastard], her teinds for 53s 4d. Set to Alexander Home for the teind sheaves of Gilstoun, 10 merks.

Kirk of Kilconquhair: Set to the laird of Sanct Monanis, the teind sheaves of Crowe, Sanct Monanis for 10 merks. Set to the lady Ardros, the teind sheaves of Carnie [Carmure, *fo 153r*], Kilbrauchmont for £20. Set to the laird of Sandfuird, the teind sheaves of Bawbuthe [Balbuthe, *fo 153r*] and Hilhous for £7. Set to the lady Kincraig [laird of Kincraig, *fo 153r*], the teind sheaves of Muircambes for £6 13s 4d. Set to Mr Alexander Wod, the teind sheaves of Lathalland for 8 merks. Set to Alexander Home, the teind sheaves of Brynt Scheillis and Bawmeill for 10 merks. Set to Mr Alexander Wod, the teind of the Newtoun [of Reres, *fo 153v*] for 15 merks. Set to John Betoun, the teind sheaves of Cawcarros [*recte*, Balcarres] with the teind peas and beans of Kilconquhair for £13 6s 8d [*in margin*, 'sould be' £14 6s 8d]. Set to John Duddingstoun, the teind sheaves of the Over Reres for £16. Set to Mr Alexander Wod, the teind sheaves of Nather Reres for 10 merks. Set to Mr Alexander Wod, the Grange and lands thereof in feu for feu mail and augmentation for 10 merks. Set to the lady Kincraig [laird of Kincraig, *fo 153v*] the teind fish of the Ferie [*i.e.* Earlsferry] for 14s. Set to Mr William Cok, the teind sheaves of Fawfeilis and Frostleis for 10 merks. 'Item, we laid in the parochin of Kilconquhair the teind of Ardros, Balclawe, Sandfuird, Kincraig'. Teinds of Ardros, 2 c. bere; 1 c. wheat; 2 c. oats; 'half chalder peis and beinis'. The teind of Kincraig, 1 c. bere; ½ c. wheat; 20 b. oats; 4 b. peas and beans. The teind of Sandefuird, 12 b. bere; 4 b. wheat; 1 c. oats; 4 b. peas and beans. [Total,] £131 17s 2d.[33]

Northberwik: Set to Alexander Home, the Manes of the Heuch with Northmedow and Law extending to 23 husbandlands and a half for the sum of £57 together with 2 c. wheat and 2 c. bere for the teind sheaves thereof. Set to Alexander Home in feu, 'ane croft callit Dixsoun croft for' 2 merks, and augmentatioun, 6s 8d. And also set in feu 'the mylnes callit Kinketh Mylnes and mylne landis, tua landis of Beinstoun for summes gers for' 10 merks, and in augmentation, 4 merks, with

---

[30] I.e. Patrick, 6th lord Lindsay of the Byres, *Scots Peerage*, v, pp. 399-400.
[31] 'Set to my lord Lindsay, the teind sheaves of Pitcurvie for yearly payment of £10', fo 152v (see above, p. 145).
[32] 'Set to John Traill, the teind sheaves of Boseymylne for £4', fo 152v (see above, p. 145).
[33] The correct calculation appears to be £112 13s 4d, including £1 in the marginal note.

24 'capones or ellis' 24s 'for thame; for the teynd of [his, *SRO, Vol. b*] cobill', 6 merks. 'Item, fourscoir akeris land set in assedatioun to Robert Lauder of the Bas for' 120 b. bere. Set to Mr Alexander Wod, 'ane croft callit Adame croft with the land occupyit be Mr Alexander Maxwell, the teind aill of North Berwik and the teynd of ane cobill for the sowme of' 50 merks. Set to Robene [Robert, *fo 154r*] Ker, the teind of Syndsorff, Luquhen and the Towir [*i.e.* Fentoun Tour, *fo 154r*] for' £10. Set to Andrew Home, 7 acres of land and a half acre with the teind sheaves of the crofts of Northberuik for 33s 4d. Set to William Horvestoun, 2 acres of land 'lyand at the tub wollis for' 2 b. malt. Set to John Spens and Isobel Lauder, 4 acres of land lying on the Hep Hill for 3 merks. Set to Adam Hepburne, Isobel Home and William Hepburne, the Mylne of Lyntoun[34] for 5 merks; 2 capons. Set to George Home, 2 acres land 'besyd the Ekkyl Welcair [Ekkylwelcare, *SRO, Vol. b*] for' 3 merks the acre. Set in feu to John Carmichell, the lands of Ademy with the pertinents for £11. Henry Kempt, for the feu lands of Munthreis, set to him in feu and heritage for £24. Set to Thomas Hammiltoun, the Grange of Breycht for £4. 'The teynd of ane cobill set to James Hammiltoun for' 6 merks. 'The teynd of ane cobill set to Thomas Cariak [Carrik, *SRO, Vol. b*] for' 6 merks. 'Item, gadderit in Northberuik yaird the teynd of Tantallon'. The teind of Bawgoun, Ester Cragy, Glegorne 'extendis in cornes to' 6 c. wheat; 6 c. bere; 2 c. peas and beans; 10 c. oats.

'The landis that we occupy our selff in Northberuik manurit with our ain guidis, that is tua fouthe plewis', £40. 'Item, we tak in teynd grys, teynd cheis, guis, lamb and woll with the rest of cobillis teynd fische extending in money altogidder to' 30 merks.

Signature: 'Ita est Robertus Lauder, notarius ex mandato domine priorisse de Northberwik.'[35]

Total: Wheat, 9 c. 12 b. Bere, 19 c. 6 b.[36] Peas and beans, 2 c. 12 b.[37] Oats, 13 c. 12 b.[38] [Money,] £48 10s 6d.[39]

Assignation of the third of the 'abacie' of North Beruick:
Third of money, £185 12s 6½d.[40] Take the kirk of Logy for £58 [*in margin*, 'fermorares, Patrik Home of Polwarth, Patrik his sone, and Patrik his oo'];

---

[34] Mylne of Lyntoun set to Isobel Home in Kingstoun, fo 154v (see above, p. 147).
[35] The text for this entry in SRO, Vol. b finishes here at the end of fo 121v, and the next entry, on fo 122r, corresponds to SRO, Vol. a, fo 182r (see below, p. 176 and n. 65).
[36] The correct calculation appears to be 19 c. 4 b.
[37] The correct calculation appears to be 3 c.
[38] The correct calculation appears to be 14 c. 4 b.
[39] The correct calculation appears to be £639 4s. Malt, salmon and cattle have been omitted from these totals.
[40] This is one third of £556 17s 8d, which bears no resemblance to the £48 10s 6d in the preceding paragraph. In the text the fraction '½' is rendered 'ob and 3 pairt ob'.

the teind sheaves of Lundie, fermorar the laird of Lundie, for £26 13s 4d; the teind sheaves of Over Prateris, Nather Prateris, Montreis, Altane and the kane of Lundie, fermorar M[r] George Lundie for £30; the feu mails of the Heuch with the North Medow and Law for £57; the teind sheaves of Over Reres for £16 [*in margin*, 'Johnne Duddingston fermarer']. 'Gif in' 40s 9⅓d.

Third of wheat, 3 c. 4 b. The teinds of Ardrois, 1 c.; the fermes of the Manes,[41] 2 c.; and the rest extending to 4 b. out of the teinds of Tantalloun.

Third of bere, 6 c. 6 b. 2 f. 2⅔ p.[42] Take the 80 acres of land set to Robert Lawder of Bas for 120 b. bere. 'Gif in' 1 c. 1 b. 1 f. 1⅓ p.

Third of peas and beans, 1 c. Out of the teinds of the Manes of Tantalloun and Gleghorne, Balgon and Eist Craig.

Third of oats, 4 c. 12 b.[43] Out of the same teinds giving 10 c.

Third of malt, 2 f. 2⅔ p. 'Fra Williame Harvesoun furth of the tua aker of land for the quhilk he payis' 2 b.

Third of salmon, 2⅓ barrels. 'Fra the laird of Largo for Levine, or' 40s 'for the barrel.'

[Third of] 'ky and oxin', 11.[44] 'Fra the laird of Bargeny for the kirk of Mayboill.'

DUNBAR, PARSONAGE OF, (SRO, Vol. a, fo 173r)

'The rentale of the personage of Dumbar pertenyng to Mr Johnne Hammiltoun, student in Pareis, gevin in be me, James Hammiltoun of Sam[u]elstoun, his brother, that is to say: the teynd schaves of Dunbar town and of Hatherwik set in assedatioun to Moreis Lauder and his mother Helene Liddale duelling in Dunbar for the sowme of' 200 merks 'be yeir.'

Calculation of third: In the whole, £133 6s 8d; third thereof, £54 8s 10⅔d.[45]

WHITEKIRK, VICARAGE OF, (SRO, Vol. a, fo 173r)

'The rentale of the vicarege of Quhytkirk set in assedatioun to Oliver Sincler for yeirlie payment of' £10 money 'allanerlie, presentit be sir Thomas Christesoun, vicar thairof, at Halyruidhous', 23 March 1573/4.

Calculation of third: In the whole, £10; third thereof, £3 6s 8d.

---

[41] Mains of the Heuch gave 2 c. wheat, fo 171r (see above, p. 167).
[42] The correct calculation appears to be 9 c. 7 b. 1 f. 1⅓ p.
[43] One third of 13 c. 12 b. is 4 c. 9 b. 1 f. 1⅓ p.
[44] The total given on fo 170r is '32 ky and oxin' (see above, p. 166).
[45] The correct calculation appears to be £44 8s 10⅔d.

ABERLADY, VICARAGE OF,   (SRO, Vol. a, fo 173v)

'The vicarege of Abirlady quhen the vicaregis wes freellie payit, set in assedatioun for four scoir merkis, it could nocht geve sua mekill now becaus of the defalcatioun.'

Signature: 'Ita subscribitur, Abrahame Chreichtoun.'

Calculation of third: In the whole, £53 6s 8d; third thereof, £17 15s 6⅔d.

HAUCH,[46] PARSONAGE OF,   (SRO, Vol. a, fo 174r)

Rental of the parsonage of the Hauche

'Item, the teynd schaves of Hailles, Trapren with the half teynd of Howesoun and the half teynd of Cowirla [i.e. Gourlaw] Bankis cummis nocht to the utilitie of the persoun bot ar takin up be the bischope of Sanctandrois[47] and his factouris.[48] The teyndis of Waichtoun and Eister Craquho[49] ar takin up be the priores[50] of Sanct Bothenis[51] and hir factouris and cummis nocht to the persounes behove. Item, thair is gevin yeirlie to a preist of the college of Dunbar out of the reddiest fruites of the said personage', £20. 'Item, thair is gevin out of the said personage yeirlie to ane vicar pensiouner', 40 merks. 'Item, the rest of the said pensionare is yeirlie' 280 merks, 'or at the leist' 240 merks 'gif guid payment wer maid thairof as it hes bene maid befoir the tyme of trubill.'

Signature: 'George Hepburne, persoun of the Hauch, with my hand.'

Calculation of third: In the whole, £233 6s 8d;[52] third thereof, £77 15s 6⅔d.

MORHAM, PARSONAGE OF,   (SRO, Vol. a, fo 174v)

'The rentale of the personage of Moram [*inserted*, 'pertenyng to sir Thomas Godrell'] set in assedatioun be umquhill Mr Robert Hoppringle to Margaret Hepburne, the relict of umquhill Johnne Hoppringill, for yeirlie payment of the

---

[46] Hauch or Linton, now Prestonkirk, cf fo 180v (see below, p. 174); Cowan, *Parishes*, pp. 81, 133.
[47] John Hamilton, archbishop of St Andrews, 1546 - 1571, Watt, *Fasti*, pp. 298-299.
[48] Cf royal warrant, fo 7v (see above, p. 7).
[49] I.e. Crauchie, cf Priory of St Bathans, SRO, Vol. a, fo 195r (see below, p. 192).
[50] Elizabeth Lamb, prioress of St Bathans, fo 195v (see below, p. 193).
[51] The rental for the priory is to be found in SRO, Vol. a, fo 195r (see below, p. 192).
[52] I.e. 280 merks plus 40 merks plus £20.

sowme of' £50 'as the said assedatioun beiris, quhairof the redar and minister of the sacramentis gettis' £16, 'sua restis' £34.

Calculation of third: In the whole, £50; third thereof, £16 13s 4d.

SPOTT, PARSONAGE AND VICARAGE OF, (SRO, Vol. a, fo 175r)

Rental of the parsonage of Spote and vicarage thereof

'The personage of Spote, ane of the prebendares of Dunbar pertenyng to Alexander Hume, takin up yeirlie the fruites thairof be Williame Hume in Dunbar, quha payis to the said Alexander Hume yeirlie the sowme of ane hundreth pundis.'

Signature: 'Ita subscribitur, Alexander Hume, with my hand.'

Calculation of third: In the whole, £100; third thereof, £33 6s 8d.

OLDHAMSTOCKS, PARSONAGE OF, (SRO, Vol. a, fo 175v)

'The personage of Auldhamstoks pertenyng to Mr Thomas Hepburne set in assedatioun to Alexander Hepburne of Quhitstoun[53] payand to him thairfoir' 280 merks.

Signature: 'Thomas Hepburne.'

Calculation of third: In the whole, £186 13s 4d; third thereof, £62 4s 5⅓d.

PINKERTON, PREBEND OF (SRO, Vol. a, fo 176r)
DUNBAR COLLEGIATE KIRK

'The rentale of the channonrie of Pinkertoun lyand within the constabularie of Hadingtoun.'

'The yeirlie availl thairof extendis to' 200 merks 'and is set in assedatioun for the said soum thir money yeiris bygane.'

Signature: 'George Hume, takisman thairof.'

Calculation of third: In the whole, £133 6s 8d; third thereof, £44 8s 10⅔d.

---

[53] Alexander Hepburn of 'Quhitsome' or 'Quhitsum', *RMS*, iv, nos 295, 482.

BOLTON, PREBEND OF
DUNBAR COLLEGIATE KIRK
(SRO, Vol. a, fo 176v)

'The rentale of the prebendarie of Beltoun, alias the personage, within the paroche kirk of Dunbar, the Eist Lowtheane, as it is set in assedatioun becaus the said fermoreris hes had it' 60 'yeiris and thair is in annuall tak for' 100 merks, 'and within this ten yeiris or thairby it is rasit to pay yeirlie dewetie the sowme of sevin scoir merkis to the haill service that wes usit for the tyme, sua this thrie yeiris bypast my ladie Boltoun[54] hes rest hir teyndis and thir tua yeiris bypast my lord Yester[55] hes tane his teyndis.'

Signature: 'M[r] Johnne Manderstoun, persoun of Beltoun.'

Calculation of third: In the whole, £93 6s 8d; third thereof, £31 2s 2⅔d.

BOTHANS COLLEGIATE KIRK, PROVOSTRY OF, (SRO, Vol. a, fo 177r)

Rental of the provostry of Bothanes

'Item, the said provestrie set to my lord Yester for' £100 'be yeir as his tak beiris. M[r] James Cambell, ane of the prebendares of his benifice', 40 merks; sir Hew Bawldis benefice, £20; sir Andrew Hayis, £10; sir William Cokburne 10 merks. 'The vicares fee of my lordis expenssis.'

Calculation of third: In the whole, £100; third thereof, £33 6s 8d.

'The prebendarie thairof. In the haill', £63 6s 8d; third thereof, £21 2s 2⅔d.

DUNBAR, ARCHPRIESTRY OF,[56]
(SRO, Vol. a, fo 177v)

'The rentale of the archiprestrie of Dunbar'

'The landis thairof gevis yeirlie' £50. 'The teynd fische of Dunbar', 40 merks. 'Item in small teyndis and utheris dewteis', £10, 'becaus the haill Pasche reknyngis, upmest claithe and corce presendis ar tane doun.'

Calculation of third: In the whole, £86 13s 4d; third thereof, £28 17s 9⅓d.

---

[54] Janet Stewart, lady Methven, gained a third of the barony of Bolton in the constabulary of Haddington in 1557, *RMS*, iv, no 1171.
[55] William, 5th lord Hay of Yester, *Scots Peerage*, viii, pp. 438-441
[56] I.e. the headship of the collegiate kirk of Dunbar.

DIRLETON COLLEGIATE KIRK, PROVOSTRY OF,
(SRO, Vol. a, fo 178r)

'The provestrie of Dirltoun quhairof sir Robert Oistler is proveist and Patrik, lord Ruthven,[57] superiour of the lordschip of Dirltoun as patrone of the samin, extendis to' £20 money 'allanerlie.'

Calculation of third: In the whole, £20; third thereof, £6 13s 4d.

DUNGLASS COLLEGIATE KIRK, PREBENDS OF,[58]
(SRO, Vol. a, fo 178v)

'The rentale of the prebendarie of Dunglas set be Frances Wolsoun to Frances Douglas in few ferme, quhilk payis yeirlie' £22 'allanerlie of the landis of Spittall.'

Calculation of third: In the whole, £22; third thereof, £7 6s 8d.

'The rentale of the prebendarie of Dunglas callit the Barnesyd and Auchincraw pertenyng sumtyme to umquhill sir Williame Mustart.'

'Item, the mailles and teyndis of the landis of Barnesyd extending to' 20 merks. The four husbandlands in Auchincraw occupied by William Paxtoun of that Ilk extending to £4. One husbandland in the said Auchincraw occupied by Robert Achesoun extending to 20s. And out of the Foulcheis and Nanewarie, 4 merks.

HERDMANSTON, CHAPLAINRY OF,
(SRO, Vol. a, fo 179r)

'The rentale of the chepellanrie besyd the place of Firdmastoun [*i.e.* Herdmanston] pertenyng to sir James Mosman gevin in be Thomas Sinclair. The haill profeittis thairof yeirlie extendis to' 21 merks money.

Calculation of third: In the whole, £14; third thereof, £4 13s 4d.

BOLTON, VICARAGE OF,
(SRO, Vol. a, fo 179v)

Rental of the vicarage of Boltoun pertaining to Andrew Symsoun, vicar thereof.

'The said vicarege set in tak and assedatioun to Laurence Cokburne yeirlie for' £35.

---

[57] Patrick, 3rd lord Ruthven, *Scots Peerage,* iv, p. 261.
[58] Cf SRO, Vol. a, fo 169r (see above, p. 166).

Calculation of third: In the whole, £35; third thereof, £11 13s 4d.

SALTOUN, VICARAGE OF, (SRO, Vol. a, fo 179v)

Vicarage of Saltoun

Calculation of third: In the whole, £33 6s 8d; third thereof, £11 2s 2⅔d.

INNERWICK, VICARAGE OF, (SRO, Vol. a, fo 179v)

Vicarage of Innerwik

Calculation of third: In the whole, £40; third thereof, £13 6s 8d.

NORTH BERWICK, VICARAGE OF, (SRO, Vol. a, fo 179v)

Vicarage of Northberuik

Calculation of third: In the whole, £11 6s 8d; third thereof, £3 15s 6⅔d.

WHITTINGEHAME, (SRO, Vol. a, fo 180r)
VICARAGE PENSIONARY OF,

'The rentale of the vicar pensionarie of Quhittinghame pertenyng to Thomas Lyle, vicar thairof, extendis to' 18 merks money 'with ane mans hous, yaird and croft extending to' 6 'fuirletis of beir sawing.'[59]

Calculation of third: 'In money', £12; third thereof, £4. Bere, 6 f. 'sawing'; third thereof, 2 f. [*In margin*, 'by the mans'.]

PRESTONKIRK OR LINTON,[60] (SRO, Vol. a, fo 180v)
CHAPLAINRY OF OUR LADY IN KIRK OF,

'The chapellanrie of Our Lady in Prestoun Kirk alias Lintoun of auld extendit to' £10 20d 'and deducit thairof be ressoun of the burnyng of Hadingtoun' 18s 6d 'and sua restis' £9 3s 2d 'and thairof' 5 merks 'wes payit to the substitute and sua restis' £5 16s 6d.

---

[59] Presumably 'sowing'.
[60] Also known as Hauch, cf fo 174r; cf above, p. 170, n. 46.

Signature: 'sir Thomas Godrell.'

Calculation of third: In the whole, £9 3s 2d; third thereof, £3 1s ⅔d.

EDINBURGH, (SRO, Vol. a, fo 180v)
TRINITY COLLEGIATE KIRK,
ORMISTON, CHAPLAINRY OF,

Chaplainry of Ormestoun

'In money', £8; third thereof, [*blank*].'

'Seik Assindane [*i.e.* Hassendean] in Selkirk[61] with foir kyis gers and thair followares and thrie akeris of land. Thrid thairof, 1 aker of land, ane kous gers and 3 pairt.'

GULLANE, VICARAGE OF, (SRO, Vol. a, fo 181r)

'Vicarege of Gulane pertenyng to George Halyburtoun and ressavit at command of the Clerk Registre.'[62]

Calculation of third: In the whole, £30; third thereof, £10.

PITCOX, PREBEND OF (SRO, Vol. a, fo 181r)
DUNBAR COLLEGIATE KIRK

'The prebendarie of Petcokis, Dunbar, pertenyng to Mr Patrik Cokburne'

Calculation of third: In the whole, £83 6s 8d; third thereof, £27 15s 6⅔d.

BOTHANS COLLEGIATE KIRK, (SRO, Vol. a, fo 181r)
CHAPLAINRY OF ST EDMUND IN,

'Chapellanrie of Sanct Edmond in the college of Bothanes'

Calculation of third: In the whole, £7 3s 4d; third thereof, 47s 9⅓d.

---

[61] Hassendean lay in the sheriffdom of Roxburgh, though the rental of the vicarage appears among rentals for Selkirk, in SRO, Vol. a, fo 259v (see below, p. 264).
[62] James McGill of Nether Rankeilour was Clerk Register at the Reformation, *Handbook of British Chronology*, p. 197.

BOTHANS COLLEGIATE KIRK, (SRO, Vol. a, fo 181r)
CHAPLAINRY OF ST COSMO AND ST DAMIAN IN,

'Chapellanrie of Sanct Cosmart and Damien in the said college'

Calculation of third: In the whole, £13 6s 8d; third thereof, £4 8s 10⅔d.

ARNISTON, PREBEND OF (SRO, Vol. a, fo 181v)
CRICHTON COLLEGIATE KIRK[63]

'The prebendarie of the college kirk of Creichtoun callit the prebendarie of Arnoldstoun.'

'The teyndis of the Manes of Arnoldstoun ar haldin and evir hes bene be the lordis of Sanct Johnne[64] be reassoun of the privilege, and payit na thing to the prebendare at ony tyme bypast thir mony yeiris as is weill knawin.'

'Item, the small teyndis of Arnoldstoun', 10 merks. The teinds of Harvestoun, £10. The teinds of Catcune, £10. The teinds of Wester Lochquhairat 'with the cotteris teyndis about Borthwik', 20 merks.[65]

HADDINGTON, PRIORY OF,[66] (SRO, Vol. a, fos 182r-184v)
(SRO, Vol. b, fos 122r-124v)

'Ane new rentale of the abbay of Hadingtoun ressavit at the queinis majesteis command.'

'Abbay of Hadingtoun'

The Manes of the Abbay of Hadingtoun, the lands of Mortounhall, Westschoipes, Eistschopes, the Wodend, Newlandis, Ryndislaw, Snawdoun, Carfray, Littill Newtoun with the teinds of the same 'set in few to the Secretar[67] for victuall, the pryces quhairof extendis to the sowme of' £187 14s. The Mylne of the Abbay, £13 10s. The lands of Quarrell Pittis and Walkmylne set for £16 4s 8d. Gymmeris Mylne, £25 0s 4d. The mails and annuals of the burgh of Hadingtoun, £4. 'In almous of the queinis custumes of Hadingtoun', 40s. The lands of Bakbie,

---

[63] Cf NLS, fo 139v (see below, p. 450).
[64] James Sandilands, preceptor of Torphichen, lord St John, *RSS*, v, nos 1396, 1547.
[65] SRO, Vol. b recommences at the start of the next paragraph, and corresponds to SRO, Vol. a, fo 182r.
[66] Cf SRO, Vol. a, fos 165v, 184v (see above, p. 161; below, p. 180).
[67] William Maitland of Lethington was Secretary at the Reformation, *Handbook of British Chronology*, pp. 193-194.

£47 6s 8d. The Nunraw with the 7 husbandlands in Garvat toun and mill, £20. The kirkland of Barro, 20s. The Mureland [Mounkland, *fo 165v*; Nunland, *fo 168v*], £5 6s 8d. The husbandland of Staintoun, 29s 8d. The husbandland of Beinstoun, 40s. The Slaid, £47 19s 4d. Garvat Grange with one husbandland in Garvalt toun, £8. Four husbandlands in Garvalt toun 'with the hauch at the kirk', £5. 'The oxingang of land of Bagbie', 20s. The teinds of the Nungait set in tack to John Young, writer, for £20. The Byres, £32. The Barnis 'riddin commonlie to' 6 b. wheat; 17 b. bere; 2 c. oats. The kirk of Garvalt, £13 6s 8d. The kirk of Elstanefuird, £50. The teinds of Bagbie set for 3 b. wheat; 4 b. bere; 18 b. oats. Harpardene set in tack for £3 6s 8d.

'Summa of silver within Lowthiane for mailles of fewlandis, annuellis, teyndis of kirkis, as is presentlie abonewrittin, is' £514 4s 8d.[68] Total wheat, 9 b. Total bere, 1 c.[69] Total oats, 3 c. 2 b.

The rental of the parish of Craill in Fyffe
The Kingis Barnes, 4 c. bere; 4 c. meal; 1 c. wheat. Randelstoun, 20 b. meal; 12 b. bere. Cammo, 1 c. meal; 28 b. bere. Wilmerstoun, 12 b. meal; 10 b. bere; 2 b. wheat. Bredleves, 2 b. bere; 2 b. meal. Drumraok, 1 b. bere; 6 b. oats; 5 f. wheat. Minarbo Hill [in Arbohill, *fo 186v*], 1 b. meal. Gawstoun, 1 b. bere; 4 b. meal. Martoun [*i.e.* Morton], 9 b. oats; 1 b. bere; 1 b. wheat. The laird of West Barnis, 3½ c. bere; 3½ c. oats. The Kingis Carne, 1 b. meal. Naikit Feild, 10 b. bere; 1 c. oats. Littill Trostrie, 2 b. meal. The teinds of Sauchop, Balcomy, Kylmynnane, 3 c. bere; £26 13s 4d 'of silver'. The ferme of Sauchope, £37 money. Newhall, 22 b. meal; 8 b. bere [8 b. bere; 12 b. meal, *fo 186r*]. Ardrie, Seipsis and the crofts of Craill, 21 b. bere; 21 b. oats; 6 b. meal. Petmillie, £23 6s 8d. The lordship of Kippo, 12 b. oats; 12 b. meal; 2 b. wheat; 6 b. bere. Pitcowy [Pitlowy, *fo 186r*], 29s 4d. Pincartoun, 49s 4d. Reidwellis, 30s 8d. Kilminane [*recte*, Kilduncan, *fo 186r*], 11 b. meal [10 b. *fo 186r*]; 3 b. bere. Cukistoun, 3 b. bere; 10 b. oats. 'Summa of the haill quheit of the kirk of Craill', 22 b. 1 f. Total bere, 17 c. 2 b. Total meal, 10 c. 13 b. Total oats, 8 c. 2 b.

The mails and annuals
'Maister George Meldrumis hous in the yeir', 2 merks. The Eist Grene, 1 merk. George Kirstorphines [Corstorphines, *fo 184r*] house, 12d. John Bissetis land, 3s. Robert Robertsoun house, 4d. Duncan Alesounes house, 4d. The laird of West Barnes house, 14s. John Kirstorphines house, 2s 6d. Mr John Arnotis house, 35s. The mails of Newtoun, £6. Litle Drostrie, 8 merks. The Furthfeild, £11 16s 8d. Petcorthy, 16 merks 8s. The Fausyd, 17s. 'Item, for the teynd fische', £18.

---

[68] The correct calculation appears to be £506 4s 8d.
[69] The correct calculation appears to be 1 c. 5 b.

'Summa of the haill rentale of the abbay of Hadingtoun': In silver, £663 9s 6d.[70] In wheat, 1c. 15 b. 1 f. In bere, 18 c. 2 b. In meal, 10 c. 13 b. In oats, 11 c. 4 b.

Signature: 'W. Maitland.'[71]

Third of money, £221 3s 2d. Third of wheat, 10 b. $1\frac{2}{3}$ f. Third of bere, 6 c. $\frac{2}{3}$ b. Third of meal, 3 c. $9\frac{2}{3}$ b. Third of oats, 3 c. 12 b.

'Aliter the rentale of Hadingtoun gevin up be the Secretar.' The Manes of the Abay of Hadingtoun, the lands [of] Mortounhall, West Hopes, Eisthopes, the Wod and [recte, Wodend, fos 182r, 186r] Newlandis, Ryndislaw, Snadoun, Carfray, Litle [Newtoun] with the teinds of the samen 'set in few to the Secretar for victuall, the pryces quhairof extendis to the sowme of' £187 14s. The Mylne of the Abay, £13 10s. The lands of Quarrell Pites and Walkmylne set for £16 4s 8d. The Gymmeris Mylne, £25 0s 4d. The mails and annuals of the burgh of Hadingtoun, 4d [recte, £4, fos 165v, 182r]. 'In almous of the queinis custoumes of Hadingtoun', 40s. The lands of Bagbie, £47 6s 8d. The Nunraw with the 7 husbandlands in Garvat toun and mill, £20. The kirkland of Barro, 20s. The Murland [Mounkland, fo 165v; Nunland, fo 168v], £5 6s 8d. The husbandland of Stayntoun, 29s 8d. The husbandland [of] Beynstoun, 40s. The Slaid, £47 19s 4d. Garvat Grange with one husbandland in Garvat toun, £8. Four husbandlands in Garvat toun 'with the hauch at the kirk', £5. 'The oxgang of land of Bagbie', 20s. The teinds of Nungait set in tack to John Young, writer, for £20. The Byres, £30 [£32, fo 182r]. The Barnis, £20.[72] The kirk of Garvat, £13 6s 8d. The kirk of Elstanefuird, £50. The teinds of Bagbie, 10 merks.[73] Harpardene set in tack for £3 6s 8d. 'Summa of silver within Lowthiane for mailles and fewlandis, annuellis, teyndis of kirkis, as is presentlie abonewrittin is' £530 18s.

The rental of the parish of Carraill in Fyffe

The Kingis Barnes, 4 c. bere; 4 c. meal; 1 c. wheat. Randelstoun, 20 b. meal; 12 b. bere. Cammo, 1 c. meal; 28 b. bere. Wilmerstoun, 12 b. oats [meal, fo 182v]; 10 b. bere; 2 b. wheat. Braidlevis, 2 b. bere; 2 b. meal. Drumraok, 1 b. bere; 6 b. oats [plus 5 f. wheat, fo 182v]. Minarbo Hill, 1 b. meal. Gastoun, 1 b. bere; 4 b. meal. Martoun [i.e. Morton], 8 b. [9 b., fo 182v] oats; 1 b. bere; 1 b. wheat. The laird of West Barnes, $3\frac{1}{2}$ c. bere; $3\frac{1}{2}$ c. oats. The Kingis Carne, 1 b. meal. Naikit Feild, 10 b. bere; 1 c. oats. Litle Trostrie, 2 b. meal. The teinds of Sauchop, Balcomy, Kylmynnane, 3 c. bere; £26 13s 4d of silver. The ferme of the Sauchope, £37 money. Newhall, 12 b. meal; 8 b. 'meill' [8 b. bere; 12 b. meal, fo 186r]. Ardrie, Seipseis and the crofts of Carraill, 21 b. bere; 21 b. oats; 6 b. meal. Petmillie, £13 6s 8d [£23 6s 8d, fo 182v]. The lordship of Kippo, 12 b.

---

[70] The correct calculation appears to be £654 2s 6d.
[71] I.e. William Maitland of Lethington who had acquired a feu of the priory's property, RSS, v, no 1881.
[72] The Barnes paid 6 b. wheat; 17 b. bere; 2 c. oats, fo 182r (see above, p. 177).
[73] The teinds of Bagbie were set for 3 b. wheat; 4 b. bere; 18 b. oats, fo 182r (see above, p. 177).

oats; 12 b. meal; 2 b. wheat; 6 b. bere. Pitcowy [Pitlowy, *fo 186r*], 29s 4d. Pincartoun, 49s 4d. Reidwellis, 30s 8d.[74] Cukistoun, 3 b. bere; 10 b. oats. 'Summa of the haill quheit of the kirk of Carraill', 21 b. Total bere, 17 c. 1 b. 'les'.[75] Total meal, 8 c. 10 b.[76] Total oats, 8 c. 13 b.[77]

The mails and annuals
    Mr George Meldrumis house in the year, 2 merks. The Eist Grene, 1 merk. George Corstorphines house, 12d. John Bissetis land, 3s. Robert Robertsoun, 4d. Duncan Alesounes house, 4d. The laird of West Barnes house, 14s. John Corstorphines house, 2s 6d. Mr John Arnotis house, 35s. The mails of Newtoun, £6. Litle Drostrie, 8 merks. The Fruitfeild [Furthfeild, *fo 183r*], £11 16s 8d. Pitcorthie, 16 merks 8s. The Fawsyd, 17s. 'Item, for the teynd fische', £18.

'Summa of the haill rentale of the abbay of Hadingtoun': In silver, £670 17s 10d. In wheat, 21 b. In bere, 17c. 1 b. 'les'. In meal, 8 c. 10 b. In oats, 8 c. 13 b.

Signature: 'W. Maitland.'

AULDHAME, PARSONAGE OF,[78]   (SRO, Vol. a, fo 184v)
    (SRO, Vol. b, fo 124v)

'The rentale of the personage of Auldhame set in assedatioun be Patrik Alschunder [*i.e.* Alexander], possessour thairof, to Jein Stewart, lady of Reidhall,[79] for' 24 merks.

Signature: 'Dumfermeling.'[80]

'James Nicolsoun,[81] ye sall ressave this rentale abonewrittin and insert the samin in your buikis and I salbe your warrand. Thir presentis subscrivit with my hand at Halyruidhous', 9 July 1573.

Signature: 'R. Dumfermlyng.'

---

[74] Included at this point on fo 182v, Kilduncan, 11 b. meal; 4 b. bere.
[75] The correct calculation appears to be 16 c. 15 b.
[76] The correct calculation appears to be 8 c. 12 b.
[77] Wheat and money have been omitted from these totals.
[78] Cf SRO, Vol. a, fo 131r (see above, p. 118). The original rental submitted also survives as a scrap inserted between the folios of SRO, Vol. b, at fo 128r (see below, p. 183, n. 1), where Auldhame is rendered 'Alden'.
[79] See above, p. 118, n. 94.
[80] At the Reformation the commendator of Dunfermline, George Dury, left for France in January 1561 and Robert Pitcairn exercised authority as commendator-designate. Cowan and Easson, *Medieval Religious Houses*, p. 58.
[81] James Nicholson, clerk of the Collectory, *Thirds of Benefices*, p. 62.

HADDINGTON, PRIORY OF,[82]

(SRO, Vol. a, fos 184v-186v)
(SRO, Vol. b, fos 124v-126v)

'The rentale of the abbay of Hadingtoun gevin in be the lord Lindesay[83] at Hadingtoun', 28 December, 1573.

'The annuellis of the said abay and the few maillis with the teyndis
    Cristiane Quhentine and hir spous' ['Cristiane Quhentyne for hir hous', *SRO, Vol. b*] 24s 9d. Hector Tait for his acres, 2s 10d. John Douglas, baxter, for his house, 10s. William Carintoun 'in stentis for his oxingang of land', 28s 4d. Bessie Kerintoun in the Fische Mercat for the annual of her house, 8s 9d. 'Item, for ane aker of land lyand in the New Gait', 3s 3d. 'James Haddowis wyff for the annuellis of hir hous', 13s 4d. John Aittoun for his house, 12d. Laurence Cokburne 'for the few mailles of his oxingait of land', 23s 4d, with 2 b. oats. James Olyphant for his mails and annuals, 39s 11d. Adam Wolsoun for his annual, 4s 8d. Janet Ogill 'for Lawsounes foirland', 7s 6d. Alexander Thomsoun for the Riglingtoun land, 20s. Henry Campbell in name of [John, *SRO, Vol. b*] Canny 'for ane aker and ane rig of land lyand in the Nunsyde and half aker in the Foull Fluires, ane aker in Sprotland with the pertinentis', 27s. William Gibsounes wife 'for ane annuell of hir hous', 28s. James Hog for the annual of Kempland, 30s. 'Douglas bakhous', 28s 8d. Patrick Lyell for the annual of his house at the Brigend, 3s. For Thomas Kerintones land, 4s. For John Kerintones land 'he payis' 5s. Lucas Tait for his house, 4s 6d. Alexander Gibsoun in Mekill Rig for the annual of his house in Hadingtoun, [*blank*]. John Gray, 'four akeris in Gallowsyd, tua aker and ane half in Sprotland and ane aker in Adameflat', 26s 8d. John Henrysoun 'for Douglas foirhous', 33s 2d. Thomas Fedler for his annual, 17s. Patrick Frenche, 28s 6d. Alexander Randie, John Creychtoun, Adam Walker, Marjorie Pillie and Janet Sandie 'for Geilles sex riggis', 11s 6d. Patrick Kerintoun 'for Geleis hous', 13s 4d. Thomas Arnot, 3s. John Mekley for Rankenis land, 4d. Robert Calbrayth, 4d. John Douglas, cordinar, for the annual of his house, 6s 8d. George Aittoun 'for Lawteis bakhous', 3s 2d. George Hepburne for his annuals, 20s 6d. 'The mylnes of Hadingtoun', £4. Cuthbert Nicolsoun for his house in Hadingtoun, 4s 6d. John Cokburne, maltman, 6s 8d. Bernard Thomesoun for Ogilbeis land, 15s. John Blak for Dalzellis land, 2s 6d. William Ogill 'for his annuellis and akeris', 25s 8d. 'To Mr Masoun' ['The maister masoun', *SRO, Vol. b*], 2s 7d. Nicol Dunlape for his house, 7s 6d. Martin Wilsoun for Pepper croft, 40s. John Clerke and Bessie Hadyngtoun, his spouse, 27s. Patrick Hepburne, 'skynner', 22s 6d. John Hepburne in the Newgait for his mails and annuals, £5 5s 6d. Archibald Kyill for the laird of Stirlingis [*recte*, Skirlingis, *SRO, Vol. b*] house, 32s. The annual of Greinlawis bakhous, 10s 6d. David Bronewod, 7d. Adam Thomsoun, 3s 6d. Cristall Galloway, 3s. Harry Cokburne, 'west the gait', 18d. Sir John Andersoun, 33d. Beatrice Hoge, 15s 9d.

---

[82] Cf SRO, Vol. a, fos 165v, 182r (see above, pp. 161, 176).
[83] See above, p. 167, n. 30.

Thomas Zeyme, 18s. Cuthbert Achesoun, 2s. John Blak, 30d. Robert Lermonth, 8s 3d. Harry Lermonth, 20s 4d. Robert Bald for his annuals, 49s. William Tait for his annuals, 14s 1d. 'And for Hectoures aker', 22d. 'Eister Nunraw, Wester Nunraw, his landis of Garwald, Garwald Mylne, ane husbandland in Benestoun', 11 'akeris of land in the Nungait with ane oustrie', £31 12s 2d. For the Slaid, £47 0s 7d. For the teinds of Garwald and Garwald Grange with the pertinents, £13 6s 8d. Laurence Patersoun for the Abbay Mylne, £13 9s 6d. David Forres for Gymmeris Mylnes, £24 4s. Patrick Home of Garwald Grange, £9 3s. Robert Diksoun 'for his hauch and hous at Garwald kirk, for his landis in Garwald', £5 5s. 'And for his duelling place and land callit the byre, loche and byre yaird', 18s. For the teinds of Quarrell Pittis, Smiddie acres, 26s 8d. 'For his akeris in the Nungait', 30s. The Nunland [Mounkland, *fo 165v*; Mureland, *fo 182r*], £5 9s 8d [*recte*, £5 6s 8d, *i.e.* 8 merks, *fo 165v*]. The Wakmylne, £8 16s 4d. The feu mails of the lands of the abbey of Hadingtoun, the lands of Mortoun, Westhopes, Eisthopes, the Wodend, Newlandis, Ryndslaw, Snadoun, Carfray, Litle Newtoun, 'payis of the haill', £195 14s [£197 14s, *fo 182r*]. The lands of Bagbie, £47 6s 8d. The lord Seitoun[84] for the teinds of Eist and West Barnes, £20. The lord Lyndesay[85] for the teind of the Byeris, £30 [£32, *fo 182r*]. George Hepburne for the teinds of Ailstounfuird, £50. John Young for the teind of Harperden and Nungait, £24 [£23 for Nungait; £3 6s 8d for Harperden, *fos 182r, 183v*]. The teind of Baikbie, 1 c. oats; 4 b. bere; 4 b. wheat [3 b. wheat; 4 b. bere; 18 b. oats, *fo 182r*; 10 merks, *fo 183v*].

Craill

The Kingis Barnes, 4 c. bere; 4 c. meal; 1 c. wheat [*in margin*, 'les be xij b.']. Cammo, 1 c. bere; 1 c. meal. Newhall, 8 b. bere; 12 b. meal [22 b. meal; 8 b. bere, *fo 182v*]. Randerstoun, 12 b. bere; 20 b. meal. Welmerstoun, 2 b. wheat; 10 b. bere; 12 b. oats [meal, *fo 182v*]. Balcomy,[86] 3 c. bere; £26 13s 4d. Nakit Feild and kirklands, 14 b. bere; 2 f. meal; 16 b. oats [10 b. bere; 1 c. oats, *fos 182v, 184r*]. Ardrie, Sipseis and the crofts of Carrell, 21 b. bere; 6 b. meal; 21 b. oats. The West Barnes, 3½ c. bere; 3½ c. oats. Rippo [*recte*, Kippo], 6 b. bere; 12 b. meal; 12 b. oats; 2 b. wheat. Kilduncane [Kilminane, *fo 182v*], 10 b. meal [11 b., *fo 182v*]; 3 b. bere. Cuikstoun, 3 b. bere; 10 b. oats. 'The ministerie' [*recte*, Morton, *fo 182v*], 1 b. wheat; 1 b. bere; 8 b. oats. Drumraok, 10 f. bere; 6 b. oats [1 b. bere; 6 b. oats; 5 f. wheat, *fo 182v*] [*in margin*, 'les mair be a f.'] Gaustone, 1 b. bere; 4 b. meal. Reedleyis [Bredleyis, *SRO, Vol. b*], 4 b. malt. Pitmillie 'for his teynd', £13 6s 8d [£23 6s 8d, *fo 182v*]. The Kingiscarne, 2 b. meal [1 b., *fos 182v, 184r*] [*in margin*, 'mair be a boll'].[87]

---

[84] George, 5th lord Seton, *Scots Peerage*, viii, pp. 585-588.
[85] See above, p. 167, n. 30.
[86] Teinds of Sauchope, Balcomie and Kilminning, fos 182v, 184r.
[87] This marginal note tallies with earlier rentals.

Feu mails and annuals of Craill
Sauchope, £40 [ferme of Sauchope, £37, *fo 182v*] [*in margin*, 'xl s mair']. 'The half teynd fische', £18. For the Newtone, £5. The Fausyd, 17s. Pitlowy and Pinkartoun, £3; and for teinds [*in margin*, 'les be viij s viij d']. [Pitcowie, 29s 4d; Pincartoun, 49s 4d., *fos 182v, 184r.*] Trostrie, £5 8s 2d. The Rudwellis, 30s. Mr George Meldrummes house, 26s 8d. William Arnotis house, 36s. John Dingwallis house, 14s. John Ramsayis house, 3s. John Corstorphines house, 30d.

'Summa of the haill silver in Lowthiane and in Fyffe', £694 15s 6d;[88] and of victual, 37 c. 12 b. 2 f.[89]

'Of this to be deducit
To ten nunis, everie ane', £20, 'summa', £200. 'Item, to Cuthbert Achesoun in pentioune', £8 13s 4d. 'Item, to Hary Cokburnes wyff in pensioun', 20 b. wheat and bere. 'Item, to James Lauder in pensioun', £240. 'Item, for the ballie fe', £40. 'Item, to the minister of Garvalt kirk', £7 13s 4d. 'Item, for the chalmerlandis fe in Craill', 36 b. victual. 'Item, mair to be deducit for the contributioun of the Lordis of the Sessioun', £21. 'Summa of silver to be deducit', £496 6s 8d,[90] and of victual, 56 b.

'Omittit in Arbohill' [Minarbo Hill, *fo 183v*], 1 b. meal.

NOTE

Rentals for the following benefices are located elsewhere in the text:

CRICHTON COLLEGIATE KIRK, PREBEND OF (NLS, fo 139v), p. 450.
DUNGLASS COLLEGIATE KIRK, PROVOSTRY OF (SRO, Vol. a, fo 188v), p. 186.

---

[88] The correct calculation appears to be £696 13s 3d.
[89] The correct calculation appears to be 37 c. 13 b. 1 f.
[90] The correct calculation appears to be £517 6s 8d.

Berwick

ECCLES, PRIORY OF, (SRO, Vol. a, fos 187r-188r)
(SRO, Vol. b, fos 127r-128r[1])

'The rentale of the abacie of Ecclis in mailles and teyndis with the vicarege and kirkis of the samin.'

'Item, the Maidlane parochin[2] extendis to' £100 'land [*recte*, 100 lands] quhilk sould pay' 200 merks 'bot the lord Home[3] withhaldis' 65 'landis and makis na payment.' In Sanct Cuthbertis parish,[4] 50 lands which extends to 100 merks. Sanct Johnnes parish[5] extends to 70 lands which extends in money to 140 merks. Lethem parish[6] extends to 88 lands 'quhilk cumis to in money to' 176 merks. 'Summa', £117 6s 8d.[7]

'The Manis teynd cumis to' £20 'led be the prioris[8] or hir factouris.'

[*In margin*, 'Striveling, Kirk of Buth Kenner.']

---

[1] Between fos 127v and 128r, a small unbound piece of text reads: 'The rentall of the personage of Alden set in assedatioun be Patrik Alexander, possessour thairof, to Jeine Stewart, ladie of Reidhall, for' 24 merks. 'R[obert Pitcairn, commendator of] Dunfermling. James Nicolsoun, ye sall ressave this rentale abonewrittin and insert the same in your bukis and I salbe your warrand be thir presentis. Subscrivit with my hand at Halyrudhous', 9 July 1573, 'R. Dunfermling.' This entry corresponds to the rental for the parsonage of Auldhame at fo 184v, (see above, p. 179); cf SRO, Vol. a, fo 131r (see above, p. 118, nn. 92, 94).
[2] I.e. the chapel of St Magdalene, situated at Birgham in the south 'quarter' of the parish of Eccles, *Fasti Ecclesiae Scoticanae*, ed. H. Scott, 10 vols. (Edinburgh, 1915 - 81), ii, p. 12.
[3] I.e. Alexander, 5th lord Home, *Scots Peerage*, iv, pp. 460-463.
[4] The parish church of Eccles was dedicated to St Cuthbert, Scott, *Fasti*, ii, p. 12.
[5] The chapel of St John was situated in the north 'quarter' of the parish of Eccles at Mersington. Scott, *Fasti*, ii, p. 12.
[6] The chapel of Our Lady was located at the Chapel Knowe of Leitholm in the east 'quarter' of the parish of Eccles. Scott, *Fasti*, ii, p. 12.
[7] This sum denotes the 176 merks of Leitholm; the total for the paragraph should read 616 merks, that is £410 13s 4d. The unit for computation throughout appears to be a double merkland.
[8] See below, p. 184, n. 11. The priory was a Cistercian nunnery.

'The kirk of Buthkener, the teyndis thairof led yeirlie be the priores extending yeirlie, a yeir with ane uther, the expenssis of the leiding, bearing tane of, cummis be guid estimatioun to' 180 merks 'or thairby.'

'The teyndis of the pendicles liand in the Mires: Rob Story, tua landis', 4 merks teind. William Lamb, 2 lands, 4 merks. Horsinpryg,[9] 1 land, 2 merks. Kettill Cheyll, 2 lands, 4 merks. The Drumscheill [*i.e.* Dronshie], 2 lands, 4 merks.

The rental of the temporal lands and annuals: The Manes of Ecclis, £28 land for [£]28 mail 'with gersum.'

Thomas Hopper 'hes tua landis in few of the place for' 34s 'in the yeir allanerlie'. Robert Story, 'tua landis in few in the Nunbank for' 40s mail in the year. Richard Egger 'quhair he duellis' for 24s in the year. In Langnewtoun, 40s land 'in the tennentis handis'. Grubbit Hauch 'quhilk the laird [of] Hunthill hes for' 6 merks mail 'and is nocht payit to the place. The guidman of Humby, thrie pund land in few in Weinsyde for thrie pund meill [mele, *SRO, Vol. b*; *recte*, maill] in the yeir'. Nicol Graden in Ecclis parish for 1 merk of mail yearly. John Diksoun in the Kamyis, 10s land for 10s mail [mele, *SRO, Vol. b*]. The laird of Mersetoun [*i.e.* Mersington], 5s land for 5s mail in the year. Robert Diksoun in the Peyll, 5s land for 5s mail in the year. George Kingorne in Darchester, 5s land for 5s mail yearly. 'The wedow in Hawssetoun [*i.e.* Hassington]', 10s land for 10s mail yearly. Arthur Wynram in Byrgone, 'a croft for' 9s in the year. The laird of Cowdun Knowis, 'ane croft for' 5s in the year. 'As for the vicarege it is worth' £40 'bot thair is na payment maid thairof this lang tyme bygane. Summa ut supra', £648 3s 8d.[10]

Signature: 'James Hammiltoun of Kincavyll.'[11]

Calculation of third: The third of the abbacy of Ecclis: In the whole, £647 13s 8d; third thereof, £215 17s 10⅔d.

Assignation of the third of the abbacy of Ecclis:
    Third of money, £215 17s 10⅔d. Take the Manes of Ecclis for £28; the kirk of Buthkenneir for £120; Sanct Johnnes parish for £93 6s 8d. 'Gif in' £25 8s 9⅓d.

---

[9] 'Horsinpryg' is possibly Horseupcleugh, near Longformacus.
[10] The correct calculation appears to be £656 13s 8d.
[11] James Hamilton of Kincavil (now Kingscavil) was sheriff of Linlithgow; his kinswoman, Marion Hamilton, was prioress of Eccles at the Reformation. *Scots Peerage*, vii, p. 550; *RSS*, v, no 3041; Cowan and Easson, *Medieval Religious Houses*, p. 146.

'Nota, all canes, custoumes, grassumes and utheris dewteis, caponis and pultrie ar omittit, thairfoir be diligent to inquire thame.'[12]

PRESTON, VICARAGE OF, (SRO, Vol. a, fo 188v)
(SRO, Vol. b, fo 128v)

Vicarage of Prestoun

Calculation of third: In the whole, £16 13s 4d; third thereof, £5 11s 1⅓d.

DUNS, KIRK OF, (SRO, Vol. a, fo 188v)
(SRO, Vol. b, fo 128v)

Kirk of Duns

Calculation of third: In the whole, £196 13s 4d; third thereof, £62 4s 5⅓d.[13]

HILTON, KIRK OF, (SRO, Vol. a, fo 188v)
(SRO, Vol. b, fo 128v)

Kirk of Hiltoun

Calculation of third: In the whole, £24; third thereof, £8.

MORDINGTON WITH LONGFORMACUS, KIRK OF, (SRO, Vol. a, fo 188v)
(SRO, Vol. b, fo 128v)

Kirk of Mordingtoun with Lochirmakhous

Calculation of third: In the whole, £21; third thereof, £7.

UPSETLINGTON, PARSONAGE OF, (SRO, Vol. a, fo 188v)
(SRO, Vol. b, fo 128v)

Parsonage of Upsettingtoun

Calculation of third: In the whole, £46; third thereof, £15 6s 8d.

---

[12] SRO, Vol. b, fo 128r, between the entries for Eccles and Preston, has an entry which seems to be deleted. It corresponds to the assumption of the third for Coldstream, SRO, Vol. a, fo 190r. In addition, SRO, Vol. b, fo 128r has a scrap of paper, consisting of the original rental submitted, which repeats the entry for the parsonage of Auldhame occuring in SRO, Vol. a, fo 184v and SRO, Vol. b, fo 124v, though 'Auldhame' is rendered 'Alden'. (See above, p. 179 and n. 78; below, p. 187.)

[13] The correct calculation appears to be £65 11s 1⅓d.

DUNGLASS COLLEGIATE KIRK, PROVOSTRY OF,[14]

(SRO, Vol. a, fo 188v)
(SRO, Vol. b, fo 128v)

'Provestrie of Dounglas'

Calculation of third: In the whole, £82; third thereof, £27 6s 8d.

CHIRNSIDE, PARSONAGE OF,

(SRO, Vol. a, fo 188v)
(SRO, Vol. b, fo 128v)

Parsonage of Chernsyd

Calculation of third: In the whole, £98; third thereof, £32 13s 4d.

COLDSTREAM, PRIORY OF,[15]

(SRO, Vol. a, fo 189r-v)
(SRO, Vol. b, fo, 129r-v)

Rental of Cauldstreme

The lands of Lanaile, £30. 'The Savok [Sawik, *fo 190r*] pertinentis of the same', £20. 'Auldhirsell with uther pertinentis of the same haldin waist be sir Andro Ker extending to' £20. The Brad Hawche and Leis, £8. The lands of Hachtnis [Hauchetnness, *fo 190r*] and Todrig, £8. Two husbandlands on Dernchester, 40s. The lands of Scaythmure, £16. The lands of Sympryn, £10. Quhyt Chester, £16.

'Mylnis
    The Mylne of Cauldstreme, the mylne callit Farburn Mylne, bayth wast and unbiggit sen the weiris.'

'Fischeingis'
    The fishing of Littill Hauche, £20. The fishing of Tylmuth Hauch, £12.

'Teyndis'
    The teinds of Lanaile kirk, 8 c. victual. The teinds of Hirsell, 3 c. The teinds of Bassinden, £40. [*In margin in SRO, Vol. b*, 'Thrid quheit, thrid beir, thrid mele.']

'Item, to be deducit heirof for the sustentatioun of nyne agit wemen sisteris in the place, [*blank*]. Item, for a ministeris fie, [*blank*]. Item for the balleis fie', £20.

---

[14] Dunglass lay in the constabulary of Haddington and sheriffdom of Edinburgh, *RMS*, v, no 1078. Its inclusion among the rentals for Berwick may be attributed to the fact that Abraham Crichton, parson of Upsetlington (in the preceding entry), was provost of Dunglass, the parsonage being annexed to the collegiate kirk. *Thirds of Benefices*, pp. 88, 283; Watt, *Fasti*, p. 356; Cowan, *Parishes*, p. 204.

[15] Cf SRO, Vol. a, fo 190r (see below, p. 187). The priory was a Cistercian nunnery.

Signature: 'Mr Robert Hoppringill.'

Calculation of third: Money, £201; third thereof, £67.
    Wheat, 3 c. 10 b. 2 f. 2⅔ p.;[16] third thereof, 1 c. 3 b. 3 f. '2 pect 3 part half pect and thrid of the thrid of half pect'.[17]
    Bere, 3 c. 10 b. 2 f. 2⅔ p.; third thereof, 1 c. 3 b. 3 f. '2 pect 3 part half pect and 3 of the 3 half pect'.[18]
    Meal, 3 c. 10 b. 2 f. 2⅔ p.;[19] third thereof, 1 c. 3 b. 3 f. '2 pect 3 part half pect and 3 of the 3 half pect.'[20]

HALLIBURTON, CHAPLAINRY OF,   (SRO, Vol. a, fo 189v)
                                                                            (SRO, Vol. b, fo 129v)

Chaplainry of Halyburtoun

Calculation of third: In the whole, £20; third thereof, £6 13s 4d.

SWINTON, VICARAGE OF,   (SRO, Vol. a, fo 189v)
                                                                            (SRO, Vol. b, fo 129v)

Vicarage of Suyntoun

Calculation of third: In the whole, £11; third thereof, £3 13s 4d.

Signature: 'Johnne Forret, vicar of Suyntoun.'[21]

COLDSTREAM, PRIORY OF,[22]   (SRO, Vol. a, fo 190r)
                                                                      (SRO, Vol. b, fo 128r)

Assumption of the third of the abbey of Cauldstreme:
    Third of money, £67. Take the lands of Lanaile for £30; the Sawik [*Savok, fo 189r*] for £20; the Braidhauch and Leyis for £8; Hauchetnnes [*Hachtnis, fo 189r*] and Todrig for £8; two husbandlands in Dernechester for, 40s. 'Gif in' 20s.

---

[16] In the text this fraction is rendered '2 peck half peck 3 half peck'.
[17] I.e. 2⅔ p., but the arithmetically correct third is 1 c. 3 b. 2 f. ⅔ p.
[18] I.e. 2⅔ p., but the arithmetically correct third is 1 c. 3 b. 2 f. ⅔ p.
[19] In the text this is rendered '2 pect half pect 3 part half pect'.
[20] I.e. 2⅔ p., but the arithmetically correct third is 1 c. 3 b. 2 f. ⅔ p.
[21] The next entry in SRO, Vol. b, fo 130r corresponds to SRO, Vol. a, fo 192r (see below, p. 189).
[22] Cf SRO, Vol. a, fo 189r (see above, p. 186).

Third of wheat, 1 c. 3 b. 3 f. 2 p. '3 part half pect 3 thairof'.[23] 'Beir alsmeikle. Meill alsmeikle. Tak thir victuallis out of the teyndis of Lanell kirk gevand' 8 c. 'quheit, beir and meill all alyk.'

['All uther thingis omittit, speir diligentlie.', *SRO, Vol. b.*]

CHANNELKIRK, VICARAGE PORTIONARY OF, (SRO, Vol. a, fo 190r)

'Produc[i]t 28 Decembris apud Hadingtoun be George Strauthachin.'

'The vicarege portionarie of Childinkirk. The woll and lamb pertenyng to the personage is set, and maill and gleib thairof in few to Helene Seytoun, the relict of Niniane Cranstoun, for' £10. 'The cheis and lynt and hay', 10 merks. Total, £16 13s 4d.

DYSART, VICARAGE PENSIONARY OF,[24] (SRO, Vol. a, fo 190r)

'The vicarege pensionar of Dysart is' 40 merks 'and hes gottin ane decreit on the persoun thairof and can get na payment, sic subscribitur, G. Strauthachin.'

COLDINGHAM, PRIORY OF,[25] (SRO, Vol. a, fo 190v)

'The rentale of the abbay of Coldinghame, bayth spiritualitie and temporalitie wer set in assedatioun to Mr Johnne Spens of Condye[26] for a thousand merkis.'

Signature: 'Johnne Maitland.'[27]

Calculation of third: 'The thrid thairof according to this rentale and yeirlie comptis extendis to' £222 4s 5½d.

---

[23] I.e. 2⅔ p., but the arithmetically correct third is 1 c. 3 b. 2 f. ⅔ p.
[24] Cf SRO, Vol. a, fo 83v (see above, p. 70). This benefice, in the sheriffdom of Fife, was included with Channelkirk in the sheriffdom of Berwick because George Strachan held both benefices.
[25] Cf SRO, Vol. a, fos 196v, 199r, 203r (see below, pp. 194, 197, 202).
[26] John Spens of Condie, Lord Advocate, 1555 - 1573, *Handbook of British Chronology*, p. 201.
[27] John Maitland, commendator of Coldingham, fo 196v (see below, p. 194).

DRYBURGH, ABBEY OF,[28]

(SRO, Vol. a, fos 192r-194v[29])
(SRO, Vol. b, fos 130r-132v)

'Rentale de Dryburcht'

The kirks set for victuals

Smalahame kirk
    'In the first, the laird of Cowdounknowis sould pay for the toun of Smalahame' 7 c. meal; 4 c. bere; 1 c. oats. 'And deducit thairof for his fee yeirlie', 1 c. meal; 1 c. bere. 'Item, the guidman of Smalahame Craiges for the teynd of Wrangholme', 4 b. bere; 14 b. oats.

Mertoun kirk
    Walter Halyburtoun for the toun of Mertoun, 3 c. 12 b. meal; 2 c. bere; 4 b. wheat. The lady of Dalcoif for the toun of Dalcoif, 24 b. meal; 24 b. bere. The laird of Coudenknowis for Quhitrig, 6 b. meal; 2 b. bere. The laird of Bennysyd,[30] 2 c. meal; 1 c. 6 b. bere; 2 b. oats. John Robsoun of Gleddiswod, 4 b. meal; 2 b. bere. Alexander Home's wife for Brotherstanis, 8 b. meal; 6 b. bere. The toun of Dryburcht 'with the New Orcheard and teynd of the akeris for the ferme and teynd', 5 c. 12 b. bere.

Maxtoun
    Item, Mark Ker for Rutherfuird 'and that ane half of the toun of Maxtoun', 3 c. meal; 2 c. bere; 1 c. wheat. The laird of Mertoun 'for that uther half of the toun of Maxtoun', 1 c. 10 b. meal; 1 c. bere; 4 b. wheat. John Halyburtoun for the Muirhouslaw, 1 c. 4 b. meal; 4 b. bere. Mark Ker for Cannestoun and Templand, 4 b. meal; 4 b. bere.

Lossuddan
    The toun of Lessaden, 3 c. 8 b. meal; 2 c. bere; 8 b. wheat. Elistoun, 1 c. meal; 8 b. bere. The Newtoun, 2 c. oats.

Total: Meal, 25 c. 12 b.[31] Bere, 21 c. 8 b. Wheat, 2 c. Oats, 4 c.[32]

The mails of the lands pertaining to Dryburcht in the year
    'Inprimis, the toun of Dryburcht with the mylne', £41 6s 8d [£31 6s 8d, *fo 198v*]. Gleddiswod, 40s. Brotherstanes, £3. The lands in Ersiltoun, 22s. Doggarflat, 20s. Sanct Johnnes chapel and Cadislie £32 [£22, *fo 198v*]. 'The

---

[28] Cf SRO, Vol a, fo 198v (see below, p. 197). See also *Liber S. Marie de Dryburgh, Registrum Cartarum Abbacie Premonstratensis de Dryburgh*, ed. W. Fraser (Edinburgh, 1847) pp. 352-361.
[29] SRO, Vol. a, fo 191r-v is blank.
[30] 'Bemyrsyd', *Liber S. Marie de Dryburgh*, p. 354.
[31] The correct calculation appears to be 26 c. (i.e. 27 c. less 1 c. in fee).
[32] This calculation is correct.

kirkland of Lauder be M[r] Andro Home', £21 6s 8d. Nather Scheilfeild, £4 6s 8d. The lands of Ugstoun, 40s. The lands of Vanglaw [Banglaw, *fos 194r, 198v*], £8. The Bernis, £6. Elbottill, £4 6s 8d. The lands of Englisberry [Inglisberry, *fo 194r*] Grange 'with the teyndis of the samin be my lord Symmervell',[33] £42 [£20, *fo 198v*]. The lands of Iflie [*i.e.* Evelaw], £6. Fouldane, 40s. Banghouswallis [Langhouswallis, *fo 194r*],[34] 46s 8d. Smalahame lands, 15s. Smalahame Spittell, £13 6s 8d. 'Summa hujus particule', £192 17s.

The annuals
    Dalcoiff, 13s 4d. Lessuden, 3s. Flemyngtoun, 6s 8d. Huntrodland, 6s 8d. Birkinsyd, 13s 4d. Aldestoun, 40s. Kersmyre, 6s 8d. Hadingtoun, 13s 4d. Dumbar, 26s 8d. The Bothanes, 6s 8d. Edinbrucht, 30s. Quhitterne, 40s. Blakbarone, 20s. Ednama, 20s. 'Summa hujus particule', £12 6s 4d.

The kirks set for money
    Mr James Cranstoun and John Farrous for the vicarage of Lauder, £60. The glebe thereof, £33 6s 8d. Cuthbert Cranstoun and sir Robert Formane for the kirk of Cheindilkirk, £66 13s 4d [£56 13s 4d, *fo 198v*]. The kirk of Saltoun 'assedatur pro' £66 13s 4d. The kirkland thereof, £7 6s 8d. The kirk of Penciatland [*sic*] 'assedatur for', £66 13s 4d. The kirkland thereof, 40s [*in margin*, 'Nota, this kirkland payis [xl, *deleted*] x li. to Den Johnne Schaw']. The kirk of Kilrennie 'assedatur for' £160 'and that payit to the pensionare of Inchemahome'. The kirk of Lanark 'assedatur to my lord Angus[35] and Sir James Hammiltoun for' £80 'and thairof payit to the preistis of the chapell, [*blank*]. The capitane of Craufuird for Pettinane' £20. The kirk of Gulen in assedation for £100. Eist Fentoun, £20. The teind of Kingistoun 'assedatur for' £13 6s 8d. The lands of Kingstoun, £11 'and that payit to the preist of Sanct Catherines chapell in Dirltoun. Summa of money for the kirkis and teynd abone mentionat', £708.[36]

'Memorandum, the personage of Lauder is in the handis of the persoun be assignatioun be provisioun of pensioun.'

'Memorandum, assignit and payit to ten channones and certan utheris pensiones of the dewteis of this rentale within writtin', 2 c. wheat; 10 c. 6 b. meal; 11 c. 12 b. bere.

'Item, in money to the saidis channones, pensionares and factouris that collectis the samin', £275 6s 8d.[37] 'Item, to the preistis of Lauder and Sanct Catherines

---

[33] James, 5th lord Somerville, *Scots Peerage*, viii, pp. 20-21.
[34] 'Bangoswallis', *RMS*, iv, no 1797.
[35] I.e. Archibald Douglas, 8th earl of Angus (who, though a minor, was retoured heir and infeft in the property of the earldom in 1559), *Scots Peerage*, i, p. 194.
[36] This total would be £707 without taking into account the £10 in the marginal note. £707 would make the calculation of the third correct.
[37] A scribal flourish above the line could indicate another £10.

chapell', £21 6s 8d. 'Item, payit of the dewteis of the kirk of Lanark be Sir James Hammiltoun to thrie preistis thair', £40. 'Item, payit thairof to the pensionare of Inchmahome', £160. 'Item, the annuellis wast and out of use of payment', £12. 'Item, inlakking of the payment of the mailles of the vicarege of Lauder be ressoun of non payment of the corspresentis, upmest claithes and Pasche finance [*recte*, fines], [*blank*]. Memorandum, the haill teynd schevis of the kirk of Lauder in the handis of Mr Andro Home, pensionar thairof.'

Assignation of the third of the abbey of Dryburcht:

Third of meal, 8 c. 10 b. 2 f. $2\tfrac{2}{3}$ p.[38] 'Tak fra Mark Ker for Rutherfuird and half toun of Maxtoun', 3 c.; the toun of Lessuden, 3 c. 8 b.; the laird of Mertoun 'for the uther half of Maxtoun', 1 c. 10 b.; Elistoun, 1 c. meal. 'Gif in' 8 b. 1 f. 1 p. '2 part of half pect' [*i.e.* $\tfrac{1}{3}$ p.].[39]

Third of bere, 7 c. 2 b. 2 f. $2\tfrac{2}{3}$ p. The lady [of] Dalcoif for the toun thereof, 24 b.; the toun of Lessuden, 2 c.; the laird of Mertoun for the half of Maxtoun, 1 c.; Mark Ker for Rutherfuird and half Maxtoun, 2 c.; the toun of Elistoun, 8 b.; Mark Ker of [*recte*, for] Cammestoun and Tempilland, 4 b. 'Gif in' 1 b. 1 f. $1\tfrac{1}{3}$ p.

Third of oats, 1 c. 5 b. 1 f. $1\tfrac{1}{3}$ p. 'Tak thir aites out of the Newtoun gevand' 2 c.

Third of wheat, 10 b. 2 f. $2\tfrac{2}{3}$ p. Take this wheat from Mark Ker for Rutherfuird and half Maxtoun giving 1 c. wheat.

Third of money, £304 1s $1\tfrac{1}{3}$d. Take the glebe of Lauder for £33 6s 8d; the kirk of Saltoun for £66 13s 4d; the kirkland thereof for £7 6s 8d; 'and Pettynane in the captane of Craufuirdis handis', £20. The mails of the lands pertaining to Dryburcht in the year: 'the toun of Dryburcht with the mylne', £41 6s 8d; Gleddiswod, 40s; Brotherstanes, £3; the lands of Merseltoun [*recte*, Ersiltoun, *fo 192v*; *i.e.* Earlston], 22s; Doggarflat, 20s; Sanct Johnnes chapel and Caldslie, £32; the kirkland of Lauder by Mr Andrew Home, £21 6s 8d; Nather Scheilfeild, £4 6s 8d; the lands in Ugstoun, 40s; the lands in Banglaw [Vanglaw, *fo 192v*], £8; the Bernis, £6; Elbottill, £4 6s 8d; the lands of Inglisberry Grange with the teinds of the same by my lord Somervell, £42; the lands of Iflie, £6; Fouldene, 40s; Langhouswallis [Banghouswallis, *fos 192v, 198v*], 46s 8d; Smalahame lands, 15s; Smalahame Spittell, £13 6s 8d. 'Gif in ' £16 2s $6\tfrac{2}{3}$d.

'Nota, the personage of Lauder omittit and thairfoir mak inhibitioun of the haill with capones, pultrie, grassumes, etc.'

---

[38] One third of 25 c. 12 b. is 8 c. 9 b. 1 f. $1\tfrac{1}{3}$ p., but if the total meal is 26 c. (see above, p. 189, n. 31), the calculation in the text is correct.

[39] These figures would give a total of 8 c. 9 b. 2 f. $2\tfrac{2}{3}$ p.

BUNKLE, VICARAGE OF, (SRO, Vol. a, fo 194v)
(SRO, Vol. b, fo 132v)

Vicarage of Boncle pertaining to sir Harry Loch, 'omittit, ungevin up.'

In the whole, 12 merks.

LANGTON, VICARAGE OF, (SRO, Vol. a, fo 194v)
(SRO, Vol. b, fo 132v)

Vicarage of Langtoun pertaining to Mr Thomas Ker.

Calculation of third: In the whole, £20; third thereof, £6 13s 4d.

ST BATHANS, PRIORY OF,[40] (SRO, Vol. a, fo 195r-v)
(SRO, Vol. b, fo 133r-v)

Rental of Sanct Bothanes

Sanct Bothanis, Frankpeta, Hardhisselles, Cornemylne and Walkmylne with the pertinents paying yearly of mail 20 merks. Blakirstoun with the pertinents paying £20 of mail yearly 'be the laird of Cadlische and his aires.' The half of Quixwod with the pertinents paying 6 merks yearly. 'Tua husband landis lyand in Stentoun, lauborit be Williame Wod of Newmylne payand' 40s yearly. Two husbandlands lying within the Manes of Kimmgery with the pertinents paying yearly 40s 'be the laird of Reidbraes'. One husbandland lying in Nanewair with the pertinents paying 20s yearly. 'Ane land in Cokburspethe with the pertinentis payand' 13s 4d 'be yeir.' Two acres of land in Duns with the pertinents paying 6s 8d in the year. 'In Bille, the mure medo with the pertinentis payand' 16s yearly. In Pople, 'ane aker of land with ane oustere' [*i.e.* hostelry] paying 5s yearly. In Buttirdane, 'half ane husband land with the pertinentis' paying 6s 8d yearly. Belhame, 'ane hous, half ane aker of land with the pertinentis' paying 18s 'of annuell be yeir'. In Edinbrucht, 30s 'be yeir of annuell.'

The teinds pertaining to Sanct Bothanis
    Cokburspethe toun and mains, 16 b wheat; 26 b. bere; 40 b. oats; 6 b. peas. The stedingis pertaining to it, 'that is to say, Fulfurdleis, tua hopriges, tua scheillis, Raquhausyd and the Clonis, 24 b. oats; 6 b. bere. Wauchtoun, 16 b. wheat; 18 b. bere; 26 b. oats; 6 b. peas. Craquha [*i.e.* Crauchie[41]] 'in aites', 2 b.; ½ b. bere; 1 f. wheat. Quixwod, 20 b. oats; 4 b. bere. Hardhessillis, 4 b. oats; 1 b. bere. Frankpet, 5 b. oats; 5 f. bere.

---

[40] Cf SRO, Vol. a fo 196r (see below, p. 194).
[41] Cf Eister Craquho, parsonage of Hauch, SRO, Vol. a, fo 174r (see above, p. 170).

Signature: 'Dame Elizabeth Lamb, prioris of Sanct Bothanes, with my hand on the pen led be the notar underwrittin at my command becaus I could nocht writ my selff.'

Assignation of the third of Sanct Bothanes:
    Third of money, £15 14s 1½d. The lands of Blakirstoun for £20. 'Gif in' £4 11s 10⅔d.[42]
    Wheat, 10 b. 3 f. 'Tak it out of Cokburnespethe payand' 16 b.
    Bere, 1 c. 2 b. 3 f. 2⅔ p.[43] Take it out of the same paying 26 b.
    Oats, 2 c. 8 b. 1 f. 1⅓ p. Take it out of the same, 40 b.; and the rest out of Wach[toun].
    Peas, 4 b. 'Tak it out of the same Cokburnespethe payand' 6 b.

'Nota efter'.[44]

FOULDEN, PARSONAGE OF,[45]   (SRO, Vol. a, fo 196r)
                                                        (SRO, Vol. b, fo 138r[46])

Rental of the parsonage of Fouldene

'Inprimis of aites', 48 b. 'Beir', 20 b. 'Quheit and ry', 8 b. 'Beinis and peis', 4 b.

Signature: 'Alexander Ramsay, persoun of Fouldene.'

Calculation of third: Third of oats, 16 b. Third of bere, 6 b. 2 f. 2⅔ p. Third of wheat and rye, 2 b. 2 f. 2⅔ p. Third of peas and beans, 1 b. 1 f. 1⅓ p.

LEGERWOOD, VICARAGE OF,   (SRO, Vol. a, fo 196r)
                                                      (SRO, Vol. b, fo 138r)

Rental of the vicarage of Lidgertwod

'The Pasche fynes, corspresandis, claithe and beist with uther sic small thingis deducit extendis in the haill to the sowme of' 18 merks.

Signature: 'sir Williame Cranstoun.'

Calculation of third: In the whole, 18 merks; third thereof, £4.

---

[42] The correct calculation appears to be £4 5s 10⅔d.
[43] This calculation is correct.
[44] The assumption of the third for the priory of St Bathans occurs at fo 196r (see below, p. 194), where the figures agree with those in this rental, though some place-names vary.
[45] Cf SRO, Vol. a, fo 197v (see below, p. 196), which repeats this entry.
[46] This text, at fos 134r-137r, has part of an entry pertaining to Melrose, which corresponds to SRO, Vol. a fos 207r-210r (see below, pp. 208-211). In SRO, Vol. b, fos 136v, 137v are blank.

ST BATHANS, PRIORY OF,⁴⁷

(SRO, Vol. a, fo 196r)
(SRO, Vol. b, fo 138r)

'The assumptioun of the priorie of Sanct Bothanes for the yeir of God' 1575:

Third of money, £15 14s 1½d. 'Tak it out of the landis of Blakstoun or fra the priores.'

Third of wheat, 10 b. 3 f. 'Tak this quheit out of Wauchtoun and Carquha [*i.e.* Crauchie] geveand be yeir' 16 b. 1 f.

Third of bere, 1 c. 2 b. 3 f. 2⅔ p.⁴⁸ 'Tak of this beir out of Wauchtoun' 18 b.; out of Carquha 2 f.; and the rest out of Hardhessillis paying 1 b.

Third of oats, 2 c. 8 b. 1 f. 1⅔ p. [1⅓ p., *fo 195v*] Take out of Wauchtoun 26 b. and the rest out of the Quickiswod 'gevand be yeir', 20 b.

Third of peas, 4 b. 'Tak thir peis out of Wachtoun gevand be yeir' 6 b.

LAUDER, PARSONAGE OF,⁴⁹

(SRO, Vol. a, fo 196v)
(SRO, Vol. b, fo 138v)

The parsonage of Lauder within the diocese of Sanctandrois set in assedation for 'tua hundreth merkis.' [*In margin,* vicarage [*sic*] of Lauder.]

Signature: 'Andro Hagye.'

Third thereof, £44 8s 10⅔d.⁵⁰

COLDINGHAM, PRIORY OF,⁵¹

(SRO, Vol. a, fo 196v)
(SRO, Vol. b, fo 138v)

Rental of the priory of Coldinghame

'The haill priorie of Coldinghame is set in tak and assedatioun for the yeirlie payment of the sowme of ane thowsand merkis money of this realme, quhairof the thrid extendis to the sowme of' 333 merks 4s 6½d⁵² 'money foirsaid.'

Signature: 'J. Maitland, commendatour of Coldinghame.'

---

⁴⁷ Cf SRO, Vol. a fo 195r (see above, p. 192).
⁴⁸ This calculation is correct.
⁴⁹ Cf SRO, Vol. a, fo 79r (see above, p. 67).
⁵⁰ This calculation is correct.
⁵¹ Cf SRO, Vol. a, fos 190v, 199r, 203r (see above, p. 188; below, pp. 197, 202).
⁵² The correct calculation appears to be 333 merks 4s 5½d.

STICHILL, VICARAGE OF, (SRO, Vol. a, fo 196v)
(SRO, Vol. b, fo 138v)

Rental of the vicarage of Stitchell lying in the sheriffdom of Beruick given in by Patrick Cokburne, vicar thereof. 'The quhilk vicarege is set of ald to George Hoppringle of Blyndley payand thairfoir yeirlie ten pundis money allanerlie.'

Signature: 'Patrik Cokburne, with my hand.'

ELLEM, KIRK OF, (SRO, Vol. a, fo 197r)
(SRO, Vol. b, fo 139r)

'The kirk of Ellem commoun to the college of Restalrig, set in assedatioun to Andro Reidpethe as the same beiris for' £76 13s 4d.

Calculation of third: 'In monie', £76 13s 4d; third thereof, £25 11s 1½d.

'Commoun to the chantorie, college kirk of Restalrig.'

POLWARTH, PARSONAGE OF, (SRO, Vol. a, fo 197r)
(SRO, Vol. b, fo 139r)

'The thrid heirof: In the haill', £10; third thereof, £3 6s 8d.

Rental of the kirk of Polwart

'Item, the persoun and his factouris ressaves the teynd schaves of Redbrais Est Manis with the toun of Polwarth and pendicles adjacent thairto extending in cornes to' 6 b. wheat; 2 c. bere; 60 b. oats; 3 b. peas 'with the vicarege thairof extending now to ten bollis.'

Signature: 'Alexander Broun [*recte*, Hume; Hwm, *SRO, Vol. b*] of Heuch, with my hand led at the pen be Robert Lauder, notar.'

Calculation of third: Wheat, 6 b.; third thereof, 2 b. Bere, 2 c.; third thereof, 10 b. 2 f. 2⅔ p. Oats, 60 b.; third thereof, 20 b. Peas, 3 b.; third thereof, 1 b.

WHITSOME, PARSONAGE OF, (SRO, Vol. a, fo 197v)
(SRO, Vol. b, fo 139v)

The parsonage of Quhitsum within the diocese of Sanctandrois pertaining to Mr James Setoun, set in assedation to the laird of Wedderburne for yearly payment of £100.

Signature: 'Presentit be me, the said M[r] James Setoun.'

Third thereof, £33 6s 8d.

FOULDEN, PARSONAGE OF,[53]   (SRO, Vol. a, fo 197v)
    (SRO, Vol. b, fo 139v)

Parsonage of Fouldeyne

'Inprimis of aites', 48 b. 'Item, of beir', 20 b. 'Item, of quheit and ry', 8 b. 'Item, of beinis and peis', 4 b.

Signature: 'Alexander Ramsay, persoun of Foulden.'

Calculation of third: Oats, 16 b. Bere, 6 b. 2 f. 2⅔ p. Wheat and rye, 2 b. 2 f. 2⅔ p. Beans and peas, 1 b. 1 f. 1⅓ p.

CRANSHAWS, PARSONAGE OF,   (SRO, Vol. a, fo 198r)
    (SRO, Vol. b, fo 140r)

Rental of the parsonage of Cranschawis within the sheriffdom of Beruik, 'quhairof that ane half is set to David Spottiswod of that Ilk and the uther half in the handis of Mr David Suyntoun, persoun thairof.'

'The said half set extending yeirlie in payment', £20, 'and the uther half of the samin in the handis of the said persoun extendis yeirlie to' £15.

'Ressavit at the command of the Clerk Register.'[54]

Calculation of third: In the whole, £35; third thereof, £11 13s 4d.

SPROUSTON, VICARAGE OF,   (SRO, Vol. a, fo 198r)
    (SRO, Vol. b, fo 140r)

Rental of the vicarage of Sproustoun pertaining to Patrick Bellenden set in assedation for £24.

Signature: 'Patrik Bellenden.'

---

[53] Cf SRO, Vol. a, fo 196r; this entry repeats the rental on fo 196r (see above, p. 193).

[54] James McGill of Nether Rankeilour, Clerk Register, 1554 - 1566, *Handbook of British Chronology*, p. 197.

Calculation of third: In the whole, £24; third thereof, £8.

DRYBURGH, ABBEY OF,[55]   (SRO, Vol. a, fo 198v)
                                                                                (SRO, Vol. b, fo 140v)

Rental of Dryburcht

'In the first, the laird of Coldenknowis sould pay for the kirk of Smalahame' [toun of Smalahame, *fo 192r*] 6 c. meal; 3 c. bere; 1 c. oats. The kirk of Mertoun 'assedat for' [*blank*]. Maxtoun kirk 'assedat for' £23 6s 8d money. Lessudden kirk 'assedat for' £66 13s 4d. Chingelkirk [Jeindilkirk, *SRO, Vol. b*] 'assedat for' £56 13s 4d [£66 13s 4d, *fo 193r*]. Saltoun kirk 'assedat for' £66 13s 4d. Pencaitland kirk 'assedat for' £66 13s 4d. Kilrenry [*recte*, Kilrenny] 'assedat for' £160. Lanrik kirk 'assedat for' £80. Pettinane kirk 'assedat for' £20. The kirk of Gulen 'in assedatioun for' 100 merks.[56]

The mails of the lands
    'In the first, the toun of Dryburcht with the mylne', £31 6s 8d [£41 6s 8d, *fo 192v*]. Glediswod, 40s. Brotherstanes, £3. The lands of Arsiltoun [*i.e.* Earlstoun], 22s. Doggarflat, 20s. Sanct Johnnes chapel and Cadeslie, £22 [£32, *fo 192v*]. The lands of Ugstoun, 40s. The lands of Banglawe [Vanglaw, *fo 192v*], £8. The Bernis, £6. Elbottill, £4 6s 8d. The lands of Inglisberry Grange, £20 [£42, *fo 192v*]. Iflie, £6. Banghouswallis [Langhouswallis, *fo 194r*], 46s 8d. The lands of Smalahame, 15s.

'Summa of this haill money by the kirk of Mertoun', £710 15s 4d; third thereof, £236 18s 5⅓d. Meal, 6 c.; third thereof, 2 c. Bere, 3 c.; third thereof, 1 c. Oats, 1 c.; third thereof, 5⅓ b.

[*In margin*, 'By Netoun [*recte*, Mertoun], £716 10s 4d; third thereof, £233 16s 9⅓d.[57]]

COLDINGHAM, PRIORY OF,[58]   (SRO, Vol. a, fos 199r-202r)
                                                                         (SRO, Vol. b, fos 141r-144r)

'The rentale of the spiritualitie of the priorie of Coldinghame and kirkis and teyndis thairof, to the quhilkis Alexander, lord Home[59] is provydit as the samen

---

[55] Cf SRO, Vol. a, fo 192r (see above, p. 189). David Erskine was commendator at the Reformation, *RMS*, iv, no 2630.
[56] In margin, in a later hand, 'Smalhame, Mertoun, Maxtoun, Lessuddan, Chingelkirk, Saltoun, Pencaitland, Kilrynnie, Lanark, Pittinane, Gulen.'
[57] The correct calculation appears to be £238 16s 9⅓d.
[58] Cf SRO, Vol. a, fos 190v, 196v, 203r (see above, pp. 188, 194; below, p. 202).
[59] Alexander, 6th lord Home, and 1st earl of Home, *Scots Peerage*, iv, pp. 463-466.

wes worthe the tyme of his provisioun thairto and presentlie geves as he may bruik the samen be the law and his provisioun gevin up be the said noble lord to the commissiouneris of Parliament undersubscryveand.'[60]

The kirk of Coldinghame
'Inprimis, the teynd schaves of the tounes and landis of Coldinghame and Eymouth collectit and led estimat yeirlie the expenssis deducit to' 6 c. oats; 6 c. bere. The teinds of the lands of Wester Lummisden, Dudinholme and their pertinents, £20. The lands of Eister Lummisden, £10. The teinds of the lands of Worthfeild 'set of auld be the erle of Bothwellis father[61] includit with the few of the landis for yeirlie payment of' £6 13s 4d. 'The teyndis of the toun landis and Mans of Rentoun', 16 b. oats; 8 b. bere. The teinds of the lands of Brokhoillis and Berrie Hill, £10. The teinds of the lands of Suanisfeild and Scheil Hopdykis, £10. The teinds of the lands of West Restoun with the teinds of the lands of Bastanerig within the parish of Aytoun 'set in tak be the said erle Bothwell to Johnne Home of West Restoun for yeirlie payment of' £40. The teinds of the lands of Eistrestoun within the said parish of Coldinghame and teinds of the lands of Prandergaist within the said parish of Aytoun 'set in tak be the said sumtyme erle Bothwell to umquhill George Auchincraw, portiouner of Eistrestoun, and his aires and for yeirlie payment of' £20. Item, the teinds of the toun and lands of Suynewod 'disponit in pensioun be the said sumtyme erle to Gilbert Pennycuik during his lyftyme for his service allanerlie worth be yeir' 24 b. oats; 8 b. bere [*in margin*, 'Assignit to Gilbert Pennycuik']. The teinds of the lands of Blakhill, 6 b. oats; 2 b. bere. The teinds of Hilend [Hilheid, *fo 204r*], 4 b. oats; 1 b. bere. The teinds of the lands of Press, Fluires, Hielawis and Benerig, 'the steill grevestyle and certane croftis and akeris possessit be Sir Alexander Home of Man[d]erstoun assignit to him self for fyve chalderis victuall as ane part of his pensioun disponit furthe of the said abacie be the said erle Bothwell for fulfilling of his majesteis decreit arbitrall quhilk pensioun is ratefeit in Parliament and declarit to be frie of all thridis, taxatiounes and utheris burdingis quhatsumevir. The saidis teyndis being worthe yeirlie incais the samin wer nocht disponit in pensioun bot wer led and gadderit', 5 c. victual; thereof 3 c. 8 b. oats; 1 c. 8 b. bere [*in margin*, 'Assignit to Sir Alexander Home of Manderstoun'].

---

[60] For commissioners, see below, p. 202, n. 71.
[61] This document is dated 1600. John Stewart, an illegitimate son of James V, was commendator of Coldingham at the Reformation; his son, Francis Stewart, born in 1563, gained the commendatorship in 1565, became earl of Bothwell in 1581, was forfeited in 1591, and attainted by act of Parliament in 1593. *Scots Peerage*, ii, pp. 168-171; *RSS*, v, no 2182; *RPC*, iv, pp. 609-610, 643-645; *APS*, iv, pp. 8-11. The commendatorship passed into the possession, first of John Maitland in 1567 and then, in 1571, of Alexander Home, younger, of Manderston, *RMS*, iv, nos 1765, 2178. In 1584 Francis Stewart protested that his claim to the priory should not be prejudiced by the election to the committee of the articles of Alexander Home as prior; and John Maitland did likewise in 1585, *APS*, iii, pp. 291, 387. In October 1589, the priory was gifted to John Stewart on the demission of Francis, earl of Bothwell, though Parliament in 1592 declared that, with Bothwell's forfeiture, the priory was annexed to the crown, SRO, E2/15, Register of Signatures in Comptrollery, fo 8; *APS*, iii, p. 561.

'Summa of the silver dewtie of the teyndis of the said kirk of Coldinghame extendis be yeir to' £116 6s 8d.[62] 'Summa of the beir', 8 c. 11 b. 'Summa of the aites', 11c. 2 b.[63]

The kirk of Auldcambuies
    The teinds of the lands of Auldcambies being the said Alexander, lord Home's heritage 'hes bene possest be him and his predicessouris in all tyme bygane for yeirlie payment of' £26 13s 4d.

The kirk of Aytoun
    'Inprimis, the teyndis of the toun and Mans of Aytoun and Fairnysyd within the said parochin togither with the teyndis of the landis of Hunwod [i.e. Houndwood] and Eister Quhitfeild extendis to' 6 c. victual: 4 c. oats; 24 b. bere; 4 b. wheat; 4 b. peas. The teinds of the lands of Nather Aytoun, half Reidhall and Gunisgrene, 24 b. 3 f. oats; 11 b. 1 f. bere; 1 b. peas; 3 b. wheat. The teinds of Nather Flemyngtoun, Fluires 'callit Broumslandis', £13 6s 8d. The teinds of Peilwallis and half Reidhall 'possessit be Jonet Auchincrawe, relict of umquhill George Auchincraw of Nether Byre', paying yearly, 4 b. bere; 16 b. oats; 2 b. peas. The teinds of Quhytrig, 16 b. oats; 3 b. bere; 1 b. rye.

'Summa of the money of the said kirk', £13 6s 8d. Total oats, 7 c. 8 b. 3 f. Total bere, 2 c. 10 b. 1 f. Total peas, 7 b. Total wheat, 7 b. Total rye, 1 b.

The kirk of Lambertoun
    The teind sheaves of the kirk of Lambertoun with the teinds of the toun of Auchincraw, £48.

The kirk of Beruik
    'Item, thair is na landis nor teyndis of the said kirk in Scotismanis handis except the teyndis of the four landis of Dringtoun [i.e. Edrington] quhilk payis' 8 b. oats; 6 b. bere; 2 b. wheat [in margin, 'Assignit to the said Sir Alexander'].

The kirk of Fischewik
    'Item, the teynd schaves of the landis of Fischeweik and toun of Paxtoun ar the haill spiritualitie of the said kirk, payis yeirlie' 5 c. oats; 2 c. bere; 1 c. wheat [in margin, 'Assignit to the said Sir Alexander Home'].

The kirk of Eddrem
    'Inprimis, the teynd schaves of the toun and Mans of Eddrem', £20. The teinds of the kirkland of Eddrem, 3 b. oats; 1 b. bere. The teinds of Bromehous, 8 b. oats; 4 b. bere. The teinds of Blakcader, Reidheuch and Quhytlaw, 5 c. oats; 2 c. bere; 1 c. wheat. The teinds of Eist Nisbit and Brume Dykis 'possest be the

---

[62] The correct calculation appears to be £116 13s 4d.
[63] The correct calculation appears to be 12 c. 10 b.

laird of Eist Nisbit assignit to the said Sir Alexander in part of payment of his pensioun payand be yeir' 6 c. victual: 4 c. oats; 24 b. bere; 4 b. wheat; 4 b. peas [*in margin*, 'Assignit to the said Sir Alexander Home']. The teinds of West Nisbit, Ryschill and Cronklie, 'possest be the laird of West Nisbit assignit lykwayis to the said Sir Alexander in pairt of payment of his pensioun', paying yearly 4 c. victual: 2 c. 7 b. oats; 1 c. 7 b. bere; 2 b. wheat. The teinds of 'the eist quarter of Eist Nisbit possessit be the guidman of Huttounhall', £13 6s 8d. The teinds of Kymmerghame toun, Kello and Kello Castell 'possest be the laird of Wedderburne for' £80. The teinds of Kymmermanis [*i.e.* Kimmerghame Mains] and Belscheill 'possest be the laird of Polwarth be rycht maid to him thairof be our soverane lord' £10. 'Summa of the money of the said kirk', £123 6s [8d]. Total oats of the said kirk, 12 c. 2 b. Total bere of the said kirk, 5 c. 4 b. Total wheat, 1 c. 6 b. Total peas, 4 b.

The kirk of Swy[n]toun
'Item, the said kirk payis' £10.

The kirk of Schichel [*i.e.* Stichill]
'Item, the said kirk of Stitchell and teyndis thairof hes ever bene ane auld possessioun of the hous of Home past memorie of man for payment of' £40.

The kirk of Ednem
'Item, the said kirk hes evir bene possest be the hous of Edmestoun, the landis thairof being thair auin propper heritage and is set be vertew of ane decreit of the Lordis of Sessioun for payment of the [auld] dewtie quhilk is' £80.

The kirk of Ersiltoun
The teinds of Ersiltoun and Coldinknowis, £40. The teinds of Mullerstanes and Fawnis £60. The teinds of Reidpethe, £10. The teinds of Carelsyde, £10.

'Summa of the haill money of the kirkis foirsaidis', £577 6s 8d.[64] 'Summa of the haill aites', 36 c. 4 b. 3 f. 'Summa of the haill beir', 18 c. 15 b. 1 f. 'Summa of the quheit', 2 c. 15 b. 'Summa of the peis', 11 b. 'Summa of the ry', 1 b.

'Thair aucht to be defasit of the money and victuall foirsaidis the sowmes and victuall efter specifeit.'

'Deductioun'
['In the first, the contributioun silver payit to the Lordis of Sessioun', £28, *SRO, Vol. b, deleted*.] The laird of Blanses pension, £66 13s 4d. 'Nicholas

---

[64] The correct calculation of the manuscript figures appears to be £577 13s 4d.

Celocht, Frensche manis, pensioun', £48. 'Summa of the defesance foirsaid', £142 13s 4d.[65]

'Item, to be defeasit of the victuall foirsaid'
Sir Alexander Homes pension extending to 30 c. victual 'quhilk is fre of all thrid thairof', 19 c. 8 b. oats; 8 c. 6 b. bere; 1 c. 12 b. wheat; 8 b. peas. 'Item, thair aucht lykwayis to be defesit the teyndis of Swynwode assignit to Gilbert Pennycuik extending to' 24 b. oats; 8 b. bere.

'Restis of fre money', £434 13s 4d; third thereof, £154 4s $5\frac{1}{3}$d.[66] 'Restis of fre aites', 15 c. 4 b. 3 f.; third of oats, 5 c. 1 b. 2 f. $1\frac{1}{3}$ p. 'Restis of fre beir', 10 c. 1 b. 1 f.; third of bere, 3 c. 5 b. 3 f. 'Restis of quheit', 1 c. 3 b.; third of wheat, 6 b. 1 f. $1\frac{1}{3}$ p. 'Restis of peis', 3 b.; third of peas, 1 b. 'Thrid of the ry' 1 f. $1\frac{1}{3}$ p.

'The fyft penny and pairt of the fre money and victuall foirsaid now pertenis to the king in place of the monkis portiounes and extendis as followis. The fyft penny of the fre money', £86 18s 8d. 'The fyft pairt aites', 3 c. 3 f. $3\frac{1}{3}$ p. 'The fyft pairt beir', 2 c. 1 f. 'The fyft pairt quheit', 3 b. 3 f. $\frac{2}{3}$ p.[67] 'The fyft pairt peis', 2 f. $1\frac{1}{3}$ p.[68] 'The fyft pairt ry', $3\frac{1}{5}$ p.

Signature: 'Alexander, lord Home.'

> 'We, the commissionares undersubscryveand contenit in the act of Parliament[69] for trying of the estait of the benifice of Coldinghame and to mak ane new just thrid thairof as the act of our said commissioun of the dait the' 15 November 'last mair fullie purportis, findis be cognitioun tane be us according thairto that the rentis of the spiritualitie of the said benifice now gevin up be Alexander, lord Home, present titular thairof, extendis to the availlis and quantiteis foirsaidis contenit in this present rentall and ordanes the same rentall to be registrat in the clerkis buikis of the Collectorie[70] and the assignatiounes to be gevin furth this yeir and in tyme cuming, conforme to the thrid thairof now set doun and allowit be us, provyding alwayis that quhen the pensioneris sall happin to deceis that the thrid of the pensiounes sall access to our soverane lord and salbe eikit to the thrid contenit in the said rentale, provyding lykwayis that quhatevir is omittit furth of this rentale sall pertein to our said soverane lord, conforme to the natour of the omissiounes in the first assumptiounes and actis maid thairanent, in witnes quhairof we have subscryvit thir presentis

---

[65] This calculation is correct when £28 from SRO, Vol. b text is included.
[66] The correct calculation appears to be £144 17s $9\frac{1}{3}$d.
[67] The correct calculation appears to be 3 b. 3 f. $\frac{2}{3}$ p.
[68] The correct calculation appears to be 2 f. $1\frac{1}{3}$ p.
[69] Cf APS, iv, p. 244.
[70] The Collectory was the department responsible for accounting and disbursing the revenues derived from the collection of the thirds of benefices.

with our handis as followis at Edinbrucht, the' 3 December 1600. 'Sic subscribitur, Thomas Ly[o]un off Auldbar, Cranstounriddell, Halyruidhous, M[r] T. Hamiltoun, Johnne Prestoun.'[71]

COLDINGHAM, PRIORY OF,[72]   (SRO, Vol. a, fos 203r-205v[73])
(SRO, Vol. b, fos 145r-147v)

The rental of Coldinghame

'The sowme of the penny maill of the' 42 'husbandland landis, ilk land payand yeirlie' 14s 4d 'with thrie capones and ane pultrie extendis to' £28.[74] 'Item, mair yeirlie in castell wairdis', 35s. Yearly in capons, 120. In poultry, 42. 'Mair in meill [*recte*, maill] yeirlie for the cotagis of the same toun', £9 9s; 'and in pultrie for the cottagis foirsaid', 59 poultry. 'Mair yeirlie in meill [*recte*, maill] for the gers landis in the same toun', 40s. 'Mair in few mailles for certane landis within the barony extending to' £69 16s 8d [*in margin*, 'Over generall']. 'Mair yeirlie for the mailles of the commountie in Coldinghame toun', 19s; and in poultry, 17. 'Mair yeirlie in annuelrent out of the said toun', £4 3s. [*In margin*, 'Sum of the money of this baronie', £116 2s 8d; 100 capons; 5 score and 18 poultry.]

Eymouthe
      The toun of Eymouthe containing 16 husbandlands, the yearly mail thereof, £9 13s 4d; 'and in castell wairdis to' 13s 4d; and in capons 48; and in poultry 16. 'Mair for the yeirlie maill of the cottagis of the same toun extending in money to' 37s; and in poultry 21. 'Mair for the annuelrent of the same toun yeirlie in maill', 33s 8d. 'Mair the yeirlie maill of the commontie of Eymouth extendis to' 51s 9d; 'and in pultrie for the said commontie' 71. [*In margin*, 'Of money', £26 [£26 9s 1d, *SRO, Vol. b*; *recte*, £16 9s 1d]; 48 capons, 108 poultry.]

The toun of Aytoun containing 14½ husbandlands extending yearly in mail to £22 5s 5d; and in capons, 44; and in poultry, 15. 'Mair the annuelrent of Aytoun extending yeirlie to' 44s 4d. 'Mair for the annualrentis of Flemyngtoun, Paxtoun, Lambertoun, Eistrestoun and Langtounes landis lyand thair extendis yeirlie' £12 18s. [*In margin*, Money, £37 7s 9d. Capons, 44. Poultry, 15.]

---

[71] This letter resembles that relating to Jedburgh included at SRO, Vol. a, fo 221r (see below, pp. 221-222). Sir Thomas Lyon of Auldbar had demitted the office of Treasurer by March 1596, and had been privy councillor till 1598; David McGill of Cranstoun Riddell, advocate (son of the former lord Advocate, who died in 1596); John Bothwell, commendator of Holyroodhouse; Thomas Hamilton of Drumcairn, lord Advocate; John Preston of Fentonbarns, Collector general. *Handbook of British Chronology*, pp. 188, 201; *Scots Peerage*, viii, p. 286; *Rotuli Scaccarii Regum Scotorum: The Exchequer Rolls of Scotland*, eds. J. Stuart and G. Burnett, *et al.*, 23 vols. (Edinburgh, 1878-1908), xxiii, pp. 148, 512; *RPC*, v, p. 288 n.; vi, pp. 98, 199.
[72] Cf SRO, Vol. a, fos 190v, 196v, 199r (see above, pp. 188, 194, 197).
[73] SRO, Vol. a, fo 202v is blank; the corresponding folio in SRO, Vol. b, fo 144v is also blank except for two lines of deleted text.
[74] The correct calculation appears to be £30 2s.

West Restoun containing 23 husbandlands, the yearly mail thereof, £15 6s 8d; 'and in castell wairdis', 15s 4d; in capons, 57; and in poultry, 39. 'Mair the yeirlie maill of the cottagis of the said toun extendis to' 13s; and in poultry, 4. 'Item, mair the annualrent of the same toun', 10s 11d. [*In margin*, Money, £17 5s 11d. Capons, 57. Poultry, 39.]

Swounwod containing 16 husbandlands, the yearly mail thereof, £11 4s; and in capons, 48; in poultry, 16. 'Mair the yeirlie maill of the cottagis of the same', 18s.; and in poultry, 6; 'and mair in annualrent', 22s. [*In margin*, Money, £13 4s. Capons, 48. Poultry, 22.]

Rentoun containing 14 husbandlands, the yearly mail thereof 'with the castell wairdis extendis to' £9 16s; and in capons, 12; in poultry, 4. The yearly mail of the cottages of the same, 11s 4d; and in annualrent yearly, 30s 6d. [*In margin*, Money, £11 17s 10d. Capons, 12. Poultry, 4.]

Auchincraw, the feu mail thereof with the annualrent of the same extends yearly to 46s 8d. [*In margin*, 'Auchincraw tenet', 46s 8d.]

Annualrents of West Lummisdan, Eist Lummisden, Blakhill, Quhytfeild, Hilheid [Hilend, *fo 199r*], Eddremtoun and mains, Fairnysyd, Bromhous and Suyntoun extends yearly [to] £55 16s. [*In margin*, 'Annuelrentis, S[umma]' £55 16s.]

'The mylnes': The Mylne of Aytoun, £10. The Mylne of West Restoun, £4; 'and the walkmylne of the same', £3 6s 8d. The Mylne of Rentoun, £3 6s 8d. The Mylne of Eymouth and Colmyln and Coldinghame, £12. [*In margin*, 'Mylnes', £32 13s 4d.]

Northfeild paying yearly in feu mail, £20. [*In margin*, 'Northfeild', £20.]

The Fewles paying yearly, 40s. [*In margin*, 'The Fewles', 40s.]

Fischweik, the feu mail thereof, £20. 'And mair for the watteris', 42 'angellis.[75] Item, mair for the augmentatioun', £10 13s 4d. [*In margin*, 'Fischweik', £30 13s 4d; 42 'angell nobillis'.]

'Augmentatiounes of certan cotagis and commounteis extending yeirlie to' £16 15s 10d.

The teinds of the 'bottis of Coldinghame' pertaining to the vicarage extending yearly to £21. 'Item, the teynd lambis and woll, teynd cheis and geis gadderit yeirlie be the place. [*In margin*, 'The boittes', £21. 'Lambis woll, geis and cheis wald be considerit upoun'.]

---

[75] An angel (noble) was an English coin worth about 10 merks.

'The small vicarege: Teynd lint, hemp and hay extending yeirlie to' £11 11s 8d. [*In margin*, 'Small Vicarege', £11 11s 8d.]

'Summa of the money of the temporalitie', £415 4s 1d.[76]

'The kirkis pertenyng to the priourie of Coldinghame set for money'
    The kirk of Ednem set in assedation for yearly payment of £106 13s 4d. 'Certan pendicles of Erssiltoun set to Johnne Hume of Carrelsyd payand thairfoir yeirlie' £60. The teinds of Fawnis and Mellerstanes set for the yearly payment of £60. The kirk of Stritchell [*i.e.* Stichill] set in tack and assedation for the yearly payment of £40. The teind of Ednem toun and Maynes, £20. The teind of Reidpethe, £10. The teind of Kymmerghame Maynes and Bedscheill,[77] £10. The teinds of West Restoun, Howburne Mylne and Basterig, £40. The teind of Northfeild, £6 13s 4d. The teind of Coldenknowis and Erssiltoun, £40. The Brokholiss and Berryhill, £10.

'The sowme of the foirsaidis kirkis in money extendis to' £403 6s 8d. 'Errat.'[78]

'Summa of the haill money, bayth kirkis and land mailles', £818 10s 9d; 'thrid of the haill money', £272 16s 11d.
    'Summa of the haill capones', 289; third thereof, $109\frac{2}{3}$.[79]
    'Summa of the pultrie', 266; third thereof, $102\frac{2}{3}$.[80]

'Followis the rentale of the kirkis of Coldinghame quhairof the teynd schaves ar in use yeirlie to be riddin as the samin wer riddin in the' 68 'yeir' [*i.e.* 1568].

Edringtoun, 2 b. wheat; 3 b bere; 8 b. oats. Fischeweik, 20 b. wheat; 4 b. rye; 3 b. peas; 26 b. bere; 92 b. oats. Swyntoun, 17 b. 3 f. 2 p. wheat; 3 f. 2 p. rye; 22 b. 2 f. peas; 55 b. 3 f. bere; 175 b. oats. Eddrem: 36 b. 3 f. wheat; 5 b. 2 p. rye; 18 b. 2 f. peas; 106 b. 2 f. 2 p. bere; 398 b. oats. Aytoun: 10 b. 2 f. wheat; 7 f. rye; 21 b. bere; 4 b. peas; 65 b. 2 f. oats. Kirk of Coldinghame, 4 b. 2 f. wheat; 2 f. rye; 1 b. 3 f. 2 p. peas; 11 b. bere; 26 b. 1 f. oats. 'By and attour the teynd schaves of the tounes of Coldinghame and Eymouth quhilkis ar yeirlie usit to be led extending yeirlie' to 121 b. 2 f. oats; 77 b. bere; 12 b. peas; 12 b. wheat; 8 b. rye.

Calculation of third: 'Summa of the quheit', 6 c. 7 b. 2 f. 2 p.; third thereof, 2 c. 2 b. 2 f. $\frac{2}{3}$ p. Bere, 19 c. 12 b. 1 f. 2 p.;[81] third thereof, 6 c. 9 b. 1 f. $3\frac{1}{3}$ p. Oats,

---

[76] The addition of the sub-totals is £414 12s 5d, but the correct total of individual entries appears to be £414 6s 1d.
[77] Belshiel (NT8149) is adjacent to Kimmerghame Mains (NT8050); Bedshiel (NT6851) is more distant.
[78] The arithmetic is correct; there seems no reason for 'Errat'.
[79] The correct calculation appears to be $96\frac{1}{3}$.
[80] The correct calculation appears to be $88\frac{2}{3}$.
[81] The correct calculation appears to be 18 c. 12 b. 1 f. 2 p.

55 c. 6 b. 1 f.; third thereof, 18 c. 6 b. 3 f.[82] Rye, 1 c. 4 b. 1 f.; third thereof, 6 b. 3 f. Peas, 3 c. 13 b. 3 f. 2 p.; third thereof, 1 c. 4 b. 2 f. 2 p. 'Summa of the raidis' 85 c. 14 b. 1 f. 2 p. Silver, £899 6s 8d. 'Summa of the cayne foullis', 52 dozen and 6 poultry.

'The landis of Coldinghame and Eymouth quhilkis ar presentlie led ar worthe ilk yeir' 20 c. 'of quheit, beir, peis and ry, as I am informit be thame quha hes maid compt of the samin to the priour[83] and hes had the hand[l]ing of the same.'

NOTE

Rental for the following benefice is located elsewhere in the text:

LAUDER, PARSONAGE OF (SRO, Vol. a, fo 79r), p. 67.

---

[82] The correct calculation appears to be 18 c. 7 b. 1½ p.
[83] See above, p. 198, n. 61.

Roxburgh (1)

MELROSE, ABBEY OF,[1] (SRO, Vol. a, fos 206r-210r[2])
(SRO, Vol. b, fos 134r-137r[3])

'This is ane just rentall of Melros as I, Mr Michaell,[4] commendatour thairof, will testifie in sua far as I knaw, extending in money fermes and teyndis with uther dew service and dewiteis as efter followis.'

Money in the year
Blainslie, £45 16s 11d. Threipwod, £32. Cummisliehill, £5. Quhitlie, £6 13s 4d. Williame Law, £5. Housbyre and Eisterfuird, £3 6s 8d. Housbyre and Westerfuird [Wester Raik and Wousbyre, *fo 255r*], £10. Langschaw, £5. Morestoun, £3 3s 4d. Reidpethe, £21. Eildoun, £26. Newtoun, £26 13s 4d. Moxpoffill, £3 6s 8d. Crangillis, £5. Freir croft, £3 6s 8d. Wouplaw, £3 6s 8d. Apletrelevis, £30. Freirschaw, £5. Moreslaw [*i.e.* Murehouslaw], £10 6s 8d. 'Custume of Melros', £6. Langlie, £12 13s 4d. Bukholme, £10. Allaneschalles, £6 13s 4d. Lassudden, £53 6s 8d. 'The annuellis of Littill Fordell nocht waist', £21. Eildoun Coit, 10s. The Mylne of Langschaw, £6 13s 4d. The Mylne of Newtoun, £8. The Mylne of Newgrange, £2. The Mylne of Hunum Grange, £2. Cartleyis, £3 6s 8d. The Femaisteris lands,[5] £3 6s 8d. Calfhill, £10. Halkburne, £3 6s 8d. Hewlabutis, 25s. Dry Grange, £22. The kirkland of Hassinden and Caveris, £6 13s 4d. The lands of Mars, £23. The mails of the lands of Eist Teviot Daill, £27 6s 8d. Lawmuirmoir, £30. Hartsyd, £40. Tweiddaill Traquair, 10s. Hairhoip, £15. Hoipcartane, £3 6s 8d. Kingildouris, £8. Wolfclid,[6] £2 13s 4d.

---

[1] Cf SRO, Vol. a, fo 255r (see below, p. 257). The rental is also printed in *Selections from the Records of the Regality of Melrose*, ed. C. S. Romanes, 3 vols. (Edinburgh, 1917), iii, pp. 133-140. See also *RMS*, iv, no 1819 which lists many of the place-names.
[2] Fo 209v is blank.
[3] SRO, Vol. b at this point does not follow the order of SRO, Vol. a; this rental is intruded between the priory of St Bathans, fo 133r-v, and the parsonage of Foulden, fo 138r.
[4] I.e. Mr Michael Balfour, *Records of the Regality of Melrose*, iii, p. 133; *RMS*, iv, no 1819.
[5] 'Semaisters landis', *Records of the Regality of Melrose*, iii, p. 134.
[6] 'Wolfurde', *Records of the Regality of Melrose*, iii, p. 134.

The teinds set for silver
 The Apletrelevis, £10. Drygrange, £20. Eildoun, £10. Calfhill teinds set to Malcolm Hoppringill, £6 13s 4d. Auld Melros set in feu to Robert Ormestoun, £6 13s 4d. The Freiris land, 33s 4d. 'Darnik Brigend and pendicles set in few to the Secretar[7] for the sowme of' £91 15s 11d. Cadeheuch,[8] 12s. 'Summa of thir precedantis being in use of payment', £720 19s 6d.[9]

Soroleisfeild, 30s. Cammestoun, £5. Plewlandis, £5. Monksfald, Thessie, Favingtoun,[10] £2. Ugingis,[11] £50. Autounburne, £10. Atrik lands, £66. Ringwodfeild, £52 6s 8d. Rodono [Powdono, *fo 257r*], £25. Eskdaillmuir, £168. 'The annuelrentis of burchtis with Beruik', £107 12s 3d. 'Blanslie beyand gevin to Hob Ormestoun be lord James',[12] £26 13s 4d.

'Summa of this last perticall nocht in use of payment', £519 2s 3d.

'Summa of the haill money precedand', £1,240 21d.[13]

'The rentale of Melros concernyng victuallis'

[*In margin*, 'Quheit':] The Abbay Milles set in tack for 4 c. 'girnell meill'. The Westoussis Myln 'of the auld', 2 c. The Newgrange, 5 c. The Braid Medow, 5 b. The Awmont Perk, 4 b. Total wheat, 11 c. 9 b.

[Meal:] The toun of Galtounsyd, 1 c. The Annay, 3 c. 1 f. The toun of Eleistoun, 3 c. The Langmedow, 1 b. 1 f. The Newgrange, 3 c. 'The Priour [Priour Wod, *following paragraph*] gevin doun be lord James to Robert Wallace', 3 b. 'be ressoun of barrennes of the grund', 6 b. The Monk Perk, 4 b. Total meal, 10 c. 13 b. 1 f.[14]

[Bere:] Dainyeltoun, 6 c. Galtunsyde, 16 c. Annay, 3 c. 1 f. Coitmedow, 3 b. 2 f. Eildoun Coit, 6 b. Quheityaird, 3 b. 'Priour Wod gevin doun as of befoir for barrennes', 3 b. 6 f. 5 p. Eleistoun, 3 c. Moshoussis, 1 c. Subcelleris land,[15] 5 b.

---

[7] William Maitland of Lethington, Secretary 1558 - 1571, *Handbook of British Chronology*, p. 193.
[8] 'Cauldheuch', *RMS*, iv, no 1819; the place has not been traced.
[9] The correct calculation appears to be £725 19s 10d.
[10] 'Farningtoun', *Records of the Regality of Melrose*, iii, p. 135; the place has not been traced.
[11] 'Wyingis', *Records of the Regality of Melrose*, iii, p. 135; the place has not been traced.
[12] I.e. James Stewart, commendator of Kelso and Melrose abbeys till his death in 1557, Cowan and Easson, *Medieval Religious Houses*, pp. 68, 77.
[13] SRO, Vol. b recommences with the start of the next paragraph on fo 134r which corresponds to SRO, Vol. a, fo 207r; cf above, p. 207, n. 3.
[14] The correct calculation appears to be 10 c. 14 b. 2 f.
[15] 'Subtelleris land', *Records of the Regality of Melrose*, iii, p. 136; 'Subsellarisland', *RMS*, iv, no 1819; the place has not been traced.

Langmedo, 3 b. 2 f. 'The Femaisteris landis[16] by Dernik etc.': Monkpark, 4 b. Newsteid, 10 c. Total bere, 40 c. 15 b. 1 f.

Oats: The Moshoussis, 3 c. The Monkis Medow, 2 b.

Butter of Cummslie: Cummslie toun and Laudopmuir, 105 stones.

Salt: 'Salt pannis of Prestoun olim thay payit', 8 c. [*In margin*, 'nota, restant salt.']

Kane Fowls: 'Cane Foullis of Melrosland with Eleistoun', 660 kane fowls; 124 capons.

Peats: Thraipwod, 400 'laid of peittis.'

'Cariages: Melrosland aucht' 615, 'by lym and peittis.'

'The estimatioun of the teyndis of the kirkis pertenyng to Melros yeirlie riddin be the maist honest men in the cuntrie and speciallie in this yeir precedand, viz., thrie scoir fyve yeiris [*i.e.* 1565].

Item in the first, the teyndis of the tua kirkis of Hassinden and Caveris quhilkis hes be[ne] led continuallie to the use of the place of the maist parte will extend in commoun yeiris as followis.' In oats, 30 c. In bere, 12 c. In wheat and peas, 3 c.

'The estimatioun of the teyndis of Eist Teviot Daill pertenyng to the hous of Melros in the yeir foirsaid.'
The teinds of Gaitschaw, Hunumgrange, Southgait[17] and Denbie [Denbray, Lounoumgrange and Southcoit, *fo 257v*]: In bere, 1 c. In oats, 2 c. 8 b.

'The estimatioun of the teyndis of the landis of Melros land of the yeir foirsaid.'
In oats, 9 c. In bere, 2 c.

'Memorandum, the kirkis of Wester [Wester Ker, *SRO, Vol. b*] and New of Aitrik[18] hes bene out of use of payment ony kynd of teyndis sen Fluddoun and as yit reducit etc. sett within.[19] Item, the teynd woll of Melrosland this yeirlie abonewrittin gaddert be honest men extendis to' 8 stones. 'Item, the lambis thairof this present yeir', 60. 'Item, the cheis of Deid Scheip[20] [*recte*, Dead Side] extendis

---

[16] 'Semaisters lands', *Records of the Regality of Melrose*, iii, p. 136; the place has not been traced.
[17] 'Southait [*sic*]', *Records of the Regality of Melrose*, iii, p. 137, now South Cote.
[18] *Records of the Regality of Melrose*, iii, p. 137 renders this 'Wester and New ... of Aitrik'. The text here should probably read 'Wester Kirk and New Kirk of Ettrick'.
[19] *Records of the Regality of Melrose*, iii, p. 137 reads 'as yit reducit ...'.
[20] *Records of the Regality of Melrose*, iii, p. 137 also renders this 'cheis of deid sheip'. 'Dead Side' is located at NT2718.

to' 8 stones. 'Item, the salmond ever[21] gaddert to the meit fische of the [*blank*] nevir being under assedatioun extending to small number. Item, the yairdis and houssis in to the handis of the abbot and convent be custome and use. Item, sum hay to small quantitie to the use of the place win upoun the expenssis thairof.'

'The rentall of the landis and kirkis in the west cuntrie pertenyng to the hous of Melros.'
The mails of lands of Kylsmure and Barmure extends in the year to £324. The lands of the Monkland in Cawrik pertaining to Melros, extends in the year to £107 13s 4d. The lands of the Monkland in Niddisdaill 'callit Dunscoir' and the kirk thereof extends in the year to £86 13s 4d. 'Summa particule', £518 6s 8d.

[*In margin*, 'Meill']
The teinds of the kirk of Uchiltrie extends in the year to 34 c. The teinds of the kirk of Mauchling extends in the year to 34 c.

'The sowme of the haill silver in use of payment as als nocht in use of payment extendis in the haill to' £1,758.[22] 'The sowme of quheit', 14 c. 9 b. 'The sowme of beir', 56 c. 5 b. 1 f.[23] 'The sowme of meill', 78 c. 13 b. 1 f. 'The sowme of aites', 44 c. 10 b. 'The sowme of capones', 104 [124, *fo 207v*]; 'thairof out of use of payment', 84. 'The sowme of pultrie', 620 [660, *fo 207v*]. 'The sowme of butyre', 105 stones. The sowme of salt', 8 c. 'all out of use of payment. The sowme of peitis', 340 laids [400 laids, *fo 207v*]. 'The sowme of cariagis', 500 [615, *fo 207v*].

'Memorandum, of thir foirsaidis sowmes of victuall alsweill as of money to be deducit and defalkit as efter followis.
Item in the first, to' 11 'religious monkis and thre portionares seculares as Williame Ormestoun, James Schaw, Patrik Hardy, extending yeirlie in money, ilk persoun as thei have had lang be assignatioun, to' 20 merks; 'the haill sowme' £186 13s 4d. 'Item, to ilk persoun of the saidis' 14 'persounes', 1 c. bere, 'cuntre met as the lordis decreit thairupoun proportis, extending in the haill in generell met to' 17 c. 3 b. 2 f. 3 p. 'or thairby. Item, to ilk ane of the saidis persounes', 4 b. wheat 'cuntre met extending in the haill in generall met to' 4 c. 14 b. 2 f. 2 p. 'or thairby with tua chalder meill to the monkis. Item, to be deducit of thir premissis be assignatioun maid to the lord Seytoun,[24] the haill mailles of Kylismure, Barmuir, the landis of Carrik and Niddisdaill extending to' £518 6s 8d 'togidder with the teynd meill of the kirkis of Mauchling and Ouchiltrie extending to' 68 c. meal 'and that in contentatioun of ane pensioun grantit to the lord Seytoun and his barnes of' 1,600 merks. 'Item, to be defalkit and deducit', 5 c. wheat; 3 c. meal, 'gevin in pensioun to the erle of Glencarne[25] and his sone and tane up fra the lard

---

[21] *Records of the Regality of Melrose*, iii, p. 138 renders this 'ever [?]'.
[22] The correct calculation appears to be £1,758 8s 5d.
[23] The correct calculation appears to be 55 c. 15 b. 1 f.
[24] George, 5th lord Seton, *Scots Peerage*, viii, pp. 585-588.
[25] Alexander Cunningham, 4th earl of Glencairn, *Scots Peerage*, iv, pp. 239-241.

of Bas of the landis of Grange. Item, to be defalkit and deducit of the teyndis of [*blank*[26]] and Caveris albeit thei be riddin and nocht in use of payment, landis of Ringwodfeild and Urdis[27] adjacent upoun' 8 c. oats 'be ressoun of devastatioun of the [*blank*]. Item, the officeris feis', £20. 'Summa of silver defaikit extendis to' [£]720.[28] 'Summa of meill defakit extend[is to]' 60[29] [*blank*].[30] 'Summa of the quheit defalkit extendis to' 14 b. 2 f.[31] 'Summa of beir defaikit', 17 c. 3 b. 3 f. 3 p.[32] 'Item, defaikit of butyre to the monkis, ilk persoun thrie stane. And sua restis de claro to be deduct betuix the thrid and the tua pairt in silver', £1,033 8s 5d; 'and heirof nocht in use of payment the sowme of' £519 2s 3d, 'sua restis in use of payment' £514 6s 2d. 'Item, restis to be devydit of the quheit', 4 c. 10 b. 1 f.[33] 'Sua restis to be devydit of beir', 39 c. 1 b. 2 f.[34] 'Sua restis to be deduct of meill', 5 c. 13 b.[35] 'Sua restis to be deduct of aites', 36 c. 10 b.

'The divisioun of this rentall of the abbay of Melros
  Summa of the money', £1,697 11s 5d;[36] third thereof, £565 17s 1½d. 'Summa of the salt', 3 c. [8 c., *fos 207v, 208v*]; third thereof, 1 c. 'Summa of butter', 50 stones [105 stones, *fos 207v, 208v*]; third thereof, 16⅔ stones. 'Summa of the peitis', 340 laids [400 laids, *fo 207v*]; third thereof, 113⅓ laids. 'Summa of cayne foullis', 380 [660, *fo 207v*; poultry, 620, *fo 208v*]; third thereof, 126.[37] 'Summa of capones', 24 [124, *fo 207v*; 104, *fo 208v*]; third thereof, 8.

BEDRULE, PARSONAGE AND VICARAGE OF,

(SRO, Vol. a, fo 212r[38])
(SRO, Vol. b, fo 148r)

The parsonage and vicarage of Redrowll extends to £20, 'pertenyng to sir Williame Tod, as the samin producit proportis.'

Calculation of third: In the whole, £20; third thereof, £6 13s 4d.

---

[26] SRO, Vol. a is blank and SRO, Vol. b is damaged at this point. Vol. a, fo 206r reads 'Hassinden and Caveris' (see above, p. 207).
[27] I.e. Ladyurd, Lochurd, Netherurd, etc. *Records of the Regality of Melrose*, iii, p. 139, renders this 'uthers'.
[28] The correct calculation appears to be £725.
[29] The correct calculation appears to be 71 c.
[30] SRO, Vol. b is damaged at this point.
[31] The correct calculation appears to be 9 c. 14 b. 2 f. 2 p.
[32] The correct calculation appears to be 17 c. 3 b. 2 f. 3 p.
[33] The correct editorial calculation appears to be 4 c. 10 b. 1 f. 2 p.
[34] The correct editorial calculation appears to be 38 c. 11 b. 1 f. 1 p.
[35] The correct editorial calculation appears to be 7 c. 13 b. 1 f.
[36] This calculation is difficult to verify as it is not clear which sub-totals have been counted.
[37] ⅔ have been omitted from this calculation.
[38] SRO, Vol. a, fos 210v-211v are blank.

WILTON, PARSONAGE OF,

(SRO, Vol. a, fo 212r)
(SRO, Vol. b, fo 148r)

Parsonage of Weltoun

Calculation of third: In the whole, £120; third thereof, £40.

LINTON, PARSONAGE OF,[39]

(SRO, Vol. a, fo 212r)
(SRO, Vol. b, fo 148r)

Parsonage of Lintoun pertaining to Mr Somerwell [*i.e.* Thomas Somerville[40]].

Calculation of third: In the whole, £26 13s 4d; third thereof, £8 17s 9⅓d.

ECKFORD, VICARAGE OF,[41]

(SRO, Vol. a, fo 212r)
(SRO, Vol. b, fo 148r)

'Inprimis in Cavertoun, tua husband landis, the yeirlie maill thairof', 40s. The teind of the same, 40s. In Haucheid, 'ane pairt of land callit Preistis Croun, payand of maill', 10s. The teind thereof, 10s. At the kirk, one croft, the rental thereof, 5s. The teind of the same, 5s. Total, £5 10s.

Signature: 'Ita est Joannes Wilsoun, vicarie de Ekfuirde.'

ASHKIRK,
PARSONAGE AND VICARAGE OF,

(SRO, Vol. a, fo 212v)
(SRO, Vol. b, fo 148v)

The parsonage and vicarage of Erskirk given in by M[r] Thomas Archibald, factor thereof, the said parson and vicar being in France. 'Quhilk personage and vicarege set to Walter Scot of Syntoun and Walter Scot of Didschaw for the sowme of' £120 yearly, 'and deducit thairof to the minister yeirlie', 20 merks.

Signature: 'Thomas Archibald, procuratour for sir Johnne Mown, persoun of Erskirk.'

Calculation of third: In the whole, £120; third thereof, £40.

---

[39] Cf SRO, Vol. a, fo 214v (see below, p. 215).
[40] See below, fo 214v (p. 215).
[41] Cf SRO, Vol. a, fo 215r (see below, p. 215).

ABBOTRULE, PARSONAGE AND VICARAGE OF,

(SRO, Vol. a, fo 212v)
(SRO, Vol. b, fo 148v)

Parsonage and vicarage of Abbotisroll pertaining to Mr Alexander Creichtoun.

Calculation of third: In the whole, £33 6s 8d; third thereof, £11 2s 2⅔d.

ANCRUM, PARSONAGE OF,[42]

(SRO, Vol. a, fo 212v)
(SRO, Vol. b, fo 148v)

Parsonage of Ancrum

Calculation of third: In the whole, £266 13s 4d; third thereof, £88 17s 9⅓d.

'Seik the deanrie of Breichen in Forfar.'[43]

SOUTHDEAN, PARSONAGE AND VICARAGE OF,

(SRO, Vol. a, fo 213r)
(SRO, Vol. b, fo 149r)

The parsonage and vicarage of Soudoun extends to 40 b. meal; 40 teind lambs.

Signature, 'Mr Hew Douglas, persoun of Soudoun.'

Calculation of third: In the whole, 40 b. meal; 40 teind lambs; third thereof, 13⅓ b. meal; 13⅓ lambs.

MAKERSTOUN, VICARAGE OF, ROBERTON, VICARAGE OF,[44]

(SRO, Vol. a, fo 213r)
(SRO, Vol. b, fo 149r)

Rental of the vicarage of McCarstoun lying in the sheriffdom of Roxburcht, and Roberttoun within the sheriffdom of Lanark.

'Vicarege of McCarstoun in the haill', 9 merks.
'Vicarege of Roberttoun in the haill', 12 merks.

Pertaining to Martin Rutherfuird.

---

[42] Cf SRO, Vol. a, fos 164v, 239v; NLS, fo 82r (see above, p. 159; below, pp. 236, 365).
[43] The rental of the deanery of Brechin is to be found in NLS, fo 82r (see below, p. 364).
[44] Makerston and Roberton, in the sheriffdom of Lanark, were both appropriated to Kelso abbey; hence their inclusion among rentals for Roxburgh.

MOREBATTLE, PARSONAGE AND VICARAGE OF,

(SRO, Vol. a, fos 213v-214v)
(SRO, Vol. b, fos 149v-150v)

'The rentale of Merbotill[45] maid be Mr Johnne Lauder, archedein of Tevi[ot]dale, anno 41 [*i.e.* 1541].'

'In the first, George Ker of Corbethous for the haill teyndis of Corbeithous and for the half teyndis of Cruikitschawis with Primsyd Myln and als for the haill teyndis of Primsyd toun with Wester Fowmerden [*i.e.* Fourmartdean] and the haill teyndis of the landis of Wester Guideltoun, for all thir teyndis abonewrittin', £32. 'Attour the teyndis abonewrittin set to the said George, the hale vicarege for' £58, 'of the quhilk the vicar pensionarie for his pensioun sould have yeirlie' £8 and sir William Young, curate of Merbottill, for his fee yearly, £10. 'Set to George Ker of Gaitschaw, the haill teyndis of the toun and pertinentis of Merbottill except the teynd of Corbethous and Kewbog with thair pertinentis respective, set to the said George Ker of Corbethous and Sir Andro Ker of Littilden and als set to the said George Ker of Gaitschaw, the hale teyndis of Gaitschaw toun and the half teynd of Crukit Schawis with Prymsyd Mylne for' £36. Andrew Ker of Schilstounbreist for the lands of Merbottill, £5. Growethe Hauch and Wyd Hoppin in the hands of James Hoppringill for £5 'and the uther half of the saidis landis Johnne Tait for uther' £5. Hallis Syd, the whole teinds thereof set to Robert Hall for 40s. Wodsyd set to Andrew Young for 46s 8d. Eister Fowmerden set to George Davidsoun for 40s. Creychtoun 'within the handis of Robert Ker, Wakkeris Burn for' £6 13s 4d. Cliftoun in the hands of George Hoppringill 'duelland in Cala for the ane half of the eister pairt of Cliftoun', £10. The other half of the easter part of Cliftoun in the hands of John Hoppringill for £4. The other half of the east part of Cliftoun in the hands of Andrew Hoppringill for £4. 'Item, the tane half of the west pairt of Cliftoun' in the hands of James Young for £5 6s 8d, 'that now Malie Davidsoun occupyis.' John Hoppringill 'of the settin of the toun', £10. Of the other half of the west part of Cliftoun in the hands of William Young for £4 13s 4d. 'The tua merkland of the west end of Cliftoun quhilkis wes occupyit be Adame Dalgleis now occupyit be David Young in Oxinsyd'. Facter Schawis in the hands of Andrew Young for £3. The whole teinds of Oxmansyd in the hands of James Ker in Mersiltoun [*i.e.* Mersington] for £3 6s 8d. The Lobricht Hous alias Louchsyd in the hands of Robert Burne for 40s. The other half thereof in the hands of Christian Hoppringill for 40s. The whole teinds of Quhiltoun in the hands of George Hoppringill 'duelland in the Merbottill' for £13 6s 8d. The Cowhog [*recte*, Cowbog], the whole teinds thereof with the pertinents in the hands of Andrew Ker of [*blank*] for' £5 3s 4d. David Ker, 'George [his] sone', of the Loycht Towir for all the teinds of the toun of Merbottill.

---

[45] The parsonage and vicarage of Morebattle formed the prebend of the archdeacon of Teviotdale, Cowan, *Parishes*, pp. 151-152. John Lauder was archdeacon of Teviotdale, 1534 - 1551; his successor (whose signature appears on p. 215 below) was John Hepburn, 1544 - 1564 x 1565. Watt, *Fasti*, pp. 178-179.

Signature: 'M[r] Johnne Hepburn.'

'Merbotle callit the archidenrie of Teviotdaill by the teindis of Merbotle salfand ij merkland of the west end of Cliftoun.'

Calculation of third: In the whole, £226 6s 8d;[46] third thereof, £73 15s $6\frac{2}{3}$d.[47]

LINTON, PARSONAGE AND VICARAGE OF,   (SRO, Vol. a, fo 214v)
       (SRO, Vol. b, fo 150v)

Rental of the parsonage and vicarage of Lintoun,[48] lying within the diocese of Glasgow and sheriffdom of Roxburcht, given up up by Thomas Somervell, parson and vicar thereof.

'The personage and vicarege extendis yeirlie to the said persoun to' £40 'money of this realme, and the 3 pairt thairof extendis to' 20 merks 'money foirsaid, and the samin wes set of auld be me and my predicessouris to umquhill Mark Ker of Litilden and umquhill Sir Andrew Ker of Hirsill, knycht, his sone, and to thair aires and assignais for' £40 'allanerlie and this I testifie be thir presentis, subscrivit with my hand at Cothlie', 24 June 1577. 'Sic subscribitur Thomas Somervell, persoun of Lintoun.'

Calculation of third: In the whole, £40; third thereof, £13 6s 8d.

ECKFORD, VICARAGE OF,[49]   (SRO, Vol. a, fo 215r)

Rental of the vicarage of Ekfuird

'In the first, tua husbandlandis liand in the toun and feild of Camertoun [*recte*, Caverton] payand yeirlie of maill', 40s. The teind of the same worth 40s. 'Certan peces of land callit the Preistis Croun and Preistis Hauches besyd the Hewcheid' paying of mail 10s. The teind thereof worth 10s. 'Thre pecis of land lyand besyd the kirk' extending in mail to 5s. The teind thereof worth 5s. Total, £5 10s.

Signature: 'Ita est Johannes Wilsoun, vicarius dicte ecclesie teste, manu propria.'

Calculation of third: In the whole, £5 10s; third thereof, 36s 8d.

---

[46] The correct calculation appears to be £220 16s 8d.
[47] The correct calculation appears to be £75 8s $10\frac{2}{3}$d.
[48] Cf SRO, Vol. a, fo 212r (see above, p. 212).
[49] Cf SRO, Vol. a, fo 212r (see above, p. 212). This rental does not appear in SRO, Vol. b.

JEDBURGH, ABBEY OF,[50]

(SRO, Vol. a, fos 216r-217v[51])
(SRO, Vol. b, fos 151r-152v)

'The rentale of mailles and annuellis, corne mylnes and walkmylnes pertenyng to the abbay of Jedburcht with the maist of the Spettell as efter followis.'

'Item, the baronie of Ustoun with the Spettall Manis of mailles, annuellis, corne mylnis, and walkmyln extending in the yeir to' £200. The barony of Wyndingtoun[52] with corn mill in the year, £36 3s 4d. The barony of Ancrome with the mill thereof in the year, £58 6s 8d. The barony of Belscheis with the mill thereof in the year, £45. The barony of Repperlaw in the year with the mill, £26. The barony of Abbotroull with the mill in the year, £40.

'The rentale of the teyndis of the abbay of Jedburcht and kirkis thairof efter the assedatiounes and estimatiounes'. The kirk of Jedburcht with the pendicles, 16 b. wheat; 8 c. 4 b. bere; 13 c. meal.

The teinds of the said kirk set for money
The teind of Over Neisbit set to Alexander Home of Howtounhall 'in the yeir for' £10. The teind of Hynnoisfeild set to Mr Alexander Home for £20. 'The half teind of Jedburcht Syd set to Williame Alysoun in the yeir', £6. 'The quarter and half quarter of the teynd of Bolsches set to Robert Carnecors of Commislie', £10 in the year.

The kirk of Ekfuird, 10 b. wheat; 5 c. bere; 7 c. meal. The kirk of Hownem, 24 b. bere; 3 c. 8 b. meal. Oxemon kirk: Oxnen, 3 c. bere; 3 c. meal. The kirk of Langnewtoun, 8 b. wheat; 1 c. 8 b. bere; 3 c. meal. The kirk of Dunmay [*i.e.* Dalmeny]: The kirk of Dunmay set for money to Thomas Hammiltoun of Preistisfeild in the year, £160. The kirk of Sibbilbe, £6 13s 4d. The kirk of Wauchope in the master [of] Maxwellis hands. The kirk of Casseltoun 'waist'.

'Item, to' 4 'brether, ane of thame ane auld man that hes double portioun, ilk ane of thame' 6 b. wheat; 16 b. bere; £20 money, 'and the auld man double'. To Patrick Hardie to his fee, 8 b. bere; 4 b. 2 f. meal; £6 money. 'To the servand cuke of the place', 8 b. bere; 4 b. 2 f. meal; £5 money. 'To the servand that keipis the place', 8 b. bere; 4 b. 2 f. meal; £5 money. 'To the gardnares fee', 8 b. meal.

'The warkmenis fee in the place of Jedburcht as followis: To the wricht in the yeir', 8 b. malt; 8 b. meal; £6 13s 4d. 'To the laremair in the yeir', 8 b. malt; 8 b. meal; £6 13s 4d money. To the bailie for his fee in the year, £10. To the officer of the barony of Abbotroull for his fee in the year, £3 6s 8d. To the officer

---

[50] Cf SRO, Vol. a, fo 219r (see below, p. 218). For the assumption of the third (with that of Restenneth), see below, p. 218.
[51] SRO, Vol. a, fo 215v is blank.
[52] 'Windingtoun', *RMS*, vi, no 1721.

of the barony of Ancrum, his fee in the year, £4 6s 8d. To the officer of the barony of Ustoun [*i.e.* Ulston] 'siclyk', £3 6s 8d. 'To the officiar of Jedburcht toun, his fee in the yeir', 40s. The Quheilland in Leddesdaill, £10 'waist'. For the vicarage of Jedburcht yearly, £20. 'To the persoun of Ancrom be yeir for ane pairt of the teyndis pertenyng to the kirk of Ancrom', £3 6s 8d. 'To the pentionare of the kirk of Jedburcht in the yeir', £13 6s 8d. 'To the pentionarie of the kirk of Crelling in the yeir', £13 6s 8d. 'To the pentionarie of the kirkis of Nesbites in the yeir', £20. 'To the pentionarie of the kirk of Prenderleth in the yeir', £13 6s 8d. 'Item, the laird of Pharinhirst haldis in his hand in the yeir', 4 c. 8 b. wheat, bere and meal. 'Item, gevin yeirlie to the College of Justice for Jedburcht', £21.

Signature: 'Nicoll Hume, chalmerlane of Jedburcht.'

RESTENNETH, PRIORY OF,[53]   (SRO, Vol. a, fos 217v-218v, 220v)
(SRO, Vol. b, fos 152v-153v, 155v)

[*In margin*, 'Restant']

Rental of Restenn[eth] lying within the diocese of Sanctandrois 'with annuellis and custoumes'.

'Item, the kirk of Restenit in victuall be yeir', 11 c. 10 b., 'tua pairt mele, 3 pairt beir [*i.e.* ⅔ meal, ⅓ bere], of the townes as efter followis, viz., Clochow, Carsburne, Muirtoun, Eilark, the croftis of Forfar, Balunschamour [*i.e.* Balmashanner], Caldrum [*i.e.* Caldhame], Petrochie [*i.e.* Pitreuchie], Auchterforfar, Burnesyd, Dod[54] and Halkartoun, Forfartoun.'

The teinds set for money of the kirks of Restenet, namely: The teind of Craignathow set for £13 6s 8d. The teind of Curbeg set for £13 6s 8d. The teind of Loure set for £20.

    Ablemno kirk: The kirk of Ablemno set for money to the laird of Bonitoun for 160 merks.

    Drummynald kirk: The teind of Drummynald kirk set to Andrew Gray for 50 merks. The vicarage of Restrint yearly, 10 merks.

Temporal lands
    The temporal lands of the Manes and feu lands with the annuals 'be yeir extending to' £163.

Total money, £168 16s 8d. Wheat, 2 c. 2 b. Bere, 23 c. 3 b. Meal, 36 c. 13 b.[55]

---

[53] Restenneth, in the sheriffdom of Forfar, was a dependent cell of Jedburgh abbey, which explains its inclusion in rentals for Roxburgh. Cf Cowan and Easson, *Medieval Religious Houses*, pp. 95-96.
[54] 'Henwode alias *lie* Dod', *RMS*, v, no 1409.
[55] These totals appear to have no relationship with the figures given above.

'Item, to the bischope of Sanctandrois[56] of ken silver in the yeir of the kirk of Ablemno', £3 6s 8d. 'Item, to the sklater for uphalding of the place of Restrint for his fie in the yeir', 1 c. meal; £6 13s 4d. 'Item, to the commoun servandis that keipis the place of Restrint and yairdis of the samin for thair feis in the yeir', £40. 'Item, gevin yeirlie to the College of Justice for Restrint', £11 4s.

Signatures: 'Thome Trotter, chalmerlane of Restrant. Presentit be M[r] Johnne Hume in my lord of Jedburchtis[57] name, M[r] Johnne Home.'

'Assumptioun and Divisioun'
Assignation of the third of the abbacy of Jedburcht with Restenneth: Jedburcht with Resteyneth, third of money, £324 16s 8d.[58] Take the barony of Wstoun with the Spittell Manes, corn mill and waukmill for £200. The temporal lands of Restenneth with the mains, feu lands and annuals for £163. 'Gif in' £38 3s 4d.

Wheat, third thereof, 11 b. 1 f. $1\frac{1}{3}$ p.[59] Bere, third thereof, 7 c. 10 b. $3\frac{2}{3}$ p.[60] Meal, third thereof, 12 c. 4 b. 1 f. $3\frac{1}{8}$ p.[61] Take the victuals out of the kirk of Jedburcht with the pendicles paying 16 b. wheat; 8 c. 4 b. bere; 13 c. meal.

'Omittit cainis, custumes, caponis, pultrie and all uther dewteis. Charge the abbot[62] to pay' £560 15s 2d; victual, 39 c. 12 b. 1 f. 3 p.

JEDBURGH, ABBEY OF,[63]

(SRO, Vol. a, fos 219r-221r)
(SRO, Vol. b, fos 154r-156r)

'The rentale of the abbacie of Jedburgh and callis [cellis, *SRO, Vol. b*] of Restenneth and Cannabie annexit thairto. To the quhilk benifice ane noble and potent lord, Alexander, lord Home,[64] is provydit, sua far as concernis the spiritualitie of the samin and gevin up be him selff to the commissiouneris of Parliament undersubscryveand.'[65]

---

[56] John Hamilton, archbishop of St Andrews, 1546 - 1571, Watt, *Fasti*, p. 298.
[57] Andrew Hume was commendator of Jedburgh at the Reformation, *RSS*, v, no 1063.
[58] This calculation is difficult to verify because of the discrepancies in the figures given for Restenneth.
[59] This is one third of 2 c. 2 b. wheat of Restenneth, not of Jedburgh and Restenneth combined.
[60] In the text, the fraction is mistakenly rendered 'half pect and 3 of the 3 of half pect'. The correct calculation appears to be 7 c. 11 b. $2\frac{2}{3}$ f., which is one third of 23 c. 3 b. bere of Restenneth, not of Jedburgh and Restenneth combined.
[61] In the text this fraction is rendered '3 of the 3 of half p.' [i.e. $\frac{1}{8}$ p.]. The correct calculation appears to be 12 c. 4 b. $1\frac{1}{3}$ f. which is one third of 36 c. 13 b. meal of Restenneth, not of Jedburgh and Restenneth combined.
[62] See above, n. 57.
[63] Cf SRO, Vol. a, fo 216r.
[64] Alexander, 6th lord and 1st earl of Home, *Scots Peerage*, iv, pp. 463-465.
[65] See below, p. 222, n. 78.

'In the first, the said lord Home is onlie provydit to the spiritualitie of the said abbay quhilk is onlie the kirkis of Jedburcht, Exnem [*i.e.* Oxnam], Hownum, Hecfuird [*i.e.* Eckford], Hopkirk and Belsis. The kirk of Dunmanie [*i.e.* Dalmeny] and patronage thairof pertenis to Mr Thomas Hammiltoun of Drumcarnie, advocat to our soverane lord, and the kirk of Langnewtoun of lang tyme bygane is evictit be Williame, erle of Mortoun,[66] be decreit gevin thairupoun aganis the last abbote.'[67]

The kirk of Jedburcht with the pendicles thereof called Nisbet and Craling

'Inprimis, the teynd schaves of Stewartfeild and Over Manis of Ulstoun', 8 b. bere; 8 b. meal. The teind sheaves of Ulstoun toun, 4 b. bere; 4 b. meal. The teind sheaves of Wollis Over and Nather, 1 b. wheat; 3 b. bere; 8 b. meal. The teind sheaves of Over Craling, 4 b. wheat; 8 b. bere; 12 b. meal. The teind sheaves of Samuelstoun, 4 b. bere; 8 b. meal. The teind sheaves of Rannestoun, 1 b. bere; 1 b. 2 f. meal. The teind sheaves of Scraisburcht and Hunthill, 4 b. bere; 4 b. meal. The teind sheaves of Edyerstoun, 4 b. bere; 4 b. meal. The teind sheaves of Auld Jedburcht, 4 b. bere; 4 b. meal. The teind sheaves of Lyntolie, 2 f. bere; 2 f. meal. The teind sheaves of Hundolie, 8 b. bere; 8 b. meal. The teind sheaves of Langtoun, 8 b. wheat; 16 b. bere; 24 b. meal. The teind sheaves of the lordship of Pherniherst, 8 b. bere; 8 b. meal. The teind sheaves of Glennislandis, 4 b. bere; 4 b. meal.

'Summa of the quheit', 13 b. 'Summa of the beir', 5 c. 2 f.[68] 'Summa of the meill', 6 c. 6 b.

Teinds of the said kirk set for silver

'Inprimis' the teind sheaves of Nather Craling, £26 13s 4d. The teind sheaves of Bonjedburcht, £26 13s 4d. The teind sheaves of Suynie, £4. The teind sheaves of Hynhousfeild, £20. The teind sheaves of Castelfeild, £20. The vicarage of the said kirk, £20. The teind sheaves of Jedburchtsyd, £12. The teind sheaves of Over and Nather Nisbitis, £20. 'Summa of the silver dewtie', £149 6s 8d.

The kirk of Oxnum

The teind sheaves of Oxnum toun, 8 b. bere; 12 b. meal. The teind sheaves of the Mans of Oxnum, 4 b. wheat; 8 b. bere; 12 b. meal. The teind sheaves of Cloiss, 4 b. bere; 4 b. meal. The teind sheaves of Suynesyd, 4 b. bere; 8 b. meal. The teind sheaves of Overtoun, 2 b. bere; 4 b. meal. The teind sheaves of Newbigging and Sykis, 8 b. bere. The teind sheaves of Do[l]phingtoun, and Fala, 12 b. meal; 12 b. bere. The teind sheaves of Prenderleyth, 1 b. bere; 2 b. meal.

---

[66] William Douglas, 5th earl of Morton, *Scots Peerage*, vi, pp. 371-375.
[67] I.e. Andrew Home, *RMS*, vi, no 501.
[68] The correct calculation appears to be 4 c. 12 b. 2 f.

Total wheat, 4 b. Total bere, 2 c. 15 b. Total meal, 3 c. 6 b.

The kirk of Hownum
The teind sheaves of 'Hownum towne', 7 b. bere; 8 b. meal. The teind sheaves of Beirhope and Philgour, 1 b. bere; 1 b. meal. The teind sheaves of Over Quhittoun, 3 b. bere; 3 b. meal. The teind sheaves of Nather Chatto, 4 b. bere; 4 b. meal. The teind sheaves of Over Chatto, 2 b. bere; 2 b. meal. The teind sheaves of Caiphopetoun, 4 b. bere; 4 b. meal. The teind sheaves of Caiphoipmans, 1 b. bere; 4 b. meal.
Total bere, 1 c. 5 b.[69] Total meal, 1 c. 10 b.

The kirk of Ekfuird
The teind sheaves of Ekfuird and Gryimschaw, 4 b. wheat; 22 b. bere; 22 b. meal. The teind sheaves of Cavertoun, Sesfuird and Ormestoun, 4 b. wheat; 2 c. bere; 3 c. meal. The teind sheaves of Cleucheid [Heuchheid, *SRO, Vol. b*], 3 b. bere; 5 b. meal.
Total wheat, 8 b. Total bere, 3 c. 9 b. Total meal, 4 c. 11 b.
The teinds of the said kirk set for silver: The teind sheaves of Mowmanis, £5 13s 4d.

The kirk of Hopkirk
The teind sheaves of the Wollis, 2 b. 1 f. bere; 2 b. 1 f. meal. The teind sheaves of Westleis, 3 f. bere; 3 f. meal. The teind sheaves of Bullerwoll [*i.e.* Billerwell], 2 b. bere; 2 b. meal. The teind sheaves of Harroull and Toun of Roull [*i.e.* Town-o'-rule], 2 b. bere; 4 b. meal. The teind sheaves of Hoppisburne and Weindis, 2 b. bere; 2 b. meal. The teind sheaves of Litle Gledstanes, 1 b. bere; 1 b. meal. The teind sheaves of Hova, 1 b. meal. The teind sheaves of Steinlethe, 1 b. bere; 1 b. meal. The teind sheaves of Appodsyd, Hathronesyd and Harwod, 4 b. bere; 4 b. bere [*recte*, meal]. The teind sheaves of Wauchop, 4 b. bere; 4 b. meal.
Total bere, 1 c. 3 b. Total meal, 1 c. 6 b.

The kirk of Belschis
The teind sheaves of the Pennakle quarter, 4 b. bere; 4 b. meal. The teind sheaves of the Myre half quarter, 4 b. bere, 4 b. meal. The teind sheaves of the Parkhill quarter, 2 b. bere; 4 b. meal.
Total bere, 10 b. Total meal, 12 b.
The teinds of the said kirk set for silver: The teind sheaves of the Mylnerig quarter, £10. The teind sheaves of Rafflat, 53s 4d. 'Summa of the silver dewtie', £12 13s 4d.

'The cell and priorie of Restennethe hes onlie thrie kirkis, to wit, Forfar, Donynauld and Abirlemno quhilkis ar all set for silver dewteis, viz. The kirk of

---

[69] The correct calculation appears to be 1 c. 6 b.

Forfar payand be yeir' £100. Donynald 'payand be yeir' £33 6s 8d. Abirlemno, 'payand be yeir' £106 13s 4d.

'Summa of the haill money of the said abbay and priorie of Restenneth is' £407 13s [4d]. 'Summa of the haill quheit', 1 c. 9 b.[70] 'Summa of the haill beir', 13 c. 5 b. 2 f.[71] 'Summa of the haill meill', 16 c. 9 b.[72]

'Defalcatioun

To be defesit of the money, the Lordis contributioun', £32 4s; 'sua restis of fre money', £375 2s 8d;[73] third thereof, £125 0s 10½d. Third of the wheat, 8 b. 1 f. 1⅓ p. Third of the bere, 4 c. 7 b. 2⅓ p. Third of the meal, 5 c. 8 b. 1 f. 1⅓ p.

'The fyft pairt of the money and victuall foirsaid now pertenis to the king in place of the monkis portiounes and extendis as followis, viz. The fyft penny of the money', £75 0s 6d 'ob'.[74] 'The fyft pairt of the quheit', 5 b. 'The fyft pairt of the beir', 2 c. 5 b. 3 f. 1 p.[75] 'The fyft pairt of the meill', 3 c. 5 b.

Priory of Cannabie

'The priorie of Cannabie hes thrie kirkis, to wit, Sibbalbie, Wachopdaill and Cassiltoun, all waist and payand na dewtie at this tyme as thei have thir monie yeiris bygane. Sic subscribitur, Alexander, lord Home.'

> 'We, the commissiouneris undersubscryveand contenit in the act of Parliament[76] for trying of the estait of the benifice of Jedburcht and to mak ane new just thrid thairof, as the act of our said commissioun of the dait the xv day of November last bypast mair fullie proportis, findis be cognitioun tane be us according thairto that the rentis of the spiritualitie of the said benifice now gevin up be Alexander Home, present titular thairof, extendis to the availlis and quantiteis foirsaidis contenit in this present rentall, and ordanes the samin rentall to be registrat in the clerkis buikis of the Collectorie[77] and the assignatiounes to be gevin furth this yeir and in tym cuming conforme to the thrid thairof now set doun and allowit be us, provyding alwayis that quhen the pensionares sall happin to deceis that the thrid of thair pensiounes sall accres to our soverane lard and salbe eikit to

---

[70] This calculation is correct.
[71] The correct calculation appears to be 14 c. 10 b. 2 f.
[72] The correct calculation appears to be 18 c. 3 b.
[73] This would be correct if the total above were £407 6s 8d.
[74] Instead of 'obol', the correct fraction is ⅔.
[75] This figure cannot be calculated from the total given in the text.
[76] In 1600, Parliament ratified the gifts and provisions to Alexander, lord Home, of Coldingham priory and Jedburgh abbey, and appointed commissioners to examine the rentals and register the thirds in the books of the Collectory, *APS*, iv, p. 244.
[77] The Collectory was the department responsible for accounting and disbursing the revenues derived from the collection of the thirds of benefices.

the thrid contenit in the said rentall, provyding lykwayis that quhatevir is omittit furth of this rentall sall pertein to our said soverane lord conforme to the natour of the omissiounes in the first assumptiounes and actis maid thairanent. In witnes quhairof we have subscryvit thir presentis with our handis as followis, at Edinbrucht' 3 December 1600. 'Sic Subscribitur: Thomas Lyoun off Auldbar; Cranstounriddell; Halyruidhous; M[r] T. Hammilton; Jo[hnne] Prestoun.'[78]

KELSO, ABBEY OF,[79] (SRO, Vol. a, fos 223r-230r[80])
(SRO, Vol. b, fos 158r-164v[81])

'The rentall of the Kelso [sic] gevin in at Hadingtown', 1 March 1573/4 'be [blank].'

'The towne and landis of Kelso with the outsettis, cottagiis, houssis, yairdis, biggingis, bairnis, killis, custumes and unis[82] [uvynes, *fo 232r*; *i.e.* ovens] thairof,[83] the Admirall landis,[84] the mylis of Kelso, the fischingis of salmond of the netheir boit upoun the Water of Tueid with the ferrie cobill of Sproustoun [Roxburcht, *fos 232r, 235r, 240r, 243r*], the fischeing of salmond of the over boit upoun the said water, the ferrie cobill of Maxwell,[85] the landis callit the West croftis, the Brome Bank and the Hoitt [Hocht, *fos 205r, 235r, 240r*; Hote, *fo 243r*], the Conyngaires and Brumland, the landis of Angreflat, the Broun [*recte*, Brome] croft,[86] the Toun croft, the Blakbalk,[87] the West Medow, the Eist

---

[78] This letter resembles that relating to Coldingham included at SRO, Vol. a, fo 202r (see above, pp. 201-202). Sir Thomas Lyon of Auldbar had demitted the office of Treasurer by March 1596, and had been privy councillor till 1598; David McGill of Cranstoun Riddell, advocate (son of the former lord Advocate, who died in 1596); John Bothwell, commendator of Holyroodhouse; Thomas Hamilton of Drumcairn, lord Advocate; John Preston of Fentonbarns, Collector general. *Handbook of British Chronology*, pp. 188, 201; *Scots Peerage*, viii, p. 286; *Rotuli Scaccarii Regum Scotorum: The Exchequer Rolls of Scotland*, eds. J. Stuart and G. Burnett, *et al.*, 23 vols. (Edinburgh, 1878-1908), xxiii, pp. 148, 512; *RPC*, v, p. 288 n.; vi, pp. 98, 199.

[79] Cf SRO, Vol. a, fos 232r, 235r, 240r, 243r (see below, pp. 230, 233, 236, 239). See also *RMS*, iv, no 1905; and *Liber S. Marie de Calchou. Registrum Cartarum Abbacie Tironensis de Kelso, 1113 - 1567*, ed. C. Innes, 2 vols. (Edinburgh, 1846), ii, pp. 489-532; the rental printed there, c. 1567, differs in detail from the five rentals in the Books of Assumption.

[80] Fos 221v-222v are blank.

[81] SRO, Vol. b, fo 156v, is blank; fo 157r-v is also blank, except for several lines of deleted text, consisting of thirds of victuals.

[82] 'Duis', *Registrum Cartarum de Kelso*, ii, p. 489.

[83] 'Outsettis, cotagias, domos, hortos, edificia, horrea et ustrina dicte ville', *RMS*, iv, no 1905.

[84] 'Amarale-landis', *RMS*, iv, no 1905; 'Almarie Lands', *Registrum Cartarum de Kelso*, ii, p. 529; i.e. the lands of the almoner.

[85] 'Maxwellheuch', *RMS*, iv, no 1905, n.

[86] 'Brome croft', SRO, Vol. a, fos 235r, 240r; SRO, Vol. b, fos 158r, 171r, 174r, 176r (see below, pp. 233, 237, 239); 'Brumecroft', *RMS*, iv, no 1905. This place name does not appear in the rental in SRO, Vol. a, fo 232r.

[87] 'Blakbalk', *RMS*, iv, no 1905; 'blak bak croft', *Registrum Cartarum de Kelso*, ii, p. 489.

Hauch,[88] the tuentie pund land of Sproustoun pertenyng in the propertie to the abay of Kelso, the landis of Reddein, the plewlandis of Halden, the four husband landis of Lyntoun, the landis of Softlawis with the tua husband landis in Heytoun,[89] the plewlandis of Hoislaw,[90] the landis of Eleischeuch,[91] the landis of Cauldscheillis, the toun and landis of Lyndene with the mylne thairof, the landis of Greneheid, the landis of Quhytmuirhall, the landis of Quhytmuirtoun, the landis of Kipperlawr, the landis of Caveris, the landis of Newhall, the landis of Auldtoun[92] besyd Hawyk, the landis of Uniscroft [Unysclos, *fo 232r*; *i.e.* Ovenscloss] with all thair pertinentis thairof lyand within the lordschip of Kelso and schirefdome of Roxburcht, the landis of Dowglen, with the pertinentis lyand within the schirefdome of Drumfreis, the landis [of] Chapethill [*recte,* Chapelhill] lyand within the schirefdome of Peblis, the kirklandis of Litill Newtoun, the kirklandis of Nanetharne, the kirklandis of Gordone, the kirklandis of Greinelaw, the kirklandis of Sympryin, the kirklandis of Hume, the landis [of] Belleschill, the landis of Gyrig,[93] the landis of Brigheid [*recte,* Bogend, *fos 232v, 236r*; Boigend, *fo 240v*], the Mylne of Fogo with all and sindrie the pertinentis lyand within the schirefdome of Beruik unitit in ane tenandrie callit Sproustoun pertenyng in few to lord Frances Stewart as air,[94] be provisioun in the infeftment maid thairof to umquhill James, erle of Murray,[95] etc., payis be yeir', £634 0s 4d.

The lands of Halyden, £10. Quhytlawhous, £10. Fawdounsyd, £10. Prestoun, £6 13s 4d. Clarelawmanes 'quhilk of auld wes set for' £210 'is set in few to young Cesfuird for' £40. Langsyd, £3. Boudentoun, £31. The Mylne of Bouden, £5. The toun of Mydlane, £31. The Mylne of Mydlane, £7. 'The girs of Midlane', £8. 'Thir tua townes with thair mylnes payis ane dosoun of capones and nyne scoir pultrie.' The kirkland of Selkirk, 40s. The kirkland of McCarstoun, £10 13s 4d.[96] Steidrig [Todrik, *fos 235r, 243r*],[97] 40s. The lands in Edinbrucht, 26s 8d. The annual of Deringtoun, 5s. Caldraa, £4. The lands in Malestanis, 40s. Bowden and Ogstounis Over and Nather, £30. Bothill, Bothelhill [*i.e.* Bushelhill] and Hareheid, £22 8s. The annuals of Duddingstoun, £8. The annual of Balintrado, £5 6s 8d.

Total money, £884 13s 4d.[98] Total capons, 12. Total poultry, 180.

---

[88] 'Eschehauch', *RMS*, iv, no 1905.
[89] 'Softlawis vith the land of hitowe', *Registrum Cartarum de Kelso*, ii, p. 492.
[90] 'Horslaw', *RMS*, iv, no 1905.
[91] 'Bolden' and 'Mydlen' are included at this point in *RMS*, iv, no 1905.
[92] 'Altoun', *RMS*, iv, no 1905; 'altowne besyde hatrik', *Registrum Cartarum de Kelso*, ii, p. 490.
[93] 'zy rig', *Registrum Cartarum de Kelso*, ii, p. 491.
[94] Francis Stewart, son of John Stewart prior of Coldingham, received a gift of Kelso from the crown in 1567, *RSS*, v, no 3212.
[95] James Stewart, earl of Moray, gained a feu of property belonging to Kelso in 1569, *RMS*, iv, no 1905.
[96] Under 'Makerstoun Kirk for Syluer', *Registrum Cartarum de Kelso*, ii, p. 507 has 'Meirden Vester ... Steidrige ... The Towne and Mains of Langtownelaw ... '.
[97] 'Stodrik', *Registrum Cartarum de Kelso*, ii, p. 491.
[98] The correct calculation appears to be £883 13s 4d.

'Humby set in few of auld to Lawsoun of Humbie for payment to the convent in pittance silver', £9 6s 8d.

'Kirkis set for silver nocht acclamit be James Cunyghame[99] nor Mungo Grahame,[100] etc.': Hopcaltzie, £10. Lyntoun, £36 13s 4d. Cawdercher [*i.e.* Calder-Clere], £44. Petircultir, £13 6s 8d. Dunsyre, £10. Symontoun, £12. Roberttoun, £20. Wystoun, £16. Kilmaris, £59 13s 4d. Stabilgortoun, £8. Trailflat, £10. Drumgrie, £5. Horneden, £20. Nanetharne and Little Newtoun, £80. McCarstoun, £100. Sanct James Kirk, 40s. 'The teynd schaves of the kirk of Sprostoun to umquhill James, erle of Murray, for the space of' 19 years, £40. 'The kirk of Kelso set alsweill to the said umquhill erle for' £10. The vicarage of Nanetharne, £13 6s 8d. The vicarage of Kelso, £10. The vicarage of McCarstoun, £10. Total of kirks set for silver, £560.[101]

'Kirkis set pairtlie for silver and pairtlie for victuall':

Lyndene parsonage[102]

Cauldscheillis, 1 b. bere; 3 b. meal. Fadounsyd, 1 b. wheat; 5 b. bere; 8 b. meal. Lyndentoun[103] with the mains thereof and Brigheucht[104] 'quhilk of auld gave' 1 c. 2 b. bere; 2 c. 4 b. meal; 2 f. wheat, 'set be umquhill abot Williame Ker[105] for the sowme of' £8. Mosilie [*i.e.* Mossilee] and Blindlie, 1 b. bere; 6 b. meal. Farnylie and Calschaw,[106] 4 b. 3 f. bere; 9 b. meal. Gallowscheillis and Boldsyd[107] 'quhilk of auld gave' 1 c. bere; 2 c. meal 'is set be the said abbot Williame Ker for' £5. Langruik,[108] 4 b. bere; 1 c. meal. 'Vicarege of Lyndene disponit to Williame Ker be the king and his Regent.[109]

Summa of the money by the vicarege', £13. Total wheat, 1 b. Total bere, 15 b. 2 f.[110] Total meal, 2 c. 9 b.[111]

---

[99] In October 1566, Queen Mary ratified the gift of a pension for life from the revenues of Kelso abbey to James Cunningham, son of Alexander, 4th earl of Glencairn, *RSS*, v, no 871; *Scots Peerage*, iv, p. 241.

[100] In August 1566, Mungo Graham, servant of the king and queen, received from the crown a gift for life of the revenues of the benefice of Selkirk (annexed to Kelso abbey) in fulfilment of his pension, *RSS*, v, no 3044.

[101] The correct calculation appears to be £530.

[102] 'Reddene [Lindene] Kirk', *Registrum Cartarum de Kelso*, ii, p. 512.

[103] 'Heyndoun Towne', *Registrum Cartarum de Kelso*, ii, p. 512.

[104] 'Brige Hauch', *Registrum Cartarum de Kelso*, ii, p. 512.

[105] William Ker was appointed abbot in 1559, *RSS*, v, no 1428.

[106] 'Ferinylie and Calfschaw', *Registrum Cartarum de Kelso*, ii, p. 512.

[107] 'Boytsyde', *Registrum Cartarum de Kelso*, ii, p. 513.

[108] 'Langreynk', *Registrum Cartarum de Kelso*, ii, p. 513.

[109] James Stewart, earl of Moray, was Regent when Ker was presented, as reader, to the vicarage of Lindean in 1569, *RSS*, vi, no 718.

[110] The correct calculation appears to be 15 b. 3 f.

[111] The correct calculation appears to be 2 c. 10 b.

Boldene [*i.e.* Bowden], 'bayth personage and vicarege'

'Boldene toun, quhilk of auld payit' 4 c. meal 'is now intromettit with be the lord of Cesfuird[112] as ballie for hors corne', 4 c. oats. Midlanetoun 'quhilk of auld gave' 3 b. wheat; 1 c. bere; 12 b. meal, 'the chalder, is set be the said umquhill abbot[113] for' 2 c. bere; 2 c. meal. Clarelew 'quhilk of auld wes worthe' 8 b. wheat; 2 c. 2 b. bere; 3 c. 8 b. meal, is set to the young laird of Cesfuird by the parson of Cleische, 'administratour', for £6 13s 4d. Halydene with the pendicles 'intromettit with be the laird of Cesfuird, ballie, paying nathing thairfoir sen the deceis of King James the fyft'. Kippellaw 'quhilk of auld gave' 1 b. wheat; 3 b. bere; 5 b. meal, 'is set be the said umquhill abbot Williame Ker for' 2 b. bere; 3 b. meal. Prestoun, 3 b. bere; 5 b. meal. Mydlame Mylne, 2 f. wheat; 3 f. bere; 2 f. meal. Quhitlawhous, 4 b. bere; 8 b. meal. Vicarage of Boldene[114] 'intromettit with be Thomas Dunkesoun [*i.e.* Duncanson], minister'.

'Summa of money by the vicarege', £6 13s 4d. Total wheat, 2 f. Total bere, 2 c. 9 b. 3 f. Total meal, 3 c. 2 b.[115] Total oats, 4 c.

Mow, parsonage and vicarage

Mowtoun Manes and Corros [Corrost, *SRO, Vol. b*][116] 'quhilk of auld wes riddin' 12 b. bere; 20 b. meal 'acclamit to be in tak and possessioun for payment of' £16. Awltoun Burne, Blakdanetreis, Halybredhoill [Halybreidholme, *SRO, Vol. b*][117] and Eleischeucht,[118] 8 b. bere; 24 b. meal. Mow Haucht, 2 b. bere; 2 b. meal. Vicarage of Mow, £13 6s 8d.[119]

Total money, £29 6s 8d. Total bere, 10 b. Total meal, 26 b.

Maxwell kirk, parsonage and vicarage

The toun of Maxwell with the mains, Wester Woddein and Howdene 'quhilk of auld gave' 6 b. wheat; 18 b. bere; 28 b. meal; and now 20 b. bere; 20 b. meal. Pendicle Hill, 1 b. wheat; 4 b. bere; 4 b. meal. 'Sanct Thomas chapell with the hauch thairof', 6 b. bere; 8 b. meal. 'The Kingis Orchart with the hauch thairof and Eister Wodend quhilk of auld payit' 5 b. wheat; 11 b. bere; 12 b. meal, 'is set be the said umquhill abbot Williame Ker for' 8 b. bere; 16 b. meal. Wester Softlaw 'quhilk of auld gave', 3 b. wheat; 6 b. bere; 8 b. meal, 'is set togither with Eister Softlaw and Midlemest Wallis quhilk alsua of auld wer led be the place and ar of the kirk of Sproustoun and wes worthe' 4 b. wheat; 10 b. bere; 12 b. meal, 'as the foirsaidis teyndis ar set be the said umquhill abbot Williame Ker for' 6 b. bere;

---

[112] I.e. Walter Ker of Cessford, *RMS*, iv, no 1988.
[113] I.e. William Ker, who died in 1566, Cowan and Easson, p. 68; see also above, p. 224, n. 105.
[114] 'Quhytlaw Hous' is included at this point in *Registrum Cartarum de Kelso*, ii, p. 514, but the vicarage of Bowden is omitted.
[115] The correct calculation appears to be 3 c. 2 f.
[116] 'Cornst', *Registrum Cartarum de Kelso*, ii, p. 511.
[117] 'Halie Burne Home', *Registrum Cartarum de Kelso*, ii, p. 511.
[118] 'Ylyscheuch', *Registrum Cartarum de Kelso*, ii, p. 511.
[119] 'Sanct James Kirk' is included at this point in *Registrum Cartarum de Kelso*, ii, p. 511, but the vicarage of Mow is omitted.

10 b. meal. Vicarage of Maxwell 'intromettit with be Katherine Ker, relict of umquhill Richard Trollope, payand nathing thairfoir, bot of auld sustenit a vicar pentionar.'

Total wheat, 1 b. Total bere, 2 c. 12 b. Total meal, 3 c. 10 b. 'By the vicarege.'

Langtoun and Sympryme

The kirks of Langtoun and Symprim are set to the laird of Langtoun for 8 c. victual, 'thairof the 3 pairt beir and the 2 pairt meill, viz.', 2 c. 10 b. 2 f. 2⅔ p. bere; 5 c. 5 b. 1 f. 3⅓ p. meal.[120] 'The vicarege of Langtoun gevin to the curat for service. The vicarege of Symprym gevin to the curat for service at the kirk.'

'Summa beir patet [*i.e.* obvious]. Summa meill patet. By the vicareges.'

Fogo kirk, parsonage and vicarage

Fogo toun, Fogo Rig and Gyrig,[121] 4 b wheat; 1 c. bere; 2 c. meal. Bogend, 2 b. wheat; 11 b. bere; 20 b. meal. Cawdray, 1 b. bere; 2 b. meal. Systerpeth,[122] the Hill, Christeris Mylne Haucht, 25 b. meal.[123] Ryslawtoun 'quhilk gave of auld' 12 b. bere; 20 b. meal, 'set in assedatioun be the said umquhill abbet Williame Ker for' £13 6s 8d. Ryslawrig, 1 b. [*blank*];[124] 3 b. meal. Hurkers [*i.e.* Harcarse], 5 b. meal. Quhincarstanes and Cauldscheilraw, 2 c. 8 b. meal. Rydpeth,[125] 8 b. meal. 'The vicarege in use to be gevin to a vicar pentionar for service at the kirk. Summa of money by the vicarege', £13 6s 8d.

Total wheat, 6 b. Total bere, 1 c. 12 b. Total meal, 8 c. 7 b.

Grenlaw kirk with the vicarage

Baitscheill and Weitfit[126] 'quhilk of auld gave' 3 b. bere; 7 b. meal, 'acclamit be Maister Thoriswod to be set to him and in possessioun thairof for' £5 6s 8d. Harelaw, 3 b. bere; 6 b. meal. Halyburtoun Eistersyd, 3 b. bere; 6 b. meal. Halyburtoun Westersyd, 3 b. bere; 6 b. meal. Grenelawdene, the Pokichaucht[127] 'and the croftis pertenand to the guidman of Grenelawden',[128] 8 b. bere; 16 b. meal. The laird of Grenelaw for his land of Westerhaw[129] 'and for his pairt of the tenandrie, Auld Greinlaw, Howlawrig, Caitmos,[130] and the Grein Syd,[131] the

---

[120] The correct calculation appears to be 5 c. 5 b. 1 f. 1⅓ p.
[121] 'Kirk land of fogo callit zy rig'; 'Fogo Towne Fogo Rige and Gayrige', *Registrum Cartarum de Kelso*, ii, pp. 491, 504.
[122] 'Fyscer Peth', *Registrum Cartarum de Kelso*, ii, p. 504.
[123] 'Cheisteris', 1 b. bere, 3 b. meal', *Registrum Cartarum de Kelso*, ii, p. 505.
[124] This 1 b. of victual is not included in any of the totals. 'Ryslawrige' gave 6 b. meal, but no bere according to *Registrum Cartarum de Kelso*, ii, p. 505.
[125] 'Reidpeth', *Registrum Cartarum de Kelso*, ii, p. 505.
[126] 'Betschule and Vitfute', *Registrum Cartarum de Kelso*, ii, p. 497.
[127] 'Polkehaugh', *RMS*, iv, no 171; 'Pokye Houich', *Registrum Cartarum de Kelso*, ii, p. 497.
[128] 'Grenladen', *Registrum Cartarum de Kelso*, ii, p. 497.
[129] 'Vesterraw', *Registrum Cartarum de Kelso*, ii, p. 498.
[130] 'Cotmos', *Registrum Cartarum de Kelso*, ii, p. 503.
[131] Grein Syd does not appear in *Registrum Cartarum de Kelso*.

Nuke',[132] 2 c. meal. Brumehill, 3 b. bere; 6 b. meal. Nicol Alexanderis croft, 1 b. meal. David Brounfeildis croft, 1 b. meal. Fluirswallis,[133] 1 b. bere; 1 b. meal. Lambden with Puddingrow, 8 b. bere; 12 b. meal. The Curogis, 4 b. bere; 6 b. meal.[134] John Trounfeildis [*recte*, Brounfeildis] 'pairt of the tenandrie', 1 b. bere; 3 b. meal. John Spens steading, 2 b. bere; 6 b. meal. Rowanstoun,[135] 8 b. bere; 24 b. meal. Eister Howlaw, 4 b. bere; 8 b. meal. Wester Howlaw, 4 b. bere; 8 b. meal. Howlawheid, 2 b. bere; 4 b. meal. William Parkis steading, 2 b. bere; 6 b. meal. Murischot 'quhilk of auld payit' 2 b. bere; 4 b. meal, 'acclamit be the laird of Greinlaw to be set for' £4 13s 4d. Elwodlaw,[136] 1 b. bere; 3 b. meal. Angelraw with Alexander Reidpethis croft in Grenelaw, 3 b. bere; 6 b. meal. Crumrig[137] with the west side of Angelraw 'quhilk of auld payit' 2 b. bere; 4 b. meal, 'acclamit be Triomour Reidpeth in tak and possessioun for' £3 13s 4d. Belleschill with Grenelaw croft, 1 b. bere; 3 b. meal. Clegdenie [*i.e.* Slegden],[138] 2 b. bere; 6 b. meal. Hardenis, 2 b. bere; 4 b. meal. Cutrig,[139] 2 f. bere; 1 b. 2 f. meal. Wodheid, 2 b. bere; 4 b. meal. The Eist Feild, 2 b. bere; 4 b. meal. The four merkland, 1 b. bere; 3 b. meal. Quhytsyd,[140] 3 b. bere; 5 b. meal. Kingis Merksworthe, 1 b. bere; 2 b. meal. Foulschotlaw,[141] 1 b. bere; 2 b. meal. 'Vicarege of Grenelaw gave of auld' £40 'and now be want of the Pace fynes and utheris dewteis', £30.

Total money, £13 13s 4d.[142] Total bere, 4 c. 11 b. 2 f. Total meal, 12 c. 3 b. 2 f.

Gordoun parsonage and vicarage

West Gordoun, 13 b. bere; 2 c. meal. Huntlie Wod, 6 b. bere; 1 c. 13 b. meal. Thornedykis 'quhilk of auld payit' 8 b. bere; 1 c. meal, 'set be the said umquhill abbot Williame' for 18 b. meal. Spottiswod 'quhilk of auld payit' 15 b. victual, 'be the said Williame, abbot', for 10 b. meal. Flas,[143] 2 b. bere; 8 b. meal. Wodderlie with the pendicles, 2 c. meal. Yifflie,[144] 3 b. meal. Fawsyd,[145] 2 b. bere; 4 b. meal. Hexpeth,[146] 3 b. bere; 6 b. meal. Rymmiltoun, 3 b. bere; 6 b.

---

[132] 'Nueke', *Registrum Cartarum de Kelso*, ii, p. 502.
[133] 'Flurislawis', *Registrum Cartarum de Kelso*, ii, p. 499.
[134] 'The Tunrik', giving 4 b. bere, 6 b. meal, follows Lambden with Puddingrow, *Registrum Cartarum de Kelso*, ii, p. 499.
[135] 'Rowistoun', *Registrum Cartarum de Kelso*, ii, p. 500.
[136] 'Alvodlaw', *Registrum Cartarum de Kelso*, ii, p. 500.
[137] 'Tuinrige', *Registrum Cartarum de Kelso*, ii, p. 500.
[138] 'Sligdene', *Registrum Cartarum de Kelso*, ii, p. 501.
[139] 'The Totrige', *Registrum Cartarum de Kelso*, ii, p. 501.
[140] 'Quheitsyde', *Registrum Cartarum de Kelso*, ii, p. 502.
[141] 'Fustalaw', *Registrum Cartarum de Kelso*, ii, p. 502.
[142] £30 paid by the vicarage of Greenlaw is not taken into account.
[143] 'Flas Johne Voddis Pairt ... Hendrie Voddis Pairt ... Rychart Edgeris Pairt?', *Registrum Cartarum de Kelso*, ii, p. 495.
[144] 'Jefflie', *Registrum Cartarum de Kelso*, ii, p. 495.
[145] 'Fenisyde', *Registrum Cartarum de Kelso*, ii, p. 496.
[146] 'Hetspeth', *Registrum Cartarum de Kelso*, ii, p. 496.

meal. Rymiltounlaw, 4 b. bere; 8 b. meal. Gordoun Manes, 2 b. bere; 2 b. meal. Eister Gordoun, 12 b. bere; 1 c. 12 b. meal. 'The Lordis Manes', 8 b. bere; 1 c. meal. Mydlethrid, 8 b. bere; 1 c. meal. Bellitaw, 4 b. bere; 8 b. meal. 'The vicarege of Gordoun of auld wes set for' 100 merks, 'is now be ressoun of the dimulitioun [*recte*, diminution] of the profeitis estemit to' £40.

Total money, £40. Total bere, 4 c. 3 b. Total meal, 14 c. 2 b.

Howme with the vicarage

The Toun Manes with Fawsyd [*i.e.* Fallsidehill][147] and Hardis Mylne, 2 c. 8 b. bere; 4 c. 8 b. meal. Todrig, 3 b. bere; 9 b. meal. Oxmure, 5 b. bere; 11 b. meal. The vicarage of Howme, 10 b. [*recte*, £10].

Total money, £10. Bere, 3 c. Meal, 5 c. 12 b.

'Summa totalis of the foirsaidis kirkis that ar pairtlie set for victuall and pairt for silver', £126.

'Summa totalis of the haill money cumand for kirkis set for silver and als for temporall landis abonewrittin pertenyng to Frances,[148] commendatour of Kelso, nocht acclamit be the said James Cunyghame and Mungo Grahame, extendis to' £1,569 13s 4d;[149] third thereof, £523 4s 5½d. Total wheat, 8 b. 2 f.; third thereof, 2 b. 3 f. 1⅓ p. Total bere, 23 c. 4 b. 6⅓ p.;[150] third thereof, 7 c. 12 b. 2 p. Total meal, 56 c. 13 b. 3 f. 1⅓ p.;[151] third thereof, 18 c. 15 b. 5 p.[152] Third of capons, 4. Third of poultry, 60.

'It is to be rememberit that the vicarege of Sproustoun is chargit be the selff in [all, *added in SRO, Vol. b*] the comptis bypast. Item, the kirk of Lyntoun is chargit be the self in the the [*sic*] comptis and thairfoir thir tua man defusit[153] [*recte*, be defaikit] of the rentall. Item, thir tua man be defaikit of the rentall the contributioun to the Lordis of Sessioun', £56. 'Item, the landis of Halyden gevin of auld for ballie fe', £10.

'Pensiounes acclamit furth of the abbacie of Kelso by and attour James Cunyghame and Mungo Grahame'

Robert Ker of Ancrom, £100. Mr William Schaw, £80; 12 c. oats 'furth of the kirk of Grenlaw'. Mr Walter Balfour has the parsonage of Lyntoun for £36 13s 4d. John Semple, £333 6s 8d. Alexander Home, brother to the laird of Aytoun, £200. William Home, brother to [the laird of] Coldenknowis, £200 'furth

---

[147] 'Fawsidhill', *Registrum Cartarum de Kelso*, ii, p. 503.

[148] Francis Stewart, see above, p. 223, n. 94.

[149] The correct calculation appears to be £1,570 13s 4d without taking into account the £9 6s 8d paid to the convent for Humbie.

[150] The correct calculation for the total of individual entries is 23 c. 5 b. 1 f. 2⅔ p.

[151] The correct calculation appears to be 56 c. 12 b. 3 f. 1⅓ p.

[152] The correct calculation appears to be 18 c. 15 b. 1 f. plus one third of 1⅓ p. (i.e. ⁴⁄₉ p.).

[153] SRO, Vol. b text also reads 'defusit'.

of the kirk of Gordoun'. The laird of Barganie, £200. Mr Alexander Machame, £20. James McCairtnay, £13 6s 8d [*in margin*, 'furth of the kirk of Caldorcleir']. Robert Scot, 'writtar', £10 'furth of the landis of Howdene'. 'The young laird of Cesfuird hes be gif of the umquhill Regent, erle of Lennox',[154] £433 6s 8d, 'and for payment thairof he intrometis with the haill landis and teyndis of Sproustoun with the pendicles, the haill mailles and teyndis of Clarelaw and baronie of Bowdenne extendis to as in the rentall foirsaid his [*recte*, is] contenit.'

'The kirkis and landis of the abbay of Kelso acclamit be James Cunyghame to pertein to him in pensioun all the dayis of his lyftyme.'

The kirk of Carlouk, £66 13s 4d. The kirkland of Carlouk, £4. The kirk of Innerlethane, £20. The kirk of Humbie, £43 6s 8d. The kirk of Duddingstoun, £66 13s 4d. The feu mails of the lands pertaining to 'umquhill' Thomas Thomsoun, £66 13s 4d. The kirk of Cloisburne, £40. The kirk of Mortoun, £40. The feu mails of the lands of Humbie pertaining to Robert Lawsoun, £10. The kirk of Drumfreis, £60. The mails of the barony of Lesmahago with Clydis Mylne, £201 6s 10d. The vicarage of Lesmahago 'set of auld to the Clerk of Register[155] for' 100 merks 'and the samin remittit to him for all the dayis of his lyff.'

The kirk of Lesmahago
'The teind schavis of the landis pertenyng to the duik[156] in few quhilkis of auld gaif tuelf chalder thrie f[irlottis] tua p[ectis], ar sett in tak to his grace for' £45. The teinds of Evindale 'quhilkis of auld gave aucht chalderis victuall ar set to the tennentis for' £40. The teinds of Corhous, 'quhilk of auld gave' £25, are set for £8 6s 8d. The teinds of Stanebyres 'quhilkis of auld gave tuenty tua bollis' are set for £11. 'Summa of money of the said kirk of Lesmahago by the vicarege' £104 6s 8d. 'Item, in victuall', 29 c. 15 b. 2 f. 2 p.; third thereof, 9 c. 15 b. 3 f. 2 p.

'It is to be rememberit that Johnne Weir of Clenedyk acclamis a pensioun furth of the samin to be gevin to him of auld for his service extending to' 4 b. bere; 12 b. meal. 'Item, thair is ane uther auld man quhilk clamis a pensioun of auld for his service extending to' 13 b. meal. 'Item, thair is ane uther auld man quhilk clamis a pentioun of auld for his service extending to' 13 b. meal.[157]

---

[154] Matthew Stewart, 4th earl of Lennox, appointed Regent in July 1570 and killed at Stirling in September 1571, *Scots Peerage*, v, p. 353.
[155] James McGill of Nether Rankeilour, Clerk Register, 1567 - 1579, *Handbook of British Chronology*, p. 197.
[156] James Hamilton, 2nd earl of Arran and duke of Chatelherault, *Scots Peerage*, iv, pp. 366-368.
[157] This rendering occurs in both texts.

'Summa of the haill money of all the saidis kirkis and landis of the abacie of Kelso acclamit be James Cunyghame, by the vicarege of Lesmahago, remittit as said is, extendis to' £723 0s 2d;[158] third thereof, £241 0s ⅔ d.

'Johnne Johnnestoun, writtar, hes furth of the kirk of Dudingstoun, a pensioun assignit to' £22 4s 5d. 'Item, Henry Killocht hes furth of the kirk of Humbie, £43 6s 8d. 'Item, Mr Johnne Stewart hes furth of the kirk of Dudingstoun', £44 8s 10⅔d.

The kirk of Selkirk,[159] 'bayth personage and vicarege, acclamit to pertein to Mungo Grahame as assignit to him for the sowme of' 400 merks 'quhairof be his gift he is astrictit to pay furth for the thrid' £88 17s 4d;[160] 'the 3 thairof extendis to as said is.'[161]

'Albeit this kirk in the auld rentale gave' 6 c. 10 f. bere; 12 c. 10 b. 2 f. meal; 1 b. wheat; 'and in silver by the vicarege', £16. 'Item, the vicarege gave' £46 13s 4d.

KELSO, ABBEY OF,[162]   (SRO, Vol. a, fos 232r-234v[163])
(SRO, Vol. b, fos 165r-167v)

Rental of the abbacy of Kelso

'The toun of Kelso: In the first, the mailles of the said toun', 100 merks. 'The cotteris [cottagiis, *fo 223r*] of Kelso', £57.[164] 'The fischeingis of Tuyde', 100 merks. 'The mylnes of Kelso', £80. 'The cobill foiranent Roxburcht' [Sprouston, *fo 223r*], £20. 'The cobill foiranent Maxwell', £10. 'The uvynes and customes of Kelso',[165] £10. The Cunyngaires, Brumebank and Hocht [Hoitt, *fo 223r*; Hote, *fo 243r*], 40s. The West croftis and Angrieflat, £3. The toun of Prestoun, £18. The toun of Reiddayne, £34. The ploughlands of Halden, £4. Heleischowcht, £5. The toun of Newtoun, 30s. The kirkland of Makerstoun, £10. The kirkland of Gordoun, £4. The kirkland of Grunlaw [*i.e.* Greenlaw], 53s 4d. Humbie, £10. The kirklands of Carlouk, £4. Downland [*recte*, Dowglen, *fos 223r, 235r, 240r, 243r*], £5. Greenheid, £5. Quhytmuirtoun, £10. Quhytmuirhall, £5. Unysclos

---

[158] This figure does not tally.
[159] Under 'Selkrik Kirk', *Registrum Cartarum de Kelso*, ii, p. 514, lists 'Quhyt Mure Hall, Quhytmure Towne, Selkrik Towne, The Schaw, Hartvodburne, Haning, Aclintour, Todrige, Phillophauch and Herheid, Carterhauch, Auld and New Vark, The Northt Bowhill, South Bowhill, Fawsyde, Bredmedow, Blakmedyngis, Myddilsteid, Harthvodmyris, Fasthucht, Crage and Yare, Sunderland Hall and Towne, Faldhoip, Grenheid, Willdunhop, Houdene.'
[160] The correct calculation appears to be £88 17s 9½d.
[161] The entry in SRO Vol. b ends here at fo 164v; 165r corresponds to SRO, Vol. a, fo 232r.
[162] Cf SRO, Vol. a, fos 223r, 235r, 240r, 243r (see above, p. 222; below, pp. 233, 236, 239).
[163] Fos 230v-231v are blank.
[164] 'The small maillis of the cottaris therof by the discedentis, £54', *Registrum Cartarum de Kelso*, ii, p. 489.
[165] 'Kelso and the borderflat', *Registrum Cartarum de Kelso*, ii, p. 489.

[*i.e.* Ovenscloss],[166] 20s. The toun of Lynden, £16. 'The fewlandis [four husbandlands, *fo 223r*; four lands, *fos 235v, 240v, 243r*] of Lyntoun', £4. Fawdounsyd, £10. Cauldschiellis, 40s. Quhytlawhous, £10. The toun of Mydlem, £30. The Mylne of Midlem 'and mylnelandis thairof', £7. The Manes of Clarelaw, £200. Keppelaw, £5.[167] Lawreis land in Clarelaw, 40s. 'Mr George Keris pairt in Clarelaw', 10 merks. Halyden, £10. Over and Nather Holdenis, £20. The Bothell, £5. The Airheid, £5. Buschelhill, 40s. The Chapelhill, £12. Dudingstones Eister and Wester, 100 merks. The ploughlands of Oislaw,[168] 40s. Fogo Rig, 20s. The Mylne of Lynden, 53s 4d. The toun of Bowden, £31. The Myln of Bowden, £5. The Preistoun £5. The kirkland of Mellerstanes, 40s. The annual of Baltraid, £5 6s 8d. The Softlaw with the two lands [husbandlands, *fo 223r*] of Hwtoun' [Heytoun, *fo 223r*; *i.e.* Heiton], £10. The Bogend, 53s 4d. The Cawdra, £3 6s 8d. The Mylne of Fogo, £4. Cliddismylne, £6 13s 4d. The Bellischill, 40s.

The mails of Lesmahago[169]
    Pownell, £6 13s 4d. Cummyir, 40s. The Skellehill,[170] 46s 8d. The Quhitsyd and Midleholm, £3. Auchelochan, 31s.[171] Aucheryn, 20s. The Auldtoun, 27s 8d. The Bawgre[172] and Bankheid, 40s. The Burdland and Dewane [*i.e.* Boreland and Devon],[173] £3 12s. Drumbekschill, 16s. Salbarry and Galowayhill, 30s.[174] Auchinlek, Thauthtis and Grenrig [*i.e.* Affleck, Teaths and Greenrig],[175] £6 13s 4d. Blakweid and Priourhill, 55s.[176] Clenoch and Gyraldweid, £3 13s 4d.[177] The Mylntoun, £3 12s. The Maynes, £14 6s. Watistoun, £5. 'The fischeing of Clyd', 6s 8d. Kype, £5. John Mures house in Lanark, £100.[178]

'Sic subscribitur' [*blank*].

'My Lordis Auditouris of this rentale, excuis me in this rentale of Lesmahago, for I am informit the just mailles of Lesmahago extendis yeirlie to' 300 merks 'and this abone is nocht for abone ane hundreth merkis, and as for the victuallis of the

---

[166] 'Vinsclos', *Registrum Cartarum de Kelso*, ii, p. 490.
[167] 'Tippilaw' paying £6, *Registrum Cartarum de Kelso*, ii, p. 491.
[168] 'Hoslaw', *Registrum Cartarum de Kelso*, ii, p. 491.
[169] Under the barony of Lesmahagow, before 'pownill', *Registrum Cartarum de Kelso*, ii, p. 492 lists 'the foullandis' paying £82.
[170] 'Skaillihill and rothart holme', 46s 8d, *Registrum Cartarum de Kelso*, ii, p. 492.
[171] 'Adflothome' paying 32s, *Registrum Cartarum de Kelso*, ii, p. 492.
[172] 'Bonegraye', *Registrum Cartarum de Kelso*, ii, p. 492.
[173] 'Borlame and dowane', *Registrum Cartarum de Kelso*, ii, p. 492.
[174] 'The gallowrig and gallow hill' paying 30s, *Registrum Cartarum de Kelso*, ii, p. 493.
[175] 'Auchinlouke to the grenerig', *Registrum Cartarum de Kelso*, ii, p. 492.
[176] 'Blekvode and forowhill', *Registrum Cartarum de Kelso*, ii, p. 493.
[177] 'Chenothe and gorvaldvode' paying £3 13s 10d, *Registrum Cartarum de Kelso*, ii, p. 493.
[178] 'Mowtons hous in lanrik' paying 20s 10d, *Registrum Cartarum de Kelso*, ii, p. 493.

same kirk, extendis to' 53 c. 'as I am informit and as the rentall beiris, the quhilk as yit I have nocht found nor gottin.'

'The set kirkis for money'

The vicarage of Selrig, 100 merks. The kirk of Sanct James, 40s. The kirk of Innerlethane, 20s. The kirk of Hopcalzo, £10. The kirk of Lintoun, £36 13s 4d. The kirk of Stabilgordoun, £8. The kirk of Symontoun, £12. The kirk of Dunsyre, £10. The kirk of Roberttoun, £20. The kirk of Wystoun, £16. The kirk of Carlouk, 100 merks. The kirk of Mortoun, £25. The kirk of Drumfreis, £40. The kirk of Trailflat, £8. The kirk of Drumgrie, £5. The kirk of Kilmaweris, £60. The kirk of Caldercleir, £44. The kirk of Humbie, £43 6s 8d. The kirk of Duddingstoun, 100 merks. Pettirculter, £24. The vicarage of Gordoun, 100 merks. The vicarage of Nenthorne, £13 6s 8d. The vicarage of Home, £10. The vicarage of Sympryn, 40s. The vicarage of Mow,[179] £13 6s 8d.

'The kirkis set for victuall and estemit half meill, half beir

Item, the kirk of Kelso, the teyndis led be the place estimat to' 20 c. victual.[180] The kirk of Sproustoun 'led be the place estimat to' 24 c. The kirk of Hume, 16 c. The kirk of McCerstoun, 8 c.[181] The kirk of Nenthorne, 16 c. The kirk of Gordoun, 14 c. The kirk of Gryinlaw, 21 c. 4 b. The kirk of Langtoun and Symprym, 8 c. The kirk of Fogo, 12 c. 12 b. The kirk of Horneden, 5 c. The kirk of Mow, 3 c. 8 b. The kirk of Bowden, 'by Clarelaw teynd', 15 c. The kirk of Lyndein, 4 c. The kirk of Selrig, 15 c.[182]

The teinds of Lesmahago

The Roswod,[183] 6 b. 3 f. 'farine' [*i.e.* meal]; 3 b. 'ordei' [*i.e.* bere]. 'Under the bank of Blakband [Blakbank, *SRO, Vol. b*]', 10 b. meal. The Hawhill, 12 b. 2 f. meal; 3 b. bere. Auchnotro, 4 b. meal; 4 b. bere. The Stalyis, Blakhilend and Clerkstoun, 21 b. meal; 3 b. 3 f. bere. Auchihache, 6 b. 1 f. meal. Auchgammyll, 22 b. meal; 8 b. 2 f. bere. Auchefardall, 8 b. meal. Auchinlek, Tathis and Nowrig[184] [*i.e.* Affleck, Teaths and Greenrig], 15 b. 2 f. 1¼ p. meal. Eylbank, 9 b. 2 f. meal. Lorrous, 1 c. 9 b. meal. Mosmyning, 1 c. 2 f. bere; 4 b. meal. The Bardyland [*i.e.* Boreland], 5 b. 1 f. 2¼ p. meal; 5 b. 1 f. 2¼ p. bere. The Bog, 3 f. bere. Aucheneill, 3 b. meal. Drumbrakischill, 3 b. meal; 1 b. bere. Awletoun, 5 b. bere; 5 b. oats. Awchron, 6 f. bere. The Bawgrein and Bawmilheid, 6 b.

---

[179] 'Vicarege of mow by the lamis', *Registrum Cartarum de Kelso*, ii, p. 494.

[180] Under 'Kelso Kirk', *Registrum Cartarum de Kelso*, ii, p. 508 includes 'Kelso Towne, Bokislaw, The Fluris, The Gallowlaw, Eister Meirdene.'

[181] '6 c. meal, 6 c. bere', fo 236v; '6 c. meal', fo 241v; '4 c. meal, 4 c. bere', fo 244v (see below, pp. 234, 238, 240).

[182] This is the only reference to the kirk of Selkirk paying victuals, except fo 230r where there is mention of the 'auld rentale'. Elsewhere the parsonage and vicarage of Selkirk pay £266 13s 4d.

[183] 'Roswod' has not been traced, but Rosehill and Woodfoot occur in the vicinity of Lesmahagow.

[184] This should probably read 'Newrig', but fo 233r reads 'Grenrig' (see above, p. 231); *Registrum Cartarum de Kelso*, ii, p. 492 has 'Auchinlouke to the grenerig'.

bere; 6 b. oats. The Dowam [i.e. Devon], 3 b. 3 f. meal; 2 b. bere. Auchlocham and Stokbrig, 4 b. meal; 11 b. 2 f. bere. Pownell, 7 b. 2 f. meal; 6 b. bere. Cummyr, 11 b. 1½ p. meal. Interaquas, 11 b. bere. Skoretholm and Skellehill, 6 b. bere; 6 b. meal. Quhytsyd and Mydleholm, 7 b. 2 f. bere. The Manes, 18 b. meal; 8 b. bere. The Clemoche and Garwelwod, 11 b. 1 f. meal; 9 b. 1 f. bere. The Kers, 8 b. 1 f. meal. The Blakwod, Kypsyd and Rogerhill, 1 c. 3 b. 1 f. meal; 1 c. 3 b. bere. The Spydhill and Candersyd, 12 b. meal. Draffyn and Southfeild, 3 c. meal; 1 c. bere. The Trypwod, 14 b. meal; 8 b. bere. Fokestoun, 2 c. 1 b. meal. The teinds of Avendaill, 5 c. 13 b. 2 p. meal; 2 c. bere.

Signature: 'Williame Scot.[185] Clericus Registri.'

KELSO, ABBEY OF,[186]   (SRO, Vol. a, fos 235r-239r[187])
                                                    (SRO, Vol. b, fos 176r-v, 168r-170r[188])

'Item, the greit mailes of Kelso toun', £66 13s 4d. 'The coil silver of Kelso', £13 6s 8d. 'The mailes of the coittares' [cottagiis, *fo 223r*], £57. 'The fischeing of Tueid', £66 13s 4d. 'The west cobill of Roxburcht' [Sproustoun, *fo 223r*], £20. 'The cobill foiranent Maxwell', £10. 'The unis [uvynes, *fo 232r*] and custumes of Kelso', £10. The Cunynghares, 26s 8d. The Bromebank and the Hocht [Hoitt, *fo 223r*; Hote, *fo 243r*], 13s 4d. 'The mylnes of Kelso', £80. The West croftis, 40s. Angriefleit, 20s. The Brome croft, 12s. The Toun croft, 10s 8d. The mails of Sproustoun, £18. The mails of Reddane 'by ane pairt of the toun that lyis to the Manes of Sproustoun', £34. The ploughlands of Haldein, £4. The Eileischeucht, £5. The kirklands of Nanetharne, 40s. The kirklands of McCarstoun, £10. 'The few of Todrik',[189] 40s. The toun of Newtoun, 30s. In Langgardoun, £4. The annual of Deringtoun, 5s. The kirkland of Grenelaw, 53s 4d. The kirkland of Sympryn, 40s. Humbie, £10. The teind lands [*recte*, kirklands, *fos 229r, 232r, 240r, 243r*] of Carlouk, £4. The Dowglen, £5. Greneheid, £5. Quhytmuirhall, £5 6s 8d. Quhytmuirtoun, £10. Unes croft [Unysclos, *fo 232r*; *i.e.* Ovenscloss], 20s. The Lyndene, £16. The four lands [husbandlands, *fo 223r*] of Lyntoun, £4. Faldensyd, £10. Cauldscheillis, 40s. Quhytlawhous, £10. The toun of Mydlane, £31. 'The gyrs of Mydlane', £8. 'The Mylne of Mydlane and mylnlandis', £7. The Clarilawmanes, £200. Kippellaw in Clarylaw [Kyplaw and Clarylaw, *fo 243v*], £5. Lawreis land in Clarylaw, 40s, 'and in the augmentatioun of the few', 20s. 'Mr George Keper [*recte*, Keris, *fos 232v, 243v, 245v*] pert in Clarylaw', £6 13s 4d. Halyden, £10. Overholden and Nethirholden, £20. The Bothell, £5. The Hairheid, £5. The Bowschelhill, 40s. The Chapelhill, £12. Duddingstoun Eister

---

[185] The text in SRO, Vol. b diverges here at the close of fo 167v; fo 168r corresponds to SRO, Vol. a, fo 236r.
[186] Cf SRO, Vol. a, fos 223r, 232r, 240r, 243r (see above, pp. 222, 230; below, pp. 236, 239).
[187] Fo 238r-v is blank.
[188] SRO, Vol. b, fo 176r-v corresponds to SRO, Vol. a, fo 235r-v; see also below, p. 234, n. 190.
[189] 'The landis of stodrik', *Registrum Cartarum de Kelso*, ii, p. 491.

and Weister, £66 13s 4d. The ploughlands of Hoislaw, 40s. Fogo Rig, 20s. The Mylne of Lynden, 53s 4d. The Bowdane toun, £31. The Mylne of Bowdane, £5. Prestoun, £5. The toun of Mellerstanes, 40s. [190]The annual of Baltrude, £5 6s 8d. Softlawis with the two lands [husbandlands, *fo 223r*] of Hevine [Heytoun, *fo 223r*; i.e. Heiton], £10. The Bogend, 53s 4d. Cadray, £3 6s 8d. The Mylne of Fogo, £4. The Clydismylne, £6 13s 4d. 'Summa of the mailles of the foirsaidis landis', £961 11s;[191] 'thairof defaikit for the erle of Glencarne',[192] £66 13s 4d; 'sua restis' £894 17s 8d.

The rental of the kirks set for money
    The kirk of Closburne, £40. The parsonage and vicarage of Selkirk, £266 13s 4d. The kirk of Sanct James, 40s. The kirk of Innerlethane with Hopcalzie, £30. The kirk of Lyntoun, £36 13s 4d. The kirk of Stablegordoun, £8. The kirk of Symontoun, £12. The kirk of Dunisie [*recte*, Dunsyre], £10. The kirk of Roberttoun, £20. The kirk of Welstoun [*i.e.* Wiston], £16. The kirk of Carlouk, £66 13s 4d. The kirk of Mortoun, £25. The kirk of Drumfreis, £60. The kirk of Trailflat, £14. The kirk of Dungrie, £5. The kirk of Kilmawres, £60. The kirk of Caldercleir, £44. The kirk of Humbie, £43 6s 8d. The kirk of Dudingstoun, £66 13s 4d. The kirk of Petirculter, 20 merks. The kirk of Maxwell, £66 13s 4d. The kirk of Newtharne, £80. The kirk of Kelso, £166 13s 4d. The teinds of Angreflat, West croft, Bordourflat, Cunynghares, [£6, *SRO, Vol. b*]. Horneden with the vicarage, £13 6s 8d. 'Item, the haill teyndis of Sproustoun parochin except the teyndis of Eister Softlaw and Mellenden[193] used to be led to the behove of the place and be commoun estimatioun we [wer, *SRO, Vol. b*] estemit to' £180. The teinds of Clarelaw 'be commoun estimatioun ar worthe' £66 13s 4d. 'Summa of all thir kirkis set for money', £1,428 13s 4d;[194] 'defalkit for the erle of Glencarne',[195] £321 13s 4d; 'sua restis' £1,070 6s 8d[196] 'by the kirk of Lyntoun.'

The kirks set for victual: The kirk of Maccarstoun, 6 c. bere; 6 c. meal. The kirk of Gordoun 'by the lord Humes[197] teyndis', 3 c. bere; 6 c. meal. The kirk of Langtoun and Symprein, 2 c. 10 b. 2 f. 2⅔ p. bere; 5 c. 5 b. 1 f. 1⅓ p. [meal]. The kirk of Mow, 2 c. meal; 24 b. bere. The kirk of Home 'halelie in the lord Humes handis togither with the said lord Humes teyndis of the landis following in the parochinis of Gordoun, Grenelaw and Fogo, viz.':
    In the parish of Gordoun: Eister Gordoun, Belletaw, Midlethrid, Wester Manes, Rymiltounlaw and the Eister Manes of Gordoun.

---

[190] SRO, Vol. b, fo 168r, which resumes here, corresponds to SRO, Vol. a, fo 236r.
[191] The correct calculation appears to be £959 11s.
[192] See above, p. 210, n. 25.
[193] 'Meltondene', *Registrum Cartarum de Kelso*, ii, p. 510.
[194] The correct calculation appears to be £1,418 13s 4d.
[195] See above, p. 210, n. 25.
[196] £1,428 13s 4d minus £321 13s 4d is £1,107.
[197] Alexander, 5th lord Home was served heir in 1551 and was forfeited for treason in 1573; the title was restored in 1578 to his son, Alexander, 6th lord Home, *Scots Peerage*, iv, pp. 462-463.

And in the parish of Grenelaw: the Hordlaw, James Spens mailing in Rochesterrig, Howlawheid, Wester Howlaw and Eister Howlow and Den [*blank*] nerles.[198]

And in the parish of Fogo: Fogo toun, John Tranentis land, Sir James Tranentis land, John Trotteris [Trollouris, *fo 241v*][199] of the Hill and the Bagend [Bogend, *fo 241v*], 'altogidder payit' 16 c. meal; 8 c. bere. [*In margin*, 'Nota, I beleve the lord Hume intrometis also with the teyndis of Rowenstounes, xxiiij b. mell & viij b. beir.'[200]]

The kirk of Bowden 'by the teyndis of Clarelaw', 3 c. 8 b. 5⅓ p. bere; 8 c. 2 f. 2⅔ p. meal [7 c. 2 f. 2 ⅔ p. meal, *fo 244v*]. The kirk of Lyndene, 3 c. meal; 24 b. bere. The kirk of Grenelaw 'by the lord Homes teyndis', 8 c. meal; 4 c. bere. The kirk of Fogo, 1 c. 9 b. 5⅓ p. bere; 3 c. 2 b. 2 f. 2⅔ p. meal. The Manes of Sproustoun 'wer set with the plenisching for' 8 c. wheat; 8 c. bere; 8 c. meal. The teinds of Melleden within the parish of Sproustoun, 4 b. meal; 2 b. bere. The teinds of Eister Softlaw within the said parish, 7 b. meal; 3 b. 2 f. bere. 'Item, in Kelso toun thair is certane fermes payit, viz., the Bank payis', 3 b. bere [4 b., *fo 242r*]; 3 b. meal. 'Item, thair the Eister Medowis in Kelso', 8 b. oats 'and ij fedderis [hay, *fo 244v*].

Summa of all thir kirkis in victuall', 9 c. 'frumenti' [*i.e.* wheat]; 40 c. 4 b. 3 f. 1⅓ p. 'beir'; 65 c. 6 b. 2 f. 2⅔ p. 'meill'; 8 b. 'avenarum' [*i.e.* oats].

'Memorandum, thir vicaregeis wer wount to be set for the pryces following': The vicarage of Nantharne, £13 6s 8d. The vicarage of Gordoun, £66 13s 4d. The vicarage of Home, £10. The vicarage of Symprine, £2. The vicarage of Mow, £13 6s 8d. 'Summa in money of thir vicaregis', £105 6s 8d.

'Summa of the haill money, by the erle of Glencarne', £2,057 4s 4d;[201] third thereof, £685 14s 9⅓d. Third of the wheat, 3 c. Third of the bere, 13 c. 6 b. 3 f. 3 p. Third of the meal, 21 c. 12 b. 3 f. 2⅓ p.[202]

The third of the abbacy of Kelso and Lesmahago, in money, £1,682 5s 6d;[203] third thereof, £560 15s 2d.

---

198 SRO, Vol. b is damaged at this point, and a blank occurs in SRO, Vol. a. Both volumes read 'Dennerles', SRO, Vol. a, fos 241v, 244v; SRO, Vol. b, fos 172v; 'Dennartes', SRO, Vol. b, fo 175v (see below, pp. 238, 241).

199 'Jane Trotter in the Heill', *Registrum Cartarum de Kelso*, ii, p. 504. John Trotter occurs in *Protocol Book of Sir William Corbet, 1529- 1555*, ed. J. Anderson (Edinburgh, 1911), no 86.

200 'xiiij b. meill, xviij b. beir', fo 241v (see below, p. 238).

201 This figure does not tally.

202 In calculating the third of the bere there was no attempt to divide ⅓ p. in the total bere by 3; but here, in calculating the third of the meal, ⅔ p. in the total meal has been divided by 3 to give ⅓ p. (which ought to read 2/9 p.). Oats have been omitted from this total and calculation of the third.

203 Lesmahagow, which belonged to Kelso, is not included in this rental as a separate entity.

'Kelso victuallis by Lesmahago': Meal, 91 c. 4 b.;[204] third thereof, 30 c. 6 b. 2 f. 2⅔ p. Bere, 91 c. 4 b.;[205] third thereof, 30 c. 6 b. 2 f. 2⅔ p.

'Lesmahago secundum [*i.e.* according to] Ker, in maill', 28 c. 1 b. 3 f. 3 p.; third thereof, 9 c. 5 b. 3 f. 3⅔ p. Bere, 12 c. 1 b. 2 f. 2¼ p.; third thereof, 4 c. 2 f. 'and half pect 3 half pect, and the 3 of the 3 part half pect' [*i.e.* 1⅜ p.; *recte*, ⅞ p.].[206]

ANCRUM, PARSONAGE OF,[207] (SRO, Vol. a, fo 239v)
(SRO, Vol. b, fo 170v)

'The rentall of the personage of Anecrame prebendrie in the kirk of Glasgw.'

'The haill teyndis and all utheris profeitis or emolumentis of the personage and vicarege of Ancrum wer set in assedatioun in the yeir of God 1559 to Mr James Thornetoun, titular of the said benifice, and George Trumbill of Barnahillis be Mr Johnne Thornetoun elder, usufructuar thairof, for the sowme of' 400 merks 'usuall money of this realme. Sen this assedatioun wes maid thir teyndis underwrittin ar reft and tane up be thir persounes, viz.': 22 husbandlands of Over Ancrum 'reft and tane up be Robert Ker, everilk husband land payand' 12 f. of victual, namely, 6 f. meal; 5 [f.] bere; 1 [f.] wheat extending in the whole to 66 b. The whole teinds of the Wodheid, the Skaw, the Gaw and the Braidlaw 'reft and tane up be the foirsaid Robert Ker' extending to 20 b. meal; 14 b. bere; 8 b. wheat, the whole extends to 42 b. The Manes of Nather Ancrum 'reft and tane up be Andro Ker quhilk payis in teynd' 4 b. meal; 7 b. bere; 4 b. wheat, the whole extends to 15 b. The whole teinds of Tronehill 'rest and tane up be Thomas Trumbill quhilkis payis of teynd' 8 b. meal; 4 b. bere, extending in the whole to 12 b.

'Summa of it that is reft and violentlie tane up extendis to' 135 b.

Signature: 'Ita est Jac[obus] Thorntoun, teste manu propria.'

KELSO, ABBEY OF,[208] (SRO, Vol. a, fos 240r-242v)
(SRO, Vol. b, fos 171r-173v)

'Ane uther rentale of Kelso'

'Item, the greit mailles of Kelso toun', £66 13s 4d. 'The coill silver of Kelso', £13 6s 8d. 'The mailles of the cottares', £57. 'The fischeing of Tueid', £66 13s 4d. 'The west cobill of Roxburcht [Sproustoun, *fo 223r*]', £20. 'The cobill foiranent

---

[204] This figure does not tally.
[205] This figure does not tally.
[206] The correct figure should be 4 c. 2 f. ⅞ p.
[207] Cf SRO, Vol. a, fos 164v, 212v; NLS, fo 82r (see above, pp. 159, 213; below, p. 365).
[208] Cf SRO, Vol. a, fos 223r, 232r, 235r, 243r (see above, pp. 222, 230, 233; below, p. 239).

Maxwell', £10. 'The unes [uvynes, *fo 232r*] and customes of Kelso', £10. The Cunynghares, 26s 8d. The Bromebank and the Hocht [Hoitt, *fo 223r*; Hote, *fo 243r*], 13s 4d. 'The mylnes of Kelso', £80. The West croftis, 40s. Angrieflett, 20s. The Brome croft, 12s. The Toun croft, 10s 8d. The mails of Sproustoun, £18. The mails of Reddane 'by ane pairt of the toun that lyis to the Manes of Sproustoun', £34. The ploughlands of Halden, £4. Eileischeucht, £5. The kirks [*recte*, kirklands] of Nanetharne, 40s. The kirklands of McKirstoun, £10. The Fewstodrik [*i.e.* Feu of Todrik, *fos 235r, 243r*], 40s. The toun of Newtoun, 30s. In Langgordoun, £4. The annual of Deringtoun, 5s. The kirkland of Grenlaw, 53s 4d. The kirkland of Sympryn, 40s. Humbie, £10. The kirklands of Carlouk, £4. The Dowglen, £5. Greinheid, £5. Quhytmuirhall, £5 6s 8d. Quhytmuirtoun, £10. Ones croft [Unysclos, *fo 232r*; *i.e.* Ovenscloss], 20s. The Lindene, £16. The four lands [husbandlands, *fos 223r*] of Lintoun, £4. Fauldounsyd, £10. Cauldscheillis, 40s. Quhitlawhous, £10. The toun of Mydlane, £31. 'The girs of Mydlane', £8. 'The Mylne of Mydlane and mylnlandis', £7. The Clarilaw Manis, £200. Kypellaw in the Clarilaw, £5. Lowreis land in Clarilaw, 40s, 'and in the augmentatioun of the few', 20s. 'Mr Georg Keper [*recte*, Keris, *fos 232v, 243v, 245v*] pairt in Clawrilaw', £6 13s 4d. Halyden, £10. Over Howden and Nethir Howden, £20. The Bothell, £5. The Hairheid, £5. The Bowschelhill, 40s. The Chapelhill, £12. Duddingstoun Ester and Wester, £66 13s 4d. The ploughlands of Hoislaw, 40s. Fogorig, 20s. The Mylne of Linden, 53s 4d. The Bowdane toun, £31. The Mylne of Bowdane, £5. Prestoun, £5. The toun of Melorstanis, 40s. The annual of Baltrude, £5 6s 8d. The Softlawis with the two lands [husbandlands, *fo 223r*] of Hevin [Heytoun, *fo 223r*; *i.e.* Heiton], £10. The Boigend, 53s 4d. Cadray, £3 6s 8d. The Mylne of Fogo, £4. The Clydismylne, £6 13s 4d. 'Summa of the mailis of the foirsaidis landis', £961 11s [*in margin*, '900.59 li. 4s 4d'[209]]; 'thairof defalkit for the erle of Glencarne', £66 13s 4d; 'sua restis' £894 17s 8d.

The rental of the kirks set for money

The kirk of Closburne, £40. The parsonage and vicarage of Selkirk, £266 13s 4d. The kirk of Sanct James, 40s. The kirk of Innerlethame [with Hopcalzie, *deleted*], £30. The kirk of Lintoun, £36 13s 4d. The kirk of Stowgordoun [*i.e.* Staplegordon], £8. The kirk of Symontoun, £12. The kirk of Dunsyre, £10 [*in margin*, 'Nota, gevin furth in a personage']. The kirk of Robertoun, £20. The kirk of Welstoun [*i.e.* Wiston], £16. The kirk of Carlouk, £66 13s 4d. The kirk of Mortoun, £26 [£25, *fo 236r*]. The kirk of Drumfreis, £60. The kirk of Trailflat, £14. The kirk of Drumgrie, £5. The kirk of Kilmaweris, £60. The kirk of Caldercleir, £44. The kirk of Humbie, £43 6s 8d. The kirk of Dudingstoun, £66 13s 4d. The kirk of Pettirculter, 20 merks. The kirk of Maxwell, £66 13s 4d. The kirk of Nantherne, £80. The kirk of Kelso, £166 13s 4d. The teinds of Angreflat, West croft, Bordourflat, Cunyngares, £6. Horneden with the vicarage, £13 6s 8d. 'Item, the haill teyndis of Sproustoun parochin except the teyndis of

---

[209] SRO, Vol. b, also has '900.59 li. 4s 4d.' The correct calculation appears to be £959 11s.

Eister Softlaw and Mellenden used to be led to the behove of the place and be commoun estimatioun we estemit to' £180. The teinds of Cleirlaw 'be commoun estimatioun ar worth' £66 13s 4d. 'Summa of all thir kirkis set for money', £1,428 13s 4d;[210] 'defalkit for the erle of Glencarne', £321 13s 4d; 'sua restis' £1,070 [*inserted*, 'richt, 1107 li.'[211]].

The kirks set for victual

The kirk of Makcarstoun, 6 c. meal. The kirk of Gordoun 'by the lord Houmes[212] teyndis', 3 c. bere; 6 c. meal. The kirk of Langtoun and Simpreay, 2 c. 10 b. 2 f. $2\frac{2}{3}$ p. bere; 5 c. 5 b. 1 f. $1\frac{1}{3}$ p. The kirk of Mow, 2 c. meal; 24 b. bere. The kirk of Hume 'halelie in the lord James handis [lord Humes handis, *fo 236v*] togidder with the said lord Homes teyndis of the landis following in the parochinis of Gordoun, Greinlaw and Fogo, viz.':

In the parish of Gordoun, Belletaw, Mydlethrid, Wester Manes, Rymmiltounlaw and the Eister Manes of Gordoun.

And in the parish of Greinlaw, the Harelaw, James Spens mailing in Rochesterik, Howlawheid, Wester Howlaw and Eister Howlow and Dennerleis.

In the parish of Fogo, Fogotoun, John Tranentis land, Sir James Tranentis land, John Trollouris [Trotteris, *fo 237r*] in the Hill and the Bogend 'altogidder payit' 16 c. meal; 8 c. bere. [*In margin*, 'Nota, I beleve the lord Home intrometis also with the teyndis of Rowenstones, xiiij b. meill; xviij b. beir'.[213]]

The kirk of Bowden 'by the teyndis of Clairlaw', 3 c. 8 b. $5\frac{1}{3}$ p. bere; 7 c. 2 f. $2\frac{2}{3}$ p. meal. The kirk of Lynden, 3 c. meal; 24 b. bere. The kirk of Greinlaw 'by the lord Humes teyndis', 8 c. meal; 4 c. bere. The kirk of Fogo, 1 c. 9 b. $5\frac{1}{3}$ p. bere; 3 c. 2 b. 2 f. $2\frac{2}{3}$ p. meal. The Manes of Sproustoun 'wer set with the plenisching for' 8 c. wheat; 8 c. bere; 8 c. meal. The teinds of Melleden within the parish of Sproustoun, 4 b. meal; 2 b. bere. The teinds of Eister Softlaw within the said parish, 7 b. meal; 3 b. 2 f. bere. 'Item, in Kelso toun thair is certane fermes payit, viz., the Bank payis', 4 b. [3 b., *fo 237r*] bere; 3 b. meal. 'Item, thair the Eister Medowis in Kelso', 8 b. oats 'and ij fedderis [hay, *fo 244v*]. Summa of all thir kirkis in victuall', 9 c. wheat; 40 c. 4 b. 3 f. $1\frac{1}{3}$ p. bere; 65 c. 6 b. 2 f. $2\frac{2}{3}$ p. meal; 8 b. oats.

'Memorandum, thir vicaregis wer wont to be set for the pryces following': The vicarage of Nantharne, £13 6s 8d. The vicarage of Gordoun, £66 13s 4d. The vicarage of Home, £10. The vicarage of Symphryn, £2. The vicarage of Mow, £13 6s 8d. 'Summa of money of the vicaregis', £105 6s 8d.

---

[210] The correct calculation appears to be £1,419 13s 4d; cf above, p. 234, n. 194.
[211] SRO, Vol. b has no equivalent correction.
[212] See above, p. 234, n. 197.
[213] 'xxiiij b. mell & viij b. beir', fo 236v (see above, p. 235).

'Summa of the haill money by the erle of Glencarne', £2,057 4s 4d;[214] third thereof, £685 14s 9¼d. Third of the wheat, 3 c. Third of the bere, 13 c. 6 b. 3 f. 3 p. Third of the meal, 21 c. 12 b. 3 f. 2¼ p.[215]

'Si per laborem honeste quippiam feceris labor abit honestum manet quod si per voluptatem turpe quippiam feceris quod suave est abit quod turpe est manet.'[216]

Signature: 'Sic subscribitur, Jacobus Hoppringill.'

[*In margin*, 'Alyk to this fra *.'[217]]

KELSO, ABBEY OF,[218]   (SRO, Vol. a, fos 243r-246v)
                                                            (SRO, Vol. b, fos 174r-178v)

Rental of Kelso

'Item, the greit mailles of Kelso toun', £66 13s 4d. 'The coil silver of Kelso', £13 6s 8d. 'The maillis of the cottaris [*cottagiis, fo 223r*]', £57. 'The fischeing of Twid', £66 13s 4d. 'The west cobill of Roxburcht [*Sproustoun, fo 223r*]', £20. 'The cobill foiranent Maxwall', £10. 'The unis [*uvynes, fo 232r*] and customes of Kelso', £10. The Cuningareis, 26s 8d. The Bromebank and the Hote, 13s 4d. 'The mylnes of Kelso', £80. The West croftis, 40s. Angrieflett, 20s. The Broun [*recte*, Brome] croft,[219] 12s. The Towne croft, 10s 8d. The mails of Sproustoun, £18. The mails of Reddane 'by ane pairt of the toune that lyis to the Manis of Sproustoun', £34. The ploughlands of Yaldan [*Halden, fos 223r, 232r, 240r*; *Haldein, fo 235r*], £4. The Elleischeucht, £5. The kirklands of Nanetharne, 40s. The kirklands of Makcarstoun, £10. The feu of Todrik, 40s. 'The towne of Newtoun', 30s. The Lange Gordoun, £4. The annual of Deringtoun, 5s. The kirkland of Grenelaw, 53s 4d. The kirkland of Symprene, 40s. Humby, £10. The kirklands of Carlowk, £4. The Dowglen, £5. The Greneheid, £5. Quhytmuirhill [*recte*, Quhytmuirhall, *fo 235v*], £5 6s 8d. Quhytmuirhill [*recte*, Quhytmuirtoun, *fo 235v*], £10. Ones Clois croft [Unysclos, *fo 232r*; *i.e.* Ovenscloss], 20s. The Lyndyne, £16. The four lands [husbandlands, *fo 223r*] of Lintoun, £4. Faldounsyd, £10. The Cauldscheillis, 40s. The Quhytlawhous, £10. 'The towne of Mydlane', £31. 'The gyrs of Mydlane', £8. 'The Mylne of Mydlane and mylnelandis', £7. The Clarylawmanes, £133 6s 8d [£200, *fos 232v, 235v,*

---

[214] This figure does not tally.
[215] Oats have been omitted from the calculation of third.
[216] I.e. 'If through honest labour something is done, the labour departs and the honesty remains; but if through dishonourable pleasure something is done, the sweetness departs and the dishonour remains.'
[217] No corresponding asterisk in the text has been traced.
[218] Cf SRO, Vol. a, fos 223r, 232r, 235r, 240r (see above, pp. 222, 230, 233, 236).
[219] Cf above, p. 222, n. 86.

240v].[220] The Kyplaw and Clarylaw [Kippellaw in Clarylaw, *fo 235v*], £5. The Lowreis land in Clarelaw 'and in the augmentatioun of the few', [*blank*].[221] 'Mr George Keris parte in Clarelaw', £6 13s 4d. Halyden, £10. Overholden and Nethirholden, £20. The Bothell, £5. The Hareheid, £5. The Bowschelhill, 40s. The Chapelhill, £12. Dudingstoun Eister and Wester, £66 13s 4d. The ploughlands of Hoisla, 40s. Fogo Rig, 20s. The Mylne of Lynden, 53s 4d. The Bowden toun, £31. The Mylne of Bowden, £5. Prestoun, £5. 'The towne of Mellerstanes', 40s. The annual of Baltrude, £5 6s 8d. Softlawis with the two lands [husbandlands, *fo 223r*] of Hewtoun [*i.e.* Heiton], £10. The Boigend, 53s 4d. Cadray, £3 6s 8d. The Mylne of Fogo, £4. The Clydismylne, £6 13s 4d. Belleischell, 40s. [*In margin*, £416 6s 8d.]

*The rental of the kirks set for money[222]

The kirk of Closburne, £40. The parsonage and vicarage of Selkirk, £266 13s 4d. The kirk of Sanct James, 40s. The kirk of Innerletheim with Hopcailzie, £30. The kirk of Lintoun, £36 13s 4d. The kirk of Stabilgordoun, £8. The kirk of Symountoun, £12. The kirk of Dunsyre, £10. The kirk of Robertoun, £20. The kirk of Wilstoun [*i.e.* Wiston], £16. The kirk of Carlowik, £66 13s 4d. The kirk of Mortoun, £25. The kirk of Drumfreis, £60. The kirk of Trainflate [*i.e.* Trailflat], £14. The kirk of Dungrie, £5. The kirk of Kilmawres, £60. The kirk of Caldircleir, £44. The kirk of Humby, £43 6s 8d. The kirk of Dudingstoun, £66 13s 4d. The kirk of Petircultir, 20 merks. The kirk of Maxwell, £66 13s 4d. The kirk of Nantharne, £80. The kirk of Kelso, £166 13s 4d. The teinds of Angreflatt, Westeroff [*recte*, West croft, *fos 232r, 236v, 241r*], Bordourflatt, Cunyngharis, £6. Horndein with the vicarage, £10 [£13 6s 8d, *fo 241r*]. 'Item, the haill teindis off Sproustoun parochin except the teindis off Eister Softlaw and Melleden usitt to be led to the behouff off the place and be commoun estimatioun war estemed' £180. 'Item, the teindis off Clarylaw be commoun estimatioun ar worth' £66 13s 4d. [*In margin*, £1,415.[223]]

The kirks set for victual

The kirk of Makcarstoun, 4 c. meal; 4 c. bere. The kirk of Gordoun 'by the lord Humes[224] teindis', 3 c. bere [and 6 c. meal, *fos 236v, 241v*].[225] The kirk of Langtoun and Semprim, 8 c., 'twa pairt meill, twa pairt beir, viz.' [*recte*, ⅔ meal, ⅓ bere], 2 c. 10 b. 2 f. 2⅔ p. bere; 5 c. 5 b. 1 f. 1⅓ p. [meal].[226] The kirk of Mow, 8 c. 'twa part meill, 3 [part] beir, pevyde [*recte*, devide, *SRO, Vol. b*] ut supra'. The kirk of Hume 'hallylie in the lordis Humes hand togider with the said lord

---

[220] 'Clarelawmanes quhilk of auld wes set for £210 is set in few to young Cesfuird for' £40, fo 223v (see above, p. 223).
[221] 'Lowreis land in Clarilaw', 40s; and in augmentation of the feu, 20s', fo 240v (above, p. 237).
[222] The text from the asterisk to n. 227 is written by a different hand.
[223] The correct calculation appears to be £1,415 6s 8d.
[224] See above, n. 197.
[225] '14 c. victual', fo 234r (see above, p. 232).
[226] The correct calculation appears to be 5 c. 5 b. 1 f. 1⅓ p.

Humes teindis in the landis following in the parechines of Gordoun, Grenlaw and Fogo, viz.':

In the parish of Gordoun: Eister Gordoun, Belitaw, Midethrid, Wester Mannis, Rymyltounlaw and the Eister Mannis of Gordoun.

And in the parish of Greinlaw: the Hordlaw, James Spensis mailing in Rochesterrig, Howlawheid, Wester Howlaw and Eister Howlaw and Dennerleis [Dennarles, *SRO, Vol. b*].

And in the parish of Fogo: Fogotoun, John Tranentis land, Sir James Tranentt land, John Trotteris off the Hill and the Boigend 'all thir togider payitt' 16 c. meal; 8 c. bere.

The Mylne [*recte*, kirk, *fos 234r, 237r*] of Bowdein 'by the teindis of Clarelaw', 10 c. 9 b., ⅔ meal, ⅓ bere, namely, 3 c. 8 b. 5⅓ p. bere; 7 c. 2 f. 2⅔ p. meal. The kirk of Lenden, 3 c. meal; 24 b. bere.[227] The kirk of Greinlaw 'by my lord Homes teyndis', 8 c. meal [and 4 c. bere, *fo 237r*]. The kirk of Fogo 'by my lord Humes teyndis', 5 c. 8 b. victual, ⅔ meal, ⅓ bere.[228] The Manes of Sproustoun 'wer set with the plenisching for' 8 c. wheat; 8 c. bere; 8 c. meal, and now the Manes of Sproustoun, 'the ane' £20 'land pertenyng to the Chartourhous and the uther to the place, and the haill being in manes of the place, the thrid of thair awin in consideratioun of the dountaking of the plenisching extendis to' 9 c. namely, 3 c. wheat; 3 c. bere; 3 c. meal. The teinds of Melleden within the parish of Sproustoun, 4 b. meal; 2 b. bere. The teinds of Eister Softlaw within the said parish, 7 b. meal; 3 b. 2 f. bere. 'Item, in Kelso toun thair is certan fermes payit, viz., the Bank', 3 b. bere; 3 b. meal. 'Item, the Eister Medowis in Kelso', 8 b. oats 'and ij fedderis hay.'[229]

'Memorandum, thir vicaragis wer wount to be sett for the prices following': The vicarage of Nantharne, £13 6s 8d. The vicarage of Gordoun, £66 13s 4d. The vicarage of Howme, £10. The vicarage of Symprene, £2. The vicarage of Mow, £13 6s 8d. 'Summa of thir vicarages', £105 6s 8d.

'Summa of beir', 33 c. 12 b. 3 f. 1⅓ p.[230] 'Summa of mell', 53 c. 14 b. 2 f. 3 p.[231] 'Summa of quheit', 3 c.[232]

Kelso 'by Lesmahago and by the kirkis of Closburne, Innerlethame,[233] Carlowk, Mortoun, Drumfres, Humby, Dudingstoun and mailles of the landis of Dudingstoun asignit to James Cunynghames pensioun and als by the personage of Lyntoun gevin out in a severall personag, the rest of the haill fruitis of Kelso

---

[227] The text at this point reverts to the hand of the original scribe.
[228] This figure does not appear in any of the previous rentals.
[229] SRO, Vol. b, fo 176r-v, here corresponds to SRO, Vol. a, fo 235r-v (see above, p. 238).
[230] The correct calculation appears to be 30 c. 11 b. 2 f.
[231] The correct calculation appears to be 56 c. 4 b. 2 f. ⅔ p.
[232] 8 b. oats have been omitted from these totals.
[233] 'Hopcalzie' is included at this point on fo 246v (see below, p. 243).

extending in money, in the haill', £1,983 17s 8d;[234] third thereof, £667 19s 2⅔d.[235] Bere in the whole, 30 c. 11 b. 2 f.[236] [30 c. 11 b. 3 f., *SRO, Vol. b*]; third thereof, 10 c. 3 b. 3 f.[237] Meal in the whole, 57 c. 14 b.;[238] third thereof, 19 c. 10⅓ b. [19 c. 1⅓ b., *SRO, Vol. b*] Wheat in the whole, 3 c.; third thereof, 1 c. Oats in the whole, 8 b.; third thereof, 2 b. 2 f. 2⅓ p.[239]

Assignation of Kelso:

Third of money, £667 19s 2⅔d. Take Faldounsyd for £10; Caldschellis 40s; Quhitlawhous, £10; 'the towne of Mydlane for' £31; 'the Mylne of Mydlane and myllandis for' £7; Kippelaw in Clarelaw, £5; 'Mr George Keris pairt in Clarelaw for' £6 13s 4d; Overholden and Natherholden for £20; Bothell for £5; Hairheid for £5; the Bowscheilhill for 40s; the Chapelhill for £22; the Mylne of Lyndane for 53s 4d; 'Bowdane towne' for £31; the Mylne of Bowdane for £5; Preistoun for £5; Softlawis with the two lands [husbandlands, *fo 223r*] of Haytoun, £10; the Mylne of Fogo for £4; the parsonage and vicarage of Selkirk for £266 13s 4d; the kirk of Symontoune for £12; the kirk of Dunsyre for £10; the kirk of Robertoun for £20; the kirk of Wolstoun [*i.e.* Wiston] for £16; the kirk of Petirculter for 20 merks; the kirk of Maxwell for £66 13s 4d; the kirk of Nantharne for £80; Quhitmuirtoun for £10. 'Giff in' £5 11s 10⅔d.[240]

Third of bere, 10 c. 3 b. 3 f. [1]⅓ p. Take the kirk of McCarstoun for 4 c.; the kirks of Langtoun and Sympriane for 2 c. 10 b. 2 f. 2⅓ p.;[241] the rest of the bere out of the kirk of Bowden 'by the teyndis of Clarelaw', 3 c. 8 b. 5⅓ p.; the kirk of Lyndene for 24 b. 'Giff in' 1 c. 7 b. 2⅔ p.

Third of meal, 19 c. 1⅓ b. Take the kirk of McCairstoun for 4 c.; the kirks of Langtoun and Sumpryne for 5 c. 5 b. 1 f. 1⅓ p.; the kirk of Bowden 'forby the teyndis of Clarelaw' for 7 c. 2 f. 2⅔ p.; the kirk of Lyndene for 3 c. 'Giff in' 4 b. ⅔ p.[242]

Third of wheat, 1 c. Out of the Manes of Sproustoun giving 3 c.

Third of oats, 2 b. 2 f. 2⅓ p.[243] Out of the Eister Medois of Kelso giving 8 b.

'The thrid of tua fadderis hay thair.'

Kelso 'by the pensioun money: in haill', £1,987 4s 4d;[244] third thereof, £662 8s 1⅓d 'assume ut supra'.

---

[234] This figure does not tally.
[235] The correct calculation appears to be £661 5s 10⅔d.
[236] This figure does not tally, but it is correct: see above, p. 241, n. 230.
[237] 1⅓ p. has been omitted from this calculation.
[238] The correct total appears to be 56 c. 4 b. 2 f. ⅔ p. (see above, p. 241, n. 231). If the total were 57 c. 4 b., the third in SRO, Vol. b would be correct.
[239] The correct calculation appears to be 2 b. 2 f. 2⅔ p.
[240] Total, £678 minus £667 19s 2⅔d is £10 0s 9⅓d.
[241] The correct calculation appears to be 2 c. 10 b. 2 f. 2⅔ p.
[242] These figures would give 19 c. 6 b., not 19 c. 1⅓ b.
[243] The correct calculation appears to be 2 b. 2 f. 2⅔ p.
[244] This figure does not tally.

The kirks of Closburne, Innerlethame 'by Hopcalzie' [with Hopcalzie, *fos 236r, 244r*], Carlouk, Mortoune, Drumfreis, Humby, Dudingstoun, 'with the mailles of baithe the Dudingstones assignit with Lesmahago in pensioun extending in the haill to' £388 6s 8d;[245] third thereof, £129 8s 10½d.

'Tak the kirk and landis of Dudingstounes gevand be yeir tua hundreth merkis. Gif in' £3 17s 9⅓d.

YETHOLM, PARSONAGE OF, (SRO, Vol. a, fo 247r)
(SRO, Vol. b, fo 179r)

'The rentale of the personage of Yettem be yeir is fourscoir pundis and of this gevin in pensiones to Mr Johnne Row' £20, 'and to David Ormestoun, sone to Robert Ormestoun,' 20 merks.

Signature: 'Thomas Crestesoun, with my hand, persone of Yettem.'

LESMAHAGOW, PRIORY OF,[246] (SRO, Vol. a, fos 247v-248v)
(SRO, Vol. b, fos 179v-180v)

'Lesmahago anno domini millesimo quingentesimo quinquagesimo sexto' [*i.e.* 1556].

'J. Weir, camerarius.'

'Computum honorabilis viri Joannis Weir'

The account of an honourable man, John Weir, chamberlain of Lesmahago, rendered at Melros 1 February 1556/7, of all his receipts and expenses in account of fermes, teinds and the annualrent of the said lordship from 23 December 1556 to the day of his account and so of the two terms within this account, namely, the feast of Pentecost[247] and of St Martin[248] 1556, and of one crop of the said year 1556, in the hearing of the the lord commendator[249] of Calco and Melros, Mr

---

[245] This calculation has not been traced.
[246] Cf *Registrum Cartarum de Kelso*, ii, p. 475, which is similar to this entry. Lesmahagow was a dependency of Kelso, Cowan and Easson, *Medieval Religious Houses*, p. 69.
[247] I.e. seven weeks after Easter.
[248] I.e. Martinmas, 11 November.
[249] James Stewart was commendator of Kelso and Melrose until his death in 1557, Cowan and Easson, *Medieval Religious Houses*, pp. 68, 77.

William Schaw, provost of Abynethys [*recte*, Abernethy], Mr Walter Balfour, parson of Lintoun, and dene Ralph.[250]

The charges of Hudsoun, the monk.

[*In margin*, 'Pecunia'] 'Memorandum, the comptar chargis withe'[251] £59 11s 4d 'of the males and money of the said lordschip be arreage and dependis of his last compt. Summa hujus onerationis patet.' [*i.e.* The total of these charges is obvious.]

Item, the same[252] charges himself with £195 4s for feufermes of the said lordship during the time of this account.

The same charges himself with £102 for fermes and teinds of the kirks of Cloisburne, Trailflat, Robertoun, Voustoun [*i.e.* Wiston][253] and Symontoun within the time of the account.

The same charges himself with £112 14d for Easter dues and teinds of the vicarage of Lesmahago in 1556.

Total of these charges, £409 5s 2d.

Item, the same charges himself with £180 for the fermes and teinds of the kirk of Drumfreis of the three years bygone still remaining, namely, of the crops of the years 1553, 1554 and 1555. With £15 for fermes and teinds of the kirk of Drumgrie of the three years abovewritten still remaining. With £30 for fermes and teinds of the kirk of Dunsyre of the three years abovewritten still remaining. With £120 for fermes and teinds of the kirk of Mortoun of the three years above mentioned still remaining. With £200 for fermes and teinds of the kirk of Kilmaweris of the three years still remaining. With £200 for fermes and teinds of the kirk of Carlouk of the three years aforesaid still remaining. Total of these charges, £745.

Total of all charges of money, £1,214 4s 6d.[254]

[*In margin*, 'Ordeum', *i.e.* Bere] Memorandum, the treasurer ('computans') charges himself with 1 c. 11 b. bere for arriages resting and remaining in the account of the crops of the years 1553 and 1554. The total of these charges is obvious.

Item, the treasurer charges himself with 13 c. 13 b. 1 f. 2 p. bere of the crop within this account for fermes and teinds of the kirk of Lesmahago. The total of these charges is obvious.

---

[250] The text has 'domino' but probably should read 'dompno'; *Registrum Cartarum de Kelso*, ii, p. 476, has 'domino rodulpho hutsoun monacho'; 'Radulphus Hudsoun [*i.e.* Ralph Hudson], monachus et registri custos manu sua signet', also occurs as a notary, *Records of the Regality of Melrose*, iii, pp. 149, 151, 154, 161.

[251] *Registrum Cartarum de Kelso*, ii, p. 476, reads 'chairgis him with'.

[252] Presumably Hudson on behalf of Weir.

[253] 'Vrmistoun', *Registrum Cartarum de Kelso*, ii, p. 476.

[254] The correct calculation appears to be £1,213 16s 6d.

Total of the charges of bere, 15 c. 8 b. 1 f. 2 p.

[*In margin*, 'Farina', *i.e.* Wheat] Memorandum, the treasurer charges himself with 1 c. 8 b. 3 f. oatmeal resting and remaining in the account of the crops of the years 1553 and 1554. The total of these charges is obvious.

Item, the treasurer charges himself with 40 c. oatmeal for fermes and teinds of the kirk of Lesmahago of the crops within this account. The total of these charges is obvious.

Total of the charges of wheat, 41 c. 8 b. 3 f.

[*In margin*, 'Avenarum', *i.e.* Oats] Memorandum, the treasurer charges himself with 8 b. oats for arriage resting and remaining in the account of the crops of the years 1553 and 1554.

[*In margin*, 'Avene', *i.e.* Oats] Item, the treasurer charges himself with 2 c. 11 b. oats for teinds of the Mains of Lesmahago within this account of 1 c. 8 b. oats of the kane within this account, namely, 1 c. oats for the lands of Auchinaich [*i.e.* Auchenheath] and 8 b. oats for the lands of Altum [*i.e.* Auldtoun].[255] Total of these charges, 4 c. 3 b. oats.

Total of the charges of oats, 4 c. 11 b.

Total of the charges: £1,214 4s 6d money; 15 c. 8 b. 1 f. 2 p. bere; 41 c. 8 b. 3 f. meal; 4 c. 3 b. oats.

Above this, the treasurer charges himself with 250 fowls counting 100 for 120 according to Scots custom '*lie* fowlis for teynd hay' of the kirk of Lesmahago within this account.

Bere. Item, the treasurer charges himself with 15 c. 8 b. 1 f. 2 p. bere, charged above of the crop of the year 1555 and balance thereof.

Total of these charges is obvious.[256]

---

[255] 'Altum', *Registrum Cartarum de Kelso*, ii, p. 478.

[256] The text in SRO, Vols. a and b ends here, though the rental for Lesmahagow in *Registrum Cartarum de Kelso*, ii, pp. 479-485, continues with further entries. Rentals for Roxburgh resume at SRO, Vol. a, fo 255r (see below, p. 257).

Peebles

GLENHOLM, PARSONAGE OF, (SRO, Vol. a, fo 249r)
(SRO, Vol. b, fo 181r)

Rental of the parsonage of Glenquhome, set in assedation by Mr William Hammiltoune, parson thereof, extends to 110 merks.

Signature: 'Mr Williame Hammiltoun, persoun of Glenquhome.'

Calculation of third: In the whole, £73 6s 8d; third thereof, £24 8s 10$\frac{2}{3}$d.

KILBUCHO, PARSONAGE AND VICARAGE OF, (SRO, Vol. a, fo 249r)
(SRO, Vol. b, fo 181r)

Parsonage and vicarage of Kilbocho

Calculation of third: In the whole, £80; third thereof, £26 13s 4d.

INNERLEITHEN, VICARAGE OF, (SRO, Vol. a, fo 249r)
(SRO, Vol. b, fo 181r)

Vicarage of Innerlethane

Calculation of third: In the whole, £66 13s 4d; third thereof, £22 4s 5$\frac{1}{3}$d.

EDDLESTON, PARSONAGE AND VICARAGE OF, (SRO, Vol. a, fo 249v)
(SRO, Vol. b, fo 181v)

Rental of the parsonage and vicarage of Eddilstoun within the diocese of Glasgow and sheriffdom of Peiblis.

'The yeirlie availl of the personage thairof extendis to' 220 b. meal. 'The lamb and woll can nocht be weill estimat to ony certane valour, bot the haill personage and vicarege being set with the haill profeitis of auld wont to be payit extendit to' 220 merks 'of the quhilkis profeitis be act of counsall the corpis presandis, upmest claithe, Paschefynes and small offerandis and utheris sik dewteis ar dischargit and thairthrow the yeirlie valour thairof gretumlie diminischit. Off the quhilkis fruites first is sustenit ane redar according to the ordour and Buik of Discipline,[1] secundlie is gevin yeirlie furth of the same to ane preist of the chore of Glasgow', 12 merks, 'togither with ane annuell of' 43s 'yeirlie payit to the chapellanes of the said chore.'

Signature: 'G. Hay.'[2]

Calculation of third: In meal, 13 c. 12 b.; third thereof, 4 c. 9 b. 1 f. $1\frac{1}{3}$ p.

'Nota, and the lamb and the woll nocht considderit on in speciall.'

RATHVEN, DUNDURCAS AND KILTARLITY, BENEFICES OF,[3]	(SRO, Vol. a, fo 249v) (SRO, Vol. b, fo 181v)

'The rentale of the kirk of Rothven and Dundurcus and Kyntallartie with the pendicles thairof togither with the mailles of the barronie of Rothven as it is [sic] frelie gefis conforme to the few set and confermit thairupoun, the beidmenis pensiounes and all uther ordinar chargis deducit and allowit, extendis to' £215 8s 10d; 2 b. victual; 2 b. 'custum aites; ij dosand vj capones.'

Signature: 'G. Hay. Tulibardin, Comptrollar.[4] James Nicolsoun,[5] ressave this rentale with the queinis grace confirmatioun of the few. J. Clericus Registri.[6] This rentall I desyr of your grace to be admittit and ressavit and commandment gevin to the clerk thairupoun.'

---

[1] See *The First Book of Discipline*, ed. J. K. Cameron (Edinburgh, 1972), pp. 105-107, 111, 162-163.
[2] The initial of the Christian name is indistinct; George Hay was parson, *RMS*, iv, no 1615.
[3] Cf NLS, fo 143v (see below, p. 456). Dundurcas and Kiltarlity were annexed to Rathven hospital, which was a prebend of Aberdeen, Cowan, *Parishes*, p. 169. These benefices, in the sheriffdoms of Banff and Inverness, are included in rentals for Peebles because George Hay, who held them, also held Eddleston.
[4] Sir William Murray of Tullibardine, Comptroller, 1565 - 1582; James Cockburn of Skirling also served as Comptroller, 1566 - 1567, *Handbook of British Chronology*, p. 191.
[5] James Nicolson, clerk of the Collectory, *Thirds of Benefices*, p. 62.
[6] James McGill of Nether Rankeilour, Clerk Register, 1554 - 1566 and 1567 - 1579; Sir James Balfour of Pittendreich, 1566 - 1567, *Handbook of British Chronology*, p. 197.

PEEBLES, MINISTRY OF THE TRINITARIANS IN,

(SRO, Vol. a, fo 250r)
(SRO, Vol. b, fo 182r)

[*In margin*, 'Ministerie of Peiblis']

'The rentale of the leving pertenyng to the ministrie of Peiblis.'

'Inprimis, the kirk of Kithins [*i.e.* Kettins], personage and vicarege thairof set in assedatioun to the tutour of Petcur[7] for' 320 merks yearly, the kirkland thereof set to James Small yearly, 12 merks. The temporal lands of Houstoun set in feu to William Wachope, younger, of Mudy and the laird of Wauchtoun yearly, 119 merks. 'Item, the akeris lyand about Dunbar set in few to James Howme, sone to Williame Hume of Lochtullo, yeirlie', £20. 6 acres of land lying beside the Croce Kirk of Peblis, £3. The Kingis Medowis, £5.

Signature: 'Gilbert Broun, minister of Peblis and persoun of Keythins.'[8]

Assignation of the third of the ministry of Peiblis:

Third of 'the ministerie', £109 11s 1½d. Take the temporal lands of Houstoun, feuars young Mudy and Wauchtoun, for £79 6s 8d [*in margin*, 'Hadingtoun']; 'the akeris lyand besyde Dunbar, fewar James Hume, set to Williame Hume of Lochtullo for' £20 [*in margin*, 'Hadingtoun']; 'the sex akeris besyd the Croce Kirk of Peiblis for' £3 [*in margin*, 'Peiblis']; the kirkland of Kytins, fermorar James Small, for 12 merks [*in margin*, 'Angus']. 'Giff in' 15s 6⅔d.

'Nota, the laird of Wauchtoun payis yeirlie for his pairt of Houstoun' £30 5s 8d money; 1 c. bere; 8 b. wheat; 8 b. rye.

STOBO, PARSONAGE OF,

(SRO, Vol. a, fos 250v-251r)
(SRO, Vol. b, fos 182v-183r)

'The rentale of the fruites of the personage of Stobo.'

'Inprimis, in the Deintoun[9] of Stobo [the toun of Stobo, *fo 251r*] extendis of victuall to' 7 c. 2 b. Hoprew Eister extends to 32 b. Hoprew Wester extends to 12 b. Drewar [*i.e.* Dreva] extends to 16 b. Brochtoun, namely, the Manes, Heuchtbray and Litilhope extends to 52 b. Burnetland extends to 6 b. Langlandhill extends to 15 b. Stirkfeild extends to 4 b. Scheilbank extends to 2 b. Clawerhill extends to 10 b. Greithetlandis extends to 10 f. Drummelzear

---

[7] James Haliburton, provost of Dundee, tutor of Pitcur, cf *RPC*, iv, p. 266.
[8] Gilbert Brown was 'minister' or head of the Trinitarian house whose church was the Cross kirk in Peebles. Kettins, in Angus, was the benefice appropriated to the Trinitarians in Peebles. Cowan, *Parishes*, pp. 93-94.
[9] 'Deintoun' has not been traced.

extends to 24 b. Dawik extends to 34 b. 'Summa of this victuall abonewrittin extendis to' 18 c. 13 b. 2 f.[10]

'Off this rentale abonewrittin in tyme of weir and waist be the thevis that is non payment', 3 c. 4 b. 'in the haill. In lambis this last yeir', 300; 'in woll', 20 stones.

Signature: 'M[r] Johnne Colquhoun, persoun of Stobo.'[11]

Assignation of the third of the parsonage of Stobo:
    Third of the parsonage of Stobo, 6 c. 13 b. $2\tfrac{2}{3}$ p.[12] Take this victual out of the toun of Stobo giving 'be yeir', 8 c. [Deintoun, 7 c. 2 b., *fo 250v.*]
    'Lambes', 100; 'Woll, 6 stanes half and thrid pairt half stane [*i.e.* $6\tfrac{2}{3}$ stones]. Tak this lambis and woll be generall charge becaus thai ar uncertanlie gevin up.'

KILBUCHO, VICARAGE PENSIONARY OF, (SRO, Vol. a, fo 251r)
(SRO, Vol. b, fo 183r)

The vicarage pensionary of Kilbocho pertaining to sir William Portuous.

Calculation of third: In the whole, £12; third thereof, £4.

Assignation:
    'Tak the gleib and kirkland for' 40s; and 40s 'out of the kirkland.'

KIRKURD, PARSONAGE OF, (SRO, Vol. a, fo 251v)
(SRO, Vol. b, fo 183v)

The parsonage of Kirkurd set to Hew Cokburne, brother to the laird of Skirling, for the yearly payment of £20 by Mr David Gibsone, possessor thereof.

'Thrid thairof', £6 13s 4d. 'Fermorar Hew Cokburne'.

---

[10] The correct calculation appears to be 20 c. 3 b. 2 f.
[11] John Colquhoun was also dean of the Christianity of Peebles, and commissary of Stobo, Watt, *Fasti*, pp. 182, 196. The benefice of Stobo was a prebend of Glasgow cathedral.
[12] This is one third of 20 c. 9 b.

LINTON, VICARAGE OF,[13]

(SRO, Vol. a, fo 251v)
(SRO, Vol. b, fo 183v)

The vicarage of Lintoun pertaining to sir Adam Kyngorne.

Calculation of third: In the whole, £43 2s; third thereof, £14 7s 4d.

NEWLANDS, PARSONAGE OF,

(SRO, Vol. a, fo 252r)
(SRO, Vol. b, fo 184r)

Rental of the parsonage of Newlandis lying in the diocese of Glasgow within the sheriffdom of Peiblis set in tack and assedation to my lord of Mortoun,[14] pays yearly 200 merks.

Signature: 'Subscryvit be me, M[r] A. Douglas.'[15]

Calculation of third: In the whole, £133 6s 8d; third thereof, £44 8s 10⅔d.

PEEBLES AND MANOR, BENEFICES OF,[16]

(SRO, Vol. a, fo 252r)
(SRO, Vol. b, fo 184r)

'The paroche kirkis of Peblis and Mennar is set to Alane Diksone for' 300 merks 'be yeir.'

Signature: 'Alane Dikesoun, with my hand.'

PEEBLES, VICARAGE OF,

(SRO, Vol. a, fo 252v)
(SRO, Vol. b, fo 184v)

Rental of the vicarage of Peiblis given in by Mr Thomas Archibald, vicar thereof.

'Quhilk vicarege is presentlie set to the tennentis and inhabitares of Peiblis, occupyeris of the akeris pertenyng thairto, for' 42 merks 'and gave yeirlie of befoir quhen corps presentis and utheris casualiteis thairof wer wont to be upliftit', £60. 'Item, heirof is to be deduce [*sic*] yeirlie to the minister', 20 merks.

Signature: 'Thomas Archibald, vicar of Peiblis, with my hand.'

---

[13] Cf SRO, Vol. a, fo 121r (see above, p. 108).
[14] I.e. James Douglas, 4th earl of Morton, *Scots Peerage*, vi, pp. 362-363.
[15] Archibald Douglas, parson, *RSS*, v, no 2740.
[16] Cf SRO, Vol. a, fo 253v (see below, p. 253). The parsonage of Peebles, and the parsonage and vicarage of Manor formed the prebend of the archdeacon of Glasgow, Cowan, *Parishes*, pp. 142-143, 162.

Calculation of third: In the whole, £28; third thereof, £9 6s 8d.

LYNE, PARSONAGE AND VICARAGE OF,   (SRO, Vol. a, fo 252v)
      (SRO, Vol. b, fo 184v)

Rental of the parsonage and vicarage of Lyne set in assedation by Mr Andrew Betoun, parson thereof, to John Wichtman, burgess of Peiblis, and William Weche [*i.e.* Waugh] of Kingis Syd, for payment yearly of £60 'of the quhilk is to be defalkit to the redar thairof', £10 money.

Signature: 'Thomas Archibald, procuratour and factour to the said Mr Andro Betoun, persoun of Lyne, with my hand.'

STOBO, VICARAGE PORTIONARY OF,   (SRO, Vol. a, fo 253r)
      (SRO, Vol. b, fo 185r)

Rental of the vicarage portionary of Stobo

'In the first', 30 b. meal 'in Tueidmure within the pendicle of Rumelzeiris [*i.e.* Drumelzier] and jurisdictioun of Stobo.' 40s 'land of auld extent with' 50 'sowmes gers gangand teynd fie quhilk payis bot' 5 merks yearly. 22 stones of cheese; $5\frac{1}{2}$ stones of butter; £22 16s 8d money.

Signature: '[*blank*] Vicarius portionarius de Stobo, manu propria.'

Calculation of third: Meal, 40 b. [30 b., *preceding paragraph*]; third thereof, 13 b. 1 f. $1\frac{1}{3}$ p. Money, £22 16s 8d; third thereof, £7 12s $2\frac{2}{3}$d. Cheese, 22 stones; third thereof, $7\frac{1}{3}$ stones. Butter, 5 stones [$5\frac{1}{2}$ stones, *preceding paragraph*]; third thereof, $1\frac{2}{3}$ stones.

[*In margin*, 'Alitar.'] 'The rentale of the vicarege of Stobo pertenyng to me, Robert Douglas, be resignatioun maid simpliciter be umquhill sir Niniane Douglas in' 1563 'extendis yeirlie in all dewteis and commoditeis to the sowme of' £60 'for the quhilk sowme the same wes set in assedatioun to Charlis Geddes, last takkisman thairof, sua the queinis thrid thairof extendis to' £20.

Signature: 'Robert Douglas, with my hand.'

GLASGOW, ARCHDEACONRY OF,[17]

(SRO, Vol. a, fo 253v)
(SRO, Vol. b, fo 185v)

'Archedeanrie of Glasgow contenand Peiblis and Mennar, Johnne Abircrumby, [*blank*] Betoune, and Allane Dikesoun,[18] takismen, payand yeirlie the sowme of' 300 merks 'iij yeiris takis enteres', 26 March 1560.

SKIRLING, PARSONAGE AND VICARAGE OF,

(SRO, Vol. a, fo 253v)
(SRO, Vol. b, fo 185v)

Rental of the parsonage and vicarage of Skraling

The said parsonage and vicarage set in tack and assedation by the parson thereof to James Cokburne of Skraling for the sum of £10 yearly.

'Ressavit at command of the Clerk of Registre.'

LINTON, PARSONAGE OF,[19]

(SRO, Vol. a, fo 253v)
(SRO, Vol. b, fo 185v)

Parsonage of Lyntoun pertaining to Mr Walter Balfour.

[*In margin*, 'personage of Lintoun reformit heirefter upoun the clxxxxvij leaf.'[20]]

PEEBLES,
CHAPLAINRY OF OUR LADY CHAPEL IN,

(SRO, Vol. a, fos 253v-254r)
(SRO, Vol. b, fos 185v-186r)

[*In margin*, 'Chapellanrie of Our Lady chaplanrie in Peiblis.']

Rental of the chaplainry of Our Lady chapel within Peiblis.

'In the first, ane corne mylne in the laird of Pirnis [Pyrmes, *SRO, Vol. b*] handis, ane corne mylne', 16 merks. 'Item, ane walk mylne with ane pairt of land thair set for the sowme of' 9 merks. 'Item, Johnne Wychtman and James Tuedyis land in Ernattis Hauche, of annuell', 40s. 'Item, the hous at the chapell bak', 9s. 'Item, ane place of Thome Tuedeis', 5s. 'Item, ane of Sanct Leonardis at the Cuinzie

---

[17] Cf SRO, fo 252r (see above, p. 251 and n. 16).
[18] John Abercrombie was recognised at the Roman curia as archdeacon of Glasgow in 1562; Andrew Betoun, who had been collated and installed as archdeacon in 1560, gained the crown's presentation in 1563; Watt, *Fasti*, p. 174. Allan Dickson, an Edinburgh burgess, was tacksman of the parsonage and vicarage of Peebles in 1561; SRO, CS7/20, Register of the Acts and Decreets of the Court of Session, fos 319r-320v.
[19] I.e. West Linton, cf SRO, Vol. a, fo 121r (see above, p. 108 and n. 63).
[20] This rental appears in SRO, Vol. b, fo 187r, not leaf 197.

Nuik', 3s 4d. 'Item, ane yaird of the Croce Kirk', 3s 4d. 'Item, Rowcastellis land', 3s. 'Item, of the laird of Pyrin land', 2 merks 'of annuell.'

Signature: 'Johnne Tuedy, cheplane of the chaplanrie of abonewrittin, with my hand. Ressavit at command of the Clerk Registre.'

Calculation of third: In the whole, £21 3s 8d; third thereof, £7 14⅔d.

PEEBLES, (SRO, Vol. a, fo 254r-v)
COLLEGIATE KIRK OF ST ANDREW, (SRO, Vol. b, fo 186r-v)
THE HOLY ROOD ALTAR IN,
CROSS KIRK, THE HOLY BLOOD ALTAR IN,

'The rentale pertenyng and gevin in be sir James Davidsoun, chaplen and prebendare of the Rude altare, situat within the college kirk of Sanct Andro in Piblis, and of the Haly Bluid altar, situat in the Croce Kirk of the said towne of Peiblis, bath unite in ane.'[21]

'Inprimis, of the Rud Mylne of Peiblis yeirlie', 17s. Melwingis land in Greistoun, 20s. The Moshoussis yearly, 4 merks. John Johnnestounes land, 2s. Robert Portuus and Done Puntis land', 4s. Robert Hoppringillis land, 6s 8d. John Mures land, 7s. John Hayis land, 2s. William Bellis land, 3s. Mage Gladdois land, 4s. William Smaillis and William Stevinsones land, 4s. William Dikesones land, 12s. William Frank and Agnes Lowis land, 18d. Andrew Wychtmanis barn at the Brig End, 6s. Michael Smythis land, 2s. William Wylieis land, 3s. John Mures land 'in the auld toun', 2s. Thomas Patersoun in the Lydgait, 2s. James Patersones land, 7s. John Bullois land in Congzie Nuik, 8s. John Stevinsones barn at the Watersyd, 18s. Thomas Inglis 'rude of land in Frankes croft', 3s. John Hogis land in Conzie, 3s. John Hoppis land, 8s. Thomas Folkertis land, 4s. Willie Dikesoun at the North Port, 4s. 'Item, of ane [croft, *deleted*] rude of land within the Deneis Park and ane barne at the syd of it', 3s. Mathesounes land, 2s. Joke Patersones land, 18d. John Coninggameis land, 3s. Willie Forsythis land, 4s. Gleidstanes land, 4s. James Thorbrandis land, 3s. Willie Robenis land, 2s. 'Summa totalis', £10 19s 2d.

'James, ressave this rentale, J. Clericus Registri.'

---

[21] The parish church in Peebles was the collegiate kirk dedicated to St Andrew. The Cross kirk in the burgh belonged to the Trinitarian house. Cowan, *Parishes*, p. 162; Cowan and Easson, *Medieval Religious Houses*, p. 224; *Charters and Documents relating to the Burgh of Peebles*, ed. W. Chambers (Edinburgh, 1872), pp. 264-268.

NOTE

Rentals for the following benefices are located elsewhere in the text:

LINTON, PARSONAGE OF (SRO, Vol. a, fo 121r), p. 108.
MANOR PARISH KIRK, ROOD ALTAR IN (EUL, fo 26r), p. 525.

Roxburgh (2)

MELROSE, ABBEY OF,[1] (SRO, Vol. a, fos 255r-258v[2])

'The rentale of the abacie of Melros pertenyng presentlie to the abbot,[3] by the landis of Kylismure, Barmure, Niddisdaill, Carrik and kirkis pertenyng thairto, and als by the fermis of Lamerm[uir][4] quhilk my lord of Glencarnes brother[5] hes in pensioun presentlie.'

The lands of Blainslie, £45 18s. Morestoun, £3 3s 4d. Langschaw, 'mylne thairof, and ester syd of Vousbyre' [*recte*, Housbyre], £18. Halkburnam [Halkburne, *fo 206r*], £3 6s 8d. Bukholme, £10. Apletrelevis, £30. Wester Raik and Wousbyre [Housbyre and Westerfuird, *fo 206r*], £10. Langlie, £12 13s 4d. Freir croft, £3 6s 8d. Meirbank, Sowtercroft, Cartleyis and New Fuird Haucht, £17 6s 8d. Drygrange, £22. Reidpethe, £21 13s 4d.

'The annuellis and custume of Littill Fordaill', £36. Newtoun Mylne, £8. Auld Melros, £6 13s 4d. Eilidoun, £26. Lossadden, £56. Murehouslaw, £12 6s 8d. Newsteid with the pendicles, £85 16s. Ladopemure, 'the nather syd', £18 6s 8d. 'Wairdis of Melros', 40s. Coulmesliehill, £5. Calfhill, £10 'or xxx woderis'.[6] Maxpoppill, £3 6s 8d. Cambestoun and Plewland, £10. Alenschawis, £6 13s 4d. Vouplaw, £3 6s 8d. Threipwode, £32. Quhitlie, £6 13s 4d. Williame Law, £6 13s 4d. Williame Law,[7] £5. Sorrowlesfeill, 30s. Newtoun, £26 13s 4d. Freirschaw, £5. Cringlis, £5. Monkfauld, 40s. Kirkland of Hassinden and Caveris, £6 13s 4d. Feymaisterlandis, £13 6s 8d. Newgrange Mylne in Lammermure, 40s. Galtrasyd with the pendicles, £119 19s 4d.

---

[1] Cf SRO, Vol. a, fo 206r (see above, p. 207). See also *Records of the Regality of Melrose*, iii, pp. 140-147; *RMS*, iv, no 1819.
[2] This rental does not appear in SRO, Vol. b.
[3] James Douglas, 2nd son of William Douglas of Lochleven, gained the commendatorship in May 1569, Cowan and Easson, *Medieval Religious Houses*, p. 77.
[4] 'Lamermure', *Records of the Regality of Melrose*, iii, p. 140.
[5] I.e. James Cunningham, *RSS*, v, no 871; cf above, p. 224, n. 99.
[6] 'Or xxxvotheris[?]', *Records of the Regality of Melrose*, iii, p. 141.
[7] There are two places with this name: William Law at NT4739, and Williamlaw at NT4739.

Feymaisterlandis,[8] £4. Anay, £43 6s. Buklawis, 16s. Walker croft, 20s. Pryourwod, £8 12s. Moshoussis, £18 13s. Urschous Mylne,[9] £12. 'Fischeing of Tueid', £6 13s 4d. Abbay Mylnes, £48. Dernik, £91. Danieltoun, £40. [*Blank*,] £37 13s.

'Summa of the haill mailles of the barronie of Melros', £1,024 16s 8d.[10]

The lands of Est Toviotdaill [*sic*]: Gaitschaw, £6 13s 4d. Hownoumgrange, £10. Cleftouncoit,[11] £6 13s 4d. Sowthcoit, £4. Autonburne,[12] £10. Hownoumgrange Mylne, 40s. Total, £39 6s 8d.

The lands of Ugginnis: Salfet [*recte*, Falset], £5. Trow,[13] £5. Coklaw, £7 10s. Enenschaw,[14] £7 10s. Bresneis,[15] £5. Sowrope, £5. Fasschaw,[16] £10. Copitrig,[17] £5. Total, £50.

The lands of the Mers: The four lands of Hassintoun Manes, Harlaw, Clerkland and Putellseucht [*i.e.* Pittlesheugh], £18 'Summa patet', £18.

The lands of Lamermure: Hertsyd, £40. Preistlaw, Panscheillis, Kingsyd, Freirdykis and Winterscheildykip, £30. Total, £70.

The lands of Tueddaill: Hairhoip, £15. Hopcarten, £3 6s 8d. Kingildoris, £8. Wolfclyd, 56s 1d. Total, £29 2s 9d.

The lands of Atrik: Glenkeyrie, £4. Migiehoipe, £6 13s 4d. Atrikhous, £6 13s 4d. Schorthope, £5. Fairhope, £6 13s 4d. Kirkhope, £5. Elspyhoipe, £6 13s 4d. Scabecleucht, £3 6s 8d. Craig, £2. Ramsecleucht,[18] £6 13s 4d. Thirlistane, £6 13s 4d. Langhope, £6 13s 4d. Total, £66.

---

[8] 'Feymaisterlandis' is listed twice. 'Femaisteris lands' paid £3 6s 8d, fo 206r (see above, p. 207). *Records of the Regality of Melrose*, iii, pp. 141, 142, has 'Seymaisterlands'; this place has not been traced.
[9] 'Ves[t]hous mylne', *Records of the Regality of Melrose*, iii, p. 142; this place has not been traced.
[10] The correct calculation appears to be £1,031.
[11] 'Elestoun Coit [or Cliftoncoat]', *Records of the Regality of Melrose*, iii, p. 142, now Cliftoncote.
[12] 'Antonburne', *Records of the Regality of Melrose*, iii, p. 142, now Attonburne.
[13] 'Trone', *Records of the Regality of Melrose*, iii, p. 142; 'Tronie', *RMS*, iv, no 1819; possibly now Trows.
[14] 'Emershaw [or Evingshaw]', *Records of the Regality of Melrose*, iii, p. 142; 'Alaneschaw', *RMS*, iv, no 1819, now Allanshaws.
[15] 'Breithchnes or [Bresmeis]', *Records of the Regality of Melrose*, iii, p. 142; 'Braicisnes', *RMS*, iv, no 1819; this place has not been traced.
[16] 'Fasschaw or [Radshaw]', *Records of the Regality of Melrose*, iii, p. 142; 'Raschaw', *RMS*, iv, no 1819; this place has not been traced.
[17] 'Capilrodrig', *RMS*, iv, no 1819; this place has not been traced.
[18] 'Ravynniscleuch', *RMS*, iv, no 1819; now Ramseycleuch.

The lands of Ringwodfeild: The Burgey, £4. Stobecoit, £6. Ringwodhaucht, £5. Bowandhill, £5. Grange, £4. Preisthauch, £5. Pennanguschoip, £5. Woisterrie, £5. Northous, £5. Sowdenrig, £5. Cauldcleuch, £3 6s 8d. Total, £52 6s 8d.

The lands of Esdaill Mure: Tulloquhair, £6 13s 4d. Crury, £5 6s 8d. Yetbyre, £6 13s 4d. Newbyre, £20. Kirkfauld, £3 6s 8d. Cristelhill, £4 6s 8d. Blakscoit, £6 13s 4d. Powdono [*recte*, Rodono], £3 6s 8d. Powmonk, £4 10s. Powcleis,[19] £3 6s 8d. Glendarge, £6 13s 4d. Cassope,[20] £5. Fynglen,[21] £5. Awnlosk,[22] £5. Mydlawheid, £5. Raburne, £10. Hairwod,[23] £10. Midleburne, £3 6s 8d. Tymmerhill,[24] £5. Johnnestoun, £5. Watterrocat, £6 13s 4d. Todschawhill,[25] £3 6s 8d. Watterrocat Grange, £40. Garualdhous, £6 13s 4d. Crukitheucht, £5. Dumfedlang,[26] £3 6s 8d. Cubeneburne, £3 6s 8d. The kirkland of Vatstirker [*i.e.* Westerkirk[27]], £3 6s 8d. Total, £164 16s 8d. 'Summa patet', £33 6s 8d.[28]

'Burrow annuellis': Sanct Johnnestoun, £3. Edinbrucht, £29. Hadingtoun, 20s. Selkirk, 40s. Peiblis, 8s. Jedburcht, 4s. Total, £35 12s.

The teinds set in assedation: The teind sheaves of Blainslie, £26 13s 4d. The teind sheaves of Apletrelevis, Meirbank, Soutercroft, Cartleyis and Norfurdhaucht', £13 6s 8d. Teind sheaves of Calfhill, £6. The teind sheaves of Eildoun, £10. The whole teinds of Atrik, £6 13s 4d. The whole teinds of Coulmesliehill, £6 13s 4d. The whole teinds of Vouplaw, 30s. The whole teinds of Langlie, £10. The whole teinds of Colmslie', £6 13s 4d. The teinds of Drygrange, £20. Total, £107 10s.

'Item, the teyndis underwrittin, viz.': Blakholme [*recte*, Buckholm], Quhitlie, Viliamlaw, Balkburne [*recte*, Halkburne], Langschaw, Woųsbyre [Housbyre, *fo 206r*], Alenschawis, Auld Melros, acres of Littill Fordaill, Vairdis of Melros, Gaitschaw, Denbray, Lounoumgrange [*i.e.* Hownam Grange] and Southcoit [Hunumgrange, Southgait and Denbie, *fo 207v*], £40 [*in margin*, 'Set in tak to the laird of Locht Levin'[29]].

---

[19] 'Powcleif', in *RMS*, iv, no 1819; this place has not been traced.
[20] 'Cassakkis', *RMS*, iv, no 1819; possibly Cassok Hill.
[21] 'Fingillen', *RMS*, iv, no 1819; now Finglen.
[22] 'Awirlosk', *Records of the Regality of Melrose*, iii, p. 144; 'Enenlesis', *RMS*, iv, no 1819; this place has not been traced.
[23] 'Halwod', *RMS*, iv, no 1819; now Harewood.
[24] 'Tynunerhill', *RMS*, iv, no 1819; this place has not been traced.
[25] 'Treshawhill', *Records of the Regality of Melrose*, iii, p. 144; now Todshawhill.
[26] 'Dumfermling', *Records of the Regality of Melrose*, iii, p. 144; now Dumfedling.
[27] 'Westerkirk', *RMS*, iv, no 1819.
[28] The correct total seems to be £195 16s 8d.
[29] I.e. William Douglas of Lochleven, *Scots Peerage*, vi, p. 371.

'The teyndis of Hassinden and Caveris referis to the tak quhilk is nocht producit as yit', 100 merks [*in margin*, 'Set to the ladie of Buccleucht'[30]].

Salt: 'Item, the four salt panes of Prestoun', 3 c. salt. 'Summa patet', 3 c.

Butter: 'Item, the oversyd of Coulmeslie payis' 50 stones of butter. 'Summa patet.'

Peats: Thraipwod, 340 laids.

Kane Fowls: 'Item, the cayne foullis', 380.

Capons: 'Item, the capones', 24.

'Summa of the haill rentale of Melros in silver', £1,697 11s 5d.[31]

'Quhairof thair is gevin furth in pensiounes: Item first, to aucht monkis portiounes yeirlie for thair pensiounes', £207 16s. To Mr James Lauder in pension, £91. To James Schaw, £16. 'Item, ane pensioun tane up be Williame Ormestoun, quhairof thair is na surtie as yit sene. Alexander Hayis pensioun', £33 6s 8d. 'Summa of the pensiounes to be deducit of the rentale, xl d. To' £347 8s.[32]

'And sua thair restis of the rentale, the sax [*sic*] pensiounes being deducit, extendis to' £1,350 3s [5d]; 'the thrid quhairof will extend to' £450 0s 3½d.[33]

> 'Rex.
> Collectour generall,[34] your depuites and clerkis. It is our will with avyse and consent of our richt traist cousing James, erle of Mortoun, lord Dalkeyth,[35] Regent to our realme and legis, and we ordane yow efter the sicht heirof to ressave fra our traist counsalour Alexander, commendatour of Culros,[36] traductour[37] and administratour to James, commendatour of Melros,[38] ane rentale of the said abacie of Melros as the samin presentlie extendis to and assume ane thrid thairof for the quhilk ye sall charge onlie in tymes cuming begynand your entrie to the yeir and crope' 1578 'nixtocum, and keip thir presentis to your warrand, subscryvit be our said

---

[30] 'Set to the laird of Buccleuch', *Records of the Regality of Melrose*, iii, p. 145. Janet Betoun, widow of Sir Walter Scott of Buccleuch, *Scots Peerage*, ii, pp. 228-230.
[31] The correct addition of the sub-totals appears to be £1,764 4s 9d.
[32] The correct calculation appears to be £348 2s 8d.
[33] The correct calculation appears to be £450 1s 1½d.
[34] Adam Erskine, commendator of Cambuskenneth, occurs as Collector general in September 1578, *RPC*, iii, pp. 29-30; Cowan and Easson, *Medieval Religious Houses*, p. 90.
[35] I.e. James Douglas, as 4th earl of Morton, was lord of Dalkeith, *RMS*, iv, no 2842.
[36] Alexander Colville, commendator of Culross, was appointed coadjutor and administrator of Melrose on 2 August 1569, *RSS*, vi, no 701.
[37] *Records of the Regality of Melrose*, iii, p. 146, reads 'tradintour'.
[38] I.e. James Douglas, see above, p. 257, n. 3.

richt traist cousing and Regent at Halyruidhous, the last day of Januar', 1577/8. Sic subscribitur: James, Regent.'

NOTE

Rentals for the following benefices are located elsewhere in the text:

ANCRUM, PARSONAGE OF (SRO, Vol. a, fo 164v), p. 159.
HASSENDEAN, VICARAGE OF (SRO, Vol. a, fo 259v), p. 264.

Selkirk

YARROW, OR ST MARY'S KIRK OF THE LOWES,[1] (SRO, Vol. a, fo 259r)
BENEFICE OF, (SRO, Vol. b, fo 187r)

'The rentale of sir Johnne Feitheis pairt of Sanct Marie kirk of the Lowis presentlie set in assedatioun be him to the laird of Cranstoun for' £60 'Scotis money be yeir, and I gat nevir penny payment fra the said laird sen his enteres quhilk wes at Lambmes[2] wes a yeir bypast and hes na uther thing to live on and thairfoir protestis for lettres for payment.'

Signature: 'sir Johnne Fithie, with my hand.'

Calculation of third: In the whole, £60; third thereof, £20.

'Nota, this is bot a pairt of the kirk.'

HAWICK, PARSONAGE OF, (SRO, Vol. a, fo 259v)
(SRO, Vol. b, fo 187v)

'The personage of Hawik instantlie set in assedatioun to Dame Jonet Betoun, lady of Bukcleucht,[3] for the yeirlie sowme of' £163 6s 8d, 'upoun the quhilk sowme I sustein a minister.'

Signature: 'Johnne Sandelandis, persone of Hawyk, with my hand.'

Calculation of third: In the whole, £163 6s 8d; third thereof, £54 8s 10$\frac{2}{3}$d.

[*In margin*, 'Lady Buccleuch, fermorar.']

---

[1] Yarrow, also known as St Mary of the Lowes, St Mary in the Forest, and St Mary of Farmainishop, was annexed to the Chapel Royal at Stirling, and its revenues were divided among the chanter, treasurer and schoolmaster. Cowan, *Parishes*, p. 211.
[2] I.e. Lammas, 1 August.
[3] Janet Betoun, widow of Sir Walter Scott of Buccleuch, *Scots Peerage*, ii, pp. 228-230.

HASSENDEAN, VICARAGE OF,[4] (SRO, Vol. a, fo 259v)
(SRO, Vol. b, fo 187v)

Vicarage of Hassinden

Calculation of third: In the whole, £20; third thereof, £6 13s 4d. 'Presentit' 28 January 1563/4.

LINDEAN, VICARAGE OF, (SRO, Vol. a, fo 259v)
(SRO, Vol. b, fo 187v)

Rental of the vicarage of Lynden within the sheriffdom of Selkirk.

'Item, the haill fruites, rentis and dewteis thairof estimat to' £40 'be yeir lyk as the samin past memorie of man hes bene sua commonlie set and estemit in all tymes bygane.'

Signature: 'Robert Ker, vicar of Lynden, with my hand. James Nicolsoun,[5] ressave this rentale. J. Clericus Registri.[6] Johnne Wyschart, Comptroller,[7] se it be of auld provydit.'

---

[4] Hassendean, appropriated to Melrose abbey, lay in the sheriffdom of Roxburgh, not Selkirk.
[5] James Nicolson, clerk of the Collectory, *Thirds of Benefices*, p. 62.
[6] James McGill of Nether Rankeilour was Clerk Register at the Reformation, *Handbook of British Chronology*, p. 197.
[7] Sir John Wishart of Pittarow, Comptroller, 1562 - 1565, *Handbook of British Chronology*, p. 191.

Annandale

CUMMERTREES, PARSONAGE OF, (SRO, Vol. a, fo 261v[1])

The parsonage of Cummertreis pertaining to me, John Tailyour, set to the lady Carlell[2] for £20 in the year.

Signature: 'Johnne Tailyeour, with my hand.'

'The thrid thairof', £6 13s 4d.

[*In margin*, 'Seik Linclouden in Kirkcudbrycht'[3]]

DRYFESDALE, VICARAGE OF, (SRO, Vol. a, fo 261v)

Vicarage of Dryvisdaill

Calculation of third: In the whole, £10; third thereof, £3 6s 8d.

MOUSWALD, PARSONAGE OF, (SRO, Vol. a, fo 262r)

Rental of the parsonage of Mouswald lying within the stewartrie of Annandaill given in by John Tailyeour, parson of Cummertreis, 'procuratour laitlie constitute be Merk Carrutheris, persoun of the samin.'

'Item, the said personage set for' 40 merks 'be yeir and now gevin doun of the samin the crospresentis [*sic*] and upmest claithes and Pasche fynes quhilk wer worthe be yeir' 10 merkis 'and better, sa the samin is now worthe be yeir' £20.

---

[1] Fos 260r-v and 261r are blank. These rentals for Annandale are not included in SRO, Vol. b.
[2] Michael, 4th lord Carlyle married, first, Janet Charteris and, secondly, Mariota Maxwell, *Scots Peerage*, ii, p. 389.
[3] Part of the lordship of Lincluden lay in Dumfries and part in Kirkcudbright, *RMS*, v, no 42. Rentals for Lincluden collegiate kirk appear in SRO, Vol. a, fo 267v (see below, p. 273), and in EUL, fos 90v, 92v, 98r-v (see below, pp. 613, 615, 620, 621).

Signature: 'Johnne Tailyeour, persoun of Cummertreis, with my hand.'

Calculation of third: In the whole, £20; third thereof £6 13s 4d.

HUTTON, PARSONAGE OF, (SRO, Vol. a, fo 262v)

Rental of the parsonage of Hutron 'quhilk wes set be my predicessouris to the laird of Gillesbie for' £20 'and in lyk maner I have set it for the said sowme to the said laird as he had it of befoir of sir Thomas Melvine [*i.e.* Melville] and I have nevir gottin ane penny as yit of the said personage.'

Signature: 'sir James Bikertoun, persone of Hutron.'

Calculation of third: In the whole, £20; third thereof, £6 13s 4d.

HODDAM, PARSONAGE AND VICARAGE OF, (SRO, Vol. a, fo 263r)

Parsonage and vicarage of Howdone

The parsonage and vicarage thereof set by Mr Andrew Layng to John Johnnestoun of that Ilk for 20 merks yearly 'unpayit thir four yeiris bygane, deduceand thairof crospresentis [*sic*], upmaist claithes and Pasche fynis, the said laird payand the curet for the tyme.'

Signature: 'M[r] Andro Laing, persone foirsaid.'

Calculation of third: In the whole, £13 6s 8d; third thereof, £4 8s 10$\frac{2}{3}$d.

MOFFAT, PARSONAGE OF, (SRO, Vol. a, fo 263v)

Rental of the parsonage of Moffet pertaining to Mr John Wardlaw set in assedation to Mr John Layng, parson of Lus, 'be the Comptroller, the abbot of Culros,[4] for' 240 merks yearly. 'Sen I wes provydit to the said personage I can get na trew and just rentale in specie be ressoun I am nocht answerit nor obeyit and hes onlie ressavit of the thriescoir yeiris', 100 merks 'and of this present yeir bot' 20 merks.

Signature: 'Mr J. Wardlaw.'

---

[4] William Colville, abbot of Culross, Comptroller, 1545 - 1555, *Handbook of British Chronology*, p. 191.

Calculation of third: In the whole, £160; third thereof, £53 6s 8d.

Dumfries

HOLYWOOD, ABBEY OF,[1]   (SRO, Vol. a, fos 264r-266r)
                                                    (SRO, Vol. b, fos 188r-189v)

[*In margin*, 'Aliter ressavit at the Comptrolleris[2] command.']

'The rentale of the teynd meill of the baronie of Halywod
     Item inprimis, Craginputtik', 1 b. meal. Thomas Smyth, 7 f. meal. Gibbe Smyth, 2 b. meal. McCullestoun [Cowestoun, *fo 265r*; Colinstoun, *EUL, fo 100r*], 6 f. meal. 'The uther half of McCollestoun', 6 f., 1 merk the boll. Ferdingrusche in assedation [*in margin*, 'to Johnne Maxwell of Hillis'], 6 f. meal, 1 merk the boll. John Maxwell, 6 f. meal. John [Thomas, *fo 268r*] Maxwell, Nather Straquhan, 5 f. meal. 'The laird of Bogrie, Kirkoch, Andro [Homer, *fo 268r*] Kirkocht thair, teynd meill set in assedatioun to the laird of Drumlangrig', £6 10s. Thomas Porter, 3 f. meal. Gibbe Geeir [*recte*, Greir] in Spadocht, 2 b. 1 f. meal. Kate Cogane, 2 b. meal. The Littill Spadocht, the teind [meal] set in assedation to John Maxwell of Kilbein, 43s 4d. The Meikill Spadocht, the teind meal set in assedation to Amer Maxwell, 40s. John Scot in Stepfuird, 14 p. meal; 2 p. bere. Sir William Edyer, David Edyer, for Thomas Scotis mailing, 1 merk. John Morrane, 9 f. 2 p. meal; 2 p. bere. Willie Scot, Luberrie, 6 p. meal. Willie Muirheid, 9 f. meal; 1 f. bere, set in assedation to John Maxwell of Hillis for 2½ merks. Adam Scot, Willie Muirheid, David Muirheid 'for the Pundland of Claychtand', 2 b. meal. Adam Scot, John Scot, Thomas Scot, 2 teinds set in assedation to John Maxwellis of the Hillis for 2½ merks. John Welsche 'under the wod', 9 f. meal; 1 f. bere. George Maxwell for Steylistoun and the 'thrie merk land [Four Merk Land,[3] *fo 270r*] set in assedatioun to him self', 6 merks 4s 5d. John Welsche for Gibbenstoun, 5 f. meal; 1 f. bere. John McByrnie, 6 f. meal; 2 p. bere. The half merk land at Brigend [of Clowdane, *fo 270r*] in assedation to John Maxwellis of Hillis, 6s 8d. The teind meal of Gribtoun set in assedation to John Maxwell of that Ilk, 24 [*blank*]. The teind meal of Baltirsoun to the lord

---

[1]   Cf SRO, Vol. a, fo 268r; EUL, fo 100r (see below, pp. 274, 622).
[2]   Bartholomew de Villemore, Comptroller, 1555 - 1562, with Thomas Grahame of Boquhaple, 1561 - 1562. They were succeeded by Sir John Wishart of Pittarow, 1562 - 1565. *Handbook of British Chronology*, p. 191.
[3]   Four Merk Land is a place beside Steilston.

Maxwell[4] 'for his ballie fie gratis.' John Hanyng, John Hawik in Mossyd, 2 b. meal. Sanct Michaellis Cros, 2 f. meal. Amer Gillesoun, 9 p. meal; 1 p. bere. Janet Harper, 9 p. meal; 1 p. bere. Kilnes, 6 f. meal. David Crosbie, 3 f. meal; 1 p. bere. Jamie Newall, 10 p. meal; 2 p. bere. The Stellintrie, 5 f. meal; 2 p. bere, 'ane half of it for 1 merk the boll' to John Maxwell of Hillis in assedation. Willie Naper, 5 p. meal. David Edyear for John McBirneis croft and sir William Edyer for Caudeis croft set in assedation, 8s 8d. George Edyer, John Edger, 7 f. 1 p. meal; 1 f. bere. Sir William Edger, David Edyer for the teind of Guliehill 'set to thame selfis for' 25s. 'Ane merk land of Mertyntoun' set to John Maxwell of Hillis, 2 merks 3s 4d. Henry Edyear, 7 f. meal; 1 p. bere. John Edyeris merkland of Mertintoun, 2 b. meal; 1 f. bere. John Narnis, 3 f. meal. Robin Carnes, 5 f. meal; 3 p. bere. Habbie Carrele, 6 f. meal. Robert Hynd, 2 f. meal; 1 f. bere. Andrew Edger, 1 f. meal. Litle Culleloung, 5 b. meal; 1 b. bere. The Myd Cuilelung, 4 b. meal; 1 b. bere. James Maxwell, his teind in assedation, 40s. The Cowhill set in assedation 'for him selff with Glengoyr, Meik[ill]feild toun, McQuhanrik for his depute fie under my lord Maxwell' [*in margin*, 5 merks 10s and 6 b. meal]. Robert Maxwell of Porterak, 'for it selff, McQuhanrik, the tene schilling land in Blairgaby the half merk land [set, *deleted*] croftis [Croftleis, *fo 265v*] set in assedatioun to him selff', £6. The Croftes, 2½ b. meal. George Harper, 1 b. meal. George Maxwell for the Newtoun in assedation, 16s 8d. Habbie Carell for the Newtoun, 5 p. meal. Robin Morane, 2 f. meal; 1 p. bere. Rob Roresoun, 10 p. meal. John Jaksoun, 3 f. meal; 2 p. bere. Mage Welche, 10 p. meal; 1 p. bere. 'The walkmylne croft', 1 f. meal. Sir William Edyer for McCalveris croft 'set in assedatioun to him selff', 20d. Rob Hanyng, 14 p. meal. Dallewodie, 3 f. meal. The lands of the Keir 'set to the maister of Maxwell[5] for maill teynd and uther dewteis in assedatioun to' £43.

'Summa of the teynd meill of the barronie of Halywod extendis to' £113. 'Summa of the landis of Keir set to the maister of Maxwell payis to me' £43 yearly. 'Summa of the milles of Clowdane set to the maister of Maxwell with the Kers for' 50 merks yearly. 'Summa totalis of the teynd meill of the Keir and Mylnes and Kers beir', £189 6s 8d.

The mails of Halywod: 'Item inprimis, Craiginputtog at the terme', £2 6s 8d. The Nather Quhytsyd, 40s. Hab Moffet 'in termino', 28s 4d. 'The thrie merkland of Cowestoun [McCullestoun, *fo 264r*; Colinstoun, *EUL, fo 100r*] in termino', 35s 9d. Ferdinrusche 'in termino', 11s 4d. 'The tua pund land of Straquhens ballie fie [*blank*].' John Greir and John Greir [*sic*] 'in termino', 18s 4d. John Chartouris of Littel Speddoche, 1 merk 10d 'in termino'. John Maxwell 'in termino' 35s. Homer Maxwell in Mekill Speddoch 'in termino', 43s 4d. The Stepfuird, Kate McFadyeane, 17s 6d. David Edyer, 1 merk land 'in termino', 16s. Cubbie

---

[4] Possibly Robert, 6th lord Maxwell, who died in 1552, *Scots Peerage*, vi, pp. 481-482, though the rental is post 1558.

[5] Presumably John, brother of 6th lord Maxwell, and heir presumptive, *Scots Peerage*, vi, p. 481. See also below, p. 624.

Morane, '1 pund land', 24s 4d 'in termino'. Willie Muirheid 'in termino', 12s 4d. Willie Scot in Luberie 'in termino', 7s 5½d. John Chartouris 'pund land', 25s. Adam Scot, 11s 4d. John Welsche 'under the wod, in termino' 35s. John McBirnie, 14s. Gibbinstoun, 23s 4d 'in termino'. Stelistoun, 23s 4d 'in termino'. 'Thrie merkland' [Four Merk Land,[6] *SRO, fo 270r*], 46s 8d 'in termino'. John Baxter, 23s 6d 'in termino'. Willie Hanyng, 12s 5d 'in termino'. John Dougane, 11s 7d 'thairof in termino'. Robert Maxwell, 11s 8d 'in termino'. John Maxwell, 22s 8d 'in termino'. Balterstane, £3 'in balye fee'. Gribtoun, 4 merks 40d 'in termino'. The Brigend [of Clowdane, *fo 270r*], 4s 4d. Sanct Michaellis Cros and the half of Killnes, David Welsche for Kilnes, 1 merk 15s. Mungo Crosbie, 19s 2d 'in termino'. John Hanyng in Mossyd, 5s 10d 'in termino'. John Hawyk, 5s 10d 'in termino'. Homer Gillesoun, 6s 4d. Janet Harper, 6s 8d. Dallewodie, 9s 2d. The Stellintrie, 6s 8d 'in termino'. Cauderis croft, Buthanes croft, McAlveris croft, 15s 11½d. 'Naper wyfe', 3s 4d. David Edyer 'at the terme', 30s. Willie Maxwell, 21s 8d 'officer fee'. Henry Edyer, 22s 7d. George Edyer, 20s 'in termino'. Daldryve, 21s 8d. Jok Edyer, 12s 5d. Thomas Scot, George Scot, 17s 4d. George Jacksoun, 8s 4d. John Narnes, 4s 8d. Robin Carnes, 17s 3d. Thomas Welsche, 40d 'officer fee'. John Bek, Croftleis, 19s 2d. Thomas Bek, 6s 4d. 'Wedow Forrest', 8s. 'Wedow McCrome', 3s 4d. 'Wedow Heslope', 2s 6d. Robin Maxwell of Potterak for Makquhanrik Mekill and Littill 10s land. Blariegabir, 'half merkland Croftleis' [crofts, *fo 264v*], 50s 8d. John Maxwell of Cullelung, 28s 7d. John Jaksoun to the abbot,[7] 40s. Rae [Rolland, *fo 269r*] Frissell, 8s. Andrew Neilsoun, 5s. David Watsoun, 5s. 'The tua pund land of Cowhill, ballie depute fee'. Robert Maxwell of Potterak, 43s 4d. Robert Maxwell of Cowhill for Mekilfeild, Muirsyd and Fischerholme, 'ballie depute fie [*blank*]'. Robin Mureheid, 22s 7d. John Maxwell of Glengoury, 8s 3d. George Harper, 14s 1d. John Jaksoun in Newtoun, 6s 8d. Rob Roresoun, 7s 10d. Habbie Carrele, 18s 3d.[8] Robie Morane, 8s 9d 'ballie fee'. Margaret Welsche, 5d. Andrew Edyer, 11s 11d. Robert Hynd, 12s 8d. Richie Edyer, 6s 6d. The Bracocht, £5. John Rig, 20s 8d. John Douglas, £4 20d. George Maxwell, 15s. Robe Edyer in Holme. 'Summa', £24 4s.

'Summa of the haill yeiris maill', £152 11s. The laird of Drumlangrig 'for his pairt of the kirk of Penpunt set in assedatioun for' £10. 'Summa of teynd meill and mailles, mylles and Kers beir and' £10 'for Drumlangrigis pairt of Penpunt extendis in the haill to' £351 17s 8d.[9] The kirk of Kirkconnell set to my lord Sancour[10] [*i.e.* Sanquhar], £20. The laird of Auchingassel 'for his pairt of the kirk of Penpunt set in assedatioun', £19. David Douglas, £4. 'Summa', £43 'quhilk is

---

[6] See above, p. 269, n. 3.
[7] Thomas Campbell was abbot or commendator of Holywood from 1550 - 1580, Cowan and Easson, *Medieval Religious Houses*, p. 102.
[8] The text for this entry in SRO, Vol. b ends here at the close of fo 189v; fo 190r corresponds to SRO, Vol. a, fo 271r.
[9] This figure is difficult to verify as insufficient information is available.
[10] Robert, 6th lord Crichton of Sanquhar, was named heir in March 1558, *Scots Peerage*, iii, p. 229.

assignit to dene Johnne Welche in pensioun and assignatioun, now minister of Halywod, confirmatur' 2 October 1558.

The kirk of Tynrun set in assedation to the laird of Lag for 45 merks 'assignit to sir Andro Mychell, yconimus and factour of Halywod, be lettres of pensioun subscryvit and selit be the abbot and convent of Halywod, confirmatur and the dait.'

Signature: 'Sic subscribitur, Thomas Campbell, commendatour of Halywod.'[11]

COLVEND, VICARAGE OF, (SRO, Vol. a, fo 266v)

Rental of the vicarage of Cowen 'registrat at command of the Lordis of the Checker' 1 July 1596 'and subscryvit be Mr Hew Fullartoun,[12] moderatour of the presbiterie of Drumfreis, and Mr Johnne [*recte*, James] Brysoun,[13] commissiouner of Nithisdaill.'

'The vicarege of Cowen presentlie possessit be sir John Tailyeour, lauchtfullie provydit thairto of new, is ane vicarege pensionarie at the kingis donatioun and is presentlie free in the said sir Johnnes handis without ony tak.'

'The profeit thairof now is thrie scoir ten lambis, thrie stanes woll and for the small teyndis' 20 merks 'and for the kirkland' 3 b. meal 'mesour of Nithe, gleib and mans in his awin possessioun. Extractit out of the buik of the rentales quhilk wes gevin up to the bretherein of the presbiterie of Nyddisdaill and subscryvit be the moderatour and commissiouner thairof at Drumfreis', 11 June 1596.

Signature: 'Hew Fulartoun, moderatour of the presbiterie of Drumfreis. Mr James Brysoun, commissiouner of Nithisdaill.'

'The warrand is daitit at Edinbrucht', 1 July 1596.

'Sic habet a tergo. Fiat ut petitur. Blantyre; Elphingstoun; Mr Almussar.'[14]

DURISDEER, PARSONAGE OF, (SRO, Vol. a, fo 267r)

The parsonage of Durisdeir pertaining to Mr John Hammiltoun, 'subchantour of Glasgow, set to umquhill Robert Douglas in Durisdeir, the takis thairof now

---

[11] See above, p. 271, n. 7.
[12] Minister of Dumfries, Scott, *Fasti*, ii, p. 264.
[13] Minister of Durisdeer, Scott, *Fasti*, ii, p. 312.
[14] Walter Stewart, commendator of Blantyre, James Elphinstone of Barnton and Peter Young, master almoner, were auditors of Exchequer in 1596, *RPC*, v, pp. 255, 500.

pertenyng to his mother and his wyff set for' 200 merks 'and now gevin downe of the same the Pasche fynes, crospresentis [sic] and upmest claith quhilkis wer worthe be yeir' £20. 'The haill is now be yow [recte, yeir]', 170 merks.

Signature: 'Hew Douglas.'

Calculation of third: In the whole, £113 6s 8d; third thereof, £37 15s 6⅔d.

SANQUHAR PARISH KIRK, ALTARS OF (SRO, Vol. a, fo 267r)
OUR LADY AND OF THE HOLY BLOOD IN,

Rental of the annuals of Our Ladie altar within the parish kirk of Sanquhar within the barony of the same and sheriffdom of Drumfreis extends to 9 merks money. Rental of the Haly Bluid altar 'situat as is abonewrittin', extends to 13 merks money foresaid, 'gevin up be me, sir Adame Frenche, cheplane thairof, at Edinbrucht', 18 December 1564, 'and subscryvit with my hand.'

Signature: 'sir Adame Frenche, with my hand.'

PENPONT, VICARAGE OF, (SRO, Vol. a, fo 267v)
LINCLUDEN COLLEGIATE KIRK,[15]
PREBEND OF LADY GALLOWAY[16]

Rental of the vicarage of Penpunt pertaining to me, John Tailyeour, set in tack and assedation to Duncan Hunter of Ballaggane and William Douglas of Parkland [? recte, Parkhead] for 80 merks 'be yeir, and the Pasche fynes, corspresentis and upmest claithes nocht payit the takismen will nocht hald the said vicarege for the foirsaid sowme abonewrittin without I allow to thame the availl of the Pasche fynes, corspresentis and upmest claithes.'

'I have siclyk ane prebendarie in Lynclowden callit Lady Galloway quhilk gevis yeirlie to me' £20, and 22 merks 'to ane redar and ane sangster.'

The assignation of the third of Penpunt: In the whole, £53 6s 8d; third thereof, £17 15s 6⅔d.

---

[15] Cf EUL, fos 90v, 92v, 98r, 98v (see below, pp. 613, 615, 620, 621).
[16] The lady Galloway was Margaret, daughter of Robert III, and wife of Archibald, 4th earl of Douglas, at whose instance, as lord of Galloway, the Benedictine nunnery at Lincluden was suppressed and a collegiate kirk erected in 1389. After her husband's death, Margaret, in 1429, endowed a chaplainry in the collegiate kirk, to which this entry relates. *RMS*, ii, no 133; W. McDowall, *Chronicles of Lincluden* (Edinburgh, 1886), pp. 50 ff.; W. Fraser, *The Douglas Book*, 4 vols. (Edinburgh, 1885), i, p. 395; *Scots Peerage*, iii, pp. 165-167; Cowan and Easson, *Medieval Religious Houses*, pp. 143, 223.

Signature: 'Mr Archibald Meinzeis.'

Memorandum: Mortoun [*recte*, Parton[17]], £4 8s 10d. Lynclowden, 10 merks. Culwen, 10 merks. 'The preceptorie of Trailtrow', £4 8s 10d. 'The vicarege vacand in the bischopis handis.'[18] Rane Patrik,[19] £4 8s 10d 'to be demittit into the kirkis handis for I get nocht this sevin yeir of it.'

'Summa of the haill', 40 merks 'quhairof gevin to the redar of Mortoun' [*sic*], £10.

HOLYWOOD, ABBEY OF,[20] (SRO, Vol. a, fos 268r-272r)
(SRO, Vol. b, fo 190r-v[21])

'The rentale of the teynd meill and beir of Halywod in anno quadragesimo quarto.' [*i.e.* 1544]

[*In margin*, 'Better and juster rentallit of befoir.']

Craginpettok, 1 b. meal. The Over Quhytsyd, 1 b. meal. Thomas Smyth, 7 f. meal. Gibbe Smyth, 2 b. meal. Malcolestoun, 3 b. meal. Ferdinrusche, 6 f. meal. John Maxwell, 6 f. meal. Thomas [John, *fo 264r*] Maxwell, 5 f. meal. The laird of Bogrie, 5 b. meal. Kirkhauch, 7 b. meal. Homer [Andro, *fo 264r*] Kirkoche, 3 b. meal. Thomas Porter, 3 f. meal. Gibbe Greir in Speddoch, 3 b. 1 f. meal. The Meikill Speddoch, 6 b. meal. The Littill Speddoche, 3 b. 1 f. meal. Kate Cogane, 2 b. meal. Comie Neilsoun, 14 p. meal; 2 p. bere. Robin Scotis mailing, 14 p. meal; 2 p. bere. John Moirane, 9 f. 2 p. meal; 2 p. bere. Richie Scotis wife, 6 p. meal. Wedes [Willie, *fos 264r, 265r*] Muirheid, 9 f. meal; 1 f. bere. John Chartouris mailing, 2 b. meal. Christian Roxburcht and Thomas Scot, 2 b. meal; 2 f. bere. Janet Neilsoun, 9 f. meal; 1 f. bere. Mungo Maxwell, 9 f. meal. John Welsche, 5 f. meal; 1 f. bere. William Neilsoun, 3 b. 3 f. meal; 1 f. bere. Jok Paull, 6 f. meal; 2 p. bere.

Brigend Clowden [22]

Brigtoun, 12 b. meal; 2 b. bere. Waltersame [*i.e.* Baltersan], 7 b. meal. Morestoun, 1 b. meal 'to the vicar.' Alexander Hannay in the Mossyd, 15 p. meal; 1 p. bere. John Hawyk, 15 p. meal; 1 p. bere. Sanct Michaellis Cros, 2 f. meal. John Gillesoun, 9 p. meal; 1 p. bere. Willie Gillesoun, 9 p. meal; 1 p. bere.

---

[17] Morton in Nithsdale was appropriated to Kelso and its dependent cell of Lesmahagow; Parton, however, was a prebend of the collegiate kirk of Lincluden, Cowan, *Parishes*, pp. 152, 162.

[18] Alexander Gordon was bishop elect of Galloway at the Reformation, Watt, *Fasti*, p. 132.

[19] Redkirk, which is also known as Raynpatrick, Cowan, *Parishes*, p. 170.

[20] Cf SRO, Vol. a fo 264r; EUL, fo 100r (see above, p. 269; below, p. 622).

[21] In SRO, Vol. b the rental for Holywood runs continuously from fo 188r (see above, p. 269). The text corresponds to that in SRO, Vol. a, which is divided into two rentals.

[22] 'Brigend of Clowdane half merk land, 6s 8d', fo 270r (see below, p. 276).

Kilnes, 6 f. meal. Andrew Corsbie, 3 f. meal; 1 p. bere. Rob Newall at Hame, 10 p. meal; 2 p. bere. The Stellintrie, 5 f. meal; 2 p. bere. Willie Naper, 5 p. meal. Jok McBirnie and Cawdeis croft, 10 p. meal. Thomas Edyer and John Edyer, 7 f. 2 p. meal; 1 p. bere. Andrew Edyeris wife, 7 f. meal; 2 p. bere. Martingtoun, 4 b. meal; 2 f. bere. Thomas Edyer elder, 7 f. meal; 1 p. bere. Jok Narnes, 3 f. meal. Robin Carnes, 5 f. meal; 3 p. bere. Luke Blak, 6 p. meal. Jok Hynd, 2 f. meal; 1 f. bere. Andrew Edyer, 1 f. meal. Litill Killelung, 5 b. meal; 1 b. bere. The Mid Colingtoun, 4 b. meal; 1 b. bere. Bessie Maitland, 4 b. meal. The Cowhill, 3 b. meal. Porterak, 5 b. meal. Hoyltone, 2 b. meal. Glengoy, 11 f. meal; 1 f. bere. The Mekill Feild, 14 p. meal. McQuharik, 3 b. meal. Jok Brechnelis mailing, 3 f. meal; 2 p. bere. Margaret Maxwell, 7 p. meal; 2 p. bere. Cubbie Roxburcht, 2 f. meal. Jok Jaksoun, 2 f. meal. Rolland [Rae, *fo 265v*] Frissell, 5 p. meal. The Croftleis, 3 b. meal. Rob Harper, 1 b. meal. Walter Newallis wife, 14 p. meal; 1 f. bere. Thomas Welsche, 1 b. meal. Jok Watsoun, 5 p. meal. John Bek, 1 f. meal. Robin Morane, 2 f. meal; 1 p. bere. Robin Roresounes mailing, 10 p. meal. Willie Jaksoun, 3 f. meal; 2 p. bere. Robin Watsoun, 10 p. meal; 1 p. bere. 'The walkmylne land with the croft', 1 f. meal. Gibbie McCalie, 2 p. meal. Nicol Hanyng, 14 p. meal. Dallavodie, 3 f. meal. The Escheholme, 2 f. meal. Herbert Edyer, 2 f. meal.

The Keirsyd
   The Capenoch, 8 b. meal. Gilbert Greir, 2 b. meal. Gibbie Greir in Penfillane, 2 b. meal. Roger Greir, 2 b. meal. Jok McCourteis mailing, 6 f. meal. Cubbie Greir in Penfillane, 6 f. meal. John Dalrumpill, 6 b. meal. Jok Greir in Broch, 6 b. meal. John Greir in Penmurtie, 2 b. meal. Andrew Porter, 3 b. meal. Gilbert Greir at Hame and for the Hilend, 6 b. 2 p. meal. Thomas Greir, 6 b. meal. Peter Greir, 6 b. meal. Kirkbryd, 3 b. meal. John Maxwell, 3 b. 1 f. meal. McClynnstoun, 5 f. meal. Cubbie Porter, 1 b. meal. Willie Edyer in the Smythtoun, 1 b. 2 p. meal. Rob Hidlsoun 'cum socio' [*i.e.* with his partner], 3 f. meal. 'The relict of George Hidlsoun', 1 b. 2 p. meal. Willie McNische, 5 f. meal. John Miller, 5 f. meal. John Muligane in the Barhill, 3 f. meal. The Blakwod, 4 b. meal; 1 b. bere. John Moffet 'and the relict of Braid Yet', 7 p. meal. 'The relict of Lowrie Horner', 3 f. meal. Willie Wauch, 2 b. 2 p. meal. Glenlauch, 2 f. meal. 'Summa', 13 c. 15 b. 3 f. meal; 9 b. 3 f. bere.

'The rentale of the mailes of Halywod in the hale yeir
   Craginputtok, thrie merk land and ane half', £2 6s 8d. The Over Quhytsyd 20s land, 20s. Nather Quhytsyd 2 'pund land', £2. McCollestoun 2 'merk land and a half', 33s 4d. The Ferdinrusche 20s land, 20s. Strowquhones 2 'pund land', £2. The Over Speddoch 20s land, 20s. The Littill Speddoch 5 merk land, £3 6s 8d. The Mekill Speddoch £4 land, £4. The Stepfuird [2 merk land, *deleted*] 'ane', 26s 8d. John Morrane 20s land, 20s. Wedes [Willie, *fos 264r, 265r*] Muirheid 20s land, 20s. Willie Scot half merk land, 6s 8d. John Chartouris 20s land, 50s. Cristiane Roxburcht 20s land, 20s. John Welsche 2 'pund land', £2. Gibbinsone 20s land, 20s. Steilstoun 20s land, 20s. 'The Four Merk Land' [thrie merk land,

fos 264r, 265r], £2 13s 4d. Aarngabyte[23] 4 'pund land', £4. Beltersane £4 land, £4. Gribtoun £4 land, £4. The Brigend of Clowdane half merk land, 6s 8d. Sanct Michaellis Cros half merk land, 6s 8d. The Mossyd 23s 4d land, 23s 4d. The Burnefuyt 10s land, 10s. Killnes, 20s. Mungo Corsbie 16s land, 16s. Dallawodie 10s land, 10s. The Stellyntrie 'ane merk land', 13s 4d. Jok McBirnie half merk land, 6s 8d. Mr George Kennadie half merk land of Caudeis croft, 6s 8d. The said Mr George 5s land of McCaleis croft, 5s. Willie Naper half merk land, 6s 8d. The Guliehill 2 pund land, £2. The Lang Muirsyd 10s land, 10s. John Edyer and George Edyer, 18s 4d. Henry Edyer 4s land, 45s 2d. Martingtoun 2 merk land, 2 merks. 'Merk land worne away with the water.' The Over Brumrig 2 merk land, 2 merks; 'and of that the officeris feis', 6s 8d. Nather Killelung £2 land, £2. Croftleis £2 land, £2. Over Killelung 5 merk land, £3 6s 8d. The Cowhill £2 land, £4 6s 8d. The Porterak £2 land, £4 6s 8d. The Meikilfeild 2 merk land, 2 merks. McQuhanrikis 4 merk land, £2 13s 4d. Glengoir £2 land, £2. Rob Harper 10s land, 10s. 'The relict of Wattie Newallis' 10s land, 10s. The Hoiltoun 20s land, 20s. Rob Newallis half merk land, 6s 8d. Robin Roresones half merk land, 6s 8d. Thomas Welchis 10s land of the Toun Heid, 10s. John Bell 5s land, 5s. Glenlauch 20s land, 20s. Marion Watsoun 5s land, 5s. Robin Moranes 7s 6d land, 7s 6d. Willie Jaksounes 10s land, 10s. Robin Watsones 10s land, 10s. Nicol Hanyng 14s land, 14s. Andrew Edyer 10s land, 10s. John Hynd 10s land, 10s. 'The relict of Luke Blakis half merk land', 6s 8d. Herbert Edger 10s land, 10s. Makqualter, £5. The Bracocht, £10, 'payis nocht'. The Beip [*recte*, Keir[24]], £2. John Riggis 'walkmylne', £2. Rob Newallis 'walkmylne', 2 merks. The Escheholme 20s land, 20s. The 12 merk land of Auchefud, £8. 'My lord Maxwellis[25] few landis callit the Keir', £88. The Mylne of the Keir, £20. The Mylne of Allantoun, 24 b. 'farine, and of that' 2 b. 'allowit for the multure of the Bracocht.'[26]

The Mylne of Gleneslen, 9 b. farine. My lord Creichtoun[27] for Kirkconnell, £20 'to be payit at Mertinmes,[28] souerteis actit Mr Rob Creichtoun, persoun of Sanquhar, and Johnne Dougall, burges of Edinbrucht, in the Officiallis buikis of Lowthiane.' The laird of Drumlangrig 'for his pairt of pensioun', £10. The laird of Auchingassell, £19, 'that is to say', £9 10s 'to be payit at Lammes'[29] and £9 10s 'to be payit at Candilmes.'[30] The laird of Lag for Tynrun, £30 'to be payit ay at Mertinmes and him selff actit in the Officialles buikis of Glasgow.' John

---

[23] 'Bllangawar', EUL, fo 100r (see below, p. 623); 'Blengaber', *RMS*, vi, no 257; 'Glengaber (vel Blengaber)', *RMS*, vii, no 1817.
[24] There is a Keip in Kirkcudbright according to *RMS*, viii, no 35.
[25] For 1544: Robert, 5th lord Maxwell, *Scots Peerage*, vi, pp. 479-480.
[26] SRO, Vol. b recommences here at fo 190r corresponding to SRO, Vol. a, fo 271.
[27] For 1544: William, 5th lord Crichton of Sanquhar, *Scots Peerage*, iii, pp. 226-227.
[28] I.e. Martinmas, 11 November.
[29] I.e. Lammas, 1 August.
[30] I.e. Candlemas, 2 February.

Maxwell of Cowhill 'for the fischeing of Nyth betuix College[31] and Herbert', 30 salmon, £3. The mills of Clowdane, 62 b. meal; 16 b. malt, 'and of this allowit' 8 b. 'farine for the uphald of the mylne.'

'This is the rentale of the mailes of [the] benefice of Halywod to' £206. 'Item, the teynd meill extendis of the haill baronie to' 183 b.; extending in silver payment to £223. The mills of Clowdane, 70 b.; in silver, £70. The Mylne of the Keir, 20 b. The My[l]ne of Allantoun, 22 b.

Kirks: Kirkconnell, £20. The kirk of Tynrun, £30. Penpunt, £10. The laird of Auchingassill 'for his pairt of Tynrun', £19. 'Summa of the haill rentale in the yeir of God 1544 in abbet Williame Kennedeis tyme quhilkis extendis, the quhilk rentale extendis to' £700.

'Thir ar the sowmes defakit of the haill benifice of Halywod of this sowme of' £700 'as this auld rentale proportis in it selff.'

'Item inprimis, to my lord Maxwellis sone', 200 merks 'in pensioun confirmatur. Item, to dene Johnne Welche, superiour and minister of Halywod in pensioun confirmat', 200 merks. Item, to sir Andrew Michell, 'yconimus', 40 merks. Item, £4 'land of Baltersane, and tua pund land of Straquhones defaikis in meill and meill [sic]', £15 12s, 'be ressoun it is my lord Maxwellis ballie fie in heritage'. John Maxwell of Brekensyd 'defaikis of' 7½ b. 'of meill and beir be a' 19 years' tack, 40s. John Maxwell of Gribtoun 'defaikis of' 14 b. 'farine and ordei [i.e. meal and bere] be ane' 19 years' tack, 7 merks. George Maxwell, Hew Maxwell, his brother, by a 19 years' tack 'defaikis of' 7 b. meal; 6 p. [bere]; 42s 6d. David Edyer 'for thrie bollis farine and ane half b. farine be ane' 19 years' tack 'defaikis' 33s 4d. James Maxwell, 4 b. meal, a 19 years' tack 'defaikis'. Robert Maxwell of Porterak by a 19 years' tack 'defaikis of' 9½ b.;[32] 4 merks 3s 4d. Robert Maxwell of Cowhill 'defaikis of' 6 b. meal by a 19 years' tack, £4. Amer Maxwell 'defaikis of' 6 b. meal by a 19 years' tack, £4. John Maxwell of Kilbane 'defaikis of' 3 b. meal by a 19 years' tack, 20s. The laird of Drumlangrig 'defaikis for' 15 b. meal by a 19 years' tack, £8 10s. The lord Maxwell of 66 b. meal 'of his awin few landis of the Keir be ane' 19 'yeir tak, defaikis' £22.[33]

'The lord Maxwell for mylnes of the Keir and of auld rentale', 31 merks. 'Summa totalis of the defesance extendis to' 622 merks 12s, 'and awand to the abbot', 422 merks. 'Item, the contributioun to the Lordis of the Sessioun yeirlie', £9 16s.

Signature: 'Ex mandati dicti domini Thome, Commendatarij subscripsit. Johannes Steinsoun, precentor Glasguensis. Ita testificor.' [I.e. By order of the

---

[31] I.e. the collegiate kirk of Lincluden.
[32] The commodity is not specified.
[33] SRO, Vol. b comes to an end here.

said sir Thomas, commendator [of Holywood].³⁴ John Steinsoun, chanter of Glasgow. So I testify.]

[*In margin*, 'mair sowmit nor we can fynd']

Calculation of third:
    In the whole, £700; third thereof, £233 6s 8d. 'Tak this money of the best of the baronie.'
        Meal, 19 c. 14 b. 3 f.; third thereof, 6 c. 10 b. 1 f. 'Tak it thair.'
        Bere, 9 b. 3 f.; third thereof, 3 b. 1 f. 'Thair.'
        Malt, 16 b.; third thereof, 5 b. 1 f. 1$\frac{1}{3}$ p.

'Nota', £206 'unlaid that I understand nocht.'

KIRKBRIDE, PARSONAGE AND VICARAGE OF, (SRO, Vol. a, fo 272r)
DURISDEER, PENSION FROM,

The parsonage and vicarage of Kirkbryd pertaining to John Douglas, parson thereof, extends to £40. The pension of Durisdeir pertaining to the said John, £26 13s 4d.

The third of Kirkbryd with this pension.

Calculation of third: In the whole, 80 merks;³⁵ third thereof, £17 15s 6d.³⁶

KIRKMICHAEL, PARSONAGE OF, (SRO, Vol. a, fo 272v)

'This is the rentale of Mr Thomas Marjoribankis, sone to M[r] Thomas Marjoribankis of that Ilk.'

The parsonage of Kirkmichaell set for 80 merks.

Calculation of third: In the whole, £53 6s 8d; third thereof, £17 15s 6$\frac{2}{3}$d.

[*In margin*, 'Aliter gevin in',³⁷ £40. 'Ressavit at the Comptrolleris³⁸ command.']

---

³⁴ Although the rental is dated 1544, it was only in 1550 that Thomas Campbell succeeded William Kennedy as commendator, Cowan and Easson, *Medieval Religious Houses*, p. 102.
³⁵ £40 plus £26 13s 4d is 100 merks.
³⁶ The correct calculation appears to be £17 15s 6$\frac{2}{3}$d.
³⁷ No other rental of the parsonage of Kirkmichael appears in the Books of Assumption.
³⁸ See above, p. 269, n. 2.

LOCHRUTTON, VICARAGE OF, (SRO, Vol. a, fo 272v)

Vicarege of Lochrutoun

Calculation of third: In the whole, £40; third thereof, £13 6s 8d.

TROQUEER, VICARAGE OF,[39] (SRO, Vol. a, fo 272v)

Vicarage of Traqueir pertaining to Mr Robert Mertyne.[40]

Calculation of third: In the whole, £20; third thereof, £6 13s 4d.

NOTE

Rentals for the following benefices are located elsewhere in the text:

CUMMERTREES, PARSONAGE OF (SRO, Vol. a, fo 261v), p. 265.
DUMFRIES, VICARAGE OF (EUL, fo 88v), p. 611.

---

[39] Cf EUL, fo 93r (see below, p. 616). This benefice, in the sheriffdom of Kirkcudbright, was appropriated to Tongland abbey. Cowan, *Parishes*, p. 200.
[40] Roger Martin in EUL, fo 93r (see below, p. 616).

Perth

PROCLAMATION (NLS,[1] fo 7r-v[2])

'The copy of the ordinance maid for the king and queinis majesties hows.'[3]

At Edinburgh, 22 December 1565, 'the king and queinis majesteis undirstanding that, thir divers yeiris bigaine and presentlie, mony and sindrye of thair subjectis hes nocht ceissit avariciouslye and without ony consideratioun to seik of thair hienes the propertye of thair croune quhilk aucht na uthir wayis be disponit bot to the sustentatioun of thair hous as alswa sic thriddis of benificis as necessarlie mon be had to the support of thair hienes propirtie forsaid and sustentatioun of thair uthir neidfull chairgeis quhairthrow oftymis thair majesteis liberalite causles schawin throw importune and schameles asking hes causit thair hienes awin propir effaris to be neglectit and want that thing quhairwith it suld haif beine performit, and thair majesteis wyth avys of thair Comptrallar[4] having calculat and considerit quhat mony and victuallis weill yeirlye furneiss [and] susteine thair majesteis hous and averye hes found that necessarilie thair mon be assignit for furnissing and susteintasioun xxxv thowsand pundis money, lxxii chalderis quheit, fiftie chalderis beir, ane hundreth and xxx chalderis aittis, for payment of the quhilkis thair majesteis with avys of the Lordis of thair Secreit Counsell hes appointit and assignit the sowmis of money following, first the money of thair hienes propirtie restand attour the payment of thair gracis ordinar servandis and uthiris deductionis, the money of the thriddis of benificeis and common kirkis wythin the schirefdomes of Forfar, Kinkardin, Pe[r]the, Fyfe, Abirdeine,

---

[1] This denotes National Library of Scotland, Adv MS 31.3.12. The title page reads 'Jac. 5.6.20. 31.3.12 Ex Libris Bibliothecae Facultatis Juridicae Edinburgi 1808.'

[2] Fo 7r was originally prepared as part of a table of contents (fos 1r-10v) for Forfar; Aberdeen; Banff; Moray, Elgin and Forres; Inverness; and (at fo 7r) 'Ylis' [blank], though rentals for Inverness and the Isles form no part of this volume. In the absence of entries for the Isles, fo 7r was utilised instead for recording the text of a proclamation of the royal household.

[3] The text of this proclamation, with minor discrepancies, occurs in *RPC*, i, pp. 412-413.

[4] Sir William Murray of Tullibardine, Comptroller, 1565 - 1566, *Handbook of British Chronology*, p. 191.

Banf, Murray,[5] Innernes, Orknay and Zetland, togithir wyth the bischoprik of Glasgow, pryorie Quihtorne [sic], abbayis of Paslay, Kilwinning, Kelso, Jedburgh and Newbottill, the rest of the beir of Orknay and Ros attour the provisioun of thair majesteis hous, the thriddis of the benefitcall landis wythin the schirefdomes of Forfar, Kincardin, Perth, Fyf, Abirdein, Banf, Murray, Innernes, Orknay, Zeitland, for quh[e]it thair will be restand undisponit of thair majesteis propirt[ie] xliiii chaldir and the rest to be takin of the thriddis of benifi[ceis], first of the archebischop[ri]e of Sanctandrois x chaldir ii boll[is], of Dumfermeling ix c[h]aldir ix bollis, of Lindoris iii chaldir xij bol[lis], of Scone v chaldir v bollis, extending in the haill to lxvii [lxxii, in RPC] chalder xii bollis the expens of thair hienessis hous in beir extending to fiftye chaldir to be takin of thair majesteis propirty in Fyfe and Styrviling schyre, the expenses of thair hienes hous in aitis extending as said is to ane hundreth and xxv [xxx, in RPC] chaldir to be takin first of the rest of thair majesteis [propirtie undisponit, added in RPC] xxiii [xxiiii, in RPC] chaldir and of the thrid of the archebischoprick of Sanctandrois xxii chaldir ix bollis, of the abbay of Dumfarmeling xliv chaldir viii bollis, of the pryorye of Sanctandrois l [i.e. 50] chaldir, of the abbay of Cowpar [blank] chalderis iii bollis, of the bischoprik off Dunkell ix chaldir vi bollis aittis, quhairfoir and to the effect that the mouths of unsaciabill askaris salbe stoppit and that thair majesteis hous may be weill and sufficientle furneist, and that nathing appoynted and assignit thairto be utherwayis distributit or gevin to ony persoun or personis for quhatsumevir caus or occasioun, thair hienes is contentit intocum [recte, in tymes to cum] to suspend thair handis fra all geving granting or dispositioun of ony pairt or porsioun of thair hienes propirtie forsaid or of the thriddis of beneficeis and common kirkis abone writtin to ony persoun or personis for ony caus or occasioun bot the semyn to be uptakin to the furnissing and sustentatioun of thair hienes hous and averye as said is, and in cais ony gift or dispositioun, tak, assigatioun or uther rycht or tytill be all readye impetrate of thir maj[e]steis hous or hir eftir sell happin to be purchest be quhatsumevir menis or cuillour, thair hienes presentlie renuncis, revokis and dischairgis the samyn simple [i.e. simpliciter] alsweill thame quhilkis ar all redye past as thame that hireftir sell happen to be punischet [recte, purchesit] and ordanis the Lordis of Counsell and Sessioun to direct and grant lettres at thair majesteis Comptrallaris instance for ansuering and obeying to hyme of the haill rentis, frutis, emolumentis and dewiteis of the landis, kirkis and possessionis, propirtye, thriddis of beneficeis and common kirkis speciallye abone expremit, nochtwythstanding ony gyft dispositioun or rycht that hes or sell happin to be grantit heireftir of the samyn or ony pairt thairof. Sick [sic] subscribitur, Marie R. Henri R.'

---

[5] Although there existed an earldom and diocese of Moray, there was no sheriffdom of that name.

PERTH, CHARTERHOUSE OF,[6]

(NLS, fos 11r-13r)
(SRO, Vol. a, fos 275r-277r[7])

'The rentale of the Charterhous gevin up [be] Adam Forman, prioure thairof.'

The annuals of houses in and about Perth
  Thomas Afflek for the house and yard lying in the Newraw, 18d. 'Pattoun Afflek for ane hous besyid Sanct Paulis chapell', 3s. John Browne for the house 'quhilk sumtyme wes William Trippis', 44s. The laird of Petkuillen 'for ane hous on the south syde of the gait without the Turat brig on the foirland', 40s. William Burry 'on his hous on the Northgait', 26s 8d. [*Blank*] Conqueriour[8] for Gilkannies land, 40s 8d. George Lathangy 'with his complices' for the land pertaining sometime to Alexander Edby, 13s 4d. John Ra for the land pertaining sometime to Andrew Huntar [William Craig, *deleted*], 13s 4d. Margaret Suentoun, 'relict of umquhile Johnne Oliphant', for the land pertaining sometime to the laird of Bowsy, 40s. James Andersoun for the land pertaining sometime to William Craig, 3s 4d. Alexander Bunche and Oliver Brok for the land pertaining sometime to Nicola Brok,[9] 15s. Thomas Crammy [Cramby, *fo 15r*] for the land pertaining sometime to Robert Carroun, 7s. Thomas Davidsoun for the land pertaining sometime to the provost of Meffen,[10] 6s 8d. Blaise Colt for his land in the Kirkgait, 24s. 'Constin Adam for the nether hous in the lugeing on the Walter of Tay in the Northgait', 40s. 'The abbote of Scone for ane laich sellar in that lugeing', 16s. Thomas Scott [Thomas Scottis wiffe, *fo 15r*] 'for the rest of that lugeing', £17. Oliver Ker 'for the land nixt thairto', 20s. 'For the land pertenand to umquhile Johnne Foular in the Watter Gait, 1 stane of walx. The relict of umquhile Robert Winrem for the land in the Wattergait, 6s 8d. Alexander Ros for the land pertaining sometime to Walter Syn [Sym, *fo 15r*], 4s. William Ade 'for the land quhilk umquhile Richard Malcum occupiit', £3 12s. Sir John Suentoun 'for ane chalmer in the said land', 15s. Alexander Moncreif for the land 'wes sumtyme' Andrew Currour, 6s. Thomas Crammy for the land pertaining sometime to the prebendary of Forgund [*i.e.* Forgandenny],[11] 3s. William Young for Thomas Lambes land, £5. William Andersoun for the land pertaining sometime to Mr James Davidsoun, 26s 8d. Hucheon Mowat 'for the land pertenand to umquhile David Cardin', 10s. Thomas Eldar for the house he occupies, 19s 8d. 'For ane uther inwert hous', 11s. Alexander Snell for the house

---

[6] Cf NLS, fo 14r, see below, p. 286.
[7] SRO, Vol. a, fos 273r-v and 274v are blank; fo 274r has the following entry: 'Perth, Forfar, Kincardin, [Abir]deine and Murray [*damaged*] 27 Ap. 1605, Thomas H [*damaged*]'; cf Thomas Hope, p. 1, n. 1.
[8] 'Alexander Conqueriour', *Protocol Book of Sir Robert Rollock, 1534 - 1552*, ed. W. Angus (Edinburgh, 1931), no 182.
[9] Oliver and Nicola Cok appear in NLS, fo 15r, see below, p. 287.
[10] David Haliburton was provost of Methven collegiate kirk, 1549 - 1573, Watt, *Fasti*, pp. 368-369.
[11] I.e. the prebend of Forgandenny which pertained to Dunkeld cathedral.

he occupies, 19s 4d. Thomas Galt 'and utheris in that foir and bak land', £3 15s 8d, 'this land hes aucht' 7s 8d annual. James Syde for the house he occupies, 26s 8d. 'Pety Fyve for the hous and yard in the southeistend off the Newraw', £6. John Lethem for the house and yard 'in the southwestend of the said Newraw', 56s 8d. John Monypenny 'for ane aiker of land', 13s 4d. Thomas Flemyng for the fishing of Bodrat, 26s 8d. Thomas Flemyng for the yard 'on the westend of the place', £9. Patrick Constine for the Tung yard, 54s 4d. The lands and fishing of the Freir toune, £46 13s 4d. The Magdalenis, £8. The kirkland of Inchemertine, 30s. The kirkland of Erroll, £7 7s 8d. Thomas Lindysay for the land in Dunkelden, 9s. James Hammiltoun for Sproustoun, £24 6s 8d. 'The pensioun of the commoun mylnis of Perth', £38 18s 8d. 'The pensioun of the greit customes of Perth', £26 13s 4d. George Johnsoun and James Moncreif for the Chartous [sic] ground, £13 6s 8d. 'The baillie of Erroll for the teindis of the Kirktoun and Erroll and Ardgath', £100.

Atholl

My lord Atholl[12] for Clunes, Cannocht and Logyrat, £16 13s 4d. Wester Denfallanceis, £6; the 'martstuik', 6s. 'Baron Fergasoun for the Myln of Pettincreif', 26s 8d. Eister Dunfallenceis 40s; the 'martstuik', 3s. 'For the tother half of Eister Dunfallencis', 40s; the 'martstuik', 3s. Pettlochery and for the glen, 14 merks; the 'martstuik', 12s. Balladolen, 40s; the 'martstuik', 6s; the glen, 16s. Drummoquhore, 40s; the 'martstuik', 6s; the glen, 16s. Letho, 40s; the 'martstuik', 6s; the glen, 16s. Ballacraig, 40s; the 'martstuik', 6s; the glen, 16s. Balladmond, £4; the 'martstuik', 6s; the glen, 16s. Dulceanis [Duncenans, *fo 16r*], £10; 'and ilk half merkland', 10d 'quhilk is' 25s. Ballachinde, Bellemenoch [Ballannoch, *fo 16r*], Belleyetoun, £9 13s 4d; the 'martstuik', 18s. Petcastell, 40s; the 'martstuik', 6s. Pettinragorie, 40s; the 'martstuik', 6s. Pettegar, 40s; the 'martstuik', 6s. Eister Tyre, £3 [4 merks, *fo 16r*]; the 'martstuik', 6s. Wester Tyre, 4 merks; the 'martstuik', 6s. Eister Balliealyenoch, 4 merks; the 'martstuik', 6s. [*In margin*, Wester Balliealyenoch, 4 merks; the 'martstuik', 6s.] Twa Balyeachanis, 10 merks; the 'martstuik', 6s. Petincreifis, £4; the 'martstuik', 6s. The Myln, 2 'mark.' 'The landis abone Lochtay quhilk the lard Glenurquhay[13] wont to have and hes yit', £21 8s 4d. The lands of Benmoir, 40s. Duncan Glas and Hevin [Evin, *fo 16r*; *i.e.* Ewen] Glas, his son, 'for uther land in tha pairtis', £9 8s 4d. 'The lard of Maknab for the landis he hes in tak and few', £22 13s 4d.

Victuals: 'The croftis aboute the toun', 62 b. bere.

---

[12] John Stewart, 4th earl of Atholl had an interest in property belonging to the Charterhouse, *RSS*, v, no 3276; *Scots Peerage*, i, pp. 444-445. In March 1563/4, Queen Mary appointed Atholl justiciar of lands including those of the Charterhouse, *RMS*, iv, no 1520.

[13] I.e. Colin Campbell, laird of Glenorchy.

Teinds of the Cars

Inchemertine, 16 b. wheat; 40 b. bere; 24 b. 'quheitt aittis'; 14 b. 'blak aittis.' Admure, 18 b. wheat; 18 b. bere; 24 b. white oats; 24 b. black oats [23 b., *SRO*]. The Myln of Admure, 3 b. wheat; 6 b. bere; 5 b. black oats; 5 b. white oats. Randelstoun, [*i.e.* Ronaldstone], 6 b. wheat; 4 b. bere; 7 b. white oats; 6 b. black oats. Leis, 2 b. wheat; 2 b. bere; 6 b. white oats; 6 b. black oats. Inchemichael, 24 b. wheat; 46 b. bere; 32 b. white oats; 32 b. black oats. Meginche, 7 b. wheat; 10 b. bere; 8 b. white oats; 8 b. black oats. Incheconane, 10 b. wheat; 20 b. bere; 20 b. white oats; 9 b. black oats. Dillalie, 2 b. wheat; 2 b. bere; 4 b. white oats; 4 b. black oats. Murei, 6 b. wheat; 13 b. bere; 11 b. white oats; 11 b. black oats. Belches, 2 b. wheat; 2 b. bere; 2 b. white oats. Muirege, 3 b. wheat; 3 b. bere; 4 b. white oats; 3 b. black oats. Hopisland, 1 b. bere; 1 b. white oats. The Myln of Erroll, 2 b. meal. The Ortchartland of Erroll, 4 b. bere. The Haltoun, 1 b. wheat; 4 b. bere. Mainis of Erroll, 7 b. wheat; 7 b. bere; 8 b. black oats. 'The ferme land of Raw', 4 b. wheat; 14 b. bere; 9 b. white oats; 9 b. black oats. Lorny, 9 b. wheat; 26 b. bere; 20 b. white oats; 26 b. black oats. Clasbany, 8 b. wheat; 26 b. bere; 18 b. white oats; 18 b. black oats. Ros and Galloflat, 1 b. wheat; 4 b. bere; 16 b. white oats [black oats, *fo 16v*]. Polkak, 1 b. wheat; 1 b. bere; 5 f. white oats; 11 f. black oats. The acres of the Kirktoun 'by the husbandrie', 4 b. wheat; 4 b. bere; 4 b. white oats.

Signature: 'A. Forman, prior.'

Calculation of third:

'Sum of the haill money of this rentall as it is gevin up heir extendis to', £509 6s 2d;[14] third thereof, £169 15s 4⅔d. Take the teinds of the kirk of Erroll and Ardgath in the hands of the bailie of Errole for £100; take the pension of the common mills of Perth for £38 18s 8d; the lands and fishing of Freirtoun for £46 13s 4d. Total, £185 12s; 'Gif in' £15 16s 7⅓d.

'Haill quheitt', 8 c. 5 b.;[15] third thereof, 2 c. 12⅓ b. Take the ferme wheat of Raw, Lorny, Clasbany, Ros, Galloflat and Polkak for 1 c. 7 b. 'as thai ar gevin up'; the teinds of Inchemichaell for 24 b. 'Gif in the rest to the priour.' [*In margin*, 'maid compt of' [*damaged*] c. 5 b.]

'Haill beir', 20 c.;[16] third thereof, 6 c. 10⅔ b. 'Tak the croftis aboute the toun for' 3 c. 14 b.; Inchemichaell for 46 b. Total, 6 c. 12 b. 'Gif in the rest.' [*In margin*, 'maid compt', 20 c.]

'Haill meill', 2 b.; third thereof, ⅔ b. Take it out of the Myln of Erroll.

'Haill quhyit aitis', 12 c. 6 b.;[17] third thereof, 4 c. 2 b. Take Inchemichaell for 2 c.; the white oats of the ferme land [of] Raw, Lorny and Clasbany as thai ar gevin up', 2 c. 15 b. 'Gif in' 13 b.

---

[14] The correct calculation appears to be £506 15s 10d.
[15] The correct calculation appears to be 8 c. 6 b.
[16] The correct calculation appears to be 19 c. 15 b.
[17] The correct calculation appears to be 13 c. 8 b. 1 f.

'Haill blak aittis', 12 c. 9 b. 3 f.;[18] third thereof, 4 c. 3 b. 1 f. Take the oats of Inchemichaell for 2 c.; take the ferme land [of] Raw and Lorny for 2 c. 3 b.; 'tak the od firlott fra the priour.'

'And nochtwithstanding this assumptioun the priour oblissis him to alter and renew the sam[e] sa oft as my lord Comptrollare[19] or Alexander Quhytlaw, his chalmerlane, sall desyre and to suffer the thrid to be tane in ony uther places gif thai sall nocht find this assumed according to thair mynd quhair better payment may be had. At Edinbrucht', 26 May 1565 [1563, *SRO*].

PERTH, CHARTERHOUSE OF,[20]

(NLS fos 14r-17r[21])
(SRO, Vol. a, fos 277v-280r)

'The rentall of the Charterhous. Apud Edinbrucht', 30 January 1569/70.

'The Lordis Auditouris of the Chekker ordanis the Collectour[22] to charge for the thrid of this priorie according to the rentall in respect of the decretis and inundatioun within mentionat and that the gevir in thairof is content that the kirk and Collectouris tak up the haill superplus, gif ony be, quhilk the auditouris ordanis the Collectour to serche out diligentlie. J. Clerk Registre, J. Bellenden, Robertus, Thesaurius; J. Spens; David Forrest.'[23]

'The rentale off the Charterhous'

The annuals of houses in and about Perth

Thomas Affleck for the house and yard lying in Newraw, 18d. 'Pawtoun Afflek for ane hous besyd Sanct Paulis chapell', 3s. John Broun for the house 'quhilk sumtyme wes William Trippis', 44s. The laird of Petcullane 'for ane hous on the south side of the gait without the Turat brig on the foirland', 40s. William Burrie for his house on the Northgait, 26s 8d. [*Blank*] Conquerour[24] for Gilkannies land, 40s 8d. George Lethangy 'with his complices' for the land sometime pertaining to Alexander Edby, 13s 4d. John Ray for the land pertaining sometime to Andrew Huntar, 13s 4d. Margaret Swentoun, 'relict of umquhile

---

[18] The correct calculation appears to be 11 c. 9 b. 3 f.
[19] Sir William Murray of Tullibardine was Comptroller in 1570, *Handbook of British Chronology*, p. 191.
[20] Cf NLS, fo 11r, see above, p. 283.
[21] NLS, fos 13v, 14v are blank.
[22] David Murray of Kerse was Collector for Perth in 1570, *Thirds of Benefices*, p. xl.
[23] James McGill of Nether Rankeilour, Clerk Register; Sir John Bellenden of Auchnoule, Justice-Clerk; Robert Richardson, Treasurer; John Spens of Condie, King's Advocate; David Forrest, general of the mint. *Handbook of British Chronology*, pp. 188, 197, 201; *RPC*, ii, p. 69; *Accounts of the Lord High Treasurer of Scotland*, eds. T. Dickson *et al.*, 12 vols. (Edinburgh, 1877-1970), xii, p. 101. These signatures do not appear in the SRO text.
[24] See above, p. 283, n. 8.

Johnne Oliphant', for the land pertaining sometime to the laird of Bowsy, 40s. James Andersoun for the land pertaining sometime to William Craig, 3s 4d. Alexander Bunche and Oliver Cok for the land pertaining sometime to Nicola Cok,[25] 15s. Thomas Cramby [Crammy, *fo 11r*] for the land pertaining sometime to Robert Carroun, 7s. Thomas Davidsoun for the land pertaining sometime to the provost of Methven, 6s 8d. Blaise Colt for his land in the Kirkgait, 24s. 'Constin Adam for the nathir hous in the ludgeing on the Wattir of Tay in the Northgait', 40s. 'The abbot of Scone for ane laich sellar in that ludgeing', 16s. 'Thomas Scottis wiffe for the rest of that ludgeing', £17. Oliver Ker 'for the land nixt thairto', 20s. 'For the land pertening to umquhile Johnne Fowlar in the Wattir Gait, 1 stane walx. The relict of umquhile Robert Wynrame for the land in the Wattir Gait', 6s 8d. Alexander Ros for the land pertaining sometime to Walter Sym, 4s. [*Blank*] Ad [William Ade, *fo 11r*] 'for the land quhilk umquhile Rechart Malcum occupiit', £3 12s. Sir John Suentoun 'for ane chalmer in the said land', 15s. Alexander Moncreif for the land pertaining sometime to Andrew Currour, 6s. Thomas Cramby for the land pertaining sometime to the prebendary of Forgoun,[26] 3s. William Young for Thomas Lambes land, £5. William Andersoun 'for the land pertening to umquhile Mr James Davidsoun', 26s 8d. Hucheoun Mowat 'for the land pertening to umquhile David Cardin', 10s. Thomas Eldar for the house he occupies, 19s 8d. 'For ane uther inwart hous', 11s. Alexander Snell for the house he occupies, 19s 4d. Thomas Galt 'for the hous and utheris in that foirland and bakland', £3 15s 8d; 'this land hes aucht' 7s 8d. James Side for the house he occupies, 26s 8d. Patrick Fyvie for the house and yard 'in the southest end of the Newraw', £6. John Lethame for a house and yard 'in the southestend of the said Newraw', 56s 8d. John Monypenny 'for ane aiker of land', 13s 4d. Thomas Fleming for the fishing of Bodrat, 26s 8d. Thomas Fleming for the yard 'on the west end of the place', £9. Patrick Constin 'for the Tung yard', 54s 4d. The lands and fishing of Freirtoun, £46 13s 4d. The Magdalannis, £8. The kirkland of Inchemertyne, 30s. The kirkland of Erroll, £7 7s 8d. Thomas Lindsay for the land in Dunkeld, 9s. James Hammiltoun for Sproustoun, £24 6s 8d. The pension of the common mills of Perth, £38 18s 8d. The pension of the great customs of Perth, £26 13s 4d. George Johnsoun and James Moncreiff for the Charterhous ground, £13 6s 8d. The bailie of Erroll for the teinds of Kirktoun, Arroll and Ardgath, £100. 'The larde of Glenorquhard'[27] for landis quhilkis he hes abone Locht Tay', £21 8s 8d. 'Summa of the money abonewrittin is' £366 2s 6d; third thereof, £122 0s 10d.

Athoill

My lord Athoill for Clunes, Cannocht and Logyrat, £16 13s 4d. Wester Dunfallances, £6; the 'martstuik', 6s. 'Barroun Fergussoun for the Miln of Pettincreif', 26s 8d. Ester Dunfallanceis 40s; the 'martstuik', 3s. 'For the uthir

---

[25] Oliver and Nicola Brok appear in NLS, fo 11r, see above, p. 283.
[26] See above, p. 283, n. 11.
[27] See above, p. 284, n. 13.

halff of Ester Dunfallances', 40s; the 'martstuik', 3s. Pettlochry and for the glen, 14 merks; the 'martstuik', 11s. Balladolen, 40s; the 'martstuik', 6s; the glen, 16s. Drumquhore, 40s; the 'martstuik', 6s; the glen, 16s. Letho, 40s; the 'martstuik', 6s; the glen, 16s. Ballacraig, 40s; the 'martstuik', 6s; the glen, 16s. Balladmound, £4; the 'martstuik', 6s; the glen, 16s. Duncenans [Dulceanis, *fo 12r*], £10; 'and ilk half markland', 10d 'quhilk is' 25s. Ballachand, Ballannoch [Bellemenoch, *fo 12r*], Balleyettoun, £10 13s 4d; the 'martstuik', 18s. Petcastell, 40s; the 'martstuik', 6s.[28] Pettegar, 40s; the 'martstuik', 6s. Eister Tyre, 4 merks [£3, *fo 12r*]; the 'martstuik', 6s.[29] Eister Balyealyenoch, 4 merks; the 'martstuik', 6s. Wester Balyealyenoch, 4 merks; the 'martstuik', 6s. Twa Balyeachanis, 10 merks; the 'martstuik', 6s. Pettincreiffis, £4; the 'martstuik', 6s. The Myln, 2 merks.[30] The lands of Benmoir, 40s. Lands occupied by Duncan Glas and Evin [Hevin, *fo 12r*] Glas, £9 8s 4d. 'For landis occupiit be lard Maknab', £22 13s 4d. 'Summa of thir evictit landis', £138 6s 8d; third thereof, £46 2s.[31]

'Item, this thrid pairt soum is to be gevin in to the priour be the Collectour or ellis allowit in his thrid of the money of the landis nocht evictit yeirly of' 1566, 1567, 1568 and 1569.

Victuals: The crofts about the toun, 62 b. bere.

Teinds of the Kers

Inchimartene, 16 b. wheat; 40 b. bere; 24 b. white oats; 14 b. black oats. Admure, 18 b. wheat; 18 b. bere; 24 b. white oats; 24 b. black oats. The Myln of Admure, 3 b. wheat; 6 b. bere; 5 b. black oats; 5 b. white oats. Rannaldsoun [*i.e.* Ronaldstone], 6 b. wheat; 4 b. bere; 7 b. white oats; 6 b. black oats. Leis, 2 b. wheat; 2 b. bere; 6 b. white oats; 6 b. black oats. Inchemichaell, 24 b. wheat; 46 b. bere; 32 b. white oats; 32 b. black oats. Incheconane, 10 b. wheat; 20 b. bere; 20 b. white oats; 9 b. black oats. Melginche, 7 b. wheat; 10 b. bere; 8 b. white oats; 8 b. black oats. Dillallie, 2 b. wheat; 2 b. bere; 4 b. white oats; 4 b. black oats. Murei, 6 b. wheat; 13 b. bere; 11 b. white oats; 11 b. black oats. Belches, 2 b. wheat; 2 b. bere; 2 b. white oats. Mureege, 3 b. wheat; 3 b. bere; 4 b. white oats; 3 b. black oats. Hopisland, 1 b. bere; 1 b. white oats. The Myln of Arroll, 2 b. meal. The Ortchartland of Arroll, 4 b. bere. The Haltoun, 1 b. wheat; 4 b. bere. Manis of Arroll, 7 b. wheat; 7 b. bere; 8 b. black oats. 'The ferme land and Raw' [The ferme land of Raw, *SRO, fo 279v*], 4 b. wheat; 14 b. bere; 9 b. white oats; 9 b. black oats. Lorny, 9 b. wheat; 26 b. bere; 20 b. white oats; 26 b. black oats. Clasbany, 8 b. wheat; 26 b. bere; 18 b. white oats; 18 b. black oats. Ros and Galloflat, 1 b. wheat; 4 b. bere; 16 b. black oats [white oats,

---

[28] At this point on fo 12r, 'Pettinragorie, 40s; the martstuik, 6s.' (See above, p. 284.)
[29] At this point on fo 12r, 'Wester Tyre, 4 merks, the martstuik, 6s.' (See above, p. 284.)
[30] At this point on fo 12r, 'The landis abone Lochtay quhilk the lard Glenurquhay wont to have and hes yit, £21 8s 4d.' (See above, p. 284.)
[31] The correct calculation appears to be £46 2s 2⅔d.

fo 12v]. Polkak, 1 b. wheat; 1 b. bere; 5 f. white oats; 11 f. black oats. The acres of the Kirktoun 'by the husbandry', 4 b. wheat; 4 b. bere; 4 b. white oats.

[*In margin*, '[*damaged*] is riddin doun of this victell that is takin away be the Wattir of [Tay], xvj bollis quheit, [x]vj b. beir and xlviij bollis quhite aittis.']

'Haill quheitt deduceand' 16 b. 'takin away be the Watter of Tay is' 7 c. 5 b.; third thereof, 2 c. 5 b. 5⅓ p.[32]

'Haill beir deduceand' 16 b. 'takin away be the Watter of Tay is', 19 c.; third thereof, 6 c. 5 b. 1⅓ p.

'Haill meill', 2 b.; third thereof, ⅔ b.

'Haill quhite aittis, deduceand' 48 b. 'takin away be the Wattir of Tay is', 9 c. 6 b.; third thereof, 3 c. 2 b.

'Haill blak aittis is' 12 c. 9 b. 3 f.; third thereof, 4 c. 3 b. 1 f.[33]

Signature: 'George Balfour, priour.'[34]

CULROSS, ABBEY OF, (NLS, fos 18r-20v[35])
(SRO, Vol. a, fos 281r-285r[36])

'The rentale of the haill benefice of the abbaye of Culros in all thingis, alsweill of fermes as of mailes, annuellis, teindis and all utheris dewiteis quhilk the said abbaye and plaice and [*sic*] hes bene in possessioun of in yeris bygane, maid the' 24 January, 1561/2.

'In the first, the yeirlie penny maile of the annuelis for the tennementis of the toun of Culros with sum certane aikaris annexat to the saidis tenementis extending to the sowme of' £26 12s 2d. The yearly mail of 11 salt pans, 'ilk salt pane payand yeirlie of sylver four poundis', extending in the whole for the said 11 salt pans to £44; 'and of salt, for ilk salt pane yeirlie twa bollis', total, 22 b. 'The haill aikeris about the toun of Culros that payis nocht ferme and by the aikeris annexat to the tenementis of the toun payis yeirlie of sylver the soum of' £40 14s 8d. Thomas Younger pays yearly for the Grene Yaird, 33s 4d. Thomas Sandis pays yearly for one half of the lands of the Sandis, £6 8s. Robert Gib pays for the other half of the lands of the Sandis, £5 5s. Robert Broun pays for the lands of the Keir, 56s 8d; 7 poultry.[37] Andrew Callender pays for the lands of Bordy, £6 20d; 14 poultry. Thomas Sandis pays for the lands of Birkinheid, 33s 4d; 4 poultry.

---

[32] The correct calculation appears to be 2 c. 7 b.
[33] These figures agree with those in the earlier rental on fos 12v. 13r (see above, pp. 285-286).
[34] In 1568 George Balfour obtained the escheat of the goods of Adam Forman, 'pretendit priour' of the Charterhouse, *RSS*, vi, no 297.
[35] Fo 17v is blank.
[36] SRO, Vol. a, fo 280v is blank.
[37] All poultry figures are in the margin in NLS, and in the text in SRO, Vol. a.

David Wrycht pays for the lands of Langside, £4 3s 4d; 10 poultry. Lady Staynhous pays for the lands of Blair, Poffillis and Weddirhill, £18 7s; 12 poultry. James Blais pays for the lands of Castailhill, £3 6s 8d; 8 poultry. John Blair pays for his lands of Westkirk, £3 10s; 7 poultry. Alexander Gaw pays for his lands 'contenit in his charter', £16; 21 poultry. Alexander Maistertoun pays for the lands of Bad, £6 13s 4d; 16 poultry. James Erskyne pays for the lands of Balgowny and Blarekry, £13 16s 8d; 14 poultry. William Aikyne pays for the lands of Burvene, £4 3s 4d; 10 poultry. John Maij [*i.e.* May] pays for the lands of Eistirbothen, 33s 4d; 4 poultry. Andrew Stewart pays for the lands of Westir Grange, £7 10s 4d; 12 poultry. Patrick Fentoun pays for the lands of Middilgrang, £6 6s 8d; 12 poultry. Adam Maistertoun pays for the lands of Estergrange, £10 16s 8d; 16 poultry. Robert Sandis pays for the lands of Overtoun, £5 16s 8d; 12 poultry. Robert Maistertoun pays for the lands of Wester Cumrie, £4 12s; 11 poultry. John Blakatter pays for the lands of Estir Cumrie and Neddir Ingzever, £15; 10 poultry. David Waid payis for the landis of Wester Ingzever, £14 3s 4d; 29 poultry. Mr Edward Bruce pays for the lands contained in his charter, £26 9s 4d; 20 poultry. Mr Robert Pont pays for the third of Schyris Myln, £4 4s 4d; 10 poultry. Cuthbert Blackatter pays for the lands of Blanhall [Blanhill, *SRO*; *recte*, Blairhall], £5 5s; 6 poultry. Robert Aikkine pays for the lands of Byregrange, £3 6s 8d; 4 poultry. Thomas Masone, Robert Broun and William Primros pay for the lands of Barhill, £5; 12 poultry.[38] James Prestoun pays for the lands of Valafeild, £13 7s 4d; 24 poultry. 'The occupiaris of the landis of Pitferris payit yeirlie' £21 19s 6s; 24 poultry. Robert Colvile of Cleische pays for the lands of Quyltis contained in his charter, £61 6s 8d. Robert Bennet pays for the lands of the Bussis, £3 6s 8d; 6 poultry. George Mill [Mylne, *SRO*] pays for the Quylt Myln, £3 6s 8d; 8 poultry. The lord Lindesay[39] pays for the third of Cassinduly, £5 2s 8d. Richard Carmychaell pays for Atherny, £16 13s 4d. 'Cowper Mylnis in Fyff payis yeirlie' £17 6s 8d, 'of the quhilk I have gottin na payment thir twa yeris bigane be ressone the Comptroller[40] wald nocht allou the same in the Chekker comptis'.[41] William Hammiltoun pays for the lands of Caraddin, £13 6s 8d. The lord Erskyne[42] pays for the lands of Gogar and Jargrayis,[43] £50.[44] Andrew Hoborne pays 'for the twa pairt of the kirk of Tuliboll', £23. John Fewthe pays for the Corne Myln of Crummye, £5 15s 8d; 12 capons. Andrew Wilsone pays for the Walkmyln of Crummy, £4 20d; 8 poultry.

---

[38] In margin in SRO, 'Nota, Robert Bennett hes ane yeirlie pensioun of' £8 'of thir landis of Byrgrange and Barhill'.
[39] John, 5th lord Lindsay of the Byres, who died in 1563, *Scots Peerage*, v, p. 398.
[40] Bartholomew de Villemore, Comptroller, 1555 - 1562; Thomas Grahame of Boquhaple also served 1561 - 1562. *Handbook of British Chronology*, p. 191.
[41] In margin in SRO, 'Nota, unpayit'.
[42] John, 6th lord Erskine, later earl of Mar, commendator of Dryburgh, Inchmahome and Cambuskenneth, *Scots Peerage*, v, pp. 612-615.
[43] 'Gogar and Gargrais', *RMS*, iii, no 1923; cf rental of the bishopric of Dunblane, NLS, fo 21r, 'Within the parish of Logie, of the lands of Gogar' (see below, p. 295).
[44] In margin in SRO, 'Nota, Robert Broun hes ane yeirlie pensioun of' £25 'of thir landis of Gogar'.

William Curry pays for Vanles Aikaris, 40s. John Callender pays for Powys, 44s, 'of the quhilk soum I have gottin na payment thir twa yeris bigane be ressone the Comptroller wald nocht allou the same in the Chekker comptis' [*in margin*, 'Nota, unpayit']. Erlysferry in Fyf pays yearly, 6s 8d. 'Ane tenement in Perth' pays yearly 22s. Dowcattis land [Dunccatis landis, *SRO, Vol. a, fo 284r*] pays yearly 20s. Craigflour pays 24s. James Colvile 'for the medowis and landis contenit in his charter,' £5 3s 4d. Lurg and Kincarne[45] [*i.e.* Kincardine] 'payis be the occupiaris in Petticommonis',[46] £4 8s. Robert Colvile of Cleische for the teind sheaves of Woddirhill, £3 5s. Euphemia Dundas for the teind sheaves of the Bussis etc., 'as is contenit in hir lettre of tak', £4 6d. Mr Alexander Colvile for the teind sheaves of the Sandis and teinds of other lands contained in his letter of tack, £32.

'Thir annuellis following hes nocht bene payit sen my interes, nevertheles the place sumtyme wes in use of payment thairof. The Pure Heuch payit be [*blank*] Merser', 32s.[47] Contill in Fyff, the laird of Torreis land, 20s. Muircammes Myln in Fyff, 26s 8d. Balgowny Myln in Fyff, 4s. 'Ane tenement in Striviling callit Haldanis land', 16s. 'Ane uther tenement thair callit Bowyis land', 2s. Arnegosk Myln pays 4s. Henry Davidsone in Innerkethine, 12s. William Blakburne 'thair', 8s. William Westwod in Dumfermling, 6s. John Pottar [Porter, *SRO*], 4s. John Andersone, alias John Myname [*i.e.* Miniman], 5s. Adam Stewart, 10s. Henry Murray, 10s. John Cowane, 6s. John Peirsone, 14s. John Keyr, 4s. 'Summa of thir annuellis that ar nocht payment', £9 3s 8d.

'The haile sylver contenit in this rentale afoir writtin extendis to the soum of' £758 16s 7d 'ob'.

[*In margin*, 'Sylver of the haill rentaele' £758 16s 7d 'ob. Errat in the haill soum be' £10 'les.']

'Off this said soum thair is yeirlie to be deducit as follois.
   In the first, thir annuellis affoirspecifiit quhilkis ar non payment as apperis befoir in this rentaele', £9 3s 8d. Cowper and the Powis 'quhilkis the Comptroller will nocht allou in the Chekker comptis', £19 10s 8d. Robert Bennetis yearly pension, £8. Robert Brownis yearly pension, £25. To the Lords of the Session yearly, £14.

'And forther thair is to be considerit nyne monkis, of the quhilkis ix monkis fyve hes recantit and the uther four wes requerit be me oft and divers tymes and in lyk maner wes requerit be the superintendent, quha wald nocht recant and swa dischargit of ony portioun or of ony uther dewitie of my plaice be

---

[45] 'Kincarne' is a common representation of Kincardine.
[46] SRO text reads 'Petticommoun'; cf *RMS*, iv, no 1632. Apart from 'Petticommone' as a place-name, 'petty-commonis' denotes an allowance of food, or sometimes of money.
[47] In margin in SRO, 'Nota, nocht payit'.

him for the present and in tyme to cum, and the five quha hes recantit wantis na maner of dewitie that thai usit to have in tyme bigane, quhilk in sylver to every ane of the five yeirlie extendis to' £20, 'swa the soum of sylver yeirlie to the v monkis wilbe' £100, 'and attour I gif yeirlie to ane minister to serf in the kirk of Culros, conforme to the Buk of Reformatioun',[48] £200, 'and als to ane uther minister in Crummy kirk', £100. 'To the erle of Argile[49] yeirlie for his bailye fee of the lordschip of Culros', £13 6s 8d. 'Summa of the haill to be deducit', £489 12d.

'This said soum befoir specifiit quhilk I want payment of and the uther thingis gevin furth of the leving yeirlie being deducit, as said is, restis fre yeirlie to me of the haill rentaele affoir mentionat of sylver the soum,' £269 15s 9d 'ob'.[50]

'Rentaele of the kane cheis': Robert Colvile of Cleische pays yearly for the lands of the Quyltis, 80 stones cheese. 'Summa of cheis, patet' [*i.e.* obvious].[51]
    'Off the quhilk cheis delyverit yeirlie to the fyve munkis quha hes recantit', 20 stones. 'Sua restis of cheis frie', 60 stones.

Rental of kane butter: Robert Colvile of Cleisch pays yearly for the said lands of Quyltis, of butter, 8 stones. 'Summa of butter, patet.'

'Rentall of the victualis that is ferme: Off quheit payit yeirlie be the tennentis and occupyeris of the landis of Lurg and Kincairne, 3 c. 3 b. 'Summa quheit, patet.'

'Caponis', 24. 'Turssis of stray', 8.

[*In margin*, 'Akeris about the toune']: Rental of ferme bere: Henry Bardnar, four acres, paying yearly for the same, 6 b. bere. Robert Bennet, two acres paying yearly 3 b. bere. William Blaikaitter, one acre paying yearly 1 b. 2 f. bere. James Chalmer, one acre paying yearly 1 b. 2 f. bere. John Blair, one acre paying yearly 2 b. bere. John Masone, one acre paying yearly 2 b. bere. Walter Sandis, one acre paying yearly 2 b. bere. David Prymros, the Wyndmilne Croce paying yearly 2 b. bere. The tenants and occupiers of the lands of Lurg and Kincairne pay yearly of ferme bere 3 c. 3 b. bere. Total, 4 c. 13 b.[52]

Rental of the ferme oats: The tenants and occupiers of the lands of Lurg and Kincairne pay yearly of ferme oats, 13 c. 13 b.

Rental of the teind bere: Bordy pays of teind bere, 10 b. Birkinheid, 4 b. Blair and Poffilis, 7 b 2 f. Castelhill, 4 b. Langsyd, 6 b. Balgony, 1 c. Blairkery, 8 b.

---

[48] I.e. the *First Book of Discipline* (1560).
[49] I.e. Archibald Campbell, 5th earl of Argyll, *Scots Peerage*, i, pp. 340-343.
[50] The correct calculation appears to be £269 15s 7d.
[51] The NLS text for this entry ends here; the rest of the entry is derived from SRO, Vol.a.
[52] The correct calculation appears to be 4 c. 7 b.

Westergrang, 11 b. Middilgrang, 10 b. Estirgrang, 1 c. 2 b. Over Grang, 10 b. Eister Cumry, 9 b. Valafeld, 1 c. 2 b. Blairhall, 6 b. Biregrang, 2 b. 2 f. Dunccatis landis [Dowcattis land, *NLS, fo 19v*], 1 b. 2 f. Andrew Callender pays for the teind sheaves of the acres 'about the toune of Culrois', 2 c. bere.

'Summa of the teind beir', 10 c. 13 b. 2 f.
'Summa of the haill beir, ferme and teind', 15 c. 10 b. 2 f.
'Summa of the haill caponis', 7 dozen.
'Summa of the haill pultrie', 26 dozen and 7 poultry.[53]
'Summa of the salt', 22 b.
'Summa of the stray', 8 'turssis'.

'Item, thair is to be consideritt the nyne monkis befoir specifeit, of the quhilkis fyve of tham that hes recantit ar answerit of thair portiounis in all thingis.' Wheat to the five monks, 1 c. 2 b. 3¼ p. Bere to the five monks, 5 c. 7 b. 3 p.

'And of aittis to be maid in, ane chalder thrie firlotis meill to the said fyve monkis allowand half meill in ilk boll of aittis becaus the use hes ay bene sua, extendand in aittis to' 2 c. 1 b. 2 f. 'and tua bollis of aittis for tua firlottis of grottis to the us[e] of thair kechin. Summa of aittis to fyv monkis', 2 c. 3 b. 2 f.

'Sua restis of quheit to me', 2 c. 3 f. 3¼ p.
'Sua restis of beir to me, 10 c. 3 b. 1 f. 3¼ p.
'Sua restis of aittis paiment to me', 11 c. 9 b. 2 f.

'And gif your wisdomes will caus the haill nyne monkis be answerit of all thair portiounis and dewteis in sylver and victuale the nyne will extend yeirlie in sylver to the sowme of' £180, 'and sua will rest payment to me of silver quhilk I gett yeirlie, all thingis beand deducit, thir uther four monkis beand payit', £189 15s 7d 'ob'; 'and of quheit to the haill nyne monkis', 2 c. 1 f. 1 p.; 'sua restis of quheit paiment to me', 1 c. 2 b. 2 f. 3 p.; 'and of beir to the haill nyne monkis', 9 c. 10 b. 2 f.; 'sua restis of beir payment to me', 6 c.; 'and of aittis to the haill nyne monkis', 4 c. 2 b.; 'sua restis of aittis payment to me', 9 c. 12 b. 2 f.; 'off cheis to the haill nyne monkis', 36 stones; 'sua restis of cheis payment to me', 44 stones; 'sua restis of salt', 22 b.; of capons, 7 dozen; of poultry, 26 dozen and 7 poultry; 'off stray', 8 'turssis'.

'Farther thair is' 11 wedders and 22 lambs, 'callit eik wodderis and eik lambis that ar payit be certane of the tennentis of evill payment, and the same wodderis and lambis wes destributit to the convent and tane up be thair provisour and the abot wes nevir in use thairof and na payment of the samyn gottin thir thrie yeiris

---

[53] The correct calculation appears to be 351; 26 dozen and 7 equals 319.

bigane, and this said rentall is subscryvit be my lord and his chalmerlane, day and yeir afoirwrittin, et sic subscribitur, William of Culros.'[54]

Assignation of Culross:

Third of money, £256 5s 6d 'ob'. Take the lands of Gogar and Jargrais, fermorar my lord Erskin, £50; the lands of Crummy, fermorar James Colvile, for £174 18s 5d; the lands of Valeyfeild, fermorar James Prestoun, for £13 7s 4d; the lands of Pitfaris for £21 19s 6d. 'Gif in', £3 18s 9d.[55]

Capons, 2 dozen and 4.

Wheat, 1 c. 1 b. Take this wheat out of the lands of Lang Geleane [*recte*, Lurg and Kincardine] yearly, 3 c. 3 b.[56]

'Tak tua turs half a turs and thrie pairt of a turs furth of the same.' [*i.e.* 2⅔].

'Lykwayis, tak' 24 capons 'furth of the same and the remanent four capones out of Pitferris. Eque, caponis, quheit and stray.'

Poultry, 8 dozen and 10.[57] 'Tak thame out of the barronie and landis about Culros set out for sylver, capones and pultrie, chairge my lord thairwith and the tennentis gefane it beis.'

Salt, 6 b. 1 f. 1¼ p.[58] 'Out of the salt panis of Culros, charge my lord.'

Cheese, 26⅔ stones. Butter, 2⅔ stones. Take this butter and cheese from Robert Colvile out of the lands of Quyltis.

Bere, 5 c. 3 b. 2 f. 'Tak the haill ferme beir of the aikeris about the toune of Culros and of the landis of Lurg and Kincardin extending to' 4 c. 13 b.; and of the teinds of Blair and Poffillis, 7 b. 2 f. 'Giff in' 1 b.

'Thre wodderis, half a wodder and 3 of half a wodder [*i.e.* 3¾]. Sevin lambis and 3 pairt of a lamb [*i.e.* 7¼].' [*In margin*, 'Chairge my lord heirwith.']

DUNBLANE, BISHOPRIC OF,[59] (NLS, fos 21r-22r)
(SRO, Vol. a, fos 285r-286r)

'Rentale episcopatus Dunblanensis'

The parish kirk of Dunblane and the teind sheaves of the touns within the same extending to 16 c. 15 b. meal; 3 c. 14 b. bere; £64 13s 4d money; 12 b. horsecorn ('avenis equorum').

The parish kirk of Muthill and the teind sheaves of the touns within the same extending to 15 c. 14 b. 2 f. meal; 2 c. 11 b. bere; 2 c. horsecorn; 2 merks money.

---

[54] I.e. William Colville, commendator, *Scots Peerage*, ii, p. 545.
[55] The arithmetic does not tally.
[56] Lurg and Kincardine gave 3 c. 3 b. wheat, fo 283v.
[57] One third of 26 dozen and 7 is 8 dozen and 10¼.
[58] The correct calculation appears to be 7 b. 1 f. 1¼ p.
[59] Cf NLS, fo 61v (see below, p. 348).

The parish kirk of Strogeycht and the teind sheaves of the touns within the same extending to 3 c. 14 b. meal; 1 c. 7 b. bere [2 c. 11 b., *fo 61v*]; 7 c. horsecorn; £60 money.

The parish kirk of Callander and the teind sheaves of the touns extending to 5 c. meal; 9 b. 2 f. bere.

The parish kirk of Monzie and the teind sheaves of the touns within the same extending to 7 c. 2 b. meal; 30 b. 1 f. bere.

The parish kirk of Fyndogask and the teind sheaves of the touns within the same extending to 20 b. meal; 20 b. bere.

The parish kirk of Kilmahug extending to 100 merks.

Within the parish of Logie, of the lands of Gogar,[60] 16 b. wheat.

A fourth part of the parish kirk of Tulliallon extending to £4; of victual, 8 b. 'quhilk is war now be ressoun of putting doune of the viccaragis and na payment gottin thairof thir v or vj yeris.'

A fourth part of the parish kirk of Glendowane. The same set to Thomas Hart for £10 'of the auld and now war be putting doune of the vicarigis.'

A fourth part of the parish kirk of Fossowy. In the old rental, £20 'and now mekle war be ressoun of the putting doune of the vicaragis.'

A fourth part of the parish kirk of Buffudder [*i.e.* Balquhidder]. In the old rental, £9, 'and now payis na thing be ressoun of putting doune of the vicaragis.'

A fourth part of the vicarage of Muthill. 'Memorandum, the auld rentale payit' 16 merks 'and now payis na thing be ressoun of putting doune of the vicaragis.'

The temporal lands of the bishopric of Dunblane

'The saidis landis extend to' 40 'merk land of auld extent and set presentlie to the tenentis and possessouris of the samyn for twa malis and ane half,[61] as the use and custume is and hes [bene] in Stratherne and the quenis landis thair sett in this maner.'

Signature: 'Maister David Gourlay, camerarius episcopatus Dunblanensis. And presentit be me, Johnne Morisoun, servand to the bischope of Dunblane.'[62]

Assignation of the bishopric of Dunblane:

Third of money, £104 6s 8d. 'Tak the haill temporall landis of the said biscoprik quhair evir thay ly for' £66 13s 4d; and £37 13s 4d out of the kirk of Kilmahug.

Meal, 16 c. 11 b. 2⅔ p. Take the parish kirk of Dunblane and teinds of the touns within the same for 16 c. 15 b. 'Gif in' 3 b. 3 f. 1⅓ p.

---

[60] See above, p. 290, n. 43.
[61] I.e. 2½ times 40 merks, which would give the correct total of £313.
[62] William Chisholm, senior, was bishop of Dunblane, 1526-1564. In 1561, his nephew, William Chisholm, was appointed coadjutor and he succeeded as bishop in his own right in 1565. Watt, *Fasti*, p. 78.

Bere, 3 c. 14 b. 2 f. 1⅓ p. Take 3 c. 14 b. of this bere out of the parish of Dunblane and the rest out of the kirk of Muthill.

Oats [*i.e.* horsecorn], 3 c. 4 b. Take these oats of the parish kirk, touns and teinds of Strogeicht giving 7 c.

Wheat, 5 b. 1 f. 1⅓ p. Take it out of the lands of Gogar in the parish of Logie giving yearly 16 b.

'Victuale undevydit': 2 b. 2 f. 2⅔ p. 'Tak it out of the ferd pairt of Tullialloun kirk gevand' 8 b.

METHVEN COLLEGIATE KIRK, PROVOSTRY OF,[63]

(NLS, fos 22r-23r)
(SRO, Vol. a, fos 286r-287r)

'Efter followis the rentale of the teynd schaves of Methvene togiddar with the few landis and vicarage thairof, giffin in be me, Maister David Halliburtoun,[64] provest and just possessour thairof be title of rycht, and sall ratifie and apprive [*sic*] this rentale trew and justlie gevin in but ony fraud.'

The teind sheaves of Tibermellocht extend to 29 b. meal; 8 b. bere. Methvene, 34 b. meal; 8 b. bere. The Abaland, 7 f. meal; 1 b. bere. Arquhilzie, 20 b. meal; 11 b. bere. Drumdevane, 20 b. meal; 8 b. bere. Bagovane [Balgowne, *fo 23v*; *i.e.* Balgowan], 22 b. meal; 13 b. bere. The Newraw, 8 b. meal; 2 b. bere. Bawquhiltoun, 16 b. meal; 4 b. bere. The Quheitbank, 5 b. 2 f. meal; 7 f. bere. Drummaoch [Drumnarrok, *fo 23v*], 2 b. 2 f. [meal]; 6 f. bere. Cassoquhy, 17 b. meal; 2 b. bere. Edberin, 8 b. 2 f. meal; 4 b. bere. Camsis [Campsie, *fo 23v*], 17 b. meal; 2 b. bere. Ardethie Ester, 21 b. meal; 6 b. bere. Ardethie Wester, 10 b. meal; 2 b. bere. Drumcarin, 10 b. meal; 3 b. 2 f. bere. Drumcruiff [Dulcrune, *fo 23v*], 16 b. meal; 10 b. bere. Lidnocht, 29 b. meal; 7 b. bere. Buspy Ester, 15 b. meal; 5 b. bere. Busspy Wester, 10 b. meal; 4 b. bere. Muirsyde [Myresyd, *fo 23v*], 22 b. meal; 4 b. bere. Buchthesyde [Lochtysyd alias Gowischawcht, *fo 23v*], 6 f. meal; 1 b. bere. 'The cottar teynd of Tibermellocht', 20 b. meal; 8 b. bere. The Bowmannis land 'callit Westerwod', 5 b. meal; 1 b. bere. Savope [Sauchop, *fo 23v*], 12 b. meal; 5 b. bere. Pittavy [Pettincrevy, *fo 23v*], 8 b. meal; 5 b. bere. The Midland [Myln land alias Multur croft, *fo 23v*], 1 b. bere. The Smyddehill [alias Pakyis land, *fo 23v*], 2 b. meal; 1 b. bere. Glook, 12 b. meal; 4 b. bere. 'My lordis Manis' [The Manys of Methven, *fo 23v*], 28 b. meal; 12 b. bere; 2 b. wheat. Lunlevin, 12 b. meal; 2 b. bere. Govis Hauch,[65] 1 b. meal; 3 f. bere.

Tullibaglis: The Barehill, 7 b. 1 f. meal; 7 f. bere. The Glak, 6 b. 2 f. meal; 7 f. bere. Gibbonstoun, 9 b. meal; 2 b. bere. Belstoun [Balstoun, *fo 24r*], 13 b. meal; 4 b. bere [12 b. meal, 3 b. bere, *fo 24r*]. Wobstaris croft, 1 b. meal;

---

[63] Cf NLS, fo 23v; see also T. Morris, *The Provosts of Methven* (Edinburgh, 1875), pp. 108-110.
[64] See above, p. 283, n. 10.
[65] Cf 'Lochtysyd alias Gowischawcht', fo 23v (see below, p. 298); see also 'Buchthesyde', fo 22v.

2 f. bere. The Miltoun, 6 b. 2 f. meal; 1 b. bere. 'The vicarage [of Methven, *fo 24r*] umquhill set in assedatioun', £40 'of the quhilkis nay payment thir thrie yeiris'. The feu lands, 50 merks. The parsonage of Auldbar, £44. 'And I will byd and stand this is the just rentale without ony fraud, and to be deducit of the fruitis abonewriting, quhilk utheris men hes. In the first, sir Henrie Mow in pensioun', £80 yearly.

Assumption of the provostry of Methven:
    Third of meal, 9 c. 14 b. 2 f. $2\frac{2}{3}$ p.[66] Take the teind sheaves of Tibbirmello for 29 b.; Methven for 24 b. [34 b., *fo 22r*]; Arquhilze for 20 b.; Habaland for 7 f.; Ballgoun for 22 b.; Drumdevane for 20 b.; Newraw for 7 b. [8 b., *fo 22r*]; Bawquhiltoun for 16 b. 'Gif in' 3 b. 2 f. 1 p. '3 part half pect' [i.e. $\frac{1}{6}$ p.].[67] [*In margin opposite the above place-names*, 'Beir, 7 b.; 8 b.; 11 b.; 1 b.; 8 b.; 13 b.; 2 b.']

'Remember to seirche out caynis, customes, caponis, carrageis, grassumes and all utheris dewiteis.'

Third of bere: 3 c. 7 b. 2 f. $1\frac{1}{3}$ p.[68] 'Tak the haill beir of thir fornamit townis as thai ar markit in the mergeane, and gif in' 1 b. 2 p. 'and 3 part p. thairof'.[69]
    Third of wheat, 2 f. $2\frac{2}{3}$ p. 'Tak this out of my lordis Manys gevand' 2 b.
    Third of money, £39 2s $2\frac{2}{3}$d. Take the feu lands of this provostry for £33 6s 8d; take the rest of the money extending to £5 15s 6d out of the parsonage of Awldbar giving 'be yeir' £44.

TULLIALLAN, PARSONAGE AND VICARAGE OF,   (NLS, fo 23r)
                                                                           (SRO, Vol. a, fo 287v)

Rental of the parsonage and vicarage of Tulliallone given in by Patrick Blakader at Hadingtoun, 12 January 1573/4.

'The haill personage and vicarage of Tulliallone pertening to Patrik Blacader, soune to the laird of Tulliallone, extendis yeirlie to the soume of' £30.

Signature: 'Patrik Blacader, with my hand.'

---

[66] The correct calculation appears to be 9 c. 15 b. $2\frac{2}{3}$ f.
[67] The arithmetic does not tally.
[68] The correct calculation appears to be 3 c. 5 b. 0 f. $2\frac{2}{3}$ p.
[69] The arithmetic does not tally.

METHVEN COLLEGIATE KIRK, PROVOSTRY OF,[70]

(NLS, fos 23v-24r)
(SRO, Vol. a, fos 287v-288r)

'The rentale[71] of the teynd schaves of the provestrie of Methven togidder with the few mailes and vicarage thairof gevin in be Mr James Hering,[72] provest and just possessour thairof be tytill of rycht, as it gevis instantly, the yeir of God', 1573, 'and hes payit thir' 11 years 'bypast, and will ratify and apprieve this rentall gevin in to be trew and just bot ony fraud.'

The teind sheaves of Tibbermallocht,' 28 b. meal; 8 b. bere. Methven, 25 b. meal; 7 b. bere. The Abbayland, 1 b. meal; 1 b. bere. Arquhilze, 22 b. meal; 8 b. bere. Drumdevane, 18 b. meal; 7 b. bere. Balgowne, 21 b. meal; 11 b. bere. Newraw, 7 b. meal; 2 b. bere. Bawchiltoun, 16 b. meal; 4 b. bere. Quhytbank, 5 b. meal; 1 b. bere. Drumnarrok [Drummaoch, *fo 22r*], 2 b. meal; 1 b. bere. Cassoquhy, 12 b. meal; 2 b. bere. Edbirne, 8 b. meal; 3 b. bere. Campsie [Camsis, *fo 22r*], 16 b. meal; 1 b. bere. Ardetye Eister, 16 b. meal; 2 b. bere. Ardetye Wester, 6 b. meal; 2 b. bere. Drumcarne, 10 b. meal; 3 b. bere. Dulcrune [Drumcruiff, *fo 22v*], 16 b. meal; 10 b. bere. Lydenocht, 25 b. meal; 6 b. bere. Busbie Eister, 14 b. meal; 5 b. bere. Busbie Wester, 9 b. meal; 5 b. bere. Myresyd [Muirsyde, *fo 22v*], 22 b. meal; 7 b. bere. Lochtysyd [Buchthesyde, *fo 22v*] alias Gowischawcht, 1 b. meal; 1 b. bere. 'The cottar teyndis of Tibbermallocht', 18 b. meal; 6 b. bere. Sauchop [Savope, *fo 22v*], 6 b. meal; 6 b. bere. Pettincrevy [Pittavy, *fo 22v*], 6 b. meal; 1 b. bere. The Myln land alias Multur croft [Midland, *fo 22v*], 1 b. bere. Glook, 7 b. meal; 4 b. bere. The Manys of Methven, Lonlavin,[73] West Wod alias the Bowmanis land, and the Smyddyhill alias Pakyis land 'quhilk my lord hes in his awin hand, payand thairfor yeirly of evil payment', 22 b. meal; 10 b. bere.

Tulybaiglis: Barehill, 6 b. meal; 1 b. bere. The Glak, 6 b. meal; 1 b. bere. Gibbinstoun, 8 b. meal; 2 b. bere. Balstoun [Belstoun, *fo 22v*], 12 b. meal; 3 b. bere [13 b. meal, 4 b. bere, *fo 22v*]. Wobstar croft, 1 b. meal; 2 f. bere. The Myltoun, 6 b. meal; 1 b. bere.

Total victual, 33 c. 3 b. 2 f.[74]

The parsonage of Auldbar set in assedation to the laird of Auldbar for, £44. The vicarage of Methven set in assedation to Robert Sinclair for £20. The mails of the feu lands of Methven, £33 6s 8d. Total, £97 6s 8d.

---

[70] Cf NLS, fo 22r (see above, p. 296).
[71] The figures for victual in this rental differ from those in the earlier rental, NLS, fo 22r (see above, p. 296).
[72] With the deposition of David Haliburton, James Herring was presented to the provostry in July 1573, and received institution and confirmation in September 1573, Watt, *Fasti*, pp. 368-369.
[73] Loanleven is omitted from SRO text.
[74] The correct calculation appears to be 33 c. 2 b. 2 f.

'And this is the just rentall quhairof[75] is to be deducit of the fruictis abonewrittin quhilkis is payit yeirly be the provest of the prebendaries of Methven as follouis: Item in the first to Robert Sinclair', 20 b. meal. To George Lermonth, £20. To Alexander Bannantyne, £20. To Alexander Young, £20. To Henry Straherny,[76] £16. To Thomas Knox, £16. To Robert Fyn, £16. To Robert Aitkyn, £16. 'Summa of the deductioun foirsaid': Meal, 20 b. Money, £124.

Signature: 'Maister James Hering, provest of Methven, with my hand.'

DUNBLANE, DEANERY OF, (NLS, fo 24v) (SRO, Vol. a, fo 288v)

'Memorandum, anent the rentale of the deanry of Dunblane, the same aucht to be reformit in respect the samyn is gevin up alanerly for thre scoir pundis quhilk is nocht the just avail thairof, becaus Williame Sinclair in Dunblane payis in few maill according to his chartour for ane rowme quhilk he hes of the said deane, thrie scoir sax bollis victuall, quhilk victuall in hurt alsweill of the kingis rentale as few foirsaid is sett be him in tak and assedatioun to George Drummond of Balloch for thre scoir pundis alanerly quhilk he hes gevin for his just rentale of his haill benefices and swa wald the thrid of the said soume wer tane[77] to the kingis behuiff quhair the fewar foirsaid payis victuall as said is. Attour, the said deane hes gottin in the vicarage gevin out in pensioun of befoir to his awin use and the proffeit being considerit with the vicarage of the haill parochin of Dunblane will extend to fourtie pundis.'

METHVEN COLLEGIATE KIRK, CHAPLAINRIES IN,[78] (NLS, fo 25r) (SRO, Vol. a, fo 289r)

'Item, the chaplanis of the chore of Methven extenden to fyve chaplanis, conforme to thair erectioun yeirlie', 106 merks 'pait to thame be the provest yeirlie of the fruitis abonewrittin.'

Signature: 'Presentit be me, Mr Johnne,[79] persoun of Cumnok, in name of the said provest.'

---

[75] 'It' is added in SRO text.
[76] Henry Strathendry had been appointed chaplain of the college of Methven on 18 July 1541, Morris, *Provosts of Methven*, p. 49.
[77] 'Away' is added in SRO text.
[78] Cf Morris, *Provosts of Methven*, p. 113.
[79] John Dunbar was parson of Cumnock, EUL, fo 49r (see below, p. 553).

DUNBLANE, ARCHDEACONRY OF,
(NLS, fo 25r)
(SRO, Vol. a, fo 289r)

Rental of the archdeaconry of Dumblane pertaining to Mr James Cheisholme.

'The said archidenrie extendis yeirlie to' 320 merks 'and of this payit furth yeirlie to sir Robert Danielstoun, persoun of Disart, in pensioun', £50; 'and siclyk to the vicar pensionar yeirlie', £8.

Signature: 'Mr James Cheisholme, archidene of Dumblane. James Nicolsoun,[80] resaiff this rentall. J. Clericus Registri.'

MONZIEVAIRD, VICARAGE OF,
(NLS, fo 25r)
(SRO, Vol. a, fo 289r)

Rental of the vicarage of Monyvarde given up by sir Thomas Cristesoun, vicar thereof, set to David Murray, brother german to Patrick Murray of Auchtirtyir, for the sum of £10, given up by the said sir Thomas at Halyrudhous, 13 March 1573/4.

Calculation of third: In the whole, £10; third thereof, £3 6s 8d.

DUNKELD, COMMON KIRKS OF,[81]
(NLS, fos 25v-26r)
(SRO, Vol. a, fos 289r-290r)

'The rentale of the rentis and commoun kirks of the chaptour of Dunkeld gevin in be Maister Robert Grahame, Collectour of the samin.'

The kirk and parsonage of Sawling set in assedation to Gabriel Merser yearly for £70. The parsonage and vicarage of Forthirgill lying in Atholl set in assedation to my said lord of Atholl[82] yearly for £96 13s 4d. The lands and toun of Mekle Moir set in feu to John Irvin paying yearly therefore, £13 6s 8d. The lands of Bawbutlir set in feu to James Scrimgeour paying therefore yearly £20. The lands of Fofardie set in feu to Alexander Piot paying therefore, £12. 'Item, payit be Mr George Cuke of his prebendarie of Creiff in pensioun yeirlie', £6 13s 4d. By the abbot of Sanct Colmis Inche[83] for his kirk of Leslie in pension yearly, £6 13s 4d.

---

[80] James Nicolson, clerk of the Collectory, *Thirds of Benefices*, p. 62.
[81] Cf *Rentale Dunkeldense, being Accounts of the Bishopric (A.D. 1505 - 1517)*, ed. R. K. Hannay (Edinburgh, 1915), p. 346.
[82] John Stewart, 4th earl of Atholl, *Scots Peerage*, i, pp. 444-445.
[83] James Stewart of Doune assumed the commendatorship of Inchcolm in 1549, *Charters of Inchcolm*, pp. 242-243; cf above, p. 63, n. 258.

Of the vicarage of Ouchterhous in pension yearly, £5. 'Summa of the haill money abonewrittin', £230 6s 8d.

'Heirof to be deduceit for the ordinar chairges
    Imprimis to the bischop of Dunkeld',[84] £8. 'Item, gevin to the beddell in his fee and utheris servandis fe', £20. 'Item, gevin to the Collectour in his fe with uther extraordinar expenssis maid be him in ryding of the teindis and inbringing of the fructis of the saidis kirkis and landis', £20. 'Item, gevin to the clerk and notar of the chaptour in his fee', £14.[85] 'Item, for the portage and cariage of' 32 c. 14 b.; '2 pairt meill, 3 pairt beir,[86] ilk boll meill', 4d, 'and ilk boll beir', 8d; 'summa', £11 14s.[87] 'Item, the personage of Auchterhous sett to the tennentis and occupyaris thairof payand thairfoir yeirlie', 16 c.; ⅔ meal, ⅓ bere. 'Item, the kirk of Megill set to the tennentis payand thairfoir yeirlie' 17 c.; ⅔ meal, ⅓ bere. Summa totalis', 33 c. 'Heirof gevin to the bot of Ylay in the boitmennis fee yeirlie', 2 b. 'Swa thair restis' 32 c. 14 b.

Signature: 'Mr Archibald Lyndesay, with my hand.'[88]

DUNKELD, DEANERY OF,[89] (NLS, fo 26v)
(SRO, Vol. a, fo 290r)

'The rentale of the denerie of Dunkeld pertening to Maister James Hepburne and produceit be Mr Archibald Lyndesay.'

The lands within the parish of Capeth, £83 15s 6d. The parsonage and vicarage of Inche Kadene[90] set in assedation for £173 6s 8d. 'Summa of money', £257 2s 4d.
    The kirk and parsonage of Cluny set to the tenants paying therefore yearly, 12 c. victual; ⅔ meal, ⅓ bere.
    'Item, ane pensioun out of the bischoprik of Dunkeld and assignit to me in the parochin of Alycht extending to' £133 6s 8d.

Signature: 'James Hepburne.'

---

[84] Robert Crichton, bishop of Dunkeld, 1543 - 1571, Watt, *Fasti*, p. 99.
[85] NLS reads '[xx, *deleted*], xiiij li.'; SRO reads 'xiiij li.'
[86] I.e. ⅔ meal and ⅓ bere.
[87] The correct calculation appears to be £11 9s 1½d.
[88] SRO, Vol. a reads 'Mr Johnne Lyndesay'. *Rentale Dunkeldense*, p. 346 has 'Subscribed by Mr Johne Lyndsay and given in by Mr Robert Grahame, collector for the chapter'.
[89] Cf *Rentale Dunkeldense*, p. 346.
[90] *Rentale Dunkeldense*, p. 346 has the additional note: 'There is a return for the kirk of Incheskadyne, which is "haldin of the dene of Dunkeld in takis payand thairfoir xiii[xx] merkis by the sustentatioun of the vicar pensionare".'

DUNKELD, BISHOPRIC OF,[91]   (NLS, fos 27r-29r)
(SRO, Vol. a, fos 290v-292v)

'Heirefter followis the rentale of the bischoprik of Dunkeld to be presentit befoir the quenis majestie and counsale conforme to the offir maid to hir grace be the prelatis for that tyme present at the last Conventioun in December anno 1561[92] quhilk wes in this effect, the kirkmen and prelatis of Scotland being restorit to thair levingis, rentis, possessionis and jurisdictiounis, thay grantit to gif hir grace for the outsetting of hir majesties honest afferis the fourt part of thair levingis for ane yeir allanerlie, protestand etc. and adherand to the protestatioun maid in name of the haill clergie and kirkmen of Scotland be ane maist reverend fadir in God, Johnne,[93] archibischope of Sanctandrois.'

The kirks of Bonkle and Prestoun in the Mers extending to' 40 c. of victual in the hands of my lord of Moirtoun,[94] paying yearly £60. The kirk and Manis of Abirlady 'reft be Archibald Douglas extending to' 40 c. victual, 'of the quhilk the Manis [of Abirlady] and teynd thairof set to Maister James Crychtoun for' 120 merks. The kirk and Manis of Cramound extending to 58 c. victual, 'of the quhilk the Manis [of Cramound] and teynd thairof extending to' 13 c. victual, 'reft be Archibald Douglas paying' 100 merks. And the rest of the teind set to Mr William Crychtoun in assedation of the said kirk paying 260 merks. The kirk and kirkland of Abircorne extending to 40 c. victual set to sir James Hamiltoun for 250 merks. The kirk of Tibbermuir extending to 18 c. victual set to Mr Gavin,[95] my lord Ruthvene[96] for £160. The kirk of Ouchtergavin extending to 18 c. victual set to Mr Gavin Hamiltoun, abbot of Kilwinging[97] for £100. The kirk of Cargill extending to 28 c. 8 b. victual set to my lord of Dunblane[98] for the same victual. 'Item, of thir kirkis foirsaidis, the said reverend fathir hes gottin nay payment thir divers yeiris bigane.' The kirk of Forgundyny extending to 20 c. victual, £53 6s

---

[91] Cf NLS, fos 55v, 57v, 58v (see below, pp. 341, 344, 345; *Rentale Dunkeldense* pp. 340-345.

[92] For ordering the 'ecclesiastical estate' Queen Mary had agreed that a 'general Convention' should meet on 15 December 1561; on 22 December, the archbishop of St Andrews and the bishops of Dunkeld, Moray and Ross then agreed to offer the crown a quarter of their revenues for one year only, though the privy council considered that it might be necessary to demand a third of the fruits of all beneficies and ordered beneficed men south of the Mounth to produce their rentals in Edinburgh for inspection on 24 January 1561/2 and those north of the Mounth on 10 February. *RPC*, i, pp. 192-210; *The Works of John Knox*, ed. D. Laing, 6 vols. (Edinburgh, 1846-1864), ii, pp. 299-309; see also R. Keith, *History of the Affairs of Church and State in Scotland*, eds. J. P. Lawson and C. J. Lyon, 3 vols. (Edinburgh, 1835-1850), iii, pp. 360 ff.

[93] John Hamilton, archbishop of St Andrews, 1546 - 1571, Watt, *Fasti*, p. 298.

[94] James Douglas, 4th earl of Morton, *Scots Peerage*, vi, pp. 362-363.

[95] I.e. Gavin Hamilton, commendator of Kilwinning. The scribe, in each text, seems to have confused this entry relating to Patrick, 3rd lord Ruthven (died 1566), with the next entry relating to Gavin Hamilton.

[96] Patrick, 3rd lord Ruthven, *Scots Peerage*, iv, p. 261.

[97] Gavin Hamilton was commendator of Kilwinning at the Reformation, Cowan and Easson, *Medieval Religious Houses*, p. 69.

[98] I.e. the bishop of Dunblane; see above, p. 295, n. 62.

8d, 'set to the lard of Awdy for the samin victuale and silver'; ⅔ meal, ⅓ bere. The kirk of Abirdaghy in my lord and Mr Oliphantis[99] hands, paying therefore yearly 7 c. victual; ⅔ meal, ⅓ bere. The kirk of Dowalie in the parishioners' hands extending to 15 b. 1 f. bere; 7 c. 8 b. horsecorn. The kirk of Alith in the parishioners' hands extending to 30 c. victual; ⅔ meal, ⅓ bere; 'in silver to' £36 'twa pairt, meil and beir'. The kirk of Letill Dunkeld in the parishioners' hands extending to 4 c. 2 b. meal; 3 c. 1 b. 2 f. bere; 15 c. 8 b. horsecorn. The kirk of Capeth in the parishioners' hands extending to 11 c. 12 b. 3 f. meal; 8 c. 12 b. bere; 5 c. 2 b. horsecorn. The kirk of Straithmeglo extending to 28 c. victual set to William Birnye for £160.

'The Temporall

Tullimulie fewit to Thomas Lyndsay and payis' £4. Wester Dulguis feued to John Stewart, 42s. Rothmell [Bot Mell, *fo 55v*] feued to the laird of Arntulie, 40s. Cragilto feued to the said laird of Arntulie, 40s. Drumboy feued with the mill to Thomas Vallantyne, £4. Kirkland Bank of Alith feued to Andrew Crychtoun, £7 6s 8d. Kirkhill of Megill feued to George Moncuir, £8 13s 4d. 'The ane half of Fordie fewit to Jonat Skrimgeour', £10. 'Ane awchland part of Fordie fewit to the said Jonat', 12s 6d. 'Ane uther auch part of the said Fordy fewit to Gilbert Bray', 12s 6d. 'Ane quarter of the said Fordie fewit to Thome the lard', 25s. Pittindyny and 'milret thairof' feued to Thomas Merschel, £12. Lycrope feued to the laird of Knokhill, £8. Foirdaill feued to John Broune, £26 13s 4d. Gawy feued to my lord of Atholl,[100] £16. Menmure feued to the parson of Menmure,[101] £3. 'Ferne with the brayis[102] fewit [*blank*] Lyndsay', £12. Kethik and Ardbraik feued to the abbot of Couper,[103] £4. Cargill feued to Andrew Abircrummy, £4. McCarsie feued to George Name, £13 6s 8d. Tibbirmuir feued to Patrick Murray, £14. Kynvaid feued to John Rattray, £3 6s 8d. Castell Campbell feued with the kirkland of Dolour to my lord of Ergill,[104] £10 13s 4d [£10 14s 4d, *SRO*]. 'Tulilum fewit to the feriris [freiris, *SRO*] of Tulilum', £4 6s 8d. Ouchtertule feued to the laird of Grange, £82. Clyntry [Clentre, *fo 56r*; Glentre, *fo 59r*] feued to [*blank*] Crychtoun 'payand thairfoir' 14 merks. Port of Logiret, 30s. Sokkoth, 40s. Kincragie, £4. Glenelvart, £4. The Myl of Glenelvart, 53s 4d. Ester Dulgus, £5 6s 8d. Kilmorich, £13 13s 4d. Dowally, £5. Drumgarthill, £6 13s 4d. Capeth, £12. Baithous of Capeth, 20s. Drummaly, £5. Logy and Birnane and Urriewall, 40s. Cluny, £16. Eister Quhitefeld, 40s.

---

[99] Laurence, master of Oliphant, son of Laurence, 3rd lord Oliphant, *Scots Peerage*, vi, pp. 545-546.

[100] See above, p. 300, n. 82.

[101] James Hamilton held the prebend of Menmuir till 1566 when John Lindsay, son of the earl of Crawford, was recognised as 'parson' of Menmuir, *RSS*, v, no 2946.

[102] *Rentale Dunkeldense* p. 341 reads 'brayis [brewhouse]'.

[103] Donald Campbell was abbot of Coupar Angus, 1526 - 1562, *Rental Book of the Cistercian Abbey of Cupar-Angus with the Breviary of the Register*, ed. C. Rogers, 2 vols. (London, 1879-1880), i, p. 100.

[104] Archibald Campbell, 5th earl of Argyll, *Scots Peerage*, i, pp. 340-343.

Cultrany Beg, Kincarny, set to my lord Erroll,[105] £6 13s 4d. Kirkland of Prestoun feued to the laird of Cokburne, 40s. The Yle of Crawmond feued to the laird of Drilaw, 5 merks. Total silver, £1,407 5s 2d.[106] Total victual, 142 c. 6 b. 2 f.[107]

'Expenssis to be deducit of the foirsaid silver and victuale
Item, to my lord Athole in yeirlie pensioun', 100 merks. 'Item, for the contributioun of the Lordis of the Seitt and College of Justice', £42. 'Item for sustening of the trene brig stene wark[108] of the same brig of Dunkeld' 300 merks. 'Item, to the maister of werk of the samin in pensioun', £40. 'Item, to the foster and twa gardianis of woddis appropriat to the brig and brig werk, in fee and boll', 100 merks; 1 c. meal. 'Item, to the portar of the place and brig and keping of the brig, in fee and boll', 20 merks; 1 c. meal. 'Item, gevin to the sustentatioun of certane puir childer callit blew freris',[109] 30 merks. 'Item, to the uphald of twa sculis and thair maister of grammer and sang', 100 merks. 'Item, to the constable that exercis [*i.e.* exercises] the jurisdictioun temporall in Dunkeld', 100 merks; 4 c. meal. 'Item, to sindrie boll men, pikmen, officiaris, auld servandis, to the nowmer of' 10 'personis, sic as wricht, masoun, sklater, gardnar, fyschar and siclyk', 5 c. meal. 'Item, to Mr Johnne Bartane in pensioun yeirlie for his service doun to me', £40. 'Item, to his[110] servandis fee, to the nowmer of' 15, 'in fee, sum' 20 merks, 'sum' £20, 'extending to the soume of' £260 6s 8d. 'Item, to the chalmerlane of Dunkeld for his expenssis and travill for the ingetting of his leving', £100. 'Item, gevin to ane grintair of his girnale of Perth and Dunkeld in fee and girnale maile inlaik of victuale', £40; 4 c. [3 c., *SRO*] victual.

'Sum of expens to be deducit', £981 13s 4d.
'Sum of the victuale to be deducit', 15 c.
'The rest of silver', £425 11s 8d.
'The rest of victuale', 127 c. 6 b. 2 f.

Signature: 'Magister Joannes Bartoun, produxit, teste chirographo.'

'Nota, this is writtin on the baksyd of the rentale.'

The rental of the bishopric of Dunkeld
Sokoth, 40s; 12 poultry; 6 b. 2 f. oats for teind. Port, 30s; 12 poultry; 2 b. 2 f. malt; 6 b. 2 f. oats for teind. Dunmacriof 'fra the burne est', 20 'sparris'; 20 'plankis of fir'; 8 b. oats; 1 b. bere 'for teynd [*blank*] in feu.' Kincragie, £4; 4 'wodderis'; 4 dozen poultry; and for the teind, 18 b. oats; 4 b. bere. Myll of

---

[105] George Hay, 7th earl of Erroll, *Scots Peerage*, iii, pp. 568-569.
[106] The correct calculation appears to be £1,408 2s.
[107] The correct calculation appears to be 141 c. 13 b. 2 f.
[108] Cf *Rentale Dunkeldense*, p. 339.
[109] The hospital of St George in Dunkeld was founded to sustain a master and seven poor folk, Cowan and Easson, *Medieval Religious Houses*, p. 175.
[110] I.e. the bishop's servants.

Kincragie, 53s 4d; 12 poultry. Glenelvart, £4; 2 wedders; 12 poultry; and for teind, 14 b. oats; 1 b. bere. Ester Dulgus and Middill Dulgus, £5 6s 8d; 8 wedders; 4 dozen poultry; 2 c. oats; 4 b. bere. Wester Dulgus, 42s; 2 wedders; 12 poultry; and for teind, 12 b. oats; 1 b. bere. Torrywald, 40s. Ladywall, £10 13s 4d; and for teind, 6 b. meal; 4 b. bere. The New Myll, £6 13s 4d; 2 dozen capons; 'ane fed beir' [*i.e.* one fed boar]. Inschevin Wester, 5 merks; the teind, 3 b. meal. Middill Inschevin, 5 merks; the teind, 3 b. meal; 1 b. bere. Dalpowy, £4; the teind, 4 b. meal; 1 b. bere. Drumboy, £4; the teind, 6 f. meal; 1 b. bere. Craighilto, 40s; the teind, 1 c. oats. Rotmell [Bot Mell, *fo 55v*] , 40s; 20 b. oats; 3 b. bere. Dowally, £5; 10 b. oats; 5 f. bere. Kilmorycht, 20 merks; the teind, 40 b. oats; 4 b. bere. Cawy, 24 merks; the teind, 2 c. 2 b. oats; 6 f. bere in feu. Cluny with the pendicles, [*blank*]. Carnies, both Ovir and Nethir, [*blank*]. Capnettis and Dungartill, [*blank*]. Kinnard in feu. Mucarsy in feu.

RUTHVEN, VICARAGE OF, (NLS, fos 29v-30r)
(SRO, Vol. a, fos 292v-293r)

The rental of the vicarage of Riffanis lying on the Wattersyd of Yle within the sheriffdom of Forfar and in the diocese of Brechin.

Of the teind sheaves of Brigtoun of Riffan pertaining to the vicarage yearly, 60 b. oats; 14 b. bere. Of the teind wool yearly 2½ stones. Of teind lambs, 25. Of teind lint, 12 'thraiffis'. 'Of teind gryssis', 7. 'Of teind geis', 7. 'Of teind hemp', [*blank*]. Of teind cheese, 5 stones. 'Item, lyand to the manis and gleib of land', 13 acres.

'And this rentaill hes beine tane up be the lairdis of Ruffanis this xiiij yeiris bygane sen the deceis of Mr Patrik Blare, viccare for the tyme, and nane part nether to the kirk nor king this fourtene yeiris past, and gif your lordschip will summound the parochinneris or possessouris of the same for the byganis and kingis thrid in tymes to cum, ye sall fund in this bill the namis of the tounis and names of the labowraris of the ground.' The Manis[111] of Ruffanis, 4 'plewis'. In the Brigtoun of Ruffanis, 4 ploughgates. The mills of Babyrnie, 2 ploughgates with 2 mills and mill lands. The Wodend, 'four oxingane of land.' Andrew Smythis land, 4 acres. 'The cotter of the Manis with the Sadilhillok', 14 acres.

'Thir ar the namis of the toun and the duellaris thairin.' The Ruffanis: John Crychtoun of Ruffanis, 'auld laird'; Adam Crychtoun, 'apperand thairof, younge laird'. The Myln of Babirnie: James Chrychtoun, John Dowglas. The Brigtoun of Ruffanis: James Crychtoun, brother to the laird, John Ductour, Pait Ducharis, John Ducharis, Patrick Smyth. The Sadilhillok: William Pypar. The Wod End: John Smyth. 'Upoun the aikeris of the Manis': Andrew Smyth, John Ducharis, John Mustard.

---

[111] SRO text reads 'names'.

'And this rentall may be provin be David Ramsay of Jurdistoun [*i.e.*, Jordanstone] and Abrahame Crychtoun in the Kirktoun of Ferne, quha wes factoris and can testefie the samyn.'

PERTH, PARISH KIRK,   (NLS, fo 30r)
CHAPLAINRY OF ST STEPHEN IN,   (SRO, Vol. a, fo 293v)

'Schir Johnne Saidlaris rentale of the chaiplanrie of Sanct Stevin cituat within the paroche kirk of Perth and pensioun granted to him be the provest[112] and baillie and counsall of Perth induring his lyftyme as followis, producit of auld pertening to his said chaplanrie of the annuellis following.'

'Item, the aikeris pertening to James Strathonis airis and now to Johnne Monypenny, liand besouth the place of sumtyme callit the Charterhous addettit in annuell yeirlie', 5 merks. 'Item, ane yarde occupyit be Johnne Pyperis yairdis liand in the Spey yarde besyd Perth, yeirlie', 36s 8d. 'Item, ane annuell rent furthcumand yeirlie of that land sumtyme pertenyng to James Blakwod liand bewest the abbot of Scounis[113] land now pertening to James Watsoun, merchand, addettit yeirlie to' 30s 8d. 'Item, beneth the Castell Gaitt Brig of the land pertenyng to Alexander Dundas airis, yeirlie addettit', 25s. 'Item, ane annuell furthcumand of the land contigue adjacente thairto addettit yeirlie', 10s. 'Item, ane yeirlie pension of auld of the said provest and baillies and counsall of the said burgh extending yeirlie to' 16 merks [15 merks, *SRO*].

DUNKELD, CHANTORY OF,[114]   (NLS, fos 30v-31r)
KINCLAVEN, BENEFICE OF,   (SRO, Vol. a, fos 293v-294r)

The rental of the chantory of Dunkeld given in by Mr William Adamsone, chanter thereof.

The kirk and parsonage of Kinclavin with the vicarage thereof set to the tenants 'ilk ane of thame for thair awin pairt extending yeirlie to' 14 c. victual, $\frac{2}{3}$ meal, $\frac{1}{3}$ bere. The vicarage thereof, £16, 'quhairof I haif gottin nay payment sen my enteres to the said benefice'. The lands of Dalrilzane set in feu to John Rattray, paying yearly £8. The lands of Athoill set in feu to George Drommound, paying

---

[112] Patrick, 3rd lord Ruthven, was provost of Perth, 1553 - 1566, M. B. Verschuur, 'Perth and the Reformation, Society and Reform: 1540 - 1560', 2 vols. (Glasgow Ph.D. thesis, 1985), ii, pp. 631-640.

[113] Patrick Hepburn, bishop of Moray, was commendator of Scone at the Reformation, Cowan and Easson, *Medieval Religious Houses*, p. 98.

[114] Cf NLS, fo 31r (see below, p. 307); see also *Rentale Dunkeldense*, p. 346.

yearly £11 10s. The lands of the Hiltoun [Haltoun, *SRO*], Linlour[115] and mill thereof set to the tenants in assedation paying yearly, £18. 'Item, the Chantour croft liand besyde the brig of Dunkeld quhilk payis yeirlie' 10s. Total: Victual, 14 c. Money, £54.

'Heirof imprimis, gevin in pensioun to Mr Johnne Dowglas, rectour of Sanctandrois, yeirlie', £26 13s 4d. To my stallar of Dunkeld, £10. To the chaplains and choir of Dunkeld, £8 13s 4d. To the minister of the said kirk, £10. To the abbey of Cambuskenneth yearly, £4. Given to the bishop of Dunkeld yearly, £4. Total, £63 6s 8d.

Signature: 'Maister Williame Adamsoun.'[116]

Calculation of third: 'In the haill of money', £54; third thereof, £18. 'In the haill of victuall', 14 c.; third thereof, 4 c. 10 b. 2 f. $2\frac{2}{3}$ p.

KILSPINDIE, VICARAGE OF, (NLS, fo 31r)
(SRO, Vol. a, fo 294r)

'The vicarage of Kinspindie pertenyng to sir Marc Jamesoun, vicar portionar of the same under the abby of Scone, in small teindis as lynt, hemp, leix, sybous, scheis, eggis and utheris small teindis and offerandis wount to be payit of the said vicarage with stirk and steng penny, extending to' £40 'and thairof to sustene the curat', £13 6s 8d.

DUNKELD, CHANTORY OF,[117] (NLS, fo 31r)
KINCLAVEN, BENEFICE OF, (SRO, Vol. a, fo 294v)

[*In margin*, 'Aliter gevin in the said chantorie of Dunkeld.']

'Item, the haill chantorie with the pertinentis set in assedatioun for payment yeirlie of' 260 merks.

'Off the quhilk to be deducet to Mr Johnne Dunglas [Dowglas, *fo 30v*] in pensioun' 40 merks. To the abbey of Cambuskynneth, 6 merks. To the stallar, 15 merks. To the chaplains of the choir of Dunkeld, 13 merks. To the vicar pensionary of Kynclavin, 13 merks. In 'procuratioun sinodallis', 6 merks.

---

[115] *Rentale Dunkeldense*, p. 347 reads 'the Haltoun Linlour'.
[116] After this signature, *Rentale Dunkeldense*, p. 347, continues with an abstract of the rental of Kinclaven; see below, NLS, fo 31r.
[117] Cf fo 30v (see above, p. 306); see also *Rentale Dunkeldense*, p. 346.

'Summa of this deductioun', 93 merks. 'And swa restis fre to the possessour', 167 merks.

Signature: 'Mr William Adamsone. James, ressaif this rentall. J. Clericus Registri.'

DUNKELD, CHANCELLORY OF,[118]
LETHENDY, PARSONAGE OF,

(NLS, fo 31r)
(SRO, Vol. a, fo 294v)

The rental of the chancellory of Dunkeld pertaining to Mr William Gordoun, produced by Mr Archibald Lyndsay, factor thereof.

'Item, the kirk and personage of Lathindene being riddin at sum tymes to' 8 c. 'Item, set to the lard of Lethindene for' £28 13s 4d. 'Item, with the fermes of the townes of Kincarneis paying yeirly' £13 6s 8d. 'Item, with the fermes of Wester Gormok set to James Hering payand thairfoir yeirlie' £10 2s 4d. 'Item, with the fermes of the toune of Lawtoun set in few to Archibald Ogilvie payand thairfoir yeirlie' £18. 'Item, with the croft in Dunkeld set in few to Thomas Lyndsay payand thairfoir yeirlie' 32s. Total, £71 14s 4d.
'Heirof to be deducit for ordinar chargis payit to the bischoipe', 40s. To the factor in his fee £10. To the stallar, £5. Total, £17, 'sua restis' £54 14s 4d. 'Item, four bollis of custome beir', 4 b. bere. 'Sum fowlis and horscorne quhilk war nocht wont to be sauld.'

Signature: 'Magister Archibald Lindsay, factour to the chancellarie of Dunkeld, with my hand.'[119]

Calculation of third: 'Money in the haill', £71 14s 4d; third thereof, £23 8s 1½d.[120] Bere, 4 b.; third thereof, 1 b. 1 f. 1⅓ p. 'Omittit fowllis and horscorne.'

DUNKELD, ARCHDEACONRY OF,[121]

(NLS, fo 32r-v)
(SRO, Vol. a, fos 295r-296r)

'Rentale fructuum et emolimentorum'

Rental of the fruits and emoluments of the archdeaconry of Dunkeld made and given in by Mr David Spens, archdeacon of the same.

---

[118] Cf *Rentale Dunkeldense*, p. 347.
[119] *Rentale Dunkeldense*, p. 347 reads 'Subscribed by Mr Archibald Lyndsay as factor on behalf of Mr William Gordon'; see above, p. 301, n. 88.
[120] The correct calculation appears to be £23 18s 1½d.
[121] Cf *Rentale Dunkeldense*, p. 348. The benefices of Lagganallachy and Tealing were annexed to the archdeaconry, Cowan, *Parishes*, pp. 126, 196.

The toun of Petpoint, 14 b. victual; 'inde, 3 pars ordeii et bina pars farina' [*i.e.* ⅓ bere, ⅔ meal]. Bakello, 36 b. similar victual ('consimulium victualium'). Ballodroun, 24 b. similar victual. Bakembo, 54 b.; 4 b. wheat ('tritici'), 50 b. similar victual. Bannwth, 34 b.; 4 b. wheat, 30 b. similar victual. Kirktoun, 44 b.; 4 b. wheat, 40 b. similar victual. The Manyis, £20. Scheilhill, 16 b. similar victual; ⅓ bere, ⅔ meal. Bagro, 38 b.; 4 b. wheat, 34 b. similar victual. Bakak, 38 b. similar victual; ⅓ bere, ⅔ meal. The Prestoun, £20 Scots money ('monete Scotie').

'Out of the saidis fruitis
    Item, tane and payit ane yeirlie pensioun of' £40 to David Balfour. The vicarage of the kirk of Logiallochie 'quhilk payis to the vicare pensionar and stallar estimat to' £20. 'And I, the said Maister David, hes payit the samin to the saidis vicare and stallar thir thre yeris and the parrochionaris will nocht answer me of ane plak of the said vicarage.'

'Ecclesia de Logialloquhy'
    Kirk of Logialloquhy annexed to the said archdeaconry: The toun of Torquhoik, 4 merks. Letill Throchtkare, £4. Ballinreicht, 26s 8d. The Croft, 13s 4d. Innercochtkill, £4. Cabelley, £3. Thomgrow, £3. Thomgarne, 40s. Bourlik, 40s. Thobane, 40s. Drumowir, 4 merks. Peitcleocht, £4. Tullach, 5 merks. Custorum, 40s. Ballyclaichtkane, £4. The Bik, 40s. Ballonalen, 26s 8d. Mekill Logye, 8 merks. Ballynathane, 26s 8d.[122] Pattoun Towne, 26s 8d. Kencailyie, 40s. Torvald, 40s. Inver, 6 merks.

Signature: 'Maister David Spens, archindene of Dunkeld.'

Calculation of third: 'Summa of this benefice: Victuale twa parte meill, 3 part beir', 17 c. 10 b.; third thereof, 5 c. 14 b. Wheat, 1 c.; third thereof, 5 b. 1 f. 1⅓ p. Money, £101 6s;[123] third thereof, £33 15s 4d. 'Omittit caynis, gressumes and utharis deweteis.'

FORGANDENNY, VICARAGE OF,[124]

(NLS, fo 33r)
(SRO, Vol. a, fo 296r)

'The vicarage of Forgindyne that Magister Johnne Bartoun hes.'

'Item, the vicarage of Forgindyne quhilk is ane prebend of Dunkeld, the valour of the quhilk the said Maister Johnne cane nocht afferme bot he belevis it wes set

---

[122] 'Mekill Logye' and 'Ballynathane' do not appear in the SRO text.
[123] The correct calculation appears to be £100, without taking into account the £40 and £20 paid to a vicar and stallar.
[124] Cf NLS, fo 43r (see below, p. 325); see also *Rentale Dunkeldense*, p. 350.

within thir few yeris for' 60 merks, 'off the quhilk to the vicare pensionar', 20 merks.

Signature: 'M[r] Johnne Bartoun.'

Calculation of third: In the whole, £40; third thereof, £13 6s 8d.

ABERNYTE, PARSONAGE AND VICARAGE OF,   (NLS, fo 33r)
(SRO, Vol. a, fo 296r)

'This is the rentall of the kirk of Abirnytie within the diocie of Dunkeld and schirefdome of Perth, quhilk kirk is devydit and assignit amangis four chaiplanis, viz., schirris Williame Stewart, Thomas Muirheid, Alexander Moncreif and James Sandesone, equall portionaris thairof as choristis of Dunkeld.'

'Item, the said kirk bayth personage and vicarage is set and payis to thame foure equalie yeirlie' £80, 'viz., ilk ane of the four' £20 'and to the vicar pensionar', 12 merks.

Signatures: 'Thomas Muirheid, Williame Stewart, Alexander Moncreif, Jacobus Sandesone, portionarij de Abirnyte.'

Calculation of third: In the whole, £88; third thereof, £29 6s 8d.

'The haill to be avisit becaus it is devydit amangis choristis in Dunkeld.'

INCHMAGRANACHAN,   (NLS, fo 33v)
PREBEND OF DUNKELD CATHEDRAL[125]   (SRO, Vol. a, fo 296v)

'The rentale of the prebendrie of Inchemagronocht within the cathedrall kirk of Dunkeld pertening to sir Robert McNeir, channon of the sameyn extendis to' £35.

Calculation of third: In the whole, £35; third thereof, £11 13s 4d.

'Presentit be Duncane McNair.'

---

[125] Cf *Rentale Dunkeldense*, p. 350.

AUCHTERGAVEN, VICARAGE OF,

(NLS, fo 33v)
(SRO, Vol. a, fo 296v)

'The vicarage of Auchtirgavin united to the said prebendarie [of Inchmagranachan] wes wont to gif in assedatioun quhane the haill fruitis of the vicarage was answerit,' £20.

Calculation of third: In the whole, £20; third thereof, £6 13s 4d.

DUNKELD, TREASURERSHIP OF,[126]

(NLS, fo 33v)
(SRO, Vol. a, fo 296v)

'The thesauraris of Dunkeld pertenyng to Maister Robert Abircrommy quhilk consistis haill in the viccarage quhilk wes wont to gif ane hundreth merkis and now be ressoun of new ordour, reformatioun and discharging of corps presentis, umest clathis, Pasche raikiningis and utharis offeringis thai quha heid[127] of befoir for the said hundreth merkis refus now for' £24.

Signature: 'Maister Robert Abircrummy, thesaurare of Dunkeld.'

Calculation of third: In the whole, £24; third thereof, £8.

DUNKELD, TREASURERSHIP OF,[128]

(NLS, fo 34r)
(SRO, Vol. a, fo 297r)

'Rentale thesaurarie cathedralis ecclesie Dunkelden'

The perpetual treasurership of the cathedral kirk of Dunkeld belonging to Mr Stephen Wilsoun extends according to common estimation to the sum of £80 annually, from which is deducted £10 for the stipend of the stallar of the choir and therefore the sum of the foresaid benefice remaining freely for the thesaurer, £70. This benefice is believed to be litigious between the said Mr Stephen and Robert Abircrommy, both contending it.[129]

Signature: 'Stephanus Wilsoun, manu sua. Presentatur et exhibitur per M. George Cok ad requisitionem specialem predicti Magistri Stephani Wilson, teste manu mea.'[130]

---

[126] Cf NLS, fo 34r; see also *Rentale Dunkeldense*, p. 347.
[127] This should probably read 'held'; *Rentale Dunkeldense* gives 'herd'.
[128] Cf NLS, fo 33v; see also *Rentale Dunkeldense*, p. 347.
[129] Stephen Culross, alias Wilson, was treasurer of Dunkeld, 1554-1561, and was latterly involved in litigation with Robert Abercrombie who also claimed the title. Watt, *Fasti*, p. 115.
[130] I.e. 'Presented and exhibited by Mr George Cok at the special request of the foresaid Mr Stephen Wilson, witnessed with my hand.'

Calculation of third: In the whole, £80; third thereof, £26 13s 4d.

CRAIGIE, PREBEND OF DUNKELD CATHEDRAL[131]

(NLS, fo 34r-v)
(SRO, Vol. a, fo 297r)

Rental of the prebend of Cragie lying in the diocese of Dunkeld 'geif befoir the yeir of God' 1558 the sum of 45 merks 12s 8d to Gilbert Banermane, prebendary of the same and canon of Dunkeld; of that sum given to the stallar, 2 merks.

Signature: 'Gilbert Banermane, prebendarie of Cragie.'

Calculation of third: In the whole, £30 12s 8d; third thereof, £10 4s $2\frac{2}{3}$d.

'The prebendarie of Cragie of Dunkeld diocie set in few to the lard of Cowtie for' 45 merks 'of this gevin to his stallar', 15 merks.

Signature: 'Johnne Bartoun, de mandato possess[oris] produxit.'

COMRIE, PARSONAGE OF,

(NLS, fo 34v)
(SRO, Vol. a, fo 297v)

The parsonage of Comry within the diocese of Dunblane pertaining to Mr Alexander Cheisholme set in assedation to John Comrie of that Ilk for the yearly payment of 40 merks.

Signature: 'Johnne Morisone, be me presentit.'

Calculation of third: In the whole, £26 13s 4d; third thereof, £8 17s $9\frac{1}{3}$d.

GLENDEVON, PARSONAGE AND VICARAGE OF,[132]

(NLS, fo 34v)
(SRO, Vol. a, fo 297v)

The parsonage and vicarage of Glendowane pertaining to Mr William Cheisholme within the said diocese set in assedation to sir John Blakwod for the yearly payment of 40 merks.

Signature: 'Presentit be me, Johnne Morisone.'

Calculation of third: In the whole, £26 13s 4d; third thereof, £8 17s $9\frac{1}{3}$d.

---

[131] Cf *Rentale Dunkeldense*, p. 349.
[132] Cf NLS, fo 53v (see below, p. 337).

ABERNETHY, VICARAGE OF,

(NLS, fo 35r)
(SRO, Vol. a, fos 297v-298r)

The vicarage of Abirnethy within the said diocese pertaining to Mr David Gourlay 'wes wount to pay yeirlie' 80 merks 'and hes payit nay thing thir foure yeris bigane.'

Signature: 'Presentit be me, Johnne Morisone.'

Calculation of third: In the whole, £53 6s 8d; third thereof, £17 15s 5½d.[133]

'Aliter gevin in be the [same, *deleted*] said Mr David him self at command of the Clerk of Registre extending to' £36 13s 4d.

LOGIE, VICARAGE OF,

(NLS, fo 35r)
(SRO, Vol. a, fo 298r)

The vicarage of Logie within the diocese of Dunblane pertaining to James Cheisholme set to Jehane [Jeane, *SRO*] Cheisholme for yearly payment of £40.

Signature: 'Presentit be me, Johnne Morisone.'

Calculation of third: In the whole, £40; third thereof, £13 6s 8d.

STROWAN, VICARAGE OF,

(NLS, fo 35r)
(SRO, Vol. a, fo 298r)

The vicarage of Strowane within the said diocese [*i.e.* Dunblane], pertaining to the said James [Chisholm] 'within the said diocie, wes wount to pay yeirlie' 20 merks 'and now payis nathing thir four yeris bigane.'

Signature: 'Presentit be me, Johnne Morison.'

Calculation of third: In the whole, £13 6s 8d; third thereof, £4 8s 10⅔d.

---

[133] The correct calculation appears to be £17 15s 6⅔d.

DUNNING, CHAPLAINRY OF,

(NLS, fo 35v)
(SRO, Vol. a, fo 298r)

Chaplainry of Donning within the said diocese [*i.e* Dunblane], pertaining to sir John Lermonth, 'quhilk hes nay propirtie bot ane pece land free for the quhilk Donald Kelty payis yeirlie' 10 merks.

Signature: 'Presentit be me, Johnne Morisone.'

Calculation of third: In the whole, £6 13s 4d; third thereof, 44s 5⅓d.

DUNBLANE CATHEDRAL, CHAPLAINRY OF ST BLAISE IN,[134]

(NLS, fo 35v)
(SRO, Vol. a, fo 298v)

The chaplainry of Sanct Blais within the kirk of Dunblane pertaining to the said sir John [Lermonth] 'wes wount to pay yeirlie' £20.

Signature: 'Presentit be me, Johnne Morisone.'

Calculation of third: In the whole, £20; third thereof, £6 13s 4d.

'Aliter gevin in and ressavit at command of the Clerk of Register, extending in the haill to' £10 money.

DUNBLANE CATHEDRAL, CHAPLAINRY OF OUR LADY IN,

(NLS, fo 35v)
(SRO, Vol. a fo 298v)

Chaplainry of [Our] Lady within the said kirk pertaining to sir William Blaikwod, 'wes wount to pay yeirlie' 30 b. meal and 6 merks money.

Signature: 'Presentit be me, Johnne Morison.'

Calculation of third: Meal, 30 b.; third thereof, 10 b. Money, £4; third thereof, 26s 8d.

DUNBLANE, SUBDEANERY OF,

(NLS, fo 36r)
(SRO, Vol. a, fo 298v)

Subdeanery of Dunblane pertaining to sir Elmond Cheisholme set to William Striveling of Ardoch for yearly payment of 10 merks.

---

[134] Cf NLS, fo 45v (see below, p. 328).

Signature: 'Presentit be me Johnne Morisone.'

Calculation of third: In the whole, £6 13s 4d; third thereof, 44s 5⅓d.

PERTH, VICARAGE PENSIONARY OF ST JOHN'S KIRK IN,[135]

(NLS, fo 36r)
(SRO, Vol. a, fo 299r)

'Vicaria pensionaria ecclesie Sancti Joannis'

The vicarage pensionary of the kirk of St John in the burgh of Perth, £20 Scots money annually paid to him by [*blank*] Ros of Cragie[136] of the fermes of the said kirk of which foresaid money he received nothing at all from the said farmer for the past term with a fourth part of the year already elapsed because of the tumult occurring.

Signature: 'Ita est M[r] G. Cok, vicarius, qui supra manu.'

Calculation of third: In the whole, £20; third thereof, £6 13s 4d.

DUNBLANE, DEANERY OF,

(NLS, fo 36r-v)
(SRO, Vol. a, fo 299r)

'Denrie of Dunblane'

'Inprimis, the vicarage of the toune and parochin of Dunblane uncertane quhat the samin wilbe be ressoun of alteratioun bot I belive it wilbe' 20 merks. 'Item, victuale of Vissurdland callit the Kir extending to' 80 b., 'the samin set in assedatioun for' £40. 'Item, ane quarter of ane kirk in Fyif callit Tullibuill set in assedatioun for' £8. 'Item, twa eckerris besyid Dunblane set in few for' 40s.

Signature: 'M[r] Roger Gordoun, den of Dunblane.'

Calculation of third: In the whole, £63 6s 8d; third thereof, £21 2s 2⅔d.

LONGFORGAN, VICARAGE OF,

(NLS, fo 36v)
(SRO, Vol. a, fo 299r-v)

Rental of the vicarage of Langforgung within Gowrie pertaining to Mr John Rattray, vicar thereof.

---

[135] Cf SRO, Vol. a, fo 87r; see also above, p. 74, nn. 294, 295.
[136] I.e. John Ross of Craigie, cf *RSS*, v, nos 3-4, 1284.

'Item, the vicarage foirsaid quhane it wes last sett in assedatioun and quhane corps presentis, viz., cowclayth and Pasche fynis and small offerandis wes payit, halelie payit in assedatioun yeirlie to my predicessouris foure scoir merkis, and the saidis corps presentis and offeringis wilbe the ane half of the said soume, viz.', 40 merks 'or neirby.'

Signature: 'Ita est Johannes Rattray, manu sua.'

Calculation of third: In the whole, £26 13s 4d; third thereof, £8 17s $9\frac{1}{3}$d.

RUFFILL, PREBEND OF DUNKELD CATHEDRAL[137] (NLS, fo 37r)
(SRO, Vol. a, fo 299v)

'Efter followis the haill just rentale of the prebendarie of Ruffill situat within the cathedrall kirk of Dunkeld, gif in be me sir George Fullourtoun, prebendar of the samin and subscrivit with my hand.'

'In the first, for the few mailis of the toune of Ruffill, brewland and eist parte of Kincarneis yeirlie, conforme to the chartouris maid thairupoun, sett in few', 34 merks 4s money; 12 b. horsecorn; 12 'turssis' straw; 4 dozen poultry 'allanerlie.'

'Off the quhilk soume to be deducit and payit yeirlie to the stallar', 5 merks.

Signature: 'Georgius Fullartoun, prebendair of Ruffill.'

Calculation of third: Money, £22 17s 4d; third thereof, £7 12s $5\frac{1}{3}$d. Horsecorn, 12 b.; third thereof, 4 b. Straw, 12 'turssis'; third thereof, 4 'turssis'. Poultry, 4 dozen; third thereof, 16.

'The uther rentall of the samin giffin in be M[r] Robert Auchmouthy [Auchinmony, *SRO*].'

'Item, the males of Ruffill and Kincarne', £22 17s 4d; 12 b. oats, 'estemit to' £6; 48 'powtre[138] [pultrie, *SRO*] fowllis', 48s; 12 'turssis' straw, 'estemit' 12s. Total, £31 17s 4d.

---

[137] Cf also *Rentale Dunkeldense*, p. 349.
[138] This may read 'powtie'; a powt is a pullet or game bird.

FORTEVIOT, PARSONAGE OF,[139]

(NLS, fo 37v)
(SRO, Vol. a, fo 300r)

Rental of the parsonage of Forteviot given in by me, Mr James Thorntoun, 'in name and behalf and as procuratour to his eme, Maister Johnne Thornetoune, usufructuar of the said personage.'

'The haill teyndis and all uthir proffitis appartening to the personage of Forteviot hes bene set in assedatioun thir xl yeiris bygane to Oliver Maxtoun of Perth and now to Patrik Murray and Thomas Monypenny, burges of Perth, for the soume of thre hundreth merkis money of this realme.'

Signature: 'Ita est Jacobus Thorntoun, manu propria.'

'The rentale of the personage of Forteviot is thre hundreth merkis to the persone yeirlie, Patrik Murray takismane one [*i.e.* on] the ane part, and Thomas Monypenny in Perth, takisman of the uthir. This is trew rentall gevin in be me to the kirk at the Conventioun[140] in Edinbrucht', 24 January 1561/2.

Signature: 'Patrik Murray.'

Calculation of third: In the whole, £200; third thereof, £66 13s 4d.

INCHTURE, VICARAGE OF,

(NLS, fo 38r)
(SRO, Vol. a, fo 300r)

Rental of the vicarage of Inchestuir with the pendicle thereof called Kynnard, lying in the Kers of Gowrie within the diocese of Sanctandrois and sheriffdom of Perth, 'gaif yeirlie befoir the moneth of Junij 1558 the sowme of thirty pundis libere to Maister Arthour Tallifer, persoun thairoff.'

Signature: 'M[r] Arthour Taillifer, with my hand.'

Calculation of third: In the whole, £30; third thereof, £10.

---

[139] Cf SRO, Vol. a, fo 164v (see above, p. 317).
[140] See above, p. 302, n. 92.

CAPUTH, PREBEND OF DUNKELD CATHEDRAL[141] (NLS, fo 38r)
(SRO, Vol. a, fo 300v)

Rental of the prebend of Capeth within the diocese of Dunkeld pertaining to Mr Archibald Lyndsay, possessor of the same.

The toun of Capeth McKarthill set in feu to the tenants thereof paying yearly £18 13s 4d. The toun of Blandarrane set in feu to my lord Gray[142] paying therefore yearly £20. The toun of Bauchland set in feu to Duncan Rattray, paying therefore yearly £9 6s 8d. Total, £48.

'Heirof to be deducit for the stallaris fee', £5. 'Sua restis' £43.

Signature: 'Maister Archibald Lyndsay.'

Calculation of third: In the whole, £48; third thereof, £16.

WEEM, PARSONAGE AND VICARAGE OF, (NLS, fo 38v)
(SRO, Vol. a, fo 300v)

The parsonage and vicarage of the kirk of Weyme extends to £60; of which £20 'to the vicare pensionar.'

Signature: 'Johnne Stewart, sone to James Stewart of Fancastell.'[143]

Calculation of third: In the whole, £60; third thereof, £20.

> 'Heir efter followis the rentailis of the benefices, chaplanries and channonreis quhilk is at the donatioun and presentatioun of Johnne, erle of Atholl and lord of Bolwanye,[144] Moraviensis [sic] diocesis.'

BLAIR ATHOLL, PARSONAGE AND VICARAGE OF, (NLS, fo 38v)
(SRO, Vol. a, fo 300v)

The parsonage and vicarage of Blair in Athol in the diocese of Dunkeld which extends to 100 merks, of which the vicar pensionary gets 20 merks.

---

[141] Cf also *Rentale Dunkeldense*, p. 349
[142] Patrick, 5th lord Gray, *Scots Peerage*, iv, pp. 283-284.
[143] I.e. Fincastle, *Scots Peerage*, ii, p. 187.
[144] John Stewart, 4th earl of Atholl, was also lord of Balvenie, cf *RMS*, iv, nos 84, 1520; *Scots Peerage*, i, pp. 444-445.

Signature: 'Presentit be Maister Leonard Leslie.'

Calculation of third: In the whole, £66 13s 4d; third thereof, £22 4s 5½d.

STRUAN, PARSONAGE AND VICARAGE OF, (NLS, fo 38v)
(SRO, Vol. a, fo 301r)

The parsonage and vicarage of Strowane in the said diocese of Dunkeld extends to 100 merks, 'the curatis to be payit thairof.'

Signature: 'Present be Maister Leonard Leslie.'

Calculation of third: In the whole, £66 13s 4d; third thairof, £22 4s 5½d.

KILMOVEONAIG,[145] PARSONAGE AND VICARAGE OF, (NLS, fo 39r)
(SRO, Vol. a, fo 301r)

The parsonage and vicarage of Kilmavernok extending to 100 merks 'diocesis eiusdem Dunkell, the curatis to be payit thairof.'

Signature: 'Presentit be Maister Leonard Leslye.'

Calculation of third: In the whole, £66 13s 4d; third thereof, £22 4s 5½d.

LUDE, PARSONAGE AND VICARAGE OF, (NLS, fo 39r)
(SRO, Vol. a, fo 301r-v)

The parsonage and vicarage of Luid in the diocese of Dunkell extends to 40 merks, 'the curatis to be payit thairof; the persoun thairof instantlie deid; the benefice in my lordis handis undisponit.'

Signature: 'Presentit be me, Maister Leonard Leslye.'

Calculation of third: In the whole, £26 13s 4d; third thereof, £8 17s 9½d.

---

[145] Cf Cowan, *Parishes*, p. 105; Scott, *Fasti*, iv, pp. 143-144.

SKIRDUSTAN[146] AND BOTRIPHNIE, PARSONAGE OF,[147]

(NLS, fo 39r-v)
(SRO, Vol. a, fo 301v)

'Item, the personage of Skeirdrostane and Pittrefeyne, twa kirkis unit togiddir and erect in ane prebendrie of the cathedrall kirk of Murray callit the prebendarie of Abirlour, of the quhilkis the teynd schaves extendis to ane hundreth merkis', 10 merks 'of the sam[in] gevin yeirlie to ane chaplane in the said cathedrall kirk of Elgin for his stall fee.'

Signature: 'Maister Leonard Leslie.'

Calculation of third: In the whole, £66 13s 4d; third thairof, £22 4s 5⅓d.

DUNKELD CATHEDRAL, CHAPLAINRY OF ST NINIAN'S ALTAR IN,

(NLS, fo 39v)
(SRO, Vol. a, fo 301v)

Chaplainry of Sanct Niniandis altar in Dunkeld

'Item, the chaiplanrie callit Sanct Nigannis [sic] alter situat within the cathedrall kirk of Dunkeld, the anuellis and landis apertening thairto extenden to' 35 merks.

Signature: 'Presentit be me, M[r] Leonard Leslie.'

Calculation of third: In the whole, £23 6s 8d; third thairof, £7 15s 6⅔d.

PERTH, ALTARAGE IN,

(NLS, fo 39v)
(SRO, Vol. a, fo 301v)

'Item, ane alterage in the kirk of Perth, the anuellis and landis apertening thairto extendis to' 14 merks.

Signature: 'Presentit be me, Maister Leonard Leslie.'

Calculation of third: In the whole, £9 6s 8d; third thairof, £3 2s 2⅔d.

---

[146] I.e. Aberlour, Cowan, *Parishes*, p. 3.

[147] Aberlour and Botriphnie, in the sheriffdom of Banff and diocese of Moray, were included with rentals for Perth because they were presented by Leonard Leslie, prebendary of Aberlour, who also presented those for the immediately preceding benefices.

KILLIN, VICARAGE PORTIONARY OF,   (NLS, fo 39v)
(SRO, Vol. a, fo 302r)

'Vicarage callit pensioun of Kellyne'

'Item, the vicarage of Kellyne callit portionar of Kelline in the diocie of Dunkeld, the haill extendis to' 22 merks, 'and that by the curatis fee.'

Signature: 'Presentit be me, Maister Leonard Leslie.'

Calculation of third: In the whole, £14 13s 4d; third thereof, £4 11s 1½d.[148]

INCHAIDEN,[149] BENEFICE OF,   (NLS, fo 40r)
(SRO, Vol. a, fo 302r)

The kirk of Incheskadyne

'Item, the kirk of Inchedene haldin of the dene of Dunkeld in tackis payand thairfoir' 260 merks 'by the sustentatioun of the vicar pentionare.'

Signature: 'Presentit be me, Maister Leonard Leslie.'

Calculation of third: In the whole, £73 6s 8d;[150] third thereof, £57 15s 6⅔d.

RHYND, VICARAGE OF,   (NLS, fo 40r)
(SRO, Vol. a, fo 302r)

'Vicaria de Rynd diocesis Sanctandrie'

The vicarage of Rynd in the diocese of St Andrews lying within the sheriffdom of Perth is set and is accustomed to pay £20, deducting procurations synodal and visitations of the dean, 'sed nunc etc.'[151]

Signature: 'Magister Johannes Logie.'
Calculation of third: In the whole, £20; third therof, £6 13s 4d.

---

[148] The correct calculation appears to be £4 17s 9½d.
[149] I.e. Kenmore, Cowan, *Parishes*, p. 84.
[150] The correct calculation appears to be £173 6s 8d.
[151] This last phrase occurs in NLS text only.

COLLACE, PARSONAGE OF,

(NLS, fo 40v)
(SRO, Vol. a, fo 302r)

'Rectoria de Colles in Gouria'

The parsonage of Colles in Gourie, parson Mr John Douglas, yearly received for the teind sheaves of the same, eight score merks.

Signature: 'M[r] Joannes Dowglas.'

Calculation of third: In the whole, £106 13s 4d; third thereof, £35 11s 1½d.

ABERNETHY COLLEGIATE KIRK, PROVOSTRY OF, (NLS, fo 40v)
(SRO, Vol. a, fo 302v)

'The rentale of the provestrie of Abirnethie is in victuale.'

The Manis, 4 c. Clasweillie, 8 b. The mill lands, 12 b. The teinds of Baiglie, 1 c. 'The saxt part teind of Carpow', 24 b. The mail silver of the mill, 10 merks 'and thrie pund.' 'Johnne Duncane croft' and 40s 'of the landis of Baiglie with' 40s 'of few maill of the landis of Knitis Pete.'[152]

'And of this rentale to be defakit' £10 'to ane priest for service makin in the kirk, and uther' 40s 'payit to the stallair in Dunblane, and this to be the haill rentale of the provestrie I appreif be this my hand writ subscrivit be me.'

Signature: 'M[r] Williame Schaw, provest of Abirnethie.'[153]

Calculation of third: Victual, ⅔ meal, ⅓ bere, 7 c. 12 b.; third thereof, 2 c. 9 b. 1 f. 1⅓ p. Money, £13 13s 4d; third thereof, £4 11s 1½d.

CRIEFF, PREBEND OF DUNKELD CATHEDRAL[154] (NLS, fo 41r)
(SRO, Vol. a, fo 302v)

Rental of the prebend of Creif lying within the cathedral kirk of Dunkeld pertaining to Mr George Cuke.[155]

---

[152] Possibly 'Wright's Pottie', which appears on the map at NO1615.
[153] William Shaw occurs as provost, 1550 - 1574, Watt, *Fasti*, p. 343.
[154] Cf *Rentale Dunkeldense*, p. 349.
[155] Cf below, p. 323, n. 156.

'Item, payit to me yeirlie be the twa personis of Creiff, prebendariis of the quenis Chapell Ryell of Striveling',[156] £26 13s 4d 'in pensioun, thairof payit be me to the dene and chaptour of Dunkeld to thair comondis' £6 13s 4d 'and yeirlie to my vicare and stallar' £5, 'and sua restis to me yeirlie of my said prebendarie', £15.

Signature: 'M[r] George Cuk, with my hand.'

Calculation of third: In the whole, £26 13s 4d; third thereof, £8 17s 9⅓d.

DUNKELD CATHEDRAL, CHAPLAINRY OF OUR LADY ALTAR IN,

(NLS, fo 41r-v)
(SRO, Vol. a, fo 303r)

Rental of the chaplainry of Our Lady altar situated within the cathedral of Dunkeld pertaining to Mr George Cuk.

£8 yearly of the lands of Eister and Wester Lowstonis, Kinvaid and the Craigheid with their pertinents with 12 b. bere and 12 b. meal of the said lands yearly. 'Item, fourte schillingis yeirlie of anuell of the mans and hous of the prebendarie of Lundeif, thairof gevin to the chaiplanis and queir of Dunkeld', 26s 8d.

Signature: 'Maister George Cuk, with my hand.'

Calculation of third: Money, £10; third thereof, £3 6s 8d. Bere, 12 b.; third thereof, 4 b. Meal, 12 b.; third thereof, 4 b.

KINNOULL, PARSONAGE AND VICARAGE OF,[157]

(NLS, fo 41v)
(SRO, Vol. a, fo 303r)

'The personage and vicarage of Kinnoule annexit thairto[158] liand within the schirefdome of Perth and diocie of Sanctandrois set in assedatioun to Thomas Monypennie, burges of Perth, and Maister Handrie Thorntoun for four hundreth and twenty merkis yeirlie be Maister Robert Carnegie, persoun thairof, as ane act in the Officiallis buikis of Sanctandrois maid thairupoun proportis, quhilkis tak

---

[156] Part of Crieff's revenues as a prebend of Dunkeld had been granted to the common fund of the cathedral. With the annexation of the prebend to the Chapel Royal at Stirling in 1501, two prebends, Crieff primo and secundo, were founded, and certain revenues were allocated to the holder of the prebend in Dunkeld cathedral who was expected to pay ten merks to the common fund. Cowan, *Parishes*, p. 39.

[157] Cf NLS, fo 47v (see below, p. 331).

[158] Kinnoull had been briefly annexed to Cambuskenneth abbey in the late fourteenth century but is said to have regained its independence. Cowan, *Parishes*, pp. 115-116.

thame furth at Lambes[159] bypast, now set of the same price to Sir Robert Carnegy of Kynnard.'

Signature: 'Sir Robert Carnegy.'

Calculation of third: In the whole, £266 13s 4d;[160] third thereof, £88 17s 9⅓d.

TULLYPOWRIE AND FYNDYNATE, CHAPLAINRY OF,   (NLS, fo 42r)
(SRO, Vol. a, fo 303v)

'The trew rentale of the chaiplanrij of Tullipourie and Fandownat pertening to sir Waltir Young, possessour of the samin, maid at Edinbrucht', 21 February 1561/2, 'subscrivit with my hand befoir thir witnes, Thomas Lindsay, Adame Lindsay.'

Fandownat feued for £6 6s 8d. Tulipowrie feued for £14 3s 4d. The Myln of Twlypourie feued for 53s 4d. Balnagard feued for £3 10s. Total, £26 13s 4d.

Signature: 'Waltir Young, chaiplane of Tulipourie, with my hand, apprevis this rentale.'

Calculation of third: In the whole, £26 13s 4d; third thereof, £8 17s 9⅓d.

LUNCARTY, PARSONAGE AND VICARAGE OF,   (NLS, fo 42r)
(SRO, Vol. a, fo 303v)

'I, Maister Waltir Balfour, persoun of Longardy, makis manifest that the personage and vicarage of Longardy is worth na mair nor fourte merkis be yeir, satifiand the curat and ridar.'

Calculation of third: In the whole, £26 13s 4d; third thereof, £8 17s 9⅓.

TRINITY, CHAPLAINRY OF,[161]   (NLS, fo 42v)
(SRO, Vol. a, fos 303v-304r)

'Item, it is worth na mair nor' 22 merks and 9 b. victual 'and this I mak manifest be this my writing,[162] and subscrivit with my hand at Edinbrucht', 27 January 1561/2.

---

[159] I.e. Lammas, 1 August.
[160] This equals 400 merks; 420 merks would be £280.
[161] Trinity chaplainry pertained to Dunblane cathedral, *Inquisitionum ad Capellam Regis Retornatarum ... Abbrevatio*, ed. T. Thomson, 3 vols. (London, 1811-1816), ii (Perth), no 1004.
[162] In the NLS text the scribe mistakenly repeated 'writing'.

Signature: 'M[r] Waltir Balfour, with my hand.'

Calculation of third: 'In the haill of money', £14 13s 4d; third thereof, £4 17s 9½d. Victual, 9 b.; third thereof, 3 b.

FORGANDENNY, CHAPLAINRY OF,[163] (NLS, fo 42v)
(SRO, Vol. a, fo 304r)

The chaplainry of Forgundine, Alexander Ruthven, brother german to Patrick, lord Ruthvene, superior and patron of the same.

'Item, the said chaplanrij set in few ferme to Waltir Ruthvene, duelland in the Few,[164] for the payment of' 40 merks yearly.

Signature: 'Alexander Ruthvene.'

Calculation of third: In the whole, £26 13s 4d; third thereof, £8 17s 9½d.

FORGANDENNY, VICARAGE OF,[165] (NLS, fo 43r)
(SRO, Vol. a, fo 304r)

Rental of the vicarage of Forgundine, which is a prebend within the kirk of Dunkeld, pertaining to Mr John Leslie, prebendary of the same.

'The said vicarage wes sett in assedatioun thir divers yeris bigane for thre scoir pundis to Johnne Browne of Fordaill quhill laitlie within thir twa yeris lastbipast that he refusis to accept it for fourte pundis and sua the fruicttis thairof restis in the parochionaris handis, off the quhilk thair is gevin furth yeirlie in ordinar chargis, awchtene pundis sax schillingis awcht d. [*i.e.* £18 6s 8d] to the vicar pensionar and stallar quha ar providit thairto, and sua the rentale thairof the saidis ordinar chargis being deducit will extend to' £21 13s 4d 'frelie be ressoun of the fruitis of the vicarage ar nocht payit as thay war wount to be.'

Signature: 'Maister Johnne Leslie, prebendare of Forgundine.'

Calculation of third: In the whole, £21 13s 4d; third thereof, £7 4s 5⅓d.

---

[163] Cf NLS, fo 46v (see below, p. 329).
[164] Cf *RMS*, iv, no 1413.
[165] Cf NLS, fo 33r (see above, p. 309).

CARGILL, VICARAGE OF,
(NLS, fo 43r-v)
(SRO, Vol. a, fo 304v)

Rental of the vicarage of Cargill within the diocese of Dunkeld.

'The said vicarage with the gleib land thairof and pertinentis thairto wes wount to be sett in assedatioun to Williame Drummound, guidman of Cargill, quhane all thingis wes thane ansurit use and wount quha geif our the same becaus of tinsale for the soum of' 90 merks 'and now, throw inlaik of the corps presentis etc., the said vicarage is worth' £40 'or thairby, and thairof to be deducit for the curatis fee' 20 merks 'and sua restis de claro' 40 merks.

Signature: 'Ita est Willelmus Drummond, vicarius de Cargill.'

Calculation of third: In the whole, £24;[166] third thereof, £8.

[*In margin*, 'Primo Maij anno' 1562. 'Cok, exhibit this rentale'.[167]]

MUCKERSIE, PARSONAGE AND VICARAGE OF,[168]
(NLS, fo 43v)
(SRO, Vol. a, fo 304v)

'My lordis, it will ples your lordschippis to undirstand as to my personage and vicarage of Mukersie, prebend in Dunkeld, wes ay set in assedatioun for ane hundreth pundis yeirlie, sene I wes promovit thairto, and laist wes set to James Johnstoun and David Andirsone, burgessis of Sanct Johnnestoun, for the said soume of' £100 'and payis thairof to the vicar pensionar' £10 'yeirlie, and to my stallar of Dunkeld' 6 merks 'and for procurage and sinage [*recte*, procurations and synodals] yeirlie', 5 merks.

Signature: 'M[r] Johnne Stevinstoun, with my hand.'

Calculation of third: In the whole, £100; third thereof, £33 6s 8d.

MUTHILL, VICARAGE OF,[169]
(NLS, fo 44r)
(SRO, Vol. a, fo 305r)

The vicarage of Muthill

Calculation of third: In the whole, £33 6s 8d; third thereof, £11 2s 2⅔d.

---

[166] 40 merks equals £26 13s 4d, not £24.
[167] This marginal note does not appear in the SRO text. George Cook was vicar pensionary of Perth; cf NLS, fos 34r, 36r (see above, pp. 311 and n. 130, 315).
[168] Cf NLS, fo 54v (see below, p. 340); see also *Rentale Dunkeldense*, p. 350.
[169] Cf NLS, fo 55r (see below, p. 340).

LUNDEIFF, PARSONAGE AND VICARAGE OF,[170]

(NLS, fo 44r)
(SRO, Vol. a, fo 305r)

Parsonage and vicarage of Lundeif

[blank]

Calculation of third: In the whole, £66 13s 4d; third thereof, £22 4s 6½d.[171]

FUNGARTH, PREBEND OF DUNKELD CATHEDRAL[172]

(NLS, fo 44v)
(SRO, Vol. a, fo 305v)

'Prebendarie of Fungard in the kirk of Dunkeld, twa clameris, McGill, Thorntoun.'[173]

Calculation of third: Money, £79; third thereof, £26 6s 8d. Bere, 2 b.; third thereof, 2 f. 2⅔ p. Horsecorn, 10 b.; third thereof, 3 b. 1 f. 1⅓ p.

DUNBLANE CATHEDRAL, CHAPLAINRY OF ST NICHOLAS IN,

(NLS, fo 44v)
(SRO, Vol. a, fo 305v)

'Heir followis the rentall of the chaplanrie of Sanct Nicholas within the kirk of Dunblane pertening to sir James Finlasoun, chaplane thairof.'

An annual yearly of 10 merks of the lands of Camscheny pertaining to my lord Drummond[174] within the parish of Dunblane. An annualrent 'of ane lugeing in Sanctjhon pertening to the lard of Kinfawnis callit Camerone land, yeirlie' 5 merks. 'Ane yeirlie annuell aucht furth of the landis and tenement liand within the toun of Dunblane pertening to Robert Lermonth, be yeir', 24s. 'Ane annuelrent aucht furth of the hous inhabit be Lucas Maxtoun within the said citie of Dunblane be yeir', 20s. 'Ane annuelrent aucht furth of the hous inhabit be Wille Morisoun within the said citie', 13s 4d.

---

[170] Cf NLS, fo 53r (see below, p. 337); see also *Rentale Dunkeldense*, p. 351.

[171] The correct calculation appears to be £22 4s 5½d.

[172] Cf NLS, fos 46v, 81v, 82r (see below, pp. 329, 365, 366); see also *Rentale Dunkeldense*, p. 349.

[173] Thomas McGill was prebendary of Fungarth in 1573, *RMS*, v, no 791; James Thornton is also on record as prebendary, see NLS, fos 46v, 81v, 82r (see below, pp. 329, 365, 366).

[174] David, 2nd lord Drummond, *Scots Peerage*, vii, pp. 46-47.

BALQUHIDDER, PREBEND OF DUNBLANE CATHEDRAL[175]

(NLS, fo 45r)
(SRO, Vol. a, fo 306r)

'Prebendarie Balquiddir'

Calculation of third: In the whole, £16; third thereof, £5 6s 8d.

FORTEVIOT, VICARAGE OF,[176]

(NLS, fo 45r)
(SRO, Vol. a, fo 306r)

The vicarage of Forteviot

Calculation of third: In the whole, £26 13s 4d; third thereof, £8 17s 9⅓d.

DUNBLANE CATHEDRAL, CHAPLAINRY OF ST BLAISE IN,[177]

(NLS, fo 45v)
(SRO, Vol. a, fo 306v)

Rental of the chaplainry of Sanct Blais within the kirk of Dunblane pertaining to sir John Lermonth 'extendis onelie in yeirlie proffite to the soum of auchtene merkis and the thrid thairof to' £4.

Signature: 'Johnne Lermonth. James, regirit [sic] this rentall. J. Clericus Registri.'

CRIEFF, CHAPLAINRY OF,

(NLS, fo 45v)
(SRO, Vol. a, fo 306v)

'Chaplanrij Chreif'

Calculation of third: In the whole, £60; third thereof, £20.

RATTRAY, PARSONAGE OF,

(NLS, fo 45v)
(SRO, Vol. a, fo 306v)

The parsonage of Rattray

Calculation of third: In the whole, £91; third thereof, £30 6s 8d.

---

[175] Cf NLS, fos 47r, 99r (see above, pp. 330, 392).
[176] Cf NLS, fo 47r (see above, p. 330).
[177] Cf NLS, fo 35v (see above, p. 314).

ABERNETHY, PARSONAGE OF,[178] (NLS, fo 46r)
(SRO, Vol. a, fo 307r)

Parsonage of Abirnethie

[blank]

Calculation of third: In the whole, £266 13s 4d; third thereof, £88 17s 9½d.

FUNGARTH, (NLS, fo 46v)
PREBEND OF DUNKELD CATHEDRAL[179] (SRO, Vol. a, fo 307v)

'The prebendrie of Foungart in the kirk of Dunkeld gevin in be Maister James Thorntoun, undoutit prebendar thairof.'

'The towne of Fungart the Hach [Hauch, *fo 81v*], the Hilheid and all uther landis bayth fewit and unfewit pertening to the said prebendrie payis in penny maill' £79 13s 4d. 'Thair is sum dewiteis of pultrie and pettis e[x]tending to ane sobir sowme [ane uther sowme, *fo 81v*] [blank]. The haill foirsaidis landis payis twa bollis beir and ten bollis hors corne', [total], 12 b.
'The prebendrie of Foungard in the said kirk of Dunkeld, twa clameris, McGill and Thornetoun.'[180]

Calculation of third: Money, £79 13s 4d; third thereof, £26 6s 8d.[181] Bere, 2 b.; third thereof, 2 f. 2⅔ p. Horsecorn, 10 b.; third thereof, 3 b. 1 f. 1⅓ p.

FORGANDENNY, CHAPLAINRY (NLS, fo 46v)
OF ST CATHERINE'S ALTAR IN,[182] (SRO, Vol. a, fo 307v)

Chaplainry of Sanct Katherine altar, situated within the kirk of Forgundeny, pertaining to sir Robert Ostlare set in feu to Alexander Ruthven.'

Calculation of third: In the whole, £4; third thereof, 26s 8d.

---

[178] Cf NLS, fo 47r (see above, p. 330).
[179] Cf NLS, fos 44v, 81v, 82r (see above, p. 327; below, pp. 365, 366); see also *Rentale Dunkeldense*, p. 349.
[180] See above, p. 327, n. 173.
[181] The correct calculation appears to be £26 11s 1¼d.
[182] Cf NLS, fo 42v (see above, p. 325).

FORTEVIOT, VICARAGE OF,[183] (NLS, fo 47r)
(SRO, Vol. a, fo 308r)

The vicarage of Forteviot set in assedation for 40 merks. 'The umest claithis, corspresentis and Paschefynes to be defalkit conforme to the rait and to my Lordis ordinance.'

Calculation of third: In the whole, £26 13s 4d; third thereof, £8 17s $9\frac{1}{3}$d.

[*In margin*, 'Thrid thairof according to this reformatioun extendis to £4 8s $10\frac{2}{3}$d.[184]]

BALQUHIDDER, (NLS, fo 47r)
PREBEND OF DUNBLANE CATHEDRAL[185] (SRO, Vol. a, fo 308r)

'Prebendrie of Balquhidder'

'Item, Magister Jacobus Rolland habet prebendam'

Mr James Rolland has the prebend of Balquhidder in the diocese of Dunblane, extending annually to 24 merks.

Calculation of third: In the whole, £16; third thereof, £5 6s 8d.

ABERNETHY, PARSONAGE OF,[186] (NLS, fo 47r)
(SRO, Vol. a, fo 308r)

Parsonage of Abirnethie pertaining to Mr Alexander Betoun.

The parsonage of Abirnethy, 400 merks.

Calculation of third: In the whole, £266 13s 4d; third thereof, £88 17s 9d.[187]

Signature: 'David Betoun of Melgund, bruther to the said Alexander.'

---

[183] Cf NLS, fo 45r (see above, p. 328).
[184] This marginal note does not appear in SRO text.
[185] Cf NLS, fos 45r, 99r (see above, p. 328; below, p. 392).
[186] Cf NLS, fo 46r (see above, p. 329).
[187] The correct calculation appears to be £88 17s $9\frac{1}{3}$d.

KINNOULL, PARSONAGE AND VICARAGE OF,[188]   (NLS, fo 47v)
(SRO, Vol. a, fo 308v)

'The personage and vicarage of Kynnoule annexit thairto[189] liand within the schirefdome of Perth and diocie of Sanctandrois, set in assedatioun to Thomas Monypenny, burges of Perth, and Maister Henry Thorntoun fore foure hundreth and tuenty merkis yeirlie be Maister Robert Carnegy, persoun thairof, as ane act of the Officialis bukis of Sanctandrois maid thairupoun proportis, quhilk tak tham furth at Lambmes bypast, now set of the same price to Sir Robert Carnegy of Kynnarde. Summa, four hundreth xx merkis [*i.e.* 420 merks], quhairof thair is to be deduceit to Maister David Methven', 80 merks 'yeirlie pensioun; and to sir Baltuer [*i.e.* Balthasar] Spens, vicar pensionar thairof', 20 merks 'yeirly; and for the corpspresentis and Pasche offeringis', £10 yearly.

Calculation of third: In the whole, £266 13s 4d;[190] third thereof, £88 17s 9⅓d.

SCONE, ABBEY OF,[191]   (NLS, fos 48r-50v)
(SRO, Vol. a, fos 309r-311v)

The rental of the abbacy of Scune

'The baronie aboun the Water of Ylay: Inprimis, the auld myln [*recte*, mail, *next paragraph*] of the said baronie extendis to' £149 3s 8d. The augmentation thereof set in feu, £18 12s 8d. The grassum silver of the same, £16 16s 8d.

The barony under the Watter of Ylay: The old mail thereof extends to £241 5s. Augmentation of the lands thereof set in feu, £17 10s. The grassum silver of the same, £11 18s.

The barony under the Bray: The old mail thereof, £92 16s 2d. Augmentation of the lands set in feu, £4 12d. The grassum of the same, £3 10s 8d.

The barony in Angus: The old mail thereof, £186 10s 8d. The augmentation of the lands thereof set in feu, £15 13s 4d. The grassum of the same, £3 16d. 'Item, for iij fed oxin silver payit be the thre tounis of Gurdeis', £7 10s.

'The secund teyndis within the baronyis foirsaidis', £11 13s 4d. 'The anuellis rent within the tounis of Perth, Dundie and Scune', £47 18s 4d. 'The fyschingis set for silver', £15 6s 8d. 'The Orcheart of Cleyne' [Cleynye, *SRO*], £6 13s 4d.

---

[188] Cf NLS, fo 41v (see above, p. 323).

[189] See above, p. 323, n. 158.

[190] This sum equals 400 merkis; 420 merks would be £280.

[191] Cf *Liber Ecclesie de Scon* (Edinburgh, 1843), pp. 215-221. Patrick Hepburn, bishop of Moray, was commendator at the Reformation, Cowan and Easson, *Medieval Religious Houses*, p. 98.

The kirks set for silver 'conforme to the auld rentale': Logyrait, £58. Blair [*i.e.* Blairgowrie], £82. Ragortoun [*i.e.* Redgorton], £60 6s 8d. Kinfawnis, £33 6s 8d. Echt, £26 13s 4d. Innergowrie, £20. The chapel of Sanct Niniane, £10. [*In margin*, 'This is the vicarage of Kynfawnis, by the woll and lamb with sum kirkland.']

'Summa of the haill money of mallis, augmentationis, gressumis, fyschingis, anuellis, secund teyndis, fed oxin silver and kirk silver befoir mentionat', £1,140 6s 8d.[192]

'The ordinar expens of money to be deducit yeirlie of the fruictis of the said abacie
    Inprimis, to the convent thairof extendand to xviij personis, the priour haiffand dowble, for thair sustentatioun', £352 3s 4d. 'In pensionis yeirlie payit to the abbayis of Dumfermeling, Cambuskinneth, Couper, Maij and utheris', £79 9s 8d. 'To divers personis be vertew of the comoune seill, the officiaris of the place and ordinaris thairof', £75 4s 8d. 'To the Lordis of Sessioun for thair yeirlie contributioun', £35. 'To be deducit of the gressumis befoir chairgit of certane landis sett in few be ressoun the possessouris had takis for yeris to rine and thairfor payis no gressum during thair saidis takis', £31 20d. 'Summa of the defalcatioun aboune writing', £572 19s 4d. 'And sua restis de claro' £567 7s 1d.[193] 'The thrid part thairof', £189 2s 4½d.

The rental of the ferme victual of the said abbacy
    The barony of Scone, 6 c. 7 b. 2 f. 2 p. bere; 1 c. meal. The barony under the Bray, 4 c. wheat; 10 c. bere; 5 c. meal. In kane wheat, 1 c. 15 b. 2 f. The barony in Angus, 2 c. 1 b. wheat; 8 c. 8 b. bere.

The rental of the teind victual of the said abbacy
    The kirk of Blair, 6 c. 5 b. 2 f. bere; 8 c. 14 b. meal. The kirk of Scone, 1 b. wheat; 13 c. 12 b. 2 f. bere; 19 c. 10 b. 2 f. meal. The kirk of Kinfawnis, 2 b. 2 f. wheat; 5 c. 15 b. bere; 4 c. meal. The kirk of Kilspindy, 3 c. 14 b. wheat; 11 c. 1 b. bere; 13 c. 1 b. 2 f. meal. The kirk of Lif, 3 c. 5 b. wheat; 11 c. 12 b. bere; 10 c. 5 b. meal. 'Summa of the haill ferme and teynd victuallis of the abbacie of Scone': Wheat, 16 c. 1 b.[194] Bere, 73 c. 13 b. 2 f. 2 p. Meal, 61 c. 15 b.

'The ordinar expens of victualis to be deducit
    Item, to the convent for thair sustentatioun', 6 c. 12 b. wheat; 22 c. 12 b. 2 f. bere; 7 c. 1 b. 3 f. meal. 'Item, in pensionis under the commond sele', 12 b. 2 f. wheat; 2 c. 8 b. bere; 12 b. 3 f. meal. 'Item, to the officiaris of the place', 3 c. 2 b. 2 f. 2 p. bere; 16 c. 12 b. meal. 'Summa of the defalcatioun foirsaid in the

---

[192] The correct calculation appears to be £1,140 7s 6d.
[193] The correct calculation appears to be £567 8s 2d.
[194] The correct calculation appears to be 14 c. 1 b.

yeir': Wheat, 7 c. 8 b. 3 f.[195] Bere, 28 c. 7 b. 2 p. Meal, 28 c. 10 b. 2 f.[196] 'And sua restis de claro', 8 c. 8 b. 2 f. wheat;[197] 45 c. 6 b. 2 f. bere; 38 c. 4 b. 2 f. meal.[198]

'Nota', 18 c. 3 b. oats horsecorn; 12 barrels of salmon 'with certane cayne caponis and pultrie nevir sauld for money bot spendit in sustentatioun of the place.'

Third of money, £380 2s 1⅔d.[199] Take the barony under the Bray for £92 16s 2d; the augmentation of the feu lands thereof for £4 12d; the grassum of the same for £3 10s 8d; the barony of Angus for £186 10s 8d; the augmentation of the feu lands thereof, £15 13s 4d; the grassum of the same for £3 16d; and £74 9s out of the mails of the barony under the Water of Ilay.

Wheat, 5 c. 5 b. 2 f. 2⅔ p. Take the barony under the Bray, 4 c.; the kane wheat thereof, 1 c. 15 b. 2 f. 'Gif in' 9 b. 1 f. 1 '3 pt half p.'[200]

Bere, 24 c. 9 b. 3 f. 2 p. The barony of Scone for 6 c.; the barony under the Bray, 10 c.; the barony of Angus for, 8 c. 8 b. 'Gif in' 5 b. 3 f.

*Meal, 20 c. 10 b. 1 f. 1⅓ p. Take the barony of Scone, 1 c.; take the barony under the Bray, 5 c.; the kirk of Lyf, 10 c. 5 b.; and out of the kirk of Kilspindie, 4 c. 5 b. 1 f. 1⅓ p.

Oats, 6 c. 1 b. Salmon, 4 barrels. 'Gar chairg my lord for thir aittis and salmond quhill he assignis sufficient men to pay the same. Omittit caynes, custumis, capones, pultrie and uther dewiteis.'[201]

'The haill money and victuallis of the abbacie of Scone extendis to the sowmes following': Money, £1,140 6s 7d [£1,140 6s 8d, *fo 48v*]. Victual, 16 c. 5 b. wheat [16 c. 1 b., *fo 49r*]; 75 c. 8 b. 2 f. 2 p. bere [73 c. 13 b. 2 f. 2 p., *fo 49r*]; 62 c. 7 b. 1 f. meal [61 c. 15 b., *fo 49r*].

'The thrid extendis in money', £380 2s 2⅓d; in victual, 5 c. 7 b. wheat; 25 c. 2 b. 3 f. 2 p. bere; 20 c. 13 b. 1⅓ p. meal.[202]

'The tutouris pensione': In money, £53 3s. In victual, 5 c. 5 b. 2 f. 2⅓ p. wheat; 15 c. 7 b. 2 f. 2 p. bere; 7 c. 12 b. 2 f. 1⅓ p. meal; total, 28 c. 10 b. 1⅓ p.[203]

'Scone hes xj kirkis: Scone, Cambusmicheall, Kinfawnis; the minister', £23 6s 8d; 2 c. bere; 3 c. meal; 'the redare at Kynfawnis', £20. Logyrait, the minister, £58.

---

[195] The correct calculation appears to be 7 c. 8 b. 2 f.
[196] The correct calculation appears to be 24 c. 10 b. 2 f.
[197] The correct calculation (from revised figures) appears to be 6 c. 8 b. 2 f.
[198] The correct calculation (from revised figures) appears to be 37 c. 4 b. 2 f.
[199] One third of £1,140 6s 8d is £380 2s 2⅔d.
[200] The correct calculation appears to be 9 b. 3 f. 1⅓ p.
[201] The section of text from * is transcribed from SRO, Vol. a, fo 310v; the NLS text for this passage is faded.
[202] These thirds are correct for the totals given immediately above.
[203] The correct calculation appears to be 28 c. 9 b. 3 f. 1⅔ p.

Blair, the minister, £60. Ragortoun [*i.e.* Redgorton], £60 6s 8d. Kilspindierait,[204] 2 c. bere; 2 c. meal. Logiedundie, Liff, Innergowrie, £20; 2 c. bere; 2 c. meal.

ABERUTHVEN, VICARAGE OF,

(NLS, fo 51r)
(SRO, Vol. a, fo 312r)

Rental of the vicarage of Arbruthvene

The vicarage of Arbruthvene pertaining to Mr Robert Herbertsoune extends to £40 yearly 'by the curattis fee, off the quhilk is to be deducit yeirirlie' [*sic*] £13 6s 8d.

Signature: 'M[r] Joannes Houstoun, manu sua.'

Calculation of third: In the whole, £40; third thereof, £13 6s 8d.

COMRIE, VICARAGE OF,

(NLS, fo 51r)
(SRO, Vol. a, fo 312r)

Rental of the vicarage of Comry pertaining to Mr John Sinclare, 'ay set for' £20 'and now gettis na thing.'

Signature: 'Thomas Sinclar, with my hand.'

Calculation of third: In the whole, £20; third thereof, £6 13s 4d.

BENVIE, PARSONAGE AND VICARAGE OF,

(NLS, fo 51v)
(SRO, Vol. a, fo 312r)

Rental of the parsonage and vicarage of Banvy

'Item, the said vicarage set in auld and now is siclike with the personage for thre scoir pundis, thairof to be payit to ane minister', 20 merks. 'Item, praxis [*recte*, procurations] and synodolis', 40s. 'Swa restis de claro' £44 13s 4d.

Signature: 'David Lindsay, rector de Banvy.'

Calculation of third: In the whole, £60; third thereof, £20.

---

[204] Kilspindie and Rait should be listed as two separate churches. For further churches appropriated to Scone, see Cowan, *Parishes*, p. 224.

CALLANDER, VICARAGE OF, (NLS, fo 51v)
(SRO, Vol. a, fo 312v)

Rental of the vicarage of Callendrech

The vicarage of Calendrech pertaining to sir John Wrycht 'wes wount to pay' £10 'now payis na thing thir thre yeiris bigane.'

Signature: 'Presentit be me, Johnne Morisoun.'

Calculation of third: In the whole, £10; third thereof, £3 6s 8d.

'Aliter gevin in be the said Johnne and ressavit at command of the Clerk of Register extending to' £6.

DUNBLANE CATHEDRAL, (NLS, fo 51v)
CHAPLAINRY OF ST MICHAEL IN, (SRO, Vol. a, fo 312v)

The chaplainry of Sanct Michael within the kirk of Dumblane pertaining to sir John Wrycht 'wes wount to pay' 24 merks.

Signature: 'Presentit be me, Johnne Morisoun.'

Calculation of third: In the whole, £16; third thereof, £5 6s 8d.

LOGIERAIT, VICARAGE OF, (NLS, fo 52r)
(SRO, Vol. a, fo 312v)

Rental of the vicarage of Logarrett

'The vicarage of Logaret had wount to [pay, *deleted*] be set in assedatioun for' £66 'and now is worth' £20 'be reassoun of wanting of umest claithis, etc.'

Signature: 'Henricus Abircrummy, manu propria.'

Calculation of third: In the whole, £20; third thereof, £6 13s 4d.

SALINE, PARSONAGE AND VICARAGE OF, (NLS, fo 52r)
(SRO, Vol. a, fo 313r)

Rental of the parsonage and vicarage of Sawling

'The haill extendis to twa hundreth merkis ij merkis half, and ane towne callit Inzefair payis' 33 b. oats and 8 b. bere 'quhair thair pa[y]it bot' 27 merks 'of this sowme forsaid of befoir that he hes reft frome us violentlie within the yeiris of vj yeiris last bipast. [*In margin*, 'Haill to the quene beaus it is commoune kirk of Dunblane.']

INVER, CHAPLAINRY OF,[205]

(NLS, fo 52r)
(SRO, Vol. a, fo 313r)

'Chaplanrie Inweir'

'In the first, the towne and Myln of Inweir', £18 yearly 'set in few with four kiddis, four geis, xx cayn fowles, ane f[irlot] of small grottis, ane yeirling hog, ane young swyne, the leiding of' 80 'leidis of peittis on my expenssis'. Dulmernocht [*i.e.* Dalmarnock] set in feu for mail 'and all dewiteis conforme to the chartour', 13 merks. The crofts in Litledon [*i.e.* Littleton],[206] £4 set for 19 years, 'swa thai be laborit be the fewaris of Inweir and Dulmernocht and failyeand thairof', £3.

Signature: 'David Morisoun, capellanus de Inveir, manu sua.'

Calculation of third: Money, £30 13s 4d; third thereof, £10 4s 5$\frac{1}{3}$d. Kids, 4; third thereof, 1$\frac{1}{3}$. Geese, 4; third thereof 1$\frac{1}{3}$. 'Small grottis', 1 f.; third thereof, 1$\frac{1}{3}$ p. 'Ane yeirling hoge, a young swyne and leiding of' 80 'leidis of peittis on his expenssis.'

MONEYDIE, PARSONAGE OR VICARAGE OF,[207]

(NLS, fo 52v)
(SRO, Vol. a, fo 313v)

'Rentale Monydie'

'Monydie parochine payis communibus annis' 14 or 13 c. victual, 'assedatur Alexandro Hepburne de Quhitsum anno 1558 pro' £100 'deductis ordinariis expenssis extenden to' £20.

Signature: 'Michaell Walcar, manu sua.'[208]

Calculation of third: In the whole, £100; third thereof, £33 6s 8d.

---

[205] The altar and chaplainry of Inver were founded in Dunkeld cathedral, *RSS*, v, no 2949.
[206] The text reads 'Litle^d'; Littleton is adjacent to Dalmarnock; Inver is adjacent to Little Dunkeld.
[207] Cf SRO, Vol.a, fo 101v (see above, p. 87); see also *Rentale Dunkeldense* p. 351. Moneydie was a prebend of Dunkeld cathedral.
[208] *Rentale Dunkeldense*, p. 351, reads 'Walcer' (indexed under 'Walker'); *RSS*, v, no 3029 reads 'Walker'.

MEIGLE, VICARAGE OF,

(NLS, fo 52v)
(SRO, Vol. a, fo 313v)

'Rentale vicarie de Megle'

This vicarage once paid 65 merks; now, however, if those dues were to be paid which are set out in the Book of Reformation[209] to this point it would be worth 10 merks.

CRIEFF, CHAPLAINRY
OF ST MICHAEL'S ALTAR IN,

(NLS, fo 53r)
(SRO, Vol. a, fo 313v)

The chaplainry of Sanct Michaellis altar in the kirk of Creif within the diocese of Dunkeld.

'Gevand be yeir to Johnne Bannantyne, alias sir Johnne', 22 merks 'allanerlie.'

'The chaiplanrie of Creif callit Sanct Michaellis'.

Calculation of third: In the whole, £13 13s 4d; third thereof, £4 17s $9\frac{1}{3}$d.[210]

LUNDEIFF, PARSONAGE AND VICARAGE OF,[211]

(NLS, fo 53r)
(SRO, Vol. a, fo 313v)

Parsonage and vicarage of Lundeiff pertaining [to] me, Alexander Creichtoun.

The parsonage and vicarage of Lundeiff in and within the diocese of Dunkeld in the parishioners' hands paying by year £66 13s 4d.

Signature: 'Mr Alexander Creichtoun.'

Calculation of third: In the whole, £66 13s 4d; third thereof, £22 4s $6\frac{2}{3}$d.[212]

GLENDEVON, VICARAGE PENSIONARY OF,[213]

(NLS, fo 53v)
(SRO, Vol. a, fo 314r)

'The rentale pensionar of Glendovan liand within the schirefdome of Perth pertening to me, Mr Johnne Hutsoun.'

---

[209] I.e. the *First Book of Discipline* (1560).
[210] 22 merks equals £14 13s 4d, which would give this third.
[211] Cf NLS, fo 44r (see above, p. 327).
[212] The correct calculation appears to be £22 4s $5\frac{1}{3}$d.
[213] Cf NLS, fo 34v (see above, p. 312).

'Off the quhilk the mans, gleib and kirkland thairof is worth the sowme of' 10 merks 'and the rest is ane pensioun out of the personage of Glendovan quhilk is worth the sowme of' 12 merks.

Signature: 'Mr Johnne Hutsoun, with my hand. Ressavit at command of the Clerk of Register.'

Calculation of third: In the whole, £14 13s 4d 'comptand thairin the mans, gleib and kirkland'; third thereof, £4 17s 9⅓d.

PITMEDDEN, (NLS, fo 53v)
PREBEND OF ABERNETHY COLLEGIATE KIRK (SRO, Vol. a, fo 314r)

Prebend of Petmaddane

'The prebendarie of Petmeddy within the college kirk of Abirnethy and schirefdome of Perth perteining to Mr Thomas Lumisden extendis in all proffettis yeirlie to' £20.

Signature: 'Mr Thomas Lumisden.'

'Prebendarie of Pittmedden in the kirk of Abirnethy', pertaining to Mr Thomas Lumisden.

Calculation of third: In the whole, £20; third thereof, £6 13s 4d.

'Ressavit at command of the Clerk Registre.'

MUCKHART, PARSONAGE OF, (NLS, fo 54r)
(SRO, Vol. a, fo 314v)

Parsonage of Mukertie pertaining to John Symple.

The rental thereof as it has been set in assedation to Andrew Abircrummye, both parsonage and vicarage, for 100 merks.

Signature: 'Subscrivit with my hand, Johnne Sympill of Mukert. Resavit at command of the Clerk of Registre.'

Parsonage of Mukart pertaining to John Sempill.

Calculation of third: In the whole, £66 13s 4d; third thereof, £22 4s 5⅓d.

AUCHTERHOUSE, VICARAGE PENSIONARY OF,

(NLS, fo 54r)
(SRO, Vol. a, fo 314v)

Rental of the vicarage pensionary of Ouchterhous within the diocese of Dunkeld.

'This vicarage, deducing the Pasche fynis, small offerandis, corpis present, umaist claithis and five pundis money yeirlie aucht to the channonis of Dunkeld in name of pensioun, is yeirlie allanerlie worth be yeir' £20.

Signature: 'Sir Duncane Gray, vicare pensionar of Ouchterhous, with my hand. Rasavit at the command of the Clerk Register.'

Calculation of third: In the whole, £20; third thereof, £6 13s 4d.

DOLLAR, VICARAGE OF,[214]

(NLS, fo 54r-v)
(SRO, Vol. a, fos 314v-315r)

Rental of the vicarage of Dolour 'be just estimatioun the umest claithis, the corpis presentis, kow deleitit, the Pasche reknyng. Item, the said rentall to extent to the sowme of' £13 6s 8d.

Signature: [blank] 'Resavit at the command of the Clerk of Register.'

Vicarage of Doler pertaining to [blank].[215]

Calculation of third: In the whole, £13 6s 8d; third thereof, £4 8s 10⅔d.

INCHTURE, VICARAGE OF,

(NLS, fo 54v)
(SRO, Vol. a, fo 315r)

The rental of Insture with the pendicle thereof called Kynnard, lying in the Cars of Gowrie and diocese of Sanctandrois.

'Gaiff yeirlie befoir the moneth of Junij [Julij, *deleted*] in the yeir of God' 1556[216] the sum of £30 'libere'.

Signature: 'Mr Arthur Taillefere, vicair thairof.'

---

[214] Cf SRO, Vol. a, fo 102r (see above, p. 88 and n. 338). This benefice, in the sheriffdom of Clackmannan and diocese of Dunkeld, was appropriated to Inchcolm abbey and is entered among rentals for both Perth and Fife.

[215] Henry Balfour was vicar in 1561, and Robert Burn was vicar pensionary, *RSS*, vi, no 2240.

[216] Both texts read '1556'.

Calculation of third: In the whole, £30; third thereof, £10.

MUCKERSIE, PARSONAGE AND VICARAGE OF,[217]

(NLS, fo 54v)
(SRO, Vol. a, fo 315r)

Rental of the parsonage and vicarage of Muckersye pertaining to Mr John Stevinsoun, 'chantour of Glasgw'.

Calculation of third: In the whole, £100; third thereof, £33 6s 8d.

PETTINBROG, (NLS, fo 54v)
PREBEND OF ABERNETHY COLLEGIATE KIRK (SRO, Vol. a, fo 315r)

Rental of the prebend of Pettinbroge pertaining to George Arnote.

'Item, the prebendarie of Pettinbroge set in assedatioun to Jonet Bruce, lady of Arnot, for' £25 6s 8d. 'Item, payit furth of the said prebendarie to ane chaplane for service in the qweir yeirlie', £12.

Signature: 'George Arnote.'

Calculation of third: In the whole, £25 6s 8d; third thereof, £8 8s 10⅔d.

MUTHILL, VICARAGE OF,[218]

(NLS, fo 55r)
(SRO, Vol. a, fo 315v)

Rental of the vicarage of Muthill set in assedation to the baron of Borland for the sum of 50 merks, 'as the lettir of tak testifeis writtin in the Officialis bukis of Glasgw' 24 April 1556, 'and als the said baroun wes oblist to pay the curattis fee and stipend.'

Signature: 'James Hammiltoun, with my hand.'

Calculation of third: In the whole, £23 6s 8d;[219] third thereof, £11 2s 2⅔d.

---

[217] Cf NLS, fo 43v (see above, p. 326).
[218] Cf NLS, fo 44r (see above, p. 326).
[219] The correct calculation appears to be £33 6s 8d.

DUNKELD, SUBDEANERY OF,[220]

(NLS, fo 55r)
(SRO, Vol. a, fo 315v)

Rental of the subdeanery of Dunkeld pertaining to Mr Richard Haldan, subdean thereof.

'In primis, the propertie of the same sett in few to the laird of Murthlie and pais to me thairfoir the soum of' 50 merks yearly, and I pay to my stallar 10 merks thereof.

Signature: 'Resavit at command of the Clerk of Registre.'

Calculation of third: In the whole, £33 6s 8d; third thereof, £11 2s $2\frac{2}{3}$d.

FOWLIS WESTER, VICARAGE OF,

(NLS, fo 55r)
(SRO, Vol. a, fo 315v)

Rental of the vicarage of Foulis pertaining to Patrick Murray extends yearly to 20 merks.

Signature: 'Patrik Murray.'

KILMADOCK, VICARAGE OF,

(NLS, fo 55r)
(SRO, Vol. a, fo 315v)

The vicarage of Kilmodok within the scheriffdom of Perth pertaining to Mr James Kennedy, chancellor of Dunblane, extending yearly to 24 merks 'deduceand of the auld rentale gevin in be corspresent and kowclayth, Pasche offeringis and uther small offeringis, dischargeit in the quenis grace lettres without prejudice of superplus and bett[er] rentall omitted.'

Signature: 'Mr James Kennedy, chancellar of Dunblane.'

DUNKELD, BISHOPRIC OF,[221]

(NLS, fos 55v-57r)
(SRO, Vol. a, fos 316r-317v)

'The rentall of the temporall landis pertening to the bischop of Dunkelden on baith the sydis of Forth in males, teindis, fermes, horscorne, presentlie pait' 4 January 1561/2.

---

[220] Cf *Rentale Dunkeldense*, p. 348.
[221] Cf NLS, fos 27r, 57v, 58v (see above, p. 302; below, pp. 344, 345).

Temporal lands

John Brownis croft, 53s 4d. Tulemulie, £4. Drumboy, £4. Craighilto, 40s. Bot Mell [Rothmell, *fos, 27v, 58v*] 40s. Dowallie, £5. Gawie, £16. Kilmoreicht, £13 6s 8d. Porte of Logirate, 30s. Sokocht, 40s. Kincragie, £4. Glenelwarte, £4. Myln of Glenelwarte, 53s 4d. Wester Dulgus, 40s. Middle Dulgus, £3 6s 8d. Eister Dulgus, £5 6s 8d. Ladywell, £10 13s 4d. Wester Incheschewin, £3 6s 8d. Eister Inschewin, 53s 4d. Dawpowe [*i.e.* Dalpowie], £4. Baithous of Capecth, 20s. Capeth, £12. Dulgarthill, £6 13s 4d. Fordy, 'the tane half', £10. 'The uther half', 50s. Adamestoun of [or, *fo 58v*] Baldornocht, 53s 4d. Boroustoun of [or, *fo 58v*] Balnavadocht, 40s. Manis of Clony, £16. Drummalie, £5. Tibbermure, £14. Pettindynie, £12. Over Petcarne, £6 13s 4d. Myln of Pitcarne, £5 6s 8d.[222] Kinvaid, £3 6s 8d. Cultranie Bege and Kincarne in Creiff, £6 13s 4d. Kirkland of Cargill, £4. Lecrop, £8. Castell Campbell with the lands of Dolour, £10 13s 4d. Fordall, £26 13s 4d. Kirkhill of Megle, £8 13s 4d. Kirkland Bank of Alicht, £7 6s 8d. Menmure, £3 6s 8d. Ferne with the brewhouse, £12. Ouchtertullie Manys and barony, £82. Clentre [Clyntry, *fo 28r*; Glentre, *fo 59r*] £9 6s 8d. Manis of Crawmond, £66 13s 4d. Crawmond Inche, £3 6s 8d. Lands of Abirladie, £80. The house in Edinburgh, £6 13s 4d. The house in Perth, 53s 4d. Kirkland of Prestoun, 46s 8d.

'The rentale of the kirkis with thair names that ar set in assedatioun of silver teindis upoun baith the syidis of Forth pertening to the bischoprik of Dunkeld, to quhome and for quhat sowme': Kirk of Auchtergawin set to Mr Gavin Hammiltoun for £100. Kirk of Tibbermure set to my lord Ruthven for £160. Kirk of Straithmeglo set to William Birny for £160. Kirk of Crawmound set to Mr William Creichtoun for £173 6s 8d. Kirk of Abircorne set to Sir James Hammiltoun for £160. Kirk of Bonkle and Preistoun set to my lord of Mortoun[223] for £60.

'Certane particular teindis of the kirkis following set for silver as followis'

Of the kirk of Alicht extending to £46 6s 8d. The kirk of Forgundynie, £53 6s 8d. The kirk of Capecht, £30. The kirk of Lyttle Dunkelden, £22 17s. The kirk of Tibbirmure, £6 13s 4d.

'The trew rentall of the kirkis as followis that pais teind victuale and horscorne upoun baith the sydes of Forth'

Kirk of Lytildunkeldensis [*sic*], 2 c. 6 b. bere; 1 c. 6 b. meal; 8 c. 6 b. 2 f. oats. Kirks of Capecht and Dowallie, 7 c. 2 b. 2 f. 1 p. bere; 11 c. 2 b. 2 f. 1½ p. meal;[224] 9 c. 2 b. 2 f. oats. Kirk of Cargill, 9 c. 5½ b. bere; 18 c. 10½ b. meal. Kirk of Alicht, 4 c. 8 b. bere; 12 c. 2 b. 2 f. meal. Kirk of Abirdagy, 2 c. 8 b. bere; 6 c.

---

[222] *Rentale Dunkeldense*, p. 344, contains the editorial note: '6 lib. 6 s. 8 d. in the succeeding "reformit" rental'.

[223] See above, p. 302, n. 94.

[224] *Rentale Dunkeldense*, p. 344, contains an editorial note that the 2 f. is in the 'reformit' rental only.

meal. Kirk of Forgundynie, 4 c. 9 b. bere; 15 c. 7 b. meal. Kirk of Abirladie, 7 c. bere; 4 c. wheat; 10 c. 9 b. oats.

'I appreif this present rentall contening twa levis writtin be my servand, Johnne Douglas, as ane trew rentall conforme to ane just calculatioun, viz.': Silver, £1,505 10s 4d. Wheat, 4 c. Bere, 37 c. 6 b. 3 f. 2⅓ p. Meal, 64 c. 12 b. 2 f. 3 p. Oats, 28 c. 2 b.

'Subscrivit with my hand at Cluny' 4 January 1561/2 'befoir thir witnessis, Maister Johnne Bartane, sir David Henrie and George Horne.'

Signature: 'R. Dunkeld. Ressavit at command of the Clerk of Registre. J. Clerk Registre.'[225]

DUNBLANE, NINE CHAPLAINRIES IN CHOIR OF, (NLS, fo 57v)
(SRO, Vol. a, fos 317v-318r)

'Thir ar the rentall of nyne chaplanis in the chore of Dunblane, pertening to sir James Forsyith, Robert Hendersoun [Henrysoun, *SRO*], Thomas Rob, Alexander Andersoun, Edmond Chisholme, Robert Sinclar, Archibald Lawder, Williame Johnnestoun, Williame Drummound, chaplanis of the qweir of Dumblane, of thair yeirlie rent as efter followis to be distribut equale amangis tham.'

'Item imprimis, of the landis of Kyirpronte be yeir', £40. The lands of Cluthybege, £10. The lands of Schavitinchame, £20. The lands of Cragaruall, 20 merks. The lands of Lundeis, 10 merks.

'The annuellis pertening to the said chaplanis
    Item, of the landis of Halychtmoir be yeir', £4. The lands of Monyward, 40s. The lands of the lard of Monyell, 10 merks. Of Andrew Toschetis lands of Monze by year, 13s 4d. Of the laird of Cultoquhais lands by year, 13s 4d. The laird of Kyltyis lands, 13s 4d. The vicarage of Kippen, £4. The priory of Inchmahomo, 4 merks. The parsonage and vicarage of Kilbryd, 40s. The parsonage of Abirfull, 2 merks. The parsonage of Comry, 40s. The parsonage of Abyrnethie, 40s. 'Item, of the fyif Quhiggis[226] pertening to the lard of Ker', £12. The lands of Culingis pertaining to Alexander Drummound, 40s. The vicarage of Kincadin of Munteithit, 4 merks.

Signature: 'James Nicolson, ressave this rentall. J. Clerk Registre.'

---

[225] The calculation of the third is misplaced after the next entry.
[226] Cf 'terras de Rathernn-Striviling alias Quaigis de Strathalloin', *RMS*, iv, no 2902.

DUNKELD, BISHOPRIC OF,[227]

(NLS, fo 57v)
(SRO, Vol. a, fo 318r)

Third of money, £501 16s 9⅓d. Take the Manis of Crawmound for 100 merks; 'tak the haill temporall landis benorth Forth for' £368 11s 8d; 'tak the rest of money furth of the silver of the kirkis of Forgundynie, Capeth and Lytle Dunkeld.'

CLUNY, VICARAGE OF,

(NLS, fo 58r)
(SRO, Vol. a, fo 318r)

The vicarage of the kirk of Clune pertaining to Mr William Salmond.

Calculation of third: In the whole, £12 13s 4d; third thereof, £4 4s 5⅓d.

AUCHTERARDER, VICARAGE OF,

(NLS, fo 58r)
(SRO, Vol. a, fo 318r)

Rental of the vicarage of Auchtirardour within the diocese of Dumblane pertaining to the dean and chapter of the same and set in assedation to sir William Blaikwode 'and wes wount to pay yeirly' £20 'and now payis nathing thir four yeiris bygane.'

Signature: 'Presentit be me, Mr Alexander Drummound.'

Calculation of third: In the whole, £20; third thereof, £6 13s 4d.

ABERFOYLE, VICARAGE OF,

(NLS, fo 58r)
(SRO, Vol. a, fo 318r)

Rental of the vicarage of Abirfule within the diocese of Dunblane pertaining to sir Stephen Sinclare.

'The quhilk vicarage hes evir bene set in assedatioun for the yeirlie payment of' £20 'and thir thre yeiris bipast payit nathing.'

Signature: 'Presentit be Alexander Setoun of Northbrig.'

---

[227] Cf NLS, fos 27r, 55v, 58v (see above, pp. 302, 341; below, p. 345); see also *Rentale Dunkeldense*, pp. 340-345. This assignation of the third of the revenues of Dunkeld, misplaced in the text, should precede, not follow, the above entry for the nine chaplainries of Dunblane, which is intrusive.

Calculation of third: In the whole, £20; third thereof, £6 13s 4d.

ST MADOES, PARSONAGE AND VICARAGE OF, (NLS, fo 58r)
(SRO, Vol. a, fo 318v)

Rental of the parsonage and vicarage of Sanctmodoce

'The haill paroche is twa townis and thair pertinentis quhilk evir wes set togedder for' 88 merks.

Signature: 'Maister David Balward, persoun of Sanct Modice, with my hand.'

DOUNE, CHAPLAINRY OF ST FILLAN, (NLS, fo 58r)
(SRO, Vol. a, fo 318v)

Rental of the chaplainry of Sanct Phillane 'besyid the castell of Doun gevin up be James Cousland, chaiplane thairof.'

Calculation of third: In the whole, £12 13s 4d [£11 18s 4d, *SRO*]; third thereof, £4 4s 5⅓d.

DUNKELD, BISHOPRIC OF,[228] (NLS, fos 58v-60v)
(SRO, Vol. a, fos 318v-320v)

'The trew rentale of the temporall landis pertening to the bischoprik of Dunkeld on baith the syidis of Forth in males, fermes, teindis, horscorne, presentlie payit' 4 January 1561/2.

[*In margin*, 'Reformit for Abirlady be our soverane lady and hir secreit counsall.']

John Brownis croft, 53s 4d. Tulemulie, £4. Drumboy, £4. Craighilto, 40s. Rothmell [Bot Mell, *fo 55v*], 40s. Dowallie, £5. Gawie, £16. Kilmoreicht, £13 6s 8d. Port of Logyrait, 30s. Sokocht, 40s. Kincragie, £4. Glenelwart, £4. Myln of Glenelwart, 53s 4d. Wester Dulgus, 40s. Eister Dulgus, £5 6s 8d. Middill Dulgus, £3 6s 8d. Ladywell, £10 13s 4d. Eister Inschewin, 53s 4d. Wester Incheschewin, £3 6s 8d. Dalpowe, £4. Baithous of Capeth, 20s. Capeth, £12. Dulgarthill, £6 13s 4d. Fordie, 'the tane half', £10. 'The uther half', 50s. Adamstoun or [of, *fo 55v*] Baldornocht, 53s 4d. Borroustoun or [of, *fo 55v*] Balnavardocht, 40s. Manis of Clony, £16. Drummalie, £5. Tibbirmure, £14. Pittindinie, £12. Over Pitcarne, £6 13s 4d. Myln of Pitcarne, £6 6s 8d [£5 6s 8d, *fo 55v*]. Kinvaid, £3 6s 8d. Cultranie Beg and Kincarne in Creif, £6 13s 4d.

---

[228] Cf NLS, fos 27r, 55v, 57v (see above, pp. 302, 341, 344); *Rentale Dunkeldense*, pp. 340-345.

Kirkland of Cargill, £4. Lecrop, £8. Castell Campell with the lands of Dolour, £10 13s 4d. Foirdaill, £26 13s 4d. Kirkhill of Megle, £8 13s 4d. Kirkland Bank of Alycht, £7 6s 8d. Menmure, £3 6s 8d. Ferne with the brewhouse, £12. Ouchtertulle Manis and barony, £82. Glentre [Clyntry, *fo 28r*; Clentre, *fo 56r*], £9 6s 8d. Manis of Crawmound, £66 13s 4d. Crawmond Inche, £3 6s 8d.[229] The house in Edinburgh, £6 13s 4d. The house in Perth, 53s 4d. Kirkland of Prestoun, 46s 8d. The Manis of Abirlady and teinds thereof with the kirk of Abirlady, £213 6s 8d.

'Nota, Abirlady wes first gevin in be the bischop for' 40 c. victual 'and thairefter reformed be him self and gevin in be him for' £80 money and 21 c. 9 b. victual 'and the quenis majestie, understandand that Abirladie teind and ferme wes sett of auld to Archibald Douglas of Kilspindie for' £200 'as his tak proportis and nevir payit mair sensyne thairfoir bot' 320 merks, 'persavand thairthrow the said bischoppis last and first rentalles to have bene calumniouslie gevin in to the hurte of the auld possessour, hes reformit that part of the said rentall and ordanit the said Abirlady teind and ferme to be rentallit for' 320 merks 'allanerly.'

'The rentall of the kirkis with thair names that ar set in assedatioun of silver teindis upoun baith the sydis of Forth pertening to the bischoprik of Dunkeld, to quhome and for quhat sowme.'

Kirk of Ouchtergawin set to Mr Gavin Hammiltoun for £100. Kirk of Tibbirmure set to my lord Ruthven for £160. Kirk of Strameglo set to William Birny for £160. Kirk of Crawmond set to Mr William Creichtoun, £173 6s 8d. Kirk of Abircorne set to Sir James Hammiltoun, £160 0s 8d. Kirks of Boncle and Preistoun set to my lord Mortoun, £60.

'Certane particulare teindis of the kirkis following set for silver as followis'

Of the kirk of Alycht extending to £46 6s 8d. The kirk of Forgundynie, £53 6s 8d. The kirk of Capeth, £30. The kirk of Lyttle Dunkeld, £22 17s. The kirk of Tibbirmure, £6 13s 4d.

'The trew rentall of the kirkis as followis that payis teind victuall and horscorne upoun baith the syidis of Forth'

Kirk of Little Dunkeld, 2 c. 7 b. bere; 1 c. 6 b. meal; 8 c. 6 b. 2 f. oats. Kirks of Capeth and Dowallie, 7 c. 2 b. 2 f. 1 p. bere; 11 c. 2 b. 2 f. $1\frac{2}{3}$ p. meal; 9 c. 2 b. 2 f. oats. Kirk of Cargill, 9 c. $5\frac{1}{3}$ b. bere; 18 c. $10\frac{2}{3}$ b. meal. Kirk [of] Alycht, 4 c. 8 b. bere; 12 c. 2 b. 2 f. meal. Kirk of Abirdagie, 2 c. 8 b. bere; 6 c. meal. Kirk of Forgundynie, 4 c. 9 b. bere; 15 c. 7 b. meal.[230]

---

[229] In the earlier rental of the bishopric of Dunkeld, at fo 56r (see above, p. 342), the 'Lands of Abirlady, £80' are included at this point.

[230] In the earlier rental of the bishopric of Dunkeld, at fo 57r (see above, p. 343), the 'Kirk of Abirladie, 7 c. bere; 4 c. wheat; 10 c. 9 b. oats' is included at this point.

'The divisioun and assumptioun of the said bischoprik maid at the quenis majesteis command in' 1564.

'Money in the haill', £1,640 17s 8d [£1,505 10s 4d, *fo 57r*]; third thereof, £546 19s 2⅔d. Take Abirlady Manis teinds and kirk for £213 6s 8d; take the Manis of Crawmound and teinds of the said mains for £66 13s 4d; take the temporal lands on the north side of Forth for £375. Total, £655. 'Gif in out of the said temporall landis on the northsyid of Forth', £108 15s.[231]

'James Nicolson, registrat this rentall and gar charge thairwith in' 1564 'and send the principall to me. Sic subscribitur, Johnne Wyischart, Comptrollare.'[232]

'Beir in the haill', 30 c. 8 b. 2 p. [37 c. 6 b. 3 f. 2⅓ p., *fo 57r*]; third thereof, 10 c. 2 b. 2 f. 3⅓ p. Take the kirk of Cargill for 6 c. 10⅓ b.;[233] take the kirk of Forgundynie for 4 c. 9 b. Total, 11 c. 3⅓ b. 'Gif in' 1 c. 2 f. 2 p.

'Meill in the haill', 64 c. 12 b. 2 f. 3 p.; third thereof, 21 c. 9 b. 2 f. 1 p. Take the kirk of Cargill for 13 c. 5⅓ b.;[234] take the rest of Forgundynie extending to 8 c. 4 b. 3⅔ p.

'Aittis or horscorne in the haill', 17 c. 9 b. [28 c. 2 b., *fo 57r*]; third thereof, 5 c. 13⅔ b. 'Tak thame out of Lytle Dunkeld.'

'Omittit canis, custumes and utheris dewiteis.'

INCHAFFRAY, ABBEY OF,[235]

(NLS, fo 61r)
(SRO, Vol. a, fo 321r)

'Regina.

Comptrollar,[236] it is oure will that incontinent efter the sight heirof ye caus lettres be direct in four formes at the instance of our lovit cousing David, lord Drummound, and Dame Lilias Ruthven, his spous,[237] as takkismen of the abbacie of Inchearffray to caus thame be answerit of the teindis, fructis, rentis, proventis, emolumentis, males, fermes and dewiteis pertening thairto of the yeiris of God' 1561 and 1562 'and siclike yeirlie and termelie in tyme cuming during the yeiris and termes contenit in the lettres of tak and lyfrent assedatioun maid to tham thairof as ye will

---

[231] These figures add up to £546 5s.
[232] See above, p. 286, n. 19.
[233] Cargill gave 9 c. 5⅓ b., fos 56v, 59v (see above, pp. 342, 346).
[234] Cargill gave 18 c. 10⅔ b., fos 56v, 59v (see above, pp. 342, 346).
[235] Cf *Liber insule missarum. Abbacie canonicorum regularium B. Virginis et S. Johannis de Inchaffery registrum vetus: premissis quibusdam comitatus antiqui de Stratherne reliquiis*, ed. C. Innes (Edinburgh, 1847), p. 83.
[236] See above, p. 286, n. 19.
[237] I.e. David, 2nd lord Drummond, and his wife Lilias, daughter of William, 2nd lord Ruthven, *Scots Peerage*, vii, pp. 46-47.

answer to us thairupoun. Subscrivit with our hand at Abirdene the xv day of October and of our regnne the twentie yeir. Sic subscribitur: Marie R.'

'The rentall of the abbay of Inchaiffray gevin in be my lord and lady Drummound assedationaris thairof as it is admittit be the quenis majestie and yeirlie charged and past in the comptis.'

'Item, the hale rentis, frutes, proventis and emolumentis of the said haill abbay ar sett be the commendater[238] and convent thairof to David, lord Drummond, and Lilias, lady Drummond, his spous, thair airis and assignais [in] lyfrent for all the dayes of the said commendataris lyftyme for payment yeirlie of the sowme of ane thowsand markis the money.'

DUNBLANE, BISHOPRIC OF,[239]

(NLS, fos 61v-63v)
(SRO, Vol. a, fos 321v-323v)

'Rentale episcopatus Dumblanensis'

The parish kirk of Dunblane and the teind sheaves of the touns within the same extending to 16 c. 15 b. meal; 3 c. 14 b. bere; £64 13s 4d money; 12 b. horsecorn.

The parish kirk of Muthill and the teind sheaves of the touns within the same extending to 15 c. 14 b. 2 f. meal; 2 c. 11 b. bere [1 c. 7 b., *fo 21r*]; 2 c. horsecorn; 2 merks money.

The parish kirk of Strogeycht and the teind sheaves of the touns within the same extending to 3 c. 14 b. meal; 2 c. 11 b. bere; 7 c. horsecorn; £60 money.

The parish kirk of Callendrech and the teind sheaves of the touns within the same extending to 5 c. meal; 9 b. 2 f. bere.

The parish kirk of Monze and the teind sheaves of the touns within the same extending to 7 c. 2 b. meal; 30 b. 1 f. bere.

The parish kirk of Findogask and the teind sheaves of the touns within the same extending to 20 b. meal; 20 b. bere.

The parish kirk of Kilmahug extending to, in money, 100 merks.

Within the parish of Logy, of the lands of Gogar, 16 b. wheat.

A fourth part of the parish kirk of Tulleallon extending to £4; 8 b. victual.

A fourth part of the parish kirk of Glendowen extending to £10, set to Thomas Hartt.

A fourth part of the parish kirk of Fossowy extending to £20.

A fourth part of the parish kirk of Buffudder extending to £9.

A fourth part of the vicarage of Muthill, 16 merks.

---

[238] Alexander Gordon, bishop elect of Galloway, was commendator at the Reformation, *RSS*, v, no 2211.

[239] Cf NLS, fo 21r (see above, p. 294).

'Item, the temporall landis of the bischoprik of Dumblane extendis to' 40 'merkis land off auld extent set presentlie to the tennentis and possessouris of the samin for twa males and ane half,[240] as the use and custume is and hes bene in Stratherne and the quenis landis thair sett in this maner.'

Assumption of the bishopric of Dumblane:
Third of money, £104 6s 8d. 'Tak: Oute of the parochin of Dumblane for the teyndis quhilk the laird of Keir hes in assedatioun', £40; the quarter of the kirk of Tullyallone, £4; the quarter of the kirk of Glendoven, £10; the quarter of the kirk of Fossowy, £20; 'the bischoppis parte of the vicarage of Muthill', 16 merks; the quarter of Buffuddir, £9; the mails of Tunruchan within the parish of Muthil, £5 6s 8d; the mails of Gannochan within the said parish, 4 merks; the mails of Drumlukocht within the said parish, 4 merks. Total, £104 6s 8d.

Meal, 16 c. 11 b. 2½ p. Bere, 3 c. 14 b. 2 f. 1½ p. 'Tak of the paroche kirk of Callender and townis as eftir followis: Item, oute of Eister Callender', 5 b. meal, 3 f. bere; Myddell Callender, 4 b. meal, 3 f. bere; Westir Callender, 5 b. meal, 3 f. bere; Kirktowne, 5 b. meal, 3 f. bere; Auchinvaik, 6 b. meal, 3 f. bere; Eister Brokland, 7 b. meal, 3 f. bere; Westir Brokland, 8 b. meal, 3 f. bere; Garth, 8 b. meal, 3 f. bere; Clasche, 10 b. meal, 3 f. bere; Grenok, 12 b. meal, 5 f. bere; Ester Gartquhone, 4 b. 2 f. meal, 3 f. bere; Wester Garquhone, 4 b. 2 f. meal, 3 f. bere; 'the tua Ibertis', 1 b. meal. 'Tak out of the parochin of Monze: Item, fra the lady Monze', 30 b. meal, 8 b. bere; Wester Cairny, 6 b. meal, 2 b. bere; Eister Cairny, 6 b. meal, 2 b. bere; the Mylnburne, 6 b. meal; Drumquhare, 5 b. meal; James Persoun, 1 b. meal; Thoumegrow, 4 b. meal, 6 f. bere; Tullemoran, 7 b. meal, 6 f. bere; Thomrandy, 5 b. meal, 2 b. bere; Classingar, 4 b. meal, 2 b. bere; Wester Growan, 4 b. meal, 1 b. bere; Myddill Growane, 4 b. meal, 2 f. bere; 'the Manis of Logy with the pertinentis' 32 b. meal, 10 b. bere. 'Tak out of the parochin of Strogeycht and barrony thairof: Item, out of Brodland Strogeycht', 15 b. meal, 4 b. bere; Kirktoun, 10 b. meal, 4 b. bere; Allycht Moir, 14 b. meal, 2 b. bere; Allingrew, 9 b. meal, 3 b. bere; Allingeroy, 2 b. meal, 1 b. bere; Drumquhare, 7 b. meal, 2 b. bere; Auchinglen, 10 b. meal, 2 b. bere; Cult Over and Nether, 7 b. meal, 1 b. bere; Brodland of Ogilbye alias Manis, 4 b. bere. Total meal, 16 c. 12 b. Total bere, 3 c. 14 b. 3 f.[241]

Third of oats [i.e. horsecorn], 3 c. 4 b. 'Tak thir aittis furth of the barrony of Ogilbeis and parochines of Strogeycht as folloues, viz.': Take out of Kinpache, 10 b. 1 f.; out of Drumcardin, 8 b.; Auchinwaik, 3 b. 1 f.; Dunduff, 2 b. 2 f.; Over Baddinheycht, 14 b.; Over Cromlikis in the parish of Dunblane, 14 b. Total, 3 c. 4 b.

Third of wheat, 5½ b. Take out of Tulleallone, 4 b.; out of Gogar, 1 b. 1 f. 1½ p.

---

[240] I.e. 2½ times 40 merks.
[241] The correct calculation appears to be 3 c. 15 b.

NOTE

Rentals for the following benefices are located elsewhere in the text:

ABERUTHVEN, VICARAGE OF (EUL, fo 10r), p. 502.
BALQUHIDDER, PREBEND OF DUNBLANE CATHEDRAL (NLS, fo 99r), p. 392.
CULROSS ABBEY, ALTARAGE OF ST MUNGO IN (EUL, fo 23v), p. 522.
DOLLAR, VICARAGE PENSIONARY OF (SRO, Vol. a, fo 102r), p. 88.
DULL, VICARAGE OF (SRO, Vol. a, fo 101v), p. 86.
FORTEVIOT, PARSONAGE OF, PENSION FROM (SRO, Vol. a, fo 164v), p. 159.
FUNGARTH, PREBEND OF DUNKELD CATHEDRAL (NLS, fos 81v, 82r), pp. 365, 366.
INCHMAHOME, PRIORY OF (EUL, fos 39r, 41v-42r), pp. 544, 548.
KILSPINDIE, VICARAGE PORTIONARY OF (SRO, Vol. a, 124r), p. 110.
MONEYDIE, PARSONAGE OR VICARAGE OF (SRO, Vol. a, fo 101v), p. 87.

Forfar (1)

ARBROATH, ABBEY OF,[1] (NLS, fo 64r-v)
(SRO, Vol. a, fo 324r-v)

The assumption of the third of the abbey of Abirbrothok:[2]

Third of the money, £851 4s 8d. Take the barony of Abirbrothok for £241 9s 8d; barony of Dynnychtin for £8 19s 8d; barony of Kyncoldrum for £170 9s 8d; barony of Athy for £108 6s 6d; barony of Newland for £123 6s 8d; barony of Torrye, £50 5s 8d; barony of Banquhory Terny £30 13s 8d; barony of Tarves, £101 0s 1d; the teind salmon of Banff, £24. 'Gif in' £7 6s 10d.[3]

Third of wheat, 10 c. 1 b. 1 f. 1½ p. Take the kirk and lordship of Arbrocht both teind and ferme for 12 c. 1 b. 3 f. 'Gif in' 2 c. 1 f. 2⅔ p.

Third of bere, 47 c. 13 b. 2 f. 3¼ p.[4] Take the same kirk and lordship of Abirbrothok both teind and ferme for 50 c. 3 b. 1 f. 'Gif in' 2 c. 5 b. 2 f. ⅔ p.

Third of meal, 65 c. 8 b. 2 f. Take the same kirk and lordship of Abirbrothok teind and ferme for 41 c. 11 b. 1 f.; the kirk of Moneyky for 24 c. 15 b. 'Gif in' 1 c. 1 b. 3 f.

Third of oats, 9 c. 3 b. 2 f. 2⅔ p. 'Remember my lord Comptrollare[5] to get particular assignatioun of this aittis becaus it is gevin in haill ouirheid', 27 c. 11 b. 'without expressing quhair it suld be liftit.'[6]

Salmon, 12½ barrels. The fishing of Banff [*i.e.* Bamff] for 12½ barrels out of the fishing of Montros.[7]

---

[1] Cf NLS, fo 77r (see below, p. 358). Cf brief extract in *Liber S. Thome de Aberbrothoc, Registrorum Abbacie de Aberbrothoc pars altera Registrum Nigrum necnon Libros Cartarum Recentiores Complectens. 1329 - 1536*, ed. C. Innes (Edinburgh, 1856), Appendix, pp. xxxiv-xxxv.

[2] This should come after the rental for Arbroath on fos 77r-81r (see below, pp. 358-364).

[3] This would add up to £851 4s 9d, which is one third of £2,553 14s 3d; the total given on fo 77v is £2,053 14s 3d; that on fo 80v is £2,553 14s; cf below, p. 359, n. 51 and p. 360, n. 57.

[4] This is correct, taking the total on fo 80v; cf below, p. 361, n. 58 and p. 363, n. 64.

[5] Bartholomew de Villemore, Comptroller, 1555 - 1562, with Thomas Grahame of Boquhaple, 1561 - 1562. They were succeeded by Sir John Wishart of Pittarow, 1562 - 1565. *Handbook of British Chronology*, p. 191.

[6] See fo 79v (below, p. 362) for the total of oats; no figures appear for individual payments.

[7] Bamff yielded 12 barrels of salmon and Montrose provided 3 barrels of salmon on fo 80r (see below, p. 362).

'Gylssis,[8] half barell and 3 pairt half barell' [*i.e.* ¾ barrel]. Out of the Northferry paying 2 barrels.

'Omittit caponis, pultre, grassumes, dawerkis and all utheris services and small dewiteis. Alswa nota that the kirkis of Abirnethy and Monyfuith ar nocht comptit heirin.'

COUPAR ANGUS, ABBEY OF,[9] (NLS, fos 65r-75r[10])
(SRO, Vol. a, fos 325r-329v)

Rental of the abbacy of Coupar

'I pray you keip this rentall weill of aventuir it speak alane.'

'Rentale of the haill temporall landis of the abbacie of Coupar as the samyn pais now, instantlie gevin up be Leonard,[11] commendatour thairof, being chergit thairto at Edinbrucht.'

The barony of Kethik from the Reidcroce west [*in margin*, 'Sett to Johnne, erle of Athoill[12] and James McBrek lyverentair'].
    Nethir Campsy with the fishings and teind sheaves of the same and of the lands of Over Campsy called the Woulfhill 'quhilk war evir set togiddir', pays of feu mail £38 6s 8d [*in margin*, £11 13s 4d 'of the malis of this soum gevin into James McBrek and his air, lyvirentaris of Campsy, in fe be umquhile Donald, abbot,[13] undir the commoun sele'].
    Woulfhill[14] of feu mail, £10 8s [*in margin*, 'Johnne Crago, fewar, pa[y]it of ald'].
    *Mr Walter Lyndesay, feuar,[15] Bruntyhill, Kemphill and Cowbyr of Kethik of feu mail, £17 16s [*in margin*, 'Archibald Ogilvy for Kemphill. Gilbert Harbert, Bruntyhill and William Alexander for Cowbyre'].
    Soutarhoussis of feu mail, £9 [*in margin*, 'William Campbell, fewar'[16]].

---

[8] This should probably read 'Grilse'.
[9] Cf NLS, fos 83v, 109v (see below, pp. 368, 409). This rental is printed in the *Rental Book of the Cistercian Abbey of Cupar-Angus with the Breviary of the Register*, ed. C. Rogers, 2 vols. (London, 1879), i, pp. 353-363; see also *Charters of the Abbey of Coupar Angus*, ed. D. E. Easson, 2 vols. (Edinburgh, 1947).
[10] NLS, fo 65v is blank.
[11] Leonard Leslie obtained a gift of the abbey on 24 August 1565, *RSS*, v, no 2284.
[12] John Stewart, 4th earl of Atholl, *Scots Peerage*, i, pp. 444-445.
[13] Donald Campbell was abbot of Coupar Angus at the Reformation, *Rental Book of the Abbey of Cupar-Angus*, i, p. 100; Cowan and Easson, *Medieval Religious Houses*, p. 74.
[14] 'Ower Camspsy alias the Wolf Hill', *Charters of Coupar Angus*, ii, p. 195.
[15] The text from * does not appear in the SRO text. 'Walter Leslie in Coubyre', *Charters of Coupar Angus*, ii, p. 238.
[16] 'Johne Campbell of the Bait', *Charters of Coupar Angus*, ii, p. 227.

Keithik and Coltward with the mills and Sanct Ninianis akir of feu mail, £74 6s 8d; and of ferme bere, 2 b. 2 f. [*in margin*, 'Mr Nicoll Campbell for vij auchtane pairtis Sanct Ninianis akir ...[17] and Thomas Campbell[18] for j auchtane pairt and the Coltward baith fewaris'].

Calsayend,[19] acres thereof, £9 13s 8d [*in margin*, 'Johnne Bell, Williame Gourlay, William Raa, William Benson and William Campbell, fewaris'].

Baitscheile, acres thereof, £17 13s 11d [*in margin*, 'Henry Thom, William Campbell, Johnne Forman, A[ndrew] Donaldsoun, Williame Wrs[20] [fewaris, *deleted*], [George Tailyour, *deleted*] and Johnne Crago'].

Neucalsay of silver mail, 41s 8d; and of ferme bere, 11 b. set in feu for 8s 4d the boll, total, £4 11s 8d [*in margin*, 'Newcalsy sett in few be Donald, abbot of Couper'].

*'Aliter the beir of Newcalsy set for silver. Lateris, money', £181; bere, 2 b. 2 f. [*in margin*, 'Lateris', £179 7s 7d; 'beir', 13 b. 2 f.].[21]

Coubyre[22] of silver mail, £5 4s [*in margin*, 'in few as it pais']; of ferme bere, 3 c. 3 b. [*in margin*, 'ferm bere'].

Balgirsche of feu mail, £15 [*in margin*, 'few as it pais'].[23]

Gallouraw of silver mail, £5 10s; of ferme bere, 1 c. 8 b. 2 p. set to the tenants in feu for 8s 4d the boll, extending in money to £9 12s 8d.[24]

Total of silver mails of the lands of Cupar from the Reid Croce west extends to £205 0s 7d.[25] Total of silver for ferme bere west of the Croce, £14 4s 4½d.[26] Total of the whole silver west of the Croce for ferme and mails extends to £219 4s 11½d. Total of ferme bere west of the Croce extends to, 3 c. 5 b. 2 f.[27]

From the Rede Croce east

Westere Denhede of feu mail, £10 17s 10d. Eister Denhede of feu mail, £9 2s 10d. Balbrogyis: Balbrogus Wester, Ester and Over, pay of few mail £59 15s 7d. Crunan of feu mail, £18 2s 4d. Airthourstane of feu mail, £18 14s. Balmyle ferme meal, 3 c. 12 b. 1 f.; of ferme bere, 3 c. 12 b. 1 f.; all this victual set in feu for 8s 4d the boll, extends in money to £50 4s 2d [*in margin*, 'lard of Ruthvennis,[28] fewar'].

---

[17] Part of the text in this marginal note is unclear.
[18] Cf *Charters of Coupar Angus*, ii, p. 231.
[19] 'Causaend', *Charters of Coupar Angus*, ii, p. 116.
[20] The name 'William Arous' occurs in *Charters of Coupar Angus*, ii, pp. 75, 76, 93, 114.
[21] The text from * to this point does not appear in the SRO volume.
[22] 'Cowbyre of Kethik', *Charters of Coupar Angus*, ii, p. 224.
[23] The marginal notes above do not appear in the SRO text.
[24] The correct calculation appears to be £10 1s ½d.
[25] This calculation is correct.
[26] £4 11s 8d plus £9 12s 8d equals £14 4s 4d.
[27] This calculation is correct.
[28] Patrick, 3rd lord Ruthven, died in 1566 and was succeeded by his second son William, 4th lord Ruthven, *Scots Peerage*, iv, pp. 261-262.

Total of silver mail from the Reid Croce east extends to £116 12s 7d. Total of silver for ferme meal and bere extends to £50 4s 2d. Total of the whole silver east of the Croce of mails and ferme victuals extends to £166 16s 9d.

'Undir the officer aboun the Waltirris of Ylay and Areicht'
Coupar Grange of silver mail, £16 13s 4d; of ferme bere, 11 c. 4 b. 3 f. set in feu for 8s 4d the boll, total of the said money, £75 6s 3d[29] [*in margin*, 'Ferm bere sett of ald in few to the lard of Kyppuithe'].

Milnhorne
Milnhorn of silver mail, £6. Ledcasse of silver mail, £8 13s 4d. Grange of Abirbothre, £41 5s. Polcak, £9 11s 8d. Blaklaw with the mill thereof 'off auld maill', £15. Ester Cotyardis 'of auld maill', £8 13s 4d. Wester Cotyardis 'of auld maill', £8 13s 4d. Tullifergus, Over Muretoun, Nethir Murtoun 'and that thrid part of the Cheppelton quhilk James Jamesone occupiis pais yeirlie of few maill' £37 6s 8d. 'Uthir twa thriddis of Cheppeltoun', £13. Eister Drymmy of feu mail, £11 6s 4d. Middill Drymmy, £5 13s 4d. Wester Drymmy with the mill and teind sheaves thereof of feu mail, £31 10s 8d. Caillies with the mill and teind sheaves thereof of feu mail, £21 4s. Wester Persey of feu mail, £8 6s 8d. Ester Persey of feu mail, £9 10s.

Total of silver mails above the Watteris extends to £252 6s 8d.[30] Total of silver for the ferme bere above the Watteris extends to £75 6s 3d. Total of the whole silver above the Watteris both for mails and ferme victual extends to £327 12s 11d.

[*In margin*, 'In Atholl']
Tullichane, Innervak, £7. Moirtullich, £11.[31] Drumfallinthie [Drumfathtie, *fo 84v*], £11. Murthlak in Mar, £11 6s 8d. Clintlaw and Auchindory, £5. Grainge of Arlie, £7 6s 8d. Blakstoun, £8. Total of the silver mails of these lands abovewritten extends to £60 13s 4d.

Grainge of Kincreich and Glenboy pays of feu mail, £55 8s. Littill Pertht, £24. Carsgrainge with the Bogmiln, Murhoussis, Westhorn and Orcheart, £111 4s. Total of these three baronies abovewritten extends to £190 12s.

Rental of Glenylay
Cambok, Over Auchinleische, Nether Auchinleische, 'exceppand the breulandis thairof', Over Ilrik, Nethir Elrik, Downy, Dalnacabok, Kirktoun, Pitlochrie, Bellite, 'v auchtan pairtis of Wester Inneraritie, thrie quartaris of Ester

---

[29] This is the total money raised from the ferme bere, excluding the silver mail.
[30] The correct calculation appears to be £252 7s 8d.
[31] There is a suggestion of a bracket around Tullichane, Innervak and Moirtullich connecting them with the marginal note.

Inneraritie, myln and mylnlandis thairof, thrie quarteris of Glenmerky quhilk the erle of Argyle[32] hes in few, pais yeirlie of few maill' £74 2s 8d. Mekill Forthir and Litill Forthir pay of feu mail, £16. 'Thre auchtane pairtis of Wester Inneraritie with the breulandis of Nether Auchinleische' of feu mail, £5 3s 4d. Quarter of Ester Inneraritie, 55s. Neutoun Freuchy, mill and Milntoun thereof, one quarter of Glenmerky, £27 17s 8d. Kirkhillokkis of feu mail, 50s. Daluany and Craigneate of silver mail, £19 8s 4d. Carnaclocht, 'the officiarris land', 20s. Wester Bogsyde, 46s 8d. 'The lap maill of Fornethie', 16s. Total of the whole silver mails of the lands of Glenylay extends to £151 19s 8d.

'Annuellis pertenyng to the abbay of Coupar and in use of payment':
Annuals within Dundie, £9 5s 8d [*in margin*, 'of denyit annuellis within the said burcht', £3 [£4, *SRO*]. Annuals within the burgh of Pertht, £28 6s 8d [*in margin*, 'of denyit annuellis within the said burcht', 44s]. Annuals within the burgh of Forfar, 26s 8d. An annual out of Scone, 33s 4d. Annual of Brunty Miln, 20s. 'Bair of Windyaige', 20s. 'Summa of annuellis confessit abon wrettin extendis to' £42 12s 4d.

'Summa of silver maillis and silver for victuell set in few of the temporall landis of the abbay of Couper extendis to' £1,159 11s 11d 'ob'.[33] 'Summa of ferm onset for silver of the temporal landis of Cupar to' 3 c. 5 b. 2 f. bere.

Rental of the kirks of Couper set for money
The kirk of Bennethie [*i.e.* Bendochy] 'quhilk extendit in the rentaile to thre scoir aucht chaldiris victuall, twa part mele and thrid part bere, set in lyverentis for' 6s 8d the boll, total, £362 13s 4d [*in margin*, 'Bennethie']. The teinds of Carsgrange which extended in the rental to 7 c. 13 b. 1½ p. wheat set in tack for 8s 4d the boll, total money, £52 2s 6d[34] [*in margin*, 'Carsgrange']. The teinds of the kirk of Alwecht both parsonage and vicarage with the lands and fishings of Awecht and Innerrychtny set for £94 [*in margin*, 'Alwecht']. The teinds of the kirk of Glenyla both parsonage and vicarage set for £80, 'and the deutie to Cambuskynnethe' [*in margin*, 'Glenyla']. The kirk of Mathie both parsonage and vicarage set for £66 13s 4d [*in margin*, 'Mathie']. Three quarters of the kirk of Fossoquhay, both parsonage and vicarage, set for £66 13s 4d [*in margin*, 'Fossoquhay']. The vicarage of Bennethie, £6 13s 4d. The vicarage of Arlie, £6 13s 4d. Total of the whole kirks of Cuper set for money extends to £735 9s 2d.

---

[32] I.e. Archibald Campbell, 5th earl of Argyll, *Scots Peerage*, i, pp. 340-343.
[33] This total is correct if for 'obol' ½d is substituted.
[34] In calculating the total, 1½ p. is counted as a tenth of a boll.

'Defalcationis off the haill rentaill of money

Item, gevin in lyvrent to James Makbrek and his airis lyverenteris of Campsy of the malis of the landis thairof be umquhile Donald, abbot of Couper[35] and convent thairof yeirlie duiring thair takkis', £6 13s 4d [*in margin*, 'James Makbrek']. To the College of Justice of yearly contribution for the said abbacy, £35 [*in margin*, 'College of Justice']. To the kirk of Arroll of yearly duty, £7 6s 8d [*in margin*, 'Errole']. To the abbot of Restennot[36] of annual, 20s [*in margin*, 'Restennot']. 'Summa of defalcatiounes extendis to', £50.

Rental of the teinds of Arlie 'as the samyn pais instantlie'

Cardene, 2 c. 14 b. meal; 1 c. 6 b. bere. Brideistoun, 3 c. meal; 1 c. 8 b. bere. Drundarne, 5 b. meal; 3 b. bere. Blakstoun, 1 c. 2 b. 2 f. 2 p. meal; 9 b. 1 f. 2 p. bere. Cukestoun, 1 c. 14 b. meal 'allanerlie.' Lunros, 20 b. meal; 14 b. bere. Litiltoun, 1 c. 14 b. meal; 14 b. bere. Bakie, 2 c. meal; 1 c. bere. Newtoun, 1 c. 8 b. meal; 12 b. bere. Grainge of Arlie, 24 b. meal ['meill tantum', *SRO*]. Manis of Arlie, 3 c. meal 'allanerlie.' Kynnalthie, 2 c. 8 b. meal; 1 c. 8 b. bere. Lundarteris, 2 c. 11 b. meal; 1 c. 5 b. bere. Redie, 2 c. 14 b. meal; 1 c. 6 b. bere. Auchindorie, 8 b. meal; 2 b. bere. Total of the teind meal of the parish of Airlie extends to 28 c. 14 b. 2 f. 2 p. Total of the teind bere of Airlie extends to 11 c. 7 b. 1 f. 2 p.

Rental of the horsecorn of the abbacy of Coupar

Campsy, 4 b. horsecorn set in feu for 3s 4d the boll, total, 13s 4d [*in margin*, 'in silver']. Woulfhill alias Over Campsie, 2 b. horsecorn. Soutarhousis, 4 b. horsecorn. Keithik, 16 b. horsecorn. Kemphill, 6 b. horsecorn set in feu for 4s 4d the boll, total, 26s. Bruntyhill, 2 b. horsecorn set in feu for 4s 4d the boll, total, 9s[37] [*in margin*, 'in silver']. Coubyre, 6 b. horsecorn. Balgirsche, 6 b. horsecorn set in feu for 3s 4d the boll, total, 20s. Gallouraw, 6 b. horsecorn set in feu for 3s 4d the boll, total, 20s [*in margin*, 'in silver']. Total of the horsecorn from the Rede Croce west with Campsy, [*blank*].

Horsecorn from the Rede Croce east

Wester Denhede, 13 b. Ester Denhede, 13 b. horsecorn set in feu for 3s 4d the boll, total, 43s 4d [*in margin*, 'few']. West side of Balbrogy, 40 b. horsecorn; 25 b. thereof set in feu for £6 13s 4d [*in margin*, 'few']; the rest which is 15 b. unset in feu. Ester Balbrogy pays of horsecorn, 1 c. 10 b.; 6 b. 2 f. thereof set in feu for 43s 4d [*in margin*, 'few']; and the rest which is 19 b. 2 f. unset in feu. Over Balbrogy, 5 b. horsecorn. Crunan, 26 b. horsecorn set in feu for 3s 4d the boll, total, £4 3s 4d[38] [*in margin*, 'few']. Airthourstaine pays of horsecorn, 20 b.; 15 b. thereof set in feu for 3s 4d the boll, total 50s [*in margin*, 'few'], 'the

---

[35] See above, p. 352, n. 13.
[36] Andrew Hume was commendator of Jedburgh and Restenneth at the Reformation, *RMS*, iv, no 1661.
[37] The correct calculation appears to be 8s 8d.
[38] The correct calculation appears to be £4 6s 8d.

remanent quhilk is' 5 b. unset in feu. Balmyle pays of horsecorn, 15 b. 3 f., the whole set in feu for 3s 4d the boll, total, 52s 6d [*in margin*, 'few']. 'Summa of hors corn under the officer fra the Rede Croce est, [*blank*].'

Horsecorn above the Wateris of Ylay and Areicht
Coupargrainge pays of horsecorn, 2 c. 4 b. set in feu for 3s 4d the boll, total, £6. Grainge of Abirbothre pays of horsecorn, 18 b.; 14 b. 1 f. thereof set in feu for 3s 4d the boll, and 3 b. 3 f. for 5s the boll, total, £3 6s 3d. Polcak, 4 b. horsecorn set in feu for 5s the boll, total, 20s. Blaklaw, 2 b. horsecorn set in feu for 3s 4d the boll, total, 6s 8d. Wester Cotyardis, 4 b. horsecorn set in feu for 3s 4d the boll, total, 13s 4d. Ester Cotyardis, 4 b. horsecorn set in feu for 3s 4d the boll, total, 13s 4d. Cheppeltoun, 6 b. horsecorn; 2 b. thereof set in feu for 3s 4d the boll, total, 6s 8d; the rest which is 4 b. unset in feu. Nethir Murtoun pays of horsecorn, 6 b. set in feu for 3s 4d the boll, total, 20s. Over Murtoun pays of horsecorn, 4 b. set in feu for 3s 4d the boll, total, 13s 4d. Tullifergus pays of horsecorn, 16 b. set in feu for 3s 4d the boll, total, 53s 4d. Total horsecorn above the Wateris of Ylay and Areicht, [*blank*].

Carsgrainge
Bogmilns, Carsgrainge, Neubiggyne, Watterybuttis, Murhoussis and Westhorne pay of horsecorn yearly 1 c. 14 b. 2 f. 2 p.
Grainge of Kincreycht, 8 b. Glenboy, 8 b. Clintlaw, 4 b. Auchindory, 4 b. Grainge of Arlie, 4 b. Littill Pertht, 8 b.

'Summa of the haill horscorn of the lordschip of Cuper conform to the auld rentaill extendis to' 23 c. 9 b. 1 f. 2 p. Total of the said horsecorn set in feu for silver, 13 c. 13 b. 1 f.; extending to in money, £41 7s 6d.[39] Total of horsecorn unset in feu extends to 9 c. 12 b. 2 p.

'Summa of the haill silver pertening to the abbacy of Cuper for the maillis of the temporal landis, fermis and hors corn set in few for money and kirkis set for silver, annuellis and all that ar in use of payment, fiftie pundis of defalcatioune being deduceit thairof, extendis to' £1,886 8s 7d.[40]

'Summa of the haill fermis now pertening to the abbacye of Cuper onset in few for silver extendis to' 3 c. 5 b. 2 f. bere.

'Summa of the haill teindis now pertening to the abbacy of Cuper onset for silvir extendis to of meill' 28 c. 14 b. 2 f. 2 p. out of the kirk of Arlie.

'Summa of the teind beir onset for silver extendis to' 11 c. 7 b. 1 f. 2 p., Airlie.

---

[39] The correct calculation from the sub-totals appears to be £41 7s 1d.
[40] This calculation is correct (apart from ½d which should be added to the total).

'Summa of the haill hors corn onset in few for silver extendis to' 9 c. 12 b. 2 p.[41]

FARNELL, BENEFICE OF,[42]   (NLS, fo 76r[43])

'Item, the rentall of Mekill Carkorie fyve chalder ane half, thrid pairt bair tuay pairt mele of this ilk plew be half ane boll quhit. Item, the rentall of Litill Carorie fourtein bollis, thrid pairt beir, tuay pairt mele. Item, the rentall of Estir Fathie, xxij bollis thrid pairt beir, tuay pairt mele. Item, the rentall of Mekill Fethie, xxviij b. thrid pairt beir, tuay pairt mele. Item, the land of Borrowfeildis Fithie ane chalder, tuay pairt mele thrid pairt beir. Item, the rentall of Farunall [*i.e.* Farnell], four chalder ane half, tuay pairt mele, thrid pairt beir. As for quhit thair is nane that payis bot Carkore onlie.'

'The rentall of the parochane of Farnall'.

ARBROATH, ABBEY OF,[44]   (NLS, fos 77r-81r[45])
(SRO, Vol. a, fos 330r-334r)

'Rentale monasterii de Abirbrothok'

Rental of the monastery of Abirbrothok containing all and sundry fermes, sums of money and victuals, of the temporality as well as the spirituality, of these lands and teinds of kirks charged according to the old rentals and also according to the accounts of the chamberlains and graniters which are sold now to the advantage and profit of the lord commendator[46] of the same as from the present accounts of this instant year 1561 [1560, *SRO*].

First, the rental of the temporality according to the accounts set for this present year
    The barony of Abirbrothok in fermes yearly and other duties according to the rental and annual accounts extending to £241 9s 8d. Barony of Dynnichtin annually extending to £8 19s 8d. Barony of Athy annually extending to £108 6s 5d. Barony of Kidcoldrum [*i.e.* Kingoldrum] annually extending to £170 9s 8d. Barony of Newlandis annually extending to £123 6s 8d. Barony of Torry annually extending to £50 5s 8d. Barony of Banquharry Terny annually, £30 13s 8d [£30 13s 4d, *SRO*]. Barony of Terves annually, £101 0s 1d. Vicarage of Fyvie

---

[41] *Rental Book of the Abbey of Cupar-Angus*, i, pp. 360-363, continues with further entries to the rental, located separately in this text at fo 83v (see below, p. 368).
[42] Farnell formed the prebend of the deanery of Brechin; see below, fo 81v (p. 364).
[43] Fo 75v is blank.
[44] Cf NLS, fo 64r (see above, p. 351).
[45] Fo 76v is blank.
[46] John Hamilton was commendator of Arbroath, at the Reformation, *RSS*, v, no 837.

annually extending to £36 13s 4d. Annualrents according to the yearly rental extending to £13 2s 6d. Total money of the said baronies with the said vicarage and annualrents yearly, £884 7s 4d.

Rental of the kirks set for money

The kirk of Innernes set for £133 6s 8d. Kirk of Gamery set for £43 6s 8d. Teind salmon of Banff set for £24. Teind white fish of Banff set for £3 3s 8d. [*In margin*, £16 13s 2d 'out of thir teindis'.[47]] Kirkland of Banff and Innerbundie set for £4. Kirk of Banff and Innerbundie set for £60. Teinds of Quhittane, Ardbrangan and Dallaquhy set for £10. Kirk of Forg set for £100. Kirk of Abirkerdouth [*i.e* Aberchirder] set for £80. Kirk of Lunglie and Fater Angus set for £50. Kirk of Coule set for £24. Kirk of Kynmarny set for £16 13s 4d. Kirk of Boithelme set for £50. Kirk of Fyvie except for the victual set for £201 6s 8d. Kirk and vicarage of Fyvie, £66 13s 4d. Kirk of Banquharryntny set for £93 6s 8d. Kirk of Terves except for the victual set for £26. Kirk of Garvok set for £80. Kirk of Newtyld set for £80. Kirk of Clovay set for £20. Kirk of Ruthven set for £10. Kirk of Kermuir set for £266 13s 4d. Kirk of Glammis set for £100. Kirk of Dunbug set for £1.

Not charged with the kirk of Abirnathy extending yearly to £293 6s 8d which is in the hands of Mr Alexander Betoun for his pension. [*In margin*, 'Non allocantur iste ecclesie.'[48]] And similarly the kirk of Monyfuith in the hands of the said Mr Alexander, vicar pensionary of the same, extending yearly to £26 13s 4d. [*In margin*, 'Nec ista vicaria computa.'[49]]

The vicarage of Carmilie set for £26 13s 4d.

Total of all the foresaid kirks and money, except for the kirks of Monyfuth and Abirnethy extends yearly to £1,669 6s 8d.[50]

Total of the sums of money of the temporality as well as of the spirituality, of lands as well as of teinds of kirks set for money except for the kirks of Monyfuth and Abirnethy which are not set but are assigned to Mr Alexander Betoun, pensioner, according to the annual accounts of the chamberlain and graniter as the foresaid in money extends to the sum of £2,053 14s.[51]

Of which total sum of money is due to be deducted according to the annual accounts of the chamberlains and graniters of the same monastery yearly allocated and accounted for the upkeep of the convent as follows. First, to the cellarer for the upkeep of the convent each year for the purchase of the kitchen

---

[47] Some of the marginal notes are in Scots, though much of the rental, with other marginal notes, is in Latin.
[48] I.e. 'These kirks are not set.'
[49] I.e. 'Nor is this vicarage accounted for.'
[50] The correct calculation appears to be £1,570 3s 8d.
[51] The correct calculation appears to be £2,454 11s; cf p. 351, n. 3, and p. 360, n. 57.

victuals extending to £663. For the upkeep of their habits yearly, £38. For their pittances, by the receipt of fermes and sums of money of Carny, Lathane, Keipny and Spittelfeild according to their accounts yearly extending to £61 10s.

Similarly they receive annualrents of Monifuth previously charged according to their annual accounts, £6 13s 4d.

Total to the said convent, £819 3s 4d.[52]

Follows the pensions in the said monastery paid annually according to the accounts.

No discharge for the kirk of Abirnethy and the vicarage of Monyfuith, intromitted by Mr Alexander Betoun, pensioner, because they were not charged above.

The see of the archbishop of St Andrews in kanes of lands and other things according to the annual accounts paying £24 13s 4d.

The bishop of Aberdeen[53] for the second teinds, 53s 4d.

Contribution to the Lords of the Council, £84. For the pension of Robert Gourlaw, £40. Mr Duncan Chalmer, £40. The vicar of Barry, £4. The vicar of St Vigeans, £21 6s 8d. The dean of Brechen, £4. The chaplain of Broth [Borth, *SRO*], £46. To two advocates before the Lords of the Council, £66 13s 4d. The chamberlain yearly in all things, £100. The graniter similarly, £100. The bailie of the regality for his office in administration of justice according to the accounts, £60 [*in margin*, 'with the mair']. Total pensions, £549 12s 8d [£549 6s 8d, *SRO*].[54]

Follows the fees of the servants of the said monastery yearly

Subprior, subchanter, sacrist, third prior, cellarer and other servants according to the annual accounts, £7 6s 8d. To the janitor of the same monastery, 40s. To the baker, brewer, architect,[55] master of works and launderer of the place according to the accounts annually extending to £37 6s 8d. Total of the servants' fees, £46 13s 4d.

Total deductions, £1,415 9s 4d,[56] so rests free yearly, £1,138 4s 8d.[57]

Rental of the spirituality of Abirbrothok and the lordship of the same as of old.

Kirk of Abirbrothok and the lordship of the same in fermes and duties from victuals of old, 12 c. 1 b. 3 f. wheat; 50 c. 3 b. 1 f. bere; 41 c. 11 b. 1 f. oatmeal.

Kirk of Ethey in teinds which the lord sets for money as above, 1 c. 14 b. wheat; 4 c. 13 b. 3 f. 2 p. [4 c. 3 f. 2 p., *SRO*] bere; 4 c. 7 b. 1 f. oatmeal.

---

[52] The correct calculation appears to be £769 3s 4d.
[53] William Gordon, bishop of Aberdeen, 1545 - 1577, Watt, *Fasti*, p. 4.
[54] The correct calculation appears to be £593 6s 8d.
[55] The text reads 'archiketto' (fo 77v), which should probably be translated as 'master builder'.
[56] This is the sum of £819 3s 4d plus £549 12s 8d plus £46 13s 4d.
[57] This totals £2,553 14s, as on fo 80v (see below p. 363); cf p. 351, n. 3, and p. 359, n. 51.

Kirk of Abirballat in teinds alone, 2 c. 15 b. wheat; 60 c. 3 b. bere; 15 c. 6 b. oatmeal.

Kirk and lordship of Dynnychthin in teinds and ferme victuals, no wheat in this kirk; 25 c. 8 b. bere; 33 c. 2 b. oatmeal.

Kirk of Moneky, 3 c. 7 b. wheat; 12 c. 3 b. bere; 24 c. 15 b. oatmeal. [*In margin*, 'Gif in' 1 c. 1 b. 3 f.]

Kirk of Panbrid in teinds, 2 c. 14 b. wheat; 10 c. 9 b. bere; 16 c. 9 b. oatmeal.

Kirk of Murehous in teinds, 1 c. 14 b. wheat; 5 c. 3 b. bere; 10 c. 10 b. oatmeal.

Kirk of Stradichin in teinds, 1 c. 8 b. wheat; 4 c. 11 b. bere; 6 c. 10 b. oatmeal.

Kirk of Innerkelour in teinds, 2 c. 13 b. 1 f. wheat; 10 c. 14 b. bere; 23 c. 13 b. oatmeal.

Kirk of Lunnane in teinds, 13 b. wheat; 6 c. 5 b. bere; 8 c. 6 b. oatmeal.

Victuals beyond the mountains ('extra montes') in teinds within the kirk of Fyvie, 4 c. bere; 11 c. oatmeal.

Total wheat in ferme as well as in teinds extends to 30 c. 4 b.

Deduction of wheat

Of which total sum of wheat there is allocated for the upkeep of the convent according to the account of the chamberlain and graniter yearly, 11 c. 11 b. Also with intromissions of the convent of the same, in victuals and teinds of the lands of Kepny [and] Carny as above, by reason of the pittances extending annually to 5 c. 6 b. In defalcation in the old rental, as by the yearly ridings of teinds of all kirks which are not now in use of payment, by the accounts of the chamberlain and graniter, 4 c. 12 b. Total deductions of wheat, 21 c. 14 b., so rests free, 8 c. 6 b.

Total bere, ferme and teind, 143 c. 7 b. 2 p.[58]

Deduction of bere out of the annual accounts for the reason foresaid.

Of which total sum of bere there is allocated for the upkeep of the convent according to the account of the chamberlain and graniter yearly, 42 c. 1 b. To the convent for their pittances intromitted with along with the victuals and the foresaid teinds of the lands of Keipny [and] Carny, as above in wheat yearly extending to 5 c. 9 b. In pensions of the place according to the accounts, namely, to the vicars and ministers of the kirks of Ethy, Barry, Lunnane and of other kirks and also ...[59] and others according to the foresaid accounts, 17 c. 4 b. In

---

[58] This total is consistent with the calculation of deductions in the next paragraph but the figure given on fo 80v is 143 c. 9 b. 2 p. which agrees with the third on fo 64r (see above, p. 351, n. 4).
[59] The text here is faded.

defalcation of the old rental according to the annual ridings of teinds of the kirks which now are not in use of payment according to the accounts of the graniter and chamberlain, 10 c. 3 b. 1 f. 1 p. Total deductions, 75 c. 1 b. 1 f. 1 p., and so rests free 68 c. 5 b. 3 f. 1 p.

Total meal, ferme as well as teind, 196 c. 9 b. 2 f.[60]

Deduction of meal out of the accounts for the reason foresaid.

Of which total sum of meal there is allocated for the upkeep of the convent according to the account of the chamberlain and graniter and also for the servants of the place. 8 c. 12 b. In pensions of the said monastery out of the annual accounts to the vicars and ministers of the kirks of the same and of other pensioners, 17 c. 5 b. In defalcation the said old rental according to the annual accounts of the said chamberlain, 16 c. 12 b. 1 f. 2 p. The almoner of the place distributes to the poor, according to the accounts, 5 c. Total deductions, 47 c. 13 b. 1 f. 2 p., and so rests free, 148 c. 12 b. 2 p.

Total sum of oats according to the account of the graniter and the old rental extends annually to 27 c. 11 b.[61]

Deduction of these oats out of the annual accounts for the reason foresaid.

To the chamberlain for his horse, 2 c. To the graniter and other servants of the place divided equally and also to the upkeep of the horses of the place for carrying fuel, 5 c. To the master of works, 2 c. Commonly for visitors yearly, 3 c. To the bailie of the regality and his deputies yearly, 3 c. Total deductions, 15 c., and so rests free, 12 c. 11 b.

Rental of the salmon of the said monastery according to the register of old and also according to the accounts of the chamberlain at present.

The salmon fishing of Banff [*i.e.* Bamff] yearly, 12 barrels of salmon. The fishing of Poldune yearly, 16 barrels of salmon. Not charged with the fishing of Monifuth because 10 barrels of salmon are set in feuferme with the lands of Northferry and so rests according to the rental two barrels of salmon and two barrels of grilse. The fishing of Montros yearly, 3 barrels of salmon. The fishing of Perthe yearly, 4 barrels of salmon. Total salmon, 37 barrels. Total grilse, 2 barrels.

Deduction of salmon

To the convent for their upkeep yearly, 24 barrels of salmon, and so rests 13 barrels of salmon and 2 barrels of grilse.

---

[60] SRO text reads '10 score and 16' instead of '9 score and 16'.
[61] Cf above, p. 351, n. 6.

Memorandum, not charged with capons, poultry and other small duties which are sold yearly for the upkeep of the place and expended for the use of strangers and others who arrive, according to the accounts.

Brought in freely from the monastery annually.
    Total of money free, £1,138 4s 8d. Total of wheat free, 8 c. 6 b. Bere, 68 c. 5 b. 3 f. 1 p. Meal, 148 c. 12 b. 2 p. Oats, 12 c. 11 b. Salmon, 13 barrels. Grilse, 2 barrels.

'Memorandum, that thair is na thing put in quhen grite lordis and utheris strangeris and siclyke the superintendent[62] cummis yeirlie twyis or thry[is] and siclyke gif my lord commendatour[63] with his servandis remanit upoun the heid of the samin thai[r] wald skantlie victell and money be gottin [to] sustene him self and utheris that wald resort unto him as eff[erit].'

    'Memorandum, that thair is nathing present in this rentall nowther in the discharg of the money and victellis quhat is gevin in to the ministeris in every kirk that is sustenit thair be my lordis grace etc. command.'

    'Memorandum, the personage of Monyfuth is nocht chargeit nor put in this rentall be reasoun it is assignit to Maister Alexander Betoun, pensionar, the rentall thairof is' 4 c. 12 b. wheat; 12 c. 9 b. bere; 15 c. 10 b. meal.

    'Memorandum, the vicarage [personage, *SRO*] of Monyfuth is nocht chargeit nor put in this rentall be reasoun that it is assignit to Maister Alexander Betoun, pensionare, and is in the yeir', £26 13s 4d.

    'Memorandum, the personage and kirk of Abirnethy is nocht put in this rentall be ressone the samin is assignit to the said pensionar and gettis for the samin yeirlie', £293 6s 8d. Total money, £320.

    'Memorandum, thair is left out amange the pensionis the pensioun grantit at the quenis grace drowrearis and Regent, derrest moder to our soverane and at hir graces command and lettres [to gif] Johnne Levingstoun, man of weir and lamit in France, yeirlie', £40. 'quhilk man be deduceit amang the haill sowme restand.'

[*In margin*, 'Memorandum, to seik the assignatioun of this rentall befoir the said rentall of it is writtin.'[64]]

Calculation of third: Money, £2,553 14s; third thereof, £851 4s 8d. Wheat, 30 c. 4 b.; third thereof, 10 c. 1 b. 1 f. 1⅓ p. Bere, 143 c. 9 b. 2 p. [143 c. 7 b. 2 p., *fo 79r*]; third thereof, 47 c. 13 b. 2 f. 3⅓ p. Meal, 196 c. 9 b. 2 f.; third thereof, 65 c. 8 b. 2 f. Oats, 27 c. 11 b.; third thereof, 9 c. 3 b. 2 f. 2⅔ p. Salmon, 37 barrels; third thereof, 12⅓ barrels. Grilse, 2 barrels; third thereof, ⅔ barrel.

---

[62] John Erskine of Dun, superintendent of Angus and the Mearns, *BUK*, i, p. 52.
[63] See above, p. 358, n. 46.
[64] This assignation is to be found in NLS, fo 64r (see above, p. 351 and n. 4).

[*In margin*, 'To be tane up quhair it salbe assignit in particular.']

[*In margin*, 'Tak j barell salmond out of Montros for the thrid parte of a barell and gylssis.']

'Pultrie, dawarkis, caponis and all utheris services and small dewiteis nocht condiscendit on in speciall. Nota, the kirkis of Abirnethy and Monyfuith ar nocht comptit heirin.'
Abirnethy, £273 [£293 6s 8d, *fos 77v, 80v*]; third thereof, £80 [*sic*].

TANNADICE, KIRK OF,   (NLS, fo 81r)
                                                  (SRO, Vol. a, fo 334r)

Tannades

Calculation of third: In the whole, £237 5s 4d; third thereof, £79 1s 9⅓d.

'Memorandum, that this thrid be payit by the pensioun of the collegis[65] becaus thay ar fre. Undir cautioun with generall restitutioun.'

TEALING, VICARAGE OF,   (NLS, fo 81r)
                                                   (SRO, Vol. a, fo 334r)

'The vicarage of Teling in dewiteis pertening to ane sympill vicarage extendit to' £40 'be yeir to thame quhilk broukit it befoir me, as tuecheing the rentall thairof sen I had it is never a penny.'

Signature: 'Maister Adame Foullis.'

Calculation of third: In the whole, £40; third thereof, £13 6s 8d.

BRECHIN, DEANERY OF,[66]   (NLS, fo 81v)
                                                   (SRO, Vol. a, fo 334v)

'The rentall of the denerie of Brichin to the quhilk Mr James Thornetoun be the quenis grace gift and nominatioun hes undoutit richt and title suppois be way of deid ane Maister David Cunynghame hes intrusit him thairin.'[67]

---

[65] Tannadice was appropriated to St Mary's College, St Andrews, cf SRO, Vol. a, fo 75r (see above, p. 64).
[66] Cf NLS, fo 76r (see above, p. 358 and n. 42).
[67] James Thornton was presented by the crown to the deanery on 3 July 1563; the previous holder was William Cunningham, *RSS*, v. no 1416; Watt, *Fasti*, p. 45. David Cunningham may be identified with the minister at Strathaven (Avondale) in 1563, *Thirds of Benefices*, p. 262.

[*In margin*, 'Money'] 'In primis the few maill of the towne of Auchdowny without ony uther service or dewtie payis' 35 merks 6s 8d. The feu mail of Stanmaquhy pays 16 merks. [*In margin*, 'Victuall'.] The lands and Manis of Fernvale, 'twa parte meill and thrid parte beir' [*i.e.* ⅔ meal, ⅓ bere] pays 5 c. The toun of Mekill Carrary, ⅔ meal and ⅓ bere pays yearly, 5 c.; the said toun pays of wheat, 8 b. The laird of Borrowfeild dwelling in Fethe pays of meal and bere, 14 b. John Carnegy dwelling in Eister Fethie pays of bere and meal, 23 b. The toun of Mekill Fethie pays of bere and meal, 20 b. The toun of Littill Carrary pays of bere and meal 17 b. Total money, 51 merks 6s 8d. Victual, 15 c.[68]

[*In margin*, 'Denrie of Brechen, in the hall' [*i.e.* whole]: Money, £34 6s 8d; third thereof, £11 8s 10⅔d. Victual, 15 c.; third thereof, 5 c. 'Tak for this victuale Mek[ill] Carrary, gif in' 8 b. 'eque. Tak the silver out of Auchdowy.']

FUNGARTH, PREBEND OF DUNKELD CATHEDRAL[69] (NLS, fo 81v)
(SRO, Vol. a, fo 334v)

[*In margin*, 'The prebendrie of Foungart, Perth'.]

'The rentall of the prebendrie of Foungart in the kirk of Dunkeld gevin in be me, Mr James Thorntoun, undoutit prebendar thairof.'

'The towne of Fowngart, the Hauch, the Hilheid and all utheris landis bayth fewit and unfewit pertening to the said prebendar payis in penny maill', £79 13s 4d. 'Thair is sum dewiteis of poultrie and pettis extending to ane uther sowme [ane sobir sowme, *fo 46v*]. The foirsaidis landis payis twa bollis beir and [tuenty, *deleted*], ten bollis hors [corne]', [total,] 12 b.

ANCRUM, PARSONAGE OF,[70] (NLS, fo 82r)
(SRO, Vol. a, fo 335r)

'The rentall of the personage of Ancrum prebendrie of the kirk of Glasgw'

'The haill teindis and all uther proffettis or emolumentis of the personage and vicarege of Ancrame wer sett in assedatioun in the yeir of God 1559 to Maister James Thorntoun, titular of the said benifice, and George Trumbill of the

---

[68] The correct calculation appears to be 15 c. 2 b.
[69] Cf NLS, fos 44v, 46v, 82r (see above, pp. 327, 329; below, p. 366); see also *Rentale Dunkeldense*, ed. R. K. Hannay (Edinburgh, 1915), p. 349. This benefice, in the sheriffdom of Perth, is included among rentals for Forfar because it was held by James Thornton, who also held the preceding benefice.
[70] Cf SRO, Vol. a, fos 164v, 212v, 239v (see above, pp. 159, 213, 236). This benefice, in the sheriffdom of Roxburgh, was a prebend of Glasgow cathedral. It is included in rentals for Forfar because James Thornton, who held the preceding benefice, possessed the title.

Barnahillis be Mr Johnne Thorntoun eldar, usufructuar thairof, for the sowme of' 400 merks 'usuale money of this realme', 400 merks.

'Sen this assedatioun wes maid thir teindis underwrittin ar reft and tane up be thir personis, viz.': 22 husbandlands of Ovir Ancrame 'reft and tane up be Robert Ker, everilk husband land payand' 12 f. of victual, namely, 6 f. meal, 5 [f.] bere 'and ane of quheit' extending in the whole to 66 b. The whole teinds of the Wodheid, the Skaw, the Gaw and the Bredlaw 'raift and tane up be the said Robert Ker' extending to 20 b. meal, 14 b. bere; 8 b. wheat, the whole extends to 42 b. The Manis of Nether Ancrame 'reft and tane up be Andro Ker, quhilk payis in teind' 4 b. meal; 7 b. bere; 4 b. wheat, the whole extends to 16 b.[71] The whole teinds of the Tronehill 'reft and tane up be Thomas Trumbill, quhilkis payit of teind' 8 b. meal; 4 b. bere, extending in the whole to 12 b.

'Summa of it that is reft and violentlie tane up extendis' 135 b. victual.

Signature: 'Ita est Ja[cobus] Thorntoun, teste manu propria.'

FUNGARTH, PREBEND OF DUNKELD CATHEDRAL[72] (NLS, fo 82r)
(SRO, Vol. a, fo 335r)

'The rentall of the prebendrie of Fongard produceit be Maister James McGill as factour for sir Thomas McGill, prebendar thairof, and now produceit be Mr James Thornetoun, ather of thame protestis that thair productionis suld nocht prejuge utheris.'

KINNELL, PARSONAGE OF, (NLS, fos 82v-83r)
(SRO, Vol. a, fos 335v-336r)

'The rentall of the fructis of the parroche kirk of Kynnell pertening to the personage with the names of the townes as followis.'

Cuggerglak, 7 f. bere; 5 b. 2 f. meal. Cokhill, 7 f. bere; 5 b. 3 f. meal. Rynmure, 4 [blank] bere; 10 b. meal. Turtestoun [i.e. Curtestoun], 2 [blank] bere; 7 b. meal. Caistoun, 2 [blank] bere; 7 b. meal. Eisterbroky, 3 b. wheat; 6 b. bere; 16 b. meal. Logyhill, 1 b. wheat; 2 b. bere; 4 b. meal. Newbiging, 2 b. bere; 8 b. meal. Westerbroky, 6 b. bere; 9 b. meal. Manys, 2 b. wheat; 4 b. bere; 12 b. meal. Haltoun, 4 b. 2 f. wheat; 10 b. 2 f. bere; 18 b. meal. Bannavis, 2 b. bere; 5 b. meal. Quhithill, 6 f. bere; 3 b. 2 f. meal. Kirktoun brewland, 1 b. wheat; 2 b. 2 f. [3 b. 2 f., *SRO*] bere; 5 b. 2 f. meal. Myltoun, 6 f. wheat; 3 b. 2 f. bere; 9 b. meal.

---

[71] The correct calculation appears to be 15 b.
[72] Cf NLS, fos 44v, 46v, 81v (see above, pp. 327, 329, 365 and n. 69).

Vicarland, 1 b. bere; 4 b. meal. Symmerhill, 2 b. bere; 6 b. meal. Auchanland, 3 b. bere; 13 b. meal. Murnyll, 6 f. bere; 2 b. meal. Sowterland, 1 f. bere; 2 f. meal. Brewlandnesse [Brewland of Rosse, *in 'Memorandum' below*], 'I haif na rentall thairof becaus it is conteinit undir the rentall of the Kirktoun'. My lord Ogilvy[73] for his Manis and lands within the par[ish of] Kynnell, 48 merks [*in margin*, 'silver'].

Totals: Wheat, 12 b.[74] Bere, 3 c. 2 f.[75] Meal, 9 c. 6 b.[76] Silver, £32.

'This is the rentall of Kinnell parroche kirk pertening to M[r] Patrik Liddell.'

Calculation of third: Wheat, 12 b.; third thereof, 4 b. Bere, 3 c. 11 b. 2 f.; third thereof, 1 c. 3 b. 3 f. 1½ p. Meal, 9 c. 6 b.; third thereof, 3 c. 2 b. Money, £32; third thereof, £10 13s 4d.

'Memorandum, to speir out the Brewland of Rosse [Brewlandnesse, *in 'Rental' above*] quhilk he allegeis to be conteinit under the Kirktoun.'

NEVAY, PARSONAGE OF,

(NLS, fo 83r-v)
(SRO, Vol. a, fo 336r-v)

'Rentale rectorie de Navay'

Rental of the parsonage of Navay in St Andrews diocese, by right of patronage annexed to the see of the archbishop by reason of the foundation, Mr George Swintoun lawfully provided to the parsonage of the same. Made at Edinbrucht, 20 January 1561/2 [1560, *SRO*].

Toun of Nava in teinds out of the east part set for money ferme yearly, 28 merks; and for grassums of three years, reckoning for each year an allocation of a sum extending to 28 merks.

The same toun, on the west part set for a money ferme yearly, £10 12s 8d; and for the grassums in the same contract, £4 14s, accounting for the three years foresaid.

Balthene: The toun of the same in teinds by tenants and occupiers, by reckoning of the contract set for 24 merks. And for grassum of the same contract by the reckoning of the three foresaid years, 30 merks.

Kirktoun of Navay and Balgrugo with the lands of Tempiltoun: In teinds according to the contract, 24 merks 'tantum'.

---

[73] James, 4th lord Ogilvy of Airlie, died before 2 January 1562 and was succeeded by his grandson James, 5th lord, who was served heir on 2 October 1563. *Scots Peerage*, i, pp. 117-119.
[74] The correct calculation appears to be 13 b.
[75] This calculation is difficult to verify owing to the omission of quantities of victual.
[76] The correct calculation appears to be 9 c. 6 b. 3 f.

Note that the reason why these teinds are leased along with the grassums on the basis of a triple lease is that thus they were accustomed to be leased by use and wont and are of greater profit.

Total annual sum in all fermes and grassums calculated together on the account of one year extending to 108 merks 2s 10d.

Signature: 'Ad hec Magister Georgius Swyntoun, rector eiusdem.'

Calculation of third: In the whole, £75 15s 1½d;[77] third thereof, £25 5s ⅓d [recte, ⅚d].

'Presentit be Maister Alexander Forrest.'

COUPAR ANGUS, ABBEY OF,[78]

(NLS, fos 83v-86v)
(SRO, Vol. a, fos 336v-339v)

Abbacy of Cowpar

'Efter followis the sowmes of the thrid parte of my lord of Couperis[79] abbacie to be intromettit with be the quenis majestie and hir Comptrollare[80] and that alsweill of fermes, teindis, hors corne as of silver and quhair the said thrid parte suld be tane up and out of quhat landis off the' 1561 'crope.'

'Item, the quenis majesteis thrid parte of silver extendis to' £412 18s 3d[81] to be taken up out of the lands following. Out of the temporal lands from the Reid Croce west with Campsay, £158 13s 4d [£158 14s 8d, *fo 84v*]; out of the lands of Kincreycht, Glenboy, Grange of Airlie, Clint Law, Auchindo[r]y, Litill Perth and half of Blakstoun, £95; out of the barony of Glenylay, £138 10s 8d; out of the lands and fishings of Awecht and Innerrychtny, £20; the annual of Gardin, 13s 4d.[82]

The queen's majesty's third of the ferme meal extends to 1 c. 4 b. 1⅓ p. to be taken up from Robert Baxter in Balmyle and Andrew Stibilis there.

---

[77] The correct calculation appears to be £104 13s 4d.
[78] Cf NLS, fos 65r, 109v (see above, p. 352; below, p. 409). In the *Rental Book of the Abbey of Cupar-Angus*, i, pp. 360-363, the following section, consisting of the assignation of the third, is rendered as an integral part of the rental, to be found in NLS, fos 65r-75r (see above, pp. 352-358). The figures do not correspond with those in that rental but relate to the rental which follows on fo 84v (see below, p. 369).
[79] See above, p. 352, n. 13.
[80] See above, p. 351, n. 5.
[81] This is one third of £1,238 14s 9d on fo 85r; cf below, p. 370, n. 91.
[82] This would total £412 17s 4d.

The queen's majesty's third part of ferme bere extends to 6 c. 13 b. 2⅓ p.,[83] to be taken up out of the lands following. From the tenants of Balmyle, 3 c. 12 b. 1 f.;[84] from the tenants of Cowbyre, 3 c. 3 f. ⅓ p.[85]

The queen's majesty's third part of teind wheat extends to 2 c. 9 b. 2 f. 3⅔ p. wheat[86] to be taken up as follows. From Thomas Turnbill of Bogmillin, 12 b.; from Robert Jaksoun, 11 b.; from Alexander Jaksoun, 11 b.; and the rest from Pautoun Henry extends to 7 b. 2 f. 3⅔ p.

The queen's majesty's third part of teind meal extends to 23 c. 2 b. 3 f. 2⅔ p.,[87] to be taken up out of the parish of Airlie.

The queen's majesty's third part of teind bere extends to 18 c. 6 b. 1 f. ⅓ p.,[88] to be taken up as follows. Out of the parish of Airlie, 11 c. 1 f. 2 p.; out of Cowty, 1 c.; Blaklaw and Wester Cotyardis, 14 b.; Ester Cotyardis, 7 b.; Nether Murtoun, 6 b.; Cheppeltoun, 13 b. 2 f.; Tullifergus, 12 b. 2 f. 2 p.; Grange of Abirbothre, 2 c. 8 b.; 9 b. 2⅓ p. bere from John Blair in Airthourstane.[89]

The queen's majesty's third of the oats and horsecorn extends to 8 c. 6 b. 3 f. 2⅔ p.,[90] to be taken up out of the lands of Airthourstane, Balmyle, Blaklaw, Cotyardis, Polcak, Murtowneis, Chapeltoun, Tullifergus and of the Grange 'quhill hir graces chamerlane be payit.'

The rental of the abbacy of Cowpar in money

The temporal lands thereof from the Reid Croce west with Campsy 'of auld penny maill', £158 14s 8d [£158 13s 4d, *fo 83v*]. The temporal lands thereof from the Reid Croce east 'of auld penny maill', £104 9s 4d. The lands thereof above the Wateris of Ylay and Areicht with Ledrassy, Perseis, Caillie, Drymmeis, Murthlie and Drumfathtie [Drumfallinthie, *fo 68v*], Tulloch and Innerwak in Atholie and Murthlie in Mar 'of auld penny maill', £274 10s 8d. The lands of Kyncreicht, Glenboy, Grange of Erlie, Clintlaw, Auchindory, Litill Perth and half Blakstoun 'of auld maill', £95. The lands of Carsgrange with the orchard thereof, £97 6s 8d. The barony of Glenylay 'of auld maill', £138 10s 8d. The lands and fishings of Alwecht and Innerrychtny, £20. The annuals of the towns of Dundie, Perth 'and uther places', £49 9s 5d.

Kirks set for silver conform to the old rental

The kirk of Alwecht, parsonage and vicarage set of old for £74. The kirk of Glenylay, parsonage and vicarage, £80. The kirk of Mathie, parsonage and

---

[83] This is one third of 20 c. 7 b. 1 f. 3 p. on fo 85v (see below, p. 370).
[84] This figure is given for meal on fo 85v (see below, p. 370).
[85] This would give 6 c. 13 b. ⅓ p.
[86] This is one third of 7 c. 13 b. 3 p. The teinds of Carsgrang are given as 7 c. 13 b. ⅓ p. wheat on fo 86r (see below, p. 370); 7 c. 13 b. 1½ p., on fo 71r (see above, p. 355).
[87] This is one third of 69 c. 8 b. 3 f. meal which occurs on fo 86r (see below, p. 370).
[88] This is one third of 55 c. 2 b. 3 f. 1 p. bere which occurs on fo 86r (see below, p. 370).
[89] This would give 18 c. 6 b. 2 f. 2⅓ p.
[90] This is one third of 25 c. 4 b. 3 f.; the figure given for oats on fo 86r is 4 c. 4 b. 3 f.; cf below, p. 371, n. 96

vicarage, 100 merks. 'Thre quarteris of the kirk of Fossoquhy, personage and vicarage', 100 merks. The teinds of Campsy and Voulschill [*recte*, Wolfhill], 20 merks. 'The thrid of the haill males, annuellis and kirkis set for silver', £1,238 14s 9d.[91] 'And this by the Medo, Cunynghar Thomesonis Park, Fergus Land for the keiping of the medo and yardis within the place.'

'The ordinare expenssis of money to be yeirlie deducit of the saidis fructis
Item, to the convent extending to nyntene bretherine for ane parte of thair sustentatioun usit and wont', £312 6s 8d. 'Item, to the baillie fie, porter and uther ordinar servandis usit of auld under the commoun seill for thair feis', £119 6s. 'Item, in pensionis yeirlie to the abbay of Restennett, chaplanis of Forfar, Carsgrange and vicarage of Errole', £10 6s 8d. 'Item, in pensioun to Johnne Scott confermet in Rome of auld', £50. 'Item, in pensioun to Maister James Thorntoun, confermit in Rome', £20. 'Item, of contributioun to the Lordis of the Sessioun', £35. 'Summa of the defalcationis of money', £546 19s 4d, 'and swa restis fre', £691 15s 5d, 'quhairof the ferd parte extendis to' £172 18s 10d.

The rental of the ferme victuals of the said abbacy
The toun of Balmyle in ferme, 3 c. 12 b. 1 f. meal. The said toun of Balmyle, Couper Grange, Cowbyre, Gallowray and 'certane akeris aboute the yett in ferme', 20 c. 7 b. 1 f. 3 p. bere. 'All the haill fermes foirsaidis sett in few for' 10 merks the chalder.

The rental of the teind victuals of the said abbacy
The kirk of Bennethy [*i.e.* Bendochy] 'by the teindis of the towne of Bennethy gevin in pensioun to sir Johnne Hummill and confirmit in Rome, and the gadderit teindis thairof', 37 c. 7 b. 3 f. 2 p. meal and 21 c. 8 b bere.[92] The kirk of Erlie [*i.e.* Airlie], 28 c. 4 b. 2 f. 2 p. meal and 11 b. 1 f. 2 p. bere. The gathered teinds of Bennethy extend to 4 c. 4 b. 3 f. oats. The teinds of the Carsgrang, 7 c. 13 b. 1⅓ p. wheat.[93] Total of the whole ferme and teind victuals of the abbacy of Cowper by year, 7 c. 13 b. 1⅓ p. wheat; 69 c. 8 b. 3 f. meal; 55 c. 2 b. 3 f. 1 p. bere; 4 c. 4 b. 3 f. oats.

'The ordinare expenssis of victuallis yeirlie to be deduceit of the haill rentall
Item, the said nyntene brethrene for thair sustentatioun brede and drink conforme to use and wont', 2 c. 6 b. 1 f. 3 p. wheat; 4 c. 2 b. 2 f. meal and 20 c. 3 b. bere. 'Item, to the portar, wrichtis and uther erand servandis for thair bollis undir the commoun seill', 12 c. 12 b. meal and 2 [c.] 8 b. bere. 'Summa of the defalcatioun of victuall': 2 c. 6 b. 1 f. 3 p. wheat; 16 c. 14 b. 2 f. meal; 22 [c.] 11 b. bere. 'And swa restis fre' 5 c. 6 b. 2 f. 2 p. wheat;[94] 52 c. 10 b. 1 f. meal;

---

[91] This is the whole, not one third; cf above, p. 368, n. 81.
[92] SRO, Vol. a, fo 338v reads '11 c. 1 f. 2 p.', which is probably a scribal slip.
[93] 7 c. 13 b. 1⅓ p., fo 71r (see above, p. 355).
[94] The correct calculation appears to be 5 c. 6 b. 2 f. 2⅓ p.

32 c. 7 b. 3 f. 1 p. bere; 'quhairof the ferd parte extendis to: off quheit, [*blank*], off meill, [*blank*], off beir, [*blank*], off aittis, [*blank*].'

'Nota', 11 c. 'hors corne, certane geis, caponis and pultre never sauld for money bot spendit to the sustentatioun of the place and hospitalitie.'

The third of the abbacy of Cowper
    The third of the silver extends to £412 18s 3d. The third of the ferme meal extends to 1 c. 4 b. 1⅓ p. to be taken from Robert Baxter and Andrew Stibbillis in Balmyle. The third of the ferme bere of landis extends to 6 c. 13 b. 2⅓ p. The third of the wheat which extends to 9 c. 9 b. 2 f. 3 p. 'and thrid parte of ane half pect' [i.e. ⅓ p.].[95] The third of the teind meal extends to 23 c. 2 b. 3 f. 2⅔ p. The third of the teind bere extends to 18 c. 6 b. 1 f. ⅓ p. The third of the oats and horsecorn extends to 8 c. 6 b. 3 f. 2⅔ p.[96]

FERN, PARSONAGE OF,[97]

(NLS, fo 87r)
(SRO, Vol. a, fo 340r)

'Unto your wisdomes presentlie convenit at this Generall Assembly humilie menis and schawis I your servitour, Mr Patrik Mure, persoun of Ferne, that quhair I obtenit be gift of our soverane lady in the yeir of God' 1561 'ane pentioun and for paiment thairof gat the haill fructis, baith of the personage and vicarage of the parochin of Ferne assignit to me, and be verteu thairof wes in possessioun of the haill fruictis of the samin sen my entrie without ony paiment of the thrid thairof to ony personis unto the yeir of God' 1567, 'in the quhilk yeir William Foulartoun, Collectour[98] of the thridis of the kirkis of Angus hes on his pretendit maner without just tryele or cognitioun in the caus put up in his comptis of the Chekker the thrid of the said benefice to the sowme of' £40 'yeirlie and thairwith hes burdanit and chargit me with the samin for the croppis and yeiris of God' 1567 and 1568 'quhan I never pait ane thrid furth of the samin befoir, and forder gif the samin aucht to be pait he hes be his estimatioun far exceidit the just thrid thairof, the haill fruictis being rychtlie considerit and that your wisdomes may understand that I am nocht prejudiciale to the rycht grantit to the sustening of the ministeris of the thrid of the benefice I am ane content that ane resonabill thrid of the said personage and vicarage be providit thairto at the sicht of the superintendent and Collectour sua that the samin be justlie estemit according to the valour of the said benefice, heirfoir I beseik your wisdomes that ye will ordane the superintendent and

---

[95] The total wheat is given as 7 c. 13 b. 1⅓ p. on fo 86r (see above, p. 370).
[96] Cf above, p. 369, n. 90.
[97] Cf NLS, fo 103r; EUL, fo 118r (see below, pp. 398, 641).
[98] William Fullarton of Ardo was sub-collector for Forfar and Kincardine, 1567 - 1569, *Thirds of Benefices*, p. xl.

Collectour to tak just triall of the haill fruictis of the said benefice and thairefter to modifie ane resonabill thrid to be tane furth of the samin according to the valour thairof and the samin salbe thankfullie pait as effeiris and your wisdomes answer humlie I beseik.'

9 March 1569/70

'Referris it to the superintendent and Collectour of Angus to try the just rentale and to asssume according thairto, sic subscribitur, James Gray subscripsit.'

11 March 1569/70

'The superintendent of Angus, with advise of Williame Foulartoun his Collectour hes, according to the desyr of the kirk, tryit the thrid of the benefice within writtin and findis the same to extend yerlie to' £20 'pundis, and this we testifie be our handis writtin. Sic subscribitur, Jhonne Erskin, William Foulartoun.'[99]

ECCLESJOHN, PARSONAGE AND VICARAGE OF, (NLS, fo 87v)
BRECHIN CATHEDRAL, (SRO, Vol. a, fo 340r)
CHAPLAINRY OF ST AGNES IN,

Rental of Eglisjohnne parsonage and vicarage in the diocese of Brechin and sheriffdom of Forfar worth 'be yeir' 20 merks with certain annuals in Newmontros extending to 5 merks.

The rental of Sanct Agnes chaplainry within the cathedral kirk of Brechen worth 'be yeir' 18 merks pertaining to John Farar. 'Ressavit at command of the Clerk of Registre.'[100]

Parsonage and vicarage of Eglisjohnne and Sanct Agnes chaplainry in the kirk of Brechin.

Calculation of third: In the whole, £16 13s 4d;[101] third thereof, £5 11s 1½d.

PANBRIDE, VICARAGE OF, (NLS, fo 87v)
(SRO, Vol. a, fo 340v)

'Neill Layng, factour to sir Williame Layng, of his benifice following.'

---

[99] John Erskine of Dun, superintendent of Angus and the Mearns. For Fullarton, see above, p. 371, n. 98; *BUK*, i, pp. 13, 52.

[100] James McGill of Nether Rankeilour was Clerk Register at the Reformation, *Handbook of British Chronology*, p. 197.

[101] This is merely the 25 merks specified in the rental for Ecclesjohn.

'Vicarage of Pambryde wes thre yeir syne promittit to Sir Robert Carnegy in assedatioun for yeirlie payment of 100 merks, 'thairoff is to be defalkit the corpresentis, umestclaythis, Paschefynes, conforme to the lait ordour takin be my Lordis of Secreit Counsale.'[102]

Signature: 'Neill Layng.'

Calculation of third: In the whole, £66 13s 4d; third thereof, £22 4s 5½d. [*In margin*, 'thrid thairof according to this reformatioun', £10.]

BRECHIN, CANONRY OF,

(NLS, fos 87v-88r)
(SRO, Vol. a, fos 340v, 342r[103])

'The channonry of Brechen callit the pensioun payis allanerlie yeirlie of maill' £16. 'This pensioun hes annexit to it furth of the vicarage of Brechen', 12 merks, 'quhairof he never gat penny nor yit possessioun sen he wes providit of the said channonrie.'

Signature: 'Neill Layng.'

Calculation of third: In the whole, £24; third thereof, £8.

LOGIE-MONTROSE, PARSONAGE OF,

(NLS, fo 88r-v)
(SRO, Vol. a, fo 342r-v)

'Rentale rectorie de Logymontros'

Rental of the parsonage of Logymontros in St Andrews diocese entirely in the collation and provision of the lord archbishop of St Andrews,[104] patron of the same, of which Maister Alexander Forrest is present parson of the same. Made at Linlythgow, 23 January 1561/2.

Teinds of the lands of Craigo set to Henry Fowlartoun, hereditary possessor, for annual payment of 28 merks. Teinds of the lands of Ardocht and Templum set to William Fowlartoun, hereditary possessor of the same, for annual payment of 13

---

[102] Such clerical exactions as the Easter offering at Communion services, the 'uppermost cloth' as a mortuary due and the 'corpse present', or funeral fine for unpaid teinds, were among the grievances presented to the provincial council of the pre-Reformation church in 1559, and were condemned as oppressive by the protestant reformers in their *Book of Discipline* in 1560. The lords who gave qualified approval to the *Book of Discipline* in January 1561 declared that exactions of this sort 'be cleane discharged, and no more taken in times comming'. *Statutes of the Scottish Church*, p. 158; *First Book of Discipline*, pp. 156-158.
[103] Fo 341r-v is blank.
[104] John Hamilton, archbishop of St Andrews, 1546 - 1571, Watt, *Fasti*, pp. 298-99.

merks 6s 8d. Teinds of Nether Craigo set to the said William Foulartoun, hereditary possessor of the same, for annual payment of 21 merks. Teinds of the kirklands of Logymontros set to my lord of Dun,[105] hereditary possessor, for annual payment of 44 merks. Teinds of the land of Galrin set to the tenants and occupiers of the same for annual payment of 80 merks. The vicarage of Logymontros, annexed to the same parsonage, in wool, lambs and certain other small teinds commonly paying 20 merks. Total, 215 merks 6s 8d.[106]

Of which annual sum is payable to Mr Patrick Liddell, pensioner of the same, by reason of provision in the Roman Curia, acquired by litigation, 50 merks, and so rests free 165 merks 6s 8d, from which sum I have to sustain the vicar and minister yearly, as I have sustained him annually, by order of the lord of Dun, administering until my present coming.

Signature: 'Alexander Forrester [sic, Forrest, SRO, Vol. a], rector hujusmodi.'

Calculation of third: In the whole, £143 13s 4d;[107] third thereof, £47 17s 9¾d.

MENMUIR, PARSONAGE OF, (NLS, fos 88v-89r)
(SRO, Vol. a, fos 342v-343r)
9 December 1550[108]

'Quo die Jacobus Hammyltoun, rector de Menmuir'

Which day James Hammyltoun, parson of Menmuir, approved and ratified as by the tenour of these presents he approves and ratifies a certain assedation of nineteen years of and concerning the teind sheaves of the parsonage of Menmuir made by himself with the consent of the dean and chapter of Dunkeld, made to the honourable man Robert Cullace de Balnomone and George Symmer, son and heir apparent of George Symmer of Balzeordie, conjointly and to their assignees and subtenants, jointly or severally as is set down at greater length in the said assedation of the date at the said chapter house, the ninth of December foresaid and insert in the register of the said chapter where it is indicated that the said James by his own proper confession for fulfilling, observing, keeping and maintaining the said assedation for the duration of the said nineteen years to the aforesaid Robert, George and their foresaid assignees and subtenants for himself and his successors under pain of excommunication with the raising of letters and the pains of infamy and of perjury on the

---

[105] I.e. John Erskine of Dun, cf above, p. 372, n. 99.
[106] The correct calculation appears to be 206 merks 6s 8d.
[107] This is equal to 215 merks 6s 8d,
[108] Both texts are dated 1550.

other hand the said Robert and George Symmer of Balzeordye were warned as the principal debtor for his said son together to pay and satisfy the said James and his successors, who for the time shall be, for the said teind sheaves, the annual sum of two hundred and sixty merks money as follows, namely, the said parson and his foresaid successors the annual sum of two hundred and forty merks of the said money at two terms in the year, namely, on the feast of St Martin[109] in winter and Pentecost[110] by equal portions, the first term's payment beginning on the feast of St Martin last and the remaining twenty merks in this way, namely, ten pounds to the stallar serving in the choir of Dunkeld for the time for the said James, four merks for the dean of Christianity in the parts of Angus and Mearns for procurations and synodals and 13s 4d for the expenses of visitations annually at the terms use and wont, paid termly under the same penalty. Witnesses, sir Andrew Philpstoun, chamberlain of Dunkeld, Walter Fairwadder and John Pypere, before which witnesses the said James of his own confession to relieve and keep free and indemnify the said Robert and George, and their assignees and subtenants from the charitable subsidy of bishops entering to the see of Dunkeld jointly or severally, of the king's taxations and pensions of the governor [i.e. the duke of Chatelherault] and other ordinary and extraordinary burdens whatsoever during the same period under the same penalty.

Signature: 'Mr James Hammyltoun, persone of Menmuir. Robert McNair, scriba, manu sua.'

Calculation of third: In the whole, £160; third thereof, £53 6s 8d.

ABERLEMNO, VICARAGE OF, (NLS, fo 89r)
(SRO, Vol. a, fo 343r)

Rental of the vicarage of Abirlemno given in by sir William Gardin, vicar thereof.

'Quhilk vicarage with teind fructis, umestclaithis, corspresentis, Paschefynes, oblationis and utheris emolumentis of the samin usit to haif bene payit of befoir wes last set in assedatioun to Johnne Alexander in Balgay for payment yeirlie to me of' £24, 'and to the curate in my name', 7 merks, 'and will now gif onlie being set', £16, 'corspresentis, umestclaithis, Paschefynes and oblationis being deduceit.'

Signature: 'Sir Williame Gardin, vicar of Abirlemno.'

Calculation of third: In the whole, £16; third thereof, £5 6s 8d.

---

[109] I.e. 11 November.
[110] I.e. seven weeks after Easter.

IDVIE, PARSONAGE OF, (NLS, fo 89r-v)
(SRO, Vol. a, fo 343r-v)

Rental of the teind sheaves of the parish of Idvy within the deanery of Angus pertaining to Mr William Hey, parson of the same.

Gardinie with the mill and pertinents, 21 b. meal; 8 b. bere; 1 b. wheat [*in margin in SRO*, 'the lard of Gardny']. Pressoik, 12 b. meal; 4 b. bere. Freok, 12 b. meal; 4 b. bere. Myddiltoun, 26 b. meal; 12 b. bere. Ligistoun, 21 b. meal; 6 b. bere. £50 money, 2 b. meal, 2 b. oats, 20 merks silver.[111] Richard Mecneile, Gask, 18 b. meal; 6 b. bere; £10 10s. Alexander Gardny, Battollo, 17 b. meal; 6 b. bere; 20 merks. James Ogilvy, Kynneris and the mill, 14 b. meal; 4 b. bere. Adscurry, 20 b. meal; 4 b. bere; £12. The Manes of Idvy, 48 b. meal; 12 b. bere; £20. Wester Petmowie, 14 b. meal; 5 b. bere [*in margin*, 'The tennentis']. Estir Petmowy, 15 b. meal; 5 b. bere [*in margin*, 'Sir James Guthrie']. The Myln of Petmowy, 3 b. meal; 1 b. bere. The Ailhoushill, 6 f. meal; 2 f. bere; £22 5s.

Signature: 'Presentit and producit', 21 February 1561/2 'be me, James Young.'

Calculation of third: In the whole, £141 8s 4d; third thereof, £47 2s 9½d.[112]

LUNDIE, PARSONAGE AND VICARAGE OF, (NLS, fos 89v-90r)
BALLUMBIE, CHAPEL OF, (SRO, Vol. a, fos 343v-344r)

Rental of the teind sheaves of the kirk of Lundy 'togidder with the vicarage and kirk gleib of the samyn lyand within the schirefdome of Anguis, gyffin in to the quenis majestie Comptrollar[113] be Marioun, lady Gray,[114] fermorar thairof, to Mr Nicoll Spittall, persoune of Lundy.'

The teind sheaves of Kirktoun of Lundy, £10. The teind sheaves of Pettermo and Dronlo, £16. The teind sheaves of Latcreiff, £6. The teind sheaves of the Nethir Smychtstoun, £7 6s 8d. The teind sheaves of Balchardo, £10. The teind sheaves of Pettendrech, £5 6s 8d. The teind sheaves of Over Smychtstoun, £12. The teind sheaves of Arguthe, £10. The teind sheaves of Wester Kethe, £6. The teind sheaves of Eister Kethe, £6. The teind sheaves of the Bowhous, £10. The teind sheaves of the Clus, £4. The teind sheaves of Petlyell and Myltree of Clus, £12. The vicarage of the parish of Lundy with the kirk glebe, £7 6s 8d.

---

[111] This portion of text is curiously arranged, with the last sentence separated from the rest by a vertical line.
[112] The various types of victual are not calculated in the third.
[113] See above, p. 351, n. 5.
[114] Patrick, 4th lord Gray, married Marion, daughter of James, 4th lord Ogilvy of Airlie, *Scots Peerage*, iv, pp. 280-281.

'Nota, the chapell of Ballumby is annexit and incorporate to the paroche kirk of Lundy and now in the handis of James Fothringtoun, persoun and vicar thairof, and suld answer for ingevin of the rentall of the samyn quhilkis extendis to' £16.

Signature: 'Presentit be me, George Gray.'

Calculation of third:
    Parsonage and vicarage of Lundy, in the whole, £134;[115] third thereof, £44 13s 4d.
    Chapel of Ballumby, in the whole, 40 merks [£16, *i.e.* 24 merks, *above*]; third thereof, 13 merks 4s 4d.

BRECHIN, TREASURERSHIP OF,   (NLS, fo 90r)
                                                               (SRO, Vol. a, fo 344r)

'The rentale of the thesaurie of Brechin'

Kethik, 40 merks. Bothers, 20 merks. Drummys, 30 merks. Auldany, 17 merks. Walkmyll of Kethik, 3 merks. Kynnell, £3.

Signature: 'James Hepburne.'

Calculation of third: In the whole, £76 6s 8d; third thereof, £25 8s 10⅔d.

DUNDEE KIRK, CHAPLAINRY IN,[116]   (NLS, fo 90v)
                                                                (SRO, Vol. a, fo 344v)

'Ane rentale pertening to Andro Gray for ane cheplanrie within the kirk of Dundie extending to' 18 merks.

Calculation of third: In the whole, £12; third thereof, £4.

Signature: 'Andro Gray.'

---

[115] The figures given above total £122.
[116] Cf NLS, fo 101r (see below, p. 395).

ABDIE, VICARAGE PORTIONARY OF,[117]

(NLS, fo 90v)
(SRO, Vol. a, fo 344v)

'The vicarage portionar of Ebdy pertenand to Johnne Simmer, quhilk vicarage wantis all teindis except the teind fruictis of the Newburgh, teind stirk, gus and grissis, the gleib thairof set in few for' 4 merks yearly.

NEWBURGH,[118] CHAPLAINRY OF ST CATHERINE IN,

(NLS, fo 90v)
(SRO, Vol. a, fo 344v)

The chaplainry of Sanct Katherinis kirk in the Newburgh pertains to the said John, extends to £16 'with the mans and yaird thairof.'

Signature: 'Johnne Symmer, with my hand.'

BRECHIN, VICARAGE OF,[119]

(NLS, fo 90v)
(SRO, Vol. a, fo 344v)

The vicarage of Brechin set in assedation to Mr John Cokburne in Brechin for payment yearly of 120 merks.

[*In margin*, 'Utherwyis gevin in and rentallit be David Watt heirefter and ressavit at the Comptrollares[120] command.']

Calculation of third: In the whole, £80; third thereof, £26 13s 4d.

KELIMORE, PARSONAGE OF, PREBEND OF BRECHIN CATHEDRAL,

(NLS, fo 90v)
(SRO, Vol. a, fo 344v)

[*In margin*, 'Prebendarie and personage of Kilmoir']

'Rentale rectorie de Kilmoir'

Rental of the parsonage of Kilmoir, prebend of Brechin, set to the lord Erskin[121] and the laird of Auldbar for £40 13s 4d, from which deducting for the fees of a stallar and a curate, £10.

---

[117] This benefice, in the sheriffdom of Fife, was appropriated to Lindores abbey, also in Fife; and was evidently submitted with rentals for Forfar.
[118] This benefice, in the sheriffdom of Fife, was evidently submitted with rentals for Forfar.
[119] Cf NLS, fo 105r (see below, p. 401).
[120] See above, p. 351, n. 5.
[121] John, 6th lord Erskine, earl of Mar, *Scots Peerage*, v, pp. 612-615.

Signature: 'Joannes Cokburne.'

Calculation of third: In the whole, £40 13s 4d; third thereof, £13 11s 1½d.

BRECHIN CATHEDRAL, CALDHAME, CHAPLAINRY OF,
(NLS, fo 91r)
(SRO, Vol. a, fo 345r)

Rental of the chaplainry of Caldhame pertaining to Matthew Hepburne, lying within the kirk of Brechin, gives yearly 40 merks.

Signature: 'James Hepburne.'

Calculation of third: In the whole, £26 13s 4d; third thereof, £8 17s 9½d.

KINGOLDRUM, VICARAGE OF,
(NLS, fo 91r)
(SRO, Vol. a, fo 345r)

The vicarage of Kincaldrowne lying within the diocese of Brechin, possessor of the same Mr David Halyburtoun, provost of Methven, gives yearly in assedation £40.

Signature: 'Presentit be me, Maister Johnne Dumbar.'

Calculation of third: In the whole, £40; third thereof, £13 6s 8d.

EDZELL, PARSONAGE OF,
(NLS, fo 91r)
(SRO, Vol. a, fo 345r)

'Rentale rectorie de Edzell'

Rental of the parsonage of Edzell pertaining to Mr John Foulis, which parsonage is set to the earl of Craufurd[122] for annual payment of £120, as is laid down in the said assedation.

Signature: 'Thomas Sinclar, de mandato dicti rectoris, manu sua.'

Calculation of third: In the whole, £120; third thereof, £40.

---

[122] David Lindsay, 10th earl of Crawford, *Scots Peerage*, iii, p. 29.

ARBROATH, CHAPLAINRY OF ST DUTHAC IN,   (NLS, fo 91r)
(SRO, Vol. a, fo 345r)

Rental of Sanct Duthois in Arbrothok extends to 20 merks, pertaining to sir William Pettillok.

Calculation of third: In the whole, 20 merks; third thereof, 6 merks 8s 8⅔d.[123]

STRACATHRO, VICARAGE OF,   (NLS, fo 91r)
(SRO, Vol. a, fo 345r)

Rental of the vicarage of Strachathro pertaining to Mr John Guthre, vicar thereof, 'quhilk payit yeirlie to me sett in assedatioun to William Huntare, Pasche fynes, umaistclathis and corps presentis beand payit', £20. 'The saidis umastclathis, Pasche fynes and corps presandis being deducit the said vicarage is' 24 merks 'and wantis v yeiris payment.'

Signature: 'Mr Johnne Guthre, vicar Strachathro, wyth my hand. Ressavit at command of the Clerk of Register.'

BRECHIN, ARCHDEACONRY OF,   (NLS, fo 91v)
(SRO, Vol. a, fo 345v)

Rental of the 'archdenrie' of Brechin

'In primis the kirk of Straquhain [*i.e.* Strachan] set in assedatioun to Robert Keith, or parsonage thairof, for' £120. The vicarage for £26 13s 4d. The steading of Drummis set for £30. 'The haill sowme', £176 13s 4d.

'Heirof gevin yeirlie to the ministeris', £26, 'sua restis fre' £100.

Signature: 'Presentit be me, Robert Pitcarne,[124] in name of the said archidene.'

Calculation of third: In the whole, £176 13s 4d; third thereof, £58 17s 9⅓d.

'Nota, this archedenrie pais victuale yeirlie and is not sett, sua of the silver by the Drummis is' £48 17s 9⅓d.

'Lettres to be direct to charge the tennentis to answer the haill of the fermes and he to haif' £20 'as for the' £30 'gevin up be him thairfoir.'

---

[123] The correct calculation appears to be 6 merks 8s 10⅔d.
[124] Robert Pitcairn had a claim to the archdeaconry of St Andrews, 1539-1584. James Pitcairn was archdeacon of Brechin at the Reformation. Watt, *Fasti*, pp. 308-309; p. 56.

DUN, VICARAGE OF,[125]

(NLS, fo 91v)
(SRO, Vol. a, fo 345v)

Rental of the vicarage of Doune 'quhen all dewties was payit in the auld maner was' £40, 'and now the Pasche fynes, umestclaithis and offerandis being deduceit extendis bot to' £20 'or thairby.'

Signature: 'Mr Robert Auchmowty.'

DUNDEE, CHAPLAINRY OF THREE KINGS' ALTAR IN,

(NLS, fo 92r)
(SRO, Vol. a, fo 346r)

[*In margin*, 'The thrie kingis of Culaneis[126] alter in Dundie.']

'Item, the thre kingis of Culaneis fundatioun scituate within the paroche kirk of Dundie extendis to' £30.

Signature: 'Mr Gilbert Oslar, with my hand.'

Calculation of third: In the whole, £30; third thereof, £10.

FINAVON, PARSONAGE OF,
INVERARITY, PARSONAGE OF,

(NLS, fo 92r)
(SRO, Vol. a, fo 346r)

'This is the rentall of the benifices underwrittin, that is to say':

'The personage of Fynnevin within the diocy of Brechin, ane lawit patronage allanerlie at the gift of the erle of Craufurde, pertening now presentlie to Maister Hew Lindesay and set in tak to the said erle for the yeirlie payment of' £40.

'The personage of Inneraritie within the diocie of Sanctandrois, at the presentatioun of the bischop of Sanctandrois[127] and the erle of Craufurde alternatis vicibus [*i.e.* alternately], pertening now presentlie to the said Maister Hew Lyndesay and sett in tak to the said erle for the yeirlie payment and proffet of' £40.

Signature: 'Erle of Crawfurde.'

---

[125] Cf NLS, fos 99r, 101v (see below, pp. 393, 396).
[126] I.e. the altar dedicated to the Three Kings of Cologne, A. Maxwell, *Old Dundee, ecclesiastical, burghal and social, prior to the Reformation* (Edinburgh, 1891), p. 33.
[127] John Hamilton, archbishop of St Andrews, 1546 - 1571, Watt, *Fasti*, pp. 298-299.

Calculation of third:
   Parsonage of Fynnevin, in the whole, £40; third thereof, £13 6s 8d.
   Parsonage of Inneraritie, in the whole, £40; third thereof, £13 6s 8d.

'But prejudice of the auld rentall and superplus becaus thair takkis ar sett sen the' 48 'yeir at the leist in diminutioun of the rentall.'

NEWDOSK, PARSONAGE OF, (NLS, fo 92v)
(SRO, Vol. a, fo 346v)

'The rentall of the benifice of Newdosk being in my possessioun, Maister Williame Chalmer, persoun thairof, for the cropeis of' 59, 60 and 61[128] 'yeiris conforme to my provisioun had thairupoun. The same suld pay yeirlie to me' £40 'monete, with the ministeris fie, deneis visitationes, procuratiounis sinodaill and wyne and walx usit for the tyme and this conforme to ane assedatioun maid thairof be sir Adame Kingorne ultimum possessorem eiusdem pro toto tempore vite sue [i.e. last possessor of the same for the whole of his life] to the erle of Crawfurd last departit and to his airis, quhairof I haif never ressavit bot' £30 'as yit and thairof hes gevin and servit thairfoir the tyme' £12, 'praying your lordschip caus my lady Crawfurd[129] as tutrix testamentare to hir sone and onlie intromettour thairwith pay me but partes of the rest for my substance will nocht serve to pley hir thairfoir.'

Signature: 'Williame Chalmer, rector de Newdosk.'

Calculation of third: In the whole, £40; third thereof, £13 6s 8d.

COOKSTON, PARSONAGE AND VICARAGE OF, (NLS, fo 92v)
(SRO, Vol. a, 346v)

Rental of the parsonage and vicarage of Cuikistoun lying in the diocese of Sanctandrois and sheriffdom of Forfar pertaining to sir Walter Grahame set in assedation to Sir Robert Carnegy of Kinnard for yearly payment of £20, 'quhairof to be deducit to the minister', £10.

Calculation of third: In the whole, £20; third thereof, £6 13s 4d.

---

[128] The SRO text reads 'lviij, lix, lx and lxi yeiris'.
[129] David Lindsay, 9th earl of Crawford, died in September 1558 and was survived by his second wife, Catherine (daughter of John Campbell of Calder) who died in 1578. Their eldest son, David (born c. 1527), who succeeded as 10th earl, was clearly not a minor in 1558, and the presumption is that the lease of Newdosk was inherited by one of the younger sons. *Scots Peerage*, iii, pp. 26-29.

BRECHIN, BISHOPRIC OF,[130]   (SRO, Vol. a, fos 347r-349v)

'Heir followis the rentall of the bischoprik [of] Brechin, of the victuall thairof justlie calculat as the samyn is now of valour and payit be the tennentis, viz., the teindis of the samin, by the temporall landis, fermes and sylver thairof, maid at Edinbrucht, the [blank].'

Teinds
Kindroquhat, 7 b. bere; 22 b. meal. Ralnabreche, [recte, Balnabriech] 1 c. bere; 2 c. 8 b. meal. Mekill Walte[r]stoun, 1 c. bere; 2 c. meal. Petforky, 5 b. 2 f. bere; 12 b. meal. Furdtoun, Littill Waterstoun and Bowhillok, 10 b. bere; 17 b. meal. Fendowrie, 4 b. bere; 8 b. meal. Braklo [Barclo, fo 352r; Bratho, fo 352v], 3 b. bere; 5 b. meal. Pendreche [i.e. Pittendreich], 6 b. bere; 12 b. meal. Petpullox, 1 c. bere; 2 c. meal. Auchnacarret, 2 b. bere; 3 b. meal. Littill Massondew, 2 b. bere; 3 b. meal. Mekill Massondew, 1 c. 6 b. bere; 1 c. 6 b. meal. Cukistoun, 4 b. 2 f. bere; 6 b. meal. Unthank, 4 b. bere; 7 b. meal. Acres of Brechen, 4 b. bere; 6 b. meal. Luchland, 1 c. 4 b. bere; 40 b. meal. Uver Kincraig, 12 b. bere; 1 c. 8 b. meal. Nether Kyncraig, 20 b. bere; 2 c. 4 b. [meal]. Arrot with the mill and Lychtounhill, 1 c. bere; 2 c. meal. Chepeltoun, 2 b. bere; 3 b. meal. Esauxtoun [i.e. Isaacstown], 4 b. bere; 6 b. meal. Newbiggyne, 6 b. bere; 14 b. meal. Tayok, 8 b. meal; 4 b. bere. Pettoskell, 14 b. bere; 26 b. meal. Talgaty [i.e. Dalgety], 2 b. bere; 4 b. meal. Kynneid and Balnamone, 10 b. bere; 18 b. meal. Acres of Montrois, 1 c. bere. Borrowfeld, 1 c. bere; 1 c. 2 b. meal. Half Hadderweik, 4 b. bere; 11 b. meal. Charltoun, 3 b. bere; 5 b. meal. Maretoun, 15 b. bere; 1 c. 14 b. meal; 3 b. wheat. Forderis [Forres, fo 351v], 1 c. 10 b. bere; 1 c. 12 b. meal; 8 b. wheat. Auld Montrois, 3 c. bere; 3 c. meal. Fullartoun, 8 b. bere; 1 c. meal. Anaine, 4 b. bere; 8 b. meal.

'Assumit in David Murrayis petitioun,[131] the sy[l]ver teindis of the bischoprik off Brechen': Kyngorny, £4. Half Hadderweik, £10. The Mainis of Catterlynge and Hiltoun [Hirltoun, fo 351r], £24. The acres of Montrois, £3 10s. Total, £41 10s.

'Assumit in David Murrayis petitioun extending to the rest of sylver to the bischope[132] and thrid.'
The rental of the temporal lands: Barrowny, Charles Dempster, £24. Newtoun, George Crawmond, 16 merks and 10s. Smyddehill, Charles Murray, 7 merks and 40s. Myldene, Walter Cullace, £8 6s 8d. Botheris, Robert Kynneir, £6. Syid, George Barclay, 19 merks. Nether Petforthie, Mr John Cokburne, 40 merks. Ardo set in feu to the Justice Clerk,[133] 25 merks and 7s 4d. Aidecat set to

---

[130] Cf SRO, Vol. a, fos 350r, 352v (see below, pp. 385, 387). This rental of the bishopric of Brechin does not appear in the NLS text.
[131] See RSS, v, nos 893, 1865; and below, p. 389.
[132] Alexander Campbell succeeded John Sinclair as bishop of Brechin in 1566, Watt, Fasti, p. 41.
[133] Sir John Bellenden of Auchnoule was Justice Clerk at the Reformation, RSS, v, no 589.

John Carnegy, £6. Egglisjohnne set to the laird of Downe [*i.e.* Dun],[134] £6. Capo set to the laird of Ullischewin, 50s. Maritoun set to my lord Argyill in feu for £49 11s 4d. Mylne of Auchdowy, George Speid, 54s 4d. Unthank, the lady Crawfurd, £3 6s 8d. Littill Dalgaty set for 26s 8d. Murtoun set for £3 3s 4d. Kethik set for £18. Kyngorny set for £18. Kirkdauche set for £6 6s 8d. Croftheddis set for £5. Wver Petforthie, £10. Littill Mylne of Brechen set for £3. Ballewny [Ballnay, *fo 353v*], one half set to John Kynneir for £3 13s 4d. The other half to William Lowsoun for £5 6s 8d. Drummy set for £10. Brauthinche set for £17 6s 8d. Walk Mylne of Stracathro set for 53s 4d. Ester Drummis set for £40 10s. The brewhouse of Stracathro, 24s. Eisauptoun [*i.e.* Isaacstown] set to my lord Argyill in feu for £10 and 7 merks.

'The haill rentall of the spiritualitie of Brechen particularlie as it standis in the buikis, off beir, meill and quheit to be deducit the 2 pairt frome the thrid, conforme to the particulare assignatioun gevin in be Mr Johnne Huttoun, chalmerlane of the said bischoprik, under his subscriptioun manuall as the said assumptioun efter followand mair fullely proportis.'

The sum of the whole bere of the said bishopric extends to 23 c. 2 b. Two thirds thereof extends to 14 c. 12 b. Third thereof extends to 7 c. 6 b.[135] 'Assignit to pay the thrid as followis of beir': Luchland, 1 c. 4 b.; Arrot with the mill and Lychtounhill, 1 c.; Newbiggin, 6 b. Fullartoun, 8 b.; Auld Montrois, 3 c.; acres of Montrois, 1 c.; Tayok, 4 b.

The whole teind meal of the bishopric of Brechen 'by the temporall landis' extends to 38 c. 2 b. Two thirds thereof extends to 25 c. 6 b. 2 f. 2$\frac{1}{3}$ p.[136] Third extends to 12 c. 11 b. 1 f. 1$\frac{1}{3}$ p. 'Assignit to pay this thrid as followis': Luchland, 2 c. 8 b.; Arrot with the mill and Lychtoun[hill], 2 c.; Newbiggin, 12 b.; Fullartoun, 1 c.; Auld Montrois, 3 c.; Tayok, 8 b.; Forderis, 1 c. 12 b.; Nether Kyncrage from George Crawmond and John Wilsoun, 1 c. 2 b.; and to be taken out of Chepeltoun, 1 b. 1 f. 1$\frac{1}{3}$ p.

The rental of the wheat of the spirituality of Brechen. The whole teind wheat extends to 11 b. Two thirds thereof extends to 7 b. 1 f. 1$\frac{1}{3}$ p. Third thereof extends to 3 b. 2 f. 2$\frac{2}{3}$ p. 'Assignit to pay this thrid as followis': To be taken out of Foderis, 3 b. 2 f. 2$\frac{2}{3}$ p.; 'The tua pairt quhiyt of Foderis, occupyit be Johnne Hill, Johnne Rany, Thomas Robertsoune, and James Smyth, extendis to' 4 b. 1 f. 1$\frac{1}{3}$ p.

The whole horsecorn[137] of the spirituality of Brechin extends to 15 b. The third thereof extends to 5 b. 'Assignit for payment thairof as followis: Furth of Stracathro', 4 b.; 'Furth of the Murtoun', 1 b.

'The names of the townes that payis mailles and teindis by David Murrayis assignatioune ar thir, viz.': Maretoun, £49 9s 4d. Croftheddis, £5. Esauxtoun, 22

[134] I.e. John Erskine of Dun, cf above, p. 372, n. 99.

[135] 14 c. 12 b. plus 7 c. 6 b. equals 22 c. 2 b.

[136] The correct calculation appears to be 25 c. 6 b. 2 f. 2$\frac{2}{3}$ p.

[137] There are no figures given above for horsecorn or oats.

merks. Acres of Montrois, £3 10s. Total, £72 14s 8d.[138] Two thirds thereof extends to £48 9s 9⅔d.[139] Third thereof extends to £24 4s 8⅓d.[140]

The Mainis of Fernell with the pendicles, mill and Croftheddis set in feu to the earl of Argill[141] for 50 merks 'quhilk is dischairgit to him in his few chartour for his bailyie fye and sua nathing to be payit of that nather to thrid nor tua pairt.'
The whole salmon fishing of Brechin extends to 3 barrels, two of them set in feu to my lord Argill for 20s the barrel; another barrel set to Captane Lawder[142] for 20s 'as the said erle and captanis chartouris maid thairupoun mair fullely proportis.'

Signature: 'Mr Johnne Hutoun, chalmerlane of Brechen, with my hand.'

BRECHIN, BISHOPRIC OF,[143] (SRO, Vol. a, fos 350r-352r)

The rental of the bishopric of Brechen pertaining to Alexander Campbell, bishop thereof.[144]

The temporal lands of the said bishopric: Barony, £24. Newtoun, £10 13s 4d; and in augmentation, 10s; 12 capons; 16 poultry; 3 geese. Smiddehill, £6 13s 4d; 6 capons; 12 poultry. Mylden, £8 6s 8d; 2 swine 'or els xx s for ilk peace thairof'; 12 capons; 12 poultry. Betharis, £6; 12 capons; 12 poultry; 3 geese. Syid, £12 13s 4d; 12 capons; 12 poultry. Nether Pitforthie, £26 13s 4d; 12 capons; 12 poultry. Ardo, £17 16d; 4 b. corn. Adecat, £6; 16 poultry. Ecclisjone, £6. Capo, 40s; 6 capons; 6 poultry. Maritoun with the fishing, 7 c. victual 'or els ten merkis for the ch[alder] and for the few aikeris' 18s; 6 dozen poultry [or] 4d the piece; 2 barrels of salmon or 20s the barell; and 6s 8d in augmentation; 'inde in the haill', £51 2s. Myladewy, 53s 4d. Unthank, £3 6s 8d. Littill Dalgaty, 26s 8d; and in augmentation, 3s 4d; 6 capons. Muirtoun, £3 3s 4d; 6 capons or 8d the piece; 10 poultry or 4d the piece; 1 b. horsecorn or 3s 4d the piece; and 3s for 'arrage and carrage addettit of auld; inde in the haill', £4 18d.[145] Kethik, one half thereof pays £9 6s 8d; and in augmentation, 13s 4d; 28 poultry; 12 capons. The other half of Kethik, £7 6s 8d; and in augmentation; 13s 4d; 12 capons; 32 poultry; 6 geese; 2 b. horsecorn. Kirkdauche, £6 13s 4d. Craftheidis, £5; and in augmentation, 3s 4d; 24 poultry or 4d the piece. Over Pitforthe, £10; 12 capons; 12 poultry. Littill

---

[138] The correct calculation appears to be £72 12s 8d.
[139] The correct calculation appears to be £48 9s 9¼d.
[140] The correct calculation appears to be £24 4s 10⅔d.
[141] See above, p. 355, n. 32.
[142] I.e. Robert Lauder, *RMS*, iv, no 1745.
[143] Cf SRO, Vol. a, fos 347r, 352v (see above, p. 383; below, p. 387). This rental of the bishopric of Brechin does not appear in the NLS text.
[144] See above, p. 383, n. 132.
[145] The correct calculation appears to be £3 17s.

Mylne of Brechen, 56s 4d; and in augmentation, 3s 4d. Balony, the one half thereof, £3 6s 8d; and 6s 8d in augmentation; 6 capons; 3 geese; 10 poultry; 1 b. horsecorn. The other half thereof, £5 16s 8d; 6 capons; 3 geese; 10 poultry; 1 b. horsecorn. Drummy, £10; 12 capons. Brathinis, £17 6s 8d; 4 b. horsecorn or 5s the boll; 12 geese or 10d the goose; 24 capons or 10d the piece; 68 poultry or 6d the piece; 'inde in the haill', £21 16s 8d.[146] Eister Drummis, 6 c. victual or 10 merks the chalder; and in augmentation, 10s; 3 dozen poultry or 4d the piece; 'inde' £10 2s.[147] Isaxtoun, £4 13s 4d; 1 c. 8 b. victual or 10 merks the chalder; 34 poultry or 4d the piece; and in augmentation, 3s 4d; 'inde in the haill', £15 8s.[148] Stracathro, £17 6s 8d; the brewhouse thereof, 24s; and in augmentation, 32d; the Walkmylne thereof, 53s 3d. 'The customes, cainis, salmound fischeing, arrage, carrage, horscorne and uther dewties payis' £9 3s 4d; 'inde in the haill', £30 10s.[149] Kingorny, £18. The annuals of the town of Brechin, £11. 'The custome firlot sylver', 26s 8d. The annuals out of the kirk of Glenyly, 53s 8d. 'The laird of Baandrowis annual out of Benhein', 40s. 'The pensioun of the toun of Montrois awant to thame yeirlie in Chekker', £12 5s 6d.

'Summa of the haill temporall landis': Money, £410 5s. Capons, 11 dozen and 6. Poultry, 15 dozen and 10. Geese, 18. Horsecorn, 8 b. 'Off the quhilk soume thair is to be deducit David Murrayis pensioun quhairunto he is provydit of auld out of the temporall landis', £333 6s 8d. To Mr William Huttoun in pension, £10. Total, £343 6s 8d. 'Sua restis of temporall landis', £66 18s 4d.

'Memorandum, the Manis of Fernewell in few to the erle of Argill with the ballerie and the dewty dischairging his ballie fie extending to' £26 13s 4d.

'The tounis and landis set for sylver.'

The parish of Catterling: Hirltoun [Hiltoun, *fos 347v, 354r*] and Catterling, £24. Kingorny, £4.

Montrois: 'The schaddow half of Hadderwik', £10. 'The sonny half thairof', £14 3s 4d. The teinds of Auld Montrois, £32. 'The new aikeris of Montrois', £3 6s 8d. 'The half fiche butis', 30s.

Brechen: Over Kincraigye, £18. Leuchland, £40. Arrot with 'the half myln', Lichtounhill and Pittindreich, £40. The acres of Brechen and Mekill Massonden, £17 13s 4d. Quarter Waterstoun, £8. Findowry, £10. Unthank, £7 6s 8d. 'Thrie barellis of salmond, thairof tua barellis set for' £11 6s 8d, so rests 1 barrel. Total of teinds set for silver, £241 6s 8d.

---

[146] The correct calculation appears to be £21 10s 8d.
[147] The correct calculation appears to be £41 2s.
[148] This calculation is correct.
[149] The correct calculation appears to be £29 9s 11d.

The rental of the teind victual.

Montrois: Borrowfeild, 1 c. bere, 2 b. meal. Acres of Montrois, 1 c. bere. Charltoun, 3 b. bere, 6 b. meal.

The parish of Dwn [*i.e.* Dun]: Forres [Forderis, *fo 352v*], 1 c. 4 b. bere; 1 c. 12 b. meal; 8 b. wheat. Tayok, 6 b. bere; 8 b. meal.

Maritoun: The teinds of the toun of Maritoun, 1 c. 2 b. bere; 1 c. 11 b. meal; 3 b. wheat. Foullartoun, 8 b. bere; 1 c. meal. Annain, 4 b. bere; 8 b. meal. Arrottis 'half mylne', 2 b. bere; 4 b. meal. Magdalen Chappeltoun, 2 b. bere; 3 b. meal. Newbiging, 6 b. bere; 14 b. meal.

Parish of Brechin: Pentoscall, 14 b. Anet, 1 c. 10 b. meal. Kynnaird, 10 b. bere; 1 c. 2 b. meal. Pitpollox, 1 c. bere; 2 c. meal. Barclo [Braklo, *fo 347r*; Bratho, *fo 352v*], 3 b. bere; 6 b. meal. Auchnacarocht, 2 b. bere; 4 b. meal. Isaxtoun, 4 b. bere; 6 b. meal. 'Haill Pitforkeis', 5 b. 2 f. bere; 12 b. meal. Nether Kincragy, 1 c. bere; 2 c. 8 b. meal. Mekill Waterstoun, 1 c. 8 b. meal; 12 b. bere. Furdtoun, Littill Waterstoun and Bowhillok, 9 b. bere; 1 c. 3 b. meal. Mekill Dalgaty, 2 b. bere; 4 b. meal. Littill Massonden, 2 b. bere; 3 b. meal. Kindrochat, 7 b. bere; 1 c. 6 b. meal. Ballnabreich, 1 c. bere; 2 c. 8 b. meal. Cuikstoun, 4 b. 2 f. bere; 6 b. meal. Dubtoun, 6 b. bere; 12 b. meal.

'Summa of the haill teind victuall foirsaid', 11 b. wheat; 14 c. 6 b. bere; 25 c. 5 b. meal; of which is given out in pension to Mr John Grahame the teinds of Kindrochat and Ba[l]nabrech extending to 1 c. 7 b. bere; 2 c. 14 b. meal; and to John Leslie in Brechin the teinds of Cuikstoun and Dubtoun[150] extending to 10 b. 2 f. bere; 1 c. 2 b. meal.

So rests free in wheat, 11 b.; in bere, 12 c. 4 b. 2 f.; meal, 20 c. 5 b.

BRECHIN, BISHOPRIC OF,[151] (SRO, Vol. a, fos 352v-354v[152])
(NLS, fos 93r-96v[153])

'Heir followis the rentall of the bischoprik of Brechin bayth spiritualitie and temporalitie as it is in use of payment sen the entres off Alexander, now bischope of Brechin, and giffin up be his brother James Campbell off Ardkynglen [*recte*,

---

[150] 'Unthank and Dubtoun giffin in pentioun of auld to Johnne Leslie out of rentall', fo 353r; cf below, p. 388, n. 155.
[151] Cf SRO, Vol. a, fos 347r, 350r (see above, pp. 383, 385).
[152] SRO, Vol. a has been used as the main text for this rental of the bishopric of Brechin; the NLS text is almost illegible here.
[153] Fo 94v is blank; fo 96v is blank except for the following 'Rentale of the bischoprik of Brechin'.

Ardkinglas], himself being in Genewy at the scullis.[154] At Edinbrucht' 28 January 1573/4.

The teind victual within the parish of Montrois: Burrowfeild, 1 c. bere; 1 c. 2 b. meal. Half Hadderwik, 4 b. bere; 11 b. meal. Acres of Montrois, 1 c. bere. Charltoun, 4 b. bere; 11 b. meal.
    The parish of Dun: Forderis, 1 c. 4 b. bere; 1 c. 8 b. meal; 8 b. wheat. Tayok, 4 b. bere; 10 b. meal.
    The parish of Maritoun: The teinds of the toun of Maritoun, 18 b. bere; 27 b. meal; 3 b. wheat. Fullertoun, 8 b. bere; 1 c. meal. Annane, 4 b. bere; 8 b. meal. Auld Montrois, 2 c. bere; 4 c. meal.
    The parish of Brechin: Leuchland, 1 c. 4 b. bere; 2 c. 8 b. meal. Arrot [Chapeltoun, *deleted*] with the mill and Lychtounhill, 1 c. bere; 2 c. meal. Chapeltoun, 2 b. bere; 4 b. meal. Newbiging, 6 b. bere; 14 b. meal. Pentoskyll, 14 b. bere; 26 b. meal. Kynnard, 10 b. bere; 18 b. meal. Pytpollox, 1 c. bere; 8 b. bere. Bratho [Braklo, *fo 347r*; Barclo, *fo 352r*], 3 b. bere; 5 b. meal. Auchnakarrat, 2 b. bere; 4 b. meal. The acres of Brechen in the bishop's hand, 4 b. bere; 6 b. meal. Cuikstoun in the bishop's hand, 4 b. 2 f. bere; 6 b. meal. Kyndrocat in the bishop's hands, 7 b. bere; 22 b. meal. Mekill Messindew in the bishop's hands, 22 b. bere; 22 b. meal. Pittindreicht, 6 b. bere; 12 b. meal. Esauxtoun, 4 b. bere; 6 b. meal. Balnabreicht, 1 c. bere; 2 c. 8 b. meal. 'Haill Pytforkeis', 5 b. 2 f. bere; 12 b. meal. Nather Kincraig, 1 c. bere; 2 c. 8 b. meal. 'Haill Mekill Walterstoun', 1 c. bere; 2 c. meal. Furdtoun, Littill Walterstoun and Bowhillok, 9 b. bere; 19 b. meal. Mekill Dagathe [*i.e.* Dalgety], 2 b. bere; 4 b. meal. Fyndoirtye, 4 b. bere; 8 b. meal. Litill Messyndew, 2 b. bere; 3 b. meal. 'Unthank and Dubtoun giffin in pentioun of auld to Johnne Leslie out of rentall', 9 b. bere; 18 b. meal.[155]

Teind wheat of Brechen
    Within the parish of Down, 8 b. out of Forderis. Within the parish of Marytoun, 3 b. out of Mar[y]toun. 'The thrid extendis to' 3 b. 2 f. 2⅔ p.

Horsecorn of Brechen, 12 b.; the third, 4 b. to be taken out of Strethcathro [*i.e.* Stracathro], 'according to the auld assumptioun.'

The rental of the temporal lands of the bishopric of Brechin
    Bawrony, £24. Newtoun, £11 3s 4d. Smyddehill, £6 13s 4d. Myldene, £8 6s 8d. Bothiris, £6. Syid, £12 13s 4d. Nether Pitforthye, £26 13s 4d. Ardo, £17 16d. Addicat, £6. Ecglisjohnne, £6. Capo, 40s. Maritoun with the fishing £48 15s 4d. Myll Audevye [Auchdowy, *fo 348r*], 54s 4d. Unthank, £3 6s 8d.

---

[154] Alexander Campbell who, as a minor, had received the bishopric of Brechin from Queen Mary in 1566, returned from Geneva with Andrew Melville in 1574, Watt, *Fasti*, p. 41; *The Autobiography and Diary of Mr James Melvill*, ed. R. Pitcairn (Ediburgh, 1842), pp. 42-43, 47.
[155] 'To Johnne Leslie in Brechin the teindis of Cuikstoun and Dubtoun', fo 352r; cf above, p. 387, n. 150.

Litill Dagathe [*i.e.* Dalgety], 26s 8d. Murtoun, £3 3s 4d. Kethik, £18. Kyngorny, £18. Kirkdawcht, £6 6s 8d. Croftheidis, £5. Wver Pitforthy, £10. Littill Mylne of Brechine, £3. Ballnay [Ballewny, *fo 348r*], £9. Drymme [Drummy, *fo 348r*], £10. Brathynsche, £17 6s 8d. Walk Mylne of Strathcathro, 53s 4d. Ester Drummis, £40 10s. Brewhouse of Strathcathro, 24s. Strethcathro, £16. Eisauxtoun, £14 13s 4d.

The Mainis of Fernvell in feu to the earl of Argyll 'with the balyarie and dischairgit in his balyie fie.' [*In margin*, 'Dischergit xl merkis.']

'Summa by Fernell', £357 10s 8d; the third extends to £119 3s 6d 'ob'; two thirds extends to £238 7s 1d 'ob'.[156]

The silver teinds within the parish of Kathirlein [*i.e.* Catterline]: Hiltoun [Hirltoun, *fo 351r*], £24. The Mainis of Kathirlein, Kyngorny, £4.
    Silver teind within the parish of Montrois: Half Haddowick, £10. [*In margin*, 'This half is the schaddow half of Haddirweik, seik Maritoun.']
    Silver teinds within the parish of Brechen: Wver Kyncraig, £18.
    Total silver teinds, £56. The third extends to £18 13s 4d. Two thirds extends to £37 6s 8d.

'Nota. Deduce of this rentall fyve hundreth merkis of pensioun tane furth off the temporall landis to David Murray, providit of auld befour the bischopis entres. Item, of pentioun to Johnne Leslie siclyk provydit', 27 b. victual. 'Item, fourtie merkis for Fernevell dischairgit in his balye fie as his infeftment beiris set in few to the erle of Argyll.'

'The auld assumptioun':
    Borrowfeild within the parish kirk of Montrois, 1 c. bere; 1 c. 2 b. meal. Half Hadderwik in the said parish, 4 b. bere; 11 b. meal. The acres of Montrois there, 1 c. bere. Charltoun, 4 b. bere; 11 b. meal.

Assumption within Dune:
    Forder, 1 c. 4 b. bere; 1 c. 8 b. meal. Tayok, 4 b. bere; 10 b. meal.

Assumption within the parish of Marytoun:
    Auld Montrois, 1 c. bere; 1 c. meal. Fullartoun, 8 b. bere; 1 c. meal.

'Assumptioun furth of the paroche of Brechin':
    Luchland, 1 c. 4 b. bere; 2 c. 8 b. meal. Arott with the mill and Lychtounhill, 2 c. meal. The quarter of Kincraig occupied by John Creychtoun, 4 b. bere; 10 b. meal.

---

[156] Instead of 'obol', the correct fractions are $\frac{1}{2}$ and $\frac{2}{3}$ respectively.

GUTHRIE COLLEGIATE KIRK, PREBENDS OF, (NLS, fo 97r)
(SRO, Vol. a, fo 355r)

'Prebendaries of Guthre'

'Maister James Straithauchan, person and provest, rentall' £47. David Arrott, vicar and readar, rental, £20. Sir Patrick Guthrie, prebendary of Langlandis in [*recte*, and] Hiltone pertaining to the same kirk, the rental, £20. 'Ane prebendarie of ten pundis out of Lytill Lowr possesst be James Guthrie.' Mr William Gardin, parson and vicar of Kirkbuddo and prebendar of Guthrie, the rental, 40 merks.

The rental of the prebend of Hiltoun given up by Alexander Guthrie of that Ilk at Hadingtoun, 13 January 1573/4. The prebend of Hiltoun and Langlandis of Guthrie set in tack by sir Patrick Guthrie, prebendary thereof, to Gabriel Guthrie for the yearly payment of £20.

Signature: 'Alexander Guthrie of that Ilk.'

At Edinbrught 17 January 1576/7.

KINNETTLES, VICARAGE OF,[157] (NLS, fo 97r)
(SRO, Vol. a, fo 355r)

The vicarage of Kennettes pertaining to George Strathauchin[158] 'is giffin up for ten merk'; third thereof, 44s 5⅓d.

Signature: 'G. Strathauchin [Strachan, *fo 98r*], vicar of Kynnettes.'

BRECHIN, PRECEPTORY[159] OF MAISON DIEU IN, (NLS, fo 98r)
(SRO, Vol. a, fo 355v)

'The preceptorie of Masondew in Brechin pertenand to Maister Robert Carnegy, preceptor thairof, set in assedatioun to the said Sir Robert Carnegy for yeirlie [pensioun, *deleted*] payment of' £40, 'quhairof is to be deduceit of yeirlie pensioun to the said sir Walter Grahame', £20.

Calculation of third: In the whole, £40; third thereof, £13 6s 8d.

---

[157] Cf NLS, fo 98r (see below, p. 391).
[158] I.e. Strachan.
[159] See Cowan and Easson, *Medieval Religious Houses*, p. 172.

KIRKBUDDO, PARSONAGE AND VICARAGE OF, (NLS, fo 98r[160])
(SRO, Vol. a, fo 355v)

Rental of the parsonage and vicarage of Kirkbodo within the diocese of Brechin pertaining to sir James Guthrie 'quhilk payit of the auld quhen Paschefynes, corspresentis and umestclaithis wes gevin', £24. 'The foirsaidis Paschefynes, umestclaithis and corspresentis being deducit extendis to' £20 'and wantis payment thairof thir five yeiris bypast.'

Signature: 'sir James Guthrie, persoun of Kirkbodo, prebendar of the college kirk of Guthrie. Ressavit at command of the Clerk of Register.'

KINNETTLES, VICARAGE OF,[161] (NLS, fo 98r)
(SRO, Vol. a, fo 355v)

'Item, the vicarage of Kinnettelis with ane hous and yard at the kirk ', 10 merks.

Signature: 'sir George Strachan. J. Clerk Register. James,[162] ressave cautioun for the thrid and caus him be obeyit of lettres. Johnne Wischart, Comptrollar.'[163]

BUTTERGILL, PARSONAGE AND VICARAGE OF, (NLS, fo 98v)
(SRO, Vol. a, fo 356r)

Rental of the parsonage of Buttergile and vicarage of the same within the diocese of Brechin

The parsonage and vicarage foresaid are set in assedation for yearly payment of 110 merks, 'heirof to be deduceit ane stallare fee in the said queir', 16 merks. So rests, 90 merks.

Signature: 'Robert Abircrumby, persoun of Buttergill, with my hand.'

Calculation of third: In the whole, £73 6s 8d; third thereof, £24 8s 10d.[164]

---

[160] Fo 97v is blank.
[161] Cf NLS fos 97r (see above, p. 390).
[162] James Nicolson, clerk of the Collectory, *Thirds of Benefices*, p. 62.
[163] Sir John Wishart of Pittarow, Comptroller, 1562 - 1565, *Handbook of British Chronology*, p. 191.
[164] The correct calculation appears to be £24 8s 10⅔d.

CORTACHY, PARSONAGE AND VICARAGE OF, (NLS, fo 98v)
(SRO, Vol. a, fo 356r)

Rental of Cortoquhy kirk, parsonage and vicarage, 'commoun kirk of the channonrie of Brechin', extending in payment to the sum of 190 merks. The temporal lands of the same parish pay yearly to the college and chapter of Brechin, £40. 'Haill to the quene.'

GLAMIS, VICARAGE OF, (NLS, fo 99r)
CLOVA, CHAPEL OF, (SRO, Vol. a, fo 356v)

'Rentale seu assedatione'

Rental or assedation of Mr James Rolland, vicar of Glammis, as follows, drawn up by him under his subscription manual, 26 January 1561/2.

First, the vicarage of Glammis with the chapel of Colvay and glebe of Glammis, now yearly extending to 80 merks, deducting fermes and oblations which used to be paid and now are not paid.

Vicarage of Glammis with the chapel of Clovay and glebe of Glammis.

Calculation of third: In the whole, £53 6s 8d; third thereof, £17 15s $6\frac{2}{3}$d.

BALQUHIDDER, PREBEND OF[165] (NLS, fo 99r)
(SRO, Vol. a, fo 356v)

The said Mr James has the prebend of Balquhidder, Dunblane diocese, extending annually to 24 merks.

Calculation of third: In the whole, £16; third thereof, £5 6s 8d.

---

[165] Cf NLS, fos 45r, 47r (see above, pp. 328, 330). This benefice, in the sheriffdom of Perth, is included in rentals for Forfar because it was held by James Rolland who also held the preceding benefice in Forfar.

DUN, VICARAGE OF,[166]

(NLS, fo 99r)
(SRO, Vol. a, fo 357r)

The kirk of Dun[167] set in assedation to the laird of Dun[168] and pertaining to the prioress of Elcho[169] for yearly payment of 107 merks.

Signature: 5 February 1561/2, 'Producit be me, Mr Johnne Wod.'

Calculation of third: In the whole, £71 6s 8d; third thereof, £23 15s 6⅔d.

MONIFIETH, PARSONAGE AND VICARAGE OF,

(NLS, fo 99v)
(SRO, Vol. a, fos 356v-357r)

Parsonage of Monifuith pertaining to Mr Alexander Betoun.

The parsonage of Monyfuith and the vicarage of the same extends to 4 c. 12 b. wheat; 12 c. 9 b. bere; 15 c. 10 b. meal; 40 merks silver, of which there is 20 merks given to the vicar pensionary; 14 barrels of salmon.

Calculation of third: Wheat, 4 c. 12 b.; third thereof, 1 c. 9 b. 1 f. 1⅓ p. Bere, 12 c. 9 b.; third thereof, 4 c. 4 b.[170] Meal, 15 c. 10 b.; third thereof, 5 c. 3 b. 1 f. 1⅓ p. Money, £26 13s 4d; third thereof, £8 17s 9⅓d. Salmon, 14 barrels; third thereof, 4⅔ barrels.

Signature: 'David Betoun of Melgund, bruder to the said Mr Alexander.'

MAINS, VICARAGE OF,

(NLS, fo 100r)
(SRO, Vol. a, fo 357r)

Rental of the vicarage of Mains pertaining to sir Patrick Grahame.

Calculation of third: In the whole, £20; third thereof, £6 3s 4d.

---

[166] Cf NLS, fos 91v, 101v (see above, p. 381; below, p. 396), where the two rentals were for £40, and were signed by Mr Robert Achmuty.
[167] The benefice of Dun was annexed to Elcho priory, Cowan, *Parishes*, p. 50.
[168] I.e. John Erskine of Dun, cf above, p. 372, n. 99.
[169] Euphemia Leslie was prioress of Elcho in October 1561; she was the last prioress and died in 1570. *RMS*, iv, no 1396; Cowan and Easson, *Medieval Religious Houses*, p. 147.
[170] The correct calculation appears to be 4 c. 3 b.

KINNETTLES, PARSONAGE OF, (NLS, fo 100r)
(SRO, Vol. a, fo 357r)

Parsonage of Kynnettis

The parsonage of Kynnettis should give yearly £124 and the lands of Brigtoun, Innerritie [*i.e.* Invereighty] and the Kirktoun, and tenants thereof 'has tane up the samin and intromittit thairwith thir four yeiris bigane.'

Calculation of third: In the whole, £124; third thereof, £41 6s 8d.

BRECHIN, SUBDEANERY OF, (NLS, fo 100r)
(SRO, Vol. a, fo 357r-v)

The subdeanery of Brechin pertaining to Mr Hercules Carnegy set in assedation to the said Sir Robert Carnegy of Kynnard for 70 merks yearly 'as it wes wont to be sett to the archidene of Brechin and utheris of befoir. Summa patet.'

Signature: 'This rentall produceit be Sir Robert Carnegy as fermorar. R. Carnegy.'

Calculation of third: In the whole, £46 13s 4d; third thereof, £15 11s 1⅓d.

DUNDEE, VICARAGE OF, (NLS, fo 100v)
(SRO, Vol. a, fo 357v)

'Item, Maister Johnne Hammiltoun is vicar of Dundie and[171] wes set in assedatioun befoir the rysing of the Congregatioun for' £40 'be yeir payit to the said Maister Johnne, and of pensioun to Maister James Thorntoun', £20. 'Summa', £60, 'and payit not ane penny thir last iij yeiris bypast.'

Calculation of third: In the whole, £60; third thereof, £20.

GUTHRIE COLLEGIATE KIRK, PROVOSTRY OF, (NLS, fo 100v)
(SRO, Vol. a, fo 357v)

'The rentale of the payment of the provestrie of Guthrie within the diocy of Brechin, the stallaris fie, vicare pensionar, and procurationis being payit to [be, *SRO, Vol. a*] the bischop', £40. 'And tuenty yeiris syne ten merkis mair and hes gottin na payment thairof thir thrie yeiris bypast.'

Signature: 'Mr Alexander Skeyne.'

---

[171] The SRO text reads 'as' instead of 'and'.

Calculation of third: In the whole, £46 13s 4d; third thereof, £15 11s 1⅓d.

ECCLESGREIG, VICARAGE OF,   (NLS, fo 101r)
                                                                                          (SRO, Vol. a, fo 357v)

Rental of the vicarage of Eglisgreige pertaining to Mr James Wilky, regent of Sanctleonardis College, 'presentlie in Sanctandrois. The rent of the vicarage of Eglisgreige was yeirlie the sowme of fourty pundis money usuall to be payit to me and tuenty merkis to be payit to the curat of the kirk in my name.'

Calculation of third: In the whole, £53 6s 8d; third thereof, £17 15s 6⅔d.

MONIKIE, VICARAGE PENSIONARY OF,   (NLS, fo 101r)
                                                                                                                   (SRO, Vol. a, fo 358r)

'This is the just rentaill of Mathow Greiff, his vicare pensionare of Moneky.'

'In primis ten pundis to be payit of the fructis of the vicarag with the croft, mans and yard extending to' £4. 'Sua in the haill extending to' £14.

Signature: 'Matheris Greif, pensionarius de Moneky, manu sua. Resavit at the command of the Clerk of Registere.'

Calculation of third: In the whole, £14; third thereof, £4 13s 4d.

DUNDEE, CHAPLAINRY IN,[172]   (NLS, fo 101r)
                                                                                   (SRO, Vol. a, fo 358r)

Rental of a chaplainry within the kirk of Dundie pertaining to Andrew Gray.

Calculation of third: In the whole, £12; third thereof, £4.

ARBROATH, VICARAGE PENSIONARY OF,   (NLS, fo 101r-v)
                                                                                 (SRO, Vol. a, fo 358r)

Rental of the vicarage pensionary of Abirbrothok

'Item, in money of the vicarage', £21 6s 8d. 'Item, the teind schavis of the burrow ruddis of Arbrothok extendis commonlie be yeir in all cornis to' 24 b. bere. 'Item,

---

[172] Cf NLS, fo 90v (see above, p. 377).

the mans and croft payis yeirlie' 40s. [*In margin*, 'xiij Martij 1573. Mr Robert Auchmowtie augmentit this rentale for the teynd fysch of Arbrocht xxx li. in the haill, extending in the thrid to ten pundis.']

Signature: 'Mr Robert Auchtmowty.'

Vicarage pensionary of Abirbrothok pertaining to Mr Robert Auchtmowty.

Calculation of third: In the whole, £23 6s 8d; third thereof, £7 15s 6d.[173] Victual, 24 b. bere; third thairof, 8 b. 'Mair for the fisching foirsaidis', £10.

NEWTYLE, VICARAGE OF, (NLS, fo 101v)
(SRO, Vol. a, fo 358v)

The vicarage of Newtyle within the diocese of Sanctandrois and sheriffdom of Forfar with the glebe and manse of the same pertaining to sir Andrew Lyndsay, set in tack to Sir Robert Carnegy for payment of £30.

Calculation of third: In the whole, £30; third thereof, £10.

KIRRIEMUIR, VICARAGE OF, (NLS, fo 101v)
(SRO, Vol. a, fo 358v)

Rental of the vicarage of Kyrmuir pertaining to Mr George Fleschar set in assedation for £60; third thereof, £20.

Signature: 'Mr George Fleschar, vicar of Kyrmuir.'

DUN, VICARAGE OF,[174] (NLS, fos 101v-102r)
(SRO, Vol. a, fo 358v)

'The rentale of [the] vicarage of Dun eftir the auld payment in all proffeittis', £40. 'And now all restis in the handis of the lard of Dun and his tennentis, I knaw nocht quhait it wilbe for I gait nathing of it thir yeiris bygan.'

Signature: 'Maister Robert Auchtmowty.'

Calculation of third: In the whole, [*blank*]; third thereof, [*blank*].

---

[173] The correct calculation appears to be £7 15s 6⅔d.
[174] Cf NLS, fos 91v, 99r (see above, pp. 381, 393).

ARBIRLOT, VICARAGE OF,

(NLS, fo 102r)
(SRO, Vol. a, fo 358v)

This is the rental of the vicarage of Arbirloth pertaining to sir James Lecprevik, vicar thereof, lying in Angus within the sheriffdom of Forfar in the years of God 1561 and 1562 extending yearly during the said two years 'be gros estimatioun to the sowme of' £14; third thereof, £4 13s 4d.

MENMUIR, VICARAGE OF,

(NLS, fo 102r)
(SRO, Vol. a, fo 359r)

Vicarage of Menmuir pertaining to Mr Robert Auchinlek.

Calculation of third: In the whole, 39 merks; third thereof, 13 merks.

KINNELL, VICARAGE OF,

(NLS, fo 102r)
(SRO, Vol. a, fo 359r)

Rental of the vicarage of Kynnell pertaining [to] sir Charles Foulartoun, vicar thereof, lying within the sheriffdom of Forfar.

Calculation of third: In the whole, £12; third thereof, £4.

CATTERLINE, VICARAGE OF,

(NLS, fo 102v)
(SRO, Vol. a, fo 359r)

Rental of the vicarage of Katerling pertaining to William Owsteane, vicar thereof.

Calculation of third: In the whole, £6; third thereof, 40s.

KINNEFF, VICARAGE PENSIONARY OF,

(NLS, fo 102v)
(SRO, Vol. a, fo 359r)

Rental of the vicarage pensionary of Kynneff pertaining to the said William Owstean extends to £13 6s 8d; and the manse and glebe to 13s 4d.

Calculation of third: In the whole, £14; third thereof, £4 13s 4d.

'Assignatioun heirof':
    Take this manse and glebe for 13s 4d 'in parte of payment off oure soverane ladyis thrid foirsaid.'

MURROES, VICARAGE OF,

(NLS, fo 102v)
(SRO, Vol. a, fo 359r)

[*In margin*, 'Vicarage of Muirhoussis']

Rental of the vicarage of Muirhous lying in Angus within the diocese of Sanctandrois pertaining to Mr Ninian Cuke [William Cuik, *SRO*], vicar thereof.

The said vicarage set in assedation 'payand thairfoir in tymes bygane yeirlie', £23 6s 8d, thereof to be paid yearly to sir David Cunnynghame in pension the sum of £10. So rests free, £13 6s 8d.

Signature: 'Ita est Ninianus Cuke, manu propria. Resavit at the command of the Clerk Register.'

ALYTH, VICARAGE OF,

(NLS, fo 103r)
(SRO, Vol. a, fo 359v)

Rental of the vicarage of Alyth pertaining to Mr Robert Grahame, 'chanoun of Dunkeld.'

This said vicarage of Alyth set in tack and assedation to William Alexandersoun and James Adame[175] for 60 merks 'in the yeiris bygane yeirlie, and now be ressoun of the excepting of the umaist clathis, Peasche fynis, corspresentis and offerandis extendis to the sowme of' 30 merks 'allanerlie.'

Signature: 'Robert Grahame of Alyth, with my hand. Rasavit at the command of the Clerk Register, etc.'

FERN, PARSONAGE AND VICARAGE OF,[176]

(NLS, fo 103r-v)
(SRO, Vol. a, fo 359v)

Rental of the personage and vicarage of Ferne pertaining to Mr Patrick Mure 'and gevin in be him self writtin with his awin hand.'

'I, Maister Patrik Mure, persone and vicare of the paroche kirk of Ferne, gevis in this rentall of the said personage and vicarage of Ferne, Pasche fynes, corspresentis and umaist claithis and small offerandis beand deducit, gevis to me instantlie' £116 12s. 'Thairfoir desyris lettres to be deliverit to me to be answerit of the' 1562 'yeris crope and this rentall to be subscrivit be your lordschip conforme to the commoun act, I findand cautioun for the thrid thairof.'

---

[175] The Christian name is omitted in the SRO text.
[176] Cf NLS, fo 87r; EUL, fo 118r (see above, pp. 371, 641).

'Rasavit at the command of the Clerk Register and subscrivit with his hand.'

Calculation of third: In the whole, £116 10s; third thereof, £38 17s 4d.[177]

DUNNICHEN, VICARAGE OF, (NLS, fo 103v)
(SRO, Vol. a, fos 359v-360r)

Rental of the vicarage of Downychin pertaining to James Cokburne, vicar of the same.

'Memorandum, the vicarage of Downichin, within the schirefdome of Angus and diocy of Brechin sett in assedatioun for' 20 merks 'of yeirlie payment only, deduceand the corspresentis, umaist clathis and Pasche fynes, is now presentlie worth the sowme of' £10 yearly.

Signature: 'James Cokburne, vicar of Dunichin. Rasavit at command of the Clerk Register.'

LOGIE-DUNDEE, VICARAGE PENSIONARY OF, (NLS, fo 103v)
(SRO, Vol. a, fo 360r)

Rental of the vicarage pensionary of Logydundy pertaining to me, William Hepburne, extends yearly to 6 merks.

Signature: 'William Hepburnne, with my hand.'

DUNDEE PARISH KIRK, ST BARBARA'S ALTAR IN, (NLS, fo 104r)
(SRO, Vol. a, fo 360r)

Rental of Sanct Barbares altar situated within the parish kirk of Dundie.

'Item, the haill rentis and emolumentis pertenand to the samyn extendis to the soume of' 12 merks money of this realm yearly, 'to be upliftit of certane annuallis of certane tennentis within the said burgh.'

Signature: 'Robert Abircrumme, with my hand. Rasavit at command of the Clerk Registre.'

---

[177] The correct calculation appears to be £38 16s 8d.

BRECHIN CATHEDRAL, ROOD ALTARAGE IN,

(NLS, fo 104r)
(SRO, Vol. a, fo 360r)

Rental of the Rud altarage situated within the cathedral kirk of Brechin now pertaining to George Wilsoun, chaplain thereof.

In the first, the lands of the Magdalane Chapell, 16 merks yearly. The lands of Tempilhill, 6 merks of mail and 40d in augmentation of feu mail and 20s of the Tempill land of Kethik and 5s in Lytill Petforth.

Signature: 'George Wilsoun, with my hand, chaplane of the Rude altarage within the kirk of Brechine. Rasavit at the command of the Clerk of Register.'

DUNDEE, CHAPLAINRY OF ST STEPHEN IN,

(NLS, fo 104r-v)
(SRO, Vol. a, fo 360v)

Rental of Sanct Stewnes chaplainry situated within the town of Dundy pertaining to sir John Burell.

'Item inprimis, ane land pertening to the airis of umquhile George Cathro quhilk payis in the yeir of few maill' £4 5s 4d. 'Item, ane land pertenand to the airis of umquhile David Schringeour [*sic*] quhilk payis in the yeir of few maill' £4. 'Item, ane land pertenand to Alexander Carnegy quhilk payis in the yeir of few maill' £5. 'Item, ane land pertenand to the airis of umquhile James Mur, the quhilk payis in the yeir of few maill' 10s 8d. 'Item, ane land pertenand to James Walsoun, the quhilk payis in the yeir of ground annuell', 5s 7d. 'Item, ane land pertenand to Johne Baxtar, the quhilk payis in the yeir of ground annuell' 3s 4d. Total, £14 4s 11d.

Signature: 'Ita est Joannes Burall, manu sua.'

BALLUMBIE, PARSONAGE OF,

(NLS, fo 104v)
(SRO, Vol. a, fo 360v)

Rental of the parsonage of Balumye pertaining to Mr James Fothringame, parson thereof, extends yearly to the sum of 40 merks; third thereof, £8 17s 9$\frac{1}{3}$d.

Signature: 'Mr James Fotheringame, with my hand.'

ARBROATH, CHAPLAINRY OF ST NICHOLAS IN, (NLS, fos 104v-105r)
(SRO, Vol. a, fo 360v)

Rental of Sanct Nicolas lying within the town of Abirbrothok pertaining to sir Thomas Meikesone.

'The said chaplanrie is worth be yeir in annuellis, rudis and houssis', £6 13s 4d.

Signature: 'Thomas Meikesone, with my hand.'

BRECHIN, VICARAGE OF,[178] (NLS, fo 105r-v)
(SRO, Vol. a, fo 361r-v)

[*In margin*, 'Aliter, the vicarage of Brechin ressavit at the Comptrollaris command.'[179]]

'The rentall of the vicarage of Brechin gevin in be David Watt, vicar thairof, as it was set in assedation, the tennour of the quhilk followis.

> Be it kend till all men be thir present lettres me James Hepburne, vicar of Brechin, to have sett and in assedatioun lattin and be thir presentis settis and in assedatioun lattis to my lovit Matho Hepburne of Caldhame and to his airis and assignais ane or ma of na griteare degre nor power than him self all and haill my said vicarage of Brechin with all and syndrie teindis, emolumentis, proffettis, commoditeis and pertinentis pertening or rychtuouslie may pertene thairto for the dayis, yeiris, space and termes of thre yeiris nixt and immediatlie following the dait heirof, quhilk dait salbe and begin the said Mathowis Hepburne, hes [*recte*, his] airis and assignais foirsaidis entre thairto and sall indure frathinefurth ay and quhill the saidis thre yeiris be fullilie and compleitlie furth run, payand thairfoir yeirlie the said Mathow, his airis and assignais foirsaidis, induring the saidis thre yeiris the sowme of fourty merkis usuall money of Scotland at twa termes in the yeir, viz., the Rude day[180] and the fest of Allhallomes,[181] be equall twa portionis allanerlie for all uther dewitie, dew service or uther charge that can be askit or requyrit thairof be ony persone or personis, and I, forsuith, the said James Hepburne, vicare foirsaid, bindis and oblissis me and my successouris vicaris of the samin to warrand, acquet and defend this my present assedatioun to the saidis Mathou Hepburne, his airis and

---

[178] Cf NLS, fo 90v (see above, p. 378).
[179] See above, p. 351, n. 5.
[180] 'The Rud day callit Beltane', NLS, fo 137v. 'Inventio Sancte Crucis' [i.e. the finding of the Holy Cross], 3 May, *Handbook of Dates for Students of English History*, ed. C. R. Cheney (London, 1978), p. 47.
[181] I.e. the feast of All Saints, 1 November.

assignais foirsaidis induring the space and termes of the saidis thre yeiris in all and be all as is abonewrittin aganis all deidlie. In witnes heirof I haif subscrivit this my assedatioun with my awin hand at Brechin' 12 March 1562/3 'befoir thir witnessis, Maister Johnne Hepburne, thesaurare of Brechin, Matho Hepburne, David Schevane, with utheris divers.

Provyding alwyis that the said James Hepburne releif and keip me skaithles of the thrid of the said vicarage at the queinis handis or utheris haifand interes thairto induring the saidis thre yeiris gif ony beis requirit thairof befoir thir witnessis foirsaidis. Sic subscribitur, James Hepburne, vicar of Brechin, with my hand.'

'James Nicolsoun, ye sall ressave fra David Watt, vicar of Brechin, the rentall of the samin conforme as the same wes sett be umquhile James Hepburne his predicessour quhilk is fourty merkis as ye may persaif be the said assedatioun sett be the said umquhile James to Mathow Hepburne his brothir, off the dait at Brechin the' 12 March 1562/3 'and failye nocht heirin keipand thir presentis to your warrand, subscrivit with my hand at Edinbrucht', 19 January 1565/6. 'Sic subscribitur, Tulibardin, Comptrollar.'[182]

---

[182] Sir William Murray of Tullibardine, Comptroller, 1565 - 1582. Rentals for Forfar resume at NLS 109v; SRO, Vol. a, 365r (see below, p. 409).

Kincardine

BENHOLM, PARSONAGE AND VICARAGE OF, (NLS, fo 106r)
(SRO, Vol. a, fo 362r)

Rental of the parsonage of Benholme

The parsonage of Benholme with the vicarage of the same annexed thereto lying within the diocese of Sanctandrois and sheriffdom of Kincardin set in assedation to Robert Carnegy of Kynnard by Mr John Thorntoun, parson thereof, for 260 merks 'as ane act of the Officialis bukis of San[c]tandrois maid thairupoun proportis.'

[*In margin*, 'Summa' 260 merks 'quhairof thair is to be deduceit' 20 merks 'of pensioun to Mr William Eldar, vicar pensioner thairof, and for the corspresentis and Pasche fynes', 20 merks 'yeirlie.']

Calculation of third: In the whole, £173 6s 8d; third thereof, £57 15s 6⅔d.

FORDOUN, VICARAGE OF, (NLS, fo 106r)
(SRO, Vol. a, fo 362r)

Rental of the vicarage of Fordoun in the Mernis within the diocese of Sanctandrois.

'In tymes bigane the said vicarage wes sett for' 100 merks 'and the curat being sustenit as it is knawin be thame quhilk had it in assedatioun thir mony yeiris bigane.'

Signature: 'James Leirmonth.'

Calculation of third: In the whole, £66 13s 4d; third thereof, £22 4s 5⅓d.

FETTERESSO, PARSONAGE AND VICARAGE OF, (NLS, fo 106v)
(SRO, Vol. a, fo 362r)

Rental of the parsonage and vicarage of Fetteresso pertaining to Mr James Broun set in tack and assedation to William, 'erle Merscheall',[1] for yearly payment of the sum of £209 3s 4d[2] 'set in nyntene yeiris tak to him and the maist parte of the yeiris thairof to rin and the vicarage of Fetteresso in commoun and gude yeiris' £20 'off the quhilk I haif ressavit na maner of payment thir tua yeiris bigane. Subscrivit with my hand at Abirdein' 12 February 1561/2.

Signature: 'M[r] Ja[mes] Broun, persoun of Fetteresso.'

Calculation of third: In the whole, £239 3s 4d;[3] third thereof, £79 13s.[4]

DUNNOTTAR, PARSONAGE AND VICARAGE OF, (NLS, fo 106v)
(SRO, Vol. a, fos 362v)

Rental of Dunnoter

'In primis, the personage and vicarage thairof yeirlie set for the sowme of foure scoir pundis being in the handis of me, Maister Johnne Eldar, and William Salmone, prebendaris of the Queinis college.'[5]

Signature: 'Joannes Eldar.'

Calculation of third: In the whole, £80; third thereof, £26 13s 4d.

DURRIS, PARSONAGE OF,[6] (NLS, fo 106v)
(SRO, Vol. a, fo 362v)

Rental of the parsonage of Durris lying in the diocese of Sanctandrois and sheriffdom of Kincardin on the north side of the Month[7] pertains to Mr John Duff, pays yearly 114 merks money freely.

---

[1] William Keith, 4th earl Marischal, *Scots Peerage*, vi, pp. 46-47.
[2] The text reads 'ten scoir nyne pundis iij s iiij d'.
[3] The correct calculation appears to be £229 3s 4d.
[4] One third of £239 3s 4d is £77 14s 5½d; one third of £229 3s 4d is £76 7s 9½d.
[5] Dunnottar was annexed to Trinity college, Edinburgh, which had been founded by Mary of Gueldres, widow of James II. Cowan and Easson, *Medieval Religious Houses*, pp. 221; Cowan, *Parishes*, p. 54.
[6] Cf NLS, fo 109r (see below, p. 409).
[7] I.e. the Mounth, the mountain chain stretching easterly from Loch Linnhe toward Stonehaven.

Signature: 'Subscrivit be me, the said Mr Johnne at Edinbrucht', 12 February 1561/2.

Signature: 'Maister Johnne Duff.'

ABERLETHNOTT, KIRK OF, (NLS, fo 107r)
(SRO, Vol. a, fo 362v)

Rental of the kirk of Abirlethnot in the Mernis annexed to the college of Abirdene.[8]

The teind sheaves extend to £200 Scots money. The vicarage extended to £20 'quhen it was weill payit. This is the griteest parte of the leving of the said college, to the nummer of fourtie and sax maisteris and studentis in divers faculteis and science, quhilk college wes never taxit be king, governour nor bischop.'

Signature: 'M[r] Alexander Skene, procurator collegij.'

GLENBERVIE, BENEFICE OF, (NLS, fos 107v-108v)
ARBUTHNOTT, BENEFICE OF, (SRO, Vol. a fos 363r-364r)

The teinds of the kirks of Glenbervy and Arbuthnot pertaining to the dean of Abirdene, the rentals thereof as follows.

Glenbervy
    The teind of Delverdis [*i.e.* Dillavaird], that is the Manis, Chappeltoun and Boggynkabyr, set for the yearly payment of £10. Inchbrok and Boggergon, the teind sheaves thereof yearly, 14 merks. Pedwethy, the teind sheaves thereof yearly, 4 merks. The Westertoun, the teind sheaves thereof, 12 merks. The Cuthill, the teind sheaves thereof, 8 merks. Annamuk, the teind sheaves thereof, 6 merks. Tannaquhy, the teind sheaves thereof, 6 merks. Peddrethy, the teind sheaves thereof, 10 merks. Nedder Kynmonth, the teind sheaves thereof, 15 merks. The Inches, the teind sheaves thereof, 4 merks. The Ailhous croft in Drumletty, the teind sheaves thereof, 6s 8d. Margyr, the teind sheaves thereof, 6 merks. Ovir Kynmonth, the teind sheaves thereof, 6 merks. The Newbyggyne, the teind sheaves thereof, 6 merks. The Coldown, Allagawynne and Blairerno, the teind sheaves thereof, 20 merks. The Manes of Glenbervy with the pendicles, 9 merks. The Myln, the Halk Hill and Murtes Hill, the teind sheaves thereof, 13 merks. 'Item, the vicaragis geif in assidatioun quhen all casualities wes ansurit,

---

[8] I.e. King's college, Aberdeen.

the sowme of' 50 merks. 'Item, thair is na temporall land except the vicaris [his, *deleted*] croft quhilk geif quhen it was sett with ane brostare hous,' 3 merks.

'This is the just rentall of the teindis and vicaragis of the benefice of Glenbervy subscrivit with my hand at Aberdene', 14 February 1561/2.

Signature: 'R. Erskin, dene of Abirdene.'[9]

Arbuthnot
    'Item, Arbuthnot is sett to the lard of Pittarro, the teind schavis thairof except the teind schavis of the toun of Drumlethy quhilk is in my awin hand extending be yeir to £20. 'The rest the lard of Pittarro hes bayth teind schavis and vicarages for the yeirlie payment of' £180. 'Sua the haill extendis to' £200 'as the rentall of the same hes bene thir hundreyth yeiris past. Thair is na temporall land except ane lytill croft that the v[i]care hes; I cane nocht geif ane particular rentall for it was nevir sene I had the samyn bot in factouris handis. This is the rentall of the benefice of Arbuthnot teindis and vicarages, subscrivit with my hand at Abirdene', 14 February 1561/2.

Signature: 'R. Erskin, den of Abirdene.'

'The ordinare chargis of Glenbervy is be yeir' 36 merks. 'The ordinare charges of Arbuthnot is be yeir' 32 merks. 'Item, I gait na thing of thir vicaragis and in speciall of Glenbervy thir foure yeiris bygane.'

The rental of two kirks pertaining to Mr Robert Erskyne, dean of Abirdene, within the Mernis. The kirk of Arbuthnot, teind sheaves and vicarage, is £200, of this Mr Andrew Patre [*i.e.* Petrie], vicar pensionary, has the manse and croft and 20 merks. William Rynd has of Arbuthnot a pension, £20. The kirk of Glenbervy, teind sheaves and vicarage, is £120; of this Mr Andrew Eldare, vicar pensionary, has the manse and glebe worth 20 merks.

Signature: 'Me, Maister Johnne Wod.'

Calculation of third:
    Kirk of Arbuthnot, in the whole, £200; third thereof, £66 13s 4d.
    Kirk of Glenbervy. 'Nota, the rentaill gevin be the dene of Abirdene, persone thairof, beiris' £149 6s 4d; third thereof, £49 15s 9d.[10]

'Ane uthire rentaile of the foirsaid kirkis.'

---

[9] Robert Erskine, dean of Aberdeen, 1540 - 1563, Watt, *Fasti*, p. 9.
[10] The correct calculation appears to be £49 15s 5½d.

The rental of the benefice of Arbuthnot pertaining to Mr Robert Erskine, parson of the same, will extend to 300 merks yearly and is set for the same sum to the laird of Pittarro. The rental of the parsonage and vicarage of Glenbervy pertaining to the said Mr Robert extends to 180 merks yearly 'and hes nocht gottin na maner of payment thairof thir foure yeiris bygane.'

Signature: 'Mr Alexander Skene, procurator, Magistri Roberti Erskyne, rectoris de Arbuthnot et Glenbervy, manu propria.'

FETTERCAIRN, PARSONAGE AND VICARAGE OF,

(NLS, fos 108v-109r)
(SRO, Vol. a, fos 364r-v)

Rental of the parsonage of Fettircarn within the diocese of Sanctandrois is 430 merks, 'and thairof yeirlie spendit in poynding or paiment culd be gotting fourty pundis and sum yeiris' 100 merks. 'Item, the rentaile of the said personage was tuenty yeiris syn the sowme of' 480 merks. 'Item, the vicarage thairof by the vicar pensionaris fie and the bischopis procurationis being payit' £10. 'Teynd woll, lambis, cheis and Pasche fynes being payit of the auld maner and hes gotting na payment thairof this foure yeiris.'

Signature: 'Alexander Skeyn, procurator Magistri Jacobi Straquhyne, rectoris de Fettercarn habens illius litteras missivas et [recte, ad] effectum suprascriptum istud rentale aprobo.'[11]

Calculation of third: In the whole, 480 merks; third thereof, £106 13s 4d.

DURRIS, VICARAGE OF,

(NLS, fo 109r)
(SRO, Vol. a, fo 364v)

The vicarage of Durris

'Item, the teind woll, lamb, milk and stirk of the vicarage of the kirk of Duris with the pertinentis is worth yeirlie' 18 merks.

Calculation of third: In the whole, £12; third thereof, £4.

---

[11] I.e. 'Alexander Skeyn, procurator of Mr James Straquhyne, parson of Fettercairn, having his missive letters to the effect above written, approves this rental.'

BARRAS, ST JOHN'S CHAPLAINRY OF, (NLS, fo 109r)
(SRO, Vol. a, fo 364v)

'The rentall of ane chaplanrie in the Memis callit Sanct Johnnes chapell of the Barrowis pertenyng to Andro Gray.'

Calculation of third: In the whole, £3 6s 8d; third thereof, 22s $2\frac{2}{3}$d.[12]

DURRIS, PARSONAGE OF,[13] (NLS, fo 109r)
(SRO, Vol. a, fo 364v)

Rental of the parsonage of Durris lying in the diocese of Sanctandrois and sheriffdom of Kincardin on the north side of the Month pertaining to Mr John Duff, pays yearly 'of money frelie' 114 merks.

'Subscrivit be the said Maister Johnne at Edinburgh', 12 February 1561/2.

Signature: 'Maister Johnne Duff.'

Calculation of third: In the whole, £76; third thereof, £25 6s 8d.

---

[12] In the text this fraction is rendered 'ob 3 ob'.
[13] Cf NLS, fo 106v (see above, p. 407).

Forfar (2)

COUPAR ANGUS, ABBEY OF,[1] (NLS, fos 109v-111r)
(SRO, Vol. a, fos 365r-367r[2])
Abbey of Coupare

'This is the just draucht and tryall of the auld rentall and registair of Cowpar of the haill silver, quheit, beir, meill, hors corne, contenit thairintill actentiklie calculat and laid with utir diligence as the said auld rentall and registar of Cowpar declaris at lenth.'

Silver: 'The haill silver of the abbay of Cowpar contenit in the said auld rentall and registar extendis in the haill de claro to the sowme of' £1,298.[3]

'Defasance: Eftir followis the silver contenit in the said rentall quhilk is out of use of payment and ungettable with the haill pensionis and infeftmentis quhilk ar to be deducit of the principall sowme abonewrittin.
    Item, imprimis Andro Blair of Balgrische denyis of his maillis contenit in the said rentall, fourtyane s, four pennyis and allegis to his few chartour', 41s 4d. 'Item, to be tane of the maillis of Beltscheill [Bedscheill, *SRO*] quhilk is ungettable be ressoun we have tryit the tennentis thairof particularlie and inlaikis', 14s 8d. 'Item, of Gleneyla the annual ongettable and furth of use of payment past memour of man extending to' 16s. 'Item, the officiaris land of Glenyla is calculat in the said rentall to tuenty s and wer ay dischargit to him for his service', 20s. 'Item, to be deducit of Litill Pertht quhilk is infet to the abbay of Restennet', 20s. 'Item, ane infeftment to ane chaplane of the Carsgrange extending to' £7 6s 8d. 'Item, Scottis pensioun provydit in Rome extending to' £66 13s 4d. 'Item, the College of Justice pensioun infet and yeirlie payit furth of Couper extending to' £36. 'Item the annuallis of Dundie ungettable be ressoun thai ar brint landis and standis waist extending to' £10 11s 8d. 'Item, the annuallis of Pertht furth of use of payment and ungettable extending to' £3 18s 8d. 'Item, annuallis in divers and

---

[1] Cf NLS, fos 65r, 83v (see above, pp. 352, 368). This rental is different in layout from the two previous rentals.
[2] Fo 366r is blank.
[3] In the previous rentals the whole silver of the abbey was given as £1,159 11s 11d 'ob' [i.e. ½d], fo 70v; £1,886 8s 7d, fo 75r (see above, pp. 355, 357).

sindry uthir places, as the auld rentall declaris at lenth, extending to' £10 14s. 'Item, the viccarage of Bennethy [*i.e.* Bendochy] furth of use and ungettable comptit in the said rentall to' £50. 'Item, the viccarage of Arlie furth of use of payment comptit in the said rentaill to' £24. 'Item, James McBraik in Campsay denyis threttysex s quhilk is contenit in the said rentall and allegis it is nocht contenit in his chartour', 36s. 'Item, the fischingis of Campsy is comptit in the said rentall', £40. 'Summa of the haill defasance befoir writin', £256 3s.[4]

'Summa of the haill silver of the abbay of Cowpar quhilk is presentlie in use of payment, the haill defalcationis befoir writin beand deducit extendis to' £1,042 10s 1d.[5] 'The just tua part heirof quhilk apertenis to my lord of Argill[6] extendis to' £695 0s 2d.[7] 'The just thrid part heirof quhilk apertenis to the quenis grace extendis to' £357 10s $\frac{1}{3}$d.[8]

Wheat: 'The haill quheit of the abbay of Cowpar contenit in the said auld rentall and registar extendis to' 7 c. 3 b. 1 f. 2 p.[9] 'The just tua part heirof quhilk apertenis to my lord of Argill extendis to' 5 c. 3 b. 1 f. 2 p.[10] 'The just thrid part heirof quhilk aperteinis to the quenis grace extendis to' 2 c. 9 b. 2 f. 3$\frac{1}{3}$ p.[11]

Bere: 'The haill beir of the abbay of Cowpar contenit in the said auld rentall [and] registar by the gatherit teyndis[12] extendis to' 53 c. 3 b. 2 f. 1 p. [55 c. 2 b. 3 f. 1 p., *fo* 86r[13]]. 'The just tua part heirof quhilk pertenis to my lord of Argill extendis to' 35 c. 7 b. 3 f. 'The just thrid part heirof quhilk apertenis to the queinis grace extendis to' 17 c. 11 b. 3 f. 1 p.[14]

Meal: 'The haill maill of the said abbay of Cowpar contenit in the said auld rentall and registar by the gatherit teyndis extendis to' 65 c. 9 b. 2 f. [65 c. 8 b. 2 f., *fo* 64r; 69 c. 8 b. 3 f., *fo* 86r]. 'The just tua part heirof quhilk pertenis to my lord of Argill extendis to' 43 c. 11 b. 2 f. 2 $\frac{2}{3}$ p. 'The just thrid part heirof quhilk pertenis to the queinis grace extendis to' 21 c. 13 b. 3 f. 1$\frac{1}{3}$ p.

Horsecorn: 'The haill hors corne of the abbay of Coupar contenit in the auld rentall and registar extendis to' 23 c. 10 b. 2 f. 2 p. [24 c. 10 b. 2 f. 2 p., *next paragraph*]. 'The just tua part heirof quhilk apertenis to my lord Argill extendis

---

[4] The correct calculation appears to be £256 12s 4d.
[5] The correct calculation appears to be £1,042 17s.
[6] I.e. Archibald Campbell, 5th earl of Argyll, *Scots Peerage*, i, pp. 340-343.
[7] The correct calculation appears to be £695 0s $\frac{2}{3}$d.
[8] The correct calculation appears to be £347 10s $\frac{1}{3}$d.
[9] 7 c. 13 b. 1$\frac{1}{3}$ p. is the figure on p. 411; cf above, p. 369, n. 86.
[10] The correct calculation appears to be 4 c. 12 b. 3 f. 2$\frac{2}{3}$ p.
[11] The correct calculation appears to be 2 c. 6 b. 1 f. 3$\frac{1}{3}$ p.
[12] The phrase 'by the gatherit teyndis' is omitted from the SRO text.
[13] See above, p. 370.
[14] $\frac{2}{3}$ p. has been omitted from this calculation.

to' 15 c. 12 b. 1 f. 2⅔ p. 'The just thrid part heirof quhilk apertenis to the quenis grace extendis to' 7 c. 14 b. 3⅓ p.

Totals: Silver, £1,298 13s 4d. Wheat, 7 c. 13 b. 1⅓ p. [7 c. 3 b. 1 f. 2 p., *fo 110r*[15]]. Bere, 53 c. 3 b. 2 f. 1 p. Meal, 65 c. 9 b. 2 f. Horsecorn, 24 c. 10 b. 2 f. 2 p. [23 c. 10 b. 2 f. 2 p., *preceding paragraph*]. 'Summa of the haill chalderis abonewrittin extendis to' 150 c. 4 b. 3 f. ⅓ p. 'by the gatherit teyndis quhilk is gevin up be the abbot of Cowpar[16] in rentall for' 2 c. 3 b. bere and 4 c. 4 b. 3 f. oats 'quhilk is to be devidit be the haill chalderis abonewritin and the just thrid thairof to be addettit to the quenis grace thrid part of the victuallis of Cowpar.'

'James Nicholsoun, ye sall resave this as the just and actentik rentall of Cowpar and keip it sua that the chalmarlane thairof be chargit thairwith and conforme thairto with the thrid of the silver and victualis befoir writtin, the haill defalcationis befoir writtin of silver beand deducit and put asyde baith fra my lord Argyllis part and the quenis grace thrid. Subscrivit with our hand at Edinbrucht', 13 January 1564/5. 'Sic subscribitur, James Nycholsoun, ressavar of this rentall; J. C[lerk] Register; Johne Wischart, Comptrollar.'[17]

Calculation of third: Third of money, £347 10s ⅓d.[18] Third of wheat, 2 c. 9 b. 2 f. 3⅓ p. Third of bere, 18 c. 7 b. 1 f. 3⅔ p.[19] Third of meal, 21 c. 13 b. 3 f. 1⅓ p. Third of horsecorn, 7 c. 14 b. 3⅓ p.[20] Third of 'Lordis aittis',[21] 1 c. 6 b. 3 f. 2⅔ p.

EASSIE, PARSONAGE AND VICARAGE OF,[22]

(NLS, fo 112r[23])
(SRO, Vol. a, fo 368r[24])

Rental of the parsonage and vicarage of the parish kirk of Esse in Angus 'augmentit be sir Hew Curry, persoun thairof, for augmentatioun of the quenis grace thrid thairof as eftir followis'.

---

[15] See above, p. 410, n. 9; p. 369, n. 86.
[16] Donald Campbell was abbot of Coupar Angus at the Reformation, *Rental Book of the Abbey of Cupar-Angus*, i, p. 100; Cowan and Easson, *Medieval Religious Houses*, p. 74. Leonard Leslie obtained a gift of the abbey on 24 August 1565, *RSS*, v, no 2284.
[17] James Nicolson, clerk of the Collectory; James McGill of Nether Rankeilour, Clerk Register, 1554 - 1566; Sir John Wishart of Pittarow, Comptroller, 1562 - 1565. *Thirds of Benefices*, p. 62; *Handbook of British Chronology*, pp. 191, 197.
[18] This is one third of £1,042 10s 1d on fo 110r (see above, p. 410).
[19] This is one third of 55 c. 6 b. 1 f. 3 p.; the total given on fo 110v (p. 411) is 53 c. 3 b. 2 f. 1 p.
[20] This is one third of 23 c. 10 b. 2 f. 2 p., which differs from the total given on fo 110v (p. 411).
[21] I.e. Earl of Argyll's oats.
[22] Cf SRO, Vol. a, fo 84v (see above, p. 72).
[23] Fo 111v is blank.
[24] Fo 367v is blank.

The teinds of the lands of Dunkany with the teinds of the cottars thereof pay yearly £13 6s 8d. The teind of the Kirktoun of Esse pays yearly £13 6s 8d. The teind of the Bagony pays yearly £20. The teinds of the Hattoun pays yearly £40. The teinds of Ynglistoun pay yearly £48. The teinds of Casseltoun pay yearly £11. The vicarage glebe and the teind of Alexander Hereis croft pays 'be estimatioun communibus annis' £6 13s 4d. Total, £152 6s 8d.

'And swa the thrid extendis to', £50 15s 8½d.[25]

NOTE

Rentals for the following benefices are located elsewhere in the text:

FERN, PARSONAGE AND VICARAGE OF (EUL, fo 118r), p. 641.
INVERKEILOR, VICARAGE OF (SRO, Vol. a, fo 99v), p. 83.
RESTENNETH, PRIORY OF (SRO, Vol. a, fo 217v), p. 217.
ROSSIE, VICARAGE OF (SRO, Vol. a, fo 92r), p. 77.
STRATHMARTINE, VICARAGE OF (SRO, Vol. a, fo 95r), p. 79.

---

[25] The correct calculation appears to be £50 15s 6⅔d.

Aberdeen

ABERDEEN, BISHOPRIC OF, (NLS, fos 113r-117v[1])
(SRO, Vol. a, fos 369r-373v)

'The rentale of the biscoprik of Abirdene as it presentlie payis.'

Auld Abirdene
   The annuals of Auld Abirdene yearly extend to the sum of £29 19s 2d. 'The ferme beir of the croftis of the said auld toun extendis to fyve chalder beir, of the quhilk is allocat and assignit to Maister Thomas Stewart tua chalder yeirlie for his fie and the uthere thre chalder for ten pundis the chalder conforme to his assedatioun maid thairupoun.' [*In margin*, £59 19s 2d.]

New Abirdene
   The teind sheaves of the half of New Abirdene extend yearly to 5 c. bere.

The barony of Abirdene
   The barony of Abirdene extends yearly of mail to the sum of £118 17s 10d; 5 c. 13 b. bere; 5 c. meal, 'of the quhilkis beir and meill assignit to Martene Hewesoun' 5 b. bere, and to Andrew Buk[2] 11 b. meal for 10s the boll conform to their feu charters; 9½ marts yearly; 22 wedders yearly; 11 b. oats;[3] 8 dozen capons yearly; 13 dozen poultry yearly; 2 c. 5 b. oats yearly. 'Of the quhilkis customes in the handis of Martene Howesoun', 1 mart for 30s; 2 wedders for 20s; 4 b. oats for 20s; 12 poultry for 4s. In Henry Multrais hands 1 mart; 3 wedders; 2 dozen poultry, 1 dozen capons 'of the same prices'. In Andrew Bukis hands, 6 poultry for 2s, conform to their feu charters. [*In margin*, £133 11s 10d.[4]]

---

[1] Fo 112v is blank.
[2] The name Andrew Buk occurs as skipper of the 'Nicolace' of Aberdeen, 1540, *Extracts from the council register of the burgh of Aberdeen*, ed. J. Stuart, 2 vols., (Aberdeen, 1844-1848), i, p. 173.
[3] This '11 b. oats' is not included in the total of either oats or wheat, but 11 swine should be included here; cf below, p. 417, nn. 15, 22.
[4] The correct calculation appears to be £134 5s 10d.

The barony of Murthill

The barony of Murthill of mail yearly extends to the sum of £30 13s 4d; 2 c. bere yearly; 2 c. meal yearly; 16 custom wedders yearly; 5 dozen capons yearly; 8 dozen poultry yearly. Of which victual and customs foresaid, 12 b. bere and 12 b. meal in Gilbert Knollis hands for 10s the boll; 2 wedders for 10s; 2 dozen poultry for 8s, conform to his assedation and set thereupon. And 4 b. bere and 4 b. meal in Andrew Bukis hands for 10s the boll; 4 wedders for 26s 8d; 2 dozen capons for 16s. In Martin Hewesonis hands, 2 wedders for 10s; 12 capons for 8s; 6 poultry for 2s, conform to their sets made thereupon. £50 14s.

The barony of Clatt

The barony of Clatt of mail yearly extends to the sum of £96 9s 10d; 8 b. bere yearly; 12 b. meal yearly; 7 marts yearly; 23 wedders yearly; 22 geese yearly; 9 dozen capons yearly; 17 dozen poultry yearly; 9 dozen moorfowl yearly; 1 c. 8 b. oats yearly; 1 swine. Of which customs before specified of the barony of Clatt in Alexander Gordounis hands, half a mart for 15s; 2 wedders for 10s; 2 dozen poultry for 8s; 2 dozen moorfowl for 12s; 2 b. oats for 10s. In the lord Forbes[5] hands, 9 poultry for 3s. In James Gordounis hands, half a mart for 15s; 2 wedders for 10s; 1 dozen capons for 8s; 2 b. oats for 10s, conform to their infeftments and sets made thereupon. In Mr Robert Lummysdane hands, 1 mart for 30s; 3 geese for 3s, conform to his set, and £13 6s 8d 'allocat yeirlie to him for procuratour fie.' £103 3s 10d.

The barony of Fetterneir

The barony of Fetternere of mail yearly extends to the sum of £17 15s 4d; 3 c. 8 b. bere yearly; 1 custom wedder yearly; 13 dozen poultry yearly, 'quhilk is haill in the lardis of Balquhannanes handis for' 10s the boll of bere, 4s the dozen of poultry, 5s for the wedder, conform to his set thereupon. £48 12s 4d.

The barony of Tulenessene

The barony of Tullenessin of mail yearly extends to the sum of £65 3s 4d; 5 b. bere yearly; 7 b. meal yearly; 6 marts yearly; 20 wedders yearly; 11 dozen capons yearly; 17 dozen poultry yearly; 10 dozen moorfowl yearly; 18 geese yearly; 1 c. 4 b. oats yearly; 1 swine yearly. 'Of the quhilkis customes in Maister William Gordonis handis', 2 marts for £3; 5 wedders for 25s; 'ane boll bere and ane uthire of meill for' 20s; 3 dozen capons for 24s; 4 dozen and 6 poultry for 18s; 1 dozen moorfowl for 6s; 5 b. oats for 25s, conform to his set made thereupon. [*In margin*, £74 16d.]

The barony of Rayne

The barony of Rayne of mail yearly extends to the sum of £98 19s 1d; 7 c. 1 b. 2 f. bere yearly; 1 c. 8 b. oats yearly; 3 c. 9 b. 2 f. meal yearly; 7 marts yearly; 21 wedders yearly; 15 geese yearly; 1 swine yearly; 18 dozen capons yearly; 23

---

[5] William, 7th lord Forbes, *Scots Peerage*, iv, p. 55.

dozen poultry yearly. 'Off the quhilkis victuall and customes foirsaidis in the laird of Balquhannanis handis', 5 c. 6 b. 2 f. 2 p. bere and 1 c. 14 b. 2 f. 2 p. meal for 10s the boll; 14 marts for £6; 9 wedders for 45s; 10 dozen capons for 8s the dozen; 14 dozen poultry for 4s the dozen; 11 b. oats for 5s the boll, conform to his set made thereupon. [*In margin*, £175 10s 1d.[6]]

The barony of Damatt
    The barony of Damate of mail yearly extends to the sum of £41; 12 c. 10 b. bere yearly; 12 c. 8 b. meal yearly; 4 b. oats yearly; 2 marts yearly; 4 wedders yearly; 2 swine yearly [1 swine, *SRO*]; 12 dozen capons yearly; 21 dozen poultry yearly. 'Off the quhilkis victuall and customes abone specifiet in the lard of Meldrumis handis', 3 c. 10 b. bere and 3 c. 10 b. meal for 10s the boll; 6 dozen poultry for 4s the dozen; 2 dozen and 4 capons for 8s the dozen, conform to his set made thereupon. [*In margin*, £101 3s 6d.[7]]

The barony of Brais [*i.e.* Birse]
    The barony of Brais of mail yearly extends to the sum of £175; of bere yearly, 3 c. 12 b.; of marts yearly, 16; of wedders yearly, 34; of swine yearly, 2; of oats yearly, 1 c. 6 b.; of capons yearly, 1 dozen and 6 capons; of poultry yearly, 7 dozen and 6 poultry. 'Off the quhilkis victuall and customes afoir specifiet, in Sir Robert Carnegyis handis', 25 b. bere for 10s the boll; 4 marts for £6; 9 wedders for 45s; 9 geese for 9s; 1 dozen and 6 poultry for 3s the dozen; 6 b. oats for 5s the boll; 6 capons for 2s; 1 swine for 8s. And in Andrew Strauchin hands 1 mart for 30s; 2 wedders for 10s; 1 b. bere for 10s; 2 geese for 2s; 6 poultry for 18d. [*In margin*, £201 2s.[8]]

The barony of Fordice
    The barony of Fordice of mail yearly extends to the sum of £12 6s 8d; 3 c. 8 b. wheat yearly.

The barony of Murtlake
    The barony of Murtlaik pays of mail yearly the sum of £111 2s 2d 'without ony custome or ony uthir thinge.'

The salmon fishing
    'Imprimis, vj half nettis upoun the Wattir of Done quhilk payis yeirlie vij lastis salmond, of the quhilkis in Walter Cowttis handis iij half nettis extending to iij lastis vj barrallis for' £4 'ilk barrall, conforme to his sett gevin thairupoun.' One half net in Andrew Bukis hands extending to 14 barrels of salmon for £4 the barrel, conform to his set made thereupon. Another half net in Mr Thomas Stewarttis hands extending to 14 barrels of salmon for £4 the barrel, conform to

---

[6] The correct calculation appears to be £175 7s 7d.
[7] The correct calculation appears to be £101 2s 8d.
[8] This calculation is correct.

his set made thereupon. Another half net in Martin Hewesonis hands extending to 14 barrels for £4 the barrel, conform to his set made thererupon. The teind salmon 'and halydayis fische of the Cruvis upoun the Wattir of Done extending to ane last of salmond in the handis of the relict of umquhile Maister Johnne Watsonis for' £4 the barrel, conform to the set foresaid 'maid lange of befoir of the quhilk halydayis fische onpayit.' One half net upon the Wattir of Dee of the Raik extending to 14 barrels of salmon in Gilbert Collesonis hands for £4 the barrel, conform to his set. The half teind salmon of the Raik upon the Watter of Dee in Gilbert Collesonis hands foresaid extending to one last salmon for £4 the barrel, conform to his set thereupon. Another of the teind salmon of the Raik in the provost of Abirdenis[9] hands extending to 8 barrels of salmon for £4 the barrel, conform to his set made thereupon. The teind salmon of the Mydchingill upon the Wattir of Dee in Mr George Myddeltonis hands extending to 10 barrels of salmon for £4 the barrel conform, to his set made thereupon. The teind salmon of the Pott and Furdis upon the Wattir of Dee in Mr Duncan Forbes hands extending to 10 barrels of salmon for £4 the barrel, conform to the set made thereupon.

'Tak the haill salmond abone writtin, teind and propertie, for the thrid of the silver and gif in' £48 14s 5d.

[*In margin in NLS; in text in SRO*] Totals: Money, £1,671 7s 1d. Wheat, 3 c. 8 b. Bere, 26 c. 5 b. 3 f. 2 p. Meal, 16 c. 15 b. 3 f. 2 p. Oats, 4 c.[10]

The second teinds called ordinary teinds
'The secund teindis callit ordinare teindis quhen thai war payit extendit in propertie to the sowme of' £88 'of the quhilkis nocht ane penny payit thir thre yeiris bypast.'
'The secund teind extraordinare, coittis of testamentis consistorie, procurationis [and] synodallis as thai accidentlie fall, of the quhilkis nathing is payit thir thre yeiris bygane.'

'The ordinare chargis to be deducit'
In the first, to Robert Gordoune, son to my lord Huntlie,[11] in yearly pension to which he is provided 500 merks. To the Lords of the Session for their yearly contribution extending to the sum of £28. To his stallar in the choir, £10.

Signature: 'Ita est Mr Alexander Setoun, cancellarius Abirdonensis, ac camerarius eiusdem de eius mandato subscripsit.'[12]

---

[9] Thomas Menzies of Petfoddellis was provost of Aberdeen in 1563, *RSS*, v, no 1350.
[10] These totals are difficult to verify because of discrepancies in the sub-totals. They are different from those which appear below in the calculation of the third.
[11] Robert Gordon, eighth son of George, 4th earl of Huntly (killed in a gun accident in 1572), *Scots Peerage*, iv, pp. 534-539.
[12] I.e. 'Thus Mr Alexander Setoun, chancellor of Aberdeen, and the chamberlain of the same by his order subscribed it.'

Calculation of third: Money 'comptand thairwith the sett salmond', £1,653 16s [9d];[13] third thereof, £551 5s 7d. Bere, 35 c. 9 b.; third thereof, 11 c. 13 b. 2 f. 2⅔ p. Meal, 24 c. 4 b. 2 f.; third thereof, 8 c. 1 b. 2 f. Marts, 47½; third thereof, 15⅚.[14] Wedders, 141; third thereof, 47. Swine, 17;[15] third thereof, 5⅔. Capons, 65 dozen and 6;[16] third thereof, 21 dozen and 10. Poultry, 119 dozen;[17] third thereof, 39 dozen and 8. Oats, 8 c. 3 b.; third thereof, 2 c. 11 b. 2 f. 2⅔ p. Moorfowl, 19 dozen; third thereof, 6 dozen and 4. Geese, 55; third thereof, 18⅓. Wheat, 3 c. 8 b.; third thereof, 1 c. 2 b. 2 f. 2⅔ p.

Third of money, £551 5s 7d. Take the six half nets upon the Watter of Done, 'thre thairof sett to Walter Cowles [Cowttis, *fo 115v*] for thre lastis, vj barellis salmound for' £4 the barrel; and to Andrew Buk for 14 barrels; and to Thomas Steuart for 14 barrels; and to Martin Hewisoun for 14 barrels; 'and the half nettis fischeingis upoun the Rak upoun the Walter of Die set to Gilbert Colesoun for' 14 barrels; the teind salmon 'and halydayis fische of the Crovis for a last salmound in the handis of Mr Johnne Watsones wyiff'; the teind salmon of the Raik in Gilbert Colesones hands 'for a last of salmound and that uther half in the provestis handis for' 8 barrels; the teind of the Midchingill in Mr George Middiltones hands for 10 barrels; and the teind of the Pott and Feudis [Furdis, *SRO*] in Mr Duncan Forbes hands for 10 barrels; 'price of ilk barell', £4, 'the haill extending to' £600. 'Giff in' £48 14s 5d 'ob'.[18]

Third of bere, 11 c. 14 b. 2 f. [11 c. 13 b. 2 f. 2⅔ p., *fo 116v*]. Take out of the barony of Abirdene, 5 c. 13 b.; out of the barony of Mu[r]thill, 2 c.; out of the barony of Clatt, 7 b.; out of the barony of Fetternere, 3 c. 8 b.; out of Tulynessill, 2 f. 2⅔ p.[19]

Third of meal, 8 c. 1 b. 2 f. Take out of the barony of Abirdene, 5 c.; out of the barony of Raine, 3 c. 9 b. 2 f. 'Gif in' 8 b.

Marts out of the barony of Byrs, 16 marts. 'Gif in the twa parte of half a mart.'[20]

Third of wedders, 47. Take out of the barony of Tulynessill, 20; out of the barony of Byrs, 34. 'Gif in' 7 wedders.

Third of swine, 5⅔.[21] Out of the barony of Abirdene paying 11.[22]

---

[13] This total is difficult to verify because of discrepancies in the sub-totals.
[14] This is rendered 15 'martis, half mart and 3 part mart'. SRO, Vol. a has '17½' marts, not '47½'.
[15] This total would include 11 swine from the barony of Aberdeen, NLS, fo 117v, and one from barony of Damatt, as in SRO text, fo 370v.
[16] The correct calculation appears to be 64 dozen and 6.
[17] The correct calculation appears to be 119 dozen and 6.
[18] 'Obol' here is unnecessary.
[19] These figures add up to 11 c. 12 b. 2 f. 2⅔ p.
[20] Third of marts is given as 15⅔ in NLS, fo 116v.
[21] In the text this fraction is rendered 'half and 3 part sow'.
[22] These are not included in the list under the barony of Aberdeen, but should be included to make a total of 17; cf above, p. 409, n. 3, and p. 417, n. 15.

Third of capons, 21 dozen and 10. Take out of Tullynessill, 11 dozen; out of Clatt, 9 dozen; take the rest extending [to] 1 dozen and 6 capons out of the barony of Abirdene.[23]

Third of poultry, 39 dozen and 8. Out of Rayne, 23 dozen; out of Tullynessill, 17 dozen. 'Gif in foure pultrie.'

Third of oats, 2 c. 11 b. 2 f. 2 p.[24] Take out of Clatt, 1 c. 8 b.; out of Tullynessill, 1 c. 4 b. 'Gif in' 6 p.

Third of moorfowl, 6 dozen and 4. Take them out of Clat giving 9 dozen.

Third of geese, 18$\frac{1}{3}$. Take them out of Clatt paying 22.

Third of wheat, 1 c. 2 b. 2 f. 2$\frac{2}{3}$ p. Take this wheat out of Fordis giving 3 c. 8 b. 'Giff in the rest.'[25]

MONYMUSK, PARSONAGE AND VICARAGE OF,[26]

(SRO, Vol. a, fo 374r)

Rental of the parsonage and vicarage of Monymusk extends as follows. 'Item, the samyn sett to me, Mr Duncane Forbes of Monymusk, for the sowme of tua hundreth merkis be yeir.'

Signature: 'Mr Duncane Forbes.'

ABERDEEN, MINISTRY OF THE TRINITARIANS IN,[27]

(NLS, fos 118r-119r)
(SRO, Vol. a, fos 374r-375r)

'The rentall of the ministerie of the Triniteis of Abirdene gevin in be me, Maister Johnne Conynghame, assignaye constitut to the fruictis thairof.'

'The annuellis undir writtin pertening thairunto'

The lands of Cromar pay yearly 10 merks. The lands of Cragttowe [i.e. Craigtove] pay yearly 40s. The Mill of Furvy pays yearly 5 merks. The lands of Alexander Sinclers of Kaithnes pay yearly 40s. ['Item, the Ile of Stromnys in Kaitnes payis yeirlie' 10 merks, *deleted*.] The lands of Lachintille pay yearly 40s. The fishing of the half net of Lipot upon the Watter of Dee pays yearly 10s. 'Item, the land of umquhill Johnne Glasinwrycht lyand in the Gren Gait of the burgh of Abirdene betuix the landis of the Carmalet Freiris thairof on the east and vost partis payis yeirlie' 30d. 'Item, furth of the place of the Carmalet Freiris of Abirdene payis yeirlie' 6s 8d. 'Item, furth of the land of Alexander Roche lyand in the Gren Gait of the said burghe betuix the land of Johnne Durlat on the southt

---

[23] This would give 21 dozen and 6.
[24] The correct calculation appears to be 2 c. 11 b. 2 f. 2$\frac{2}{3}$ p.
[25] Third of wheat appears in SRO text only.
[26] Cf NLS, fo 137v (see below, p. 446).
[27] Cf NLS, fo 120r; EUL, fo 48r (see below, pp. 420, 558).

pairt and the Quenis Streit on the north pairt payis yeirlie' 6s 8d. Furth of the land of John Berryhill lying in the Grein Gait between the burn of the said burgh on the east and the Commowne Vennall of the same burgh on the west parts pays yearly 4s. Furth of the land lying at the foot of Sanct Katrenis Hill between the land of Sanct Thomas alter the Evangelist on the north 'and the vennall quhilk gangis to Sanct Katrenis chappell on the southt pairt' pays yearly 24s. Furth of the land of David Colp lying in the Schip Gait between the land of Alexander Gray on the east, the land of Alexander Malysoun on the south pays yearly 5s. Furth of John Kintors land yearly 30s [30d, *deleted*]. Furth of Andrew Bowsounis land in the Schip Gait pays yearly 40d. Furth of the land of John Ramsay lying in the Schip Gait between the land of Alexander Malysoun on the east and the Commowne Gait on the west pays yearly 6s 8d. Out of the land of John Crawfurd lying in the Schip Gait between the land of Robert Blindscheild on the north 'and the vennall that gangis to Sanct Katreinis chappell on the southe' pays yearly 4s. 'Item, the waistland pertening to the said Triniteis in parte lyand in the Gaistraw of the said burghe betuix the land of Mathow Wrycht on the northe and the land of David Giffurd on the southe pairtis, yeirlie' 6s 8d. Furth of the land of David Goldsmycht lying in the Castellgait of the said burgh between the lands of James Leslie on the east part and the land of Richard Greg on the west part pays 6s 10d. Furth of the lands of Andrew Stratoun lying in the Nedder Gait of the kirk of the said burgh between the lands of Alexander Gray on the east and Robert Kenarthy on the west parts pays yearly 8s 4d. Furth of the land of Thomas Culane lying in the Theif Gait[28] pays yearly 6s 8d. Furth of the lands of William Young lying in the Theif Gait of the said burgh between Mr Alexander Culane on the north and Gilbert Lethe on the south parts pays yearly 5s. Furth of the land of John Hill lying in the Gren Gait between the lands of the Carmalite Friars of the said burgh on the south and the King Street on the north parts pays yearly 6s 8d. Furth of the lands of George Hervy in Futy pays yearly 4s 4d. Furth of the land of Andrew Fyff lying in Castell Gait of the said burgh between the lands of Thomas Prate on the west and Sanct Peters altar on the east parts pays yearly 21s. Furth of the lands of William Blindscheill lying in Sanct Trinyanis [Ninianis, *fo 121v*] Streit between the lands of Robert Fudes on the west and John Mair on the east parts pays yearly 13s 4d. 'Furth of the waistland' pertaining to Alexander Blindscheill in the Theif Gait pays yearly 5s. Furth of John Blakis land yearly, 20s. The Mekill Yard occupied by John Chalmeris lying beside the place of the Triniteis pays yearly 21s. 'Item, the kariour boit in Gilbert Menzeis handis payis yeirlie' 8 merks.

'And attour the anuellis abone writtin thair is the malis of the landis of Ferryhill quhilkis ar sett in few ferme to Patrik Menzeis, burges of Abirdene, for payment yeirlie of' £20.

---

[28] I.e. The Justice Port or Robbers' Gate, Porta Latronum, *Cartularium Ecclesiae Sancti Nicholai Aberdonensis*, ed. J. Cooper, 2 vols. (Aberdeen, 1888-1892), ii, p. 153.

Signature: 'Mr Johannes Cunyghame. James Nicolsoun,[29] resaiff this rentale. J. Clericus Registri.'[30]

Calculation of third: In the whole, £54 14d; third thereof, £18 0s 4⅔d.

COLMONNEL, VICARAGE OF,[31] (NLS, fo 119v)
NIGG, VICARAGE OF, (SRO, Vol. a, fo 375r)
KINKELL, PENSION FROM THE PARSONAGE OF,[32]
GLASGOW, PENSION FROM THE PARSONAGE OF,[33]

Rental of the vicarage of Commenell pertaining to Mr John Davidsoun, £20. The vicarage of Nyg pertaining to Mr John, £10. The rental of the pension of the parsonage of Kinkell pertaining to the said Mr John, £40. The rental of the pension of the parsonage of Glasgow pertaining to the said Mr John, £33 6s 8d.

Signature: 'Ita est Johannes Davidsoun.'

ABERDEEN, TRINITARIANS IN,[34] (NLS, fo 120r-121v)
(SRO, Vol. a, fos 375v-376v)

'The draucht and copie of the rentall of the yeirlie annuelrentis of the Trinitie Freiris of Abirdene auchtand thairto within the burcht of the samyn continualie sen the yeir of God' 1557 'extending to xviij yeiris as eftir followis, viz.'

'The thesaurare in name of the toun of Abirdene of ane halff nettis fischeing of the Pot yeirlie' 10s, 'inde' £9. Patrick Handibellis in the Grene yearly, 30d, 'inde' 27s.[35] William Menzeis 'for the Quheitt Freiris place of Abirdene' yearly, 6s 8d, 'inde' £6. John Routh and Catherine Crawfurd 'of the land occupeitt be hir and hir tennentis pertening to him in the Grene yeirlie', 6s 8d, 'inde' £6. Gilbert Collesoun for the land of Walter Hay occupied by him in the Geistraw yearly, 6s 8d, 'inde' £6. Robert Forbes for his land in the Schipraw occupied by him yearly, 6s 8d, 'inde' £6. Patrick Gray for his land in the Castellgaitt occupied by him yearly, 6s 8d, 'inde' £6. Agnes Lowsoun 'sumtyme pertening to Andro Lowsoun

---

[29] James Nicolson, clerk of the Collectory, *Thirds of Benefices*, p. 62.
[30] James McGill of Nether Rankeilour was Clerk Register at the Reformation, *Handbook of British Chronology*, p. 197.
[31] Cf EUL, fos 56r, 56v (see below, pp. 569, 570). This benefice, in the sheriffdom of Ayr, was included here with Nigg and Kinkell because John Davidson held the benefices.
[32] Cf NLS, fos 137r, 139r (see below, pp. 445, 448).
[33] Cf EUL, fo 17v. This benefice, in the sheriffdom of Lanark, was included with Nigg and Kinkell in the sheriffdom of Aberdeen because John Davidson held the benefices.
[34] Cf NLS, fo 118r; EUL, fo 48r (see above, p. 418; below, p. 558).
[35] The correct calculation appears to be 45s.

for David Allasoune land and occupeitt be thame in the Schipraw yeirlie' 40d, 'inde' £3. Gilbert Kintor at the Keyheid yearly 'for his landis that James Ewin and [*blank*] Kintour his spous yeirlie' 30s, 'inde' £27. Alexander Fullertoun and [*blank*] for the land occupied by them in the Nethir Kirkgaitt yearly, 8s 8d, 'inde' £7 16s. Marion Patersoun in the Gaistraw of the land occupied by her yearly, 40d, 'inde' £3. Janet Ailhous for the [*blank*] crofts occupied by her yearly, £3, 'inde' £54. Mr Robert Chalmer 'for the tennentis yaird and waist forland occupeitt be him', yearly 25s, 'inde' £22 10s. Thomas Schand for the land occupied by him in the Schip Raw yearly, 13s 4d, 'inde' £12. 'Malie Walker in Futty, relict of umquhile Williame Cadonheid, yeirlie', 7s, 'inde' £6 2s.[36] Andrew Wysman, cordiner, for the land occupied by him yearly 6s 8d, 'inde' £12.[37] The land of John Berryhill in the Grene between the burn of the said burgh on the east and the Commoun Vennill of the same on the west parts pays yearly 4s, 'inde' £3 12s. Out of the land of David Colp lying in the Schip Raw between the lands of Alexander Gray on the east and the land of Alexander Malysoun on the south pays yearly 5s, 'inde' £4 10s. Furth of the land at the foot of Sanct Katherinis Hill between the lands of Sanct Thomas altaire the Evangelist on the north 'and the vennall that gangis to Sanct Katherinis chappell at the south', yearly 24s, 'inde' £21 12s. Furth of John Kintorres land yearly 30d, 'inde' 45s; 'I beleiff this be the same land that ye call Gilbert Kintorris land and that thir' 30d 'suld be' 30s, 'thairfoir gar speir it out'. Alexander Leslie of Kincragie yearly the said years for the lands thereof yearly, 40s, 'inde' £37.[38] Andrew, earl of Erroll,[39] for the Myln of Furvie yearly the said years, 5 merks, 'inde' £60. The earl of Cathnes[40] yearly the said years, 10 merks, 'inde' £120. The laird of Dunbeth for the Thriddis croft yearly, 40s, 'inde' £37.[41] [*blank*] Andersoun for the lands of [*blank*] Watsoun of Cromar occupied by them yearly the said years, 10 merks, 'inde' £120. [*blank*] for the lands of Craigtowy yearly, 40s, 'inde' £36. Alexander Sincleir for his lands in Cathnes yearly, 40s, 'inde' £36. Alexander Forbes of Petslego for the lands of Leichintullie yearly, 40s, 'inde' £36. Patrick Menzeis, burgess of Abirdene, for the yearly feu mails of the lands of Ferry Hill each year the said years, £20, 'inde' £360. Furth of the land of John Ramsay lying in the Schip Gaitt between the lands of Alexander Malysoun on the east and the Commoun Gaitt on the west yearly, 6s 8d, 'inde' £6. Furth of the land of John Crawfurd lying in the Schip Gaitt between the land of Robert Blindscheill on the north 'and the vennell that gangis to Sanct Katherinis chappel on the south' pays yearly 4s, 'inde' £3 12s. The 'waist land' pertaining to the Triniteis in perpertuity lying in the Gaistraw between the land of Matthew Wrycht on the north and the land of David Giffuird on the south yearly 6s 8d, 'inde' £6. Furth of the land of David Goldsmycht in the

---

[36] The correct calculation appears to be £6 6s.
[37] The correct calculation appears to be £6.
[38] The correct calculation appears to be £36.
[39] This 'copy' of the rental for 1557 was used in assessing valuations for 18 years. Andrew Hay, 8th earl of Erroll, succeeded his father, George, 7th earl, in 1574, *Scots Peerage*, iii, pp. 569-572.
[40] I.e. George Sinclair, 4th earl of Caithness, *Scots Peerage*, ii, pp. 338-340.
[41] The correct calculation appears to be £36.

Castell Gaitt of the said burgh between the lands of James Leslie on the east part and the land of Richard Greg on the west pays yearly 6s 10d, 'inde' £6 4s.[42] Furth of the land of Thomas Cullane lying in the Gaillowgaitt yearly 6s 4d. Furth of the land of William Young lying in the Theif Gaitt of the said burgh, now called the Gallowgaitt,[43] between the land of Mr Alexander Cullane in the north, Gilbert Leithe on the south yearly £5,[44] 'inde' £4 10s. Furth of the land of John Hill lying in the Grene Gaitt between the lands of the Carmalite Friars on the south and the Kingis Street on the north yearly 6s 8d, 'inde' £6. Furth of the land of George Hervie in Futty yearly 4s 4d, 'inde' £3 18s. Furth of the lands of Andrew Fyff lying in the Castell Gaitt of the said burgh between the lands of Thomas Prattis on the west and Sanct Peteris altar on the east yearly 21s, 'inde' £18 18s. Furth of the lands of William Blindscheill lying in Sanct Ninianis [Trinyanis, *fo 119r*] Strett between the lands of Robert Feddes [Fudes, *fo 119r*] on the west and John Mair on the east yearly 13s 4d, 'inde' £12. 'Furth of the waist land' pertaining to Alexander Blindscheill in the Theif Gaitt or Gallowitt [*sic*] yearly 5s, 'inde' £4 10s. Furth of John Blakis land yearly, 20s, 'inde' £18. 'Item, the karior boit or the keill in Gilbert Menzeis handis' yearly 8 merks, 'inde' £140.[45]

ABERDEEN, DEANERY OF,[46]

(NLS, fos 122v-124r[47])
(SRO, Vol. a, fos 377r-378v)

'Dene of Abirdene'

Third of money, £110 7s 6d 'ob 3 ob'. Take the teinds of Stralocht for 33 merks; the teinds of Monycabok and Swaillend for 25 merks; the teinds of Elryk for 10 merks; the teinds of Kingis Seitt for 4 merks; the teinds of half Kynmund for 10 merks; Ald Govill for 10 merks; Boghoill for 5 merks; Clubbisgovill for 17 merks; Litill Govill for 6 merks; Perslie for 6 merks 6s 8d; Granden for 6 merks; Dilspro for 8 merks; Deynstoun for 15 merks; Scottistoun for 12 merks. Total, £111 13s 4d. 'Gif in' 25s 9d '2 part ob'.

Third of bere, 1 c. 10 b. Take it out of the Grene End of Abirdene giving 3 c. 'Gif in the rest.'

Third of meal, 1 c. 1 b. Out of Mamewlaycht giving 29 b. meal and 8 b. bere. 'Gif in the rest of the meal.'

Third of malt, 1 b. 'Out of Kethokis meill [Catholris Myln, *fo 123v*] payand' 3 b. malt.

Third of 'victuall undividit' 4 b. 2 f. 2⅔ p. Out of the Cottoun paying yearly 14 b. victual.

---

[42] The correct calculation appears to be £6 3s.
[43] See above, p. 419, n. 28.
[44] To give a total of £4 10s, this should read 5s.
[45] The correct calculation appears to be 144 merks or £96.
[46] Cf NLS, fo 131r (see below, p. 436).
[47] Fo 122r is blank.

Third of salmon, 1 barrel. Out of Rutherstoun paying 3 barrels.
'By the vicarage [i.e. of St Machar] quhilk wes wont as he affirmis to giff' £80 'and now na thing, caus thairfoir my lord Comptrollare[48] avyis thairwith.'

'The rentale of the denrie of Abirdene, teind schavis, maillis of few landis and vicaragis.'

The lands of Straloycht set to the laird of the same, the teind sheaves thereof extends 'be yeir', 33 merks. Monycabok and Swayll End, the teind sheaves thereof, 25 merks. Elryk, the teind sheaves thereof, 10 merks. The Kingis Sett, the teind sheaves thereof, 4 merks. The half of Kynmownd, the teind sheaves thereof, 10 merks. Auld Govyll, the teind sheaves thereof, 10 merks. The Boghoill, the teind sheaves thereof, 5 merks. Clubbisgovill, the teind sheaves thereof, 17 merks. Lytill Govyll, the teind sheaves thereof, 6 merks. Perslie, the teind sheaves thereof, 6 merks 6s 8d. Grandene, the teind sheaves thereof, 6 merks. Dilspro, the teind sheaves thereof, 8 merks. The Deinstoun, the teind sheaves thereof, 15 merks. Carnfeild and Cordebuk, the teind sheaves thereof, 6 merks. The Scottistoun, the teind sheaves thereof, 12 merks. The West Feild, the teind sheaves thereof, 6 merks 8s. The Brigtoun of Balgowny, the teind sheaves thereof, 20 merks. Murcowr, the teind sheaves thereof, 30 merks. Mondowrnocht, the teind sheaves thereof, 10 merks. The Berry Hillok, the teind sheaves thereof, 3 merks. The Ward of Kynmonde, the teind sheaves thereof, 14 merks. The Setoun, the teind sheaves thereof, 7 merks. The New Myln crofts, the teind sheaves thereof, 24s. 'The toun of the teind of Crwis',[49] the teind thereof, 7 merks. Auchmull, the teind sheaves thereof, 12 merks. The Bischoppis Clintirte,[50] the teind sheaves thereof, 4 merks 6s 8d [4 merks 4s 8d, *SRO*]. The lands of Stanywod, the teinds thereof set to the laird for the sum of 43 merks 6s 8d [43 merks 6s 4d, *SRO*]. Crabbistoun, the teind sheaves thereof, 10 merks. The Tullacht, the teind sheaves thereof, 5 merks. Schetaucht Ley, the teind sheaves thereof, 9 merks. Kingis Wallis, the teind thereof, 2 merks. Heslie Hed, the teind sheaves thereof, 5 merks 6s 8d. Ribbislaw, the teind sheaves thereof, 5 merks. The Ferry Hill, the teind sheaves thereof, 10 merks. Bogferlow, the teind sheaves thereof, 6 merks. Gilcamstoun, the teind sheaves thereof, 15 merks. Rutherstoun, the teind sheaves thereof, 8 merks, 'and thre barraill of salmond'. Petmukkistoun, the teind thairof, 7 merks[51] 6s 8d. Capristoun, the teind thereof, 3 merks. The Froster Sett, the teind thereof, 13s 4d. Mamewlacht, the teind sheaves thereof set for 37 b. victual, 8 b. thereof being bere. The half of Kynmundy set for 18 b. victual, 4 b. thereof being bere. The teind sheaves of

---

[48] Bartholomew de Villemore, Comptroller, 1555 - 1562, with Thomas Grahame of Boquhaple, 1561 - 1562. They were succeeded by Sir John Wishart of Pittarow, 1562 - 1565. *Handbook of British Chronology*, p. 191.
[49] SRO text reads 'the toune of the Crowes, the teind thairof'; 'The Cruives', W. M. Alexander, *The Place-Names of Aberdeenshire* (Aberdeen, 1952), pp. 44-45.
[50] Cf *Place-Names of Aberdeenshire*, p. 215.
[51] NLS text reads 'vii [vi, *deleted*]'.

Catholris Myln set for 3 b. malt. Slatyr, the teind sheaves thereof, 9 b. meal and 5 b. bere, with 40s money. 'The Cottowne usit to be leid and is' 14 b. victual. The teinds of the Greyne End 'yeirlie leid and ar commonly worth' 3 c. 12 b. bere. The mail of the feu lands is 36 merks 10s. 'Item, the ordinare chargis is be yeir' 62 merks. 'Item, the vicaragis was wont to be' 120 merks 'bot now thair is na payment gotting thairof.'

'This is the trew rentaill of the denrie of Abirdene that it gevis presentlie, subscrivit with my hand at Abirdene', 14 February 1561/2.

Signature: 'R. Erskin, dene of Abirdene.'

Calculation of third: Money, £331 2s 8d; third thereof, £110 7s 6d 'ob 3 ob' [i.e. $\frac{7}{8}$d]. Bere, 4 c. 14 b.;[52] third thereof, 1 c. 10 b. Meal, 3 c. 3 b.;[53] third thereof, 1 c. 1 b. Malt, 3 b.; third thereof, 1 b. 'Victuall in general', 14 b.; third thereof, 4 b. 2 f. 2$\frac{2}{3}$ p. Salmon, 3 barrels; third thereof, 1 barrel.

'By the vicarage [*i.e.* of St Machar] quhilk he affermis to geif nathing bot was wount to giff' £80. 'Seik the assumptioun of this [rentall, *deleted*] benefice immediatlie befoir this rentall.'[54]

ABERDOUR, PARSONAGE AND VICARAGE OF, (NLS, fo 124r)
(SRO, Vol. a, fo 378v)

Rental of the parsonage and vicarage of Abirdoure pertaining to Mr Robert Carnegy, parson thereof.

The parsonage and vicarage of Abirdour set in assedation to Sir Robert Carnegy of Kynnard, 'knycht, lyik as it was sett of befoir be Maister James Leislie, thane persone of the samyn, to the lard of Phillorth for' 220 merks, 'quhairof is to be deducit to sir Alexander Ramsay, vicare pensionare thairof', 20 merks; 'and for the corpspresentis, umaist clathyhs and Pasche offerandis', 20 merks. 'Sua thair restis frie to the persoun foirsaid' 180 merks.

Calculation of third: *see below*, Parsonage and vicarage of Tough.

---

[52] The correct calculation appears to be 4 c. 13 b.
[53] The correct calculation appears to be 3 c. 4 b.
[54] I.e. on fo 122v (see above, pp. 422-423).

TOUGH, PARSONAGE AND VICARAGE OF, (NLS, fo 124v)
(SRO, Vol. a, fo 379r)

Rental of the parsonage and vicarage of Towch pertaining to William Carnegy, parson thereof.

The parsonage and vicarage of Touch set in assedation to Sir Robert Carnegy of Kynnaird, knight, 'lyik as it was sett of befoir be Maister Andro Arnote, thane persone thairof, to Maister Williame Gordone and utheris for ane hundreyth merkis yeirlie, quhairof thair is to be deducit to sir Petir Murray, vicare pensionare thairof, the sowme of' 20 merks; 'and for the corps presentis, umaist clathis and Pasche offerandis', 10 merks. 'Sua thair restis to the persoun foirsaid the sowme of' 70 merks.

Signature: 'R. Carnegy, etc.'

Calculation of third:
   Parsonage and vicarage of Abirdowir, in the whole, £146 13s 4d; third thereof, £48 17s 9⅓d.
   Parsonage and vicarage of Touch, in the whole, £66 13s 4d; third thereof, £22 4s 5⅓d.

ABERDEEN, COLLEGE OF, (NLS, fo 125r)
(SRO, Vol. a, fo 379r)

The college of Abirdene within the diocese thereof

The kirks of Glenmowk, Abirgardin, Slanis annexed to the college extend to £60 Scots money and 4 c. victual. And the kirk of Snaw annexed to the said college extends to 30 b. bere, 'thairof payit to the vicare' 10 merks, 'and to the maister of grammer scowll' 5 merks.

[*In margin*, 'Gevin in and insert be the rentall of the college self.']

ABERDEEN CATHEDRAL, COMMON KIRKS OF,[55] (NLS, fo 125r)
(SRO, Vol. a, fo 379r)

'The commoun kirkis of the chaptoure of Abirdene': The fruits of the kirk of Fordyce extend yearly to £200. The fruits of Logybuchane set in assedation for £120. The fruits of the kirk of Rathin set in assedation for £120. The fruits of the kirks of Kildrymmee, Glenbuchatt and Logymare set in assedation for £100. The

---

[55] Cf NLS, fo 132r (see below, p. 437).

fruits of Dunmeth set in assedation for £20. 'Off the quhilk payit yeirlie in pensionis and charges' £200 'as thair yeirlie comptis proportis.'

[*In margin*, 'Bettir rentallit and insert be the dene of Abirdeneis ingevinge.']

SPITTAL, PARSONAGE OF,[56]

(NLS, fo 125r)
(SRO, Vol. a, fo 379v)

The parsonage of Spittaill pertaining to Mr Alexander Kyd, 'subshantour', extends yearly to £50 'of the quhilk payit yeirlie to tua chaplanes of the cathedrall kirk' £20.

[*In margin* 'Rentallit and insert be his awin ingevinge.']

KINAIRNEY, VICARAGE OF,[57]

(NLS, fo 125v)
(SRO, Vol. a, fo 379v)

The vicarage of Kynarny pertaining to him [*i.e.* Mr Alexander Kyd] extends to 40 merks.

COULL, VICARAGE OF,

(NLS, fo 125v)
(SRO, Vol. a, fo 379v)

The vicarage of Cowll pertaining to Mr Andrew Leslie extends yearly to 40 merks.

Calculation of third: In the whole, £26 13s 4d; third thereof, £8 17s $9\frac{1}{3}$d.

DYKE, VICARAGE OF,[58]

(NLS, fo 125v)
(SRO, Vol. a, fo 379v)

The vicarage of Dyik within the diocese of Murray pertaining to Mr John Leslie 'was wont to be sett in assedatioun syndre yeiris bygane for' 40 merks, 'bot nathing payit thir thre yeiris bygane.'

Calculation of third: In the whole, £26 13s 4d; third thereof, £8 17s $9\frac{1}{3}$d.

---

[56] Cf NLS, fo 129v (see below, p. 434). Spittal was a prebend of Aberdeen cathedral, belonging to the subchanter; Alexander Kyd was subchanter, 1533 x 1534 - 1563, Watt, *Fasti*, p. 17.
[57] Cf NLS, fo 130r (see below, p. 434).
[58] Dyke lay in the sheriffdom of Elgin and Forres, cf, *RMS*, v, no 1727. The Books of Assumption contain no separate section for the sheriffdom of Elgin and Forres.

TULLYNESSLE, PARSONAGE OF,[59] (NLS, fo 125v)
(SRO, Vol. a, fo 379v)

The parsonage of Tullinessill pertaining to Mr William Cabell set in assedation for £40 'thir money [*i.e.* many] yeiris bygane.'

[*In margin*, 'Bettir rentalit be his awin hand wratis.']

INVERURIE, VICARAGE OF,[60] (NLS, fo 125v)
(SRO, Vol. a, fo 379v)

The vicarage of Innerowry pertaining to him [*i.e.* Mr William Cabell] set in assedation yearly for 40 merks.

PETERCULTER, VICARAGE OF, (NLS, fo 125v)
(SRO, Vol. a, fo 379v)

The vicarage of Petircultir pertaining to sir William Meldrum extends yearly to 40 merks 'bot onpayit thir thre yeiris bygane.'

Calculation of third: In the whole, £26 13s 4d; third thereof, £8 17s 9⅓d.

LONMAY, PARSONAGE OF, (NLS, fo 126r)
(SRO, Vol. a, fo 380r)

The parsonage of Lunmey pertaining to Mr James Gordone set in assedation for 100 merks 'thir mony yeiris bygane'.

'Personage of Lunmey, Dynmur,[61] takkisman.'

Calculation of third: In the whole, £66 13s 4d; third thereof, £22 4s 5⅓d.

---

[59] Cf NLS, fo 131v (see below, p. 436).
[60] Cf NLS, fo 131v (see below, p. 437).
[61] Cf 'David Dynmur', *RMS*, iv, no 1131; G. F. Black, *The Surnames of Scotland* (New York, 1946), p. 230.

ABERDEEN, CHANCELLORY OF,[62]   (NLS, fo 126r)
                                                        (SRO, Vol. a, fo 380r)

The chancellory of Abirdene pertaining to Mr Alexander Setoun set in assedation for £160 'thir mony yeiris bygane.'

[*In margin,* 'Gevin in be him self.']

Signature: 'Ita est Magister Joannes Leslie de Owyne, manu sua.'

RUTHVEN, VICARAGE OF,   (NLS, fo 126r)
                                                        (SRO, Vol. a, fo 380r)

The vicarage of Ruthven in Badyenocht pertaining to Mr Hercules Carnegy set in assedation to Mr George Gordoun for yearly payment of 40 merks, 'quhairof to be deducit for the corps presentis and Pasche offerandis yeirlie the sowme of' £10.

Calculation of third: In the whole, £26 13s 4d; third thereof, £8 17s 9⅓d.

Signature: 'Robert Carnegy.'

ROTHIEMAY, VICARAGE OF,   (NLS, fo 126v)
                                                        (SRO, Vol. a, fo 380r)

The vicarage of Rothemay, 'quhairof Maister Henrie Lummysden is vicare, liand within the schirefdome of Abirdene, extendis yeirlie in all proffeittis to' 20 merks.

Signature: 'Maister Henrie Lummysden. Rasavit at the command of the Clerk Register.'

Calculation of third: In the whole, £13 6s 8d; third thereof, £4 8s 10⅔d.

OYNE, PARSONAGE AND VICARAGE OF,   (NLS, fo 126v)
                                                                  (SRO, Vol. a, fo 380v)

Rental of the parsonage and vicarage of Owyn pertaining to Mr John Leslie, parson of the same.

The said parsonage and vicarage has been set in assedation to Patrick Leycht 'thir dyveris yeiris bygane for' 260 merks 'bot presentlie gevis nocht sameklie be

---

[62] Cf NLS, fo 129r (see below, p. 432), where the rental is given as £260, not £160 as here.

ressoun of the decaiy of the vicarage in' £20, 'off the quhilk he payis to the pensionar and stallare' 32 merks.

Signature: 'Mr Joannes Leislie de Owyn.'

Calculation of third: In the whole, £173 6s 8d; third thereof, £57 15s 6⅔d.

FORGUE, VICARAGE OF, (NLS, fos 126v-127r)
(SRO, Vol. a, fo 380v)

The vicarage of Forge pertaining to the said Mr John Leislie, set in assedation to the laird of Frendracht for £40, of which there is paid to the curate £10.

Signature: 'Joannes Leislie de Owyn.'

Calculation of third: In the whole, £40; third thereof, £13 6s 8d.

'Omittit the mans quhilk is ane plewcht of land quhilk presentlie is in the Officialis awin hand and the quene to be answerit.'

BANCHORY-DEVENICK, VICARAGE OF, (NLS, fo 127r)
(SRO, Vol. a, fo 380v)

[*In margin*, 'Vicarage of Banchorie'.]

Rental of the vicarage of Banchorie-Devynik within the diocese of Abirdene pertains to Mr Robert Merser, vicar thereof, extends to the sum of £24 yearly.

Calculation of third: In the whole, £24; third thereof, £6 13s 4d [*recte*, £8].

TYRIE, PARSONAGE AND VICARAGE OF, (NLS, fo 127r)
(SRO, Vol. a, fo 381r)

Rental of the kirk of Tyrye within the diocese of Abirdene pertaining to Alexander Andersoun, parson thereof.

The teind sheaves of the said kirk extend to £40; and the vicarage of the same was set for £20 'quhen it was payit, quhilk was nevir taxit be kynge nor bischope.'

Signature: 'Alexander Andersoun, persoun of Tyrye.'

Calculation of third: In the whole, £60; third thereof, £20.

'Nota, this pertenis nocht to the college.'

AUCHTERLESS, PARSONAGE AND VICARAGE OF,

(NLS, fo 127v)
(SRO, Vol. a, fo 381r)

Parsonage and vicarage of Ochtirles

'To the quhilk Johnne, lord Innermayth,[63] is takkisman and subtennent ontill him the lard of Fyvie and his tutouris, is sett in assedatioun for' 240 merks. 'Off the quhilk mone be deducit the stallare fie, viz.', 12 merks; 'the vicare pensionaris fie', £10. 'By procurationis, synodullis and visitationis sylver in patria non reformata[64] extendinge to' £4. 'Item, ane fourty pund pensioun to Maister David Dyschintoun togidder with ane resonable fie to ane minister. Item, the fermoraris hes intakit hithirtill of compleit payment be ressoun of want nowadayis of the small teindis and umaist claith, siclyik.'

Signature: 'Archibald Betoun. Presentit be my lord Innermeth in name of Mr Archibald Betoun. Innermeth.'

Calculation of third: In the whole, £160; third thereof, £53 6s 8d.

KINCARDINE O'NEIL, VICARAGE OF,

(NLS, fo 127v)
(SRO, Vol. a, fo 381r)

The vicarage of Kincardine Onyll pertaining to Mr Thomas Levingstoun set by his predecessors to sir Alexander Ostaige[65] and Thomas Edy for yearly payment of £20.

Signature: 'Alexander Levingstoun.'

Calculation of third: In the whole, £20; third thereof, £6 13s 4d.

INSCH, VICARAGE OF,

(NLS, fo 128r)
(SRO, Vol. a, fo 381v)

Rental of the vicarage of Inche lying within the diocese of Abirdene pertaining to William Ballingall, vicar of the same, 'was of auld' £40 'and now fure pleuchis

---

[63] John Stewart, 4th lord Innermeath, *Scots Peerage*, v, p. 5.
[64] I.e. 'in the unreformed country; before the Reformation.'
[65] Alexander Ewstache is on record as reader at Kincardine O'Neil by 1563, *Thirds of Benefices*, p. 223.

teind pertenand to the said vicarage', 24 merks; and the glebe, 4 merks; 'and I want the haill uther teindis thir thre yeiris quhilk wald have maid out the' £40.

Signature: 'William Ballingall, vicar of Inche, with my hand.'

Calculation of third: In the whole, £40; third thereof, £13 6s 8d.

CRIMOND, PARSONAGE OF,[66] (NLS, fo 128r)
(SRO, Vol. a, fo 381v)

Rental of the parsonage of Crechmond lying in the diocese of Abirdene gives yearly the sum of £100, 'fyve pundis thairof gevin to the stallaris of Abirdene yeirlie befoir the yeir of God' 1558.

Signature: 'Arthoure Taillefere, vicar thairof.'

Calculation of third: In the whole, £100; third thereof, £33 6s 8d.

ABOYNE, VICARAGE OF,[67] (NLS, fo 128r)
(SRO, Vol. a, fo 381v)

Rental of the vicarage of Awboyne lying in the diocese of Abirdene gives yearly 80 merks, 'twenty pundis[68] thairof to Maister William Hay, persoun of Torreff, be provisioun of the paipe, sua restis to the said Maister Arthoure bot fyftie merkis yeirlie befoir the yeir of God' 1558.

Signature: 'Mr Arthoure Tallifere, vicar thairof.'

Calculation of third: In the whole, 80 merks; third thereof, 26 merks 8s $10\frac{2}{3}$d.

PETERUGIE, VICARAGE OF, (NLS, fo 128v)
(SRO, Vol. a, fo 382r)

Rental of the vicarage of Peitirugy

100 lambs, £12. 20 stones of wool, £10. 40 stones of cheese, £9. Teind fish, £9. Total, £40.

---

[66] Cf NLS, fo 134r (see below, p. 440).
[67] Cf NLS, fos 134r, 140v (see below, pp. 440, 451).
[68] Fo 140v reads 'xx merkj'.

'Item, the curattis fie to be takin of this', £10. 'Item, the biscope and archidane for thair procurationis', £4. Total, £14. 'And sua restis fre', £26.

Signature: 'Ita est Dominus Patricius Ogstoun, vicarius de Petirugy, manu propria. Presentit be me, Mr Alexander Skeyne.'

Calculation of third: In the whole, £40; third thereof, £13 6s 8d.

RAYNE, PARSONAGE AND VICARAGE OF,[69] (NLS, fo 128v)
(SRO, Vol. a, fo 382r)

Rental of the parish of Rayne within the diocese of Abirdene

The parsonage and vicarage set for 290 merks 'heirof to be deducit for the ordinare chargis, viz., for the vicare pensionaris fie and stallares with procurationis' £30. 'Item, by this rentaill, the corps present, umaist claith with the utheris oblationis and Pasche fynes' £24 'quhilkis ar nocht payit now. Item, the visitationis pertenynge to the archidenrie siclyk nocht payit now extendinge to' £32.

'And this rentall presentit be Mr Johnne Stewart.'

'Archidenrie thairof, personage and vicarage of Rayne.'

Calculation of third: In the whole, £193 6s 8d; third thereof, £64 8s 10⅔d.

ABERDEEN, CHANCELLORY OF,[70] (NLS, fo 129r)
(SRO, Vol. a, fo 382r-v)

Rental of the chancellory of Abirdene pertaining to Mr Alexander Setoun, chancellor.

The parsonage and vicarage of the parish kirk of Birs pertaining to the said chancellory set in assedation 'thir money yeiris bigane' to Alexander Gordone of Stradowyne yearly for the payment of £160. The teind salmon of the Waltir of Done set in assedation 'thir money yeiris bygane for the payment of, pertenand to the said chancellarie', £70. The vicarage of Bothelny [*i.e.* Bethelnie] set in

---

[69] Rayne was a prebend of Aberdeen cathedral pertaining to the archdeacon, John Stewart, who held office from 1551 to 1563, Watt, *Fasti*, p. 21.

[70] Cf NLS, fo 126r (see above, p. 428), where the rental is given as £160, not the £260 (i.e. £160 plus £70 plus £30) as specified here.

assedation 'thir mony yeiris bygane' for £30 pertaining to the said Mr Alexander Setoun. 'Item, of the chancellarie to be deducit to the vicar of the kirk', 20 merks. 'Item, to the stallar of the kirk', £9. 'Item, of the vicarage of Bothelny to be deducit to the vicar of the kirk', 20 merks.

Signature: 'Mr Alexander Setoun, chancellar of Abirdene, etc.'

Calculation of third: In the whole, £260; third thereof, £86 13s 4d.

TURRIFF, PARSONAGE AND VICARAGE OF,[71]   (NLS, fo 129r-v)
                                                  (SRO, Vol. a, fo 382v)

Rental of the parsonage of Turreff

The parsonage of Turreff, lying within the diocese of Abirdene pertaining to Mr William Hay, gives 93 b. victual; in silver, £140. The vicarage of the same pertaining to the said Mr William Hay gives yearly £100. Total, £240 argent; 93 b. victual.

'Item, out of this gevin furth to the vicare pensionar yeirlie' 40 merks. 'Item, to the chaplane stallar in the cathedrall kirk' 40 merks. 'Item, in small uthire chargis' 53s 4d.

Parsonage and vicarage of Turreff

Calculation of third: 'In the haill of money', £240; third thereof, £80. Victual, 5 c. 13 b.; third thereof, 1 c. 14 b. 1 f. 1½ p.[72]

WESTHALL, CHAPLAINRY OF,[73]   (NLS, fo 129v)
                                        (SRO, Vol. a, fo 383r)

Rental of the chaplainry called the Waisthaill

The chaplainry called the Westhall in the diocese of Abirdene pertaining to sir William Capell is set for £40, 'heirof yeirlie pensioun is gevin to the chaplanis in Abirdene extendinge to' 50s.

Calculation of third: In the whole, £40; third thereof, £13 6s 8d.

---

[71] Cf NLS, fo 136v (see below, p. 445).
[72] The correct calculation appears to be 1 c. 15 b.
[73] Westhall is situated in the neighbourhood of the parish church of Oyne.

SPITTAL, PARSONAGE OF,[74]

(NLS, fos 129v-130r)
(SRO, Vol. a, fo 383r)

'The kirk of Spittal callit the hospitall'

'The rentall of the hospitall kyrk besyd Abirdene pertenand to the subschantour of Abirdene.'

'Imprimis of penny maile payit be the fewaris and induellaris of the landis of the said hospitall' £20 'be yeir'. The teind sheaves of the said hospital extend in the year to £26 13s 4d. The fermes of the same 'be yeir' extend to 30 b. bere.

'Item, payit furth of this foirsaid rentale be yeir to tua chaplanis of the first and premive [sic] fundatioun of the hospitall', £20. 'Item, for the servinge of the cure', £10.

Signature: 'Alexander Kyd, succentor Abirdonensis.'

'The kirk of Spittall quhilk heir is callit hospitall betuix the Aberdenes.'

Calculation of third: Money, £46 13s 4d; third thereof, £15 11s 1½d. Bere, 30 b.; third thereof, 10 b.

KINAIRNEY, VICARAGE OF,[75]

(NLS, fo 130r)
(SRO, Vol. a, fo 383v)

Rental of the vicarage of Kynarny within the diocese of Abirdene.

'Item in primis, for the teind schavis pertenand to the said vicarage be yeir payit be the tennentis and induellaris in the samyn drawis to' £40 'be yeir. Item, the lambis woll and cheis of the same be yeir extendis to the sowme of' £6 'and eftir as the cours of the yeir happines, quhyllis les, quhyllis mair. Item, payit furth yeirlie of the said vicarage to ane chaplane and minister to serve the cuir of the samyn', £13 6s 8d.

Signature: 'Alexander Kyd, succentor Abirdonensis.'

Calculation of third: In the whole, £46; third thereof, £15 6s 8d.

---

[74] Cf NLS, fo 125r (see above, p. 426).
[75] Cf NLS, fo 125v (see above, p. 426).

ECHT, VICARAGE OF, (NLS, fo 130r-v)
(SRO, Vol. a, fo 383v)

The vicarage of Echt in Mare set for 40 merks 'quhen corps present and umaist clayth was payit to the vicare pensionar of the samyn, quha quhen corps present and umaist clayth was layit done geif our the samyn.'

Signature: 'Mr James Gray, with my hand.'

Calculation of third: In the whole, £26 13s 4d; third thereof, £8 17s 9½d.

METHLICK, PARSONAGE AND VICARAGE OF,[76] (NLS, fo 130v)
(SRO, Vol. a, fos 383v-384r)

Rental of Methlyk within the diocese of Abirdene

'Imprimis, the lard of Haddowkis haill teind schavis within the said paroche extendis to' £80. The laird of Tullegowneis whole teind sheaves extend to £13 6s 8d. The laird of Schethinis whole teind sheaves extend to £29 6s 8d. The teind sheaves of Lytill Drumquhendill, £16. The teind sheaves of Lytill Methlyk, £3 6s 8d. The vicarage 'quhen it was payit in the auld manere wald haif extendit to' £20 'or thairby bot nathinge heirof payit thir thre yeiris bypast. Summa of the haill personage and vicarage beinge payit extendis to' £162.

Signature: 'Thomas Burnett of Methlik.'

'And payis of the foirsaid sowme to the minister' £20 'by uther ordinar charges.'

Calculation of third: In the whole, £162; third thereof, £54.

CLATT, PARSONAGE OF, (NLS, fo 131r)
(SRO, Vol. a, fo 384r)

The parsonage of Clett within the diocese of Abirdene pertains to Mr James Gordone, son to the earl of Huntlie,[77] set in assedation with the vicarage of the same for £100. Thereof to my lord Kilwyning[78] in yearly pension, £40; and to the curate, £16. 'Sua restis to the possessoure' £44.

---

[76] Cf NLS, fo 137r (see below, p. 445).

[77] James Gordon, fifth son of George, 4th earl of Huntly, subsequently spent much of his life abroad as a Jesuit in Paris, Bordeaux and Rome, *Scots Peerage*, iv, pp. 534-537. This rental is to be dated before the 4th earl's death at Corrichie in October 1562. Cf below, p. 436, n. 80.

[78] I.e. Gavin Hamilton, commendator of Kilwinning, Cowan and Easson, *Medieval Religious Houses*, p. 69; Watt, *Fasti*, p. 299.

Signature: 'Mr Thomas Keir, manu sua, and presentit be him at the command of the erle Huntlie in the name of the foirsaid possessouris.'

Calculation of third: In the whole, £100; third thereof, £33 6s 8d.

ABERDEEN, DEANERY OF,[79]

(NLS, fo 131r)
(SRO, Vol. a, fo 384r)

'The denrie of Abirdene pertening to Maister Robert Erskin.'

'Item, sett in assedatioun for' 500 merks. 'Thairof ane hundreyth merkis in vicarages quhilkis is nocht payit thir thre yeiris. Sua the possessoure gaittis bott' 400 merks.

Signature: 'Mr Thomas Keir, manu sua. Presentit be him in name of the foirsaid possessour.'

BANCHORY-DEVENICK, PARSONAGE OF,

(NLS, fo 131r-v)
(SRO, Vol. a, fo 384v)

The parsonage of Banchorie within the diocese of Abirdene pertaining to Mr James Gordoun, brother to the earl of Huntlie.[80]

'Item, sett in assedatioun for the sowme of' £60.

Signature: 'Mr Thomas Keir, manu sua, and presentit be him in the name of the said possessour.'

Calculation of third: In the whole, £60; third thereof, £20.

TULLYNESSLE, PARSONAGE OF,[81]

(NLS, fo 131v)
(SRO, Vol. a, fo 384v)

Rental of the teind sheaves of the parsonage of Tullenessill.

The teind sheaves of Mongarye, £21 15s. Fowlislie, £10 13s 4d. Quhothorny and Curemyll, £16. The Bray, £3. Dulpersie and Warakstoun, £8.

---

[79] Cf NLS, fo 122v (see above, p. 422).

[80] In the rental of the parsonage of Clatt, above, James Gordon is described as the son of the earl of Huntly. With the death of George, 4th earl, in 1562 and his subsequent forfeiture by 1563, the earldom passed to the second son, George, 5th earl, who gained the office of Chancellor in 1565, though the sentence of forfeiture was repealed only in 1567. *Scots Peerage*, iv, pp. 534-540. This entry was evidently composed at a later date than that for Clatt, see above, p. 435, n. 77.

[81] Cf NLS, fo 125v (see above, p. 427).

'Off the quhilk thair is to be payit furth in ordinare' £5 6s 8d.

Calculation of third: *see below*, Vicarage of Inverurie

INVERURIE, VICARAGE OF,[82] (NLS, fo 131v)
(SRO, Vol. a, fo 384v)

Vicarage of Innerowry

'It was wount to pay to me quhen kirk richtis and all uthir dewities usit and wont was payit' 50 merks Scots money, 'off the quhilk I haif nocht rasavit nor gottin payment of the thrid parte thairof thir thre yeiris bypast. Abuf the foirsaid vicarage I haif four pleuche teind schavis of Cokkinglas pertenand to the said vicarage extending till' £22 'in assedatioun to the lard of Balhaggertie within the paroche of Innerowrie, of the quhilkis' £22 'I haif nocht rasavit ane grott thir haill thre yeiris bypast.'

Signature: 'Mr William Cabell, persoune of Tullinessill and vicar of Innerowry, with my hand.'

Calculation of third:
  Parsonage of Tullynessill, in the whole, £59 8s 10d;[83] third thereof, £19 16s 1½d.
  Vicarage of Innerowry, in the whole, £55 6s 8d; third thereof, £18 8s 10d.[84]

ABERDEEN CATHEDRAL, COMMON KIRKS OF,[85] (NLS, fos 132r-133r)
(SRO, Vol. a, fo 385r-v)

[*In margin*, 'Commoun kirkis of the chaptour of Abirdene']

The kirk of Fordice: For the teind sheaves of Findlatter, Derskfurd, Cullane, Draquhadlie, Knonkburne [*i.e.* Knockdurn], Tulmacht, the half of Baddinspink, New Myln, the half of Clawnacht and barony of Deskfurd; total, £122 13s 4d. For Tordiquhill [*i.e.* Ordiquhill], 'beand pendiclie of Fordice', £30. For the teind sheaves of Durne, Auchmillir, Randintracht, Fordice, Petchadlie, Bogmowkillis within the said parish of Fordice, £30 16s 8d. For the teind sheaves of the 'thaindour of Boyn and Auchanachy' and their pertinents within the said parish of Fordice; total, £39. For the teind sheaves of Ordinhuissis [Ordehuissis, *fo 133r*] with the pertinents within the said parish of Fordice; total, £33 6s 8d. For the

---

[82] Cf NLS, fo 125v (see above, p. 427).
[83] The correct calculation appears to be £59 8s 4d.
[84] The correct calculation appears to be £18 6s 2⅔d.
[85] Cf NLS, fo 125r (see above, p. 425).

teind sheaves of Lies, Dettambrage and Classindammir with the pertinents; total, £20 13s 4d within Fordice. For Glassaucht 'cum pertinentiis' within the said Fordice, £9 13s 4d. For the teind sheaves of Muraik, Rochtnakenzeis and Brakanhillis within the said Fordice; total, £22. For the 'uthir half landis teind schavis of Baddinspink within the said Fordice; total, £4 13s 4d. For the mail of Channons croft of the kirk of Fordice; total, 13s. For the vicarage of Fordice 'quhen payment was haid thairof with all dewities uisit and wontt'; total, £80 'of the quhilk thair is litill or nathinge gottin presentlie.'

'The ordinare chargis of the said kirk of Fordice with the pendicles extendis till' 80 merks 'or thairby quhilk suld be defalcat of the principall sowme.'

Logybuchane: for the whole teind sheaves and vicarage of Logybuchane, total, £126 13s 4d. Rathin, for the teind sheaves and vicarage of Rathin, total, £126 13s 4d. The kirk of Cabrawch: for the teind sheaves and vicarage thereof, £26 13s 4d. Twnmeth [*i.e.* Dalmeath]: for the teind sheaves and vicarage of Tunmeth, £20. Kyldrymme, Glenbuchet and Logymar: for the teind sheaves and vicarage of Kildrummye, Glenbuchett and Logymar, total, £113 6s 8d.

Signature: 'R. Erskin, dene of Abirdene.'

'The ordinare charges to be deducit
    In the first to the curattis of Fordice, Cullane, Desfurd and Ordefull [*i.e.* Ordiquhill] to thair feis and chargis as is abone rehersit', 80 merks. 'Item, for the uphaldinge of the cathedrale kirk and all necessaris thairin' £80. 'Item, to tua chaplanis and sex barnis', £56. 'Item, to Maister Robert Lumsden for his procuratouris fie', 20 merks. 'Item, to Maister Alexander Skeyne for his procuratouris fie', 10 merks. 'Item, to Maister Nicoll Hay, scrib of the chaptour', 20 merks. 'Item, deduct yeirlie of the saidis teind schavis of Fyndatour [*i.e.* Findlater] in all tymes bypast', £8. 'Item, deduct yeirlie of the teindis of Ordehuissis' [Ordinhuissis, *fo 132r*], £6 13s 4d. 'Item, of Rathin', £6 13s 4d. 'Item, of Logybuchane', £13 6s 8d. 'Becaus the foirsaidis teindis was nevir in use of payment of samekle as is contenit in the foirsaid rentall.'

Common kirks of Abirdene
    Fordice, £393 10s. Logybuchane, £126 13s 4d. Rathin, £126 13s 4d. Cabrawch, £26 13s 4d. Vicarage of Tunneth [*i.e.* Dalmeath], £20. Kyldrymme, Glenbuchett and Logymar, £113 6s 8d. Total of the common kirks is £806 16s 8d. 'Nota, thir commoun kirkis resavis na divisioun bot ar halelie appliet to the quene.'

KINCARDINE O'NEIL, PARSONAGE OF, (NLS, fo 133r-v)
(SRO, Vol. a, fo 386r)

Rental of the benefice of Kincardin Oneill extending to five kirks, namely, the kirk of Kyncardene 'wantand the vicarage', the kirk of Lowfannan with the vicarage, the kirk of Cluny with the vicarage, the kirk of Mydmar with the vicarage, the kirk of Glentannar with the vicarage.

The parsonage of the kirk of Kincardin with the temporal lands extending yearly to 300 merks; 'the temporall extendis to thre score pundis; ane myln', 22 b. malt of the sum foresaid. The kirk of Lonfannan, the teind sheaves thereof extending to £60, 'and the vicarage in tyme of peax sett for' 40 merks. The kirk of Cluny, the teind sheaves thereof extending to £50, and the vicarage in time of peace gave in assedation £16. The kirk of Mydmar, the teind sheaves thereof extending to £50, and the vicarage in time of peace gave in assedation £16. The kirk of Glentannar, the teind sheaves, parsonage and vicarage thereof is set and has been set 'thir diveris yeiris bygane' for the sum of 50 merks. 'Nochtwythstanding this foirsaid rentall is sett in assedatioun be umquhile Maister Robert Hammyltoun, my predecessour, with my consent to James Wetherspoune of Brighous and Elizabeth for the sowme of' 300 merks 'to be payit yeirlie.'

Signature: 'James Wetherspoune of Brighoussis.'

Calculation of third: In the whole, £452; third thereof, £150 13s 4d.

[*In margin*, 'Memorandum'] 'Nota, James Wytherspune of Bryghous allegis assedatioun of this personage for thre huntreyth merkis.'

BELHELVIE, PARSONAGE AND VICARAGE OF,[86] (NLS, fos 133v-134r)
(SRO, Vol. a, fo 386r)

Rental of the parsonage and vicarage of Balhelvy within the diocese of Abirdene.

'All teindis beinge payit of the auld maner concernynge the vicarage is' 300 merks.

'And thairof to be deducit to the vicar pensionar, stallaris fie in Abirdene, bischoppis procurationis' 40 merks yearly.

Signature: 'Mr Alexander Skeyne.'

Calculation of third: In the whole, 300 merks; third thereof, 100 merks.

---

[86] Cf NLS, fo 135v (see below, p. 442).

'Nota this rentall is gevin in also be Johnne Carnegy bot diminissed.'[87]

CRIMOND, PARSONAGE OF,[88] (NLS, fo 134r)
(SRO, Vol. a, fo 386v)

Rental of the parsonage of Creychtmond lying within the diocese and sheriffdom of Abirdene.

'Gaif yeirlie befoir the moneth of Junij 1558 yeiris the sowme of' £100, 'thairof to the stalleris of Abirdene and the remanent to me, Maister Arthour Talliefer, persoun of Creichmond, subscrivit with my hand.'

Signature: 'Mr Arthure Talliefere.'

Calculation of third: In the whole, £100; third thereof, £33 6s 8d.

ABOYNE, VICARAGE OF,[89] (NLS, fo 134r)
(SRO, Vol. a, fo 386v)

Rental of the vicarage of Aboyne lying within the same diocese and sheriffdom.

'Gaif yeirlie befoir the moneyth of Junij 1558 the sowme of' 80 merks; thereof £20 to Mr William Hey, parson of Turreff, 'as ane pensioun be provisioun of the paip, and the sowme of fyftie merkis to me the said Maister Arthour, vicar thairof, subscrivit wyth my hand.'

Signature: 'Mr Arthour Talliefer.'

Calculation of third: In the whole, £53 6s 8d; third thereof, £17 15s $6\frac{2}{3}$d.

ABERGAIRN, BENEFICE OF, (NLS, fo 134v)
(SRO, Vol. a, fo 386v)

Rental of the kirk of Abirgardein within the diocese of Abirdene, annexed to the said college thereof.

---

[87] This note is in the margin in SRO, but in the text in NLS.
[88] Cf NLS, fo 128r (see above, p. 431).
[89] Cf NLS, fos 128r, 140v (see above, p. 431; below, p. 451).

The teind sheaves of the said parish extends to £33 Scots money, and the vicarage extended to £20 'quhen it was payit, quhilk was gevin to the curat and uthir necessar ordinar charges.'

Signature: 'Alexander Andersoun, subprincipalis.'

Calculation of third: In the whole, £53; third thereof, £17 13s 4d.

GLENMUICK, BENEFICE OF, (NLS, fo 134v)
(SRO, Vol. a, fo 387r)

Rental of the kirk of Glenmowik annexed to the said college of Abirdene within the diocese thereof.

The teind sheaves and vicarage of Glenmuik is set for the sum of £55 'bot the vicarage was nevir weill payit.'

Calculation of third: In the whole, £55; third thereof, £18 6s 8d.

SLAINS, BENEFICE OF, (NLS, fos 134v-135r)
(SRO, Vol. a, fo 387r)

Rental of the kirk of Slanis annexed to the said college of Abirdene within the diocese of the same.

The teind sheaves extend to 5 c. of victual and £8 of money, and the vicarage of the said kirk was set for £10 of money 'quhen it was payit.'

'The kirkis foirsaid [*i.e.* Glenmuick and Slains] war nevir taxit be king, governour nor quene, nor yitt be biscope for thair is fourty and viij maisteris and studentes to be sustenit on the samyn.'

Signature: 'Alexander Andersoun, subprincipalis.'

Calculation of third: 'In the haill of money', £18; third thereof, £6. Victual, 5 c.; third thereof, 1 c. 10 b. 2 f. $2\frac{2}{3}$ p.[90]

---

[90] In the text this $\frac{2}{3}$ is rendered 'half pect 3 part half pect'.

SNOW, PARSONAGE OF,
(NLS, fo 135r)
(SRO, Vol. a, fo 387r)

The parsonage of Snaw, 'unite to the said college' [*i.e.* of Aberdeen], pertaining to Mr John Leislie, extends yearly to 40 merks, 'off the quhilk payit to the curat thairof', 10 merks, 'and to the maister of the grammar scuill', 5 merks.

Signature: 'Mr Joannes Leslie subscripsit.'

Calculation of third: In the whole, £26 13s 4d. 'Deducit thairof' 5 merks 'to the maister of the grammer skull, restis to devyd' £23 6s 8d; third thereof, £7 15s $6\frac{2}{3}$d.[91]

INVERNOCHTY, PARSONAGE AND VICARAGE OF,
(NLS, fo 135r)
(SRO, Vol. a, fo 387v)

Rental of the parsonage of Innerochty pertaining to Mr John Elphingstoun with the vicarage thereof.

The whole parsonage and vicarage foresaid extends in rental to £100, 'of the quhilk thair is ane pensioun providit to James Elphinstoun, sumtyme persoun thairof, of' 40 merks, 'and sua restis ane hundreyth and x merks.'

Total, 110 merks.

Signature: 'Mr Henrie Kynros, presentar of the samyn, with my hand.'

Calculation of third: In the whole, £100; third thereof, £33 6s 8d.[92]

BELHELVIE, PARSONAGE AND VICARAGE OF,[93]
(NLS, fo 135v)
(SRO, Vol. a, fo 387v)

The parsonage and vicarage of Balhelvy within the diocese of Abirdene pertaining to Mr James Strathauchin, parson thereof.

'Item, sett in assedatioun to Johne Carnegy, sone and appeirand air to Robert Carnegy of Kynnaird, knycht, for' 260 merks 'lyik as it was sett of befoir be Maister Gilbert Strathauchin, than persone thairof, to William Lyoun and Gilbert

---

[91] Deductions in this case seem to have been allowed.
[92] Deductions here were not allowed.
[93] Cf NLS, fo 133v (see above, p. 439).

Menzeis, quhairof is to be deducit to Maister William Strathauchin, vicar pensionar thairof', 20 merks, 'and for the corps presentis and umaist clathis and Pasche offerandis' 20 merks. 'Sua restis to the persoun', 220 merks.

Signature: 'Johnne Carnegy, with my hand.'

GAMRIE, VICARAGE OF, (NLS, fo 135v)
(SRO, Vol. a, fo 387v)

Rental of the vicarage of Gemrye within the diocese of Abirdene.

'Item, sett to sir Thomas Cristysoun, pensionar and minister of the same for the sowme of' 80 merks yearly 'and hes nocht gottin payment thairof this thre yeiris bigane and was sett to him' 24 'yeiris syne quhen the profeittis was graittar thairof nor it is now be Pasche fynes, corps presentis and umaist claithis togidder with small offerandis.'

Signature: 'Johnne Cokburne.'

Calculation of third: In the whole, £53 6s 8d; third thereof, £17 15s $6\frac{2}{3}$d.

AUCHINDOIR, PARSONAGE OF, (NLS, fo 136r)
(SRO, Vol. a, fo 388r)

The parsonage of Auchindoir extends to 140 merks. 'Maister Johnne Stewarte, in name and behalf of Maister Johnne Sinclar, dene of Restalrige,[94] and Maister George Lauder.'

Signature: 'Mr Johnne Stewarte.'

Calculation of third: In the whole, £93 6s 8d; third thereof, £31 2s $2\frac{2}{3}$d.

LESLIE, VICARAGE OF, (NLS, fo 136r)
(SRO, Vol. a, fo 388r)

'Rentale domini Thome Rayth.'

The vicarage of Leslie in Garioche within the diocese of Aberdeen set in past times yearly, £20.

---

[94] John Sinclair, who held the parsonage at the Reformation (and demitted it in October 1561, in favour of George Lauder) was also dean of Restalrig, 1542 - 1566. *RSS*, v, no 876; Watt, *Fasti*, p. 370.

Signature: 'Thomas Raith, manu propria.'

Calculation of third: In the whole, £20; third thereof, £6 13s 4d.

[*In margin*, 'Aliter gevin in and ressaved at the Clerk of Registeris command for' £13 6s 8d 'in the haill.']

CUSHNIE, PARSONAGE AND VICARAGE OF, (NLS, fo 136r)
(SRO, Vol. a, fo 388r)

The parsonage and vicarage of Cuschnye pertaining to Mr John Sandelandis, parson thereof, pays yearly £60; thereof the vicarage, £20; 'and na payment thairof thir thre yeiris last bypast.'

Signature: 'Magister Robertus Lineynden [*recte*, Lumsden], fermorarius eiusdem. Presentit be me, Maister Alexander Skeyn, subscripsit.'

Calculation of third In the whole, £60; third thereof, £20.

MORTLACH, PARSONAGE AND VICARAGE OF, (NLS, fo 136v)
(SRO, Vol. a, fo 388r)

Rental of the kirk of Morthlak within the diocese of Abirdene.

The teind sheaves of the said kirk extend to £100, 'and the vicarage was sett thir yeiris bygane for' £40.

Signature: 'Alexander Andersoun [of Morthlak, *deleted*], persoun of Morthlik.'

Calculation of third: In the whole, £140; third thereof, £46 13s 4d.

Murthlaik
    The rental of Murthlak pertaining to Mr Alexander Andersoun extending to £100.

ABERDEEN, TREASURERSHIP OF, (NLS, fo 136v)
(SRO, Vol. a, fo 388v)

'The thesaurarie of Abirdene'

The 'thesaurarie' of Abirdene pertaining to Mr Patrick Myrtoun extends to 300 merks.

Calculation of third: In the whole, £200; third thereof, £66 13s 4d.

TURRIFF, PARSONAGE AND VICARAGE OF,[95]   (NLS, fo 136v)
(SRO, Vol. a, fo 388v)

The parsonage and vicarage of Turreff pertaining to Mr William Hay extends yearly to 400 merks, 'thairof the vicarage restis onpayit thir tua yeiris bygane.'

Calculation of third: In the whole, £266 13s 4d; third thereof, £88 17s 9$\frac{1}{3}$d.

METHLICK, PARSONAGE AND VICARAGE OF,[96]   (NLS, fo 137r)
(SRO, Vol. a, fo 388v)

The parsonage and vicarage of Methlik pertaining to Mr Thomas Burnet extends yearly to £100, 'thairof the vicarage onpayit thir diveris yeiris bygane.'

Signature: 'Mr Alexander Skeyn.'

KINKELL, PARSONAGE OF,[97]   (NLS, fo 137r-v)
(SRO, Vol. a, fos 388v-389r)

Rental of the parsonage of Kinkell with the pendicles annexed thereto.

The parish kirk of Kynkell, £52 13s 4d. The parish of Kintor extends to £107. The parish of Kynnellar, £64 13s 4d. The parish of Dys, £58 3s 4d. The parish of Kemany [i.e. Kemnay], £46 6s 8d. The parish of Skeyn, £104. The parish of Drumblaitt, £100. 'Summa totalis of the personage of Kynkell presentlie payit to the persoun', £532 16s 8d.

Thereof to be deduced to Mr John Davidsoun and Mr Andrew Galloway in pensions,[98] £80. To the stallar, £18. 'And sua restis to the personis parte', £434 6s 8d.[99]

Calculation of third: 'In the haill vij kirkis', £532 16s 8d; third thereof, £177 12s 2$\frac{2}{3}$d.

---

[95] Cf NLS, fo 129r (see above, p. 433).
[96] Cf NLS, fo 130v (see above, p. 435).
[97] Cf NLS, fos 119v, 139r (see above, p. 420; below, p. 448).
[98] John Davidson's pension is recorded on fo 119v (see above, p. 420), that of Andrew Galloway has not been traced.
[99] The correct calculation appears to be £434 16s 8d.

MONYMUSK, PRIORY AND PARSONAGE OF,[100] (NLS, fos 137v-138v)
(SRO, Vol. a, fos 389r-390r)

'The rentall of the priorie [*inserted*, and personage] of Monymusk set in assedatioun to Mr Duncan Forbes for the space of thre yeiris nixt followinge the fest of Candylmes'[101] 1560/1 for payment yearly of the sum of £400 usual money of this realm 'in this maner and at the termis following, that is to say, begynand the first termis of payment thairof quhilk salbe ane hundreyth pundis togidder in ane sowme at the said fest of Candylmes nixt[tocum] of his and thair entrie thairto; at the Rud day callit Beltane[102] nixt thaireftir the yeir of God' 1561 'ane uther hundreyth pundis; at the fest of Lambes[103] nixt thaireftir in the said yeir of of [*sic*] God' 1561, 'ane hundreyth pundis; and at the fest of All Sanctis callit Hallomes[104] in the said yeir of God' 1561 'the last ane hundreyth pundis in compleit payment of the foirsaidis four hundreyth pundis and sua furth yeirlie and termelie togidder in ane sowme during the saidis thre yeiris. Attour the said Maister Duncane sall upoun his expenssis furnes ane sufficient precheour as salbe fund qualifiit at the sych and jugment of Adam Heriote, precheour in the Newtoun of Abirdene, or sic as salbe deputt in his place in caice of his absence, to remane and duell at the said plaice of Monymusk or at ony ane of the kirkis pertening to the said priorie and caus him preche upoun ilk festuall day at ane of the said fyve kirkis pertening to the saidis priorie and personage be ordoure, and syclyik sall furnes at ilk ane of the four kirkis pertening to the said priory and at the paroche kirk thairof ane sufficient minister to remane and duell thairat for ministratioun of the sacramenttis, doing and usinge of sic service as is requirit for the tyme, and als sall pay all utheris ordinar charges quhilk hes bene in use of payment of befoir in caice he be requirit and compellit thairto. And becaus thair is certane vicar pensionaris of the saidis kirkis quha suld and aucht to furnes ministeris for serving of the cuir thairin or at the lest remane thameself for doing thairof in caice that ony of the saidis vicar pensioneris remane and byd at the saidis kirkis or furneis qualifiet men under thame for service to be maid thairin, the said Maister Duncan sall redount, content and pay to the said Maister Johnne [Hay], his airis and assignayis or factouris yeirlie during the said space of thre yeiris samekle as the said pentioun and fie of that minister or ministeris could extend to yeirlie by the foirsaid soume of four hundreyth pundis, and the said Maister Duncan sall entir and imputt the said precheour and ministeris betuix the dait heirof and

---

[100] SRO text reads 'priorie and personage'; NLS text has 'and personage' inserted with a caret. The parsonage formed a prebend of Aberdeen cathedral and was separate from the priory, Cowan, *Parishes*, pp. 150-151. The rental of the parsonage and vicarage of Monymusk occurs in SRO, Vol. a, fo 374r (see above, p. 418). For Hay, see above, p. 118, n. 91.
[101] I.e. 2 February.
[102] 'Inventio Sancte Crucis' [i.e. the finding of the Holy Cross], 3 May, *Handbook of Dates for Students of English History*, p. 47
[103] I.e. Lammas, 1 August.
[104] I.e. the feast of All Saints, 1 November.

Mertymes[105] nixttocum or soner as he may guidlie, and in caice the said Maister Johnne heppin to permute, resigne, or gif oure thir benefeceis, this present contract to be expyrit in the self and to be of nane availl thaireftir, he advertisand the said Maister Duncan thairof and mak him sure of samekle as he hes in his awin handis pertening to the said priory conforme to the rentall, and in caice thair be ane generall reformatioun maid be the estaittis of this realme and ressavit universalie throw the samyn quhat beis defalkit and takin doun of the saidis benefices the samyn to be allowit to the said Maister Duncan, his airis and assignayis, in the sowmes aboun specifiit or ellis to cum to new appoyntment with the said Maister Johnne and gif ony of thir ministaris to be put to the saidis kirkis at the tyme aboun expremit be fund unhabill and nocht of gud conversatioun to the contentment of the parochenaris, in that caice the said Maister Johnne gevis and grantis to the said Maister Duncan be this present contract full powir to output and imput utheris in thair places, qualifiet and habill thairfoir, at the sycht of the said Adam Heriot or sic uthir as salbe present precheour within the said burgh of Abirdene for the tyme. Attour, the said Maister Johnne byndis and oblissis him, his airis, assignayis and successouris, to warrand and keip the said Maister Duncan, his airris and assignayis, skaythles anent the fruitis and dewities off the said beneficeis at all handis quhome it effeiris as law will. And siclyik ordanis the said Maister Duncan to convene, call, fallow and persew [blank], lord Forbes, and Arthour Forbes befoir the Lordis of Counsall and Cessioun at this nixt cessioun for reductioun of thair pretendit assedationis quhilk thai allege tham to haif of the teindis of the landis of Argathun and Cunlie, and in caice the samyn beis fundin valeud befoir the saidis Lordis the said Maister Johnne to defaik to the said Maister Duncan yeirlie samekle of the sowme abonwrittin as the samyn extendis to conforme to the rentall thairof.'

Calculation of third: In the whole, £400; third thereof, £133 6s 8d.

ABERDEEN, CHAPLAINRY OF COCKLARACHY IN ST NICHOLAS KIRK IN,[106]

(NLS, fo 138v)
(SRO, Vol. a, fo 390r)

The chaplainry of Cowclarachquhy pertaining to Mr Robert Carnegy lying within the diocese of Aberdeen, set in feu to the earl of Huntlie for £20.

Calculation of third: In the whole, £20; third thereof, £6 13s 4d.

---

[105] I.e. Martinmas, 11 November.
[106] I.e. the chaplainry of St Mary of Cocklarachy, *The Place-Names of Aberdeenshire*, p. 218.

LOGIE DURNO, VICARAGE OF, (NLS, fos 138v-139r)
(SRO, Vol. a, fo 390r)

'Rentale Magistri Joannes Philp ut infra' [*i.e.* Rental of Mr John Philp as below.]

The vicarage of Logydornocht in Garioche within the sheriffdom of Aberdeen belonging to the foresaid John Philp extends in the year, deducting a pension payable from it and other things to be deducted according to common order, to the sum free for the said vicar, £26 13s 4d.

Calculation of third: In the whole, £26 13s 4d; third thereof, £8 17s 9⅓d.

KINKELL, PARSONAGE OF, PENSION FROM,[107] (NLS, fo 139r)
(SRO, Vol. a, fo 390r)

[*In margin*, 'Pensioun out of the personage of [Ki]nkell']

Rental of a pension pertaining to Mr John Davidsoun, 'maister of the pedagoge of Glasgw, furth of the personage of Kinkell liand within the diocy of Abirdene', which extends yearly to £40; third thereof, £13 6s 8d.

TARVES, VICARAGE OF,[108] (NLS, fo 139r)
(SRO, Vol. a, fo 390r)

Rental of the vicarage pertaining to Mr Alexander Ogilby 'and presentit be me, Mertene Foirman, becaus of his absence, extending in the haill to yeirlie in all proffeittis to the soume of' £26 13s 4d; third thereof, £8 17s 9⅓d.

---

[107] Cf NLS, fos 119v, 137r (see above, pp. 420, 445); the rentals on fos 119v and 137r are for the parsonage.

[108] The name of this vicarage is omitted from the text, but is identified both from the name of the benefice-holder, Alexander Ogilvy (who also held Duthill), and from the third (which approximates that for the vicarage of Tarves, £7 16s 8d, in *Thirds of Benefices*, pp. 8, 227).

ROTHES, PARSONAGE OF,[109] (NLS, fo 139r[110])

[*In margin*, 'M[u]rray']

Rental of the parsonage of Rothos

The Manys, 29 b. 'of awld and this yeir bot' 16 b. The Fischertoun, 6 b. 'of auld and now bot' 2 b.[111]

DRUMOAK, PARSONAGE AND VICARAGE OF, (NLS, fo 139v)
COLLYHILL, CHAPLAINRY OF,[112] (SRO, Vol. a, fo 390v)

Rental of the benefice pertaining to Thomas Hay, parson of Drummaok and vicar of the same.

The parsonage of Drummaok yearly, 50 merks. The vicarage of the same yearly, 24 merks. The chaplainry of Collyhill yearly, 36 merks. Total, 110 merks.

'CHAPELDEN OF KINDRIMY',[113] CHAPLAINRY OF, (NLS, fo 139v)
(SRO, Vol. a, fo 390v)

The chaplainry of the Chapelden of Kindrimy containing the lands of Ovir Kilbacho with the mill and mill lands thereof, namely, Rapoche and croft, Murryell and their pertinents pertaining to sir George Broune, chaplain thereof, set in feuferme to John Forbes of Towyis for payment of 25 merks.

Signature: 'sir George Broune,[114] chaplen of the cheplanrie of Chapellden. James Nicolsoun, resaiff this rentall. J. Clerk Registre.'

---

[109] Cf NLS, fo 162r, 174r (see below, pp. 475, 491). As the marginal note indicates, the benefice belonged to the diocese of Moray; it also lay in the sheriffdom of Elgin and Forres; cf *RMS*, iv, no 149.
[110] SRO, Vol.a, fo 390v contains the heading only.
[111] This entry ends abruptly here and is deleted.
[112] I.e. the chaplainry of Collyhill in the chapel of Garioch, *RSS*, v, no 2313, where Thomas Hay, last possessor, is described as deceased (September, 1565).
[113] Possibly Chapelton of Glenkindie in Kildrummy parish, W. Macfarlane, *Geographical Collections relating to Scotland*, ed. A. Mitchell, 3 vols. (Edinburgh, 1905-1908), i, p. 30.
[114] 'Sir' occurs in NLS text only.

ARNISTON, (NLS, fo 139v)
PREBEND OF CRICHTON COLLEGIATE KIRK[115]

'The prebendarie of the college kirk of Creichtoun callit the prebendarie of Arnoldstoun.' [*In margin*, 'This is registrat in Hadingtoun.']

'The teyndis of the Manes of Arnoldstoun ar haldin and ever hes [bene be] the lordis of Sanct Johnne[116] be reasoun of his privilege, and payit [na thing] to the prebendair at ony tyme bypast thir mony yeiris as is [weill knawin].'
'Item, the small teindis of Arnoldstoun', 10 merks. The teinds of Harvestoun, £10. The teinds of Katkune, £10. The teinds of Wester Lochtquharrut with the cottaris teindis aboute Borthuike', 20 merks.

LONGLEY AND FETTERANGUS, VICARAGE OF, (NLS, fo 140r)
(SRO, Vol. a, fo 390v)

Rental of the vicarage of Lungley and Feterangus 'unite', lying within the diocese of Abirdene pertaining to Mr John Eldar 'extending be yeir to fourty merkis, furth of the quhilk is gevin in pensioun be yeir to sir Johnne Masoune', 20 merks.

Signature: 'Maister Johnne Eldar, with my hand. James Nicolson, ressave this rentale. Sic subscribitur, J. Clerk Registre.'

ALFORD, VICARAGE OF, (NLS, fo 140r)
(SRO, Vol. a, fo 390v)

Rental of the vicarage of Aufurd pertaining to sir John Colysoun, vicar thereof.

'Item, the said vicarage of Aufurd lyand in the diocie of Abirdene pertenis to the said sir Johnne of ald and wes wont to gif in assedatioun' 40 merks 'and now be resoun of the doun taking of the corps presentis, umaist claythis, Pasche fynes and offerandis, is nocht worth bot' 30 merks 'yeirlie allanerlie.'

Signature: 'Sir Johnne Colysoun, vicar of Aufurd, with my hand.'

---

[115] Cf SRO, Vol. a, fo 181v (see above, p. 176). Crichton collegiate kirk and its prebend of Arniston lay in the constabulary of of Haddington.
[116] James Sandilands, preceptor of Torphichen, otherwise lord St John, Cowan and Easson, *Medieval Religious Houses*, p. 161.

ABOYNE, PENSION FROM THE VICARAGE OF,[117]   (NLS, fo 140v)
(SRO, Vol. a, fo 391r)

'The rentale of ane pensioune of the vicarage of Aboyne, provydit of ald to me, Maister Williame Hay, persoun of Turreff, and in peciable possessioun thairof this' 43 years, 'per' £20.

Signature: 'W. Hay of Turreff. Presentit be Mr George Hay, xx Martij, anno lxxiij' [*i.e.* 1573/4].

NOTE

Rentals for the following benefices are located elsewhere in the text:

ABERDEEN, MINISTRY OF THE TRINITARIANS IN (EUL, fo 48r), p. 558.
DEER, ABBEY OF (NLS, fo 146v), p. 457.

---

[117] Cf NLS, fos 128r, 134r (see above, pp. 431, 440).

Moray (1)

ELGIN, BLACK FRIARS OF,[1] (NLS, fo 141r-v)
  (SRO, Vol. a, fo 392r[2])
'Jesus Maria'

'Ane rentaill of the Freiris Predicatouris of Elgin' in 1555.

The lands of Straithspey of Cardell with the pertinents, £20. Pottinfar, 20 merks. The Over Monbennis which Jarum Spens has in feu, 9 merks. The Hauch of Nethir Monbennis, the feuar James Innes, 20½ merks. The barony of Abircheirdour, 10 merks.

'This is the annuellis of the towne of Elgin.'
 'Item, of Johnne Youngis landis at the Over Port of his awin landis', 6s 8d. John Tailyouris lands, 6s. William Smithis lands, 12d. Of Watsonis land, 2s. Of James Robbis lands, 16s. Of Thomas Youngis 'benorth the Cors', 2s. Of Makconnes lands, 2s. Of Thomas Umphrais lands 'umquhill Quarteris land', 6s. Of George Cruikschankis lands, 2s. Of Andrew Alies lands 'umquhill Willmannis land', 2s. Of Alexander Winsisteris [*i.e.* Winchester's] land, 2s. 'Of glaissin wrycht land', 6s. 'Of the maister cuikis land', 2s. Of Thomas Airis lands 'umquhill Thome Youngis', 8s. Of the lands outside the Eist Port occupied by sir William Gibsone, sir Andrew Pantar and sir John Russell, 40d. Total annualrents of Elgin, 59s 4d.[3]

The fermes of Massindew: The lands of Massindew[4] pay yearly 2 c. 4 b. meal. The Boig Syid pays 24b. The Freiris Hauch, 1 c. 2 b. [*In margin*, 'Disponit to

---

[1] I.e. Dominicans.
[2] Fo 391v is blank.
[3] The correct calculation appears to be 67s.
[4] The hospital of Maison Dieu was granted to the Black Friars of Elgin in 1520, Cowan and Easson, *Medieval Religious Houses*, p. 179.

James Innes, servitour to my lord Regentis grace.'⁵]

Signature: 'F[rater] James Christisoun, manu propria.   F[rater] Jo[hnne] Adamesoun, supprior, manu sua.'⁶

---

⁵ James Innes of Drainie was bailie of the bishopric of Moray and had an interest in the fruits of the vicarage of Elgin, *RMS*, iv, no 2760; *Thirds of Benefices*, p. 215. The marginal note to the Regent, presumably of a date later than that of the rental, is evidently a reference to James Stewart, earl of Moray, Regent, 1567 - 1570.

⁶ Rentals for Moray resume at EUL, fo 149r; SRO, Vol. a, fo 396r (see below, p. 461).

Banff

CULLEN COLLEGIATE KIRK, PROVOSTRY OF,   (NLS, fo 143r[1])
RATHVEN, VICARAGE OF,   (SRO, Vol. a, fo 393r[2])

'The provestrie of the college kirk of Cullane fundat onlie upoun the fructis of the vicarage of Rothven pertenyng to sir George Duff, provest, liand in the diocy of Abirdene within the schirefdome of Banff, being at all tymes collectit be the said sir George him self sen his entres thairto, the yeirlie availl estimat to' 50 merks, 'quhairof he payis to sir Andro Hay, vicar pensionar of Rothven', 20 merks yearly.

Signature: 'sir George Duff, provest of Cullane.'

'Provestrie of Cullane quhilk is the vicarag of Rathven.'

Calculation of third: In the whole, £33 6s 8d; third thereof, £11 2s $2\frac{2}{3}$d.

BANFF AND INVERBOYNDIE, VICARAGE OF,   (NLS, fo 143r)
      (SRO, Vol. a, fo 393r)

The vicarage of Banff and Innerbundy within the diocese of Abirdene pertaining to Mr William Carnegy, extending 'be yeir' and set in assedation to sir William Smycht for £40.

Calculation of third: In the whole, £40; third thereof, £13 6s 8d.

---

[1] Fo 142r-v is blank.
[2] Fo 392v is blank.

FORDYCE, PARISH CLERKSHIP OF,[3]

(NLS, fo 143r-v)
(SRO, Vol. a, fo 393r)

Diocese of Abirdene and sheriffdom of Banff

'Paroche clerkschip of Fordice'

The parish clerkship of the kirk of Fordice pertaining to Alexander Hay, was worth yearly, 'quhen the same was compleitlie payit', 60 b. meal or bere 'in the optionar and chois of the parochenaris.'

Calculation of third: In the whole, 60 b. victual; third thereof, 20 b.

RATHVEN, PARSONAGE OF,[4]

(NLS, fos 143v, 146r[5])
(SRO, Vol. a, fo 393v)

'The rentall of the personage of Rothven with the annexis thairof, liand within the diocesis of Abirdene and Murray, respective, schirefdomes of Banff and Murray[6] respective.'

The teind silver of the parish of Rothven, £146. The mails of the barony of Rothven, 31 merks 7s 4d. The fermes of the Loynheid acres 'and myln multuris' extend to 100 b. The kirk of Dundurcus set for £40. The kirk of Kintallarlie [*i.e.* Kiltarlity], £24. The lands of Mullen lying in the parish of Dundurcus, 16 merks. 'Summa off the haill', [*blank*].

'Heirof deducit of ordinar charges to sex bedmen', 42 merks. To their habits, £7 4s. To the stallar in Abirdene, [*blank*]. 'Item, gevin furth of Dundurcus to the abbay of Kinlos', £5.[7]

Signature: 'G. Hay, with my hand.'[8]

'Personage of Rathven, the bedmenis pensiounis and claith deducit, in the haill', £206 10s; third thereof, £68 16s 8d. Bere, 6 c. 4 b.; third thereof, 2 c. 1 b. 1 f. $1\frac{1}{3}$ p.

---

[3] Cf NLS, fo 146r (see below, p. 457), which repeats this rental.
[4] Cf SRO, Vol. a, fo 249v (see above, p. 248). The revenues of Rathven, with those of Dundurcas and Kiltarlity, were annexed to the hospital of Rathven, Cowan, *Parishes*, p. 169.
[5] Fos 144r-v, 145r-v are blank.
[6] Although there existed an earldom and diocese of Moray, there was no sheriffdom of that name. The annexed kirks of Dundurcas and Kiltarlity lay in the sheriffdoms of Elgin and Inverness respectively. *RMS*, iii, no 2810; ix, no 1260.
[7] There is a marginal note here which has been deleted.
[8] George Hay was parson of Rathven at the Reformation; his brother, Andrew, was coadjutor and successor, *RMS*, iv, no 1615.

ARNDILLY, VICARAGE OF, (NLS, fo 146r)
(SRO, Vol. a, fos 393v-394r)

Rental of the vicarage of Artilduill, lying in the diocese of Murray and within the sheriffdom of Banff pertaining to sir John Robertsoun, is yearly worth £10.

Signature: 'Presentit be me, Alexander Hay, in name of the said vicar. Alexander Hay.'

Calculation of third: In the whole, £10; third thereof, £3 6s 8d.

FORDYCE, PARISH CLERKSHIP OF,[9] (NLS, fo 146r)
(SRO, Vol. a, fo 394r)

Diocese of Aberdene and sheriffdom of Banff

The parish clerkship of the kirk of Fordice, pertaining to Alexander Hay, was worth yearly, 'quhen the same was completlie payit', 60 b. victual 'meill or beir, in the optioun and chois of the parochinaris.'

Calculation of third: In the whole, 60 b. victual; third thereof, 20 b.

DEER, ABBEY OF,[10] (NLS, fos 146v-148v)
(SRO, Vol. a, fos 394r-395v)

'The rentale of the mailles and custumis of the temporall landis of the abbacie of Deir set to my lord erle Merschell.'[11]

    The feu mails and custom silver of the temporal lands of the barony of Deir set in feu to my lord earl Merschell extending to £161 10s 7d. The feu mails and custom silver of Glacrewch and Lytill Elrik set in feu to Alexander Keith of Glacreuch extending to £12 19s. [*In margin*, 'Glakreauch'.] The feu mails and custom silver of Mekle Auchtchrydie within the said barony and parish of Deir set in feu to Mr Robert Lumisdane extending to £9 9s 4d. The feu mails and custom silver of the one half of Fischirtoun of Peterheid and Grange[hill, *deleted*] of Rahill within the barony of Deir and parish of Peiterugy set to Monane Hog and his heirs extending to £5 15d. *The feu mails of Feychill and Monkishill [set to] my lord Forbes[12] within the parishes of Foverne and Ellen extending to £5 17s 2d.[13] The annualrents of Tulliocht, Towkis, Litill Creiche [*i.e.* Crichie] and

---

[9] Cf NLS, fo 143r (see above, p. 456); this repeats the rental on fo 143r.
[10] Deer Abbey lay in the sheriffdom of Aberdeen, though its property extended into the sheriffdom of Banff, *RMS*, v, no 1309.
[11] William Keith, 4th earl Marischal, *Scots Peerage*, vi, pp. 46-48.
[12] William, 7th lord Forbes, *Scots Peerage*, iv, pp. 55-56.
[13] The text from * to the end of this sentence is omitted from the SRO text.

Fethok extending to £9. 'Summa of the haill mailles and custum silver of the temporale landis foirsaid extending to' £203 13s 4d.[14]

The teind silver of the parish of Deir set to my lord earl Merschell extending to £120 7s 6d. The vicarage of Deir extending to £66 13s 4d. The teind silver of the parish of Peterugy, my lord earl Merschell, tacksman, extending to £61 2s. The teind silver of the parish of Foverne, my lord earl Merschell, tacksman, extending to £40 13s 4d. The teind silver within the said parish of Foverne, my lord Forbes, tacksman, extending to £44. The vicarage of the parish of Foverne extending to £13 6s 8d. The teind silver of the parish of Kynedwart, the lord Drum,[15] tacksman, extending to £82. The vicarage of Kynedwart extending to £20.
The rental of the fermes of the temporal lands of the abbey of Deir.

The feu fermes of the one half of the Grange of Rahill within the parish of Peiterugy set to my lord Merschell extending to 4 c. bere. The other half of the feu fermes of the said Grange of Rahill set to Monan Hog and his heirs extending to 4 c. bere. The multures of the Myln of Boupie and set to my lord Merschell and lying within the parish of Deir extending to 40 b. meal. The feu fermes of Feychill and Monkishill, my lord Forbes, tacksman, 'bayth stok and teynd' lying within the parishes of Foverne and Ellen extending to 7 c. 8 b. victual, 'heirof' 2 c. bere; 6 b. wheat; 5 c. 2 b. meal. The Manis of Deir, 'bayth stok and teynd' extending to 10 c. victual, 'heirof' 3 c. bere; 8 b. wheat; 6 c. 8 b. meal. The teind ferme meal of the parish of Deir, my lord Merschell, tacksman to the abbey extending to 25 c. 11 b. 2 f. meal. The custom of the said teind fermes set to my lord Merschell for silver extending to £15 18s. The ferme mail of the teinds of the parish of Peterugy, my lord Merschell, tacksman, extending to 10 c. 3 b. 2 f. meal. The custom of the said teind ferme set to my lord Merschell for silver extending to £4 12s. The ferme meal of the parish of Foverne for the teinds of the same, whereof my lord Merschell is tacksman, extending to 9 c. 10 b. meal. The customs of the said teind ferme set to my lord Merschell for silver extending to 13s 2d. The teind ferme bere within the parish of Fovarne, my lord Merschell, tacksman, extending to 26 b.

'Summa of the haill silver abonewrittin extending to' £672 8s 6d.[16] 'Thairoff few mailles, custum silver and annuell rentis to be pait at Witsonday[17] and Mertymes[18] yeirlie', £203 13s 6d. 'Item, of the said' £672 8s 6d 'of teynd silver and vicarage silver to be pait at Peasche[19] and Lambmes[20] conforme to the use of pament extending to' £468 15s.

---

[14] The correct calculation appears to be £203 17s 4d.
[15] I.e. Alexander Irvine of Drum, *RMS*, iv, no 2158.
[16] The correct calculation appears to be £673 3s 4d.
[17] I.e. seven weeks after Easter.
[18] I.e. Martinmas, 11 November.
[19] I.e. Easter.
[20] I.e. Lammas, 1 August.

'The third of the foirsaid few mailles and annuell rentis to be pait at Witsonday and Mertymes yeirlie extending to' £67 17s 10d.

'The third of the teynd silver and vicarage silver to be pait at Peasche and Lambmes, conforme to the use of pament extending to' £156 5s.

'Summa of the haill third of silver abonewrittin extending to' £224 2s 10d.[21]

'Summa of the beir foirsaid abonewrittin', 14 c. 10 b.; third thereof, 4 c. 14 b.

'Summa of the quhyit abonewrittin', 14 b. wheat; third thereof, 4 b. 2 f. 2⅔ p.

'Summa of the haill meill foirsaid', 59 c. 11 b.; third thereof, 19 c. 14 b. 1 f. 1⅓ p.

The parsonage of Phillorcht, £70; third thereof, £23 6s 8d.

The parsonage and vicarage of Dunnet extending to 200 merks; third thereof, £44 7s 10d.[22]

Signature: 'Deir.'[23]

'Apud Halyrudhous', 21 December 1573. 'Productum.'

'Apud Edinburgh', 22 December 1573. 'The Lordis[24] findis this rentale unsufficient, being gevin in les nor hes bene maid compt of the yeiris preceding.'

'Apud Edinburgh', 23 December 1573. 'My lord of Deir is content of his awin consent that the thrid of the meill of the abbay of Deir be chargit for according to the charges of the comptis preceding, nochtwithstanding that the haill rentale will nocht extend to samekill, viz., the thrid of the meill', 21 c. 3 b. 1⅓ p.[25] 'Sic subscribitur, Deir.'

---

[21] This is one third of £672 8s 6d.
[22] The correct calculation is £44 8s 10⅔d.
[23] Robert Keith, son of William, 4th earl Marischal, was provided to the commendatorship of Deer, 19 October 1552, Cowan and Easson, *Medieval Religious Houses*, p. 74.
[24] I.e. the Lords of Exchequer.
[25] The third of the meal on fo 148r is 19 c. 14 b. 1 f. 1⅓ p.

NOTE

Rentals for the following benefices are located elsewhere in the text:

RATHVEN, DUNDURCAS AND KILTARLITY, BENEFICES OF (SRO, Vol. a, fo 249v), p. 248.
SKIRDUSTAN AND BOTRIPHNIE, PARSONAGE OF (NLS, fo 39r), p. 320.

Moray[1] (2)

KINLOSS, ABBEY OF,[2] (NLS, fos 149r-152v)
(SRO, Vol. a, fos 396r-398v)

'The haill rentale of the abbay of Kynlos in all maillis, teindis, kirkis and dewities as eftir followis.'

The parsonage and the kirk of Allane [*i.e.* Ellon] extends in teind silver and gives yearly £135 5s. The same kirk gives in victual yearly, all meal, 33 c. 1 b.

The barony of Straithylay
    The Manis and lands of Straithylay with 'tour, fortalice and orchard' of the same, the Clerk Sett, Boglugy, Thornitoun, Hauches, Murifald, Brakhaw, Carnihillis, Craigleithe, Auchindanery, Ovir Myln, Nether Myln and mill lands of the same set for £122 15s 1d. The lands underwrittin, namely, the lands of Millegin, Garwotwod, Eistir Cranokis, Newland thereof, Westir Cranokis, Eister Croylettis, Westir Croylettis, Echres, the half lands of Ballnamene, Fortrie, Newland of Fortrie, the Ovir Sett and Nethir Sett of Kilmanitie, the lands of the Clerk Saitt 'of the west syd of the burne', the lands of the 19 oxgangs of the Knok set for £162 4s. 'Item, the remanent of the kirk extendinge to xiij oxingange' set to the tenants for yearly payment of £12 18d; 3 b. 3 f. custom meal; 3 b. 3 f. custom oats; $3\frac{1}{4}$ wedders; 8 capons. The lands of Auchinhovis with the pendicles, Glengarrok and Mengreowis set for £73 5s 2d. The lands of the Hauches of Kilmyntutie set for 5 merks 6s 8d; 2 f. custom meal; 1 b. custom oats; 1 wedder; 1 goose; 3 capons; 3 poultry. The lands of Kelliesmonth, Toirmoir [*i.e.* Tarmore] and Nethir Kylmanedy set for £38 18s. The lands of Pethnik set for £7 11s; 8 capons; 16 poultry; 2 geese. The lands of Edingeicht with the pendicles set for £19 7s 8d; 11s 'ryne mart silver'; 2 geese; 8 capons; 16 poultry. The lands of Ovir and Nether Cantlie set for £8 11s; 2 b. custom meal; 2 b. custom oats; 2 wedders; 8 capons; 16 poultry; 11s 'ryne mart sylver'. The lands of Fluris and the lands of

---

[1] The parishes listed below lay in the diocese of Moray, and were located in the sheriffdoms of Elgin and Forres, Nairn, Banff, Aberdeen, and Inverness.
[2] A printed rental for Kinloss and Beauly occurs in *Records of the Monastery of Kinloss*, ed. J. Stuart (Edinburgh, 1872) pp. 154-163.

the 'auld toun of Ballamene' set for yearly payment of £9 9s 5d; 2 f. custom meal; 2 f. custom oats; ½ wedder; 13 geese; 3 capons; 3 poultry; 2s 9d 'in ryne mart silver'. The lands of Wynde Hillis called the 'Sauchy town' set for the yearly payment of £7 14s 8d; 5s 6d 'in ryne marte silver'; 2 b. custom meal; 2 b. custom oats; 2 wedders; 12 geese; 12 capons; 12 poultry. The lands of Ovir Hauches of Kelleismonth set for the yearly payment of £4 20d; 1 b. meal; 1 b. oats; 1 wedder; 1 goose; 3 capons; 6 poultry. The lands of Lynnache set for payment of £6 13s 4d. The Newlandis of Millegin 'callit Jonettis Scheill', with Straib Know set for the yearly payment of 20s. The Lady land set for yearly payment of £3 6s 8d.

The kirk of Awach [i.e. Avoch]

The parsonage of the said kirk pays in victual yearly, 190 b. 2 f. bere; in teind silver with the lands and mains thereof, £22. 'Item, tuenty tua wedderis quhilkis of auld ar assignit to the tennentis and payis for the pece of evirilk ane thairof', 3s; total, £3 6s.

The barony of Kinlos

The lands of Kilboyok set for the yearly payment of £7 silver; 10 b. bere; 3 b. meal; dry multure; 10 geese; 3 dozen poultry; 24s 'ryne marte silver'. The lands of Langcoit set for yearly payment of £6 13s 4d. The lands of Tannoquhy set for yearly payment of £9 17s 4d; 19 b. victual. The lands of Stanyiris set for the yearly payment of £3 9s 7d; 3 b. 3 p. victual. The lands called the Ordeyis set for yearly payment of 21s.

The annuals of Forres pay yearly 40s. The feu lands of Elgin pay yearly 33s 4d. The feu lands of Narne pay yearly 26s 8d. The annuals in Innernes pay yearly 32s. 'Item, the fresche Wattir of Findorne sett to the communietie of Forres' for the yearly payment of £10. The fishing called the Evinstell, 'viz., the ferd pairt thairof and ane xvj day of the fresche Waltir of Findorne' set yearly for £8 2s. 'Item, [*blank*] day of the fischeing of the fresche Wattir of Findorne' set yearly for the payment of £27 10s paid by Alexander Urquhat, burgess of Forres. The Schirefstell set for the yearly payment of £5 10s. 'Item, the remanent landis of the said baronie with the manis and woddis thairof and the remanent fischeingis of the yardis and stellis upoun the Watter of Findorne and mouth thairof and odmaill of the said toun sett for the yeirlie payment of' £414 11s. [*In margin*, 'Nota, thir ar mekle better of.']

'And sua the haill abbacy of Kinlos extendis yeirlie in silver, victuall, aittis, wedderis, geis, caponis and pultries as eftir followis, viz. Summa of the haill victuall sylver', £1,152 12d 'oblo' [*sic*].[3] 'Summa of the haill victuall', 47 c. 11 b. 1 f. 3 p.[4] 'beir and male as said is'. Oats, 10 b. 3 f.[5] Geese, 41.[6] Capons, 60.[7] Poultry, 124.[8] Wedders, 34½.[9]

---

[3] The correct calculation appears to be £1,143 13s.
[4] The correct calculation appears to be 47 c. 10 b. 1 f. 3 p.

'Thir ar the thingis that ar to be deducit of the money and victuallis abone specifiet, payit as eftir followis.'

'Inprimis, to the baillie of Kinlos for his fie', £13 6s 8d. 'Item, to the baillie of Straithylay for his fie', £10. 'Item, of the Chapel croft of Udnie, quhilk is ane parte of the kirk of Allane abone specifiit, the lard of Udnie refussis lik as he hes refusit this mony yeiris bygane of the teind silver of the said croft quhilk is putt in the rentall of the auld extending yeirlie to' £16. 'Item, to the officiar of Kinlos for his fie quhilk he hes in lyfrentt of auld fra the abbott and convent thairof', 26s 8d. 'Item, to the officiar of Straythylay for his fie', 40s. 'Item, to Maister Johnne Ferrerwus[10] [Ferarius, *SRO*] for his pensioun quhilk he hes under the commoun seill of the place for his lyftyme yeirlie', £40. 'Item, the Lordis of the Saitt[11] contribusioun yeirlie', £18 4s. 'Item, to the stallar of the kirk cathederall of Ros quhilk stallar is for the personage of Avache', £6 13s 4d. 'Item, for the prebendar and stallaris pensioun within the kirk cathederall of Abirdene for the kirk of Allane', £28. 'Item, to the vicar pensionar of Allane for his yeirlie pensioun as his provisioun beiris', £13 6s 8d. 'Item, to the redare of the commoun prayaris in the kirk of Allane', £26 13s 4d. 'Item, thair is to be deducit of the said soum of silver for xiiij monkis habite silver, ilk monk haiffand' 50s 'be yeir, quhilk extendis to' £35. 'Item, for thair fische and flesche be yeir, ilk ane of the said xiiij monkis haiffand' 8d 'in the day for thair flesche and' 2d 'in the day for thair fische, extending in the haill to' £93. 'Item, for thair fyre, buttir, candill, spice and Lentroun meitt', £12. 'Item, the annuallis of Innernes afoir specifiit quhilkis ar nocht payit as apperis befoir in this rentall becaus the provest and ballies of Innernes hes maid actis in thair bukis that annuallis be nocht payit concernynge kirk men', 32s. 'Item, in lyik maner the annuallis of Forres', 40s.

'Summa of annuellis nocht [*sic*] payment with the uther sowmes gevin furth and the uthir sowmes to be deducit, as is abone specifiit, extendis to '£319 2s 8d. 'Et sic restat frie to the abbottis partt', £832 18s 'oblo.'

'Item, thair is to be deducit to the said xiiij monkis for breid and drynk, ilk ane of thame haiffand in the yeir' 19 b. 1 f. 2 p. 'extending in the haill to' 16 c. 15 b. 1 f. 'Item, to be deducit for the officiar of Kinlos [Kinros, *SRO*] breid and drink during his lyftyme haveand in the moneyth' 1 b. 'extending in the yeir to' £13. 'Item, in the lard of Wemes handis for his landis of Chapeltoun, quhilk no payment is maid of', 4 b.

---

[5] The correct calculation appears to be 10 b. 1 f.
[6] This calculation is correct.
[7] The correct calculation appears to be 53.
[8] The correct calculation appears to be 108.
[9] The correct calculation appears to be 32¼.
[10] This is a reference to the Piedmontese scholar Johannes Ferrerius or Ferrerio who was recruited by abbot Robert Reid to teach the monks of Kinloss, and who wrote lives of abbots Thomas Crystall and Robert Reid, *Records of the Monastery of Kinloss*, pp. xiii ff.
[11] I.e. Lords of the Court of Session.

'Summa of the victuallis deducit as is abonewrittin extendis to' 18 c. 1 f. victual 'and sua restis to the abbot', 29 c. 11 b. 3 p. victual; 10 b. 3 f. oats; 34¼ wedders; 41 geese; 60 capons; 124 poultry.

'Memorandum, that the vicarage of Allane was wount to pay to me teind buttur, teind cheis, teind lambis woll and utheris dewities quhilk payis nathing in thir dayis.'

Signature: 'W., abbot of Kinlos.'[12]

Calculation of third: Money, £1,152 12d; third thereof, £384 0s 4d. Victual, 48 c. 14 b. 1 f. 3 p. [47 c. 11 b. 1 f. 3 p., *fo 150v*]; third thereof, 15 c. 15 b. 1 f. 3⅔ p.[13] Oats, 12 b. 1 f. [10 b. 3 f., *fo 151r*]; third thereof, 4 b. 1⅓ p. Geese, 41; third thereof, 13⅔. Capons, 60; third thereof, 20. Poultry, 124; third thereof, 41⅓. Wedders, 34¼; third thereof, 11 wedders '3 parte wedder and 3 of ane quarter' [i.e. five twelfths].

Third of money, £384 0s 4d. Take the lands, mains, woods, fishings of the barony of Kinlos 'quhilkis ar in the abbottis handis unsett for' £414 11s. 'And giff in agane' £30 10s 8d.

Third of victual, 15 c. 15 b. 1 f. 3⅔ p. Take the kirk of Awache for 11 c. 14 b. 2 f.; and 4 c. 3 f. 3⅔ p. out of the kirk of Ellein, 'et eque.' [*In margin*, 'The kirk of Ellon tane be Pitarro[14] for this haill victuale.']

Third of oats, 4 b. 1⅓ p. Take out of the Knok, 3 b. 3 f. 'and out of the same, thre wedderis, and the rest out of the Hauches of Killeismonth, and out of the same, a wedder'; out of Pethnik, 2 geese, 16 poultry, 8 capons; out of Edingeicht, 8 capons, 16 poultry; Over and Nether Cantlie, 2 wedders, 4 capons, 10 poultry;[15] Over Kilmenedy, 2 wedders; Windehillis called Sauchquhytoun, 2 wedders, 12 geese, 1 f. 1⅓ p. oats; Fluris, ½ wedder; Over Hauches of Killeismonth, 1 wedder. 'Eque for aittis, wodderis, caponis, pultrie and geis.'

BEAULY, PRIORY OF,[16] (SRO, Vol. a, fos 398v-399r)

Third of money, £45 11s 1⅓d. Take 'this thrid of the baronie of Bewlie gevand be the rentall' £61. 'Gif in' £16 8s 10⅓d.

Third of victual, 4 c. 11 b. 2 f. 2⅔ p. Take the barony of Bewlie for 4 c.; 'tak the rest of the tua mylnis of Bewlie et sic eque.'

---

[12] Walter Reid, nephew of Robert Reid, was abbot of Kinloss at the Reformation, Cowan and Easson, *Medieval Religious Houses*, p. 76.
[13] This is one third of 47 c. 14 b. 1 f. 3 p.
[14] I.e. Sir John Wishart of Pittarow, Comptroller, 1562 - 1565, who was appointed Collector general on 1 March 1562, *Handbook of British Chronology*, p. 191; *RSS*, v, no 998.
[15] The NLS text ends abruptly here at fo 152v; SRO, Vol. a continues at fo 398v.
[16] The entry for Beauly, a dependent cell of Kinloss, is engrossed in the rental for Kinloss.

'Aittis, martis, muttoun, pultrie and wodderis ar gevin in in generall, thairfoir gar seik thame out.'
Third of salmon, 10 barrels. Out of the fishing of Bewlie.

'Remember my lord Comptrollare[17] to speir the rentall of thir tua Kynlos and Bewlie better for thay ar suspicious anent the fischeingis.'

ROTHIEMURCHUS, VICARAGE OF, KINCARDINE,[18] VICARAGE OF,

(SRO, Vol. a, fo 399r)

Rental of the vicarage[19] of Rothemurcus and Kincarn given up by Robert Hepburne, vicar thereof, 12 March 1574/5, extending yearly to £10.

Signature: 'Robert Hepburne.'

MORAY, BISHOPRIC OF,

(NLS, fos 153r-154v [part])
(SRO, Vol. a, fos 399v-401v)

Rental of the bishopric of Murray

'Money'
    The barony of Spyne: The mails, feuferme, annuals and dry multures within the said barony extend to £97 3s 8d.
    The barony of Kynnedour: The mails, feufermes, annuals and dry multures within the same extend to £76 12s 5d.
    The barony of Birnethe: The mails, feufermes, annuals and dry multures thereof extend to £101 11s 6d.
    The barony of Rafforte: The mails, feufermes, annuals and dry multures thereof extend to £92 9s 8d 'obl' [i.e. ½d].
    The barony of Ardilat: The mails, feufermes, annuals and dry multures thereof, £143 15s 4d 'obl' [i.e. ½d]. [£148 15s 4d 'obl', *SRO, Vol. a, fo 401r.*]
    The barony of Keyth: The mails, feufermes, annuals and dry multures thereof extend to £92 3s 7d.
    The barony of Kilmyles: The mails, feufermes, annuals and dry multures thereof extend to £97 5s 3d 'obl' [i.e. ½d].
    The barony of Straithspey: The mails and duties thereof, £187 3s 9d.
    The barony of Moye: The mails thereof, £20 15s 4d.

---

[17] Bartholomew de Villemore, Comptroller, 1555 - 1562, with Thomas Grahame of Boquhaple, 1561 - 1562. They were succeeded by Sir John Wishart of Pittarow, 1562 -1565. *Handbook of British Chronology*, p. 191.
[18] I.e. Kincardine-on-Spey.
[19] The two vicarages of Rothiemurcus and Kincardine were held by the same vicar, Robert Hepburn.

The kirks set for silver

Dyik, £16. Rothemay, £40. Keyth, £333 6s 8d. Grantulye [i.e. Gartly], £193 6s 8d.[20] Wardlaw, £40. Rothymurcus, £13 6s 8d. David [i.e. Daviot], £38 13s 4d. Tallaracie [i.e. Dalarossie], £30. Innerallane, £30 13s 4d.

'The teynd salmond of thre cobillis in Spey extending to sevin lastis aucht lastis yeirlie quhylis les quhyllis mair as the fische rynnes, pryce of the last', £18. 'Lay viij lastis inde summa', £184.[21]

Total, £2,033 7s 3d 'ob' [i.e. ½d].

'Summa of the haill maillis, few fermis, annuellis, dry multuris, kirk sylver, fischeingis, teind salmond and wther dewties', [blank].

'The ordinar expenssis to be deducit thairof':
'Imprimis, the fruittis of the kirk of David and Tullaracie, quhilkis ar waist and nathing payit thir thrie yeiris bypast', £68 13s 4d. 'Item, the landis of Over Blarie liand waist', £8 13s 4d.[22] 'Item, Conyeocht Durris pertenand to the lady Murrestoun, waist', £5 5s. 'Item, the officiaris feis of the said baronies', £10 10s. 'Item, in pensionis to certane chaplanis in Elgin, the vicar of Ogstoun and officiaris of the cathederall kirk', £48 8s 8d. 'Item, to thre childir sangstarris in the queir', £12. 'Item, to the maister of grammar scull in Elgin', £3 6s 8d. 'Item, to the ballie of Mwrray and officiaris feis of the said place', £30 8s 8d. 'Item, to the Lordis of Cessioun for thair contributioun in the yeir', £28. 'Summa of the money to be deducit of the fructis foirsaidis', £215 5s 8d.

'And sua restis de claro', £1,626 19d 'ob' [i.e. ½d]; third thereof, £542 0s 6d 'tantum'.

The rental of the ferme victual of the said bishopric of bere, meal and rye.
The barony of Spyne, 1 c. 1 b. The barony of Kynnedour, 13 c. The barony of Byrneth, 3 c.

The teind victual of the said bishopric.
The kirks of Elgin and Sanctandros,[23] 41 c. 4 b. 3 f. 2 p. The kirk of Dyk, 15 c. 1 b. The kirk of Ugstoun, 4 c. 'Item, within the said parochianis of quheit', 10 b.

'Summa of the haill ferme and teind victuallis', 77 c. 6 b. 3 f. 2 p. victual and 10 b. wheat.

---

[20] The text seems to read 'j^{c}iiij^{xx} li. xxiij li. vj s. viij d.', but presumably should read '£193 6s 8d', if the first 'x' of 'xxiij' is read as an ampersand.

[21] The correct calculation appears to be £144.

[22] The NLS text recommences at the next sentence, which corresponds to SRO, Vol. a, fo 400r.

[23] I.e. the church of St Andrew at Kilmalemnock, subsequently united with Lhanbryde; Scott, Fasti, vi, p. 397.

'Tharof to be deducit in pensionis and officiaris feis extending to' 8 c. 13 b. victual 'and sua restis' 68 c. 9 b. 3 f. 2 p. victual, 'quhairof the thrid pairt extendis to' 22 c. 13 b. 3 f. 3⅓ p. 'Off quheit', 10 b.; third thereof, 3 b. 1 f. 1⅓ p.

'Nota, the auld rentall of the ferme and teind victuallis of the said bischoprik extendit in tymes bypast to' 94 c. 'victuall or thairby, quhairof' 16 c. 'or thairby is defalkit for inundatioun and sandinge of the landis be watteris and wound and povertie of tennentis and truble of this tyme.'

'Nota, tua chalder viij b. hors corne with certane custume mertis, muttoun, caponis and pultrie nevir sauld for money bot spendit in the plaice.'

'Nota, the procurationis and synodallis was wont to extend yeirlie to' £80 'and nathing gotting thairof thir thre yeiris bygane.'

Calculation of third: Money, 'the salmond comptit thairwith', £2,033 7s 3d 'ob' [i.e. ½d]; third thereof, £677 15s 9d '3 ob' [i.e. ⅜d]. 'Money, by salmond', £1,649 7s 3d 'ob' [i.e. ½d]; third thereof, £549 15s 9d.[24] Victual, 77 c. 6 b. 3 f. 2 p.; third thereof, 25 c. 12 b. 3 f. 3⅓ p. Wheat, 10 b.; third thereof, 3 b. 1 f. 1⅓ p. Horsecorn, 8 b.; third thereof, 2 b. 2 f. 2⅔ p.

'Nota, custoume martis, muttoun, pultrie, caponis and all utheris dewities omittit.'

'Item, he affirmes this rentall of the victuallis to have extendit of auld to' 94 c. 'ut infra.'

The third of the money 'by the salmound' extends to £549 15s 9d. 'Lift this money in this maner'. Take the barony of Spyne for £97 3s 8d; the barony of Kynedwart for £76 12s 5d; the barony of Birneth for £101 11s 6d; the barony of Rafferd for £92 9s 8d 'ob' [i.e. ½d]; the barony of Ardclath for £148 15s 4d 'obl' [i.e. ½d] [£143, *SRO, Vol. a, fo 399v*]; the barony of Keith for £92 3s 7d. 'Giff in agane of this', £59 0s 6d.[25]

'Be remembrit to gar my lord Comptrolare[26] speir out and tak wp the hail thrid of the salmound quhilk the bischope[27] in his rentale sayis ar the teindis of the coblis and esteims thame to viij lastis be yeir and the last to' £48 [£18, *SRO, Vol. a, fo 400r*] 'bot thair ar better. Remember my lord to tak ordour thairanent.'

---

[24] '3 ob' [i.e. ⅜d] has been omitted here; see preceding sentence.
[25] This calculation is correct, with 'obol' denoting ½d.
[26] See above, p. 465, n. 17.
[27] Patrick Hepburn, bishop of Moray, 1538 - 1573, Watt, *Fasti*, p. 217.

Third of victual, 25 c. 12 b. 3 f. 3⅓ p. Take the barony of Spyne for 1 c. 1 b.; the barony of Kynnedward for 13 c.; the barony of the Birnett for 3 c.; and out of the kirks of Elgin and Sanctandrois, 8 c. 1[1 b. 3 f. 3⅓ p.].

Third of wheat, 3 b. 1 f. 1⅓ p. 'Tak this quhyte out of the parochin of Elgin and Sanctandrois quhilk payis in the haill' 10 b.

'The horscorne he estemis to' 2 c. 8 b. 'quhairoff the thrid wilbe' 13⅓ b.

'Omittit grassumes, canes, custumes, martis, pultrie and all dewiteis.'

DUFFUS, PARSONAGE OF,[28]   (NLS, fo 155v[29])
(SRO, Vol. a, fo 402r)

'The tennentis in Duffois'

Keme

    John Rob, 8 b. Richard Thomesoun, 10 b. Thomas Rob, 8 b. Thomas Watsoun, 12 b. Alexander Falconar, 8 b. Alexander Thomesoun, 8 b. John Suthirland, 8 b.

Rossill

    John Brabnir, 5 b. Thomas Innes, 15 b. 2 f. David Stevin, 12 b. 2 f. John Archebald, 6 b. 1 f. John Forsy 'eldar', 12 b. John Forsy 'youngar', 5 b. Andrew Wod, 5 b. John Thomesoun, 5 b. Alexander Andersone, 5 b. William Chayne, 10 b. James Nicholl, 8 b. 3 f. Nicol Archebald, 5 b. John Inchekeill, 12 b.

MORAY, DEANERY OF,   (NLS, fos 156r-157r)
(SRO, Vol. a, fos 402r-403r)

Rental of the deanery of Mwrray

'I, Johnne Lamb, burges of Edinburgh, takkismane and fermorar to Maister Alexander Campbell, dene of Murray, for the space of thre yeiris, my entres thairto begynand at [blank] lastwas, hes the said benefice as my assedatioun fullellie proportis as is notifiet in the bukis of Consall quhat dewitie I aucht thairfoir quhilk extendis to the some of [blank] to be payit to the said Maister yeirlie and ilk yeir of' 400 merks 'and just calcalatioun of the rest to be furthcummand to me till I be payit of the some of' 855 merks 6s 8d 'yitt nocht to defraud the just rentall bot that I sall geif it eftir my intelligence. It is worth as eftir fallowis.'

---

[28] Cf NLS, fo 168v (see below, p. 482). This entry lacks a title, apart from the heading 'The tennentis of Duffois', and is incomplete. It appears to relate to the parsonage of Duffus, a prebend of Elgin cathedral.

[29] Fo 155r is blank.

Mekill Pennyk set for 2 c. 14 b. Lytill Pennyk set for 1 c. 14 b. Golfurd set to the sheriff for 2 c. 13 b. 2 f. Mynes set to the sheriff for 3 c. 10 b. 3 f. 3 p. 'Leythynbar (1 chalder)[30] be the rentall', 24 b. 2 f. Brychtmony set to John Ros for 2 c. 1 b. Kinstarie, 3 c. Brodland, 1 c. 1 b. Kynnowdir, 1 c. 2 f. Petquhiwin, 8 b. Park Auld, 3 c. Eistir Geddis, 2 c. 8 b. Westir Geddis, 14 b. 'In Johnne Ros of Brychtmony', 9 b. Narne set for 3 c. 11 b. Boychthill set for 4 b. Total victual, 31 c. 5 [b.] 1 f. 3 p. 'conforme to my knawledge [and] the rentale gevin to me.'

'Item, for the martis ane price', 26s 8d. 'Item, for the teindis sett for money', £114 13s 4d. 'Item, for the temporal landis maillis', £14 0s 10d. 'Item, of wedderis callit cayn wadderis', 110 wedders. 'Item, of cayne aittis', 6 b. 'Item, of caponis', 24.

'Off the quhilk thair is to be deducit to me ane chalder victuall, the wedderis and caponis'. To Andrew Mairsar for his fee, 'thre quarteris of Lytill Pynnik', 22 b. 2 f. To George Dumbar for his fee, 7 b. 2 f. To Alexander Campbell, 20 b. 'Item, to the gardnare of the yarde', 6 b. 'Item, to the chalmerlane for his fie', £24.

'I, Johnne Lamb apprevis this be my just calcolatioun.'

'Memorandum, that my assedatioun beiris that I sall haif the fermis and victuallis abone expressit as the price extendis and is sauld betuix Mertymes[31] and Youll[32] ilk yeir of my thre yeiris takkis and gif thair be ony thing omittit in the rentall I will testifie it is nocht in my knawlege.'

Signature: 'Johnne Lamb, with my hand.'

Calculation of third: Victual, 31 c. 5 b. 1 f. 3 p.; third thereof, 10 c. 7 b. $2\frac{1}{3}$ p. Money, £130 0s 10d; third thereof, £43 6s $11\frac{1}{3}$d. Wedders, 110; third thereof, $36\frac{2}{3}$. Oats, 6 b.; third thereof, 2 b. Capons, 24; third thereof, 8.

Third of victual, 10 c. 7 b. $2\frac{2}{3}$ p.[33] Take Mekle Penik for 2 c. 14 b.; Litil Penik for 1 c. 14 b.; Kinstarry, 3 c.; Park Auld, 3 c. 'And gif in agane' 4 b. 3 f. $1\frac{1}{3}$ p.

Third of money, £43 6s $11\frac{1}{3}$d. Take the mart for 26s 8d; the temporal lands for £14 0s 10d; and the rest of money extending to £29 6s $1\frac{1}{3}$d 'out of the rediest of the teinds quhilkis giffis' £114 13s 4d.

'Tak the wedderis, aittis, capones quhair thai ar to be upliftit and remember my lord Comptrolare[34] to tak partic[ulare] assignatioun of this benefice becaus it is gevin in in generall.'

---

[30] These brackets are in the text.
[31] I.e. Martinmas, 11 November.
[32] I.e. Christmas.
[33] The correct calculation appears to be 10 c. 7 b. $2\frac{1}{3}$ p.
[34] See above, p. 465, n. 17.

MORAY, SUBCHANTORY OF,

(NLS, fo 158r[35])
(SRO, Vol. a, fo 404r[36])

'Rentale succentoris Moraviensis'

[*In margin*, 'Subchantorie of Murray']

The kirk of Raforte set to Alexander Dunbare of Kilboyok for 335 merks. The kirk of Arclaich set to Marjorie Dunbar of Bellewot for £40.

'Item, payit to the subchantouris stallar of the soum abonewrittin', £10. 'Item, to the vicar of Raforth', 12 merks. 'And sua restis frie to the subchantour', 388 merks.[37]

Signature: 'Alexander Dunber, subchantour of Murray.'

Calculation of third: In the whole, £263 6s 8d;[38] third thereof, £87 15s 6⅓d.[39]

MOY, PARSONAGE OF,

(NLS, fo 158r-v)
(SRO, Vol. a, fo 404r-v)

The rental of the parsonage of Moy within the diocese of Murray of the crop of the year of God 1560 'pertening to ane discreit man, sir Williame Sutherland, persone of the samyn, and sett in assedatioun to ane honorabill man, Robert Dunbar of Durris, fewar of the landis of Grante Grein [Grangegreen, *next paragraph*], Boggis and ane parte of the Kirktoun of Moye as eftir fallowis.'

'In primis, the towne and landis teindis of Grangegren [Grante Grein, *previous paragraph*] and Boggis sett to the said Robert and his subtennentis, lauboraris of the ground, payand thairfoir yeirlie at tua termis, Pasche[40] and Lambes[41] usit and wount', 40 merks. 'Item, the Kirktoun and landis of Moiye in the said Robertis handis and his subtennentis, as said is, conforme to ane assedatioun gevin to the said Roberte, payment yeirlie, as said is', 40 merks. Total, 80 merks.

'Off this sowme defalcat payit to the stallar that makis dalie residence in the cathederall kirk of Murray[42] for his staill fie', 10 merks. 'Item, defalcat for thekin of the queir of Murray and uther reformatioun within the samyn', 10 merks. 'Item,

---

[35] Fo 157v is blank.
[36] Fo 403v is blank.
[37] The correct calculation appears to be 368 merks.
[38] This is the total of 335 merks and £40.
[39] The correct calculation appears to be £87 15s 6⅔d.
[40] I.e. Easter.
[41] I.e. Lammas, 1 August.
[42] I.e. Elgin cathedral.

becaus the vicarage of Moy may nocht sustene ane minister of the tendis of the said townis to supporte his fie, yeirlie to him', 8 merks. 'Summa defalcat', 18 merks.[43]

'Quarta pars hujus beneficij'. The fourth part of this benefice which falls to the queen contains the sum of 13 merks.[44]

Signature: 'Ita est dominus Willelmus Swtherland, rector de Moy, manu propria.'

Calculation of third: In the whole, £53 6s 8d;[45] third thereof, £17 15s $6\frac{2}{3}$d.

PLUSCARDEN, PRIORY OF,

(NLS, fos 159r-161r)
(SRO, Vol. a, fos 404v-406v)

'The rentale of the priorie of Pluscardin conteninge all and syndre the fermis, mallis, sowmes of money, alsweill for the temporall landis as for the fischeingis that ar in use of payment and of the victuallis alsweill of the temporall landis as of the teindis of kirkis thairof, eldar sett for mony [sic] or victuallis, extract furth of the register and chalmerlane comptis of the samyn for this present yeir of God' 1561.

'The lordschipe of Pluscardin and barronies Farmen and Urquherde conforme to the auld rentallis, registeris and the yeirlie comptis of the chalmerlane now present calculat togidder, and first the said thre baronies in money by the fischeingis and cobillis extendis to' £257 16s 2d. And in wheat to 1 c. 1 b. 2 f. And in dry multures to 9 c. 11 b. And in victuals, malt, bere and meal 'ourheid', 14 c. 15 b. 2 f. And in custom oats, 5 c. 13 b. 'Summa, patet.'

'The fischeingis pertening to the said priorie eftir the rentallis and chalmerlane comptis presentlie.'
    'Ane cobill, half cobill and thrid parte cobill of lange tyme bypast in the places handis fischeit be thair awin servandis and be gud estimatioun yeirlie gud yeir and evill yeir ourheid wald extend to' 30 'lastis salmond, bott now intromettit with be the schiref of Murray allegand him to haif the samyn in fewferme to the grait hurte of the said place and standinge under reductioun. Ane uthir half cobill sett for yeirlie payment of' £33 6s 8d. 'Ane uthir half cobill sett for' £66 13s 4d. 'Summa of the fischeingis by the schireff of Murrayis intromissioun,' £100.

---

[43] The correct calculation appears to be 28 merks.
[44] 80 merks minus 28 merks equals 52 merks, divided by 4 equals 13 merks.
[45] I.e. 80 merks.

'The rentall of the spritualitie and kirkis perteninge to the said priorie sett for money and victuallis conforme to the auld rentallis, registeris and chalmerlane comptis as eftir fallowis.'

'The kirk of Pluscardin conforme to the rentall, raid and chalmerlane comptis extendis yeirlie commownlie in victuallis, meill and beir to' 7 c. 11 b. The kirks of Urquharde and Bellie 'conforme to the rentall, rydinge thairof, and chalmerlane comptis extendis yeirlie commownlie in victuallis, mele and beir oureheid to' 28 c. 10 b. 1 f. 1 p. 'And in money with the vicaragis of the samyn bot out of use of payment during this instant contraversie and truble', £122 0s 8d. The kirk of Durris [*i.e.* Dores] and Dacrus [*i.e.* Dalcross] set in assedation for £46 13s 4d. 'Summa totalis of the haill money befoir writtin, by the fischeingis occupyit and tane up be the schiref of Murray and the vicaragis out of use errat[46] of payment', £404 9s 6d.

'And thair is to be deducit and defalcat thairof the ordinar charges as eftir fallowis, viz.'

'To the sustentatioun of v monkis, ilk ane of thame in keching and habit silver', £16; total, £80. 'And to the sustentatioun of the chalmerlane and his tua servandis contractit to gif him yeirlie the sowme of' £133 6s 8d. 'And in almous to the puir folkis and puir tennentis yeirlie' £40. 'Summa deductorum in pecunia', £253 6s 8d.

'And sua restis free of silver', £151 2s 10d.

'Summa of quheit free, conforme to the rentall and chalmerlanis comptis', 1 c. 1 b. 2 f. 'frumenti' [*i.e.* wheat].

'Summa of drie mullturis free, conforme to the rentall and chalmerlanis comptis', 9 c. 11 b.

'Summa of victuallis bayth of the temporalitie and spiritualitie as is befoir writtin', 51 c. 4 b. 3 f. 1 p.

'Quhairof to be deducit the ordinar chargis as eftir fallowis.'

'Item, to the said fyve monkis of the convent, ilk ane of thame yeirlie having ane chalder fyve bollis, summa in the haill', 6 c. 9 b. 'To the maister cuk', 14 b. 'To the portare', 14 b. 'To the maister baxtar', 14 b. 'To the gardnar', 14 b. 'To the puir folkis and utheris puir tennentis and passingerris', $2\frac{1}{2}$ c. 'For makinge of the malt', 14 b. 'Summa deductorum in victualibus', 13 c. 7 b.[47]

'Sua restis of frie victuall', 37 c. 13 b. 3 f. 1 p.

'Nota the schiref of Murray intromettis with the milnis of Forres extendinge to' 4 c. 6 b. 'sed nihil solvit' [*i.e.* but nothing is paid].[48]

---

[46] 'Errat' appears to have been inserted; the arithmetic, however, is correct.
[47] The correct calculation appears to be 12 c. 15 b.
[48] This note does not appear in the SRO text.

'Memorandum, thair is na charge of casualities and dewities sic as fallis be accidentis nor yitt of custome muttones, kyddis, nor pultrie becaus the samyn servis bot the place and strangeris as efferis off honour and honestie.'

'Summa of the charge of aittis', 5 c. 13 b. 'And to be deducit for the chalmerlanis hors maitt', 2 c. 'Sua restis fre of aittis', 3 c. 13 b.

'Summa totalis frie as is befoir notit', £151 2s 10d money; 1 c. 1 b. 2 f. wheat; 9 c. 11 b. dry multures; 37 c. 13 b. 3 f. 1 p. victuals; 3 c. 13 b. oats.

Signature: 'Oeconomus, etc.'

[*In margin*, 'Nota, quhairof the thre baronies to be nemmit in particular.']

Calculation of third: Money, £525 10s 2d; third thereof, £175 3s 3⅔d.[49] Wheat, 1 c. 1 b. 2 f.; third thereof, 5 b. 3 f. 1⅓ p. Victual, dry multures, 9 c. 11 b.; third thereof, 3 c. 3 b. 2 f. 2⅔ p. Oats, 5 c. 13 b.; third thereof, 1 c. 15 b. Victual, malt, bere and meal, 51 c. 4 b. 3 f. 1 p.; third thereof, 17 c. 1 b. 2 f. 1⅔ p.[50]

'Nota, all gressowmes, martis, muttounes, caponis, kiddis, pultreis and sic small dewities omittit.'

'Memorandum, tweching the schiref of Murayis salmond.'

*Third of money, £175 3s 4⅔d. Take this money of the three baronies, namely, Pluscardin, Farnen and Urquhard, 'bot remember my lord Comptrollare[51] to ressave particulare assignatioun becaus thay ar gevin in ourheid in the rentall.'

Take the wheat out of the same baronies also which extend to 5 b. 3 f. 1⅓ p.

Take the victual and dry multures thereof, 3 c. 3 b. 2 f. 2⅔ p.

Victual malt, beir and meill. Take the victual malt, bere and meal of the three baronies extending to 14 c. 15 b. 2 f.; the kirk of Pluscardin, 8 c. 11 b.; total, 22 c. 10 b. 2 f. 'Gif in' 5 c. 8 b. 3 f. 2⅔ p.

Third of oats, 1 c. 15 b. Take the oats of the three baronies 'and remember my lord Comptrollare to ressaif particulare assignatiounis thairof.'

'Nota, to tak ordour withe the salmond fyscheingis, viz., a cobill and half a cobill and 3 parte cobill quhilkis ar in the schireff of Murrayis handis and estimit to' 30 'last salmond be yeir.'

---

[49] The correct calculation appears to be £175 3s 4⅔d.
[50] In the SRO text, the calculation of the third, together with the 'Nota' and 'Memorandum', is separated from the rest of the entry, and is located at fo 407v.
[51] See above, p. 465, n. 17.

'Item, to serche out the gressumes, canis, custumes, pultrie, caponis and all utheris dewties omittit.'[52]

DIPPLE, PARSONAGE OF,
(NLS, fos 161v-162r)
(SRO, Vol. a, fos 407v-408r)

Rental of the parsonage of Dyppil pertaining to Adam Hepburne, parson thereof.

'Item, with the personage and kirk of Ruthven pertening to the said personage and ane of the kirkis thairof beand within the diocy of Murray sett in assedatioun to my lord erle of Huntlie,[53] payand thairfoir yeirlie the sowme of' £80. 'Item, the kirk of Dippill, ane half thairof sett to James Innes of Aichthintowill payinge thairfoir yeirlie the sowme of' £18 3s 4d. 'Summa of the money of bayth the kirkis extendis to' £98 3s 4d.

'Heirof to be deducit for ordinar chargis of baith the saidis kirkis, yeirlie payit to the bischope of Murray[54] the sowme of' £4. To the factor of the said kirks for his fee, £10. Paid to the stallar for his fee, £10. Total, £24.

'Sua restis all chargis beand deducit the sowme of' £74 3s 4d.

'Item, for the rest of the kirk of Dippill sett to the tennentis for victuall, payand thairfor yeirlie' 2 c. 4 b. 'Summa of victuall', 2 c. 4 b.

Signature: 'Adamus Hepburne, rector de Dyppill, manu sua.'[55]

'The personage of Dippill contenand the kirkis of Dippill and Ruthven in Baidyenache'.

In the whole, £98 13s 4d [£98 3s 4d, *fo 161v*]; third thereof, £32 14s 6d 'ob'.[56] Victual, 2 c. 4 b.; third thereof, 12 b.

---

[52] The text of the assignation of the third from the second asterisk on p. 473 appears only in the SRO text, fo 406r-v. It is followed by the rental of the parsonage of Rothes which corresponds to NLS, fo 162r and by the calculation of the third of Pluscarden, in SRO, Vol. a, fo 407v, which corresponds to NLS, fo 174r (see below, pp. 475, 491).
[53] George Gordon, 4th earl of Huntly, *Scots Peerage*, iv, pp. 534-536.
[54] See above, p. 467, n. 27.
[55] From the next sentence, the NLS text, fos 162r-163v, is badly damaged. The deficiencies are made good from the SRO text.
[56] One third of £98 13s 4d is £32 17s 9½d; one third of £98 3s 4d is £32 14s 5½d.

ROTHES, PARSONAGE AND VICARAGE OF,[57]

(NLS, fo 162r)
(SRO, Vol. a, fo 408r)

[*In margin in SRO*, 'Personage and vicarage of Rothos, aliter in fine'.]

Rental of the parsonage of Rothos

The teind silver, 23½ merks. Of feu lands and annuals, 26½ merks. Of teind victuals, 124 b., half malt, half meal. 'The vicarage be estimatioun as I cane esteme is now' 10 merks. 'Subscrivit with my hand.'

Signature: 'Peter Leslie, persoune of Rothos.'

Calculation of third: In the whole, £40; third thereof, £13 6s 8d. Victual, half malt, half meal, 124 b.; third thereof, 41 b. 1 p. 1 f. ⅓ p. [*recte*, 41 b. 1 f. 1⅓ p.].

ESSIE, PARSONAGE OF,

(NLS, fo 162r-v)
(SRO, Vol. a, fo 408v)

Rental of the parsonage of Essye pertaining to Mr George Arnot.

The parsonage of Essye lying within the diocese of Murray set in assedation to George Gordoun of Lesmoir for £22.

Signature: 'George Arnote.'

Calculation of third: In the whole, £22; third thereof, £7 6s 8d.

DUFFUS, CHAPLAINRY OF,
UNTHANK, PARSONAGE OF,

(NLS, fo 162v)
(SRO, Vol. a, fo 408v)

Rental of the chaplainry of Duffois 'callit uther wayis of now the personage of Unthank', pertaining to John Gibsoun, parson thereof, and set in assedation by him to Alexander Sudderland of Duffeus for the sum of 40 merks yearly.

Calculation of third: In the whole, £26 13s 4d; third thereof, £8 17s 9⅓d.

---

[57] Cf SRO, Vol. a, fo 406v (see below, p. 491). NLS, fo 139r contains an incomplete rental entitled 'Rothes' which differs from those in NLS, fo 162r and SRO, fo 406v (see above, p. 449).

ELGIN CATHEDRAL,
CHAPLAINRY OF ST DUTHAC IN,

(NLS, fo 162v)
(SRO, Vol. a, fo 408v)

Rental of Sanct Duthois chaplainry within the parish kirk of Elgine pertaining to John Gibsoun, being yearly worth 12 merks.

Calculation of third: In the whole, £8; third thereof, 54s 4d.

ELGIN CATHEDRAL,
CHAPLAINRY OF OUR LADY IN,

(NLS, fo 162v)
(SRO, Vol. a, fo 408v)

Rental of the chaplainry of Oure Lady within the said kirk pertaining to the said John Gibsoun, being worth yearly 9 merks.

Calculation of third: In the whole, £6; third thereof, 40s.

'Thir rentallis was gevin in be the said Alexander Sudderland of Duffous', 24 February 1561/2.

'Quhairof the quenes haill thrid extendis of all the saidis benefices[58] to' 20 merks.[59]

Signature: 'Joannes Gibsoun, manu propria, and ressavit at the command of the Clerk Registere.'[60]

INVERKEITHNY, PARSONAGE OF,

(NLS, fo 163r)
(SRO, Vol. a, fo 409r)

Rental of the parsonage of Innerkethnie 'quhilk is ane prebendarie in the kirk of Murray[61] gevin be me, Mr James Thornetoun, in name and behalf and as procuratour to Maister Hew Cragy, possessour and titular of the said personage.'

'The haill teindis and all uther proffeit of the personage of Innerkethnie quhilk is ane prebendre in the kirke of [Mur]ray, hes bene thir mony yeiris and as yitt ar sett in assedatioun to Alexander Dunbar of Kellugrie for the sowme and yeirlie payment of' £80 money of this realm.

Signature: 'Ita est Jacobus Thorntoun, procurator, teste manu propria.'

---

[58] I.e. the chaplainries of Duffus, of St Duthac and of Our Lady.
[59] The correct calculation appears to be 20 merks plus 5s 5½d.
[60] James McGill of Nether Rankeilour was Clerk Register at the Reformation, *Handbook of British Chronology*, p. 197.
[61] I.e. Elgin cathedral.

Calculation of third: In the whole, £80; third thereof, £26 13s 4d.

MORAY, ELEVEN CHAPLAINRIES IN,[62] (NLS, fos 163r-166r[63])
(SRO, Vol. a, fos 409r-410r)

'The rentall of levin [i.e. eleven] capellanies within the diocie of Murray of the quhilk sir Alexander Douglas, chaplane in the kirk of Murray, is possessoure, gevin in be Maister James Thornetoun, in name and behalf and as procuratour to the said sir Alexander.'

INVERNESS, (NLS, fo 163r-v)
CHAPLAINRY OF OUR LADY IN THE GREEN, (SRO, Vol. a, fo 409r)

[*In margin*, 'Chaplanrie of Our Lady in the Grein.']

'Ane capellanie callit Our Lady of the Grene besyd Innernes in the diocy of Murray, of the quhilk the said sir Alexander is possessour, all thingis being deduct, frelie payis to the possessour thairof', 25 merks.

Calculation of third: *see* Chaplainry of St Thomas, *below*.

ELGIN CATHEDRAL, (NLS, fo 163v)
CHAPLAINRY OF ST THOMAS'S ALTAR IN, (SRO, Vol. a, fo 409v)

[*In margin*, 'Chaplanrie of Sanct Thomas altar in Murray kirk.']

'Ane capellanie callit at Sanct Thomas alter in the kirk of Murray foundat be biscope Gavane Dumbar,[64] of the quhilk the said sir Alexander is possessour, all chairges being deducit, payis frelie to the possessour thairof' 20 merks.

Signature: 'Ita est James Thornetoun, procurator, teste manu propria.'

Calculation of third:
  'Chaplanrie of the Lady in the Grene besyd Innernes. In the haill frie', £16 13s 4d; third thereof, £5 11s 10½d.[65]
  'Chaplanrie att Sanct Thomas altar. In the haill free', £13 6s 8d; third thereof, £4 8s 10⅔d.

---

[62] Cf NLS, fo 168r (see below, p. 481).
[63] Fos 164r-v, 165r-v are blank.
[64] Gavin Dunbar, junior, was dean of Moray, 1517 - 1525, bishop of Glasgow, 1523 - 1547, Watt, *Fasti*, pp. 149, 221.
[65] The correct calculation appears to be £5 11s 1⅓d.

ELGIN CATHEDRAL, CLYNE, CHAPLAINRY OF,

(NLS, fo 163v)
(SRO, Vol. a, fo 409v)

'Ane chaplanie callit at Sanct Androis altar in the kirk of Murray callit alias the chaplanrie of Clyn, of the quhilk the said sir Alexander is possessour, all chargis being deducit payis frelie to the possessour thairof' 15 merks.

Signature: 'Ita est Jacobus Thornetoun, procurator, teste manu propria.'
Calculation of third: In the whole, £10; third thereof, £3 6s 8d.

ELGIN CATHEDRAL, KNOCKANDO, CHAPLAINRY OF,

(NLS, fo 166r)
(SRO, Vol. a, fo 409v)

'Ane chaplanie at the Magdalanes altar in the foirsaid kirk of Murray callit alias the Knokandoche, of the quhilk the said sir Alexander is possessour, all charges beinge deducit payis frelie to the possessour thairof' 13 merks.

'Summa of the haill chaplanries abone expremit extendis to' 73 merks.[66]

Signature: 'Ita est James Thornetoun, procurator, teste manu propria.'

Calculation of third: 'In the haill free', £8 13s 4d; third thereof, 57s 9⅓d.

BOTARY, VICARAGE OF,

(NLS, fo 166r)
(SRO, Vol. a, fo 410r)

20 January 1561/2

Rental of the vicarage of Botarie with the pendicle of the same called the kirk of Petruifnye[67] [*i.e.* Botriphnie] with manse, yard and croft within the diocese of Murray and deanery of Strathboggie in James Innes hands of Dranyes 'be sett payand wont' 40 merks money 'to my [*blank*], Johne Hepburne, vicar of the samyn, quhairof justlie now is deducit' £10 money 'for the umaist claith, corps present, Pasche penny and uther oblationis conforme to the act maid of laitt, the haill', 25 merks. 'The quenes grace thrid thairof is', 8 merks 4s 5⅓d.

Signature: 'Johnne Hepburne, vicar of Bottarrie, with my hand. Rasavit at the command of the Clerk Registere.'

Calculation of third: In the whole, 25 merks; third thereof, 8 merks 4s 5d.

---

[66] Four chaplainries only (not eleven) are listed here.
[67] The vicarage perpetual of Botriphnie, in the sheriffdom of Banff, was linked with that of Botary to form a single cure, Cowan, *Parishes*, p. 21.

ARDCLACH, VICARAGE OF,[68] (NLS, fo 166v)
(SRO, Vol. a, fo 410r-v)

Rental of the vicarage of Ardclach lying within the diocese of Murray given up by Thomas Ros in Daltuly, 'fermorar thairof, and subscrivit with his hand at the pen in maner underwrittin.'

'The said Thomas Ros, quha is fermorar of the said vicarage with the croft and mans thairof and utheris fructis, rentis and pertinentis of the samyn, hes brukit it be the space of xv yeiris lastbypast payand thairfor yeirlie to Patrik Browne, vicar thairof, the sowme of' £20 'allanerlie. Item, to defalk thairof the Pasche fynes, offerandis, corps presentis, umaist clathis, walx, yerding silver and utheris sic as teind allis [sic] and small offerandis at mariages and cristynning of barnes quhilk wald half extendit allanerlie to the valour of the half of the said' £20, 'viz.' to £10. 'And sua the thrid parte now of the said vicarage will extend to the soum of' £3 6s 8d 'allanerlie.'

Signature: 'Thomas Ros of of [sic] Daltulie with my hand at the pen led be the notar underwrittin at my request becaus I cane nocht write. At Edinburgh', 4 June, 1562 'befoir thir witnes, Johnne Falconar, Alexander Wache, Johnne Wat, Johnne Andro and Nicol Wischart with utheris dyveris. Ita est Magister Gilbertus Grote, notarius publicus, ac testis in premissis manu propria.'

Calculation of third: In the whole, £10; third thereof, £3 6s 8d.

MORAY, CHANTORY OF, (NLS fo 167r)
ALVES, PARSONAGE AND VICARAGE OF, (SRO, Vol. a, fo 410v)
LHANBRYDE, PARSONAGE AND VICARAGE OF,

Rental of the chantory of Murray given in by Mr James Thornetoun 'in name and behalf and as procurar [recte, procurator] of his eme, Maister Johnne Thornetoun eldar, quha is in titill of the said chantorie.'

'The haill teindis and all uder proiffeittis of the kirkis of Alves and Lambryd[69] personage and vicarage quhilkis ar annexit to the said chantorie was sett in assedatioun to [blank] Spens and Alexander Andersoun viij yeiris syne for the sowme of' 440 merks.

'Now the said Maister Johnne Thorntoun takkis up him self the haill tendis [sic] foirsaid quhilk in victualibus, the maist parte beir, the remanent quheit, will extend to' 18 c.

---

[68] Cf NLS, fo 172r (see below, p. 488).
[69] Alves and Lhanbryde were assigned to the chanter of Moray, Cowan, *Parishes*, pp. 6, 132.

'The mony that is payit of the said chantorie will extend to the sowme of' 180 merks.

'Off this for the feis of the vicar pensionaris of Alves and Lambrid quhilk is payit to thame is yeirlie deducit extending to' 40 merks.

'Summa in victualibus', 18 c. In money ('deductis oneribus'[70]), 140 merks.

Signature: 'Ita est Jacobus Thorntoun, procurator, teste manu propria.'

Calculation of third: Money, £120; third thereof, £40. Victual, bere and wheat, 18 c.; third thereof, 6 c.

'Nota, he grantis it was sett befoir for' 440 merks.

ADVIE AND CROMDALE, PARSONAGE OF,   (NLS, fo 167v)
    (SRO, Vol. a, fo 411r)

'The rentall of the personage of Advy and Cromdell, prebendre in the kirk of Murray, gevin in be me, James Thorntoun, in name and behalf and as procurator to his brother, Maister Johnne Thorntoun younger, possessour and titular of the samyn.'

'The haill proffeittis and all teindis of the personage of Advy and Crowmdell, prebendrie in the kirk of Murray liand in tressory[71] and ar sett in assedatioun for the space of xix yeiris to Johnne Grant of Fruquhy for the yeirlie payment of the sowme of' 40 merks, 'quhilk suld be frelie payit to the said Maister Johnne.'

Signature: 'Ita est Jacobus Thornetoun, procurator, teste manu propria.'

Calculation of third: In the whole, £26 13s 4d; third thereof, £8 17s $9\tfrac{1}{3}$d.

ABERCHIRDER, VICARAGE OF,[72]   (NLS, fos 167v-168r)
    (SRO, Vol. a, fo 411r)

Rental of the vicarage of Abirkardur in the diocese of Murray, 'gevin in be the said me [sic], James Thornetoun, in name and as procuratour for his said brother Mr Johnne Thorntoun younger, possessour and titular of this vicarage.'

---

[70] I.e. 'deducting the charges'; but in the calculation of the third, the payments to the vicars pensionary have not been allowed.
[71] I.e. treasury.
[72] Cf SRO, Vol. a, fo 164v (see above, p. 159).

'The haill teindis, proffeittis and all uther casualities of the vicarage of Abirkardoure within the diocy of Murray quhen all teindis and casualities war payit was sett and as yitt is sett in assedatioun to James Innes of Towchyis for the sowme of' £86 13s 4d, 'and now lytill or nothing gottin thairof be ressoun no teyndis of vicaragis ar payit.'

'Off this gevin to the vicar pensionare for his susten[tatioun] the sowme of' 20 merks.

'Summa, oneribus deductis' [*i.e.* total, charges deducted], 110 merks.

Signature: 'Ita est Jacobus Thornetoun, procurator, teste manu propria.'[73]

Calculation of third: In the whole, £86 13s 4d; third thereof, £28 17s 9¼d.

MORAY, ELEVEN CHAPLAINRIES IN,[74]

(NLS, fo 168r)
(SRO, Vol. a, fo 411v)

'The rentall of ane vicarage and levin lytill capellanies in the diocy of Murray off the quhilkis sir James Spens, vicar of the queir of Murray, is possessour, gevin in be me, James Thornetoun, in name and behalf and as procurour to the said sir James.'

DALLAS, VICARAGE OF,

(NLS, fo 168r-v)
(SRO, Vol. a, fo 411v)

The vicarage of Doles in the diocese of Murray 'quhen corpspresent, umaist claith and Pasche penny was payit, gaif in assedatioun to sir James Spens, possessoure thairof, the sowme of' £10 money of this realm 'and now nothing thairof is payit'

Calculation of third: *see* Chaplainry of Artillich, *below*.

ELGIN CATHEDRAL,
CLYNE, CHAPLAINRY OF,[75]

(NLS, fo 168r-v)
(SRO, Vol. a, fo 411v)

'Ane capellanie callit Clynis in Ros in the diocy of Murray of the quhilk the said sir James is possessour payis in assedatioun to the said sir James'[76] £10.

---

[73] I.e. 'Thus, John Thornton testifies with his own hand.'
[74] Cf NLS, fo 163r (see above, p. 477).
[75] Cf NLS, fo 163v (see above, p. 478).
[76] Sir Alexander Douglas is named as possessor, fo 163v (see above, p. 478).

Calculation of third: *see* Chaplainry of Artillich, *below*.

ARTILLICH,[77] CHAPLAINRY OF,   (NLS, fo 168r-v)
(SRO, Vol. a, fo 411v)

'Ane capellanie callit of Artillich in the diocy of Murray, of the quhilk the said sir James is possessour, payis to him in assedatioun' 8 merks.

'Summa of the haill', 38 merks.[78]

Signature: 'Ita est James Thornetoun, procurator, teste manu propria.'

Calculation of third:
    Vicarage of Dolas, in the whole, £10; third thereof, £3 6s 8d.
    Chaplainry of Clynis in Ros, in the whole, £10; third thereof, £3 6s 8d.
    Chaplainry of Artillich, in the whole, £5 6s 8d; third thereof, 35s $6\frac{2}{3}$d.

[*In margin*, 'The haill the quenis.']

DUFFUS, PARSONAGE OF,[79]   (NLS, fos 168v-169r)
(SRO, Vol. a, fos 411v-412r)

Rental of the parsonage of Duffows

Rossyill, 10 c. victual. Burnesyd and the Bagra, 3 c. victual. Ynchekyll and the Starwod, 3 c. victual. Kayme, 7 c. victual set to James Innes of Drany for £10 the chalder extending to £70. Kirktoun of Duffows 'with the manys thairof with the towne of Saltcottis', 8 c. victual 'or thairby' set to the laird of Duffows for £10 the chalder extending to £82 10s.
Total victual, 16 c. Total money, £152 10s.

'Quhilkis victuallis and mony fallis to the last personis executouris be ressoun of his annat to pay the dettis awand be him at his deces and to fulfill the gudwill of the ded conforme to the tennour of the testament.'

Signature: 'Ita est Andreas Creif,[80] manu propia [*sic*]'.

'Personage and vicarage of Duffows'.

---

[77] Presumably a rendering of Arndilly, a prebend of Elgin cathedral, Scott, *Fasti*, vi, p. 336.
[78] This figure includes the rental of the vicarage of Dallas and the chaplainry of Clyne, as well as that of Artillich.
[79] Cf NLS, fos 155v, 170r (see above, p. 468; below, p. 485).
[80] This is rendered 'Andrew Moncreif' on fo 169v, (see below, p. 483).

KINNOIR, PARSONAGE OF,

(NLS, fo 169r)
(SRO, Vol. a, fo 412r)

The parsonage of Kynnour within the diocese of Murray.

Set in assedation to my lord earl of Huntlie[81] for £100 yearly 'and the persoun to pay the stallar, quhilk is' 20 merks; the third part extending to £28 17s 9½d.

Calculation of third: In the whole, £100; third thereof, £33 6s 8d.

ELGIN, VICARAGE OF,

(NLS, fo 169r)
(SRO, Vol. a, fo 412r)

Rental of the vicarage of Elgin

'The vicarage of Elgin payis now allanerlie bot' 2 c. victual 'be ressoun thair is na payment maid nowthir of woll, lamb nor utheris dewities payit to the vicaris in tymes bypast quhilk had wont to be sett in assedatioun for' 80 merks.

Calculation of third: In the whole, 2 c. victual; third thereof, 10 b. 2 f. 2⅔ p.

'Nota, it was wount to be sett for' 80 merks.

INVERNESS, VICARAGE OF,[82]

(NLS, fo 169v)
(SRO, Vol. a, fo 412v)

Rental of the vicarage of Innernes 'had wontt to pay in assedatioun yeirlie' 100 merks, 'of the quhilk thair is na payment maid be the parochenaris nothir of woll, lamb, Pasche pennies, umaist clathis nor corps presentis thir thre yeiris bygane.'

Signature: 'Ita est Andreas Moncreiff,[83] manu propria.'

Calculation of third: In the whole, £66 13s 4d; third thereof, £22 4s 5⅓d.

---

[81] See above, p. 474, n. 53.
[82] Cf NLS, fo 77r (see above, p. 359).
[83] See above, p. 482, n. 80.

KINNEDDAR, VICARAGE OF,[84]

(NLS, fo 169v)
(SRO, Vol. a, fo 412v)

Rental of the vicarage of Kynnedoir within the diocese of Murray.

'In primis, the samyn is sett be sir James Douglas, vicare thairof, yeirlie to James Innes of Drany for the sowme of' 40 merks.

Signature: 'Ita est Joannes Douglas, vicarius de Galstoun, manu propria.'

Calculation of third: In the whole, £26 13s 4d; third thereof, £8 17s $6\frac{2}{3}$d.[85]

ST LEONARD'S, CHAPLAINRY OF,

(NLS, fos 169v-170r)
(SRO, Vol. a, fo 412v)

'Ane cheplanrie callit Sanct Leonardis chapell pertenyng to Maister Johnne Dumbar, to the quhilk cheplanrie the taxman of the erldome of Murray aucht to pay' £5 yearly 'and I gat never payment thairof thre yeris bigane.'

Calculation of third: In the whole, £5; third thereof, 33s 4d.

Signature: 'Maister Johnne Dumbar, persoun of Cumnok.'

BOTARY, PARSONAGE OF,[86]

(NLS, fo 170r)
(SRO, Vol. a, fo 412v)

Rental of the parsonage of Botary within the diocese of Murray is 100 merks 'and thairof ane pensioun and stallaris fie in Elgin yeirlie extending to' 20 merks.

Signature: 'Mr Alexander Skeyne.'

Calculation of third: In the whole, £80;[87] third thereof, £26 13s 4d.

---

[84] Cf NLS, fo 173v (see below, p. 490).
[85] The correct calculation appears to be £8 17s $9\frac{1}{3}$d.
[86] Cf NLS, fo 172v (see below, p. 489).
[87] The total should be 100 merks, or 80 merks if the pension and stallar's fee are deducted.

DUFFUS, VICARAGE OF,

(NLS, fo 170r)
(SRO, Vol. a, fo 413r)

Rental of the vicarage of Duffows pertaining to John Cocburne, lying within the diocese of Murray and sheriffdom of the same pays yearly £20, 'subscrivit with my hand.'

Signature: 'Johnne Cokburne, vicar of Duffois, with my hand.'

Calculation of third: In the whole, £20; third thereof, £6 13s 4d.

GLASS, PARSONAGE AND VICARAGE OF,

(NLS, fo 170r-v)
(SRO, Vol. a, fo 413r)

Rental of the parsonage and vicarage of Glas pertaining to Maister James Gartlie,[88] 'extendis of auld and usit and wount to' 105 merks, 'off the quhilk to be deducit yeirlie to the vicar pensionar the sowme of' 12 merks.

Signature: 'I, Maister James Gartlie, presentis the samyn.'

Calculation of third: In the whole, £70; third thereof, £23 6s 8d.

KIRKMICHAEL, VICARAGE OF,

(NLS, fo 170v)
(SRO, Vol. a, fo 413r)

Rental of the vicarage of Kirkmichell in Stradoun within the diocese of Murray pertaining to James Gartlie 'and use and wount being payit extendis to' 60 merks.
    'Of this to be deducit yeirlie to the curat thairof' £10, 'the casualities nocht being payit it wilbe' £30 'be yeir.'

Calculation of third: In the whole, £40; third thereof £13 6s 8d.

CAWDOR, VICARAGE OF,[89]

(NLS, fo 170v)
(SRO, Vol. a, fo 413r)

Rental of the vicarage of Kwyd lying within the sheriffdom of Narne[90] and diocese of Murray pertaining to Mr George Strang.

---

[88] The surname appears as 'Grantuly' and 'Garnetullie', *Thirds of Benefices*, pp. 168, 226.
[89] Also known as Braaven, Cowan, *Parishes*, p. 22; Scott, *Fasti*, vi, p. 438.
[90] The Books of Assumption contain no separate list of entries for the sheriffdom of Nairn.

Calculation of third: In the whole, £13 6s 8d; third thereof, £4 8s 10⅔d.

LAGGAN, VICARAGE OF, (NLS, fo 171r)
(SRO, Vol. a, fo 413v)

Rental of the vicarage of Logan in Murray pertaining to sir John Nicolsoun.

'Item, the said vicarage of Logan gaif of auld' £20, 'and now gevis nothing quhil generall ordoure be tane.'

Signature: 'Mr Henrie Kynrois, presentar, with my hand.'

Calculation of third: In the whole, £20; third thereof, £6 13s 4d.

RHYNIE, PARSONAGE OF, (NLS, fo 171r)
(SRO, Vol. a, fo 413v)

Rental of the parsonage of Ryne within the diocese of Murray pertaining to Mr James Gordoun,[91] son to the earl of Huntlie, set in assedation for 80 merks. 'Thairof deducit to the queir of Elgin', £8. 'Sua restis to the possessoure', 68 merks.

Calculation of third: In the whole, £53 6s 8d; third thereof, £17 15s 6⅔d.

MORAY, CHANCELLORY OF, (NLS, fo 171r)
(SRO, Vol. a, fo 413v)

Rental of the chancellory of Murray pertaining to the said Mr James Gordoun set in assedation for £100.

Calculation of third: In the whole, £100; third thereof, £33 6s 8d.

'Bettir rentalit in Mr Robert Gordones tyme for' 240 merks, 'and sua maid ... '.[92]

---

[91] See above, p. 435, n. 77.
[92] NLS, fo 171r is torn at the right-hand corner; the phrase 'and sua maid' is missing from the SRO text. Robert Gordon succeeded James Gordon as chancellor in 1566, Watt, *Fasti*, p. 228.

MORAY, COMMON KIRKS OF: (NLS, fo 171v)
(SRO, Vol. a, fo 413v)

BIRNIE, PARSONAGE AND VICARAGE OF, (NLS, fo 171v)
(SRO, Vol. a, fo 413v)

The kirk of Burneyth, the parsonage and vicarage of the same in the diocese of Murray extending yearly to 4 c. 12 b. victual. John Stanis, vicar and minister of the said kirk and reader and exhorter in the said kirk; the vicarage of the said kirk extending to 10 merks 'allanerlie.'

Signature: 'Joannes Stanis, vicarius de Burneyth, manu propria.'

Calculation of third: Victual, 4 c. 12 b.; third thereof, 1 c. 9 b. 1 f. $1\frac{1}{4}$ p. Money, £6 13s 4d; third thereof, 44s $5\frac{1}{3}$d.

KINGUSSIE, PARSONAGE OF, (NLS, fo 171v)
(SRO, Vol. a, fo 414r)

The rental of the parsonage of Kyngusy pertaining to Mr George Hepburne, 'productit be me, Archibald Lyndesay.'

The kirks and parsonage of Kingusye set to Mr George Gordoun for £80. 'Heirof to be deducit for the stallaris fie', £6 13s 4d. 'Sua restis' £73 6s 4d. 'And sua the thrid parte thairof to the quenis grace extendis to' £18 6s 8d.[93]

Signature: 'George Hepburne.'

Calculation of third: In the whole, £80; third thereof, £26 13s 4d.

MORAY, ARCHDEACONRY OF,[94] (NLS, fo 172r)
(SRO, Vol. a, fo 414r)

Rental of the archdeaconry of Murray 'producit be me [Mr, *added in SRO*] Archibald Lyndesay.'

The kirks and parsonages of Fores and [Enkethe, *deleted*] Edinkellie[95] set in assedation to Thomas Tullache and William Wauche for £146 13s 4d.

---

[93] The correct calculation appears to be £24 8s $9\frac{1}{3}$d.
[94] Cf SRO, Vol. a, fo 417v (see below, p. 494).
[95] This place-name is blank in the SRO text.

Signature: 'Maister Archibald Lyndesay, with my hand.'

Calculation of third: In the whole, £146 13s 4d; third thereof, £48 17s 9⅓d.

'Nota, this archidenrie is allegit to be vacand and to extend to' £160; third thereof, 80 merks.

SPYNIE, PARSONAGE OF,   (NLS, fo 172r)   (SRO, Vol. a, fo 414r)

Rental of the parsonage of Spynye

The parsonage of Spynnie, 'ane channownrie of Murray', pertaining to Mr Thomas Hay set for 200 merks money yearly 'deduceand thairof yeirlie to the stallar of the queir', £10 money.

Signature; 'Thomas Hay.'

Calculation of third: In the whole, £133 6s 8d; third thereof, £44 8s 10⅔d.

ABERTARFF, VICARAGE OF,   (SRO, Vol. a, fo 414v)   (NLS, fo 172r[96])

'The rentall of the vicarage of Abirtarff, sir Patrik Broun possessor thairoff, at the hed of Lochnes, payit in assedatioun' 40 merks. 'Off the quhilk na payment is gottin this thrie yeiris bygane of ane grot, for your lordschip knawis it lyis [*blank*[97]] Clane Ranald and utheris sik clanis quha usis to mak na payment.'

Calculation of third: *see* Vicarage of Ardclach, *below*.

ARDCLACH, VICARAGE OF,[98]   (NLS, fo 172v)   (SRO, Vol. a, fo 414v)

Rental of the vicarage of Ardclacht within the diocese of Murray pertaining to sir Patrick Browne, vicar thereof, which paid in assedation £20. 'And na maner of payment gotting thairof thir thre yeiris bygane. Sucscrivit with my hand at Aberdene', 12 February 1561/2.

---

[96] The SRO text here supplies the transcription, for the NLS text is damaged.

[97] The blank space here indicates that the scribe was utilising the text in the NLS version which is torn at the corner of the folio.

[98] Cf NLS, fo 166v (see above, p. 479).

Signature: 'sir Patrik Browne, vicar of Abirtarf and Arclauch'

Calculation of third:
   Vicarage of Abirtarff 'in the hed of Lochnes', in the whole, £26 13s 4d; third thereof, £8 17s 6d.[99]
   Vicarage of Ardclauch, in the whole, £20; third thereof, £6 13s 4d.

BOTARY[100] AND ELCHIES, PARSONAGE OF,   (NLS, fo 172v)
(SRO, Vol. a, fo 414v)

Rental of the parish kirk of Botarra and Elchis pertaining to Mr James Strathquhene, parson of the same, within the diocese of Murray.

'And set in assedatioun for' 80 merks.

'And this gevin in be Johnne Gordoun in Karnbarrow, factour of the samyn.

Signature: 'Johnne Gordoun.'

ABERLOUR, VICARAGE OF,   (NLS, fo 173r)
(SRO, Vol. a, fo 415r)

Rental of the vicarage of Abirloche within the diocese of Murray, 'quhilk wes settand in the lard of Grantis handis thir yeiris bypast, payand' £18 money, 'quhair may be justlie deducit for Pasche penny, oblationis, corps presentis, umaist clathis out of use of payment presentlie ten merkis. Sua restis free', 17 merks. 'Pars regine [*i.e.* the queen's part]', 5 merks 8s 10⅔d.

Signature: 'William Douglas, with my hand.'

DUNDURCAS, VICARAGE OF,   (NLS, fo 173r)
(SRO, Vol. a, fo 415r)

Rental of the vicarage of Dundurcus within the said diocese pertaining to Mr William Wismoun [*i.e.* Wiseman], vicar of the same, in William Ogilbeyis hands 'and his executouris' paying yearly 23 merks money. 'Thairof deducit ar 8 merks 'for the Pasche penny, oblatioun, corpspresent and umaist claith be just rentall, sua restis, £10; third thereof, £3 6s 8d.

---

[99] The correct calculation appears to be £8 17s 9½d.
[100] Cf NLS, fo 170r (see above, p. 484) for Botary; there is no further entry for Elchies.

'This rentall subscrivit be me the said William Douglas at the said Maister Williames command conforme to his hand writt send me.'

KINNEDDAR, VICARAGE OF,[101]

(NLS, fo 173v)
(SRO, Vol. a, fo 415v)

Rental of the vicarage of Kynnedoure within the diocese of Murray set in assedation to James Innes of Drany for £20. The possessor of the said vicarage is named James Douglas.

ELGIN CATHEDRAL, CHAPLAINRY OF ST CATHERINE IN,[102]

(NLS, fo 173v)
(SRO, Vol. a, fo 415v)

'The said James Douglas hes ane chaplanrie fundit within the cathedrall kirk of Murray, callit Sanct Katherines chaplanrie, extending yeirlie to' 8 b. victual 'payit of certane croftis of the burgh of Elgin and fyve pundis money payit of the patromine of the bischoprik of Murray and' 22s money 'payit of certane annuallis of certane houssis of the said burch.'

'This rentall subscrivit be me, William Douglas, vicar of Abirlocht, brother to the said James Douglas, at his command.'

Signature: 'James Douglas, with my hand. William Douglas, with my hand.'

ELGIN CATHEDRAL, CHAPLAINRY OF ST CATHERINE IN,[103]

(NLS, fo 173v)
(SRO, Vol. a, fo 415v)

The chaplainry of Sanct Katherine in the cathedral kirk of Murray pertaining to sir Alexander Sinclare, extends in yearly rent as follows.

'Item, furth of the croftis aboute the towne of Elgin', 8 b. bere; 'and furth of the byschoprik of Murray', £5 money; and of annuals within the town of Elgin, 22s. 'Sua the haill extendis to' 8 b. bere and £6 2s money.

Signature: 'Jacobus Makgill.'

---

[101] Cf NLS, fo 169v (see above, p. 484).
[102] Cf next rental.
[103] Cf preceding rental.

ROTHES, PARSONAGE OF,[104]   (SRO, Vol. a, fos 406v-407r)
(NLS, fo 174r[105])

'The rentall of the personage of Rothes of the victuallis.'

Boighaid, 4 b. victual. Blakhall, 9 b. victual. The Ylis, 10 p. victual. In Kendalie, 7 b. victual. The Croftis, 7 f. victual. Enchebowoquhy, 5 b. victual. The Ovir Glen, 4 b. victual. Petcraigy, 3 b. victual. The Manis, 24 b. victual. Auchinroith, 2 b. 1 f. victual. Auld Yardis, 2 f. victual. Ovir Petteddedrie, 17 b. victual. The Courocht, 4 b. victual. The acres, 4 b. victual. The Brochahill, 2 b. victual. Total, 88 b. 8 f. 2 p. victual.[106]

The teind silver of the parish of Rothes

The Nether Glene, 50s. Blaklawacht, 7 merks. Ardcannie, 40s. Smychttie Grene, 13s 4d. The Fischartoun 40s.

The mails of the feu lands and annuals of the parsonage of Rothes

Akynwaiy, £8. The Brig, £4. The Lingestoun, 52s in the sheriff's handis. The Kirkhill of Rothes, 40s 'and liand west [i.e. waste] the space of thir thrie yeiris bygane'. The mills of Narrine, 52s. The vicarage of Rothes extends in money to £9 10s 8d. Total, £38 11s 4d.[107]

Signature: 'James Leslie,[108] persoune of Rothes, with my hand. Apud Edinbrucht', 27 September, 1566. 'The Lordis Auditouris of Chekker ordanis this rentall to be ressavit and chairgit yeirlie in tyme to cum. Sic subscribitur: Adam Orchadensis; Johnne Rossensis; J. Bellendin.'[109]

KINGUSSIE, VICARAGE OF,[110]   (SRO, Vol. a, fo 407r)

'We, George, erle of Huntlie,[111] etc., forsamekle as inrespect that Mr Williame Carnegy hes set to us the viccarage of Kynguyse for the space of thre yeiris for tuentie merkis yeirlie as in the assedatioun is conteinit, we ar content to pay the said tuentie merkis yeirlie in compleit payment of the

---

[104] Cf NLS, fo 162r (see above, p. 475; see also, p. 474, n. 53).

[105] NLS, fo 174r is badly damaged; it appears to form part of the rental for Rothes in SRO, fo 406v; a marginal note in a later hand reads: 'See the rental of Pluscardin'. The NLS text ends at fo 174v which is also badly damaged. SRO, Vol. a continues for several folios.

[106] The correct calculation appears to be 88 b. 2 p.

[107] The correct calculation appears to be £40 11s 4d.

[108] The rental at fo 162r (see above, p. 475) is signed by Peter Leslie, who was parson at the Reformation; James Leslie was parson in 1564, *Thirds of Benefices*, pp. 108, 168-169, 216; *RMS*, iv, no 1787.

[109] Adam Bothwell, bishop of Orkney; John Leslie, bishop of Ross; Sir John Bellenden, Justice Clerk. Watt, *Fasti*, pp. 254, 270; *RSS*, v, no 589.

[110] Cf SRO, Vol. a, fo 416r (see below, p. 492).

[111] George, 5th earl of Huntly, *Scots Peerage*, iv, pp. 539-541.

dewtie of the said vicarage, tua parte and third at the feist of Bartholmes[112] ilk yeir respective provyding that the said Williame find me sufficient dischairge of the thrid at the Collectouris[113] handis gif neid beis and siclyk gif that the tennentis hes intromettit with the fructis of this present yeir of threscoir fyftene [*i.e.* 1575] inrespect that na inhibitioun hes proceidit thairon the said Mr Williame sall rais lettres and sene ane [*sic*][114] upoun the tennentis for payment thairof, and for mair verificatioun of the premissis we have subscrivit this [*sic*] presentis with our hand at Huntlie the' 9 June 1575 'beffoir thir witnessis, Robert Bruce of Inches, Antone Bruce, James Curle and Mr Thomas Austoun. Sic subscribitur, Huntlie.'

KINGUSSIE, VICARAGE OF,[115] (SRO, Vol. a, fo 416r)

Rental of the vicarage of Kyngusy within the diocese of Murray pertaining to Mr Hercules Carnegy, set in assedation to Mr George Gordoun for 40 merks yearly; third thereof, £8 17s 9½d.

DALLAS, PARSONAGE OF,[116] (SRO, Vol. a, fos 416r-v)

'The xxiiij day of August the yeir of God ane thousand fyve hundreth and sextie ane yeiris, the cornes of the paroche of Dolles riddin the teindis of the same estimat be James Runsimane in Innerlochtie, Robert Andersoune in Barmwcutty, Andro Garnar, Theophilus Johnnestoun with sir James Johnnestoun, notar publict.'

'The rentall of the saidis teind schavis of Doles as eftir followis.'

The Congtoun 'riddin and set for' 4 b. The Littill Brunquallis, 9 b. The Mekill Ranquallis, 9 b. The Lewin aucht to' 4 b. Auchynes [*i.e.* Auchness] to 5 b. Kirktoun to 2 b. 'Ballavrauche with the out saittis', 5 b. 'Makinrais set', 7 f. 'Macawnis satt', 7 f. 'Jokkie Tailyouris satt', 7 f. The Sochowche, 6 f. Runynnour, 2 b. Rotmethe, 1 b. Wester Killis, 12 b. The Mainis and Haltoun of Dolles, 14 b. 'Summa of the haill teindis extending to' 5 c. 2 b. 3 f. victual.[117]

Signature: 'Ita est Joannes Johnnestoune, notarius publicus in premissis requisitus, manu propria.'[118]

---

[112] I.e. the feast of St Bartholomew, 24 August.
[113] Robert, lord Boyd was Collector general, *RPC*, ii, pp. 313, 504.
[114] This should possibly read 'letters and summons' or a similar legal phrase.
[115] Cf SRO, Vol. a, fo 407r (see above, p. 491).
[116] See next rental.
[117] The correct calculation appears to be 4 c. 9 b. 3 f.
[118] I.e. 'Thus, John Johnnestoune, notary public, required to be present in the aforesaid [transaction], with his own hand.'

DALLAS, PARSONAGE OF,[119] (SRO, Vol. a, fo 416v)

Rental of the parsonage of Doles pertaining to Mr William Patersoun, subdean of Murray, set in assedation to James Innes of Drany, sir Nicholas Tulloche, vicar of Ruthven, and Archibald Dunbar of Penneik 'for the space of thre yeiris nixt following the feist of Lambmes callit Peter Advincula'[120] 1561, paying therefore yearly 5 c. 2 b. 3 f. victual;[121] third thereof, 2 c. 6 b. 2 f.

RUTHVEN, VICARAGE OF, (SRO, Vol. a, fo 416v)

Rental of the vicarage of Ruthven, pertaining to sir Nicol Tullach, set in assedation to James Innes of Drany 'for the space of thrie yeiris nixt following, the' 14 December 1561.

Calculation of third: In the whole, £24; third thereof, £8.

AULDEARN, VICARAGE OF,[122] (SRO, Vol. a, fo 417r)

'Payand thairfoir yeirlie the sowme of' £30 'usuall money of this realme at the feist and terme of Lambes yeirlie during the said space of thrie yeiris, with ten pundis money foirsaid to the pensionar of Aulderne allanerlie.'

The vicarage of Aulderne pertaining to Mr William Patersoun, subdean of Murray.

Calculation of third: In the whole, £40; third thereof, £13 6s 8d.

CONVETH,
PENSION FROM THE PARSONAGE OF, (SRO, Vol. a, fo 417r)

'The rentall of ane yeirlie pensioun perteining to Alexander McKenzeis quhilk he hes yeirlie to be liftit of the first and reddiest fruictis of the said personage of Conwath annexit to the priorie of Bewlie, extendis in the haill yeirlie to' 20 merks.

Signature: 'Mr Alexander McKenze, with my hand. Rasavit at the command of the Clerk Registre.'

---

[119] See preceding rental.
[120] Lammas, 1 August, was the feast of St Peter ad Vincula.
[121] This figure may have been copied in error from the same figure in the preceding rental.
[122] The entry in the SRO text lacks a heading and starts abruptly. In this rental, the vicar pensionary's stipend of £10 ought to be subtracted from the total of £30, not added to it.

MORAY, ARCHDEACONRY OF,[123] (SRO, Vol. a, fo 417v)

Rental of the archdeaconry of Murray pertaining to Mr Michael Walker.

'Item, the haill archidenrie in all dewties extendis to' £165 9s.

'Produced be Johnne Ogstoun in name of the said Mr Michaell Walker.'

Signature: 'Johnne Ogstoun, with my hand.'

DUTHIL, PARSONAGE AND VICARAGE OF, (SRO, Vol. a, fo 417v[124])

Rental of the parsonage and vicarage of Duthell pertaining to Mr Alexander Ogilby, parson thereof, 'and presentit be me, Merteine Forman, becaus of his absence.'

'Item, the said personage and vicarage gevis in lamb woll, butter and cheis and utheris proffeitis pertening thairto' £48.

Calculation of third: In the whole, £48; third thereof, £16.

'Resavit at command of the Clerk Registre.'

NOTE

Rentals for the following benefices are located elsewhere in the text:

ABERCHIRDER, VICARAGE OF, PENSION FROM (SRO, Vol. a, fo 164v), p. 159.
ROTHES, PARSONAGE OF (NLS, fo 139r), p. 449.

---

[123] Cf NLS, fo 172r (see above, p. 487).
[124] In SRO, Vol. a, fo 418r-v is blank. An index follows in a more modern hand.

Lanark

GLASGOW, ARCHBISHOPRIC OF, (EUL,[1] fos 1r-3r)

'Rentall of the archbischoprik of Glasgow' [*in another hand*, 'taken up about 1561.']

The penny mail of the barony of Glasgow yearly extends to the sum of £268 13s 4d; of ferme [*blank*] extends in the year to [*blank*] c. 8 b.; of ferme malt, [*blank*]; of kane in the year, [*blank*]; of horse[corn] in the year, [*blank*].
 Barony of Carstairis: 'the haill penny maill' of the barony of Carstairis in the year, £47; the Mylne of Carstairis in the year, [*blank*]; of horse[corn] in the year, [*blank*].
 Barony of Ancrum: the penny mail of the barony of Ancrum in the year 'alanerlie', [*blank*].
 Barony of Lillisleiff: the penny mail of the barony of [*blank*] in the year 'alanerlie', [*blank*].
 Barony of Askirk, in penny mail in the year 'alanerlie', [*blank*].
 Barony of Stobo, in penny mail in the year, £27; the Mylne of Stobo in the year, £20; of kane bere in the year, 40 b. (at 10s per boll);[2] kane wedders in the

---

[1] This denotes Edinburgh University Library, Dc.4.32, 'Rental of Assumptions, 1561 - 66 etc.' The description of the volume reads: 'A manuscript of the 16th and 17th centuries, of which the principal part is a copy or record of the different rent-rolls given up to Government by the different Popish Beneficiaries after the Reformation in Scotland, and principally in the year 1561 - 66, the thirds of which were to be set apart for the provision of the clergy. This record has been kept by Murray of Tullibardin, the comptroller. The volume also contains copies of rentals of several bishopricks, &c. as Argyle, Orkney, Galloway, St Andrews, &c. written about the year 1695; and an historical account of several of the Scots bishops, *in the handwriting of Robert Mylne, writer in Edinburgh, to whom the volume appears to have belonged. Folio.*' Above this printed description is the information: 'This buik is begun the thrid day of November 1624. Hic liber pertinet ad Johannem [*damaged*]. Hunc mihi si forte contingat perdere librum. Compertor ut proprio restituetur hero. [*I.e.* This book belongs to John [*damaged*]. If it should happen by chance that I should lose this book, the finder should restore it to its proper heir.] Constable [*i.e.* the bookseller], Edinburgh 1801.' Before fo 1 there is a list of contents, followed by an index apparently in the hand of Robert Mylne.
[2] These brackets appear in the text.

year, 40 wedders and [blank] yearly in the tenants' hands, for 4s 'ilk peice'; the [blank] of Drava, with the 'steilbo guidis', £60.

Barony of Edlistoun, the penny mail of the barony of Edlistoun in the year, £13 18s 8d; the Mylne of Edlistoun in the year, £3 6s 4d; the Maines of the Lang Coit within the barony in the year, £6 13s 4d; 'pro kane wedderis' yearly, in the tenants' hands for 4s 'the peice', [blank]; of kane bere in the year, 8 b.; the penny mail of the kirkland of Cambusnethane in the year, 30s; penny mail of the Halfpenny land of Carrik in the year, £5; the penny mail of Nudry Foster in Louthian in the year, £6 13s 4d; the penny mail of the Bischopis Forrest in Nidisdaill in the year, £13 6s 8d; 'sum capounis and pultrie'.

The kirk of Cambusnetham set in assedation to Sir James Hamiltoun yearly for the sum of £16 13s 4d. The kirk of Deymyne [recte, Drymen] set to John Schaw in the year for the sum of £160. The kirk of Dryfdaill set to Cuthbert Hudstoun and Mungo Jonstoun in the year for £30. The kirk of Traquhair set to Patrick Murray of Hangitschaw in the year for £50. 'Item, sum casualities for rentallyng of tenentis and confearmyng of tenentis.'

'Heirof is to be defalkit to Mr Michell Cheisholme, viccar of Calder and Munkland for the teind of Choismylne', 6 b. meal; for the teind meal of the Mylne of Partik to the parson of Glasgow, 24 b. meal; 'for my lordis contributioun to the Coledge of Justice', £42.

Signature: 'Thomas,[3] heritabill chalmerlane of Glasgow, with my hand.'

Calculation of third: Money, £987 8s 7d; third thereof, £329 2s 10d.[4] Meal, 32 c. 8 b.; third thereof, 10 c. 13 b. 1 f. 1½ p. Malt, 28 c. 6 b.; third thereof, 9 c. 7 b. 1 f. 1½ p. Bere, 8 b.; third thereof, 2 b. 2 f. 2 'pectis half pect 3 parte pect' [i.e. 2⅝ p.; recte, '2 pectis half pect 3 half pect', i.e. 2⅔ p.]. Horsecorn, 12 c. 13 b. 3 f.; third thereof, 4 c. 4 b. 2 f. 1½ p. Salmon, 14 dozen; third thereof, 56.

Assignation of the archbishopric of Glasgow:
Third of the money, £329 2s 10½d. Take the penny mail of the barony of Glasgow extending in the year to £268 13s 4d; take the penny mail of the barony of Carstairis extending in the year to £47 13s 4d [£47, *fo 1r*]; the Mylne of Carstairis, £26 13s 4d. 'Gif in' £13 17s 4½d.[5]

Meal, 10 c. 13 b. 1 f. 1½ p. 'Tak it up of the ferm meill of the barronie of Glasgow geivand in the haill' 32 c. 8 b.; 'and tak speciall assignatioun of the best payeris.' Take the third of the malt extending to 9 c. 8 b. 1 f. 1½ p.[6] out of the same barony giving in the whole 28 c. 6 b.

---

[3] I.e. Thomas Archibald.
[4] ½d has been omitted from this calculation.
[5] The correct calculation appears to be £13 17s 1⅔d.
[6] The correct calculation appears to be 9 c. 7 b. 1 f. 1½ p.

Take the third of the horsecorn extending to 4 c. 4 b. 2 f. 1⅓ p. out of the same barony giving in the whole 11 c. 3 f.

Salmon, the third extending to 4 dozen and 8 out of the same barony 'and tak particular assignatioun as of befoir.'

Kane bere, third thereof, 2 b. 2 f. 2⅔ p. Out of the barony of Edlistoun giving 8 b.

'Memorandum, to chairge for the haill grassumes, entrie silver, capounis, pultrie, caynes, deywarkis, cariages and all uther services omittit.'

GLASGOW, SUBDEANERY OF, (EUL, fos 3v-7r)

Rental of the subdeanery of Glasgow as it pays yearly in meal and bere.

The parish of Calder: The town of Auchinairne, 16 b. meal; 4 b. bere. Auchinloch, 16 b. meal; 2 b. bere. Bogtoun, 10 b. meal; 1 b. bere. Buchlay, 12 b. meal; 6 f. bere. Balmyldie, 15 b. meal; 2 b. bere. Badlay and Molinis, 12 b. meal. Cadderquhult, 10 b. meal; 2 b. bere. Chrystoun, 14 b. meal; 2 b. bere. Davidstoun, 16 b. meal; 2 b. bere. Drumcawillhill, 4 b. meal. Edingecht, 3 b. 3 f. meal; 2 f. bere. Eister Cadder, 11 b. meal, 1 b. bere. Gartinquhen, 10 b. meal; 2 b. bere. Gartinquhennour, 5 b. meal. Gardarocht, 12 b. meal; 3 b. bere. Gardein kirk, 12 b. meal; 6 f. bere. Gartferrie, 8 b. meal; 1 b. bere. Hiltoun, 11 b. meal; 2 b. bere. Hartsyd 'parte'.[7] Johnstoun, 8 b. meal; 1 b. bere. Kirktoun, 10 b. meal; 2 b. bere. Lumocht, 13 b. meal; 2 b. bere. Mukray Eister, 4 b. meal; 2 b. bere. Mukray Waster, 4 b. meal; 1 b. bere. Robrestoun, 13 b. meal; 6 f. bere.

The parish of Munkland:[8] Ardrie, 12 b. meal. Arbukill, 5 b. meal. Auchingray, 4 b. meal. Auchinlonying, 10 f. meal; 2 f. bere. Barachany, 6 b. meal; 1 b. bere. Blairthomas, 4 b. meal; 1 b. bere. Brointbrowme, 4 b. meal. Blaklandis, 10 f. meal. Bartonehill, 5 b. meal; 1 b. bere. Brumesyd, 4 b. meal. Bartebeithe, 2 b. meal. Blakyairdis, 4 b. meal; 1 b. bere. Brumeleis, 6 f. meal. Braidisholme, 5 b. meal; 1 b. bere. Balgadie Over, 8 b. meal; 1 b. bere. Balgadie Nether, 10 b. meal; 2 b. bere. Blainelyne Eistir, 4 b. 2 f. meal. Blainelyne Wastir, 3 b. meal. Blainelyne Middill, 2 b. meal. Bredenhill, 6 b. meal. Carmyldie Over, 20 b. meal; 8 b. bere. Carmyldie Nether, 20 b. meal; 4 b. bere. Conflattis, 30 b. meal; 6 b. bere. Cuillhill, 4 b. meal; 1 b. bere. Cattis [Coittes, *SRO, Vol. a, fo 116r*], 'pariche Petegrew', 6 b. meal; 'paris Jacobi Hamiltoun', 4 b. meal; 'pars quondam Walteri Petegrew', 4 b. meal, 2 f. bere; 'pars Margrete Boyd', 4 b. meal, 2 f. bere. Crumlotis, 4 b. meal. Caddercruikis, 4 b. meal. Daldowe Eistir, 8 b. meal. Daldowe Waster, 16 b. meal; 1 b. bere. Denbank, 2 b. 2 f. meal. Dundivane, 4 b.

---

[7] 'Hartsyd parte' has been inserted by a later hand.
[8] Many of these place names also appear in the rental for the abbey of Newbattle, barony of Monkland, SRO, Vol. a, fo 116r (see above, p. 102).

meal. Dikie [*i.e.* Dykes], 5 f. meal. Drumplear, 16 b. meal. Fiskane, 14 b. meal. 'Gartscharie Tenandrie', 11 b. meal. Gartyngavok, 4 b. meal. Glenhuif, 4 b. meal. Glenter Eister, 4 b. meal. Glenter Waster, 4 b. meal. Gartluskane, 5 b. meal. Gartmylane, 3 b. 2 f. meal. Gayne, 4 b. meal. Gairtlie, 9 f. meal. Hallhill, 8 b. meal. Haggis, 24 b. meal. Inschurt, 4 b. meal. Killyart [Kilgarth, *SRO, Vol. a fo 116r*], 2 b. 2 f. meal. Kypchaplane, 4 b. meal; 2 f. bere. Kypbyre, 12 b. meal. Kenmure, 20 b. meal; 5 b. bere. Lughill, 2 b. meal. Milcrist, 1 b. meal. Magiscrist, 10 f. meal. Maynhill, 7 b. meal; 1 b. bere. Meddrawis and Mywetis, 16 b. meal. Patequhen, 4 b. meal. Ruchinving, 6 f. meal. Ruscholis, 11 b. meal. Ruchsellocht, 6 b. meal. Riyardis, 2 b. meal. Scheariscroft, 2 f. meal. Wyndehege, 4 b. meal; 2 f. bere.

'The steiding of the subdene, craigis and mylne thairof', 80 merks.

'Item, the ten pund [land, *deleted*] of annuell in the Rattownraw and Densyd croft to be allowit to schir James Stewart, the subdene viccar in the queir of Glasgow, for his fie in pairte of payment.'

Signature: 'Subdene of Glasgow.'[9]

Calculation of third: Meal, 39 c. 11 b.; third thereof, 13 c. 3 b. 2 f. $2\frac{2}{3}$ p. Bere, 4 c. 9 b. 2 f.; third thereof, 1 c. 8 b. 2 f. Money, £63 6s 8d;[10] third thereof, £21 2s 2d.[11]

Assignation of the subdeanery of Glasgow:
　　Third of the meal, 13 c. 3 b. 2 f. $2\frac{2}{3}$ p.; third of the bere, 1 c. 8 b. 2 f. Take the 'town' of Auchnairne for 6 b. meal, 4 b. bere; Auchinloch for 16 b. meal, 2 b. bere.; Bogtoun, 10 b. meal, 1 b. bere; Buchlay, 3 b. meal, 6 f. bere; Balmyldie, 15 b. meal, 2 b. bere; Badlay and Molnys, 12 b. meal; Cadderquhult, 10 b. meal, 2 b. bere; Christoun, 14 b. meal, 2 b. bere; Davidstoun, 16 b. meal, 2 b. bere; Drumcavillhill, 4 b. meal; Edingeycht, 3 b. 3 f. meal, 2 f. bere; Eister Cadder, 11 b. meal, 1 b. bere; Gartenquheyn, 10 b. meal, 2 b. bere; Garteynquhermuire, 5 b. meal; Gartynquhosche,[12] 4 b. meal; Gardarocht, 12 b. meal, 2 b. bere[13] [3 b. bere, *fo 3v*]; Gartynkirk, 12 b. meal, 6 f. bere; Gartforrie, 8 b. meal; Hiltoun, 11 b. meal; Johnstoun, 8 b. meal; Kirktoun, 10 b. meal; Lunlocht, 14 b. meal. 'Gif in' 5 b. $1\frac{2}{3}$ p. meal,[14] 'et sic eque for meill and beir'.
　　Third of the money, £21 2s $2\frac{2}{3}$d. 'Tak this money out of the steiding of the subdene, craigis and mylne thairof, geivand be yeir' 80 merks.

---

[9] James Hamilton was subdean of Glasgow 1550 - 1580, Watt, *Fasti*, p. 168.
[10] The correct calculation appears to be £53 6s 8d.
[11] $\frac{2}{3}$d has been omitted from this calculation.
[12] 'Gartynquhosche' does not appear in the entry for Cadder in EUL, fo 3v (see above, p. 497). 'Gartyngavok', in the parish of Monkland, gave 4 b. meal in the list on fo 5r (p. 498).
[13] Gardarrocht gave 3 b. bere, fo 3v; this would give the correct total.
[14] The correct calculation appears to be 3 b. $1\frac{2}{3}$ p.

GLASGOW, COMMON KIRKS OF, (EUL, fo 7r-v)

'The rentall of the comoun kirkis of Glasgow, perteining to the kirk and chaptour thairof.'

'Item, inprimis, the comoun kirk of Commonell quhilk was sett to the laird of Bargeny and to the guidman of Adromillane for' 460 merks yearly, 'quhairof we have gottin na payment thir four yeiris last bypast'. The kirk of Glencairne was set to William Fergusoun of Bardarrocht 'and to uthir thrie his caligis paroschineris thairof' yearly for 400 merks, 'quhairof siklyk na payment to us of thir four yeiris bygain'. The kirk of Lilisleif was set to the laird of Cesnod [*recte*, Cesnok] for 120 merks yearly, 'quhairof he hes maid na payment thir fyve or sax yeiris bygane howbeit the said kirk and fructis thairof be yeirlie worth' 400 merks. The kirk of Dalzell 'sumtyme sett to James Tailfeir and laitlie to the viccaris of the queir of Glasgow' yearly for 20 merks. [*In margin*, 'na payment maid heirof thir fyve yeiris last bygaine.'] The kirk of Wolstoun set to the parishioners thereof yearly for £40 'payit'.

Signature: 'Joanes Stevinstoun, precentor Glesguensis, pro decano et capitulo Glesguensi, manu sua' [*i.e.* chanter of Glasgow, on behalf of the dean and chapter of Glasgow, with his own hand.][15]

Comonell, 'haill', £240;[16] Glencairne, 'haill', £266 13s 4d; Lelileiff, 'haill', £80; Dalzeall 'haill', £13 6s 8d; Welstoun 'haill', £40.

KILBRIDE, PARSONAGE AND VICARAGE OF, (EUL, fo 8r)

Rental of the parsonage and vicarage of Kilbryd pertaining to Mr John Stevinsoun, 'chanter of Glasgow.'

Calculation of third: In the whole, £266 13s 4d; third thereof, £88 12s 9⅓d.[17]

THANKERTON, PARSONAGE AND VICARAGE OF, (EUL, fo 8r)

Rental of the parsonage and vicarage of Thankertoun pertaining to the 'chantarie' [of Glasgow].

Calculation of third: In the whole, £26 13s 4d; third thereof, £8 12s 9⅓d.[18]

---

[15] John Stevenson was chanter of Glasgow, 1544 - 1564. Henry Sinclair was dean, 1550 - 1561; James Balfour, December 1561 - 1589. Watt, *Fasti*, pp. 156, 160.
[16] £240 equals 360 merks; the figure given above for Colmonell is 460 merks.
[17] The correct calculation appears to be £88 17s 9⅓d.
[18] The correct calculation appears to be £8 17s 9⅓d.

GLASGOW CATHEDRAL, ALTAR OF ST STEPHEN AND ST LAURENCE IN,

(EUL, fo 9r-v)

'The rentale of Sanct Stevinis and Sanct Laurence altare cituat in the metrapolitane kirk of Glasgow perteining to sir Thomas Knox, possessor of the samin, and foundit be umquhill Mr James Lyndsay.'

'Item in primis, the ane half of the Scroggis in Tweddell within the schireffdome of Peblis and baronie of Leyne, the maill thairof', 12 merks yearly. In the Borrow Muire of Edinbrucht 'callit Sanct Geiligrainge of the Seinis', of annual yearly, 10 merks.

Small annuals in Glasgow: the tenement of Catherine Wilsoun, the spouse of umquhile John Rankein in Ratownray, 2 merks; the tenement of sir John Measoun, 6s; the tenement of sir John Smyth, 7s; the tenement of John Muire, 2s; the tenement of Matthew Heriot, 2s; the tenement of John Winyet, 4s; the tenement of Henry Burrell, 14s; 'of ane aker of land callit the Berrieaker' of annual, 4s.

'Of the quhilk rentall abonewrittin thair is ordanit yeirlie to be distribute to the tennour of the fundatioun: to the canons of the kirk, 40s yearly; to the vicars of the said kirk, 2 merks yearly; 'to the puire folkis in the said town', 2 merks yearly; 'and for uther small service', 8s. 'Sua restis to the chaplane frie', 19 merks 4s 3d.[19]

Calculation of third: In the whole, £17 18s;[20] third thereof, £5 19s 4d.

GLASGOW CATHEDRAL, ALTAR OF THE HOLY ROOD IN,[21]

(EUL, fo 9v)

'Rud altare in Glasgow'

Rental of the Ruid altar situated in the said kirk pertaining to the said sir Thomas Knox and sir James Stewart, 10 merks annual [*i.e.* annualrent] yearly, 'to be taine of Lochebochtsyd and payit be Robert Muntgumrey of Skalmorlie.'

Signature: 'sir Thomas Knox, with my hand.'

Calculation of third: In the whole, £6 13s 4d; third thereof, 44s 5½d.

---

[19] I.e. £12 17s 7d; the correct calculation appears to be £17 19s minus £5 1s 4d, i.e. £12 17s 8d.
[20] The correct calculation appears to be £17 19s.
[21] Cf EUL, fo 10r (see below, p. 502).

GLASGOW CATHEDRAL, CHAPLAINRIES OF ST MICHAEL[22] AND ST MUNGO IN,[23]
(EUL, fo 10r)

[*In margin*, 'Chaplanrie[s] in Glasgow and besyd.']

Sanct Michallis chaplainry in Glasgow kirk pertaining to sir William Wilkie, chaplain, 'payand yeirlie now to the chaplane' £12 'or thairby'. Sanct Mungois chaplainry pertaining to the said sir William paying £16 'or thairby'.

'Presentit be sir Mark Johnnestoun.'[24]

GLASGOW CATHEDRAL, CHAPLAINRY OF ST MUNGO IN,[25]
(EUL, fo 10r)

Sanct Mungois chaplainry in Glasgow kirk pertaining to sir David Kirke [*recte*, Kirkland[26]], paying to the said sir David, £12 'or thairby'.

'Presentit be sir Mark Johnnestoun [*recte*, Jamieson].'[27]

GLASGOW CATHEDRAL, CHAPLAINRY OF ST PETER IN,
(EUL, fo 10r)

Sanct Petiris chaplainry in Glasgow kirk pertaining to Mr Robert Harbertsoun, chaplain of the same, yearly paying £14.

DARNLEY, CHAPLAINRY OF,
(EUL, fo 10r)

'Darnly chappell,[28] the chaplanrie tharof perteinand to the said Mr Robert payand yeirlie to him' £20 'or thairby'.

---

[22] Cf EUL, fos 19r, 92r (see below, pp. 515, 615).
[23] Cf EUL, fo 25r (see below, p. 523); see also next entry.
[24] This is a scribal error; it should read Mark Jamieson, cf *Abstracts of Protocols of the Town Clerks of Glasgow*, ed. R. Renwick, 11 vols. (Glasgow, 1894-1900), iv, 1318.
[25] Cf EUL, fo 25r (see below, p. 523); see also preceding entry.
[26] See fo 25r (see below, p. 523); see also *Protocols of the Town Clerks of Glasgow*, ii, no 273; iv, no 1318.
[27] See above, n. 24.
[28] I.e. the chapel of St Ninian at Darnley.

ABERUTHVEN, VICARAGE OF, (EUL, fo 10r)

Vicarage of Abritane pertaining to the said Mr Robert paying yearly £16 13s 4d[29] 'or thairby'.

Calculation of third: *See* Chaplainry of St Mungo outside the town, *below, p. 503.*

GLASGOW CATHEDRAL, (EUL, fo 10r)
CHAPLAINRY OF ST JAMES IN,

Sanct James chaplainry in Glasgow kirk pertaining to sir Thomas Knox, chaplain of the same, paying yearly £24 'or thairby'.

GLASGOW CATHEDRAL, (EUL, fo 10r)
CHAPLAINRY OF ST STEPHEN IN,[30]

Sanct Stevinis chaplainry pertaining to sir Thomas Knox, 'chaslere [*sic*] in Glasgow kirk' paying yearly £16 'or thairby'.

GLASGOW CATHEDRAL, (EUL, fo 10r)
CHAPLAINRY OF THE HOLY BLOOD IN,

Haly Bluid chaplainry pertaining to sir [John Knox, *deleted*] Richard Harbertsoun paying yearly £14 'or thairby'.

GLASGOW CATHEDRAL, (EUL, fo 10r)
CHAPLAINRY OF THE HOLY ROOD IN,[31]

The Ruid chaplainry pertaining to sir John [*recte*, Thomas] Knox[32] paying yearly to the said John £16 'or thairby'.

---

[29] This is given as £26 13s 4d in the calculation of the third, after the entry for the chaplainry of St Mungo outside the town; see below, p. 503, n. 35. Aberuthven vicarage, which lay in the sheriffdom of Perth, is included among the rentals for Lanark as the holder possessed chaplainries in Lanarkshire.

[30] Cf EUL, fo 9r (see above, p. 500).

[31] Cf EUL, fo 9v (see above, p. 500).

[32] Thomas Knox was a stallar in the choir of Glasgow cathedral, *RMS*, iv, no 1629; see also fo 9v (see above, p. 500).

GLASGOW, CHAPLAINRY OF ST MUNGO OUTSIDE THE TOWN

(EUL, fo 10r)

Sanct Mungowis chaplainry 'of the chappell outwith the town' pertaining to the said sir John paying yearly £16 'or thairby'.

Signature: 'Be Mark Johnstoun [*recte*, Jamieson].'

Calculation of third:
'Chaplanries in Glasgow and besyd. The haill extending to' £198 13s 4d;[33] third thereof, £66 4s 8d.[34]
'The vicerage of Abritane, haill', £26 13s 4d;[35] third thereof, £8 17s 9½d.

GLASGOW, VICARAGE PORTIONARY OF,[36]

(EUL, fos 10v-11r)

'Rentale viccarrie portionarie callit Glasgow Secundo perteining to Mr Johne Howstoun, viccar thairof, the haill emolumentis of the said viccerage was sett to William Davidsoun, burges of Glasgow for' 103 merks, 'and to be deducit thairof the viccar pensioner fie and pensioun togidder with the stallar fie extending togidder' 83 merks 'and of the said assedatioun I want unpayit to me twa yeiris and 3, beseikand your [lordschippis] that I may have lettres for payment according to my assedatioun etc.'

'The speciall rentall of the viccerag foirsaid consistis in corpspresentis, umesclaithes, teynd lint and hempe, teind of the yeardis of Glasgow, the thrid pairt of the teind of the boatis that aryves to the brig of Glasgow, Paschmes[37] teind of the browsteris, the oblatiounis at Pasche. Anentis uther dewties of the viccerage, viz., lamiswoll, hay, the teind of the ky, the twa parte teind of the watter, that is the teind of boatis pertenis onto the persoun. Memorandum, I haif na mans nor glyb, I haif the said viccerag now in my awin handis unset sen Hallow day[38] and can gett nothing of any dewtie thairof, beseiking your lordschippis for helpe.'

Signature: 'Mr Johne Johnstoun, with my hand.' [Mr Johne Howstoun, *above*.]

Calculation of third: In the whole, £68 13s 4d; third thereof, £22 12s 10½d.[39]

---

[33] The correct calculation appears to be £160.
[34] The correct calculation appears to be £66 4s 5⅓d.
[35] This is given as £16 13s 4d on fo 10r; see above, p. 502, n. 29.
[36] The vicarage of Glasgow was erected into a prebend which was known as Glasgow Secundo; the parsonage was known as Glasgow Primo, Cowan, *Parishes*, p. 74.
[37] I.e. Easter.
[38] I.e. the feast of All Saints, 1 November.
[39] The correct calculation appears to be £22 17s 9⅓d.

DOLPHINTON, PARSONAGE OF, (EUL, fo 11r)

Rental of the parsonage of Dolphingtoun made by John Cokburne, parson thereof, at Dolphingtoun, 19 February 1561/2.

'Oneracio': The said parsonage is set in assedation for the sum of £50 in the year.

'Exonera[tio] humana: In primis givin in the yeir to the minister' £13 8s 8d. The archbishop of Glasgow,[40] 'procuratiounis synodalis' in the year, £3 6s 8d. 'Summa exonerationis', £15 6s 8d.[41] 'And sua restis frie' £32 13s 4d;[42] 'and sua the thrid that cumis to the queinis grace is' £10 12s 9d.

Signature: 'Johne Cokburne, with my hand.'

Calculation of third: In the whole, £50; third thereof, £16 13s 4d.

GLASGOW, DEANERY OF, (EUL, fo 12r[43])

Rental of the deanery of Glasgow pertaining to Mr James Balfour 'as it is payis presentlie and hes done thir mony yeiris bygaine.'

Silver, £359;[44] meal, 16 b.; oats, 24 b.; capons, 24; 'by his pairt of the comounis blank'.

'Item, the said Mr James hes the viccerage of Scuny sett in assedatioun quhen all dewties was payit for' 80 merks 'and now not sa guid payment as of befoir.' [*In margin*, 'Nota, this v[iccerage l]yis in Fyff.'[45]]

Signature: 'Henrie Livingstoun, with my hand, in name and behalf of the foirsaid dene.'

Calculation of third: Money, £349; third thereof, £116 6s 8d. Meal, 16 b.; third thereof, 5 b. 1 f. 1⅓ p. Oats, 24 b.; third thereof, 8 b. Capons, 24; third thereof, 8.

---

[40] James Betoun was archbishop of Glasgow, 1550 - 1570, Watt, *Fasti*, p. 149.
[41] The correct calculation appears to be £16 15s 4d.
[42] £50 minus £15 6s 8d is £34 13s 4d.
[43] Fo 11v is blank.
[44] This is given as £349 in the calculation of the third below.
[45] Cf SRO, Vol.a, fo 83v (see above, p. 71).

CARNWATH, VICARAGE PENSIONARY OF,[46] (EUL, fo 12v)

'Viccerage pentionarie of Carinveth'

Rental of the vicarage pensionary of Karinvath pertaining to sir John Cuninghame.

'Item in primis, the glyb land in the handis of Hew Sumervell extending to' £8. 'And in the handis of [blank],[47] persoun thairof, in name of pensioun', £8. Total, 24 merks.

Calculation of third: In the whole, £16; third thereof, £5 6s 8d.

BLANTYRE, PARSONAGE OR PRIORY OF, (EUL, fo 12v)

'Rentale rectorie sive prioratus Blantyre'[48]

'In primis, the haill personage viccerag with the annuellis of the kirkland with mans and gleib thairof with' 25 merks 'of pensioun of Quhithorne sett in assedatioun thir mony yeiris be the predicessouris and pryour instant, now presentlie in lykwayis to David Hamiltoun of Boithwellhauch for the sowme of' 197 merks 'as the said tak proportis', and of this there is 40 merks 'payit to ane minister' and 20 merks in pension to Mr William Salmond, 'conform to his provisioun papistolik', 13 merks given to Robert Lyndsay of Dinrod [*i.e.* Dunrod] in bailie fee; 'the rest is the comendatouris,[49] that is to say' 120 merks.[50]

Signature: 'Ita est per me Magistrum Willelmum Chirnsyd, quondam retroscriptum.[51] Presentit be Patrik Hepburne.'

Calculation of third: In the whole, £131 6s 8d;[52] third thereof, £43 15s 6⅔d.

---

[46] Cf EUL, fo 16v, the treasurership of Glasgow (see below, p. 510). The benefice of Carnwath, a prebend of Glasgow cathedral, was annexed to the treasurership. Cowan, *Parishes*, p. 28.
[47] I.e. Thomas Livingston, cf fo 16v (see below, p. 510).
[48] Blantyre was the only benefice appropriated to the priory of Blantyre, Cowan, *Parishes*, pp. 19, 214.
[49] William Chirnside was prior of Blantyre in 1558, *RMS*, iv, no 2473.
[50] 197 merks minus 73 merks is 124 merks.
[51] I.e. 'formerly the aforewritten'.
[52] I.e. 197 merks.

MONYABROCH, VICARAGE PENSIONARY OF, (EUL, fo 13r)

'Rentale viccerrie [sic] pensionarie de Miniabracht pro domino Gilberto Law.'[53]

'Viccaria pensionaria ecclesie'

Vicarage pensionary of the kirk of Minabrocht extending to two arable acres of land with a small house for room and small offerings extending in total to the sum of 20 merks.

Signature: 'Ita est ut supra de speciali mandato predicti viccari [sic] pensionarij de Meneabracht attestor. M[r] G. Cok, notarius, manu propria sua.'[54]

DOUGLAS, PARSONAGE OF, (EUL, fo 13r)

Rental of the parsonage of Douglas

The parsonage of Douglas pertaining to Mr Archibald Douglas, parson thereof, lying in the sheriffdom of Lanark and diocese of Glasgow, set in tack and assedation gives yearly 300 merks 'libere'.

Signature: 'Subscryvit be me, the said Mr Archbald, at Edinbrucht', 15 January 1561/2.

Calculation of third: In the whole, £200; third thereof, £66 13s 4d.

BOTHWELL, PARSONAGE AND VICARAGE OF, (EUL, fo 13v)

Rental of the parsonage of Bothwell

'Quhilk personage and viccerage ar sett in assedatioun to the laird Carinphin and Kleland town sen the Feild of Flowdoun for the sowme of' 300 merks, 'thay havand nyntein yeiris takis thairof, of the quhilk thair ar nyn yeiris to rin. Item, the gleib of the said proveistrie',[55] £10 'land of auld extent sett in few and heritage of auld for' £22 yearly annual. Total, £212.[56]

---

[53] I.e. 'Rental of the vicarage pensionary of Miniabracht pertaining to sir Gilbert Law'.

[54] I.e. 'I am witness that it is as above by the special mandate of the foresaid vicar pensionary of Meneabracht. M[r] G[eorge] Cok, notary, with his own hand.'

[55] I.e provostry of Bothwell collegiate kirk.

[56] The correct calculation appears to be £222.

'Off the quhilk sowme Mr Alexander Hepburne,[57] sumtym proveist of Bothwell, is provydit be the court of Roome till ane hundreth pundis yeirlie pensioun and hes bein in use and possessioun thairof be the space of' 18 years 'or thairby, siklyk as he is yett.'

'I, Mr Williame Chirinsyd,[58] last provest of Bothwell, is provydit siklyk be the court of Rome till the sowme of fourtie merks yeirlie pensioun of the said provestrie and hes bein in continuall possessioun of the sam in the space of [blank] and yeiris or thairby siklyk as he is yett. The quhilk pensioun beand tain of the said sowme of' £222, 'than restis to the said provest yeirlie to his sustentatioun the sowme of the sowme of [sic]' £96 6s 8d.[59]

Signature: 'Joanes Hamiltoun, prepositus de Bothwell, manu propria.'

Calculation of third: In the whole, £222; third thereof, £74.

BALDERNOCK, PARSONAGE AND VICARAGE OF, (EUL, fo 14r)

'This is the nest [sic] rentale that I, Mr Johne Weir, persoun of the parosch kirk of Bethernok and viccar of the samyne and als viccar of the paroch kirk of Wistoun and Lanark presentis and gyvis, subscryvit with my hand as followis.'

Bothernok: 'In the first, the haill teind schavis of the personage of Bothernok extendis yeirlie to' 419 b. meal, 'off the quhilk James Stirvilling of Keir hes sett in tak and assedatioun of me' 61 b. meal for 13s 4d the boll. The kirkland pays 7 merks of mail yearly. The laird of Badowy [i.e. Bardowie] pays for his Maynes, £5 money yearly. 'And the haill viccerage was sett quhen all maner of casualities was payit and upliftit, usit and wont for' £10 'be yeir and now be estimatioun is worth' 10 merks.

'Off this foirsaid rentall the curet hes' £16 'of fie'. And the bishop of Glasgow 'procurage and sinodall' extends to 16s 'be yeir.'

Signature: 'James Weir, re[c]tor de Bothernok et viccarium [sic] de Lanark et Wilstoun, manu propria, etc.'

Calculation of third: 'In the haill', [blank]; third thereof, [blank].

---

[57] Alexander Hepburn, provost of Bothwell, 1525 - 1534, Watt, Fasti, pp. 345-346.
[58] William Chirnside, provost of Bothwell, 1534 - 1552; John Hamilton, provost, 1552-1594, Watt, Fasti, p. 346.
[59] The correct calculation appears to be £95 6s 8d.

WISTON, VICARAGE OF, (EUL, fo 14r-v)

'Item, the viccerage of Wistoun was sett for' 50 merks yearly 'quhen all maner of auld dewties was upliftit, quhairof the curat gat at that tyme' £10 'of fie yeirlie. And to the bischopis patronage putt to the sett of Glesgow extendis to' 37s 'be yeir. And now the said viccerage of Wostoun be reasoun of the deminutioun is estimat yeirlie' 40 merks.

Signature: 'James Weir, re[c]tor de Bothernok et viccarius de Lanark et Wistoun, manu propria.'

Calculation of third: 'In the haill', [blank]; third thereof, [blank].

LANARK, VICARAGE OF, (EUL, fo 14v)

'The viccerage of Lanark with the kirkland and glyb of the samen, the teind schaves of the beir yeardis extendis yeirlie to' 28 b. meal and bere and 6s 8d money. 'Of the quhilk victuall' [to] Andrew Livingstoun in pension, 14 b. meal and bere and 13s 4d money, and John Bannatyne of Carhous[60] has the other 14 b. victual 'with the houss, mans and yaird' for 7 merks by year.

'Item, the remanent of the haill viccerage quhen all maner of dewties was payit of auld was worth be estimatioun' 40 merks 'be yeir and now be estimatioun worth' 20 merks. 'And of this the procurage of the bischopis sinodall extendis yeirlie to' 5 merks 10s 8d. 'And the curatis fie' £10 'with ane pairt of the small offerandis.'

Signature: 'Joanes Weir, re[c]tor de Bothernok et viccarius de Lanark et Wistoun, manu propria.'

Calculation of third: 'In the haill', [blank]; third thereof, [blank].

CAMBUSLANG, PARSONAGE OF, (EUL, fo 15r)

Rental of the parsonage of Cambuslayng

The 'town of Cambuslyng', 32 b. meal; 8 b. bere. Fleymingtoun, 32 b. meal; 4 b. bere. Newtown, 14 b. meal; 4 b. bere. Lettirik, 40 b. meal. Clydsmyll, 2 b. meal. Worestoun, 8 b. meal; 1 b. bere. Bellcattis, 9 b. meal. Estir Greinleis, 8 b. meal; 1 b. bere. Wastir Greinleis, 8 b. meal; 1 b. bere. Spittell, 1 b. 2 f. meal. Crokascheill, 2 b. meal. Turnlay, £5. Gilbitfeild, 8 b. meal. Unnertown, 8 b.

---

[60] John Bannatyne of Corhous, *RMS*, iv, no 1274.

meal. Bellillsyd, 2 b. meal. Mylkinwood, 1 b. 2 f. meal. 'The Chapell', 2 f. bere. Total: Meal, 11 c. 11 b. 2 f.[61] Money, £5.

'Item, my lord dukis grace hes in pensioun of me' 403 c. [sic]. [*In margin*, 'Thir soumes to be the duke of Chastelherault']. 'My viccar pensioun hes' 22 merks and 10 acres of lands 'with mans and coillheucht, he will be in proffeit worth' £40. Sir David Christisum has in pension £26 13s 4d. 'Thir thingis being considderit I will have lytill to leive upon.'

Calculation of third: Meal, 11 c. 11 [b.] 2 f.; third thereof, 3 c. 14 b. 2 f. Bere, 1 c. 3 b. 2 f.; third thereof, 6 b. 2 f. Money, £5; third thereof, 33s 4d.

Signature: 'Mr William Hayttoun [*recte*, Hamilton], persoun of Cambuslaying.'

DUNSYRE, VICARAGE OF, (EUL, fo 15v)

'Memorandum, this is the first rentale of the viccerage of Dunsyre perteining to maister schir James Greg.'

The said vicarage is set for £20 'of new and of auld' within the diocese of Glasgow and sheriffdom of Lanark.

Signature: 'sir James Greg, with my hand.'

Calculation of third: In the whole, £20; third thereof, £6 13s 4d.

CARSTAIRS, VICARAGE OF, (EUL, fo 16r)

Rental of the vicarage of Carstairis pertaining to John Scott, vicar thereof.

'Memorandum, the viccerage of Carstairis gaif in assedatioun in tymes bypast' £40 yearly 'and is in my handis, quhilk viccerag lyis in the diocie of Glesgow. Item, in my lord Cowperis[62] landis [*recte*, hands]' £50 'of pensioun, and thairof the abbott defalkis to me in tak as occuris in his pairt', and 22s to the College of Justice.

---

[61] The correct calculation appears to be 11 c. meal; 1 c. 3 b. 2 f. bere has been omitted from the total.
[62] I.e. the abbot of Coupar Angus; Donald Campbell was abbot at the Reformation, *RMS*, iv, no 1387.

CARMUNNOCK, VICARAGE OF, (EUL, fo 16r)

Rental of the vicarage of Carmannoch pertaining to Mr James Hamiltoun.

The said vicarage set in assedation to Robert Hamiltoun, chaplain of Kilwining, for £40.

Signature: 'Mr James Hamiltoun, with my hand.'

Calculation of third: In the whole, £40; third thereof, £13 6s 8d.

NEWTOWN, (EUL, fo 16r)
PREBEND OF BOTHWELL COLLEGIATE KIRK

'Prebendenrie of Bothuill callit New-town'

'Prebendria de Newtoun ex ecclesia collegiata de Bothwill'

Prebend of Newtoun in the collegiate kirk of Bothwill in the diocese of Glasgow, of which the possessor is Mr John Robert[s]oun, prebendary, now in fact set in assedation to Matthew Hamiltoun of Mylburn for payment yearly of £20 Scots money.

Signature: 'Mr Joanes Robertsoun.'

Calculation of third: In the whole, £20; third thereof, £6 13s 4d.

GLASGOW, TREASURERSHIP OF,[63] (EUL, fo 16v)

Rental of the 'thesaurarie of Glesgow' pertaining to Mr Thomas Livingstoun.

'Quhilk personage and viccerag[64] was sett in assedatioun of auld for' 260 merks 'and now is sett for the yeirlie payment of' £200 'by his pairt of the commounis'.

Signature: 'Henrie Livingistoun, in name and behalf the foirsaid thesawrer.'

Calculation of third: In the whole, £200; third thereof, £86 13s 4d.[65]

---

[63] Cf EUL, fo 12v; see above, p. 505 and n. 46.
[64] I.e. Carnwath, see above, p.505, n. 46.
[65] £86 13s 4d is one third of £260.

CARMICHAEL, PARSONAGE AND VICARAGE OF, (EUL, fos 16v-17r)

Rental of the parsonage and vicarage of Carmichall given in by George Douglas, parson thereof.

'In primis, the teind schavis of the towne of Carsrig and brewhous extending to ten pund land', the teind sheaves thereof extends to 36 b. victual. The Ovirtown of Carmichall 'uther tend pund land' and duty of the teind sheaves thereof extends to 36 b. victual. The Nethertown of Carmichall 'tend pund land', the teind sheaves thereof extends to 36 b. victual. 'The town of Laklayok, uther tend pund land', the teind sheaves thereof, 36 b. victual.

'The viccerage the foirsaid sett in assedatioun in tymes bygain for' 6 merks, 'the auld offringis and uther oblatiounis being dischargit. And the said haill personag and viccerag in tymis bygain hes bein sett togidder for' 100 merks.

Signature: 'George Douglas, persoun of Carmichaell.'

Calculation of third:
    Parsonage of Carmichall, in the whole, 10 c. 1 b.;[66] third thereof, 3 c. 1 f. $1\frac{1}{3}$ p.[67]
    Vicarage thereof, in the whole, £4; third thereof, 26s 8d.

STONEHOUSE, (EUL, fo 17r-v)
PREBEND OF BOTHWELL COLLEGIATE KIRK

'Rentale pro Wilielmo Tailyifeir, prebendario eclesie [sic] collegiate de Bothuill.'

Prebend of Stenhous in the collegiate kirk of Buthuill in the diocese of Glasgow pertaining to William Tailyifeir set to James Hamiltoun of Stenhous paying for the teinds of the same £24 yearly. And to Andrew Hamiltoun of Ardoch paying yearly for the lands of the same prebend called the forty-shilling land of Cathkin yearly 5 merks in name of feuferme ('nomine feudiferme'), and paying for three and a half oxgangs of land of Netherurd with an eighth part of the mill of the same yearly 5 merks in name of feuferme. Total of the said prebend, £30 13s 4d. Thereof the substitute deducting, 24 merks. And rests free for paying the said William Teilfeir, £14 13s 4d.

---

[66] The correct calculation appears to be 9 c.
[67] The correct calculation appears to be 3 c. 5 b. 2 f. $2\frac{2}{3}$ p.

Signature: 'Ita est domino Thomas Godrell, factorio seu procur[ator]io nomine predicti Wilielmi Teilfeir de speciali mandato eiusdem testavi manu mea, domino Thomas Godrell.'[68]

Prebend of Bothuill called Stainhous

Calculation of third: In the whole, £30 13s 4d; third thereof, £10 4s 5⅓d.

CRAWFORD LINDSAY, VICARAGE OF,[69] (EUL, fo 17v)

Rental given in by Mr George Strange 'of his small benefeice and prebendaries of the yeirlie rent thairof.'

'Item, in the first, his viccerage of Crawfurd Lindsay within the schireffdome of Lanark and diocie of Glesgow extending be year to' £32 10s.

Calculation of third: In the whole, £32 10s; third thereof, £10 16s 8d.

GLASGOW, PARSONAGE OF,[70] (EUL, fos 17v-18r)

Rental of the parsonage of Glesgow pertaining to Henry, bishop of Ros,[71] 'usufructuare of the said personage'.

Meal, 32 c. 8 b. Bere, 9 c. 3 b. Silver, £60 4s 8d. Herring, 3 barrels. 'Small dewties of ane pairt of the viccerage quhilkis quhen thay war gottin comonlie was not worth' 10 merks by year.

Mr John Davidsoun has 100 merks pension of the said parsonage 'for the quhilk is contentit and satisfiet yeirlie thir' 18 years 'bygaine or thairby, and siklyk man be yeirlie dureing his lyftyme.'
    The said Mr Henry has a pension of 400 merks of the abbey of Kilwining.

Signature: 'Henricus Rossensis.'

Calculation of third: Meal, 32 c. 8 b.; third thereof, 10 c. 13 b. 1 f. 1⅓ p. Bere, 9 c. 3 f. [9 c. 3 b., *above*]; third thereof, 3 c. 1 f. Money, £60 4s 8d; third thereof, £20 1s 6⅔d. Herring, 3 barrels; third thereof, 1 barrel.

---

[68] I.e. 'Thus, sir Thomas Godrell, factor or procurator, in name of the foresaid William Teilfeir by special order of the same, witnessed with my hand, sir Thomas Godrell.'
[69] I.e. Crawford, Scott, *Fasti*, iii, p. 295; known as Crawford Lindsay or Crawford Douglas; cf chaplainry of the Trinity in St Andrews, SRO, Vol. a, fo 83v (see above, p. 71).
[70] Cf NLS, fo 119v. This parsonage was also known as Glasgow Primo, see above, p. 503, n. 36.
[71] Henry Sinclair was bishop of Ross, 1558 - 1565, Watt, *Fasti*, p. 270.

The pension of Kilwyning, £266 13s 4d; third thereof, £88 17s 9½d.

WALSTON, VICARAGE OF, (EUL, fo 18r)

Vicarage of Wailstoun in Cliddisdaill pertaining to sir David Dalgleisch set in assedation to Michael Lescheman, John Leschman 'and utheris thair coligis' for 50 merks 'frie to the said schir David yeirlie, by and abone the sowme of' 20 merks 'quhilk the saidis fermoureris payis to Laurence Leschman, minister at the said kirk.'

Signature: 'David Dalgleish.'

Calculation of third: In the whole, £46 13s 4d; third thereof, £15 10s 1⅓d. [72]

STONEHOUSE, VICARAGE OF, (EUL, fo 18v)

Rental of the vicarage of Stainhous given by the provost of Bothwell.[73]

The glebe and manse thereof set in assedation to John Hamiltoun of Brumhill, 'he payand for it yeirlie' 4 merks. 'The rest consistis in lambwoll and small teindis extending yeirlie to' £4. Total, 10 merks.

'Ressavit at command of the Clerk Registre.'

Calculation of third: In the whole £6 13s 4d; third thereof, 44s 5⅓d.

HARTSIDE, PARSONAGE OF, (EUL, fo 18v)

Rental of Hartsyd within the diocese of Lanark.[74]

Ten merkland of Hilhous, 9 b. meal. Six merkland of Eistir Hartsyd, 10 b. meal. Six merkland of Wastir Hartsyd, 8 b. meal. Nine merkland of the 'eist syd of Wandellburne', 9 b. meal. Six merkland of Lytillgell, 8 b. meal. Eight merkland of the 'wast syd of Wandelburne', 9 b. meal. 'The Cald Chappell at Appilgirth [*i.e.* Cold Chapel] takis up' 14 b. meal. 'The steid of Bunok', 1 b. meal.

---

[72] The correct calculation appears to be £15 11s 1½d.
[73] John Hamilton, provost of Bothwell collegiate kirk, 1552-1594, Watt, *Fasti*, p. 346.
[74] This should read 'diocese of Glasgow, deanery of Lanark'.

'This is the trew rentall that I, Mr Nicoll Craufuird, sould haif payment of my lordis.'

Calculation of third: In the whole, 4 c. 6 b.;[75] third thereof, 1 c. 8 b.[76]

'Aliter, the said personage sett in assedatioun to the laird of Leffnoreis for' £66 13s 4d; third thereof, £22 4s 5½d.

LANARK, CHAPLAINRY OF ST NICHOLAS IN,[77] (EUL, fo 19r)

'Sanct Nicolas chaplanrie, Glasgow' [*recte*, Lanark]

'Rentale domini Thome Godrall'

'The chaplanrie of Sanct Nicolas chapell in Lanark', £40, 'thairof the substitut deducit' £10, 'sua restis' £30. Total, £30. 'Off the quhilk nathing payit thir last thrie yeiris.'

Calculation of third: In the whole, £40; third thereof, £13 6s 8d.

GLASGOW, CHAPLAINRY OF ST ROCH, (EUL, fo 19r)

'Chaplanrie [of] Sanct Roch [besy]d Glesgow'

'I, schir Thomas Fleyming, curate of the ceitie of Glesgow, grantis me to have ane chaplanrie in the kirk of Sanct Roche besyd the town of Glesgow extending yeirlie in valour to the sowme of' £7 12s money, etc., 'and this mak knawin be this my hand writt writtin and subscryvit with my hand at Glasgow, 1561.'

Signature: 'Thomas Fleyming, qui supra propria manu ita scripsi.'

Calculation of third: In the whole, £7 12s. [*Third not given.*]

---

[75] The correct calculation appears to be 4 c. 4 b.
[76] One third of 4 c. 6 b. is 1 c. 7 b. 1½ f.
[77] Cf EUL, fo 23r, 25v (see below, pp. 521, 524).

GLASGOW CATHEDRAL, CHAPLAINRY OF ST MICHAEL IN,[78]
(EUL, fo 19r)

[*In margin*, 'Prebendrie of Bothuill', *deleted*]

'Sanct Michallis chaplanrie'

Sanct Michalis chaplainry within the kirk of Glesgow, extending yearly commonly to £26 money in fermes, mails and annuals 'unpayit thir thrie yeiris bygain, chaplan and possessor thairof, M[r] David Gibsoun.'

Signature: 'David Gibsoun, manu sua.'

Calculation of third: In the whole, £26; third thereof, £8 13s 4d.

TORRANCE, PARSONAGE OF,
(EUL, fo 19v)

Parsonage of Torrence lying within the parish of Kilbryd pertaining to Mr Robert Hamiltoun gives yearly in assedation 'for all proffeitis, alsweill cropis [*recte*, corpse], umest, as small offerandis, the sowme of' 40 merks.

Calculation of third: *See* Vicarage of Cambusnethan, *below*.

HAZELDEAN, PREBEND OF BOTHWELL COLLEGIATE KIRK
(EUL, fo 19v)

'Item, ane prebendrie of Bothuill callit Hisileden' pertaining to the said Mr Robert Hamiltoun 'quhilk payis yeirlie in assedatioun', 50 b. meal. [*In margin*, 'Nota, this rentall is givin in sensyne be the said Mr Robert for' 40 b. meal at 13s 4d the boll 'alanerlie.']

Calculation of third: *See* Vicarage of Cambusnethan, *below*.

CAMBUSNETHAN, VICARAGE OF,
(EUL, fo 19v)

'Item, ane viccerage of Cambusnetham payis yeirlie in assedatioun perteining to Johne Hamiltoun the sowme of' 30 merks 'for all proffeitis and dewties perteining thairto.'

Signature: 'Joanes Hamiltoun, vicarius de Cambusnethame.'

---

[78] Cf EUL, fo 10r, 92r (see above, p. 501; below, p. 615).

Calculation of third:
 Parsonage of Torrens, in the whole, £26 13s 4d; third thereof, £8 12s 9¼d.[79]
 Prebend of Hessilden, in the whole, 50 b. meal. [*Third not given.*]
 Vicarage of Cambusnetham, in the whole, £20; third thereof, £8 13s 4d.[80]

GLASGOW, PREBEND OF THE NEW KIRK OF,[81]   (EUL, fo 20r-v)

Rental of Mr James Hamiltoun of the mails and annuals of the New Kirk of Glesgow, 'quhairof he was prebendar'.

'Item, ane tenement quhilk David Lyoun and Robert Fortis inhabitis, bak and foirland', £11. A tenement which Alexander Park of Bagra and John Blair inhabit, £8 14s. A tenement of James Wilsoun, £5 4s 6d. A tenement which James Law inhabits, 5 merks 9s 2d. A tenement 'quhilk was callit the auld chappell', 5 merks 6s 8d. David Craufurd of Fairme for his lands 'lyand in wodsett', £8. George Harbesoun for a tenement 'callit Patiooup',[82] of annuals, 15s. Margaret Rankein of the annuals of the lands 'callit Pacakis', 15s. Elizabeth Hamiltoun for 'vj ruid of land lyand in the Lang croft', 30s. Alexander Leggat for two acres of land lying in the Kinclaith, 2 b. bere. John Stark for an acre of land lying in the Densyd, 12s. Margaret Muire for an acre of land lying on the Burrow Mure of Glasgow, 6s 8d. Alexander Stene for an acre of land lying in the Burrow Muire of Glasgow, 6s 8d. Henry Berall for an acre of land lying in the Burrow Muire of Glasgow, 6s 8d. David Landellis for an acre of land lying in the Burrow Muire of Glasgow, 6s 8d. Robert Rankin for the annual of his house, 30s. Catherine Laying for the annual of her house, 30s. William Lowdiane for an acre of land lying in the Burrow Muire of Glasgow 'and ane buith', 16s 8d. Thomas Spankis for two and a half acres lying in the Burrow Muire of Glesgow, 16s 8d. John Campbell for four acres of land lying in the Burrow Muire of Glasgow, 2 merks.

[*In margin in a later hand*, 'In all', £51 6s 4d & 2 b. 'of Bear.']

Signature: 'Joanes [*recte*, James] Hamiltoun, with my hand.'

'Prebendre of the New College of Glasgow quhairof Mr James Hamiltoun was prebendar.'

---

[79] The correct calculation appears to be £8 17s 9¼d.
[80] The correct calculation appears to be £6 13s 4d.
[81] I.e. the collegiate kirk of St Mary and St Anne, see fos 22r, 24v, 25v (see below, pp. 519, 523).
[82] This should probably read 'pretorii loci' [*i.e.* the Tolbooth], *Liber Collegii Nostre Domine. Registrum Ecclesie B. V. Marie et S. Anne infra muros civitatis Glasguensis, MDXLIX*, ed. J. Robertson (Glasgow, 1846), p. 26; *Protocols of the Town Clerks of Glasgow*, xi, no 3728.

Calculation of third: Money, £51 6s 4d; third thereof, £17 2s 1½d.

NETHERFIELD, PREBEND OF BOTHWELL COLLEGIATE KIRK
(EUL, fo 20v)

'Prebendre of Netherfeild with the pertinentis, viz.'

'The Netherfeild with the kirkland and maillis thairof; Goyslintoun; Oylandis; Vithank; the persounis mansioun with the maillis thairof; the aucht pairt of the fewmaillis and grassumes with the augmentatioun of Caythkin and Netherfeild extending to' 40 merks.

Signature: 'William Strutheris.'

Calculation of third: In the whole, £26 13s 4d; third thereof, £8 17s 9½d.

SYMINGTON, VICARAGE OF,
(EUL, fo 21r)

Vicarage of Symontown in Cliddisdaill set to William Symontoun of Hardingtoun for yearly payment of £30 in assedation, 'he findand the service dewlie done as effeiris conforme to his assedatioun maid thairupon.'

Signature: 'William Ruth[e]rfuird, vicarius de Rutherfuird [recte, Symington].'[83]

Calculation of third: In the whole, £30; third thereof, £10.

CARSTAIRS, PARSONAGE OF,
(EUL, fo 21r)

Rental of the parsonage of Carstairis pertaining to Mr James Kenneddis, 8 c. victual, 'the twa pairt thairof meill and thrid pairt beir, sett to the tenentis, occupyeris and laboureris of the grund sen my entres and for yeiris to cum for' 16s the boll 'as thair assedatioun maid be me mair fullilie proportis. Summa of silver', £105 12s.[84]

Signature: 'Ita est Thomas Westoun, de mandato Jacobi Yung, principalis procuratoris prefati rectoris.'[85]

Calculation of third: In the whole, £105 12s; third thereof, £35 4s.

---

[83] Nicol (not William) Rutherford was vicar of Symington in 1575, RSS, vii, no 360.
[84] The correct calculation appears to be £102 8s.
[85] I.e. 'by order of James Yung, principal procurator of the foresaid parson.'

THANKERTON, PARSONAGE AND VICARAGE OF, (EUL, fo 21r)

[*In margin in a later hand*, 'Tankertoun']

Rental of the kirk of Tanbutoun

'Memorandum, the haill personage and viccerage of the said kirk was sett of the auld and payis now instantlie yeirlie for the haill fructis and dewties thairof to the persoun the sowme of' 100 merks 'for all casualities of the said parosch kirk.'

Signature: [*blank*].

Calculation of third: In the whole, £66 13s 4d; third thereof, £22 4s 5⅓d.

CADDER AND MONKLAND, VICARAGE OF, (EUL, fo 21v)

Rental of the vicarage of Cadder and Munkland 'producit be' Mr Michael Coscholme [*recte*, Chisholm], vicar thereof.

'The yeirlie availl of the said viccerage extendis as followis, that is to say', 8 b. meal; 60 teind lambs; 8 stones of wool; 'togidder with corps presentis, umest claithis and Pasche fynes extending yeirlie to' £40, 'quhilk viccerage with the haill proffeitis thairof, as said is, being sett in assedatioun payit yeirlie' £54, 'and now the corps presentis, um[e]st claithes, Pasch fyines, and uther small dewties being be act of counsell dischargit, the yeirlie availl of the said viccerage is demin[i]schit.'

Signature: 'Mr Michell Chisholme.'

Calculation of third: In the whole, £54; third thereof, £18.

HAMILTON, VICARAGE PENSIONARY OF, (EUL, fo 21v)

Rental of the 'viccerage personarie' of Hamiltoun

'The haill [viccerage, *deleted*] pensioun of this viccerage appoyntit for the service extendis to the sowme of' 20 merks 'givin be principall provost[86] for the caus foirsaid, off the quhilk thair be' 12 merks 'givin be the provost, and the rest thairof dois consist in hay and siklyk dewties conserneing ane viccerage personarie.'

---

[86] I.e the provost of Hamilton collegiate kirk. Arthur Hamilton was provost in 1561, *RMS*, iv, no 2585; Watt, *Fasti*, p. 361.

Signature: 'Mr Archbald Karray [*recte*, Barry], viccar pensionar of Hamiltoun.'

Calculation of third: In the whole, £13 6s 8d; third thereof, £4 8s 10⅔d.

RUTHERGLEN, VICARAGE OF, (EUL, fo 22r)

Rental of the vicarage of Rutherglen 'extendis in yeirlie valour to' 40 merks, 'the quhilk perteinis to Mr Mathow Fleyming'.

Signature: 'Mr Joanes Houstoun, manu sua.'

Calculation of third: In the whole, £26 13s 4d; third thereof, £8 17s 9⅓d.

DALZIEL, PARSONAGE AND VICARAGE OF, (EUL, fo 22r)

Parsonage and vicarage of Dalzell, 'quhairof schir Malcum Fleyming, Alexander Bane, James Stewart, David Pitcarnis, Bartilmo Simpsoun, David Andirsoun, Archibald Dikie, Christell Knox, Johne Nesmyth, and Duncan Stevin, viccaris of the queir of Glesgow, ar persounis and viccaris', extends yearly to 10 merks money; 68 b. oatmeal ('thrie scoir aucht bollis aitmeill').

Calculation of third: In the whole, 10 merks; third thereof, 3 merks 4s 5⅓d. Meal, 68 b.; third thereof, 22 b. 2 f. 2⅓ p.[87]

GLASGOW, PREBEND OF THE NEW KIRK OF,[88] (EUL, fo 22r)

'The prebendrie of the New College of Glasgow.'

'Ane prebendrie of the New [Colege, *deleted*] Kirk of Glesgow perteining to Mr James Kennedie, the rentall quhairof is' 80 merks 'as his institutioun beiris.'

'The principall of this rentall is among my warrandis becaus Mr Michell Cheisholmeis ressait of' 1561 'is writtin thairon.'

Calculation of third: In the whole, 80 merks; third thereof, £17 15s 2⅔d.[89]

---

[87] The correct calculation appears to be 22 b. 2 f. 2⅔ p.
[88] Cf fos 20r, 20v, 24v, 25v (see above, p. 516; below, p. 523).
[89] The correct calculation appears to be £17 15 6⅔d.

CATHCART, VICARAGE OF, (EUL, fo 22v)

'Memorandum, the viccerage of Cathcart within Cleddisdaill sett to Mr David Gibsoun and sir Bernard Pebillis for' £48 yearly by Mr John Rettray, vicar thereof, 'as his tak heir reddie to produce proportis, of the quhilk your lordschippis man tak cognitioun of the inlaik of payment of corpspresentis and Paschefynis extending as your lordschippis will guidlie modifie be [the] viccaris aith or his takisman, givin be Mr David Gibsoun in the viccaris name.'

Signature: 'David Gibsoun, takisman.'

Calculation of third: In the whole, £48; third thereof, £16.

MUIRHALL, CHAPLAINRY OF, (EUL, fo 22v)

'Item, this is the just rentall of the chappell of Muirhall within the barronie of Carinveth, schireffdome of Lanark and diocie of Glasgow extending yeirlie to' 16 merks 5s money, 'occupyit be me, sir Thomas King, cheplane of the same chaplanrie.'

'Ressavit at command of the Clerk of Registre.'

Signature: 'Ita est Thomas King, prebendarius dicte cheplane [sic], manu propria.'

GLASGOW, PENSION FROM THE PARSONAGE OF, (EUL, fo 22v)

'Pensioun furth of the personage of Glasgow'

'The rentale of ane pensioun perteining to Mr Johne Davidsoun, maister of the pedagog of Glasgow, quhilk sould be payit yeirlie furth of the personage of Glesgow extendis to' 50 merks.

Calculation of third: In the whole, 50 merks; third thereof, £11 2s $2\frac{2}{3}$d.

LANARK PARISH KIRK, ALTARAGES IN, (EUL, fo 23r)
LANARK, ALTARAGES IN ST NICHOLAS'S KIRK,[90]

'Alterage in Lanark kirk [and] Sanct Nicolas kirk'

'Thir ar the rentallis justlie underwrittin perteinand to me, Thomas Zetoun [*recte*, Hetoun], cheplene in Lanark of thir alterage underwrittin.'

The Ruid altar within the parish kirk of Lanark worth yearly £7. Our Lady altar there worth £5. The Halie Bluid altar in Sanct Nicolas's kirk worth £4. Sanct Michellis altar there worth £3.

Signature: 'Thomas Hetene, with my hand. Ressavit at comand of the Clerk Registre.'

BIGGAR, PARSONAGE AND VICARAGE OF, (EUL, fo 23r)

Rental of the parsonage and vicarage of the kirk of Biggar given in by John Jaesoun, 'factour to the lord Fleyming,[91] as it hes payit thir mony yeiris past, the sowme of, £100.

'Ressavit at comand of the Clerk Registre.'

CARNWATH COLLEGIATE KIRK, (EUL, fo 23r-v)
PREBEND OF THE AISLE OF CARNWATH,[92]

'The rentale of the prebendrie of the Ile of Carinvath perteining to sir Duncan Aikman, prebendar thairof, extendis yeirlie in the haill to' 24 merks 'to be payit to the chaplane thairof'.

Signature: 'sir Duncan Aikman, prebendar of the Ille of Carinvath , with my hand. Ressavit at command of the Clerk Registre.'

'The Ile of Carnvath pertaining to schir Duncan [per, *deleted*] Carnweyth [*recte*, Aikman].'

Calculation of third: In the whole, £16; third thereof, £5 6s 8d.

---

[90] Cf EUL, fo 19r, 25v (see above, p. 514; below, p. 524).
[91] John, 5th lord Fleming, *Scots Peerage*, viii, pp. 543-544.
[92] For Carnwath collegiate kirk, see Cowan and Easson, *Medieval Religious Houses*, p. 216.

GLASGOW CATHEDRAL, ALTARS OF THE NAME OF JESUS AND OUR LADY OF PITY IN, CULROSS ABBEY, ALTAR OF ST MUNGO IN,[93]

(EUL, fo 23v)

Rental of the altars of 'In Nomine Jesu and Lady Pety' within the cathedral kirk of Glesgow and the altar of Sanct Mungo within the abbey kirk of Culros pertaining to sirs John Nasmyth, Christopher Knox and John Brown in Culros, chaplains of the same, extends yearly in all profits to 83 merks money of this realm 'and that be the maillis and dewties of the land of Craigrossy lyand within the schireffdome of Pertht annexat and mortyfiet to the saidis altaris, quhilkis landis ar sett in assedatioun to Andro Rollok of Duncrube for yeirlie payment of the foirsaid sowme'.

Signature: 'sir Johne Nasmyth, with my hand; sir Christall Knox, with my hand. Ressavit at command of the Clerk Registre.'

GLASGOW, BLACK FRIARS OF,[94]

(EUL, fo 24r)

'The rentall of the freiris of Glesgow quhilk was not givin up in compt.'

'Item, the annuellis of the town of Glesgow, quhairof thair is ane pairt not in use of payment', £12. 'The fruit within the greit yeard payit yeirlie' £4. Kilmacolme, Kilbryd in Argyll and McClychtan lands, £3.

'Item, the yeard occupyit be Alexander Lindsay payit yeirlie of malt', 20 b. malt. 'Sum akeris besyd the toun payis' 14 b. malt. Total, 2 c. 2 b. malt.

LANARK, FRANCISCAN FRIARS OF,[95]

(EUL, fo 24v)

'The rentall of the freiris of Lanark'

'Item, by the annuellis of the town quhilk was givin up befoir for the castell wairdis of the schireffdome of Lanark', £13 6s 8d. 'Thair is fyve ruid of land and ane lytill yeard quhilkis the freiris occupyit thameselfis quhilk is now occupyit be Mr David Cughame [recte, Cunningham].'

---

[93] This altar in Culross abbey, in the sheriffdom of Perth, is included in rentals for Lanark because the holders also held altarages in Glasgow cathedral.
[94] I.e. Dominicans.
[95] I.e. Friars Minor Conventual.

GLASGOW, PREBEND OF SACRISTA MAJOR IN THE NEW KIRK OF,[96]

(EUL, fo 24v)

The rental of the prebend called Sacrista Major in the New Kirk of Glesgow pertaining to me, Patrick Salmond.

'The half viccerage of Mayboill extending' to 10 merks [sic] with 40s of annuals in Glesgow.

Signature: 'Patrik Salmond, with my hand'

Calculation of third: In the whole, £10; third thereof, £3 6s 8d.

CAMBUSLANG, LADY CHAPEL OF KIRKBURN IN,

(EUL, fo 25r)

Rental of the Lady Chapell of Kirkburne in Carbuslaying [sic] pertaining to sir John Millar within the sheriffdom of Lanark extends yearly to 7 merks.

Signature: 'sir Johne Millar, with my hand.'

GLASGOW CATHEDRAL, CHAPLAINRIES OF ST MUNGO'S ALTAR IN,[97]

(EUL, fo 25r)

'This is the rentall of the annuellis perteinand yeirlie to the chaplaneris of Sanct Mungois altar in the laich kirk of Glesgow, and now to ane schir [David] Kirkland chaplan perteinis, ane tenement of land lyand in Glesgow occupyit be me [sic] Adame Myllar, payand yeirlie to the chaplanrie' £16 'usuall money of this realme.'

Signature: 'Subscryvit with my hand at Edinbrucht', 22 January 1565/6, 'be his hand writt sic subscribitur, sir David Kirkland.'

GLASGOW, NEW KIRK OF,[98]

(EUL, fo 25v)

'The rentall of my levin [inserted, in a later hand, 'living'] in the New Kirk of Glesgow quhilk is ane feall givin in be the town thairof.'

'In the first, James Wilsounis place, umquhill Johne Clemis of Nunbothis', £6. 'Item, of Kathrein Langis place, umquhill Jonat Mcalyionis in the Briggait', £3.

---

[96] Cf fos 20r, 20v, 22r (see above, pp. 516, 519).
[97] Cf EUL, fo 10r (see above, p. 501).
[98] Cf fos 20r, 20v, 22r, 24v (see above, pp. 516, 519, 523).

'Item, tua akeris and ane half of land lyand in the Gallow Muir off Marian Gainis and Johne Takettis', 33s 8d.

Signature: 'James,[99] ressave this rentall, Georgius Maxwell. Sic subscribitur, J. Clerk Registre.'

LANARK, ST NICHOLAS'S KIRK, CHAPLAINRY OF OUR LADY ALTAR IN,[100]

(EUL, fo 25v)

Lady altar pertaining to sir William Stewart.

'This is the rentall of Our Lady altar cituat within Sant Nicolas kirk within the burcht of Lanark givin in be William Stewart, chaplane of the samyn', 15 b. victual, 'ilk boll extending to' 13s 4d; 40s of annuals 'the quhilk was wont to be givin to ane preist to say mes twys in the week and now the town takis the sam to help to say the comoun prayeris within the kirk. This is the just rentall subscryvit with my hand, sua that the haill yeirlie rent of the said chaplanrie that I gett payment of extendis bot to the saidis' 15 b. meal 'at the pryce foirsaid'. Total, 15 merks.

Signature: 'Williame Stewart, chaplane of Our Lady altar in Lanark.'

'Aliter'

'Lady alterag of Lanark, aliter'

'This Lady alterage in Lanark pertining to schir William Stewart extendis yeirlie to' £10 'alanerlie givin up be me, schir William Stewart at Edinbrucht' 12 January 1565/6.

Signature: 'William Stewart with my hand. James, ressave this rentall and chairge thairwith in tyme cuming. Sic subscribitur, Pattireu [*i.e.* Pettigrew], capellanus. J. Clerk Registre.'

---

[99] James Nicolson, clerk of the Collectory, *Thirds of Benefices*, p. 62.
[100] Cf EUL, fos 19r, 23r (see above, pp. 514, 521).

GLASGOW CATHEDRAL, ALTAR OF ST NICHOLAS IN,
(EUL, fo 26r)

'Sanct Michalis [sic] alter'

Rentall of Sanct Nicolas altar within the cathedral kirk of Glesgow pertaining to Mr John Davidsoun, 'maister of the pedagog of Glesgow.'

In William Lowdounis house, 4 merks. In Nicol Andrewis house, 40s. In Thomas Hogis house, 30s. In Wilsounis house, 10s.

MANOR PARISH KIRK,[101] ROOD ALTAR IN,
(EUL, fo 26r)

Rental of the Ruid altar situated within the parish kirk in Maner within the diocese of Glesgow and sheriffdom of Peblis.

A tenement of land on the south side of the Corce Gait of Peblis pertaining to James Gowan extending to 30s. An annualrent of a tenement of land pertaining to Thomas Lowis, 10s. An annualrent of a tenement of land pertaining to Arthur Jaksoun, 5s.

Signature: 'Idem asserit Adamus Lowis, notarius publicus, attestante chirographo.'[102]

GLASGOW CATHEDRAL, PREBEND OF 'TUA OF THE THRIE CHILDREN' IN,
(EUL, fo 26r)

'The rentall of the prebendrie of Tua of the Thrie Children situat within the metropolitan kirk of Glesgow, givin up be Mr Archbald Craufurd, persoun of Egilshame', 5 March 1567/8 'extending in all proffeitis, rebus habentibus ut nunc [i.e. as things stand at the moment] to' 10 merks.

Signature: 'Ita est Archbaldus Craufurd, de mandato domini Roberti Watsoun, prebendi.'

---

[101] This kirk, in the sheriffdom of Peebles, lay within the diocese of Glasgow and formed part of the prebend of the archdeacon of Glasgow, which explains its inclusion in rentals for Lanark, Cowan, *Parishes*, pp. 142-143.

[102] I.e. 'Adam Lowis, notary public, states the same, as his signature testifies.'

GOVAN, CHAPLAINRY OF THE LADY ALTAR OF,

(EUL, fo 26v)

'The chaplanrie of the Lady altar of Gevane' [*in margin in a later hand*, 'Govan'].

'Item, the yeirlie valour thairof being colectit and gatherit out of paroschineris handis extending' 12 b. oats. A tenement of houses annexed thereto extending yearly to 3 b. meal 'with tua yeardis lyand thairto'. Of annualrent of the tenement of William Wilsoun yearly to be paid, 14s. An other annual of the tenement of Robert Bronsyd in Glesgow extending yearly to 12s.

Signature: 'Ita est Jacobus Hill, chaplanus [*recte*, capellanus], manu ipsa.'

NOTE

Rentals for the following benefices are located elsewhere in the text:

ANCRUM, PARSONAGE OF, PENSION FROM (SRO, Vol. a, fo 164v), p. 159; (SRO, Vol. a, fos 212v, 239v), pp. 213, 236; (NLS, fo 82r), p. 365.
CULTER, PARSONAGE AND VICARAGE OF (EUL, fo 45v), p. 554.
GLASGOW CATHEDRAL, CHAPLAINRY OF ST MICHAEL IN (EUL, fo 92r), p. 615.
GLASGOW, PARSONAGE OF, PENSION FROM (NLS, fo 119v), p. 420.
KITTYMUIR, PREBEND OF BOTHWELL COLLEGIATE KIRK (SRO, Vol. a, fo 133v), p. 121.
MAKERSTOUN, VICARAGE OF (SRO, Vol. a, fo 213r), p. 213.
ROBERTON, VICARAGE OF (SRO, Vol, fo 213r), p. 213.
STRATHAVEN, VICARAGE PENSIONARY OF (EUL, fo 83v), p. 603

Renfrew

PAISLEY, ABBEY OF,[1] (EUL, fos 27r-30r)

'Rentale monasterij Pasleti'

Rental of the monastery of Pasley of the most reverend lord John,[2] present archbishop of St Andrews and perpetual abbot of the same [*i.e.* Paisley], containing all fermes and victuals and sums of money both for augmentations and grassums, by reason of certain lands [held] in feuferme and on perpetual lease ('perpetua empheteosi'), and of kirks likewise set in feu, briefly comes from the registers and accounts of the graniter according to the present annual rent of the same monastery as it is at present to be paid for the third part of all fruits of the same, both of the temporality and spirituality of the aforesaid monastery, payable to our lady the queen, compiled in January 1561/2.

The lordships of Paislay, Kilpatrik, Glen and Munktoun according to the annual accounts of the graniter and the rental in total calculated ('calentatis'; *recte*, calculatis) in money and augmentations of lands set in feuferme extending to £569 7s 1d. Total oatmeal of these lordships according to the foresaid rental and accounts, 7 c. 9 b.

Total ferme bere of these lordships, 15 c. 6 b. 1 f. 2 p. Total ferme oats of small measure of these lordships counting 3 f. to the Paisley boll[3] according to the custom and use of the foresaid place, 53 c. 3 b.

Rental of property of the spirituality[4] and the kirks of the said monastery set for teind victuals with the defalcations of the old rental and others things unpaid according to the annual accounts of the graniters of the said monastery.

---

[1] For an earlier, more detailed rental, see J. C. Lees, *The Abbey of Paisley from its Foundation to its Dissolution* (Paisley, 1878), pp. lvi-clxxviii.

[2] I.e. John Hamilton, who, as commendator in 1553, had resigned the abbacy in favour of his nephew, Claud Hamilton, while reserving its fruits for life and its administration to himself; cf above, p. 6, n. 24.

[3] There are usually 4 firlots to a boll.

[4] The Latin text here reads 'summalitatis', but should probably read 'spiritualitatis'.

Kirk of Paislay according to the annual accounts and rental in meal, 5 c. 1 f. 2⅓ p.; and in bere of the same kirk, 6 c. 9 b. Kirk of Kilpatrik according to the rental and annual accounts in meal, 28 c. 15 b. 2 f.; and in bere of the same kirk, 7 c. 3 b. 3 f. 2 p. Kirk of Dundonald according to the rental and annual accounts, 2 c. 8 b. [meal]. Kirk of Ruglen according to the rental and annual accounts in meal, 2 c. 'fearm'; and in bere of the same, 2 c. 3 b.; and in oats of the same, 3 c. 10 b. Kirk of Eistwood according to the rental and annual accounts in meal, 1 c. 7 b. 3 f.; and in bere of the same, 1 c. 3 b. 2 f. Kirk of Mernis according to the rental and annual accounts in meal, 6 c. 10 b. 3 f.; this [kirk] has no bere. Kirk of Houstoun according to the rental and annual accounts in meal, 2 c. 2 b. 1 f.; and in bere, 7 b. 1 f. Kirk of Killelane according to the rental and annual accounts in meal, 1 c. meal; and in bere of the same, 8 b. Kirk of Cumray according to the rental and annual accounts in bere, 2 c. 8 b. bere. Kirk of Rilkertoun [*i.e.* Riccarton] according to the rental and annual accounts in meal, 17 c. 6 b. 1 f.

Memorandum, the kirk of Craigy with which the graniter is not charged as it is set for the pension of the vicar pensionary for the same payment, and in fee to the laird of Craigy, bailie of the said monastery in that part, for 6 c. 15 b. meal, but charged in bere of the same, 11 b. bere.

Sum total of the teind meal of the kirks and fermes abovewritten, 72 c. 3 b. 3 f. 2⅓ p.[5]

    Of which total there is deducted out of the ferme meal for the upkeep of the convent. Item, of annual payment to the number of 18 persons, 5 c. 11 b. meal.

    And also for the fee of the lord of Ros[6] of ferme meal of his lands within the foresaid lordship, 1 c. 14 b.

    And also for the alms to the poor distributed weekly by the almoner of the place, 7 c.

    Total deductions, 14 c. 9 b. Rests free in meal, 57 c. 10 b. 3 f. 2⅓ p.

Total bere, ferme and teind, 40 c. 12 b.[7]

    Of which total ferme deducted for the upkeep of the convent of the said monastery according to the accounts of the graniter, 14 c. 1 b. 2 f. So rests free, 26 c. 10 b. 2 f.

Total horsecorn, ferme and teind reckoning three chalders from the teinds of Ruglen extending in great measure to 43 c. 1 b. 1 f. 1 p.[8]

---

[5] The correct calculation appears to be 74 c. 1 b. 3 f. 2⅓ p.
[6] I.e. Henry Sinclair, bishop of Ross, 1558-1565, Watt, *Fasti*, p. 270 (and not James, 4th lord Ross, *Scots Peerage*, vii, pp. 252-254).
[7] The correct calculation appears to be 36 c. 1 b.
[8] The figures given for oats and oatmeal total 64 c. 6 b.

Of which total there is deducted according to the accounts of the graniter, by use and wont annually extending to 7 c. 8 b. And so rests free, 35 c. 9 b. 1 f. 1 p.

Rental of the kirks set for money according to the assedations of the place made above.

Kirks of Largis, Innerkip and Lochvinock, £460. Kirk of Kilmacalme, £133 6s 8d. One part of the kirk of Mernis and the teinds set for £48. Another part of the said kirk, £36. Also another part of the said kirk, £20. Kirk of Neilstoun, £66 13s 4d. Kirk of Cathcart, £40. Kirk of Carmanok, £20. Kirk of Rasneth, £146 13s 4d. Kirk of Killmane in Arcadia [*i.e.* Kilmun in Argyll], £52. Kirk of Kilkerrane in Kintyr, £13 6s 8d. Kirk of Calmanle [*i.e.* Kilcalmonell], £12. Kirk of Kilbarchan, £65 13s 4d. Kirk of Kililane, £19 6s 4d.[9] Kirk of Dundonald, £140. Kirk of Auchinlek, £66 13s 4d. Kirk of Munktoun with one part of the kirk of Sancti Kevoci [*i.e.* St Quivox], £108. Another part of the kirk of Sancti Kevoci, £30. A third part of the same kirk, £33 6s 8d. Teinds of Munktoun, £14. Teinds of Over Manis of Munktoun, £4. Kirk of Prestuik, £13 6s 8d. Kirk of Innerweck, £80. Kirk of Lydgerwod £50. Teinds of Quhytfuird and Railstoun, £10. Teinds of the town of Paislay £26. Teinds of Cothmay within the parish of Kilpatrik, £7. Teinds of Drumry, £33 6s 8d. One part of the teinds of Drumnerbek, Calektoun and Auchintorlie, £5. Teinds of Over and Nether Dalmeynis and Garskadden, £33 6s 8d. Vicarage of Paislay and Lochv[i]nok, £100. Total money in teinds of kirks set for money, £1,887.

Total sum of the sums of money, both of the temporality and spirituality with the teinds of the kirks of this monastery set for money, reckoned according to the foresaid annual accounts and rental likewise extending to £2,467 19s.[10]

Of which total sum there have to be deducted the sums as follows below.

For the upkeep of the convent in the kitchen and for their habits yearly according to the accounts of graniter and cellarer of the same place £473 8s 4d.

For the fees of the graniter and appointed servants, and the cellarer of this place annually extending to the sum of £38.

For the archbishop of Glasgow[11] for procurations, £13 6s 8d.

For the Lords of the Council for their contribution yearly, and in pensions by reason of our letters and in fees, namely, to the lord duke,[12] Mr John Maquhyn and James Jonstoun, preachers,[13] and sir John Skleater and John Nasmyth and David Hamiltoun of Bothwellhaucht, £550 2s 8d.

---

[9] This should probably read 8d, which would give the correct total.

[10] £1,887 plus £569 7s 1d, from the lordships, totals £2,456 7s 1d.

[11] James Betoun was archbishop of Glasgow at the Reformation, Watt, *Fasti*, p. 149.

[12] I.e. James Hamilton, 2nd earl of Arran and duke of Chatelherault, *Scots Peerage*, iv, pp. 366-368.

Total deductions, £1,074 13s 7d.[14] And so rests free in money, £1,393 16d.

Total of bere, ferme and teind free, 26 c. 10 b. 2 f. Total of meal, ferme and teind free, 57 c. 10 b. 3 f. $2\frac{1}{3}$ p. Total horsecorn, ferme and teind, 35 c. 9 b. 1 f. 1 p.

Memorandum, Crukitschot does not pay salmon because of the sterility of the river and few are caught these three and four years. Lymbarne[15] likewise not paid. 'Sicklyk' Balquhair and Wollscheit not paid. The burns of Paislay and Blakstoun do not pay because they are barren and few are caught. In fact this year were caught only 32 salmon. Not charged with capons and poultry because they are sold for the upkeep of the place.[16]

Calculation of third: Money, £2,467 19s; third thereof, £822 13s. Meal, 72 c. 3 b. 3 f. $2\frac{1}{3}$ p.; third thereof, 24 c. 9 b. 1 f. 'half pect 3 part' [i.e. $\frac{5}{6}$ p.].[17] Bere, 40 c. 11 b. [40 c. 12 b., *fo 28r*]; third thereof, 14 c. 9 b. 1 f. $1\frac{1}{3}$ p.[18] Horsecorn, 43 c. 1 b. 1 f. 1 p. 'gret met'; third thereof, 14 c. 5 b. 3 f. $\frac{1}{3}$ p. Cheese, 576 stones; third thereof, $235\frac{1}{3}$ stones.[19]

Assignation:

Third of money, £822 13s. Take the penny mails of the lordship of Paislay, Kilpatrik, Glen and Munktoun for £569 7s 1d; the kirk of Lydgertwod for £50; the kirk of Rosneth, £146 13s 4d; the kirk of Kilbarchan for £66 13s 4d [£65 13s 4d, *fo 28v*]. 'Gif in' £10 0s 9d.

Third of meal, 24 c. 1 b. $1\frac{2}{3}$ f. [24 c. 9 b. $1\frac{2}{3}$, *above*]. The kirk of Kilpatrik for 28 c. 15 b. 2 f. 'Gif in' 4 c. 14 b. 3 p.[20]

Third of bere, 13 c. 9 b. 1 f. $1\frac{1}{3}$ p.[21] 'That is the fearme beir of the saidis lordschipis of Paislay, Kilpatrik, Glen and Mun[k]toun givand be yeir' 15 c. 6 b. 1 f. 2 p. 'Gif in' 1 c. 13 b. 2 p.[22]

---

13 John McQueen, schoolmaster, and James Johnston, a Dominican friar, were appointed special preachers according to the Provincial Council's decision of 1549. *Statutes of the Scottish Church, 1225 - 1529*, ed. D. Patrick (Edinburgh, 1907), pp. 88, 108; *Essays on the Scottish Reformation*, ed. D. McRoberts (Glasgow, 1962), pp. 161, 351.

14 The correct calculation appears to be £1,074 17s 8d.

15 Cf 'Linbren', 'Lenbreyn', 'Lymbrane', *Registrum Monasterii de Passelet 1163 - 1529*, ed. C. Innes, (Paisley, 1882), p. 220; *RMS*, v, no 2070; possibly now Linburn, NS4570.

16 The Latin text reads 'Memorandum, Crukitschot non solut salmonis que fluminis ster[i]les et pauci predicitur [*recte*, prendentur] his tribus et quatuor annis. Lymbarne similiter non soluit. Sicklyk Balquhair et Wollscheit non solvet. Rivole de Paislay et Blakstoun non solvent quia steriles et pauci prenduntur. Veram hoc anno precepti [*recte*, percepti] sunt tantundo [*recte*, tantumodo] xxxij salmonis. Non onerat tam caponibus et pulteris quia veniunt ad sustentationem loci.'

17 The correct calculation appears to be 24 c. 1 b. 1 f. plus one third of $2\frac{1}{3}$ p. (i.e. $\frac{7}{9}$ p.).

18 The correct calculation appears to be 13 c. 9 b.

19 No figure for cheese is given in the rental above.

20 These figures total 24 c. 1 b. 1 f. 1 p.

21 See above, n. 18.

22 These figures total 13 c. 9 b. 1 f.

Third of horsecorn, 13 c. 6 b. 3 f. ⅓ p. [14 c. 5 b. 3 f. ⅓ p., *fo 29v*] 'of gret met. Tak thir aitis of the best payment and quhair thay can be best gottin.'

'Tak the haill cheis of Glen and Paislay, becaus we knaw thair is superplus and gif in the tua parte.' [*In margin*, 'Cheis', 11 score and 15 [stane]s 3 pt. stane'.]

GLASSFORD, PARSONAGE OF, (EUL, fo 30r)

Rental of the parsonage of Glasfuird given up by John Sympill, provost of Sempill.

'The said personage sett of auld for yeirlie payment in langtymis past tua chalderis aittis and fourty pund money quhairof I have ressavit nathing sen my provisioun thairto.'

Signature: 'Johne Sempill of Bultries.'

RICCARTON, (EUL, fo 30r-v)

'Rentale de Richertoun' [*in a later hand*, 'Kyle']

Thripwod, 1 b. meal. Coilgoist [*i.e.* Cowgove], 3 b. meal; 1 b. bere. Brontwod, 2 c. 2 b. meal; 4 b. bere. Milrig, 4 b. meal; 1 b. bere. Soumertait,[23] 4 b. meal. Smedan hous, 5 b. meal. 1 b. bere. Sornhill, 4 b. meal; 1 b. bere. Drumdraet [*i.e.* Drumdroch], 5 b. meal; 1 b. bere. Brussrydoun, 3 b. meal; 1 b. bere. Wraichis [*i.e.* Wraes], 5 b. meal; 1 b. [bere]. Overlond, 10 b. meal; 2 b. bere. Neth[e]rland, 16 b. meal; 4 b. bere. Maynis of Hayning, 20 b. meal; 4 [b.] bere. Prok [*i.e.* Purroch], 2 b. meal. Auchindonan, 14 b. meal; 1 b. bere. Craigo, 8 b. meal. Baith the Airdis, 7 b. meal; 1 b. bere. Auchincros Manis, 1 c. meal; 1 b. bere. Ladyaird, 3 b. meal; 2 f. bere. Hunthall and Rassallis croft, 6 b. meal; 6 f. bere. Dullar Fullartoun, 4 b. meal; 1 b. bere. Windihall, 4 b. meal; 1 b. bere. Hawestok, 7 f. meal; 1 f. [bere]. Ladyland, 4 b. meal; 1 b. bere. Auchten and Boghous, 2 b. meal; 1 b. bere. Cattacroch, 6 f. meal. Ester Balgray, 9 f. meal; 2 f. bere. Wester Balgray, 6 f. meal; 2 f. bere. Brighous, 10 f. meal; 2 f. bere. Staucherlat [Quhitrighill, *deleted*], 2 b. meal; 2 f. bere. Quh[i]trigmerschell, 4 b. 2 [f.] meal; 2 f. bere. Barlethtoun Lethill, 11 b. meal; 2 b. bere. Sessnak Glesfuird, 5 b. meal; 1 b. bere. Sempill, 2 f. meal. Schawhill, 6 f. meal. Quhirlfuird, 7 b. meal; 1 b. bere. Blair, 3 b. meal; 1 b. bere. Cansheill, 4 b. meal; 1 b. bere. Ritchartoun with the common, 7 b. [meal]; 1 b. bere. Richartoun Holms, 5 b. meal. Dullarhill, 14 b. meal. Dullarschaw, 2 b. meal. New Comonis, 10 b. meal; 1 b. bere. Burnbank, 4 b. meal; 2 f. bere. Schortles, 1 b. meal; 1 f. bere. Inschgottrig Eister, 4 b. 2 f. meal; 5 f. bere. Middill Inschgotrig, 2 b. meal;

---

[23] This is possibly a scribal mistranscription of Sornbeg, *i.e.* Little Sorn, NS4933.

2 f. bere. Waster Inschgottrig, 2 b. meal; 2 f. bere. Husland Tak Harlaw, 8 f. [meal]; 1 f. bere.[24]

RENFREW, PARSONAGE OF, (EUL, fo 31r)

Rental of the parsonage of Ranfrew

'This haill of this personage in victuall with the pensioun of Paislay annexit thairto extendis to' 19 c. victual, 'the quhilk was sett ever for' 240 merks 'as I am abill to prove and as the matir is notour. The viccerage thairof, the haill dewties thairof almost being dischargit be[ing, *deleted*] act of Parliament, is sett for' £12.

'Off the quhilkis to be deducit the ordinar charges of my reidar quha is satisfiet by my [*recte*, be me] instantlie nochtwithstanding my awin residence and service is [*recte*, as] the superintendencie[25] in Glesgow will verifie according to the Buik of Dissiplein.[26] Item, the said benefice is astrictit to ane yeirlie payment of ane pensioun to the hospitall of Glesgow' 6½ merks. 'Item, the said personage is obleissid to the interteinment of ane chaplanrie yeirlie', 12 merks.

Signature: 'Andro Hay.'

Calculation of third: In the whole, £172;[27] third thereof, £57 6s 8d.

KILMACOLM, VICARAGE OF, (EUL, fo 31r)

'The rentale of Umphray Cunghams benefice.'

The vicarage of Kilmacholm lying within the sheriffdom of Ranfrew worth by year 15 merks, 'nothing payit this thrie yeir bygain.'

Signature: 'Umphra Cuninghame.'

Calculation of third: In the whole, £33 8s 8d;[28] third thereof, £9 2s 2d '3 ob' [*recte*, 'ob 3 ob', i.e. $\frac{2}{3}$d].[29]

---

[24] This entry ends abruptly at this point.
[25] John Willock, superintendent of Glasgow; see *First Book of Discipline*, p. 121.
[26] I.e. the *First Book of Discipline* (1560).
[27] I.e. 240 merks plus £12.
[28] £33 6s 8d equals 50 merks, although the figure in the preceding paragraph is 15 merks.
[29] The correct calculation appears to be £11 2s 10⅔d.

GOVAN, PARSONAGE AND VICARAGE OF, (EUL, fo 31v)

Rental of the parsonage and vicarage of Govane pertaining to Mr Stephen Betoun which are set in assedation by him to his kinsman, Thomas Forret, burgess of Glesgow, 'lyk as he fand the samin sett of befoir be utheris his predicessouris for the sowme of' 300 merks usual money, of which sum foresaid his mi[ni]ster has 40 merks, 'sua remainis to the persounis use' 240 merks.[30] 'In witnes quhairof becaus he may not travell hes subscryvit this present in verificatioun of his leving conforme to the queinis majesties generall proclamatiounis[31] and his takisman in lyk maner.'

Signature: 'Thomas Forret. S. Betoun. Presentit be me, Mr Thomas Archbald, for for [sic] Mr Stevin Betoun'.

Calculation of third: In the whole, £200; third thereof, £66 13s 4d.

INVERKIP, VICARAGE OF, (EUL, fo 31v)

Rental of the vicarage of Innerkip pertaining to sir David Crystysoun and set by him to the laird of Greinok and Mr William Schaw for 100 merks yearly, 'and this is the uttermest and haill proffeit I haif thairof.'

Signature: 'sir David Crystisoun, viccar of Innerkip. M[r] William Schaw, takisman, with Johne Schaw of Greinok.'

Calculation of third: In the whole, £66 13s 4d; third thereof, £22 4s 5⅓d.

KILLELLAN, VICARAGE OF, (EUL, fo 32r)

'Rentale pro M[agistro] Roberto Maxwell, viccario de Kilelane.'

The vicarage of the kirk of Killelan in the diocese of Glasgow is set to William Fleyming of Barochane for the annual sum of £40, of which sum nothing at all has been paid to the said vicar for the space of three years.

Signature: 'Mr Robertus Maxwell, vicarius Killelan.'

Calculation of third: In the whole, £40; third thereof, £13 6s 8d.

---

[30] 40 merks plus 240 merks is 280 merks, not 300.
[31] See *RPC*, i, pp. 193-194, 196-197, 199-200, 202, 204-206.

KILBARCHAN, VICARAGE OF, (EUL, fo 32r)

'Rantale [sic] pro Magistro Joane Maquhyn'[32]

Vicarage of the parish kirk of Kilbarchan in the diocese of Glasgow is set to William Wallace of Johnstoun for 40 merks Scots money yearly payable to him, of which nothing was paid to him for the payment for the past term as it is assigned to the aforesaid commendator.[33]

Signature: 'Ita est ut supra prout asseruit prefatus Magister Joanes McQuyn. M[agister] G. Cok, notarius, de speciali mandato eiusdem.'[34]

Calculation of third: In the whole, £26 13s 4d; third thereof, £8 17s 9⅓d.

ERSKINE, PARSONAGE OF, (EUL, fo 32v)

Rental of the parsonage of Erskein 'quhilk is ane channoun of Glesgow' lying within the diocese of the same.

'Item, the said personage quhilk pertenis presentlie to David Stewart, persoun thairof, havand na do with the viccarage thairof, payit in assedatioun or he obtenit the sam' 200 merks. 'And as yett is sett to David Hamiltoun of Bothwellhaucht for the said sowm of' 200 merks.

Signature: 'David Stewart, persoun of Erskein.'

Calculation of third: In the whole, £133 6s 8d; third thereof, £44 8s 10⅔d.

INCHINNAN, VICARAGE OF, (EUL, fo 32v)

Rental of the vicarage of Inschinane pertaining to sir Bernard Peblis.

'Item, the said viccerage extendis yeirlie with all proffeitis and dewteis to' £60.

Signature: 'Ressavit at command of the Clerk Registre.'[35]

---

[32] I.e. 'Rental for John Maquhyn'.

[33] Kilbarchan was appropriated to Paisley, whose commendator at the Reformation was Claud Hamilton. Cowan, *Parishes*, p. 94; Cowan and Easson, *Medieval Religious Houses*, p. 65; see also above, p. 6, n. 24; p. 527, n. 2.

[34] I.e. 'Thus it is as above, as the aforesaid Mr John McQueen has asserted. Mr George Cook, notary, by his special mandate'.

[35] James McGill of Nether Rankeilour, Clerk Register, 1554 - 1566, *Handbook of British Chronology*, p. 197.

Calculation of third: In the whole, £60; third thereof, £20.

PAISLEY, PARISH KIRK OF, (EUL, fo 33r)
CHAPLAINRY OF THE ALTAR OF ST MIRIN AND ST COLM IN,

'The rentall of the chaplanrie of Sanct Mernis and Sanct Colmis alterage foundit and situat within the parosch kirk of Paislay perteining to sir Johne Urie, chaplan thairof, extending yeirlie in all commodities and proffeitis to' 15 b. victual, 'tua pairt meill and 3 pairt beir yeirlie,[36] to be liftit of the lands of Leidhill[37] perteining thairto, lyand within the paroschin of Paislay and schireffdom of Ranfrew togidder with' 50s 'for the maillis of the Welmedo perteining to the samin chaplanrie lyand within the paroschin and schireffdom foirsaid.'

Calculation of third: In the whole, 15 b. victual; third thereof, 5 b. Third of money, 50s; third thereof, 16s 8d.

Signature: 'James, ressave this rentill. J. N. registrat.'[38]

---

[36] I.e. ⅔ meal, ⅓ bere.
[37] 'Sedhill', *Registrum Monasterii de Passelet*, p. 265; 'Seidhill', *RMS*, v, no 2070.
[38] I.e. James Nicolson, clerk of the Collectory, *Thirds of Benefices*, p. 62.

Stirling (1)

CAMBUSKENNETH, ABBEY OF,[1] (EUL, fo 34r-v)

'Regina.
    Comptrollare,[2] we greit yow weill. Forsameikill as we understand that the rentall givin up to yow of our abay of Cambuskenneth is far abone the just valour present and thairthrow it is not possible [to] our weilbelovit clark, Adame, now comendatour[3] of our said abay, to pay the thrid thairof, according to the said rentall, quhairfoir it is our will and we comand and chairge yow that ye ressave fra our said comendatour sic rentall as he sall give in to yow subscryvit with his hand and conform thairto tak assumptioun for our third of the said abay without prejudice of the superplus gif ony thing be omittit quhilk being done that ye deleit the former rentallis extending abone the just valour, as said is, as ye will answer to us thairupon, keepand thir presentis for your warrand, sub[scryvit] with our hand at [*blank*], the [*blank*] day of [*blank*], the yeir of God' 1567. Sic sub[scribitur], Marie R.'

Rental of Cambuskenneth

[*In another hand*, 'vide page 39, 40, etc.']

Victual: Sanct Meanis [*in margin in another hand*, 'St Ninians'] kirk, 16 c. meal; 16 c. bere, 'by pensiounis givin furth, and of money', £78. Grainges set in feu 'for ane hundreth pund.' The Ruid [*recte*, Hood[4]] acres and Craig set in feu for £90. Bothkenner set in feu for 40 merks. The Cambus, £16. The Powis, £20. Kintulloch, 40 merks. Culray [Colwy, *fo 39r*], £20. Baddindrech, £20. Mortoun,

---

[1] Cf EUL, fos 38v, 39r, 40r (see below, pp. 543, 544, 546); cf also *Registrum monasterii S. Marie de Cambuskenneth A.D. 1147 - 1535*, ed. W. Fraser (Edinburgh, 1872).

[2] Sir William Murray of Tullibardine, Comptroller, 1565 - 1582; James Cockburn of Skirling also served as Comptroller, 1566 - 1567, *Handbook of British Chronology*, p. 191.

[3] Adam Erskine was commendator of Cambuskenneth, 1562 - 1608, Cowan and Easson, *Medieval Religious Houses*, p. 91.

[4] See *Registrum Monasterii S. Marie de Cambuskenneth*, pp. cxvii-cxviii, cxxii; cf below, fo 38v (see below, p. 544).

20 merks. Kettestoun [*i.e.* Kettleston], £35 [£25, *fo 39r*]. The kirk of Cla[k]manan set for 100 merks. The kirk of Tullibody, £20. The kirk of Lacorpe [*i.e.* Lecropt], £20. The kirk of Lenzie, £80. The kirks of Larbeth and Dunpais, £80. The kirk of Aringosk, £44 [£14, *fo 39r*]. The kirk of Crothe [*i.e.* Crathie], £22. The kirk of Kilmaronnok, 100 merks. The kirk of Alvecht [*i.e.* Alva], 50 merks. The kirk of Tulycultie [*i.e.* Tillicoultry], 50 merks. The kirks of Kippone and Kincairdein, £80 [£60, *fo 39r*]. Annuals in Edinbrucht, £8 13s 4d. The customs of Stirvilling, £15 6s 8d. Annuals of Perth, 50s. The pension of Glenluce, £10. The pension of Linclouden, £4. Pension of Blair in Gourie, £5. The lands of Kerssie and Throsk, £30.

'Item, of the foirsaid sowme is to be deducit first be Alexander Grahame givin be the quenis majestie in pensioun' 500 merks. 'Item, to xv channounis, ilk ane of them four scoir merkis, conform to ane decreit quhilk they obtenit thairupon', total, 1,200 merks. The contribution to the Lords of the Session, £36.

Signature: 'Adam, comendatour of Cambuskenneth.'[5]

---

[5] Rentals for Stirling resume at NLS, fo 38v (see below, p. 543).

Dunbarton

DUMBARTON COLLEGIATE KIRK, (EUL, fo 36r-v)
PROVOSTRY OF,

'The rental of the yeirlie proffeit of the proveistrie of Dumbartane.'

The kirk of Fintray pays yearly 80 merks, set to James Stewart of Cardonald. The kirk of Strablane, 200 merks, set to James Edamsoun and Lucas Stirvling. The kirk of Bullill [*i.e.* Bonhill], 5 c. meal. The temporal lands of Strablane, £22. The lands of Ladytoun, 16 merks. The lands of Bularnok, 6 merks. The lands of Knokdoron, 6 merks. The lands of Stakithort, 6 merks. The lands of Mariland, 40s.

'Item, of this to be deducit and payit to sax prebendaris yeirlie ilk prebendar to his fie ane chalder meill and nyn merkis money.'

Signature: 'Johne Kennedie, presentar of the samin, in name of my lord bischope of Cathnes.[1] Johne Kennedie.'

Calculation of third: In the whole, £233 6s 8d; third thereof, £77 15s $6\frac{2}{3}$d. Meal, 5 c.; third thereof, 1 c. 10 b. 2 f. 2 p. 'half pect 3 part pect' [*recte*, 'half pect, 3 half pect', i.e. $\frac{2}{3}$ p.]. 'Omittit capinis, custumis, canys, grassumis and all uther dewties.'

Assignation:
    Third of money, £77 15s $6\frac{2}{3}$d. Take the kirk of Strablane paying £133 6s 8d. [*In margin*, 'Not[a], this kirk givis x merkis mair.'] 'Gif in' £55 11s 1d '3 part ob' [i.e. $\frac{1}{6}$d; *recte*, $\frac{1}{3}$d].
    Third of meal, 1 c. 10 b. 2 f. $2\frac{2}{3}$ p. Take this meal out of the kirk of Bullule paying yearly 5 c. 'Omittit grassumis, entrie silver, caponis, caynis, custumis and small dewties, ye may speir out and bring in haill.'

---

[1] Robert Stewart, provost of Dumbarton collegiate kirk, c. 1530 - 1570, was also bishop of Caithness, 1541 -1586, Watt, *Fasti*, pp. 61, 353.

CARDROSS, PARSONAGE AND VICARAGE OF, (EUL, fo 36v)

Rental of the parsonage of Cardros set in assedation to Mr John Wood of Gillistoun for payment yearly of 100 merks 'for the space of fyve yeiris nixt following Lambmes'[2] 1558.

Calculation of third: In the whole, £66 13s 4d; third thereof, £22 4s 5½d.

INCHCAILLOCH, PARSONAGE OF, (EUL, fo 36v)
STEVENSTON, VICARAGE OF,

'I, James Walker of Inschecalocht and vicar of Stevinstoun, grantis the said personage to be sett to Umphra Cuninghame, burges of Dumbartane, for payment of four scoir merkis yeirlie, the said viccerage to be worth the valour of' 80 merks 'or thairby. I can not tell certanlie becaus it is in my awin my [sic] hand and the samin is payit to me and my under reader. In witnes heirof I have subscrivit thir presentis with my hand at Glesgow', 21 January 1561/2.

[*In margin*, 'To be payit to me [and] my under reider and siklyk grantis.']

[*In margin*, 'Personage of Inchelyocht, viccerage of Stevinstoun.']

Signature: 'James Walker, with my hand.'

Calculation of third: In the whole, £106 13s 4d; third thereof, £36 11s 1½d.[3]

DUMBARTON KIRK, (EUL, fo 37r)
CHAPLAINRY OF THE HOLY ROOD IN,

Rental of the Ruid chaplainry founded in the kirk of Dumbartane pertaining to sir Robert Watsoun extends yearly to £22.

Calculation of third: In the whole, £22; third thereof, £7 6s 8d.

DUMBARTON CASTLE, CHAPLAINRY OF, (EUL, fo 37r)

'The rentall of ane chaplanrie foundit in the castell of Dumbartane perteining to sir Johne Cuik quhilk extendis yeirlie to' 10 merks.

---

[2] I.e. Lammas, 1 August.
[3] The correct calculation appears to be £35 11s 1½d.

Calculation of third: In the whole, £6 13s 4d; third thereof, 44s 5⅓d.

BONHILL, VICARAGE OF, (EUL, fo 37r)

'Patrik Reid, viccar of Bullull, affirmis be this my writting, my viccarage to be of the valour of ten merk and ane chalmer with ane aker of land. As for my offeringis that was wont to be had, thai ar decayit sa as I have na mair lefte. The persoun heirof is the proveist of Dumbartane.[4] Sic subscribitur be me Patrik Reid, the viccar of puire Bullill.'

Calculation of third: In the whole, £6 13s 4d; third thereof, 44s 5⅓d.

LUSS, PARSONAGE AND VICARAGE OF, (EUL, fo 37r)

Rental of the parsonage and vicarage of Lus set by Mr John Layng, parson thereof, to John Culquhoun of Kilmerdeny for payment of £160 money yearly, [plus] £13 6s 8d;[5] 'the persoun payand the viccar pensiounis fe ten[6] merk'; to the stall of Glesgow', 5 merks 'procurage and sinodall, and als the said persoun sall releif the said John, factour foirsaid, of the quenis taxatioun and all uther ordinar and extraordinar charges.'

Calculation of third: In the whole, £173 6s 8d; third thereof, £7 15s 6⅔d [*recte*, £57 15s 6⅔d].

KILPATRICK, VICARAGE OF, (EUL, fo 37v)

'Rentale pro Magistro Archbaldo Barry, viccario de Kilpatrik prope Dumbartan.'

The vicarage perpetual of the kirk of Kilpatrik near Dumbartane in the diocese of Glasgow pertaining to Mr Archibald Barry, according to the total estimation and assedation of the same to the sum of 80 merks. Deducted from this for the stipend of the curate of the kirk, 24 merks. And so the sum of the foresaid vicarage free for the foresaid vicar, 56 merks.

Signature: 'Magister Archbaldus Barry, vicarius ut supra manu propria.'

Calculation of third: In the whole, £53 6s 8d; third thereof, £17 15s 6⅔d.

---

[4] See above, p. 539, n. 1.
[5] These two figures match the total below.
[6] This may read 'ane'.

LUSS, VICARAGE PENSIONARY OF, (EUL, fo 37v)

Rental of the vicarage pensionary of Lus

'Malcum Stevinstoun, viccar pensionar of Lus, affirmis the fructis thairof to be worth be yeir' 24 merks 'payit yeirlie out of the personage.'

Signature: 'Malcum Stevinsoun.'

Calculation of third: In the whole, £16; third thereof, £5 6s 8d.

LUSS, CHAPLAINRY OF, (EUL, fo 37v)

Rental of the chaplainry of Lus pertaining to sir Thomas Henrisoun, chaplain thereof.

The said chaplainry extends yearly to 20 merks, 'haifand the landis of Craigtuly[7] et lie muer with the multuris of the tua millis of Lus and Fynlawis sett in few fearme to Adam Colquhoun be the said sir Thomas Henrisoun of the dait' 6 February 1556/7 'givin up be the said Adam in name and behalf of the said sir Thomas.'

ROSSDHU, LADY CHAPLAINRY OF, (EUL, fo 38r)

Rental of the Lady chaplainry of Rosdew extends yearly [to] the sum of 10 merks 'of certan anuellis lyand within the toun of Dumbartan, perteining to the said sir Thomas and givin in be the said Adame.'

Signature: 'Adam Colquhoun, in name and behalf of the said sir Thomas Henrysoun.'

CARDROSS, VICARAGE PENSIONARY OF, (EUL, fo 38r)

Rental of the vicarage pensionary of Cardros pertaining to Robert Cuik, lying within the sheriffdom of Dumbartan extends yearly to £10; third thereof, £3 6s 8d.

Signature: 'Robert Cuik, with my hand at the pen.'

---

[7] 'Craigynthoye' or 'Craigentui', W. Fraser, *The Chiefs of Colquhoun and their Country*, 2 vols. (Edinburgh, 1869), ii, p. 48.

Stirling (2)

CAMBUSKENNETH, ABBEY OF,[1] (EUL, fo 38v)

'Regina.
    Comptrollare,[2] we greit yow weill. Forsameikill as the rentallis of the benefices quhilkis ar in the possessioun of our trest cousing and counsallour Johne lord Erskein and his consignnes[3] ar givin up to yow far abone the extent and valour that thai presentlie pay, thairfoir it is our will and we comand yow that ye ressave the saidis according as the said benefices payis at this present and ressave our thrid of the samin accordinglie for sua is our pleasour. Subscryvit with our hand at Striviling, the tuenty day of Appryl 1565. Sic subscribitur, Marie R.

Rental of Cambuskenneth

The kirk of Kilmaronoch set in assedation for 100 merks. The kirks of Teinzie [*recte*, Leinzie] in my lord Fleymeinis[4] hands, £80. The kirks of Cipane and Kincar[d]ine set in assedation for £73 6s 8d. The kirks of Larber and Dunipace, £80. The kirk of Auldwecht [*i.e.* Alva] set for £33 6s 8d. The kirk of Airngosk set in assedation for £43 [£44, *fo 34v*]. The kirk of Dillycoutry set in assedatioun for £33. 6s 8d. The kirk of Crathie set in assedation for £22 13s 4d. The kirk of Clakmanan set in assedation for 100 merks. The kirk of Tulibodie set in assedatioun for £20. The kirk of Licrept set in assedation for £20; and of victual bere and meal, 18 c. The lands of Crassid [Carsyd, *fo 40v*], Cowie, Bandeth, Murtoun extending to £83 [£84, *fo 40v*]. Ketilstoun, £35. The feu lands of

---

[1] Cf EUL, fos 34r, 39r, 40r (see above, p. 537; below, pp. 544, 546).
[2] Sir John Wishart of Pittarow, Comptroller, 1562 - 1565, *Handbook of British Chronology*, p. 191.
[3] With David Panter's death in 1558, as commendator of Cambuskenneth, John, 6th lord Erskine, was appointed by the crown to Cambuskenneth; he was in posssession by 1561; and in his name, Queen Mary nominated Adam Erskine as commendator of Cambuskenneth in 1562. *Scots Peerage*, v, pp. 612-614; *Thirds of Benefices*, pp. 113-114; *RSS*, v, no 1066; Cowan and Easson, *Medieval Religious Houses*, pp. 90, 91, 101.
[4] John, 5th lord Fleming, *Scots Peerage*, viii, pp. 543-544.

Dunypaice, £8 6s 8d. The customs of Striviling, £16 6s 8d [£15 6s 8d, *fo 34v*]. Pitcrathie, 26s 8d. The Palhous, £22. The Camous [*i.e.* Cambus], £16. The Hudis acres set in feu for £117. The fermes of Brachkenneth set in feu for £56. The Grange is in Arthur Erskinis hands 'givin be the quenis majestie. Item, of this foirsaid money givin in pensioun to Alexander Grahame be the quenis majestie', 500 merks.

Money in the whole, £973;[5] third thereof, £324 6s 8d. Victual, 18 c., namely 9 c. bere, 9 c. meal, 'thairof 3 chalderis of ilk ane.'

INCHMAHOME, PRIORY OF,[6] (EUL, fo 39r)

Rental of Inschmahomo

The kirk of Kilmadok should pay 10 c. victual; £40 money. The kirk of the Porte, 5 c. victual; £6 money. The kirk of Lany set in assedation to the laird of Touch for £53 6s 8d. The kirk of Lamtrachen [*i.e.* Lintrathen] in assedation to my lady Craufuird[7] for 100 merks.

Calculation of third: In the whole, £166; third thereof, £55 6s 8d. Victual, 15 c.; third thereof, 5 c.

CAMBUSKENNETH, ABBEY OF,[8] (EUL, fo 39r)

'Regina.
Comptrollar, we greit yow weill. Forsameikill as we understand that the rentall givin up to yow of our abay of Cambuskeneth is far abone the just valour present and thairthrow it is not possible to our weilbelovit clark, Adam,[9] comendatour of our said abay, to pay the 3 thairof, according to the said rental, quhairfoir it is our will and we comand and chairge yow that ye ressave fra the said comendatour sic rentall as he sall give in to yow, subscryvit with his hand, and conform thairto tak assedatioun [assumptioun, *fo 34r*] for the third of our said abay without prejudice of the superplus, gif ony thing be omittit, quhilk being done that

---

[5] The correct calculation appears to be £894, excluding the pension of 500 merks paid to Alexander Graham.
[6] Cf EUL, fo 41v (see below, p. 548 and n. 25).
[7] David Lindsay, 9th earl of Crawford, died in September 1558 and was survived by his second wife, Catherine who died in 1578. *Scots Peerage*, iii, pp. 27-29.
[8] Cf EUL, fos 34r, 38v, 40r (see above, pp. 537, 543; below, p. 546); this rental repeats that on fo 34r.
[9] Adam Erskine, commendator of Cambuskenneth, 1562 - 1608, Cowan and Easson, *Medieval Religious Houses*, p. 91.

ye deleit the former rentallis extending abone the just valour, as said is, as ye will answer to us thairupon, keipand thir presentis for your warand, subscryvit with our hand at [blank], the [blank] day of [blank], the yeir of God' 1567. Sic subscribitur, Marie R.'

Rental of Cambuskenneth

Victual: Sanct Neinianis kirk, 16 c. meal; 6 c. [16 c., *fo 34r*] bere, 'by pensiounis givin [furth, and] of money', £78. Graingis set in feu for £100. The Heid [*recte*, Hood[10]] acres and Craig set in feu for £90. Bothkenner set in feu for 40 merks. The Cambus, £16. The Powis, £20. Baddinderth, £20. Colwy [Culray, *fo 34r*], £20. Kintuleoch, 40 merks. Mortoun, 20 merks. Kettilstoun, £25 [£35, *fo* 34r]. The kirk of Cla[k]manan set for 100 merks. The kirk of Tullebody, £20. The kirk of Lecrope, £20. The kirk of Lenzie, £80. The kirks of Larbeth and Dunpaice, £80. The kirk of Aringosk, £14 [£44, *fo 34v*]. The kirk of Creth [*i.e.* Crathie], £22. The kirk of Kilmaronok, 100 merks. The kirks of Alwecht [*i.e.* Alva], 50 merks. The kirk of Tilleculter, 50 merks. The kirks of Kippin and Kincair[di]n, £60 [£80, *fo 34v*]. Annuals in Edinbrucht, £8 13s 4d. The customs of Striviling, £15 6s 8d. Annuals of Perth, 50s. The pension of Glenluce, £10. The pension of Kynclydyn [*i.e.* Lincluden], £4. Pension of Blair in Gourye, £5. The lands of Kersye and Throsk, £30.

'Item, of the foirsaid sowme is to be deducit first be Alexander Grahame givin be the quein in pensioun' 500 merks. 'Item, to xv channounis, ilk ane of thame lx [merkis] ['four scoir merkis', *fo 34v*] conform to ane decreit quhilk they obteinit thairupon', total, 1,200 merks.[11] The contribution to the Lords [of the] Session, £36.

Signature: 'Adame, comendator of Cambuskenneth.'

STIRLING PARISH KIRK, (EUL, fo 39v)
CHAPLAINRY OF ST LAURENCE'S ALTAR IN,

Rental of Sanct Laurence altar within the parish kirk of Stirvilling given by Mr William Gulen, chaplain of the same 'and maister of the gramer schole in the said burcht.'

Sanct Laurence croft occupied by the laird of Randfurd and David Kar yearly, £4 lying beside the Brig of Stirviling. 'Item, the foir bowchis of the laird of Keiris land' lying within the burgh of Stirviling occupied by Gavin Drumond, William

---

[10] See above, p. 537, n. 4.
[11] This total is accurate if 'four scoir merkis' is the correct reading of the amount paid to the 15 canons.

Bourie and Janet Costland yearly, £6 6s 8d. The land of Robert Cosland 'with the turnpyk' lying in the said burgh yearly, 10s. The land of Agnes Ballquhaip lying on the west side of the said Robert's land yearly, 6s. Mr John Stewartis land, 8s. James Watsounis land, 5s. The land of 'umquhill William Main lyand besyd the James Watsounis landis', 5s. Of the customs of Stirviling, 20s. Total, £13 0s 9d.[12]

ST NINIANS, CHAPEL OF, (EUL, fo 39v)

'The rentall of Sanct Neinianis chapell besyd Stirviling'

Out of the Hillis of Towch paid by William Adam and Thomas Adam, tenants of the same, yearly 8 merks. Sanct Neinianis croft lying beside the chapel, by William Bell and Sibilla Drumond, his spouse, 40s. The land of John Thomsoun 'male [sic] lyand besyd John Gibis land, cuitlar, within the said burcht', 6s 8d. The said Gibbis land, 6s 8d. 'Item, Johne Neilsoun wyffis land lyand besyd the mid port', 4s 8d. The Guis croft and Clay croft, 5s. The land of Robert Drumond, 30d. 'Item, the buith that William Lawrie occupyis of the laird of Keir', 6s 8d. Total, £8 19s 10d.[13]

Signature: 'Ita est Mr Gulielmus Gulan, manu propria.'[14]

CAMBUSKENNETH, ABBEY OF,[15] (EUL, fos 40r-41r)

Rental of Cambuskenneth

The kirk of Kilmaronok set to the laird of Drumquhassel for 100 merks. The kirks of Lenzie in my lord Fleymingis hands extending to 33 c. meal; 3 c. bere, 'payand to abay thairfoir', £80. The kirks of Kippane and Kincairdein set to the laird of Torwodheid, £73 6s 8d. The kirks of Lethbert [i.e. Larbert] and Dunipaice set to the laird of Dunipaice for £80; 4 c. bere [in margin, 'money, beir and mony', sic]. The kirk of Alwech [i.e. Alva] set to Robert Monteith for 50 merks. The kirk of Tulliculie [i.e. Tillicoultry] set to the lady Lochlevin[16] for 50 merks. The kirk of Aringosk in the laird of Balvairdis hands for £44 [£14, fo 39r; £43, fo 38v]. The kirk of Crathie set to Thomas Burnet for 50 merks [£22 13s 4d, fo 38v]. The kirk of Clakmanan set to my lord Erskin for 200 merks [100 merks, fo 38v]. The kirk of Tullybody set to David Balfour for 2 c. bere; 8 b. wheat; and

---

[12] The correct calculation appears to be £13 0s 8d.
[13] The correct calculation appears to be £8 18s 10d.
[14] I.e. 'Thus, Mr William Gulan with his own hand'.
[15] Cf EUL, fos 34r, 38v, 39r (see above, pp. 537, 543, 544).
[16] I.e. Margaret Erskine, widow of Sir Robert Douglas, Scots Peerage, vi, p. 369.

of money, £20. The kirk of Lucrope, 4 c. meal, 26 b. bere.[17] The kirk of Sanct Neinian should pay yearly 42 c. meal; 10 c. bere, 'and the samin in the laird of Touchis handis', 11 c. meal; 12 c. bere for £40 money. [*In margin*, 'Remember to devyd.'] And of the same kirk in James Cuninghamis hand, 41 b. meal. [*In margin*, 'Restis of this meill by it that is sett for silver' 27 c. 6 b.] And in James Schawis hands 18 b. meal, 14 b. bere for 40 merks money. [*In margin*, 'Restis lykwys of this beir by it that is sett for silver' 7 c. 2 b.] [Total,] £674.[18]

The lands of Cambuskenneth with the Hude, 10 c. bere; 8 b. wheat; £37 money. [*In margin*, 'mony, quheit, beir.']

The Eist Grainge and Wast Grainge in the hands of Arthur Ersken paying yearly therefore, 'conforme to the quenis grace gift payit of auld', 6 c. wheat; 4 c. [bere]; 20 c. oats.

The lands of Cowie, Carsyd, Bandeth, Muirtoun, Throsk, £84 [£83, *fo 38v*]. The lands of Dough and Kintuloch, £86 13s 4d. The lands of Kittilstoun, £35. The lands of Dunypaice, £8 6s 8d. The lands of Pitcarthie, 26s 8d. The customs of Striviling, 33 merks. The lands of Bothkenner, 4 c. 11 b. 2 f. wheat.

'Item, givin furth of pensiounis of rentall foirsaid as follows. In the first, to Alexander Grahame givin be the quenis grace gift in mony yeirlie', 500 merks. 'Item, to xvij channounis', 17 c. bere, 4 c. 11 b. wheat, 5 c. meal and £293 money. 'Item, to the officiaris of the place', 5 c. meal. 'Item, to the puris', 3 c. 8 b. meal. 'Item, to the bailyies fie of Cambuskenneth', £100.

Signature: 'Andro Hagy.'

Calculation of third: Money, £930 13s 4d;[19] third thereof, £310 4s 5¼d. Bere, 'the Graingis comptit', 28 c. 13 b.;[20] third thereof, 9 c. 9 b. 2 f. 2 p. 'half pect 3 part pek' [*recte*, 'half pect 3 half pect', i.e. ⅔ p.]. Wheat, with the Graingis, 11 c. 11 b. 2 f.; third thereof, 3 c. 14 b. 2 f. Meal, 31 c. 8 b.;[21] third thereof, 10 c. 7 b. 2 f.[22] Oats of the Graingis, 20 c.; third thereof, 6 c. 10 b. 2 f. 2⅔ p.

'Omittit all capounis, pultrie, grassumis, deywarkis, dewties.'

The assignation of Cambuskenneth:
    Third of money, £310 4s 5¼d. Take the kirks of Lethbert [*i.e.* Larbert] and Dunpaice, the laird of Dunipais fermorar, for £80; the lands of Cowye, Carsyd,

---

[17] Lecropt gave £20 and 18 c. victual, fo 38v (see above, p. 543).
[18] The correct calculation appears to be £664.
[19] The correct calculation from individual entries appears to be £938 6s 8d.
[20] The correct calculation appears to be 27 c. 10 b.
[21] The correct calculation appears to be 48 c. 9 b.
[22] A third of 31 c. 8 b. is 10 c. 8 b.

Bendeth, Murtoun and Throsk for £83; the lands of Dawglie [*i.e.* Dough] and Kintulocht, £86 13s 4d; the lands of Kittilstoun for £35; the lands of Dunipais for £8 6s 8d; and the kirk of Tullibody for £20. 'Gif in' 55s $6\frac{1}{3}$d.[23]

Third of bere, 9 c. 9 b. 2 f. $2\frac{2}{3}$ p. Take the 'town' and lands of Cambuskenneth with the Hude by year, 10 c. bere. 'Gif in' 6 b. 1 f. $1\frac{1}{3}$ p.

Third of wheat, 3 c. 14 b. 2 f. Take the lands of Bothkenner giving by year 4 c. 11 b. 2 f. 'Gif in' 13 b.

Third of meal, 10 c. 7 b. 2 f. $2\frac{2}{3}$ p.[24]

Third of oats, 6 c. 10 b. 2 f. $2\frac{2}{3}$ p. Take these oats and meal out of the Graingis, Arthur Erskein fermorar.

'Omittit grassumes, cainis, capounis and uther dewties.'

INCHMAHOME, PRIORY OF,[25] (EUL, fos 41v-42r)

Rental of Inschmahomo

The kirk of Kilmadok, 38 c. meal; 5 c. 8 b. bere. The kirk of the Porte, 16 c. meal; 14 b. bere. The kirk of Lanzie set in assedation to the laird of Touch for 80 merks. The kirk of Lunkathen [*i.e.* Lintrathen] set to my lady Crafurd for 200 merks. The lands of Cardrous, 4 c. meal; £44 money. The vicarage of the Porte, £20.

'Item, the pensiounis, assedatiounis and lyfrentis': 'Item in the first to the auld priour[26] in pensioun', 22 c. meal. 'Item, to ix channounis', 18 c. meal; 180 merks money. Set in assedation to the laird of Drumquhassill, 2 c. meal; 8 b. bere for £10 [*in margin*, 'layd']. Set in assedation to William Sinklar, 24 b. meal for £8 [*in margin*, 'layd']. Set in assedation to George Home, 2 c. meal, 8 b. bere for £16 [*in margin*, 'layd']. To James Erskein 'for his bailyie fie', 3 c. meal; 12 b. [bere]. [*In margin*, 'Not layd'.]

Calculation of third: Meal, 59 c.;[27] third thereof, 19 c. 10 b. 2 f. 2 p. 'half pect [*recte*, 'half pect 3 thairof', i.e. $\frac{2}{3}$ p.]. Bere, 7 c.;[28] third thereof, 2 c. 5 b. 1 f. $1\frac{1}{3}$ p. Money, £234;[29] third thereof, £78.

---

[23] The correct calculation appears to be 55s $6\frac{2}{3}$d.
[24] See above, p. 547, n. 22.
[25] Inchmahome priory, in the sheriffdom of Perth, was held by the Erskine family, as was Cambuskenneth abbey, which explains its inclusion in rentals for Stirling. Cf EUL, fo 39r (see above, p. 544).
[26] David, an illegitimate son of Robert, master of Erskine, succeeded Johne Erskine (later 6th lord Erskine) as commendator of Inchmahome in 1556, *Scots Peerage*, v, pp. 610-612.
[27] The correct calculation appears to be 58 c.
[28] The correct calculation appears to be 6 c. 6 b.
[29] The correct calculation appears to be £250 13s 4d.

'Omittit gressumis and custumis, carragis, capounis, pultrie and all uther dewties.'

Assignation:
Third of meal, 19 c. 10 b. [2 f.] 2 p. 'half pect' [*recte*, 'half pect 3 thairof', i.e. ⅔ p.]. Take this meal out of the said kirk of Kilmadok giving by year 38 c.
Third of bere, 2 c. 5 b. 1 f. 1⅓ p. Take it out of the same kirk giving 5 c. 8 b.
Third of money, £78. Take the lands of Cardros for £60 [£44, *fo 41v*]; and the rest, £18 out of the kirk of Lany, [*i.e.* Lenzie], fermorar the laird of Touch, paying by year, 80 merks. 'Omittit caynis, custumis, gressumis, caponis, pultrie and all uther dewties.'

MANUEL, PRIORY OF, (EUL, fos 42v-43r)

Rental of Manwell [*in another hand*, 'pryorie']

The Maynis of Manuell and the Myre Heidis pay yearly of mail £20, 'and now plinissed with the prioris plinissing payis yeirlie' 7 c. meal; 3 c. bere, 'all rynnan in ebb'. The Wa[l]kmyln toun pays yearly of mail 28 merks 6s 8d. 'The Williame Craigis payis of meall' 4 merks. The Burrow Myln and Lytill Myln pay of mail 20 merks. The Mongale Milln pays 8 merks. The Grenis pays of mail 40s. The augmentation, 20s. The fishing in the Wast Kers, 24 salmon. The lands of Almont pay of annual, 50s. In Linlithgow of annuals, 33s 4d. The Walkmylntoun pays 'of plewing dargis', 13; 'and of harrowing dargis', 14. 'Making and leiding in hervest of ilk ane', 14 'dargis. Item, of collis', 22 'laid at' 8d 'the laid. Item, four dussane of capounis.'

Signature: 'M[r] Alexander Livingstoun.'

Assignation:
Third of meal, 2 c. 5 b. 1 f. 1⅓ p. Third of bere, 1 c. Take this bere and meal out of the Manis of Manwell giving by year 8 c. meal; 3 c. bere.
Third of money, £17 11s 6d 'ob'. Take the Walkmylntoun for 28 merks 6s 8d. 'Gif in' 28s 5d 'ob'. [*In margin*, 'The last rental of Manuell'.]

'Salmond, dawarkis, capones, conform to the roll of the ad, speir and chairge the principall fermourar thairwith. Inprimis in pennymaill', £21 13s 4d. 'Item, of fermis', 7 c. 12 b. 'extending to' 124 b. 'aittmeill; pryce of the boll', 6s 8d; 'summa' £41 6s 8d. 'Item, of beir', 3 c., 'extenting to' 48 b.; 'pryce of the boll', 8s 4d; 'summa', £20. 'Item, of malt', 12 b.; 'pryce of the boll', 12s 8d; 'summa', £7 12s. 'Item, for dusane of capones' 22 'leid coillis, pryce of thir', 50s; 'thrie pund wax, the leading of v doson laid of coillis'. [*In margin*, 'The pryce of thir within the scoir' 50s.] 'Item, the augmentatioun of the few chartour', 20s. 'Summa totalis', £94 2s;[30] third thereof, £32 7s 4d.[31]

'Item, of annuellis, the tua mylnes of Linlithgow' £13 6s 8d. Of Mongall Myln, £5 6s 8d. 'Item, of quheit blolkis [recte, blocks]', 50s. 'Furth of Achway', £5. 'Furth of West Kers', 24 salmon, 'pryce of the peice' 5s, total £6. Of annuals in Linlithgow, 33s 4d. Total, £33 16s 8d. 'Summa of the thrid' £11 5s 6⅔d ('ob the thrid part ob').

'Summa of the haill rentall' £127 18s 8d; 'summa of the haill thrid' £42 12s 10⅔d ('ob the thrid part of ane ob').

Signature: 'M. Hawns,[32] de mandato domine promissa [recte, priorisse].[33] Sic habet a tergo,[34] James Nicolsoun, ressave this rental and tak cautioun for the thrid of this benefice conforme thairto. Sic subscribitur with my hand at Edinbrucht', 31 January 1565/6.

STIRLING, VICARAGE OF, (EUL, fo 43v)

Rental of the vicarage of Stirviling 'efter the auld'.

'Item, in Paschfynis, houklie of the brousteris, umestclaithis, small offrenis, teindis of gressis, gryssess, hemp and lynt, of the quhilk thair is nothing gottin now', £70. 'Item, the fisching, [of] Cambus quhilk the reader hes', £16. 'Item, the teind schavis of the Burrow Rigis', £20. 'Summa efter the auld payment', £106. 'And efter the new alanerlie', £36. 'Of the quhilk the reader hes' £16. 'And sua restis onlie to my sustentatioun of all the foirsaidis sowmes', £20.

Calculation of third: In the whole, £36; third thereof, £12.

BANNOCKBURN, CHAPEL OF, (EUL, fo 43v)

Rental of the chapel of Bannaburne, 1 c. bere 'with ane croft and sum small houssis, of the quhilk chalder I gett stope and impediment maid to me be our soveran ladyis Comptrollar and obtenis na payment.'

---

[30] The 50s in the marginal note has been omitted from this calculation.
[31] The correct calculation appears to be £31 7s 4d.
[32] This name has been corrupted in copying, with at least the loss of the initial following 'M[aister]'. The original signature might have been 'M. [N. T]hownis', for Mr Nicol Thounis, a notary in the Linlithgow area. *Protocol Book of Nicol Thounis, 1559 - 1564*, ed. J. Beveridge and J. Russell (Edinburgh, 1927).
[33] Janet Livingston was prioress of Manuel at the Reformation, *RSS*, iii, no 332; *RMS*, v, no 16.
[34] I.e. 'by order of the prioress. Thus it is endorsed.'

Signature: 'Mr Robert Auchtmuty.'

Calculation of third: In the whole, 1 c. bere; third thereof, 5 b. 1 f. 1⅓ p.

CAMBUSKENNETH, (EUL, fo 44r)
CHAPLAINRY OF OUR LADY ALTAR IN,
ST NINIANS KIRK, CHAPLAINRY IN,
STIRLING, CHAPLAINRY OF ST THOMAS IN,

Rental of the benefices pertaining to Mr Alexander Chalmer.

'Ane chaplanrie of the Lady altar in Cambuskenneth', £13 6s 8d.

'Ane chaplanrie in Sanct Nenianis kirk', £10; third thereof, £3 6s 8d.

'Ane chaplanrie of Sanct Thomas with the almous in Stirviling', £13 6s 8d, 'and tenementis pertening thairto', £12.

'3 of the haill', £8 8s 10⅔d.[35]

Signature: 'sir Alexander Chalmer.'

STIRLING PARISH KIRK, ST ANDREW'S ALTAR IN, (EUL, fo 44r)

'The rental of Sanct Anderis altar within the kirk of Stirvilling payit at Witsonday[36] and Martymes[37] 1563, 1564, 1565.'

'In the first, the hous of Banarage in the Cownycht Wynd under and abone', 31s. Alexander Bowmannis lands, 6s 6d. Jonet Cowanis lands 'bak and foir' 40s. The land of John Allen, baxter, 2s 6d. The lands of John Madrell, 14s. Walter Watsounis lands, 10s. John Chalmeris 'bak land', 40s. 'Item, the foirland pertening to the laird Geden', 10s. Watsounis land behind 'McKnalis smyde yard', [blank]. 'Summa the haill', £8 4s; third thereof, 54s 8d.

'Surtie [i.e. surety], David Foster, burges of Edinbrucht for the thrid.'

---

[35] I.e. of £23 6s 8d for St Thomas chaplainry.
[36] I.e. Seven weeks after Easter.
[37] I.e. 11 November.

MONYABROCH,[38] PARSONAGE AND VICARAGE OF, (EUL, fo 44v)

Rental of the parsonage and vicarage of Monyabrik extends to 10 c. meal, given in by Ninian Aytoun, factor thereof.

Signature: 'Subscryvit with my hand, Neinian Aytoun.'

Calculation of third: In the whole, 10 c. meal; third thereof, 3 c. 5 b. 2 f. $1\frac{1}{3}$ p.[39]

SLAMANNAN, PARSONAGE AND VICARAGE OF, (EUL, fo 44v)

Parsonage and vicarage of Slamananmure

Rental of the parsonage and vicarage of Slenannamure extends to 80 merks by year given up by Ninian Aytoun, factor thereof.

Signature: 'Subscryvit with my hand, Nenian Aytoun.'

Calculation of third: In the whole, £53 6s 8d; third thereof, £17 15s $6\frac{2}{3}$d.

SLAMANNAN, VICARAGE PENSIONARY OF, (EUL, fo 44v)

Rental of the vicarage pensionary of Sleimanane

'Item, the gleib and mans thairof occupyit be me, schir James Arthour, pensionar thairof, be the space of 26 years, extends yearly to 40s and £8, to be paid yearly four times in the year. Total, £10.

Signature: 'Ita est Dominus Jacobus Arthour, vicarius pensionarius, manu propria.'

Calculation of third: In the whole, £10; third thereof, £3 6s 8d.

Assignation: Take this glebe and manse for 40s; take the rest, 25s 8d[40] 'out of the best payment of the paroschin.'

'Ressavit at comand of the Clerk Registre.'[41]

---

[38] I.e. Kilsyth, Cowan, *Parishes*, p. 150.
[39] The correct calculation appears to be 3 c. 5 b. 1 f. $1\frac{1}{3}$ p.
[40] £3 6s 8d minus 40s is 26s 8d.
[41] James McGill of Nether Rankeilour, Clerk Register, 1554 - 1566, *Handbook of British Chronology*, p. 197.

KIRK O' MUIR, CHAPLAINRY OF,[42] (EUL, fo 45r)

Rental of the 'vicarage of the chaplanrie' of the Kirk of Mure pertaining to William Kinros, 'being ane lawit patronage at my lord Grahamis[43] gift.'

'Inprimis', 1 c. meal and 40s money.

Signature: 'Mr Henrie Kinros, presentar thairof, with my hand.'

Calculation of third: In the whole, 1 c. meal; third thereof, 5 b. 2 f. $1\frac{1}{3}$ p. Money, 40s; third thereof, 13s 4d.

FALKIRK, VICARAGE OF, (EUL, fo 45r)

Vicarage of Fawkirk

Calculation of third: In the whole, £67 13s 4d; third thereof, £23 8s $10\frac{2}{3}$d.

MUIRTON, VICARAGE OF,[44] (EUL, fo 45r)

Vicarage of Muirtoun

Calculation of third: In the whole, £13 6s 8d; third thereof, £4 8s $10\frac{2}{3}$d.

BERWICK, (EUL, fo 45r)
CHAPLAINRY OF ST MUNGO'S ALTAR

'This is the rentall of the anuellis perteinand to the chaplandis of Sanct Mungowis altar in the.'[45]

---

[42] I.e. the chapel of St Mary of Garvald (near Fintry), a pendicle of the parish of St Ninians (Kirkton), Cowan, *Parishes*, p. 123.

[43] I.e. William, 2nd earl of Montrose, who was served heir in 1513 and died in 1571, *Scots Peerage*, vi, pp. 226-228.

[44] The parochial chapel of Muirton in St Ninians (Kirkton) parish is possibly to be identified with the village of Muirton (Muiralehouse), at NS8189, near Bannockburn. *Ordnance Gazetteer of Scotland*, ed. F. H. Groome (Edinburgh, 1901), p. 1209; *Registrum monasterii S. Marie de Cambuskenneth A.D. 1147 - 1535*, ed. W. Fraser (Edinburgh, 1872), pp. 274-277.

[45] This intrusive entry abruptly ends at this point.

CULTER, PARSONAGE AND VICARAGE OF,[46] (EUL, fo 45v)

Rental of the parsonage and vicarage of Culter pertaining to Mr Archibald Livingstoun, which parsonage and vicarage is set in assedation for yearly payment of 160 merks.

Signature: 'Henry Livingstoun, in name and behalf of the foirsaid persoun.'

Calculation of third: In the whole, £106 13s 4d; third thereof, £35 11s $1\frac{1}{3}$d.

KILLEARN, PARSONAGE AND VICARAGE OF, (EUL, fo 45v)

Rental of the parsonage and vicarage of Killerne

The parsonage and vicarage of Killerne lying within the sheriffdom of Stirling set in tack and assedation by William Graham, parson and vicar thereof, to John Broch 'the yeiris thairof not yett run furth', for the sum yearly of 160 merks.

Signature: 'William Graham, persoun of Killern.'

Calculation of third: In the whole, £106 13 4d; third thereof, £35 11s $1\frac{1}{3}$d.

STIRLING PARISH KIRK, (EUL, fo 46r)
ALTARAGE OF ST MICHAEL IN,

Rental of the altarage of Sanct Michell, situated within the parish kirk of Striviling, pertaining to Alexander Fergy 'extendis yeirlie to be upliftit of the tenementis of William Bell and James Robertsoun, off the haill', 24 merks 10s.

Signature: 'Alexander Fergy, with my hand. Ressavit at comand of the Clerk Registre.'

STIRLING, BLACK FRIARS OF, (EUL, fo 46r-v)

Rental of the Blakfreiris of Stirvling

'Item, the M[aison]dew occupyit be my lord Graham',[47] 50s. 'Item, in mony in the Row of Menteth intromettit with be the quenis chalmerland thairof and givin in

---

[46] This benefice, which was unappropriated, lay in the sheriffdom of Lanark, Cowan, *Parishes*, p. 41.
[47] I.e. William Graham, 2nd earl of Montrose, *Scots Peerage*, vi, pp. 226-228.

compt be the chalmer in Cheker', £10. The annuals of Auchindrain, £7 6s 8d. The Garwell beside Duning, 20s. The lands of Stirling [*in margin*, 'besyd the toun'], £3. James Ednisteis lands of Dunterth 'quhilk he denyis alanerlie', £7 6s 8d. 'Siklyk', 40s. 'Item, the fisching of ane cowbill', 20s. The small annuals of the 'toun, quhilk will never be weill payit, extendis to' £7 15s 8d. Total money, £41 19s.

'Victuallis of the said freiris': The Burrow Myln in Alexander Erskenis hands pays 36 b. malt. In the said Alexander's hands 'for the croft adjacent to the said freiris place', 20 b. malt. In the said Alexander's hands for the acres in Dumfermling, 7 b. 2 f. malt. The lands called Akeynis lands 'at the Calwy and besyd the freiris', 3 b. malt. The Myln of Kincairn, 1 b. malt. Total, 5 c. 8 b. 2 f.[48]

'The maill of the saidis freiris': The Calsay End, 4 b. meal. The lands of Fintrie, 40 b. meal. Bartholomew Balfouris steading in Menteth, 15 b. meal. Total, 3 c. 12 b.[49]

STIRLING PARISH KIRK, (EUL, fos 46v-47r)
ALTARAGE OF ST JOHN THE BAPTIST IN,

'Rentale aikeris [*recte*, altaris] Beate [*recte*, Beati] Johanes [*recte*, Johannis] Baptiste situat infra ecclesiam parochialem de Striveling.'[50]

'In primis, de terris Kinloch', 14s. The lands of Alexander Brown, 13s 4d. Of the tenement of Mariota Broun, 15s. Of the upper tenement of David Forestar, 6s. Of the tenement of Edward Forest, 20s. 'De Croft Lytilltoun', 2s. Of the Burrow Rudis alias Brady, 4s. Of the tenement of Richard Davidsoun and Janet Moffat, 10s. Of the tenement of Smart, 3s. To the founder, 5s. Of the tenement of Alexander Livingstoun in which 'Magistro Prestoun' lives, 8d. Of the lands of William Brown, 12d. Of the garden of Craigingalt, 4s. Of the tenement of William Lamb, 6s. Of the tenement Edmistoun, 20s. Of the tenement of Catherine Greg, 16s 8d. Of the land of Richard Narne, 3s 4d. 'Item, pro sumba' [*recte*, cumba, *i.e.* a small boat], 17s. 'Hattelar Spens.'[51]

---

[48] The correct calculation appears to be 4 c. 3 b. 2 f.
[49] The correct calculation appears to be 3 c. 11 b.
[50] I.e. 'Rental of the altar of St John the Baptist, situated within the parish kirk of Stirling.'
[51] This appears as a signature. I.e. Balthasar Spens, presumably chaplain.

STIRLING PARISH KIRK, HOLY ROOD ALTAR IN, (EUL, fo 47r)

Rental of Ruid altar situated within the parish kirk of Strivilling pertaining to me, John Ardhill, extending yearly 'togidder with my pairt of the comounis to' £12 6s 8d 'to be upliftit out of the landis and tenementis underwrittin.'

The Rud croft lying beside Stirvilling occupied by Malcolm Kinros and Isobel Broun extending to 12 merks. 'The subdenis land', 2s. The land of Kippen Ros, 2s. Sanct Thomas land 2s. James Fairneis land, 2s. John Schawis land, 2s. Marion Noklaus land, 2s. Agnes McHoweis land, 40d. David Balfouris, 5s. Robert Drumondis land, 30d. William Makesounis, 7s. The laird of Gairdenis land, 5s. Alyman Guidnychtis, 12d. William Leschmanis land, 6s 8d. Walter Douglas land, 5s. John Akins land, 7s. Thomas Comes land, 6s 8d. The lord Grahamis land, 30d. John Alexanderis land, 12d. Norwallis land, 3s 4d. [The Almis Hous, 12d, *deleted*.] Maister cuik land, 2s 2d. John Robisoun, 'fleschour', 2s. Restis land,[52] 19s. The Almis Hous, 12d. Alexander Bonaris house, 8s. My lord Elphingstounis[53] land, 26s 8d.

Signature: 'Joanes Muray. Resaiff this rentall. J. Clericus Registri.'

NOTE

Rental for the following benefice is located elsewhere in the text:

AIRTH, VICARAGE OF (SRO, Vol. a, fo 163r), p. 157.

---

[52] The scribal transcription is indistinct here.
[53] Robert, 3rd lord Elphinstone, *Scots Peerage*, iii, pp. 534-536.

Ayr : Kyle

AYR PARISH KIRK, (EUL, fo 47v)
CHAPLAINRY OF ST PETER'S ALTAR IN,

Rental of Sanct Petiris altar situated within the burgh of Air in the parish kirk of the same, pertaining to sir Richard Millar, chaplain, paying 'be yeir' 8 merks money 'of the comoun purs, and uther' 8 merks 'of thir tenementis underwrittin. In the first':

'Item, to be upliftit of the tenement of Johne Lokart lyand in the north part of the yett of the said Johne', 4s. Of the tenement occupied by John Campbell, 18d. The tenement occupied by William Fergushill, 32d. The tenement of Archibald Osburne, 6s 8d. Of the tenement of William Aird, 32d. Of the tenement of John Law, 3s 4d. The tenement pertaining to Archibald Fergushill, 3s 4d. Of the tenement pertaining to James Bannatyne, 20d. The tenement of George Blair, 8s. Of the tenement of Thomas Mure, 4s. The tenement of William Dalrumpill, 12d. The tenement of Alan and John Huntaris, 12d. Of the yard of Thomas Broun in the Sandgaitt, 4s. In the yard of David Campbell in the Foul Vennell, 12d. The tenement of David Reid, 12s 8d. Of the tenement 'of umquhill Gilbert Turnbill lyand betuixt Johne Walker on the south pairt and Jonat Beynan on the north', 12s. Of the tenement of Alexander Kennedie, 4s. The tenement of Thomas Aichter, 4s. Of the tenement of John Cuninghan [sic], 'flescher', 4s. Of the tenement 'of umquhill Jonat Smycht lyand betuixt the tenement of Thomas Broun in Sandgait on the south pairt and the tenement of Henrie Prestoun in the north pairt', 26s 8d.

Signature: 'sir Richard [Millar], with my hand.'

Calculation of third: In the whole, £10 15s 9d; third thereof, £4.[1]

---

[1] The correct calculation appears to be £3 11s 11d.

FAIL, or FAILFORD, MINISTRY OF THE TRINITARIANS AT,[2] ABERDEEN, MINISTRY OF THE TRINITARIANS IN,[3]

(EUL, fo 48r-v)

'The rentall of the place of Failfuird in victuall and silver.'

'In primis in the paroschinaris of Barmweill in victuall', 9 c. 6 b. 1 f. in meal and bere. In Barnweill of silver, £8. In the parish of Galstoun that pertains to the parson, 9 c. 2 b. 'les'. To the vicarage, 5 c. which is set for 50 merks. The parishioners of Symontoun, 9 b. bere 'alanerlie to the persoun' and 6 c. to the vicarage, set in assedation for 50 merks.

'This is the warst payment of victuall in Scotland.'

'Item, the xx pund land[4] of Carbello payis yeirlie' £20 'and ilk mark land of the foirsaid xx', 40d in augmentation. The kirk of Tertelell [*i.e.* Torthorwald] in Niddisdaill should pay 40 merks. 'Item, ane kirk in Argyll[5] sould pay' 26 merks in the laird of Laumundis [*i.e.* Lamont's] hand 'quha hes not payit this vj yeir ane penny.' $2\frac{1}{2}$ merks for the Croce land. 5 merks for the Wyffand [Waiffuird, *fo 52v*].

The 'ministrie' of Abirdein pertaining to me: £20 of feu mail of the lands of Ferrehill. 40s of the lands of Craigtove [Craigtoun, *fo 48v*]. 12 merks 'of the quenis graces landis of Bray in Air, the quhilk is unpayit this langtym bygain.'

'Off the quhilk ministrie givin to tua puir men in the place to sustein them', £22.

'Extraordinar that the minister gives out to four ald men of the convent evrie ane of the four', 11 b. meal and also 13 b. malt. 'Item, to evrie ane' 8 merks 'of habeitis silver and for thair kiching silver'. To Rankin Davidsoun, minister in Galstoun, £40. To the minister of Symontoun, 50 merks. To the minister of Barnweill [*i.e.* Barnwell], 42 merks. To the 'Lordis of the Seat' [*i.e.* Lords of Session], £7. The minister of Torthoraill, 24 merks. To the minister of Interquhallan [*i.e.* Inverchaolain] in Argil, £10.

Signature: 'Robert, minister of Failfuird.'[6]

---

[2]   Cf EUL, fo 51v (see below, p. 562); see also Cowan and Easson, *Medieval Religious Houses*, p. 109.
[3]   Cf NLS, fos 118r, 120r (see above, p. 418, 420); see also Cowan and Easson, *Medieval Religious Houses*, p. 108.
[4]   I.e. Twenty-pound land.
[5]   I.e. Inverchaolain, appropriated to Fail, Cowan, *Parishes*, p. 88.
[6]   Robert Cunningham, 'last minister' of Failford, EUL, fo 55v (see below, p. 569).

Calculation of third: In the whole, £174 6s 8d;[7] third thereof, £58 2s 2⅔d.
Victual, bere and meal, 18 c. 4 b.;[8] third thereof, 6 c. 1 b. 1 f. 1⅓ p.

Assignation of Failfuird and Abirdeine:
Third of money, £58 2s 2⅔d. The £20 land of Carbello, £36 13s 4d;[9] the feu mails of Ferryhill, £20; annual of Craigtoun [Craigtove, *fo 48r*], 40s. 'Gif in the gift of the anuell extending to' 11s 1⅓d.
Victual, bere and meal, third, 6 c. 1 b. 1 f. 1⅓ p. Take this victual out of the parish of Barnweill giving 9 c. 6 b. 1 f.

CUMNOCK, PARSONAGE AND VICARAGE OF,[10] (EUL, fo 49r)

'Parsonage and vicarage of Cumnok'

Rental of the parsonage of Cumnok within the diocese of Glesgow.

'The said personage is worth be yeirlie rent togidder with the viccerage of the samin be comoun estimatioun the sowme of' 500 merks 'and is sett in assedatioun to [*recte*, be] me, Mr Johne Dumbar, persoun thairof, to Patrik Dumbar, fear of Cumnok, for the sowme of' £40 'to be payit to me yeirlie at the tearmis specifeit in the said assedatioun, of quhilk I want tua tearmis payment last bypast, and suirlie I could never obtein ane just rentall heirof for any labouris I could gett, and the said assedatioun and appoyntment is insert in the buikis of Counsall afoir the lordis thairof, and the said Patrik is oblist thairintill under the paine of horneing to pay all taxatiounis or exactiounis to be imput be the queinis grace or any uther lauchtfull contributioun aucht of the said personage any maner of way and to releif me thairof without any stop or impedement as the samin beiris. I beseik your lordschip to provyd sum ordour quhairby I may be payit of the foirsaid' £40 'considdering the sowme is but small and be releivit of ony taxatioun or exactioun may be imput be the quenis grace to be payit of the foirsaid benefice according to the foirsaid appoiyntment and justice.'

Signature: 'Mr Johne Dumbar, persoun of Cumnok.'

Calculation of third: In the whole, £40; third thereof, £13 6s 8d.

---

[7] The correct calculation appears to be £173 13 4d without taking into account the augmentation.
[8] The correct calculation appears to be 18 c. 8 b. 1 f.
[9] The Twenty-pound land of Carbello yielded £20, plus an augmentation of 40d for each merkland, on fo 48r (see above, p. 558).
[10] Cf EUL, fo 51r (see below, p. 561), which repeats this rental.

DUNDONALD, VICARAGE OF, (EUL, fo 49v)

The just rental of the vicarage of Dundonald within the diocese of Glesgow pertaining to Mr Hew Muntgumry.

'The said viccerage extendis in small teindis, corspresentis and all uther dewties and proffeitis usit and wont to the sowme of' £60 'quhilk was frie to me by the dewties payit to tua sindrie curatis extending to' 40 merks 'in thair fies.'

Signature: 'Ita est Hugo Mungumrie, vicarius de Dundonald, manu propria.'

Calculation of third: In the whole, £86 13s 4d; third thereof, £28 17s 9⅓d.

COYLTON, BENEFICE OF, (EUL, fo 49v)

Rental of the kirk of Queltoun pertaining to George Ros and William Angus, 'channounis of our soverane ladyis chappell royall of Stirvilling.' 'Item, the said kirk payis, lyk as it payit of befoir in assedatioun to ilk ane of us', £40, 'ilk ane of us payand' £6 'pensioun and sua restis of fie to ilk ane of us', £34.

Signature: 'George Ros, with my hand.'

Calculation of third: In the whole, £80; third thereof, £26 13s 4d.

AYR, HOLY ROOD CHAPLAINRY OF, (EUL, fo 50r)

'Rentale domini Thome Rycht' [*In margin*, 'chaplan, Rud of Air']. In the parish kirk of Air, the chaplainry of the Holy Rood[11] extending in annualrents to 20 merks.

Calculation of third: In the whole, £13 6s 8d; third thereof, £4 8s 10⅔d.

DALMELLINGTON, VICARAGE OF, (EUL, fo 50r)

Rental of the vicarage of Dalmellingtoun within Kyle, pertaining to sir John Donlope, set in assedation to the lord of Cathcart[12] for £20, 'payand by this' £12 'to the curetis fiell.'

---

[11] The copyist has corrupted the Latin text here, which should probably read 'capellania Sancte Crucis'. Thomas Raith: 'sangster', *Ayr Burgh Accounts 1534 - 1624*, ed. G.S. Pryde (Edinburgh, 1937), p. 27; and chaplain of St Leonard in Ayr, *RSS*, viii, no 715.

[12] Alan, 4th lord Cathcart, *Scots Peerage*, ii, pp. 514-516.

Signature: 'Mr Johne Houstoun, viccar viccar [sic] of Glasgow, with my hand.'

Calculation of third: In the whole, £32; third thereof, £10 13s 4d.

AYR, PARSONAGE AND VICARAGE OF, (EUL, fo 50v)

Rental of the kirk of Air, both parsonage and vicarage thereof, set to the said Thomas Kennedie of Bargeny by sir Robert Dennestoun and Mr David Gibsoun, the parsonage and vicarage for 153 merks yearly payment, the vicarage thereof pertaining to sir Robert Legatt. At Edinburcht, 14 February 1561/2 'subscryvit with my hand'.

Signature: 'Thomas Kennedie of Bargeny. Presentit be Mr David Gibsoun in name of Bargeny. David Gibsoun.'

Calculation of third: In the haill, £102 13s 4d;[13] third thereof, £34 4s 5⅓d.[14]

CUMNOCK, PARSONAGE AND VICARAGE OF,[15] (EUL, fo 51r)

Rental of the parsonage and vicarage of Cumnok within the diocese of Glesgow.

'The said personage is worth be yeir of yeirlie rent with the viccerage of the samin be comoun estimatioun the sowme of' 500 merks 'and is sett in assedatioun be me, Mr Johne Dumbar, persoun thairof, to Patrik Dumbar, fear of Cumnok, for the sowm of' £40 'to be payt to me yeirlie at the tearmis specifeit in the said assedatioun, of quhilk I want tua tearmis payment last bypast and suirlie I could never obtein ane just rentall heirof for any labouris I could mak, and the said assedatioun and appoyntment is insert in the buikis of Counsall afoir the lordis thairof and the said Patrik is oblist thairintill under the pain of horneing to pay all taxatiounis or exactiounis to be imput be the queinis grace or any uther lauchtfull contributioun aucht of the said personage any maner of way and to releif me thairof without any stop or impedement as the said appoyntment beiris, beseikand heirfoir your lordschip to provyd sum ordour quhairby I may be pay[i]t of the foirsaid' £15 [recte, £40] 'considdering the sowm is but small and to be releivit of ony taxatioun or exactioun may be imput be the quenis grace to be payit of the foirsaid benefice according to the foirsaid appoyntment and justice [recte, mister]'.

---

[13] 153 merks is £102.

[14] This rental is repeated immediately below, on fo 50v; 'the said', in square brackets, is from the second version, and the date in the first version is given as '1661' [recte, 1561]; the figure given for the third in the first version is £34 4s 4⅓d.

[15] Cf EUL, fo 49r (see below, p. 559); this entry repeats that on fo 49r, with few significant differences.

Signature: 'Mr Johne Dumbar, persoun of Cumnok, with small proffeit.'

Calculation of third: In the whole, £40; third thereof, £13 6s 8d.

MONKTON, VICARAGE OF, (EUL, fo 51v)

Rental of the vicarage of Munktoun pertaining to Matthew Forestar, vicar thereof.

Calculation of third: In the whole, 6 c. meal; third thereof, 2 c. meal.

FAIL, or FAILFORD, 'MINISTRY' OF,[16] (EUL, fos 51v-53v)

'Ane new rentall of the minstrie of Failfuird wtherwayis givin in.'

'The rentall of the plaice of Failfuird in victuall and silver.'

'In primis in the paroschineris of Barmweill in victuall', 9 c. 1 b. 1 f. bere and meal; 'of the quhilk victuall thair is' 18 b. meal and 3 b. bere in the hands of William Wallace of Barmweill for payment of 10s the boll 'alanerlie'. Sum of the victual of Barmweill 'ungivin for silver' 7 c. 12 b. 1 f. 'Sum of silver of the victuall givin for silver' £10 10s. The silver mail of Brounhill [*recte*, Barnwell] with the augmentation extends yearly to £8 16s 8d. In the parish of Galstoun that pertains to the parson, 7 c. 11 b.; 'off the quhilk victuall thair is in the handis of the laird of Bar', 24 b. meal and 1 b. bere for 10s the boll; and in the hands of the laird of Somebege 4 b. meal for 10s the boll; and in the hands of the lairds of Galstoun and Cessnok, 4 b. meal for 10s the boll. 'Summa of the victuall of Galstoun ungivin for silver', 5 c. 10 b. 'Summa of the silver of victuall givin for silver', £16 10s. The parishioners of Symontoun, 11 b. bere to the parson.

'Item, xx li. land[17] of Carbello payis yeirlie' £20, 'xxx stanis cheis, ten hoggis, iij stirkis, xl d. in augmentatioun of evrie merk land of the said xx li. land.'

The kirk of Tarthorrall [*i.e.* Torthorwald] in Niddsdaill should pay yearly the sum of £40. 'Item, ane kirk in Ergyll callit Interchallen [*i.e.* Inverchaolain] sould pay' 26 merks in the laird of Laumondis hands 'quha hes not payit ane penny this vj yeir. Item, thrie pund land in Corsby in the laird of Cuninghamheidis hand payand yeirlie', 5 merks. 'Item, the fisching of the toun extendis to' 2 dozen grilse and salmon in the laird of Cuninghamheidis hands. For the Waiffuird [Wyffand, *fo 48r*], 40s 40d. 'Item, tua merk land in Symontoun callit the Countrie land payis yeirlie' 2 merks. The annual out of Mauchlein, 10 merks. The Spittalhill within

---

[16] Cf EUL, fo 48r (see above, p. 558).
[17] I.e. Twenty-pound land.

the parish of Symontoun 'extending to tua merk land' in the laird of Carnellis hand pays yearly, 2 merks 6s 8d. 'Item, ten schilling land of the Maynes of Failfuird with the myln thairof', 12 merks 10s.

'Extraordinar that the minister gives out to four auld beidmen of the convent evrie ane of the four', 11 b. meal, 12 b. malt. 'Item, to evrie ane' 8 merks 'of habit silver and kichein silver.' To the Lords of Session, £7. To Mr John Douglas in pension, 50 merks, 'for the quhilk I am actit in the Buikis of Counsall for payment thairof furth of the rentis of my place of Failfuird.'

Minsters' stipends: To John Millar, exhorter at Barnweill, £40. To Thomas Keringtoun, reader at Symontoun, £20. To Rankin Davidsoun, exhorter at Galstoun, £50. To John Wallace, readar at Tarthorall, £20.

Calculation of third: *see* Vicarage of Symington, *below*.

GALSTON, VICARAGE OF, (EUL, fo 52v)

Rental of the vicarage of Galstoun pertaining to Mr John Stevinsoun extends to 5 c. and is set in assedation [to] the minister of Failfuird for 50 merks. [*In margin*, 'For the Colege of Justice.']

Calculation of third: *see* Vicarage of Symington, *below*.

SYMINGTON, VICARAGE OF,[18] (EUL, fo 53r)

Rental of the vicarage of Symontoun pertaining to sir John Millar extends to 7 c. and is set in assedation to Adam Cuningham for 40 merks.

Signature: 'Robert, minister of Failfuird.'[19]

Calculation of third:
    Ministry of Failford: 'Victuall unsett for silver, in the haill', 14 c. 1 f.; third thereof, 4 c. 10 b. 3 f. Money, in the whole, £141 16s 8d; third thereof, £47 6s 6⅔d.[20] Salmon and grilse, in the whole, 2 dozen; third thereof, 8. Cheese, 30 stones; third thereof, 10 stones. Hogs, 10; third thereof, 3⅓. Stirks, 3; third thereof, 1.
    Vicarage of Symontoun, in the whole, £26 13s 4d; third thereof, £8 17s 9⅓d.

---

[18] Cf EUL, fo 55v (see below, p. 569).

[19] I.e. Robert Cunningham, see above, p. 558, n. 6.

[20] The correct calculation appears to be £47 5s 6⅔d.

Vicarage of Galstoun pertaining to Mr John Stevinsoun, in the whole, £33 6s 8d; third thereof, £11 2s 2⅔d.

Assignation of the ministry of Failfuird:
Third of victual, 4 c. 10 b. 3 f. Take this victual as follows in the parish of Barnweill: in the Foultoun, 26 b. meal.[21] In the parish of Symontoun: the Maynis of Ilingtoun pertaining to the laird of El[d]irslie, 18 b.; in the Hilhous in James Blairis hands, 8 b. In the parish of Galstoun: the Greinholme, 4 b. 2 p. 'les victuall'; Eschard, 6 b. meal, 6 f. bere; [*in margin*, 'Johne'] Gordoun in Lufnoris, 6 f. meal, 3 p. bere. John Roxburcht in Luffnoris, 10 f. meal, 6 p. bere. 'Summa of this victauall pertenand' 74 b. 3 f.

The third of the ministry of Failfuird extends to £46 5s 6⅔d.[22] The kirk of Interchallan for £17 6s 8d; the annual of Mauchlyne for £6 13s 4d; the laird of Baris victual extending [to] 24 b. meal, 1 b. bere, the laird of Sornibegis victual extends to 4 b. meal and the lairdis of Galstoun and Cessnakis victuals, 4 b. 'all thir for x s. the boll, montis in mony to' £16 10s; 'the tua markland of Symontoun callit the Countreland payis yeirlie', 26s 8d; 'the Spittal, ij merkland in the laird of Carnelis handis' 33s 4d; and the rest extending to £4 15s 6⅔d out of the hands [*recte*, lands] of William Wallace of Barnweill giving £10 10s for 21 b. victual.

'Agriance for the lxj yeiris'
Third of victual of Faill, 4 c. 10 b. [4 c. 10 b. 3 f., *fo 53r*] at 2 merks the boll, total, £99 13s 4d.[23] Third of money, £47 5s 6⅔d. 'Summa both of mony and victuall extendis to' £141 18s 10⅔d.[24] 'Of this debursit to the ministeris' £130 'and tak of the kirk of Interchallane quhilk is tain up in your assignatioun extending to' £17 6s 8d. 'Restis', 54s 10⅔d 'quhilk is defeist for viij salmond and grylssis, x stanis cheis, ane stirk, iij hoggis and a 3 part hog.'

Third of the vicarage of Symontoun which extends to £8 17s 9⅓d 'quhilk salbe payit to the minister as taksman of the said vicerage to my lard Comptrollar[25] et sic eque, eque, eque.'

'Chalmerland of Kyle, ye sall ressave this new assignatioun tain of the ministrie of Failfuird and according thairto tak this assignatioun for the lxj[26] bot as tuiching the lxj yeir it is halilie payit as ye sie be this former agrianc[e] befoir, ye have nathing to tak thairof. Tak only fra him the 3[27] of the vicarage of Symontoun as taksman thairof and certifie be your writtin the chalmerland of Argyll that the kirk

---

[21] There is no record of Foultoun or any other place paying 26 b. meal in the rental.
[22] The correct calculation appears to be £47 5s 6⅔d.
[23] The correct calculation appears to be £98 13s 4d.
[24] £99 13s 4d plus £47 5s 6⅔d is £146 18s 10⅔d.
[25] Sir John Wishart of Pittarow, Comptroller, 1562 - 1565, *Handbook of British Chronology*, p. 191.
[26] I.e. 1561, but it presumably should read 1562.
[27] I.e. third.

of Interchallan is haill tain up in the quenis 3[28] and thairfoir gar him charge the laird of Laumond for the haill, bayth for the lxj and lxij yeiris. Keepe this present for your warand, subscryvit with my hand at Edinbrucht', 24 February 1562/3.

Signature: 'Johnne Wischart, chaplane' [sic].

NOTE

The following rental is located elsewhere in the text:

AYR PARISH KIRK, OUR LADY ALTAR AND CHAPLAINRIES IN (EUL, fo 57r), p. 571.

---

[28] I.e. the queen's third.

Ayr : Carrick

CROSSRAGUEL, ABBEY OF, (EUL, fo 54r-54v)

'Thir ar the ferme landis of Corsraguell quhilk ar teind free and multure frie': The 26 merkland of Drumgirlo[c]h pays 160 b. bere and meal. The 8 merkland of Balchristen pays 70 b. bere and meal. The 6 merkland of Dynein pays 30 b. bere and meal. The 6 merkland of Mochrumtill, 80 b. oats. The Manis of the place pays 36 b. bere and meal. The two mills pay 48 b. bere and meal.

'The landis of Corsraguell quhilk payis of penny maill': David Kennedie of Baltarsyd 9 merkland pays £12. Niven McClassill 20s land pays 40s. Gilbert Kennedie 20s land pays 40s. John Kennedie in Deinyn 16 merkland pays £26 13s 8d. John Kirkpatrik 40s land pays £3. Gilbert Ecclis 4 merkland pays £4. The laird of Cors 3 merkland pays £5. John Fergusoun 6 merkland pays £8. Robert Chalmour £3 land pays £5. Matthew Ramsayis 20s land pays £3. William Kennedie in Brounstoun, 10 merkland, 18 merks. Adam Bydis 40s land pays 5 merks. James Kennedie in Kirkdoun 6 merkland pays £10. John Smyth in Straitoun 3 merkland pays 20s. 'Ane uther 3 merk land payis' 20s. John Kennedie, Ardmillan, 5 merkland pays £5. In Kirkoswell, 300 b. bere and meal. Girvane kirk, 260 b. bere and meal. The kirk of Daylie, 260 merks. The kirk of Straitoun pays £60. The kirk of Kirkcudbrycht[1] pays 40 merks.

Calculation of third: 'Meill and beir in the haill', 55 c. 8 b.;[2] third thereof, 18 c. 8 b. Oats, 80 b.; third thereof, 26 b. 2 f. 2⅔ p. Money in the whole, £383;[3] third thereof, £127 13s 4d.

'Item, the Steding of Drumusquhen occupyit be the abbottis[4] awne gudis estimat worth' 40 merks by year.

---

[1] Kirkcudbright-Innertig (now Ballantrae) was annexed to Crossraguel abbey, Cowan, *Parishes*, p. 120.
[2] The correct calculation appears to be 56 c. 8 b.
[3] The correct calculation appears to be £363 0s 4d.
[4] Quentin Kennedy was abbot of Crossraguel at the Reformation, Cowan and Easson, *Medieval Religious Houses*, p. 64.

'Apud Haddingtoun', 28 December, 'givin in be my lord of Cassallis.'[5]

'My lard erle of Cassallis being chargit to produce a rentall of the abacie of Corsraguell and Glenluce affirms that he hes na rycht nor tytill to the saidis benefeices bot bruikis the fructes and proffeitis thairof as taksman thairunto as his predicessour bruikit the samin this long tyme bypast for payment of the deuties respective eftir specifeit that is to say': For the fruits of the abbey of Glenluce yearly, £666 13s 4d; third thereof, £222 4s 5½d, 'quhairof the said erle hes maid compleit payment of all yeris preceding to the Colectouris.' For the fruits of the abbey of Crosraguell yearly, the sum of 700 merks 'quhilkis he payit to umquhill Queintein, comendatour of Crosraguell,[6] unto his deceis and thairefter yeirlie to Mr George Buquhanon be decreit of the lordis obtenit at his instance thairupon.'

MAYBOLE COLLEGIATE KIRK, PREBENDS OF, (EUL, fo 55r)

'Givin in be sir Johne Kennedie.'

'Sir Johne Kennedy being chargit to produce a rentall of his prebendrie of Mayboill, the haill deutie yeirlie of his prebendrie is' £25; 'the thrid thairof is' £8 6s 8d.

'This rentall of the prebendrie of Mayboill givin in be Androw Gray.'

'The prebendry of Mayboill callit the thrid stall thairof, pertenand to Andro Gray extending in the yeir to fyftie four merkis.'

Signature: 'Andro Gray, with my hand.'

MAYBOLE COLLEGIATE KIRK, PROVOSTRY OF, (EUL, fo 55r)

Rental of the provostry of Mayboill

The 'town' of Barcly yearly, 30 b. meal. 'Ane uther toun callit Barcly of the said provestry giffis yeirlie' 20 b. meal. 'Ane uther toun callit Auchinnaicht gifis yeirlie' 16 b. 'Ane uther toun callit Waustoun gifis yeirlie' 6 b. meal; and of silver yearly, £20.

---

[5] Gilbert Kennedy, 4th earl of Cassillis, was served heir in October 1562, *Scots Peerage*, ii, pp. 471-472.
[6] Quentin Kennedy died in 1564, Cowan and Easson, *Medieval Religious Houses*, p. 64.

SYMINGTON, VICARAGE OF,[7] (EUL, fo 55v)

Rental of the vicarage of Symontoun given up by John Millar, vicar thereof.

In the whole, 8 c. oatmeal, set in tack to the parishioners thereof, namely, 6 c. 14 b. for 13s 4d the boll, 'and the remnant' 18 b. are set in tack to William Wallace of El[d]irslie for 10s the boll 'lyk as he and his predicessouris hes bruikit the samin tyms bygaine past memorie of man. And farder the gleibland of the said viccerage extendis yeirlie to' £10 6s 8d 'and was sett in few forsamekill thrie scoir yeiris syne or thairby.' The teind hay of the 'midow callit Ellingtoun Boig' set to William Wallace of Eldirslie, pays yearly 10s money. The smaller teinds extends yearly to 25s 8d.

Signature: 'This rentall givin up be Johne Miller, viccar of Symontoun, and subscryvit with his awin hand at Edinbrucht', 23 May 1580.

[*In margin*, 'Nota'] 'This first rentall of the viccerage was givin up be umquhill Robert Cuninghame, last minister of Failfuird.'

COLMONELL, VICARAGE OF,[8] (EUL, fo 56r)

Rental of the vicarage of Commonell pertaining to Mr John Davidsoun, 'minster [*recte*, maister] of the pedagog of Glesgow', extends yearly to £40.

Calculation of third: In the whole, £40; third thereof, £13 6s 8d.

STRAITON, VICARAGE OF,[9] (EUL, fo 56r)

This is the just rental of the vicarage of Straitoun, given up by Mr William Bosuell, vicar thereof, 16 June 1562, set to the earl of Cassillis for £46 'havand the gleib extending to half ane merkland of auld extent, the valour thairof be my polecy' 20 merks.

'Defalkait thairof', 20 merks to the m[in]ister. 'Sua the thrid will extend to the sowme of [*blank*]'.

Signature: 'Mr William Bothwell, viccar of Straitoun, with my hand.'

---

[7] Cf EUL, fo 53r (see above, p. 563).
[8] Cf NLS, fo 119v (see above, p. 420); EUL, fo 56v (see below, p. 570).
[9] Cf EUL, fo 65y (see below, p. 582).

Calculation of third: In the whole, £59;[10] third thereof, £19 5s 6⅔d.[11]

KIRKCUDBRIGHT-INNERTIG, PARSONAGE AND VICARAGE OF,[12] (EUL, fo 56v)

[*In margin*, 'Personage and vicarage of Kirkcudbrycht alias Innertig.']

The rental of the kirk of Mayboill, parsonage thereof, set to Thomas Kennedy of Bargeny and his assignees for the sum of £22, 'thairof quhilk I haif in few I refar to the priores[13] rentale of Kirkcudbrycht alias Innertig be personage and vicerage for the personage yeirlie' 40 merks. To the vicar yearly, £30. To the vicar pensionary yearly, £4. The same is a kirk of the abbacy of Corsraguell set by the abbot to the said Thomas.

[*In margin*, 'Haill to the queine becaus it was pittance of Corsraguell.']

Calculation of third: In the whole, £60 13s 4d;[14] third thereof, £20 4s 5⅓d.

COLMONELL, PARSONAGE OF,[15] (EUL, fo 56v)

'The rentall of that ane half of the personage of Colmonell sett be the chaptour of Glesgow to Thomas Kennedie of Bargeny for the yeirlie payment sett of au[l]d for' £80 'and now thai desyre for the samin' 180 merks yearly.

COLMONELL, VICARAGE OF,[16] (EUL, fo 56v)

The vicarage of Colmonell set for £20 yearly to Mr Gilbert Kennedie; third thereof, £6 13s 4d.

---

[10] The correct calculation appears to be £59 6s 8d.
[11] One third of £59 is £19 13s 4d; one third of £59 6s 8d is £19 15s 6⅔d.
[12] Cf EUL, fo 90r (see below, p. 612).
[13] I.e. the abbot of Crossraguel; cf above, pp. 567, n. 4, 568, n. 6.
[14] The figure of £22 has been omitted from this calculation.
[15] Cf EUL, fo 7r (see above, p. 499), where Colmonell appears as one of the common kirks of Glasgow.
[16] Cf NLS, fo 119v; EUL, fo 56r (see above, pp. 420, 569).

AYR PARISH KIRK, OUR LADY ALTAR AND CHAPLAINRIES IN,[17]
(EUL, fo 57r)

Sir Alexander Kerris rental of Our Lady altar situated within the parish kirk of Air and of the chaplainries of the said kirk.

'Item', £10 'of the commoun guidis of the toun of Air yeirlie for the setting and raising of the said mare landis of my said alterage in few and heritage and hes beine in use and possessioun of payment of the samyn' 30 'yeris bygaine.' Upon John Wallace tenement and lands 'in the said gaitt of the said toun' 4s of yearly annual. Upon Robert Galt tenement and lands in the said burgh, 3s yearly annual. 'On the lard of Bargenyis tua tenementis of land within the said burcht', 6s yearly. Of Hew Munfoid tenement and lands within the said burgh, 6s. On Alexander Kennedies tenement and lands in the said burgh, 11s. Upon sir Robert Kennedies tenement which William Blair occupies, 4s. On Huchoun Wallace tenement and lands, 4s. Upon Adam Measouns tenement 'quhilk was umquhill the laird of Carnellis tenement', 3s. On Robert Rankeins tenement lying in the Meill Marcat of the said burgh, 6s. Upon Archibald Fergussouns tenement 4s lying in the Meill Marcatt. Upon Stephen Tenandis father's tenement lying in the Leik Yeardis, 40d. Total, £12 17s 4d.[18]

'The quenis grace hes dischargit hir thrid of my small benefice of Bargeny and to cum.'

KIRKMICHAEL, VICARAGE OF,
(EUL, fo 57v)

'The rentall of the half viccerage of Kirkmichaell in Carrik perteining to me, sir Thomas Muntgumry, viccar pensionar thairof, extendis to' £15. 'And in the handis of Alexander Kennedy of Smeithstoun for the maillis of the gleib landis with the manssis and places thairof', £5. Extending in the whole, the said vicarage yearly to £20; third thereof, £6 13s 4d.

Signature: 'sir Thomas Mungumry, with my hand. James Nicolsoun,[19] resaiff this rentall, J. N. registrat.'

---

[17] The town of Ayr is situated in Kyle, not Carrick.
[18] The correct calculation appears to be £12 14s 4d.
[19] James Nicolson, clerk of the Collectory, *Thirds of Benefices*, p. 62.

NOTE

Rentals for the following benefices are located elsewhere in the text:

COLMONNEL, VICARAGE OF (NLS, fo 119v), p. 420.
COLMONNEL, PARSONAGE OF (EUL, fo 7r), p. 499.

Ayr : Cunninghame

KILWINNING, ABBEY OF,[1] (EUL, fos 58r-63v)

'The last assumptioun of Kilwining tain be Sir Williame Murray of Tullibairdin, knycht, Comptrollar[2] to our soveran lady for the tyme, and affearmit be the Lordis Auditouris of our saidis soveran Checker the [*blank*] day of [*blank*], the yere of God' 1566.

'The thrid of mony conforme to the former assumptioun extendis to' £293 7s 9d. The kirk of Dalry, £100; the teind of the Manis of Rowallone, £10; the kirk of Dumbartane, £66 13s 4d; the teinds of the Mayns of Dene pertaining to lord Boyd,[3] £8; the Mayns of Threipwod in John Hamiltonis hands of Stainhous, £10 13s 4d; the kirk of McHarmik [*i.e.* Kilmacocharmik] in the hands of the laird of Auchinlek, £16; the prior of Quhitarne[4] for the kirk of Kirkmichell, £8 [*in margin*, 20 b. 'farine', 4 b. 'ordei']; 'N[ig]eli Muntgumry for the 3 of Langschaw', £13 6s 8d [*in margin*, '14 b. 'farine', 1 b. 1 p. 'ordei']; in William Muirs hands for the teinds of the Boghill beside the Bigholme, £10 3s 8d [*in margin*, 9 b. 'farine']; the teinds of Wodsyd in the hands of the laird of Ranalstoun in Beith, £6 [*in margin*, 9 b. 'farine']; from John Boyd in Bonschaw teinds within the parish of Stewartoun, £6; the teinds pertaining to Grissall Hamiltoun, lady Robertland within the parish of Stewartoun, [£6, *deleted*] £10 13s 4d [*in margin*, 14 b. 'farine', 2 f. 'ordei']; the teinds of Schoulderflatt pertaining to the lady Sempill[5] in Beitht, £5 [*in margin*, 10 b. 'farine'; 1 b. 'ordei'; 4 b. 'frumenti']; Hugh Huntlie for his teinds of Brumhillis in Beitht, 53s 4d [*in margin*, 2 b. 'farine', 2 b. 'ordei']; the laird of

---

[1] This composite entry contains an assumption of the third dated 1566, followed by a rental dated 1561, and another 'assumption' dated 1564 (see below, pp. 575, 579).
[2] Sir William Murray of Tullibardine, Comptroller, 1565 - 1582, *Handbook of British Chronology*, p. 191.
[3] Robert, 5th lord Boyd, *Scots Peerage*, v, pp. 155-161.
[4] Malcolm Fleming, dean of Dunblane, was prior of the Premonstratensian house at Whithorn in 1566. G. Donaldson, 'The Bishops and Priors of Whithorn', in *Transactions of the Dumfriesshire and Galloway Natural History and Antiquarian Society*, 3rd ser., xxvii (1950), pp. 127-152, at pp. 146-147.
[5] Robert, 3rd lord Sempill, married Isabel, daughter of Sir William Hamilton of Sanquhar, *Scots Peerage*, vii, pp. 538-543.

Heislat for his Maynis of Heislat within the parish of Beith, £9 6s 8d; for Robert Hamiltoun younger of Dalsuff [i.e. Dalserf] for his teinds within Kilwyning, 40s.[6]

'Item, deducit for the teindis of the Lordis contributioun', £9 6s 8d [in margin, '3 of 28 lib.'[7]]; 'to be taine fra the chalmerlane of Kilwyning' 13d. 'Et sic eque.'

'The 3 of maill conforme to the said assumptioun', 22 c. 6 b. 1 f. 1 p. Take the kirk of Kilmarnak 'as salbe givin in be the chalmerlane of Kilwining particularlie and in use of payment', 21 c. 10 b. 1 f. 1 p.; 'for the teindis of Hegatt and Begatt perteining to the lard of Cauldwell within the paroschin of Beith quhairof thair is tua f. ordei', 13 b. 'Et sic eque.'[8]

'The 3 of beir conforme to the last assumptioun', 4 c. 11 b. 1 f. 1 p. Take the kirk of Kilmarnok 'as salbe givin in particularlie be the chalmerlane and in use of payment' 1 c. 5 b. 1 f. 1 p.; the teinds of Dubbis within the parish of Steinsoun, 1 b.; in Kilbryd 'as salbe givin in particularlie be the chalmerlane and in use of payment', 3 c. 5 b. 'Et sic eque.'

'Apud Edinburcht', 26 September 1566

'The Lordis Auditouris of the Chelker with advys of the Comptroller ordanis the assumptioun of Kilwyning to be tain and ressavit of this yeir thriescoir sax yeiris and in tyme to cum with command to the Clerk Collectour[9] to delait this former assumptioun and this assumptioun to be insert in the buikis of the Colectorie to the effect that the Collectour of Kilwynning[10] may have lettres for inbringing of the sam. Sic subscribitur, Cathness, Ep[iscopus]; Jo[hnne], E[piscopus] Rossensis; Ad[am] Orchadensis; J. Spens; Tullibardin, Comptrollar; Robert, Thesaurer.'[11]

'We, Sir Williame Murray of Tullibardin, knycht Comptrollar to our soverane lord and lady, and Colectour universall of thair hienes thridis of the benefices within this realme with advyse, consent and assent of the Lordis Auditouris of our said soveranis Cheker be thir presentis settis and in assedatioun lattis to our weilbelovit William Hyltoun of Torschaw, his factouris ane or mae all and haill the thrid of Kilwining abay for the present yeir thriescoir sax yeiris and sik lyk for all yeiris heirefter sa lang

---

[6] This money totals £284 10s 4d.
[7] I.e. one third of £28.
[8] These figures do not tally.
[9] I.e. James Nicolson, clerk of Collectory, *Thirds of Benefices*, p. 62.
[10] The sub-collectors for Ayrshire were Robert Campbell of Kinyeancleuch, 1561-2, and Eustace Crichton, 1565-6, *Thirds of Benefices*, p. xl.
[11] Robert Stewart, bishop of Caithness; John Leslie, bishop of Ross; Adam Bothwell, bishop of Orkney; John Spens of Condie, Lord Advocate; Sir William Murray of Tullibardine, Comptrollar; Robert Richardson, Treasurer. Watt, *Fasti*, pp. 61; 270; 254; *RSS*, v, no 2913; *Handbook of British Chronology*, pp. 191; 188.

as it sall pleis our said soveranis to tak up and intromet with the said thrid during Gavin, comendatour of Kilwining[12] now presentlie, his lyftyme, to be bruikit, joysit, upliftit and disponit be the said William and his foirsaidis at thair pleasour dureing all the haill space mentionat and sall caus lettres to be derect in our lord soveranis names and ouris to the said William and his factouris foirsaid ane or mae for the ingetting of payment of the samin siclyk as Eusage Crechtoun, now Collectour of Cunighame,[13] and utheris Colectouris gettis for inbringing of the restis of our said soveranis thridis the said William and his factouris ane or mae payand yeirlie to our successouris Comptrollaris or universall Collectouris of our soveranis thrid for the tyme and our factouris the sowme of sevin hundreth merkis sevin hundreth merkis [sic] yeirlie at [blank] and [blank] be [blank] equall portiounis the first tearmis payment beginand at [blank] in the yeir of God' 1567 'and sua furth tearmlie dureing the space abonementionat and we forsuith, the said Sir William, Comptrollar and universall Collectour, and our successouris Comptrolleris for the tyme be the avys of the Lordis Auditouris foirsaid, sall warrand, acquyet and defend this present tak and assedatioun dureing the space abonementionat to the said William and his foirsaidis contrair all deidlie as law will and for suire payment of the said' 700 merks 'yeirlie al the tearmis abonementionat Arthour Grainger, burges of Edinbrucht, is becum cautioun and souertie togidder with the said William self. In verificatioun heirof we have subscrivit the samin present tak at Edinbrucht the [blank] day of [blank].

'At Edinburcht, the [blank] day of October the yeir of God' 1566, 'Arthour Grainger is becum souertie for payment of the sevin hundreth merkis yeirlie and tearmlie as is within writtin and the said William to releif him and ar content the samin be registrat in the Colectour buikis and dyellis [recte, dyettis] to be raisit thairupon in forme as effeiris. In witnes heirof baith the said souertie and principall hes subscryvit the samin, day, place and yeir abonewrittin.'

'The rentale of the abycie of Kilwynning baith of the temporale landis, maillis and fructis as of the kirkis sett for mony and victuallis and uther dewties yeirlie conforme to the chalmerlanis comptis givin up to Gavin, commendatour of Kilwinning foirsaid and extractit be him breiflie to be givin in befoir our soverane lady and my lordis commissaris, ressaveris of the saidis rentallis, anno 1561.'[14]

---

[12] I.e. Gavin Hamilton, commendator of Kilwinning, Cowan and Easson, *Medieval Religious Houses*, p. 69.
[13] See above, p. 574, n. 10.
[14] In January 1562, the queen had commissioned James McGill of Nether Rankeilour, Clerk Register; Sir John Bellenden of Auchnoule, Justice Clerk; Robert Richardson, Treasurer; William Maitland of Lethington, Secretary; John Spens of Condie, Lord Advocate; and Sir John Wishart of Pittarow to receive rentals of benefices. *RPC*, i, pp. 196-197.

The rental of the mails of the temporal lands conform to the 'chaplanis' [*recte*, chamberlain's] accounts. The yearly mails of the lands within the parishes of Kilwynning, Kilmarnok, Dalry and Beith extend yearly to £232 15s.

'The rentall of the kirkis sett for mony yeirlie conforme to the chalmerlanis comptis thairof': The kirk of Dalry, £100. The kirk of Lowdoun, £100. The kirk of Dumbartane, £66 13s 4d. The kirk of Kilmaharmull [*i.e.* Kilmacocharmik] within Argyll, £16. The pension of Quhithorne, £16. The kirk of Kilbirny, £8. The kirk of Dunlope set for £40. Total of the kirks set for money, £446 13s 4d.[15]

'The rentall of the kirkis sett for victuallis and mony yeirlie conforme to the chaplanis [*recte*, chamberlain's] comptis thairof.'

Kirk of Kylwinning: 'Kylwining, the defalcatiounis and attour the ordinar chairges conforme to the chaplanis [*recte*, chamberlain's] comptis extendis in teindis yeirlie to' 6 c. 8 b. meal; 4 c. bere; 8 b. 1 f. wheat and 40 stones of cheese; 'and teindis thairof, stirkis and cheis of the samyn sett for mony yeirlie', £51 10s.

Kirk of Ardrossane: 'Ardrossane attour the defalcatioun and ordinar charges conforme to the chalmerlandis comptis extendis in teindis to' 3 c. 6 b. meal; 9 b. bere; 'and na teindis thairof sett [for] mony.'

Kirk of Kilbryd:[16] 'Kylbryd conforme to the chaplanis [*recte*, chamberlain's] comptis abone mentionat extendis in teindis yeirlie to' 4 c. 15 b. 2 f. meal; 3 c. 5 b. bere; teinds set for money, £8.

Kirk of Irvyn: 'Irvyn conforme to the chalmerlanis comptis abone mentionat extendis in teindis yeirlie to' 2 c. 7 b. meal; 9 b. 2 f. bere; teinds set for money, £7 6s 8d; 'and als iiij huggutis of wyn.' [*In margin*, 'Nota, this sould be' £17 6s 8d.]

Kirks of Perstoun and Dreghorne: 'Peirstoun and Dreghorn conforme to the chalmerlanis comptis yeirlie in teindis', 1 c. 12 b. meal; 1 c. 14 b. bere; teinds set for money, £75; 'and als sett to my lord Glencairne',[17] 11 c. 8 b. oats for money extending yearly, £38 17s.

Kirk of Stewartoun: 'Stewartoun conforme to the chalmerlanis [comptis] abone specifeit in teindis yeirlie', 8 c. 5 b. meal; 1 b. bere; 15 c. 14 b. oats; teinds set for money, £34 6s 8d.

---

[15] The correct calculation appears to be £346 13s 4d.
[16] Later West Kilbride, Cowan, *Parishes*, p. 95.
[17] Alexander Cunningham, 4th earl of Glencairn, 1548 - 1574, *Scots Peerage*, iv, pp. 239-241.

Kirk of Kilmernok: 'Kilmernok conforme to the chalmerlanis comptis as said is yeirlie in teindis', 21 c. 11 b. 2 f. 1 p. meal; 1 c. 5 b. 2 f. 1 p. bere; teinds set for money, £33 6s 8d.

Kirk of Stevinsoun: 'Stevinsoun conforme to the chalmerlanis comptis as said is in teindis yeirlie and sen syne is na meill by the viccaris pensioun and conform to the saidis comptis', 13 b. bere; teinds set for money, £6 13s 4d.

Kirk of Baeth: 'Baeth conforme to the chalmerlandis comptis as said is in teindis yeirlie', 10 c. 9 b. 2 p. meal; 9 b. 3 f. 2 p. bere; teinds set for money, £43 5s.

Total meal of the said kirks extends to 59 c. 10 b. 3 p. 'And thairof thair is deducit to the yeirlie sustentatioun of sax [7, *next paragraph*] munkis, ilk ane of thame haifing in the yeir' 9 b. 2 f. 2 p.; total, 4 c. 4 b. 2 f.[18] 'deducit. 'Sua restis of frie meill', 55 c. 5 b. 2 f.[19] 'And attour this' 15 c. 14 b. oats 'resulting in meill to' 7 c. 15 b. Total meal, 63 c. 4 b. 2 f.

Total bere of the said kirks, 14 c. 1 b. 3 f. 3 p.[20] 'And thairof is deducit to the yeirlie sustentatioun of the sevin munkis [six, *preceding paragraph*], ilk ane haifing yeirlie' 8 b. 2 f. 2 p.; total 3 c. 13 b. 2 p.[21] 'And sua restis of frie beir de claro', 10 c. 4 b. 3 f. 1 p.

'Suma of the haill quheit quhilk is givin to the sustentatioun of the munkis yeirlie', 8 b. 1 f. 'Suma of cheis unsett for mony that is givin to the sustentatioun of the munkis yeirlie', 40 stones. Total horsecorn extends to 15 c. 8 b. oats. Total wine, 3 hogsheads [4, *fos 62v, 63r*]. Total money for the teind stirks and cheese is £300 15s 4d.

'Suma totalis of maillis, kirksilver and teindis stirkis and cheis altogidder sowmit extendis to' £880 3s 8d.[22] 'And thairof is to be deducit for the servandis fies standing yeirlie in the said place and uther ordinar charges conforme to the comptis foirsaid usit and wont' £76 17s. 'And als for the kiching rowis of the saidis munkis and uther habite silver, ilk ane having in the yeir', £16 3s 4d; total thereof £112 3s 4d. 'And siklyk yeirlie to pay my lord Rossis[23] pensioun' £216 13s 4d. 'And als to pay Mr Johne Layng, persoun of Lus, for his yeirlie pensioun under the comoun seill', £13 6s 8d. 'And alsua to Gavin Hamiltoun for his yeirlie pensioun quhilk is provydit be assegnatioun of Andro Hamiltoun, quha had the sam cum potestate transferrendi[24] and the Lardis decreit obtenit heirupon', £100.

---

[18] These figures do not tally.
[19] 59 c. 10 b. 3 p. minus 4 c. 4 b. 2 f. is 55 c. 5 b. 2 f. 3 p.
[20] The correct calculation appears to be 13 c. 2 b. 3 f. 3 p.
[21] These figures do not tally.
[22] This is the total of £232 15s plus £346 13s 4d plus £300 15s 4d.
[23] I.e. the bishop of Ross, see below, p. 579, n. 32.
[24] I.e. 'with power of transferring'.

'Item, to the Lardis of the Seat for thair yeirlie contributioun extending to' £28. 'And for advocatis fiellis to persew and defend the actiounis and caussis of the said abay yeirlie', £12. 'Suma of the haill deductis' £607 0s 4d.[25]

'Sua restis de claro of frie mony', £273 3s 2d.[26] And so rests of free meal, 63 c. 4 b. 2 f. And of free bere, 10 c. 4 b. 3 f. 1 p. And of wine, 4 hogsheads.

'Memorandum, as to the viccerages of Kilmarnok, Kilwynning and Beith, thay ar na sett rentall of them and thay depend in pley and never ane penny of them gottin at the leist thir thrie yeiris bygaine and sasone as I may haif ony certitude and payment of the samin I sall produce the yeirlie proffeit thairof.'

'Item, gif it pleassis my saidis Lardis Auditouris and comisseris to confer this draucht with the yeirlie comptis that I haif tain of my chalmerlandis in divers yeiris bygaine and als my predicessouris tyme I sall produce the samin that the ordinar charges may be considderit quhilk is not heir befoir particularlie mentionat.'

Signature: 'Gawan, comendatour of Kilwyning.'

Calculation of third: Money, £830 3s 8d [£880 3s 8d, *fo 62r*]; third thereof, £293 7s 9⅓d.[27] Meal, 67 c. 9 b. 3 p.; third thereof, 22 c. 8 b. 1 f. 2⅓ p. Bere, 14 c. 1 b. 3 f. 3 p.; third thereof, 4 c. 11 b. 1 f. 1 p. Wheat, 8 b. 1 f.; third thereof, 2 b. 3 f. Cheese, 40 stones; third thereof, 13 stones 5⅓ lb.[28] Wine, 4 hogsheads; third thereof, 1⅓ hogsheads.

Assignation thereof:

Third of money, £293 7s 9⅓d. Take the temporal lands of Kilwyning within the parishes of Kilwyning, Kilmarnok, Dalry and Beith for £232 15s; and the rest of the money extending to £60 12s 10⅓d[29] out of the kirk of Dalry giving by year £100.

Third of meal, 22 c. 8 b. 1 f. 2⅓ p. The kirk of Stewartoun for 8 c. 5 b.; the kirk of Beith for 10 c. 9 b. 2 p.; the kirk of Kilbryd for 4 c. 15 b. 2 f. 'Gif in' 1 c. 5 b. 3⅓ p.[30]

Third of bere, 4 c. 11 b. 1 f. 1 p. The kirk of Kilbryd for 3 c. 5 b.; Peirstoun and Dreghorne for 1 c. 14 b. 'Gif in' 7 b. 2 f. 3 p.

'Tak the thridis of quheit and cheis quhair it may be gottin and wyne out of Irvyne.'

---

25 The correct calculation appears to be £559 0s 4d.

26 The correct calculation from the figures cited in the text appears to be £273 3s 4d.

27 Three times £293 7s 9⅓d is £880 3s 4d.

28 Cheese was commonly measured by the stone of 15 pounds. This calculation is therefore correct. In modern reckoning, a third of 40 stones is 13 stones 4⅔ lb.

29 The correct calculation appears to be £60 12s 9⅓d.

30 The correct calculation appears to be 1 c. 5 b. 3⅔ p.

'Omittit, canis, custumis, pultrie, capounis, entrie silver, grassumis, carrages and all uther dewties quhilk ye man diligentlie speir and bring tham haill in.'

'The last assumptioun of the abay of Kilwyning takin be Sir Johne Wischart of Pittarro, knycht, Comptrollar[31] to our soverane lady', 16 January 1563/4.

The third of the money of the foresaid abbey extends to £292 7s 9½d [£293 7s 9d, *fo 58r*]. Take out of the kirk of Dalry, £100; out of the kirk of Dumbartan, £66 13s 4d; the mails of the Maynis of Rowalane, £10; the feu mails of the lord Boydis Maynis, £8; the feu mails of the Threipwod in John Hamiltounis hands, £10 13s 4d; 'Item, the 3 of the bischop of Rossis[32] pensioun givin frie to him', £88 15s 5½d;[33] 'Item, the 3 of the contributioun', £9 6s 8d. 'Gif in to the comendatour' 12d 'et sic eque for silver.'[34]

Third of meal, 22 c. Third of bere, 4 c. 11 b. 1 f. 1 p. Take these victuals out of the parishes of Beith, Ardrossane, Irvyn and Kilbryd 'as it was of befoir.'

'Apud Edinbrucht' 25 De[cember], 1566.

'Thir lettres ar sene and admittit be Sir William Murray of Tullibardin, knycht, Comptrollar to the king and quenis majestie, and ordanit [*in margin*, to haif] farder exemtioun reservand the assignatioun within mentionat. Sic subscribitur, Tulibardin, comptonar [*sic*], and sensyn affearmet be Sir William Murray of Tullibardin, knycht, in the Chekker be the avys of the Auditouris', 21 January 1565/6.

DUNLOP, VICARAGE OF, (EUL, fo 64r)

Rental of the vicarage of Dunlope pertaining to Mr John Houstoun, vicar thereof, which is set to William Cuninghame of Akett for £78.

Signature: 'Mr Johne Houstoun, with my hand.'

Calculation of third: In the whole, £78; third thereof, £26.

---

[31] Sir John Wishart of Pittarow, Comptroller, 1562 - 1565, *Handbook of British Chronology*, p. 191.
[32] John Leslie, bishop of Ross, 1566 - 1592, Watt, *Fasti*, p. 270.
[33] Three times £88 15s 5½d is £266 6s 4d; the figure given for the bishop's pension on fo 62r is £216 6s 4d (see above, p. 577).
[34] This calculation does not tally.

EAGLESHAM, PARSONAGE OF, (EUL, fo 64r)

'The persoun of Egilshame rentale'

'Item, iij hundreth merkland of Egilshame payis in teind meill and na beir' 14 c. 13½ b. 'Sett communiter' [*i.e.* commonly] for 240 merks.

Calculation of third: In the whole, £186 13s 4d; third thereof, £82 4s 5⅓d.[35]

KILBRIDE, VICARAGE OF, (EUL, fo 64r)

The vicarage of Kilbryd, £40, 'at sumtyme, v merk or thairby better, or sumtyme war.'

Calculation of third: In the whole, £40; third thereof, £13 6s 8d.

KILMAURS, VICARAGE OF, (EUL, fo 64v)

'Memorandum, the viccerage of Kilmawris within Cuninghame perteining to Mr Andro Layng, viccar thairof, consisting in woll, lamb, Paschefynis and kirkland. The kirkland sett in few to the laird of Robertland xx yeir bygaine for viij merkis be yeir quhilk was better nor the rest of the viccerage or it is sett in few and now it beand sett the viccerage is na better nor instantlie nor' £30 yearly. 'To be tain of the said sowme for non payment of corspresentis and umest clothis and Paschfynis yeirlie', 10 merks, 'and the curatis fe', 20 merks.

Signature: 'Presentit be me, Mr Andro Layng.'

Calculation of third: In the whole, £30; third thereof, £10.

TARBOLTON, PARSONAGE OF, (EUL, fo 64v)

'Channonrie of Carboltoun'

'Rentale rectorie seu canonicatus de Tarboltoun perteining to James Cheisholme.'

'In the said kirk sett in assedatioun to the laird of Caprintoun for the yeirlie payment of' 240 merks. 'Item, deduct of the said sowme to the curet for his fie', £20. 'Item, to the chaplane for his stall within the queir of Glesgow yeirlie' 11 merks.

---

[35] The correct calculation appears to be £62 4s 5⅓d.

'Your lordschip may have consideratioun of this rentall becaus [the] viccerage payis not conforme to use and wont.'

Calculation of third: In the whole, £160; third thereof, £53 6s 8d.

'But prejudice of the rycht rentall.'

DREGHORN, VICARAGE OF,[36] (EUL, fo 65r)

'Memorandum, the viccerage of Dreghorne within Cuninghame, the ane half thairof set to Mr James Broun, burges of Irvyne, be Mr Andro Layng, viccar thairof, for fyfty merkis mony yeirlie and the uther half in the said viccaris handis extending to uther' 50 merks money extending in the whole to 100 merks money yearly, 'of the quhilk mon be deducit and payit to the viccer pensionare or curratte' £20 yearly and to the archbischop of Glasgow[37] 'for procurage and sinodage' £4 'and sua restis' 64 merks money, 'the nopayment of corspresentis, umest clothis and Paschfynis beand unteane of nor deducis quhilk was the thrid pairt of the proffeit of the said viccerage in yeiris bygain, referand to your lordschippis discretioun quhat ye will pleis to deduce for thair nopayment.'

Signature: 'Mr Andro Layng, viccar foirsaid.'

Calculation of third: In the whole, £66 13s 4d; third thereof, £22 4s 5⅓d.

ARDROSSAN, VICARAGE OF, (EUL, fo 65r)

Rental of the vicarage of Ardrossane

'I, William Portarfeild, viccar of Ardrossane, grant and testifies my viccarage to be of yeirlie proffeit,' 2 c. 'as pensioun meill, togidder with' 24s 'mony of few fearme for my kirkland. As to uther small teindis usit to be payit they ar of small valour and of quantitie uncertan. In witnes heirof I have subscrivit thir presentis with my hand at Glesgow the' 15 January 1551/2 [sic].

Signature: 'Williame Portirfeild, with my hand.'

Calculation of third: Meal, 2 c.; third thereof, 10 b. 2 f. 2 p. '3 p. 3 part' [recte, 2⅔ p.]. Money, 24s; third thereof, 8s.

---

[36] Cf EUL, fos 65x, 65y (see below, pp. 582 and n. 38, 583).
[37] James Betoun was archbishop of Glasgow, 1550 - 1570, Watt, *Fasti*, p. 149.

IRVINE, VICARAGE OF, (EUL, fo 65x[38])

Rental of the vicarage of Irvyne

'Memorandum, that the viccerage of Irvyn quhen it was sett in assedatioun as it hes beine thir fourty yeiris bypast quhen all thingis and oblatiounis and corspresendis was weill answerit for', 40 merks 'be yeir, and this we testifie treulie befoir God be our handwritt writtin and subscryvit with our hand at Irvyn', 3 March 1561/2.

Signature: 'Ita est Thomas Andreas, viccarius de Irvyn, manu propria.'

Calculation of third: In the whole, £26 13s 4d; third thereof, £8 17s $9\frac{1}{3}$d.

DREGHORN, VICARAGE OF,[39] (EUL, fo 65x[40])

[*In margin*, 'Aliter givin in Vicarage of Dreghorn.']

Rental of the vicarage of Dreghorne within Cuninghame.

The 'toun' of Dreghorne yearly of teind meal, 2 c. Henry Speir and Janet Auld in the 'toun' of Dreghorne for the fermes of the kirkland, 8 b. bere. Patrick Mowat for his kirkland, 6 b. meal. For the Cranschaw Myln yearly, 2 b. meal. The yard of Warkwik Hill yearly, 1 b. meal. Patoun Small yearly for the yards and kirkland 1 b. bere. Patrick Mowatis kirkland 'ilk thre yeir of grassum', 40s money. 'Item, for the samin yeirlie', 2 dozen capons. The kirkland at the kirk, 12 dozen capons. The rest of the teinds pertaining to the said vicarage 'callit the alterage' were set yearly in assedation for 40 merks. Total victual, 73 b. meal;[41] 19 b. bere.[42]

STRAITON, VICARAGE OF,[43] (EUL, fo 65y[44])

The vicarage of Straitoun pertaining to Mr Andrew Mongummry extends yearly to £36.

---

[38] A discrepancy in the numbering of the foliation occurs at this point.
[39] Cf EUL, fos 65r, 65y (see above, p. 581; below, p. 583).
[40] See above, n. 38.
[41] The correct calculation appears to be 41 b.
[42] The correct calculation appears to be 9 b.
[43] Cf EUL, fo 56r (see above, p. 569).
[44] See above, n. 38.

DREGHORN, VICARAGE OF,[45] (EUL, fo 65y[46])

'Memorandum, in victuall beir, meill and ferme teind', 80 b. 'viz., within the toun of Dreghorne of teind quhilk extendis to' 24 merks 'land and ilk pund land thairof payand of teind meill', 4 b.

James Patoun for a 20s land, 4 b. meal. William Galt for a 20s land, 4 b. meal. Margaret Ros for a 20s land, 4 b. meal. Alexander Galt for 40s land, 8 b. meal. John Speir, 20s land, 4 b. meal. Janet Bogart for 20s land, 4 b. meal. John Rutherfuird, 20s land, 4 b. meal. John Caldwell, 40s 30d land paying of teind meal, 8 b. 2 f. George Brydein for 1 merk land, 2 b. 10 p. 1 f. Adam Galt younger, 24s 4d land paying 4 b. 12 p. meal. Nicol Galt for 20s land, 4 b. meal 'x pectis ij xij pectis'. John Galt in Foirsyd, £3 land paying 11 b. 2 f. 2 p. 'ij xij pectis'. Janet Auld, 15s land paying yearly 3 f. meal.

'Gleib kirkland': Janet Auld of ferme, 4 b. bere and 3 capons. Elizabeth Reid, 'relict of Henrie Speir', for the same, 4 b. bere of ferme and 3 capons.

Kirkland feued by Charles Mowat paying yearly of ferme meal, 6 b.; the kirkland of money, 40s; of poultry, 12 capons, 12 hens; 'and ilk yeir of grassum', 40s. The Warwikhill yards yearly, 1 b. bere. 'Item, the gleib, kirkland, yardis in the toun of Dreghorn', 1 b. bere. The Cramschaw Myln yearly of teind meal, 2 b. meal. 'Item, the rest of the small teindis perteining to the viccerage war aftymis and comounlie reknit to of befoir to' 40 merks, 'and now estemit to' £20.

---

[45] Cf EUL, fos 65r, 65x (see above, pp. 581, 582).
[46] See above, p. 582, n. 38.

Wigtown

GALLOWAY, BISHOPRIC OF,[1] (EUL, fos 65v-68v)

'Vide aliud heirefter uptane into Pat[rik] Goveanis hand.'[2]

Rental of the bishopric of Galloway and abbacy of Tungland. [*In margin*, 'aliter Galloway'.][3]

Nethir Bartcappill, Alexander Gordoun of Troquhoun [*i.e.* Troquhain] tacksman, in mail in the year, £7 10s; the teinds, £20; 'the gersum beir and bollis' [*recte*, the grassum bere, 10 b., *fo 69r*]; 1 b. multure bere; 5 b. multure meal. Over Bartcapill, David Arnot 'sellar', in mail in the year, £6; the teinds, £15; 'the gersum beir and bollis' [*recte*, the grassum bere, 10 b., *fo 69r*]; 1 b. multure bere; 5 b. multure meal. Kirkconnell, Roger Gordoun tenant, in mail in the year, £3; the teinds, £8; 2 b. multure meal. Larmanoch, Roger Gordoun tenant, in mail in the year, £13 6s 8d; the teinds, £13 6s 8d.[4] Beoch, Roger Gordoun in Slograry[5] tacksman, in mail in the year, 49s 4d; the teinds, £5.[6] Over Calquha, Mr Roger Martyn, tenant, in mail in the year, £10 13s 4d;[7] the teinds, £20; the grassum bere, 10 b.; 100 loads of peats; 1 b. multure bere; 6 b. multure meal. Nethir Calquha, Robert Aschman,[8] tenant to the half of the same, in mail in the year, £5; the teind, £10; the grassum bere, 5 b.; 'tua greyt pectis multir beir'; 3 b. multure meal; 60 loads of peats. John Lachlisoun, for a quarter of the same, in mail in the year, 50s; the teind, £5; 2 b. 2 'gryt pectis grassum beir'; 1 'gryt pect multir beir'; 1 b. 2 'gryt pectis multir maill'; 30 loads of peats. Ninian Lachlisoun for a quarter of the same, in mail in the year, 50s; the teind, £5; 2 b. 2 'grytt pectis' grassum bere; 1 'gryt pect' multure bere; 1 'gryt boll' 2 'grytt pectis' multure meal; 30 loads of

---

[1] Cf EUL, fos 69r, 86r (see below, pp. 589, 606).
[2] This sentence has been inserted in another hand.
[3] The following paragraph is headed 'under Cree' on fo 69r (see below, p. 589).
[4] Included at this point on fo 69r, 'the multure, 50s' (see below, p. 589).
[5] Cf 'Slogarie', *RMS*, v, no 1401.
[6] Included at this point on fo 69r, 'the multure, 20s' (see below, p. 589).
[7] Included at this point on fo 69r, 'the Stalling croftis, 2 merks' (see below, p. 589).
[8] Cf 'Aschennen of Culquha', *RMS*, v, no 782.

peats. Balanan, John Aschnanen[9] of Duniop, tacksman to the half of the same, in mail in the year, £5; the teind, £10; 5 b. grassum bere; 5 b. multure meal; 60 loads of peats. Henry Arnott, tacksman to a quarter of the same, in mail in the year, 50s; the teind, £5; the grassum bere, 23 b. 2 'gryt pectis'; 2 b. 2 'gryt pectis' multure meal; 30 loads of peats. Roger Gordoun in Kirkconnell, tenant to a quarter of the same, in mail in the year, 50s; the teinds, £5; 2 b. 1 'gryt pect' grassum bere; 2 b. 2 'gryt pectis' multure meal; 30 loads of peats. Over Bancrosche [Over Balmerosche, *fo 69v*; *i.e.* Barncrosh], James Aschman, tacksman to a half of the same, in mail in the year, 8 merks; the teinds, £10; 5 b. grassum bere; 6 b. multure meal; 2 'gryt pectis' multure bere; 60 loads of peats. Edward Carnis, tacksman to a quarter of the same, in mail in the year, 4 merks; the teind, £5; 2 b. 2 'gryt pectis' grassum bere; 1 'gryt pect' multure bere; 1 b. 2 'grytt pectis' multure meal; 30 loads of peats. Thomas Maclellan, tacksman to a quarter of the same, in mail in the year, 4 merks; the teinds, £5; 2 b. 2 'gryt pectis' grassum bere; 1 'gryt pectis' multure bere; 1 b. 2 'gryt pectis' multure meall; 30 loads 'of meill' [*recte*, peats]. Nether Bancrosche [Nether Balmerosche, *fo 69v*; *i.e.* Barncrosh], Margaret Crechtoun, lady Lochinvar,[10] tackswoman, in the year of mail, £5; the teinds, £15; 1 b. 2 'gryt pectis' multure bere; 6 b. multure meal. Duniop, John Aschman feuar, in mail in the year, £12; the teinds, £10; 4 b. multure meal. Lytill Park, John Gordoun of Lochinvar, tacksman, in mail in the year, 4 merks; the teinds, £7 10s; 5 b. grassum bere; 1 b. multure bere; 5 b. multure meal; 60 loads of peats. Argrenan, John Gordoun of Lochinvar, tacksman, in mail in the year, £4 13s 4d; the teinds, £7 10s; 1 b. multure bere; 3 b. multure meal; 60 loads of peats.[11]

The mails of the Clachan: Henry Ansterthrope [Anthrotheris, *fo 69v*; Anstruthir, *fo 86r*], Few croftis in mail in the year £3 16d. The Walkmyln set to Andrew McKerry for £10. The Cuikis croft [Kirkis croft, *fo 69v*] set to William Muirheid for 20s. The Lymkill boat and crofts of the same set to John MacGuffok[12] for 40s. The fishing of the Water of Dee set to George Anstrother for £16 and 160 salmon. The Corn Myln set to George Anstruther with the crofts pertaining thereto for £20. The multure bere of the Clachan extends to 1 b. 3 'gryt pectis' 3 'auchlettis'. The Mainis of Tungland 'ten pund land', £20.

Total silver within the barony of Tungland of mails and teinds, £351 7s 4d. Total grassum and multure bere of the barony of Tungland, [*blank*]. Total multure meal of the barony of Tungland, 3 c. 13 b.[13]

---

[9] Cf 'Aschennane of Dunjop', *RMS*, v, no 501.
[10] In 1520 Margaret Crichton, daughter of Robert Crichton of Kirkpatrick, married James Gordon of Lochinvar, who was killed at Pinkie in 1547. *Scots Peerage*, v, pp. 106-108.
[11] The details here vary from those below, fo 69r-v (see below, p. 589).
[12] Cf 'Macguffog', Black, *Surnames*, p. 507.
[13] In these calculations 1 great peck equals 1 firlot, though the 'gryt boll' paid by Ninian Lachlisoun still measures 1 boll; an auchlett is $\frac{1}{8}$ of a boll.

The barony of Kirkcryst: Nuntoun, James McKlellane feuar, in mail in the year, £20. New Myln, James McKlellane feuar, in mail in the year, £20. Newtoun, Roger Gordoun in Kirkconnell tacksman of a half of the same, in mail in the year, £3. William Schaw tacksman to the other half of the same, in mail in the year, £3. Kirkoch, James McClellane feuar, in mail in the year, 40s. The augmentation of James McClellan feus yearly, £5. Cudoig [Kudag, *fo 70r*], John Gordoun of Lochinvar tacksman, in mail in the year, £20. James McClellane, feuar to the fishing of Kirkcryst, paying yearly 108 salmon. Kirkcryst in the year, £10. Total of the mails of Kirkcryst, £88.[14]

Kirks set for money beneath Cree: The vicarage of Tungland set to David Arnott for £40; 'thairof deducit for Paschfynis, capounis [*recte*, corpse] presentis and umest clothis', £10; so rests, £30. The parsonage and vicarage of Troweir [*i.e.* Troqueer] set to John Maxwell of the Hills for £160; 'thairof to be deducit for Paschfynis, corpspresentis and umestclothis', £26 13s 4d; so rests, £93 6s 8d.[15] The parsonage and vicarage of Girthtoun [*i.e.* Girthon] set to John Gordoun of Lochinvar for £113 6s 8d; 'thairof to be deducit for Paschfynis, corpspresentis and umest clothis', £30; so rests £83 6s 8d. The parsonage and vicarage of Mingaff to John Gordoun of Lochinvar for £100; 'deducit thairof for Paschfynis, corpis presentis and umest clothis', £20; so rests, £80. Total silver set in assedation under Cree extends to £287 13s 4d.[16]

The kirk of Sannyk set for victual: Norman Muirheid for Over Sannik, 6 b. bere; 7 b. meal. Nether Sannik, the half of the same in James McClellanis hands ridden to 3 b. 2 'grit pectis' bere; 4 b. meal. The other half of the same in the laird of Lochinvaris hands ridden to 2 b. bere; 4 b. meal. Esserttoun in the laird of Lochinvaris hands ridden to 4 b. 3 'gritt pectis' bere; 5 b. meal. Meikill Dunrod in George Schartcrowis hand ridden to 11 b. bere; 16 b. meal. George Mure 'for the Myln croft', 1 'auchleit beir'. Total of the teind bere of the parish of Sannik, 1 c. 12 b. 1 'grit pect' 1 'auchtles'. Total of the teind meal of the parish of Sannik, 2 c. 4 b.

Rental of the bishopric of Galloway and abbacy of Tungland above Crie: The Mans, £5 land in Patrick Durlis hand paying yearly £10. The Clary 5 merkland; the Grange, 5 merkland; the Nether Bar 5 merkland; Barlaching 5 merkland, extending to 20 merkland in the laird of Garleis hands, £40. Balsallocht [*i.e.* Barsalloch] 5 merkland in the laird of Lachinvaris hands, £13 6s 8d. Casnestik and Balquhilly in Mr John Stewartis hands, £6 13s 4d. Barquharrand 5 merkland in Allan McKrisknes hand, £6 13s 4d. The Over Bar 5 merkland in the laird of Largis hands, £6 13s 4d. Baltarssin and Barvernocht 8 merkland 'in the landis [*recte*, laird] of Barinbarochis handis', £11 3s 4d. The Myln of Peninghame, 24 b.

---

[14] The correct calculation appears to be £83.
[15] The text reads 'iiij xiii li. vj s. viij d.'; the correct calculation appears to be £133 6s 8d.
[16] The correct calculation from the figures cited in the text appears to be £286 13s 4d.

meal, 8 b. bere. Total of mails of the barony of Peninghame, £94 10s. Total in malt [*recte*, meal], 1 c. 8 b. Total in bere, 8 b.

The barony of Glasnycht: Nethir Glasnycht and Over Glasnycht 'and multir extending to' 12 merkland in William Gordoun of Threthrevis[17] hands paying yearly £16. The Grange 4 merkland, Barchrachand 4 merkland, the Cruif 5 merkland in the hands of Alexander Gordoun in feu, £25 6s 8d. Kirkcryst, Kilyenar [*i.e.* Killiemore] and Bartrostan, 7 merkland in Robert Gordounis hand, £10 6s 8d. Total, £50 13s 4d.[18]

The barony of Quhiterne: Bischoptoun and Balaschweir in the young laird of Garleis hands, 10 merkland paying yearly £18. Respyn and Craig in Robert Maxwellis hands, 10 merkland, £25s. Total, £43.

The 'Mans of Insch'[19] called Cullarby [*i.e.* Cullurpattie] £5 land, the 40s land of Innermessan, the 10s land of Kinaroche [*i.e.* Kirminnoch], all in my lord of Cassillis[20] hands paying therefore yearly £20. In Ninian Boydis hands, £5 land paying therefore yearly £10 10s. In the laird of Garthlandis hands, £10 land paying yearly £18. The Myln of Insch paying yearly £16. Total, £69 10s.

Kirks above Cree set for money: The parsonage and vicarage of the kirks of Insch and Leswatt set to the earl of Cassillis for 260 merks; 'deducit thairof for the corpspresentis, Paschfynis and umest clothis of the viccerage of the Insch', £20; so rests, £153 6s 8d. Total, £153 6s 8d.

Total of the whole silver of the bishopric of Galoway and abbacy of Tungland and of the kirks set for silver, 'the deductiounis beand deducit for corps presentis, umest clothis and Paschfynis extendis to' £1,138 0s 8d. 'Summa of the haill beir abone and beneth Cree extendis to' 6 c. 15 b. 3 'gryt pectis'. 'Summa of the haill maillis [*recte*, meill] abone and beneth Cree', 8 c. 10 b.[21]

Signature: 'Mr Alexander, bischop of Galoway.'[22]

---

[17] I.e. High Threave, Middle Threave and Low Threave.
[18] The correct calculation appears to be £51 13s 4d.
[19] The following entries appear under the heading of 'the barony of Insch' on fo 71r; 'barony of Innermessane' on fo 86v (see below, pp. 590, 607).
[20] I.e. Gilbert Kennedy, 4th earl of Cassillis, *Scots Peerage*, ii, pp. 471-473.
[21] The correct calculation appears to be 7 c. 9 b.
[22] Alexander Gordon, bishop of Galloway, 1559 - 1575, Watt, *Fasti*, p. 132.

GALLOWAY, BISHOPRIC OF,[23] (EUL, fos 69r-73r)

Rental of the bishopric of Galoway and abbey of Tungland under Cree: Nether Bartcapill in mail in the year, £7 10s; the teinds, £20; the grassum bere, 10 b.; 1 b. multure bere; 5 b. multure meal. Over Bartcappill in mail in the year, £6; the teinds, £15; the grassum bere, 10 b.; 1 b. multure bere; 5 b. multure meal. Kirkconnell in mail in the year, £3; the teinds, £16 [£8, *fo 65v*]; 2 b. multure meal. Larmanoch in mail in the year, 20 merks; the teinds, 20 merks; the multure, 50s. Beech [*i.e.* Beoch] in mail in the year, 49s 4d; the teinds, £5; the multure, 20s. Over Calquha in mail in the year, 16 merks; the Stalling croftis, 2 merks; the teinds £20; the grassum bere, 10 b.; 1 b. multure bere; 6 b. multure meal; 100 loads of peats, 'payand be the byd ij d. for ilk laid, ij gallainis of aill, ane gryt pect of meill and tua stain of cheis, iiij s., iiij dos of pultrie, iiij geis and iiij penneis allowit for ilk guis and siklyk of all them that payis peittis and geis.' Nether Calquha in mail in the year, £5; the teinds, £20; the grassum bere, 10 b.; 1 b. multure bere; 6 b. multure meal; 100 loads of peats; 4 dozen poultry; 4 geese. Ballanan in mail in the year, £5; the teinds, £20; the grassum bere, 10 b.; 1 b. multure bere; 6 b. multure meal; 100 loads of peats; 4 dozen poultry; 4 geese. Over Balmerosche [Over Bancrosche, *fo 66r*; *i.e.* Barncrosh] in mail in the year, 16 merks; the teinds, £20; the grassum bere, 10 b.; 1 b. meal; ½ b. multure bere; 6 b. multure meal; 100 loads of peats; 4 dozen poultry; 4 geese. Nether Balmerosche [Nether Bancrosche, *fo 66r*; *i.e.* Barncrosh] in mail in the year, £5; the teinds, £15; 1½ b. multure bere; 6 b. multure meal. Dunyop in mail in the year, £12; the teinds, £10; 4 b. multure meal. Lytill Park in mail in the year, 4 merks; the teinds, £7 10s; 5 b. grassum bere; 1 b. multure bere; 5 b. multure meal; 60 loads of peats. Argrenan in mail in the year, £4 13s 4d; the teinds, £7 10s; 1 b. multure bere; [*in margin*, 4 b. 'multir meill']; 60 loads of peats.[24]

Henry Anthrotheris [Ansterthrope, *fo 66v*; Anstruthir, *fo 86r*], Few croftis in mail in the year, £3 16d.[25] The Walkmyln in the year, £10. The Kirkis croft [Cuikis croft, *fo 66v*] in the year, 20s. The Lymekill boat in the year, 40s. 'The dry fisch yard' in the year, 40s. The fishing of the Water of Dee in the year, £16; 160 salmon. The Corne Myln in the year, £20. The Mainis of Tungland £10 land, £20. The kirk of Troqueir in the year, £160. The kirk of Girthtoun [*i.e.* Girthon] in the year, £113 6s 8d. The kirk of Sennikin [*i.e.* Senwick] in the year, 'quhylis mair victuall, quhylis les, conforme to the rait, bot will be ilk yeir worth' 40 b. meal [malt, *next paragraph*] 'with the better' and 30 b. bere 'with the better.' Newtoun in the year, £6. Kirkcryst in the year, £11. Nuntoun in the year, £20. Newmyln in the year, £20. Bischoptoun in the year, £20. Kirkeich in the year, £3. Angaschill, 40s. The augmentation of James McClellanis feu, £5. Kudag

---

[23] Cf EUL, fos 65v, 86r (see above, p. 585; below, p. 606).
[24] The details here vary from those above, on fos 65v-66v (see above, pp. 585-586).
[25] This paragraph is headed 'The maillis of the Clachan' on fo 66v (see above, p. 586).

[Cudoig, *fo 67r*] in the year, £20. The fishing of Kirkcryst in the year, 168 salmon [108, *fos 67r, 73r*].

'Summa of the haill mony under Cree', £697 10s 8d.[26] Total of grassum bere of the barony of Tungland, 65 b. Total of multure bere in the barony of Tungland and of the crofts of Insch Clachan extends to 10 b. The teind bere of Sannik by year 'conforme to the raid', 30 b. 'The haill sowme of beir', 105 b. The teind malt [meal, *preceding paragraph*] of Sannik in the year, 40 b. The teind meal of the crofts of Clachin 'and the delwingis' extends to 8 b. Total meal, 103 b.[27]

The rental above Crie

The barony of Peninghame: The Barthland in the year, £10. The Clarie in the year, 10 merks. The Grange in the year, £20. Balsallosch [*i.e.* Barsalloch] in the year, 30 merks. Balterssan in the year, £4 10s. The Over Bar in the year, 10 merks. The Nether Bar in the year, 10 merks. Barlachlelan in the year, 10 merks. Barvennen in the year, 10 merks. The Myln of the same in the year, 24 b. meal; 8 b. malt. Total, £81 3s 4d[28] 'by meill and malt.'

The barony of Quhitterne: The Bischoptoun and Balyiequhair in the year, £18. Respine and Craig in the year, £25. Balcry in the year, 20 merks. Total, £56 6s 8d.

The barony of Insch:[29] 'The schireff[30] in the yeir', £10 10s. The laird of Garthland in the year, £10 4s. Quentin McDowell in the year, £20. Cullartby [*i.e.* Cullurpattie] in the year, £20. Ninian Boyd in the year, £4 6s 8d. Balyett in the year, 40s. Kirmenoch [*i.e.* Kirminnoch] in the year, 40s. The Myln of the same in the year, £16 4s. Total, £74 6s 8d.[31]

The Glasnych: The barony of Glasnycht in the year, £40. Alexander Gordounis augmentation, £4.

The kirk of Monygaff: The kirk of Monygaff in the year, £100. The kirks of the Insche and Leswane [*i.e.* Leswalt] in the year, £173 6s 8d. Total, £317 6s 8d.

Total of silver above Crie £529 3s 4d.

Total of the whole silver above and under Crie, £1,226 14s.

---

[26] The correct calculation appears to be £743 17s 4d.
[27] The correct total appears to be 96 b. and 1 great peck.
[28] The correct calculation appears to be £87 16s 8d.
[29] The following entries begin with 'the Mans of Insch' on fo 68v (see above, p. 588), and appear under the heading of 'barony of Innermessane' on fo 86v (see below, p. 607).
[30] Patrick Agnew was sheriff of Wigtown at the Reformation, *RMS*, iv, no 1492.
[31] The correct calculation appears to be £85 4s 8d.

Deduction of silver: To the master [of] Maxwell[32] in pension, £400. To the earl of Cassillis bailie fee, £26 13s 4d. To the lord Maxwellis[33] bailie fee,[34] £10. To Simon McQuharrie 'for his portioun givin be bischop Dury[35] under the comoun seill of the place', £40. To William Dury and his wife 'or the langest levar of them tua for his portioun yeirlie givin be the said bischope under the comoun seill of the place', £40. To 'vij brethir of the place that had haill portioun to ilk brother for his kiching silver', £9 3s; 'and for his habit silver', 4 merks, 'extending in the haill to' £102 13s 4d.[36] 'Item, to iiij noveissis, tua haill portiounis of silver', £23 12s 8d. For casting of peats above Crie, £10. For casting of the convent's peats of Tungland, £10. To the abbot of Halyrudhous[37] for an annualrent of Tungland, £10. 'To the Seit for contributioun', £22 8s. To Mr David Borthuik, procurator, £10. To the procurator's fee in Kirkcudbrycht, £4. To the procurator's fee in Wigtoun, £4. To the chamberlain's fee, 'his hors and man ordinar', £40. Total deduction of silver, £753 7s 4d.

Deduction of meal: 'Item, vij brether and iiij noveissis maid ix portiounis, ilk portioun ix b. meill, summa of the haill' 81 b. 'Item, for thair grottis', 4 b. 'Item, to thair baxter', 12 b. 'Item, to thair brouster', 12 b. 'Item, to thair portar', 4 b. 'Item, to thair cuik', 6 b. 'Item, to the skleater', 4 b. 'Item, to the officiar', 4 b. 'Summa in meill', 127 b.[38] 'Sic eque.'

Deduction of bere: 'Item, for ix portiounis, ilk portioun viij b.', 72 b. 'Sua restis of' 105 b. 'to the possessour', 33 b.

'Item, of the haill bischoprik of Galloway and abacie of Tungland, the foirsaidis beand deducit, restis in money' £473 6s 8d; 'and in maill nathing'; in bere, 14 b.; in malt, 8 b.; in fish, 160 salmon. The fishing of Kirkcryst, 160 salmon[39] 'unpayit.'

Signature: 'Alexander, bischop of Galloway, with my hand.'

Bishopric of Galloway and abbacy of Tungland

Calculation of third: Money, £1,226 14s; third thereof, £408 18s. Bere, 8 c. 6 b.;[40] third thereof, 2 c. 13 b. Meal, 10 c. 7 b.; third thereof, 3 c. 7 b. 2 f. 2⅔ p.

---

[32] With the death in 1552 of Robert, 6th lord Maxwell, Robert, his eldest son, became 7th lord but died aged four in 1555; he was succeeded by his brother, John, 8th lord Maxwell, a posthumous child, born in 1553, married 1572, and died 1593. *Scots Peerage*, vi, pp. 481-483.
[33] See above, n. 32.
[34] The text reads 'for', probably a scribal slip.
[35] Andrew Dury, bishop of Galloway, 1541 - 1558, Watt, *Fasti*, p. 132.
[36] The correct calculation appears to be £82 14s 4d.
[37] I.e. Robert Stewart, commendator of Holyroodhouse, Cowan and Easson, *Medieval Religious Houses*, p. 90.
[38] Cf above, p. 590, n. 27.
[39] The fishing of Kirkchrist paid 108 salmon, fos 67r, 73r (see above, p. 587; below, p. 592).
[40] The total bere given above is 105 b., i.e. 6 c. 9 b.

[2 p. 'half thairof'; *recte*, 'half pect 3 thairof', i.e. ⅔ p.]. Malt, 8 b.; third thereof, 2 b. 2⅔ p. [2 p. '3 parte half pecte'; *recte*, 'half pect 3 thairof', i.e. ⅔ p.].⁴¹ Salmon, 268; third thereof, 89⅓.

'Nota, the peitis, the geis, the cheis, the pultrie ar sua givin up that they cannot be understand. Tak this thrid without ony deductioun of the expenssis.'

Assignation of Galloway and Tungland:
    Third of money, £408 18s. Take the barony of Peninghame for £81 3s 4d; take the barony of Quhitarne, 'conteinand' Bischoptoun, Respyn, Craig and Bartry [*i.e.* Balcray], for £56 6s 8d; the barony of Insche for £74 6s 8d; the Glasnyth for £44; the kirk of Insche and Leswaid [*i.e.* Leswalt] for £173 6s 8d. 'Gif in' £20 5s 4d.
    Third of bere, 2 c. 13 b. Take the kirk of Sannik for 30 b.; the multure bere of Tungland and c[r]oftis of the Clachan, 10 b.; and 5 b. out of Nether Bartcapill.
    Third of meal, 3 c. 7 b. 2 f. 2⅔ p. Take the kirk of Sannik for 2 c. 8 b.; Over Balmorsch for 6 b. multure meal; Dunyop for 4 b. multure meal. 'Gif in' 1 f. 1⅔ p.
    Third of malt, 2 b. 2⅔ p. Out of the Myln of Peninghame giving 8 b.
    Third of salmon, 89⅓. Take the fishing of Kirkcryst for 108 salmon. 'Gif in' 18⅓.⁴²

WHITHORN, PRIORY OF,   (EUL, fos 73v-77v)

Rental of the 'abbay' of Quhittern of lands, fermes and land mails with kirks set in assedation.

Quhitorne parish: Bischeptoun 'wast in yung Garleis handis', 16 b. meal; 6 b. bere. Balyequhair, 12 b. meal; 4 b. bere, in Andrew McKenlaris hands. Wiggingairne occupied by Thomas Gibsoun, 6 b. meal; 2 b. bere. Meidwig in Walter Douglas hands, 6 b. meal; 2 b. bere. Meikillwig in the Ladie Wigis' 6 b. meal; 2 b. bere. Outtoun Gallous occupied by John Barbour, 6 b. meal; 2 b. bere. Lytill Guttoun occupied by John McCurreir, 5 b. meal; 2 b. bere. Outoun Chapell occupied by Duncan Murray, 5 b. meal; 2 b. bere. Brachtoun Wall, 18 b. meal; 6 b. bere. The Skyrcht, 10 b. meal; 4 b. bere. Stannuk Carbot,⁴³ 6 b. meal; 2 b. bere. Outoun Burges, 8 b. meal; 2 b. bere. Drumanstann [*i.e.* Drummaston], 5 b. meal; 1 b. bere. The croft of Drumanstoun, 2 f. meal. Olbrek,⁴⁴ 9 b. meal; 3 b. bere. [*In*

---

⁴¹ 2 f. has been omitted from this calculation.
⁴² The correct calculation is 18⅔.
⁴³ 'Stennok-Corbak', 'Stennok-Corbat', *RMS*, iv, no 2823.
⁴⁴ Cf 'Howreik', fo 80r; 'Yowrek', fo 80v; 'Bowraik', fo 81r (see below, pp. 598-599).

margin, Chappell] Ahbirquharne [*i.e.* Chapelheron], 6 b. meal; 2 b. bere. The Ile, 11 b. meal; 3 b. bere. Portcarrik, 5 b. meal; 1 b. bere. Nether Ersik, 11 b. meal; 3 b. bere. The Crage, 18 b. meal. Respyne, 13 b. meal; 5 b. bere. Cotreoch, 13 b. meal; 6 b. bere.

Glastertoun: The Reuchtwen, 10 b. meal; 5 b. bere. Lytill Arrow, 5 b. meal; 2 b. bere. Schalowchtbroun, 5 b. meal; 2 b. bere, 'in auld Garleis handis wast.' Craigdon, 13 b. meal; 6 b. bere. Craigemain [*i.e.* Craiglemine], 8 b. meal; 3 b. bere. The Hillis, 8 b. meal; 2 b. bere, 'in auld Garleis handis wast.' Kiddisdaill, 20 b. meal; 7 b. bere. Over Ersslis [*i.e.* High Ersock], 10 b. meal; 3 b. bere. The Greinand, 7 b. meal. Lochtoun and Drumra, 8 b. meal. Carltoun, 34 b. meal; 16 b. bere. Careindunie [*i.e.* Cairndoon], 20 b. meal; 10 b. bere.

Kirkmaidin parish: Blairballie, 14 b. meal; 4 b. bere. Balcraig, 20 b. meal; 6 b. bere. Barwinnok, 6 b. meal; 2 b. bere. Drumaddie, 12 b. meal. The Knok, 16 b. bere.

Sorbie parish: The Blair, 6 b. meal; 2 b. bere. The Ereychtlyoun, 1 b. [meal]; 1 b. bere. The Mylne croft of Ralveinstoun [*i.e.* Ravenstone], 2 f. bere. Sanct Johnis croft, 1 f. bere. [*In margin*, 'Barliezow'] Barkdow [*i.e.* Barledziew], 5 b. meal; 2 b. bere. Culmoker, 6 b. meal. Donelstoun, 16 b. meal. Barinoking [*i.e* Barvernochan], 8 b. meal. The Clounsche [*i.e.* Claunch], 7 b. meal; 3 b. bere. The Corver, 7 b. meal; 2 b. bere. The Quhythillis, 13 b. meal; 3 b. bere.

Crugiltoun parish: [*in margin*, 'Quhitorne'] Crugiltoun Cawnis,[45] 15 b. meal; 5 b. bere. Balteir, 4 b. meal; 2 b. bere. Crugiltoun Castell, 80 b. meal; 30 b. bere. Apilbie, 30 b. meal. Prestrie, 60 b. meal; 32 b. bere.

[*In margin*, 'Mylnes'] 'The Milwe [*i.e.* Mill] fearmes': The Milwe [*i.e.* Mill] of Apillbie, 21 b. meal; 10 b. bere. The Mylne of Quhythillis, 52 b. meal. The Mylne of Poltoun, 40 b. meal; 6 b. bere, 'in Mochrowmis handis wast'. The Myln of Portcarrik, 52 b. meal. The Mylne croft of Poltoun, 4 b. bere.

[*In margin*, 'Omittit, the paroschin[46] of Erssillis [Erssips, *fo 76r*; *i.e.* Ersock]', 6 b. meal. The Walkmyln of Powtoun, £3 6s 8d. The Mill of Lytill Arisch, £5 6s 8d.[47]

The kirks set in assedation: The kirk of Mowchtroum [*i.e.* Mochrum] yearly, 340 merks at three terms, namely, 'Martymes,[48] Yuile[49] and Pasche'[50] set to Robert

---

[45] Now Kevans, NX4642.
[46] 'Paroschin' should read 'Mylne'; cf fo 76r (see below, p. 595).
[47] These details differ from those in a similar note on fo 76r; cf below, p. 595, n. 54.
[48] I.e. Martinmas, 11 November.
[49] I.e. 25 December.
[50] I.e. Easter.

Gordoun. Kirkmichaell in Karrik set to Janet Mure for £100 yearly 'at Martymes and Midsomer'. Borg set to William McClellane, Bawangaid,[51] for £20 'payit at Midsumer'. Gelstoun set to the laird of Lochinvar for £30 'payit at Sanct Lawrence day'.[52] Kirkdaill set to Alexander Mure of Kers for £24. Toscartoun and Clanschanacht set to the laird of Kilarstair for £80 at two terms, 'Androsmes[53] and Midsumer'. Kirkmarrow in the Ile of Man 'pro deleto.' [*In margin*, 'Nota, omittit kirk and' £100 'in Mr Thomas McClellanis handis.']

Land mails: The 25 merkland of Kintyre in Archibald McConnis hands for 25 merk mail, 'inde nihill'. Powtoun £10 land paying £32. Ardmillane £20 [land] yearly paying, £63. Deswintoun, £5 land paying £15 yearly. Crugiltoun 10 merkland paying £16 yearly. Meikillwig 10 merkland paying yearly £20. Outoun Burges 5 merkland paying yearly 20 merks. Quhythillis 5 merkland paying yearly 10 merks. Balsmycht 5 merkland paying yearly £10. The Ile 5 merkland paying yearly £22. Wiggingairne, 4 merkland paying yearly £10. Medwig, 4 merkland paying yearly 10 merks. Barwarnokun [*i.e.* Barvernochan], 4 merkland paying yearly 8 merks. Arles and Culmalzo, 10 merkland paying yearly 20 merks. Lytill Areis, 4 merkland paying yearly 8 merks. Meikill Areis, 5 merkland paying yearly 9 merks. Outoun Gallous, 4 merkland paying yearly 8 merks. Outoun Chappell, 4 merkland paying yearly 10 merks. Lytill Outoun, 4 merkland, yearly £7 [£8, *fo 77r*]. Sedoch and Scamnoch, 5 merkland paying yearly 10 merks. Creachdaw [*i.e.* Craigdhu], 5 merkland paying yearly 10 merks. Craigilmain, 5 merkland paying 20 merks [10 merks, *fo 76r*]. 'Nota heir is omittit out of this rentall the pryouris manis of the toun of Quhittarne occupyit with his awin guidis and plewis extending to a ten pund land.'

'Sett teindis in assedatioun: Meikilwig assedatur Michalj McClelane et Kathrine Spottis pro' 12 merks 'monete.' Balnab set to John Braidfutte for £5. Lytill Scamnocht set to John McCrory, for £6. Dunance set to Thomas McDowell for £10. Sedocht set to Arthur Braidfutte for £5. Balsmycht set to John McCleirlie for £8. Baltray [*i.e.* Balcray] set to 'Fargiltun' [Fergus, *fo 77r*] McDowell for £10. Merwicht set to Thomas McCulloch for 10 merks. Drumarell set to the laird of Arduell for £10. The Lagand set to Mr Robert Stewart for 8 merks. Fisgill set to Michael McCrakin for £10. Meikill Arrow set to the laird of Arduell for £6. Arbrok set to Alexander McKie for 10 merks. The Laroch set to Thomas Dunmell for 12 merks. Skellocht set to Rudolph Peirsoun for £5. Garrorie and Balmaill set to Robert Maxwell for £10. The Mawr set to the laird of Murhycht for 10 merks. Balseir [set to] James McCulocht for £8. Kilstuire set to John Gordoun for £8. Culkaye set to Thomas Diksoun for 8 merks. The Insche set to John Hanna for 10 merks. Inglistoun set to the laird of Sorbie for £10. The

---

[51] Presumably 'Balmangane', *RMS*, vi, no 1249.
[52] I.e. 10 August.
[53] I.e. the feast of St Andrew, 30 November.

Cawtis set to the laird of Blakster for £10. Polmowert set to William Hamiltoun for 6 merks. Powtoun set in the hands of the inhabitants for 16 merks.

The Few crofts lying about the toun of Quhittorne: Duncan McKies 2 crofts paying £3. Francis Murray, 1 croft, 30s. James Cuninghame, 2 crofts, £3. Andrew Dumbar, 1 croft, 26s. Peter McKilweynd, 13s 4d. Catherine Spottis croft, 20s. Neill Adair, 1 croft, 30s. James Addair, 1 croft, 24s [23s, *fo 77v*]. James Telfer, 1 croft, 26s 8d.

'My lord, this is the just rentall of my benefice of the pryorie of Quhittarne, quhairof at your lordschippis discretioun is to be considdirit deducit and tain thairof the conventis sustentatioun quhilk extendis yeirlie to' 900 merks.

'Nota, mair omittit out of this rentall the Mylne of Erssips [Erssillis, *fo 74v*; i.e. Ersock]', 6 b. meal. The Walk Mylne of Poltoun, £3 6s 8d. The Myln of Lytill Areis, £5 6s 8d. The toun and lands of Clamadie, 20 b. meill, 8 b. bere.[54]

The contribution to the Lords of the College of Justice and Session yearly the sum of £42. The lands 'haldin waist be the auld and young lairdis of Garleis and the laird of Mochrum quhilkis extendis yeirlie as your lordschip man considder, to wit,' in meal yearly, 75 b.; and in bere, 18 b. 'This rentall send to me quhilk I beleif trew and subscryvis in the sendaris name.'

Signature: 'Johanes Stevinsoun, precentor Glasguensis, ex mandato et nomine dictie [*recte*, dicti] Malcomii, comendatarij prioratus antedicti.'[55]

Calculation of third: Money, £1,016 3s 4d;[56] third thereof, £338 14s 5½d. Bere, 15 c. 14 b. 3 f.;[57] third thereof, 5 c. 4 b. 3 f. 2⅔ p. Meal, 51 c. 15 b. 2 f.;[58] third thereof, 17 c. 5 b. 2⅔ p.

'Omittit caynis, custumis and all uther dewties.'

9 March 1584/5

'Nota, the kirk of Kirkcandreis [*i.e.* Kirkandrews], Myln of Ersik, Wakmylne of Poltoun, Myln of Lytill Aris, with the toun and landis of Clamadie, quhilk was

---

[54] These details differ from those in the similar note on fo 74v; cf above, p. 593, n. 47.
[55] I.e. 'by order of and in the name of the said Malcolm, commendator of the foresaid priory'. Malcolm Fleming, dean of Dunblane, was prior of the Premonstratensian house at Whithorn in 1566 (this rental is dated 1585). G. Donaldson, 'The Bishops and Priors of Whithorn', *Transactions of the Dumfriesshire and Galloway Natural History and Antiquarian Society*, 3rd ser., xxvii (1950), pp. 127-152, at pp. 146-147.
[56] This total does not tally.
[57] This calculation is correct (omitting the valuation for 'wasted lands').
[58] The correct calculation appears to be 52 c. 12 b. 2 f.

omittit of befoir was givin up be the laird of Barinbarro in name of the pryour of Quhittorne, and ordaine the Chelker to be rentallit with the rest of the fructis of the samin besyd, extending to the particular quantitie underwrittin, viz'., Mylne of Essik, 6 b. meal; Claimadie, 20 b. meal; 8 b. bere; Walkmylne of Poltoun, £3 6s 8d. 'My land of Lytill Abay', £5 6s 8d. The kirk of Kirkcaindelis [*i.e.* Kirkandrews], £100.

Assumption of the 'abbay' of Quhitterne:
Third of money, £338 14s 5½d. Take the teinds of Dinance, Thomas McDuell fermorar, for £10; Fysgill, Michaell McCraken for £10; the teinds of Baltray [*i.e.* Balcray], Fergus McDowell fermorar, for £10; Meikill Arow the laird of Arduell for £6. Take the 'landis maillis and Few croftis undermentionate, viz.,' Poltoun, £32; Ardmillan, £63; Daswintoun, £15; Crugiltoun, £16; Mikillwig, £20; Outtoun Burges, 20 merks; Balsmycht, £10; and the Ile, £22; Wiggingairne, £10; Midwig, 10 merks; Barwarnok, 8 merks; Arles and Culmalzo, 20 merks; Lytill Areis, 8 merks; Meikeill Areis, 9 merks; Outtoun Gallous, 8 merks; Outtoun Chappell, 10 merks; Lytill Outtoun, £8 [£7, *fo 75v*]; Sedacht and Scamnocht, 10 merks; Creocht Dow [*i.e.* Craigdhu], 10 merks; Craigilmain, 10 merks [20 merks, *fo 75v*]. The Few [croftis, *inserted*] lying above the toun of Quhittorne: Duncan McKies, 2 crofts, £3; Francis Murray, 1 croft, 30s; James [Cuninghame, *fo 76r*], 2 crofts, £3; Andrew Dumbar, 1 croft, 26s; Peter McKilweynd, 13s 4d; Catherine Spotis croft, 20s; James Adar, 1 croft, 23s [24s, *fo 76r*]; James Tailfeir 1 croft, 23s 8d [26s 8d, *fo 76r*], extending to £335 3s 4d. 'Tak the rest out of the best and reddiest of the haill priorie.'

Third of bere, 5 c. 4 b. 3 f. 2⅔ p. Third of meal, 17 c. 5 b. 2⅔ p. Take this bere and meal out of the parishes of Kirkmaidin and Crugiltoun, the whole meal thereof, 15 c. 1 b. and the rest thereof out of Quhittorn parish extending [to] 2 c. 4 b. 2⅔ p. and the whole bere out of Crugiltoun giving 7 c. 3 b. 1 f. 1 p. 'Nota, Crugiltoun payis' 3 c. bere 'alenerlie and thairfor tak the rest out of Quhittorne paroschin.'

'Omittit, caponis, henis, gressis entressis.'

'Nota, Kirkmadin and Crugiltoun paroschinis gives' 18 c. 8 b. meal [15 c. 1 b., *preceding paragraph*] 'by and attour Prestrie, Apilbie and myln thairof quhilkis ar worng [*recte*, wrongly] givin up to be of the saidis paroschinis. Nota this 3 is eikit and reformit as followis', third of the bere, 5 c. 7 b. 2 f. 1⅓ p. Take this bere out of Crugiltoun, 3 c. and the rest thereof out of Quhittarne parish. Third of the meal, 17 c. 13 b. 3 f. 1⅓ p. Take this meal out of Kirkmaidin and Crugiltoun parishes giving by year 18 c. 8 b. meal.

WIGTOWN, PARSONAGE AND VICARAGE OF, (EUL, fo 78r-v)

'Copia rentalis rectorie de Wigtoun'

'Item, of silver for the maillis of the landis' £26. 'Item, the teind beir of the akeris of Wigtoun', 16 b. 'Item, of teind meill': The Coitland, 12 b. Crevane, 6 b. Clachran, 10 b. The Braidfeild, 8 b. Tartyis McCulloch, 14 b. Tartus McKer, 7 b. Tartyus Mur,[59] 8 b. The Blakmark,[60] 1 b. 2 f. The Merkbradan, 1 b. The Hillis, 11 b. The Husbandtoun, 8 b. The Wod, 6 b. The Glenturk, 8 b. The Chepeltoun, 6 f. The Carsgoun, 2 b. The Carslie, 5 b. The Burrow muirs,[61] 18 b. The Kirkland teind, 12 merks. The Culquhork, 8 b. The Maidland teind, 8 merks.

Total meal, 135 b. Total bere, 16 b. Total silver, 53 merks.[62]
'Item, of Paschefynis commonlie', 20 merks.

'Item, of this givin to James Woodis, minister', 40 merks. 'Item, to the viccar pentionar, reidar', 30 merks.

Signature: 'Presentit the' 5 February 1561/2 'be me, Mr Johne Wod, in name of Patrik Vaus.'

Calculation of third: Bere, 16 b.; third thereof, 5 b. 1 f. 1⅓ p. Meal, 8 c. 7 b.; third thereof, 2 c. 13 b. Money, £48 13s 4d;[63] third thereof, £16 4s 5⅓d.

MOCHRUM, VICARAGE OF, (EUL, fo 78v)

Rental of the vicarage of Mochrum within the diocese of Galloway 'was ay sett with gleib and kirkland to Johne Ramsay for' £80, 'quhairof thir thrie yeiris bygaine I have not gottin ane penny and haldin be the lard of Mochrum, prayand your lordschippis to considder thir premissis quhilkis ar of veritie and to provyd remeid that I may be answered in tymis to cum.'

Signature: 'Ita est Stevinsoun, C. G.,[64] manu sua.'

Calculation of third: In the whole, £80; third thereof, £26 13s 4d.

---

[59] Cf Torhousekie, NX3756; Torhousemuir, NX3957.
[60] Probably Blackpark, NX3757.
[61] The text here is abbreviated as 'mrs'; Borrow Moss occurs at NX4357.
[62] The correct total appears to be 59 merks.
[63] £48 13s 4d equals 73 merks, which includes the Easter dues.
[64] 'Cantor Glasguensis', i.e. chanter of Glasgow.

SOULSEAT, ABBEY OF, (EUL, fos 79r-81r)

Rental of Salseitt in mails, teinds, multures and fowls.

'In primis' 40 'merk land of auld extent lyand within the baronie of Saullseitt occupyit be the tenementis [*recte*, tenants] as followis, and meill that followis is for the personage and teindis of the samin landis. James Kennedie occupyer of the v merkland of Auchtirl[u]ir', £10 [mail]; 7½ b. meal. 'Harbert George, Nyvin Acculattane[65] and thair caligis, occupyeris of the v merkisland of Airdis payis' £10 [mail]; 40 b. meal; 2 dozen capons. Thomas Mak, Mr Gavin McMytin and their colleagues, occupiers of the 2½ merkland of Coilhorne pays 10 merks mail; 40 b. meal; 2 dozen capons. Thomas McCarlie, William McMaister and their colleagues, occupiers of the two merkland of Glenquhappill pays 10 merks mail; 12 b. 3 p. meal; 2 dozen capons. Thomas Hoge, occupier of the 'tua merkland, ane half of Balmulo, payis', £10 [mail]; 16 b. meal; 1 f. 'mayne'; 2 dozen capons. Andrew McCaulie 'and his marrow, occupyeris of the merkland of Drumdocht payis' £4 mail; 15 b. meal; 1 dozen capons.

Gilchrist McMaister, occupier of the 20s land of Barsollis pays £4 [mail]; 10 b. meal; 1 dozen capons. John McMikin, occupier of the merkland of Merkgain [McGowin, *fo 81r*] pays 40s mail; 5 b. meal; 1 dozen capons. James McClanochtane,[66] 'occupyer of the iij merkland of Culmanan, payis' £8 mail; 18 b. meal 'allannate'. James McClanochan, 'occupyeris' [*sic*] of the 2½ merkland of Kurmanocht payis' £4 mail; 3 b. meal; 1 dozen capons. The lady Barnbarrocht, occupier of the 2½ merkland of the Gannocht, pays 40s mail; 3 b. meal. Michael Umfray, occupier of the 20s [land] of Kirkmering and the merkland of the Manis of Saulseitt, £4 mail; 9 b. meal; 1 dozen capons. Thomas Jonsoun, 'fewar of the v merkland of the Homland of Saulseitt and the meill of Sant Johne and the meill of Barsollis payis' 16 merks mail; 24 b. meal. Sanct Johnis croft occupied by James Kennedie pays 40s. In Quhittorn, 20 merkland occupied as follows: Duncan McGain, 'occupyer of the v merkland of Skyocht', 21 merks mail. William Annas 'wifit', occupier of the two merkland of Balnaige [Balmachee, *fo 81r*] pays 5 merks [£5, *fo 81r*]. John McDowellis wife, occupier of Drumasson 2½ merkland, pays 5 merks. Howreik [Olbrek, *fo 73v*; *i.e.* Auldbreck], 2½ merkland occupied by the sheriff of Galloway,[67] 'the saidis landis for his fie.' The laird of Barnebarrocht, occupier of the 'tua merkland of Chappellquhair and payis' 40s. 'The Drumnanis v merkland and the meill occupyit be Johne McDowell payis' 25 merks mail. As to the kirk of Saulseitt, the teindis thairof is estimatt with the maillis of the landis, the viccerage thairof payis the viccar his' 20 merk pension, in the hands of William Adair, laird of Kinhill, 'quha sould pay to the place yeirlie of new the sowme of' 300 merks; 100 b. bere; 'sum almous'.

---

[65] I.e. 'A'Cultan', Black, *Surnames*, p. 6.
[66] I.e. 'MacClannachan', Black, *Surnames*, p. 469.
[67] Patrick Agnew was sheriff of Wigtown at the Reformation, *RMS*, iv, no 1492.

The croft of Mainport occupied by the wife of Harry McCulloch pays 40s. 'The croft of the kirk occupyit be airis of the laird of Freuch payis ane pund of wax. The croft of Bollinga occupied by James Chalmer, 20s.

Signature: 'Joanis, camerarius, Sedis Animarum.'[68]

Calculation of third: Money, £343 13s 4d; third thereof, £114 11s 1½d. Meal, 13 c. 4 b. 2 f.; third thereof, 4 c. 6 b. 3 f. 1¼ p. Bere, 7 c. 8 b.; third thereof, 2 c. 8 b. Capons, 13 dozen and 6; third thereof, 4 dozen and 6. 'Thrid pairt of ane pund of wax.'

'Bye the the [sic] thrid of the twa merkland and ane half of Yowrek [Olbrek, *fo 73v*; i.e. Auldbreck] occupyit be the schireffis fie.'

'The said abacie sett be sir Johne Johnsoun, abbott thairof, to umquhill James Johnsoun of Wamfray and his assignayis to quhom was sumtyme umquhill Gilbert, erle of Cassillis, and we as air to him for the yeirlie payment of' 380 merks.

In the whole, £253 6s 8d; third thereof, £84 8s 10⅔d.

'Memorandum, to derect lettres to answer the thrid pairt as it is givin in, the tua pairte to him and na mair bot as it is givin in heir, and the remanent superplus to the quein gif the balkis be maid sen the inhibitioun.'

Assumption of the abbey of Sauldseitt:
 Third of money, £111 11s 1½d.[69] Take the temporal lands and mails thereof of the touns and lands after following, namely, 'The fyve merkland [of] Auchirlair', £10; 5 merkland of Ardis, £10; 2½ merkland of Cehorne [i.e. Culhorn], 10 merks; 2½ merkland of Glenquhappill, 10 merks; 2½ merkland of Barmullo, £10; the lands of Drumdocht' £4; the lands of Barsollis, £4; the lands of McGowin [Merkgain, *fo 79v*], 40s; the lands of McQuhair, £4; the lands of Culman, £8; the lands of Gannoch, £4; the lands of Kilmuraig, [*blank*].

The lands of Drumanston in Quhittorne pertaining to the foresaid abbey: The lands of Skeoch, 21 merks; the lands of Balmachee [Balnaige, *fo 79v*], £5 [5 merks, *fo 79v*]; Bowraik [Olbrek, *fo 73v*; i.e. Auldbreck], £5; the lands of Drumanstane, 5 merks; the lands of Chappallquhairan, 40s; the lands of Drumnance and mill thereof, 25 merks; Neill Aidairs croft, 50s. Total of the mails extends to £121 11s 8d. 'Gif in' £10 9s 5½d. 'Eque, eque, eque.'[70]

---

[68] I.e. 'John, chamberlain of Soulseat.' The chamberlain's identity has not been traced. John Johnston was commendator of Soulseat at the Reformation, Cowan and Easson, *Medieval Religious Houses*, p. 102; cf *RMS*, vi, no 785.
[69] In the calculation of third above, the figure given is £114 11s 1½d.
[70] These figures do not tally.

Signature: 'Presentit primo Januarij be Harie Smycht, Colectour of Galloway. Sic subscribitur, Clerk Registre.'[71]

GLENLUCE, ABBEY OF,   (EUL, fo 81r)

'Rentale albacie [sic] de Glenluce'

The same set by Thomas,[72] abbott, to Gilbert, earl of Cassillis,[73] for the yearly payment of 1,000 merks money conform to the assedation made thereupon at Edinbrucht, 14 February 1561/2 'quhilk was sett affoir on the samin pryce be Galtir[74] [i.e. Walter], abott, to William,[75] abbott[76] of Corsraguell and to the laird of Lochinvar be James,[77] albott [sic], his broder, 1561. Presentit be the persoun of Air in the erle of Cassillis name. David Gibsoun.'

Calculation of third: In the whole, £666 13s 4d; third thereof, £222 4s 5⅓d.

MINNIGAFF, VICARAGE OF,   (EUL, fo 81v)

Rental of the vicarage of Minigaff pertaining to George Arnold, set to the laird of Larg in assedation for £50.

Calculation of third: In the whole, £50; third thereof, £16 13s 4d.

KIRKMAIDEN, VICARAGE OF,   (EUL, fo 81v)

Rental of the vicarage of the parish kirk of Kirkmaidin in Ferines [i.e. Kirkmaiden-in-Farines], lying within the diocese of Galloway and sheriffdom of Wigtoun 'gevin be Rauff Peirsoun, viccar thairof'.

---

[71] James McGill of Nether Rankeilour was Clerk Register at the Reformation, *Handbook of British Chronology*, p. 197.
[72] Thomas Hay was appointed abbot in 1560, Cowan and Easson, *Medieval Religious Houses*, p. 75.
[73] I.e. Gilbert, 4th earl of Cassillis, *Scots Peerage*, ii, pp. 471-473.
[74] Walter Mallen was an earlier abbot of Glenluce, Cowan and Easson, *Medieval Religious Houses*, p. 75.
[75] William Kennedy, brother of Gilbert, 2nd earl of Cassillis, was commendator of Crossraguel till 1547, Cowan and Easson, *Medieval Religious Houses*, p. 64.
[76] The phrase 'to William, abbott', is repeated in the text.
[77] In 1547, Walter Mallen resigned the abbacy in favour of James Gordon, brother of the laird of Lochinvar, Cowan and Easson, *Medieval Religious Houses*, p. 75.

'Quhilk viccerage with the haill te[i]ndis, fructis and gleib thairof, corppresentis, umest clathis, oblatiounis and offreingis being deducit now comounlie will gif yeirlie in proffeit' 34 merks 'alanerlie.'

Signature: 'Raff Peirsoun, viccar of Kirkmadin, with my hand.'

Calculation of third: In the whole, £22 13s 4d; third thereof, £7 11s 1½d.

CRUGGLETON, VICARAGE OF, (EUL, fo 81v)

Rental of the vicarage of Crugiltoun lying within the diocese of Galloway and sheriffdom of Wigtoun given in by William Telfeir, vicar thereof, 'quhilk viccerage with the teindis, fructis and gleib of the samin hes gevin tymis bygain and yett will gif comownlie in yeiris to cum in proffeit, corpspresentis, umast clethis and oblatiounis being deducit', 24 merks 'alanerlie.'

Signature: 'I, William Tailfer, with my hand.'

Calculation of third: In the whole, £16; third thereof, £5 6s 8d.

KIRKMADRINE, VICARAGE OF, (EUL, fo 82r)

Rental of Kirkmad[r]in pertaining to sir Nicol McBlaine, vicar thereof, which lies within the sheriffdom of Wigttoun in the barony of Yettoun within the diocese of Galloway of the patrimony of Sanct Maries Ile is worth yearly to be set in assedation of money, £10.

Signature: 'Thomas Browne, burges of Edinburcht.'

Calculation of third: In the whole, £10; third thereof, £3 6s 8d.

PENNINGHAME, BENEFICE OF,[78] (EUL, fo 82r-v)

Rental of Peninghame in Galloway

The kirk of Penninghame [pertaining] to the archdeacon of Galloway[79] set in tack and assedation to Alexander Stewart of Garleis paying for the same yearly to the

---

[78] Penninghame was a prebendal church of Whithorn cathedral, Cowan, *Parishes*, p. 225.
[79] Andrew Arnot was archdeacon of Galloway, 1543 - 1575, Watt, *Fasti*, p. 138.

archdeacon 276 merks. To the vicar[80] of the said kirk, 24 merks. Total, 100 merks [*recte*, 300 merks].

Signature: 'Garleis.'

Calculation of third: In the whole, £200; third thereof, £73 6s 8d.[81]

KELTON, PARSONAGE OF, (EUL, fo 82v)

Rental of the parsonage of Keltoun lying within the diocese of Galloway and stewartry of Kirkcudbrycht set in assedation by Robert, commendator of Halyrudhous,[82] to James McCleland of the Muirtoun [Muntros, *below*] for the sum of £18 usual money of this realm.

Signature: 'James McCleland of Muntros' [Muirtoun, *above*].

'Of the quhilk personage the toun of Slegna [*i.e.* Slognaw] pertenis to the abay of Halyrudhous, quhairof he takis up the proffeit extending to [*blank*]. Sic subscribitur ut supra.'

Calculation of third: In the whole, £16 [*recte*, £18]; third thereof, £6.

KIRKINNER AND KIRKCOWAN, BENEFICES OF, (EUL, fos 82v-83r)

Kirks of Kirkkineir and Kirkcowane

'The rentall of the haill kirk [*sic*] of Kirkkynneir and Kirkcowan in Galloway perteinand to sir George Clappertoun, subdene, and sir James Patirsoun, sacristane of the Chappell Ryall of Stirvilling.'

'Item, the said kirkis payis as they payit of befoir in assedatioun to ilk ane of us' 340 merks. 'Ilk ane of us now payand to the mi[ni]ster and precher be yeir' 50 merks. 'And sua restis frie to everilk ane of us', 280 merks.[83]

Signature: 'James Patirsoun, sacristane.'

Calculation of third: In the whole, £453 6s 8d; third thereof, £151 2s 2⅔d.

---

[80] Martin Gib was vicar portionary and reader at Penninghame, *Thirds of Benefices*, p. 93.
[81] Three times £73 6s 8d is £220.
[82] See above, p. 591, n. 37.
[83] The correct calculation appears to be 290 merks.

KELLS, PARSONAGE AND VICARAGE OF, (EUL, fo 83r)

Rental of the parsonage and vicarage of Kellis pertaining to 'the quenis grace Chappall Royall of Strivilling' set to John Gordoun of Barskeocht by Andrew Gray, parson and stallar of the queen's Chapel of Stirvilling for the yearly payment of 100 merks.

Calculation of third: In the whole, £66 13s 4d; third thereof, £22 4s 5$\frac{1}{3}$d.

BALMACLELLAN, PARSONAGE AND VICARAGE OF, (EUL, fo 83r)

Rental of the parsonage and vicarage of Balmaclellane pertaining to the Chapel Royal of Stirvilling set to Robert Gordoun of Chirmis by sir George Gray, parson and stallar thereof, for yearly payment of 50 merks; third thereof, £12 2s 2$\frac{2}{3}$d.[84]

Signature: 'Andreas Gray.'

SORBIE, VICARAGE OF, (EUL, fo 83v)

Rental of the vicarage of Sorbie within the diocese of Galloway which pertains to sir Gilbert Oslay [Ostan, *below*] set in assedation yearly to William Haynyis wife in Wigtoun for £20 'as he writtis and send me ward quhilk I beleif trew.'

Signature: 'Mr Gilbert Ostan[85] [Oslay, *above*]. Steinstoun, C. G.,[86] manu propria.'

Calculation of third: In the whole, £20; third thereof, £6 13s 4d.

STRATHAVEN, VICARAGE PENSIONARY OF,[87] (EUL, fo 83v)

'The rentall of the viccerage pensionar of Strathabane quhilk sir Johne Andirsoun is viccar pensionar thairof' which pays yearly 24 merks, 'and ix akeris of land quhilkis is in the Hamiltounis handis quhairof he gettis in [*blank*] merks weill payit as he writtis to me quhilk I beleif of veritie.'

Signature: 'Steinsoun, C. G.,[88] manu sua.'

Calculation of third: In the whole, £22 13s 4d; third thereof, £7 11s 1$\frac{1}{3}$d.

---

[84] The correct calculation appears to be £11 2s 2$\frac{2}{3}$d.
[85] 'Oislar', *RSS*, v, no 3270.
[86] 'Cantor Glasguensis', i.e. chanter of Glasgow.
[87] Strathaven, in the sheriffdom of Lanark, was appropriated to Bothwell collegiate kirk.
[88] I.e. John Stevenson or Steinstoun, Watt, *Fasti*, p. 160.

KIRKCORMACK, VICARAGE OF,[89] (EUL, fo 83v)

Rental of the vicarage of Kirkcarmok within the diocese of Galoway extending to the sum of £16; third thereof, £5 6s 8d.

Signature: 'sir Henry Dun of Kilcormok.'[90]

KIRKGUNZEON, BENEFICE OF, (EUL, fo 84r)

Rental of the kirk of Gargonzoun pertaining to Mr Alexander Home within the diocese of Glasgow.

The said kirk set in assedation to John Maxwell of Brakinsyd by year for £93 6s 8d. 'Item, deduct to the pensionar yeirlie for his fie' £20. 'The rentall presentit be me, Mr Johne Howme, in Mr Alexander Howme name.'

Signature: 'Maister Johne Howme.'

Calculation of third: In the whole, £93 6s 8d; third thereof, £31 2s $2\frac{2}{3}$d.

SENWICK, VICARAGE OF, (EUL, fo 84r-v)

Rental of the vicarage of Sawilk 'producit be me, Mr Williame McClellan, taksman and fermourar thairof, befoir the comissioneris constitute be the quenis majestie for resseving of the rentallis of the benefeices of this realme.'

'The viccerage of Sanwik was sett in all tymes bygaine for the sowme of' £40. 'And becaus the corpspresent, umest clothis and Pasche rekiningis quhilk wald have extendit yeirlie to' 20 merks 'dischargit, the proffeitt of the said viccerage will extend yeirlie to now' 40 merks, 'quhairupon ane reader within the said kirk may be sustenit to reid the commoun prayeris.'

Signature: 'Williame McClelland, with my hand.'

Calculation of third: In the whole, £26 13s 4d; third thereof, £8 17s $10\frac{2}{3}$d.[91]

---

[89] Cf EUL, fo 95r (see below, p. 618).
[90] Sir Herbert (not Henry) Dun was vicar of Kirkcormack at the Reformation, *Thirds of Benefices*, pp. 148, 289.
[91] The correct calculation appears to be £8 17s $9\frac{1}{3}$d.

CLAYSHANT, VICARAGE OF, (EUL, fo 84v)

Vicarage of Glauschant

Rental of the vicarage of Gleuschart in Rynnis of Galloway set in assedation to Uthreid McDowell for yearly payment of £20 'and that quhen umest clothis and all small offerandis was payit.'

Signature: 'Robertus Watsoun.'

Calculation of third: In the whole, £20; third thereof, £6 13s 4d.

DALRY, PARSONAGE AND VICARAGE OF, (EUL, fo 84v)

Rental of the parsonage and vicarage of Dalry pertaining to Mr John Hepbroun, parson thereof, set in assedation for £220, that is to say, John Gordoun of Barskeocht, £120; Fergus Accaiman of Killochrig, £100.

Calculation of third: In the whole, £220; third thereof, £73 6s 8d.

TOSKERTON, VICARAGE OF, (EUL, fo 85r)

Rental of the vicarage of Toskartoun lying within the diocese of Galloway set by sir Michael Hathern, vicar thereof, to Guthrie McCulocht of Arduell for the sum of 20 merks.

Calculation of third: In the whole, £13 6s 8d; third thereof, £4 8s 10⅔d.

GLASSERTON, VICARAGE OF, (EUL, fo 85r)

Rental of the vicarage of Glasshertoun given in by the said sir Michael, vicar thereof, lying with the said diocese, yearly worth £20.

Signature: 'Ita est Michaell Huthron, viccarius predictarum terrarum, manu propria.'

Calculation of third: In the whole, £20; third thereof, £6 13s 4d.

LESWALT, VICARAGE OF, INCH, VICARAGE PENSIONARY OF,

(EUL, fo 85r)

Vicarage of Laswalt and vicarage pensionary of the Insche.

Rental of the vicarage of Lasualt in Galloway pertaining to sir William McDowell set in assedation for the sum of 40 merks yearly. 'And my viccerage pensionare of the Insche extendis yeirlie to the sowme of' £10.

Signature: [*in another hand*] 'Makdowell.'

WIGTOWN, BLACK FRIARS OF,

(EUL, fo 85v)

'The rentall of the freiris of Wigtoun'

'Item, in money', £20 3s 8d. 'Item, of dustmeill', 10 c. 14 b. 'Item, the cruik besyd the peir', 2 b. malt. 'Item, thair was ane fisching that payit them sum salmond bot the haill is intromettit with be the auld laird of Garleis quha aledgis him self to haif ane gift and seasing of the haill of the quein Regent.'[92]
[*In margin*, 'Nota, the boll will not gif' 10s 'quhen it was darest.']

DUNDRENNAN, ABBEY OF,

(EUL, fo 85v)

Rental of the abbey of Drumdrannan[93] as the same is presently set in assedation 'and was lykwayis sett of auld' extends yearly to the sum of £500; third thereof, £166 13s 4d. Thereof to the minister of Drundrannan 'or' [*sic*] £20 6s 8d. To the reader thereof, £20. To the minister of Kirkmabrek, £30. To the reader there, £20. Total, £90 6s 8d. 'Sua restis of superplus', £76 6s 8d.

GALLOWAY, BISHOPRIC OF,[94]

(EUL, fos 86r-87r)

'The new rentall of the bischoprik of Galloway ressevit at comand of Sir William Murray of Tulibarand, knycht, Comptroller', 8 September, 1566 'and ordanit to be registrat.'

---

[92] I.e. Mary of Guise, who died in June 1560.
[93] Adam Blackadder, commendator at the Reformation, had leased the abbacy in 1551 for three years to Sir John Maxwell of Terregles, who retained possession of the property at the Reformation. His son, Edward Maxwell, was appointed commendator in 1562. *Acts of the Lords of Council in Public Affairs, 1501-1554*, ed. R. K. Hannay (Edinburgh, 1932), pp. 607-608; *Calendar of State Papers relating to Scotland and Mary, Queen of Scots, 1547-1603*, eds. J. Bain, *et al.*, 13 vols. (Edinburgh, 1898-1969), i, no 533; *RSS*, v, no 1101.
[94] Cf EUL, fos 65v, 69r (see above, pp. 585, 589).

Rental of the bishopric of Galloway and abbacy of Tungland

Barony of Tungland: The laird of Lochinvar conform to his charter, £166 15d. The laird of Schennan conform to his charter, £12. David Arnot conform to his charter, £6. Henry Anstruthir conform to his charter, £3 16d. Total, £187 16s 4d.[95]

Barony of Kirkcryst: The laird of Blairquhan conform to his charter, £11. James McClellane conform to his charter, £72 6s 8d. The laird of Lochinvar conform to his charter, £20 6s 8d. Roger Gordoun of Kirkconnell and William Schaw for the Newtoun conform to their charters, £11 6s 8d. Total, £110.[96]

Barony of Penninghame: The laird of Garleis conform to his charter, £48. William Gordoun conform to his charter, £10 6s 8d. The laird of Barinbaroch conform to his charter, £12 17s 8d. The laird of Larg conform to his charter, £10 6s 8d. The laird of Lochinvar conform to his charter, £13 13s 4d. Total, £95 4s 4d.

Barony of Glasnycht: The laird of Craichlaw conform to his charter, £30 4s 5d. Alexander Gordoun conform to his charter, £25 6s 8d. Total, £55 11s 1½d.[97]

Barony of Innermessane:[98] The earl of Cassillis conform to his charter, £30. Quentin Boyd conform to his charter, £30 7s 4d. The sheriff conform to his charter, £13 13s 4d. The laird of Garthland conform to his charter, £33 15s 1d. John McDowell conform to his charter, 46s 8d. Total, £99 12s 5d.[99]

Total of the whole temporality of the bishopric of Galloway and abbacy of Tungland, £125 4s 2½d.[100]

Kirks set for money pertaining to the bishopric of Galloway and abbacy of Tungland: Parsonage of Traqueir set to John Maxwell of Hillis for £120. Parsonage and vicarage of Tungland set for £225 6s 8d. Parsonage of Sannyk set to the laird of Lochinvar for £20. Parsonage of Girthtoun set to the laird of Lochinvar for £113 6s 8d. Parsonage of Manygaff set to the said laird for £100. Parsonage and vicarage of the kirk of Insche and Leswalt set to the earl of Cassillis for £173 6s 8d. Total of the spirituality of the bishopric of Galloway and abbacy of Tungland, £752.

Total of the spirituality and temporality of the bishopric of Galloway and abbacy of Tungland extends to £1,057 4s 2½d.[101]

---

[95] The correct calculation appears to be £187 2s 7d.
[96] The correct calculation appears to be £115.
[97] The correct calculation appears to be £55 11s 1d.
[98] The following entries begin with 'the Mans of Insch', on fo 68v; and come under the heading of 'the barony of Insch' on fo 71r (see above, pp. 588, 590).
[99] The correct calculation appears to be £110 2s 5d.
[100] The correct addition of the sub-totals in the text is £548 4s 2½d.
[101] This total does not tally.

Kirkcudbright

KIRKBEAN, VICARAGE OF,[1] (EUL, fos 87v-88r)

Assignation for the third of the vicarage of Kirkbeyne

'My lord Regent,[2] unto your grace and Lordis Auditouris of this present Chelker. Unto your lordschip humblie meanis and schawis I your servitour, Mr Williame Sumervell, viccar of Kirkbyne, that quhair sen that Assumptioun of the Thridis of all Benefices within this realme to our soverane I payit the thrid of my said viccerage yeirlie to the Comptrollar[3] and his colectiounis [*sic*] for the tyme be the space of tua yeiris extending yeirlie to the sowme of fyftie merkis albeit that I sensyne ressavit na payment nether of the tua pairt nor thrid of the fructis thairof, thairfoir I be my suplicatioun meant me [*sic*] to the quenis majestie and Lordis of Secreit [Council] for the tyme upon the sevint day of Marche the yeir of God' 1565/6 'and be thair delyvrance obtenit ane comand to the Lordis of the Chekker for the tyme ordaineing me to compeir befoir them and mak the assignatioun following or farder at [*recte*, as] they thocht expedient for satisfactioun of the thrid of my said benefice and ordanit them to ressave and doe the samin, viz.':

The toun and lands of Lytill Ardres [*i.e.* Airdrie] extending to 5 merks 'and occupyit be Robert Hewat, Richard Wod and the relict of umquhill Thomas McCon' paying yearly, 5 b. meal 'mesour of Neyth, with teind lamb woll and uther small teindis'. The toun and lands of Mekle Airdrie extending to ten merkland occupied by Walter Colter, William Myligane, Thomas Tailyour, John Diksoun, John Gregane, Patrick Wilsoun, John Cand, James Glessinwrycht, paying yearly, 10 b. meal 'mesour of Neyth, with teind lamb woll and uther dewties etc.' The 40s land of Mylerlandis, the 10s land of Cavanis occupied by Herbert Maxwell in Cavanis and his

---

[1] Cf EUL, fo 91v (see below, p. 614).
[2] James Stewart, earl of Moray, was Regent 1567 - 1570, *Scots Peerage*, vi, pp. 313-316.
[3] Sir William Murray of Tullibardine, Comptroller, 1565 - 1582, although James Cockburn of Skirling also served as Comptroller, 1566 - 1567, *Handbook of British Chronology*, p. 191.

subtenants paying therefore yearly 4 b. meal 'mesour of Neyth, with teynd lamb woll of all lyand within the paroschin of Kirkbyne and stewartrie of Kirkcudbrycht. And in the meintyme suspendit all lettres and proces of horne upon me raisit or to be raisit aganis me for non payment of the said thrid of my said viccerage for all yeiris bygane and in tyme cuming during my lyftyme notwithstanding the quhilk ordinance the saidis Lordis of Chelker did nathing thairintill and I daylie sensyne awaiting upon the ansuer thairof the tyme of evrie Chekker howbeit I was ay lyk as I am yett reddie to have obeyit and fullfillit the foirsaid ordinance and delyvrance according to the tennour of the samin and thairfoir all lettres purchest aganis me for not payment of the said thrid aucht and sould be suspendit simpliciter upon me in tyme cuming. Heirfoir I beseik your grace that in considderatioun of the premissis and that I ressavit na payment nether of the tua pairt nor of the thrid of my said viccerage quhilk gif I obtenit I wald glaidlie satisfie for the said thrid that ye will caus my lord Comptroller accept my foirsaid assignatioun and farder as ye think resonable in payment and full satisfactioun of the said thrid of my said viccerage for all yeiris bygaine restand awand unpayit and siklyk in tyme cuming dureing my lyftyme according to the foirsaid ordinance and delyverance, and in the mein tyme to provyd sum remeid quhairby I sall be payit of my said vicerag suspending all lettres purchest aganis me for non payment of the said thrid for ony yeiris bygaine and siklyk in tyme cuming dureing my lyftyme, discharging all officiaris of putting of me to the horne be vertew thairof and of thair offices in that pairt and gif I be putt to the horne to relax me thairfra for the caussis foirsaidis and your grace answer humblie I beseik.

Sic habet a tergo [*i.e.* Thus it is endorsed].

Apud Edinbrucht, secundo Decembris anno lxvij [*i.e.* 2 December 1567].

My lord Regent with advyse of the Lordis of Secreit Consall ordanis the Comptrollar and Auditouris of the Chelker to ressave this compleineris assignatioun within mentionat of all yeiris bygane and to cum and to gif and grant to him lettres in the four formis to be answerit and payit of the rest, conforme to the desyre of the said compleinere.'

Signatures: 'James, Regent. James Nicolsoun, ye sall ressave this bill and doe conforme to the delyvrance of the samin. W. Maitland. Tullibardyn, Comptrollar. J. Ballantyne. J. Spens.'[4]

---

[4] See above, p. 176, n. 67; p. 7, n. 28; p. 8, n. 30; p. 286, n. 23.

SWEETHEART, ABBEY OF,[5] (EUL, fo 88v)

Assumption of the abbey of Sueithartis: In the whole, £682; third thereof, £227 6s 8d.

Take the kirks of Buthill and Corsmichall 'sett of auld for' 400 merks and now for £213 6s 8d. The kirk of Kirkpatrik Durand [*i.e.* Durham] for £24. 'Giff in' £10 out of Kirkpatrik Durand.

Signature: 'J. Clericus Registri.[6] Receavit this assumptioun' 13 July 1570.

DUMFRIES, VICARAGE OF, (EUL, fo 88v)

Rental of the third of the vicarage of Dumfreis

The third of the vicarage of Drumfreis extends to £13 6s 8d. 'Givin up be me, sir Johne Bryce, vicar thairof, subscryvit with my hand at Drumfreis', 29 April 1576.

Signature: 'Johne Bryce, vicar of Drumfreis, with my hand.'

'Producit apud Edinburcht 3 of May 1575 [*sic*] and ordanit to be chargit in this maner in tyme cuming. Sic subscribitur, J. Hay.'

SWEETHEART, ABBEY OF,[7] (EUL, fos 89r-90r)

Rental of the abbey of Sueitharte

'The fyftie tua markland of Kirkpatrik Durane of the quhilkis landis sett to the lord Maxuell[8] in the yeir of God' 1541, 'xlix markland and xx d of the samin in feufearme for double meall. Item, of the said land thrie markland xx d less in tak and assedatioun to James Gordoun for double maill. Item, the personage of Kirkpatrik Durane sett in assedatioun the space of aucht scoir yeir syne for the yeirlie payment of' £24. 'Item, the ten merkland of Laichis sett in fewfearme the space of xij yeiris bypast for the yeirlie payment of' 20 merks. The parsonage and vicarage of Kirkcum set to the laird of Corswall for the sum yearly of 100 merks. The kirks of Buttill and Corsmichaell 'sett in auld tymis bypast' for the yearly payment of 400 merks, and now set for 320 merks 'and that evill payment be

---

[5] Cf EUL, fos 89r, 99v (see below, pp. 611, 622).
[6] James McGill of Nether Rankeilour was Clerk Register, 1567 - 1579, *Handbook of British Chronology*, p. 197.
[7] Cf EUL, fos 88v, 99v (see above, p. 611; below, p. 622).
[8] In 1541, Robert, 5th lord Maxwell, *Scots Peerage*, vi, pp. 479-480.

ressoun of non payment of corpspresent, umest claithis and uther small dewties, quhilk sowme will not sustein tua qualifiet ministeris.'

[*In margin*, 'New Abay']

'The barrounie of Lochindillocht, fourscoir markland sett in fewfearme to the inhabitaris of the maist pairt for dowble maile and of thir landis in our awin hand vj mearkland quhen it was sett to the tenentis payit yeirlie' 32 merks for 16 b. mail of the mesour of Nyth.'

5 merklands of the foresaid 80 merklands set in feuferme to my lord Maxuell for his bailie fee 'fyftie yeir syne or thairby.' The Mylne of the barony of New Abay set in assedation for the yearly payment of 33 merks. The parsonage and vicarage of Lochindilloch set in assedation for 80 b. victual for the sum of 160 merks. 'Summa pecunie', £682.

Signature: 'Johne, abott of Sueithart,[9] with my hand.'

Calculation of third: In the whole, £682; third thereof, £227 6s 8d.

'Omittit, grassumis, canis, custumis, etc.'

'This rentall was presentit be Petir Thomsoun, herald, in name of abott of New Abay.'

Signature: 'Petir Thomsoun, Ilay Herald.'

KIRKCUDBRIGHT-INNERTIG, VICARAGE PORTIONARY OF,[10]

(EUL, fo 90r)

'Vicaria portionaria de Kirkcudbrycht alias Inertig'

Vicarage portionary of Kirkcudbrycht alias Inertig in Carrik, in the diocese of Glasgow, whose possessor is Mr Andrew Oliphant, for divers years past was set to a noble man, Thomas Kennedie of Bargenie, his heirs and assignees for the sum of £30 money Scots, which vicarage with its fruits the said Thomas detains its profits as usufructuary.

Signature: 'Magister Andreas Oliphant, manu sua propria.'

---

[9] John Brown, the last pre-Reformation abbot of Sweetheart, resigned in 1565. Cowan and Easson, *Medieval Religious Houses*, p. 78.
[10] Cf EUL, fo 56v (see above, p. 570).

Calculation of third: In the whole, £30; third thereof, £10.

LINCLUDEN COLLEGIATE KIRK, PROVOSTRY OF,[11]   (EUL, fo 90v)

Rental of the provostry of Lencluden, assedation of the lands of the baronies of Corsmichall and Drumsleitt.

The mails of the barony of Corsmichaell extends to, £113 6s 8d. The mails of the barony of Drumsleit, £36 6s 8d. The grassum of both the said baronies 'in there yeiris' extends to £120 0s 7d. The whole multures and fermes of both the said baronies and Maynis of the place extends by year to 13 c. 8 b. victual of the measure of Nythsdaill, 'of the quhilk thair is to be defalkit' 24 b. 'throw inlaik of ane mylne.' The salmon fishing of Lyncluden 'quhairof the valour is uncertane.' The poultry extends yearly to 13 dozen.

The five kirks: The kirk of Carlaverok paying 380 merks. The kirk of Kirkbeane paying 26 b. bere set for 20s the boll. The kirk of Culven paying 38 b. meal set for 20s the boll. The kirk of Tereglis paying 10 merks. The kirk of Lochrutoun paying £20.

'This is the haill rentall that sustenis the proveist and aucht preistis and xxiiij beidmen. Off the quhilk is assignit to the saidis beidmen to thair sustentatioun, fyre and cleithingis, 12 c. victual. 'and siklyk to everrie ane of the saidis viij preistis alanerlie quhilk in the haill extendis to' 360 merks.

Signature: 'Robert Douglas.'[12]

Calculation of third: In money, £423 6s 8d;[13] third thereof, £141 2s 2⅔d. Victual, 15 c. 2 b. 'ob 3 ob' [i.e. ⅔ b.];[14] third thereof, 5 c. 2 f. 2 p. 'half pect and 3 part half pect' [*recte*, 5 c. 3 f. 2⅔ p.].

'Omittit etc.'

---

[11] Cf SRO, Vol. a, fo 267v; EUL, fo 98v (see above, p. 273; below, p. 620).
[12] Robert Douglas, provost of Lincluden, 1546-1609, Watt, *Fasti*, p. 365.
[13] These figures do not tally.
[14] These figures do not tally.

KIRKBEAN, VICARAGE OF,[15] (EUL, fos 91v-92r)

Rental of the vicarage of Kirkbene lying in the diocese of Glasgow and stewartry of Kirkcudbrycht pertaining to Mr William Somervell, vicar thereof, 'quha was provydit thairto in' 1547.[16]

'Quhilk vicarage with the haill teindis thairof, gleib, mans and kirklandis with corspresentis, umast claithis, Paschfynis and offerandis, war sett be umquhill Hew, lord Sumervell,[17] father to the said vicar, to Harbert Maxwell in Calvennis [*i.e.* Cavens], in the said vicaris minoritie, being at the schullis, in the xlviij yeir of God for the space of sax yeiris thairefter following, for payment of the sowme of' 150 merks 'yeirlie and continuallie; sen the ische of the said tak the said Harbert and umquhill William, his sone, hes violentlie withhaldin the said vicarage fra the said vicar, lyk as the said Harbert as yett withhaldis the samyn, quhairthrow the said vicar nor na utheris in his name hes gottin proffeitt thairof, and hes the said Harbert instantlie in proces befoir the Lordis of Cessioun thairfoir, and swa the said vicar can gett na entres thairto and thairthrow can have na uther knowledge quhat the rentall extendis to, and to verifie thir premissis the said vicar is reddie to depone his aith thairupon. Subscryvit with his hand at Edinburcht' 19 February 1561/2.

Signature: 'William Somervell.'

Calculation of third: In the whole, £100; third thereof, £33 6s 8d.

ANWOTH, VICARAGE OF, (EUL, fo 92r)

Rental of the vicarage of Anwach lying within the stewartry of Kirkcudbrycht 'in lamb woll and cheis with kirk gleib, consistis yeirlie to' 54 merks 'or thairby gif payment war maid. Givin in by Mr Malcum McCullocht, vicar of the said kirk.'

Signature: 'I, Mr Malcum McCuloch, with my hand.'

Calculation of third: In the whole, 54 merks; third thereof, £12.

---

[15] Cf EUL, fo 87v (see above, p. 609).
[16] William Somerville, sixth son of Hugh, 4th lord Somerville, *Scots Peerage*, viii, pp. 15-19.
[17] Hugh, 4th lord Somerville, *Scots Peerage*, viii, pp. 15-19.

GLASGOW CATHEDRAL, CHAPLAINRY OF ST MICHAEL IN,[18]

(EUL, fo 92r-92v)

Rental of Sanct Michallis chaplainry within the kirk of Glesgow extending yearly commonly to £25 [£26, *below*] in fermes, mails and annuals 'unpayit this thrie yeir bygane'. Chaplain and possessor thereof, Mr David Gibsoun.

Signature: 'David Gibsoun, manu sua.'

Calculation of third: In the whole, £26 [£25, *above*]; third thereof, £8 13s 4d.

LINCLUDEN COLLEGIATE KIRK, PREBEND OF,[19]

(EUL, fo 92v)

Rental of the prebend of Lyncluden extending to 10 merks 'servand the fees. Apud Edinburcht', 14 February 1551/2.[20]

Signature: 'David Gibsoun.'

Calculation of third: In the whole, £6 13s 4d; third thereof, 44s 6½d.[21]

[*In margin*, 'Haill to the queine.']

KIRKPATRICK-JUXTA, PARSONAGE AND VICARAGE OF,

(EUL, fos 92v-93r)

Rental of the benefice of Kirkpatrik Juxta, both parsonage and vicarage thereof, set by sir Humphrey Calquhoun, parson of the same, to Robert Johnstoun and Adam Johnstoun of Beithach, 'brether german, hes the personage thairof in tak for' £40. And James Johnstoun, 'sone to umquhill Gawan Johnstoun in Kirkdone', has the vicarage in assedation for £24 yearly. The parson paying yearly the vicar pensionary 'and uther ordinar and extra ordinar charges.'

Signature: 'sir Umphra Calquhoun, persoun of Kirkpatrik Juxta.'

Calculation of third: In the whole, £84;[22] third thereof, £28.

---

[18] Cf EUL, fos 10r, 19r (see above, pp. 501, 515).
[19] Cf EUL, fo 98r (see below, p. 620).
[20] This is presumably a scribal slip for February 1561/2.
[21] The correct calculation appears to be 44s 5½d.
[22] The correct calculation appears to be £64.

TROQUEER, VICARAGE OF,[23] (EUL, fo 93r)

Rental of Troqueir within the diocese of Glasgow, stewartry of Kircudbrycht and of the kirk glebes of the same, 'corspresent, umast claithis, Paschfynis, servandis offerandis being exceptit'. Given by Mr Roger Marteine, possessor of the said vicarage, extending yearly to £20. 'Subscryvit with my hand as followis.'

Signature: 'I, Maister George Marteine,[24] vicar of Troquair, with my hand.'

'Ressavit at command of the Clerk Registre.'

Calculation of third: In the whole, £20; third thereof, £6 13s 4d.

PARTON, PARSONAGE AND VICARAGE OF, (EUL, fo 93v)

Rental of the kirk of Partoun, both parsonage and vicarage, set in assedation to John Glendining of Drumrasche by Mr Charles Geddis, parson of Partoun, for the sum of 53 merks, 'lyk as it hes beine sett thir fourtie yeiris bygaine be me and my predicessouris.'

Signature: 'I, Mr Charlis Geddes, persoun of Partoun, with my hand. Presentit be me, Petir Thomsoun, herald, in name of Mr Charlis Geddes, persoun of the said benefice. Petir Thomsoun, Ilay Herald.'

Calculation of third: In the whole, £35 6s 8d; third thereof, £11 15s 6⅔d.

'Guid chaip vicarage rentallit.'

KIRKCHRIST, PARSONAGE AND VICARAGE OF, (EUL, fo 93v)

Rental of the parsonage and vicarage of Kirkcryst within the diocese of Galloway and stewartry of Kirkcudbrycht set in assedation to James McClellan of Muntoun [*i.e.* Nunton] for the sume of £40; third thereof, £13 6s 8d.

Signature: 'Mr Robert Balfour, persoun of Kirkcryst.'

[*In margin*, 'Nota, ane better rentall of this givin utherwayis.'[25]]

---

[23] Cf SRO, Vol. a, fo 272v (see above, p. 279).
[24] 'Roger Marteine', in preceding paragraph; Roger Martin appears as a witness at Tongland in 1566, *RMS*, iv, no 1573.
[25] No other rental for the benefice of Kirkchrist has been traced.

ST MARY'S ISLE, PRIORY OF,[26]   (EUL, fo 94r-94v)

The rental of the priory of Sanct Marie Iyle pertaining to Mr Robert Richartsoun, prior thereof.

The lands of Torris extending to ten merkland of auld extent pay yearly conform to the old rental the sum of £20. 'The sevin merkland and ane half of Lytill Galtua' pays yearly in the old rental £10. 'Baith the saidis landis being sett in few payis the sowmis foirsaidis togidder with' 13s 4d in augmentation of the rental extending in the whole to £30 13s 4d. The ten merkland of Grange set in feu pays yearly the sum of 30 merks. 'The tua markland and ane half of the ile of Sanct Marie Iyle with the maner place, woodis and fisch thairof' set in feu for the yearly payment of 7 merks 6s 8d. 'And the saidis landis of Great [*i.e.* Great Cross] and iyle of Sanct Marie Iyle payis alsua the sowme of' 13s 4d in name of augmentation. The Mylne of the Grange pays yearly set in feuferme the sum of 10 merks. The lands of Eitoun set in feu to the laird of Bombie pay yearly the sum of £48 8s. The lands of Bankhill set in feu to the laird of Mochrum pay yearly the sum of £19 10s. The teinds of the kirk of Kirkmadyne set in tack for the yearly payment of the sum of £46 13s 4d. The kirk of Anwath set in tack to the laird of Bomby for the yearly payment of the sum of £50. The kirk of Sanct Marie Iyle called Lytill Galtua, both parsonage and vicarage thereof set in assedation for yearly payment of £80.

Signature: 'Maister Robert Richisoun, pryour of Sanct Marie Iyle.'

Calculation of third: In the whole, £307 11s 4d; third thereof, £102 10s 5⅓d.

[*In margin*, 'Nota, thair is ane uther better rentall givin in of this in nixt leafe.']

KIRKCUDBRIGHT, VICARAGE OF,[27]   (EUL, fos 94v-95r)

The vicarage of Kirkcudbrycht, possessed by George Crychtoun, 'corspresentis, umast claithis and Paschfynis of servandis lauboris exceptit', extends in the year to the sum of £40.

Signature: 'Dene George Crychtoun, vicar.'

Calculation of third: In the whole, £40; third thereof, £13 6s 8d.

---

[26] Cf EUL, fo 95v (see below, p. 618).
[27] Kirkcudbright was appropriated to Holyrood abbey, Cowan, *Parishes*, p. 119.

BALMAGHIE, VICARAGE OF, (EUL, fo 95r)

Rental of the vicarage of Balmegie

'Item, the said vicarage of Balmegie, possest be the said George in maner abonewrittin, is worth be yeir the sowme of' £40.

Signature: 'Dene George Crychtoun, vicar.'

Calculation of third: £40; third thereof, £13 6s 8d.

KIRKCORMACK, VICARAGE OF,[28] (EUL, fo 95r)

Rental of the vicarage of Kirkcrumok

The said vicarage possessed by Henry Down, vicar thereof, 'corspresentis, umast claithis and Paschfynis of servandis laboure exceptit', is worth by year the sum of £16.

Signature: 'Henrie Down, vicar of Kirkcormok.'

Calculation of third: In the whole, £16; third thereof, £5 6s 8d.

KIRKMAHOE, PARSONAGE AND VICARAGE OF, (EUL, fo 95v)

Parsonage and kirk of Kirkmaho

The said parsonage pertaining to the laird of Garleis in patrimony set in tack and assedation to him, John Stewart his son, for the sum of 180 merks. To the vicar thereof, 20 merks.

Signature: 'Garleis.'

Calculation of third: In the whole, £133 6s 8d; third thereof, £44 8s $10\frac{2}{3}$d.

ST MARY'S ISLE, PRIORY OF,[29] (EUL, fos 95v-97v)

'Rentale Sancte Marie de Traill callit Sanct Marie Iyle.'

---

[28] Cf EUL, fo 83v (see above, p. 604).
[29] Cf EUL, fo 94r (see above, p. 617).

The kirk of Awecht [*i.e.* Anwoth] in the laird of Bombies hands pays yearly £50 'to be payit at Midsumer and Candlemes'.[30] The lands above Crie,[31] namely, Balfarrine, Etoun, Stelary [*i.e.* Skellarie], Orcharttoun and Stewartoun in the hands of the laird of Bombie pay yearly £48 8s. Bankill in the laird of Mochrumis hands pays yearly the sum of £20 conform to his feu, namely, [£]19 10s for augmentation and 10s for 2 dozen capons. The teind of Orchartoun and Bankhill and Eigirnes in Mochrumis hands, £30. The teind of Cuddrurie, Yettoun Orchart, £10. The teind of Culschaddan, 10 merks.

The lands of Torris. One quarter occupied by John McKnelland for mail and teind paid £10, 'and tua bollis beir callit multir beir'. One quarter occupied by Thomas Tailyiefeir for mail and teind, £10 and 2 b. bere called multure bere. Another quarter occupied by Edward Maxwall for mail and teind, £10 and 2 b. multure bere. One quarter occupied by Elspeth Maklellan for teind and mails, £10 and 2 b. multure bere 'and sa money capounis as is contenit in thair takis and not in use of payment.'

The Galtway: William Stirviling occupies three quarters and pays yearly £7 10s; 9 capons; 18 b. meal; 6 b. bere for his teind. Thomas Johnstoun, tenant to William Bowgred, 'and the ferd quartar' pays yearly 50s; 3 capons; 6 b. meal; 2 b. bere for his teind. The Myln and Myln croft in William Quhytheidis hand pay yearly £20 mail; 2 b. teind bere; 2 dozen capons. The said William pays for the Quhytfald of mail 20s; and for the teind of the same, [*blank*]; and 12s in augmentation.

The crofts of the Grange: William Quhytheid, subtenant to William of Bawmonthe, pays for the 'multir hill croft', 40s and 2 b. grassum bere. 'Item, in use of payment for the teind', 40s. John Comling for the Quhytcroft for teind and mail, 22s. Hobbie Edward for his croft, 40s. Sir Adam Magony for two crofts pays yearly for teind and mail 40s. Total money, £235 4s 4d.[32]

'Rentale victualium': William Stirvilling for three quarters of Galtway pays yearly 18 b. meal; 6 b. bere. Thomas Johnstoun, one quarter of the same, pays yearly 6 b. meal; 2 b. bere.

The Meikill Galtway: Edward Martine, 3 b. 2 f. meal; 1 b. bere for his teind. John Martyne, 3½ b. meal; 1 b. bere. William Schennand, 3½ b. [meal]; 1 b. bere. William Boyd and Henry Martyne, 3½ b. meal; 1 b. bere.

Gilgowrie Galtway: Patrick Schennand, 6 b. meal; 2 b. bere. Alexander McClelland, 3 b. meal; 1 b. bere. John Wilsoun, 1 b. 2 f. meal; 2 f. bere. Edward Martynis wife, Janet Makmochinee pays yearly 1 b. 2 f. meal; 2 f. bere.

---

[30] I.e. 2 February.
[31] These lands are located to the west of Wigtown Bay and the River Cree.
[32] The correct calculation appears to be £245 15s 4d.

The Grange: Patrick Schennand pays for the half of it 20 b. meal; 20 b. bere; 40 b. oats; 1 dozen capons and 1 dozen poultry. John Edwarte occupies the other half paying yearly therefor 20 b. meal; 20 b. bere; 40 b. oats; 1 dozen capons; 1 dozen poultry. The Myln and Myln croft pay 2 b. teind bere. William Quhytheid pays for the Muthill, 2 b. grassum bere. John Elwand pays for his croft, 1 b. bere and 4 capons. John Comlinge for his croft, 1 b. bere and 3 capons. John McKarnell, 1 b. bere; 3 capons. John McKyrnell, 1 b. bere; 3 capons. James Camlinge, 1 b. bere and 3 capons. John Spart for his croft, 1 b. bere; 3 capons, Thomas Jewark for his croft, 1 b. bere; 3 capons. [blank] Quhytheid for his croft, 1 b. bere, and for his teind, 1 dozen capons.

Total meal, 90 b.[33] Total oats of the Grange, 80 b. Total bere, 77 b.[34] Total meal and bere and [blank], 247 b. 'preter decimas eristarum [recte, aristarum].'[35]

'Item, thir sommis ar by the land of the Iyle and proffeitis thairof.'

LINCLUDEN COLLEGIATE KIRK, PREBEND OF,[36]   (EUL, fo 98r)

The rental of the prebend of Lyncluden pertaining to Mr Archibald Meinzies extends yearly to £20.

Signature: 'Mr Archibald Meinzies.'

Calculation of third: In the whole, £20; third thereof, £6 13s 4d.

LINCLUDEN COLLEGIATE KIRK, PROVOSTRY OF,[37]   (EUL, fo 98v)

'The rentall of the provestrie of Lyncluden by the beidmen and prebendaris extendis yeirlie to sax hundreth tuelff merkis [i.e. 612 merks], quhairof I am obleist to susteine yeirlie curaitis and uphalding of fyve kirkis, denis visitatiounis, procuratiounis, sinodallis and utheris exactiounis by corpspresandis, umast claithis, Paschfynis and offerandis quhilkis ar of laitt taine downe quhairthorow my said rentall will not extend to the foirsaid sowme be ane greit way quhilk I desyre to be modifiet and deducit be your wisdomis.'

---

[33] This figure is correct, if the repeated entries for William Stirling and Thomas Johnston are excluded.

[34] The correct calculation appears to be 80 b., excluding the repeated entries for William Stirling and Thomas Johnston.

[35] I.e. 'except the teinds of corn'; *arista*, an ear of corn.

[36] Cf EUL, fo 92v (see above, p. 615).

[37] Cf EUL, fo 90v (see above, p. 613). For provost, see below, p. 621.

LINCLUDEN COLLEGIATE KIRK, PREBENDS OF, (EUL, fo 98v)

'The rentall of the aucht prebendaris of Lynclowden.'

'Item, thair perteinis to the saidis prebendaris assignit to thame be the provest the mailis of the barronie of Drumsleit extending yeirlie to' 80 merks. 'Item, the teind schaves of the kirk of Carlavrok assignit lykwayis be the said provest for payment of the saidis prebendaris extending yeirlie to fyve scoir four bollis meill of the measour of Nyth[38] sett of auld for twa mark the boll.' Total, 208 merks. Total, 288 merks.

Calculation of third: 'The provestrie of Lynclowdan by the prebendaris and beidmen. In the haill', £408;[39] third thereof, £136.

'Tak this out of the temporall landis and best payment of the kirkis of the provestrie.'

'The aucht prebendaris of Lynclowdan by the provestrie, the beidmen and sir Johne Tailyauris chaplanrie, in the haill, £192; third thereof, £64.

Assignation of the third of these prebends:
    Take the mails of the barony of Drumsleit extending to yearly £53 6s 8d; and the rest extending to £10 13s out of the kirk of Carlavrok.

'The coppie of my lord Comptrolleris assignatioun for payment of the provest of Lynclowdan his pensioun.'
    'James Nicolsoun, forsameikill as it hes pleisit the queinis majestie to gif ane yeirlie pensioun to Robert Douglas, provest of Lynclowdan, of the sowme of tua hundreth pundis and hes chargit me to mak him payment thairof of thridis of the benefices, thairfoir we have assignit and be the tennour heirof with his awin consent assignis to him for compleit payment of his said pensioun the thrid of the said provestrie with the thrid of the chaplanries and prebandries perteining thairto of the cropis and yeiris of God' 1561, 1562 and 1563 'and siklyk yeirlie in tymecuming indureing hir majesties will, thairfoir answer him of lettres upon the said haill provestrie and thridis of the chaplanries and prebendries foirsaidis of the yeiris abonewrittin, ye keepand this precept for your warrand. Subscryvit with our hand at Edinburcht', [blank] June 1563.

---

[38] The measure of Nith was about one tenth larger than the standard Scots measure, *Concise Scots Dictionary*, ed. M. Robinson (Aberdeen, 1985), p. 442.
[39] 288 merks is £192.

SWEETHEART, ABBEY OF,[40] (EUL, fo 99v)

Rental of the abbacy and benefice of Sweithart called New Abay, 'baith the maillis, teindis, grassumis, pultrie, service and thair dewties.'

The barony of Lokcindalocht[41] extending to 80 merkland set in feu for the yearly payment of £164 16s 8d. The corn mill of the said barony set in feu for £22. The Walk Mylne of the said barony set in feu for 26s 8d. The teinds of the said barony of Lokindilloch, both parsonage and vicarage, set for 19 years for yearly payment of £106 13s 4d. The 50 merkland of Kirkpatrik Durane with the corn mill thereof set in feu to my lord Maxwell for £87 2s. The 40s land of Marcaitnay set in feu for £4 10s. The parsonage of the kirk of Kirkpatrik Durane set for 19 years for the yearly payment of £24. The 10 merkland of Laichis set in feu for £13 6s 8d. The 20s land of Airdis set in feu for 30s. The teinds, both parsonage and vicarage, of the kirks of Buttill and Corsmichall with kirkland set for 20 years for the yearly payment of £213 6s 8d. The parsonage and vicarage and kirkland of Kirkcum set for 10 years for the yearly payment of £66 13s 4d. Out of the lands of Orchartoun a yearly annualrent of £3 6s 8d. Total, [blank].

'Gevin out of this to the brether of the place' £117 6s 8d. To James Marcaitnay, burgess of Edinbrucht, £13 6s 8d. To the 'Lordis of the Seat', of contribution, £14.

Signature: 'Mr Gilbert Broun, abbott of Sweithart,[42] with my hand.'

HOLYWOOD, ABBEY OF,[43] (EUL, fos 100r-100v)

'The Witsounday[44] maill of Halywood'

'Item, in primis, in Craygyn Puttok', 46s 8d. In Overquhytsyd, 16s 3d. The Nether Quhytsyd, 40s. The Colinstoun, 35s. The Ferdunruische, 11s. The Skuyfuird [Skynfuird, *fo 100v*; *i.e.* Skinford], 15s 4d. The Chartetoun, 28s. The Lytill Speddoche, 35s. The Meikill Spedoche, 43s. The Steipfuird, 17s. David Edyar 'for the uther merkland' 16s. The Mornstoun, 34s. Willie Muirheid for the Clachan, 11s. The Libberie, 7s. 'Under the wood', 35s. Gibbonestoun, 20s 4d. Stealstoun, 22s. The 'thrie merkland' [Four Merk Land,[45] *SRO, Vol. a, fo 270r*],

---

[40] Cf EUL, fos 88v, 89r (see above, p. 611).
[41] New Abbey parish was formerly known as Lochkindeloch or Kinderloch, Cowan, *Parishes*, p. 135.
[42] Gilbert Brown became abbot of Sweetheart on the resignation of John Brown in 1565. Cowan and Easson, *Medieval Religious Houses*, p. 78.
[43] Cf SRO, Vol. a, fos 264r, 268r (see above, pp. 269, 274).
[44] I.e. seven weeks after Easter.
[45] Four Merk Land is a place beside Steilston.

46s 8d. Bllangawar[46] [*i.e.* Glengaber], 42s 7d. 'The half markland in the Brigend', 4s. Makbrunstoun, 16s 5d. Sanct Michallis Cros, 9s 4d. Kilnes, 26s. Brogsyd [*i.e.* Bogside], 19s. The Mosseyde, 5s. Amer Gilstoun, 6s. The Stelling Trie, 6s. Dalwoody, 9[s]. Naperis Cros, 3s. Gulyhill and Dalryve, 41s 2d. Brumrig, 24d. The Furd, 20s. Wrynns croft, 4s. Mertingtoun, 17s 8d. Cairinshome, 16s. The Cors Lies, 15[s] 6d. The Forrest croft [Forestanis croft, *fo 100v*], 7s. The Partarrak 'with all his [*sic*] landis extendis' £9. In Mydekilliboung, 28s. The merkland of Glengory, 8s. Muirheidis 20s land of McQuhinrik, 22s. Hayperis croft, 6s. Carowdis croft 'in the samin place with the Nentoun', 16s. Welschis croft, 5s. The Wod Nuk, 12s. Richard Edyar, 20s. The Walkmyln croft and Walk Mylne, 15s. The Home, £3. David Douglas, £4. The Breckoche, £5.

'The teindis of Halywoodis as they war in use thir xxx yeiris.'

Craigyn Puttok, 24d. The Overquhytsyd, 13s 4d. The Nether Quhytsyd, 24s. The Colinstoun, 40s. The Ferdingousche, 13s 4d. The Nether Strauchane, 13s 4d. The Steipfuird, 2 merks. The Lytill Speddoche, 40s. The Meikill Speddoch, 40s. Mornstoun, 51s 4d. The Skynfuird, 13s 4d. Barefregane, 2 merks. The Lybrie, 6s. 'Under the wood', 3 merks. The Thrie Merkland, 4 merks. McBurnistoun, 2 merks. Gribtoun, 14 merks. Gillestounis croft, 16s. Killnes, 23s. The Bogsyd, 15s. The Stelling Trie, 13s 4d. The Gulliehill, 25s. Mertingtoun, 2 merks. The Brumrig, 40s. Wrynnes croft, 9s. The Dowblahill, 6s. The Cors Lies, 24s. Mydkillelong, 40s. Forestanis croft [Forrest croft, *fo 100r*], 4s. Caudies croft, 20. 'The mylnis and Cars', 53 merks. The Keir, 50 merks.

The kirks: The kirk of Tynram, £30, 'quhilk was in pensioun to sir Andro Michall.' The laird of Auchingaschill, £19. The kirk of Kirkconnell, £20, 'the quhilk the hous of Sanquhar hes haldin fra me this xiiiij yeiris, and siklyk the kirk of Penponait quhilk payis' £10 'be yeir, quhilk the laird of Drumlagnerk [*i.e.* Drumlanrig] hes haldin fra me thir xij yeiris, quhilk I was faine to agrie with him for xl merkis. Beseiking the Regentis grace and the Lordis of the Chelker Counsall to haif consideratioun how that I am ane singill man and far frae my awin frendis and am constand and beir with thame quhill the Regentis grace and the Counsall caus thame to mak me thankfull payment quhairthrow I may mak guid and thankfull payment to my soverane, his factouris and Colectouris, and siklyk compleinis to my lord Regentis grace and the Counsall that Robert Maxwell of Cowhill haldis' 18 or 20 merkland 'baith maill and teind fra me thir xiiij yeiris, beseiking the Regentis gracis honour and the Counsall for remeid for Godis caus.'

---

[46] Cf above, p. 276, n. 23.

[*In margin*, 'Sic subscribitur, Thomas Cambell, comendatour of Halywood,[47] with my hand.']

'Thir ar the channonis pentionis that I am constranit to pay yeirlie and hes payit it sen the lviij yeir. Item in primis, Johne Maxwell, sone to my lord Heres',[48] 200 merks. To William Crechtoun in Librie, £20. To Gilbert Greir, £20. To David Welsche, £20. To Mr James Betoun, minister at Halywood, £30.

NOTE

Rental for the following benefice is located elsewhere in the text:

TROQUEER, VICARAGE OF (SRO, Vol. a, fo 272v), p. 279.

---

[47] Thomas Campbell was commendator, 1550 - 1580, Cowan and Easson, *Medieval Religious Houses*, p. 102.

[48] John, master of Maxwell, 4th lord Herries of Terregles by reason of his marriage to Agnes, lady Herries, eldest daughter of William, 3rd lord Herries, was the second son of Robert, 5th lord Maxwell (who died in 1546), and heir presumptive to his brother, Robert, 6th lord Maxwell, who died in 1552. Lord Herries' fifth son, John, of Newlaw, is the person mentioned here. *Scots Peerage*, iv, pp. 409-412; vi, pp. 479-482. See also above, p. 270, and nn. 4-5.

Inverness

ROSS, BISHOPRIC OF, (EUL, fos 101r-103v)

Assumption of the bishopric of Ros:
    Money, the third of the silver thereof, £217 5s 4d 'quondam'. Take the teind silver of the parish of Nig extending to £15; the mairdom of Nyg, £161 18s[1] 3d; take the rest extending to £40 9s 1d 'quondam, furth of the laird of Fowlis handis and his brethir out of the mairdome of Fairnedonald.'[2]
    Victual: Third of victual extends to 24 c. 12 b. 3 p. Take the teinds of the parish of Nyg for 18 c. 12 b.; the Myln of Morocht in George Moreis hands for 2 c.; the mills of Tarbat and Kincardin in the laird of Balnagowne and Walter Innes of Terbacks hands for 20 b.; take from the abbot of Fearnis[3] teinds, 2 c. 14 b. 3 p.[4]
    Oats: Third of oats extends to 2 c. $10\frac{1}{4}$ b. The mairdom of Nyg for 1 c. 6 b. 2 f.; the mairdom of Ferindonald, 11 b. Total, 2 c. 1 b. 2 f., and so rests to be taken from the chamberlain of the bishopric of Ros, 9 b. $2\frac{1}{4}$ p.
    Marts: Third of custom marts, $13\frac{1}{4}$ marts. Take the mairdom of Nyg for $8\frac{1}{4}$ marts; the mairdom of Fernedonald, $8\frac{1}{4}$ marts. Total, $13\frac{3}{4}$.[5] 'Gif in ane quarter mairt.'
    Mutton: Third of mutton extends to $82\frac{3}{4}$ muttons. Take the teind mutton of the parish of Nyg for 29 muttons; take the mairdom of Nyg for 42 muttons; take the mairdom of Fairnedonald for 25 muttons. Total, 96 muttons. 'Gif in to the cheplane [*recte*, chalmerlane] of the bischoprik', $13\frac{1}{4}$jp muttons.
    Kids: Third of kids extends to $54\frac{2}{3}$ kids. Take in the mairdom of Nyg for 36 kids; take in the mairdom of Fairnedonald, 24 kids. Total, 60 kids. 'Gif in againe to the chalmerland', 6 kids.

---

[1] In the text this appears to read 'xliij' shillings, possibly a scribal error for 'xviij'.
[2] These figures total £217 7s 4d.
[3] Nicholas Ross, commendator of Fearn, fo 112v (see below, p. 634).
[4] These figures total 24 c. 14 b. 3 p.
[5] This figure does not tally.

Capons: Third of capons, 16 dozen and 4. Take the mairdom of Nyg for 12 dozen and 6 capons; take out of Fairnedonald, 6 dozen and 10 capons. 'Gif in againe to the chalmerland' 2 dozen capons.[6]

Poultry: Third of poultry, 15 dozen. Take the barony of Nyg for 11 dozen and 6 poultry; take in Fernedonald for 5 dozen and 9 poultry. Total 17[7] dozen and 3. 'Gif in againe' 2 dozen and 3.

'The rentall of the bischoprik of Ros as it payis now salvo justo calculo [*i.e.* saving a just calculation] sua far as I can gett wit presentlie.'

'In primis, the Witsounday[8] and Martymes[9] tearmis maillis of the landis of the mairdomis of Allane, Ardmanocht, Fyrndonald, Nyg, Ardrosscher [*i.e.* Ardersier] and uther landis perteining to the said bischoprik of Ross yeirlie extendis to the sowme of' £462 4s 2d.

Ferme of the temporality thereof, 20 c. 8 b. victual; custom oats, 7 c. 4 b.; custom marts, $39\frac{1}{4}$ marts;[10] custom muttons, 140; custom capons, 10 dozen; custom poultry, 57 dozen; custom kids, 134, 'and with ilk kyd xxx aggis, paying thairfoir' 4d.

'And in salmound tua or thrie last and sum yeiris not sa many, quhairof is to be deducit salt, taes and uther expenssis, quhilk the thrid pairt of the fische beis gottin will nocht outred yeirlie for commoun, and sum yeiris not tua last, and sua uncertain quhilk is lesse quhilk is mair.'

'Item, ane pairt of the teindis of the paroschinis of Kilmowir and Kilernane yeirlie riddin quhylis les, quhylis mair, estimat to' 5 c. 8 b. 'comounly.'

The parish kirk of Tarbat, 33 c. 1 p. [victual]; and in teind silver, £26 8s; and 40 muttons. In the parish kirk of Nig, 19 c. 5 b. victual; and in teind silver, £15 0s 9d; 29 muttons.

'The sowmes of money and victuall givin out of the bischoprik of Ros in Ardmar now yeirlie.'

To the Lords of the College of Justice, £16 16s. To the chamberlain in fee, £40. 'To the grintar men of Nyg and Terbat', 18 b. victual and £10 money. To the chaplain of Allane, 12 b. victual. 'To the fischar bottis of Rosmerkin', 2 b. 2 f. victual. 'To the salmond fischaris of the Nes of the Channonrie', 19 b. victual and £4. To the 'kenar of the Nes', 12 b. victual and £10. For the officer's fees, £10. To the curates of Nyg and Terbat, £40. To the 'kenar of Kincairne' 4 b.

---

[6] These figures total 17 dozen and 4.
[7] In the text this appears to read 'xlii', possibly a scribal error for 'xvii'.
[8] I.e. Seven weeks after Easter.
[9] I.e. Martinmas, 11 November.
[10] The text reads 'xxxix mairtis, ij quarteris'.

victual and 40s money. 'To the vicar of chore in the Channonry', £20. To the preacher of the kirks of Nyg and Terbat in the year, £50.

'Item, for the expenssis and fies of men to keepe the houssis and place of the Chanounrie quhen I am furth of it in the Cessioun or wtherwayis in the queinis grace service, quhilk lyis in ane far heland cuntrie and ellis stollin this tyme twa yeir fra my servandis be brokin men as is notarlie knawin quha withheld it fra me nyne monthis or thairby to my great skaith and opprest nocht alanerlie the landis pertening to that kirk bot sindrie utheris of the queinis grace tennentis and utheris perteining to uther landit men of the cuntrie thairabout quhairthrow it is force to me to haif ane guid cumpany of men in my absence in the said place lyk as I haif now presentlie in the samin quhairof the expenssis extendis as efter followis': Victual, 12 c.; 20 marts; mutton, 80; poultry, 20 dozen; 'Item, for fleschis and utheris necessaris and the saidis menis fies', £300.

'Quhilk is to be considderit for on force I am constrainit to caus keepe that place as said is and wtherwayis gif brokin men mycht have it, it sould not only stop me to be answerit of my leiving in thai pairtis bot als be ane instrument to truble the cuntrie thairabout.'

Signature: 'Hen[ricus] Rossan[sis] [*recte*, Rossensis]. [*In a later hand*, 'Henry Stclair, bishop of Ross.']

Calculation of third: Money, £504 1s 4d;[11] third thereof, £168 0s 4⅔d.[12] Victual, 78 c. 5 b. 1 p.; third thereof, 26 c. 1 b. 2 f. 3 p. Custom oats, 7 c. 4 b.; third thereof, 2 c. 6 b. 2 f. 2⅔ p. Martis, 39¾;[13] third thereof, 13 '3 of half mart' [i.e. 13⅙]. Mutton, 169;[14] third thereof, 66⅓.[15] Capons, 10 dozen; third thereof, 3 dozen and 4. Poultry, 57 dozen; third thereof, 19 dozen. Kids, 134; third thereof, 44⅔. 'Thrid of the haill salmond.'

CAITHNESS, BISHOPRIC OF, (EUL, fos 104r-106v)

Rental of the bishopric of Cathnes given in by [*blank*].

'Item, the barronie of Ardurmes callit xv davoch land with the salmond fisching of the samyne, pendicles and pertinentis thairof, with the teind schavis of the samin sett in few and payis yeirlie in all dewtie' £81 6s 8d. The 'townis' of Skaill and Regeboll pay yearly in all duty £6. The barony of Skebo with pendicles and pertinents thereof set in feu and pays yearly in all duty the sum of £54 19s 8d.

---

11 The correct calculation appears to be £503 12s 11d, if no account is taken of payment for eggs.

12 One third of £504 1s 4d is £168 0s 5⅓d.

13 I.e. 39½.

14 The correct calculation appears to be 209.

15 One third of 169 is 56⅓.

Stoirdaill, Nygdaill and Lytill Creicht set in feu for the yearly payment of £20. Killmallie, Kirktoun and Reard pay yearly the sum in feu mail 20 merks money. The crofts and tenants in Dornoche pay yearly the sum of £10. 'Lateris', £185 13s.

The feu lands within Cathnes: The barony of May with pendicles and pertinents thereof set in feu to the earl of Cathnes[16] paying yearly in all duty, £84. The toun of Dorare pays yearly in all duty, the sum of £15 18s 5d. Lytill Ullagrahame yearly in all [duty], £7. Subunster yearly in all duty, 11s 3d. Halieerik 'with myln, cowff and salmond fisching' pays yearly in all duty, £16 17s 11d. Wnsterdaill yearly in all duty, £4. Kisterdaill yearly in all duty, £3. Thornisdaill yearly in all duty, 20s. Mereinethalis yearly in all duty, 13s 4d. Stansall yearly in all duty, £9. Lycht within the parish of Bowar in all duty, £9. Atterdaill [Alterwaill, fo 110r] yearly in all duty, £5 8s 8d. [Lateris,] £160 4s 11d.[17]

Derane yearly in all duty, £28 5s. 'Item, thir particular townis fewit and payis mair in augmentatioun of the rentall' £3 5s. Brymis yearly in all duty, £46. Forss with the mill and salmon fishing thereof pays yearly in all duty, £28 4s 6d. Bailyie pays yearly in all duty, 56s. 'The twa pairt of Lochmair with the twa pairt of Albist and tua penny land mair nor the saidis tua pairtis payis yeirlie in all dewtie', £21 8s 4d. Stainunsteir in all duty, £9 16s 8d. Strabusteir in all duty, £34 15s. The crofts of Strabuster yearly 10s. The quarter of the Watter of Thurseth pays yearly in all duty, £13 6s 8d. 'Ten pennyland in Weik with Bischopis Quyis[18] and Canis Quyis yeirlie in all' £16 5s 4d. North Kilmster yearly in all, £19 16s 8d. The Myln of Vindleis[19] yearly in all, £5 'Thrie ottounis in Netherland Norne[20] yeirlie', 5s. Lateris, £249 15s 6d.[21]

The Mylne of Lychtmoir in all, £6 5s. 'Item, thir particullar townis fewit and payis mair in augmentatioun of rentall', £6. For the annuals of Thursocht of the tenants thereof yearly, 6 dozen geese. [*In margin*, 'Particularlie', £12 5s; 6 dozen 'geis'.]

Total of this whole temporality, £607 18s 5d.[22]

The rental of the teinds of the bishopric foresaid: The teind sheaves of the parish of Ray within Cathnes set in assedation yearly for the sum of £79 6s 8d. The teind sheaves of the parish of Thursoche within Cathnes set in assedation for the

---

[16] I.e. George Sinclair, 4th earl of Caithness, *Scots Peerage*, ii, pp. 338-340.
[17] The correct calculation appears to be £156 9s 7d.
[18] 'Bischopisquyis', *RMS*, iv, no 1669; the place-name 'Quoy' is common in Caithness and Orkney.
[19] 'Windeleis', *RMS*, iv, no 1669.
[20] Cf 'Myrelandnorne', *RMS*, vii, no 766; presumably Myrelandhorn, ND2758.
[21] The correct calculation appears to be £229 14s 2d.
[22] This tallies with the MS sub-totals.

yearly sum of £126 16s 8d. The teind sheaves of the parish of Weik within Cathnes set in assedation for the yearly payment of £196 13s 4d. The teind sheaves of the parish of Lothrin [*i.e.* Latheron] within Cathnes set in assedation for the yearly payment of £91 11s 8d. The teind sheaves of the parish of Lothe within Sutherland set in assedation for the yearly payment of £75 17s 4d. The teind sheaves of the parish of Kilmaly within Suthirland set in assedation for the yearly payment of £105 15s.

'Deducit: Item, thair is to be deducit of this prenominat rentall that is givin in yeirlie pensioun to Mr Alexander Gordoun, bischope of Galloway etc., to the quhilk he is providit and cautioun actit for yeirlie payment thairof, viz., the sowme of' 500 merks money. 'Item, mair to be deducit yeirlie for contributioun to be payit to the Lordis of Counsall', £14. 'Item, siklyk to be deducit in heritable bailyie fie to my lord of Sutherland',[23] £100 money, 'conforme to his infeftment maid thairupon.'

'The rentall presentit be Johne Kennedie, subscryvit with my hand.'

Signature: 'Johne Kennedy, with my hand.'

Calculation of third: Money, £1,283 18s 9d;[24] third thereof, £426 19s 7d.[25] Geese, 6 dozen; third thereof, 2 dozen.

'All uther thingis omittit.'

'Memorandum, that this be tain up but prejudice of the auld rentall quhill thir takis and fewis be producit to sie the tyme of thair giving.'

ASSYNT, PARSONAGE AND VICARAGE OF, (EUL, fo 108r)

Rental of the parsonage and vicarage of Assent set in assedation for the yearly payment of £40.

Signature: 'Johne Kennedy.'

Calculation of third: In the whole, £40; third thereof, £13 6s 8d.

---

[23] John Gordon, earl of Sutherland, *Scots Peerage*, viii, pp. 339-342.
[24] The correct calculation appears to be £1,283 19s 1d.
[25] The correct calculation appears to be £427 19s 7d.

CAITHNESS, DEANERY OF, (EUL, fo 108r)

Rental of the deanery of Cathnes pertaining to Mr William Hepburne, dean thereof, is 10 c. bere and 40 merks for the vicarage of Kirktoun of Clyne and Denes feild, 'heirof the vicarage onpayit.'

Calculation of third: Money, £35;[26] third thereof, £11 13s 4d. Victual, 20 c. [10 c., *preceding paragraph*]; third thereof, 6 c.[27]

CAITHNESS, CHANTORY OF, (EUL, fo 108r-108v)

Rental of the chantory of Cathnes pertaining to Robert Stewart, 'chantour thairof', set in assedation for yearly payment of the sum of £100. For the Chantouris feild in feu yearly, 40s.

Signature: 'Johne Kennedie, with my hand.'

Calculation of third: In the whole, £102; third thereof, £34.

CAITHNESS, TREASURERSHIP OF, (EUL, fo 108v)

Rental of the 'thesaurarie of Cathnes' pertaining to Mr William Gordoun, pays yearly $3\frac{1}{2}$ c. bere and 100 merks money. The Thesaureris feild yearly, 40s.

Calculation of third: Money, £68 13s 4d; third thereof, £22 17s $9\frac{1}{3}$d. Bere, 3 c. 8 b.; third thereof, 1 c. 2 b. 2 f. $2\frac{2}{3}$ p.

CAITHNESS, CHANCELLORY OF, (EUL, fo 108v)

Rental of the chancellery of Cathnes pertaining to Mr John Jaksoun, chancellor thereof, pays yearly 6 c. bere and 100 hundred merks money for the parsonage and vicarage. The Chancellaris feild yearly, 40s.

'Heirof thair is to be deducit of the chansellarie that is givin in yeirlie pensioun to Thomas Mathesoun', £40.

Calculation of third: Money, £68 13s 4d; third thereof, £22 17s $9\frac{1}{3}$d. Bere, 6 c.; third thereof, 2 c.

---

[26] 40 merks is £26 13s 4d.
[27] The correct calculation appears to be $6\frac{2}{3}$ c.

KILDONAN, PARSONAGE AND VICARAGE OF, (EUL, fo 109r)

Rental of the parsonage and vicarage of Kildonane pertaining to 'Dene Henrie Abircrumby'[28] set in assedation for the yearly payment of 80 merks.

Calculation of third: £53 6s 8d; third thereof, £17 15s 6⅔d.

FARR, PARSONAGE AND VICARAGE OF, (EUL, fo 109r)

Rental of the parsonage and vicarage of Fard pertaining to sir Alexander Gray set in assedation for yearly payment of 80 merks.

Calculation of third: £53 6s 8d; third thereof, £17 15s 6⅔d.

SPITTAL, PARSONAGE OF, (EUL, fo 109r)

Parsonage of Sipittall [sic]

Rental of the parsonage of Spittall pertaining to Mr William Gordoun set in assedation for the yearly payment of £80.

Calculation of third: In the whole, £80; third thereof, £26 13s 4d.

HALKIRK AND SKINNET, BENEFICES OF, (EUL, fo 109r-109v)

Rental of the kirks of Halkirk and Skenand called the common kirks of Cathnes 'duty' pays yearly 8 c. bere.

Signature: 'Johne Kennedie, with my hand.'

Calculation of third: In the whole, 8 c. bere; third thereof, 2 c. 10 b. 2 f. 2 p 'half p.'[29]

THURSO, VICARAGE OF, (EUL, fo 109v)

Rental of the vicarage of Thurso 'only pertening to the bischope', set yearly for £16.

---

[28] Kildonan, a prebend of Dornoch cathedral, was held by Henry Abercrombie, 'prior of Scone' abbey, *RSS*, vii, no 1032.

[29] The correct calculation appears to be 2 c. 10 b. 2 f. 2⅔ p.

Calculation of third: In the whole, £16; third thereof, £5 6s 8d.

HELMSDALE, CHAPLAINRY OF, (EUL, fo 109v)

Rental of the chaplainry of Helmisdaill in Mr Thomas Braidis hands set in assedation for the yearly payment of £20 'callit Sanct Johnis chaplanrie'.

Calculation of third: In the whole, £20; third thereof, £6 13s 4d.

[*In margin*, 'Haill to the queine.']

ST ANDREW, CHAPLAINRY OF, (EUL, fo 109v)

Rental of the chaplainry of Sanct Androw in the hands of sir Robert McCrayth, vicar of Kilmaly,[30] pays yearly in assedation £10.

Signature: 'Johne Kennedy, with my hand.'

Calculation of third: In the whole, £10; third thereof, £3 6s 8d.

[*In margin*, 'Haill to the queine.']

CAITHNESS, ARCHDEACONRY OF, (EUL, fos 110r-111r)

'The rentall of the archdenrie of Cathnes, the estimatioun of the teind schavis of the paroschinis of Bewar and Vactin [*i.e.* Watten] perteining to the said archdenrie of the crope of the saxtie ane yeiris as efter followis.'

Parish of Bewar: Claok, 6 b. Scarmalat, 2 c. Guilschfeild and Laroll, 3 b. Scampstar, 22 b. Tusbust with the pertinents, 20 b. Brabustar, 2 c. Bowartour, 18 b. Halbero, 7 b. The Lwne and Bowar, 12 b. 'in the curatis fie of Bowar'. Bowar Madin, 24 b. Alterwaill, 9 b. Lyth with the pendicles, 37 b. Stanestall, 18 b. Kirk, 24 b.

The parish of Vactin: Vactin 28 b. 'in the pensionaris handis for serving of the cure.'

Strathnawer: Cagill and Gers, 11 b. Lynekirk, 12 b. Bilbustar [*i.e.* Bilbster], 20 b. Nedder Staddaill, 18 b. Over Sculdaill with the pendicles, 16 b. Monsarne, Kansarne and the Rowenes, 6 b. Toftingaill, 8 b. Waistbust, 20 b. Dune, 40 b.

---

[30] Kilmalie, now Golspie, Cowan, *Parishes*, p. 103.

'The dewties and rentall of the said archdene, his corporall landis efter followis': Scarneclati with the pertinents, namely, Larillis, Galthie Feild, Cloak [*i.e.* Clayock] and Camster extending to 18d land 'ilk penny land' 14s 'money alanerlie; inde suma', £12 12s.

The Mylne of Scarmlat yearly, £16. Ballinknok pays yearly 5 merks. 'Of thir foirsaidis to be deducit yeirlie to the chirrister' [*sic*], £16. The vicarages of Bauar and Vactin extending yearly to £40.

Calculation of third: Money, £55;[31] third thereof, £18 12s 10⅔d. Victual, 28 c. 15 b.;[32] third thereof, 9 c. 10 b. 1 f. 1⅓ p.

'Ane uther rentall of the archdenrie of Cathnes'

'The archdenrie of Cathnes sett for takis to ryn to David Sinclar of Dune, his airis and assignayis, for the sowme of' 240 merks yearly 'with the payment of the stallaris fie of Dornocht and curatis fie of Bowair with all uther ordinar chargis. Subscryvit with my hand.'

Signature: 'Williame Baine.'

Calculation of third: In the whole, 240 merks; third thereof, 80 merks.

FEARN, ABBEY OF, (EUL, fos 111v-112v)

Rental of Pherne

'First, the landis contenit in the laird of Ballangownis few chartour': Innercarroun, Westis Ferme, Downy, Westray, Muldarg, Knokydaff, Myltoun, Balmoch, Midilgany, Pitkory, the Manes of Fearnie, Eistir Gany, Wastir Gany, Meikill Rany, Ballie Blair [*i.e.* Balblair], The Dow croft, Brighous, Mylcroft and Weitland and the fishing of Bonath.

'Quhilkis givis in maillis, fearmis, girsum, bonage silver, mairtis, muttoun, capounis, hennis and in augmentatioun as his chartour proportis with sic as efter followis.'
    'Item, in maillis, girsum and bonage silver and augmentatioun, the sowme of' £89 12s 8d '1 ob. iij f.'
    'Item, mairtis, muttoun, capoun, hen and fre silver, the sowme of' £20 10s 8d; 23 c. 2 f. 2 p. victual; 16 b. oats.

---

[31] The correct calculation appears to be £55 18s 8d.
[32] The correct calculation appears to be 27 c. 11 b.

'And heirof of the said money allocat to the said laird as his chartour proportis in bailyie fie', £40.

'Item, the landis of Catboll Fischar, Lawch Clawethe, Tulloch [*i.e.* Tullich], Lytill Rany, Amot, Eistir Ferne, Beloun, 'sett in few to the Dunnunis, quhilkis payis conforme to thair [chartour] as efter followis: Item, in maill, girsum, bonage silver and augmentatioun, the sowme of' £31. 'Item, mairtis, mutoun and capoun silver', £6 14s; in victual, 2 b.; in oats, 6 b.

'The mylne and uth[e]ris landis quhilkis ar not sett in few payis as efter followis': The Ferne in victual, 7 c. 'Item, four ailhoussis in Ferne with the croftis', £4 6s 8d. The Smithis Landis, Barne croft, the croft called Roresounis croft,[33] the croft called Ballanascharach, 'utherwayis callit the cottaris delwingis, payis' 54s. 'Item, the fischaris aucht akeris of land quhilk never payit ane penny bot givin to them to dwell upon for furnisching of fische to the place and cuntrie upon the cuntries expenssis. Item, the place and yeardis with the waird for feding of hors never payit ane penny nor can never in rentall.'

'Deducit heirof': To the laird of Balnagowne conform to his charter as said in bailie fee, £40. In contribution to the College of Justice, £5 12s. 'Item, to the sustentatioun of the channounis', 4 c. 12 b. victual and £24 money. A pension to John Nicolsoun 'quhairwith he is provydit of auld', 24 b. victual.

Signature: 'Nicolas Ros, comendatour of Ferne.'

Calculation of third: Money, £165 7s '1 ob. 3 f.' [i.e. $1\frac{1}{4}$d];[34] third thereof, £55 2s 4d '3 ob. 1 f.' [i.e. $4\frac{5}{12}$d]. Victual, 30 c. 2 f. 2 p.;[35] third thereof, 10 c. 1 f. $3\frac{1}{3}$ p.[36] Oats, 22 b.; third thereof, 7 b. 1 f. $1\frac{1}{3}$ p.

ROSS, CHANCELLORY OF, (EUL, fo 113r)

Rental of the chancellory of Ros pertaining to Mr Duncan Chalmere, 'usufructuare'.

The parsonage of Suddye, the parsonage and vicarage of Kennetis 'with the foure pairt of the teind schavis of Cormarte and Rosmerkie perteinand to the said chanslerie sett in assedatioun to Mr David Chalmer, titular of the samin, and Richard Urwing [*i.e.* Irving] his factour, for the sowme of' 460 merks.

---

[33] 'Waltir-McRoreis-croftis', *RMS*, iv, no 2220.
[34] The correct calculation appears to be £154 18s $1\frac{1}{4}$d.
[35] The correct calculation appears to be 30 c. 2 b. 2 f. 2 p. minus 6 c. 4 b., which equals 23 c. 14 b. 2 f. 2 p.
[36] One third of 30 c. 2 f. 2 p. is 10 c. $3\frac{1}{3}$ p.

'Of the quhilk thair is to be deducit to the vicaris and chaplandis, ministaris of the samin, the sowme of' 50 merks.

'Presentit be me, Mr David Chalmer, with my hand.'

Calculation of third: In the whole, £173 6s 8d; third thereof, £57 15s 6⅔d.[37]

KILTEARN, PARSONAGE OF,   (EUL, fo 113r-113v)

Rental of Mr John Sandelandis, parson of Kittarne.

The said parsonage is set in tack and assedation to the laird of Fowlis for the yearly payment of the sum of £96 13s 4d.

Of which I pay in pension to sir Donald Scherar yearly the sum of £12. To John Sandilandis, son to the laird of Sanct Monanis, 42 merks yearly 'with the yeirlie dewties alsua to the minister.'

Signature: 'Johne Sandilandis with my hand.'

Calculation of third: In the whole, £96 13s 4d; third thereof, £32 4s 2¼d.[38]

URRAY, PARSONAGE OF,   (EUL, fo 113v)

Rental of the parsonage of Urray, 'ane prebendary within the diocy of Ros' pertaining to Mr David Halyburtoun.

Bram extending yearly to 20 merks. Kingis Urray extending yearly to 11 merks. Kirk Urray extending yearly to 2 merks. Arokyne extending yearly to 26 merks. Eistir Farbrowne extending yearly to 5 merks. Androquhenren extending yearly to 40s. Killquhillie Drum extending yearly to 7 merks. Mekill Moy with Murdo McCenzies feu lands extending yearly to £50.

Signature: 'Subscryvit be me, Mr Johne Dumbar.'

Calculation of third: In the whole, £102 13s 4d;[39] third thereof, £34 4s 5¼d.

---

[37] 460 merks is £306 13s 4d, but £173 6s 8d is 260 merks.
[38] The correct calculation appears to be £32 4s 5¼d.
[39] The correct calculation appears to be £99 6s 8d.

DINGWALL, (EUL, fo 114r)
CHAPLAINRIES OF ST LAURENCE AND ARFAILL IN,

'Chaplanries of Sant Larent and Ardsoull[40] [*recte*, Arfaill] in Dingwell'

Rental of the chaplainries of Sanct Lawrent and Arsaill lying within the diocese of Ros pertaining to sir David Bartan[41] 'sett in few be chartour and seasing' paying yearly to the chaplain £15.

Signature: 'Presentit be me, sir David Barquhan.'

Calculation of third: In the whole, £15; third thereof, £5.

[*In margin*, 'Haill to the queine.']

LOGIE-EASTER, VICARAGE OF, (EUL, fo 114r)

Rental of the vicarage of Logy lying within the sheriffdom of Innernes and diocese of Ros pertaining to sir Donald Reid, vicar of the same.

The said vicarage extends yearly to the sum of £12 yearly; third thereof, £4.

KILMORACK, VICARAGE OF,[42] (EUL, fo 114r-114v)

Rental of the vicarage of Kilmorak in Ros pertaining to sir John Nicolsoun.

The vicarage of Kilmorak gave of old £20 in assedation 'and now nothing payit thir tua yeiris quhill universall ordour be tane.'

Signature: 'Maister Henrie Kinros, presentar, with my hand.'

Calculation of third: In the whole, £20; third thereof, £6 13s 4d.

TAIN COLLEGIATE KIRK, PREBEND OF, (EUL, fo 114v)

Rental of the prebend of Than pertaining to Simon Blyth which should pay yearly to him £4 'and serve the self albeit I haif gottin na payment thairof this vj yeiris bygaine.'

---

[40] 'Ardafaillie', *RMS*, v, 1225; presumably Ardeville, NH7363.
[41] I.e. Barton or Barchan, cf *RSS*, vi, nos 545, 769.
[42] Cf EUL, fo 118r (see below, p. 640).

Calculation of third: In the whole, £4; third thereof, 26s 8d.

[*In margin*, 'Haill to the queine.']

LOGIE-EASTER, PARSONAGE OF, (EUL, fo 114v)

Rental of the parsonage of Logy 'ane channonry in Ros' set yearly for 100 merks, deducing thereof yearly to the stallar, £10.

Signature: 'Thomas Hay.'

Calculation of third: In the whole, £66 13s 4d; third thereof, £22 4s 5½d.

ROSS, TREASURERSHIP OF, (EUL, fo 115r)

'The rentall of the thesaurarie of Ros'

The fruits of the said treasurership and teind sheaves of the same, namely, the kirks of Logy and Urquhart and the quarter of Crumarty and Rosmerky set in assedation to Matthew Hamiltoun of Mylneburne for the sum of 300 merks.

'Of the quhilk thair is to be deducit for the uphold of the kirkis and to the ministaris', 100 merks, and so rests 200 merks.

Signature: 'Ita est Jacobus Lamb, notarius publicus, ex mandato dicti Mathei, factoris dicte thesawrarie, scribere nescien[tis], manu propria.'[43]

Calculation of third: In the whole, £200; third thereof, £66 13s 4d.

OBSTULE, CHAPLAINRY OF,[44] (EUL, fo 115r)

Rental of the chaplainry callit Obstull within the diocese of Ros pertaining to Mr John Dumbar, parson of Cumnok, which is set in assedation by the said possessor to George Monro of Dalcarty for £20.

Signature: 'Mr Johne Dumbar, persoun of Cumnok.'

Calculation of third: In the whole, £20; third thereof, £6 13s 4d.

---

[43] I.e. 'Thus James Lamb, notary public, by order of the said Matthew, factor of the said treasury, who cannot write, by his own hand.'
[44] Obstule was a chaplainry in Fortrose cathedral, *RSS*, iii, no 2530.

[*In margin*, 'Haill to the queine.']

ROSS, ARCHDEACONRY OF, (EUL, fo 115v)

Rental of the 'archdenrie' of Ros pertaining to Mr Duncan Frascher,[45] archdeacon thereof.

Calculation of third: Money, £12 13s 4d; third thereof, £4 4s 5$\frac{1}{3}$d. Victual, 20 c.; third thereof, 6 c. 10 b. 2 f. 2$\frac{2}{3}$ p.

CANISBAY, PARSONAGE AND VICARAGE OF, (EUL, fos 115v-116r)

Rental of the parsonage and vicarage of Cannesby within the diocese of Cathnes.

'The teind schavis of the parsonage thairof sett dyveris yeiris for money extendand yeirlie to the sowme of' 175 merks 'except within thir thrie or four yeiris bygaine or thairby, the foirsaidis teind schavis ar intromettit with for the maist pairt be the erle of Cathnes[46] servandis and tenantis and withhaldin be thame respective unpayit as yett throw default of justice and the victuall thairof for the maist restand in thair handis.'

'The vicarage thairof quhilk payit the vicar only worth' 20 merks 'to the stallar, intromettit with for the maist pairt thir thrie or four yeir bygaine be the wayis of the said erlis servandis and occupyit, beseikand your lordschip for remeid in all tymis bygaine and to cum according to equitie and justice.'

Signature: 'H. Barclay, with my hand.'[47]

Calculation of third: In the whole, £116 13s 4d;[48] third thereof, £38 17s 9$\frac{1}{3}$d.

'Nota, this personage gives' 17 c. bere 'or thairby, and lettres to answer the quein of the haill and the possessour of the twa pairt of thir sowme.'

'Vicarage thairof. Nota, thair is na rentall thairof givin up but only to the stallar' 20 merks, 'the rest alegit givin to susteine the vicar.'

Calculation of third: In the whole, £13 6s 8d; third thereof, £4 8s 10$\frac{2}{3}$d.

---

[45] 'Donald Fraser', archdeacon of Ross, *RSS*, vi, no 1512; cf Watt, *Fasti*, p. 287.
[46] See above, p. 628, n. 16.
[47] 'Hercules Barclay', *RSS*, vi, no 1551.
[48] This is 175 merks.

CROY AND MOY, KIRKS OF, (EUL, fo 116r-116v)

'The rentall of Croy and Moy lyand amang the Clan Chattan, persoun thairof, M[r] Patrik Lyddall, quha hes not gottin payment thairof thir four yeiris.'

Croy 'in anno', 5 merks. Cantnaleis in the year, 5 merks. Cantrafreis and Drumoyer in the year, 5 merks. Holme in the year, 4 merks. Galmantna in the year, 5 merks. Dalgrauch in the year, 40s. Contra Downe in the year, 4 merks. Eistir and Wastir Cantna in the year, £4. Clawak in the year, 2 merks. Dalry and Drunmotoune, 40s. Daldaicht, 52s. Wastir Daldolicht in the year, 4 merks. Leamicht in the year, 46s.

Calculation of third: £59 16s;[49] third thereof, £19 18s 8d.

SUDDIE AND KILMUIR-WESTER, VICARAGES OF, (EUL, fo 116v)

Rental of the vicarage of Suddy and Kilmowr within the diocese of Ros pertaining to sir David Barthane,[50] vicar of the same.

'The saidis twa vicaragis in tyme bygaine quhen all dewties and teindis was dewlie payit sic as lamb, woll, stirk, buttir, cheis, teind aill, corspresentis and Paschfynis and teind lynt and hempt and teind fisches of the Stell of Kissok was worth in thes yeiris' 20 merks 'and now thir twa yeiris bygaine nothing gottin.'

Signature: 'David Barthan, vicar of Suddy, with my hand, and Kilmowr. Presentit be me, sir Alexander Pedder, procurator.'

Calculation of third: In the whole, £13 6s 8d; third thereof, £4 8s 10⅔d.

ST MONANCE AND APPLECROSS, CHAPLAINRIES OF, (EUL, fo 117r)

Rental of the chaplainries of Sanct Monanis and Apilcorce lying within the diocese of Ros extending by yearly profit and annual to 36 merks, 'presentit be me, Mr William Monro, minister and vicar at Dingel.'

'And becaus I gett nathing of the said vicarag except' 5 merks 'of twa cobillis fisching, I am content the quenis grace dispone thairof becaus I can gett na lyf thairin without hir grace caus me have ane lyf lyk ane minister and be speciallie exertit.'

---

[49] The figures above total £35 11s 4d.
[50] See above, p. 636, n. 41.

Signature: 'Ita est Willielmus Monro, vicarius de Dingel, manu sua, subscripsit.'

Calculation of third: In the whole, 36 merks; third thereof, 8 merks.[51]

URRAY, VICARAGE OF,   (EUL, fo 117r-117v)

Rental of the vicarage of Urray within the diocese of Ros pertaining to sir Alexander Peddir, vicar of the same, 'the space of thir four yeiris bygaine or thairby.'

'The said vicarage of Vurray was quhen guid payment was maid worth yeirlie' £20, 'and now nothing gottin thir thrie yeiris.'

Signature: 'Alexander Peddir, vicarius ut supra, manu sua.'

Calculation of third: In the whole, £20, third thereof, £6 13s 4d.

AVOCH, VICARAGE OF,   (EUL, fo 117v)

Rental of the vicarage of Awoche within the diocese of Ros pertaining to the said sir Alexander Peddir, vicar of the same, 'the space of thir four yeiris bygaine or thairby.'

'The vicarage of Awoch was, quhen guid payment was maid, worth be yeir' 20 merks, 'and now not worth' 10 merks.

Signature: 'Alexander Pedder, vicarius ut supra, manu sua, subscripsit.'

Calculation of third: In the whole, £6 13s 4d; third thereof, 44s 5⅓d.

'Nota, it was worth' 20 merks.

TAIN COLLEGIATE KIRK, PROVOSTRY OF,   (EUL, fos 117v-118r)

Rental of the provostry of Tayne pertaining to Nicolas Ros, provost thereof.

'The said haill provestrie consistit in offrandis and the vicarage of Tayne, of the quhilk vicarage the kirk kow and cl[a]yth with the Pasch offrandis ceissis and only restis teind lamb and teind lynt quhilk will not extend to' £20 'or thairby.'

---

[51] The correct calculation appears to be £8.

Signature: 'Mr Henrie Kinros, presentar of this present rentall.'

Calculation of third: In the whole, £20; third thereof, £6 13s 4d.

FERN, PARSONAGE AND VICARAGE OF,[52] (EUL, fo 118r)

[*In margin*, 'This personage and vicarage lyis in Forffar.']

Rental of the parsonage and vicarage of Ferne pertaining to Mr Patrick Mowre, parson thereof.

The said parsonage of Pherne is £120 12s 4d. The vicarage gave 40 merks 'and now geiffis nathing.'

'Item, of this givin to the stellar [*recte*, stallar] in Dunkeld and to the man that servis the kirk of Pherne' £20.

'Writtin be me, Mr Patrik Mowr, persoun of Ferne.'

Signature: 'Mr Patrik Mowr, with my hand.'

Calculation of third: In the whole, £147 5s 8d; third thereof, £49 1s 10⅔d.

KILMORACK, VICARAGE OF,[53] (EUL, fo 118r-118v)

Rental of the vicarage of Kilmorak in the diocese of Ros pertaining to sir John Nicolsoun, vicar of the same.

The said vicarage of Kilmorak gave yearly in assedation 'quhen teindis and oblatiounis was payit' the sum of 40 merks.

'Presentit be me, sir Alexander Pedder, in name of the said sir Johne Nicolsoun.'

Calculation of third: In the whole, £26 13s 4d; third thereof, £8 17s 9¼d.

---

[52] Cf NLS, fos 87r, 103r (see above, pp. 371, 398). This benefice, in the sheriffdom of Forfar, was a prebend of Dunkeld cathedral, Cowan, *Parishes*, p. 65.

[53] Cf EUL, fo 114r (see above, p. 636).

ROSS, DEANERY OF, (EUL, fo 118v)

Rental of the deanery of Ros pertaining to Mr Mungo Monypenny, dean of the same.

The kirk of Ardrosscher [*i.e.* Ardersier], 5 c. victual. The quarter teind of Rosmerkin and Crumbathy [*i.e.* Rosemarkie and Cromarty], 6 c. 1 b. victual; of silver, £35. 'And to be deducit of the rentall' 20 merks yearly 'to the choristare for his fie.'

'Presentit be the said, Alexander Pedder, procuratour for the said dene.'

Calculation of third: Victual, 20 c.;[54] third thereof, 6 c. 10 b. 2 f. $2\frac{2}{3}$ p. Money, £35; third thereof, £11 13s 4d.

ROSS, CHANTORY OF, (EUL, fos 118v-119r)

Rental of the chantory of Ross pertaining to Mr John Cairncorce, 'chantour of the samin'.

In victual, 8 c. In silver, 100 merks.

Mr William Cranstoun has of the said chantory in yearly pension the sum of £30. John Gibiesoun, 'choristar in the channonry of Ross', has yearly for his fee, 21 merks. Total of the money paid yearly out of the chantory of Ross extending to the sum of 66 merks.

Signature: 'Mr Johne Cairncorce, chantour of Ros.'

Calculation of third: Victual, 8 c.; third thereof, 2 c. 10 b. 2 f. $\frac{2}{3}$ p.[55] Money, £66 13s 4d; third thereof, £22 4s $5\frac{1}{3}$d.

CROMARTY, VICARAGE OF, (EUL, fo 119r-119v)

Rental of the vicarage of Crumarty within the diocese of Ros 'quhen all dewtie pertening thairto was answerit.'

Paid yearly 22 merks, 'the curat being sustenit, and now lytill thairof answerit except the teind of the yeardis within the town quhilk wilbe' 6 b. victual or thereby; pertains to sir John Andirsoun, chamberlain of Murray.

---

[54] The correct calculation appears to be 11 c. 1 b.
[55] 2 p. have been omitted from this calculation.

Signature: 'Mr Thomas Ker at the command of sir Johne Hendirsoun, possessour of the foirsaid benefice.'

Calculation of third: In the whole, 6 b. victual; third thereof, 2 b.

LEMLAIR, PARSONAGE OF, (EUL, fo 119v)

Rental of the parsonage of Lymmlar lying within the diocese of Ros pertaining to Henry Kincaid gives to him by year the sum of 100 merks 'alanerlie.'

Signature: 'Henrie Kincaid, with my hand.'

Calculation of third: In the whole, £66 13s 4d; third thereof, £22 4s 5⅓d.

KILMUIR-EASTER, PARSONAGE OF, (EUL, fos 119v-120r)

Rental of the parsonage of Kilmowr within the diocese of Ros pertaining to George Dumbar, parson thereof.

The parsonage of Kilmowre yearly in rental 'quhen it was in use of payment', 100 merks.

Signature: 'Georg Dumbar, persoun of Kilmowr and vicar of Ramsky [i.e. Rosemarkie].'

'Personage of Kilmowr and vicarage thairof.'

Calculation of third: In the whole, £66 13s 4d; third thereof, £22 4s 5⅓d.

ROSEMARKIE, VICARAGE OF, (EUL, fo 120r)

Rental of the vicarage of Rosmaky [sic] pertaining to the said George Dumbar, 'quhen all teindis and small offrandis was in use of payment, and nathing payit now be the space of thrie yeiris extending in use and wont to' £20. Total, £20.

Signature: 'George Dumbar, persoun of Kilmowr, and vicar of Ramskye. Presentit be me, Alexander Pedder, procuratour for the said persoun of Kilmowr.'

Calculation of third: In the whole, £20; third thereof, £6 13s 4d.

ROSSKEEN, PARSONAGE OF, (EUL, fo 120r-120v)

Rental of the parsonage of Roskyne within the diocese of Ros pertaining to Mr Gavin Dumbar, parson of the same.

The said parish of Reskyne set all for money to the tenants of the same extending in the whole to the sum of £101 6s 8d.

'And of that deducit givin to the choristaris yeirlie', £10.

'Presentit be sir Alexander Pedder, procuratour for the said persoun of Roskyne.'

Calculation of third: In the whole, £101 6s 8d; third thereof, £33 15s 5⅓d.[56]

FODDERTY, VICARAGE OF, (EUL, fo 120v)

Rental of the vicarage of Feddirdy, lying within the diocese of Ros pertaining to sir Andrew Robertsoun.

'The said vicarage of Fedirdy, quhen all dewties usit and wont was payit, gaiff be yeir in assedatioun to the vicar' 20 merks.

'Presentit be sir [Alexander] Pedder, procuratour for the said sir Andro Robertsoun.'

Calculation of third: In the whole, £13 6s 8d; third thereof, £4 8s 10⅔d.

INCHRORY AND ST RULE, CHAPLAINRIES OF, (EUL, fos 120v-121r)

Rental of the chaplainries of Inschrory and Sanct Regule lying within the diocese of Ros pertaining to sir Andrew Robertsoun.

The said chaplainry of Inscherory yearly £5. The said chaplainry of Sanct Regule, £10.

'Presentit be sir Alexander Pedder, procuratour for the said sir Andro Robertsoun.'

Calculation of third:
  Chaplainry of Inschrory: In the whole, £5; third thereof, 33s 4d.
  Chaplainry of Sanct Regule: In the whole, £10; third thereof, £3 6s 8d.

---

[56] The correct calculation appears to be £33 15s 6⅔d.

[*In margin*, 'Haill to the queine etc.']

CONTIN, PARSONAGE AND VICARAGE OF, (EUL, fo 121r)

Rental of the parsonage and vicarage of Contane within the diocese of Ros pertaining to Mr Robert Burnet, parson of the same.

The said parsonage and vicarage of Burnett [*sic*] set in assedation for the yearly payment of the sum of £40.

Calculation of third: In the whole, £40; third thereof, £13 6s 8d.

KINCARDINE, PARSONAGE AND VICARAGE OF, (EUL, fo 121r-121v)

Rental of the parsonage and vicarage of Kincairdein within the diocese of Ros [pertaining] to Mr Thomas Denowne, parson thereof.

The said parsonage and vicarage set in assedation for 120 merks 'and now in falt of payment of the vicarage can not be maid ane perfyt rentall on the samin.'

Signature: 'Presentit be me, Mr Thomas Ros, persoun of Alnes.'

Calculation of third: In the whole, £80; third thereof, £26 13s 4d.

ALNESS, PARSONAGE OF, (EUL, fo 121v)

Rental of the parsonage of Alnes within the diocese of Ros 'devydit amang four portionaris.'

'That is to say, Mr Thomas Ros, principall persoun of the samin, quha gettis for his pairt set in assedatioun' 100 merks 'and thairof giffis furth of yeirlie pensioun to sir Andro Robertsoun', 20 merks. 'Item, the uther thrie portionaris, viz., sir James Buschart, Mr Alexander McKeinzie and Johne Robertsoune hes evrie ane of thame' 36 b. victual 'or thairby of the said personage be yeir.'

Signature: 'Maister Thomas Ros, persoun of Alnes, with my hand.'

Calculation of third: Money, £66 13s 4d; third thereof, £22 4s 5½d. Victual, 6 c. 12 b.; third thereof, 2 c. 4 b.

BALNAGOWN, CHAPLAINRY OF, (EUL, fo 122r)

Rental of the chaplainry of Balnagowne pertaining to the said Mr Thomas gives by year the sum of £4.

Signature: 'Mr Thomas Ros.'

Calculation of third: In the whole, £4; third thereof, 26s 8d.

KILTEARN, VICARAGE OF, (EUL, fo 122r)

Rental of the vicarage of Kilterne lying within the diocese of Ros pertaining to John Saidserff.

'The quhilk vicarage gaiff yeirlie in assedatioun to the said sir Johne the sowme of' 40 merks 'and now be yeir nathing be reasoun the paroschineris will not pay quhill farder ordour be put to the kirk of the samin.'

'Presentit be me, Mr Alexander Pedder, procuratour for the said sir Johne Saidserff.'

Calculation of third: In the whole, £26 13s 4d; third thereof, £8 17s 9$\frac{1}{3}$d.

CULLICUDDEN, PARSONAGE OF, (EUL, fo 122r-122v)

Rental of the parsonage of Cullicudden pertaining to David Dumbar, parson of the same.

The 'towne' of Cullicudden, 42 b. Drummecudyne, 22 b. Sanct Martenis, 9 b. Kynebarch, 10 b. The Craighouse, 4 b. Eistir Culbell, 14 b. Wastir Culboll, 20 b. The Wodheid, 4 b. The 'towne' of Braire, 16 merks.

Total victual, 165 b.,[57] 7 c. 13 b.; and in money, 16 merks.

Signature: 'David Dumbar, persoun of Cullicudyne, with my hand. Presentit be sir Alexander Pedder, procuratour for the said persoun, etc.'

Calculation of third: Victual, 7 c. 13 b.; third thereof, 2 c. 9 b. 2 f. 2$\frac{2}{3}$ p. Money, £10 13s 4d; third thereof, £3 11s 1$\frac{1}{3}$d.

---

[57] The correct calculation appears to be 125 b., which is 7 c. 13 b.

TAIN COLLEGIATE KIRK, NEWMORE, CHAPLAINRY OF,
(EUL, fo 123r)

Rental of the chaplainry of Newmoir[58] within the diocese of Ros given by George Munro, feuar[59] thereof.

Newmoir extends to 12 b. bere; 12 b. oatmeal; 12 b. small custom oats; four marts; four muttons; 4 dozen poultry; 16 merks; 16 merks [sic]. 'The quhilk rentall was sett to the said George in assedatioun be umquhill Johne Bissatt, chaplane thairof, and now be Hectour Monro, chaplane for this tyme present, for the sowme of' £30 money 'allanerlie.'

Signature: 'George Munro, with my hand.'

Calculation of third: In the whole, £30; third thereof, £10.

TAIN COLLEGIATE KIRK, TARLOGIE, CHAPLAINRY OF,
(EUL, fo 123r)

Chaplainry of Talaquhy

Rental of the chaplainry of Carlaquhy [sic] set to George Munro in feu by sir Robert Melvill, chaplain thereof.[60]

'Item, Tarlaquhy in few maill', £20.

Signature: 'George Munro, with my hand.'

Calculation of third: In the whole, £20; third thereof, £6 13s 4d.

[*In margin*, 'Haill to the queine.']

ALNESS, VICARAGE OF,
(EUL, fo 123v)

Rental of the vicarage of Alnes pertaining to Mr John Davidsoun, 'maister of the peddagog of Glasgow'.

---

[58] See *RSS*, vi, no 2533.
[59] Cf *RMS*, iv, no 2792.
[60] See *RSS*, v, no 3061; vi, no 2520.

'The quhilk gaiff quhen payment was maid the sowme of' 20 c. [*recte*, £20, *next paragraph*] 'be yeir with ane plaid bot the laird of Fowlis and his freindis hes not lattin me gett ane penny thairof this fyve yeiris bygaine.'

Calculation of third: In the whole, £20 and a plaid; third thereof, £6 13s 4d, 'plaid[61] with the pleneissing, etc.'

OLRIG, PARSONAGE AND VICARAGE OF, (EUL, fo 123v)

Rental of the parsonage and vicarage of Alrig pertaining to Mr William Sincklar lying within the sheriffdom of Innernes extending in the whole to the sum of £134 9s; third thereof, £44 9s 8d.[62]

LATHERON, VICARAGE OF, (EUL, fo 123v)

Rental of the vicarage of Ladroun pertaining to Mr William Sincklar.

Calculation of third: In the whole, £40; third thereof, £13 6s 8d.

REAY, VICARAGE OF, (EUL, fo 123v)

Rental of the vicarage of Ray pertaining to sir William Reid, vicar thereof.

Calculation of third: In the whole, £20; third thereof, £6 13s 4d.

THURSO, VICARAGE OF, (EUL, fo 124r)

Rental of the vicarage of Thursoch pertaining to Mr Walter Innes, lying within the sheriffdom of Innernes.

Calculation of third: In the whole, £16; third thereof, £5 6s 8d.

WICK, VICARAGE OF, (EUL, fo 124r)

Rental of the vicarage of Weik lying within the sheriffdom of Innernes and diocese of Cathnes pertaining to Mr Andrew Grahame, vicar thereof.

---

[61] In the MS the scribe has placed a diagonal cross before 'plaid'.
[62] The correct calculation appears to be £44 16s 4d.

The said vicarage is set yearly to John Keith, 'capitane of Akergile', for £40, 'and of grassum ilk thrid yeir', £20.

Signature: 'Androw Grahame. Ressavit at comand of the Clerk Registre.'[63]

DOUNIE, CHAPLAINRY OF, (EUL, fo 124r)

Rental of the chaplainry of Dunny extending in the whole to £10; third thereof, £3 6s 8d.

Signature: 'James,[64] resaiff this rentall, J. Clerk Registre.'

BEAULY, PRIORY OF, (EUL, fos 124v-126r)

Rental of the priory of Bowlyne, 'baith of the maillis, silver, fearmis, teindis, martis, wedderis and utheris dewties as efter followis.'

The rental of silver: The silver mail of the barony of Bewlyne with the Maynis of the same extends to £61. The kirks of Convith and Cumer should pay in silver in the year the sum of £33. The kirk of Abirtarfe should pay in silver the sum of £42. 'Summa of the haill silver in maillis and teindis extendis to' £136 13s 4d.[65]

The rental of the victual of the said priory: The whole victual of the barony of Bowlyne with the Maynis of the same extends to 4 c. victual. The kirks of Conveith and Cummer in victual extends to 7 c. 11 b. The two mills of Bowlyne set for 2 c. 8 b. meal and malt. Total of the whole victual extends to 13 c. 4 b.[66]

The rental of oats: The whole oats of the said barony, 8 b. The marts: In marts, 10 marts. The mutton: In wedders, 20 wedders. In poultry, 21 dozen.

'As for the fisching of Bowlyne, it is uncertane sumtymis les sumtymis mair and uther tymis verie lytill, and thir twa yeiris bygaine hes scarslie givin' 2 last 6 barrels.

[*In margin*, 'Beaulyne']

'And sua the haill priorie of Bewlyn extendis yeirlie in silver, victuale and wedderis, aitis, mairtis, pultrie and salmond as efter followis': £136 13s 4d; 14 c.

---

[63] James McGill of Nether Rankeilour was Clerk Register at the Reformation, *Handbook of British Chronology*, p. 197.
[64] James Nicolson, clerk of the Collectory, *Thirds of Benefices*, p. 62.
[65] The correct calculation appears to be £136.
[66] The correct calculation appears to be 14 c. 3 b.

3 b. victual; 8 b. oats; 10 marts; 20 wedders; 21 dozen poultry; 2 last 6 barrels salmon.

'Thir ar the thingis that ar to be deducit of the money, salmond and victuallis abone specifeit, payit as efter followis: Item in primis, thair is to be deducit be payment maid to the aucht brethir for thair habit silver, ilk breder havand in the yeir' 40s; which extends to £16.

'Item, thair is to be deducit for the said viij brether for thair flesch and fisch in the yeir, ilk brother havand for thair flesch', 3d 'in the day, for thair fisch ilk day' 2d, extending in the whole in the year to £29 14s 8d.[67]

For the Lords of the 'Seit' contribution yearly, £4 4s. Mr Alexander McKenzie for his yearly pension, 'quhilk he hes of the said pryorie and provydit thairof in Roome' £13 6s 8d. To the officer of Bowlyne yearly for his fee 'quhilk he hes dureing his lyftyme' 26s 8d. 'Item, thair is to be deducit for the said aucht bretheris drink in the yeir', 112 b. victual. 'Item, for thair breid in the yeir', 57 b. 3 f. 1 p. 'Item, for the officiaris fie', 1 b. 'Item, thair is to be deducit for the officiaris fie', 1 b. [*sic*] 'Item, thair is to be deducit for the teind fisch of the kirk of Warlaw', $3\frac{1}{3}$ barrels of salmon. 'Summa of the haill silver, victuallis and salmond deducit as is abonewrittin extendis to' £64 13s[68] of silver; 10 c. 10 b. 3 f. 1 p.[69] victual; $3\frac{1}{3}$ barrels of salmon.

'And sua restis to the prior' £72 16d;[70] 3 c. 8 b. 3 p. victual; 2 last $2\frac{1}{3}$ barrels[71] of salmon.

'Memorandum, that the kirk of Conveith was wont to pay for the vicarage thairof the sowme of' £27 13s 4d 'and now gettis na payment of the samin.'

Signature: 'W[alter], abbot of Kinlos.'[72]

Calculation of third: Money, £136 13s 4d; third thereof, £45 11s $1\frac{1}{3}$d. Victual, 14 c. 3 b.; third thereof, 4 c. 11 b. 2 f. 2 'half p.'[73] Oats, 8 b.; third thereof, 2 b. 2 f. $2\frac{2}{3}$ p. Marts, 10; third thereof, $3\frac{1}{3}$. Wedders, 20; third thereof, $6\frac{2}{3}$. Poultry, 21 dozen; third thereof, 7 dozen. Salmon, 2 last 6 barrels; third thereof, 10 barrels.

---

67 The correct calculation appears to be £60 16s 8d.

68 The correct calculation appears to be £64 12s.

69 This is correct if the 1 b. for the officer's fee is added once only.

70 This is correct if £64 12s is deducted.

71 The correct figure should be '$2\frac{2}{3}$ barrels'.

72 Walter Reid was abbot of Kinloss and the dependent priory of Beauly at the Reformation, Cowan and Easson, *Medieval Religious Houses*, pp. 76, 80.

73 The correct calculation appears to be 4 c. 11 b. 2 f. $2\frac{2}{3}$ p.

'Memorandum, to tak the salmond the thrid not as it is rentallid bot as it givis, for this rentall is mauchlitt.'

'Assumptioun as it is in this rentall':
    Third of money, £45 11s 1½d. Take the barony of Bewlyne with the Manis giving by year according to this rental £61. 'Give in' £15 8s 10½d.[74]
    Third of victual, 4 c. 11⅔ b. Take the Manis and barony of Bewlie according to this rental for 4 c.; take the rest out of the two mills of Bewlyn 'et sic eque.'
    'Aitis, mairtis, muttoun, powder [*recte*, poultry] and wedderis ar givin up in generall, thairfoir gar the Collectour[75] seik thame out.'
    Third of salmon, 10 b. out of the fishing of Bewlyne.

'Remember my lord Comptrollar[76] to speir the rentall of thir twa, Kinlos and Bewlyne, for they ar suspitious anent the fisching.'

ROSS, SUBDEANERY OF,   (EUL, fos 126v-127v)
TAIN AND EDDERTON, PARSONAGES OF,

Rental of the kirks of Tayne and Eddirtane 'annexit to the subdenrie of Ros as thay pay'.

The parish of Thayne: The 'towne' of Tayne payis in rental to the hands of sir Nicholas Ros, commendator of Ferne, Thomas Fiddes, Andrew Ros and his mother, Nicholas Ros and 'Mitchell' Furde, 104 merks. Mornichie, half a davach of land pays by Nicholas Ros, 12 merks. Talraquhy, one davach of land pays by Alexander Ros, laird of Balnagowne, 22 merks. Cambuscurry, three quarters of a davach of land pays by Adam Hay, 21 merks. Plaiddes, three quarters of a davach of land pays by Alexander Innes of Catboll, 8 merks. Pettogarty, half a davach of land pays by the said Alexander Innes of Catboll, 8 merks. Ballecherye, one quarter of a davach of land pays by the said Alexander Innes, 4 merks. Innerrartie and Balna Touch, half a davach of land pays by Michael Furde, 6 merks. Petgerello, half a davach of land pays by the hands of John Drumond, 8 merks. Balnagaw, half a davach of land pays by Walter Innes, 4 merks. Lochiskyne [*i.e.* Lochslin] and Newtown, one davach of land, 'the ane half pertenis to the bischope of Ros, the uther half payis subdeane of Ros', 6 merks. Skardy with the pendicles, occupied by Agnes Ros, pays yearly 6 merks.

---

[74] The correct calculation appears to be £15 8s 10⅔d.
[75] Presumably Sir John Wishart of Pittarow, who was appointed Collector general on 1 March 1562, *RSS*, v, no 998.
[76] Bartholomew de Villemore, Comptroller, 1555 - 1562, with Thomas Grahame of Boquhaple, 1561 - 1562. They were succeeded by Sir John Wishart of Pittarow, 1562 - 1565. *Handbook of British Chronology*, p. 191.

Kerskeith, occupied by Andrew Ros, pays 3 merks. Anley, occupied by Agnes Ros, pays 2 merks.

The parish of Eddirthane: Iddirthane, half a davach of land occupied by the said Alexander Ros of Balnagown, pays 10 merks 6s 8d. Westrey and Meltoun, half a davach land occupied by the said laird', 4 merks. Rowny [Downe, *fo 127v*], one davach of land occupied by the said Alexander Ros of Balnagown, pays 14 merks. Mekle Doles and Bechestoun [Leichstoun, *fo 127v*], two quarters of a davach of land occupied by the said Alexander Ros of Balnagown, pays 10 merks. Lytill Doles, three quarters of a davach of land occupied by John McCalmestoun, alias Ros, pays 10 merks. Wastir Farnie, one davach of land occupied by William Ros, pays 18 merks. Estir Ferne, one davach of land occupied by Thomas Ros, pays 8 merks. Dathan Mekle, three quarters of a davach of land occupied by Walter Ros, alias Alexander Waltersoun, his mother and Thomas Ros, pays 6 merks. Dathan Lytle, one quarter of a davach of land occupied by William Ros and his brother, pays 3 merks. 'The subdeanis craft [*sic*] and maynis within the channonrie of Ros occupyit be Andro Wilgrief wyff payis' 3 merks.

'Suma of the haill rentall of the subdenrie of Ros as it payis in silver on the uther syd extendis to' 300 merks 6s 8d;[77] 'ditto', 3 merks.

The assumption of the third of the subdeanery of Ros:
Thyne parish: 'Ane quarter of the town of Thane led be sir Nicolas Ros, comendatour of Ferne, payis' 26 merks; 'and half quarter of the said town of Thane led be Thomas Fiddes of Thane payis' 13 merks; Talrogy, one davach of land occupied by the laird of Balnagown, pays 22 merks; Kerskeith, occupied by Andrew Ros, 3 merks.

Accherthane [*i.e.* Edderton] parish: Eddirthane occupied by the laird of Balnagown, half a davach of land pays 10 merks 6s 8d; Westray and Myltoun occupied by the said laird, half a davach of land pays 4 merks; Mekle Doles and Leichstoun [Bechestoun, *fo 127r*], three quarters [$\frac{2}{4}$ *above*] of a davach of land occupied by the said laird pays 10 merks; Downe, one davach of land occupied by the said laird pays 14 merks.

'Sum[m]a of the haill,' 102 merks 6s 8d.

Signature: 'Mr James Thorntoun, subdeane of Ros.'

---

[77] This calculation is correct.

NOTE

Rentals for the following benefices are located elsewhere in the text:

BEAULY, PRIORY OF (SRO, Vol. a, fo 398v), p. 464.
ROSS, SUBDEANERY OF, PENSION FROM (SRO, Vol. a, fo 164v), p. 159.

Orkney

ORKNEY, BISHOPRIC OF,[1]   (EUL, fos 128r-139r)

'Rentale fructuum episcopatus Orchadie'

Rental of the fruits of the bishopric of Orkney, within Caithness and Shetland and Orkney.[2]

The lands of Wastray and Papea Westray of the said bishopric set in feu to Gilbert Balfour for yearly payment of the sum of £113 'quhilk is lyand waist in hieland[3] menis punisoun this last yeir'. The parsonage here is adjoined to the bishopric and the teinds[4] set with the lands. The lands of Barsay [Birsay, *fo 144r*] and Marwik and Birsay 'be eist and wast is lyk wayis sett in few to the said Gilbert' for yearly payment, £173 18s 8d argent.[5] Also the parsonage here is adjoined 'as said is' to the bishopric. The lands of Ewe[6] and Sansyd [Iwey Sandis, *fo 138v*; Evey and Sandis in Deirnes, *fo 147v*] with Stoknes [*i.e.* Scockness] in Rowsay, together with the teinds thereof, are set to Edward Sinclar for yearly payment of £102. Also the parsonage here is adjoined 'as said is' to the bishopric. The isle of Ethie set in feu to Oliver Sinclar of Quh[y]tkirk for payment of 23 merks argent yearly; 5 lasts flesh. Also the parsonage here is adjoined to the bishopric. The isle of Egilsay set to Duncan Scolay for £20 argent. Also the parsonage adjoined 'as said is.'

---

[1] Cf EUL, fo 144r (see below, p. 668). Two earlier and different rentals occur in *Records of the Earldom of Orkney 1299 - 1614*, ed. J. Storer Clouston (Edinburgh, 1914), pp. 404-419.
[2] In the rentals for the bishopric of Orkney, the measurements differ from those elsewhere; 24 marks = 1 settin; 6 settins (sts.) or 6 lispunds = 1 meil (m.); 24 meils = 1 last; 12 barrels = 1 last; 1 last or 24 meils = 1 chalder. *Thirds of Benefices*, p. 42, n. 1; *Old-Lore Miscellany of Orkney, Shetland, Caithness and Sutherland*, ed. A. W. Johnston, ix (London, 1933), pp. 242-243.
[3] The entry originally read 'his handis' and was altered to 'hieland'; cf entries for Westray on fos 140v, 141v (see below, pp. 666-667).
[4] 'Of Mary Kirk' is included at this point on fo 144r (see below, p. 668).
[5] 'Argent' was the term used for Scots money in the rentals to distinguish it from Orkney currency, *Old-Lore Miscellany*, p. 71.
[6] Presumably the parish of Evie (and not Eves Howe in Deerness), despite the distance of Evie from Sandside in Deerness.

Total of the foresaid lands in argent extends to £404 5s 4d;[7] 6 lasts victual;[8] 5 lasts fish [flesh, *preceding paragraph*].

Sanday: Sanday in land mails extends to 6 lasts 2 sts. [5 lasts 2 sts., *fo 144v*]; £3 11s 5d [£3 6s 5d, *fo 144v*; £3 11s 6d, *fo 147v*] silver; 2 barrels of butter. 'Item, the auld rentale hes' 12 'meillis fisch [flesh, *fo 147v*], not exprimit in the new nor in use of payment. Suma [*sic*] patet.' [*i.e.* The total is obvious.]

Rowsay: The lands of Rowsay set to Mr Magnus Halcro of Brugh for yearly payment of 5 lasts [5 lasts 5 m. 3 sts., *fo 144v*; 5 lasts 5 m. 4 sts., *fo 147r*]; £10 argent; 15 barrels of butter. The teinds of Rowsay, the parsonage annexed therewith, extends to 5 lasts 29 m.[9] 4 sts. cost.[10] Total of this isle in victual is 11 lasts 1 m. 1 st.; 15 barrels of butter; £20 argent.[11]

Stronsay: The lands of Stronsay extend to 7 lasts 14 m. 3 sts. bere; 7 barrels of butter; 4 sts. [4 m., *fo 144v*] of flesh. The teinds thereof, because the parsonage of 'S P K' [*i.e.* St Peter's kirk] is adjoined to the bishopric, extend yearly to 7 lasts 16 m. bere. Extending in the whole in cost to 10 lasts 4 m. 2 sts. Total of this isle of Stronsay in victual extends 'in cost to' 10 lasts 4 m. 2 sts.;[12] 7 barrels of butter; 4 sts. flesh.

Schaipnysay: The lands of Scarpinsay, with the teinds thereof because they are adjoined, extends [to] 14 lasts 19 m. 3 sts. [14 lasts 19 m. 4 sts., *fo 144v*] cost; 5 lasts 8 m. 4 sts. flesh [fish, *below*]; 2 barrels of butter; 1 barrel of oil; £4 3s 5d silver. Total of this isle in victual, 14 lasts 19 m. 3 sts.; 2 barrels of butter; 1 barrel of oil; £4 3s 5d silver; 5 lasts 8 m. 4 sts. fish [flesh, *above*].

Deirnes: The lands of Deirnes with the teinds thereof are set for yearly payment of 10 lasts 8 m. 5 sts. cost; 3 lasts 17 m. 2 sts. flesh with 46s of 'skeat silver'; 2 barrels of butter; 158 kane fowls; 2 oxen; 'and twa bairris' [*i.e.* boars]; 46 'fouodome'[13] peats. 'And heirof yeirlie defalkit for our blawin landis,[14] ane last victuale to the commounte.'

Burray, Flattay [*i.e.* Flotta], Futhanne [*i.e.* Switha; Schewsay, *fo 138v*], Suaay [*i.e.* Swona; Sallway, *fo 138v*] set to Barbara Stewart for the sum of £60 argent and with 24 'mais[15] skray[16] and heir alsua for our blawin landis yeirlie defalkit of the foirsaid sowme', 10 merks. 'Item, schoe sould pay' 80 'cuple of cunningis or ellis for everilk cuple', 4d argent.

---

[7] The correct calculation appears to be £424 5s 4d.
[8] No figures for victual are given in the preceding paragraph.
[9] '5 lasts 19 m. 4 sts.', fo 144v (see below, p. 669). As there were 24 meils to a last, '29 m.' is probably a scribal slip.
[10] The term 'cost' was used in Orkney to mean victual paid for rent, *Thirds of Benefices*, p. 42.
[11] These figures do not tally.
[12] 7 lasts 14 m. 3 sts. plus 7 lasts 16 m. equals 15 lasts 6 m. 3 sts.
[13] I.e. fathoms, a measurement of peats.
[14] I.e. lands overblown with sand, see fo 132r (see below, p. 658).
[15] Maise were 'heavy nets or net-like bags used for carrying peats or fish', *Thirds of Benefices*, p. 44, n. 11.
[16] I.e. 'scrafish', young saithe cured by drying in the sun, *Thirds of Benefices*, p. 44, n. 10.

Sanct Awlaw [*i.e.* St Ola's] parish: For the lands [landis maillis, *fo 145r*] thereof, 1 last 6 m. cost; 20 [nyntein, *fo 145r*] barrels of butter; 1 cow; £5 9s 8d [£5 19s 8d, *fo 145r*] argent; 'and xx lib. of walkis [*i.e.* 20 lb. of wax]'. For the teinds, 4 lasts cost. 'Suma of victuale', 5 lasts 6 m.

Holme: The land mails of the said parish extend to 6 lasts 12 m. 3 sts. 6 marks; with 1 last 18 m. 5 sts. 20 marks flesh. 'Item, for the teindis quhilk ar in maner foirsaid annexit', 13 m. [23 m., *fo 145r*] cost; 28s silver; 10 'fowdom' peats [1 fathom, *fo 145r*]; 60 fowls. Total victuals, 7 lasts 11 m. 3 sts. 6 marks.[17]

Rannaldsay Besouth: The teinds thereof extend to 18 lasts 14 m. 4 sts. bere; and in land mails, 15 m. bere; extending in cost to 12 lasts 19 m. 4 sts. 16 marks.[18]

Wawis [*i.e.* Walls]: 'Item, Wawis payis half ane barrell oyle.'

Sanctandros parish: The teinds of the foresaid extend to 2 lasts 14 m. cost; 2 barrels of butter; 10s silver. 'Patet summa.'

Orphir: Orphir pays yearly 2 barrels of butter; 9 sts. cost; 9 sts. flesh.

Firthe: The teinds of Firth extend to 3 lasts 22 m. cost; 1 barrel of butter.

Stromnes: Stromnes pays yearly ½ barrel of butter; 8 sts. cost.

Harray: The teinds thereof extend yearly to 5 lasts 16 m. 4 sts. cost. 'Suma patet etc.'

Ethallow [*i.e.* Eynhallow]: The same pays 1 barrel of butter.

Sandwike: The teinds thereof extend to 7 lasts 2 m. 1 st. cost; in land mails, 1¼ barrels of butter; 10s silver.

Stenhous [*i.e.* Stenness]: In teinds, 1 last 17 m. cost; 4 barrels of butter; and in flesh, 1 mart.[19]

Randall [*i.e.* Rendall] and Garsay [*i.e.* Gairsay]: The teinds thereof extend to 4 lasts 11 m. 3 sts. cost.

Farray: Farray pays 2 barrels of butter.

The fruits of the bishopric of Orkney within parts of Caithness: The lands of Strybustar [*i.e.* Scrabster] with the 8d land of Dourne [*i.e.* Dounreay] with the mill thereof set in feu to Mr William Mudie for yearly payment of £11 6s. The lands of Halcro in Cathnes with the mill thereof set in feu to James McBath for the yearly payment of £4 argent and 6 b. meal. The 16d land of Drumray [*i.e.* Dounreay] with Brubister, Furro [Furay, *fo 146r*] and Lumay [Lunay, *fo 146r*] set in feu to the earl of Sutherland[20] [to William Sinclar of Dumbaith, *fo 146r*] for yearly payment of £30 argent. Total of the duties of the lands of Cathnes extends to £45 6s; 6 b. meal.

---

[17] The correct calculation appears to be 8 lasts 20 m. 3 sts. 2 marks.
[18] The correct calculation appears to be 19 lasts 5 m. 4 sts.
[19] These teinds are set for £10 'argent' on fo 145v (see below, p. 670).
[20] John Gordon, 10th earl of Sutherland, died in 1567, and was succeeded by his son, Alexander, 11th earl. *Scots Peerage*, viii, pp. 339-344.

Total of the whole rental in silver is £537 19s 10d.[21] 'Item, for the cuningis', £1 6s 8d. 'Inde in money' £539 6s 8d.[22]

'Of the quhilk thair is yeirlie deducit for landis overblawin with sand in Burray' £7 13s 4d. 'And sua de claro the sowme totall is' £531 6s 6d.[23] Total of the whole victual extends to 109 lasts 13 m. 3 sts. 22 marks cost; 6 b. meal.[24]

'Nota, thair is to be defalkit of the blawin land with sand in Derenes' 1 last cost. And of butter in Orknay, 5 lasts 2 barrels 6 lispunds; 'and ane barrell oyle and ane half';[25] and of flesh, 13 lasts 14 m. 2 sts. 20 marks and 4 marts; 2 swine; 218 poultry; 56 'foudoume' peats; 20 lb. wax; 'and of skray', 24 'mais'.

Rental of the fruit of the bishopric of Orkney within parts of Shetland: 'In primis in Done Rosness' 16 'pece';[26] 'inde' 8 barrels of butter, 8 barrels of oil. The Fair Yle, 2 'pece'; 'inde' 1 barrel of butter, 1 barrel of oil. Brasay and Burray, 4 'pece'; 'inde' 2 barrels of butter, 2 barrels of oil. Sandsting and Nesting[27] [*recte*, Aithsting], 6 'pece'; 'inde' 3 barrels of butter, 3 barrels of oil. Vannis [*recte*, Walls; Waynis, *fo 135r*; Wais, *fo 146v*], 6 'pece'; 'inde' 3 barrels of butter, 3 barrels of oil. Skatstay and Delting, 6 'pece'; 'inde' 3 barrels of butter, 3 barrels of oil. Northmawing [*i.e.* Northmavine], 8 'pece'; 'inde' 3 barrels of butter, 5 barrels of oil. Yell set to Mr William Lowder for 10½ [*recte*, 10½] 'pece'; 'inde' 4½ barrels of butter, 6 barrels of oil. Fetlair, 7½ 'pece'; 'inde' 3 barrels butter, 4½ barrels of oil. Onst [*i.e.* Unst], 13½ 'pece'; 'inde' 6 barrels of butter, 7½ barrels of oil. Nestinge, Lunay, Quhalsay and the Skarris, 8 'pece'; 'inde' 4 barrels of butter, 4 barrels of oil.

'Inde summa buteri' 3 lasts 8 barrels.[28] 'In oyle', 3 lasts 11 barrels.[29] 'Extending in integro to' 7 lasts 3½ barrels of butter and oil.[30]

In the North Iles 'of scatt', 1 barrel of butter; 11 barrels of oil. 'Item, mair in Zetland in land maill: Occlositter' [Collessetter, *fo 139r*; Facctosetter, *fo 146v*], 10s Zetland payment. The land mail [of] Grindsulle [Grinsett, *fo 139r*;

---

[21] This total is difficult to verify owing to discrepancies in the sub-totals.
[22] The correct calculation appears to be £539 6s 6d.
[23] The correct calculation appears to be £531 13s 4d.
[24] These totals are difficult to verify owing to the discrepancies in the sub-totals.
[25] Walls gave ½ barrel of oil, fo 130v (see above, p. 657); no other oil is recorded in this sub-section.
[26] I.e. a unit of any commodity, cf *Court Book of Shetland*, ed. G. Donaldson (Lerwick, 1991), p. 178.
[27] Sandsting was united to Aithsting in the sixteenth century, Cowan, *Parishes*, p. 180. The adjacent parish of Nesting lies to the east of Sandsting.
[28] The correct calculation appears to be 3 lasts 4½ barrels.
[29] The correct calculation appears to be 3 lasts 10¾ barrels.
[30] These figures do not tally.

Greinsulle, *fo 147r*], 8s Zetland payment. The land mail of Dypdall, 5s Zetland payment. The land mail of Normaven [*i.e.* Northmavine], 12s Zetland payment.[31] [*In margin*, 'Nota, the schilling of Zetland payment is bot four d Scottis.']

'Thir ar the ordinar charges and pensiounis of the bischoprik of Orknay: In the first, for the haill taksmenis fies', 5 lasts 10 c. 4½ sts. cost. 'Item, to the cowpar for his fie', 100 merks. To sir Magnus Murray, 'ane auld man', for his duty, £13 6s 8d. To the constable and bailie for his fee, 100 merks. To sir Robert Sinclar 'siklyk for his fie quhilk is ane ordinar dewtie', 1 last victual. To my lord Caldanghame[32] of yearly pension, £400. To my lord Ruthvenis[33] son of yearly pension, £200. 'Item, to the Justice Clerkis[34] sonnes of yeirlie pensioun', £400. To Mr James Thorntoun of yearly pension, £20.

'Item, to the maisteris' [*recte*, ministeris], [*blank*].

'Heir followis the names of the kirkis of Orknay that hes neid of ministeris to serve the people, to minister the sacramentis, to instruct and teach thame in the knawlege of the word of God.

In primis, the ile of Ralaldsay [*i.e.* South Ronaldsay], thair is in it twa parosch kirkis, the ane calit Sant Peteris kirk perteining to the bischope[35] and the uther callit Our Lady kirk pertening to the provest.[36] Borray with the Holmes thairof hes ane parosch kirk and neidis in speciall ane minister. Flottay, Sounay [*i.e.* Swona] and Skynthy [*i.e.* Switha], distant four or fyve myle of sett [*recte*, sea] sindrie haifand ane kirk in Flattay and sua nedis ane minister. The ile of Wawis [*i.e.* Walls] and of Hoy, ten myle distant, neidis twa ministeris bot may skears susteine ane of the haill fruitis thairof. Gramsay and Stormnes perteining to sir Maggus [*sic*] Strang as persoun thairof and neidis a minister bot payis na mair bot iiij last of beir. Sandwik neidis ane minister. Bresay and Marway[37] [*i.e.* Birsay and Marwick] neidis ane minister. Harray neidis ane minister, quhilk is distand thairfra fyve myle. Orpher, Firth and Stanehous lyand within sax mylis neidis ane minister. Randell and Eway neidis ane minister. Papplay[38] and Holme

---

[31] 'Item, fyve pund out of the custumis of Abirdeine' is included at this point on fo 147r (see below, p. 670).
[32] John Stewart was prior of Coldingham at the Reformation, and his son, Francis Stewart gained the commendatorship in 1565, *RSS*, v, no 2182; cf above, p. 198, n. 61.
[33] This rental, signed by Tullibardine as Comptroller, is therefore to be dated no earlier than 1565. William, 4th lord Ruthven, succeeded his father in June 1566, was created earl of Gowrie in 1581, and was executed in 1584. *Handbook of British Chronology*, p. 191; *Scots Peerage*, iv, pp. 261-263.
[34] Cf below, p. 670, n. 86.
[35] Adam Bothwell was bishop of Orkney, 1559 - 1593, Watt, *Fasti*, p. 254.
[36] In the reconstitution of 1544, the vicarage of South Ronaldsay was assigned to the provost of Orkney, Cowan, *Parishes*, p. 186. Alexander Dick was provost, 1553-1576 x 1584, Watt, *Fasti*, p. 255.
[37] See H. Marwick, *Orkney Farm-Names* (Kirkwall, 1952), pp. 130, 136-137.
[38] Paplay denotes the eastern end of the parish of Holm, Marwick, *Orkney Farm-Names*, p. 89.

neidis ane minister. Sanct Androis parosch and Dernenes, sax myle sindrie, neidis ane minister. Edday, ane ile of sax myle lang, and Farray, ane uther ile, neidis ane minister. Schailpinsay, ane ile of sax myle land [sic], neidis ane minister. Stromsay hes twa parosch kirkis,[39] Sanct Petteris quhilk perteinis to the bischope of [Orkney],[40] Sanct Michallis [recte, Nicholas] quhilk belangis to Mr Francis Bothwell[41] and neidis ministeris. The ile of Papy Sonsay [i.e. Papa Stronsay] inhabitant [sic], thrie myle wyd, neidis ane minister. Sanday, twelf myle lang, havand twa paroschinaris [recte, parishes], ane belangand to Mr Johne Maxwell quhilk is callit Our Lady parosch, and the Corce parosch quhilk Mr William Persoun hes, and ilk ane of thame neidis ane minister.[42] North Ronaldsay, distant from ony ile four myle, neidis ane minister. Item, Westray and Papy [i.e. Papa Westray], nyne or ten myle wyde, haifand twa paroschinis, Our Lady parosch, perteining to the bischope, and the Corce kirk, perteinand to Mr James Annand,[43] quhilkis twa kirkis neidis twa ministeris. The ile of Rousay, fyve myle of lenth, and neidis ane minister. The ile of Eglisay neidis ane minister.'

'Heir followis the kirkis of Zetland quhilkis neidis ministeris, of the quhilkis kirkis halylie the bischope is persoun, and neidis ilk ane of thame ane minister be reasoun of thair far distance fra utheris.

In primis, the Fair Ile, distant from all landis xl myle of wicked seas, neidis ane minister. Item, the parosch of Drumrosnes, xxiiij myle long, neidis twa ministeris. Item, the ile of Brasin neidis ane minister. The ile of Burray, Troundray [i.e. Trondra] neidis ane minister. The parosch of Sandsting and Eistinge[44] [i.e. Aithsting] neidis ane minister. Wanyis [i.e. Walls; Vannis, fo 132v; Wais, fo 146v], viij myle, neidis ane minister. Foulay, xviij myle from any landis, neidis ane minister. Skatsay and Delting, xij myle wyd, neidis ane minister. Northtmavin, xiiij or xvj myle wyd, neidis ane minister. The ile of Yell, xvj myle lang, neidis ane minister. Fettlar, fyve myle lang, neidis ane minister. Wnster [i.e. Unst], xij or xvj myle lang, neidis ane minister. Nefang [i.e. Nesting] and Lunay in the new land [i.e. Mainland] neidis ane minister. Qualsay [i.e. Whalsay], from ony land [blank] myle, neidis ane minister. The kirk of Tingwall, quhilk is the archdeanis, neidis a minister.'

'Memorandum, that the landis sett in few to Gilbert Balfour ar sett for silver in the tyme of the inhibitioun and thairfoir null, albeit in the rentale befoir they payit victuale, ule [i.e. oil], flesche, quhilk thing man be tane ordour with that the victuale be payit conforme to the rentall [blank], and sua the sam to be cravit and

---

[39] Stronsay had three parishes, St Peter, St Nicholas and St Mary, Cowan, *Parishes*, pp. 192-193.
[40] St Peter's parish on Stronsay belonged to the treasurer of Orkney, not to the bishop, Cowan, *Parishes*, p. 193.
[41] Francis Bothwell was treasurer of Orkney, 1561 - 1574, Watt, *Fasti*, pp. 257-258.
[42] Sanday had three parishes, the third being St Colme, Cowan, *Parishes*, p. 180.
[43] James Annand was parson of Westray and minister there, *Thirds of Benefices*, p. 151.
[44] See above, p. 658, n. 27.

na silver to be tane of thir few landis, bot to defalk the silver of this rentall, and to crave the victuall and thir dewties usit and wont. Item, the lyk is to be done of the fewis sett to Oliver Sinclar, to Mr Magnus Hetar [*recte*, Halcro] and to Mr Williame Lawder.'

'Memorandum, that the haill rentall as it is givin in be the bischope of Orknay extendis in silver to the sowme of' £529 18s 2d [£537 19s 10d, *fo 131v*].

'Defalcatio: Thairof to be defalkit for the landis sett to Gilbert Balfour in few for silver quhairof the queine mane crave victuale, fische, butter thairof as of auld', £266 18s 8d. 'Item, for the teindis of Ethay sett to Oliver Sinclar in few for the samin reasoun', £15 6s 8d. 'Item, for over blawin land with sand in Burray, the sowme of' £6 13s 4d. 'Suma of the haill defalcatioun of the money foirsaid', £288 18s 8d. 'Sua restis of silver to be cravit payment of', £250 19s 6d;[45] third thereof, £84 13s 2d[46] [*in margin*, '3a argenti'].

'Victuale: The haill victuall givin in be the bischope in rentall extendis to' 109 lasts 21 m. 3 sts. 22 marks cost [109 lasts 13 m. 3 sts. 22 marks, *fo 132r*].

'Defalcatio: Thairof to be deducit for the ovir blawin landis in Dernis', 1 last victual. 'Sua restis of victuall to be cravit payment de claro', 108 lasts 21 m. 3 sts. 22 marks; third thereof, 26 lasts 8 m. 1 sts. 7 marks.[47] 'Nota, that the victuall of the few landis sett for silver is not comprehendit heirintill bot to be cravit conforme to the rentall.'

'The haill buttir givin in be the bischope in rentall extendis to' 8 lasts 6⅓ barrels 6 lispunds of butter;[48] third thereof, 2 lasts 10⅓ barrels 2 lispunds of butter.

'Oleum: Suma of the oyle givin in be the bischope in rentall extendis to' 4 lasts; third thereof, 1 last 4⅓ barrels.[49] [*In margin*, 'j brell. 3 b. oyle; 3a olei', *i.e.* 1⅓ barrels of oil; third of the oil.]

'Flesche: The haill flesch givin in be the bischope in rentall extendis to' 13 lasts 13 m. 2 sts. 20 marks; third thereof, 4 lasts 12 m. 4 sts. 20⅓ marks.[50] [*In margin*, '3 flesche'.]

Marts: Total, 4 marts;[51] third thereof, 1⅓ marts. Swine, 2; third thereof, [*blank*]. Poultry: 218; third thereof, 72⅓.[52] Peats, 56 fathoms; third thereof, 18⅔ fathoms. Wax, 20 lb.; third thereof, 6⅔ lb.

'Seray fische: Seray fische' [*recte*, skray, *fos 130r, 132r*], 24 'mais'; third thereof, 8 'mais.'

---

45 The correct calculation appears to be £240 19s 6d.

46 The correct calculation from the figures cited appears to be £83 13s 2d.

47 The correct calculation appears to be 36 lasts 7 m. 1 st. 7⅓ marks.

48 The correct calculation appears to be 8 lasts 6⅛ barrels of butter.

49 The correct calculation appears to be 1 last 4 barrels.

50 The correct calculation appears to be 4 lasts 12 m. 2 sts. 22⅔ marks.

51 The text here reads 'marks'.

52 The correct calculation appears to be 72⅔.

'Memorandum, that the butter, oyle, fische, martis, pultre, pettis and uther dewties of the landis sett in few ar not comprehendit heirintill.'

'The pryce of the last [of] victuall callit cost' £16; 'inde' £580 16s 3d. 'The pryce of the last of flesche be neyth [*recte*, weycht] of the cuntre, suma', £3 12s; 'inde' £16 6s 6d. The price of the barrel of butter, £4 15s; 'inde' £164 7s 6d. The price of the barrel of oil, £3 5s 8d; 'inde' £53 12s 6d.[53]

'Memorandum, the prycis of the thridis of martis, swyne, pultre, petis, wax and scrayis ar not modifiet bot to be modifiet at the chalmerlandis comptis'.

'Suma of the haill silver to be payit to the Comtrollar,[54] by the dewties and victualis of the few landis that ar not allowit, was comprehendit heirintill is altogidder' £598 15s 11d.

[*In margin*, 'Nota, 8818 [*sic*] lib. 5s jd.]

'Divisioun and assumptioun of the bischoprik of Orknay according to the last rentall gottin thairof be Sir William Murray of Tullibardin, knycht, Comptrollar.'
    Money. Third of money, £214 8s 3d. Take Deirenes for 16s 4d; Holme, 24s; 'Sanct Awlayis [*i.e.* St Ola's] parochis'; £5 3s 8d; Schalpinschaw, £4 3s 5d; Sanday, £3 11s 5d; Burray, Flattay, Sallway [*i.e.* Swona; Suaay, *fo 130r*] and Schewsay [*i.e.* Switha; Futhanne, *fo 130r*], £60; Iwey,[55] Sandis [Sansyd, *fo 128r*] and [in, *fo 147v*] Deirnes', £100; Egilsay, £20; Ethay, £14; and the rest extending [to] £4 out of Ralsay [*i.e.* Rousay].
    Victual. Third of victual, 'baith of cost and beir, extending to' 37 c. 16 c. [*sic*]. Take Deirnes for 10 c.; Holme for 6½ c.; Sanct Alawis parish for 1 c. 6[56] 'eque' [*sic*]; Schalpinschaw for 14 c. 1 m. 3 sts.; the rest extending to 2 m. 3 sts. out of Sanda.
    Flesh. Third of flesh, 5 lasts 19 m. 1 st. Take Deirnes up for 3 lasts 17 m. 5 sts.; Holme for 1 last 17 m. 5 sts.; take the rest extending to 6 m. 5 sts. 4 marks out of Scalpinschaw.[57]
    Butter. Third of butter, both of Orknay and Zetland extends to 3 lasts 6 barrels. Take St Allauis [*i.e.* St Ola's] parish, 1 last 8 barrels [*blank*] lispund; Sandrowis [*i.e.* St Andrew's] parish, 2 barrels; Salpinschaw [*i.e.* Shapinsay], 2 barrels; Fara, 1 last 2 barrels; Ralsay [*i.e.* Rousay], 1 last 2 barrels; out of Sanda, 1 barrel. 'Gif' 2 lispunds.[58] 'Or tak this haill butter out of Zetland.'

---

[53] The figures in *Thirds of Benefices*, p. 144, for the sale of Orkney cost are £5 a last; butter, £18 a last; oil, £1 a barrel; and flesh, £3 a last.
[54] Sir William Murray of Tullibardine, Comptroller, 1565 - 1582, *Handbook of British Chronology*, p. 191.
[55] See above, p. 655, n. 6.
[56] St Ola paid 1 last 6 m. cost, fo 130r (see above, p. 657).
[57] These figures total 5 lasts 17 m. 3 sts. 4 marks.
[58] These figures do not tally.

Oil. Third of oil, both out of Orknay and Zetland, 10½ barrels. Take it out of Zetland.

'Cain fowlis, oxin and bairris' [*i.e.* boars]. Third of kane fowls, 67. Take out of Deirnes and Holme 'togidder with the pairt of ane fed ox and ij parte of ane bair.'[59]

'Paittis' [*recte*, skray fish, *fos 130r, 132r*]. Third thereof, 8 'mais'. Take them out of Sowna [*i.e.* Swona], Burra, Flatta and Sowtha [*i.e.* Switha].

'Gild kow.'[60] The third of a geld kow out of Sowlowis [*i.e.* St Ola's] paroschin.'

Wax. Third of wax, 5 lb. out of Lawrok.[61]

'Boit teindis in Zetland. Foulsilver', the third thereof, 14. 'Tak thame togidder with the third of Soutsilver of Zetland, ilk sout payand ane lycht gudling'.

[*In margin*, 'Land mails of Zetland'] And the third of the land mails of Northmovin [*i.e.* Northmavine], Collessetter [Occlositter, *fo 133r*; Facctosetter, *fo 146v*], Grinsett [Grindsulle, *fo 133r*; Greinsulle, fo 147r], and Dipdell extending to 10s 8d Zetland payment.

Sic subscribitur: 'Tulibardin, Comptrollar, etc.'

KIRKWALL CATHEDRAL, ST CATHERINE'S ALTAR IN,[62]

(EUL, fo 139r)

Rental of Sanct Kathrenis altar within the cathedral kirk of Orknay set in feu to Archibald Balfour, son to Gilbert Balfour of Wastrey, by sir Thomas Richartsoun, prebendary thereof, for £17 2s. 'Givin up' 5 December 1566.

'Item, in silver', £11 12s. 'Item, in malt, Orknay meillis, saxteine meilis, parte [*recte*, pryce] of the meile five s, inde' £4. 'Item, half ane barrell butter as Archibald Balfour hes the few', 30s. Total, £17 2s; third thereof, £5 14s.

Signature: 'sir Thomas Richartsoun, with my hand.'[63]

---

[59] I.e. one third of an ox and two thirds of a boar.
[60] I.e. one without defect, and not over eight years of age, Marwick, *Orkney Farm-Names*, p. 203.
[61] St Ola is probably intended here; see above, fo 130r (see above, p. 657).
[62] Cf fos 139v, 140r (see below, pp. 664, 665).
[63] The rental for St Catherine's altarage is interrupted at this point by the insertion of the rental for St Colme's parsonage on Sanday; St Catherine's altarage resumes on fo 139v (see below, p. 664).

ORKNEY, SUBCHANTORY OF, SANDAY, ST COLME, PARSONAGE OF,[64]

(EUL, fo 139v)

'Followis the just rentall of Sanct Cowme callit the subchantorie of Orknay perteinand instantlie to Mr Jerome Tullocht.'

In Stromnes, 4 lasts of malt. In Shabray, 18 meils victual, half malt, half meal. In 'land meillis', 1 lispund of butter; £13 silver. In Schabray, $\frac{1}{2}$ barrell of butter; 9s silver; 3 meils of malt; 5 meils of flesh. 'Item, the Burgh, Birsay and Bequy of beir', 2 meils 'les.'

This benefice is set in assedation by the said Mr Jerome Tullocht, parson of the same, to Duncan Scola, David Scola and Nicol Tullocht for the sum of 130 merks.

Signature: 'Mr Jerome Tullocht, persone of Sanct Colme, with my hand.'

'James Nicolsoun, ye sall give lettres in largest forme to Mr Jerome Tullocht upon the subchantorie of Orknay, he findand cautioun for his thrid conforme conforme [*sic*] to his assedatioun givin in thairupon, subscryvit with our handis at Edinbrucht', 26 March 1585.[65] 'Sic subscribitur, Tulibardin, Comptrollar.'

['Personage of Skemesay, callit Sanct Nicolas paroch', *deleted*]

KIRKWALL CATHEDRAL, ST CATHERINE'S ALTAR IN,[66]

(EUL, fo 139v)

'Item, in silver', £11 12s. 'Item, in malt', 16 meils Orknay, 'the pryce of the meill as Archibald Balfour hes in few is' 5s; the sum is, £4. 'Item, half ane barrell butter as the few is sett', 30s. Total, £17 2s.

Signature: 'sir Thomas Richartsoun. Tulibardin, Comptrollar.'

DUNROSSNESS, VICARAGE OF,

(EUL, fo 140r)

The vicarage of Dumrosnes in Zetland pertaining to Mr George Ballanden, 'sett of auld and of new to the Fold [*i.e.* Foud] of Zetland for' £80. 'And now quhen the maist pairt of the fruitis of the vicarage ar not payit be reasoun of the actis of

---

[64] The parsonage of St Colme on Sanday formed the prebend of the subchanter of Orkney; Jerome Tulloch was subchanter, 1562 - 1592. Cowan, *Parishes*, p. 180; Watt, *Fasti*, p. 259.

[65] This should probably read '1581'. Tullibardine's tenure of office had ended by November 1582, *Handbook of British Chronology*, p. 191.

[66] Cf fos 139r, 140r (see above, p. 663; below, p. 665).

Parliament[67] thair man be defalkit of the maill according to the rait of the demonitioun, etc.'

SHETLAND, ARCHDEACONRY OF, (EUL, fo 140r)

The archdeaconry of Zetland pertaining to Mr Jerome Schen [*i.e.* Cheyne] set to the Foud of Zetland and to sir George Strange for 360 merks.

Calculation of third: In the whole, 360 merks; third thereof, £80.

FETLAR, VICARAGE OF, (EUL, fo 140r)

The vicarage of Fotlar pertaining to sir John Reid, in wool, lamb and fish, set to sir George Strange for 24 merks 'of auld quhen the dewties of vicaragis custumit was healilie payit, etc.'

Calculation of third: In the whole, £16; third thereof, £5 6s 8d.

KIRKWALL CATHEDRAL, (EUL, fo 140r)
PREBEND OF ST CATHERINE IN,[68]

'Prebendrie of Sanct Kathreine, habetur pro deleto' [*i.e.* understood to be deleted].

The prebend of Sanct Kathreines, $3\frac{1}{2}$ barrels of butter; 16 meils of malt, extending to in the whole, £20 money, 'for fructis sett in tak and hes beine divers yeiris bygaine.'

'The prebendrie of Sanct Kathreine, sir Thomas Richarsoun.'

Calculation of third: In the whole, £20; third thereof, £6 13s 4d.

---

[67] In 1560, Parliament is said to have ordained that all possessors of teinds should retain payment until the Privy Council issued further instructions; and by 1562 the Privy Council ordered that no one should obey any title to beneficies preceding 1 March, or make payments, pending a settlement of the 'thirds'. Keith, *History*, i, p. 325; *RPC*, i, pp. 204-206.

[68] Cf fo 139r-v (see above, pp. 663, 664).

WESTRAY, CROSS KIRK, PARSONAGE OF,[69] (EUL, fo 140v)

Rental of Mr James Annandis benefice in Orknay

'The personage of Wastrey callit the Cors Kirk was wont to give ten lastis and ane half beir and half barell butter, sett to Johne Culland and James Alexander for the sowme of' 80 merks. 'Heirof I pay in yeirlie pensioun to James Alexander', £20. 'This parosch is wastit be the Lewis men [hiland men, *fo 141v*].'

Calculation of third: In the whole, £53 6s 8d; third thereof, £17 15s $6\frac{2}{3}$d.

SANDAY, VICARAGE OF,[70] (EUL, fo 140v)

Rental of the vicarage of Sanday pertaining to the said Mr James [Annand[71]] 'was wont to give in land mailis and boit teind', 4 lasts of bere; 'in bow teind', 4 barrels of butter with teind lamb and teind wool set to Henry Annand for £30.

Calculation of third: In the whole, £30; third thereof, £10.

STRONSAY, ST NICHOLAS, PARSONAGE OF, (EUL, fos 140v-141r)

Rental of Mr Francis Bothuell benefice in Orknay

Parsonage of Stromsay called Sanct Nicolas parish

'The personage of St[r]onsay callit Sanct Michallis [*recte*, Nicholas'] parosche was wont to give' 12 lasts bere, set to James and Duncan Darley for 100 merks.

Calculation of third: In the whole, £66 13s 4d; third thereof, £22 4s $5\frac{1}{3}$d.

KIRKWALL CATHEDRAL, PREBEND OF ST MAGNUS IN, (EUL, fo 141r)

The prebend of Sanct Magnus pertaining to Mr Robert Chyne, set to John Brown for £30.

Calculation of third: In the whole, £20 [£30, *above*]; third thereof, £6 13s 4d.

---

[69] Cf EUL, fo 141r (see below, p. 667).

[70] Cf EUL, fo 141v (see below, p. 667). Sanday had three parishes, Cross Kirk, Lady or Mary Kirk, and St Colme; all three vicarages formed the prebend of the chancellor of Orkney, then held by James Annand. Cowan, *Parishes*, pp. 179-180; Watt, *Fasti*, p. 257.

[71] *Thirds of Benefices*, pp. 93, 151, 204; Watt, *Fasti*, p. 257.

KIRKWALL CATHEDRAL, PREBEND OF ST JOHN IN, (EUL, fo 141r)

Rental of Mr Gilbert Fouldis benefice in Orknay.

'Sanct Johnes prebendarie sett to James Alexander was wont to pay' 3 lasts 18 m. victual, with one barrel of butter, set for £32.

Calculation of third: In the whole, £32; third thereof, £10 13s 4d.

WESTRAY, CROSS KIRK, PARSONAGE OF,[72] (EUL, fo 141r-141v)

Rental of Mr James Annandis benefice in Orknay.

'The personage of the Cors Kirk in Wastrey was wont to give x lastis beir and half barrell butter, now sett to Johne Cullane and James Alexander, thrie yeiris takis payment thairof yeirlie' 80 merks. 'Furth thairof is payit to the said James Alexander in pensioun', £20.

'The rowmes ar devastat be the hiland men [Lewis men, *fo 140v*] this last yeir, so presentlie no payment can be had. In yeiris of payment the third will be' 26 merks 7s 9d [*recte*, 26 merks 8s 10⅔d, *see next paragraph*].

Calculation of third: In the whole, 80 merks; third thereof, 26 merks 8s 10⅔d.

SANDAY, VICARAGE OF,[73] (EUL, fo 141v)

Rental of the vicarage of Sanday pertaining to the said Mr James [Annand] 'was wont to give four lastis of beir in bow teind and four barellis butter, by teind woll and teind lamb sett to Harie Annand for' £30 'be yeir. Of this as yett na payment can be had conforme to the auld use and wont be reasoun they war wont to pay beir for thair teind fische quhilk now they refuis. Payment beand maid, the third will be yeirlie' 15 merks.

'Suma [*sic*] of the thridis foirsaidis wilbe' 41 merks 8s 11d.[74]

Calculation of third: In the whole, £30; third thereof, £10.

---

[72] Cf EUL, fo 140v (see above, p. 666).
[73] Cf EUL, fo 140v (see above, p. 666).
[74] 26 merks 8s 10⅔d (from Cross Kirk, Westray) plus 15 merks is 41 merks 8s 10⅔d.

SANDWICK AND STROMNESS, VICARAGE OF, (EUL, fo 141v)

Vicarage of Sandwik

Rental of the vicarage of Sandwik and Stromsay [*recte*, Stromness] in Orknay 'was wont to be sett in assedatioun for' £40, 'now deducing the corpspresent, umest clyth, Paschfynis and small offerandisis bot worth' £20.

Signature: 'Mr Johne Bonkle, fermourar of the said vicarage, with my hand.'

KIRKWALL CATHEDRAL, (EUL, fo 142r)
WOODWICK, PREBEND OF,

'The rentall of the stallrie of Wedwik extendis to ten merkis money and to ane last victuall, quhilk last is sett for' £5 6s 8d money. 'Givin in be Mr William Lawder.'[75]

Calculation of third: In the whole £12;[76] third thereof, £4.[77]

ORKNEY, BISHOPRIC OF,[78] (EUL, fos 144r-148r)

'Rentale fructuum episcopatus'

Rental of the fruits of Orkney within parts of Orkney.

The lands of Wastray and Papa Wastray set in feuferme to Gilbert Balfour for yearly payment of the sum of £122 0s 8d. The parsonage here is adjoined to the bishopric and the teinds of Mary Kirk set with the lands. The lands of Birsay [Barsay, *fo 128r*] and Marwik and Birsay 'be eist and be wast' are set in feu for yearly payment of the sum of £182. Also the parsonage here is adjoined to the bishopric. And the teinds of Marvik pays 35 m. cost; 1 m. flesh. The lands of Evie, Stanhous, Enhallow and Birstane are set in feu to Patrick Balenden for payment of £95. The ile of Ethay set for in feu to Oliver Sinclar [for] payment of £14; 5 lasts flesh. Also the parsonage here is adjoined to the bishopric. The Burray, Flattay, Sunay [*i.e.* Swona] are set in feu for payment of the sum of £60 argent; 'and twentie scoir mais skra with lxxx cupill cunningis or for evrie cupill,

---

[75] William Lauder was prebendary of Woodwick, *Thirds of Benefices*, p. 149.
[76] The figure given in the preceding paragraph is £5 6s 8d.
[77] At this point there follows, in a later hand, an entry entitled 'Information to the King', which is placed in Appendix 2 (see below, p. 677).
[78] Cf EUL, fo 128r (see above, p. 655).

iiij d'. The ile of Egilsay is set for yearly payment of £20 argent; 9 lasts bere. Also the parsonage is here adjoined 'as said is' to the bishopric.[79]

The lands of Sanday pertaining to the bishopric pay yearly 5 lasts 2 sts. [6 lasts 2 sts., *fos 128v, 147r*] cost; £3 6s 5d [£3 11s 5d, *fo 128v*; £3 11s 6d, fo 147v] silver; 2 barrels of butter.

Rowsay: The lands of Rowsay are set for yearly payment of 5 lasts 5 m. 3 sts. [5 lasts 5 m. 4 sts., *fo 147r*] cost; 15 barrels of butter. The teinds of Rowsay, the parsonage being annexed with the bishopric extends to 5 lasts 19 m. 4 sts. cost.[80]

Stronsay: The lands of Stronsay extend to 7 lasts 14 m. 3 sts. bere; 7 barrels of butter; 4 m. [4 sts., *fo 129r*] of flesh. The teinds thereof because the parsonage of Sanct Peteris kirk is adjoined to the bishopric extends to 7 lasts 16 m. bere. Extending altogether in cost to 10 lasts 4 m. 2 sts.[81]

Scalpinsay: The lands of Schalpinsay with the teinds thereof because they are adjoined to the bishopric extending to 14 lasts 19 m. 4 sts. [14 lasts 19 m. 3 sts., *fo 129v*] cost; 5 lasts 8 m. 4 sts. flesh; 2 barrels of butter; 1 barrel of oil; £4 3s 5d silver.

Darnes: The lands of Deirnes with the teinds thereof are set for yearly payment of 10 lasts 8 m. 5 sts. cost; 46s argent ['skeat silver', *fo 129v*]; 3 lasts 17 m. 2 sts. flesh; 2 barrels of butter; 158 kane fowls; with two oxen; 'and twa bewis [*i.e.* boars]'; and 46 fathoms of peats.

Sanct Awlawis [*i.e.* St Ola's] parish: For the land mails thereof 1 last 6 m. cost; 19 [20, fo 130r] barrels of butter; one cow; £5 19s 8d [£5 9s 8d, *fo 130r*] argent; 20 lbs of wax. And for the teinds thereof, 4 lasts cost.

Holme: The land mails of the same parish extend to 6 lasts 12 m. 3 sts. 6 marks; 1 last 18 m. 5 sts. 20 marks flesh. 'Item, for the teindis quhilk ar in maner foirsaid annexit to the bishoprik', 23 [13, *fo 130r*] m. cost; 28s silver; 1 fathom of peats [10, *fo 130r*]; 60 fowls.

Rannaldsay Besouth: The teinds thereof extending yearly to 18 lasts 14 m. 4 sts. bere; and in land mails, 15 m. bere; extending in cost to 12 lasts 19 m. 4 sts. 16 marks.

Wais [*i.e.* Walls; Wawis, *fo 130v, 134r*]: 'Item, Wais payis half ane barrell oyle.'

Sanct Andros parish: The lands [teindis, *fo 130v*] thereof extend yearly to 2 lasts 14 m. cost; 2 barrels of butter; 10s silver.

Orphar: Orphar pays yearly 2 barrels of butter; 9 sts. flesh; 9 sts. cost.

Firth: The teinds of Firth extend to 3 lasts 22 m. cost; 1 barrel of butter.

Stromnes: Stromnes pays yearly ½ a barrel of butter; 8 sts. cost.[82]

Harray: The teinds thereof extend to yearly 5 lasts 16 m. 4 sts. cost.

---

[79] The details in the above paragraph differ from those in the previous rental on fo 128r (see above, p. 655).

[80] £10 'argent' is included in this assessment on fos 128v, 147r (see above, p. 656; below, p. 670).

[81] Cf above, p. 656, n. 12.

[82] Eynhallow and Sandwick are included at this point on fo 131r (see above, p. 657).

Stenhous: The teinds of Stenhous set to Patrick Balenden for payment of £10 argent.[83]

Sandwill [*recte*, Sandwick]: The teinds thereof extend yearly to 7 lasts 2 m. 1 sts. cost; in land mails, 1½ barrels of butter; 10s silver.

Randall and Garsay: The teinds of Randall and Garsay yearly extend to 4 lasts 11 m. 3 sts. cost.

Farray: Farray pays 2 barrels of butter.

Fruits of the bishopric of Orkney within parts of Caithness: The lands of Skubster [Strybustar, *fo 131v*; *i.e.* Scrabster] with 8d lands of Drumray' [*i.e.* Dounreay] and the mill thereof set in feu for yearly payment of £11 6s. The lands of Halcro in Cathnes with the mill thereof set in feu for yearly payment of £4 argent; 6 b. meal. The 16d land of Drumray with Bombstar [*i.e.* Broubster], Furay [Furro, *fo 131v*] and Lunay [Lumay, *fo 131v*] set in feu to William Sinclar of Dumbaith [to the earl of Sutherland, *fo 131v*] for yearly payment of £30 argent.

Rental of the fruits of the bishopric of Orkney within parts of Shetland: 'Item in primis, in Drumrosnes' 16 'pece'; 'inde' 8 barrels of butter, 8 barrels of oil. The Sey Ile [Fair Yle, *fo 132r*], 2 'pece'; 'inde' 2 [*recte*, 1] barrel of butter, 2 [*recte*, 1] barrel of oil.[84] Standsting [*i.e.* Sandsting] and Nesting,[85] 6 'pece'; 'inde' 3 barrels of butter, 3 barrels of oil. Wais [*recte*, Walls; Vannis, *fo 132v*; Waynis, *fo 135r*], 6 'pece'; 'inde' 3 barrels of butter, 3 barrels of oil. Northmoving [*i.e.* Northmavine], 8 'pece'; 'inde' 4 barrels of butter; 4 barrels of oil. Skattay and Delting, 6 'pece'; 'inde' 3 barrels of butter, 3 barrels of oil. Yell 10½ 'pece'; 'inde' 4½ barrels of butter, 6 barrels of oil. Fetlar, 7½ 'pece'; 'inde' 3 barrels butter, 4½ barrels of oil. Unst, 13½ 'pece'; 'inde' 6 barrels of butter, 7½ barrels of oil. Nestinge, Lunay, Quhalsay, Skeirreis, 7 'pece'; 'inde' 3 barrels of butter, 4 barrels of oil. 'Item, mair in Zetland in land maill; Facctosetter [Occlositter, *fo 133r*; Collessetter, *fo 139r*], ten schilling Zetland payment. Item, the land mailis of Greinsulle [Grindsulle, *fo 133r*; Grinsett, *fo 139r*], payis aucht schilling Zetland payment. Item, the landis of Deipdaill lyand waist and was wont to pay' 5s 'Zetland payment. Item, the land maillis of Northmoving, twelf schilling Zetland payment. Item, fyve pund out of the custumis of Abirdeine.'

'Heir followis of the foirsaid rentall quhat is not of the bischopis teindis, and the assignatioun takin up be the Justice Clerk[86] and his chalmerlandis, quhairof the bischope aucht na thrid. Item, the haill dewties of Egilsay payand' 9 lasts bere; £20 money. 'Item, the haill land mailis of Rowsay extending to' 5 lasts 5 m. 4 sts. [5 lasts 5 m. 3 sts., *fo 144v*] cost; £10 argent; 15 barrels of butter.

---

[83] On fo 131r, the teinds of Stenness were given as 1 last 17 m. cost; 4 barrels of butter; 1 mart flesh (see above, p. 657).
[84] 'Brasay and Burray' are included at this point on fo 132v (see above, p. 658).
[85] See above, p. 658, n. 27.
[86] Sir John Bellenden of Auchnoule, Justice Clerk, received a gift of the temporality of the bishopric of Orkney from the crown in 1559, *RSS*, v, no 589.

Of Deirnes, the whole victual extending to 9 lasts 20 m. 5 sts. cost; 46s 'argent' 4d 'silver'; 3 lasts 17 m. 3 sts. flesh [10 lasts 8 m. 5 sts. cost; 46s argent; 3 lasts 17 m. 2 sts. flesh, *fo 145r*].

The land mails and teinds of Sanday extending to 6 lasts 2 sts. [5 lasts 2 sts., *fo 144v*] cost; 26 lispunds of butter, 'inde twa barelis buter; 12 m. flesh [fish, *fo 128v*]; £3 11s 6d [£3 11s 5d, *fo 128v*; £3 6s 5d, *fo 144v*] silver.

The yearly duties of his lands of Holme extend to 3 lasts 4 m. cost. 'Item, assignit to the Justice Clerk upon the lands of Evey[87] and Sandis [*i.e.* Sandside; Sansyd, *fo 128r*] in Deirnes, £102, 'and of the reddiest of the fruitis of Birsay', £98.

'The uther ordinar charges to be defalkit: Of the money within writtin to be defalkit thrie hundreth merkis for a pensioun thairof givin to Archibald Ruthven', 300 merks. 'Item, to be defalkit for a pensioun to Mr James Thorntoun', £20. 'Item, to my lordis lectour in the cathedral kirk', £20. 'Item, to be defalkit for overblawin landis in Burray', 10 merks. 'Item, for barrelis to aucht last butter, pryce of ilk barrell', 10s; in the whole, £64.[88] 'Item for barelis to iij lastis vij barrels, pryce of ilk barell', 1 merk; total, £28 6s 8d.[89] 'Item, to the constable keepar of the place of yeardis in Kirkwall and baillie thairof for thair fies', 100 merks. 'Item, to the chalmerland of Zetland for his fie', 100 merks. 'Item, to the Lardis of the Saitt[90] for his pairt of the contributioun', £14. 'Item, to be defalkit of the victualis foirsaid for the taksmenis fies, collectouris of the victuall, for ilk last' 1 m.; total, 3 lasts 8 m. 'Item, for the cowpar fies that collectis the buter', 1 last cost. 'Item, defalkit for overblawin landis in the dewties', 1 last victual. 'Item, to be defalkit for the teindis of overblawin landis in Renaldsay', 2 lasts victual. 'Item, to twa servandis', 3 lasts victual. 'Item, to the weyar of the victuall', 1 last cost.

---

[87] See above, p. 655, n. 6.
[88] Counting 12 barrels to 1 last, the correct figure should be £48.
[89] The correct calculation appears to be £28 13s 4d.
[90] I.e. Lords of the Court of Session.

Appendix 1

ISLES, BISHOPRIC OF,[1]
IONA, ABBEY OF,[2]

'Rentale of the Bishoprick of the Ilis and Abbacie of Ecolmkill

Imprimis, the twentie pund landis of Ecolmkill, 20 lib.
    Item, of Rosse, 20 lib.
    Item, in Brolos ane pennie land, callit Torrinichtrache, 1d.
    Item, the pennie land of Cairsage, 1d.
    Item, the pennie land of Carvalge, 1d.
    Item, the half-pennie land of Glasveildirie, ½d.
    Item, ane pennie land of Kilphubbill, 1d.
    Item, the pennie land of the Keallinne, 1d.
    Item, the pennie land of Kilbrandane, 1d.
    Item, the pennie land of Kilneoning, 1d.
    Item, the half-pennie land of Cengarwgerrie, ½d.
    Item, the pennie land of Kilmorie, 1d.
The foirsaid nyne pennie land, all lyand within the Isle of Mulle.

The Abbatis landis within the Isle of Teirie

    Item, Baillephuille, 4 lib.
    Item, the Wyle, 13s. 4d.

The Abbatis landis within the Clanrannaldis boundis

    Item, the Ile of Cannay, 20 lib.
    Item, Ballenamanniche, lyand within the Ile of Weist.

---

[1] The survey of rentals recorded in the Books of Assumption does not extend to Argyll and the Western Isles; but an early rental for the bishopric of the Isles and Iona abbey, printed in *Collectanea de Rebus Albanicis* (Edinburgh, 1847), pp. 1-4, is reproduced here as an appendix.
[2] The abbacy of Iona was annexed to the bishopric of the Isles, Cowan and Easson, *Medieval Religious Houses*, p. 59.

The Abbatis landis within Donald Gormis boundis

Item, [in] the Ile of Weist the tuentie-four pennie land, callit Unganab, 24d.
Item, Baillenakill in Eillera.
Item, Kirkapost in Eillera.
Item, Cairenische thair.
Item, in Trouterneiss ane half *teirunge*,[3] callit Keilbakstar.
Item, in Sleatt the tua Airmadillis.

The landis that McAen[4] hes of the Abbatis

Item, Geirgadeill in Ardnamurchan.

The Abbatis landis possest be McCloid of Heries

Item, the Ards of Glenelge.

The landis quhilk the Clandonald of the West Illis haldis of the Abbatt

Item, the tuentie pundis of Laintymanniche and Mwicheleische in Illa, 20 lib.
Item, Ardneiv in Illa, 8 lib. 13s. 4d.
Item, the fourtie merk landis of Skeirkenzie in Kintyre, 26 lib. 13s. 4d.
Item, the sax merkis land of Camasnanesserin in Melphort, 4 lib.
Item, the 1^c pennie land of Muckarn.[5]

The Kirkis and Personagis perteining the Abbatt of Ecolmkill

Item, the teindis of Ecolmkill, callit the personaige of Tempill-Ronaige.
Item, the personag of Kilviceowin in Rosse.
Item, the personag of Keilfeinchen in Mulle.
Item, the personag of Keilnoening in Mulle.
Item, the personag of Keilchallumkill in Quyneise in Mulle.
Item, the personag of Keillean in Toirrasa in Mulle.
Item, the personag of Soiribie in Teirie.
Item, the personage of Keilpedder in Veist.
Item, the personage of Howmoir thair.
Item, the personage of Sand thair.
Item, the personage of Cannay.

---

[3] The meaning of the term is unclear.
[4] I.e. MacIan of Ardnamurchan.
[5] The six merk lands of Camusnanesrin and the lands of Muckairn seem erroneously to be included.

Item, the personage of Sleatt.
Item, the personage of Mwidort.
Item, the personage of Skeirkenze in Kintyre.
Item, the personage of Keilcheirran thair.
Item, the personage of Keilchrist in Strathawradall.[6]

Rentale of Bischopis landis within the Illis

Imprimis, Keilvennie in Illa.
Item, the Ille callit Ellanamwk, possest be McAen of Ardnamurchane.
Item, the Ille of Rasay.
Item, the fyve Illis of Barray.
Item, Skeirachnaheie in Loise.[7]
Item, Rona na nav.
Item, in Orknay.
Item, Snisport in Troutirneise.
Item, Kirkapost in Teirie.

The Teindis and Personagis perteining the Bischop

Item, the Teindis of Buitt.
Item, the Teindis of Arran.
Item, the personage of Kirkapost in Teirie.
Item, the personage of Elie in Loise.
Item, the personage of Roidill in Hereis.
Item, the personage of Snisport in Troutirneise.
Item, the thrid pairt of all personagis perteining to the Abbatt, the personag of Ecomkill and Rosse onlie exceptit.'

---

[6] I.e. in Skye.
[7] I.e. Lewis.

Appendix 2

BISHOP OF ORKNEY TO KING JAMES VI   (EUL, fos 142r-143v)

'Information[1] to the king, his most excellent majestie, from the bishop of Orkney[2] anent the yearly rent of the landis designed to him and his successors.

May it please your sacred and most gracious majestie to consider these few and short articles which I have upon my knowledge and tryall set doun treulie and laite direction.

1. First, it is of verity that the rentall which I delyvered to your majesties officers is the last, best and most perfect that ever wes in Orkney, conforme to quhich all the tacksmen of heall [*i.e.* whole] Isles and paroches had thir particular rentalls given to them subscryved be the [clerk, *deleted*] earle.[3]

2. It wes tryet and reported be the Clerk Register,[4] Sir Henry Wardlaw, Archibald Prymrose and James Baillie (who had directions from the lords) that all the rent of the lands designed to me wes not equivalent to the thrids and eight thousand merks contracted to be payed to me yearly be the earle of Orkney.

3. Whereas it is objected be my lord Thesaurer depute[5] for the faithfull discharge of his service that the rent of these lands being counted according to the tacksmens counts and payment will exceid the proportion of the 3ds [*i.e.* thirds] and money adebted to me yet thair is no materiall difference so farr as I can try betuixt the rentall and the compts except in three [*blank*] tuo chalders teind in the paroch of Stromness quhich belongs justly to the subchanter[6] and came laitlie to the earle his possession more be usurpation and violence than good right.

---

[1] The 'Information' and 'Rentall of the frie rent' are in a later hand, and are dated 1614.
[2] James Law, bishop of Orkney, 1605 - 1615, Watt, *Fasti*, p. 254.
[3] Robert Stewart, 1st earl of Orkney, who died in 1593, was succeeded by Patrick, 2nd earl, who was executed in January 1615, *Scots Peerage*, vi, pp. 573-577.
[4] Sir Alexander Hay of Whitburgh, Clerk Register 1612 - 1616, *Handbook of British Chronology*, p. 197.
[5] I.e. Sir Gideon Murray of Elibank, *RMS*, vii, no 913.
[6] No subchanter of Orkney is recorded between 1592 and 1620 in Watt, *Fasti*, p. 259.

4. And thairfoire howsoever the rent of thes lands shall be compted ath[e]r according to the rentall or according to the tacksmens charge and exhoneration, the summa thairof with all the few maills of the lands of Orkney reserved to me (and no deduction of lyferent pensions being made) shall not surmount 8,000 merks besyde the 3ds [*i.e.* thirds] as I have tryed by just calculation.

5. Bot it is alleat [*i.e.* alledged] 2dly [*i.e.* secondly] that the garsoms [*i.e.* grassums] of the lands assigned will every year be worth' £546 'besyde the rent. God knows that I know not nor can I to this day learne the trewth thairof. This much I am informed that it is very credible that the earle did menfullie impose and vigorouslie exact these thrie years garsoms [*i.e.* grassums] upon every pennyland attour the ferms to the great damnage of the poor and grinding of thair faces, whose paterne I purpose not to imitate.

6. Bot let it be granted that the rent of the lands with thir garsomes [*i.e.* grassums] be 5 or 600 lib. more nor may counter value the 3ds [*i.e.* thirds] and 8,000 merks.

I hope that your most gracious majestie who is now to lay ane new foundation for the bishoprick of Orkney, who hes bein so liberall and bountifull to other bishops, will favourablie consider my travells, troubles and great losse which I have sustained thir nyne years bygone and provyde so for my successors, they in these remotest places may be habile both to live honourablie at home and serve your majestie abroad as shall be fitting for thair dignity especiallie for these persons.

First, seing all the archbishops and bishopes in Scotland have, by the properties of thair patrimonie, many casualities I hope non shall be able to move your majestie to allow to me and my successors the casualities of poor garsomes [*i.e.* grassums] for ane part of our set rent.

Secondly, the rents, ferms and teinds of the lands designed must be subject to the burdings of taxations and other impositions without releife of fewars (for thair is not abone four in my knowledge) for tacksmen of teinds (for the teinds are disponed as ane. part of the rent) and all other bishops besyde thir properties and casualities hes releife of thir vassalls and tacksmen.

Thirdly, your majestie would be pleased to consider that the rent of the lands will never be payed compleitly in one year yett many years hes bein and maybe wherin the halfe or tuo part may not be payed.

And if your majestie pleases to impar my condition I had rather 7,000 merks of good payment and securitie for it then the rent and garsomes of all the land designed and thairfoir equitie requires that the garsomes and casualities be allowed to supplie and support the defect of yearly payment and burdings of the rent and service.

Further, I will humblie beseich your gracious majestie to ponder how much the benefeice it selfe, I and my successors are hurt and damnified by this new fundation and the hard conditions therof.

I must now resigne the superioritie of all lands fewed to the haill [recte, earl] and some others with the casualities thereof.

The teinds of all the set lands which be tyme might have bein recovered.

The patronages of threttie or moe benefeics great or small besyde the viccarages I think most hurtfull to me and hinderfull for planting desolate and waist kirks.

I will not mention to your majestie the regalitie quhairof I have tuo or thrie charters given be your majesties predicessors from King James the 3d.

These priveledges and prerogatives belonged be all good right to the bishopes of Orkney, were disponed to me be your majestie, ratified in Parliament, and I in possession of them, and now I shall have nothing in recompence of them but ane vaiked rent unable to be payed yearly and garsomes assigned for ane pairt therof.

As for the lands of Greinweill which are esteemit to be worth £26 mor than thay pay presentlie.

They are the kyndly possession of Edmond St Clair, his son, for the old rent and teynd quhich is rentalled; they are clamed as propertie be Edward Stuart,[7] base son to the last earle of Orkney; they ly in the heart of the lands helin and disponed to me and if no better may be I ame content to ex[c]amb them with land in St Olais paroch; ther is no other thing objected which neideth my answer.

If it shall please your most gracious majestie upon the consideratione of thes articles (which I have set doun trewlie and so breifely as I could, fearring to weary your majestie) to dispone the lands with thir rents, garsomes and casualities as is contracted alreadie and to command the samen to be sent to your majestie to be signed I shall glaidly continow.

Bot if your majestie shall be moved to put me in worse estate and condition than wes contracted to me be the earle and to give me no comfort nor help nor setleing for my losses and troubles quhich I have patiently and constantly endured in yor majesties service thir 9 years bygone then I will beg most humble pardon and favour rather to resigne my office and malefice [recte, benefice], yea and my native soyle than with such discredit, trouble, hurt and uncertaintie to continow one haife year longer in it.

---

[7] I.e. Edward Stewart of Brugh, illegitimate son of Robert, 1st earl of Orkney, *Scots Peerage*, vi, p. 575.

Bot bearing my selfe in the best hope that your majestie will now after so long tyme aither setle me in the land with all the casualities thairof according to the contract agried [blank] upon or else cause some securitie to be made to me for payment of 8,000 or 7,000 merk with the 3ds without prejudice to my first gift of the superiorities and patronages, I shall ever make prayers and thanksgiving for your majesties sacred person, royall progenie and prosperous government.

Your majesties most humble servant, sic subscribitur Ja[mes], b[ishop] Orcaden[sis]', Edinburgh, 22 June 1614.

'Rentall of the frie rent of the lands designed to the bishop of Orkney according to the accompts taken up be the bishop and me, John Finlason, chamberlane depute to Sir James Stuart, of the cropes 1611 and 1612.

The malt of the lands conteined in the said designation comptand ane last or [blank] to ane chalder of malt, is' 72 c. 12 b. whereof there is to be 'deduced' [i.e. deducted] for the thirds, 15 c., so rests 57 c. 12 b., price of the chalder or last, £64; 'inde' £3,696.

The bere of the lands contained in the said designation counting one and a half lasts to one chalder of bere is 16 c. 14 b. or thereby, whereof there is to be deducted for the thirds, 15 c., so rests 1 c. 14 b., price of the chalder, £48; 'inde' £90.

The meal of the lands contained in the said designation, whereof there is nothing to be deducted for the thirds, is $7\frac{1}{2}$ lasts, price of the last is £72; 'inde' £540.

The butter of the said lands contained in the said designation is 4 lasts $10\frac{2}{3}$ barrels, whereof there is deducted for the thirds, 3 lasts $7\frac{1}{3}$ barrels, so rests 1 last $3\frac{1}{3}$ barrels, price of the barrel, £24; 'inde' £368.

The flesh paid furth of the lands contained in the said designation is $36\frac{2}{3}$ lasts, whereof is deducted for the thirds, $5\frac{2}{3}$ lasts, so rests 31 lasts at £20 the last is £600.[8]

The feu mails paid to the bishop furth of the 'last' [recte, lands] contained in the designation, with the feu mails of the lands of Caithnes, is £340 1s 6d, whereof there is to be deducted for the thirds, £214 8s 4d, so rests £125 13s 2d.

The lands of Greenwall [in] Paplay were set in tack by the earl of Orkney for £8 of yearly rent, which tack is expired; before the setting of the said tack, the earl might first [have] got for the said lands beside the feu duty yearly as much malt

---

[8] The correct calculation appears to be £620.

and flesh as will extend at the prices foresaid to the sum of £126, so the said lands should be allowed to the bishop for the said sum of £126.

The lands contained in the designation pay of grassum every three years £1,638, which being divided in three parts will augment the yearly duty each year, £546.

'There doeth lye yearly overhead within the bounds afoirsaid of fisher boothes 15, quhich booths payes of ground lieve, ilke ane of them 2 dozen of fish and ane barrell salt, estimat to' £6 'for the boll, inde' £90.

'The kayne fowlls of the whole lands designed to the bishop are' 1,086 fowl, 'compting 6 stone to the 100th, 'quhairof deduced [*i.e.* deducted] for the 3d', 87, so rests 1,019 fowl at 1s 4d the piece, 'inde' £81 5s 4d.[9]

'The peats payed furth of the said lands are 31 faddome are [*sic*] 53s 4d the faddom, inde' £82 13s 4d.

'Summa', £6,363 11s 10d.[10]

SIR GIDEON MURRAY OF ELIBANK (EUL, fo 143r-v)
TO KING JAMES VI

'The kings most sacred majestie. Most sacred soveraigne. According to the direction of your majesties letter sent to the Lords of Privy Councill I have set doun heire under my hand the difference betuixt the bishop of Orkney and me for the yearly duties of the lands designed to him quhich sould not exceid in yearly rent the soume of 8,000 merks be the 3ds appointed for the ministers stipends according to the agriement made betuixt the earle of Orkney and him quhich differences are comprehended under these articles following.

First, ther was omitted in the reackoning made with the bishope the pensions given out be the earle and his father[11] quhich are bot a temporarie right to these that possess them and will expyre with thair death and so remaine with the bishop and his successors in all tyme comeing and so he can desyre no more bot that thair may be compensation allocat to him during the [*blank*] thair lyftymes.

2. The lands are omitted that are lyeing ley quhich for the most part hes proceidit from the negligence of the chamberlane and tacksmen that had charges of the earles affairs, the said lands being set for the dutie that they are worth and hes

---

[9] This arithmetic does not tally.
[10] The correct addition of the sub-totals in the text appears to be £6,345 11s 10d.
[11] See above, p. 677, n. 3.

payd in tyme bygone will increase to the yearly dutie and so ought to be accepted be the bishop for a pairt of his rent.

3. There are some of the lands set in tacke for a small dutie quhairof the tacks are expyred and so the bishop ought to accept them according as they are worth and as the possessors were contented to have payed for them befoir they were set in tacke.

4. The bishope craves deduction for lands quhich he alleadges were injustlie purchassed be the earle and set at over high a raite, notwithstanding that they have payd the full dutie all the years preceiding.

5. The bishope craves deduction for lands designed be him to ministers for there gleybs surpassing the act of Parliament to the triple availl.

6. The few mailles in the reackoning made with the bishop are for the most pairt omitted.

7. In the said reackoning there was no consideration had of the grassomes payd furth of the saids lands ilk and 3 year extending to' £1,638 'quhich being divydit in 3 parts will make the increase of' £546 yearly.

'8. The dutie payed be the fishers for ground lieve within the bounds designed to the bishop with the pultrie and peats payed furth of the saids lands were not compted.

It is true that direction being given by the Lords of Exchequer the last summer to contrair persons to select furth of the rentall of Orkney such paroches and lands as might countervaile the yearly duty of 8,000 merks for the bishops rent and so much victuall and other commodities as might answer to the quantitie of the 3ds according to the contract made betuixt the earle and him it wes then found by these that were imployed being strangers to the bussiness and not weill acquainted with the nature of Orkney rents and trusting withall to ane old rentall baith produced be the bishop (quhich is farr short of the rent that it is now put too) that all the rent of the lands contained in his designatione, the 3ds being deduced, were found not to exceid the sowme of 6,700 merks, in consideration whereof the gressomes, pultrie, peats and other casualities of that kynd were not reckonid to him but the same allouit for the proportion of' £800 'with the rents of the lands designed to him were found be the said reackoning to inlaick bot fra I tryed that the said reackoning wes not weill made and that the rents of thes lands were better nor 8,000 merks be the 3ds it seimed to be agrieable to reason that all these commodities sould be reckoned with the rest of the duties and no more of all given to the bishop nor might make up the yearly rent of 8,000 merks with the 3ds by quhich reackoning thair will be of overplus according to the particular rentall sent herewith' £1,032 5s; 'the rentall is for the most part made up be these compts taken in be the bishop himselfe according to the quhich the possessors have made payment both befoir and since the informations for the gressomes ground lieve of the bolls and the rent quhich may be had for Greinwall in Paplay are made to me by these who have particular knowledge of the rents of Orkney and I think that the bishop will not make great question thairanent. The desyre I have to give your majestie particular informatione of all the differences concerning this bussines hes moved me to draw thes differences to more lenth nor can be agrieable which your majesties serious and princlie affaires, but I hope the caire I have to make your majestie clear and true informationes will plead your gracious favour and acceptance and so I end with a most humble and fervent prayer for your majesties long preservation in all health and happiness.

Your Majesties most humble and faithfull servant, sic subscribitur, Murray.[12]

Edinburgh, the 24 June 1614.'

---

[12] I.e. Murray of Elibank, see above, p. 677, n. 5.

Glossary

3 pairt one third
4 pairt one fourth
8 pairt one eighth

abacie, abbacy the dignity, rights and privileges of an abbot
abone above
actentik authentic
ad addition
addebtit indebted
aggis eggs
agriance agreement
aires heirs
aires and assignais, aires and assignaris heirs and assignees
aites oats
aitmeill oatmeal
aittes, aittis oats
albeit although
aldar elder
aliter, alitter otherwise, else
allanerlie, allanerly alone, only
almois, almous, almus alms
almussar almoner
als also
als as
alsmeikill, alsmeikle as much
alswa also
alsweill as well
amangis among
ane one
anent concerning
anes once
angell, angell nobillis English coin worth about 10 merks Scots
annat, annates originally the revenues of a benefice in the first year of a vacancy which went to the heirs or executors of the deceased benefice-holder; but by the later middle ages, the papacy had established the practice of exacting half a year's fruits of all benefices provided at the apostolic see within the limits of 24 to 100 florins *auri de camera*; this was not intended to exclude the claims of executors but the papal demand had priority.
annuals, annuellis annual rent
appeirand, apperand heir apparent
aragies *see* arriage and carriage
archdeanrie, archidenrie archdeaconry
archedein, archideane, archidein archdeacon
argent term used in Orkney for Scots money as distinct from local currency
arreage *see* arriage and carriage
arriage and carriage feudal services owed by tenants, in men and horses, to their landlords
as efferis as is appropriate
asedation lease
assignais, assignaris assignees; those to whom a right or property is formally assigned
assumit collected

assumption collection
ather either
attour above, beyond; in addition, besides
auchleit, auchlet one eighth of a boll
aucht eight
aucht ought
aucht owed, owing
aucht to have a claim to, to possess
aucht pairt one eighth
auchtane eighth
augmentation increase in the amount of a payment; increase in rent as a result of feuing
auin own
auld old
auld extent ancient valuation of land for the purpose of taxation
avail, availl value
avena oats
avys advice
awand, awant owing
awin own
ay always

b. *see* boll
bailie officer of an administrative area; a magistrate
bailyie *see* bailie
bair boar
bairn barn
baithe both
balleis *see* bailie
ballie *see* bailie
barnes children
barrel measure equal to one twelfth of a last (in the Northern Isles)
baxter baker
bayth both
be by
be the se by the sea
beand being
bedmen, beidsmen residents of an almshouse

behouff obligation
behove to be obliged; to be dependent on
beir *see* bere
beiris testifies
beist beast, animal
bere a type of barley
beseik beseech
betuix between
bigging building
biggit built
bipast past
birnyng burning
black oats oats used as horse fodder
blawin lands lands damaged by weather
block quantity of goods sold at one time; an exhange; an agreement; a bargain
boll dry measure; one sixteenth of a chalder; equal to four firlots
boll men men in charge of cattle
bollman a cottager
bonage feudal service of labour; harvest work
bonage silver money paid in lieu of service
bot but
bot only
bot without
boundis boundaries, borders
bow places grazing pastures
bow teind teind of livestock
bowchis enclosures for livestock
brewster brewer
brig bridge
brint burnt
broder sone nephew
brokin men outlaws
brostare hous brewhouse
brouster, browster brewer
bruik and joyse to use and enjoy
brunt burnt
buith, buithe booth
bund bound
burcht burgh

burdingis burdens
burgess freeman of a burgh
burgh acres land belonging to a burgh
burrow mail duties payable within burghs
burrow rudis cultivated land belonging to a burgh
bursar, bursare holder of a scholarship or endowment
but without
buyth booth
by except
bygane bygone
byggin building
byre cowshed

c. *see* chalder
cain, canis *see* kane
callit called
caragies, cariages, cariagis *see* arriage and carriage
carrage *see* arriage and carriage
castell wairdis payments in commutation of the feudal service of guarding a castle
casting digging or cutting of peats
casualitie incidental item of income or revenue
cave roum or volt cellar or vault
caynes *see* kane
Cessioun *see* Session
chaipland chaplain
chalder dry measure; equal to sixteen bolls
chalmer chamber
chalmerland, chalmerlane chamberlain; the chief financial officer of an estate
channon canon
chantorie the benefice of a chanter
Checker Exchequer
cheik side
childir children
chore choir

claithe, clayth *see* uppermost cloth
claithes clothes, cloth
clois courtyard
cobill, coblis flat-bottomed boat used for salmon-fishing
cognitioun judicial knowledge, cognizance
coil coal
coil silver commuted cash payments in place of customary coal
coilheucht coal pit
coill pot coal pit
coittares *see* cottar
coittis (quots) dues payable for the confirmation of testaments
colden coal pit
College of Justice the supreme civil court
commendator holder of the revenues of a monastic house
commonis food; community of goods
commontie common possession or enjoyment of pasture land
commoun common
commounte community
communibus annis in common or normal years
comprehendit included
compt account
comptar accountant
condiscend specify, state one's case; acquiesce in, agree upon
contigue, contiguous adjacent
corce presendis *see* corpse offerings
corpes presentis, corpis presandis, corpresentis *see* corpse offerings
corpse offerings mortuary or funeral gifts to the church
cost duty paid in kind
cottar, cotter occupier of a cottage
cowclayth *see* uppermost cloth
cravit requested
croce presend *see* corpse offerings
croft small piece of land enclosed for tillage and pasture;

smallholding
crospresentis *see* corpse offerings
crovis fish traps
cryis summonses
cultellar cutler
cuningis, cunningis rabbits
cuntre met local measure
custumit accustomed

daill an allocated share of land
darg a day's work
davach, davoch measure of land used for assessment of tax
dawark, dawerk a day's work
deane dean, a title also assigned to canons regular
debursit disbursed
de claro free
decraipit decrepit
decreit decree
decreit arbitrall decree pronounced by an arbitrator
deduceand deducting
deducit deducted
dedus deduct
defaikis deducts
defakit deducted
defalcatioun authorised deduction from income in accounts
defalkand deducting
defalkit deducted
defasance acquittance, discharge
defese allow as a deduction; acquit or discharge from an obligation
delwingis pieces of ground dug
den, dene *see* deane
deywarkis *see* dawark
diocie diocese
dispone convey in legal form
disponit conveyed in legal form
dochter daughter
doun taking, dountaking reducing
draucht draft
drowrear dowager
dustmeill particles of meal and husk produced in grinding corn
dykis walls

Easter dues payments made to the church at Easter
eckerris acres
effaris affairs
effeiris *see* as efferis
eik also; in addition to
eikit added
eismentis advantages
eldar elder
eldar either
elemosinarie almoner
ellis else
ellis, els otherwise
eme uncle
enteres, entres entrance
enteres, entres interest
entrie silver money paid by an heir on taking possession of his property
erde earth
erding silver payment for burial
everilk every
excamb exchange
exprimit expressed

f. *see* firlot
fadder cart-load
failyeit failed
failyie fail
farina meal
fathom measurement of peats
fe fee
feall fee
fear owner of the fee-simple of a property
fed oxin silver, feid oxin silver commuted cash payment in place of custom oxen
fedder *see* fadder
feft put in legal possession
ferd fourth

GLOSSARY

fermarar, fermarer, fermorar, fermorer, fermourer farmer
ferme rent
ferrie cobill ferry boat
feu heritable and perpetual lease of property
feu mail feu duty
feuar holder of a feu
feudaris feuars
feued disponed on a perpetual heritable tenure for an annual, fixed money rent
feuferme, feu ferme form of land tenure in which a vassal held property heritably on payment on an annual feu duty to a superior, in place of military service
few *see* feu
few annuell annual payment of feu duty
fie fee
fiell fee
firlot dry measure; one quarter of a boll; equal to four pecks
flesche meat
foddome, fouodome, fowdom, foudoume, foddone *see* fathom
foir in front
foiranent in front of
foirfurthe from now on
followar calf, young animal
for far
forby besides
force consequence
forder further
forrow kow cow not in calf
forsamekill for as much
Foud of Shetland chief official and judge in Shetland
fouodome *see* fathom
fouthe abundant, prosperous
fowdom *see* fathom
fra from
freiris friars
frumentum wheat
fuirlet *see* firlot
fure four
furnish furnish
furth, furthe out of
fyft fifth
fyft pairt one fifth

gadderit gathered
gait road, way
gar cause
garsom *see* grassum
gefis gives
geif gave
generell met usual measure
gers grass
gersum, girsum *see* grassum
gif, giff give
gif, giff if
giff in return
gild kow one without defect, and not over eight years of age
girnell granary
girs grass
glasinwreycht, glasynwrycht glazier
glebe, gleib the land assigned to a parish priest or minister
gleissing glazing
glieb *see* glebe
glyb *see* glebe
gottin received
granitar officer in charge of collecting grain rent
grassoum, grassum sum paid by a tenant at the renewal of a lease
grilse young salmon
grintar man one in charge of a lord's granary
grissis grass, pasture
groats hulled grain
grot, grott *see* groats
grott groat (money)
gryis, grys, gryssess grass, pasture
gudling guilder
guidlie easily
guidman yeoman, small proprietor

farming his own land
guidsone son-in-law
guidwife, guidwyif wife of a proprietor
guis, gus goose
gyff give
gylssis *see* grilse

habit silver money assigned to members of religious houses for clothing
haif have
haill whole
hald hold
haldin held
hale whole
halelie wholly
harrowing darg a day's work at harrowing
hauch, hauche flat land usually beside a river
haveand having
healilie wholely
heavin haven, harbour
hebbet *see* habit
heid head
heireftire hereafter
heland cuntrie the Highlands
helin highland
heritour proprietor of heritable property
hithirtill hitherto
holmes low-lying land beside a river
hoprig sloping hollow between ridges
horneing process of proclaiming an outlaw
horsecorn, hors corne oats for horse fodder
hors gers grass for feeding a horse
hospitalitie hospital
houklie weekly
huggut hogshead
humest claithes *see* uppermost cloth
humilie humbly

husbandland measure of land, originally 26 Scots acres

ilk every
Ilk, of that Ilk proprietor of the land of the same name
ilk ane every one
impetrate obtained by petition or formal application to an authority
in pley subject to litigation
inche island
incontinent immediately
inde thence
infeft legally granted
infeftment legal possession of heritable property
ingeving, ingiveing formal submission of a document
ingifar one who submits a document
ingros engross
initit united
inlakking lacking
inpreving improving
inserche search
intakit taken in
interes entrance
intromet, intromett interfere with the goods of another
inundatioun flooding
inwithe within
ische ending
ische issue

joyse enjoy
joysit enjoyed

kane cain, customary payments in kind by a tenant to his landlord
keching and habit silver money allocated to members of religious houses for the upkeep of their kitchen and clothing
ken silver payment on entry to

property
kenar water bailiff
kicheing, kiching kitchen
kild cow *see* gild cow
kill kiln
kirk kow mortuary or funeral fine or present of a cow, given to the parson or vicar on a parishioner's death
kous gers grass for feeding a cow
kow *see* kirk kow
ky cattle
kyis gers grass for feeding cattle

laid of peittis load of peat
laird lesser landed proprietor
lait let
landwart inland
laremair lorimer
last measure of quantity equal to 12 barrels
lateris pertaining to the side
lattin, lattis let
lauborit worked
lauchfullie lawfully
lauender launderer
led to cause to be done in form of legal proceedings
legis lieges
leid load
leix leeks
lemitur licensed beggar
Lentroun Lenten
lesch pund *see* lispund
levand living
levander launderer
liand lying
lipper leper
lispund measure of weight; one sixth of a meil (Northern Isles)
litster dyer
Lords of the Seat judges of the Court of Session
los be less by

luckin buthes covered stalls which could be locked
ludgeing lodging
luiging lodging
lyand lying
lycht light
lymekill limekiln
lynt lint
lyverent liferent

m. *see* meil
mail rent
mailing ground for which rent is paid; a rented farm; a holding
mains home farm
mair executive officer of the law
mair more
mairt *see* mart
mais nets used for carrying fish
maister master; Master of Arts; the heir apparent or heir presumptive of a peer
male *see* mail
mallis *see* mail
man, mane must
manes *see* mains
manifestit made evident
manis *see* mains
mans *see* manse
manse house for the use of the parish priest or minister
mark measure of weight; one twenty-fourth of a settin (Northern Isles)
marrow companion
mart cow or ox fattened for slaughter; an animal due rendered by a tenant at Martinmas
mart silver money paid instead of marts as rent
martstuik beasts paid as rent, often commuted for a money payment
mauchlitt botched
maynes *see* mains

meil measure of weight, equal to 6 lispunds; one twenty-fourth of a last
meill *see* meal
meinis and schawis complains and demonstrates
meit measure
meit fische fish for food
mekill much
mekle much
menfullie manfully
menis and schawis complains and demonstrates
menit complained; sought redress
merk 13s 4d Scots
merkland measure of land originally worth one merk
mert *see* mart
met measure
mickale, mikall large
milwe mill
mister need or requirement
moe more
mon must
monie, mony many
multure tax on grain or meal ground at the barony mill and payable to the owner or tenant of the mill
mure moor
mylis mills
myll mill
mylne mill

na no
nane none
nest next
nichel, nichell nothing
nocht not
nor nor
nor than
notar notary public
nothir, nowthir neither
notour commonly known
novatioun innovation
nuckald, nukald newly calved

oblatioun offering
obleis oblige
obleist obliged
oblissis obliges
oblist obliged
offerand, offrenis offering
Official bishop's deputy, with judicial functions
oftymis often
onnawyse in no way
onpayit unpaid
ontill unto
onto unto
oo grandson
or before
ordeum bere
ottounis out touns, outlying fields
oustere hostelry
out saittis *see* outsettis
outred to settle a payment or discharge a debt
outrynnyng expiry, termination
outsettis outlying land belonging to a larger property
overblawin lands lands damaged by weather

p. *see* peck
Pace fynes *see* Easter dues
parechines parishes
paroche, parochin parish
parochiner parishioner
Pasch fines, Pasche fynes *see* Easter dues
Pasch offrandis, Pasche offering *see* Easter dues
Pasche, Peasche Easter
Pasche penneis, Pasche pennis *see* Easter dues
Pasche raikiningis, Pasche reknyngis *see* Easter dues
Peasche fynis *see* Easter dues
Peax Easter
peax peace
pece a unit of any commodity

peck dry measure; one quarter of a firlot
pecunia money
pedagog teacher
pedagogie school, university
pendicle dependency
penny mail rent paid in money
pentionarie pensionary
personal teinds tithes derived from the profits of personal labour
pertinents anything connected with a piece of land or heritable property that is not specially reserved from the grant
petty-commonis an allowance of food or money
pikmen miller's servants
pittance an allowance of money given to a member of a religious community
place property belonging to a religious community
plak small coin worth 4d Scots
planting (of kirks) providing churches with ministers
plenisching provision; goods; stock
pleuch, plew plough
plew ploughgate
plewing darg a day's work at ploughing
plewland measure of arable land, variable
pley *see* in pley
porportand purporting
porte gate
portion monk's pension allocated from monastic revenues
portionar owner or holder of part of a property
portionares seculares secular residents in a religious house
poynding distraint; appropriating goods for sale to defray debts
prebendar prebendary
prebendarie, prebendreis prebend, prebends

precentor chanter; leader of the choir in a cathedral
precept legal injunction; a written order from a superior; an order to pay money
prejudgit prejudiced
premive first
premissis that which is before mentioned; the early portion of a legal document where the subject-matter is stated in full
prepositive relating to provost
prepositus provost
presentis present letter or document
priorassie prioress
priores, prioris prioress
procurage *see* procurations
procurations sums paid as commutation for the duty to entertain the bishop, archdeacon and dean on their visitations
procuratour advocate, agent, representative
proport purport
provost, proveist chief magistrate of a town
provost, proveist head of collegiate kirk
provostry, provestrie office of provost
pundland measure of land originally worth one pound
punisoun attack by way of punishment

queir choir
querrell yaird quarry yard
quha who
quhair where
quhairin wherein
quhairof of which
quhairthrow through which
quhairunto to which
quhat what
quhatsumevir whatsoever

quheit wheat
quheit aites, quheitt aittis white oats
quhich which
quhilk, quhilkis which
quhill until
quhill while
quhit what
quhome whom
quhylis, quhyllis sometimes
quhylis, quhyllis until
quhyt aites white oats
quots dues payable for the confirmation of testaments
quoy township or farm

raid *see* riding
raid, rait rate
raift *see* reft
recantit recanted
redar reader
reddin *see* riding
redount fell to be paid
reft plundered, stolen
regent teacher in a college
relict widow
rentaleit rentalled
restand remaining
rests, restis arrears of rent or other payments; remainder, remains
richteuslie righteously
riddin *see* riding
riding determining the boundaries of land by riding around the perimeter
rig ridge made by ploughing
riggis *see* rig
rin run
rowmes lands
rude rood of land
rycht right
rychtin writing
ryddin, rydin, ryding *see* riding
ryne mart sylver commuted cash payment as rent in place of an animal due at Martinmas

sa so
salbe shall be
salfand saving, except
sall shall
salt girsis, salt gres pasture near salt water
sameikill, samekill as much
samen, samin, samyn same
sangschole song school
sangster chorister
sasine act of giving possession of feudal property
sawing sowing
scat, scatt tax or tribute
schawis shows
scheill summer pasture
schir sir, the title given to a priest
scho she
scray young saithe cured by drying in the sun
se sea
seasing *see* sasine
Seat Court of Session, supreme civil court
seggis gelded oxen
selit sealed
sen since
sensyne since then
servand serving
servitour servant
Sessioun Court of Session, supreme civil court
set in assedatioun to give in lease
set in tak to give in lease
settin setting; a measure of weight in Orkney, equal to 24 marks
severall individual
sex pairt one sixth
sic such
sicht sight
siclyk likewise
sict such
sik such

sindrie sundry, separate, distinct
sinodage, sinodall *see* synodals
sir *see* schir
skant scarce
skeat *see* scat
sklater slater
skray *see* scray
small teinds vicarage teinds
smiddie, smyde smithy
socht sought
soud a quantity
sould should
sowme sum
sowmes gers horseloads of grass; as much grass as will pasture one cow or five sheep
sowmis sums
sowmit calculated
specie special
speir ask, inquire
stallar cleric occupying the stall of a canon in a cathedral choir; usually a synonym for vicar choral
steading farm buildings
steding *see* steading
steilbo guidis the stock and seed on a farm which remained the property of the landlord and not to be removed by an outgoing tenant
stellis places in a river over which nets are drawn to catch salmon
stene stone
stent tax
stewartrie, stewartry administrative district
stirk steer, young bullock or heifer
sts. *see* settin
sua so
suld should
summes gers horseloads of grass
sumtyme formerly
superplus surplus
surtie surety
sustenit sustained

sustentatioun upkeep
swa so
sybous spring onions
syne since
syne then
synodals payments to the bishop from each church annually, as fixed by a canon in the Decretal of Gratian

tack lease
tack take
tacksman leaseholder; tenant farmer
tae salmon spear
taill riggis narrow strips of land jutting out from a larger piece
tak lease
tak take
takisman, takkismane *see* tacksman
tane taken
taxman *see* tacksman
teind tithe; one tenth of the produce of the land (and water) in a parish, paid for the support of the church in the form of a stipend for the parson and vicar (but often leased); *see also* small teinds
teind fish tithes leviable on fish
teind sheaves sheaves of corn payable as tithes
teind stirk, gus and grissis cattle, geese and grass or hay payable as tithes
tenandrie tenancy; a tenant's holding
tenement land held in tenure and built on
testamentore testamentary
teynd *see* teind
teynd fische tithes leviable on fish
teynd grys tithes leviable on grass
teynd schaves, teynd schevis *see* teind sheaves
thairintill thereto
than then

thei they
thekin thatching
thesaurer treasurer
thesaurerie treasury
thir, thire these, those
thirds a tax, equivalent to a third of the income, levied on benefices from 1562 to support the crown and reformed ministry
threttie thirty
threttie ane thirty-one
thrid pairt one third
thridis see thirds
throucht through
till to
till until
toft land for cultivation, attached to a house
tolbooth town hall
toun rural settlement; area of arable land on an estate, occupied by a number of farmers as co-tenants
traductour leader, guide, curator
traist trusted
trans narrow passageway
transumpt copy, transcript
trene wooden
tressory treasury
tronn instrument for weighing heavy goods
truittis trout
tua two
tua pairt two thirds
tuelf twelve
turs, turssis truss, bundle, load
tutour tutor; legal guardian
twa part two thirds
tyne lose
tynsall loss

umaist claithes, umestclaythis see uppermost cloth
umquhile, umquhill former; deceased
unes, unis ovens
unstabill irregular
unvitiat unimpaired, valid
upgeving act of delivering up
upmaist claithes see uppermost cloth
uppermost cloth mortuary due, sometimes the uppermost cloth on the bed, sometimes the outermost garment of the deceased, exacted by the pre-Reformation church
uptakin, uptaking act of collecting or uplifting dues or taxes
usis in the habit of
usit and wont use and wont, custom
usitt used, accustomed, tried
usufructuar possessor of the fruits or revenues
utermaist utmost
uvynes ovens

vaik vacant
valour value
vennale, vennall alley; narrow street
victuall victual; grain
videlicet namely
visie visit
vitiat rendered null and void
volt vault

wairdis see castell wairdis
wald would
walk mylne, walkmyln, walkmylne mill for fulling cloth
walx wax
wantand lacking
war worse
wark work
waukmill mill for fulling cloth
wedder wether; castrated ram
wedder gangis right of pasturage for wethers
weyar weigher
weycht weight

win to harvest, to gain by labour
wit know
wit knowledge
without unless
woder *see* wedder
wodsett wadset; grant of land in pledge for a debt, with a reserved power to the debtor to recover his lands on payment of the debt; mortgage
woll wool
wont custom
wount *see* wont
wricht wright
writtar writer; lawyer or lawyer's clerk
wrycht wright
wynd narrow street

yconimus oeconomus, manager of a monastic house, a term often used synomymously with commendator
yerding silver payment for burial-ground
yet gate
youll *see* yule
yule Christmas

Bibliography

1 Manuscript Sources

Edinburgh University Library:
   MS Dc.4.32, Book of Assumptions

National Library of Scotland:
   Adv MS 31.3.12, Book of Assumptions

Scottish Record Office:
   CS7/20, Register of the Acts and Decreets of the Court of Session
   E2/15, Register of Signatures in Comptrollery
   E48/1/1, Book of Assumptions
   E48/1/2, Book of Assumptions

2 Primary Printed Sources

Abstracts of Protocols of the Town Clerks of Glasgow, ed. R. Renwick, 11 vols. (Glasgow, 1894-1900)
Accounts of the Collectors of Thirds of Benefices, 1561-1572, ed. G. Donaldson (Edinburgh, 1949)
Accounts of the Lord High Treasurer of Scotland, eds. T. Dickson, *et al.*, 12 vols. (Edinburgh, 1877-1970)
Accounts of the Masters of Works, eds. H. M. Paton, J. Imrie and J. G. Dunbar, 2 vols. (Edinburgh, 1957-82)
Acts of the Lords of Council in Public Affairs, 1501-1554, ed. R. K. Hannay (Edinburgh, 1932)
Acts of the Parliament of Scotland, eds. T. Thomson and C. Innes, 12 vols. (Edinburgh, 1814-1875)
The Autobiography and Diary of Mr James Melvill, ed. R. Pitcairn (Ediburgh, 1842)
Ayr Burgh Accounts 1534-1624, ed. G. S. Pryde (Edinburgh, 1937)
The Booke of the Universall Kirk of Scotland: Acts and Proceedings of the General Assemblies of the Kirk of Scotland, ed. T. Thomson, 3 vols. and appendix vol. (Edinburgh, 1839-1845)

Calendar of State Papers relating to Scotland and Mary, Queen of Scots, 1547-1603, eds. J. Bain, et al., 13 vols. (Edinburgh, 1898-1969)
Carte Monialium de Northberwic, ed. C. Innes (Edinburgh, 1847)
Cartularium Ecclesiae Sancti Nicholai Aberdonensis, ed. J. Cooper, 2 vols. (Aberdeen, 1888-1892)
Charters and Documents relating to the Burgh of Peebles, ed. W. Chambers (Edinburgh, 1872)
Charters of the Abbey of Coupar Angus, ed. D. E. Easson, 2 vols. (Edinburgh, 1947)
Charters of the Abbey of Inchcolm, eds. D. E. Easson and A. Macdonald
Chartulary of Lindores Abbey 1195-1479, ed. J. Dowden (Edinburgh, 1903)
Collectanea de Rebus Albanicis (Edinburgh, 1847)
Concilia Scotiae: Ecclesiae Scoticanae Statuta tam provincialia quam synodalia quae supersunt, MCCXXV-MCLIX, ed. J. Robertson (Edinburgh, 1866)
Court Book of Shetland, 1615-1629, ed. G. Donaldson (Lerwick, 1991)
Extracts from the council register of the burgh of Aberdeen, 2 vols., ed. J. Stuart (Aberdeen, 1844-1848)
The First Book of Discipline, ed. J. K. Cameron (Edinburgh, 1972)
An Historical Atlas of Scotland c.400 - c.1600, eds. P. McNeill and R. Nicholson, (St Andrews 1975)
Inquisitionum ad Capellam Regis Retornatarum ... Abbrevatio, ed. T. Thomson, 3 vols. (London, 1811-1816)
John Knox's History of the Reformation in Scotland, ed. W. C. Dickinson, 2 vols. (Edinburgh, 1949)
The Knights of St John of Jerusalem in Scotland, eds. I. B. Cowan, P. H. R. Mackay and A. Macquarrie (Edinburgh, 1983)
Liber Cartarum Sancte Crucis. Munimenta Ecclesie Sancte Crucis de Edwinesburg, ed. C. Innes (Edinburgh, 1840)
Liber Collegii Nostre Domine. Registrum Ecclesie B. V. Marie et S. Anne infra muros civitatis Glasguensis, MDXLIX, ed. J. Robertson (Glasgow, 1846)
Liber Conventus S. Katherine Senensis Prope Edinburgum, ed. J. Maidment (Edinburgh, 1841)
Liber insule missarum. Abbacie canonicorum regularium B. Virginis et S. Johannis de Inchaffery registrum vetus: premissis quibusdam comitatus antiqui de Stratherne reliquiis, ed. C. Innes (Edinburgh, 1847)
Liber S. Marie de Calchou. Registrum Cartarum Abbacie Tironensis de Kelso, 1113-1567, ed. C. Innes, 2 vols. (Edinburgh, 1846)
Liber S. Marie de Dryburgh, Registrum Cartarum Abbacie Premonstratensis de Dryburgh, ed. W. Fraser (Edinburgh, 1847)
Liber S. Thome de Aberbrothoc, Registrorum Abbacie de Aberbrothoc pars altera Registrum Nigrum necnon Libros Cartarum Recentiores Complectens. 1329-1536, ed. C. Innes (Edinburgh, 1856)
Miscellany of the Wodrow Society, ed. D. Laing (Edinburgh, 1844)
Papal Negotiations with Mary Queen of Scots during her reign in Scotland, 1561-1567, ed. J. H. Pollen (Edinburgh, 1901)
Protocol Book of Mr Gilbert Grote, 1552-1573, ed. W. Angus (Edinburgh, 1914)

Protocol Book of Nicol Thounis, 1559-1564, ed. J. Beveridge and J. Russell (Edinburgh, 1927)
Protocol Book of Sir William Corbet, 1529-1555, ed. J. Anderson (Edinburgh, 1911)
Protocol Books of Dominus Thomas Johnsoun, 1528-1578, eds. J. Beveridge and J. Russell (Edinburgh, 1920)
Records of the Earldom of Orkney 1299-1614, ed. J. Storer Clouston (Edinburgh, 1914)
Records of the Monastery of Kinloss, ed. J. Stuart (Edinburgh, 1872)
Register of the Great Seal of Scotland. Registrum Magni Sigilli Regum Scotorum, eds. J. M. Thomson, *et al.*, 11 vols. (Edinburgh, 1882-1914)
Register of Ministers, Exhorters and Readers, ed. A. Macdonald (Edinburgh, 1830)
Register of the Privy Council of Scotland, eds. J. H. Burton, *et al.*, 1st ser., 14 vols. (Edinburgh, 1877-1898)
Register of the Privy Seal of Scotland. Registrum Secreti Sigilli Regum Scotorum, eds. M. Livingstone, *et al.*, 8 vols. (Edinburgh, 1908-1982)
Registrum Cartarum Ecclesie Sancti Egidii de Edinburgh, ed. D. Laing (Edinburgh, 1859)
Registrum de Dunfermelyn, ed. C. Innes (Edinburgh, 1842)
Registrum Monasterii de Passelet 1163-1529, ed. C. Innes, (Paisley, 1882)
Registrum Monasterii S. Marie de Cambuskenneth A.D. 1147-1535, ed. W. Fraser (Edinburgh, 1872)
Rental Book of the Cistercian Abbey of Cupar-Angus with the Breviary of the Register, ed. C. Rogers, 2 vols. (London, 1879-1880)
Rentale Dunkeldense, being Accounts of the Bishopric (A.D. 1505-1517) with Myln's 'Lives of the Bishops' (A.D. 1483-1517), ed. R. K. Hannay (Edinburgh, 1915)
Rotuli Scaccarii Regum Scotorum: The Exchequer Rolls of Scotland, eds. J. Stuart and G. Burnett, *et al.*, 23 vols. (Edinburgh, 1878-1908)
The Scots Peerage, ed. J. B. Paul, 8 vols. (Edinburgh, 1904-1914)
Selections from the Records of the Regality of Melrose, ed. C. S. Romanes, 3 vols. (Edinburgh, 1917)
Statutes of the Scottish Church, 1225-1529, ed. D. Patrick (Edinburgh, 1907)
Valor Ecclesiasticus, eds. J. Caley and J. Hunter, 6 vols. (London, 1810-1834).
The Works of John Knox, ed. D. Laing, 6 vols. (Edinburgh, 1846-1864)

3 Secondary Works

Alexander, W. M., *The Place-Names of Aberdeenshire* (Aberdeen, 1952)
Black, G. F., *The Surnames of Scotland* (New York, 1946)
Campbell, J., *Balmerino and its Abbey* (Edinburgh, 1899)
Cowan, I. B. and Easson, D. E., *Medieval Religious Houses, Scotland* (London, 1976)
Cowan, I. B., *The Parishes of Medieval Scotland* (Edinburgh, 1967)

Dickinson, W. C., 'The *Toschederach*', *Juridical Review*, liii (1941)
Dilworth, M., 'The Commendator System in Scotland', *Innes Review*, xxxvii (1986)
Donaldson, G., 'The Bishops and Priors of Whithorn', in *Transactions of the Dumfriesshire and Galloway Natural History and Antiquarian Society*, 3rd ser., xxvii (1950)
Donaldson, G., 'The "New Enterit Benefices", 1573-1586', *Scottish Historical Review*, xxxii (1953)
Essays on the Scottish Reformation, ed. D. McRoberts (Glasgow, 1962)
Fasti Ecclesiae Scoticanae, ed. H. Scott, 10 vols. (Edinburgh, 1915-1981)
Fraser, W., *The Chiefs of Colquhoun and their Country*, 2 vols. (Edinburgh, 1869)
Fraser, W., *The Douglas Book*, 4 vols. (Edinburgh, 1885)
Handbook of British Chronology, eds. E. B. Fryde, *et al.* (London, 1986)
Handbook of Dates for Students of English History, ed. C. R. Cheney (London, 1978)
Hannay, R. K., *The College of Justice* (Edinburgh, 1933; reprinted 1990)
Haws, C. H., *Scottish Parish Clergy at the Reformation, 1540-1574* (Edinburgh, 1972)
Hay, G., 'The late medieval development of the High Kirk of St Giles, Edinburgh', *Proceedings of the Society of Antiquaries of Scotland*, cvii (1975-76)
Keith, R., *History of the Affairs of Church and State in Scotland*, eds. J. P. Lawson and C. J. Lyon, 3 vols. (Edinburgh, 1835-1850)
Laing, A., *Lindores Abbey and its Burgh of Newburgh* (Edinburgh, 1876)
Lees, J. C., *The Abbey of Paisley from its Foundation to its Dissolution* (Paisley, 1878)
Macfarlane, W., *Geographical Collections relating to Scotland*, ed. A. Mitchell, 3 vols. (Edinburgh, 1905-1908)
Marwick, H., *Orkney Farm-Names* (Kirkwall, 1952)
Maxwell, A., *Old Dundee, ecclesiastical, burghal and social, prior to the Reformation* (Edinburgh, 1891)
McDowall, W., *Chronicles of Lincluden* (Edinburgh, 1886)
Moir Bryce, W., *The Black Friars of Edinburgh* (Edinburgh, 1911)
Morris, T., *The Provosts of Methven* (Edinburgh, 1875)
Old-Lore Miscellany of Orkney, Shetland, Caithness and Sutherland, ed. A. W. Johnston, ix (London, 1933)
Ordnance Gazetteer of Scotland, ed. F. H. Groome (Edinburgh, 1901)
Orkney and Shetland Records, eds. A. W. Johnston and A. Johnston, i (London, 1907-1913)
Rankin, W. E. K., *The Parish Church of the Holy Trinity, St Andrews* (Edinburgh, 1955)
Verschuur, M. B., 'Perth and the Reformation, Society and Reform: 1540-1560', 2 vols. (Glasgow Ph.D. thesis, 1985)
Watt, D. E. R., *Fasti Ecclesiae Scoticanae Medii Aevi ad annum 1638* (Edinburgh, 1969)
Zupko, R. E., 'The weights and measures of Scotland before the Union', *Scottish Historical Review*, lvi (1977)

Index of Persons

Abercrombie (Abircrumby), Andrew, 303 (Cargill)
——, *dene* Henry, parson of Kildonan, prior of Scone abbey, vicar of Logierait, 332, 335, 631, 631n.
——, James, 92 (Holyroodhouse abbey)
——, John (Glasgow archdeaconry), 253, 253n.
——, John, 62 (Inchcolm abbey)
——, Richard, abbot of Inchcolm, 62, 62n.
——, Mr Robert, treasurer of Dunkeld, 311, 311n.
——, Robert, parson of Buttergill, 391
——, Robert, chaplain of St Barbara's altar, Dundee, 399
Accaiman, *see* Aikman
Acheson, Alexander, 101 (Newbattle abbey)
——, Cuthbert, 181, 182 (Haddington priory)
——, John, parson and vicar of Tarvit, 73, 88
——, Robert, 173 (Dunglass collegiate kirk)
A'Cultan (Acculattane), Niven, 598 (Soulseat abbey)
Adair, James, 595, 596 (Whithorn priory)
——, Neil, 595 (Whithorn priory)
——, William, 598 (Soulseat abbey)
Adam, David, 158 (Airth)
——, James, 398 (Alyth)
——, Thomas, elder, 158 (Airth)
——, Thomas, younger, 158 (Airth)
——, Thomas, 546 (St Ninians)
——, William, 546 (St Ninians)
Adamson, George, 102 (Newbattle abbey)
——, James, in Edinburgh, 102

——, James, 101, 102 (Newbattle abbey)
——, John, friar of Elgin, 454
——, John, in Linlithgow, 155, 157
——, Sanders, in Edinburgh, 127
——, Mr William, chanter of Dunkeld, 306, 307, 308
——, William, in Edinburgh, 127, 130
Addie (Ade), William, 283, (Perth)
Ade, *see* Addie
Affleck, Pattoun, 283, 286 (Perth)
——, Thomas, 283, 286 (Perth)
Agnew, Patrick, sheriff of Wigtown, 590, 590n., 607 (Galloway bishopric); 598 (Soulseat abbey)
Aichter, Thomas, 557 (Ayr)
Aiken (Akin, Aikkine), John, 556 (Stirling)
——, Robert, 290 (Culross)
——, William, in Edinburgh, 122
——, William, 290 (Culross abbey)
Aikenhead (Akinheid), Marjorie, in Edinburgh, 137
Aikman (Accaiman) of Killochrig, Fergus, 605
——, Donald, 105 (Kirknewton), 105
——, Duncan, prebendary of the aisle of Carnwath collegiate kirk, 521
——, James, lister, in Edinburgh, 137
——, James, in Edinburgh, 124, 125
Ailhous, *see* Alehouse
Air, Thomas, in Elgin, 453
Aird, William, 557 (Ayr)
Airdrie (Ardree), laird of, 18
Airth, laird of, 92
Aitken (Aitkyn), Robert, prebendary of Methven collegiate kirk, 299
Aiton (Aittoun), *see also* Ayton
——, George, 180 (Haddington priory)
——, John, 180 (Haddington priory)

Aldoch (Aldoke), Robert, 114 (Leith hospital)
Alehouse (Ailhous), Janet, 421 (Aberdeen friary)
Alexander (Alschunder), James, 667 (Kirkwall)
——, James, 666, 667 (Westray)
——, John, 375 (Aberlemno)
——, John, 556 (Stirling)
——, Nicol, 227 (Greenlaw)
——, Patrick, parson of Auldhame, 179, 183n.
——, William, 352 (Coupar Angus abbey)
Alexanderson, William, 398 (Alyth)
Alie, Andrew, in Elgin, 453
Alison (Alesoun, Alysoun), Duncan, 177, 179 (Haddington priory)
——, William, 216 (Jedburgh abbey)
Allan (Allane, Allen), Adam, in Edinburgh, 137
——, Alexander, 158 (Airth)
——, Charles, 158 (Airth)
——, Giles, 104 (Kirknewton)
——, John, baxter, 551 (Stirling)
——, John, 158 (Airth)
——, Robert, 155 (Dalmeny)
——, William, 100 (Newbattle)
Allanson (Allansone), David, 420 (Aberdeen friary)
Alschunder, see Alexander
Altrie, lord, see Keith, Robert
Anderson, [blank], 421 (Aberdeen friary)
——, Mr Alexander, subprincipal, King's college, Aberdeen, 441; parson of Mortlach, 444
——, Alexander, chaplain, Dunblane cathedral, 343.
——, Alexander, tenant in Duffus, 468
——, Alexander, parson of Tyrie, 429
——, Alexander, 479 (Moray chantory)
——, David, burgess of Perth, 326
——, David, vicar choral in Glasgow cathedral, parson and vicar of Dalziel, 519
——, James, 283, 287 (Perth)
——, John, chamberlain of Moray, 642
——, John, vicar pensionary of Strathaven, 603
——, John, in Edinburgh, 139
——, John, 291 (Culross abbey)
——, John, 181 (Haddington priory)
——, Robert, 492 (Dallas)
——, William, in Edinburgh, 127
——, William, 283, 287 (Perth)
Anderston, Gilbert, in Linlithgow, 11
Andrew, John, 479 (Ardclach)
Angus, Andrew, vicar of Leslie, 82
——, William, canon of Chapel Royal, Stirling, 560 (Coylton)
Annand, Henry, 667, 667 (Sanday)
——, Mr James, parson of Papa Westray - Cross kirk, 660, 660n., 666, 667; vicar of Sanday, 666, 666n., 667
Annas, William, 598 (Soulseat abbey)
Anstruther (Ansterthrope, Anstrother, Anthrother), George, 586 (Galloway bishopric)
——, Henry, 586, 589, 607 (Galloway bishopric)
Anton (Antane), 'Franche', in Edinburgh, 140
Archibald, John, tenant in Duffus, 468
——, Nicol, tenant in Duffus, 468
——, Mr Thomas, chamberlain of archbishopric of Glasgow, 212, 496, 496n., 533
——, Mr Thomas, vicar of Peebles, 251
——, Thomas, factor, 252 (Lyne)
Ardhill, John, chaplain of Holy Rood altar, Stirling parish kirk, 556
Ardros, lady, 167; see also Crichton, Elizabeth
Arduell, laird of, 594, 596 (Whithorn priory)
Argyll, earls of, see Campbell
Arnold, George, vicar of Minnigaff, 600
Arnot (Arnocht, Arnott), Mr Andrew, parson of Tough, 425
——, Andrew, archdeacon of Galloway, 601n.
——, David, 585, 587, 607 (Galloway bishopric)
——, Mr George, parson of Essie, 475
——, George, prebendary of Pettinbrog, Abernethy collegiate kirk, 340
——, Henry, 586 (Galloway bishopric)
——, Mr John, 177, 179 (Haddington priory)
——, John, 18 (St Andrews priory)
——, Thomas, 180 (Haddington priory)
——, William, 182 (Crail)
Arntully (Arntulie), laird of, 303
Arous (Wrs), William, 353n. (Coupar Angus abbey)

INDEX OF PERSONS

Arran, earls of, *see* Hamilton
Arrois, James, 101 (Newbattle)
——, Robert, 101 (Newbattle)
Arrott, David, vicar and reader at Guthrie, 390
Arthur, James, vicar pensionary of Slamannan, 552
Aschennan (Aschman, Aschnanen) of Dunjop, John, 586, 586n.
——, James, 586 (Galloway bishopric)
——, Robert, 585 (Galloway bishopric)
Atholl, earls of *see* Stewart
Auchendinny, Over, laird of, 142
Auchincraw of Nether Byre, George, 199
——, George, portioner of East Restoun, 198 (Coldingham priory)
Auchingassell (Auchingaschill), laird of, 271, 276, 277, 623
Auchinleck, laird of, 573
——, Mr Robert, vicar of Menmuir, 397
Auchmuty (Auchmowty), Mr Robert, vicar of Dun, 381, 393n., 396; vicar pensionary of Arbroath, 396; vicar of Stirling, chaplain of Bannockburn, 550; 316 (Dunkeld)
Auld, Janet, 582, 583 (Dreghorn)
Auldbar, laird of, 298, 378
Auldcorn, James, 158 (Airth)
Austin (Austeand, Austoun, Oustane, Owsteane), Mr Thomas, 492 (Huntly)
——, William, vicar of Catterline, 397
——, William, in Linlithgow, 11
——, William, 66 (St Andrews archdeaconry)
Awdy, laird of, 303
Ayton (Aytoun), *see also* Aiton
——, laird of, 228
——, Ninian, factor, 552 (Kilsyth)
——, Robert, in Edinburgh, 132

Baillie, James, 677 (Orkney bishopric)
Bain (Baine, Bane), Alexander, vicar choral in Glasgow cathedral, parson and vicar of Dalziel, 519
——, William, 633 (Caithness archdeaconry)
Baird, Duncan, 158 (Airth)
——, William, 158 (Airth)
Balcanquhal (Ballincanquall), John, chaplain of Holy Blood altar, Kirkcaldy, 81, 81n.
Bald (Bawld), Hew, prebendary of Bothans collegiate kirk, 172
——, Robert, 181 (Haddington priory)
Balderston (Balderstoun), Richard, 156 (Linlithgow priory)
Baldrany (Baldreny), Thomas, 101 (Newbattle abbey)
Balfour of Montquhanny, Andrew, 60
—— of Pittendreich, James, 248n.
—— of Westray, Gilbert, 663
——, Archibald, son of Gilbert Balfour of Westray, 663 (Kirkwall)
——, Bartholomew, 555 (Menteith)
——, Mr David, pension to, 309 (Dunkeld archdeacony)
——, David, 14, 18 (Forgan)
——, David, 556 (Stirling)
——, David, 546 (Tullibody)
——, George, prior of Charterhouse, Perth, 288, 289
——, Gilbert, 655, 660, 661, 668 (Orkney bishopric)
——, Henry, vicar of Dollar, 339n.
——, Mr James, dean of Glasgow, vicar of Scoonie, 71n.
——, Mr James, parson of Flisk, 60
——, John, 114 (Leith hospital)
——, Mr Michael, commendator of Melrose abbey, 207
——, Mr Robert, parson of Kirkchrist, 616
——, Mr Walter, chaplain of Trinity chaplainry, Dunblane cathedral, 325
——, Mr Walter, parson of Linton, chaplain of Newhaven, 108, 228, 244, 253
——, Mr Walter, vicar pensionary of Kinross and Orwell, 88
——, Mr Walter, pension to, 228 (Kelso abbey)
——, Walter, parson of Luncarty, 324
Balhaggertie, laird of, 437
Ballantyne (Bellentyne), *see also* Bellenden
——, James, in Edinburgh, 127
——, John, 115 (Lasswade)
Ballingall, William, vicar of Insch, 430-431
Ballquhaip, Agnes, 546 (Stirling)
Balmowto, laird of, 72
Balnagown (Ballangown, Balnagowne), laird of, bailie of Fearn abbey, 625,

633, 634, 652
Balquhannane, laird of, 414, 415
Balvaird, (Balward), laird of, 546
——, Mr David, parson of St Madoes, 345
Bannatyne of Carhouse, John, pension to, 508 (Lanark)
——, Alexander, prebendary of Methven collegiate kirk, 299
——, James, 557 (Ayr)
——, John, chaplain of St Michael's altar, Crieff, 337
——, John, 100 (Newbattle)
——, John, 141 (Restalrig)
——, Margaret, in Edinburgh, 135
——, Mr Thomas, prebendary of Vogrie, Crichton collegiate kirk, 109
——, William, chaplain of St Laurence's altar, St Giles' collegiate kirk, Edinburgh, 125
Bannerman, Gilbert, prebendary of Craigie, canon of Dunkeld cathedral, 312
Bar, *see* Barr
Barber (Barbour), John, 592 (Whithorn priory)
——, William, prebendary of Bute, Restalrig collegiate kirk, 143
Barchan, *see* Barton
Barclay, George, 383 (Brechin bishopric)
——, Hercules, parson of Canisbay, 638, 638n.
——, John, 113 (Leith hospital)
——, Thomas, 18 (Forgan)
Bardner (Bardnar), Henry, 292 (Culross abbey)
Bardoy, laird of, 92 (Holyroodhouse abbey)
Bargany, laird of, *see* Kennedy
Barker (Barcare), Alexander, in Edinburgh, 133
——, John, in Edinburgh, 137, 138
Barnbarroch (Barinbaroch, Barnebarrocht), lady, 598
——, laird of, 587, 596, 598, 607
Baron, James, in Edinburgh, 140
Barquhan, *see* Barton
Barr (Bar), laird of, 562
——, Hew, in Edinburgh, 133
Barry, Mr Archibald, vicar pensionary of Hamilton, 519
——, Mr Archibald, vicar of Kilpatrick, 541
Barthane, *see* Barton
Barthren, Thomas, in Edinburgh, 136
Barton (Barquhan, Bartan, Bartane, Barthane), David, chaplain of chaplainries of St Laurence and Arfaill, Dingwall, 636
——, David, vicar of Suddie and Kilmuir-Wester, 639, 639n.
——, Mr John, prebendary of Forgandenny, Dunkeld cathedral, portioner of Abernyte, 309-310; pension to, 304 (Dunkeld); 312, 343 (Dunkeld)
——, Robert, comptroller, 114, 114n.
——, Walter, chaplain of St Laurence's altar, St Giles' collegiate kirk, Edinburgh, 132
——, William, 114 (Leith hospital)
Bass, laird of, 147, 211
Bassendine, James, in Edinburgh, 127, 128
Bathgate (Bathcat), Mr Thomas, prebendary of St Vincent, St Mary of the Fields' collegiate kirk, Edinburgh; song schoolmaster, 124
Bathok, John, 158 (Airth)
Bawmonthe, *see* Beaumont
Baxter (Baxtar), David, vicar of Newburn, reader at Newburn, 69
——, John, 400 (Dundee)
——, John, 271 (Holywood abbey)
——, Robert, 368, 371 (Coupar Angus abbey)
Beath, laird of, 63n.
Beaton, *see* Betoun
Beattie (Batie, Betty), burgess of Edinburgh, 138, 139
Beaumont (Bawmonthe), William of, 619 (St Mary's Isle priory)
Beck, John, 271, 275 (Holywood abbey)
——, Thomas, 271 (Holywood)
Bell, Adam, in Edinburgh, 133
——, Alexander, 155 (Dalmeny)
——, Andrew, in Edinburgh, 141
——, John, 353 (Coupar Angus abbey)
——, John, 276 (Holywood abbey)
——, William, 102 (Newbattle abbey)
——, William, 254 (Peebles)
——, William, 546 (St Ninians)
——, William, 554 (Stirling)
Bellenden (Ballanden, Ballantyne), *see also* Ballantyne

—— of Auchnoule, John, justice clerk, 8, 8n., 100, 100n., 286, 286n., 383, 610, 659, 659n., 670, 670n.
——, Christine, prioress of Sciennes nunnery, Edinburgh, 94, 94n.
——, Mr George, vicar of Dunrossness, Shetland, 664
——, John, justice clerk, 491, 491n., 575n.
——, Patrick, vicar of Sprouston, 196
——, Patrick, 92 (Holyroodhouse abbey)
——, Patrick, 668, 670 (Orkney bishopric)
Bennet, David, 19 (Markinch)
——, Robert, pension to, 291; 290, 292 (Culross abbey)
Bennie (Benny), James, 158 (Airth)
Benson, William, 353 (Coupar Angus abbey)
Berryhill, John, 419, 421 (Aberdeen friary)
Bertram (Bertrem), Walter, in Edinburgh, 128, 133
Best, John, in Edinburgh, 127
Betoun of Melgund, David, 330, 393
—— of Pitlochy, John, 146
——, Mr Alexander, parson of Abernethy, 330; pension to, 359, 360, 363 (Arbroath abbey, Monifieth); parson of Monifieth, 393
——, Mr Andrew, parson of Lyne, 252
——, Andrew, tacksman, Glasgow archdeaconry, 253, 253n.
——, Archibald, parson and vicar of Auchterless, 430
——, Mr James, archbishop of Glasgow, 124, 124n., 504, 504n., 529, 529n., 581, 581n.
——, Mr James, minister at Holywood, pension to, 624
——, Janet, lady of Buccleuch, 260, 263, 263n.
——, John, 146, 167 (Kilconquhar)
——, Mr Stephen, parson and vicar of Govan, 533
Beveridge (Bawerage), James, 64 (Aberdour)
Beynan, John, 557 (Ayr)
Bickerton (Bikertoun), James, parson of Hutton, 266
Biggar, James, 104 (Kirknewton)
Binnie (Bynnie), Libra, in Linlithgow, 156

Binning, Walter, in Edinburgh, 127
Birnie, William, 303, 346 (Strathmiglo)
Birrell (Berall, Burell), Henry, 500, 516 (Glasgow)
——, John, chaplain of St Stephen's chaplainry in Dundee, 400
Bishop (Bischope, Byschope), James, vicar pensionary of Ratho, reader and minister, 123, 131, 131n.
——, William, 101 (Newbattle abbey)
Bisset (Bissat), John, chaplain of Newmore chaplainry, Tain collegiate kirk, 647
——, John, 177, 179 (Haddington priory)
——, Robert, 158 (Airth)
Black, John, 419, 422 (Aberdeen friary)
——, John, 180, 181 (Haddington priory)
——, Luke, 275, 276 (Holywood abbey), 276
——, Walter, in Dundee, 12
Blackadder (Blaikaitter), Adam, commendator of Dundrennan, 606n.
——, Cuthbert, 290 (Culross abbey)
——, John, 290 (Culross abbey)
——, Patrick, parson and vicar of Tulliallan, 297
——, William, 292 (Culross abbey)
Blackburn, John, elder, in Leith, 122
——, John, younger, in Leith, 122
——, William, in Inverkeithing, 291
Blackhall, Andrew, 92 (Holyroodhouse abbey)
Blacklock (Blacklaik, Blaiklok), Richard, in Edinburgh, 124
——, Walter, in Edinburgh, 132
Blackstock, John, in Edinburgh, 127
Blackstone, David, in Edinburgh, 122
Blackwood, James, 306 (Perth)
——, John, 312 (Glendevon)
——, Peter, 92 (Holyroodhouse abbey)
——, William, chaplain of Our Lady, Dunblane cathedral, 314
——, William, vicar of Duddingston, 99
——, William, 344 (Auchterarder)
Blair of Balgrische, Andrew, 409
——, Andrew, 114 (Leith hospital)
——, George, 557 (Ayr)
——, James, 564 (Failford)
——, John, 369 (Coupar Angus abbey)
——, John, 290, 292 (Culross abbey)
——, John, 516 (Glasgow)

——, Mr Patrick, vicar of Ruthven, 305
——, William, 571 (Ayr)
Blairquhan, laird of, 607
Blais, James, 290 (Culross abbey)
Blanerne, laird of, 93
Blindseil (Blindscheill), Alexander, 419, 422 (Aberdeen friary)
——, Robert, 419, 421 (Aberdeen friary)
——, William, 419 (Aberdeen friary)
Blyth, Ninian, 32 (Abdie)
——, Simon, in Edinburgh, 127; chaplain of Trinity altar, St Cuthbert's kirk, Edinburgh, 128, 129; prebendary of St Salvator's altar, St Giles' collegiate kirk, Edinburgh, 129
——, Simon, prebendary of Tain collegiate kirk, 636
Bogart, Janet, 583 (Dreghorn)
Bogrie, laird of, 269, 274
Bolton, lady, 172
Bomby, laird of, 617, 619
Bonar, Alexander, 556 (Stirling)
Bonitoun, laird of, 217
Bonkle, *see* Buncle
Borland, baron of, 340
Borrowfield, laird of, 365
Borthwick, John, 6th lord, 3, 3n.
——, Alexander, 104 (Kirknewton)
——, Mr David, 92 (Holyroodhouse abbey); 591 (Galloway bishopric)
——, John, 102 (Newbattle abbey)
——, Nicol, in Edinburgh, 128
——, Mr Ninian, prebendary of Half Gogar and Half Addiston, Corstorphine collegiate kirk, 107, 113
——, William, in Edinburgh, 141
Boswell, Mr William, vicar of Straiton, 569
——, William, vicar of Inverkeithing, 76
Bothwell, earls of, *see* Hepburn; Stewart
——, countess of, *see* Sinclair, Agnes
——, Adam, bishop of Orkney, 491, 491n., 574, 574n., 659, 659n., 660, 660n., 661
——, Mr David, in Edinburgh, 134
——, Mr Francis, treasurer of Orkney, parson of Stronsay - St Nicholas, 660, 660n., 666
——, Mr Francis, in Edinburgh, 130
——, John, commendator of Holyroodhouse abbey, 202, 202n., 222, 222n.
Bourie, William, 545-546 (Stirling)
Boward, laird, 128
Bowden (Baudoun), Elizabeth, in Edinburgh, 136
Bower (Bowar), John, 158 (Airth)
Bowgred, William, 619 (St Mary's Isle priory), 619
Bowie (Bowyie), Archibald, 158 (Airth)
Bowman, Alexander, 551 (Stirling)
Bowsie (Bowsy), goodwife of, 167
——, laird of, 283, 287
——, William, prebendary of St John the Evangelist, Crail, 66
Bowsoun, Andrew, 419 (Aberdeen friary)
Boyd, Robert, 5th lord, collector general, 149, 149n., 150, 150n., 492, 492n., 573, 573n., 579
——, John, 573 (Stewarton)
——, Margaret, 497 (Monkland), 497
——, Ninian, 588 (Galloway bishopric)
——, Quentin, 607 (Galloway bishopric)
——, William, 619 (St Mary's Isle priory)
Boyman, John, 114 (Leith hospital)
Brabner, John, tenant in Duffus, 468
Braid, Mr Thomas, chaplain of Helmsdale, 632
Braidfoot (Braidfutte), Arthur, 594 (Whithorn priory)
——, John, 594 (Whithorn priory)
Brechnel, John, 275 (Holyroodhouse abbey)
Brethocht, Thomas, in Edinburgh, 136
Briscot (Brissat), Anthony, in Edinburgh, 134
Broch, John, 554 (Killearn)
Brock (Brok), Nicola, 283, 287n. (Perth)
——, Oliver, 283, 287n. (Perth)
——, William, in Edinburgh, 132
Broman, David, 19 (Crail)
Bronewod, David, 180 (Haddington priory)
Bronsyd, *see* Brownside
Brony, James, 158 (Airth)
Brotherstone, laird of, 92, 93
Brounfield, *see* Brownfield
Brown (Broune) of Fordell, John, 325
—— of Heuch, Alexander, 195
——, Alexander, 555 (Stirling)
——, George, chaplain of 'Chapelden of Kindrimy', 449, 449n.

——, Gilbert, abbot of Sweetheart abbey, 622, 622n.
——, Gilbert, 'minister' of Trinitarians, Peebles, parson of Kettins, 249
——, Helen, 158 (Airth)
——, Hew, in Edinburgh, 134
——, Isobel, 556 (Stirling)
——, Mr James, burgess of Irvine, 581
——, Mr James, parson and vicar of Fetteresso, 404
——, Mr James, parson of Kirknewton, 104-105
——, John, abbot of Sweetheart abbey, 612, 612n.
——, John, chaplain of altarage of St Mungo, Culross abbey kirk, 522
——, John, chaplain of altars of St Ninian and St Patrick, St Giles' collegiate kirk, Edinburgh, 134
——, John, prebendary of Our Lady altar, Crail collegiate kirk, 65
——, John, in Edinburgh, 129
——, John, 342, 345 (Dunkeld)
——, John, 283, 286 (Perth)
——, Mariota, 555 (Stirling)
——, Mark, in Edinburgh, 126
——, Mary, 114 (Leith hospital)
——, Nicol, in Edinburgh, 138
——, Patrick, vicar of Ardclach and Abertarff, 479, 488, 489
——, Robert, pension to, 291; 289, 290n. (Culross abbey)
——, Thomas, burgess of Edinburgh, 601
——, Thomas, 557 (Ayr)
——, William, 158 (Airth)
——, William, 555 (Stirling)
Brownfield (Brounfeild), David, 227 (Greenlaw)
——, John, 227 (Greenlaw)
Brownlea, 'widow', 101
Brownside (Bronsyd), Robert, in Glasgow, 526
Bruce of Inches, Robert, witness at Huntly, 492
——, Alexander, 117n., 118 (Edinburgh)
——, Anthony, witness at Huntly, 492
——, Mr Edward, parson of Torrie, 73, 73n.; 290 (Culross abbey)
——, Janet, lady of Arnot, 340
——, John, vicar of Airth, 158
——, Robert, vicar pensionary of Dollar, 88

——, Robert, in Edinburgh, 134
——, Robert, 19 (St Andrews priory)
Bryce, John, vicar of Dumfries, 611
Bryden (Brydein, Brydin), George (Dreghorn), 583
——, Ninian, chaplain of St John's aisle, St Giles' collegiate kirk, Edinburgh, 140
Bryson, James, commissioner of Nithsdale, minister of Dumfries, 272, 272n.
Buchanan, Mr George, pension to, 568 (Crossraguel abbey)
Buck (Buk), Andrew, 413, 414, 415, 417 (Aberdeen bishopric)
Bulloch (Bullo), John, 254 (Peebles)
Bunch (Bunche), Alexander, 283, 287 (Perth)
Buncle (Bonkle), Mr John, tacksman of Sandwick vicarage, 668
Burn, Robert, vicar pensionary of Dollar, 88, 339n.
——, Robert, 214 (Morebattle)
Burnet, Mr Robert, parson of Contin, 645
——, Mr Thomas, parson and vicar of Methlick, 435, 445
Burrie (Burry), William, 283, 286 (Perth)
Buschart, see Butchart
Butchart (Buschart), James, parson portionary of Alness, 645
Buthill (Buthile), Mr John, prebendary of Our Lady altar, Crail collegiate kirk, 65
Butler, Thomas, in Edinburgh, 141

Cabell, Mr William, parson of Tullynestle, vicar of Inverurie, 437
Cadenhead (Cadonheid), William, 421 (Aberdeen friary)
Cady (Cadie), William, in Edinburgh, 127
Cairncross (Carncerce) of Colmslie, Robert, 216
——, Mr Robert, chanter of Ross, 642
——, Robert, 99 (Holyroodhouse abbey)
——, William, in Edinburgh, 138
Cairns (Carnes, Carnis, Kernis), Adam, in Edinburgh, 137
——, Edward, 586 (Galloway bishopric)
——, Henry, 18 (St Andrews priory)

——, Robert, 270, 271, 275 (Holywood abbey)
——, Thomas, 101 (Newbattle abbey)
Calder, laird of, 18, 94
Caldwell, John, 583 (Dreghorn)
Calein, Ellen, 155 (Dalmeny)
Caling, James, 156 (Linlithgow friary)
——, William, 156 (Linlithgow friary)
Callander (Callendar), Andrew, 289, 293 (Culross abbey)
——, John, 291 (Culross abbey)
——, William, 158 (Airth)
Cameron (Cambroun), William, 62 (Inchcolm abbey)
Camlinge, see Combline
Campbell, Archibald, 5th earl of Argyll, bailie of bishopric of Brechin, bailie of Culross abbey, 7, 7n., 40, 40n., 292, 303, 303n., 355, 355n., 384, 385, 385n., 386, 389, 410, 410n.
—— of Ardkinglas, James, 387
—— of Glenorchy, Colin, 284, 284n., 287
—— of Kinyeancleuch, Robert, sub-collector, 574n.
—— of Lawers, James, 58
——, Alexander, bishop of Brechin, 383, 385, 387, 388, 388n.
——, Alexander, dean of Moray, 468
——, Alexander, 469 (Moray deanery)
——, Catherine, lady Crawford, 382, 382n., 384
——, David, 557 (Ayr)
——, Donald, abbot of Coupar Angus, 303, 352, 352n., 353, 356, 368, 411, 509, 509n.
——, Henry, 180 (Haddington priory)
——, Mr James, prebendary of Bothans collegiate kirk, 172
——, John, 557 (Ayr)
——, John, 516 (Glasgow)
——, Mr Nicol (Coupar Angus abbey), 353
——, Thomas, commendator of Holywood abbey, 271, 271n., 272, 277-278, 278n., 624, 624n.
——, Thomas, 353 (Coupar Angus abbey)
——, Mr William, parson of Tullynessle, vicar of Inverurie, 427
——, William, feuar, 352, 353 (Coupar Angus abbey)
Cand, John, 609 (Kirkbean)

Cannes, William, 101 (Newbattle abbey)
Canny, George, 101 (Newbattle abbey)
Canoch (Kanocht), Allan, in Edinburgh, 124, 124n.
Cant, Henry, in Edinburgh, 124
——, James, in Edinburgh, 125
——, John, 114 (Leith hospital)
——, Margaret, in Edinburgh, 131
——, Thomas, burgess of Edinburgh, 132, 133
Capell, William, chaplain of Westhall, Aberdeenshire, 433, 433n.
Caprington, laird of, 580 (Tarbolton)
Carden, see Garden
Cardin, David, 283, 287 (Perth)
Cargill, the goodman of, 326
Carinphin and Kleland, laird of, 506
Carkettle (Carkettill), Adam, in Edinburgh, 136, 141
——, John, in Edinburgh, 125, 128
——, John, 102 (Newbattle abbey)
——, Thomas, in Edinburgh, 94
Carlyle, Michael, 4th lord, 265n.
——, lady, 265, 265n.; see also Charteris
Carmichael, John, 168 (North Berwick)
——, Richard, 290 (Culross abbey)
——, Richard, 145 (Largo)
——, Robert, in Edinburgh, 136
——, Robert, in Linlithgow, 11
Carnegie (Carnegy) of Colluthie, David, 21n.
—— of Kinnaird, Robert, 331, 373, 382, 390, 394, 403, 415, 424, 425
——, Alexander, 400 (Dundee)
——, Mr Hercules, vicar of Kingussie, 492; vicar of Ruthven in Badenoch, 428
——, Hercules, subdean of Brechin, 394
——, John, son of Robert Carnegie of Kinnaird, 440, 442, 443 (Belhelvie)
——, John, 365 (Brechin deanery); 384 (Brechin bishopric)
——, Mr Robert, chaplain of chaplainry of St Mary of Cocklarachy, St Nicholas' kirk, Aberdeen, 447, 447n.
——, Mr Robert, parson of Aberdour, 424
——, Mr Robert, parson of Kinnoull, 323, 324, 331
——, Mr Robert, preceptor of Maison Dieu, Brechin, 390

INDEX OF PERSONS

——, Mr Robert, vicar of Leuchars, 83
——, Robert, 396 (Newtyle)
——, William, parson of Tough, 245
——, William, vicar of Banff and Inverboyndie, 455
——, William, vicar of Kingussie, 491-492
Carnell, laird of, 563, 571
Carr (Cor), *see also* Ker
——, Andrew, in Edinburgh, 134
——, James, chaplain of St Salvator's altar, St Giles' collegiate kirk, Edinburgh, 96
Carrick (Cariak, Carrik), Thomas, 168 (North Berwick)
Carrington (Keringtoun, Kerintone, Kerintoun), Elizabeth, 180 (Haddington priory)
——, John, 180 (Haddington priory)
——, Patrick, 180 (Haddington priory)
——, Thomas, reader at Symington, 563
——, Thomas, 180 (Haddington priory)
——, William, 180 (Haddington priory)
Carrol (Carell, Carrele), Habbie (Robert), 270, 271 (Holywood abbey)
Carron (Carroun), Robert, 283, 287 (Perth)
Carruthers, Mark, parson of Mouswald, 265
Cassillis, earls of, *see* Kennedy
Castlehill, Thomas, in Edinburgh, 126
Cathcart, Alan, 4th lord, 560, 560n.
Cathro, George, 400 (Dundee)
Cauldwell, laird of, 574
Celocht, Nicholas, Frenchman, 200-201 (Coldingham priory)
Cessford, laird of, *see* Ker
Cessnock, laird of, 499, 562, 564
Chalmers (Chalmour), Mr Alexander, chaplain of Our Lady altar, Cambuskenneth; chaplain in St Ninians; chaplain of St Thomas' altar, Stirling, 551
——, Mr Alexander, prebendary of St Matthew's prebend, St Mary of the Fields' collegiate kirk, Edinburgh, 139
——, Mr Alexander, vicar of Liberton, 107
——, Alexander, 92 (Holyroodhouse abbey)
——, Mr Duncan, chancellor of Ross, 634, 635
——, Mr Duncan, 360 (Coupar Angus abbey)
——, James, 292 (Culross abbey)
——, James, 599 (Soulseat abbey)
——, John, 419 (Aberdeen friary)
——, John, 551 (Stirling)
——, Mr Robert, 421 (Aberdeen friary)
——, Robert, 567 (Crossraguel abbey)
——, Mr William, parson of Newdosk, 382
Chapman, David, in Edinburgh, 127
——, Mr John, in Edinburgh, 138
——, Robert, in Edinburgh, 138
——, Walter, in Edinburgh, 140
Charteris (Chartouris), Janet, lady Carlyle, 265, 265n.
——, John, 271, 274, 275 (Holywood abbey)
Chatelherault, duke of, *see* Hamilton, James
Chattan, clan, 639
Chatto, John, vicar of Pencaitland, 165
Cheyne (Chayne, Chyne, Schen), Mr Jerome, archdeacon of Shetland, 665
——, Mr Robert, prebendary of St Magnus, Kirkwall cathedral, 666
——, William, tenant in Duffus, 468
Chirnside, Mr William, provost of Bothwell collegiate kirk, 507, 507n.
——, William, prior of Blantyre, 505, 505n.
Chisholm (Cheisholme), Mr Alexander, parson of Comrie, 312
——, Edmund (Elmond), subdean of Dunblane cathedral, 314; chaplain, Dunblane cathedral, 343
——, Mr James, archdeacon of Dunblane, 300
——, James, parson of Tarbolton, 580
——, James, vicar of Logie, 313
——, James, vicar of Strowan, 313
——, Jean, 313 (Logie)
——, Mr Michael, vicar of Cadder and Monkland, 496, 518
——, Mr William, parson and vicar of Glendevon, 312
——, William, bishop of Dunblane, 295, 295n., 302
Christison (Crystisoun), David, vicar of Inverkip, 533
——, David, pension to, 509 (Cambuslang)

——, James, friar of Elgin, 454
——, Thomas, parson of Yetholm, 243
——, Thomas, vicar of Monzievaird, 300
——, Thomas, vicar pensionary and minister of Gamrie, 443
Clanranald, 488 (Abertarff)
Clapperton, George, provost of Trinity college, Edinburgh, 118, 118n.
——, George, subdean of Chapel Royal, Stirling, 602
——, Laurence, provost of Trinity college, Edinburgh, 118, 118n.
——, William, in Edinburgh, 128
Cleghorn, Archibald, in Linlithgow, 156
Clerk, John, 180 (Haddington priory)
——, Robert, 101 (Newbattle abbey)
Clerkington, laird of, 18
Clydesdale (Cliddisdale), James, in Edinburgh, 133
Cockburn of Skirling, James, comptroller, 7n., 76, 76n., 537n., 609n.
——, Alison, in Edinburgh, 127
——, Henry, 180, 182 (Haddington priory)
——, Hew, brother of laird of Skirling, 250 (Kirkurd)
——, James, vicar of Dunnichen, 399
——, Mr John, vicar of Brechin, 378, 379
——, Mr John, 383 (Brechin bishopric)
——, John, maltman, 180 (Haddington priory)
——, John, parson of Dolphinton, 504
——, John, vicar of Duffus, 485
——, John, vicar of Gamrie, 443
——, Laurence, 173 (Bolton)
——, Mr Patrick, prebendary of Pitcox, Dunbar collegiate kirk, 175
——, Patrick, vicar of Stichill, 195
——, William, prebendary, Bothans collegiate kirk, 172
——, William, in Haddington, 11
Codynnar, see Cordiner
Cogan (Cogane), Catherine (Kate), 269, 274 (Holywood abbey)
Coill, Mr Alexander, prebendary of Half Gogar and Half Addiston, Corstorphine collegiate kirk, 107, 112, 115
——, Patrick, 155 (Linlithgow friary)
Coldenknowes (Cowdenknowes), laird of, 184, 189, 197, 228
Collace (Cullace) of Balnamoon, Robert, 374, 375
——, Walter, 383 (Brechin bishopric)
Collier (Colyeir), William, 64 (Aberdour)
Collison (Collesoun, Colysoun), vicar of Alford, 450
——, Gilbert, 416, 417, 420 (Aberdeen)
Colp, David, 419, 421 (Aberdeen friary)
Colquhoun (Calquhoun, Culquhoun) of Kilmardinny, John, 541
——, Adam, 542 (Luss)
——, Humphrey, parson of Kirkpatrick-Juxta, 615
——, Mr John, parson of Stobo, 250, 250n.
Colt, Blaise, 283, 287 (Perth)
Colter, Walter, 609 (Kirkbean)
Colville of Cleish, Robert, 290, 291, 292
——, Mr Alexander, 291 (Culross abbey)
——, Alexander, commendator of Culross abbey, administrator of Melrose abbey, 260, 260n.
——, James, 291 (Culross abbey); 294 (Crombie)
——, James, 18 (St Andrews priory)
——, William, commendator of Culross abbey, comptroller, 266, 294, 294n.
Combe, Thomas, 556 (Stirling)
Combline (Camlinge, Comlinge), James, 620 (St Mary's Isle priory)
——, John, 619 (St Mary's Isle priory)
Comrie of that Ilk, John, 312
Congalton (Congiltoun, Ungiltoun), laird of, 163
——, Mr Hew, chaplain of St Triduana's aisle, Restalrig collegiate kirk, 143
Conqueror, [blank], 283, 283n., 286, 286n.
Constane, Adam, 283, 287 (Perth)
——, Patrick, 284, 287 (Perth)
Constin, see Constane
Cook (Cok, Cuik, Cuke), Mr George, vicar of Monimail, 74
——, Mr George, vicar pensionary, Perth, 74, 74n., 326, 326n.
——, Mr George, 300, 311 (Dunkeld)
——, George, prebendary of Crieff, Dunkeld cathedral, 322, 323; chaplain of Our Lady altar, Dunkeld cathedral, 323

——, John, chaplain, Dumbarton castle, 540
——, Nicola, 283n., 287 (Perth)
——, Mr Ninian, vicar of Murroes, 398
——, Oliver, 283n., 287 (Perth)
——, Robert, vicar pensionary of Cardross, 542
——, Mr William, notary, 506, 506n., 534
——, Mr William, 146, 167 (Kilconquhar)
——, William, vicar pensionary of Perth, 315
Cookston (Cokstoun), Robert, in Edinburgh, 127
Cooper, *see* Cowper
Copland (Coipland, Coupland), John, in Edinburgh, 139
Cor, *see* Carr
Cordiner (Codynnar), Thomas, in Edinburgh, 130
Cornwall, Begge, in Linlithgow, 156
Cors, laird of, 567 (Crossraguel abbey)
Corsbie, *see* Crosbie
Corstorphine, laird of, 100, 127
——, George, 177, 179 (Haddington priory)
——, John, 182 (Crail)
——, William, prebendary of Our Lady at high altar, Crail, 66
Corswall, laird of, 611
Cossour, George, 101 (Newbattle)
Cousland (Cosland, Costland), James, chaplain of the chaplainry of St Fillan, Doune, 345
——, Janet, 546 (Stirling)
——, Robert, 546 (Stirling)
Coutts (Coutes, Couttis, Cowttis), Allan, chamberlain of Dunfermline abbey, 23, 40, 45, 51, 73, 73n.
——, John, in Edinburgh, 126
——, Walter, 415 (Aberdeen bishopric)
Cowan, Janet, 551 (Stirling)
——, John, 291 (Culross abbey)
Cowdenknowes, *see* Coldenknowes
Cowie (Cowye), John, 158 (Airth)
——, Marion, 158 (Airth)
Cowper, George, in Edinburgh, 127
——, John, in Dundee, 12
Cowtie, laird of, 312
Crab, [*blank*], in Aberdeen, 12
Crago, John, feuar, 352, 353 (Coupar Angus abbey)

Craichlaw, laird of, 607
Craig, Andrew, in Edinburgh, 127
——, Mr John, 117, 117n. (Edinburgh)
——, John, 102 (Newbattle abbey)
——, William, 283, 287 (Perth)
Craighall, laird of, 19
Craigie (Cragy), laird of, 26, 48
——, Mr Hew, parson of Inverkeithny, 476
Craigmillar, laird of, 25, 47, 93
Crail, James, in Dundee, 12
Crambie, Thomas, 283, 287 (Perth)
Crammy, *see* Crambie
Cramond (Crawmond), George, 383, 384 (Brechin bishopric)
Cranston, laird of, 263
——, Cuthbert, 190 (Channelkirk)
——, Mr James, 190 (Lauder)
——, John, in Edinburgh, 136
——, John, 18 (St Andrews priory)
——, Ninian, 188 (Channelkirk)
——, Thomas, in Edinburgh, 136
——, Mr William, pension to, 642 (Ross chantory)
——, William, vicar of Legerwood, 193
Crawford (Craufuird), earls of, *see* Lindsay
——, Catherine, countess of, 544, 544n., 548
—— of Ferme (Fairme), David, 516
——, Captain of, 190
——, Mr Archibald, parson of Eaglesham, 525
——, Catherine, 420 (Aberdeen friars)
——, John, bailie of Monkland, 103 (Newbattle abbey)
——, John, 419, 421 (Aberdeen friary)
——, Margaret, in North Berwick priory, 148
——, Nicol, parson of Hartside, 514
——, Robert, 100 (Newbattle)
——, Thomas, in Edinburgh, 136
Creich (Creycht), laird of, 58
Crichton (Crechtoun, Creychtoun, Crychtoun) of Sanquhar, Robert, 6th lord, 271, 271n.
—— of Sanquhar, William, 5th lord, 276, 276n.
—— of Cranston Riddell, Martin, 124
—— of Kirkpatrick, Robert, 586n.
—— of Ruthven (Ruffanis), Adam, younger, 305
—— of Ruthven, John, elder, 305

——, Mr Abraham, parson of Upsetlington, provost of Dunglass, 186, 186n.
——, Abraham, prebendary of Aberlady, 170
——, Abraham, in Kirktoun of Fern, 306 (Dunkeld)
——, Mr Alexander, parson of Abbotrule, 213
——, Alexander, parson and vicar of Lundeiff, 337
——, Andrew, 303 (Alyth)
——, Elizabeth, lady of Ardros, 64, 146, 167
——, Eustace, sub-collector for Cunningham, 574n., 575, 575n.
——, George, vicar of Kirkcudbright, vicar of Balmaghie, 617, 618
——, Mr James, 302 (Aberlady)
——, James, 305 (Ruthven)
——, Janet, in North Berwick priory, 148
——, John, 389 (Brechin bishopric)
——, John, 180 (Haddington priory)
——, Margaret, lady Lochinvar, 586, 586n.
——, Mr Robert, parson of Sanquhar, 276
——, Mr Robert, 18 (St Andrews priory)
——, Robert, bishop of Dunkeld, 301, 301n., 307, 308, 341, 343, 346
——, Thomas, in Edinburgh, 125
——, Mr William, in Edinburgh, 122
——, Mr William, in Linlithgow, 156
——, Mr William, 302, 346 (Cramond)
——, Mr William, 17 (Linlithgow)
——, William, pension to, 624 (Holywood abbey)
Crockatt (Crokat), John, 114 (Leith hospital)
Crosbie (Corsbie), Andrew, 275 (Holywood abbey)
——, David, in Edinburgh, 140
——, David, 270 (Holywood abbey)
——, Mungo, 271, 276 (Holywood abbey)
Cruickshanks, George, in Elgin, 453
Crunzane (Crounzeong, Crunzeome), John, in Edinburgh, 136
Cuby, 'widow', 102 (Newbattle abbey)
Cuikland, Martin, in Edinburgh, 122
Cuke, see Cook
Cullace, see Collace

Cullen (Cullane, Culland), Mr Alexander, 419, 422 (Aberdeen friary)
——, John, 666, 667 (Westray)
——, Thomas, 419, 422 (Aberdeen friary)
Cultoquhay, laird of, 343
Cunieson (Kwnisoun), Andrew, 158 (Airth)
Cunningham, Alexander, 4th earl of Glencairn, 148, 210, 210n., 224n., 235, 237, 238, 239, 576, 576n.
——, William, 5th earl of Glencairn, 257
——, Mr David, 364, 364n. (Brechin deanery)
——, David, pension to, 398 (Murroes); 522 (Lanark friary)
——, Humphrey (Umphra), burgess of Dumbarton, 540 (Inchcailloch)
——, Humphrey, vicar of Kilmacolm, 532
——, James, son of 4th earl of Glencairn, 224, 224n., 228, 229, 230, 257, 257n.
——, James, 547 (St Ninians)
——, James, 595, 596 (Whithorn priory)
——, Mr John, 418, 420 (Aberdeen friary)
——, John, flesher, 557 (Ayr)
——, John, vicar of Carnwath, 505
——, John, 104 (Kirknewton)
——, John, 254 (Peebles)
——, Robert, 'minister' of Failford, 558, 563, 569
——, Thomas, 104 (Kirknewton)
——, William, dean of Brechin, 367n.
——, William, nephew to earl of Glencairn, 148 (Maybole); pension to, 149 (Maybole)
Cunninghamhead, laird of, 562
Curle, James, in Edinburgh, 131
——, James, witness at Huntly, 492
Currell, David, 114 (Leith hospital)
Currie (Curry), Alexander, 113 (Leith hospital)
——, Hew, parson of Eassie, 411
——, John, procurator in Ayr, 114
——, William, 291 (Culross abbey)
Curror (Currour), Mr Alexander, in Edinburgh, 126
——, Andrew, 283, 287 (Perth)
Cuthbertson (Cudbertson), James, 104 (Kirknewton)

INDEX OF PERSONS

Dalcoif, lady, 189, 191
Dalgleish, Adam, 214 (Morebattle)
—, David, vicar of Walston, 513
—, John, in Edinburgh, 157
—, Robert, in Edinburgh, 128
Dalmahoy, John, in Leith, 143
Dalrymple, John, 158 (Airth)
—, John, 275 (Holywood abbey)
—, William, 557 (Ayr)
Dalziel, James, 156 (Linlithgow friary)
—, John, friar in Scotlandwell, 56
Danielston, Robert, parson of Dysart, 70, 300
Darling (Darley, Derling), Duncan, 666 (Stronsay)
—, Helen, in North Berwick priory, 148
—, James, 666 (Stronsay)
Davidson, Mr Andrew, vicar of Dalkeith, 110, 110n.
—, Mr Andrew, 4, 90 (St Andrews archbishopric)
—, George, 214 (Morebattle)
—, Henry, chaplain of St James' altar [Fife], 85
—, Henry, in Inverkeithing, 291
—, Mr James, 283, 287 (Perth)
—, James, chaplain of Holy Rood altar, St Andrew's kirk, Peebles; chaplain of Holy Blood altar, Cross kirk, Peebles, 254
—, Mr John, principal of Glasgow university, pension to, 420, 445, 448 (Kinkell); 512, 520 (Glasgow); chaplain of St Nicholas' altar, Glasgow cathedral, 525; vicar of Alness, 647; vicar of Colmonell, 420, 569; vicar of Nigg, 420
—, John, prebendary of St James' altar, Crail collegiate kirk, 66
—, John, in Edinburgh, 132
—, Malie, 214 (Morebattle)
—, Rankin, exhorter, minister at Galston, 558, 563
—, Richard, 555 (Stirling)
—, Thomas, 283, 287 (Perth)
—, William, burgess of Glasgow, 503
—, William, in Edinburgh, 126
Dawlein, Robert, 155 (Dalmeny)
Dawson, Archibald, 155 (Dalmeny)
—, John, 155 (Dalmeny)
—, William, in Leith, 132
Dee, John, in Edinburgh, 127

Dennistoun (Dennestoun), James, in Linlithgow, 156
—, Robert, parson and vicar of Ayr Tertio, 561
Denoon (Dennun, Denowne), Robert, in Edinburgh, 127
—, Mr Thomas, parson of Kincardine (Ross), 645
—, family of, 634
Dewar, John, 18 (St Andrews priory)
Dick, Alexander, provost of Orkney, 659, 659n.
Dickie (Dikie), Archibald, vicar choral in Glasgow cathedral, parson and vicar of Dalziel, 519
Dickson (Diksoun), Alan, 251, 253, 253n. (Peebles, Manor)
—, Bartholomew, 101 (Newbattle abbey)
—, John, 184 (Eccles priory)
—, John, 609 (Kirkbean)
—, John, 114 (Leith hospital)
—, Margaret, in Edinburgh, 131
—, Robert, in Garvald, 181
—, Robert, 184 (Eccles priory)
—, Robert, 101 (Newbattle abbey)
—, Mr Thomas, 143 (Restalrig)
—, Thomas, vicar of Torphichen, 152
—, Thomas, in Edinburgh, 127
—, Thomas, 19 (St Andrews priory)
—, Thomas, 594 (Whithorn priory)
—, William, 101 (Newbattle abbey)
—, William, 254 (Peebles)
Dingwall, John, 182 (Crail)
Dishington, Mr David, pension to, 430 (Auchterless)
Donald, David, vicar pensionary of Kirkforfar, 84
Donaldson, Andrew, 353 (Coupar Angus abbey)
—, Margaret, in North Berwick priory, 148
Dougall, John, burgess of Edinburgh, 276
Dougan, see Dugan
Douglas, Archibald, 8th earl of Angus, 190, 190n.
—, Archibald, 4th earl of Douglas, 273n.
—, James, 4th earl of Morton, 64, 103, 103n., 122, 122n., 141, 147, 147n. 148, 251, 260, 260n., 342, 346
—, William, 5th earl of Morton, 219

—— of Kilspindie, Archibald, 346
—— of Lincluden, Robert, collector general, provost of Lincluden collegiate kirk, 10, 10n., 21, 21n., 54, 97, 97n., 98, 613, 613n.
—— of Lochleven, Robert, 17n., 26n.
—— of Lochleven, William, 9, 9n., 17n., 257n.
—— of Parkland [? Parkhead], William, 273
—— of Whittingehame, William, 97
——, Alexander, chaplain in Elgin cathedral, 477; chaplain of St Andrew's altar *alias* Clyne, Elgin cathedral, 478; chaplain at St Magdalene's altar *alias* Knockando, Elgin cathedral, 478; chaplain of St Thomas' altar, Elgin cathedral, 477; chaplain of Our Lady of the Green chaplainry, near Inverness, 477
——, Mr Archibald, parson of Douglas, 506
——, Archibald, parson of Newlands, 251
——, Archibald, 302 (Aberlady, Cramond)
——, Barbara, 102 (Newbattle abbey)
——, David, 271, 623 (Holywood abbey)
——, Florence, chorister, Corstorphine collegiate kirk, 130
——, Francis, 173 (Dunglass collegiate kirk)
——, George, brother to laird of Lochleven, 9 (Markinch)
——, George, parson and vicar of Carmichael, 511
——, George, son of laird of Whittingehame, 97, 98
——, Hew, parson of Southdean, 213
——, James, son of Douglas of Lochleven, commendator of Melrose abbey, 257, 257n., 260, 260n.
——, James, vicar of Kinneddar, chaplain of St Catherine's chaplainry, Elgin cathedral, 484, 490
——, Mr John, parson of Collace, 322
——, Mr John, pension to, 563 (Failford)
——, Mr John, provost of St Mary's college, rector of St Andrews university, 64, 65; pension to, 307 (Kinclaven)
——, John, baxter, 180 (Haddington priory)
——, John, cordiner, 180 (Haddington priory)
——, John, parson of Kirkbride, 278
——, John, servant of Robert Crichton, bishop of Dunkeld, 343
——, John, vicar of Galston, 484
——, John, 271 (Holywood abbey)
——, John, 305 (Ruthven)
——, Ninian, vicar of Stobo, 252
——, Robert, provost of Corstorphine collegiate kirk and prebendary of Half Bonnington and Half Platt; provost of Dalkeith collegiate kirk, 100, 100n., 107, 115, 116
——, Robert, vicar of Stobo, 252
——, Robert, in Durisdeer, 272
——, Thomas, in Dalmahoy, 97
——, Walter, 536 (Stirling)
——, Walter, 592 (Whithorn priory)
——, William, vicar of Aberlour, 489, 490
——, William, vicar of Aberclach, 490
Down, *see* Dun
Downie, Andrew, 32 (Abdie); 34 (Lindores abbey)
Drumlanrig, laird of, 269, 271, 276, 277, 623
Drummelzier, laird of, 150
Drummond, David, 2nd lord, 347, 347n., 348
—— of Balloch, George, 299
——, Mr Alexander, 344 (Auchterarder)
——, Alexander, 343 (Dunblane)
——, Catherine, 148 (North Berwick priory)
——, Gavin, 545 (Stirling)
——, George, 306 (Atholl)
——, John, 651 (Tain)
——, Robert, 546 (St Ninians)
——, Robert, 556 (Stirling)
——, Sibilla, 546 (St Ninians)
——, William, chaplain, Dunblane cathedral, 343
——, William, goodman of Cargill, 326
——, William, vicar of Cargill, 326
Drumquhassill, laird of, 548
Drylaw, laird of, 304 (Cramond)
——, Alexander, 158 (Airth)
Duchars, John, 305 (Ruthven)
——, Peter, 305 (Ruthven)

Ductour, John, 305 (Ruthven)
Duddingston, John, 146n., 167 (Kilconquhar); 169 (North Berwick)
Duff, George, provost of Cullen collegiate kirk and vicar of Rathven, 455
——, Mr John, parson of Durris, 404, 405, 408
——, William, in Leith, 132
Duffus, laird of, 482
Dugan (Dougane), John, 271 (Holywood abbey)
Dun (Down), Henry [*recte*, Herbert], vicar of Kirkcormack, 604, 618
——, Herbert, vicar of Kirkcormack, 604n.
Dunbar (Dumbar) of Cumnock, Patrick, 559, 561, 562
—— of Durris, Robert, 470
—— of Kellugrie, Alexander, 476
—— of Kilboyack, Alexander, 470 (Rafford)
—— of Pennick, Archibald, 493
——, Alexander, subchanter of Moray, 470
——, Andrew, 595, 596 (Whithorn priory)
——, David, parson of Cullicudden, 646
——, Mr Gavin, parson of Roskeen, 644
——, Gavin, dean of Moray, bishop of Glasgow, 477, 477n.
——, George, parson of Kilmuir-Easter, vicar of Rosemarkie, 643
——, George, 469 (Moray deanery)
——, Mr John, parson of Cumnock, chaplain of St Leonard's chapel (Moray), chaplain of Obstule, Fortrose cathedral, 299, 299n., 484, 559, 561, 637
——, Mr John, 379 (Kingoldrum)
——, Mr John, 635 (Urray)
——, Marjory, 470 (Ardclach)
Duncan, Alexander, in Edinburgh, 132
——, James, 158 (Airth)
——, John, 322 (Abernethy)
Duncanson, Mr Thomas, minister, 225 (Kelso abbey)
Dundas of Newliston, James, 3
——, Alexander, in Perth, 306
——, Euphemia, 291 (Culross abbey)
Dunipace, laird of, 546, 547
Dunkany, John, in Edinburgh, 136
Dunlop, John, vicar of Dalmellington, 560
——, Nicol, 180 (Haddington priory)
Dunmell, Thomas, 594 (Whithorn priory)
Dunmore (Dynmur), David, 427, 427n. (Lonmay)
Dunsire (Dynsyre), Thomas, in Edinburgh, 127
Durie, *see also* Dury
—— of that Ilk, Henry, 90
——, lady, 145 (Largo)
Durl, Patrick, 587 (Galloway bishopric)
Durlat, John, 418 (Aberdeen friary)
Dury, *see also* Durie
——, Andrew, bishop of Galloway, 591
——, George, archdeacon of St Andrews, 67n.
——, George, commendator of Dunfermline abbey, 40, 40n., 44, 72, 73, 73n., 179n.
——, Peter, 40 (Dunfermline abbey)
——, William, 591 (Galloway bishopric)
Dyke, Thomas, 158 (Airth)
Dynmur, *see* Dunmore

East Craig, laird of, 147 (North Berwick)
East Nisbet, laird of, 200
Eccles (Ecclis), Gilbert, 567 (Crossraguel abbey)
Eclynsoun, John, parson of Tarvit, 73
Edby, Alexander, 283, 286 (Perth)
Edgar (Edger, Edyar, Edyear), *see also* Eggar
——, Andrew, 270, 271, 275, 276 (Holywood abbey)
——, David, 269, 270, 271, 277, 622 (Holywood abbey)
——, George, 270, 271, 276 (Holywood abbey)
——, Henry, 270, 271, 276 (Holywood abbey)
——, Herbert, 275, 276 (Holywood abbey)
——, John, 270, 271, 275, 276 (Holywood abbey)
——, Richard, 271, 623 (Holywood abbey)
——, Robert, 271 (Holywood abbey)
——, Thomas, elder, 275 (Holywood abbey)
——, William, 269, 270, 275 (Holywood

abbey)
Edmeston (Ednistie, Edmistoun, Edmondston) of Duntreath, 555
——, [blank], 555 (Stirling)
——, Archibald, in Leith, 132
Edmondston, see Edmeston
Ednam (Ednem), [blank], in Edinburgh, 136
Ednistie, see Edmeston
Edward (Edwarte), Hobbie (Robert), 619 (St Mary's Isle priory)
——, John, 620 (St Mary's Isle priory)
Eggar (Egger), see also Edgar
——, Richard, 184 (Eccles priory)
——, Richard, 102 (Newbattle abbey)
Elder (Eldar), Mr Andrew, vicar pensionary of Glenbervie, 406
——, Mr John, parson of Dunnottar, prebendary of Trinity college, Edinburgh, 404
——, Mr John, vicar of Longley and Fetterangus, 450
——, Thomas, 283, 287 (Perth)
——, Mr William, vicar pensionary of Benholm, 403
Elderslie, laird of, see Wallace,
Elliot (Elwand, Elwart), James, in Edinburgh, 122
——, John, 620 (St Mary's Isle priory)
Elphinstone, Robert, 3rd lord, 556
—— of Barnton, James, 272, 272n.
——, Euphemia, 92n. (Holyroodhouse)
——, James, parson of Invernochty, 442
——, Mr Nicol, 18, 92, 93 (St Andrews priory)
——, William, in Edinburgh, 128
Elwand, see Elliot
Elwart see Elliot
Erroll, earls of, see Hay
Erskine, John, 5th lord, 26n.
——, John, 6th lord; earl of Mar; commendator of Dryburgh, Inchmahome and Cambuskenneth, 290, 290n., 378, 378n., 543, 543n., 546, 548n.
——, Robert, master of (son of John 5th lord Erkine), 548n.
—— of Dun, John, superintendent of Angus and the Mearns, 45, 45n., 363, 363n., 371, 372, 372n., 374, 374n., 384, 384n., 393
——, Adam, commendator of Cambuskenneth abbey, collector general, 260, 260n., 537, 537n., 538, 544, 544n.
——, Alexander, 555 (Stirling)
——, Arthur, 544, 547, 548 (Cambuskenneth abbey)
——, David, commendator of Dryburgh abbey, 197n.; commendator of Inchmahome priory, 548n.
——, James, bailie of Inchmahome priory, 548
——, James, 290 (Culross abbey)
——, Margaret, lady Lochleven, 26, 26n., 546
——, Mr Robert, parson of Arbuthnott and Glenbervie, dean of Aberdeen cathedral, 406, 407, 424, 426, 436, 438
Esplin, John, 158 (Airth)
Eustace (Ewstache, Ostaige), Alexander, reader at Kincardine O'Neil, 430, 430n.
Ewan (Ewein), Archibald, prebendary, Dunglass collegiate kirk, 166
——, James, 421 (Aberdeen friary)

Fairlie, James, burgess of Edinburgh, 131
Fairnie, James, 556 (Stirling)
Fairweather, Walter, 375 (Menmuir)
Falcon, David, in Edinburgh, 136, 137
——, James, in Edinburgh, 136
Falconer, Alexander, tenant in Duffus, 468
——, John, witness, 479 (Ardclach)
——, 'widow', 101 (Newbattle)
Farrar (Farar, Pharar), John, parson and vicar of Ecclesjohn, chaplain of St Agnes chaplainry, Brechin cathedral, 372
——, Margaret, 105 (Kirknewton)
——, Thomas, in Linlithgow, 11
Farrous, John, 190 (Lauder)
Fawsyde (Fawsyd), laird of, 25, 47
Feddes, see Fiddes
Fedlar, see Fiddler
Feithie, see Fithie
Fender (Fendar), John, in Edinburgh, 124
Fenton, laird of, 127
——, Margaret, in Edinburgh, 132
——, Patrick, 290 (Culross abbey)
Fergushill, Archibald, 557 (Ayr)

Ferguson (Fergussoun) of Bardarroch, William, 499
——, baron, 284, 287 (Atholl)
——, Archibald, 571 (Ayr)
——, John, 567 (Crossraguel abbey)
——, William, 557 (Ayr)
Fergy, Alexander, chaplain of St Michael's altar, Stirling parish kirk, 554
Ferrerio, Johannes, Piedmontese scholar at Kinloss abbey, 463, 463n.
Ferrie, William, in Edinburgh, 135
Fewthie, *see* Fithie
Fiddes (Feddes, Fudes), Robert, 419, 422 (Aberdeen friary)
——, Thomas, 651 (Tain)
Fiddler (Fedlar), Thomas, 180 (Haddington priory)
Fife (Fyff), Andrew, 419, 422 (Aberdeen friary)
Fillan (Fillane), Robert, 64 (Aberdour and Dalgetty)
Fin (Fyn), Robert, prebendary of Methven collegiate kirk, 299
Fingland, James, in Edinburgh, 129
Finlayson (Finlason), John, chamberlain depute, bishopric of Orkney, 680
Fisher (Fischer), Andrew, in Edinburgh, 133
——, John, in Edinburgh, 129, 136
Fithie (Feithie, Fewthie), John, chanter of Chapel Royal, prebendary of Yarrow, 263
——, John, 290 (Crombie)
——, William, in Edinburgh, 124
Fleming, John, 5th lord, 92, 92n., 521, 521n., 543, 543n., 546
—— of Barrochan, William, 533
——, Malcolm, prior of Whithorn priory, dean of Dunblane cathedral, 573, 573n., 595, 595n.
——, Malcolm, vicar choral in Glasgow cathedral, parson and vicar of Dalziel, 519
——, Matthew, vicar of Rutherglen, 519
——, Patrick, in Edinburgh, 127, 137, 157
——, Thomas, curate of Glasgow, chaplain in kirk of St Roche, beside Glasgow, 514
——, Thomas, 284, 287 (Perth)
——, Thomas, 58 (Balmerino abbey)
——, William, in Edinburgh, 131

Flesher, Mr George, vicar of Kirriemuir, 396
Flint (Flent), Janet, 104 (Kirknewton)
Folkard (Folkert), Thomas, 254 (Peebles)
Forbes, William, 7th lord, bailie of Lindores abbey, 32, 32n., 36, 414, 457, 457n., 458
—— of Monymusk, Mr Duncan, 416, 418, 446, 447
—— of Pitsligo, Alexander, 421, 447
—— of Reres, Arthur, 146
—— of Towyis, John, 449
——, Arthur, 447 (Monymusk priory)
——, Robert, 240 (Aberdeen friary)
Fordie, laird of, 303
Forman (Foirman), Adam, prior of Charterhouse, Perth, 285
——, John, vicar of Kilrenny, 75
——, John, 353 (Coupar Angus abbey)
——, Martin, 448 (Tarves)
——, Robert, 190 (Channelkirk)
Forres, Mr Alexander, vicar of Clatt, 129n.
——, David, 181 (Haddington priory)
Forrest, Mr Alexander, parson of Logie-Montrose, 373, 374
——, Mr Alexander, provost of Fowlis Easter, vicar of Inverkeilor, 83, 83n., 84n.
——, Mr Alexander, provost of St Mary of the Fields' collegiate kirk, Edinburgh, 105, 105n., 120, 120n.
——, Mr Alexander, 368 (Nevay)
——, David, master of the mint, 286, 286n.
——, Edward, 555 (Stirling)
——, John, 156 (Linlithgow friary)
——, 'widow', 271 (Holywood abbey)
Forrester (Forestar) of Corstorphine, James, 112, 115
——, Alexander, 92, 93 (Holyroodhouse abbey)
——, David, 555 (Stirling)
——, John, 113 (Leith hospital)
——, Matthew, vicar of Monkton, 562
——, Matthew, 114 (Leith hospital)
——, Robert, 58 (Balmerino abbey)
——, William, in Auchendinny, 142 (Restalrig)
Forret, of that Ilk, John, 79
——, laird of, 59
——, George, prebendary of St Mary of

the Fields' collegiate kirk, Edinburgh, 134
——, John, vicar of Swinton, 187
——, John, 18 (St Andrews priory)
——, Mr Thomas, vicar pensionary of Logie-Murdoch, 79
——, Thomas, burgess of Glasgow, 533
Forsy, see Forsyth
Forsyth, [blank], 158 (Airth)
——, James, chaplain, Dunblane cathedral, 343
——, John, elder, tenant in Duffus, 468
——, John, younger, tenant in Duffus, 468
——, William, 254 (Peebles)
Fortis, Robert, 516 (Glasgow)
Foster, David, burgess of Edinburgh, 551
——, Robert, abbot of Balmerino, 58, 58n., 60
——, William, 142 (Restalrig)
Fotheringham (Fothringane, Fothringtoun), Mr James, parson of Ballumbie, 377, 400
Foular, see Fowler
Fould, Mr Gilbert, prebendary of St John, Kirkwall cathedral, 667
Fouleit, John, in Leith, 133
Foulis, Mr Adam, prebendary of Lambeletham, St Andrews, 86
——, Mr Adam, vicar of Tealing, 364
——, Mr John, parson of Edzell, 379
——, Mr Henry, 26, 48 (Hailes)
Fourhous, John, 102 (Newbattle abbey)
Fowler (Foular), John, 283, 287 (Perth)
——, William, keeper of place and yards of Scotlandwell priory, 56
——, William, in Edinburgh, 126
Fowlis, laird of, 625 (Ross bishopric); 635 (Kilteam); 645 (Alness)
Frank, William, 254 (Peebles)
Fraser (Frascher), Mr Duncan, archdeacon of Ross, 638
French, Adam, chaplain of the altars of Our Lady and Holy Blood, Sanquhar parish kirk, 273
——, Patrick, vicar of Linlithgow, 153
——, Patrick, 180 (Haddington priory)
Frissell, Rae, 271, 275 (Holywood abbey)
——, Rolland, 271, 275 (Holywood abbey)
Frog, Elizabeth, 47 (Dunfermline abbey)

Frost, John, prebendary, Dunglass collegiate kirk, 166
Froster, John, in Linlithgow, 156
Fudes, see Fiddes
Fullerton (Fowlartoun, Fullarton, Fullourtoun) of Ardo, William, collector for Forfar and Kincardine, 371, 371n., 372
——, Alexander, 421 (Aberdeen friary)
——, George, prebendary of Ruffell, Dunkeld cathedral, 316
——, Henry, 373 (Logie-Montrose)
——, Mr Hew, moderator of Dumfries presbytery, minister of Dumfries, 272, 272n.
——, William, 373, 374 (Logie-Montrose)
Furde, Michael, 651 (Tain)
Fyn, see Fin
Fyvie, laird of, 430
——, Patrick, 284, 287 (Perth)

Gadie (Gady), John, 92 (Holyroodhouse abbey)
Gain, Marian, 524 (Glasgow)
Gairdner, see Gardener
Gairlie, Mr David, 73 (Torry)
Galbraith (Galbrayth), Mr Robert, 180 (Haddington priory)
Galloway, Alexander, 66 (St Andrews archdeaconry)
——, Mr Andrew, pension to, 445 (Kinkell)
——, Cristall, 180 (Haddington priory)
——, John, 114 (Leith hospital)
Galston, laird of, 562, 564
Galt, Adam, 583 (Dreghorn)
——, John, 583 (Dreghorn)
——, Nicol, 583 (Dreghorn)
——, Robert, 571 (Ayr)
——, Thomas, 284, 287 (Perth)
——, William, 583 (Dreghorn)
Garden (Gairden, Geden), laird of, 26, 48, 551, 556
——, Mr William, parson and vicar of Kirkbuddo and prebendary of Guthrie collegiate kirk, 390
——, William, vicar of Aberlemno, 375
Gardener (Gairdner, Gardinar, Gardner, Gairtner), Andrew, 158 (Airth)
——, John, 158 (Airth)
——, John, 155 (Dalmeny)

——, Robert, in Linlithgow, 156, 157
——, Rolland, in Edinburgh, 127
——, William, 158 (Airth)
Gardyne (Gardny), laird of, 376 (Idvie)
——, Alexander, 376 (Idvie)
Garlies, laird of, 587, 606, 607, 618
——, laird of, younger, 588, 595
Garnar, Andrew, 492 (Dallas)
Garthland, laird of, 588, 590, 607
Gartlie (or Grantullie), Mr James, vicar of Kirkmichael, parson and vicar of Glass, 485, 485n.
Gaw, Alexander, 290 (Culross abbey)
Geden, laird of, *see* Garden
Geddes, Mr Charles, parson of Parton, 616
——, Charles, tacksman, Stobo, 252
Gemmell, Richard, in Edinburgh, 136
George, Herbert, 598 (Soulseat abbey)
Gibb (Gib), Adam, in Edinburgh, 127
——, John, cutler, 546 (St Ninians)
——, Martin, vicar and reader at Penninghame, 602n.
——, Robert, 289 (Culross abbey), 289
Gibson, Alexander, 180 (Haddington priory)
——, Mr David, chaplain of St Michael's chaplainry, Glasgow cathedral, 515, 615
——, Mr David, parson and vicar of Ayr, 561, 600
——, Mr David, tacksman of vicarage of Cathcart, 520
——, David, 615 (Lincluden)
——, George, in Edinburgh, 128
——, John, bower in Edinburgh, 139
——, John, chaplain of Duffus, parson of Unthank, 475
——, John, chaplain of Our Lady chaplainry, Elgin cathedral; chaplain of St Duthac's chaplainry, Elgin cathedral, 476
——, John, in Linlithgow, 11, 156
——, Michael, in Linlithgow, 156
——, Robert, 155 (Dalmeny)
——, Thomas, 592 (Whithorn priory)
——, William, priest in Elgin, 453
——, William, 180 (Haddington)
Gifford (Giffurd, Gifhert), David, 419, 421 (Aberdeen friary)
——, James, 100 (Newbattle abbey)
Gilbert, Michael, in Edinburgh, 128
Gillesbie, laird of, 266

Gilleson, Aymer, 270 (Holywood abbey)
——, Homer, 271 (Holywood abbey)
——, John, 274 (Holywood abbey)
——, William, 274 (Holywood abbey)
Gillespie, David, in Edinburgh, 141
Gilmore, John, writer to the signet, 13n.
Gilmour, Robert, 158 (Airth)
Gilson, Aymer, 623 (Holywood abbey)
Gilston, Lancelot, prebendary of Bute, Restalrig collegiate kirk, 143
Gladdo, Mage, 254 (Peebles)
Glamis, lord, *see* Lyon
Glass, Duncan, 284, 288 (Atholl)
——, Ewen, 284, 288 (Atholl)
Glassenwright (Glasinwrycht, Glessinwrycht), James, 609 (Kirkbean)
——, John, 418 (Aberdeen friary)
Glassford (Glasfuird), William, in Edinburgh, 133
Gleghorn, *see* Cleghorn
Glen (Glenn), Mr Robert, in Edinburgh, 135
——, Thomas, in Linlithgow, 156
Glenbervie, laird of, 94
Glencairn, earls of, *see* Cunningham
Glendinning of Drumrasche, John, 616
Godrell, Thomas, chaplain of Our Lady, Prestonkirk (Linton), 175
——, Thomas, chaplain of St Nicholas' chaplainry, Lanark, 514; factor, 512, 512n.
——, Thomas, parson of Morham, 170
Goldsmith, David, 419, 421 (Aberdeen friary)
Goodlad, Thomas, in Linlithgow, 156
Goodnight (Guidnycht), Alyman, 556 (Stirling)
Goodson (Goodsoun), David, 92, 93 (Holyroodhouse abbey)
Gordon, Alexander, 11th earl of Sutherland, 657n.
——, George, 4th earl of Huntly, 416, 416n., 435, 435n., 436n., 474, 474n., 483, 486
——, George, 5th earl of Huntly, 7, 7n., 8, 8n., 436, 436n., 491
——, John, 10th earl of Sutherland, 629, 629n., 657, 657n., 670
—— of Barskeocht, John, 603, 605
—— of Kirkconnell, Robert, 607
—— of Lochinvar, *see also* Lochinvar,

laird of
—— of Lochinvar, James, 586n.
—— of Lochinvar, John, 586, 587
—— of Shirmers (Chirmis), Robert, 603
—— of Strathdon, Walter, 432
—— of Three Threaves, William, 588
—— of Troquhain, Alexander, 585
——, Mr Alexander, bishop elect of Galloway, commendator of Inchaffray, 7, 7n. 26, 92, 92n., 348, 348n., 588, 588n.; pension to, 629 (Caithness bishopric)
——, Alexander, 414 (Aberdeen bishopric)
——, Alexander, 588, 590, 591, 607 (Galloway bishopric)
——, David, 113-114 (Leith hospital)
——, Mr George, parson of Kingussie, 487, 492
——, Mr James, chancellor of Moray, 486, 486n.
——, Mr James, parson of Clatt, 435
——, Mr James, parson of Lonmay, 427
——, Mr James, parson of Rhynie, 486
——, James, abbot of Glenluce, 600
——, James, parson of Banchory-Devenick, 436
——, James, 414 (Aberdeen bishopric)
——, James, 611 (Sweetheart abbey)
——, John, factor, 489 (Botary)
——, John, in Leiffnoris, 564 (Failford)
——, John, 594 (Whithorn priory)
——, Martin, in Edinburgh, 133
——, Robert, chancellor of Moray, 486, 486n.
——, Robert, son of 4th earl of Huntly, 416, 416n.
——, Robert, 588 (Galloway bishopric)
——, Roger, dean of Dunblane cathedral, 315
——, Roger, 585, 586, 587 (Galloway bishopric)
——, Mr William, bishop of Aberdeen, 360, 360n.
——, Mr William, chancellor of Dunkeld cathedral, 308, 308n.
——, Mr William, parson of Spittal, 631
——, Mr William, treasurer of Caithness, 630
——, Mr William, 425 (Tough); 414 (Aberdeen bishopric)
——, William, 607 (Galloway bishopric)
Gourlay, Mr David, vicar of Abernethy, 313; chamberlain of bishopric of Dunblane, 295
——, Robert, pension to, 360 (Coupar Angus abbey)
——, William, feuar, 353 (Coupar Angus abbey)
Govan (Govean), James, in Peebles, 525
——, Patrick, 585 (Galloway bishopric)
Graden, Nicol, in Eccles parish, 184,
Graham (Grahame), William, 2nd earl of Montrose, 553, 552n., 554, 556,
—— of Alyth, Robert, 398
—— of Boquhaple, Thomas, comptroller, 73n., 94n., 152n., 290, 291, 351, 351n., 423n., 465n., 651, 651n.
——, Alexander, pension to, 538, 544, 545, 547 (Cambuskenneth abbey)
——, Mr Andrew, vicar of Wick, 648, 649
——, Archibald, 94 (Edinburgh)
——, Mr John, pension to, 387 (Brechin bishopric)
——, Mungo, servant to Mary, Queen of Scots, 224, 224n., 228, 230
——, Patrick, vicar of Mains, 393
——, Mr Robert, vicar of Alyth, canon of Dunkeld, 398; collector to the chapter of Dunkeld, 300, 301n.
——, Walter, parson and vicar of Cookston, 382; pension to, 390 (Maison Dieu, Brechin)
——, William, parson and vicar of Killearn, 554
Grainger, Arthur, burgess of Edinburgh, 575
Grandtully, laird of, 17
Grange, laird of, 303 (Dunkeld)
Grant of Freuchie, John, 480
——, laird of, 489
Grantullie (or Gartlie), Mr James, vicar of Kirkmichael, parson and vicar of Glass, 485, 485n.
Gray, Patrick, 5th lord, 18, 18n., 54, 54n., 318, 318n.
——, Patrick, 6th lord, commendator of Dunfermline abbey, 54, 54n.
——, Alexander, parson of Farr, 631
——, Alexander, 419, 421 (Aberdeen friary)
——, Andrew, chaplain in Dundee, 377, 395; chaplain of St John's chaplainry, Barras, Mearns, 408
——, Andrew, parson of Kells, stallar in

INDEX OF PERSONS

Chapel Royal, Stirling, 603
——, Andrew, 603 (Balmaclellan)
——, Andrew, 114 (Leith hospital)
——, Andrew, 568 (Maybole)
——, Andrew, 217 (Restenneth priory)
——, Arthur, 124 (Edinburgh)
——, David, 156 (Linlithgow friary)
——, Duncan, vicar pensionary of Auchterhouse, 339
——, George, parson and vicar of Balmaclellan, stallar in Chapel Royal, Stirling, 603
——, George, 377 (Lundie)
——, Mr James, prebendary of Half Norton and Half Rathobyres, Corstorphine collegiate kirk, 107, 109, 130
——, James, scribe to general assembly, 372
——, James, vicar of Echt, 435
——, John, 180 (Haddington priory)
——, Patrick, 420 (Aberdeen friars)
——, Patrick, 114 (Leith hospital)
——, Richie, in Edinburgh, 127
——, Robert, in Edinburgh, 132, 137
——, Robert, in Linlithgow, 156
——, Thomas, chaplain of St Michael's altar, St Giles' collegiate kirk, Edinburgh, 140
Greenlaw (Greinlaw), laird of, 226
——, John, prebendary of Half Hatton and Half Dalmahoy, Corstorphine collegiate kirk, 107
Greenock, laird of, *see* Shaw
Greer (Greir), Cubbie, 275 (Holywood abbey)
——, Gibbe, in Speddoch, 269, 274 (Holywood abbey)
——, Gilbert, 275; pension to, 624 (Holywood abbey)
——, John, in Broch, 275 (Holywood abbey)
——, John, in Penmurtie, 275 (Holywood abbey)
——, Peter, 275 (Holywood abbey)
——, Roger, 275 (Holywood abbey)
——, Thomas, 275 (Holywood abbey)
Gregan (Gregane), John, 609 (Kirkbean)
Greiff, *see* Grieve
Greig (Greg), Catherine, 555 (Stirling)
——, James, vicar of Dunsyre, 509
——, James, 105 (Kirknewton)
——, Richard, 419, 422 (Aberdeen friary)
Grenelawdene, goodman of (Kelso abbey), 226
Grieve, Matthew, vicar pensionary of Monikie, 395
Grote, Mr Gilbert, notary public, 156n., 479
Guild, James, chaplain in Edinburgh, 136
Gullan (Gulen), David, 19 (St Andrews priory)
——, William, chaplain of St Laurence's altar, Stirling parish kirk, grammar schoolmaster, 545
Guthrie of that Ilk, Alexander, 390
——, Alexander, in Edinburgh, 127
——, Mr David, vicar of Dull, 86
——, David, 18 (St Andrews priory)
——, Gabriel, 390 (Guthrie collegiate kirk)
——, James, parson of Kirkbuddo and prebendary of Guthrie collegiate kirk, 390, 391
——, James, 376 (Idvie)
——, Mr John, vicar of Stracathro, 380
——, Patrick, prebendary of Langlands and Hilton, Guthrie collegiate kirk, 390

Hacket, laird of, 11
Haddington (Hadyngtoun), Elizabeth, 180 (Haddington priory)
Haddow, James, 180 (Haddington priory)
——, Robert, in Edinburgh, 123
Haddowk, laird of, 435
Hagye, *see* Heggie
Hain, John, in Edinburgh, 129
Haining (Hanyng, Hayny), John, in Mosside, 270, 271 (Holywood abbey)
——, Nicol, 275, 276 (Holywood abbey)
——, Robert, 270 (Holywood abbey)
——, William, 271 (Holywood abbey)
——, William, 603 (Sorbie)
Halcro of Brugh, Mr Magnus, 656, 661 (Orkney bishopric)
Haldane, Mr Richard, subdean of Dunkeld, 341
Haliburton (Halyburtoun), Mr David, parson of Urray, 635
——, David, provost of Methven

collegiate kirk, 283, 283n., 296, 298n., 379
——, George, vicar of Gullane, 175
——, James, tutor of Pitcur, 249
——, John, 189 (Dryburgh abbey)
——, Laurence, in Edinburgh, 126
Halkerston (Halkarstoun), Adam, in Edinburgh, 133
——, George, in Edinburgh, 136, 138
——, James, in Edinburgh, 138
Hall, Robert, 158 (Airth)
——, William, 158 (Airth)
Haltoun, laird of, 100
Hamill (Hummill), John, pension to, 370 (Coupar Angus abbey)
Hamilton, James, 2nd earl of Arran, duke of Chatelherault, Governor, 39, 39n., 58, 92, 92n., 129n., 229, 229n., 375, 405, 509, 529, 529n.
——, James, 3rd earl of Arran, 40, 40n.
—— of Ardoch, Andrew, 511
—— of Bothwellhaugh, David, 505, 529, 534
—— of Briggs, Robert, 3
—— of Broomhill, John, 513
—— of Dalserf, Robert, younger, 574
—— of Drumcairn, Thomas, king's advocate, 202, 202n., 219, 222, 222n.
—— of Kingscavil, James, sheriff of Linlithgow, 184n.
—— of Mylburn, Matthew, 510; factor to treasurer of Ross, 637
—— of Priestfield, Thomas, 216
—— of Samuelston, James, 169
—— of Samuelston, Mr John, vicar of Kilconquhar, 77
—— of Sandhill, James, 155
—— of Stanehouse (Stenhous), James, 511
—— of Stanehouse, John, 573
——, Alexander, vicar of Carriden, 157
——, Alexander, 101 (Newbattle abbey)
——, Anne, in Edinburgh, 128
——, Mr Archibald, vicar of St Cuthbert's kirk, Edinburgh, 105
——, Arthur, provost of Hamilton collegiate kirk, 518
——, Arthur, in Edinburgh, 124
——, Claud, commendator of Paisley abbey, 527n., 534n.
——, Elizabeth, 516 (Glasgow)
——, Mr Gavin, commendator of Kilwinning abbey, 4, 4n., 302, 302n., 342, 346, 435, 435n., 575, 575n., 578
——, Gavin, pension to, 577 (Kilwinning abbey)
——, Grissall, lady Robertland, 573
——, James, parson of Menmuir, 303, 303n., 374, 375
——, James, prebendary of New college [St Mary and St Anne], Glasgow, 516
——, James, subdean of Glasgow, 498, 498n.
——, James, vicar of Carmunnock, 510
——, James, in Edinburgh, 131
——, James, in Linlithgow, 156, 157
——, James, 302, 342, 346 (Abercorn)
——, James, 496 (Cambusnethan); 497 (Monkland)
——, James, 190, 191 (Lanark)
——, James, 340 (Muthill)
——, James, 102 (Newbattle abbey)
——, James, 168 (North Berwick)
——, James, 284, 287 (Perth)
——, Mr John, archbishop of St Andrews, commendator of Paisley, 6, 6n., 7, 90, 104, 104n., 105, 156, 156n., 170, 170n., 218, 218n., 302, 367, 373, 373n., 381, 381n., 527
——, Mr John, student in Paris, parson of Dunbar, 169
——, Mr John, subchanter of Glasgow cathedral, and parson of Durisdeer, 272
——, Mr John, vicar of Dundee, 394
——, John, baxter, in Edinburgh, 135
——, John, commendator of Arbroath abbey, 83n., 114, 114n., 358, 358n.
——, John, provost of Bothwell collegiate kirk, 507, 507n., 513n.
——, John, vicar of Cambusnethan, 515
——, John, in Linlithgow, 11
——, John, 579 (Kilwinning abbey)
——, Marion, prioress of Eccles nunnery, 184n.
——, Mr Ninian, prebendary of Ravelston, St Giles' collegiate kirk, Edinburgh, 112
——, Mr Robert, parson of Carrington, 121
——, Mr Robert, parson of Kincardine O'Neil, 439
——, Mr Robert, parson of Torrance,

prebendary of Hazeldean, Bothwell collegiate kirk, 515
——, Robert, chaplain of Kilwinning, 510
——, Robert, in Linlithgow, 11, 156
——, Thomas, in Edinburgh, 124
——, Thomas, 168 (North Berwick), 168
——, Mr William, parson of Cambuslang, 509
——, Mr William, parson of Glenholm, 247
——, William, 290 (Culross abbey)
——, William, 11 (Linlithgow)
——, William, 18 (St Andrews priory)
——, William, 595 (Whithorn priory)
Handibellis, Patrick, 420 (Aberdeen friary)
Hanna (Hannay), Alexander, in Mosside, 274 (Holywood abbey)
——, John, 594 (Whithorn priory)
Harcarse (Harcas, Harkers, Herkes), [blank], in Linlithgow, 11
——, Adam, 94 (Sciennes priory, Edinburgh)
——, Alexander, 92, 93 (Holyroodhouse abbey)
Hardie (Hardy), David, 158 (Airth)
——, Patrick, 216 (Jedburgh abbey)
——, Patrick, 210 (Melrose)
Harkers, see Harcarse
Harlaw, James, in Edinburgh, 126
——, William, in Edinburgh, 136
Harley (Harlie), Giles, in Edinburgh, 137
Harper, George, 270 (Holywood abbey)
——, Janet, 270 (Holywood abbey)
——, John, 271 (Holywood abbey)
——, Robert, 275, 276 (Holywood abbey)
Harrot, George, 101 (Newbattle abbey)
——, William, 101 (Newbattle abbey)
Hart, Thomas, 295 (Glendevon)
Harvey, see Hervey
Harvieston (Horvestoun), William, 168, 169 (North Berwick)
Hawick (Hawyk), John, 270, 271, 274 (Holywood abbey)
Hawns, M., 550 (Manuel priory)
Hawthorn (Hathern, Huthron), Michael, vicar of Glasserton, 605
——, Michael, vicar of Toskerton, 605
Hawthornden, laird of, 142
Hay, Andrew, 8th earl of Erroll, 421, 421n.
——, George, 7th earl of Erroll, 421, 421n., 304, 304n.
——, William, 5th lord Yester, 172, 172n.
—— of Easter Kennet, Alexander, clerk register, 21n.; clerk to privy council, 149, 149n.
—— of Tallo, William, 122, 122n.
—— of Whitburgh, Alexander, clerk register, 677, 677n.
——, Adam, 651 (Ross subdeanery)
——, Alexander, parish clerk of Fordyce, 456, 457; 457 (Arndilly)
——, Alexander, pension to, 260 (Melrose abbey)
——, Andrew, parson of Rathven, 456n.
——, Andrew, parson of Renfrew, 532
——, Andrew, prebendary of Bothans collegiate kirk, 172
——, Andrew, vicar pensionary of Rathven, 455
——, Mr George, parson of Kingussie, 487; parson of Rathven, Dundurcas and Kiltarlity, 248, 456, 456n.; 451 (Aboyne); parson of Eddleston, 248, 248nn.
——, Gilbert, in Edinburgh, 127
——, J., 611 (Dumfries)
——, Mr John, parson of Monymusk, commendator of Balmerino abbey, 19, 19n., 59, 59n., 61, 118, 118n., 446, 446n., 447
——, Mr John, in Peebles, 128
——, John, 254 (Peebles)
——, Nicholas, vicar portionary of Borthwick, 111
——, Mr Nicol, scribe to chapter of Aberdeen, 438
——, Mr Thomas, parson of Spynie, canon of Elgin cathedral, 488; abbot of Glenluce abbey, 600, 600n.
——, Thomas, parson of Logie-Easter, 637
——, Walter, 420 (Aberdeen friary)
——, Mr William, parson of Idvie, 376
——, Mr William, parson of Turriff, 431, 433, 440, 445, 451
Heggie (Hagye), Andrew [vicar of Lauder], 194
——, Andrew, 547 (Cambuskenneth abbey)

Heggin (Hegeyn), John, 158 (Airth)
——, Patrick, 158 (Airth)
Heislat, laird of, 574
Henderson (Henryson) of Fordaill, James, 126
——, Mr David, vicar of Rossie, 77
——, James, prebendary of Kinkell, 80
——, James, in Edinburgh, 129
——, James, 101 (Newbattle abbey)
——, Mr John, 101 (Newbattle abbey)
——, John, vicar of Cromarty, 643
——, John, in Edinburgh, 127
——, John, 180 (Haddington priory)
——, Richard, 102 (Newbattle abbey)
——, Robert, chaplain, Dunblane cathedral, 343
——, Simon, 113 (Leith hospital)
——, Thomas, chaplain of Luss, chaplain of Rossdhu, Dumbarton, 542
——, Thomas, 102 (Newbattle abbey)
——, Mr William, chaplain, Restalrig collegiate kirk, 143
Henry (Henrie), Adam, 158
——, David, 343 (Dunkeld)
——, John, 158 (Airth)
——, John, 101 (Newbattle)
——, Pautoun, 369 (Coupar Angus abbey)
——, William, 158 (Airth)
Henryson, *see* Henderson
Hepburn (Hepbroun), James, 4th earl of Bothwell, bailie of Haddington priory, 7, 7n., 163, 163n.
—— of Beanstoun, Patrick, bailie depute of Haddington priory, 163
—— of Caldhame, Matthew, 401, 402
—— of Whitsome, Alexander, 171, 171n., 336
——, Captain, pension to, 58 (Balmerino abbey)
——, Adam, parson of Dipple, 474
——, Adam, 168 (North Berwick)
——, Mr Alexander, provost of Bothwell collegiate church, pension to, 507 (Bothwell)
——, Elizabeth, prioress of Haddington nunnery, 161, 164, 164n.
——, George, parson of Hauch (Prestonkirk or Linton), 170
——, George, 180 (Haddington priory)
——, James, vicar of Botary, 478
——, James, vicar of Brechin, 401, 402

——, James, 377, 379 (Brechin)
——, Mr John, parson of Dalry (Galloway), 605
——, Mr John, 214n., 215 (Morebattle)
——, John, treasurer of Brechin, 402
——, John, 180 (Haddington priory)
——, Margaret, 170 (Morham)
——, Matthew, chaplain of Caldhame, Brechin cathedral, 379
——, Matthew, 402 (Brechin)
——, Patrick, bishop of Moray, commendator of Scone abbey, 283, 287, 306, 331n., 467, 467n., 474
——, Patrick, skinner, 180 (Haddington priory)
——, Patrick, 505 (Blantyre)
——, Patrick, 65 (Tyninghame)
——, Robert, vicar of Rothiemurcus and Kincardine, 465
——, Mr Thomas, parson of Oldhamstocks, 171
——, Mr William, dean of Caithness, 630
——, William, vicar pensionary of Logie-Dundee, 399
——, William, in Dunbar, 171
——, William, 168 (North Berwick)
Herbert (Harbert), Gilbert, 352 (Coupar Angus)
Herbertson (Harbesoun), George, 516 (Glasgow)
——, Richard, chaplain of Holy Blood chaplainry, Glasgow cathedral, 502
——, Mr Robert, vicar of Aberuthven, chaplain of St Peter's chaplainry, Glasgow cathedral, 334, 501, 502; chaplain of St Ninian's chaplainry, Darnley, 501
Hering, Mr James, provost of Methven collegiate kirk, 298, 298n., 299
——, James, 308 (Lethendy)
Heriot of Trabroun, James, 153
——, Adam, preacher in New Aberdeen, 446, 447
——, John, in Edinburgh, 111
——, Matthew, in Glasgow, 500
——, Patrick, prebendary, Dunglass collegiate kirk, 166
Herkes, *see* Harcarse
Herries (Hereis) of Terregles, *see also* Maxwell
—— of Terregles, lady Agnes, 624n.
—— of Terregles, John, 4th lord, 624,

INDEX OF PERSONS

624n.
—— of Terregles, William, 3rd lord, 624n.
——, Alexander, 412 (Eassie)
——, John, grammar schoolmaster, prebendary of St John the Baptist and Holy Rood prebends, Crail collegiate kirk, 66
Hervey (Hervy), Andrew, in Edinburgh, 126
——, George, 419, 422 (Aberdeen friary)
Hetoun, Thomas, chaplain of altars of Holy Rood, Our Lady, and Holy Blood in Lanark, 521
Hewat, Peter, prebendary, Dunglass collegiate kirk, 166
——, Robert, 609 (Kirkbean)
Hewison (Hewesoun), Martin, 414, 416, 417 (Aberdeen bishopric)
Hiddleston, George, 275 (Holywood abbey)
——, Robert, 275 (Holywood abbey)
Hill, laird of, 47
——, George, 155 (Dalmeny)
——, James, chaplain of Lady altar, Govan, 526
——, John, 419, 422 (Aberdeen friary)
——, John, 384 (Brechin bishopric)
——, William, in Edinburgh, 128
Hislop (Heslope), William, 92 (Holyroodhouse abbey)
——, 'widow', 271 (Holywood abbey)
Hoborne, Andrew, 290 (Culross)
Hodge (Hoge), [blank], in Edinburgh, 135
——, Alexander, 158 (Airth)
——, Beatrice, 180 (Haddington priory)
——, Patrick, 158 (Airth)
——, Thomas, 598 (Soulseat abbey)
Hogg (Hog), Andrew, 158 (Airth)
——, George, in Scotlandwell, 56
——, James, 180 (Haddington priory)
——, John, 158 (Airth)
——, John, 254 (Peebles)
——, Monane, 457, 458 (Deer abbey)
——, Robert, 158 (Airth)
Hoggarth, John, prebendary of King's Work, Leith, Restalrig collegiate kirk, 143
Home, Alexander, 5th lord, 234, 234n., 235, 238
——, Alexander, 6th lord, 1st earl of, 197, 197n., 199, 201, 218, 218n., 219, 221, 221n., 234n.
—— of Carrelsyd, John, 204
—— of Garvald Grange, Patrick, 181 (Haddington priory)
—— of Heuch, Alexander, 146, 195
—— of Heuch, Robert, 147
—— of Huttonhall, Alexander, 216
—— of Lochtullo, William, 249
—— of Manderston, Alexander, 166, 198, 199, 200, 201
—— of Manderston, Alexander, younger, commendator of Coldingham priory, 198, 198n.
—— of North Berwick Mains, Alexander, 147 (North Berwick)
—— of Polwarth, Patrick, 146, 147, 148, 166, 168
—— of West Restoun, John, 198
——, [blank], cutler, in Edinburgh, 114
——, Mr Adam, parson of Polwarth, pension to, 149 (North Berwick priory)
——, Mr Adam, 147 (North Berwick)
——, Mr Alexander, [parson of Kirkgunzeon], 604
——, Mr Alexander, 216 (Jedburgh abbey)
——, Alexander, brother of laird of Ayton, 228
——, Alexander, prebendary of Spott, Dunbar collegiate kirk, 171
——, Alexander, son of Patrick Home of Polwarth, 147 (North Berwick)
——, Alexander, 189 (Dryburgh abbey)
——, Alexander, 167 (Largo; North Berwick)
——, Alexander, 145 (North Berwick priory)
——, Mr Andrew, 190, 191 (Lauder)
——, Andrew, commendator of Jedburgh abbey, 218, 218n.
——, Andrew, 147, 168 (North Berwick)
——, George, 548 (Inchmahome priory)
——, George, 168 (North Berwick)
——, George, 171 (Pinkerton)
——, Isobel, in Kingstoun, 147, 168n. (North Berwick priory)
——, Isobel, 168 (North Berwick priory)
——, James, son of William Home of Lochtullo, 249
——, James, in Leith, 122
——, Jean, in North Berwick priory, 148

——, Mr John, 604 (Kirkgunzeon)
——, Mr John, 218 (Restenneth priory)
——, John, 114 (Leith hospital)
——, Margaret, prioress of North Berwick nunnery, 166, 166n., 168
——, Nicol, chamberlain of Jedburgh abbey, 217
——, Patrick, son of Patrick Home of Polwarth, 166, 168 (Logie)
——, Patrick, 166, 168 (Logie)
——, William, brother of laird of Coldenknowes, pension to, 228 (Kelso abbey)
Hope (Hopp), John, in Edinburgh, 130
——, John, 254 (Peebles)
Hopper (Hoppar), James, prior of Carmelite friary, Linlithgow, 156
——, John, in Edinburgh, 127
——, Thomas, 184 (Eccles priory)
Hoppringle (Hoppringill) of Blyndley, George, 195
——, Andrew, 214 (Morebattle)
——, Christian, 214 (Morebattle)
——, George, 214 (Morebattle)
——, James, 214
——, John, 239 (Kelso abbey)
——, John, 214 (Morebattle)
——, John, 170 (Morham)
——, Malcolm, 208 (Melrose abbey)
——, Mr Robert, parson of Morham, 170
——, Mr Robert, 187 (Coldstream priory)
——, Robert, 254 (Peebles)
Horner, Lowrie, 275 (Holywood abbey)
Horvestoun, *see* Harviestoun
Hothe, John, 94 (Sciennes priory, Edinburgh)
Houston (Howstoun), Mr John, vicar of Dunlop, 579
——, Mr John, vicar of Glasgow Secundo, 503, 561
——, Mr John, 334 (Aberuthven)
Hucheon (Houchoun), [*blank*], 84 (Kirkcaldy)
Hudson (Hudstoun), Cuthbert, 496 (Dryfesdale)
——, Hew, prebendary of Holywell, Dunglass collegiate kirk, 166
——, Mr John, vicar pensionary of Glendevon, 337-338
——, *dene* Ralph, monk, 244, 244n. (Lesmahagow priory)

Humbie (Humby), the goodman of, 184
Hume, *see* Home
Humphrey (Umphray), Michael, 598 (Soulseat abbey)
——, Thomas, in Elgin, 453
Hunter (Huntar) of Ballaggane, David, 273
——, Alan, 557 (Ayr)
——, Alexander, 101 (Newbattle abbey)
——, Andrew, 283, 286 (Perth)
——, John, 557 (Ayr)
——, John, 100 (Newbattle abbey)
——, Martin, in Edinburgh, 136
——, Stephen, 101 (Newbattle abbey)
——, William, 380 (Stracathro)
Hunthill, laird of, 184
Huntly, earls of, *see* Gordon
——, Hugh, 573 (Kilwinning abbey)
Hutcheson, Nicol, prebendary of Our Lady, St Mary of the Fields' collegiate kirk, Edinburgh, 125
Huthron, *see* Hawthorn
Hutson, *see* Hudson
Hutton, Mr John, chamberlain of Brechin, 384, 385
——, Mr William, pension to, 386 (Brechin bishopric)
Hynd, John, 275, 276 (Holywood abbey)
——, Robert, 270, 271 (Holywood abbey)
Hyntrod, Robert, in Edinburgh, 129

Inchekeill, John, tenant in Duffus, 468
Inglis, James, prebendary of Half Hatton and Half Dalmahoy, Corstorphine collegiate kirk, 97
——, Thomas, portioner of Aldlistoun, tutor to James Inglis, prebendary, 97
——, Thomas, 254 (Peebles)
——, William, vicar of Cupar, 87
Innes of Aichthintowil, James, 474 (Dipple)
—— of Catboll, Alexander, 651 (Tain)
—— of Drainie, James, servitor to the Regent Moray, 454, 454n., 484, 490, 493; 478 (Botary)
—— of Terbacks, 625
—— of Towchyis, James, 481
——, James, 454, 454n. (Elgin friary)
——, Thomas, tenant in Duffus, 468
——, Mr Walter, vicar of Thurso, 648

—, Walter, 651 (Tain)
Inverleith (Innerleith), laird of, 93, 127
Irvine (Urwing) of Drum, Alexander, 458
—, John, 300 (Dunkeld)
—, Richard, factor to chancellor of Ross, 634
—, Robert, 17 (Migvie; Tarland)

Jackson (Jaksoun), Alexander, 369 (Coupar Angus abbey)
—, Arthur, 525 (Manor)
—, George, 271 (Holywood abbey)
—, Mr John, chancellor of Caithness, 630
—, John, 270, 271, 275 (Holywood abbey)
—, Robert, 369 (Coupar Angus abbey)
—, Thomas, in Edinburgh, 128
—, William, 275, 276 (Holywood abbey)
James II, King of Scots, 404n.
James III, King of Scots, 679
James V, King of Scots, 225
James VI, King of Scots, 10, 54, 97, 98, 198n., 224, 260, 623, 677-683
Jameson, George, 101 (Newbattle abbey)
—, Mark, vicar pensionary of Currie, 106, 110, 110n.; vicar portionary of Kilspindie, 110, 110n., 307; 501, 503 (Glasgow)
—, Robert, in Linlithgow, 11
Jamieson, *see* Jameson
Jardin (Jerdin), Alexander, in Edinburgh, 127
—, John, 101 (Newbattle abbey)
Jason (Jaeson), John, factor to lord Fleming, 521
Jewark, Thomas, 620 (St Mary's Isle priory)
Johnson (Johnsoun), Thomas, 598 (Soulseat abbey)
Johnston (Johnnestoun) of that Ilk, John, 266
— of Beithach, Adam, 615
— of Wamphray, James, 599
—, [blank], 155 (Linlithgow friary)
—, Andrew, 104 (Kirknewton)
—, Gavin, 615 (Kirkpatrick-Juxta)
—, George, 284, 287 (Perth)
—, James, burgess of Perth, 326

—, James, notary, 492, 492n. (Dallas)
—, James, 'preacher', 529, 529n. (Paisley)
—, James, in Edinburgh, 125
—, James, 615 (Kirkpatrick-Juxta)
—, Mr John, 503 (Glasgow)
—, John, commendator of Soulseat, 599, 599n.
—, John, vicar of Kinneil, 154
—, John, writer, 230 (Duddingston)
—, John, in Edinburgh, 128
—, John, 113 (Leith hospital)
—, John, 254 (Peebles)
—, Mark, 503 (Glasgow)
—, Mungo, 496 (Dryfesdale)
—, Patrick,101 (Newbattle abbey)
—, Robert, 615 (Kirkpatrick-Juxta)
—, Theophilus, 492 (Dallas)
—, Thomas, in Linlithgow, 11
—, Thomas, 619 (St Mary's Isle)
—, William, chaplain, Dunblane cathedral, 343
—, William, chaplain of Holy Blood altar, St Giles' collegiate kirk, Edinburgh, 131
Jolly (Joly), John, in Edinburgh, 133
Joy, John, in Edinburgh, 127

Karray [*recte* Barry], Mr Archibald, vicar pensionary of Hamilton, 519
Kay, William, 101 (Newbattle abbey)
Keir (Ker, Keyr), laird of, 19, 343, 349, 545, 546
—, James, in Dundee, 12
—, John, 291 (Culross abbey)
—, Mr Thomas, 436 (Clatt; Aberdeen deanery; Banchory-Devenick)
Keith, William, 4th earl of Marischal, 17n., 153, 153n., 404, 459n.
—, Robert, 1st lord Altrie (second son of William, 4th earl Marischal), commendator of Deer abbey, 17, 17n., 457, 457n., 458, 459, 459n.
— of Drumtochy, James, 82
— of Glackreoch, Alexander, 457
—, Mr Gilbert, pension to, 153 (Strathbrock)
—, Gilbert, 'capitane of Akergile', 649 (Wick)
—, Robert, 18 (St Andrews priory)
—, Robert, 380 (Strachan)
Kello, Mr John, 156 (Linlithgow friary)

Kelty, Donald, 314 (Dunning)
Kempt, Henry, 168 (North Berwick)
——, Robert, chaplain of Markle, 161
Kennardy (Kenarthy), Robert, 419 (Aberdeen friary)
Kennedy, Gilbert, 2nd earl of Cassillis, 600, 600n.
——, Gilbert, 3rd earl of Cassillis, 599
——, Gilbert, 4th earl of Cassillis, 568, 568n., 569, 588, 591, 600, 600n., 607
—— of Baltarsyd, David, 567 (Crossraguel abbey)
—— of Bargany, Thomas, 148, 166, 169, 229, 499, 561, 570, 571, 612
—— of Smithston, Alexander, 571
——, Alexander, 557, 571 (Ayr),
——, Mr George, 276 (Holywood abbey)
——, Mr Gilbert, vicar of Colmonell, 570
——, Gilbert, 567 (Crossraguel abbey)
——, Mr James, parson of Carstairs, 517
——, Mr James, prebendary of the New kirk [college of St Mary and St Anne], Glasgow, 519
——, Mr James, vicar of Kilmadock, chancellor of Dunblane, 341
——, James, 567 (Crossraguel abbey)
——, James, 598 (Soulseat abbey)
——, John, prebendary of Maybole, 568
——, John, in Ardmillan, 567 (Crossraguel abbey)
——, John, in Deinyn, 567 (Crossraguel abbey)
——, John, 629, 630 (Caithness bishopric; chantory)
——, John, 539 (Dumbarton)
——, John, 631 (Halkirk and Skinnet)
——, John, 632 (St Andrew's chaplainry, Caithness)
——, Quentin, abbot of Crossraguel, 568, 568n., 570n.
——, Robert, 571 (Ayr)
——, William, abbot of Crossraguel, 600, 600n.
——, William, abbot of Holywood abbey, 278n.
——, William, 567 (Crossraguel abbey)
Kent, Thomas, chaplain, Linlithgow, 156, 157, 157n.
Ker (Kar, Kerr), see also Carr
—— of Ancrum, Robert, 147, 228
—— of Caverton, son and heir apparent of Walter Ker of Cessford, 229
—— of Cessford, Walter, bailie of Newbattle, 102, 103; of Kelso, 225
—— of Corbethous, George, 214
—— of Gaitschaw, George, 214
—— of Hirsell, Andrew, 215
—— of Littleden, Andrew, 214
—— of Littleden, Mark, 215
—— of Schilstounbreist, Andrew, 214 (Morebattle)
——, [blank], 236
——, Alexander, chaplain of Our Lady altar, Ayr parish kirk, 571
——, Alexander, notary at Balmerino, 58, 58n.
——, Andrew, 236, 366 (Ancrum)
——, Andrew, 186 (Coldstream)
——, Andrew, 214 (Morebattle)
——, Catherine (Katherine), relict of Richard Trollope, 226 (Maxwell)
——, David, 214 (Morebattle)
——, David, 545 (Stirling)
——, Mr George, 231, 237, 240 (Kelso abbey)
——, George, 214 (Morebattle)
——, James, in Mersington, 214
——, James, 100 (Newbattle abbey)
——, John, 114 (Leith hospital)
——, Mark, commendator of Newbattle abbey, 25, 25n., 103, 103n.
——, Mark, 189, 191 (Dryburgh abbey)
——, Oliver, 283, 287 (Perth)
——, Robert, vicar of Lindean, 264
——, Robert, 236 (Ancrum)
——, Mr Thomas, vicar of Langton, 192
——, Mr Thomas, 643 (Cromarty)
——, Robert, 366 (Ancrum)
——, Robert, 214 (Morebattle)
——, Robert, 168 (North Berwick)
——, Walter, in Edinburgh, 129
——, Walter, 114 (Leith hospital)
——, William, abbot of Kelso, 224, 224n., 225, 225n., 226, 227
——, William, in Edinburgh, 127
Keringtoun (Kerintoun, Kerintone), see Carrington
Kernis, see Cairns
Kidd (Kyd), Alexander, parson of Spittal, vicar of Kinairney, subchanter of Aberdeen, 426, 434
Kilarstair, laird of, 594
Killock (Killocht), Henry, 230 (Humbie)

Kincaid, David, in Edinburgh, 136
——, Henry, parson of Lemlair, 643
——, Muriel, in Edinburgh, 128
——, Thomas, in Edinburgh, 137
Kincraig, lady, 146, 167
——, laird of, 167; *see also* Leslie
King, Alexander, 18 (Bourtie)
——, Thomas, chaplain of Muirhall, Carnwath barony, 520
Kinghorn, Adam, vicar of Kinglassie, 67
——, Adam, vicar of Linton, 251
——, Adam, parson of Newdosk, 382
——, Adam, pension to, 39 (Dunfermline abbey)
——, David, in Kirkcaldy, 48
——, George, 184 (Eccles priory)
——, John, in Kirkcaldy, 26
Kinhill, laird of, 598
Kinloch, Andrew, 58 (Balmerino abbey)
——, David, in Edinburgh, 127
Kinnear of that Ilk, John, 57, 58, 59, 60, 61
——, John, 384 (Brechin bishopric)
——, Robert, 383 (Brechin bishopric)
——, Thomas, chaplain of Magdalene altar, St Andrews, 1n.
——, Thomas, prebendary of St Nicholas' altar, Crail, 66
Kinross, Mr Henry, 85 (Aberdour)
——, Mr Henry, 442 (Invernochty)
——, Mr Henry, 636 (Kilmorack)
——, Mr Henry, 553 (Kirk O' Muir)
——, Mr Henry (Laggan), 486
——, Mr Henry, 641 (Tain)
——, Malcolm, 556 (Stirling)
——, William, chaplain of Kirk O' Muir, 553
Kintore, Gilbert, 421 (Aberdeen friary)
——, John, 419, 421 (Aberdeen friary)
Kirkcaldy, David, 101 (Newbattle abbey)
Kirkhaugh (Kirkocht), Andrew, 269, 274 (Holywood abbey)
——, Homer, 269, 274 (Holywood abbey)
Kirkland, David, chaplain of St Mungo's chaplainry, Glasgow cathedral, 501, 523
Kirkpatrick, John, 567 (Crossraguel abbey)
——, John, 102 (Newbattle abbey)
Knightson (Knychtsone), George, 114 (Leith hospital)

Knokhill, laird of, 303
Knowles (Knollis, Knowis), Gilbert, 414 (Aberdeen bishopric)
——, James, 101 (Newbattle abbey)
——, Thomas, chaplain of St Mungo's altar, Currie, 139
Knox (Knokis), Christopher, vicar choral in Glasgow cathedral, parson and vicar of Dalziel, 519; chaplain of altar of Our Lady of Pity, Glasgow cathedral, 522
——, Gilbert, in Edinburgh, 126
——, Thomas, chaplain of the altars of St Stephen and St Laurence, Glasgow cathedral, 500, 502; chaplain of Holy Rood chaplainry, Glasgow cathedral, and stallar in choir, 500, 502, 502n.; chaplain of St James' chaplainry, Glasgow cathedral, 502
——, Thomas, prebendary of Methven collegiate kirk, 299
Kyle (Kyill), Archibald, 180 (Haddington priory)
——, Donald, in Edinburgh, 127
Kylty, laird of, 343
Kyppuithe, laird of, 354

Lachlieson (Lachlisoun), Ninian, 585 (Galloway bishopric)
Lag, laird of, 272, 276
Laing (Layng), Mr Andrew, vicar of Dreghorn, 581
——, Mr Andrew, parson of Hoddam, 266
——, Mr Andrew, vicar of Kilmaurs, 580
——, Bartholomew, pension to, 103 (Newbattle abbey)
——, Catherine, 516 (Glasgow)
——, James, in Edinburgh, 137
——, Mr John, parson and vicar of Luss, 266, 541; pension to, 577 (Kilwinning)
——, Mr John, vicar of Bathgate, 152
——, Neil, canon of Brechin cathedral, 373
——, William, vicar of Panbride, 372
Lamb, Alexander, 114 (Leith hospital)
——, Edward, in Edinburgh, 139
——, Elizabeth, prioress of St Bathans, 170, 170n., 193

——, James, notary public, 637 (Ross treasurership)
——, James, younger, 158 (Airth)
——, John, burgess of Edinburgh, tacksman to the dean of Moray, 468, 469
——, Thomas, in Perth, 12
——, Thomas, 283, 287 (Perth)
——, William, 184 (Eccles priory)
——, William, 555 (Stirling)
Lamont (Lamonthe, Lawmonthe), laird of, 558, 562, 565
——, Mr Allan, 18 (St Andrews priory)
——, Henry, 18 (St Andrews priory)
Landells, David, 516 (Glasgow)
Lang, Catherine, 523 (Glasgow)
Langland, Janet, in Edinburgh, 131
Lanton, *see* Linton
Larg, laird of, 600, 607
Largo, laird of, 146, 169
Lathangy, George, 283, 286 (Perth)
Lauder (Lawder, Lowder) of Bass, Robert, 168, 169
——, Archibald, chaplain, Dunblane cathedral, 343
——, Mr George, parson of Auchindoir, 443, 443n.
——, Mr George, prebendary of Half Norton and Half Rathobyres, Corstorphine collegiate kirk, 107
——, Gilbert, in Edinburgh, 129
——, Isobel, 168 (North Berwick)
——, Mr James, pension to, 260 (Melrose abbey)
——, James, 182 (Haddington priory)
——, Mr John, archdeacon of Teviotdale, 214, 214n.
——, Captain Robert, 385, 385n.
——, Robert, notary, 168 (North Berwick); 195 (Polwarth)
——, Mr William, prebendary of Woodwick, Kirkwall cathedral, 668, 668n.
——, Mr William, 658, 661 (Orkney bishopric, Shetland)
——, William, in Edinburgh, 127-128, 136
Laverock, Thomas, in Edinburgh, 136
Law, Gilbert, vicar pensionary of Monyabroch [Kilsyth], 506
——, James, bishop of Orkney, 677, 677n., 680, 681, 682, 683
——, James, 516 (Glasgow)

——, Janet, in Leith, 133
——, John, 557 (Ayr)
——, Stephen, in Edinburgh, 138
——, William, in Edinburgh, 126
Lawmonth, *see* Lamont
Lawrie, William, 155 (Dalmeny)
——, William, 546 (St Ninians)
Lawson (Lowsoun), Agnes, 420 (Aberdeen friary)
——, Andrew, 420 (Aberdeen friary)
——, James, in Edinburgh, 122, 124
——, Robert, 229 (Humbie)
——, William, bonnetmaker in Edinburgh, 137
——, William, in Edinburgh, 127, 157
——, William, 384 (Brechin bishopric)
Learmonth (Lermonth), George, prebendary of Methven collegiate kirk, 299
——, Henry, 181 (Haddington priory)
——, James, provost of St Mary on the Rock, St Andrews, 55, 55n.
——, James, 403 (Fordoun)
——, John, chaplain of St Blaise chaplainry, Dunblane cathedral, 314, 328; chaplain of Dunning, 314
——, John, 117 (Gogar)
——, Patrick, 18, 19 (St Andrews priory)
——, Robert, 181 (Haddington priory)
Leggat, Alexander, 516 (Glasgow)
Leiffnoris, laird of, 514
Leighton (Lychtoun), John, in Edinburgh, 136
Leishman (Leschman), John, 513 (Walston)
——, Laurence, minister at Walston, 513
——, Michael, 513 (Walston)
——, William, 556 (Stirling)
Leith, Archibald, in Edinburgh, 127
——, Gilbert, 422 (Aberdeen friary)
Lekprevik (Lecprevik), James, vicar of Arbirlot, 397
Lennox, dukes of, *see* Stewart
Leslie, Andrew, 5th earl of Rothes, 19, 19n., 36
——, George, 4th earl of Rothes, 17n.
—— of Kincraig, Alexander, 421; *see also* Kincraig, laird of
——, Agnes, lady Lochleven, 17, 17n.
——, Mr Andrew, vicar of Coull, 426
——, Euphemia, prioress of Elcho nunnery, 393, 393n.

——, Mr James, parson of Aberdour, 424
——, James, parson of Rothes, 491
——, James, 419, 422 (Aberdeen friary)
——, Mr John, bishop of Ross, 7, 7n., 8, 8n. 30, 491, 491n., 574, 574n., 577, 577n., 579, 579n.
——, Mr John, parson of Oyne, 428, 429
——, Mr John, parson of Snow, 442
——, Mr John, prebendary of Forgandenny, Dunkeld cathedral, 325
——, Mr John, vicar of Dyke, 426
——, Mr John, vicar of Forgue, 429
——, John, in Brechin, 387, 388, 388n.
——, John, 389 (Brechin bishopric)
——, John, 284, 287 (Perth)
——, Mr Leonard, parson and vicar of Blair Atholl, 319; parson of Skirdustan and Botriphnie, 320; 319 (Lude); 319 (Kilmoveonaig); 320 (Perth); 320 (Dunkeld); 321 (Inchaiden); 321 (Killin)
——, Leonard, commendator of Coupar Angus abbey, 352, 411
——, Peter, parson of Rothes, 475, 491n.
Lethendy (Lethindene), laird of, 308
Leycht, Patrick, 428 (Oyne)
Leyne, Edward, chaplain of the Holy Rood altar, Kirkcaldy, 84
Liddel (Liddell, Lyddall, Lyddell), Helen, in Dunbar, 169
——, John, 101 (Newbattle abbey)
——, Morris, in Dunbar, 169
——, Mr Patrick, parson of Croy and Moy, 639
——, Mr Patrick, parson of Kinnell, 367; pension to, 374 (Logie-Montrose)
——, Robert, in Edinburgh, 127
Lillie (Lily), Matthew, in Edinburgh, 125
Lindsay, David, 9th earl of Crawford, 382, 382n., 544n.
——, David, 10th earl of Crawford, 30, 30n., 379, 379n., 381
—— of the Byres, John, 5th lord, 40n., 290, 290n.
—— of the Byres, Patrick, 6th lord, 18, 18n., 145, 145n., 180, 181; (as master of) 40, 40n., 145n. 167, 167n.
—— of Dowhill, James, 62, 62n.
—— of Dunrod, Robert, bailie of Blantyre priory, 505

——, [blank], 146 (Largo)
——, Adam, 324 (Tullypowrie)
——, Alexander, 522 (Glasgow friary)
——, Andrew, vicar of Newtyle, 396
——, Andrew, in Edinburgh, 137
——, Mr Archibald, prebendary of Caputh, Dunkeld cathedral, 318; factor to chancellor of Dunkeld, 308, 308n.; 301 (Dunkeld)
——, Archibald, 487 (Kingussie); 487, 488 (Moray archdeaconry)
——, David, parson of Benvie, 334
——, Mr Hew, parson of Finavon, parson of Inverarity, 381
——, Mr James, founder of altarage of St Stephen and St Laurence, Glasgow cathedral, 500
——, Mr John, 301n. (Dunkeld)
——, John, parson of Menmuir, 303, 303n.
——, Thomas, Snowdon herald, 146 (Largo)
——, Thomas, in Dunkeld, 308
——, Thomas, 303 (Dunkeld)
——, Thomas, 284, 287 (Perth); 324 (Tullypowrie)
——, Mr Walter, feuar, 352 (Coupar Angus abbey)
——, William, in Edinburgh, 128
Lineynden, see Lumsden
Linlithgow, John, vicar of Abercorn, 153
Linton (Lanton, Lyntoun), Francis, 117n., 118 (Edinburgh)
——, 'widow', 101 (Newbattle abbey)
Lister (Litster, Lytstar), Elizabeth, 155 (Linlithgow friary)
——, Stephen, 92, 93 (Holyroodhouse abbey)
Liston (Listoun), Ninian, 101 (Newbattle abbey)
Litster, see Lister
Little, Clement, in Edinburgh, 127, 129
——, Edward, in Edinburgh, 126
——, Helen, in Edinburgh, 125
——, John, 155 (Dalmeny)
——, Robert, prebendary of St Mary of the Fields' collegiate kirk, Edinburgh, 134
Littlejohn, George, chaplain of Nomine Jesu, Holy Blood and St Blaise altars, St Giles' collegiate kirk, Edinburgh, 117, 117n., 118

Liverance (Leverance), Andrew, in Linlithgow, 156
Livingston (Levingstoun), Mr Alexander, 549 (Manuel priory)
——, Alexander, burgess of Edinburgh, 139
——, Alexander, 555 (Stirling)
——, Andrew, pension to, 508 (Lanark)
——, Mr Archibald, parson and vicar of Culter, 554
——, George, 92 (Holyroodhouse abbey)
——, Henry, in Livingston, 11
——, Henry, 554 (Culter)
——, Henry, 504 (Glasgow)
——, Janet, prioress of Manuel nunnery, 550, 550n.
——, John, 'man of weir and lamit in France', 363
——, Mr Thomas, treasurer of Glasgow cathedral, parson and vicar of Carnwath, 505, 505n., 510
——, Mr Thomas, vicar of Kincardine O'Neil, 430
——, William, 92 (Holyroodhouse abbey)
Loch, Henry, vicar of Bunkle, 192
——, John, burgess of Edinburgh, 131
Lochinvar, lady, 586, 586n.
——, laird of, 594, 600, 600n., 607; see also Gordon
Lochleven, lady, 17, 26; see also Erskine, Margaret
Lockburn, Laurence, 180 (Haddington priory)
Lockhart (Lokart), John, chaplain of Holy Blood altar, St Giles' collegiate kirk, Edinburgh, 130
——, John, 557 (Ayr)
——, William, in Edinburgh, 127
Logan of Flures, Walter, in Leith, 133
——, Alexander, 158 (Airth)
——, Mr Andrew, pension to, 119 (Restalrig)
——, Mr John, parson of Restalrig, 119
——, John, bailie of regality of Holyroodhouse, 92
——, John, 158 (Airth)
——, John, 114 (Leith hospital)
——, William, 158 (Airth)
Logie (Logy), Catherine, 155 (Dalmeny)
——, Effie, 155 (Dalmeny)
——, Mr John, vicar of Rhynd, 321

——, Peter, 155 (Dalmeny)
Lothian (Lowdiane), William, 516 (Glasgow)
Low, Agnes, 254 (Peebles)
Lowder, see Lauder
Lowes (Lowis), Adam, notary, 525 (Manor)
——, Thomas, 525 (Manor)
Lowrie, James, in Edinburgh, 137
Lowson, see Lawson
Lugtoun, laird of, 118, 158
Lumisden, see Lumsden
Lumsdale (Lummisdale), see also Lumsden
——, Mr Henry, 64 (Tannadice)
Lumsden (Lumisden), see also Lumsdale
——, Mr Henry, parson and vicar of Kirkforthar, 86
——, Mr Henry, vicar of Rothiemay, 428
——, Mr Robert, 414 (Aberdeen bishopric); procurator, 438 (Aberdeen); 444 (Cushnie)
——, Robert, 58 (Balmerino abbey)
——, Robert, 457 (Deer)
——, Mr Thomas, prebendary of Pitmedden, Abernethy collegiate kirk, 338
——, William, parson and vicar of Cleish, administrator of Kelso abbey, sacristan of Dunfermline, 72, 225
Lundie, laird of, 17, 145, 146, 148, 167, 169
—— of Baldastard, Janet, 145, 167
——, Mr George, brother to laird of Lundie, 148 (North Berwick priory)
——, Mr George, 145, 167, 169 (North Berwick priory)
Lyall (Lyell), see also Lyle
——, Alexander, 114 (Leith hospital)
——, Patrick, 180 (Haddington priory)
Lychtoun, see Leighton
Lyle, see also Lyall
——, Thomas, vicar of Whittingehame, 174
Lyon (Lyoun), John, 8th lord Glamis, 17n.
—— of Auldbar, Thomas, 202, 202n., 222, 222n.
—— of Baldukie, Thomas, master of Glamis, treasurer, 21, 21n., 54

——, David, 516 (Glasgow)
——, Jean, daughter of John, 8th lord Glamis, 17n.
——, William, 442 (Belhelvie)

MacAulay (McCalie, McCaulie), Andrew, 598 (Soulseat abbey)
——, Gibbie, 275 (Holywood abbey)
MacBeath (McBath), James, 657 (Orkney bishopric, Caithness)
McBirnie (McByrnie), John, 269, 270, 271, 275, 276 (Holywood abbey)
MacBlain (McBlaine), Nicol, vicar of Kirkmadrine, 601
McBreck (McBraik, Makbrek), James, 352, 356, 410 (Coupar Angus)
McCalie, McCaulie, *see* MacAulay
McCalmestoun, John, 652 (Edderton)
McCalzean (Mcalyion), Janet, 523 (Glasgow)
MacCarlich (McCarlie), Thomas, 598 (Soulseat abbey)
MacCartney (McCairtnay), James, burgess of Edinburgh, 622 (Sweetheart abbey)
——, James, 229 (Kelso abbey)
MacClannachen (McClanochtane), James, 598 (Soulseat abbey)
MacCleary (McCleirlie), John, 594 (Whithorn priory)
MacClellan (McCleland, McClellane, McKlellane, McLilland, Maklellan) of Muirtoun, James, 602
—— of Nunton, James, 587, 616
——, Alexander, 619 (St Mary's Isle priory)
——, Elspeth, 619 (St Mary's Isle priory)
——, James, 587, 607 (Galloway bishopric)
——, Michael, 594 (Whithorn priory)
——, Mr Thomas, 594 (Whithorn priory)
——, Thomas, in Edinburgh, 136
——, Thomas, 586 (Galloway bishopric)
——, Mr William, 604 (Senwick)
——, William, 594 (Whithorn priory)
McCoard (McCourte), John, 275 (Holywood abbey)
MacCone (Makconne, McCon, McConn, Magony), [*blank*], in Elgin, 453
——, Adam, 619 (St Mary's Isle priory)
——, Archibald, 594 (Whithorn priory)
——, Thomas, 609 (Kirkbean)
McCourt, John, 275 (Holywood abbey)
MacCrackan (McCrakin), Michael, 594, 596 (Whithorn priory)
MacCraith (McCrayth), Robert, chaplain of St Andrew (Caithness), vicar of Kilmalie, 632
MacCristin (McKrisknie), Allan, 587 (Galloway bishopric)
MacCrom (McCrome), 'widow', 271 (Holywood abbey)
MacCrorie (McCrory), John, 594 (Whithorn priory)
MacCulloch (McCulocht) of Arduell, Guthrie, 605 (Toskerton)
——, Henry, 599 (Soulseat abbey)
——, James, 594 (Whithorn priory)
——, Mr Malcolm, vicar of Anwoth, 614
——, Thomas, 594 (Whithorn priory)
MacCurrie (McCurreir), John, 592 (Whithorn priory)
McDowall (McDowell, McDowyll, McDuell), Fergus, 594, 596 (Whithorn priory)
——, John, 607 (Galloway bishopric)
——, John, 598 (Soulseat abbey)
——, Quentin, 590 (Galloway bishopric)
——, Thomas, 594, 596 (Whithorn priory)
——, Uthred, 605 (Clayshant)
——, William, vicar of Dalmeny, 152
——, William, 'vicar' of Holyroodhouse; chaplain of St Giles' collegiate kirk, Edinburgh, 95; master of hospital of St Paul's work, Leith, 95, 95n.
——, William, vicar of Leswalt, vicar pensionary of Inch, 606
MacGaw, John, in Canongate, Edinburgh, 132
McGill of Cranstoun Riddell, David, 92, 93, 202, 202n., 222, 222n.
—— of Nether Rankeilour, Mr James, clerk register, 61, 71, 71n., 77, 82, 84, 88, 90, 92, 93, 93n., 96, 155, 175, 196, 229, 248, 248n., 264, 264n., 286, 286n., 372, 372n., 391, 411, 411n., 420, 420n., 476, 478, 513, 534, 534n., 575, 575n., 600, 600n., 611, 611n., 649, 649n.
——, Mr James, factor, 366 (Fungarth);

490 (Elgin)
——, Thomas, prebendary of Fungarth, Dunkeld cathedral, 327, 327n., 329, 366
McGuffog (MacGuffok), John, 586 (Galloway bishopric)
McHowie, Agnes, 556 (Stirling)
Mack (Mak), Thomas, 598 (Soulseat abbey)
MacKellar (McKenlar), Andrew, 592 (Whithorn priory)
MacKenzie (McCenzie), Mr Alexander, parson portionary of Alness, 645; pension to, 650 (Beauly priory)
——, Alexander, pension to, 493 (Conveth)
——, Murdo, 635 (Urray)
MacKerral (McKarnell, McKyrnell), John, 620 (St Mary's Isle priory)
Mackerrow (McKerry), Andrew, 586 (Galloway bishopric)
Mackie (McKie, Makkye), Alexander, 158 (Airth)
——, Alexander, 594 (Whithorn priory)
——, Duncan, 595, 596 (Whithorn priory)
——, Robert, 158 (Airth)
——, William, chaplain of St Salvator's aisle, St Giles' collegiate kirk, Edinburgh, 135
Mackieson, William, 556 (Stirling)
MacKilvain (McKilweynd), Peter, 595, 596
McKnellane, John, 158 (Airth)
——, John, 619 (St Mary's Isle priory)
McMaster (McMaister), Gilchrist, 598 (Soulseat abbey)
——, William, 598 (Soulseat abbey)
MacMeeken (McMikin, McMytin), Mr Gavin, 598 (Soulseat abbey)
——, John, 598 (Soulseat abbey)
MacMonies (Makmochinee), Janet, 619 (St Mary's Isle priory)
MacNab, laird of, 284, 288
McNair, Duncan, 310 (Inchmagranachan, Dunkeld)
——, Robert, prebendary of Inchmagranachan, canon of Dunkeld, 310
——, Robert, scribe, 375 (Menmuir)
McNeil, Richard, 376 (Idvie)
MacNeillie (McKnalie), [blank], 551 (Stirling)

McNeish (McNische), William, 275 (Holywood abbey)
McQueen (Maquhyn), Mr John, vicar of Kilbarchan, 534
——, Mr John, 'preacher', 529, 529n. (Paisley)
——, Michael, 130 (Edinburgh)
MacQuharrie, Simon, 591 (Galloway bishopric)
Madrell, John, 551 (Stirling)
Magony, see MacCone
Maily, Marion, 158 (Airth)
Main (Maine, Mayn), George, 137
——, Jasper, 141
——, William, 546 (Stirling)
Mair, John, 422 (Aberdeen friary)
Maitland of Lethington, William, 176, 176n., 178, 178n., 208, 575n., 610
—— of Thirlestane, John, chancellor, 21, 21n., 54
——, John, commendator of Coldingham priory, 194, 198n.
——, Elizabeth, 275 (Holywood abbey)
Makmochinee, see MacMonies
Malcolm, Alexander, 157 (Airth)
——, James, 158 (Airth)
——, Richard, 283, 287 (Perth)
——, William, burgess of Edinburgh, 139
Malison (Malesoun, Malysoun), Alexander, 419, 421 (Aberdeen friary)
——, John, in Edinburgh, 136
Mallace (Mailles), William, in Edinburgh, 128
Mallen, Walter, abbot of Glenluce, 600, 600n.
Maltmaker, Thomas, in Edinburgh, 127
Manderston, Mr John, parson of Bolton, 172
——, Mr John, vicar of Lasswade, 115
——, Thomas, curate and reader at Tyninghame, 65
Mar, earls of, see Erskine
——, James, 18 (St Andrews priory)
——, John, in Edinburgh, 127
Marcaitnay, see MacCartney
March (Marche), Peter, in Edinburgh, 127
Marischal, earls of, see Keith
Marjoribanks of that Ilk, Mr Thomas, 278
——, [blank], in Edinburgh, 133

INDEX OF PERSONS

——, James, chaplain of the Barras, Edinburgh, 96-97
——, James, in Edinburgh, 123
——, Mr Thomas, prebendary of Half Bonnington and Half Platt, Corstorphine collegiate kirk, 107, 121
——, Mr Thomas, parson of Kirkmichael, 278
——, Mr Thomas, in Edinburgh, 123, 127, 128
Marshall, Thomas, 303 (Dunkeld)
Martin (Martyn, Martyne, Mertyne), Edward, 619 (St Mary's Isle priory)
——, George, 616 (Troqueer)
——, Henry, 619 (St Mary's Isle priory)
——, John, 619 (St Mary's Isle priory)
——, Mr Roger, vicar of Troqueer, 279, 279n., 616, 616n.
——, Roger, tenant, Over Calquha, 585 (Galloway bishopric)
Mary of Guelders, 408n.
Mary of Guise, 163, 606, 606n.
Mary, Queen of Scots, 6, 41, 176, 281, 282, 284n., 336, 341, 348, 368, 369, 388n., 392, 397, 411, 429, 476, 489, 537, 543, 543n., 544, 545, 547, 550, 559, 561, 565, 570, 603, 604, 615, 621, 627, 645, 647, 661
Mason (Measoun), Adam, 571 (Ayr)
——, Alexander, in Edinburgh, 137
——, John, in Glasgow, 500
——, John, pension to, 450 (Longley, Fetterangus)
——, John, 292 (Culross abbey)
——, Thomas, 290 (Culross abbey)
Masterton (Maistertoun), Adam, 290 (Culross abbey)
——, Alexander, 290 (Culross abbey)
——, Robert, 290 (Culross abbey)
Matheson, Andrew, 136 (Edinburgh)
——, John, 114 (Leith hospital)
——, Thomas, pension to, 630 (Caithness chancellory)
Matthew (Mathow), John, 158 (Airth)
Mauchan (Machame, Mauchane), Mr Alexander, 229 (Kelso abbey)
——, Isobel, 94 (Sciennes priory, Edinburgh)
Maxton, Oliver, in Perth, 317
Maxwell, John, 8th lord, bailie of Galloway bishopric, 591, 591n.; bailie of Sweetheart abbey, 612, 622
——, Robert, 5th lord, bailie of Holywood abbey, 276, 276n., 277, 611, 611n., 624n.
——, Robert, 6th lord, 591n.; bailie of Holywood abbey, 270, 270n., 624n.
——, Robert, 7th lord, 591n.
——, John, master of (later 4th lord Herries of Terregles), 270, 270n., 606n., 624, 624n.
——, John, master of, 216, 270
——, Robert, master of, 591, 591n.
—— of that Ilk, John, 269
—— of Brekensyd, John, 277, 604
—— of Cowhill, John, 276-277
—— of Cowhill, Robert, bailie depute of Holywood abbey, 271, 277, 623
—— of Glengoury, John, 271
—— of Gribtoun, John, 277
—— of Hills, John, 269, 270, 587, 607
—— of Kilbane, John, 269, 277
—— of Killilung (Celelung), John, 271
—— of Newlaw, John, 624, 624n.
—— of Portrack (Porterak), Robert, 270, 277
——, Mr Alexander, 168 (North Berwick)
——, Aymer, 269, 277 (Holywood abbey)
——, Donald, in Edinburgh, 132
——, Edward, commendator of Dundrennan abbey, 606n.
——, Edward, 619 (St Mary's Isle priory)
——, George, 524 (Glasgow)
——, George, 269, 270, 271, 277 (Holywood abbey)
——, Herbert, 609, 614 (Kirkbean)
——, Homer, 270 (Holyroodhouse abbey)
——, Hugh, 277 (Holywood abbey), 277
——, James, 270, 277 (Holywood abbey)
——, Mr John, parson of Our Lady kirk, Sanday, 660
——, John, 269, 274, 271, 275 (Holywood abbey)
——, Margaret, 275 (Holywood abbey)
——, Mariota, lady Carlyle, 265n.
——, Mungo, 274 (Holywood abbey)
——, Robert, vicar of Killellan, 533
——, Robert, 588 (Galloway bishopric)
——, Robert, 271 (Holywood abbey)
——, Robert, 594 (Whithorn priory)

——, Thomas, 92 (Holyroodhouse abbey)
——, Thomas, 269, 274 (Holywood abbey)
——, William, 271 (Holywood abbey)
May, John, 290 (Culross abbey)
Measoun, *see* Mason
Meggat, George, 100 (Newbattle)
Meikeson, Thomas, chaplain of St Nicholas' chaplainry, Arbroath, 401
Mekley, John, 180 (Haddington priory)
Meldrum, laird of, 415
——, Mr George, 177, 179 (Haddington priory)
——, George, 182 (Crail)
——, William, vicar of Peterculter, 427
Melrose, David, in Edinburgh, 128
Melville (Malvile, Melvin), Andrew, 388n.
——, Robert, chaplain of Tarlogie, Tain collegiate kirk, 647
——, Thomas, parson of Hutton, 266
——, Walter, 147 (North Berwick)
——, Walter, 18 (St Andrews priory)
——, William, in Edinburgh, 137
Menteith, *see also* Monteith
——, Patrick, 92 (Holyroodhouse abbey)
Menzies of Pitfoddels, Thomas, provost of Aberdeen, 416, 416n.
——, lady, 349
——, Mr Archibald, prebendary, Lincluden collegiate kirk, 620
——, Mr Archibald, 274 (Penpont)
——, Gilbert, 419, 422 (Aberdeen friary); 443 (Belhelvie)
——, Patrick, burgess of Aberdeen, 419, 421
——, William, 420 (Aberdeen friary)
Merser (Mairsar), Andrew, 469 (Moray deanery)
——, Gabriel, 300 (Saline)
——, Mr Robert, vicar of Banchory-Devenick, 429
Mersington, laird of, 184
Merton, laird of, 189, 191
Methven, Mr David, vicar of Forgan, 75
——, Mr David, vicar of Lathrisk, 75
——, Mr David, pension to, 331 (Kinnoull)
——, John, 19 (St Andrews priory)
——, Thomas, prebendary of Kingask and Kinglassie, 65, 65n., 77
Michelson, John, 155 (Dalmeny)

——, Nicol, prebendary, Dunglass collegiate kirk, 166
——, Thomas, 114 (Hailes)
Middleton, Mr George, 416, 417 (Aberdeen bishopric)
Mill, *see* Miln
Millar (Myllar), Adam, 523 (Glasgow)
——, John, chaplain of the Lady chapel of Kirkburn, Cambuslang, 523
——, John, exhorter at Barnwell, 563
——, John, vicar of Symington, 563, 569
——, Richard, chaplain of St Peter's altar, Ayr parish kirk, 557
Miller (Myller), Ebby, in Linlithgow, 156
——, James, 102 (Newbattle abbey)
——, John (Holywood abbey), 275
Milligan (Muligan, Myligane), John, 275 (Holywood abbey)
——, William, 609 (Kirkbean)
Miln (Mill, Mylne), George, 290 (Culross abbey)
Miniman (Myname), John, 291 (Culross abbey)
Mitchell (Michall, Michell, Mychell), Andrew, economus and factor of Holywood abbey, 277
——, Andrew, pension to, 623 (Tynron)
Mochrum, laird of, 595, 597, 619
Moffat (Moffet), Hob (Robert), 270 (Holywood abbey)
——, Janet, 555 (Stirling)
——, John, 275 (Holywood abbey)
——, John, 101 (Newbattle)
——, Robert (Hob), 270 (Holywood abbey)
——, Robert, 101 (Newbattle)
Moncrief, Alexander, 283, 287 (Perth)
——, Alexander, chaplain and chorister of Dunkeld, portionary of Abernyte, 310
——, Andrew, 482 (Duffus); 483 (Inverness)
——, James, 284, 287 (Perth)
Moncur, laird of, 19
——, George, 303 (Meigle)
——, John, prebendary, Dunglass collegiate kirk, 166
Monfode, *see* Munfoid
Monro, *see* Munro
Monteith, *see also* Menteith
——, Robert, 546 (Alva)
Montgomery of Skelmorlie, Robert, 500

INDEX OF PERSONS 739

——, Mr Andrew, vicar of Straiton, 582
——, Mr Hew, vicar of Dundonald, 560
——, Neil, 573 (Kilwinning abbey)
——, Thomas, vicar pensionary of Kirkmichael, 571
Month, *see* Mount
Monyell, laird of, 343
Monypenny, John, 284, 287 (Perth)
——, Mr Mungo, dean of Ross, 642
——, Thomas, burgess of Perth, tacksman of Forteviot parsonage, 317, 323, 331
Monzie (Monze), lady, 349
Moodie (Mudie), Mr William, 657 (Orkney bishopric, Caithness)
Moray, earls of, *see* Stewart
Morie, James, in Edinburgh, 133
——, Thomas, in Edinburgh, 133
Morison (Moresoun), Alexander, in Edinburgh, 133
——, David, chaplain of Inver, 336
——, John, servant to bishop of Dunblane, 295; 312, 313, 314, 315 (Comrie); 335 (Callander); 335 (Dunblane)
Morran (Moirane, Morane), John, 269, 274, 275 (Holywood abbey)
——, Robert, 270, 271, 276 (Holywood)
——, Thomas, 269 (Holywood abbey)
Morris (Moreis), George, 625 (Ross bishopric)
——, James, in Crail, 12
Morrison, *see* Morison
Morton (Marton), earls of, *see* Douglas
——, Adam, 158 (Airth)
——, John, prebendary of St Michael's altar, Crail, 65
——, Thomas, 114 (Leith hospital)
Mossman (Mosman), James, chaplain of Herdmanston, 173
Mot, George, 102 (Newbattle abbey)
Moultray (Multray), Mr James, vicar of Kirkcaldy, 67
Mount (Month), James, 18 (St Andrews priory)
Mow, Thomas, in Edinburgh, 137
Mowat, Charles, 583 (Dreghorn)
——, Hucheon, 283, 287 (Perth Charterhouse)
——, Patrick, 582 (Dreghorn)
Mowbray, Andrew, in Edinburgh, 127, 128, 135
——, Andrew, in Leith, 133

——, Mr James, pension to, 39 (Dunfermline abbey)
——, John, parson of Ecclesmachan, 151
——, Patrick, chaplain of St Cuthbert's altar, Dalmeny kirk; chaplain of St Thomas' altar, Cramond kirk, 149, 150
——, Robert, in Edinburgh, 135
Mown, *see* Mun
Mows, John, vicar of Kinross and Orwell, 88
Mudie, *see* Moodie
Muir (Mowre, Muire, Mur, Mure) of Kers, Alexander, 594
——, Adam, in Edinburgh, 125
——, George, 587 (Galloway bishopric)
——, James, 400 (Dundee)
——, Janet, 594 (Whithorn priory)
——, John, in Edinburgh, 125
——, John, in Glasgow, 500
——, John, in Lanark, 231
——, John, in Linlithgow, 11
——, John, 104 (Kirknewton)
——, John, 254 (Peebles)
——, Margaret, 516 (Glasgow)
——, Mr Patrick, parson of Fern, 371, 641
——, Thomas, 557 (Ayr)
Muirhead, David, 269 (Holywood abbey)
——, Robert, 271 (Holywood abbey)
——, Thomas, chaplain and chorister of Dunkeld, portionary of Abernyte, 310
——, William, 586 (Galloway bishopric)
——, William, 269, 271, 274, 275, 622 (Holywood abbey)
Muirton, Robert, 64 (Aberdour)
Mun (Mown), John, parson of Ashkirk, 212
Munfoid, Hew, 571 (Ayr)
Munro (Monro), George, 647 (Tain)
——, Hector, 647 (Tain)
——, Mr William, minister and vicar of Dingwall, 639, 640
Murdy (Murdie), John, in Edinburgh, 127
Murhycht, laird of, 594 (Whithorn priory)
Murray of Blackbarony, Andrew, 140
—— of Elibank, Gideon, treasurer depute, 677, 677n., 681, 683, 683n.

—— of Hangitschaw, Patrick, 496
—— of Kerse, collector for Perth, 286n.
—— of Ochtertyre, Patrick, 300
—— of Tullibardine, William, collector general, comptroller, 5n., 7, 7n., 76, 76n., 248, 248n., 281, 281n., 286, 286n., 402, 402n., 537, 537n., 573, 573n., 574, 574n., 575, 579, 606, 609, 609n., 610, 662, 662n., 663, 664
——, Alexander, 94 (Sciennes priory, Edinburgh)
——, Andrew, in Edinburgh, 128
——, Andrew, 102 (Newbattle abbey)
——, Charles, 383 (Brechin bishopric)
——, David, pension to, 386; 383, 384, 389 (Brechin bishopric)
——, David, 300 (Monzievaird),
——, Duncan, 592 (Whithorn priory)
——, Francis, 595 (Whithorn priory)
——, Henry, 291 (Culross abbey)
——, John, in Edinburgh, 122
——, Magnus, pension to, 659 (Orkney bishopric)
——, Patrick, vicar of Fowlis-Wester, 341
——, Patrick, in Perth, tacksman of Fortevoit parsonage, 317
——, Patrick, 158 (Airth)
——, Patrick, 303 (Tibbermore)
——, Peter, vicar pensionary of Tough, 425
——, Robert, 19 (St Andrews priory)
——, Mr William, 'archdeacon of St Andrews', treasurer of Dunblane, 65, 65n.
——, William, chaplain of St Roche's altar, St Giles' collegiate kirk, Edinburgh, 138, 139
Murthly (Murthlie), laird of, 341
Mustard, John, 305 (Ruthven)
——, William, prebendary, Dunglass collegiate kirk, 173
Mylne, *see* Miln
Myname, *see* Miniman
Myreton (Myretoun, Myrtoun), Mr David, prebendary of St Catherine's altar, Crail, 66
——, Mr Patrick, treasurer of Aberdeen cathedral, 444
——, Patrick, provost of Crail collegiate kirk, 82
——, William, 18 (St Andrews priory)

Nairn (Narne), George, 303 (Muckersie)
——, John, 270, 271, 275 (Holywood abbey)
——, Richard, 555 (Stirling)
Naismith, John, prebendary of Bute, Restalrig collegiate kirk, 143
——, John, vicar choral in Glasgow cathedral, parson and vicar of Dalziel, 519; chaplain of altar of Name of Jesus, Glasgow cathedral, 522
——, John, in Linlithgow, 11
——, John, 529 (Paisley)
——, Michael, in Edinburgh, 135
Napier, Andrew, chaplain of St Salvator's altar, St Giles' collegiate kirk, Edinburgh, 141
——, William, 270, 275, 276 (Holywood abbey)
Neilson, Andrew, 271 (Holywood abbey)
——, Comie, 274 (Holywood abbey)
——, Janet, 274 (Holywood abbey)
——, John, in Edinburgh, 122
——, John, 546 (St Ninians)
——, 'widow', 101 (Newbattle)
——, William, 274 (Holywood abbey)
Nesbit, *see* Nisbet
Newall, James, 270 (Holywood abbey)
——, Robert, 275, 276 (Holywood abbey)
——, Walter, 275 (Holywood abbey)
Newlands, John, in Edinburgh, 137
Newman, James, in Dundee, 12
Nicol, James, tenant in Duffus, 468
Nicolson, Cuthbert, in Haddington, 180
——, James, clerk to the Collectory, 57, 57n., 61, 71, 81, 82, 100, 109, 129, 133, 134, 135, 137, 140, 149, 150, 157, 157n., 179, 183n., 248, 248n., 254, 264, 264n., 300, 308, 328, 343, 347, 391, 391n., 402, 411, 411n., 420, 420n., 449, 450, 524, 524n., 535, 535n., 550, 571, 571n., 610, 649, 649n., 664, 664n.
——, John, vicar of Laggan, 486
——, Martin, in Edinburgh, 122
——, Patrick, 101 (Newbattle)
——, Thomas, prebendary of Middleton, Crichton collegiate kirk, 108
Niddrie (Nudry), laird of, 93
Nisbet (Nesbit), William, in Edinburgh,

128
Noklaus, Marion, 556 (Stirling)

Ogill, see Ogle
Ogilvie (Ogilby, Ogilvy) of Airlie, James, 4th lord, 367, 367n.
——, Mr Alexander, vicar of Tarves, 448, 448n.
——, Alexander, parson and vicar of Duthil, 494
——, Archibald, 308 (Lethendy); 352 (Coupar Angus abbey)
——, James, 376 (Idvie)
——, Marion, lady Gray, 376, 376n.
——, William, 489 (Dundurcas)
Ogilvy, see Ogilvie
Ogle (Ogill), Janet, 180 (Haddington priory)
——, William, 108 (Haddington priory)
——, William, 101 (Newbattle abbey)
Ogston, John, 494 (Moray archdeaconry)
——, Patrick, vicar of Peterugie, 432
——, Patrick, vicar of Strathbrock, 152
Oisler, see Ostlar
Oliphant, Laurence, 3rd lord, 112n., 383n.
——, Laurence, master of, 112, 112n., 303, 303n.
——, Mr Andrew, vicar portionary of Kirkcudbright-Innertig, 612
——, Henry, 58 (Balmerino abbey)
——, James, 180 (Haddington priory)
——, John, 283, 287 (Perth)
Orkney, John, 114 (Leith hospital)
Orm (Orame, Orme), David, 18, 19 (St Andrews priory)
Ormiston (Ormestoun), David, pension to, 243 (Yetholm)
——, Robert, 92, 93 (Holyroodhouse abbey)
——, Robert, 208 (Melrose abbey)
——, Robert, 243 (Yetholm)
——, William, pension to, 210, 260 (Melrose abbey)
Osborne (Osburne), Archibald, 557
Ostaige, see Eustace
Ostlar (Oisler, Oslar, Oslay, Ostan, Ostlare), Mr Gilbert, 381 (Dundee)
——, Gilbert, vicar of Sorbie, 603
——, Robert, provost of Dirlton collegiate kirk, 173

——, Robert, chaplain of St Catherine's altar, Forgandenny, 329
Otterburn, Elizabeth, in Edinburgh, 137
Oustane, see Austin
Owsteane, see Austin

Panter (Pantar), Andrew, priest in Elgin, 453
——, David, commendator of Cambuskenneth abbey, 543n.
Park of Bagra, Alexander, 516
——, Alexander, in Edinburgh, 127, 131
——, Elizabeth, in Edinburgh, 131
——, William, 227 (Greenlaw)
Pate (Payet), [blank], in Edinburgh, 135
Paterson (Patirsoun), Alexander, 149 (North Berwick priory)
——, James, sacristan of Chapel Royal, Stirling, 602
——, James, 254 (Peebles)
——, John, 254 (Peebles)
——, Laurence, 181 (Haddington priory)
——, Marion, 421 (Aberdeen friary)
——, Mr Robert, vicar pensionary of Moonzie, 83
——, Robert, 254 (Peebles)
——, Thomas, 254 (Peebles)
——, Mr William, subdean of Moray, parson of Dallas, vicar of Auldearn, 493
——, William, in Edinburgh, 137
Patillo (Pettillok), William, chaplain of St Duthac's chaplainry, Arbroath, 380
Paton (Patoun), James, 583 (Dreghorn)
Patrick, see Petrie
Paul, John, 274 (Holywood abbey)
Pawtoun, Alexander, 158 (Airth)
Paxton of that Ilk, William, 166, 173
Payet, see Pate
Peacock (Pecok, Porok), George, in Edinburgh, 128, 128n.
Pearson (Persoun), James, 349 (Dunblane)
——, John, 291 (Culross abbey)
——, Ralph, vicar of Kirkmaiden, 600, 601
——, Rudolph, 594 (Whithorn priory)
——, Mr William, parson of Sanday - Cross kirk, 660
Pedder (Peddir), Alexander, vicar of Avoch, 640; vicar of Urray, 640;

procurator, 639, 641, 643, 644, 646
Peebles, Bernard, vicar of Inchinnan
——, John de, in Edinburgh, 129
Pendreich, John, in Edinburgh, 135
Pennycuik, Gilbert, 198, 201 (Coldingham priory)
——, William, provost of St Mary of the Fields' collegiate kirk, Edinburgh, 106-107, 120, 120n.
Petcullane, laird of, 283, 286
Petrie (Patre), Mr Andrew, vicar pensionary of Arbuthnott, 406
Pettie, Alexander, 113 (Leith hospital)
Pettigrew (Petegrew), [blank], chaplain, 524 (Lanark)
——, Walter, 497 (Monkland)
Pettillok, see Patillo
Philip (Phillip), Henry, 32 (Abdie)
Philipston (Philpstoun), Andrew, chamberlain of Dunkeld, 375
Philorth, laird of, 424 (Aberdeen)
Philp, Mr John, abbot of Lindores, 29n.; vicar of Logie Durno, 448
Pillie, Marjorie, 180 (Haddington priory)
Piot, see Pyott
Piper (Pypar), John, in Perth, 11-12, 306
——, John, 375 (Menmuir)
——, William, 305 (Ruthven)
Pitcairn, David, vicar choral in Glasgow cathedral, parson and vicar of Dalziel, 519
——, James, archdeacon of Brechin, 380n., 394
——, Robert, archdeacon of St Andrews, 67, 67n., 380n.
——, Robert, commendator of Dunfermline abbey, 40n., 44, 45, 73, 73n., 159, 179, 179n., 183n.
——, Robert, parson of Strathbrock, 153
——, Robert, 380 (Brechin archdeaconry)
Pittarow, laird of, see, Wishart
Pollock, David, 156 (Linlithgow friary)
Polton, laird of, 142
Polwarth, laird, 200
Pont (Punt), Donald, 254 (Peebles)
——, Mr Robert, 290 (Culross abbey); provost of Trinity college, Edinburgh, 119, 119n.
Porok, see Peacock
Porteous (Portius, Portuous), James, in Edinburgh, 133

——, Oswald, 101 (Newbattle)
——, Robert, 254 (Peebles)
——, William, vicar pensionary of Kilbucho, 250
——, William, 101 (Newbattle)
Porter, Andrew, 275 (Holywood abbey)
——, Cubbie, 275 (Holywood abbey)
——, John, 291 (Culross abbey)
——, Nicol, in Edinburgh, 127
——, Thomas, 269, 274 (Holywood abbey)
Porterfield (Portarfeild), William, vicar of Ardrossan, 581
Potter (Pottar), John, 291 (Culross abbey)
Pratt (Prattis), Thomas, 419, 422 (Aberdeen friary)
——, Thomas, 92 (Holyroodhouse abbey)
Prest, see Priest
Preston of Fentonbarnes, John, 202, 202n., 222, 222n.
——, Archibald, 24, 47 (Dunfermline abbey)
——, Henry, 557 (Ayr)
——, James, 290 (Culross abbey); 294 (Valleyfield)
——, Mr John, in Edinburgh, 128, 135
——, 'Magister', 555 (Stirling)
——, Richard, 39, 40 (Dunfermline abbey)
——, Simon, in Edinburgh, 137
Priest (Prest), Francis, in Edinburgh, 133
Primrose (Prymros), Archibald, 677 (Orkney bishopric)
——, David, 292 (Culross abbey)
——, John, 100 (Newbattle)
——, William, 290 (Culross abbey)
Punton (Puntoun), Alison, in North Berwick priory, 148
Purves, Adam, in Edinburgh, 127
——, Janet, 104 (Kirknewton)
——, John, 105 (Kirknewton)
——, Thomas, in Edinburgh, 133
Pyott (Piot), Alexander, 300 (Dunkeld)
Pyrin, laird of, 254

Quentin (Quhentine), Christian, 180 (Haddington priory)
Quhyt, see White
Quhytheid, see Whitehead

INDEX OF PERSONS

Quhytlaw, *see* Whitelaw

Ra, *see* Ray
Rae, *see* Ray
Raith (Rathe, Rayth, Rycht), laird of, 41
——, Mr Thomas, provost of Seton collegiate kirk, 99-100
——, Thomas, vicar of Leslie, 443, 444
——, Thomas, 'sangster'; chaplain of St Leonard, Ayr, 560, 560n.
Ramsay of Jordanstone, David, 306
——, Alexander, parson of Foulden, 97, 193, 195
——, Alexander, vicar pensionary of Aberdour, 424
——, George, in Edinburgh, 136
——, Henry, in Edinburgh, 127
——, John, in Linlithgow, 157
——, John, 419, 421 (Aberdeen friary)
——, John, 182 (Crail)
——, John, 597 (Mochrum)
——, Matthew, 567 (Crossraguel abbey)
——, Robert, 61 (Balmerino abbey)
Ranalstoun, laird of, 573
Randfurd, laird of, 545
Randie, Alexander, 180 (Haddington priory)
Rankeilour, laird of, 78 (Creich)
Rankin, John, in Glasgow, 500
——, Margaret, 516 (Glasgow)
——, Robert, 571 (Ayr)
——, Robert, 516 (Glasgow)
Rannald, Lawrie, 158 (Airth)
Rannell, Richard, 158 (Airth)
Rannie (Rany), John, 384 (Brechin bishopric)
Rattray (Rettray), Mr John, vicar of Cathcart, 520
——, Mr John, vicar of Longforgan, 315, 316
——, John, 303 (Dunkeld); 306 (Kinclaven)
Ray (Ra, Raa), Mr John, vicar of Tranent, 144
——, John, 283, 286 (Perth)
——, William, feuar, 353 (Coupar Angus abbey)
——, William, in Edinburgh, 127, 138
——, William, 158 (Airth)
Reddein, Robert, in Linlithgow, 156
Redpath (Reidpeth), Alexander, 227 (Greenlaw)

——, Andrew, 195 (Ellem)
——, Triomour, 227 (Greenlaw)
Redwell (Ridwall), William, in Edinburgh, 135
Reid, Alexander, 102 (Newbattle abbey)
——, David, 557 (Ayr)
——, Donald, vicar of Logie-Easter, 636
——, Elizabeth, 583 (Dreghorn)
——, George, 61 (Balmerino abbey)
——, Helen, 128
——, James, 19 (Balmerino abbey)
——, John, vicar of Fetlar, 665
——, John, 101, 102 (Newbattle abbey)
——, Patrick, vicar of Bonhill, 541
——, Robert, 101 (Newbattle abbey)
——, Walter, abbot of Kinloss, 464, 464n., 650, 650n.
——, William, vicar of Reay, 648
Reidhall, lady *see* Stewart, Jean
Renton (Rentoun), Elizabeth, in North Berwick priory, 148
Rhind (Rynd), Alexander, in Edinburgh, 141
——, Henry, in Edinburgh, 129
——, James, in Edinburgh, 96, 128, 129
——, William, pension to, 406 (Arbuthnot)
Richardson (Richartsoun), David, 100 (Newbattle)
——, George, 100 (Newbattle)
——, Mr Robert, prior of St Mary's Isle, 617
——, Robert, treasurer, 286, 286n., 574, 574n., 575n.
——, Thomas, prebendary of St Catherine's altar, Kirkwall cathedral, 663, 664, 665
Richie, John, prebendary of St Mary of the Fields' collegiate kirk, Edinburgh; prebendary of St Colm's altar, St Giles' collegiate kirk, Edinburgh, 123, 129n.
Ridwall, *see* Redwell
Rigg, James, in Edinburgh, 137
——, John, 271, 276 (Holywood abbey)
Robb, James, tenant in Duffus, 468
——, James, in Elgin, 453
——, Thomas, chaplain, Dunblane cathedral, 343
——, Thomas, tenant in Duffus, 468
Robert III, King of Scots, 273n.
Robertland, lady, *see* Hamilton, Grissall
——, laird of, 580

Robertson, Andrew, vicar of Fodderty, chaplain of Inchrory and St Rule, diocese of Ross, 644
——, Andrew, pension to, 645 (Alness)
——, James, 554 (Stirling)
——, Mr John, prebendary of Newtown, Bothwell collegiate kirk, 510
——, John, parson portionary of Alness, 645
——, John, vicar of Arndilly, 457
——, John, 129n. (Edinburgh)
——, Robert, 177, 179 (Haddington priory)
——, Thomas, 384 (Brechin bishopric)
——, Walter, vicar of Aberdour, 85
Robeson (Robesoun, Robisoun), Adam, in Edinburgh, 136
——, George, in Edinburgh, 136
——, John, cutler, in Edinburgh, 138
——, John, flesher, 556 (Stirling)
——, John, provost of Roslin collegiate kirk, 117
——, John, 102 (Newbattle abbey)
——, Mr Richard, 132 (Edinburgh)
Robin (Roben), William, 254 (Peebles)
Robison, see Robeson
Robson (Robsoun) of Gleddiswod, John, 189 (Dryburgh)
Rolland, Mr James, vicar of Glamis, prebendary of Balquhidder, 330, 392, 392n.
——, Thomas, witness, 61 (Balmerino abbey)
Rollock of Duncrub, Andrew, 522
——, David, in Dundee, 12
Roreson, Robert, 270, 271, 275, 276 (Holywood abbey)
Roslin, laird of, 141
Ross of Balnagown, see also Balnagown, laird of
—— of Balnagown, Alexander, 651, 652
—— of Brychtmony, John, 469 (Moray deanery)
—— of Craigie, John, 74, 74n., 315, 315n.
——, Agnes, 651, 652 (Tain)
——, Alexander, 283, 287 (Perth)
——, Andrew, 652 (Edderton)
——, Andrew, 651, 652 (Tain)
——, George, canon of Chapel Royal, Stirling, 560 (Coylton)
——, Helen, in Edinburgh, 127
——, John, 652 (Edderton)

——, Nicholas, commendator of Fearn abbey, 625, 625n., 634, 651, 652; provost of Tain collegiate kirk, 640
——, Nicholas, 651 (Tain)
——, Mr Thomas, parson of Alness, 645; chaplain of Balnagown, 646
——, Thomas, 479 (Ardclach)
——, Thomas, 652 (Edderton)
——, Walter, 652 (Edderton)
——, William, 652 (Edderton)
Roth, Andrew, in Linlithgow, 11
——, John, 420 (Aberdeen friary)
Rothes, earls of, see Leslie
Row, Mr John, vicar of Kennoway, 83; pension to, 243 (Yetholm)
Roxburgh, Christine, 274, 275 (Holywood abbey)
——, Cubbie, 275 (Holywood abbey)
——, John, in Leiffnoris, 564 (Failford)
Runcieman (Runsimane), James, in Innerlochtie, 492 (Dallas)
Russell, Allan, 158 (Airth)
——, Christian, in Linlithgow, 11
——, James, in Edinburgh, 124
——, John, priest in Elgin, 453
——, Thomas, in Edinburgh, 127
Rutherford, John, 583 (Dreghorn)
——, Martin, vicar of Makerstoun and Robertson, 213
——, Nicol, vicar of Symington, 517n.
——, William, vicar of Symington, 517, 517n.
Ruthven, Patrick, 3rd lord, 325, 342, 346, 659n.; provost of Perth, 173, 173n., 302, 302n., 306, 306n., 353, 353n.
——, William, 2nd lord, 347n.
——, William, 4th lord, 353n., 659, 659n.
——, Alexander, brother to Patrick lord Ruthven, superior and patron of chaplainry of Forgandenny, 325, 329
——, Archibald, pension to, 671 (Orkney bishopric)
——, Lilias, 347, 347n., 348
——, Walter, 325 (Forgandenny)
Rycht, see Raith
Rynd, see Rhind

Saddler (Saidler), Luke, in Edinburgh, 136
Saidserff, see Sydserf

INDEX OF PERSONS

St Monance, laird of, 17, 92, 635
Salmond, Patrick, prebendary of Sacrista Major, New kirk [college of St Mary and St Anne], Glasgow, 523
———, Mr William, vicar of Cluny, 344
———, Mr William, pension to, 505 (Blantyre)
———, William, vicar of Dunottar, prebendary of Trinity college, Edinburgh, 404
———, William, 102 (Newbattle abbey)
Sanderson, James, chaplain and chorister of Dunkeld, portionary of Abernyte, 310
Sandfuird, laird of, 146, 148, 167
Sandie, Janet, 180 (Haddington priory)
Sandilands, James, preceptor of Torphichen, 110, 176, 176n., 450, 450n.
———, James, 37 (Lindores abbey)
———, James, 19 (St Andrews priory)
———, Mr John, parson of Cushnie, 444
———, Mr John, parson of Kiltearn, 635
———, John, parson of Hawick, 263
———, John, son of laird of St Monance, pension to, 635 (Kiltearn)
Sands, Robert, 290 (Culross abbey)
———, Thomas, 289 (Culross abbey)
———, Walter, 292 (Culross abbey)
Schartcrow, George, 587 (Galloway bishopric)
Schaw, see Shaw
Schen, see Cheyne
Schennan, laird of, 607
Schennand, Patrick, 619, 620 (St Mary's Isle priory)
———, William, 619 (St Mary's Isle priory)
Scherar, see Shearer
Schethin, laird of, 435
Schevane, see Shewan
Scollay (Scola, Scolay), David, 664 (Sanday)
———, Duncan, 655 (Orkney bishopric); 664 (Sanday)
Scott (Scot) of Buccleuch, Walter, 260n.
——— of Didschaw, Walter, 212
——— of Syntoun, Walter, 212
———, Adam, 269, 271 (Holywood)
———, David, chaplain of St John the Evangelist's chaplainry, St Giles' collegiate kirk, Edinburgh, 138
———, Mr George, 78 (Rossie)
———, George, 271 (Holywood abbey)
———, Mr James, provost of Corstorphine collegiate kirk, 100, 100n.
———, James, 78 (Rossie)
———, John, chaplain of altars of St Ninian and St Eloy, St Giles' collegiate kirk, Edinburgh, 134
———, John, vicar of Carstairs, 509
———, John, pension to, 370 (Coupar Angus abbey)
———, John, 269 (Holywood abbey)
———, Marion, in Edinburgh, 128
———, Nicol, in Canongate, Edinburgh, 132, 137
———, Robert, writer, 229, 233 (Kelso abbey)
———, Robert, 274 (Holywood abbey)
———, Mr Thomas, vicar of Collessie, 81
———, Mr Thomas, vicar of Cramond, 126
———, Thomas, 269, 271, 274 (Holywood abbey)
———, Thomas, 104 (Kirknewton)
———, Thomas, 283, 287 (Perth)
———, Mr William, 70 (Strathmiglo)
———, William, vicar of Auchtermuchty, 76, 85
———, William, 269, 271, 275 (Holywood abbey)
Scougal (Skowgall), David, 100 (Newbattle)
Scrimgeour, David, 400 (Dundee)
———, James, 300 (Dunkeld)
———, Janet, 303 (Dunkeld)
———, Mr John, in Edinburgh, 125
Sellar, John, in Edinburgh, 136
Semple (Sympill), Robert, 3rd lord, 163, 163n.
——— of Beltries, John, provost of Semple collegiate kirk, 531 (Glassford)
———, Isabel, lady, 573, 573n.
———, John, parson of Muckhart, 338
Seton (Seytoun), George, 5th lord, 3, 3n., 4, 181, 181n., 210, 210n.
——— of Northbeg, Alexander, 344
——— of Parbroath, David, comptroller, 21, 21n.
———, Mr Alexander, chancellor of Aberdeen, 416, 416n., 428, 432, 433
———, Helen, 188 (Channelkirk)
———, Mr James, parson of Whitsome, 195, 196

——, Mr John, vicar of Creich, 78
——, John, pension to, 228 (Kelso abbey)
Shand, Thomas, 421 (Aberdeen friary)
Sharp (Scharp), Alexander, in Edinburgh, 137
——, Marion, 155 (Dalmeny)
Shaw (Schaw) of Greenock, John, 533
——, Helen, in North Berwick priory, 148
——, James, pension to, 210, 260 (Melrose abbey)
——, *dene* John, 190 (Dryburgh abbey)
——, John, 496 (Drymen)
——, John, 556 (Stirling)
——, Mr William, tacksman, 533 (Inverkip)
——, Mr William, provost of Abernethy collegiate kirk, 244, 322, 322n.
——, Mr William, pension to, 228 (Kelso abbey)
——, William, 587, 607 (Galloway bishopric)
Shearer, Donald, pension to, 635 (Kiltearn)
Shewan, David, 402 (Brechin)
Shoreswood (Schoriswod, Thoriswod), Mr Richard, 66 (St Andrews archdeaconry)
——, Mr, 226 (Kelso abbey)
Sibbald, David, 78 (Creich)
——, Henry, 92 (Holyroodhouse abbey)
Side, *see* Syde
Sim (Sym, Syme), Adam, 122, 123 (Dunbar)
——, John, in Edinburgh, 123
——, Thomas, in Leith, 122
——, Walter, 283, 287 (Perth)
Simmers (Simmer, Symmer), of Balzeordie, George, 374, 375
——, Adam, in Edinburgh, 124
——, George, son of George Simmers of Balzeordie, 374
——, James, 82 (Garvock)
——, John, vicar portionary of Abdie, chaplain of St Catherine's chaplainry, Newburgh, 378
——, William, vicar portionary of Abdie, 68
Simson (Symsoun), Andrew, vicar of Bolton, 173
——, Andrew, in Edinburgh, 126
——, Bartholomew, vicar choral in Glasgow cathedral, parson and vicar of Dalziel, 519
——, Laurence, in Edinburgh, 130
——, Thomas, 158 (Airth)
——, William, 111, 123 (Edinburgh)
——, William, 114 (Leith hospital)
Sinclair (St Clair, Sincklar), George, 4th earl of Caithness, 421, 421n., 628, 628n., 638, 638n.
——, Agnes, countess of Bothwell, 7
—— of Dun, David, 633
—— of Dunbeath, William, 657, 670
—— of Whitekirk, Oliver, 655
——, Alexander, chaplain of St Catherine's chaplainry, Elgin cathedral, 490
——, Alexander, 418, 421 (Caithness; Aberdeen friary)
——, Edmond, 679 (Orkney bishopric)
——, Edward, 655 (Orkney bishopric)
——, George, in Linlithgow, 11
——, Henry, dean of Glasgow, 499, 499n.; bishop of Ross, 528, 528n., 627, 651; pension to, 512 (Kilwinning abbey); usufructary of Glasgow parsonage, 512, 512n.
——, Henry, 92 (Holyroodhouse abbey)
——, Mr John, dean of Restalrig, parson of Auchindoir, 116, 116n., 141, 443, 443n.
——, Mr John, vicar of Comrie, 334
——, John, in Linlithgow, 11
——, Margaret, in North Berwick priory, 148
——, Oliver, 91 (Holyroodhouse abbey)
——, Oliver, 661, 668 (Orkney bishopric)
——, Oliver, 169 (Whitekirk)
——, Robert, chaplain, Dunblane cathedral, 343
——, Robert, prebendary of Methven collegiate kirk, 298, 299
——, Robert, 659 (Orkney bishopric)
——, Stephen, vicar of Aberfoyle, 344
——, Thomas, 379 (Edzell)
——, Thomas, 173 (Hermanston)
——, Mr William, parson of Olrig, 148
——, Mr William, vicar of Latheron, 648
——, William, in Dunblane, 299
——, William, in Lasswade, 141
——, William, 548 (Inchmahome priory)
Skaithmuir (Skarthmure), Robert, in

INDEX OF PERSONS 747

Edinburgh, 127
Skene, Mr Alexander, 394 (Guthrie), 405, 407, 407n.; 438 (Aberdeen); 439 (Belhelvie); 484 (Botary); 444 (Cushnie); 445 (Methlick)
Skirling, laird of, 150, 180, 250
Slater (Skleater), John, 529 (Paisley)
Sloan (Slowane), John, in Edinburgh, 122
Smailhome Craigs, laird of, 189
Small (Smaill), James, 249 (Kettins)
——, John, 158 (Airth)
——, Patoun, 582 (Dreghorn)
——, William, 254 (Peebles)
Smart, [blank], 555 (Stirling)
Smetoun, laird of, 147
Smith (Smyth), Alexander, 101 (Newbattle)
——, Andrew, 305 (Ruthven)
——, Gibbe, 274 (Holywood abbey)
——, Henry, collector for Galloway, 600
——, James, 158 (Airth)
——, James, 384 (Brechin bishopric)
——, Janet, 557 (Ayr)
——, John, in Straiton, 567 (Crossraguel abbey)
——, John, 305 (Ruthven)
——, Michael, 254 (Peebles)
——, Patrick, 305 (Ruthven), 305
——, Robert, vicar of Dunino, 90
——, Robert, 157 (Airth)
——, Robert, 105 (Kirknewton)
——, Thomas, 158 (Airth)
——, Thomas, 269, 274 (Holywood abbey)
——, 'widow', 101 (Newbattle)
——, William, servant to provost of Trinity college, Edinburgh, 118
——, William, in Elgin, 453
——, William, 455 (Banff and Inverboyndie)
Snell, Alexander, 283, 287 (Perth)
Somerville (Sumervell), Hew, 4th lord, 614, 614n.
——, James, 5th lord, 190, 191
——, Alexander, pension to, 4 (St Andrews archbishopric)
——, Hew, 505 (Carnwath)
——, Thomas, parson and vicar of Linton, 212, 215
——, Mr William, vicar of Kirkbean, 609, 614, 614n.
Sommers (Somers), William, 114 (Leith hospital)
Sorbie, laird of, 594
Sornbeg, laird, 562
Spankis, Thomas, 516 (Glasgow)
Spart, John, 620 (St Mary's Isle priory)
Speir, Henry, 582, 583 (Dreghorn)
——, John, 583 (Dreghorn)
Spens of Condie, Mr John, king's advocate, 40, 92, 93, 188, 286, 286n., 574, 574n., 575n., 610, 610n.
——, [blank], 479 (Moray chantory)
——, Adam, in Edinburgh, 127
——, Balthasar, vicar pensionary of Kinnoull, 331
——, Balthasar (Hattelar), 555, 555n. (Stirling)
——, Mr David, archdeacon of Dunkeld, 308, 309
——, James, chaplain of Arndilly (Artillich), Moray, 482, 482n.; chaplain of Clyne, Moray, 481; chaplain and vicar choral, Elgin cathedral, 481; vicar of Dallas, 481
——, James, 235, 238, 241 (Greenlaw)
——, Jarum, 453
——, John, in Edinburgh, 96, 141
——, John, 227 (Greenlaw)
——, John, 168 (North Berwick)
——, John, 18 (St Andrews priory)
——, Mr Peter, in Edinburgh, 139
——, Thomas, in Linlithgow, 11
Spittal (Spittall), Mr Henry, in Edinburgh, 94
——, Nicol, parson of Lundie, 376
Spott (Spot), Catherine, 595, 596 (Whithorn priory)
Spottiswoode of that Ilk, David, 196
——, John, parson of Calder [Mid-Calder], 99
——, John, in Edinburgh, 127
Spreull, Matthew, 114 (Hailes)
——, Patrick, 111 (Edinburgh)
Sprott (Sprot), John, in Edinburgh, 128, 157
Stables (Stibilis, Stibbillis), Andrew, 368, 371 (Coupar Angus abbey)
Stanehouse, lady, 137, 290
——, laird of, see Hamilton
Stanes, see Stones
Stark, John, 516 (Glasgow)
Steel (Steill), George, 63 (Auchtertool and Dollar)

—, John, 62, 63 (Auchtertool and Dollar)
—, Thomas, 101 (Newbattle)
Stein (Stene), Alexander, 516 (Glasgow)
Steinson, *see* Stevenson
Steven (Stevin), David, tenant in Duffus, 468
—, Duncan, vicar choral in Glasgow cathedral, parson and vicar of Dalziel, 519
—, John, 101 (Newbattle)
Stevenson, (Steinsoun, Stevinsoun), Mr John, parson and vicar of Muckersie, chanter of Glasgow cathedral, 277-278, 326, 340, 499, 499n., 595, 597, 597n.; 603, 603nn.; parson and vicar of Kilbride, 499; vicar of Mochrum, 597
—, Mr John, vicar of Galston, 563, 564
—, John, 254 (Peebles)
—, Malcolm, vicar pensionary of Luss, 542
—, Thomas, 101 (Newbattle)
—, William, 254 (Peebles)
Stewart (Stuart), Henry, lord Darnley, King of Scots, 281, 282
—, Ludovic, 2nd duke of Lennox, 20, 20n.
—, Francis, earl of Bothwell, commendator of Coldingham priory, 198, 198n.
—, James, earl of Moray, Regent, commendator of St Andrews priory, 9, 9n., 57, 57n., 223n., 224, 224n., 454, 454n., 609, 610
—, John, 4th earl of Atholl, 7, 7n., 284, 284n., 300, 303, 304, 318, 318n.
—, Matthew, 4th earl of Lennox, 229, 229n.
—, Patrick, 2nd earl of Orkney, 677, 677n., 679, 680, 681, 682
—, Robert, 1st earl of Orkney, 677n., 679n.
—, John, 4th lord Innermeath, 430, 430n.
— of Doune, James, commendator of Inchcolm abbey, 63, 63n., 64, 64n., 85, 300, 300n.
— of Garlies, *see also* Garlies, laird of
— of Garlies, Alexander, 601, 602
— of Minto, John, 20n.

— of Rosyth, Robert, 62, 62n.
—, Adam, 291 (Culross abbey)
—, Alexander, brother of laird of Grandtully, 17 (Dull)
—, Andrew, 290 (Culross abbey)
—, Bernard, prior of the Friars Preachers (Dominicans) in Edinburgh, 128
—, David, parson of Erskine, 534
—, Francis, commmendator of Kelso abbey, 223, 223n., 228, 228n.
—, James, chaplain of Holy Rood altar, Glasgow cathedral, 500; vicar choral in Glasgow cathedral, parson and vicar of Dalziel, 519
—, James, commendator of Kelso and Melrose abbeys, 208, 208n, 243, 243n.
—, James, subdean of Glasgow, vicar in the choir of Glasgow, 498
—, James, 680 (Orkney bishopric)
—, Janet, lady Methven, 172n.
—, Jean, lady Reidhall, 117n., 118, 118n., 179
—, Mr John, parson and vicar of Rayne, archdeacon of Aberdeen, 432, 432n.
—, Mr John, 443 (Auchindoir)
—, Mr John, pension from Duddingston kirk, 230
—, Mr John, 587 (Galloway bishopric)
—, Mr John, 92 (Holyroodhouse abbey)
—, Mr John, 546 (Stirling)
—, John, commendator of Coldingham, 198n., 223n.
—, John, son of James Stewart of Fincastle, parson and vicar of Weem, 318
—, John, son of laird of Garlies, 618 (Kirkmahoe)
—, John, 303 (Dunkeld)
—, John, 92, 93 (Holyroodhouse abbey)
—, Margaret, lady Galloway, 273, 273n.
—, Mr Robert, 594 (Whithorn priory)
—, Robert, bishop of Caithness, provost of Dumbarton collegiate kirk, 539, 539n., 574, 574n.
—, Robert, chanter of Dornoch cathedral, Caithness, 630

INDEX OF PERSONS 749

——, Robert, commendator of Holyroodhouse abbey, 92n., 93, 93n., 591, 591n., 602, 602n.
——, Mr Thomas, 413, 415 (Aberdeen bishopric)
——, Thomas, 417 (Aberdeen bishopric)
——, Walter, commendator of Blantyre priory, 20, 20n., 21, 272, 272n.
——, William, chaplain and chorister of Dunkeld cathedral, portioner of Abernyte, 310
——, William, chaplain of our Lady altar, St Nicholas' kirk, Lanark, 524
Stibilis, see Stables
Stirling (Stirviling, Striviling) of Ardoch, William, 314
——, Euphemia, in Edinburgh, 140
——, William, 619 (St Mary's Isle priory)
Stoddart, John, burgess of Edinburgh, 125
——, John, in Edinburgh, 138
Stones (Stanis), James, 77 (Ballingry)
——, John, vicar and minister of Birnie, 487
——, John of, 114 (Leith hospital)
Storie (Sturry), Alexander, 149 (North Berwick priory)
——, James, in Edinburgh, 136
——, Robert, 184 (Eccles priory)
Strachan (Strathquhene, Strauthauchin), Andrew, 415 (Aberdeen bishopric)
——, George, vicar of Kinnettles, 390
——, George, vicar pensionary of Dysart, 71, 188
——, Mr James, parson and vicar of Belhelvie, 442
——, Mr James, parson of Botary and Elchies, 489
——, Mr James, parson of Fettercairn, 407, 407n.
——, Mr James, provost of Guthrie collegiate kirk, 390
——, Mr William, vicar pensionary of Belhelvie, 443
Strang, Mr George, vicar of Cawdor (Braaven), 485
——, Mr George, vicar of Crawford, 92, 512
——, Mr George, 58 (Balmerino abbey)
——, Magnus ('Maggus'), parson of Stromness, 659
Strathendry (Straherny), Henry, prebendary of Methven collegiate kirk, 299, 299n.
Strathmartine, laird of, 19
Strathon, James, 306 (Perth)
Straton, Andrew, 419 (Aberdeen friary)
——, Arthur, 19 (Ecclesgreig)
Struthers, William, prebendary of Netherfield, Bothwell collegiate kirk, 517
Sutherland of Duffus, Alexander, 475, 476
——, John, tenant in Duffus, 468
——, William, parson of Moy, 470, 471
Swine, George, 82 (Methil)
Swinton (Swenton), Cuthbert, chaplain of Ross, Restalrig collegiate kirk, 143
——, Mr David, parson of Cranshaws, 196
——, Mr George, parson of Nevay, 367, 368
——, John, 26, 48 (Inverkeithing)
——, John, 283, 287 (Perth)
——, Margaret, 283, 286, (Perth)
Syde, James, 284, 287 (Perth)
Sydserf (Saidserff), John, vicar of Kiltearn, 646
Sym, see Sim
Symington (Symontoun) of Hardington, William, 517
Symple, see Semple

Tacket (Takett), John, 524 (Glasgow)
Tait, Hector, 180 (Haddington priory)
——, John, 414 (Morebattle)
——, John, 100 (Newbattle)
——, Lucas, 180 (Haddington priory)
——, William, 181 (Haddington priory)
Takett, see Tacket
Tannahill (Tonnohill), John, in Edinburgh, 128
Taylor (Tailyour), George, 353 (Coupar Angus abbey)
——, John, chaplain, Lincluden collegiate kirk, 621
——, John, parson of Cummertrees, 265, 266
——, John, vicar of Penpont, prebendary of Lady Galloway prebend, Lincluden collegiate kirk, 273
——, John, vicar pensionary of Colvend, 272

——, John, in Elgin, 453
——, Thomas, 609 (Kirkbean)
Telfer (Taillefere, Tailyefeir, Tallifere),
    Mr Arthur, vicar of Aboyne, 431,
    440; parson of Crimond, 431, 440;
    parson and vicar of Inchture, 317,
    339
——, George, in Edinburgh, 136
——, James, 499 (Dalziel)
——, James, 595, 596 (Whithorn priory)
——, Thomas, 619 (St Mary's Isle
    priory), 619
——, William, prebendary of
    Stonehouse, Bothwell collegiate
    kirk, 511, 512, 512n.
——, William, vicar of Cruggleton, 601
Tenant (Tenand, Tennent), Francis, 141,
    143 (Lasswade)
——, Patrick, in Edinburgh, 127
——, Stephen, 571 (Ayr)
Teviotdale, John, in Canongate, 136
Thomas, Alexander, 180 (Haddington
    priory)
——, Andrew, vicar of Irvine, 582
Thomson, Adam, 180 (Haddington
    priory)
——, Alexander, tenant in Duffus, 468
——, Bernard, 180 (Haddington priory)
——, Edward, in Edinburgh, 135
——, Hector, in Edinburgh, 136
——, James, in Edinburgh, 122
——, John, tenant in Duffus, 468
——, John, in Cramond, 150
——, John, in Edinburgh, 124
——, John, elder, in Linlithgow, 156
——, John, 158 (Airth)
——, John, 101 (Newbattle)
——, John, 546 (St Ninians)
——, Peter, Islay herald, 612, 616
——, Philip, 104 (Kirknewton)
——, Thomas, 229 (Duddingston)
——, Richard, tenant in Duffus, 468
——, William, in Edinburgh, 140
Thorburn (Thurbrand), James, 254
    (Peebles)
——, Robert, 100 (Newbattle)
Thoriswod, *see* Shoreswood
Thornton, laird of, 18
—— of Benholm, Mr John, 403
——, Gilbert, pension to, 159
    (Forteviot)
——, Mr Henry, 323, 331 (Kinnoull);
    pensions to, 159 (Aberchirder,
    Ancrum, subdeanery of Ross)
——, Mr James, dean of Brechin, 364,
    364n.; 317 (Forteviot); pension to,
    394 (Dundee)
——, Mr James, prebendary of Fungarth,
    Dunkeld cathedral, 327, 327n., 329,
    365, 365n.
——, Mr James, subdean of Ross, 652
——, Mr James, titular of parsonage of
    Ancrum, 236, 365, 365n., 366
——, Mr James, pension to, 659, 671
    (Orkney bishopric)
——, Mr James, 477, (Elgin cathedral);
    476 (Inverkeithny); 479, 480 (Moray
    chantory)
——, James, 481 (Aberchirder); 480
    (Advie and Cromdale); 482
    (Artillich); 478, 481 (Elgin
    cathedral)
——, Mr John, elder, parson of Ancrum,
    parson of Forteviot, subdean of
    Ross, 159; chanter of Moray, 479;
    usufructary of Ancrum, 236, 366
——, Mr John, usufructary of Forteviot
    parsonage, 317; pension to, 370
    (Coupar Angus abbey)
——, Mr John, younger, vicar of
    Aberchirder, 159, 480; parson of
    Advie and Cromdale, 480
——, John, in Edinburgh, 126
Thounis, Mr Nicol, notary public, 550n.
Thurbrand, *see* Thorburn
Tod, William, parson and vicar of
    Bedrule, 211
——, William, in Edinburgh, 141
Toddoch (Toddog), Thomas, in
    Linlithgow, 156, 156n.
Todrick, George, in Edinburgh, 127
Torrance, John, 101 (Newbattle)
Torwoodheid, laird of, 546
Touch, laird of, 544, 547, 548, 549
Town, James, in Edinburgh, 126
Trail, John, 145 (Largo); 148 (North
    Berwick)
Tranent, James, 238, 240 (Fogo)
——, John, 235, 238, 241 (Fogo)
Trent, John, 100 (Newbattle)
Tripp, William, 283, 286 (Perth)
Trollope, Richard, 226 (Kelso abbey)
Trotter, Jane, 235n. (Fogo)
——, John, 235, 235n., 238, 241 (Fogo)
——, Thomas, chamberlain of
    Restenneth priory, 218

Tullegowne, laird of, 435
Tulloch (Tullo, Tullocht), Mr Jerome, subchanter of Orkney, parson of Sanday - St Colme, 664
——, Michael, in Edinburgh, 127
——, Nicol, 664 (Sanday)
——, Nicholas, vicar of Ruthven, 493
——, Thomas (Moray archdeaconry), 487
Turing, Thomas, 101 (Newbattle)
Turnbull (Trumbill, Turnbill) of Barnahillis, George, 236, 365-366
—— of Bogmillin, Thomas, 369 (Coupar Angus abbey)
——, Gilbert, 557 (Ayr)
——, Thomas, 236 (Ancrum)
Turner (Tornour), Thomas, burgess of Edinburgh, 139
Tweedie (Twedie, Twedy), Alexander, in Edinburgh, 132
——, James, 253 (Peebles)
——, John, chaplain of chaplainry of Our Lady, Peebles, 254
——, Thomas, 253 (Peebles)

Umphray, *see* Humphrey
Ungiltoun, laird of, *see* Congalton
Ure, William, goldsmith, in Edinburgh, 122
Urie, John, chaplain of the altar of St Mirin and St Colm, Paisley parish kirk, 535
Urquhart, Alexander, burgess of Forres, 462 (Kinloss abbey)
Urr (Ur), Alexander, 92 (Holyroodhouse abbey)
Urwing, *see* Irving

Vallantyne, Thomas, 303 (Dunkeld)
Vaus, Patrick, parson of Wigtown, 597
——, Robert, in Edinburgh, 132
Verner (Vernour), John, in Edinburgh, 128
Villemore, Bartholomew de, comptroller, 73n., 94n., 152n., 290, 291, 351, 351n., 423, 423n., 465n., 651, 651n.
Vocat, Mr David, 106 (Livingston)

Waddie (Wody), Neil, 113 (Leith hospital)
Wade (Waid), David, 290 (Culross abbey)
Walker (Walcar) of Inchcailloch, James, parson and vicar of Inchcailloch, vicar of Stevenston, 540
——, Adam, 180 (Haddington priory)
——, David, 19 (St Andrews priory)
——, John, 557 (Ayr)
——, Malie, 421 (Aberdeen friary)
——, Mr Michael, archdeacon of Moray, 494
——, Michael [prebendary of Moneydie, Dunkeld cathedral], 336
Wallace of Barnwell, William, 562, 564
—— of Elderslie, William, 564, 569 (Symington)
—— of Johnston, William, 534
——, John, reader at Torthorwald, 563
——, John, 571 (Ayr)
——, Margaret, 129n. (Kilmaurs)
——, Robert, 208 (Melrose abbey)
——, William, 158 (Airth)
Wallane, John, 156 (Linlithgow friary)
Walson, James, 400 (Dundee)
Walterson, Alexander, 652 (Edderton)
Wardlaw, Mr Alexander, parson and vicar of Ballingry, 76, 86
——, Henry, 677 (Orkney bishopric)
——, Mr John, parson of Moffat, 266
——, Mr John, vicar of Garvock, chaplain of St Andrew's chaplainry, Garvock, 82
——, John, in Leith, 86, 150
Waterson, Mr William, notary, 164 (Haddington priory)
Watson of Cromar, [*blank*], 421 (Aberdeen friary)
——, [*blank*], in Elgin, 453
——, Archibald, in Edinburgh, 124
——, David, 271 (Holywood abbey)
——, James, merchant, 306 (Perth)
——, James, in Edinburgh, 132
——, James, 546 (Stirling)
——, Mr John, 416, 417 (Aberdeen bishopric)
——, John, in Edinburgh, 127, 136, 140
——, John, 275 (Holywood abbey)
——, Marion, 276 (Holywood abbey)
——, Robert, chaplain of Holy Rood chaplainry, Dumbarton kirk, 540
——, Robert, prebendary of Two of the Three Children, Glasgow cathedral,

525
—, Robert, vicar of Clayshant, 605
—, Robert, 275, 276 (Holywood abbey)
—, Thomas, tenant in Duffus, 468
—, Thomas, in Edinburgh, 137
—, Walter, 551 (Stirling)
Watt (Wat), David, vicar of Brechin, 378, 401, 402
—, David, witness, 61 (Balmerino abbey)
—, John, witness, 479 (Ardclach)
—, Patie, 158 (Airth)
—, Thomas, in Linlithgow, 156
Wauch, see Waugh
Wauchope (Wachope) of Mudy, William, younger, 249 (Peebles friary)
Wauchton, laird of, 249
Waugh (Wache, Wauch), Alexander, witness, 479 (Ardclach)
—, John, in Edinburgh, 114, 127
—, William, 252 (Lyne)
—, William, 275 (Holywood abbey)
—, William, 487 (Moray archdeaconry)
Weatherspoon (Wetherspune, Wooderspune, Wydderspyne), of Brighouses, James and Elizabeth, 439 (Kincardine O'Neil)
—, Archibald, vicar of Carriden, 156, 157
—, Robert, 11 (Linlithgow)
Weddell, John, 102 (Newbattle abbey)
Wedderburn, laird of, 195
Weir of Clenedyk, John, pension to, 229 (Kelso abbey)
—, James, parson of Baldernock, vicar of Lanark and Wiston, 508
—, John, chamberlain, Lesmahagow priory, 243
—, John, 155 (Dalmeny)
Welsh (Welsche), David, 271 (Holywood abbey)
—, David, pension to, 624 (Holywood abbey)
—, John, 'superior and minister of Holywood abbey', 272, 277
—, John, 269, 271, 274, 275 (Holywood abbey)
—, Margaret, 270, 271 (Holywood abbey)
—, Thomas, 271, 275, 276 (Holywood abbey)
Wemes, laird of, 463 (Kinloss abbey)
Wemyss, Mr John, 26, 48 (Kinghorn Wester)
West, George, 92 (Holyroodhouse abbey)
—, John, 100 (Newbattle)
West Barns, laird of, 177, 178, 179
West Nisbet, laird of, 200
Weston (Waston), Mr John, in Edinburgh, 126
—, Mr Thomas, chaplain of St Duthac's altar, St Giles' collegiate kirk, Edinburgh, 124
—, Thomas, 517 (Carstairs)
Westwood, William, in Dunfermline, 291
White (Quhyt), David, 114 (Leith hospital)
—, Mr Henry, 64 (St Mary's college, St Andrews)
Whitehead (Quhytheid), [blank], 620 (St Mary's Isle priory)
—, David, 113 (Leith hospital)
—, William, 619, 620 (St Mary's Isle priory)
Whitelaw (Quhytlaw), Alexander, chamberlain to comptroller, 286
—, Janet, 100 (Newbattle)
Wight (Wicht), Walter, in Edinburgh, 128
Wightman (Wychtman), Andrew, 254 (Peebles)
—, John, burgess of Peebles, 252, 253
Wightmuir (Wychtmuir), James, chaplain of Our Lady altar, Dalmeny, 154, 154n., 155
Wilgrief, Andrew, 652 (Edderton)
Wilkie, Mr James, prebendary of Half Dalmahoy and Half Hatton, Corstorphine collegiate kirk, 107, 121
—, Mr James, vicar of Ecclesgreig, regent in St Leonard's college, St Andrews, 395
—, James, 102 (Newbattle abbey)
—, James, 18 (St Andrews priory)
—, William, chaplain of chaplainries of St Michael and St Mungo, Glasgow cathedral, 501
William, John, 158 (Airth)
Willmann, [blank], in Elgin, 453
Wilson (Wolsoun), [blank] in

Edinburgh, 133
—, Adam, 180 (Haddington priory)
—, Andrew, 290 (Crombie)
—, Archibald, in Linlithgow, 156
—, Catherine, in Glasgow, 500
—, Elizabeth, 155 (Linlithgow friary)
—, Francis, chaplain in Edinburgh, 120, 120n.
—, Francis, prebendary of Spittal, Dunglass collegiate kirk, 166, 173
—, James, 523 (Glasgow)
—, John, vicar of Eckford, 212, 215
—, John, vicar of Kinghorn Easter, 80, 80n.
—, John, 384 (Brechin bishopric)
—, John, 516 (Glasgow)
—, John, 92 (Holyroodhouse abbey)
—, John, 619 (St Mary's Isle priory)
—, Martin, 180 (Haddington priory)
—, Patrick, in Edinburgh, 133
—, Patrick, 609 (Kirkbean)
—, Stephen, treasurer of Dunkeld cathedral, 311, 311n.
—, Stephen, in Edinburgh, 124
—, Thomas, 101 (Newbattle abbey)
—, William, 526 (Govan)
Winchester, Alexander, in Elgin, 453
Winram (Wynram, Wynrame), Arthur, 184 (Eccles priory)
—, Mr John, prior of Portmoak, Lochleven; superintendent of Fife and Strathearn, 45, 45n., 56-57
—, Mr John, 87 (Dull)
—, Mr Robert, vicar of Dairsie, 79
—, Mr Robert, 12, 18 (Linlithgow; Binning); 100 (Ratho)
—, Robert, sub-collector, Fife, 44n.
—, Robert, 283, 287 (Perth)
Winyet, John, in Glasgow, 500
Wiseman (Wismoun, Wysman), Andrew, cordiner, 421 (Aberdeen friary)
—, William, vicar of Dundurcas, 489
Wishart of Pittarow, John, comptroller, 5n., 61, 61n., 71, 73, 73n., 75, 88, 94, 94n., 129, 152, 152n., 264, 264n., 347, 351, 351n., 376, 378, 391, 391n., 406, 407, 411, 423, 423n., 464, 464n., 465, 465n., 467, 543, 543n., 564, 564n., 575n., 579, 579n., 651, 651n.
—, John, chaplain, 564 (Symington)
—, John, in Edinburgh, 129

—, Nicol, witness, 479 (Ardclach)
Woddell, see Weddell
Wody, see Waddie
Wolsoun, see Wilson
Wood (Wod) of Gillistoun, Mr John, 540
— of Grange, Mr Alexander, 146,
— of Largo, Andrew, comptroller, 53, 167
— of Newmylne, William, 192
—, [blank], 146 (Largo)
—, Mr Alexander, vicar of Largo, 78
—, Mr Alexander, 167 (Kilconquhar); 168 (North Berwick)
—, Andrew, tenant in Duffus, 468
—, Andrew, 18 (St Andrews priory)
—, George, chaplain of the Holy Rood altarage, Brechin cathedral, 400
—, James, son of Andrew Wood of Largo, 167 (Largo)
—, James, 145 (Largo)
—, Mr John, 87 (Dull); 393 (Dun); 406 (Glenbervie)
—, Mr John, 597 (Wigtown)
—, Marion, in North Berwick priory, 148
—, Richard, 609 (Kirkbean)
—, Thomas, 122 (Dunbar)
Woods (Woodis), James, minister in Wigtown, 597
Wright (Wrycht), David, 290 (Culross abbey)
—, James, 100 (Newbattle)
—, John, chaplain of St Michael's chaplainry, Dunblane cathedral, 335; vicar of Callander, 335
—, John, in Edinburgh, 126
—, Matthew, 419, 421 (Aberdeen friary)
Wrs, William, 353, 353n. (Coupar Angus abbey)
Wylie, William, 254 (Peebles)

Yair, Thomas, 157 (Airth)
Yellowley, Duncan, vicar of Crawford, 71n.
Yeoman (Yeman, Zeyme), James, 19 (St Andrews priory)
—, Thomas, 181 (Haddington priory)
Young, Alexander, prebendary of Methven collegiate kirk, 299
—, Andrew, 214 (Morebattle)

——, David, 214 (Morebattle)
——, Henry, chaplain of Holy Rood altar, Kirkcaldy, 85
——, James, procurator for Mr James Kennedy, parson of Carstairs, 517, 517n.
——, James, 376 (Idvie)
——, James, 214 (Morebattle)
——, Mr John, scribe, Musselburgh, 40 (Dunfermline abbey)
——, John, writer, 177, 178 (Haddington priory)
——, John, in Edinburgh, 140
——, John, in Elgin, 453
——, John, in Lasswade, 118
——, John, 33 (Collessie)
——, Peter, master almoner, 272n.
——, Richard, 114 (Leith hospital)
——, Robert, maltman in Linlithgow, 155, 156, 157
——, Robert, 58 (Balmerino abbey)
——, Mr Thomas, vicar of Abercrombie, 69
——, Thomas, in Elgin, 453
——, Thomas, 158 (Airth)
——, Walter, chaplain of Tullypowrie and Fyndynate, 324
——, William, curate of Morebattle, 214
——, William, 422 (Aberdeen friary)
——, William, 283, 287 (Perth)
Younger, Edward, 104 (Kirkliston)
——, James, 101 (Newbattle)
——, Thomas, 289 (Culross abbey)
——, William, 114 (Hailes)
Yut, James, 101 (Newbattle)

Zeyme, *see* Yeoman

Index of Places

Aarngabyte, *see* Glengaber
Abaland, Abbayland, Habaland, *Methven collegiate kirk*, 296, 297, 298
Abay Croft, *St Andrews priory*, 18
Abay Myln, *Dunfermline abbey*, 39
Abbatis Medo, *Holyroodhouse abbey*, 92
Abbay croft, *Crail*, 162
Abbay kirk, *see* St Andrews
Abbay Milles, Abbay Mylnes, *Melrose abbey*, 208, 258
Abbay Mylne, *Haddington priory*, *see* Abbey Mill
Abbey Mains (Manes of the Abay, Maynes of the Abbay), NT5375; 162, 165, 176, 178
Abbey Mill (Abbay Mylne, Mylne of the Abay, Mylne of the Abbay), *Haddington priory*, NT5374; 162, 176, 178, 181
Abbey Mylnes, *St Andrews priory*, 18
Abbey St Bathans, *see* St Bathans
Abbotrule (Abbotisroll, Abbotroull), NT6012; 213, 216; parson of, 213; parsonage of, 213; vicarage of, 213
Abbottis Hall, *Kirkcaldy*, 24, 28, 46, 50
Abdie (Ebdy) [Newburgh, NO2318], kirk of, 32, 36; parish of, 34; vicars portionary of, 68, 90, 378; vicarage portionary of, 378; *see also* Auld Lindores, Lindores, Newburgh
Aberbothrie (Abirbothre), Grange of, NO2344; 354, 357, 369
Aberbrothok, *see* Arbroath
Aberchirder (Abirkerdour, Abirkerdouth) [Marnoch, NJ5950], 159, 359; kirk of, 359; vicarage of, 159, 159n.
Abercorn (Abircorne), NT0878; 153-154, 302, 342, 346; kirk of, 302, 342, 346; vicar of, 153; vicarage of, 153-154
Abercrombie (Abercrommy, Abircrommy, Abircrummie, Abircrummy), NO5102; 16, 17, 69; kirk of, 16, 17; vicar of, 69; vicarage of, 69
Aberdalgie (Abirdaghy, Abirdagie, Abirdagy), NO0720; 303, 342, 346; kirk of, 342, 346
Aberdeen (Aberdene, Abirdeane, Abirdein, Abirdeine, Abirdene, Abirdeneis, Abirdenis, Abirdonensis), NJ9206; 12, 19, 114, 248n., 348, 360, 360n., 404, 405, 405n., 406, 406n., 413, 413n., 416, 416n., 417, 417n., 418-426, 428-444, 447, 448, 450, 451, 455, 456, 457, 463, 488, 558, 559; archdeacon of, 432; bishop of, 360, 360n., 432; bishopric of, 413; burgh treasurer of, 420; Castle Gait, 419, 420, 422; Common Gait, 419, 421; Common Vennel, 419, 421; East End of, 422; Gaistraw (Geistraw), 419, 420, 421; Gallow Gait, 419, 419n., 422; grammar schoolmaster, 425, 442; Green Gait, 418, 419, 422; Green, 420, 421; hospital beside, 434; King Street, 419, 422; Nether Gait, 419; Official of, 429; provost of burgh of, 416, 416n.; Queen Street, 419; St Catherine's chapel in, 419, 421; St Catherine's Hill in, 419, 421; St Ninian's Street, 419, 422; St Peter's altar in, 419, 422; St Thomas the Evangelist's altar in, 419, 421; Ship Gait, 419; Ship Raw, 421; Thief Gait, 419, 422; vicarage of, 423,

424; see also New Aberdeen, Old Aberdeen
—— cathedral, 248n., 425, 426, 426n., 430, 431, 432n., 433, 437, 438, 440, 446n., 456, 463; chancellor of, 416, 416n., 428, 432-433; chancellory of, 428, 432-433; chaplains of, 438; common kirks of chapter of, 437, 438; dean of, 406, 406n., 424, 426, 435, 438; deanery of, 422, 423, 424, 435; prebends of, 248, 426n., 432n., 446n.; prebendary in, 463; scribe to chapter of, 438; stallars in, 429, 431, 432, 433, 439, 440, 456, 463; subchanter of, 426, 426n., 434; treasurer of, 444; vicarage of St Machar's kirk, 423, 424; see also Rathven
—— college, see below Aberdeen, King's college
—— university, 405, 425, 430, 440, 441, 442; sub-principal of, 440, 441
——, Carmelite friars (White friars) in, 418, 420, 422
——, King's college, 405, 405n., 425, 441, 442
——, St Nicholas' kirk, 447; chaplainry of Cocklarachy (St Mary) in, 447, 447n.
——, Trinitarians in, 418, 420, 421, 558, 559; 'friary' of, 558, 559; ministry of, 558-559
——, *sheriffdom*, 1nn., 31, 36, 281, 281n., 282, 283n., 420n., 457n., 428, 431, 432, 433, 434, 440, 448, 461n.
Aberdour (Abirdour), *Fife*, NT1985; 24, 46, 62, 63, 64, 64n., 84, 85; kirk of, 62; mains of, 63, 64; mill of, 62; nuns of, 63; parish of, 63; parson of, 64; vicarage of, 84-85
Aberdour (Abirdour, Abirdoure, Abirdowir, Abirdowr), *Aberdeenshire*, NJ8460; 424, 435; parsonage of, 424; vicarage of, 424
Aberfoyle (Abirfule, Abirfull), NN5200; 343, 344; parsonage of, 343; vicar of, 344; vicarage of, 344
Abergairn (Abirgardein, Abirgardin), NO3597; 425, 440; kirk of, 440-441
Aberlady (Abirladie, Abirlady), NT4679; 94, 170, 342, 343, 345, 346, 346n.; kirk of, 302, 342, 346, 346n.; vicar of, 170; vicarage of, 170

Aberlady Mains (Abirlady Manis, Mains of Abirlady), NT4779; 302, 346, 347
Aberlemno (Abirlemno, Ablemno), NO5255; 217, 218, 220, 221, 375; kirk, 217, 218, 220, 221; vicar of, 375; vicarage of, 375
Aberlethnott (Abirlethnot) [Marykirk, NO6865], 405; kirk of, 405
Aberlour (Abirloche, Abirlocht, Abirlour) [Charlestown of Aberlour, NJ2642], 320, 320n., 489, 490; prebend of Elgin cathedral, 320, 320n.; vicar of, 489, 490; vicarage of, 489; see also Skirdustan
Abernethy (Abirnathy, Abernethie, Abirnethy, Abyrnethie, Abynethys), NO1916; 30, 30n., 35, 313, 322, 329, 330, 338, 340, 343, 352, 359, 360, 363, 364; kirk of, 352, 359, 363, 364; parson of, 330; parsonage of, 329, 330, 343, 363; provost of, 322n.; vicar of, 313; vicarage of, 312, 343
—— collegiate kirk, 30, 35, 244; prebend of, 340; prebend of Pitmedden in, 338; prebendary of, 338; provost of, 244, 322; provostry of, 322; see also Pettinbrog, Pitmedden
Abernyte (Abirnyte, Abirnytie), NO2531; 310; kirk of, 310; parsonage of, 310; vicar of, 310
Abertarff (Abirtarf, Abirtarfe, Abirtarff) [Boleskine and Abertarff parish, NH3710], 488, 489; kirk of, 649; vicar of, 488, 489; vicarage of, 488
Aberuthven (Abritane, Arbruthvene), NN9715; 334, 350, 502, 503; curate of, 334; vicar of, 503; vicarage of, 334, 502, 503
Abirballat, see Arbirlot
Abirbothre, see Aberbothrie
Abirbrothe, Abirbrothock, Abirbrothok, see Arbroath
Abircheirdour, see Aberchirder
Abircorne, see Abercorn
Abircrommy, Abircrummie, Abircrummy, see Abercrombie
Abirdaghy, Abirdagie, Abirdagy, see Aberdalgie
Abirdeane, Abirdein, Abirdeine, Abirdene, Abirdenis, see Aberdeen
Abirdour, Abirdoure, Abirdowir, see

INDEX OF PLACES

Aberdour
Abirfule, Abirfull, see Aberfoyle
Abirgardein, Abirgardin, see Abergairn
Abirkardoure, Abirkardur, Abirkerdour, Abirkerdouth, see Aberchirder
Abirladie, Abirlady, see Aberlady
Abirlady Manis, see Aberlady Mains
Abirlemno, see Aberlemno
Abirlethnot, see Aberlethnott
Abirloche, Abirlocht, see Aberlour
Abirlour, see Aberlour
Abirnathy, Abirnethie, Abirnethy, see Abernethy
Abirnyte, Abirnytie, see Abernyte
Abirtarf, Abirtarff, see Abertarff
Ablemno, see Aberlemno
Aboyne (Awboyne), NJ5398; 431, 440, 451; vicar of, 431; vicarage of, 431, 440, 451
Abritane, see Aberuthven
Abynethys, see Abernethy
Abyrnethie, see Abernethy
Accherthane, see Edderton
Accorne Waird, Actcorward, *Dunfermline*, 27, 49
Achway, Almont, *Manuel priory*, 549, 550
Aclintour, *Selkirk parish*, 230n.
Actcorward, see Accorne Waird
Adame croft, Adames croft, *North Berwick priory*, 147, 168
Adameflat, *Haddington priory*, 180
Adamestoun of Baldornocht, Adamstoun or Baldornocht, *Dunkeld bishopric*, 342, 345
Addeweill, see Addiewell
Addicate (Addicat, Adecat, Aidecat), NO6362; 383, 385, 388
Addiewell (Addeweill), NT9962; 98
Addiston (Adestoun, Alderstoun, Aldstoun) [Mains, NT1569], prebend (with Gogar) of Corstorphine collegiate kirk, 107, 107n., 112, 113, 115
Adecat, see Addicate
Aderny, see Aithernie
Adflothome, see Auchelochan
Admirall lands (Almarie lands, Amarale-landis), *Kelso abbey*, 222, 222n.
Admure, see Aithmuir
Adnistoun, see Edmonstone
Adorny, see Aithernie

Adromillane, *Colmonell*, 499
Adscurry, see Ascurry
Advie (Advy), NJ1234; 480
—— and Cromdale, parson of, 480; parsonage of, 480; see also Cromdale
Adyistoun Nether, *Calder*, 98
Adyistoun Over, *Calder*, 98
Affleck (Auchinlek, Auchinlouke), *Lesmahagow* [Nether Affleck, NS8442], 231, 231n., 232, 232n.
Aglisgrig, see Ecclesgreig
Aichthintowill [*Auchintoul, NJ6151], 474
Aidecat, see Addicate
Aikenway (Akynwaiy), NJ2949; 491
Ailhous croft, *Drumlithie, Aberdeen deanery*, 405
Ailhoushill, *Idvie*, 376
Ailstounfuird, see Athelstaneford
Air, see Ayr
Airdis, *Riccarton*, 531
Airdit (Ardat), NO4120; 13, 20
Airdrie (Ardree, Ardrie), *Crail, St Andrews priory*, NO5608; 18, 177, 178, 181
Airdrie (Ardrie, Ardry), *Monkland, Newbattle abbey*, NS7665; 102, 497
Airheid, see Harehead
Airlie (Arlie, Erlie) [Kirkton of, NO3151], 355, 356, 357, 367n., 369, 370, 376n., 410; kirk of, 355, 357, 370; vicarage of, 410; see also Grange of Airlie, Mains of Airlie
Airlies (Arles), NX3651; 594, 596
Airmadillis, see Armadale
Airngosk, see Arngask
Airntully (Arntulie), NO0935; 303
Airth (Airthe, Arthe), NS8987; 91, 92, 157-158; kirk of, 91; vicarage of, 157-158
Airthourstaine, Airthourstane, see Arthurstone
Aithernie (Aderny, Adorny, Atherny) [Castle, NO3703], 15, 21, 145, 168, 290
Aithmuir (Admure), NO2725; 285, 288; mill of, 285, 288
Aithsting (Eistinge), HU3355; 658n., 660; see also Sandsting
Aitrik, see Ettrick
Akergile, *Wick*, 649
Akynwaiy, see Aikenway

Alaneschaw, *see* Allanshaws
Albist, *Caithness bishopric*, 628
Ald Govill, *see* Old Goval
Aldbar (Auldbar, Awldbar) [Muirside of, NO5756], 21n., 202, 202n., 222, 222n., 297, 298, 378; parsonage of, 297, 298
Aldcambuies, *see* Old Cambus
Alden, *see* Auldhame
Alderstoun, *see* Addiston
Aldestoun, *Dryburgh abbey*, 190
Aldhamestok, *see* Oldhamstocks
Aldincraw, *see* Auchencrow
Aldlistoun, *see* Old Liston
Aldstoun, *see* Addiston
Alenschawis, *see* Allanshaws
Alford (Aufurd), NJ5716; 450; vicar of, 450; vicarage of, 450
Alichmore (Allycht Moir, Halychtmoir), NN8420; 343, 349
Alicht, *see* Alyth
Alith, *see* Alyth
Allagawynne, *Aberdeen deanery*, 405
Allan (Allane), *Inverness*, NH8177; 626
Allane, *Kinloss abbey*, *see* Ellon
Allanhill (Elenenhill), NO5214; 10, 13
Allanshaws (Alaneschaw, Alenschawis, Allaneschalles, Emershaw, Enenschaw, Evingshaw), NT4943; 207, 257, 258, 258n., 259
Allanton (Allantoun), NX9184; mill of, 276, 277
Allhallow altar, *Linlithgow kirk*, 154
Allingeroy, *Strageath*, 349
Allingrew, *Strageath*, 349
Allycht Moir, *see* Alichmore
Almarie lands, *see* Admirall lands
Almerie Cruik, Almery Cruik, *Lindores abbey*, 30, 35
Almont, *see* Achway
Alness (Alnes), NH6569; 645, 647; parson of, 645; parsonage of, 645; vicar of, 647; vicarage of, 647-648
Alrig, *see* Olrig
Altane, *North Berwick priory*, 167, 168
Alterwall (Alterwaill, Atterdaill), ND2865; 628, 632
Altoun, Altowne, *see* Auldtoun
Altum, *see* Auldtoun
Alva (Alvecht, Alwech, Alwecht, Auldwecht), NS8896; 538, 543, 545, 546; kirk of, 538, 543, 545, 546
Alvah (Alwecht, Awecht) [Kirktown of, NJ6760], 355, 368, 369; kirk of, 355, 369; parsonage of, 369; vicarage of, 369
Alvecht, *see* Alva
Alves, NJ1362; parsonage of, 479; vicar pensionary of, 480; vicarage of, 479
Alvodlaw, *see* Elwodlaw
Alwech, Alwecht, *Cambuskenneth abbey*, *see* Alva
Alwecht, *Coupar Angus abbey*, *see* Alvah
Alycht, *see* Alyth
Alyth (Alicht, Alith, Alycht), NO2448; 301, 303, 342, 346, 398; kirk of, 303, 342, 346; parish of, 301; vicar of, 398; vicarage of, 398
Amarale-landis, *see* Admirall lands
Amot, *Fearn abbey*, 634
Anaine, Annain, Annane, *Maryton*, 383, 387, 388
Anay, Annay, *Melrose*, 208, 258
Ancrum (Ancrame, Ancrom, Ancrome, Anecrame), NT6224; 147, 213, 216, 217, 228, 236, 261, 365, 495, 526; kirk of, 217; parson of, 217, 365; parsonage of, 159, 159n., 213, 236, 365; prebend of Glasgow cathedral, 159n.; vicarage of, 236, 365
Ancrumlaw, *see* Ankrielaw
Androquhenren, *Urray*, 635
Anecrame, *see* Ancrum
Anet, *Brechin parish*, 387
Angaschill, *near Kirkeoch, Galloway bishopric*, 589
Angelrow (Angelraw), NT7445; 227
Angreflat, Angreflatt, Angrieflat, Angriefleit, Angrieflett, *Kelso abbey*, 222, 230, 233, 234, 237, 239, 240
Angus (Anguis), barony, 30, 31, 331, 332, 333; deanery of, 376; lordship of, 2
——, *sheriffdom*, 35, 45n., 66, 67, 72, 77, 249, 249n., 363n., 371, 372, 372n., 375, 376, 397, 398, 399, 411; *see also* Forfar
Ankrielaw (Ancrumlaw), NT2758; 142, 143
Anley, *Tain*, 652
Annain, Annane, *see* Anaine
Annamuick (Annamwk), NO7984; 405
Annandale (Annandaill, Annandardaill), NY0994, NY1583, NY1679,

NS3938; 1n., 113, 265, 265n.
Annatis Croce, *Calder*, 98
Annay, *see* Anay
Anstruther (Anstruder) [Easter, Wester, NO5703, NO5603], 22, 74; kirk of, 22; vicarage of, 22; vicarage pensionary of, 74
Antonburne, *see* Attonburn
Anwach, Anwath, *see* Anwoth
Anwoth (Anwach, Anwath, Awecht), NX5856; 614, 617, 619; glebe of, 614; kirk of, 617, 619; vicar of, 614; vicarage of, 614
Apilbie, Apillbie, *see* Appleby
Apilcorce, *see* Applecross
Apletrelevis [*Appletreehall, NT5217], 207, 208, 257, 259
Appilgirth, *Hartside*, *see* Applegarth
Appleby (Apilbie, Apillbie), NX4140; 593, 596; mill of 593
Applecross (Apilcorce), NG7144; 639; chaplainry of, 639
Applegarth, Cold Chapel at, NS9324; *Hartside*, 513
Appodsyd, *Hopkirk*, 220
Arbirlot (Abirballat, Arbirloth), NO6040; 361, 397; kirk of, 361; vicar of, 397; vicarage of, 397
Arbohill, Minarbo Hill, *Crail*, 177, 178, 182
Arbroath (Aberbrothok, Abirbrothe, Abirbrothock, Abirbrothok, Arbrocht, Arbrothok), NO6441; 7, 79, 83, 84n., 114, 114n., 351, 351n., 358, 358n., 360, 380, 395, 396, 401; chaplain of St Duthac in, 380; chaplain of St Nicholas in kirk of, 400; chaplainry of St Nicholas in kirk of, 400; grammar schoolmaster of, 83; school in, 79; vicar pensionary of, 396; vicarage pensionary of, 395-396; *see also* Broth
—— abbey, 7, 83, 84n., 351, 351n., 352, 358-364; abbot of, 114n.; commendator of, 358, 358n., 363; 'third prior' of, 360; sacrist of, 360; subchanter of, 360; subprior of, 360
Arbrocht, *see* Arbroath
Arbrock (Arbrok), NX4537; 594
Arbrothok, *see* Arbroath
Arbruthvene, *see* Aberuthven
Arbukill, Arnbukill, *Monkland*, 497

Arbuthnott (Arbuthnot), NO7975; 405, 406, 407; benefice of, 405, 406; kirk of, 405; parson of, 407; vicar of, 406; vicar pensionary of, 406; vicarage of, 406
Arcadia, *see* Argyll
Arcan (Arokyne) [Mains, NH5053], 635
Arclaich, Arclauch, *see* Ardclach
Ardafaillie, *see* Ardeville
Ardat, *see* Airdit
Ardbraik, *with Keithick, Dunkeld bishopric*, 303
Ardbrangan [Brangan, NJ6164], 359
Ardcanny (Ardcannie), NJ2649; 491
Ardclach (Arclaich, Arclauch, Ardclacht, Ardclath, Ardclauch), NH9545; 467, 470, 479, 488, 489; vicar of, 479, 488, 489; vicarage of, 479, 488
Ardersier (Ardrosscher), NH7855; 626, 642; kirk of, 642
Ardethie Easter, Ardetye Easter [South Arditie, NO0128], 296, 298
Ardethie Wester, Ardetye Wester [North Arditie, NO0029], 296, 298
Ardetye Easter, *see* Ardethie Easter
Ardetye Wester, *see* Ardethie Wester
Ardeville (Ardafaillie, Ardsoull, Arfaill, Arsaill), NH7363; 636, 636n.; chaplainry of, 636; *see also* Dingwall
Ardgaith (Ardgath), NO2222; 284, 285, 287
Ardgath (Arguth), NO2837; 376
Ardgathen (Argathun), NJ5616; 447
Ardis, *see* Aird
Ardkinglas (Ardkynglen) [Ho, NN1710], 387
Ardkynglen, *see* Ardkinglas
Ardmanocht, *Ross bishopric*, 626
Ardmar, *Ross bishopric*, 626
Ardmillan (Ardmillane) [Ho, NX1694], 567, 594, 596
Ardnamurchan (Ardnamurchane), NM5267; 674, 675
Ardneiv, *Islay*, 674
Ardo, NO6262; 371n., 383, 385, 388
Ardoch (Ardoche, Ardoiche), *Mearns, Lindores abbey*, 31, 35
Ardoch, *Bothwell collegiate kirk*, 511
Ardoch, *Dunblane subdeanery*, 314
Ardocht, *Logie-Montrose*, 373
Ardree, Ardrie, Ardry, *see* Airdrie

Ardross (Ardrois, Ardros), NO5000; 64, 66, 89, 94, 146, 167, 169
Ardrossan (Ardrossane), NS2343; 576, 579, 581; kirk of, 576; parish of, 579; vicar of, 581; vicarage of, 581
Ardrosscher, *see* Ardersier
Ards of Glenelge, *Iona abbey*, 674
Ardsoull, *see* Ardeville
Ardurness (Ardurmes), *Caithness bishopric*, 627
Ardwell (Arduell) [Mains, NX1045], 594, 596, 605
Areicht, *see* River Ericht
Arfaill, *see* Ardeville
Argathun, *see* Ardgathen
Argell, *see* Argyll
Argil, Argile, Argill, *see* Argyll
Argownzie, *Monkland*, 102
Argrennan (Argrenan) [Mains, NX7056], 586, 589
Arguthe, *see* Ardgath
Argyll (Arcadia, Argell, Argil, Argile, Argill, Argyill, Argyl, Argyle, Ergill), NN0407, NR9399; 495n., 529, 558, 562, 564, 576; chamberlain of, 564
Aringosk, *see* Arngask
Arles, *see* Airlies
Arlie, *see* Airlie
Armadale (Airmadillis), NG6404; 674
Arnbukill, *Monkland*, 102
Arndilly (Artilduill), NJ2947; vicarage of, 457
Arnegosk Myln [Arngask, NO1410], 291
Arngask (Airngosk, Aringosk), NO1410; 538, 543, 545, 546; kirk of, 538, 543, 545, 546
Arniston (Arnoldstoun), NT3259; 176, 450; prebend of Crichton collegiate kirk, 176, 450, 450n.
Arnoldstoun, *see* Arniston
Arnot, *Abernethy collegiate kirk*, 340
Arntulie, *see* Airntully
Arokyne, *see* Arcan
Arott, *see* Arrat
Arquhilze, Arquhilzie, *Methven collegiate kirk*, 296, 297, 298
Arran, NR9536, NR9540; 675
Arrat (Arott, Arrot, Arrott), NO6358; 383, 384, 386, 387, 388, 389
Arroll, *see* Errol
Arrot, Arrott, *see* Arrat

Arsaill, *see* Ardeville
Arsiltoun, *see* Earlston
Arthe, *see* Airth
Arthmur, *Portmoak priory*, 56
Arthurstone (Airthourstaine, Airthourstane), NO2642; 353, 356, 369
Artilduill, *see* Arndilly
Ascurry (Adscurry), NO5446; 376
Ashkirk (Askirk, Erskirk), NT4722; 212, 495; parson of, 212; parsonage of, 212; vicarage of, 212
Ashyard (Eschard), NS4735; 564
Askirk, *see* Ashkirk
Assent, *see* Assynt
Assindane, *see* Hassendean
Assynt (Assent) [Ho, NH5967], 629; parsonage of, 629; vicarage of, 629
Athelstaneford (Ailstounfuird, Elstanefuird, Elstanfuird), NT5377; 162, 163, 177, 178, 181; kirk of, 162, 163, 177
Atherny, *see* Aithernie
Atholl (Athoill, Athol, Athole), NN6871, NN8571; 8, 16, 86, 86n., 284, 287, 300, 306, 318, 354, 369
Athy, *see* Ethie
Atrik, *see* Ettrick
Atrikhous, *see* Ettrick Ho
Atterdaill, *see* Alterwall
Attonburn (Antonburne, Autounburne), NT8122; 208, 258, 258n.
Auchanachie (Auchanachy), NJ5962; 437
Auchanland, *Kinnell parish*, 367
Auchcarroch (Auchnacarocht, Auchnacarret, Auchnakarrocht), NO3356; 383, 387, 388
Auchdowny, Auchdowy, *Brechin deanery*, 365
Auchefardall, *see* Auchtyfardle
Auchefud [*Auchenfad, NX8150; *Auchenfad Hill, NX9469], 276
Auchelochan, *see* Auchlochan
Auchencrow (Aldincraw, Auchincraw), NT8560; 166; prebend of Dunglass collegiate kirk, 173
Auchencrow (Auchincraw), *Lamberton*, NT8560; 199, 203
Aucheneill, *Lesmahagow*, 232
Auchengassell (Auchingaschill, Auchingassel, Auchingassell, Auchingassill), NX8299; 271, 276,

INDEX OF PLACES 761

277, 623
Auchenheath (Auchihache, Auchinaich), NS8043; 232, 245
Auchenlaiche (Auchinvaik, Auchinwaik), NN6407; 349
Auchentorlie (Auchintorlie) [Ho, NS4374], 529
Aucheryn, *see* Auchren
Auchgammyll, *see* Auchtygemmel
Auchihache, *see* Auchenheath
Auchinaich, *see* Auchenheath
Auchinairn (Auchinairne, Auchnairne), NS6169; 497, 498
Auchincross, NS5713; mains of, 531
Auchindanery, *Kinloss abbey*, 461
Auchindoir, NJ4822; 443; parsonage of, 443
Auchindonan, *Riccarton*, 531
Auchindorie, Auchindory, *near Clintlaw, Coupar Angus abbey*, 354, 356, 357, 368, 369
Auchindrain, *Stirling Black friars*, 555
Auchinglen, NN9017; 349
Auchingray (Auchengray), *Monkland*, *NS9954; 102, 497
Auchinhove (Auchinhovis), NJ4551; 461
Auchinleck (Auchinlek), NS5521; 529; kirk of, 529
Auchinlek, Auchinlouke, *Lesmahagow*, *see* Affleck
Auchinloch, NS6570; 497, 498
Auchinlonying, *Monkland*, 497
Auchinnaicht, *Maybole* [*Aucheninch, NS2514], 568
Auchinroath (Auchinroith), NJ2651; 491
Auchinvaik, Auchinwaik, *see* Auchenlaiche
Auchirlair, *see* Ochtrelure
Auchlochan (Adflothome, Auchelochan, Auchlocham), NS8037; 231, 231n., 233
Auchmillie (Auchmillir), NJ5963; 437
Auchmull [*Westerton of Auchmill, NJ8908], 423
Auchmuty (Awchmawtie), NO2700; 15
Auchnacarocht, Auchnacarret, Auchnakarrat, *see* Auchcarroch
Auchnairne, *see* Auchinairn
Auchness (Auchynes), NJ1149; 492
Auchnotroch (Auchnotro), NS8243; 232
Auchnoule, 8n., 100n., 286n., 383n.,
670n.
Auchren (Aucheryn, Awchron), NS8238; 231, 232
Auchten, *Riccarton*, 531
Auchterarder (Auchtirardour), NN9412; 344; vicarage of, 344
Auchterderran (Ochterderay, Ochtirtdery), NT2195; 72; parsonage of, 72
Auchterforfar, NO4749; 217
Auchtergaven (Auchtirgavin, Auchtergawin, Ouchtergavin, Ouchtergawin), NO0235; 302, 311, 342, 346; kirk of, 302, 342, 346; vicarage of, 311
Auchterhouse (Auchterhous, Ouchterhous), NO3337; 301, 339; parsonage of, 301; vicarage of, 301; vicarage pensionary of, 339
Auchterless (Ochtirles) [Howe of, NJ7242], 430; minister of, 430; parsonage of, 430; vicarage of, 430
Auchtermairnie (Ochtermerny) [Fm, NO3403], 15
Auchtermonsie, Auchtirmonsy, *see* Moonzie
Auchtermouthie, Auchtirmothie, *see* Auchtermuchty
Auchtermuchty (Auchtermouthie, Auchtermuchtie, Auchtirmothie, Auchtirmuchtie, Auchtirmuktie, Ochtermouthie, Ouchtermuchtie), NO2311; 30, 33, 33n., 35, 76, 85; parish of, 33; vicarage of, 76, 85; *see also* Bondhalff
Auchterstruder [Struthers, NO3709], 55
Auchtertool (Auchtertule, Auchtertwill, Ouchtertule, Ouchtertulle, Ouchtertullie), NT2290; 62, 63, 63n., 303, 342, 346; kirk of, 62, 63; mains of, 342, 346
Auchtirardour, *see* Auchterarder
Auchtirgavin, *see* Auchtergaven
Auchtirluir, *see* Ochtrelure
Auchtirmousall, *see* Moonzie
Auchtirtyir, *Monzievaird, see* Ochtertyre
Auchtyfardle (Auchefardall), NS8141; 232
Auchtygemmel (Auchgammyll) [Fm, NS8142], 232
Auchynes, *see* Auchness
Aufurd, *see* Alford
Auld Abirdene, *see* Old Aberdeen

Auld and New Vark, *Selkirk parish*, 230n.
Auld Govyll, *see* Old Goval
Auld Greinlaw, *see* Old Greenlaw
Auld Jedburcht, *see* Old Jeddart
Auld Lindores, Auld Lundores, Auld Lundoris [Lindores, NO2617], 30, 35; *see also* Abdie, Lindores, Newburgh
Auld Melros, *see* Old Melrose
Auld Montrois, *see* Old Montrose
Auld Yardis, *Rothes*, 491
Auldany, *Brechin treasurership*, 377
Auldbar, *see* Aldbar
Auldbreck (Bowraik, Howreik, Olbrek, Yowrek), NX4640; 592, 592n., 598, 599
Auldcambies, Auldcambuies, *see* Old Cambus
Auldearn (Aulderne), NH9155; 493; vicar of, 493; vicarage of, 493
Auldhame (Alden, Haldain, Haldame, Haldan), NT5984; 117n., 118, 118n., 183n., 185n.; parish of, 118n.; parson of, 179, 183n.; parsonage of, 118, 179, 179n., 183n., 185n.; vicarage of, 118
Auldhamstoks, *see* Oldhamstocks
Auldhirsell [Hirsel, NT8240], 186
Auldmore, *Leuchars*, 14
Auldtoun (Altoun, Altowne), *Kelso abbey*, 223, 223n.
Auldtoun (Altum, Awletoun), *Lesmahagow*, NS8239; 231, 232, 245, 245n.
Auldwecht, *see* Alva
Autonburne, Autounburne, *see* Attonburn
Avache, *see* Avoch
Avendaill, *see* Strathaven
Avoch (Avache, Awach, Awache, Awoch, Awoche), NH7055; 462, 463, 464, 640; kirk of, 462, 464; parsonage of, 463; vicar of, 640; vicarage of, 640
Avondale, *see* Strathaven
Awach, Awache, *see* Avoch
Awboyne, *see* Aboyne
Awchmawtie, *see* Auchmuty
Awchron, *see* Auchren
Awdy, *Forgandenny*, 303
Awdy, *with Lalathan, St Andrews priory*, 15

Awccht, *Coupar Angus abbey, see* Alvah
Awecht, *St Mary's Isle priory, see* Anwoth
Awgmowty, *Portmoak priory*, 56
Awirlosk, *see* Awnlosk
Awldbar, *see* Aldbar
Awletoun, *see* Auldtoun
Awltoun Burne, *Mow*, 225
Awmont Perk, *Melrose abbey*, 208
Awnlosk, Awirlosk, Enenlesis, Eskdalemuir, *Melrose abbey*, 259, 259n.
Awoch, Awoche, *see* Avoch
Ayr (Air), NS3422; 114; chaplainry of the Holy Rood in, 560; Foul Vennell, 557; Leik Yeardis, 571; Meal Market, 571; parsonage of, 561; St Peter's altar in, 557; Sandgaitt, 557; town of, 571, 571n.; vicarage of, 561
—— parish kirk, altar of Our Lady in, 571; chaplain in, 571; chaplain of St Peter's altar in, 557; chaplainries in, 571; chaplainry of St Peter's altar in, 557
——, *sheriffdom*, 420n.
Ayton (Aytoun), NT9261; 198, 199, 202, 204, 228; kirk of, 199, 204; mains of, 199; mill of, 203; parish of, 198

Baandrowis, *Brechin bishopric*, 386
Babey, *Dysart*, 70
Babirnie, Babyrnie, *see* Balbirnie
Bacclero, *Inchcolm abbey*, 62
Bachilton (Bawchiltoun, Bawquhiltoun), NO0023; 296, 297, 298
Bad, *Culross abbey*, 290
Baddinderth, Baddindrech, *see* Bandeath
Baddinspink, *see* Badenspink
Baddis, *Calder* [*Baads Mains, NT0061], 98
Baddyforrow, Badyforrow, Baldiforrow, *Fintray, Lindores abbey*, 32, 36
Badenoch (Badyenocht, Baidyenache), NN4985, NN6291; 428, 474
Badenspink (Baddinspink), NJ5961; 437, 438
Badlay, *see* Bedlay
Badnormy, *Moorfoot, Newbattle abbey*,

INDEX OF PLACES 763

101
Badowy, *see* Bardowie
Badyenocht, *see* Badenoch
Badyforrow, *see* Baddyforrow
Baeth, *see* Beith
Bagbe, Bagbie, *see* Begbie
Bagend, *see* Bogend
Bagony, *see* Balgownie
Bagovane, *see* Balgowan
Bagra, *Duffus*, *see* Begrow
Bagra, *Glasgow*, 516
Bagro, *Dunkeld archdeaconry* [*Balgray, NN4038], 309
Baidyenache, *see* Badenoch
Baiglie, NO1515; 322
Baikbie, *see* Begbie
Baikie (Bakie), NO3149; 356
Baiklandis, *Monkland*, 102
Baillenakill, *bishopric of the Isles*, 674
Baillephuille, *see* Balephull
Baillie (Bailyie), ND0465; 628
Bailliewhir (Balaschweir, Balyiequhair), NX4242; 588, 590, 592
Bair of Windyaige, *Coupar Angus abbey*, 355
Bait, *Coupar Angus abbey*, 352n.
Baith Bell, Bayth Bell, *Dunfermline*, 27, 49
Baith Bonella, Baith Bonolla, Bayth Bonalla, Bayth Bownalla, *Dunfermline*, 27, 49
Baith Danyell, Bayth Danyell, *Dunfermline*, 27, 27, 49
Baith Keir, *see* Keirsbeath
Baith Maistertoun, Baithmaisterstoun, Bayithe Maistertoun, Bayth Maistertoun, Bayth Maisterstoun, *Dunfermline*, 27, 49
Baith Persone, Baith Persoun, Bayth Persoun, *Dunfermline*, 27, 49
Baith Steinstoun, Baith Stenestoun, Bayth Stevinstoun, *Dunfermline*, 27, 49
Baith Stewart, Bayith Stewart, Bayth Stewart, *Dunfermline*, 27, 49
Baith Trumbill, Bayth Trumbill, *Dunfermline*, 27, 49
Baithcat, *see* Bathgate
Baithous *of Caputh*, 303, 342, 345
Baitscheile, Bedscheill, Beltscheill, *Coupar Angus abbey*, 353, 409
Baitscheill, Betschule, *Greenlaw* [*Bedshiel, NT6851], 226, 226n.

Bakak, *see* Balcalk
Bakbie, *see* Begbie
Bakello, *see* Balkello
Bakembo, *see* Balkemback
Bakie, *see* Baikie
Bakran, *Aberdour*, 63
Balannan (Balanan), NX7059; 586
Balaschweir, *see* Bailliewhir
Balbaird (Balbard, Balbardy, Ballbaird, Ballbard, Bawbaird, Bawbard), NO4305; 24, 28, 29, 46, 50, 51
Balbardy, *see* Balbaird
Balbirnie (Babirnie, Babyrnie), *Ruthven*, NO2948; 305; mill of, 305
Balbirnie (Ballbirny), *St Andrews priory* [Mill, NO2801], 15
Balbithan (Balbuchein, Balbuthine), NJ7917; 31, 36
Balblair (Ballie Blair), NH8277; 633
Balbow Mylne, *see* Ballomill
Balbreiche, *see* Ballinbreich
Balbrogie (Balbrogy, Balbrogus Wester, Easter and Over, Balbrogyis), NO2442; 353, 356
Balbrogyis, *see* Balbrogie
Balbuchein, *see* Balbithan
Balbuthie (Balbuthy, Bawbuthe), NO5002; 146, 148, 167
Balbuthine, *see* Balbithan
Balcalk (Bakak), NN3939; 309
Balcanquhal (Belcanquell), NO1609; 69
Balcarres (Cawcarros) [Ho, NO4704], 146, 167
Balchardo, *Lundie*, 376
Balcherry (Ballecherye), NH8282; 651
Balchriston (Balchristen), NS2411; 567
Balchrystie (Balcristie, Ballchrist, Ballcrist, Ballcriste), *Newburn*, NO4503; 24, 28, 29, 46, 50, 51
Balchrystie, *Portmoak*, 56
Balclawe, *Kilconquhar*, *North Berwick priory*, 167
Balcomie (Balcomy), NO6209; 66, 177, 178, 181, 181n.
Balcomy Croft, Balcomy's Croft, *Crail*, 11, 18
Balcormo (Balcarmo), NO5104; 145, 166
Balcraig [Moor, NX3745], 593
Balcray (Balcry, Baltray) [High, Low, NX4538], 590, 592, 594, 596
Balcristie, *see* Balchrystie
Balcurvehauch [Balcurvie, NO3400], 15

Baldastard (Bandastet), NO4106; 145, 167
Balderane, see Bandirran
Baldernock (Bethernok, Bothernok), NS5774; 507, 508; parson of, 507, 508; parsonage of, 507; vicar of, 507; vicarage of, 507
Baldiforrow, see Baddyforrow
Baldinnie (Balduny), NO4211; 13, 20, 21
Baldornocht, *Dunkeld bishopric*, 342, 345
Baldridge [Easter, Wester, NT0888], Halbaudrik, Hoill Baudrik, Hoill Bawidrik, 27, 49
Baldukie, 21n.
Balduny, see Baldinnie
Baledmund (Balladmond, Balladmound), NO0051; 284, 288
Balephull (Baillephuille), NL9640; 673
Balfarrine [North, South Balfern, NX4350, NX4450], 619
Balfour, NO3200; 15
Balgadie Nether, *Monkland*, 497
Balgadie Over, *Monkland*, 497
Balgarvie (Ballgarvy) [Easter, Wester, NO3515, NO3415], 14
Balgay, NO2727; 375
Balgersho (Balgirsche, Balgrische), NO2238; 353, 356, 409
Balgirsche, see Balgersho
Balgoiff, see Balgove
Balgone (Balgon, Balgonye, Bawgoun) [Ho, NT5682], 147, 168, 169
Balgonie (Ballgony) [Cott, NO3000], 15
Balgony, see Balgownie
Balgove (Balgoiff), NO4817; 10, 12, 18, 20
Balgowan (Bagovane, Balgowane, Balgowne, Ballgoun) [Home Fm, NN9823], 296, 297, 298
Balgowne, see Balgowan
Balgownie (Bagony), *Eassie*, NO3546; 412
Balgownie (Balgony, Balgowny), *Culross abbey* [Mains, NN9888], 290, 292
Balgownie (Balgowny), *Aberdeen deanery* [Balgownie Ho, NJ9309], Brigtoun of, 423
Balgowny Myln [Balgownie, NY9888], 291
Balgrische, see Balgersho

Balgrugo, *Nevay*, 367
Balgrummo (Ballgrummo), NO3703; 15, 21
Balhalgardy (Balhaggertie, Ballaggarthy, Bathalgarthy), NJ7623; 32, 36, 437
Balhelvy, see Belhelvie
Balintrado, Baltraid, Baltrude, *Kelso abbey*, 223, 231; see also Temple
Balkburne, see Halkburne
Balkello (Bakello), NN3637; 309
Balkemback (Bakembo), NN3938; 309
Ballachand, Ballachinde [*Upper Ballachandy, NN9756], 284, 288
Ballacraig, *Atholl, Perth Charterhouse*, 284, 288
Balladmond, Balladmound, see Baledmund
Balladolen, *Atholl, Perth Charterhouse*, 284, 288
Ballaggane, *Penpont*, 273
Ballaggarthy, see Balhalgardy
Ballamene, Ballnamene, *Kinloss abbey* [*Balnamoon, NJ4855], 461, 462
Ballanan, *Galloway bishopric*, 589
Ballanascharach, *Fearn abbey*, 634
Ballangown, see Balnagown
Ballannoch, Bellemenoch, *Atholl, Perth Charterhouse*, 284, 288
Ballantrae, see Kirkcudbright-Innertig
Ballavrauche, *Dallas*, 492
Ballbaird, Ballbard, see Balbaird
Ballbirny, see Balbirnie
Ballbroky Myll, *Kennoway*, 15
Ballbroky, *Kennoway*, 15
Ballchrist, Ballcrist, Ballcriste, see Balchrystie
Ballcreiff, see Balchrystie
Ballcurvie Major [Balcurvie, NO3400], 15
Ballcurvie Minor [Balcurvie, NO3400], 15
Ballecherye, see Balcherry
Ballenamanniche, *Uist, Iona abbey*, 673
Ballenkirk (Ballinkirk), NO3204; 15
Ballewny, see Ballnay
Balleyettoun, Belleyetoun, *Atholl, Perth Charterhouse*, 284, 288
Ballgony, see Balgonie
Ballgoun, see Balgowan
Ballgrummo, see Balgrummo
Ballie Blair, see Balblair
Ballinbreich (Balbreiche, Ballinbrach,

INDEX OF PLACES 765

Ballinbrocht), NO2720; 11, 15, 19
Ballingall, *Lathrisk, St Andrews priory*, 14, 21
Ballingry (Ballingery), NT1797; 64n., 76, 86; glebe of, 76, 86; parsonage of, 86; vicar pensionary of, 76; vicarage of, 86; vicarage pensionary of, 76
Ballinhard, *see* Bonhard
Ballinkirk, *see* Ballenkirk
Ballinknok, *Caithness archdeaconry*, 633
Ballinreicht, *Lagganallachy*, 309
Ballmongy, Ballmoungy, *see* Balmungo
Ballmullo, *see* Balmullo
Ballnabreich, *see* Balnabriech
Ballnamene, *see* Ballamene
Ballnay, Ballewny, *Brechin bishopric*, 384, 389
Ballo, *with Glasslie, St Andrews priory*, 15
Balloch, *Dunblane*, 299
Ballodmounth, *Forgan*, 14
Ballodroun, *see* Balluderon
Ballomill (Balbow Mylne, Bellow Mylne), NO3210; 33, 37
Ballonalen, *Lagganallachy* [*Ballinloan, NN9740], 309
Ballone, Ballonen, Ballonie, *see* Balone
Ballqhany, *see* Mountquhanie
Balluderon (Ballodroun) [North, South, NN3738], 309
Ballumbie (Ballumby, Balumbie, Balumye), NO4433; 58, 376, 400; chapel of, 376, 377; parson of, 400; parsonage of, 400
Ballyclaichtkane, *Lagganallachy* [*Ballachraggan, NN9338], 309
Ballynathane, *Lagganallachy*, 309, 309n.
Balmachee, Balnaige, *Soulseat abbey*, 598, 599
Balmaclellan (Balmaclellane), NX6579; 603; parson of, 603; parsonage of, 603; vicarage of, 603
Balmadysyd, *see* Balmeadowside
Balmaghie (Balmagy, Balmegie) [Ho, NX7163], 91, 618; vicar of, 618; vicarage of, 618
Balmaill, *see* Barmeal
Balmallo, *see* Belmont
Balmangane, Bawangaid, *Borgue*, 594, 594n.

Balmashanner (Balunschamour), NO4649; 217, 234, 237, 240
Balmaw, *see* Belmont
Balmeadowside (Balmadysyd), NO3218; 34
Balmerino (Balmerinoch, Balmerinocht, Balmeryinoch, Balmerynoch, Balmerynoche, Balmerynocht), NO3524; 10, 57, 58, 58n., 59, 60, 61; kirk of, 58; minister of, 58; reader of, 58
—— abbey, 19, 19n., 57-61, 118; abbot of, 59, 59nn.; commendator of, 57, 59
Balmoch [*Balmuchy, NH8678], 633
Balmowto, *see* Balmuto
Balmule (Balmullis), NT2088; 63
Balmullis, *see* Balmule
Balmullo (Ballmullo), *St Andrews priory*, NO4220; 13, 20
Balmulo, Barmullo, *Soulseat abbey*, 598, 599
Balmungo (Ballmongy, Ballmoungy), NO5214; 13, 20, 21
Balmuto (Balmowto), NT2289; 72
Balmyldie, *Calder*, 497, 498
Balmyle, NO2744; 353, 357, 368, 369, 370, 371
Balmyll Mylne, *with Balmule, Aberdour, Inchcolm abbey*, 63
Balna Touch, *Tain*, 651
Balnab, NX4639; 594
Balnabriech (Ballnabreich, Balnabrech, Balnabreicht, Ralnabreche), NO5458; 383, 387, 388
Balnagall (Balnagaw), NH8382; 651
Balnagown (Ballangown, Balnagowne), *NH8154; 625, 633, 634, 646, 651, 652; chaplain of, 646; chaplainry of, 646
Balnaguard (Balnagard), NN9451; 324
Balnaige, *see* Balmachee
Balnamoon (Balnamone, Balnomone), NO5563; 374, 383
Balnavadocht, Balnavardocht, *Dunkeld bishopric*, 342, 345; *see also* Boroustoun
Balniel (Banneill, Bawmeill), NO4705; 146, 167
Balone (Ballone, Ballonen, Ballonie, Balonie), NO4815; 10, 12, 19, 20, 21
Balony, *Brechin bishopric*, 386
Balquhain (Balquhannanes,

Balquhannanis) [Mains of, NJ7323], 414, 415
Balquhair, *Paisley abbey*, 530, 530n.
Balquhany, *see* Mountquhanie
Balquhidder (Balquiddir, Buffudder, Buffuddir), NN5320; 295, 328, 330, 348, 349, 350, 392; kirk of, 295, 348, 349; prebend of Dunblane cathedral, 328, 329, 392
Balquhilly, *see* Palwhilly
Balrymonth Easter, Balrymounthe Easter, *see* Easter Balrymonth
Balrymonthe Wester, Balrymounth Wester, Balrymounthe Wester, *see* Wester Balrymonth
Balsallocht, Balsallosch, *see* Barsalloch
Balsier (Balseir), NX4346; 594
Balsmycht [Meikle Balsmith, NX4639], 594, 596
Balstoun, *see* Belston
Baltarssin, *see* Baltersan
Baltarsyd, *Crossraguel abbey* [*Baltersan Mains, NS2809], 567
Balteir, *see* Baltier
Baltersan (Baltersane, Balterstane, Baltirsoun, Beltersane, Waltersame), *Holywood abbey*, NX4261; 269, 271, 274, 276, 277
Balthene, *Nevay*, 367
Baltier (Balteir), NX4643; 593
Baltilly (Baltulye), NO3911; 55
Baltraid, Baltrude, *see* Balintrado
Baltray, *see* Balcray
Baltulye, *see* Baltilly
Balumbie, *see* Ballumbie
Balumye, *see* Ballumbie
Balunschamour, *see* Balmashanner
Balvaird, NO1712; 546
Balvenie (Bolwanye) [Castle, NJ3240], 318, 318n.
Balwearie (Balwery), NT2590; 24, 28, 46, 50
Balyequhair, *see* Bailliewhir
Balyett [High Balyett, NX0862], 590
Balyiequhair, *see* Bailliewhir
Balzeordie (Balzeordye), NO5564; 374, 375
Bamff (Banff), *Arbroath abbey*, NO2251; 351, 351n., 359, 362
Banchory-Devenick (Banchorie-Devynik), NJ9102; 429, 436; parsonage of, 436; vicar of, 429; vicarage of, 429

Banchory-Ternan (Banquharry Terny, Banquharryntny, Banquhory Terny), NO6998; 351, 358, 359; kirk of, 359
Bandastet, *see* Baldastard
Bandeath (Baddinderth, Baddindrech, Bendeth), NS8592; 537, 543, 545, 547, 548
Bandirran (Balderane), NO4010; 55
Bandon (Bandonen), NO2704; 15
Bandrum, NT0491; 24, 46
Banff (Banf), NJ6664; 359, 455, 456; kirk of, 359
—— and Inverboyndie, vicarage of, 455; *see also* Inverboyndie
——, *sheriffdom*, 1n., 248n., 281n., 282, 320n., 457, 457n., 478n.
Banff, *Arbroath abbey*, *see* Bamff
Banghouswallis (Bangoswallis, Langhouswallis), *Dryburgh abbey*, 190, 190n., 191, 197
Bangly (Banglaw, Banglawe, Vanglaw) [Hill, *NT4875], 190, 191, 197
Bank, *Dollar*, *see* Dollarbank
Bank, *Kelso abbey*, 235, 238, 241
Bankhead (Bankheid, Bawmilheid), *Lesmahagow*, 231, 232
Bankhill, Bankill, *St Mary's Isle priory*, 617, 619
Bannaburne, *see* Bannockburn
Bannafield (Bannafeild, Banofeild), NO5211; 11, 19
Bannaty (Bonnat), NO1608; 69
Bannaty Mill (Bannate Myll), NO1708; 69
Bannavis, *Kinnell parish* [Balneavis Cottage, NO6049], 366
Banneill, *see* Balniel
Bannockburn (Bannaburne), NS8190; 553n.; chapel, 550; chaplain in chapel of, 550
Bannwth, *Dunkeld archdeaconry*, 309
Banofeild, *see* Bannafield
Banquharry Terny, Banquharryntny, Banquhory Terny, *see* Banchory-Ternan
Banvy, *see* Benvie
Bar, *Galston*, 562, 564
Bara (Barro), NT5669; 162, 177, 178
Barachany, *see* Barrachnie
Barclay (Barcly), NS3208; 568
Barclo, Braklo, Bratho, *Brechin bishopric*, 383, 387, 388
Bardarroch (Bardarrocht) [Fm,

NS4718], 499
Barden, *Balmerino abbey*, 58
Bardowie (Badowy, Bardowie), NS5873; 507; mains of, 505
Bardoy, *Holyroodhouse abbey*, 92
Bardyland, *see* Boreland
Barefregane, *Holywood abbey*, 623
Barehill, *Methven collegiate kirk*, 296, 298
Bargany (Barganie, Bargeny), NS2400; 148, 166, 169, 229, 499
Barhill, *Holywood abbey* [Fm, NX1044], 275
Barhill, *Culross abbey*, 290, 290n.
Barinbaroch, Barinbarro, *see* Barnbarroch
Barinoking, *see* Barvernochan
Barkdow, *see* Barledziew
Barlaching, *see* Barlauchlin
Barlachlelan, *see* Barlauchlin
Barlauchlin (Barlaching, Barlachlelan), NX3961; 587, 590
Barledziew (Barkdow, Barliezow), NX4245; 593
Barlethtoun Lethill, *Riccarton*, 531
Barmeal (Balmaill), *Whithorn priory*, NX3841; 594
Barmuckity (Barmwcutty) [Moss of, NJ2461], 492
Barmuir (Barmure), NS4428; 210, 257
Barmullo, *see* Balmulo
Barmweill, *see* Barnwell
Barnbarroch (Barinbaroch, Barinbarro, Barnbarrocht, Barnebarrocht), NX4051; 587, 596, 598, 607
Barne Croft, *Fearn abbey*, 634
Barnebarrocht, *see* Barnbarroch
Barnes, Barnis, East and West Barnes, *Haddington priory*, 162, 177, 178, 178n.
Barnford (Barnfuird, Barnefurd), NS3513; 102, 103
Barnhill, Barnehill, *Aberdour*, 63, 64
Barnhills (Barnahillis), *Ancrum*, NT5921; 236, 366
Barnside (Barnesyd), NT7462; 166; prebend of Dunglass collegiate kirk, 173
Barnton, 272n.
Barnwell (Barmweill, Barnweill, Brounhill), NS4130; 558, 559, 562, 563, 564; exhorter at, 563; minister of, 558; parish of, 558, 559, 562

Barochan (Barochane) [Moss, NS4268], 533
Barony, Barrowny, Bawrony, *Brechin bishopric*, 383, 385, 388
Barquharrand [*Barwhirran, NX4061], 587
Barra (Barray), NF6700; 675
Barrachan (Barchrachand), NX3657, NX3649; 588
Barrachnie (Barachany), NS6664; 497
Barras (Barrowes) [Chapel of, NO8378], 408; chaplain of, 408; St John's chaplainry of, 408
Barras, chapel of the Blessed Virgin Mary, *Edinburgh*, 96, 96n.
Barro, *Lothian* [*Bara, NT5669], 91, 92; kirk of, 91; minister of, 92
Barony of Ogilbeis, *see* Ogilvie
Barrowny, *see* Barony
Barry (Barrie), NO5334; 58, 59, 360, 361; kirk of, 58; minister of, 59, 361; vicar of, 360, 361
Barsalloch (Balsallocht, Balsallosch), NX4558; 587, 590
Barsay, *Orkney bishopric*, 655, 668
Barskeocht, *Kells*, 603, 605
Barsolus (Barsollis), *Soulseat abbey*, NX1056; 598, 599
Bartebeithe, *Monkland*, 497
Barthland, *Penninghame*, 590
Bartonehill, *Monkland*, 497
Bartrostan, NX3859; 588
Bartry, *see* Balcray
Barvennan (Barvennen), NX3860; 590
Barvernochan (Barinoking, Barvernocht, Barwarnok, Barwarnokun), NX3852; 587, 593, 594, 596
Barvy, *see* Bervie
Barwarnok, Barwarnokun, *see* Barvernochan
Barwinnock (Barwinnok), NX3843; 593
Bass (Bas), *North Berwick priory*, 168, 169, 211
Bassendean (Bassinden), NT6245; 186
Bassinden, *see* Bassendean
Bastanerig, *see* Bastleridge
Basterig, *see* Bastleridge
Bastleridge (Bastanerig, Basterig), NT9359; 198, 204
Bateis land, Bateislandis, *Restalrig collegiate kirk*, 142, 143
Bathalgarthy, *see* Balhalgardy

Bathedloskis, *see* Nathertoun of Bathedloskis
Bathgate (Baithcat, Bathcat), NT9769; 92n., 101, 102; kirk of, 102; vicar of, 152; vicarage of, 152
Battollo, *Idvie* [*Bractullo, NO5247], 376
Bauar, *see* Bower
Bauchland, *Caputh*, 318
Bawangaid, *see* Balmangane
Bawbaird, Bawbard, *see* Balbaird
Bawbuthe, *see* Balbuthie
Bawbutlir, *Dunkeld*, 300
Bawchiltoun, *see* Bachilton
Bawenevy Hauch, *Markinch*, 15
Bawenzie, *Leuchars*, 13
Bawgoun, *see* Balgone
Bawgre, Bawgrein, Bonegraye, *Lesmahagow*, 231, 231n., 232
Bawmeill, Bawmell, Bawmill, Bawmyll, *Dunfermline abbey*, 23, 27, 27n., 45, 49
Bawmeill, *see* Balniel
Bawmilheid, *see* Bankhead
Bawmonthe, 619
Bawquhiltoun, *see* Bachilton
Bawrony, *see* Barony
Bayith Stewart, *see* Baith Stewart
Bayithe Maistertoun, *see* Baith Maistertoun
Bayth Bell, *see* Baith Bell
Bayth Bonalla, Bayth Bownalla, *see* Baith Bonella
Bayth Creir, *see* Keirsbeath
Bayth Danyell, *see* Baith Danyell
Bayth Ker, *see* Keirsbeath
Bayth Maisterstoun, *see* Baith Maistertoun
Bayth Persoun, *see* Baith Persone
Bayth Stevinstoun, *see* Baith Steinstoun
Bayth Stewart, *see* Baith Stewart
Bayth Trumbill, *see* Baith Trumbill
Bayth under the Hill, *Dunfermline abbey*, 23, 45
Bayth, *see* Baith, Beath
Beanstoun (Beinstoun, Benestoun, Beynstoun), NT5476; 147, 162, 163, 167, 177, 178, 181
Beath (Bayth), NT1392; 62, 63n., 64, 64n.; chapel of, 62, 63; kirk of, 64; *see also* Baith, Bayth under the Hill
Beattock (Beithach), NT0802; 615
Beauly (Beaulyne, Bewlie, Bewlyn, Bewlyne, Bowlyne), NH5246; 464, 464n., 465, 493, 649, 650, 651; mains of, 649, 651
—— priory, 461n., 464-465, 493, 649-651, 653; commendator of, 650, 650n.; monks of, 650
Bechestoun, Leichstoun, *Edderton*, 652
Bedlay (Badlay) [Cas, NS6970], 497
Bedrule (Redrowll), NT6018; 211; parson of, 211; parsonage of, 210; vicarage of, 210
Bedscheill, *Coupar Angus abbey, see* Baitscheile
Bedscheill, *Edrom, see* Belshiel
Bedshiel, NT6851; 204n.
Beech, *see* Beoch
Begall, *Beith*, 574
Begbie (Bagbe, Bagbie, Baikbie, Bakbie), NT4970; 162, 165, 177, 178, 178n., 181
Begrow (Bagra), NJ1668; 482
Beirhope, *Hownam*, 220
Beith (Baeth, Beith, Beitht), NS3453; 573, 574, 576, 577, 578, 579; kirk of, 577, 578; parish of, 574, 576, 578, 579; vicarage of, 578
Beithach, *see* Beattock
Belcanquell, *see* Balcanquhal
Belches, *Carse of Gowrie, Perth Charterhouse*, 285, 288
Belhame, *St Bathans priory*, 192
Belhelvie (Balhelvy), NJ9417; 439, 442; parson of, 442; parsonage of, 439, 442; vicar pensionary of, 442; vicarage of, 439, 442
Belitaw, *see* Bellitaw
Bellaty (Bellite), NO2359; 354
Bellcattis, *Cambuslang*, 508
Bellemenoch, *see* Ballannoch
Bellewot [Loch Belivat, NH9547], 470
Belleyetoun, *see* Balleyettoun
Bellie, NJ3659; 472; kirk of, 472
Bellillsyd, *Cambuslang*, 509
Bellitaw (Belitaw, Belletaw), NT6943; 228, 234, 238, 241
Bellite, *see* Bellaty
Bellow Mylne, *see* Ballomill
Bellshiell (Belleischell, Belleschill, Bellischill), *NT8149; 223, 227, 231, 240
Belmont (Balmallo, Balmaw), *Newtyle*, 30, 30n., 34, 35
Beloquhy (Bequy), HY6239; 664

INDEX OF PLACES 769

Beloun, *Fearn abbey*, 634
Belscheis, Belschis, *see* Belses
Belses (Belscheis, Belschis, Belsis, Bolsches), NT5725; 216, 219, 220; kirk of, 219, 220
Belshiel (Bedscheill, Belscheill), *Edrom*, NT8149; 200, 204, 204n.
Belston (Balstoun, Belstoun), NO0335; 296, 298
Beltersane, *see* Baltersan
Beltoun, *see* Bolton
Beltscheill, *Coupar Angus abbey, see* Baitscheile
Bemersyde (Bemyrsyd, Bennysyd), NT5933; 189, 189n.
Bendeth, *see* Bandeath
Bendochy (Bennethie, Bennethy), NO2244; 355, 370, 410; kirk of, 355, 370; vicarage of, 410
Bene, Benie, *Strathearn*, 31, 32, 35, 36
Benerig, *Coldingham priory*, 198
Benestoun, *see* Beanstoun
Bengour Law [East Bangour, NT0471], 151
Benhein, *Brechin bishopric*, 386
Benholm (Benholme), NO8069; 403; parson of, 403; parsonage of, 403; vicar pensionary of, 403; vicarage of, 403
Benie, *see* Bene
Benmoir, *Atholl, Perth Charterhouse*, 284, 288
Bennethie, Bennethy, *see* Bendochy
Bennysyd, *see* Bemersyde
Benvie (Banvy), NO3231; 334; minister of, 334; parsonage of, 334; vicar of, 334; vicarage of, 334
Beoch (Beech), NX6861; 585, 589
Bequy, *see* Beloquhy
Beriehoill, Berieholl, *Lindores abbey*, 29, 33, 35, 37
Bernis, *Dryburgh*, 190, 191, 197
Berrie Hill, Berryhill, *Coldingham priory*, 198, 204
Berrieaker, *Glasgow*, 500
Berry Hillok [*Berryhill, NJ9512], 423
Berstane (Birstane), HY4610; 668
Beruick, Beruik, *see* Berwick
Bervie (Barvy, Bervy), *Mearns, Lindores abbey*, 31, 34, 35
Berwick (Beruick, Beruik, Berwyck), 553; chaplainry of St Mungo's altar in, 553; *see also* Berwick-Upon-Tweed, North Berwick
——, *sheriffdom*, 1n., 67, 67n., 179n., 186n., 188n., 195, 196, 223
Berwick-Upon-Tweed, NT9952; kirk of, 199, 208
Betharis, Botheris, Bothiris, *Brechin treasurership*, 377, 383, 385, 388
Bethelnie (Boithelme, Bothelny), NJ7830; 359, 432, 433; kirk of, 359; vicar of, 432; vicarage of, 432
Bethernok, *see* Baldernock
Betschule, *see* Baitscheill
Bewar, *see* Bower
Bewlie, Bewlyn, Bewlyne, *see* Beauly
Beynstoun, *see* Beanstoun
Biggar, NT0437; 521; parsonage of, 521; vicarage of, 521
Bigholm (Bigholme), NS3654; 573
Bik, *Lagganallachy*, 309
Bilbster (Bilbustar), ND2852; 632
Billerwoll (Bullerwoll), NT5915; 220
Billie (Bille) [Mains, NT8558], 192
Binning, kirk of 17, *see also* West Binny
Biregrang, Byregrange, Byrgrange, *Culross abbey*, 290, 290n., 293
Birgham, NT7839; 183n.
Birkenside (Birkinsyd), NT5642; 190
Birkinheid, *Culross abbey*, 289, 292
Birnam (Birnane), NO0341; 303
Birnie (Birneth, Birnethe, Birnett, Burneyth, Byrneth), NJ2154; 465, 466, 467, 468, 487; exhorter at, 487; minister of, 487; parsonage of, 487; reader at, 487; vicar of, 487; vicarage of, 487
Birsay (Bresay), *with Marwick, Orkney bishopric*, *HY3022; 655, 659, 668; kirk of, 659; minister of, 659; parsonage of, 655, 668
Birsay, *with Beloquhy, Orkney subchantory*, 664
Birsay, *with Deerness, Orkney bishopric*, 671
Birse (Birs, Brais, Byrs), NO5579; 415, 417; parsonage and vicarage of, 432
Birstane, *see* Berstane
Bischeptoun, Bischoptoun, *see* Bishopton
Bischopis Forrest, *Nithsdale, Glasgow archbishopric*, 496
Bischopis Quyis, Bischopisquyis, *Wick*, 628, 628n.

Bischoppis Clintirte [*Bishopston, NJ8412; *Clinterty Cottage, NJ8311], 423
Bischopschyre, *St Andrews archbishopric*, 2, 5
Bishopton (Bischoptoun), *Kirkchrist*, NX6650; 589
Bishopton (Bischeptoun, Bischoptoun), *Whithorn, Galloway bishopric*, NX4440; 588, 590, 592
Black Barony (Blakbarrony), NT2347; 140
Blackadder (Blakcader) [Mount, NT8553], 199
Blackcraig, NT0371; Mydblak Craigis, 151; Wester Blak Craigis, 151
Blackhall (Blakhall), NJ2648; 491
Blacklands (Blaklandis), NS7266; 497
Blacklaw (Blaklaw), *Millhorn, Coupar Angus abbey*, NO2245; 354, 357, 369
Blacklaw (Blaklaw, Blaxlaw), *Dunfermline abbey*, NO1186; 23, 27, 45, 49
Blackpark (Blakmark), NX3757; 597, 597n.
Blackstone (Blakstoun) [Mains, NS4666], 530, 530n.
Blackwood (Blakwod), NX9087; 275
Blainelyne Easter, Blainelyne Middle, Blainelyne Wester, *Monkland*, 497; see also Blairlinn, 102
Blainslie (Blanslie) [New Blainslie, NT5444], 207, 208, 257, 259
Blair Atholl (Blair in Atholl), NN8765; 318; parsonage of, 318; vicar pensionary of, 318; vicarage of, 318
Blair, *Culross abbey*, *NO0298 [*Blair Mains, NN9686]; 290, 292, 294
Blair, kirk of, *see* Blairgowrie
Blair, *Riccarton*, 531
Blair, *Sorbie* [High, Low Blair, NX4347], 593
Blair, *in Gowrie*, *see* Blairgowrie
Blairathie, Blairathy, Blairrathie, Blairrathy, *with Blairenbathie, Dunfermline abbey*, 23, 27, 45, 46, 49
Blairballie, *Kirkmaiden*, 593
Blairboth Norther, Blairbothy Norther [Blairenbathie, NT1194] 27, 49
Blairbothy Souther, Blairinbother Souther [Blairenbathie, NT1194], 27, 49
Blairenbathie (Blairnbothie), NT1194; 23, 46
Blairerno, see Blererno
Blairgowrie (Blair), NO1745; 332, 334; kirk of, 332, 538, 545; minister of, 334
Blairhall (Blanhall, Blanhill), NS9987; 290, 293
Blairkery, *see* Blinkeerie
Blairlinn (Blairlynis), NS7572; 102; *see also* Blainelyne Easter, Middle, Wester, 497
Blairquhan, 607
Blairrathie, Blairrathy, *see* Blairathie
Blairthomas, *Monkland*, 497
Blakbalk, Blak Bak croft, *Kelso abbey*, 222, 222n.
Blakbank, Blakband, *Lesmahagow*, 232
Blakbarone, *Dryburgh abbey*, 190
Blakbarrony, *see* Black Barony, 140
Blakcader, *see* Blackadder
Blakdanetreis [*Blakedean, NT8222], 225
Blakfreir Wynd, *see* Edinburgh
Blakhall, *see* Blackhall
Blakhilend, *Lesmahagow*, 232
Blakhill, *Coldingham priory*, 198, 203
Blakholme, *see* Buckholm
Blakirstoun, Blakstoun, *St Bathans priory*, 192, 193, 194
Blaklandis, *see* Blacklands
Blaklaw, *see* Blacklaw
Blaklawacht, *Rothes parish*, 491
Blakmark, *see* Blackpark
Blakmedyngis, *Selkirk*, 230n.
Blakscoit, *Eskdalemuir, Melrose abbey*, 259
Blakstoun, *near Grange of Airlie*, 354, 356, 368, 369
Blakstoun, *Paisley, see* Blackstone
Blakstoun, *St Bathans priory, see* Blakirstoun
Blakweid, *see* Blakwod
Blakwod, Blakweid, Blekvode, *Lesmahagow*, 231, 231n., 233
Blakwod, *Holywood abbey, see* Blackwood
Blakyairdis, *Monkland*, 497
Blandarrane, *Caputh*, 318
Blanerne [Berwickshire], 93
Blanhall, Blanhill, *Culross abbey, see* Blairhall

Blanses, 200
Blanslie, *see* Blainslie
Blantyre, NS6857; 20, 21, 272, 272n., 505, 505n.; glebe of, 505; manse of, 505; minister of, 505; parsonage of, 505
Blantyre priory, 272, 272n., 503, 505, 505n.; bailie of, 505; commendator of, 272; prior of, 20, 20n., 21, 505, 505n.
Blarekry, *see* Blinkeerie
Blariegabir, *see* Blairgaby
Blaverne, *see* Blanerne
Blaxlaw, *see* Blacklaw
Blekvode, *see* Blakwod
Blengaber, *see* Glengaber
Blererno (Blairerno), NO7783; 405
Blessed Virgin Mary, chapel of, *see* Barras
Blindlie, *with Mossilee, Kelso abbey*, 224
Blinkeerie (Blairkery, Blarekry), NN9789; 290, 292
Bllangawar, *see* Glengaber
Blyndley, *Stichill*, 195
Boarhills (Byirhillis, Byrehillis, Byrhillis), NO5614; 2, 5, 13, 13n., 20, 21
Bodrat, *Perth Charterhouse*, 284, 287
Bog, *Lesmahagow*, NS7653; 232
Bogend (Bagend, Boigend, Brigheid), NT7949; 223, 226, 231, 234, 235, 237, 238, 240, 241
Bogferlow [Bogfur, NJ7618], 423
Boggergon, *see* Bogjurgan
Boggis, *Moy*, 470
Boggynkabyr, *see* Bogincaber
Boghill, *with Bigholm, Kilwinning abbey* [*Bog Hall, NS3654], 573
Boghoill, *with Old Goval, Aberdeen deanery*, 422, 423
Boghouse (Boghous), *Riccarton*, 531
Bogincaber (Boggynkabyr), NO7482; 405
Bogjurgan (Boggergon), NO7584; 405
Boglugy, *Kinloss abbey*, 461
Bogmillin, Bogmiln, Bogmilns, *Coupar Angus abbey*, 354, 357, 369
Bogmowkillis, *see* Bogmochals
Bogmuchals (Bogmowkillis), NJ5559; 437
Bogrie, NX8184; 269, 274
Bogside (Bogsyd, Brogsyd), *Holywood abbey*, 623
Bogtoun, *Cadder*, 497, 498
Bogy, *Kirkcaldy*, 24, 28, 46, 50
Boig Syid, *Elgin*, 453
Boigend, *see* Bogend
Boighaid, *Rothes*, 491
Boithelme, *see* Bethelnie
Boithwellhauch, *see* Bothwellhaugh
Bokislaw, *Kelso*, 232n.
Boldene, *see* Bowden
Boldside (Boldsyd, Boytsyde), NT4933; 224, 224n.
Boleskine, *see* Abertarff
Bollinga, *Soulseat abbey*, 599
Bolsches, *see* Belses
Bolton (Beltoun, Boltoun), NT5070; 91; barony of, 172n.; kirk of, 91; parson of, 172; prebend of Dunbar collegiate kirk, 172; vicar of, 173; vicarage of, 173-174
Bolwanye, *see* Balvenie
Bombie (Bomby), NX7150; 617, 619
Bombstar, *see* Broubster
Bonath, *Fearn abbey*, 633
Boncle, Bonkle, *see* Bunkle
Bondhalff ('Bond-half of Auchtermuchty called Jerveselands'), *Lindores abbey*, 33, 33n.
Bonegraye, *see* Bawgre
Bonhard (Ballinhard) [East Bonhard Farm, NT0279], *Carriden*, 157
Bonhill (Bullill, Bullule, Bullull), NS3979; 539, 541; kirk of, 539; vicar of, 541; vicarage of, 541
Bonitoun, 217; *see also* Bonnington
Bonjedward (Bonjedburcht), NT6522; 219
Bonnat, *see* Bannaty
Bonnington (Bonitoun, Bonyntoun, Bonytoun, Bruntoun), NT1169; prebend (with Platt) of Corstorphine collegiate kirk, 107, 107n., 116, 121
Bonnyton, Bonytoun, *Trinity parish*, 13, 20
Bonshaw (Bonschaw), NS3744; 573
Bonyntoun, Bonytoun, prebend of Corstorphine collegiate kirk, *see* Bonnington
Bonytoun, *St Andrews parish*, *see* Bonnyton
Boquhapple (Boquhaple), NN6500; 73n., 94n., 152n., 269n., 290n., 351n., 423n., 465n., 651n.

Bordeaux, *France*, 435n.
Borderflat, Bordourflat, Bordourflatt, *Kelso abbey*, 230n., 234, 237, 240
Borders (Bordouris), 118
Bordie (Bordy), NS9586; 289, 292
Boreland (Bardyland, Borlame, Burdland), *Lesmahagow*, NS8340; 231, 231n., 232
Boreland (Borland), *Muthill*, NN8517; 340
Borgue (Borg), NX6348; 594; kirk of, 594
Borlame, Borland, *see* Boreland
Borlick (Bourlik), NN9539; 309
Boroustoun of Balnavadocht, Borroustoun or Balnavardocht, *Dunkeld bishopric* [Borestone Cottage, NN9718], 342, 345
Borray, *see* Burray
Borrow Muire, *Edinburgh*, 500
Borrowfield (Borrowfeild, Borrowfeld, Burrowfeild), NO7059; 358, 365, 383, 387, 388, 389
Borth, *see* Broth
Borthwick (Borthuik, Borthuike, Borthwik), NT3659; 111, 176, 450; vicar of, 111; vicarage of, 111
Borthy, *see* Bourtie
Boseymylne, Bowsie, Bussiemyln, *Largo*, 145, 145n., 148, 167, 167n.
Bosmerkin, *see* Rosemarkie
Bot Mell, *see* Rotmell
Botary (Botarie, Botarra, Bottarrie) [Mains, NJ4745], 478, 484, 489, 489n.; parsonage of, 484; vicar of, 478; vicarage of, 478, 478n.
— and Elchies, parsonage of, 489, 489n.; *see also* Elchies
Bothans (Bothanes) collegiate kirk [Yester, NT5467], 122, 172, 175, 176, 190; chaplainries of St Cosmo and St Damien, 176; prebend of St Edmund, 175; prebendary of, 172; provostry of, 172
Bothedillach, Bothedlach, *see* Nathertoun of Bathedloskis
Bothelhill, *see* Bushelhill
Bothell, Bothill, *Kelso abbey*, 223, 231, 233, 237, 240, 242
Bothelny, *see* Bethelnie
Botheris, Bothers, Bothiris, *Brechin treasurership*, *see* Betharis
Bothernok, *see* Baldernock

Bothkennar (Bothkenner, Brachkenneth, Buth Kenner, Buthkener, Buthkenneir), NS9083; 183, 184, 537, 544, 545, 547, 548; kirk of, 183-184
Bothuill, *see* Bothwell
Bothwell (Bothuill, Bothwill, Buthuill), NS7058; parsonage of, 506; vicarage of, 506
— collegiate kirk, 121, 506n., 507n., 510, 511, 515, 517, 526, 603n.; prebends of, 122, 510, 511-512, 515, 517; prebendaries of, 510, 511-512, 515, 517; provost of, 507, 507nn., 513; provostry of, 506, 506n.; *see also* Hazeldean, Kittymuir, Netherfield, Newtown, Stonehouse
Bothwellhaugh (Boithwellhauch, Bothwellhaucht), NS7157; 505, 529, 534
Botriphnie (Petruifnye, Pittrefeyne), NJ3844; 320, 320n., 460, 478, 478n.; kirk of, 478, 478n.
— and Skirdustan, parsonage of, 320, 320n.; *see also* Skirdustan
Bottarrie, *see* Botary
Bouden, Boudentoun, *see* Bowden
Bouprie (Bouipre, Bowpre) [Banks, NT1785], 62, 63
Bourlik, *see* Borlick
Bourtie (Borthy, Bourthy, Bourty, Burthy) [Kirktown of], NJ8024; 10, 16, 17, 18; kirk of, 16, 17
Bowair, Bowar, *see* Bower
Bowanhill (Bowandhill), NT4105; 259
Bowar Madin, *see* Bowermadden
Boward, *Duddingston*, 128
Bowartour, *see* Bowertower
Bowden (Boldene, Bouden, Boudentoun, Bowdane, Bowdein, Bowden, Bowdenne), NT5530; 223, 223n., 225, 225n., 229, 231, 232, 234, 235, 237, 238, 240, 241, 242; kirk of, 232, 235, 238, 241, 242; mill of, 223, 231, 234, 237, 240, 242; minister of, 225; parsonage of, 225; vicarage of, 225
Bower (Bauar, Bewar, Bowair, Bowar) [Sta, ND2058], 628, 632, 633; curate of, 633; parish of, 628, 632; vicarage of, 633
Bowermadden (Bowar Madin), ND2464; 632

Bowertower (Bowertower), ND2261; 632
Bowhillok, *Brechin parish*, 383, 387, 388
Bowhouse (Bowhous), *NO1829, *NO2001; 376
Bowis, *Collessie*, 33
Bowlyne, *see* Beauly
Bowmannis land ('callit Westerwod', West Wod alias the Bowmanis land), *Methven collegiate kirk*, 296, 298
Bowndis, West Boundis, *with Monkegy, Lindores abbey*, 32, 36
Bowpre, *see* Bouprie
Bowraik, *see* Auldbreck
Bowscheilhill, Bowschelhill, *see* Bushelhill
Bowsie, *see* Boseymylne
Bowsy, *Perth*, 283, 287
Bowyis land, *Stirling*, 291
Boychthill, *Moray deanery* [*Boath Ho, NH9255], 469
Boyne (Boyn) [Castle, NJ6165], 437
Boytsyde, *see* Boldside
Braaven [Cawdor, NH8449], vicar of, 485; vicarage of, 485, 485n.
Brabster (Brabustar) [Moss, ND2359], 632
Brachkenneth, *see* Bothkennar
Brachtoun Wall, *Whithorn priory* [Broughton Mains, NX4544], 592
Brackenhills (Brakanhillis), NJ5859; 438
Brackmont (Brakmounth), NO4222; 14
Braco (Bracocht, Breckoche), NX8775; 271, 276, 623
Brad Hawche, *see* Broadhaugh
Brahan (Bram) [Ho, NH5154], 635
Braicisnes, *see* Bresneis
Braid Medow, *Melrose abbey*, 208
Braid Yet, *Holywood abbey*, 275
Braidenhill (Bredenhill, Brydenhill), NS7467; 102, 497
Braideston (Brideistoun), NO3147; 356
Braidfeild, *see* Broadfield
Braidhauch, *see* Broadhaugh
Braidisholme, *Monkland*, 497
Braidlaw, Bredlaw, *Ancrum*, 236, 366
Braidlevis, Bredleves, Reedleyis, *Crail*, 177, 177, 181
Braidschaw, *see* Broadshaw
Braire, *Cullicudden*, 646
Brais, *see* Birse

Brakanhillis, *see* Brackenhills
Brakhaw [*Braco, NJ4951], 461
Braklo, *see* Barclo
Brakmounth, *see* Brackmont
Bram, *see* Brahan
Brasay, Brasin, *see* Bressay
Braterstoun, *see* Brotherstone
Brathinch (Brathinis, Brathynsche, Brauthinche), NO5964; 384, 386, 389
Bratho, *see* Barclo
Brathynsche, *see* Brathinch
Brauthinche, *see* Brathinch
Bray, *Ayr* [*Brae, NS3747], 558
Bray, *Scone abbey*, 331, 332, 333
Bray, *Tullynessle*, 436
Brechin (Brechen, Brechine, Breichen, Brichen, Brichin), NO6060; 7, 255, 305, 358n., 360, 364, 365, 372, 373, 377, 378, 379, 380, 381, 383, 383n., 384, 385, 385n., 386, 387, 387n., 388, 388n., 389, 390, 391, 392, 394, 399, 400, 401, 402; archdeacon of, 394; archdeaconry of, 380; bishop of, 383, 383n., 385, 387, 388, 388n., 394; bishopric of, 7, 383-389; parish of, 387, 388, 389; preceptory of Maison Dieu in, 390; vicar of, 401-402; vicarage of, 373, 378, 401-402
— cathedral, canonry of, 373; chaplain of Holy Rood altarage in, 400; chaplainry of Caldhame in, 379; chaplainry of St Agnes in, 372; dean of, 360, 364, 364n.; deanery of, 213, 213n., 364-365; Holy Rood altarage in, 400; prebend of, 378; subdeanery of, 394; treasurer of, 402; treasurership of, 377; *see also* Farnell
Breckoche, *see* Braco
Bredenhill, *see* Braidenhill
Bredlaw, *see* Braidlaw
Bredleves, *see* Braidlevis
Bredmedow, *Selkirk*, 230n.
Brego, *with Croftgary, Inchcolm abbey*, 62, 63
Breich (Breycht), NS9660; Grange of, 147, 168
Breichen, *see* Brechin
Breichmylne [*Breich, NS9660], 98
Breithchnes, *see* Bresneis
Bresay, *with Marwick, Orkney bishopric, see* Birsay

Bresneis, Braicisnes, Breithchnes, Bresmeis, *Ugginnis, Melrose abbey*, 258
Bressay (Brasay, Brasin), HU5040; 658, 659, 660, 670n.; kirk of, 660; minister of, 660
Brethertoun, *see* Brotherton
Brewland of Rosse, Brewlandnesse, *Kinnell parish*, 367
Brewland, Broister land, Browistane Land, *Kinglassie*, 24, 29, 46, 51
Brewland, *Kinneddar*, 24, 47
Brewlandnesse, *see* Brewland of Rosse
Breycht, *see* Breich
Brichen, Brichin, *see* Brechin
Brideistoun, *see* Braideston
Bridgehouse (Brighous) [Fm, NS4334]; 531
Bridgend (Brigend), *Melrose abbey*, NT5235; 208
Brig, *Rothes* [Boat o' Brig, NJ3251], 491
Brige Hauch, Brigheucht, *Lindean*, 224, 224n.
Brigend, Brigend Clowden, Brigend of Clowdane, *Holywood abbey* [Cluden, NX9379], 269, 271, 274, 274n., 276, 623
Brigend, *Haddington*, 180
Brigend, *Leuchars* [*Bigend, NO4822], 14, 20
Brigend, *Melrose abbey*, *see* Bridgend
Brigheid, *see* Bogend
Brigheucht, *see* Brige Hauch
Brighous, Brighoussis, Bryghous, *Kincardine O'Neil*, 439
Brighous, *Fearn abbey*, 633
Brighous, *Riccarton*, *see* Bridgehouse
Brightmony (Brychtmony), NH9253; 469
Brigis, *Kirkliston*, 3
Brigton (Brigtoun), *Kinnettles* [Ho, NO4146], 394
Brigton (Brigtoun), *Holywood abbey*, NX3674; 274
Brigton (Brigtoun), *Ruthven*, NO2848; 305
Brigtoun of Balgowny, *Aberdeen deanery* [Balgownie Ho, NJ9309], 423
Brims (Brymis), ND0470; 628
Brintscheillis, *see* Bruntshiels
Broadfield (Braidfeild), NX4256; 597

Broadhaugh (Brad Hawche, Braidhauch), NT8654; 186, 187
Broadshaw (Braidschaw), NT0462; 98
Broch [*Brochloch, NX5396], 275
Brochahill, *Rothes*, 491
Brochtoun, *Holyroodhouse abbey*, 91, 93
Brochtoun, *Stobo*, *see* Broughton
Brockholes (Brokhoillis, Brokholiss), NT8263; 198, 204
Brodland, *near Penick, Moray deanery*, 469
Brodland, Brodlandis, *Lindores abbey*, 30, 32, 34, 35, 36
Brodland, *Leuchars*, 13, 20
Brodland of Ogilbye, *Dunblane bishopric*, *see* Ogilvie
Brogsyd, *see* Bogside
Brointbrowme, *Monkland*, 497
Broister Land, *see* Brewland
Brokhoillis, Brokholiss, *see* Brockholes
Brolas (Brolos), NM4723; 673
Brome Bank, Bromebank, Brumebank, *Kelso abbey*, 222, 230, 233, 237, 239
Brome croft, Broun croft, Brumecroft, *Kelso abbey*, 222, 222n., 233, 237, 239
Bromehous, Bromhous, *Edrom*, 199, 203
Bromepark, *Balmerino abbey*, 58
Brontwod, *see* Bruntwood
Broomdykes (Brume Dykis), NT8753; 199
Broomhill (Brumehill), *Greenlaw*, *NT7046; 227
Broomhill (Brumhillis), *Beith*, *NS3733; 573
Broomrigg (Brumrig), NX9779; 623
Broth, Borth, *Arbroath abbey*, 360; chaplain of, 360; *see also* Arbroath
Brotherstone (Braterstoun, Bruterstoun, Brotherstanes, Brotherstanis), NT6136; 92, 93, 189, 191, 197
Brotherton (Brethertoun), NS0364; 98
Broubster (Bombstar, Brubister), ND0360; 657, 670
Broughton (Brochtoun), NT1136; 249; mains of, 249. *See also* Bruchtoun
Broumslandis, ('Fluires callit Broumslandis'), *see* Fleurs
Broun croft, *see* Brome croft
Brounhill, *Fail*, *see* Barnwell
Brounhill, Brumhill, *Dunfermline*, 27,

INDEX OF PLACES

49
Brounstoun, *see* Brunston
Brountoun, *see* Brunton
Browistane Land, *see* Brewland
Brownside (Brownsyd, Brumesyd), *Monkland*, 102, 497
Brubister, *see* Broubster
Bruchtoun, *Restalrig collegiate kirk*, 143
Brugh, *Rousay*, 656
Brume Dykis, *see* Broomdykes
Brumebank, *see* Brome Bank
Brumecroft, *see* Brome croft
Brumehill, *see* Broomhill
Brumeleis, *Monkland*, 497
Brumesyd, *Monkland*, *see* Brownside
Brumhill, *Dunfermline*, *see* Brounhill
Brumhill, *Stonehouse*, 513
Brumhillis, *Beith*, *see* Broomhill
Brumland, *Kelso abbey*, 222,
Brumrig, *see* Broomrigg
Brunston (Brounstoun) [Castle, NS2601], 567
Brunton (Brountoun), *with Dalginch, St Andrews priory* [Ho, NO3002], 15
Bruntoun, prebend of Corstorphine collegiate kirk, *see* Bonnington
Bruntshiels (Brintscheillis, Brynt Scheillis), NO4310; 146, 167
Bruntwood (Brontwd), NS5032; 531
Brunty Miln [Brunty, NO1938], 355
Bruntyhill [Brunty, NO1938], 352, 356
Brussrydoun, *Riccarton*, 531
Bruterstoun, *see* Brotherstone
Brychtmony, *see* Brightmony
Brydenhill, *see* Braidenhill
Bryghous, *see* Brighous
Brymis, *see* Brims
Buchley (Buchlay), NS5972; 497, 498
Buchthesyde, *see* Lochtysyd
Buckholm (Blakholme, Bukholme), NT4838; 207, 257, 259
Buffudder, Buffuddir, *see* Balquhidder
Buit, Buite, Buitt, *see* Bute
Buittle (Buthill, Buttill), NX7859; 611, 622; kirk of, 611; parsonage of, 622; vicarage of, 622
Bukholme, *see* Buckholm
Buklawis, *Melrose abbey*, 258
Bularnok, *Dumbarton collegiate kirk*, 539
Bullerwoll, *see* Billerwell
Bullill, Bullule, Bullull, *see* Bonhill

Bultries, *Glassford*, 531
Bunkle (Boncle, Bonkle) [Bunkle and Preston parish, NT8057], 192, 302, 342, 346; kirk of, 302; vicar of, 192; vicarage of, 192
—— and Preston, kirk of, 342, 346; *see also* Preston
Bunok, *Hartside*, 513
Burdland, *see* Boreland
Burgess Outon (Outoun Burges, Outtoun Burges), NX4540; 592, 594, 596
Burgey, *Ringwodfeild, Melrose abbey*, 259
Burgh [Broughtown, *Sanday*, HY6641], 664
Burn Mouth, Burne Mouth, Burne Moutht, Burne Mowthe, Burne Mowtht, Burnemouth, *Dunfermline*, 23, 27, 45, 49
Burnbank, *Riccarton*, NS3931; 531
Burnefute, *Leith*, 128
Burnegranis, *Calder*, 98
Burnemill, Burnmylne, *Scoonie*, 15, 21
Burnesyd, *see* Burnside
Burnetland, NT1036; 249
Burneyth, *see* Birnie
Burnfoot (Burnefuyt), *Holywood abbey*, 276
Burngrenis, *Lindores abbey*, 33
Burnmylne, *see* Burnemill
Burnside (Burnesyd), *Duffus*, NJ1669; 482
Burnside (Burnesyd), *Restenneth*, NO5050; 217
Burnturk (Burneturk), NO3308; 14, 21
Burra (Burray), *Shetland*, HU3732; 658, 660
Burra, *with Flotta, Orkney, see* Burray
Burray (Borray, Burra), *Orkney*, ND4896; 656, 658, 659, 661, 662, 663, 668, 670n., 671; kirk of, 659; minister of, 659
Burray, *with Nesting, Shetland, see* Burra
Burrow Muire, Burrow Mure *of Glasgow*, 516
Burrow muirs, *Wigtown* [Borrow Moss, NX4357], 597, 597n.
Burrow Myln, *Manuel priory*, 549
Burrow Myln, *Stirling*, 555
Burrow Rigis, *Stirling*, 550
Burrow Rudis alias Brady, *Stirling*, 555

Burrowfeild, see Borrowfield
Burrowine (Burvene), NS9789; 290
Burthy, see Bourtie
Burvene, see Burrowine
Busbie Easter, Buspy Easter [Busby, NO0326], 296, 298
Busbie Wester, Busspy Wester [Busby, NO0326], 296, 298
Bushelhill (Bothelhill, Bowscheilhill, Bowschelhill, Buschelhill), *NT7263; 223, 231, 233, 237, 240, 242
Buspy Easter, see Busbie Easter
Bussie Myln, see Boseymylne
Bussis, *Culross abbey*, 290, 291
Busspy Wester, see Busbie Wester
Bute (Buit, Buite, Buitt), NR9971, NS0665, NS1052; 143, 675
Buth Kenner, Buthkener, Buthkenneir, see Bothkennar
Buthanes croft, *Holywood abbey*, 271
Buthill, see Buittle
Buthuill, see Bothwell
Butterdean (Buttirdane), NT8064; 192
Buttergill (Buttergile), parson of, 391; parsonage of, 391; vicarage of, 391
Buttill, see Buittle
Byeris, see Byres
Byirhillis, see Boarhills
Bynnie, see West Binny
Byregrange, Byrgrange, see Biregrange
Byrehillis, see Boarhills
Byres (Byeris), *Haddington priory*, NT4976; 145, 145n., 162, 167, 167n., 177, 178, 181
Byres, prebend of Corstorphine collegiate kirk, see Rathobyres
Byrgone, *Eccles priory*, 184
Byrhillis, see Boarhills
Byris Orchard [Byres, NT4976], 11
Byrneth, see Birneth
Byrs, see Birse
Bythlie, *Markinch*, 15

Cabelley, see Cablea
Cablea (Cabelley), NN9138; 309
Cabrach (Cabrawch), NJ3827; 438; kirk of, 438
Cadboll (Catboll), NH8877; 651
Cadder (Calder), NS6172; 496, 497, 518; parish of, 497, 498n.; vicar of, 496

— and Monkland, vicar of, 518; vicarage of, 518; see also Monkland
Caddercruikis, see Caldercruix
Cadderquhult, see Cawder Cuilt
Cadeheuch, Cauldheuch, *Melrose abbey*, 208, 208n.
Cadeslie, Cadislie, see Kedslie
Cadlische, see Kedslie
Cadray, see Caldra
Caerlaverock (Carlaverok, Carlavrok), NY0067; 613, 621; kirk of, 613, 621
Cagill, see Cogle
Caillie, Caillies, *Millhorn, Coupar Angus abbey*, 354, 369
Caiphopetoun, *Hownam*, 220
Cairenische, see Carinish
Cairnborrow (Karnbarrow) [Lodge, NJ4540], 489
Cairndoon (Careindunie), NX3739; 593
Cairnfield (Carnfeild), NJ8418; 423
Cairnishome, *Holywood abbey*, 623
Cairns (Kernis), prebend of St Mary on the Rock collegiate kirk, St Andrews, 87
Cairsage, *Mull*, see Carsaig
Caistoun, *Kinnell*, 366
Caithness (Caithnes, Cathenis, Cathnes, Kaithnes, Kaitnes), 114, 418, 421, 539, 539n., 627, 628, 628n., 629, 630, 631, 632, 633, 638, 648, 655, 657, 670, 680; archdeaconry of, 632-633; bishop of, 9n., 574, 574n., 631; bishopric of, 627-629
—, cathedral and dignitaries of, see Dornoch
Caitmos, see Catmoss
Cakinche, *Inchcolm abbey*, 62
Cala [Burn, NT4716], 214
Calco, see Kelso
Caldangham, see Coldingham
Caldcoites, Caldcoittis, Caldcottis, see Cauldcoats
Calder, 382n.
Calder [Mid Calder, NT0767], 3, 18, 94, 98, 99; parsonage of, 98-99; vicarage of, 99
Calder, *Lanarkshire*, see Cadder
Calder-Clere, Caldircleir, Caldorcleir, Cawdercher [East Calder, NT0867], 224, 229, 232, 234, 237, 240; kirk of, 224; see also East Calder
Caldercruix (Caddercruikis, Caldercruikis), NS8268; 102, 497

INDEX OF PLACES 777

Caldermure, *Calder*, 98
Caldhame (Caldrum), NO4748; 217, 379, 401; chaplainry, Brechin cathedral, 379
Caldircleir, Caldorcleir, *see* Calder-Clere *and* East Calder
Caldra (Cadray, Caldraa, Cawdra, Cawdray), NT7057 [*Caldra Fm, NT7749]; 223, 226, 231, 234, 237, 240
Caldrum, *see* Caldhame
Caldschellis, *see* Cauldshiel
Caldslie, *see* Kedslie
Caldwell (Cauldwell), NS4254; 574
Caldycoites, *Markinch*, 15
Calektoun, *Paisley abbey*, 529
Calendrech, *see* Callander
Calfhill, NT5138; 207, 208, 257, 259
Calfshaw (Calfschaw, Calschaw), NT4633; 224, 224n.
Callander (Calendrech, Callender, Callendrech, Callendreth), NN6307; 295, 335, 348, 349; kirk of, 295, 348; vicar of 335; vicarage of, 335
Calludye, *St Mary on the Rock collegiate kirk, St Andrews*, 55
Calmanle, *see* Kilcalmonell
Calroust (Cornst, Corros, Corrost), NT8219; 225, 225n.
Calsay End, *Stirling Black friars*, 555
Calsayend, Causaend, *Coupar Angus abbey*, 353, 353n.
Calschaw, *see* Calfshaw
Calvennis, *see* Cavens
Calwy, *Stirling Black friars*, 555
Calyhillis, Cannyhillis, *Aberdour*, 63, 64
Cambestoun, *see* Camieston
Cambo (Cammo), NO6011; 65, 66, 177, 178, 181
Cambok [Easter Camock, NO2458], 354
Cambus (Camous), NS8594; 537, 544, 545, 550
Cambuscurrie (Cambuscurry) [Bay, NH7285], 651
Cambuskenneth (Cambuskeneth, Cambuskinneth, Cambuskynneth, Cambuskynnethe), NS8094; 290n., 307, 323n., 332, 355, 547, 548, 551; chaplain of Our Lady altar in, 551; chaplainry of Our Lady altar in, 551
— abbey, 307, 332, 537-538, 543-548, 543n.; bailie of, 547; canons of, 538, 545, 547; commendator of, 260n.,
290n., 537, 537n., 538, 543, 543n., 544, 544n.
Cambuslang (Cambuslaying, Cambuslayng, Cambuslyng, Carbuslaying), NS6459; 508, 509, 523; chapel, 509; manse of, 509; parson of, 509; parsonage of, 508-509; vicar pensionary of, 509; *see also* Kirkburn
Cambusmichael (Cambusmicheall), NO1132; 333; kirk of, 333
Cambusnethan (Cambusnetham, Cambusnethame, Cambusnethane), NS8155; 496, 515, 516; kirk of, 496; vicar of, 515; vicarage of, 515, 516
Cameron (Camerone), NT3499; 15
Camertoun, *see* Caverton
Camieston (Cambestoun, Cammestoun, Cannestoun), NT5729; 189, 191, 208, 257
Camilty (Camiltie), NT0660; 98
Cammestoun, *see* Camieston
Cammo, *see* Cambo
Camous, *see* Cambus
Campsay, *see* Campsie
Campsie (Campsay, Campsy), *Coupar Angus abbey*, NO1223; 352, 356, 368, 369, 370, 410
Campsie (Camsis), *Methven collegiate kirk*, NN9828; 296, 298
Camscheny, *Dunblane parish*, 327
Camsis, *see* Campsie
Camster, *Caithness archdeaconry*, *ND2160; 633
Camusnanesrin (Camasnanesserin), *Melfort*, 674
Canderside (Candersyd), NS7647; 233
Canis Quyis, *Wick*, 628
Canisbay (Cannesby), ND3472; 638; parson of, 638; parsonage of, 638; vicarage of, 638
Canna (Cannay), NG2405, 673, 674; parsonage of, 674
Cannabie, *see* Canonbie
Cannesby, *see* Canisbay
Cannestoun, *see* Cammestoun
Cannocht [*Cammach Hill, NN8959], 284, 287
Cannogait, *see* Canongate
Cannyhillis, *see* Calyhillis
Canonbie (Cannabie), NY3976; priory, 218, 221
Canongate (Cannogait), 92, 114, 126,

129, 132, 133, 137, 143; *see also*
  Holyroodhouse, Edinburgh
Canours land, *see* Cavouris land
Cansheill, *Riccarton*, 531
Canterland (Cantiland), NO7165; 16
Cantnaleis, *Croy* [Cantraybruich,
  NH7746], 639
Cantrafreis, *Croy* [Cantraybruich,
  NH7746], 639
Cantraydoune (Contra Downe),
  NH7946; 639
Cantward, *see* Cartward
Capecht, Capecth, Capeth, *see* Caputh
Capenoch, *Holywood*, 275
Capeth McKarthill, *see* Caputh
Capilrodrig, *see* Copitrig
Capnettis, *with Dungarthill, Dunkeld
  bishopric*, 305
Capo, NO6266; 384, 385, 388
Caprington (Caprintoun) [Castle,
  NS4036], 580
Capristoun, *see* Craibstone
Caputh (Capecht, Capecth, Capeth),
  NO0840; 301, 303, 318, 342, 344,
  345, 346; kirk of, 303, 342, 344,
  346; parish of, 301; prebend of
  Dunkeld cathedral, 318; prebendary
  of, 318; Baithous of, 303, 342, 345;
  Capeth McKarthill, 318
Caraddin, *Culross abbey*, 290
Carbellow (Carbello), NS6122; 558,
  559, 562
Carberry (Carbarie, Carbarrie, Carbarry,
  Carberie, Carbery), NT2894; 24, 25,
  26, 38, 40, 41, 42, 43, 47, 48, 53
Carboltoun, *see* Tarbolton
Carbuslaying, *see* Cambuslang
Carcary (Carkore), NO6455; 358
Cardell, *Strathspey, Elgin Black friars*
  [*Mains of Kirdells, NJ1839], 453
Carden, *Stirling*, NT0498; 26, 48
Cardene, *Airlie*, 356
Cardonald, NS5364; 539
Cardross (Cardros, Cardrous), NS3477;
  540, 542, 548, 549; parsonage of,
  540; vicar pensionary of, 542;
  vicarage of, 540; vicarage
  pensionary of, 542
Careindunie, *see* Cairndoon
Carelsyde, *see* Carolside
Carfin (Carinphin), *NS7758, *NS8346;
  506
Carfrae (Carfra, Carfray), NT5769; 162,
165, 176, 178, 181
Cargill, NO1536; 302, 303, 326, 342,
  346, 347, 347n.; kirk of, 342, 346,
  347; vicar of, 326; vicarage of, 326
Carhous, Corhous, *Lanark*, 508, 508n.
Caringtoun, *see* Carrington
Carinish (Cairenische), NF8160; 674
Carinphin, *see* Carfin
Carinvath, Carinveth, *see* Carnwath
Carkettill, *see* Kirkettle
Carkore, *see* Carcary
Carlaquhy, *see* Tarlogie
Carlaverok, Carlavrok, *see* Caerlaverock
Carleton (Carltoun), NX3937; 593
Carling croft, *Monkland*, 102
Carlinglippes, *Moorfoot, Newbattle
  abbey*, 101
Carluke (Carlouk, Carlowik, Carlowk),
  NS8450; 229, 230, 232, 233, 234,
  237, 239, 240, 241, 243, 244; kirk
  of, 229, 232, 234, 237, 240, 241,
  243, 244
Carmannoch, Carmanocht, *see*
  Carmunnock
Carmanok, *see* Carmunnock
Carmichael (Carmichaell, Carmichall),
  NS8937; 511; parson of, 511;
  parsonage of, 511; vicarage of, 511
Carmunnock (Carmannoch,
  Carmanocht, Carmanok), NS5957;
  510, 529; kirk of, 529; vicar of, 510;
  vicarage, 510
Carmure, Carnie, *with Kilbrackmont,
  Kilconquhar*, 146, 167
Carmyldie Nether, *Monkland*, 497
Carmyldie Over, *Monkland*, 497
Carmyllie (Carmilie), *Forfar*, NO5442;
  359
Carnaclocht, *Coupar Angus abbey*, 355
Carnbee (Carnbie, Carnebie, Carneby),
  NO5306; 29, 38, 42, 51, 52, 69;
  parish of, 29, 38, 51, 52; vicarage of,
  69
Carnel, Carnell, 563, 564, 571
Carnfeild, *see* Cairnfield
Carnie, *see* Carmure
Carnies (Carny) *Arbroath abbey* [Fm,
  NO9628] 360, 361
Carnies ('both Ovir and Nethir'),
  Dunkeld bishopric, 305
Carnihillis, *Kinloss abbey* [*Cairnhill,
  NJ5051], 461
Carnis Easter, *Calder*, 98

Carnis Wester, *Calder*, 98
Carnock, NT0488; kirk of, 56
Carnurrie, *St Mary's college, St Andrews*, 64
Carnwath (Carinvath, Carinveth, Carnvath, Carnweyth, Karinvath), NS9846; 520; benefice of, 505n.; parsonage of, 510, 510n.; prebend of Glasgow cathedral, 505, 505n.; vicar pensionary of, 505; vicarage of, 510, 510n.; vicarage pensionary of, 505; *see also* Muirhall
—— collegiate kirk, 521, 521n.; prebend of the aisle of Carnwath, 521; prebendary of, 521
Carolside (Carelsyde, Carrelsyd), NT5639; 200, 204
Carowdis croft, *Holywood abbey*, 623
Carpow (Carpowy, Carpowye) [Ho, NO2017], 32, 36, 322
Carquha, *see* Crauchie
Carraill, Carrell, *see* Crail
Carrelsyd, *see* Carolside
Carrick (Carrik, Cawrik, Karrik), NX5750, NY0491, NR9087, NX2794, NX3594; 113, 210, 257, 496, 571, 594, 612
Carriden (Carriddin, Carridein), NT0181; 91, 157, 159n.; kirk of, 91; kirkland of, 157; vicar of, 157; vicarage of, 157
Carrill, Cawill, *Dunfermline abbey*, 27, 49
Carrington (Caringtoun), NT3160; 121; parson of, 121; parsonage of, 121
Cars and Pikin, Pinkie et Cars, Pynkin and Cars, Pynkin and Kers, *Musselburgh*, 24, 24n., 47; *see also* Pikin, Pinkie
Carsaig, NM5421; *Mull*, 673
Carse of Gowrie (Cars of Gowrie, Kers of Gowrie), NO2322, NO2524; 285, 288, 317, 339; *see also* Gowrie
Carseburn (Carsburne), NO4751; 217
Carsegown (Carsgoun), NX4258; 597
Carsenestock (Casnestik), NX4462; 587
Carsgoun, *see* Carsegown
Carsgrainge, Carsgrang, Carsgrange, *Coupar Angus abbey*, 354, 355, 357, 369, 369n., 370, 409
Carskerdo, NO3908; 55
Carslae (Carslie), NX4358; 597
Carslogie (Carslogy), NO3514; 14

Carsrig, *Carmichael*, 511
Carstairs (Carstairis), NS9346; 495, 496, 509, 517; mill of, 495, 496; parson of, 517; parsonage of, 517; vicar of, 509; vicarage of, 509
Carsyd, Crassid, *with Bandeath, Cambuskenneth abbey*, 543, 547
Carterhauch, *Selkirk*, 230n.
Cartleyis, *Melrose abbey*, 207, 257, 259
Cartward, Cantward, *Lindores abbey*, 33, 37
Carvalge, *Mull*, 673
Caskieberran (Cassebarean, Cassebarrian), NO2500; 24, 29, 46, 51
Casnestik, *see* Carsenestock
Cassebarean, Cassebarrian, *see* Caskieberran
Casseltoun, Cassiltoun, *see* Castleton
Cassindilly (Cassindellie, Cassindulie, Cassinduly), NO3909; 55, 290
Cassindonald (Cassindonat), NO4612; 10, 13, 18, 21
Cassope, Cassakkis [Cassok Hill, NT2204], 259, 259n.
Cassoquhy, *Methven collegiate kirk*, 296, 298
Castailhill, Castelhill, *Culross abbey*, 290, 292
Castelfeild, *Jedburgh abbey*, 219
Castell of Dumbartane, *see* Dumbarton Castle
Castle Campbell (Castell Campbell, Castell Campell), NS9699; 303, 342, 346
Castlelaw (Castellaw), NT2363; 142
Castleton (Casseltoun), *Eassie*, NO3346; 412
Castleton (Casseltoun, Cassiltoun), *parish*, NY5092; 216, 221; kirk of, 216
Catboll Fischar [Cadboll, NH8877], 634
Catboll, *see* Cadboll
Catcune (Katkune), NT3559; 176, 450
Cathcart, NS5860; 520, 529, 560; kirk of, 529; vicar of, 520; vicarage of, 520
Cathenis, Cathnes, *see* Caithness
Cathkin (Caythkin) [Ho, NS6258], 511, 517
Cathnes, *see* Caithness
Catholris Myln, Kethokis, *Aberdeen deanery*, 422, 424

Catmoss (Caitmos, Cotmos), NT7145; 226, 226n.
Cattacroch, *Riccarton*, 531
Catterline (Catterling, Catterlynge, Katerling, Kathirlein), NO8678; 386, 389, 397; mains of, 383, 389; parish of, 386, 389; vicar of, 397; vicarage of, 397
Cattis, *see* Coatbridge
Caudeis croft, Cauderis croft, Caudies croft, Cawdeis croft, *Holywood abbey*, 270, 271, 275, 276, 623
Cauldcleuch [Head, NT4500], 259
Cauldcoats (Caldcoites, Caldcoittis, Caldcottis, Cauldcoit, Cauldcoites, Cauldcot), NT3070; 24, 26, 43, 47, 49, 53, 54
Cauldhame, *Markinch*, 15
Cauldheuch, *see* Cadeheuch
Cauldscheilraw, *Fogo*, 226
Cauldshiel (Caldschellis, Cauldscheillis, Cauldschiellis) , *Lindean* [Hill, NT5131], 223, 224, 231, 233, 237, 239, 242
Cauldstreme, *see* Coldstream
Cauldwell, *see* Caldwell
Cauldyhame, *Forgan*, 14
Causaend, *see* Calsayend
Cavanis, *see* Cavens
Cavens (Cavanis, Calvennis), NX9758; 609, 614
Cavernours land, *see* Tavernouris land
Cavers (Caveris), NT5315; 207, 209, 211, 211n., 223, 257, 260; kirk of, 209, 211, 211n.
Caverton (Cavertoun), NT7427; 212, 215, 220
Cavillis Myln, Cavillis Mylne, *with Easter Disblair, Lindores abbey*, 32, 36
Cawcarros, *see* Balcarres
Cawdeis croft, *see* Caudeis croft
Cawder Cuilt (Cadderquhult), NS5670; 497, 498
Cawdercher, *see* Calder-Clere *and* East Calder
Cawdor (Kwyd), NH8449; 485; vicar of, 485; vicarage of, 485
Cawdra, Cawdray, *see* Caldra
Cawill, *see* Carrill
Cawrik, *see* Carrick
Cawy, Gawie, Gawy, *Dunkeld bishopric*, 303, 305, 342, 345

Caythkin, *see* Cathkin
Cehorne, *see* Culhorn
Cengarwgerrie, *Mull*, 673
Ceres (Seres), NO4011; 55
Cessford (Cesfuird, Cosfuird, Sesfuird), NT7323;102, 220, 225, 225n., 229
Cessnock (Cesnok, Cessnak, Cessnok, Sessnak) [Castle, NS5135], 499, 531, 562, 564
Cestertoun, *see* Costerton
Chancellaris feild, *Caithness chancellory*, 630
Channelkirk (Cheindilkirk, Childinkirk, Chingelkirk, Jeindilkirk) [New Channelkirk, NT4855], 188, 188n., 190, 197, 197n.; kirk of, 190, 197, 197n.; parson of, 188, 188n.; parsonage of, 188; vicarage portionary of, 188
Channonbank, *see* Shannabank
Channons croft, *Fordyce*, 438
Chantour croft ('besyde the brig of Dunkeld'), 307
Chantouris feild, *Caithness chantory*, 630
Chapel croft, *Udny*, 463
Chapel of Garioch, NJ7223; chaplainry of Collyhill in, 449, 449n.
Chapel Outon (Outoun Chapell, Outoun Chappell, Outtoun Chappell), NX4441; 592, 594, 596
Chapel Royal (Chapell Ryell), *Stirling*, 263n., 323, 323n., 560, 602, 603
Chapelden of Kindrimy [*Chapelton of Glenkindie, NJ4216], 449, 449n.; chaplainry of, 449
Chapelheron (Chappallquhairan, Chappell Ahbirquharne, Chappellquhair), NX4541; 593, 598, 599
Chapelhill (Chapethill), *Peebles*, 223, 231, 233, 237, 240, 242
Chapelkettill, Chapell Kettill, *Forgan*, 10, 18
Chapelton (Chapeltoun), *Mearns, St Andrews priory*, 10, 18
Chapelton (Chappeltoun), *Glenbervie*, NO7382; 405
Chapelton (Chepeltoun), *Wigtown*, NX4959; 597
Chapeltoun, *Brechin parish*, *see* Magdalene Chappeltoun
Chapeltoun, Cheppeltoun, *Millhorn*,

INDEX OF PLACES 781

Coupar Angus abbey [*Chapelhill, NO2445], 354, 357, 369
Chapeltoun, *Kinloss abbey*, 463
Chapelyaird, *Leuchars*, 14
Chappallquhairan, *see* Chapelheron
Chappell Ahbirquharne, *see* Chapelheron
Chappellquhair, *see* Chapelheron
Charleton (Charltoun), NO7160; 383, 387, 388, 389
Charterhouse (Chartourhous), *Kelso abbey*, 241
Charterhouse (Charterhous, Chartous), *Perth*, 241, 283-289, 284n., 289n., 306; prior of, 283, 285, 286, 289, 289n.
Chartetoun, *Holywood abbey*, 622
Chartous, *see* Perth
Cheindilkirk, *see* Channelkirk
Cheisteris, *see* Christeris Mylne Haucht
Cheles, *see* Schelis
Chenothe, *see* Clenoch
Chepeltoun, *near Isaacstown*, *see* Magdalene Chappeltoun
Chepeltoun, *Wigtown*, *see* Chapelton
Cheppeltoun, *Millhorn, Coupar Angus abbey, see* Chapeltoun
Chernsyd, *see* Chirnside
Childinkirk, *see* Channelkirk
Chingelkirk, *see* Channelkirk
Chirmis, *see* Shirmers
Chirnside (Chernsyd, Chrinesyd, Chyrnsyd), NT8756; 166, 186; parsonage of, 186
Choismylne, *Cadder and Monkland*, 496
Chreif, *see* Crieff
Chrinesyd, *see* Chirnside
Christeris Mylne Haucht, Cheisteris, *Fogo* [*Christies Bog, NT8300], 226, 226n.
Christoun, *see* Chryston
Christskirk (Christis Kirk, Chrystis Kirk), NJ6026; 31, 32, 35, 36
Chryston (Christoun, Chrystoun), NS6870; 497, 498
Chyrnsyd, *see* Chirnside
Cill mo Charmaig, kirk of, *see* Kilmacocharmik
Cipane, *see* Kippen
Clachan (Clachin), *Tongland, Galloway bishopric* [High Clachan, NX6955], 586, 589n., 590, 592; corn mill of, 586, 589; feu crofts of, 596; waulk mill of, 586, 589, 596
Clachan (Claychtand, Clayland), *Holywood abbey*, NX8880; 269, 622
Clachran, *see* Clauchrie
Clackmannan (Clakmanan, Clakmannan, Clamanan), NS9191; 538, 543, 545, 546; kirk of, 538, 543, 545, 546, *sheriffdom*, 88, 339n.
Clackriach (Glacreuch, Glacrewch, Glakreauch) [Mains of, NJ9347], 457
Claimadie, Clamadie, *see* Claymoddie
Clairbastoun, *Inchcolm abbey*, 62
Clairlaw, *see* Clarilaw
Claivag, *see* Clevage
Clakmanan, Clakmannan, *see* Clackmannan
Clamanan, *see* Clackmannan
Clannochdyke (Clenedyk), NS8040; 229
Clanschanacht, *see* Clayshant
Claok, *see* Clayock
Claremont (Claremounthe, Clarmonthe), NO4514; 10, 18
Clarilaw (Clairlaw, Clarelaw, Clarelew, Clarilaw, Clarylaw, Clawrilaw, Cleirlaw), NT5527; 225, 229, 231, 232, 233, 234, 235, 237, 238, 240, 241, 242
Clarilaw Mains (Clarelawmanes, Clarilaw Manis, Clarilawmanes, Clarylawmanes) [Clarilaw, NT5527], 223, 231, 233, 237, 239, 240n.
Clarkston (Clerkstoun), NS8342; 232
Clary (Clarie) [Carse of, NX4260], 587, 590
Clasbany, *see* Clashbenny
Clash (Clasche), NN6306; 349
Clashbenny (Clasbany), NO2121; 285, 288
Clashendamer (Classindamir), NJ5461; 438
Clashwilly (Clasweillie), *Abernethy collegiate kirk*, 322
Classindammir, *see* Clashendamer
Classingar, *Monzie*, 349
Clatt (Clat, Clett), NJ5325; 129n., 414, 417, 418, 435, 436n.; parsonage of, 435; vicarage of, 129n.
Clatto (Cletty), *Lathrisk*, NO3507; 15, 21
Clatto, *St Andrews parish*, NO4315; 12,

20, 21
Clauchrie (Clachran), *NX4056,
    *NX4058; 597
Claunch (Clounsche), NX4248; 593
Clavege, see Clevage
Clawak, Croy and Moy, 639
Clawerhill, see Cloverhill
Clawnacht, Fordyce, 437
Clawrilaw, see Clarilaw
Clay croft, Stirling, 546
Claychtand, Clayland, see Clachan
Clayis, Lindores abbey, 30, 35
Claymoddie (Claimadie, Clamadie),
    NX4236; 595, 596
Clayock (Claok, Cloak), Bowar,
    ND1759; 632, 633
Claypotts (Clay Pottis, Claypottis)
    [Castle, NO4531], 30, 34, 35
Clayshant (Clanschanacht, Glauschant,
    Gleuschart), NX1152; 605; kirk of,
    594; vicar of, 605; vicarage of, 605
Cleddisdaill, see Clydesdale
Cleftouncoit, see Cliftoncote
Clegdenie, see Slegden
Cleirlaw, see Clarilaw
Cleish (Cleisch, Cleische), NT0998; 72,
    225, 290, 291, 292; kirk of, 225;
    parsonage of, 72
Cleland (Kleland), *NS7859, *NS7958;
    506
Clemoche, see Clenoch
Clenedyk, see Clannochdyke
Clenoch, Clemoche, Chenothe,
    Cheisteris, Lesmahagow,
    [Clannochdyke, NS8040], 231,
    231n., 233
Clentre, Clyntry, Glentre, Dunkeld
    bishopric, 303, 342, 346
Clerkington (Clerkingtoun), NT5072; 8,
    10, 16, 18, 100; parsonage of, 100;
    mill, 101; see also Corstorphine
Clerkland, Merse, Melrose abbey, 258
Clerkseat (Clerk Sait, Clerk Sett),
    NJ4752; 461
Clerkstoun, see Clarkston
Clett, see Clatt
Cletty, Lathrisk kirk, see Clatto
Cleucheid, Calder, 98
Cleucheid, Heuchheid, Eckford, 220
Clevage (Claivag, Clavege), NO0514;
    32, 36; 'vicarage of the toun of', 32
Cleyne, Cleynye, Scone abbey, 331
Cliddisdaill, see Clydesdale

Cliddismylne, see Clydesmill
Cliedisdaill, see Clydesdale
Clifton (Cliftoun), Morebattle, NT8126;
    214, 215
Clifton Hall (Cliftounhall), NT1070; 3
Cliftoncote (Cleftouncoit, Cliftoncoat,
    Elistoun Coit), NT8123; 258, 258n.
Cliftoun, with Clifton Hall, St Andrews
    archbishopric, 3
Clintlaw (Clint Law), NO2953; 354,
    357, 368, 369
Cloak, see Clayock
Clochridgestone (Clochrie Stane,
    Cloichrie Steyne), NO1413; 89
Clochtow (Clochow), *NO4852; 217
Cloichrie Steyne, see Clochridgestone
Cloisburne, see Closeburn
Cloiss, Oxnam, 219
Clonis, St Bathans priory, 192
Clony, see Clunie
Closeburn (Cloisburne, Closburne),
    NX8992; 229, 234, 237, 240, 241,
    243, 244; kirk of, 229, 234, 237,
    240, 241, 243, 244
Clounsche, see Claunch
Clova (Clovay, Colvay), NO3273; 359,
    392; chapel of, 392; kirk of, 359
Cloverhill (Clawerhill), NT1138; 249
Clowdane, Clowden, see Cluden
Clubbisgovill, with Old Goval,
    Aberdeen deanery, 422, 423
Cluden (Clowdane, Clowden), NX9379;
    270, 277
Cluikhill, Dalgety, 63
Clune, Dunkeld deanery, see Clunie
Clunes, Atholl [Wester Clunie,
    Cluniemore, NN9258], 284, 287
Clunie (Clony, Clune, Cluny), Dunkeld
    deanery, NO1043; 301, 303, 305,
    343, 344; mains of, 342, 345;
    parsonage of, 301; vicar of, 344;
    vicarage of, 344
Clunie Mylne, Cluniemylne [Clunie,
    NT2495], 29, 51
Clunie, Dunfermline abbey, see Cluny
Cluny (Clunie), Dunfermline abbey,
    NT2495; 24, 27, 46, 49
Cluny Easter, see Easter Clunie
Cluny, parish, NJ6911; 439; kirk of, 439
Clus, Lundie, 376
Cluthybege, with Lundeis, Dunblane,
    343
Clyd, see River Clyde

INDEX OF PLACES

Clydesdale (Cleddisdaill, Cliddisdaill, Cliedisdaill), NS7951, NS8645, NS8744; 113, 513, 517, 520
Clydesmill (Cliddismylne, Clydis Mylne, Clydismylne, Clydsmyll), 229, 231, 234, 237, 240, 508
Clyne (Clyn, Clynis), chaplainry, *Elgin cathedral*, *NJ8251; 478, 481, 482, 482n.
Clyne, *see* Kirkton of Clyne
Clyntes, *Leith*, 128
Clyntry, *see* Clentry
Coatbridge (Cattis, Coittes), NS7265; 497
Coates (Coites, Coittes), NT2161; 101, 102, 103
Cobinschaw Nethir [South Cobbinshaw, NT0157], 98
Cobinschaw Ovir [North Cobbinshaw, NT0157], 98
Cock Rig, NT1559; Cokrig Easter, 98; Cokrig Wester, 98
Cockairnie (Cowcarny), NO1785; 63
Cockburn (Cokburne), *NT7658, *NT1465; 304
Cockburnspath (Cokburnespethe, Cokburspethe), NT7770; 120n., 192, 193
Cockhill (Cokhill), NO5244; 366
Cocklarachy (Cowclarachquhy), chaplainry of St Mary, *St Nicholas kirk, Aberdeen*, NJ5337; 447, 447n.
Cocklaw, Coklaw, *Dunfermline*, 23, 27, 45, 49
Cockpen (Cokpen, Cowpen) [Fm, NT3263], 102, 103; kirk of, 102, 103
Cogle (Cagill), ND2657; 632
Coilgoist, *see* Cowgove
Coilheuche, Kelleuch, *Inchcolm abbey*, 62, 64
Coilhorne, *see* Culhorn
Coit, Cot, Cott [West Coates, East Coates, NO4404], 24, 25, 28, 29, 46, 47, 50, 51
Coitcoit, *Moorfoot*, 101
Coites, *Newbattle abbey*, *see* Coates
Coitland, *see* Cotland
Coitlaw, *Moorfoot*, 101
Coitmedow, *Melrose abbey*, 208
Coittes, *Newbattle abbey*, *see* Coates
Cokburne, *see* Cockburn
Cokburnespethe, Cokburspethe, *see* Cockburnspath

Cokhill, *see* Cockhill
Coklaw, *Dunfermline abbey*, *see* Cocklaw
Coklaw, *Melrose abbey*, 258
Cokpen, *see* Cockpen
Cokrig Easter, *see* Cock Rig
Cokrig Wester, *see* Cock Rig
Cokside Nether, *Moorfoot*, 101
Cokstoun, *see* Cookston
Coksydholl, *Moorfoot*, 101
Colassie, *see* Collessie
Cold Chapel, NS9324; *Hartside*,
Coldenknowis, Coldinknowis, *see* Cowdenknowes
Coldingham (Caldanghame, Coldinghame), NT9066; 198, 199, 201, 202, 203, 204, 205, 221n., 222n., 223n.; kirk of, 198, 204; parish of, 198; vicarage of, 203, 204
—— priory, 188, 188n., 194, 197, 198n., 202, 204, 221n. 222n.; commendator of, 188, 188n., 194, 197, 198n., 201, 223n., 659, 659n.
Coldown, *with Blererno, Aberdeen deanery* [*Cowden, NO7582], 405
Coldstream (Cauldstreme), NT8439; mill of, 186; minister of, 186; *see also* Lennel
—— priory, 185n., 186, 187; nuns of, 186
Colferie, *Inchcolm abbey*, 62
Colinstoun, *see* Collieston
Collace (Colles), NO1832; 322; parson of, 322; parsonage of, 322
Collessetter, Facctosetter, Occlositter, *with Northmavine, Orkney bishopric*, 658, 663, 670
Collessie (Colassie, Culessie, Cullesse, Cullessie, Cullessy), NO2813; 30, 33, 34, 35, 37, 81, 89; kirk of, 37; mill croft of, 37; vicar of, 89; vicarage of, 81, 89
Collieston (Colinstoun, Cowestoun, Malcolestoun, McClynnstoun, McCollestoun, McCullestoun) [Hill, Moor, NX8182], 269, 270, 274, 275, 622, 623
Colluthie (Culluthie), NO3319; 21, 21n.
Collyhill, chaplainry of, *see* Chapel of Garioch
Colmonell (Commenell, Commonell, Comonell), NS1484; 499, 499n., 569, 570, 572; kirk of, 499, 499n.,

570n.; parsonage of, 570; vicar of, 420; vicarage of, 420, 570
Colmslie (Commislie, Coulmeslie, Cummslie), NT5139; 209, 216, 259, 260
Colmsliehill (Coulmesliehill, Cummisliehill), NT5141; 207, 257, 259
Colmyln, *Coldingham priory*, 203
Cologne, altar of the Three Kings of, *Dundee*, 381, 381n.
Coltward, NO2137; 353
Colvay, *see* Clova
Colvend (Cowen, Culven, Culwen), NX8654; 272, 274, 613; glebe of, 272; kirk of, 613; manse of, 272; vicarage of, 272
Colwy, *see* Cowie
Colzeame Easter, *Calder*, 98
Colzeame Wester, *Calder*, 98
Comar (Cumer, Cummer), NH3331; 649; kirk of, 649
Commenell, Commonell, Comonell, *see* Colmonell
Commislie, *see* Colmslie
Commonis of Dunbar, *see* Dunbar Common
Commoun of Caldermure, *Calder*, 98
Comrie (Comry), *parish*, NN7722; 312, 334, 343; parson of, 312; parsonage of, 312, 343; vicar of, 334; vicarage of, 334, 343
Condeland [East, West Conland, NO2604, NO2504], 15
Condie (Condy, Condye), 40, 188, 286n.
Conflattis, *Monkland*, 497
Congtoun, *Dallas*, 492
Congzie Nuik, Cuinzie Nuik, *Peebles* [*Cunzierton Fm, NT7418], 253, 254
Contill, *Fife*, 291
Contin (Contane), NH4555; 645; parson of, 645; parsonage of, 645; vicarage of, 645
Contra Downe, *see* Cantraydoun
Conveth (Conveith, Convith), *parish*, *Inverness-shire*, 649, 650; kirk of, 649, 650
Conveth (Conwath), *New college, St Andrews*, NO7272; 493; curate at, 64; parson of, 493; parsonage of, 64, 493; reader at, 64; vicarage of, 64
Conwath, *see* Conveth

Conyeocht Durris, *Moray bishopric* [*Dores, NH6034], 466
Conyngaires, Cuningareis, Cunyngaires, Cunyngares, Cunynghares, Cunyngharis, *Kelso abbey*, 222, 230, 233, 234, 237, 239, 240
Conzie, *Peebles* [*Cunzierton Fm, NT7418], 254
Cookston (Cokstoun, Gokstoun, Goukstoun, Jokstoun, Sastoun), NO5810; 10, 12, 18, 20, 21
Cookston (Cuikistoun), *parish* [Kinnaird, NO2428], 382; minister of, 382; parsonage of, 382; vicarage of, 382; *see also* Kinnaird
Cookston (Cuikstoun, Cukistoun), *Brechin bishopric*, NO5961; 383, 387, 388, 388n.
Cookston (Cuikstoun, Cukistoun), *Crail*, NO5810; 177, 179, 181
Cookston (Cukestoun), *Airlie*, NO3348; 356
Copitrig, Capilrodrig, *Ugginnis, Melrose abbey*, 258, 258n.
Corbeithous, Corbethous, *Morebattle*, 214
Cordebuk, *Aberdeen deanery*, 423
Corhous, *Lesmahagow*, 229
Cormarte, *see* Cromarty
Cornst, *see* Calroust
Correllpittis, *see* Quarrypits
Corrichie, 435n.
Corros, Corrost, *see* Calroust
Cors Lies, *see* Crossleys
Cors, *Crossraguel abbey*, 567
Corsby, *see* Crosbie
Corshous, *see* Crosshouse
Corsmichaell, Corsmichall, *see* Crossmichael
Corsraguell, *see* Crossraguel
Corston (Corstoune) [Mill, NO2009], 69
Corstorphine (Corstorphin, Corstorphing), NT1972; 91, 100, 107, 109, 112, 115, 127; hospital of, 100; kirk of, 91; minister of, 98; parsonage of, 100; *see also* Clerkington, Edinburgh
—— collegiate kirk, 97, 100, 112, 113, 115, 116, 121, 130; choir of, 130; prebendaries of, 100, 107, 109, 112, 113; prebends of, 97, 100, 107, 107n., 109, 112, 113, 115, 116, 121, 130, 131; provost of, 100; provostry

INDEX OF PLACES 785

of, 100; *see also* Addiston,
Bonnington, Dalmahoy, Gogar,
Hatton, Norton, Platt, Rathobyres
Corswall, *Sweetheart abbey*, 611
Corswod Middle, *see* Mid Crosswood
Corswodburne, *see* Crosswoodburn
Corswodhill, *see* Crosswoodhill
Cortachy (Cortoquhy), NO3959; 392;
parsonage of, 391; vicarage of, 392
Corver, *Sorbie*, 593
Cosfuird, *see* Cessford
Cosland, *see* Cousland
Costerton (Cestertoun, Costertoun),
NT4363; 24, 47
Cot, Cott, *see* Coit
Cot-town (Cottoun, Cottowne), NJ8423;
422, 424
Cothlie, *Linton*, 215
Cothmay, *Kilpatrick*, 529
Cotland (Coitland), NX4154; 597
Cotlandis, *Auchtermuchty*, 33
Cotlaw, *Kirkliston*, 3
Cotmos, *see* Catmoss
Cotreoch, *see* Cutreoch
Cotyardis, *Airlie, Coupar Angus abbey*,
369
Coubyre, *see* Cowbyre
Couch Myln, *see* Touch Myln
Coudenknowis, *see* Cowdenknowes
Coul (Coull), *Markinch*, NO2703; 16
Coull (Coule, Cowll), *Aberdeenshire*,
NJ5102; 359, 426; kirk of, 359; vicar
of, 426; vicarage of, 426
Coulmeslie, *see* Colmslie
Coulmesliehill, Coulmesliehill, *see*
Colmsliehill
Countreland, Countrie land, *Symington*,
562, 564
Coupar Angus (Coupare, Couper,
Cowpar, Cowper, Cupar, Cuper),
NO2240; 303, 303n., 332, 352,
352n., 353, 355, 356, 357, 368, 369,
370, 371, 409, 410, 411, 411n.
— abbey, 282, 332, 352-364, 368-371,
409-412; abbot of, 303, 303n., 356,
368, 509, 509n.; commendator of,
352, 352n.
Coupar Grange (Coupargrainge, Couper
Grange), NO2342; 354, 357, 370
Couper, *Fife*, *see* Cupar
Courocht, *Rothes*, 491
Cousland (Cosland, Cowisland),
NT3768; 26, 38, 40, 42, 48, 53

Couston (Coustoun), NO3239; 63
Couttie (Cowty), NO2140; 369
Cowbakie (Cowbaikye, Cowbaky) [Hill,
NO4425], 13, 20
Cowbog (Cowhog, Kewbog), NT7625;
214
Cowbyre (Coubyre, Cowbyr) *of
Keithick, Coupar Angus abbey*, 352,
352n., 353, 353n., 356, 369, 370
Cowcarny, *see* Cockairnie
Cowclarachquhy, *see* Cocklarachy
Cowdenknowes (Coldenknowis,
Coldinknowis, Coudenknowis,
Cowdounknowis, Cowdun Knowis),
NT5836; 184, 189, 197, 200, 204,
228
Cowen, *see* Colvend
Cowestoun, *see* Collieston
Cowgove (Coilgoist), NS5134; 531
Cowhill [Tower, NX9582], 270, 271,
275, 276, 277, 623
Cowhog, *see* Cowbog
Cowie (Colwy, Cowye, Culray),
NS8489; 537, 543, 545, 547
Cowirla Bankis, *see* Gourlaw
Cowisland, *see* Cousland
Cowll, *see* Coull
Cownycht Wynd, *Stirling*, 551
Cowpar, Cowper, *abbey*, *see* Coupar
Angus
Cowpar, Cowper, *Fife*, *see* Cupar
Cowpen, *see* Cockpen
Cowper Mylnis, *Fife* [Cupar, NO3714],
290
Cowthrople, *Newbattle abbey*, 102
Cowtie, *Craigie*, 312
Cowty, *see* Couttie
Cowye, *see* Cowie
Coylton (Queltoun, Quiltoun), NS4119;
560; benefice of, 560
Crabbistoun, *see* Craibstone
Craftheidis, *see* Crofthead
Crafurd, *see* Crawford
Cragaruall, *with Lundeis, Dunblane*, 343
Cragduky Wester, *see* Craigduckie
Crage, *see also* Craig
Crage, *Selkirk parish*, 230n.
Crage, *Whithorn parish*, *see* Craig
Cragie, *prebend of Dunkeld cathedral*,
see Craigie
Cragie, *Perth*, *see* Craigie
Cragieluscur, *see* Craigluscar
Cragilto, Craighilto, *Dunblane*

bishopric, 303, 305, 342, 345
Craginpettok, Craginputtik,
   Craginputtok, *see* Craigenputtock
Cragttowe, *see* Craigtove
Cragy, *Ecclesgreig* [*Craigo, NO6965;
   *Criggie, NO7367], 16
Cragy, *Leuchars, see* Craigie
Cragy, *Perth, see* Craigie
Cragy, *with Claypotts, see* Craigie
Craibstone (Capristoun, Crabbistoun),
   NJ8710; 423
Craichlaw, *see* Craighlaw
Craig (Crage), *Whithorn parish*,
   NX4142; 588, 590, 592, 593
Craig, *Ettrick* [Hill, NT2515], 258
Craig End, Craigend, *Lindores abbey*,
   30, 33, 35, 37
Craig Mylne, Craigmyln, Craigmylne,
   Lindores abbey, 30, 32, 35, 37
Craig Nether, *Calder*, 98
Craig Over, *Calder*, 98
Craig Rossie (Craigrossy), NN9812;
   522
Craig, *Cambuskenneth abbey*, 537, 545
Craigbayth, *Inchcolm abbey*, 62, 64
Craigdhu (Creachdaw, Creocht Dow),
   NX3940; 594, 596
Craigdon, *near Craiglemine, Whithorn
   priory*, 593
Craigduckie Easter (Craigduky Easter,
   Easter Craigduky) [Craigduckie,
   NT1091], 23, 27, 45, 49
Craigduckie Wester (Cragduky Wester,
   Craigduky Yester, Craigenky
   Wester, Wester Craigduky)
   [Craigduckie, NT1091], 23, 23n., 27,
   45, 49
Craigemain, *see* Craiglemine
Craigenputtock (Craginpettok,
   Craginputtik, Craginputtok,
   Craiginputtog, Craigyn Puttok,
   Craygyn Puttok), NX7782; 269, 270,
   274, 275, 622, 623
Craigflour, *Culross abbey*, 291
Craigfoodie (Craigfudie, Craigfudy,
   Fudeis), NO4017; 2, 5, 14,
Craigforthie (Craigforthy), NJ8019; 32,
   36
Craigfud, *Strathmiglo*, 69
Craighall, NO4010; 19, 55
Craigheid, NO0330; 323
Craighilto, *see* Craghilto
Craighlaw (Craichlaw), NX3061; 607

Craighouse, *Cullicudden*, 646
Craigie (Cragie), prebend of Dunkeld
   cathedral, 312; prebendary of, 312
Craigie (Cragie, Cragy), *Perth*, 26, 48,
   74, 74n., 315, 315n.
Craigie (Cragy), *Leuchars*, NO4524; 13,
   20
Craigie (Craigy), *parish*, 528; kirk of,
   528; vicar pensionary of, 528
Craigie (Craigy, Cragy), *with Claypotts*,
   NO4231; 34, 35
Craigilmain, *see* Craiglemine
Craiglemine (Craigemain, Craigilmain),
   NX4039; 594, 593, 596
Craiglethie (Craigleithe), NJ4952; 461
Craigluscar (Cragieluscur, Craigluscer,
   Craigluscour), NT0690; 23, 27, 45,
   49
Craigmillar (Craigmiller), NT2871; 25,
   47, 93
Craignathro (Craignathow), NO4648;
   217
Craigneate, *Coupar Angus abbey*, 355
Craigo, *Logie-Montrose*, NO6764; 373
Craigo, *Riccarton*, 531
Craigrothie (Craigrothe), NO3710; 55
Craigsyd, Craigsyid, *Kinglassie*, 29, 51
Craigtoun, *St Andrews archbishopric*
   [Park, NO4714], 10, 12, 13, 19, 20,
   21
Craigtoun, *Aberdeen Trinitarians*, *see*
   Craigtove
Craigtoun, *Wranghame, Aberdeenshire*,
   Lindores abbey, 31, 36
Craigtove (Cragttowe, Craigtoun,
   Craigtowy), 418, 421, 558, 559
Craigtuly, Craigynthoye, Craigentui,
   chaplainry of Luss, 542, 542n.
Crail (Carraill, Carrell, Craill), NO6107;
   11, 12, 18, 19, 58, 65, 66, 162, 163,
   164, 165, 177, 178, 179, 181, 182;
   grammar schoolmaster of, 66; kirk
   of, 162, 163, 164, 165, 179; minister
   of, 66; parish of, 177, 178; reader at,
   82
—— collegiate kirk, 65-66, 82; Holy
   Rood altar, 66; Our Lady altar, 65;
   prebenbary of St John the
   Evangelist's altar, 66; prebend of
   'Our Lady servant', 66; prebend of St
   John the Evangelist, 66;
   prebendaries of, 65; prebendary of
   Holy Rood altar, 66; prebendary of

Our Lady altar, 65; prebendary of Our Lady at the High altar, 66; prebendary of St Catherine's altar, 66; prebendary of St James' altar, 66; prebendary of St John the Baptist's altar, 66; prebendary of St Nicholas' altar, 66; provost of, 82; provostry of, 82
Crailing (Craling, Crelling), NT6824; 219; kirk of, 217, 219
Craling, *see* Crailing
Crammond Riggis [Cramond, NT1876], 150
Cramond (Crammond, Cramound, Crawmond, Crawmound), NT1876; 126, 144, 149; altar of St Thomas in, 144; chaplain of St Thomas' altar in kirk of, 149-150; kirk of, 149, 149n., 150, 342, 346, 302, 342, 346; mains of, 302, 342, 344, 346, 347; vicar of, 126; vicarage of, 126
Cramond Island (Crawmond Inche, Yle of Crawmond), NT1978; 304, 342, 346
Cramschaw Myln, *see* Cranschaw Myln
Crangillis, Cringlis [Cringie Law, NT4418], 207, 257
Cranschaw Myln, Cramschaw Myln, *Dreghorn*, 582, 583
Cranshaws (Cranschawis), NT6861; 196; parsonage of, 196
Cranston (Cranstoun), *Moorfoot, Newbattle abbey*, 101
Cranston (Cranstoun), *parish*, NT3866; kirk of, 3; vicarage of, 112
Cranstoun Riddel (Cranstoun Riddell, Cranstoun Ryddell, Cranstounriddell), NT3865; 96, 124, 202, 202n., 222, 222n.; chaplainry of, *St Salvator's college, St Andrews*, 96
Cranstoun, *Selkirk*, 263
Craquha, *see* Crauchie
Crassid, *see* Carsyd
Crathie (Creth, Crothe), NO2694; 538, 543, 545, 546; kirk of 538, 543, 545, 546
Crauchie (Carquha, Craquha), NT5678; 192, 194; *see also* Easter Craquho
Craufurdmure, *see* Crawfordmuir
Crawford Douglas, *see* Crawford Lindsay
Crawford Lindsay (Craufurd Lindesay, Crawford Lindsay, Crawfurd Lindsay, Crawfurd Lyndesay) [Crawford, NS9520], 71, 71n., 91, 91n., 92; parsonage of, 91, 91n.; vicar of, 71n., 512; vicar portionary of, 92; vicarage of, 71, 71n., 91, 91n., 512, 512n.
Crawfordmuir (Craufurdmure, Crawfuirdmure) [Crawford, NS9520], 102, 103
Crawmond Inche, *see* Cramond Island
Crawmond, Crawmound, *see* Cramond
Crawnarland, Crownarland, *Dunfermline*, 27, 49
Craygyn Puttok, *see* Craigenputtock
Creachdaw, *see* Craigdhu
Crechmond, *see* Crimond
Cree, *see* River Cree
Creich (Creiche, Creycht), NO3221; 30, 33, 35, 58, 78; parish of, 33; vicar of, 78; vicarage of, 78; Derache land, Derachland, Deratland of, 30, 30n., 35
Creichmond, *see* Crimond
Creichtoun, see Crichton
Creif, Creiff, *see* Crieff
Crelling, *see* Crailing
Creocht Dow, *see* Craigdhu,
Creth, *see* Crathie
Crevane, *see* Kirvennie
Creycht, *see Creich*
Creychtmond, *see* Crimond
Creychtoun, *see* Crichton
Crichton (Creichtoun), NT3862; 450
—— collegiate kirk, 108, 109, 119, 128, 176, 182, 450, 450n.; prebendaries of, 108, 109; prebends of, 108, 109, 182; provostry of, 119; *see also* Arniston, Middleton, Vogrie
Crichton (Creychtoun), *Morebattle*, 214
Crie, *see* River Cree
Crieff (Chreif, Creif, Creiff), NN8721; 300, 322, 323, 328, 337; chaplainry of St Michael's altar in, 337; chaplainry of, 328; prebendaries of Dunkeld cathedral, 322-323, 323n.
Crimond (Crechmond, Creichmond, Creychtmond), NJ6214; 431, 440; parson of, 440; parsonage of, 431, 439, 440; vicar of, 431
Cringlis, *see* Crangillis
Cristelhill, *Eskdalemuir, Melrose abbey*, 259

Croce land, *Fail*, 558
Croce, *see* Rede Croce
Croft Lytilltoun, *Stirling*, 555
Croft, *Lagganallachy*, 309
Croftes, *Holywood abbey*, 270
Croftgary (Crogary), NT1886; 62, 63
Crofthead (Craftheidis, Croftheddis, Croftheidis), *Brechin bishopric*, 384, 385, 389
Croftleis, *Holywood abbey*, 270, 271, 275, 276
Crofts (Croftis), *Rothes*, NJ2850; 491
Crogary, *see Croftgary*
Crokascheill, *Cambuslang*, 508
Cromar, *NJ4404, *NJ5104; 418, 421
Cromarty (Cormarte, Crumarty, Crumbathy), NH7967; 634, 637, 642; curate of, 642; kirk of, 634, 637, 642; minister of, 635; vicar of, 635, 643; vicarage of, 642-643
Crombie (Crummy, Crummye), NT0485; 23, 46, 292, 294; kirk of, 292; mills of, 290; minister of, 292
Cromdale (Cromdall, Cromdell, Crowmdell), NH2728; 480; *see also* Advie
Cromlet (Crumlotis), NS7367; 497
Cronklie, *Edrom*, 200
Cronystoun, *Ecclesgreig*, 16
Crookedshaws (Cruikitschawis, Crukit Schawis), NT8025; 214
Crosbie (Corsby), NS2149; 562
Cross kirk (Corce), *parish*, *Sanday*, HY6238; 660, 666n.; minister of, 660
Cross kirk (Corce, Cors), *parish*, *Westray*, 660, 666, 667, 667n.; parson of, 666, 667; parsonage of, 666, 667
Cross kirk (Croce Kirk), *Peebles*, 249, 249n., 254
Crosshouse (Corshous), NT2363; 142
Crossleys (Cars Lies), NX9580; 623
Crossmichael (Corsmichaell, Corsmichall, St Michael's Cross), NX7366; 270, 271, 274, 276, 611, 613, 622, 623; kirk of, 611; parsonage of, 622; vicarage of, 622
Crossraguel (Corsraguell, Crosraguell) Abbey, NS2708; 567, 568, 570, 600, 600n.; abbot of, 567, 567n., 568, 568n., 600, 600n.; mains of, 567
Crosswoodburn (Corswodburne), NT0557; 98
Crosswoodhill (Corswodhill), NT0456; 98
Crothe, *see* Crathie
Crovis, *see* Cruives
Crowe Sanct Monanis, *see* St Monance
Crowes, *see* Cruives
Crowmdell, *see* Cromdale
Crownarland, *see* Crawnarland
Croy, NH7949; 639
—— and Moy, kirks of, 639; parson of, 639; *see also* Moy
Cruggleton (Crugiltoun), NX4843; 593, 594, 596, 601; vicar of, 601; vicarage of, 601; parish of, 593, 596
—— Castle (Crugiltoun Castell), 593
Crugiltoun Cawnis [*Kevans, NX4642], 593, 593n.
Cruif, *Glasnick, Galloway bishopric*, 588
Cruik [*Crook of Devon, NO0300], 58
Cruikitschawis, *see* Crookedshaws
Cruives (Crovis, Crowes, Cruvis, Crwis), *near Persley, Aberdeen bishopric*, 416, 417, 423, 423n.
Crukit Schawis, *see* Crookedshaws
Crukitheucht, *Eskdalemuir, Melrose abbey*, 259,
Crukitschot, *Paisley abbey*, 530, 530n.
Crumarty, *see* Cromarty
Crumbathy, *see* Cromarty
Crumlat, *Monkland*, 102
Crumlotis, *see* Cromlet
Crummy, Crummye, *see* Crombie
Crumrig (Tuinrige), NT7344; 227, 227n.
Crumzeane's land, *Holyroodhouse abbey*, 92
Crunan, *near Arthurstone*, 353, 356
Crurie (Crury), NY2594; 259
Cruvis, *see* Cruives
Crwis, *see* Cruives
Cubeneburne, *Eskdalemuir, Melrose abbey*, 259
Cudbertstoun, *Linlithgow*, 156
Cuddrurie, *near Culscadden*, 619
Cudoig, Kudag, *Kirkchrist*, 587, 589
Cuggerglak, *Kinnell parish*, 366
Cuiken (Cuikkin, Cwickin), NT2361; 142, 143
Cuikis croft, Kirkis croft, *Clachan, Galloway bishopric*, 586, 589
Cuikistoun, *parish, see* Cookston
Cuikstoun, *see* Cookston

INDEX OF PLACES 789

Cuillhill, *Monkland*, 497
Cuinzie Nuik, *see* Congzie Nuik
Cukestoun, Cukistoun, *see* Cookston
Culdrachy, *near Linlithgow*, 11
Culessie, *see* Collessie
Culhorn (Cehorne, Coilhorne) [Mains, NX0858], 598, 599
Culilow, *see* Cullalloe
Culingis, *see* Cullings
Culkaye [East Culkae, NX4245], 594
Cullalloe (Culilow), NT1888; 63
Cullarby, Cullartby, *see* Cullurpattie
Cullelung, *see* Killylung
Cullen (Cullane), NJ5167; 437, 438, 455; curate of, 438
Cullen collegiate kirk, 455; provostry of, 455
Cullesse, Cullessie, Cullessy, *see* Collessie
Cullicudden (Cullicudyne), NH6664; 646; parson of, 646; parsonage of, 646
Cullings (Culingis), NN7606; 343
Cullurpattie (Cullarby, Cullarthy), NX1062; 588, 590
Culluthie, *see* Colluthie
Culmalzie (Culmalzo), NX3753; 594, 596
Culman, Culmanan, *Soulseat abbey*, 598, 599
Culmoker, *with Barledziew, Sorbie*, 593
Culquha, NX6958; 585n.
Culquhirk (Culquhork), NX4256; 597
Culray, *see* Cowie
Culross (Culrois, Culros), NS9885; 41n., 260, 260n., 266, 266n., 289, 292, 293, 294, 522; altar of St Mungo in abbey kirk of, 522; chaplain of altar of St Mungo in abbey kirk of, 522; kirk of, 292; minister of, 292; saltpans of, 294; town of, 289, 294
—— abbey, 289-294, 350; commendator of, 260, 294n.; monks of, 290-293
Culsalmond (Culsalmonth, Culsalmound), *Lindores abbey*, kirk of, 32, 36. *See also* Kirkton of Culsalmond
Culscadden (Culschaddan), NX4649; 619
Cult Over and Nether, *Strageath*, 349
Culter [Ho, NT0234], 526, parson of, 554; parsonage of, 554; vicarage of, 554
Cultoquhey (Cultoquhais), NN8923; 343
Cultranie Beg, Cultranie Bege, Cultrany Beg [*Coulterenny, NO0635], 304, 342, 345
Culven, Culwen, *see* Colvend
Cumbrae (Cumray), NS1655; 528; kirk of, 528
Cumer, Cummer, *see* Comar
Cummertrees (Cummertreis), NY1466; 265, 266, 279; parson of, 265; parsonage of, 265
Cummisliehill, *see* Colmsliehill
Cummslie, *see* Colmslie
Cummyir, Cummyr, *Lesmahagow*, 231, 233
Cumnock (Cumnok), NS5820; 299, 299n., 484, 637; parson of, 299, 299n., 484, 559, 561, 562, 637; parsonage of, 559, 561; vicarage of, 559, 561
Cumray, *see* Cumbrae
Cunighame, Cuninghame, *see* Cunninghame
Cuningareis, *see* Conyngaires
Cuninghamheid, *see* Cunninghamhead
Cunlie, *with Ardgathen, Monymusk*, 447
Cunninghame (Cunighame, Cuninghame, Cunyghame), NS3146, NS4046; 113, 575, 580, 581, 582
Cunninghamhead (Cuninghamheid), NS3741; 562
Cunyghame, *see* Cunninghame
Cunyngaires, Cunyngares, Cunynghares, Cunyngharis, *see* Conyngaires
Cunynghar Thomesonis Park, *Coupar Angus abbey*, 370
Cupar (Couper, Cowpar, Cowper), *Fife*, NO3714; 10, 14, 17, 18, 81, 87, 89, 291; chaplainry of Our Lady in St Catherine's kirk, 81; kirk of, 14, 17; St Catherine's kirk, 81; vicarage of, 87; Cowper Mylnis, 290
Cupar, Cuper, *abbey*, *see* Coupar Angus
Curbeg, *Restenneth*, 217
Curemyll, *Tullynessle*, 436
Curling, *NT5126; 129n.
Curogis, *Greenlaw*, 227
Currie (Curry), NT1867; 101, 106, 110, 117, 139; chaplain of St Mungo's altar in, 139; parsonage of, 117; St

Mungo's altar in, 139; vicar of, 106; vicarage of, 106
Curtestoun (Turtestoun), *Kinnell*, 366
Cuschnye, see Cushnie
Cushnie (Cuschnye) [Leochel-Cushnie, NJ5210], 444; parson of, 444; parsonage of, 444; vicarage of, 444
Custorum, *Lagganallachy*, 309
Cuthilhill, see Cuttlehill
Cuthill, see Quithel
Cutilhill, see Cuttlehill
Cutreoch (Cotreoch), NX4635; 593
Cutrig, see Todrig
Cuttlehill (Cuthilhill, Cutilhill), NT1589; 62, 63
Cwickin, see Cuiken

Dachy, see Dallachy
Dacrus, see Dalcross
Daftmill (Daftmyll, Daftmylne), NO3112; 33, 37
Dailly (Daylie), NS2701; 567; kirk of, 567
Dainyeltoun, see Danieltoun
Dairsie (Darsie, Dersy, Dersye), NO4117; 2, 5, 10, 14, 17, 18, 79; kirk of, 14; vicar of, 79; vicarage of, 79; waulk mill of, 10, 18
Dalarossie (Tallaracie, Tullaracie) [Moy and Dalarossie parish, NH7630], 466; kirk of, 466
Dalcarty, *Obstule*, 637
Dalcove (Dalcoif, Dalcoiff) [Mains, NT6532], 189, 190, 191
Dalcross (Dacrus), NH7650; 472; kirk of, 472
Daldaicht, *Croy*, 639
Daldowe Easter [Wester Daldowie, NS6662], 497
Daldowe Wester, see Wester Daldowie
Daldryve, Dalryve, *with Guillyhill, Holywood abbey*, 271, 623
Daleally (Dillalie, Dillallie), NN2521; 285, 288
Dalgady, Dalgathe, Dalgatie, Dalgaty, see Dalgety
Dalgarno [*Dalgairn, NO3715], 91; parsonage of, 91; vicarage of, 91
Dalgety (Dalgathe, Dalgady, Dalgatie, Dalgaty), *Inchcolm abbey, Fife* [Bay, NT1683], 62, 63, 64, 64n.; kirk of, 62; parish of, 63; parson of, 64; vicarage of, 64
Dalgety (Talgaty), *Brechin*, NO6159; 383
Dalginch (Dallginche), NO3102; 15
Dalgrauch, *Croy*, 639
Dalhousie [Grange, NT3163], 101
Dalkeith (Dalkeyth), NT3367; 103, 103n., 110, 115, 115n., 116, 116n., 141, 142; vicar of, 110; vicarage of, 110
— collegiate kirk, 115, 115n.; provost of, 115, 115n.
Dallachy (Deachy, Dachy), *Aberdour and Dalgety*, NT2086; 63, 64
Dallaquhy, *Arbroath abbey* [Upper, Nether Dallachy, NJ6365, NJ6464], 359
Dallas (Dolas, Doles, Dolles), NJ1252; 481, 482, 482n., 492, 493; mains of, 492; parson of, 493; parsonage of, 492-493; vicar of, 481; vicarage of, 481, 482n.
Dallawoodie (Dallavodie, Dallawodie, Dallewodie, Dalwoody), NX9579; 270, 275, 276, 623
Dallginche, see Dalginch
Dalmahoy [Mains, NT1467], prebend (with Hatton) of Corstorphine collegiate kirk, 97, 107, 107n., 112, 121
Dalmarnock (Dulmernocht), NN9945; 336, 336n.
Dalmeath (Dunmeth, Tunmeth, Twnmeth), *parish* [*Dunmeath, NJ4237], 426, 438; kirk of, 426, 438; vicarage of, 438
Dalmellington (Dalmellingtoun), NS4806; 560; curate of, 560; vicar of, 560; vicarage of, 560-561
Dalmeny (Dumany, Dummanie, Dummany, Dunmay, Dunnany), NT1477; 216, 219; altarage of Our Lady in kirk of, 154-155; chaplain of Our Lady altar in kirk of, 154-155; chaplain of St Cuthbert's altar in kirk of, 149-150, 159; chaplainry of St Cuthbert's altar in kirk of, 149-150; kirk of, 216, 219; vicar of, 152; vicarage of, 152
Dalnacabok, *Coupar Angus abbey*, 354
Dalpowie (Dalpowe, Dalpowy, Dawpowe), NO0539; 305, 342, 345
Dalrilzane, *Kinclaven*, 306

INDEX OF PLACES 791

Dalroy (Dalry), NH7644; 639
Dalry, *Croy and Moy*, see Dalroy
Dalry, *Cunninghame*, NS2949; kirk of, 573, 576, 578, 579; parish of, 576, 578
Dalry, *Wigtown*, NX6281; parsonage and vicarage of, 605
Dalryve, *see* Daldryve
Dalserf (Dalsuff), NS7950; 574
Dalsuff, *see* Dalserf
Dalswinton (Daswintoun, Deswintoun), NX9385; 594, 596
Daltulich (Daltulie, Daltuly), NJ9848; 479
Daluany, *Coupar Angus abbey*, 355
Dalwoody, *see* Dallawoodie
Dalziel (Dalzeall, Dalzell), NS7556; 499, 519; parsonage of, 519; kirk of, 499; vicarage of, 519
Damate, Damatt, *barony, Aberdeen bishopric*, 415, 417n.
Damberny, *see* Dunbarney
Dambray, *see* Denbrae
Damheid, *see* Denhead
Danestone (Deinstoun, Deynstoun), NJ9210; 422, 423
Danieltoun, Dainyeltoun, *Melrose abbey*, 208, 258
Darchester, *see* Darnchester
Darnchester (Darchester, Dernchester, Dernechester), NT8142; 184, 186, 187
Darness, *see* Deerness
Darnick (Dernik), NT5334; 208, 209, 258
Darnley (Darnly) [Mains, NS5359], 501; chapel of St Ninian, 501, 501n.; chaplainry of chapel of St Ninian at, 501, 501n.
Darsie, *see* Dairsie
Daswintoun, *see* Dalswinton
Dathan Lytle, *Edderton*, 652
Dathan Mekle, *Edderton*, 652
David, *see* Daviot
Davidston (Davidstoun), NS6770; 497, 498
Daviot (David), NH7239; 466; kirk of, 466
Dawglie, *see* Dough
Dawik, *see* Dawyck
Dawmoirmyln, Dawmoirmylne, *Restalrig collegiate kirk*, 142, 143
Dawpowe, *see* Dalpowie

Dawyck (Dawik) [Ho, NT1735], 250
Daylie, *see* Dailly
Deachy, *see* Dallachy
Dead Side (Deid Scheip), NT2718; 209, 209n.
Dee, *see* River Dee
Deer (Deir) [Old Deer, NJ9747], 457, 458, 459; mains of, 458; parish of 457, 458; vicarage of, 458
Deer abbey, 451, 457, 457n., 458, 459, 459n.; commendator of, 17, 17n., 459n.
Deerness (Darness, Deirenes, Deirnes, Derenes, Dernenes, Derneness, Dernis), HY5606; 655n., 656, 658, 660, 661, 662, 663, 669, 671; kirk of, 659; minister of, 660; parish of, 662
Deid Scheip, *see* Dead Side
Deinstoun, *see* Danestone
Deintoun *of Stobo*, 249, 249n.
Deinyn, Dynein, *Crossraguel abbey*, 567
Deipdaill, Dipdel, Dypdall, *with Northmavine, Orkney bishopric*, 659, 663, 670
Deir, *see* Deer
Deirenes, Deirnes, *see* Deerness
Delting, HU4067; 658, 660, 670; kirk of, 660; minister of, 660
Delverdis, *see* Dillavaird
Demperston (Dempstertoun), NO2211; 33
Den Mylne, Denemylne, *Lindores abbey*, 32, 35
Denbank, *Monkland*, 497
Denbie (Denbray), NY1072; 209, 259
Denbrae (Dambray, Denbray, Denebray, Tenenbray), NO3918; 10, 12, 18, 20, 21
Denbray, *see* Denbie
Dene Bank, *Monkland*, 102
Dene, mains of, *Kilwinning abbey*, 573
Denebray, *see* Denbrae
Deneheid, *see* Denhead
Deneis Perk, *Peebles*, 254,
Denemylne, *see* Den Mylne
Denes feild, *Caithness deanery*, 630
Denhead (Damheid, Deneheid, Denneheid, Tenenheid), NO3810; 10, 13, 18, 20, 21
Dennarles, Dennartes, Dennerles, Dennerleis, *Greenlaw* [*Dunterlee,

NT7355], 235, 235n., 238, 241
Denneheid, see Denhead
Denork (Dunmork, Dunnork, Dunnorke, Innerork), NO4513; 10, 13, 18, 20, 21
Densyd croft, *Glasgow*, 498
Densyd, *Glasgow*, 516
Derache land, Derachland, Deratland, *of Creich, Lindores abbey*, 30, 30n., 35
Derane, see Durran
Derenes, see Deerness
Deringtoun, *Kelso abbey*, 223, 233, 237, 239
Dernchester, Dernechester, see Darnchester
Derneness, see Deerness
Dernik, see Darnick
Dernis, see Deerness
Derskfurd, see Deskford
Dersy, Dersye, see Dairsie
Desfurd, see Deskfurd
Deskford (Derskfurd, Desfurd, Deskfurd) [Kirktown of, NJ5061], 437, 438; curate of, 438
Deswintoun, see Dalswinton
Dettambrage, *with Clashendamer, Fordyce*, 438
Dewane, Dowam, Dowane [*Glendevon, NS8139], 231, 231n., 233
Deymyne, see Drymen
Deynstoun, see Danestone
Didschaw, 212
Die, see River Dee
Dikheid, *Calder*, 98
Dikie, see Dykes
Dillalie, Dillallie, see Daleally
Dillavaird (Delverdis) [Mains of, NO7482], 405
Dillycowtry, see Tillicoultry
Dilspro, *near Persley, Aberdeen bishopric*, 422, 423
Dingwall (Dingel, Dingwell), NH5458; 636, 639, 640; chaplain of the chaplainries of St Laurence and Ardeville in, 636; chaplainries of St Laurence and Ardeville in, 636; minister of, 639; vicar of, 639, 640; see also Ardeville
Dinnans (Drumnance, Drumnanis, Dunance), NX4740; 594, 596, 598, 599
Dinrod, see Dunrod

Dipdel, see Deipdaill
Dipple (Dippill, Dyppil, Dyppill), NJ3258; 474; parson of, 474; parsonage of, 474
Dirleton (Dirltoun), NT5183; 173, 190; St Catherine's chapel in, 190-191
—— collegiate kirk, 173; provost of, 173; provostry of, 173
Disart, see Dysart
Dixsoun croft, *North Berwick priory*, 167
Dodd (Dod, Henwode alias Dod), NO4539; 217, 217n.
Doggarflat, *Dryburgh abbey*, 189, 191, 197
Dolas, see Dallas
Doler, see Dollar
Doles, see Dallas
Dollar (Doler, Dolour, Doloure), NS9698; 24, 25, 26, 39, 46, 47, 48, 51, 52, 53, 62, 63, 63n., 88, 303, 339, 342, 346, 350; kirk of, 62, 63; mill of, 24, 46; parish of, 26, 48, 52; vicar of, 339n.; vicar pensionary of, 88, 339n.; vicarage of, 339; vicarage pensionary of, 88, 88n.; *see also* Dollarshire
Dollarbank (Bank), NS9598; 24, 24n., 25, 46, 47, 51
Dollarbeg (Dolour Beig, Dolour Brig), NS9796; 24, 46; mill of, 24, 46
Dollarshire, 26, 37, 38, 41, 42, 44, 48, 53; *see also* Dollar
Dolles, see Dallas
Dolour Beig, Dolour Brig, see Dollarbeg
Dolour, Doloure, see Dollar
Dolourschyre, see Dollarshire
Dolphingston (Dolphingtoun), *Oxnam*, NT3872; 219
Dolphinton (Dolphingtoun), *Lanark*, NT1046; 504; parson of, 504; parsonage of, 504
Done Rossness, see Dunrossness
Done, see River Don
Donelstoun, *with Barledziew, Whithorn priory*, 593
Donibristle (Donibirsall, Dunnybirsall, Dynbirsall, Dynibirssall, Dynybirsall), NT1688; 62, 63, 64
Donning, see Dunning
Donyface, see Duniface
Donynald, Donynauld, see Dunninald
Dorare, see Dorrery

INDEX OF PLACES 793

Dores (Durris), kirk of, 472
Dornoch (Dornoche, Dornocht), NH7989; 628, 633
—— cathedral, 631n.; chancellor of, 630; chancellory of, 630; chanter of, 630; chantory of, 630; common kirks of, 631; dean of, 630; deanery of, 630; stallar in, 633, 638; treasurer of, 630; treasurership of, 630; *see also* Kildonan
Dorrery (Dorare), ND0755; 628
Dough, Dawglie, *Cambuskenneth abbey*, 547, 548
Douglas (Dowglas), NS8330; 506; parson of, 506; parsonage of, 506
Doune (Doun), *Perthshire*, NN7301; 63n., 64n., 300n., 345; St Fillan's chaplainry, 345
—— Castle (Castell of Doun), 345
Doune, *Forfar, see* Dun
Dounfeild, *see* Downfield
Dounglas, *see* Dunglass
Dounie (Downe, Rowny), *Edderton*, NH6986; 633, 652
Dounie (Dunny), *Inverness-shire*, NH5142; 649
Dounie, Douyne, *Lathrisk*, 14, 21
Dounreay (Dourne, Drumray), NC9867; 657, 670
Dourne, *see* Dounreay
Douyne, *see* Dounie
Dow Croft, *Fearn abbey*, 633
Dow, *see* Dull
Dowally (Dowalie, Dowallie), NO0047; 303, 305, 342, 345, 346; kirk of, 303, 342, 346
Dowam, Dowane, *see* Dewane
Dowblahill, *Holywood abbey*, 623
Dowcattis land, Dunccatis landis, *Culross abbey*, 291, 293
Dowglas, *see* Douglas
Dowglen (Downland) [Hill, NY3389], 223, 230, 233, 237, 239
Dowhill, NT1197; 62, 62n.
Down, Downe, *see* Dun
Downfield (Dounfeild, Tounfeild), NO3407; 15, 21
Downichin, Downychin, *see* Dunnichen
Downingis [Downing Point, NT1582], 63
Downland, *see* Dowglen
Downy, *Fearn, see* Dounie, *Edderton*
Downy, *Coupar Angus abbey*

[*Donnies, NO1962], 354
Draffan (Draffyn), NS7945; 233
Drainie (Drany, Dranyes), NJ2168; 453n., 478, 482, 484, 490, 493; *see also* Ogston
Drany, Dranyes, *see* Drainie
Draquhadlie, *see* Drochedlie
Drava, *see* Dreva
Dreghorn (Dreghorne), NS3538; 576, 578, 581, 582, 583; curate of, 581; glebe of, 583; kirk of, 576, 578; vicar of, 581, 582; vicar pensionary, 581; vicarage of, 581, 582, 583
Drem (Trem), NT5079; 10, 18
Dreva (Drava, Drewar), NT1435; 249, 496
Drewar, *see* Dreva
Drilaw, *see* Drylaw
Dringtoun, *see* Edrington
Drochedlie (Draquhadlie), NJ5561; 437
Dron (Drone) [South, Wester, NO4217, NO4116], 13, 20
Dronlo, *with Pitermo, Lundie*, 376
Dronshiel (Drumscheill), NT7055; 184
Drum, *King Edward*, 458, 458n.
Drumaddie, *see* Drummoddie
Drumanstane, Drumanstann, Drumanston, Drumanstoun, *see* Drummaston
Drumarell, *see* Drummorral
Drumasson, *see* Drummaston
Drumbekschill, Drumbrakischill, *see* Dumbraxhill
Drumblade (Drwmblaitt), NJ5840; 445; parish of, 445
Drumboy, NS6138; 303, 305, 342, 345
Drumcairn (Drumcarnie), 202n., 219, 222n.
Drumcardin, *Dunblane bishopric*, 349
Drumcarin, Drumcarne, *Methven collegiate kirk*, 296, 298
Drumcarro (Trumcarro), NO4512; 10, 13, 18, 20, 21
Drumcavillhill, Drumcawillhill, *Cadder*, 497, 498
Drumcruiff, *see* Dulcrune
Drumdevane, *Methven collegiate kirk*, 296, 297, 298
Drumdoch (Drumdocht), NX0957; 598, 599
Drumdraet, *Riccarton* [*Drumdroch, NS4934], 531
Drumdrannan, *see* Dundrennan

Drumdreel, NO2008; 69
Drumeldrie (Drummeldre, Drummeldrie, Drumneldrie), NO4403; 24, 25, 28, 29, 46, 47, 50, 51
Drumelzier (Drummelzear, Drummelzer, Rumelzeiris), NT1334; 150, 249, 252
Drumfallinthie, Drumfathtie, *Coupar Angus abbey*, 354, 369
Drumfathtie, *see* Drumfallinthie
Drumfreis, Drumfres, *see* Dumfries
Drumgarthill, *see* Dungarthill
Drumgirloch, *Crossraguel abbey*, 567
Drumgray [Ho, NS7770], 102
Drumgrie, *see* Dumgree
Drumharne, *see* Dunearn
Drumkappie, Drumkeppe, *Kinneddar*, 24, 46
Drumlagnerk, *see* Drumlanrig
Drumlaken (Drumlukocht), NN8517; 349
Drumlanrig (Drumlagnerk, Drumlangrig) [Castle, NX8599], 269, 271, 276, 277, 623
Drumlethy, Drumletty, *see* Drumlithie
Drumlithie (Drumlethy, Drumletty), NO7880; 405, 406
Drumlukocht, *see* Drumlaken
Drummaird (Drummard), NO3603; 15
Drummalie, Drummaly, *Dunkeld bishopric*, 303, 342, 345
Drummaoch, Drumnarrok, *Methven collegiate kirk*, 296, 298
Drummaston (Drumanstane, Drumanstann, Drumanston, Drumanstoun, Drumasson), NX4640; 592, 598, 599
Drummecudyne, *Cullicudden*, 646
Drummis, *Brechin archdeaconry*, 380
Drummoddie (Drumaddie), NX3944; 593
Drummond (Drummond, Drummound) [Castle, NN8418], 327, 347, 348
Drummoquhore, *see* Drumquhore
Drummore (Drumoyer), NH7643; 639
Drummorral (Drumarell), NX4636; 594
Drummy, Drymme, *Brechin bishopric*, 384, 386
Drummynald, *see* Dunninald
Drummys, *Brechin treasurership*, 377
Drumnance, Drumnanis *see* Dinnans
Drumnarrok, *see* Drummaoch

Drumnerbek, *Paisley abbey*, 529
Drumoak (Drummaok), NO7898; 449; parsonage of, 449; vicarage of, 449
Drumour (Drumowir), NN9640; 309
Drumoyer, *see* Drummore
Drumpellier (Drumpender, Drumplear) [Ho, NS7165], 102, 498
Drumpender, *see* Drumpellier
Drumquhar (Drumquhare), *Monzie*, *NO0133; 349
Drumquhassle (Drumquhassel, Drumquhassill) [Park of, NS4886], 546, 548
Drumquhore, Drummoquhore, *Atholl, Perth Charterhouse*, 284, 288
Drumrack (Drumraok), NO5408; 177, 178, 181
Drumrae (Drumra), NX4043; 593
Drumrash (Drumrasche), NX6671; 616
Drumray, *see* Dounreay
Drumrosnes, *see* Dunrossness
Drumry, NS5070; 529
Drums, NO2606; 15
Drumscheill, *see* Dronshiel,
Drumsleet (Drumsleit, Drumsleitt), NX9474; 613, 621
Drumtenant (Drumtennent), NO2909; 33, 37
Drumthuill, Drumthuthill, Druthuill, *Dunfermline*, 27, 49
Drumtiktie, *see* Drumtochty
Drumtochty (Drumtiktie) [Castle, NO6980], 82
Drumusquhen, *Crossraguel abbey*, 567
Drundarne, *Airlie*, 356
Drundrannan, *see* Dundrennan
Drunmotoune, *Croy and Moy*, 639
Druthuill, *see* Drumthuill
Drwmblaitt, *see* Drumblade
Dryburgh (Dryburcht), NT5932; 189, 191, 197; St John's chapel of, 189, 191, 197
— abbey, 189-191, 197; commendator of, 197n., 290n.
Dryden [Tower, NT2764], 141
Dryfesdale (Dryfdaill, Dryvisdaill) [Ho, NY1283], 265, 496; vicarage of, 265
Drygrange (Dry Grange), NT5735; 207, 208, 257, 259
Drylaw (Drilaw), NT2175; 304
Drymen (Deymyne), NS4788; kirk of, 496
Drymme, *Brechin bishopric, see*

Drummy
Drymmeis, *Coupar Angus abbey*, 369
Dryvisdaill, *see* Dryfesdale
Dubbs (Dubbis), NS2842; 574
Dubton (Dubtoun), NO5860; 387, 387n., 388, 388n.
Duddingston (Duddingstoun), *Inchcolm abbey*, *NT1077; 62
Duddingston (Duddingstoun, Dudingstone, Dudingstoun), *parish*, NT2972; 99, 128, 223, 229, 230, 232, 234, 237, 240, 241, 243; kirk of, 229, 232, 234, 237, 240, 241, 243; vicar of, 99; vicarage of, 99; Duddingston Easter and Wester, 231, 233, 237, 240
Dudingstone, Dudingstoun, *see* Duddingston
Dudinholme, *Coldingham priory*, 198
Duffus (Duffeus, Duffois, Duffous, Duffows), NJ1768; 468, 475, 476, 482, 485; prebend of Elgin cathedral, 468n., 476n.; parson of, 482; parsonage of, 468, 468n., 482; vicar of, 485; vicarage of, 485; *see also* Unthank
Dulceanis, Duncenanis, *Atholl, Perth Charterhouse*, 284, 288
Dulcrune, Drumcruiff, *Methven collegiate kirk* [*Dalcrue, NO0427], 296, 298
Dulgarthill, *see* Dungarthill
Dull (Dow), NN8049; 8, 16, 17, 86, 350; kirk of, 16, 17; vicar of, 86; vicarage of, 86-87
Dullar Fullartoun, *Riccarton* [Law Dollars, NS4533], 531
Dullarhill, *Riccarton* [Law Dollars, NS4533], 531
Dullarschaw, *Riccarton* [Law Dollars, NS4533], 531
Dulmernocht, *see* Dalmarnock
Dulpersie [*Terpersie Castle, NJ5420], 436
Dumany, *see* Dalmeny
Dumbaith [*Dunbeath, ND1629], 657, 670
Dumbar, *see* Dunbar
Dumbarny, *see* Dunbarny
Dumbarton (Dumbartan, Dumbartane), NS3876; 539, 540, 541, 542 573, 576, 579; burgess of, 540; kirk of, 573, 576, 579

—— Castle (Castell of Dumbartane), NS4074; 540; chaplain in, 540; chaplainry in, 540-541
—— collegiate kirk, 539, 539n.; prebendaries of, 539; provost of, 541; provostry of, 539
—— parish kirk, chaplain of Holy Rood chaplainry in, 540; chaplainry of Holy Rood in, 540
Dumblane, *see* Dunblane
Dumbraxhill (Drumbekschill, Drumbrakischill), NS8240; 231, 232
Dumbug, *see* Dunbog
Dumfarmeling, Dumfermeline, Dumfermeling, Dumfermelyng, Dumfermling, Dumfermlyng, *see* Dunfermline
Dumfarmlingschyre, Dumfermelyneschyre, *see* Dunfermlineshire
Dumfedling (Dumfedlang, Dumfermling), NT2401; 259, 259n.
Dumfermling, *Melrose abbey*, *see* Dumfedling
Dumfries (Dumfreis, Drumfreis, Drumfres), NX9776; 229, 232, 234, 237, 240, 241, 243, 244, 272, 272n., 279, 611; kirk of, 229, 232, 234, 237, 240, 241, 243, 244; presbytery of, 272; vicar of, 611; vicarage of, 611
——, *sheriffdom*, 1n., 223, 265n., 273
Dumgree (Drumgrie), NY0696; kirk of, 232, 234, 240
Dummanie, Dummany, *see* Dalmeny
Dumrosnes, *see* Dunrossness
Dun (Doune, Down, Downe, Dune, Dwn), NO6660; 363n., 372n., 374, 374n., 381, 384, 384n., 387, 388, 389, 393, 393n., 396; parish of, 387, 388; vicar of, 396; vicarage of, 381, 393, 396
Dunance, *see* Dinnans
Dunbar (Dumbar), NT6879; 169, 171, 172, 172n., 190, 249; parson of, 169; parsonage of, 169,
—— collegiate kirk, 170, 171, 172, 172n., 175; archpriestry of, 172; prebend of, 171; *see also* Bolton, Pinkerton, Pitcox, Spott
Dunbar Common (Commonis of Dunbar), NT6778; 122
Dunbarney (Damberny, Dumbarny)

[Ho, NO1118], revenues of the kirk of, 111, 111n., 113, 113n.
Dunbeth, *Aberdeen Trinitarians*, 421
Dunblane (Dumblane), NN7801; 65n., 290n., 294, 295, 295n., 296, 299, 300, 302, 302n., 312, 313, 314, 315, 322, 324n., 327, 328, 330, 335, 336, 341, 343, 344, 344n., 348, 349, 350; archdeacon of, 300; archdeaconry of, 300; bishop of, 295, 295n.; bishopric of, 290n., 294-296, 348-349; chamberlain of bishopric of, 295; diocese of, 148n.; kirk of, 294, 295; parish of, 349; town of, 327; vicarage of, 315
— cathedral, 336; chancellor of, 341; chaplain of Our Lady in, 314; chaplain of St Blaise in, 314, 328; chaplain of St Michael's chaplainry in, 335; chaplain of St Nicholas in, 327; chaplainries in, 343; chaplainry of Our Lady in, 314; chaplainry of St Blaise in, 314, 328; chaplainry of St Michael in, 335; chaplainry of St Nicholas in, 327; chaplains in, 343; chapter of, 344; dean of, 65n., 299, 315, 344, 595n.; deanery of, 299, 315; prebendary in, 316; prebends of, 328, 330 392; prebend of Ruffill, 316; subdean of, 314; subdeanery of, 314; Trinity chaplainry in, 324, 324n.; *see also* Balquhidder, Ruffill
Dunbog (Dumbug), NO2817; 68, 359; kirk of, 359; vicarage of, 68
Dunccatis landis, *see* Domcattis land
Duncenans, *see* Dulceanis
Duncrub (Duncrube), NO0014; 522
Dundas [Mains, NT1177]; 3
Dundee (Dundie, Dundy), NO3632; 12, 19, 30, 32, 35, 36, 58, 331, 355, 369, 377, 381, 394, 395, 399, 400, 409; Magdalene chapel in, 400; Murraygait, 12; town of, 31, 331; provost of, 249n.; vicar of, 394; vicarage of, 394
— kirk, 32, 36, 377; altar of the Three Kings of Cologne, 381, 381n.; chaplain of St Stephen in, 400; chaplainry in, 377, 395; chaplainry of St Stephen in, 400; chaplainry of Three Kings' altar in, 381; St Barbara's altar in kirk of, 399
Dundivane, Dundyvane, *Monkland*, 102,
497
Dundonald, NS3634; 528, 529, 560; kirk of, 528, 529; vicar of, 560; vicarage of, 560
Dundrennan (Drumdrannan, Drundrannan), NX7447; 606; minister of, 606; reader of, 606
— abbey, 606
Dunduff, *Dunblane bishopric*, NN8211; 349
Dunduff, *Dunfermline*, NT0891; 27, 49
Dundurcas (Dundurcus), NJ2950; 456, 456n., 460, 489; benefice of, 248, 248n.; kirk of, 456, 456n.; vicar of, 489, 490; vicarage of, 489
Dundyvane, Dundivane, *Monkland*, 102, 497
Dune, *Forfar*, *see* Dun
Dune, *Inverness-shire*, *see* Dunn
Dunearn (Drumhame, Dunhair), NT2187; 28, 50
Dunekeir, *see* Dunnikier
Dunfermline (Dumfarmling, Dumfermeline, Dumfermeling, Dumfermelyng, Dumfermling, Dunfarmeling, Dunfarmline, Dunfarmling, Dunfermeling, Dunfermelyng, Dunfermling, Dunfermlyne, Dunfermlyng), NT1087; 23, 25, 26, 37, 38, 39, 26, 27, 29, 47, 48, 49, 51, 52, 53, 282, 332, 291, 555; customs of burgh, 23, 45; vicarage of, 68, 72; *see also* Dunfermlineshire
— abbey, 23-29, 37-54, 83n., 103-104, 104n., 150, 150n., 179, 179n., 282, 332; abbot of, 54; commendator of, 40, 40n., 44, 45, 45n., 54, 54n., 72, 73, 73n., 150, 150n., 179, 179n., 183n.; sacristan of, 72
Dunfermlineshire, 23, 44; *see also* Dunfermline
Dungar Drumgarthill, 303
Dungarthill (Drumgarthill, Dulgarthill, Dungartill), NO0541; 305, 342, 345
Dunglass (Dounglas, Dunglas), NT7671; 166, 173, 186n.
— collegiate kirk, 120n., 166, 173, 182, 186; prebendaries of, 120n., 166, 173; prebends of, 166, 173; provost of, 186n.; provostry of, 186; *see also* Auchencrow, Barnside, Haliewelll, Spittall

INDEX OF PLACES 797

Dunhair, see Dunearn
Dunichin, see Dunnichen
Duniface (Donyface, Dunyface), NO3501; 15, 66
Dunino (Dunnennio), NO5311; 90; vicar of, 90; vicarage of, 90
Duniop, see Dunjop
Dunipace (Dunipaice, Dunipais, Dunpaice, Dunpais, Dunypaice), NS8083; 538, 543, 544, 545, 546, 547, 548; kirk of, 538, 543, 545, 546, 547
Dunisie, see Dunsyre
Dunjop (Duniop, Dunyop), NX7160; 586, 586n., 589, 592
Dunkany, see Dunkenny
Dunkeld (Dunkelden, Dunkell), NO0242; 35 282, 283n., 284, 287, 300, 301, 301n., 302, 302n., 304, 306, 307, 308, 309, 310, 311, 311n., 312, 316, 318, 319, 320, 321, 322, 323, 323n., 325, 326, 327, 329, 336n., 337, 339, 339n., 341, 342, 343, 344, 344n., 345, 346, 346n., 350, 365, 366, 374, 375, 398; archdeacon of, 308, 309; archdeaconry of, 308-309, 309n.; bishop of, 301, 301n., 307, 308, 341, 346, 375; bishopric of, 69, 282, 301-305, 341-344, 344, 345-347; chapel of St Ninian in, 332; dean of the Christianity of, 375; diocese of, 88n.; grammar schoolmaster of, 303; hospital of St George in, 304n.; minister of, 307; song schoolmaster of, 303
—— cathedral, 87n., 283n., 375, 641n.; chapter of, 374; beadle in, 301; canons of, 398, 339; chancellory of, 308; chanter of, 306; chantory of, 306-307, 307-308; chaplainry of Inver in, 336, 336n.; chaplainry of Our Lady altar in, 323; chaplainry of St Ninian's altar in, 320; chaplains in, 307, 323; chaplain of Inver in, 336; choir of, 307; choristers of, 310; clerk of chapter of, 301; common kirks of cathedral of, 300; dean of, 301, 374; deanery of, 301; prebend of Inchmagranachan, 310, 311; prebends of, 283n., 309, 310, 312, 318, 325, 326, 327, 329, 365; prebendaries of, 300, 329, 365, 366;
stallars in, 307, 309, 641; subdean of, 341; subdeanery of, 341; treasurer of, 311, 311n.; treasurership of, 311; see also Caputh, Craigie, Crieff, Fern, Forgandenny, Fungarth, Moneydie
Dunkenny (Dunkany), NO3547; 412
Dunlop (Dunlope), NS4049; 576, 579; kirk of, 576; vicar of, 579; vicarage of, 579
Dunmacriof, Dunkeld bishopric, 304
Dunmanie, see Dalmeny
Dunmay, see Dalmeny
Dunmeth, see Dalmeath
Dunmork, see Denork
Dunmure, Abdie [Denmuir, NO3018], 32, 34, 36
Dunn (Dune), ND1956; 632, 633
Dunnany, see Dalmeny
Dunnennio, see Dunino
Dunnet, ND2171; 459; parsonage of, 459; vicarage of, 459
Dunnichen (Downichin, Downychin, Dunichin, Dynnichtin, Dynnychthin, Dynnychtin), NO5048; 351, 358, 361, 399; kirk of, 361; vicar of, 399; vicarage of, 399
Dunnikier (Dunekeir) [Ho, NT2894], 24, 46
Dunninald (Donynald, Donynauld, Drummynald) [Mains, NO7054], 217, 220, 221; kirk of, 217, 220, 221
Dunning (Donning, Duning), NO0214; 314, 555; chaplain of, 314; chaplainry of, 314
Dunnork, Dunnorke, see Denork
Dunnottar (Dunnoter) [Mains, NO8783], 404, 404n.; parsonage of, 404; vicarage of, 404
Dunny, see Dounie
Dunnybirsall, see Donibristle
Dunnygask (Tonigask, Tunegask), NT0592; 23, 27, 28, 46, 49, 50
Dunpaice, Dunpais, see Dunipace
Dunrod (Dinrod, Dunrode, Dunroid), NS2273; 91, 93, 505; kirk of, 91
Dunrossness (Done Rosness, Drumrosnes, Dumrosmes), HU3917; 658, 660, 664, 670; ministers of, 660; parish of, 660; vicar of, 664; vicarage of, 664-665
Dunruchan (Tunruchan), NN8016; 349
Duns, NT7853; 185, 192; kirk of, 185

Dunscore ('Monkland in Niddisdaill callit Dunscoir'), NX8684; 210; kirk of, 210
Dunsyre, NT0748; 224, 232, 237, 240, 242, 244, 509; kirk of, 224, 232, 234, 237, 240, 242, 244; vicar of, 509; vicarage of, 509
Dunterth [*Duntreath Castle, NS5381], 555
Dunyface, *see* Duniface
Dunyop, *see* Dunjop
Dunypaice, *see* Dunipace
Dura (Dury) [Mains, NO4014], prebend (with Rumgally) of St Mary on the Rock collegiate kirk, St Andrews, 71
Durie (Dury) [Ho, NO3702], 15, 18, 21, 90, 145
Duris, *see* Durris
Durisdeer (Durisdeir), NS8903; 272, 272n., 278; parsonage of, 272-273
Durn (Durne) [Ho, NJ5865], 437
Durran (Derane), ND1963; 628
Durris (Duris) [Kirkton of, NJ7796], 404, 407, 408, 470; parson of, 404, 408; parsonage of, 404, 408; vicarage of, 407
Dury, *near Kilmux, Scoonie, see* Durie
Dury, *with Rumgally, see* Dura
Duthil (Duthell), NH9324; 494; parson of, 494; parsonage of, 494; vicar of, 448n.; vicarage of, 448n., 494
Dwn, *see* Dun
Dyce (Dys), NJ8812; 445; parish of, 445
Dyik, *see* Dyke
Dyke (Dyik, Dyk), NH9858; 426, 426n., 466; kirk of, 466; vicar of, 426; vicarage of, 426
Dykes (Dikie), *Monkland*, 498
Dynbirsall, *see* Donibristle
Dynein, *see* Deinyn
Dynibirssall, *see* Donibristle
Dynnichtin, Dynnychthin, Dynnychtin, *see* Dunnichen
Dynybirsall, *see* Donibristle
Dypdall, *see* Deipdaill
Dyppil, Dyppill, *see* Dipple
Dys, *see* Dyce
Dysart (Disart, Dysert), NT3093; 70, 71, 90, 188, 300; parson of, 70, 188, 300; parsonage of, 70; vicar of, 188, 188n.; vicar pensionary of, 71; vicarage of, 70; vicarage portionary of, 188

Eaglesham (Egilshame), NS5752; 525, 580; parson of, 525, 580; parsonage of, 580
Earlsferry (Erlisferry, Erlysferry, Ferie), NT4899; 146, 167, 291
Earlston (Arsiltoun, Ersiltoun, Erssiltoun), NT5738; 189, 191, 197, 200, 204; kirk of, 200, 204
Eassie (Esse, Essy) [Church, NO3547], 72, 411; glebe of, 412; parson of, 475; parsonage of, 72, 411-412; vicarage of, 72, 411-412
East Bangour (Eister Bengour), NT0471; 151
East Barns (Eist Bairnis, Barnes, Barnis), *Dunfermline*, 23, 27, 45, 49
East Calder, NT0867; kirk of, 229, 232, 234, 237, 240; *see also* Calder-Clere
East Craig (Eist Craig, Eistcraig, Ester Cragy), NT5882; 147, 168, 169
East Drimmie (Eister Drymmy), NO1749; 354
East Drums (Eister Drummis, Ester Drummis), NO5857; 384, 386, 389
East Feddellis, Easter Feddellis [Wester Feddal, NN8208], 31, 36, 32, 35
East Fenton (Eist Fentoun), NT5281; 190
East Field (Eist Feild), *Kirkliston*, 3
East Field (Eist Feild), *Greenlaw*, 227
East Gordon (Eister Gordoun), NT6643; 228, 234, 241
East Grange (Eist Grainge), *Cambuskenneth abbey* [West Grange, NS8194], 547
East Grange (Eistergrang, Estergrange, Estirgrang), *Culross abbey*, NT9989; 290, 293
East Green (Eist Grene), *Crail*, 177, 179
East Hauch, Eschehauch, *Kelso abbey* [*East Haugh, NT1164], 222, 222n.
East Hope (Eisthopes, Eistschopes) [*Hope, NT4062], 162, 165, 176, 178, 181
East Hous, East Houssis, *see* Easthouses
East Kilbride, *see* Kilbride
East Lothian (Eist Lowtheane), 172
East Lumsdaine (Eist Lummisden, Eister Lummisden) [Lumsdaine, NT8769], 198, 203
East Luscar (Luscer Ewat, Luscerewert, Luscour Evert, Luscour Ewart), NT0589; 23, 27, 45, 49, 50

East Mathers (Eister Mathouris), NO7766; 16
East Mill (Eist Mylne) *of Kirkcaldy*, 24, 46
East Nisbet (Eist Nisbit), NT6726; 199, 200
East Quarter, Easter Quarter, *Kinghorn*, 24, 28, 46, 50
East Reston (Eistrestoun), NT9061; 198, 202
East Teviotdale (Eist Teviot Daill, Eist Toviotdaill [Teviotdale, NT4815, NT5822, NT6123], 207, 209, 258
East Whitefield (Eister Quhitefeld), NO1734; 303
East Yard, *Lindores abbey, see* Eist Yaird
East Wood (Eist Wod), *Lindores abbey*, 30, 35
Easter and Wester Bochlaweis [Easter Bucklyvie, NT1788], 62
Easter and Wester Cantna, *Croy and Moy*, 639
Easter and Wester Lowstonis, *with Kinvaid, Dunkeld*, 323
Easter Balbrogy [Balbrogie, NO2442], 356
Easter Baldridge (Bawdrick, Bawdrik), NT0888; 23, 45
Easter Balgray [Midton of Balgray, NS4434], 531
Easter Balliealyenoch, Balyealyenoch [Ballyalnach, NN9453], 284, 288
Easter Balrymonth (Balrymonth Eister, Balrymounthe Eister), NO5314; 10, 18
Easter Bawdrick, Bawdrik, *see* Easter Baldridge
Easter Bengour, *see* East Bangour
Easter Bothylokis, *Inchcolm abbey*, 62; *see also* Nathertoun of Bathedloskis
Easter Brackland (Eister Brokland), NN6608; 349
Easter Braikie (Eisterbroky), NO6351; 366
Easter Brokland, *see* Easter Brackland
Easter Cadder [Cadder, NS6172], 497, 498
Easter Cairny, *Monzie*, 349
Easter Callander (Eister Callender) [Callander, NN6307], 349
Easter Cash (Eister Casche), NO2309; 69

Easter Clunie (Cluny Eister, Cluny Ester), NO2217; 30, 35
Easter Comrie (Cumrie, Cumry) [Comrie, NO0189], 290, 293
Easter Cotyardis, *Millhorn, Coupar Angus abbey*, 354, 357, 369
Easter Cragy, *see* East Craig
Easter Craigduky, *see* Craigduckie Easter
Easter Cranokis [Crannach, NJ4954], 461
Easter Craquho [Crauchie, NT5678], 170, 170n., 192n.; *see also* Crauchie
Easter Croylettis [Croylet, NJ5056], 461
Easter Culbell, *see* Easter Culbo
Easter Culbo (Eistir Culbell), NH6461; 646
Easter Cumrie, Cumry, *see* Easter Comrie
Easter Dalguise (Eister Dulgus, Ester Dulgus), NN9947; 303, 305, 342, 345
Easter Denhead (Eister Denhede, Ester Denhede), NO2441; 353, 356
Easter Denshoussis [Wester Deanshouses, NT2151], 101
Easter Disblair [Disblair, NJ8619], 32, 36
Easter Drummis, *see* East Drums
Easter Drymmy, *see* East Drimmie
Easter Dulgus, *see* Easter Dalguise
Easter Dunfallanceis, Dunfallances, Dunfallenceis, Dunfallencis [Dunfallandy Ho, NO9556], 284, 287, 288
Easter Farbrowne [Fairburn Mains, NH4753], 635
Easter Fathie, *see* Easter Fithie
Easter Fearn [Fearn, ND8377], 634, 652
Easter Feddellis, *see* East Feddellis
Easter Ferne, *see* Easter Fearn
Easter Fethie, *see* Easter Fathie
Easter Fithie (Fathie, Fethie) [Fithie, NO6354], 358, 365
Easter Fotteris, *Leuchars* [*Fetterdale, NO4725], 14
Easter Fowmerden [Fourmartdean, NT7927], 214
Easter Gany [Geanies Ho, NC8979], 633
Easter Gartquhone [Gartchonzie, NN6007], 349
Easter Geddis [Meikle Geddes,

NH8752], 469
Easter Gellet (Eister Gellet, Eister Gullet), NT0985; 26, 27, 49; *see also* Gellets
Easter Glentore (Glenter Eister), NS7872; 498
Easter Gordoun, *see* East Gordon
Easter Greinleis [West Greenlees, NS6359], 508
Easter Gullet, *see* Easter Gellet
Easter Hailles, Halis, Hallis, *with Traprain, St Andrews archbishopric* [Nether, Over Hailes, NT5775, NT5776], 3, 6
Easter Happrew (Hoprew Eister), NT1939; 249
Easter Hartsyd, *Hartside parish*, 513
Easter Hartsyde, *Dunbar*, 122
Easter Howlaws (Eister Howlaw, Eister Howlow), NT7242; 227, 235, 238, 241
Easter Inneraritie, *Coupar Angus abbey* [Inverharity, NO1963], 354, 355
Easter Inschewin, [Middle Inchewan, NO0341], 342, 345
Easter Keith (Eister Kethe), NO2937; 376
Easter Kennet, 21n.
Easter Kilwhiss (Kylquhis Eister), NO2810; 33
Easter Kinsleith (Kynsleif Eister), NO3318; 33
Easter Lassodie (Lasoiddy, Lasoidey) [Lassodie, NT1292], 23, 45
Easter Lathrisk (Eister Lauthrisk, Lawthrisk Eister), NO2808; 14, 20
Easter Mains of Gordon, *see* Gordon East Mains
Easter Mathouris, *see* East Mathers
Easter Medois, Medowis, *Kelso abbey*, 235, 238, 241, 242
Easter Meirdene, *Kelso abbey*, 232n.
Easter Newtoun, *Kirknewton* [*Easter Newton, NT1267], 104
Easter Nunraw [Nunraw, NT5970], 181
Easter Persey [Wester Pearsie, NO3458], 354
Easter Petmowy [*Pitmuies Mill Fm, NO5849], 376
Easter Petscote, Petscottie, *see* Easter Pitscottie
Easter Pitlour [Pitlour, NO2111], 69
Easter Pitscottie (Eister Petscote, Eister Petscottie), NO4113; 55
Easter Quarter, *see* East Quarter
Easter Quhitefeld, *Dunkeld*, *see* East Whitefield
Easter Quhitfeild, *Old Cambus*, 199
Easter Softlaw (Eister Softlaw), NT7532; 225, 234, 235, 238, 240, 241; *see also* Softlaw
Easter Toun, *Beath*, 64
Easter Tyre, *see* Eastertyre
Easter Wodend, *see* Easter Wooden
Easter Wooden (Eister Wodend), NT7225; 225
Eastertyre (Eister Tyre), NN9552; 284, 288
Easthouses (Eist Hous, Eist Houssis), NT3465; 101, 103
Eastwood (Eistwood), NS8239; 528; kirk of, 528
Ebdy, *see* Abdie
Eccles (Ecclis, St Cuthbert's), NT7641; mains of, 183, 184; Our Lady chapel in parish of, 183, 183n.; parish of, 183, 183n., 184; St John's chapel in parish of, 183, 183n.; St Magdalene's chapel in parish of, 183, 183n.; vicarage of, 184
—— priory, 183, 184, 184n., 185n.; prioress of, 184, 184n.
Ecclesgreig (Aglisgrig, Eglisgreig, Eglisgreige, Eglisgrig), NO7365; 12, 16, 17, 19, 395; kirk of, 16, 17; vicarage of, 395
Ecclesia Trinitatis, *see* St Andrews, Trinity kirk
Ecclesiamagirdle (Egleismagreltoun, Eglismagall, Eglismagreill, Eglismagrill) [Ho, NO1016], 31, 32, 34, 35, 36; kirk of, 32, 36
Ecclesjohn (Ecclisjone, Ecglisjohnne, Egglisjohnne, Eglisjohnne), *kirk, Brechin bishopric*, 372, 384, 385, 388; parsonage of, 372; vicarage of, 372
Ecclesmachan (Eglismachane), NT0573; glebe of, 151; parson of, 151; vicarage of, 151
Ecglisjohnne, *see* Ecclesjohn
Echres, *see* Euchries
Echt, NJ7305; 332, 435; kirk of, 332; vicar pensionary of, 435; vicarage of, 435
Eckford (Ekfuird, Ekfuirde, Hecfuird),

NT7026; 212, 215, 216, 219, 220; kirk of, 216, 219, 220; vicar of, 212, 215; vicarage of, 212, 215
Ecolmkill, Ecomkill, *see* Iona
Eday (Edday, Ethay, Ethie), HY5531; 655, 660, 661, 662, 668; kirk of, 660; minister of, 660; parsonage of, 655, 668
Edberin, Edbirne, *Methven collegiate kirk*, 296, 298
Edday, *see* Eday
Edderlick (Hedelik, Hedelyk), NJ6226; 31, 36
Edderton (Eddirtane, Eddirthane), NH7184; 651, 652; parish of, 652; parson of, 651; parsonage of, 651
Eddilstoun, *see* Eddleston
Eddirtane, Eddirthane, *see* Edderton
Eddleston (Eddilstoun, Edlistoun), NT2447; 247, 248n., 496, 497; mill of, 496; parson of, 248, 248n.; parsonage of, 247-248; vicarage of, 247-248, 248n.
Eddrem, Eddremtoun, *see* Edrom
Edgerston (Edyerstoun), NT6822; 219
Edinburgh (Edinbrucht, Edinbrught, Edinburcht), NT2773; 7, 8, 25, 42, 47, 53, 57, 61, 92, 94, 95, 96, 99, 102, 105, 106, 111, 113, 114, 117, 118, 120, 122, 123, 124, 125, 126, 127, 128, 129, 130, 131, 132, 133, 134, 135, 137, 138, 139, 140, 141, 149, 157, 161, 163, 166n., 175, 190, 192, 202, 222, 223, 253n., 259, 272, 273, 276, 281, 286, 302n., 317, 324, 342, 346, 352, 367, 383, 388, 390, 402, 404n., 405, 408, 411, 459, 468, 479, 491, 495n., 506, 523, 524, 538, 545, 550, 551, 561, 565, 569, 574, 575, 579, 600, 601, 610, 611, 614, 615, 621, 622, 680; altarage in, 120; Auldoucht land, 128; Baith's Wynd, 123; Bakraw, 124; Bell's land, 127; Bell's Wynd, 127, 130, 133; Bissatt's land, 124; Blak Freir Wynd, 131, 137; Booth Row, 137; Bowquann's yard, 127; burgess of, 276, 575, 601, 622; Burgh muir, 500; Butlar's land, 128; Castlehill, 140; Cavour's land (Canour's land), 127, 127n.; chaplain in, 120; chaplain of the Barras (St Mary's chapel) in, 96-97; chaplainry of the Barras (St Mary's chapel) in, 96-97; Cochren's land, 127; Conn's Close, 143; Cowgate, 96, 123, 127, 130, 132, 133, 134, 135, 136, 140, 141, 157; Cuikland, 122; Dundas land, 127; Fairnlie land, 126; Fawside's (Fallsydis) land, 127, 128; Fish Market, 137; Flesh Market Close, 140; Forrester's (Foster's) Wynd, 129, 135; Freir Wynd, 126, 139, 140; Galloway's land, 127; Gilbert Lauder's Close, 129; Gillespie's Close, 129, 130; Greyfriars' Port, 94; Halkerstoun's Wynd, 137; Henderson's land, 126; High Street, 96, 135; James Aikman's Close, 124; James McGill's Close, 96; John Barcar's Close, 138; John Fisher's Close, 129; King Street, 138, 139; King's Stables Road, 96n.; Kirk o' Field Wynd, 125; Lamb's land, 127, 128; Lawson's land, 127; Leith Wynd, 127, 133; Lowch's land, 128; Lucken Booths, 129; Magdalen chapel, 127; Magdalene land, 127; Market Cross, 123, 132, 140; Meal Market, 132, 137, 139; Melrose land, 127; Merlin's Wynd, 94, 104; Methestoun land, 128; Muiswall, 127; Muncurr's (alias John Thornton's) land, 126; Neilson's land, 128; Nether Bow, 134, 140; Niddry's Wynd, 94, 124, 132, 133, 136; Over Bow, 124, 127, 131, 140; Patok's land, 127; Peebles Wynd, 124, 135; Pinkerton's alias Nory's land, 127; Purves' land 128; Pyott's land, 127; Queen Street, 96, 125, 132, 133, 134, 136, 137, 141; reader in, 119; Robert Bruce's Close, 134; Rodd's land, 127; St Catherine's altar, 134; St James' land, 131; St Leonard's hospital, 92; St Mary's Wynd, 131, 157; Salt tron, 140; Slater's Close, 140; Soltray's Wynd, 135; song school in, 124; Spens's land, 130; Stinkand Style, 129, 137; Straw Market, 129n.; Tavernour's land (Cavernour's land), 128, 128n.; Todd's Close, 132, 134; Tweedie's land, 127; Watson's land, 127; West Port, 129; Wyntrepp's now Stevinsoun's land, 127; *see also* Canongate, Corstorphine,

Holyroodhouse
—— Castle, 95n., 96, 131, 136
——, Friars Preachers (Dominican or Black friars) of, 126-128
——, priory of Sciennes (Seinis, Sennis), 94; prioress of, 94, 94n.; St Giles' grange (Sanct Geiligrainge) of, 500
——, St Cuthbert's kirk, 91, 93; chaplain of Trinity altar in, 128-129; minister of, 92, 93; vicar of, 105; vicarage of, 105
——, St Giles' collegiate kirk, altars in, 95; chaplain of Holy Blood altar, 130, 131; chaplain of St Denis' altar, 132; chaplain of St Duthac's altar, 124; chaplain of St Eloy's altar, 134, 135; chaplain of St Francis' altar, 132; chaplain of St John the Evangelist's chaplainry, 138; chaplain of St John's aisle, 140; chaplain of St Laurence's altar, 125-126, 132; chaplain of St Patrick's altar, 134, 135; chaplain of St Roche's altar, 138-139; chaplain of St Salvator's altar, 96, 129; chaplainry at Holy Blood altar, 131; chaplainry in Holy Blood aisle, 135; chaplainry of Holy Blood altar, 130; chaplainry of Nomine Jesu and Holy Blood, 117-118; chaplainry of St Denis' altar, 132-133; chaplainry of St Duthac's altar, 124; chaplainry of St Francis' altar, 132-133; chaplainry of St John the Evangelist, 138; chaplainry of St Laurence's altar, 125-126; chaplainry of St John's aisle, 140; chaplainry of St Laurence's altar, 132; chaplainry of St Roche's altar, 138-139; chaplainry of St Salvator in Holy Blood aisle, 135-137; chaplainry of St Salvator's altar, 96; prebend of, 113; prebendaries of, 104, 113; St Colme's altar, 123; St Eloy's altar, 134-135; St Michael's altarage, 140; St Nicholas' altar, 137; St Patrick's altar, 134-135; St Salvator's altar, 129, 130; St Salvator's altarage, 140; see also Ravelston
——, St Mary of the Fields' collegiate kirk (Kirk a Feild, Kirk Feild, Kirk of Feild, Kirkafeild), 83n., 105, 105n., 106, 119, 120, 120n., 123, 124, 125, 134, 139; prebend of Our Lady, 125; prebend of St Matthew, 139; prebend of St Vincent or songschool, 124; prebend of, 123, 134; prebendaries of, 106, 134; prebendary of Our Lady, 125; prebendary of St Matthew, 139; prebendary of St Vincent, 124; provost of, 83n., 105, 105n., 106-107, 120, 120n.; provostry of, 105-107, 119-120
——, Trinity collegiate kirk (Queen's college), 94, 111, 118, 119n., 175, 404; chaplainry of Ormiston in, 175; glebe of the provostry of, 119; prebendaries of, 404, 404n.; provost of, 118, 118n., 119, 119n.; provostry of, 118-119
——, *sheriffdom*, 1n., 68n., 108n., 110n., 118n., 121n., 126, 130, 149n., 179n., 186
Edingecht, Edingeycht, *Cadder* [*Auchengeich, NS6871], 497, 498
Edingight (Edingeicht) [Ho, NJ5155], 461, 464
Edinkillie (Edinkellie), NJ0047; 487; kirk of, 487; parsonage of, 487
Edlistoun, *see* Eddleston
Edmestoun, *Ednam*, 200
Edmestoun, *Musselburgh*, *see* Edmonstone
Edmonstone (Adnistoun, Edmestoun), NT2970; 25, 26, 41, 42, 47, 48
Ednam (Ednama, Ednem), NT7337; 190, 200, 204; kirk of, 200, 204; mains of, 204
Ednem, *see* Ednam
Edrington (Dringtoun, Edringtoun) [Mains, NT9454], 199, 204
Edrom (Eddrem, Eddremtoun, Eddren), NT8356; 199, 203, 204; kirk of, 199, 204; mains of, 199
Edyerstoun, *see* Edgerston
Edzell, NO5969; 379; parson of, 379; parsonage of, 379
Eggerness (Eigirnes), NX4947; 619; *see also* Kirkmadrine, *St Mary's Isle priory*
Egglisjohnne, *see* Ecclesjohn
Egilsay (Eglisay), HY4729; 655, 660, 662, 669, 670; kirk of, 660; minister of, 660; parsonage of, 669
Egilshame, *see* Eaglesham

INDEX OF PLACES 803

Egleismagreltoun, *see* Ecclesiamagirdle
Eglisay, *see* Egilsay
Eglisgreig, Eglisgreige, Eglisgrig, *see* Ecclesgreig
Eglisjohnne, *see* Ecclesjohn
Eglismachane, *see* Ecclesmachan
Eglismagall, Eglismagreill, Eglismagrill, *see* Ecclesiamagirdle
Eigirnes, *see* Eggerness
Eilark, *Restenneth priory*, 217
Eildon (Eildoun, Eilidoun), NT5732; 207, 208, 257, 259; Eildoun Coit, 207, 208
Eileischeucht, *see* Elisheugh
Eilidoun, *see* Eildon
Eillera, *see* Illeray
Eisauptoun, Eisauxtoun, *see* Isaacstown
Eist Yaird, Eistyaird, *Lindores abbey*, 30, 35
Eistcraig, *see* East Craig
Eisterbroky, *see* Easter Braikie
Eisterfuird, *Melrose abbey*, 207
Eisterfuird, *with Housbyre, Melrose abbey*, 207
Eistergrang, *Culross abbey*, see East Grange
Eisthopes, Eistschopes, *see* East Hope
Eistinge, *see* Aithsting
Eistirbothen, *Culross abbey*, 290
Eistraw, Preistlaw, *Restalrig collegiate kirk*, 142, 143
Eistrestoun, *see* East Reston
Eistwood, *see* Eastwood
Eitoun, Etoun, *St Mary's Isle priory*, 617, 619
Ekfuird, Ekfuirde, *see* Eckford
Ekkyl Welcair, Ekkylwelcare, *North Berwick priory*, 168
Elbottill, *Dryburgh abbey*, 190, 191, 197
Elchies (Elchis) [Forest, NJ2046], 489, 489n.; *see also* Botary
Elcho, NO1620; 393, 393n.; prioress of, 393, 393n.
Elderslie (Eldirslie), NS4462; 564, 569
Eleischeuch, Eleischeucht, *see* Elisheugh
Eleistoun, *Melrose abbey* [*Elliston, NT5628], 208, 209
Eleistoun, *see* Elliston
Elenenhill, *see* Allanhill
Elgin (Elgine), NJ2162; 453, 453n., 454, 462, 466, 468, 476, 483, 484, 486, 490; Cross, 453; East Port, 453; grammar schoolmaster of, 466; hospital of Maison Dieu in, 454, 454n.; Over Port, 453; parish of, 468; town of, 490; vicar of, 483; vicarage of, 454n., 483
—— cathedral, 320, 466, 468n., 470, 470n., 476, 476n., 477, 478, 481, 482n., 490; chaplain of Clyne in, 478; chaplain of Our Lady in, 476; chaplain of St Catherine's chaplainry in, 490; chaplain of St Duthac in, 476; chaplainries in, 477-478; chaplainry of Clyne in, 478, 481; chaplainry of Knockando in, 478; chaplainry of Our Lady in, 476, 476n.; chaplainry of St Catherine in, 490; chaplainry of St Duthac in, 476, 476n.; chaplainry of St Thomas' altar in, 477; chaplains in, 466, 477-478; choir of, 486; choristers in, 466; common kirks of, 487; prebends of, 320, 476, 480; stallars in, 470, 483, 484, 487; vicar choral in, 481; *see also* Aberlour, Clyne, Duffus, Knockando *and* Moray
——, Friars Preachers (Dominican or Black friars) in, 453-454
——, *sheriffdom*, 281n., 426n., 449n., 456n., 461n.
Elie, *Lewis, see* Eye
Elisheugh (Eileischeucht, Eleischeuch, Eleischeucht, Elleischeucht, Heleischowcht, Ylyscheuch) [Hill, NT8121], 223, 225, 225n., 230, 233, 237, 239
Elistoun Coit, *see* Cliftoncote
Elistoun, *Kirkliston*, 3
Elistoun, *Lessudden* [*Elliston, NT5628], 189, 191
Elistoun, *see* Elliston
Ellanamwk, *bishopric of the Isles*, 675
Ellein, Ellen, *see* Ellon
Elleischeucht, *see* Elisheugh
Ellem [*Ellemford, NT7260], 143, 195; kirk of, 195
Ellingtoun Boig [*Helenton, NS3930], 569; *see also* Ilingtoun
Ellon (Allane, Ellein, Ellen), NJ9530; 457, 458, 461, 463, 464; kirk of, 461, 463; parish of, 457, 458; reader at, 463; vicar pensionary of, 463; vicarage of, 464
Elrick (Elryk) [Ho, NJ8818], 422, 423

Elryk, *see* Elrick
Elspyhoipe, *Ettrick*, 258
Elstanefuird, Elstanfuird, *see* Athelstaneford
Elwodlaw, Alvodlaw, *Greenlaw*, 227, 227n.
Emershaw, *see* Allanshaws
Enchebowoquhy, *Rothes*, 491
Enenlesis, *see* Awnlosk
Enenschaw, *see* Allanshaws
England (Ingland), 143
Englisberry Grange, Inglisberry Grange, *Dryburgh abbey*, 190, 191, 197
Enhallow, *see* Eynhallow
Ennerawrie, *see* Inverurie
Ennervarit, *see* Inneraritie
Ereychtlyoun, *Sorbie*, 593
Ergill, *see* Argyll
Erlie, *see* Airlie
Erlisferry, Erlysferry, *see* Earlsferry
Ernattis Hauche, *Peebles*, 253
Errol (Arroll, Errole, Erroll), NO2522; 284, 285, 287, 304, 356, 370; bailie of, 287; kirk of, 356; mill of, 285, 288; orchardland of, 285, 288; vicarage of, 370
Ersik, *see* Ersock
Ersiltoun, *see* Earlston
Erskine (Erskein, Erskin, Erskyne), NS4571; 534; parson of, 534; parsonage of, 534
Erskirk, see Ashkirk
Ersock (Ersik, Essik, Erssillis, Erssips) [High, Low, NX4437], mill of, 593, 595, 596; waulk mill of,
Erssillis, *see* Ersock
Erssiltoun, *see* Earlston
Erssips, *see* Ersock
Esauxtoun, *see* Isaacstown
Eschard, *see* Ashyard
Eschehauch, *see* Eist Hauch
Escheholme, *Holywood abbey*, 275, 276
Esdaill Mure, *see* Eskdalemuir
Esdaill, *see* Eskdale
Eskdale (Esdaill), NY3489; 113
Eskdalemuir (Esdaill Mure, Eskdaillmuir), NY2597; 208, 259
Esse, *see* Eassie
Essertoun, *Senwick*, 587
Essie (Essye), NJ4627; 475; parsonage of, 475
Essik, *see* Ersock
Essy, *see* Eassie

Essye, *see* Essie
Estergrange, Estirgrang, *Culross abbey*, *see* East Grange
Ethallow, *see* Eynhallow
Ethay, Ethie, *Orkney*, *see* Eday
Ethey, *Forfar*, *see* Ethie
Ethie (Athy, Ethey, Ethy) [Castle, NO6846], 351, 358, 360, 361; kirk of, 360; minister of, 361
Ethy, *see* Ethie
Etoun, *see* Eitoun
Ettrick (Aitrik, Atrik), NT2714; 208, 258, 259; kirk of, 209, 209n.; New Kirk of, 209, 209n.
Ettrick Ho (Atrikhous), NT2514; 258
Euchries (Echres), NJ4955; 461
Evelaw (Iflie), NT6652; 190, 191, 197
Eves Howe, *Orkney*, 655n.
Evey, *see* Evie
Evie (Evey, Eway, Ewe, Iwey), HY3621; 655, 655n., 659, 662, 668, 671; kirk of, 659; parsonage of, 655
Evindale, *see* Strathaven
Evingshaw, *see* Allanshaws
Evinstell, *fishing on Findhorn, Kinloss abbey*, 462
Eway, *see* Evie
Ewe, *see* Evie
Ewesdale (Ewisdaill) [*Ewes Water, NT3845], 113
Exnem, *see* Oxnam
Eye (Elie), *Lewis*, NB5032; parsonage of, 675
Eyemouth (Eymouth, Eymouthe), NT9464; 198, 202, 204, 205; mill of, 203
Eylbank, *Lesmahagow*, 232
Eymouth, Eymouthe, *see* Eyemouth
Eynhallow (Enhallow, Ethallow), HY3529; 657, 668, 669n.

Facctosetter, *see* Collessetter
Facter Schawis, *Morebattle*, 214
Faddinche, *see* Feddinch
Fadonch, *see* Feddinch
Fadounsyd, *see* Faldonside
Fail (Failford, Failfuird, Faill), NS4228, NS4626; 558, 559, 562, 563, 564, 569; mains of, 563
——, Trinitarians, 'minister' of, 558, 563, 569; ministry of, 558-559, 562-563

INDEX OF PLACES 805

Fair Isle (Fair Ile, Fair Yle, Sey Ile), HZ2271; 658, 660, 662, 670; kirk of, 660; minister of, 660
Fairhillis, *Restalrig collegiate kirk*, 142, 143
Fairhope, *Ettrick*, 258
Fairme, *Glasgow*, 516
Fairnedonald, Ferindonald, Fernedonald, Fyrndonald, *Ross*, 625, 626
Fairnieside (Fairnysyd), NT9461; 199, 203
Fairnilee (Farnylie, Ferinylie) [Ho, NT4533], 224, 224n.
Fala, NT4361; teinds of, 219
Faldhoip, *Selkirk*, 230n.
Faldonside (Fadounsyd, Faldensyd, Faldounsyd, Fauldonsyd, Fawdounsyd), NT5032; 223, 224, 231, 233, 237, 239, 242
Falkirk (Falkirktoun, Fawkirk), NS8880; 91, 92, 93, 553; kirk of, 91; minister of, 92; vicarage of, 553
Falkland, NO2507; 15, 21, 30, 35; queen's chapel in, 30, 35
Fallsidehill (Fawsyd), *Hume*, NT6841; 228, 228n.
Falside (Fausyd, Fawsyd), NO5405; 177, 179, 182
Fancastell [*Fincastle Ho, NN8662], 318, 318n.
Fandownat, *see* Fyndynate
Fans (Fawnis), NT6140; 200, 204
Fara, *see* Faray
Faray (Fara, Farray), HY5236; 657, 660, 662, 670; kirk of, 660; minister of, 660
Farburn Mylne, *see* Fireburnmill
Fard, *see* Farr
Farmen, Farnen [*Fearnan, NN7244], 471, 473
Farnell (Farnall, Farunall, Fernell, Fernevell, Fernewell, Fernvale, Fernvell), NO6255; 358, 389; benefice of, 358; prebend of Brechin cathedral, 358n.; *see also* Mains of Farnell
Farnen, *see* Farmen
Farningtoun, *see* Favingtoun
Farnylie, *see* Fairnilee
Farr (Fard), ND7263; 631; parson of, 631; parsonage of, 631; vicarage of, 631
Farray, *see* Faray

Fasschaw, Radshaw, Raschaw, *Melrose abbey*, 258, 258n.
Fasthucht, *Selkirk*, 230n.
Fater Angus, *see* Fetterangus
Fauldounsyd, *see* Faldonside
Fausyd, Fawsyd, *Musselburgh* [Falside Hill, NT3771], 25, 47
Fausyd, *Haddington priory*, *see* Falside
Favingtoun, Farningtoun, *Melrose abbey*, 208, 208n.
Fawdounsyd, *see* Faldonside
Fawfeildis, Fawfeilis [*Falfield, NO4408], 146, 167
Fawkirk, *see* Falkirk
Fawnis, *see* Fans
Fawside (Fawsyd, Fenisyde), *Gordon*, NT6445; 227, 227n.
Fawsidhill, *Hume*, *see* Fallsidehill
Fawsyd, *Haddington priory*, *see* Falside
Fawsyd, *Hume*, *see* Fallsidehill
Fawsyd, *Musselburgh*, *see* Fausyd
Fawsyde, *Selkirk*, 230n.
Fearn (Fearnie, Ferne, Pherne), ND8377; 625, 625n., 633, 634, 651, 652; mains of, 633
— abbey, 633-634, 651, 652; abbot of, 625, 625n.; bailie of, 634; commendator of, 634, 651, 652
Fechil (Feychill), NJ9629; 457, 458
Feddal (Feddellis) [Ho, NN8208], 31; mill of, 31, 35
Feddinch (Faddinche, Fadonch, Feddinche, Fedinche), NO4813; 11, 13, 19, 20, 21
Feddirdy, *see* Fodderty
Feddonch, *see* Feddinch
Fedinche, *see* Feddinch
Fedirdy, *see* Fodderty
Feild of Flowdoun, *see* Flodden Field
Feistertoun, *see* Fosterton
Feldye, Feyldie, *St Andrews Black friars*, 89
Femaisteris lands, Feymaisterlandis, Semaisters lands, Seymaisterlands, *Melrose abbey*, 207, 207n., 209, 209n., 257, 258, 258n.
Fendowrie, *see* Findowrie
Fenisyde, *see* Fawside
Fenton (Fentoun) [Barns, NT5181], 127
Fenton Tower (Fentoun Tour, The Towir), NT5482; 147, 168
Fentonbarns, NT5181; 202n., 222n.
Fentrey, *see* Fintray

Ferdingousche, Ferdingrusche,
  Ferdinrusche, Ferdunruische, *near
  Craigenputtock, Holywood abbey*
  [*Fardingjames, NX8787]; 269, 270,
  274, 275, 622, 623
Fergus Land [*Fergus, NO1968], 370
Ferie, *see* Earlsferry
Ferietoun, Ferritoun [West Ferry,
  NO4431], 20, 30, 34, 35
Ferindonald, *see* Fairnedonald
Ferinylie, *see* Fairnilee
Ferme (Fairme), *Glasgow*, 516
Fern (Ferne, Pherne), NO4681; 303,
  342, 346, 371, 398, 412, 641; parson
  of, 371, 398, 641; parsonage of, 371-
  372, 398, 641, 641n.; prebend of
  Dunkeld cathedral, 641n.; vicar of,
  398; vicarage of, 371, 398, 641,
  641n.
Ferne, *abbey*, *see* Fearn
Ferne, *parish*, *see* Fern
Fernedonald, *see* Fairnedonald
Fernell, Fernevell, Fernewell, Fernvale,
  Fernvell, *see* Farnell
Ferniehirst (Pharinhirst, Pherniherst),
  NT4441; 217, 219
Ferritoun, *see* Ferietoun
Ferryhill (Ferrehill, Ferry Hill),
  Aberdeen, NJ9304; 419, 421, 423,
  558, 559
Ferryport-on-Craig, parish of, 90n.
Ferrytoun, Ferytoun, *Leuchars* [Tayport,
  NO4628], 14, 20
Feterangus, *see* Fetterangus
Fethe, *see* Fithie
Fethok, *with Toux, Deer abbey*, 458
Fetlar (Fetlair, Fettlar, Fotlar), HU6291;
  658, 660, 665, 670; kirk of, 660;
  minister of, 660; vicar of, 665;
  vicarage of, 665
Fetterangus (Fater Angus, Feterangus),
  NJ9850; 359, 450; kirk of, 359; *see
  also* Longley
—— (and Longley), vicarage of, 450; *see
  also* Longley
Fettercairn (Fettercarn, Fettircarn),
  NO6473; 407, 407n.; parson of, 407,
  407n.; parsonage of, 407, 407n.;
  vicarage of, 407
Fetteresso [Castle, NO8485], 404;
  parson of, 404; parsonage of, 404;
  vicarage of, 404
Fetternear (Fetterneir, Fetternere) [Ho,
  NJ7217], 414, 417
Fettlar, *see* Fetlar
Feu crofts, *Clachan, Galloway
  bishopric*, 586, 589
Feu crofts, *Whithorn*, 595, 596
Feudis, *see* Furdis
Few, *Forgandenny*, 325
Fewles, *Coldingham priory*, 203
Feychill, *see* Fechil
Feyldie, *see* Feldye
Feymaisterlandis, *see* Femaisteris lands
Fife (Fyf, Fyfe, Fyff, Fyffe, Fyif, Fyiff),
  sheriffdom, 1n., 2, 5, 9, 12, 16, 17,
  20, 22, 29, 32, 34, 36, 44n., 45n., 66,
  67n., 68n., 69, 72n., 74n., 75, 77n.,
  79, 79n., 83n., 86n., 88n., 89, 90,
  113, 162, 177, 178, 182, 188n., 281,
  282, 290, 291, 315, 339n., 378n.,
  504
Finavon (Finaven, Fynnevin), NO4957;
  381, 382; parsonage of, 381-382
Findlater (Findlatter, Fyndatour) [Castle,
  NJ5467], 437, 438
Findo Gask (Findogask, Fyndogask),
  NO0020; 295, 348; kirk of, 295, 348
Findowrie (Fendowrie, Findowry,
  Fyndoirtye), NO5561; 383, 386, 388
Fingask, NO3918; 14
Finglassie (Fynglassie, Fynglassy)
  [East, West, NT2699, NT2599], 24,
  29, 46, 51
Finglen (Fingillen, Fynglen) [Rig,
  NT1332], 259, 259n.
Fingling [*Nether, Upper Fingland,
  NS9310, NS9209], 102
Finmont (Fynmont, Fynmonth,
  Fynnocht) [Fm, NT2399], 24, 29,
  46, 51
Fintray (Fentrey, Fyntray), NJ8371; 31,
  32, 34, 36; kirk of, 32, 36
Fintry (Fintray, Fintrie), NS6186; 539,
  553n., 555; kirk of, 539
Firdmastoun, *see* Herdmanston
Fireburnmill (Farburn Mylne), NT8239;
  186
Firth (Firthe), *parish*, HY3514; 657,
  659, 669; kirk of, 659
Firth (Forthe), *Restalrig collegiate kirk*
  [Mains, NT2660], 142
Fischartoun, Fischertoun, *Rothes*, 449,
  491
Fischer Hill [*Fisherhills, NO7262], 31,
  35

INDEX OF PLACES 807

Fischerholme, *Cowhill, Holywood abbey*, 271
Fischeweik, Fischewik, Fischweik, *see* Fishwick
Fischirtoun, *Peterhead*, 457
Fisgill, *see* Physgill
Fishwick (Fischeweik, Fischewik, Fischweik), NT9151; 199, 203, 204; kirk of, 199, 204
Fiskane, *Monkland*, 498
Fithie (Fethe), NO6354; 358, 365
Flanschill, *Forgan*, 14
Flass (Flas), NT6251; 227, 227n.
Flatta, Flattay, *see* Flotta
Flemington (Flemyngtoun), *Dryburgh abbey*, NT9460; 190, 202
Flemington (Fleymingtoun), *Cambuslang*, NS6559; 508
Flendaris, Flynderis [Old Flinder, NJ5927], 31, 36
Fleurs ('Fluires callit Broumslandis'), *Ayton*, 199
Fleymingtoun, *see* Flemington
Flisk, NO3322; 60; parson of, 60
Flodden Field (Feild of Flowdoun, Fluddoun), NT8937; 209, 506
Floors (Fluris), *Kelso abbey* [Castle, NT7134], 232n.
Floors (Fluris), *Kinloss abbey*, NJ4952; 461, 464
Flotta (Flattay, Flottay, Flatta), ND3594, ND3796; 656, 659, 662, 663, 668; kirk of, 659; minister of, 659
Fluddoun, *see* Flodden
Fluires, 'callit Broumslandis', *see* Fleurs
Fluires, *Coldingham*, NT9165; 198
Fluirswallis, Flurislawis, *Greenlaw*, 227, 227n.
Flures, *Restalrig barony*, 133
Fluris, *Kelso abbey*, *see* Floors
Fluris, *Kinloss abbey*, *see* Floors
Flynderis, *see* Flendaris
Fodderty (Feddirdy, Fedirdy), NH5159; 644; vicarage of, 644
Foderis, Forderis, Forres, *near Old Montrose* [*Fordhouse, NO6660], 383, 384, 387, 388, 389
Fofardie, *Dunkeld*, 300
Fogo (Fogotoun), NT7748; 226, 232, 234, 235, 238, 241; kirk of, 220, 221, 226, 232, 238, 241; mill of, 223, 231, 234, 237, 240, 242; parish of, 234, 235, 238, 241; parsonage of, 226; vicarage of, 226
Fogorig (Fogo Rig, Fogo Rige), NT7748; 226, 231, 234, 237, 240
Foirdaill, Fordaill, Fordall, *Dunkeld bishopric* [*Fordel, NO1312, NO4423; *Fordell, NT1588], 303, 325, 342, 346
Foirrdis South and North, *see* Northfod, Southfod
Foirsyd, *Dreghorn*, 583
Foistertoun, *see* Fosterton
Fokestoun, *Lesmahagow*, 233
Fongard, *see* Fungarth
Ford (Furd), NX9578; 623
Fordaill, *Edinburgh*, 126
Fordel (Fordell, Fyrdaill), *Leuchars*, NO4428; 13, 20, 63
Forder, Forderis, *near Old Montrose*, *see* Foderis
Fordie (Fordy), *Dunkeld bishopric*, *NN7922 [*Mains of Fordie, NO0941], 303, 342, 345
Fordis, *see* Fordyce
Fordorno, *see* Fordoun
Fordoun (Fordorno), NO7475; 16, 17, 403; kirk of, 16, 17; vicarage of, 403
Fordyce (Fordice, Fordis), NJ5563; 415, 418, 425, 437, 438, 456, 457; curate of, 438; kirk of, 425, 437, 456; parish clerkship of, 456, 457
Fores, *see* Forres
Forestanis croft, Forrest croft, 623
Foresterseat (Frostar Sait, Froster Sett), *NJ8616; 31, 36, 423
Forfar (Forfartoun), NO4550; 217, 220. 221, 355, 370
——, *sheriffdom*, 1n., 36, 72n., 77n., 79, 79n., 83n., 213, 217n., 281, 281n., 282, 283n., 305, 365n., 371n., 372, 378n., 382, 392n., 396, 397, 402n., 641, 641n.; *see also* Angus
Forgan (Forgone, Forgoun, Forgound, Forgund) [Church, NO4425], 9, 14, 17, 20, 75, 76; Kirktoun of, 18, 19; kirk of, 9, 14, 17, 20; parish of, 90n.; vicar of, 75; vicarage of, 75-76
Forgandenny (Forgindyne, Forgoun, Forgund, Forgundeny, Forgundine, Forgundynie, Forgundyny), NO0818; 283, 287, 302, 309, 325, 329, 342, 343, 344, 346, 347; benefice of, 283n.; chaplainry of, 325; chaplainry of St Catherine's

altar in, 329; kirk of, 302, 342, 343, 344, 346, 346, 347; prebend of Dunkeld cathedral, 283n.; vicar pensionary of, 310, 325; vicarage of, 309, 325
Forgoun, *see* Forgandenny
Forgue (Forg, Forge), NJ6140; 359, 429; kirk of, 359; manse of, 429; vicar of, 429; vicarage of, 429
Forgund, *see* Forgandenny
Fornethy (Fornethie) [Ho, NO2455], 355
Forowhill, *see* Priourhill
Forres, *near Old Montrose, see* Foderis
Forres (Fores), NJ0358; 462, 463, 472, 487; kirk of, 487; parsonage of, 487
——, *sheriffdom*, 281n., 426n., 449n., 461n.
Forrest croft, *see* Forestanis croft
Forret [Mill, NO3921], 32, 35, 59, 79
Forss [Ho, ND0368], 628
Forteviot, NO0517; 317, 328, 330, 350; parsonage of, 159, 317; vicarage of, 328, 328, 330
Forth, *see* River Forth
Forthe, *see* Firth
Forther Ramsay, Forthir Ramsay, 15, 21; *see also* Kirkforthar
Forthirgill, *see* Fortingall
Fortingall (Forthirgill), NN7347; 300; parsonage of, 300; vicarage of, 300
Fortrose cathedral, NH7256; 463, 637; chaplainry of Obstule in, 637, 637n.; chanonry of, 652; chorister in, 642; stallar in, 463; vicar choral in, 627; *see also* Obstule
Fortry (Fortrie), NJ4953; 461
Fosken, *Monkland*, 102
Fossoway (Fossoquhay, Fossoquhy, Fossowy), NO0499; 295, 348, 349, 355, 370; kirk of, 295, 348, 349, 355, 370; parsonage of, 355, 370; vicarage of, 355, 370; *see also* Tullibole
Fosterton (Feistertoun, Foistertoun, Fostertoun), *Kinglassie*, NT2596; 29, 51
Fotlar, *see* Fetlar
Fouddeis South and North, *see* Northfod, Southfod
Fouddis, Foudeis, *see* Northfod, Southfod
Foula (Foulay), HT9539; 660; kirk of, 660; minister of, 660
Foulcheis, Foulches, *with Manderston and Ninewar, Dunglass collegiate kirk*, 166, 173
Foulden (Fouldane, Fouldene, Fouldeyne, Fulden), NT9255; 97, 190, 191, 193, 196, 207n.; parson of, 97, 193, 196; parsonage of, 193, 196
Foulford (Fowilfuird, Fowilfurd, Fowillfuird, Fowlisfuird) [Inn, NN8926], 27, 49
Foulis, *see* Fowlis Wester
Foull Fluires, *Haddington priory*, 180
Foullandis, *Lesmahagow*, 231n.
Foullartoun, *see* Fullerton
Foullis, *see* Fowlis
Foulshot Law (Foulschotlaw, Fustalaw), NT7400; 227, 227n.
Foultoun, *Barnwell*, 564
Foungard, Foungart, *see* Fungarth
Fourmerkland (Four Merk Land), *Holywood abbey*, NX9080; 269, 269n., 271, 275, 622, 622n.
Foveran (Fovarne, Foverne), NJ9723; 457, 458; parish of, 457, 458; vicarage of, 458
Fowilfuird, Fowilfurd, Fowillfuird, *see* Foulford
Fowlfurde, *Restalrig collegiate kirk*, 142
Fowlisfuird, *Dunfermline, see* Foulford
Fowlis (Foullis), NO3233; 8, 16, 18
Fowlis, *Alness and Kiltearn*, 625, 635, 648
Fowlis Easter, NO2934; kirk of, 8, 16, 18
—— collegiate kirk, 83n.; provost of, 83n.
Fowlis Wester (Foulis), NN9323; 341; vicar of, 341; vicarage of, 341
Fowlislie, *Tullynessle*, 436
Fowngart, *see* Fungarth
Foxton (Foxtoun), NO3916; 14
France, 40n., 73n., 124n., 169, 179n., 212, 363
Frankes croft, *Peebles*, 254
Frankpet, Frankpeta, *St Bathans priory*, 192
Freir croft, *Melrose abbey*, 207, 257
Freir toune, Freirtoun, *see* Friarton
Freirdykis [*Friarsdykes Dods, NT6668], 258
Freiris Hauch, *Elgin*, 453
Freiris land, *Melrose abbey*, 208

INDEX OF PLACES 809

Freirmyll, *Strathmiglo*, 69
Freirschaw, *see* Friarshaw
Freirsflat, *with Kedlock, St Andrews priory*, 18
Freirtoun, *Lindores, see* Ferietoun
Freirtoun, *St Andrews priory* [South Friarton, NO4325], 10, 12, 18
Freland, *Lindores abbey*, 32, 37
Frendraught (Frendracht) [Ho, NJ6141], 429
Freok, *see* Friock
Freris Croft, *Crail*, 19
Freuchie (Freuchy, Frouchie), NO2806; 15, 21
Freugh (Freuch) [East, NX1155], 599
Friarshaw (Freirschaw), NT5225; 207, 257
Friarton (Freir toune, Freirtoun), NO1121; 284, 285, 287
Friock (Freok) [Mains, NO5849], 376
Frithfield (Fruitfeild, Furthfeild, Furdefeild), NO5507; 162, 177, 179
Frostar Sait, Froster Sett, *see* Foresterseat
Frostleis, Frostleyis, *Kilconquhar*, 146, 167
Frouchie, *see* Freuchie
Fruitfeild, *see* Frithfield
Fruquhy, *Advie*, 480
Fudeis, *see* Craigfoodie
Fulden, *see* Foulden
Fulfordlees (Fulfurdleis), NT7669; 192
Fullerton (Foullartoun, Fullartoun, Fullertoun), NO6756; 383, 384, 387, 388, 389
Fungarth (Fongard, Foungard, Foungart, Fowngart, Fungard, Fungart), NO0343; 327, 327n., 329, 350, 365, 366; prebend of Dunkeld cathedral, 327, 329, 365, 366; prebendary of, 365, 366
Furay, Furro, *with Dounreay, Orkney bishopric*, 657, 670
Furd, *see* Ford
Furdefeild, *see* Frithfield
Furdis, Feudis, *fishing on the Dee, Aberdeen bishopric*, 416, 417
Furdtoun, *Brechin*, 383
Furro, *see* Furay
Furthfeild, *see* Frithfield
Fustalaw, *see* Foulshot Law
Futhanne, *see* Switha
Futty, Futy, *Aberdeen*, 419, 421, 422

Futy, *see* Futty
Fyf, Fyfe, Fyff, Fyffe, Fyif, Fyiff, *see* Fife
Fynave, Fynnave, *Scotlandwell priory*, 56
Fyndatour, *see* Findlater
Fyndogask, *see* Findo Gask
Fyndoirtye, *see* Findowrie
Fyndynate (Fandownat), NN8953; 324; chaplain of (with Tullypowrie), 324; chaplainry of (with Tullypowrie), 324
Fynglassie, Fynglassy, *see* Finglassie
Fynlawis, *Luss*, 542
Fynmont, Fynmonth, *see* Finmont
Fynnave, *see* Fynave
Fynnevin, *see* Finavon
Fynnocht, *see* Finmont
Fyntray, *see* Fintray
Fyrdaill, *see* Fordel
Fyrndonald, *see* Fairnedonald
Fysgill, *see* Physgill
Fyvie, NJ7637; 358, 359, 361, 430; kirk of, 359, 361

Gadden, *Balmerino abbey*, 58
Gain (Gayne), NS7370; 102, 498
Gairden, *see* Garden
Gairsay (Garsay), HY4422; 657, 670
Gairtlie, *see* Gartlea
Gaitmilk, Gaitmylk, *see* Goatmilk
Gaitschaw, *see* Gateshaw
Galashiels (Gallowscheillis), NT4836; 224
Galbesyd, *Monkland*, 98
Gallouraw, Gallowray, *Coupar Angus abbey*, 353, 356, 370
Gallow Hill, *see* Galowahill
Galloway (Galoway), NX3466, NX4866, NX6566; 113, 273, 274n., 585, 587, 588, 588n., 589, 591, 591n., 592, 597, 598, 600, 601, 601n., 602, 603, 604, 605, 606, 607, 616; archdeacon of, 601; bailie of bishopric of, 591; bishop of, 7, 7n., 92, 92n., 113, 274, 274n., 348n., 588, 588n., 591, 591n., 629; bishopric of, 495n., 586-592, 606-607; chamberlain of bishopric of, 591; diocese of, 616; minister of, 92
Gallowflat (Galloflat), NO2121; 285, 288

Gallowlaw, *Kelso abbey*, 232n.
Gallowrig, *Lesmahagow*, 231n.
Gallows Outon (Outoun Gallous, Outtoun Gallous), NX4542; 592, 594, 596
Gallowscheillis, *see* Galashiels
Galmantna, *Croy and Moy*, 639
Galoway, *see* Galloway
Galowayhill, Gallow Hill, *Lesmahagow*, 231, 231n.
Galrik, *Dunfermline*, 27, 49
Galrin [*Gallery, NO6765], 374
Galston (Galstoun), NS4936; 484, 558, 562, 563, 564; exhorter at, 563; minister of, 558; parish of, 558, 562, 564; vicar of, 484, 563, 564; vicarage of, 563, 564
Galstoun, *Balmerino abbey* [Gilston Ho, NO4406], 58
Galthie Feild, *with Clayock, Caithness archdeaconry*, 633
Galtounsyd, Galtrasyd, Galtunsyde, *see* Gattonside
Galtway, NX7148; 619; parsonage of, 617; vicarage of, 617; mill and mill croft of, 619, 620
Gamrie (Gamery, Gemrye), NJ7962; 359, 443; minister of, 443; vicar pensionary of, 443; vicarage of, 443
Gannoch, *see* Genoch
Gannochan (Ganochan), NN8509; 349
Ganochan, *see* Gannochan
Gardarocht, Gardarrocht, *Cadder*, 497, 498, 498n.
Gardein kirk, *see* Garnkirk
Garden (Gairdin, Geden), *Stirling*, 26, 48, 551, 556
Gardin, Gardinie, *see* Gardyne
Gardyne (Gardin, Gardinie, Gardny) [Castle, NO5748], 368, 376
Gargonzoun, *see* Kirkgunzeon
Gargrais, *see* Jargrayis
Garioch (Garioche), NJ6924; 443, 448; *see also* Chapel of Garioch
Garlies (Garleis) [Castle, NX4269], 587, 588, 592, 593, 595, 601, 602, 606, 607, 618
Garnkirk (Gardein kirk, Gartynkirk), NS6768; 497, 498
Garpit (Garpet, Gorpot), NO4627; 14, 20
Garrerie (Garrorie), NX3840; 594
Garrowood (Garwotwod), NJ5050; 461

Garsay, *see* Gairsay
Garscadden (Garskadden), NS5268; 529
Gart (Garth), NN6406; 349
Gartchery, *see* Gartsherrie
Gartenkeir (Gartinfallo, Gartinkeir), NS9394; 23, 46
Gartenquheyn, Gartinquhen, *Cadder*, 497, 498
Garteynquhermuire, Gartinquhennour, *Cadder*, 497, 498
Gartferry (Gartferrie, Gartforrie), *Cadder*, 497, 498
Gartforrie, *see* Gartferry
Garth, *Callander*, *see* Gart
Garthe (and Stok), *fishing near Perth, Balmerino abbey*, 58
Garthery, *see* Gartsherrie
Garthland [Mains, NX0755], 588, 590, 607
Gartinfallo, *see* Gartenkeir
Gartinquhen, *see* Gartenquheyn
Gartinquhennour, *see* Garteynquhermuire
Gartlea (Gairtlie, Gartlie), *NS4583, *NS7664; 102, 498
Gartloskan (Gartluskane, Gartlusken) [Hill, *NR7013], 102, 498
Gartly (Grantulye), NJ5232; 466; kirk of, 466
Gartmillan (Gartmylane, Gartmyllane), NS7469; 102, 498
Gartscharie, *see* Gartsherrie
Gartsherrie (Gartchery, Garthery, Gartscharie), NS7166; 102, 498
Gartturk, *Monkland, Newbattle abbey*, 102
Gartwery, *Monkland, Newbattle abbey*, 102
Gartyngailboik, *Monkland, Newbattle abbey*, 102
Gartyngavok, *Monkland, Glasgow subdeanery*, 498, 498n.
Gartynkirk, *see* Garnkirk
Gartynquhosche, *Cadder*, 498, 498n.
Garualdhous, *see* Garwald Ho
Garvald (Garvak, Garvalt, Garvat, Garvet, Garvok, Garvoll, Garwald, Garwat), NT5870; 181; hauch at the kirk of, 162, 165, 177, 178, 181; husbandlands of, 162, 165, 177, 178; kirk of, 162, 163, 165, 177, 178, 182; mill of, 162, 163, 165, 177, 178, 181, 182; minister of kirk of,

182; toun of, 162, 163, 165, 177, 178, 181, 182
Garvald Grange (Garvat, Garvet, Garwald, Garwat Grange), NT5871; 162, 165, 177, 178, 181
Garvald, chapel of St Mary of, *near Fintry, Stirlingshire*, 553n.; *see also* Kirk O' Muir, Kirkton, *St Ninians*
Garvock (Garwell), *near Dunning*, NO0314; 555
Garvock (Garvok), *St Andrews diocese*, NO7470; 82, 359; chaplainry of St Andrew in, 82; kirk of, 359; vicar of, 82; vicarage of, 82
Garwald Grange, Garwat Grange, *see* Garvald Grange
Garwald Ho (Garualdhous), NY2300; 259
Garwald Mylne, *see* Garvald
Garwald, Garwat, *see* Garvald
Garwell, *near Dunning, see* Garvock
Garwelwod, Gorvaldwode, Gyraldweid, *Lesmahagow*, 231, 231n., 233
Garwotwod, *see* Garrowood
Gask, *Dunfermline abbey* [Gask Ho, NN9918; Gask, NO5347], 23, 27, 46, 49, 376
Gaston (Gastoun, Gaustone, Gawstoun), NO5408; 177, 178, 181
Gateshaw (Gaitschaw), NT7722; 209, 214, 258, 259
Gattonside (Galtounsyd, Galtrasyd, Galtunsyde), NT5434; 208, 257
Gaustone, *see* Gaston
Gaw, *Ancrum*, 236, 366
Gawie, Gawy, *see* Cawy
Gawstoun, *see* Gaston
Gayne, *see* Gain
Gayrige, *see* Gyrig
Geden, *see* Garden
Geilles sex riggis, *Haddington priory*, 180
Geirgadeill, *Ardnamurchan*, 674
Gellets (Gulatis, Gullates, Gullatis, Gullattis) [Easter, Wester Gellet, NT0985], 25, 26, 47, 48
Gelshfield (Guilschfeild), ND1859; 632
Gelston (Gelstoun), NX7658; 594; kirk of, 594
Gemrye, *see* Gamrie
Geneva (Genewy), *Switzerland*, 388, 388n.
Genoch (Gannoch, Gannocht) [Mains, NX1356], 599
Gers, *see* Gersa
Gersa (Gers), ND2758; 632
Gerusland, *Lindores abbey*, 33
Gevane, *see* Govan
Gibbenstoun, Gibbinsone, Gibbinstoun, Gibbonestoun, *Holywood*, 269, 271, 275, 622
Gibbinstoun, Gibbonstoun, *Methven collegiate kirk* [*Gibbeston, NO0236], 296, 298
Gilbertfield (Gilbitfeild), NS6558; 508
Gilcamstoun, *near Ruthrieston, Aberdeen deanery*, 423
Gilgowrie Galtway [Galtway, NX7148], 619
Gillespie (Gillesbie) [Ho, NX2451], 266
Gillestounis croft, *Holywood abbey*, 623
Gillistoun, *Cardross*, 540
Gilmerstoun, *St Andrews priory* [*Gilmerton Ho, NO5111], 13, 20, 21
Gilmerton (Gilmertoun), NT2968; 94
Gilmertoun Grange, *Moorfoot, Newbattle abbey*, 101
Gilmestoun, *with Kinaldy, St Mary on the Rock collegiate kirk, St Andrews* [*Gilmerton Ho, NO5111], 55
Gilston (Gilstoun) [Ho, NO4406], 145, 167
Girs Muirland, Girsmuir Land, Girsmuirland, *Dunfermline*, 26, 27, 48, 49
Girthon (Girthtoun), NX6053; 587, 589, 607; kirk of, 589; parsonage of, 587; vicarage of, 587
Girvan (Girvane), NX1897; 567; kirk of, 567
Glacreuch, Glacrewch, *see* Clackriach
Gladhouse (Gledhous), NT2951, NT2954; 101
Gladswood (Gleddiswod, Glediswod), *NT5934; 189, 191, 197
Glak, *Tullybeagles, Methven collegiate kirk*, 296, 298
Glakreauch, *see* Clackriach
Glamis (Glammis), NO3846; 359, 392; glebe of, 392; kirk of, 359; vicar of, 392; vicarage of, 392
Glammis, *see* Glamis
Glasfuird, *see* Glassford
Glasgow (Glasgw, Glesgow), NS5965; 71, 71n., 94, 102n., 124, 124n., 215,

236, 247, 248, 250n., 251, 251n., 253, 253n., 272, 276, 277, 278, 282, 340, 365, 365n., 420, 448, 477n., 495, 496, 497, 498, 498n., 499, 499n., 500, 501, 502, 503, 504, 504n., 506, 507, 508, 509, 510, 511, 512, 514, 516, 519, 520, 522, 523, 525, 526, 529, 532, 533, 534, 540, 541, 559, 560, 561, 569, 570, 580, 581, 595, 597n., 603n., 604, 612, 614, 615, 616; archbishop of, 124, 124n., 504, 504n., 507, 508, 529, 529n., 581, 581n.; archbishopric of, 282, 495-497; archdeacon of, 251n., 253n., 525n.; archdeaconry of, 253; Brig, 503; Bridegate, 523; burgesses of, 503, 533; burghmuir of, 516; chamberlain of archbishopric of, 496; chaplain in, 94; chaplain of St Mungo's chaplainry outside the town of, 503; chaplain of St Roche's chaplainry beside, 514; diocese of, 513n., 525n.; chaplainry of St Mungo outside the town of, 503; chaplainry of St Roche beside, 514; curate of, 514; Gallow Muir, 524; hospital of, 532; Official of, 276, 340; parson of, 496, 512; parsonage of, 420, 512, 512n. 520; Rottenrow, 498, 500; superintendent of, 532, 532n.; Tolbooth (tenement 'callit Patiooup'), 516, 516n.; town of, 500; vicar of, 503, 561; vicarage portionary of, 503, 503n., 561
—— cathedral, 500, 501, 502, 504n., 515, 522, 522n., 523, 525, 526; chanter of, 277, 278, 340, 595, 597, 597n., 603, 603nn.; chantory of, 499; chaplain of altar of Name of Jesus in, 522; chaplain of altar of St Stephen and St Laurence in, 500; chaplain of altar of Our Lady of Pity in, 522; chaplain of Holy Blood chaplainry in, 502; chaplain of Holy Rood chaplainry in, 502; chaplain of St James' chaplainry in, 503; chaplain of St Michael's chaplainry in, 501, 515; chaplain of St Nicholas' altar in, 525, 615; chaplain of St Peter's chaplainry in, 501; chaplain of St Stephen's chaplainry in, 502; chaplainry of Holy Blood in, 502; chaplainry of Holy Rood in, 502; chaplainry of St James in, 503; chaplainry of St Michael in, 501, 515, 615; chaplainry of St Mungo in, 501, 502n.; chaplainry of St Peter in, 501; chaplainry of St Stephen in, 502; chaplains in, 248, 498, 499, 580; chaplains of altar of Holy Rood in, 500; ; chaplains of St Mungo's altar in, 523; chaplains of St Mungo's chaplainry in, 501; chapter of, 570; common kirks of, 499; dean of, 71, 71n., 499, 499n., 504; deanery of, 504; prebend of, 236, 250n.; stallar in choir of, 502n.; subchanter of, 272; subdean of, 498; subdeanery of, 102n., 497-498; treasurer of, 510; treasurership of, 505n., 510; vicars choral in, 498, 499; *see also* Ancrum, Manor, Stobo *and* Teviotdale

——, Friars Preachers (Dominican or Black friars) of, 522

——, New kirk (collegiate kirk of St Mary and St Anne) of, 516, 519, 523-524; prebend of, 516-517; prebendary of, 516, 519, 523; prebend of Sacrista Major in, 523

—— Primo, parsonage of, 512, 512n.

—— Secundo, vicarage of, 503, 503n.

—— university, 569; principal of, 448, 520, 569, 647

Glasslie (Glaslie) [Easter, Wester, NO2305, NO2304], 15

Glasnick (Glasnych, Glasnycht, Glasnyth) [High, Low, NX3562, NX3461], 588, 590, 592, 607

Glasnych, Glasnycht, Glasnyth, *see* Glasnick

Glass (Glas) [Haugh of, NJ4239], 485; parson of, 485; parsonage of, 485; vicarage of, 485

Glassaugh (Glassaucht) [Mains of, NJ5564], 438

Glasserton (Glasshertoun, Glastertoun), NX4237; 593, 605; parish of, 593; vicar of, 605; vicarage of, 605

Glassford (Glasfuird, Glesfuird), NS7247; 531; parsonage of, 531

Glastertoun, *see* Glasserton

Glasveildirie, *Mull*, 673

Glauschant, *see* Clayshant

Gleddiswod, Glediswod, *see* Gladswood

Gledhous, *see* Gladhouse

INDEX OF PLACES 813

Gleghornie (Gleghorne, Glegorne), NT5983; 147, 168, 169
Glen Muick (Glenmowik, Glenmowk, Glenmuik), NO3188; 425, 441; kirk of, 425, 440
Glen Tanar (Glentannar) [Ho, NO4795], 439; kirk of, 439
Glen, NS6514; 527, 530, 531
Glenbervie (Glenbervy), NO7680; 405, 406, 407; mains of, 323, 405; benefice of, 405; glebe of, 406; kirk of, 405; manse of, 406; parson of, 407; parsonage of, 407; vicar pensionary of, 406; vicarage of, 405, 406, 407
Glenbervie, *Sciennes priory, Edinburgh*, 94
Glenboy, *near Kincreich, Coupar Angus abbey*, 354, 357, 368, 369
Glenbuchat (Glenbuchatt, Glenbuchet, Glenbuchett) [Kirkton of, NJ3715], 425, 438; kirk of, 425, 438
Glencairn (Glencairne), NX7589; kirk of, 499
Glencaple, NS9221; 102
Glencors, *Monkland*, 102
Glencorse (Glencors), *Restalrig collegiate kirk* [Mains, NT2462], parish of, 141, 142, 143
Glendearg (Glendarge), NT5227; 259
Glendevon (Glendovan, Glendoven, Glendowane, Glendowen), NN9904; 295, 312, 337, 338, 348, 349; glebe of, 338; kirk of, 295, 348, 349; manse of, 338; parson of, 312; parsonage of, 312, 338; vicar pensionary of, 337; vicarage of, 312
Glenelvart, Glenelwart, Glenelwarte, *with Kincraigie, Dunkeld bishopric*, 303, 305, 342, 345; mill of, 303, 342, 345
Glenesslin (Gleneslen), NX8284; mill of, 276, 277
Gleneyla, Glenyla, Glenylay, *see* Glenisla
Glengaber (Aarngabyte, Blengaber, Bllangawar) [Burn, NX9080], 276, 276n., 623
Glengeith, NS9416; 102
Glengerrack (Glengarrok) [Mains of, NJ4552], 461
Glengower (Glengoir, Glengory, Glengoy, Glengoyr, Glengoury), NX9382; 270, 271, 275, 276, 623
Glenholm (Glenquhome), NT1032; 247; parson of, 248; parsonage of, 248
Glenhove (Glenhuif), NS7772; 102, 498
Glenhuif, *see* Glenhove
Glenisla (Gleneyla, Glenyla, Glenylay, Glenyly) [Kirkton of, NO2160], 354, 355, 368, 369, 386, 409; kirk of, 355, 369; parsonage of, 355, 369; vicarage of, 355, 369
Glenkerry (Glenkeyrie), NT2710; 258
Glenlaugh (Glenlauch), NX8688; 275, 276
Glenluce, NX1957; 538, 545
—— abbey, 538, 545, 568, 600, 600n.; abbot of, 600, 600nn.
Glenmarkie (Glenmerky) [Lodge, NO2364], 355
Glenmowik, Glenmowk, *see* Glen Muick
Glenmuik, *see* Glen Muick
Glennislandis, *Jedburgh abbey*, 219
Glenorchy (Glenorquhard, Glenurquhay), NN2433; 284, 284n., 287
Glenquhapple (Glenquhappill) [Moor, NX3570], 598, 599
Glentannar, *see* Glen Tanar
Glenter Easter, *see* Easter Glentore
Glenter Wester, *see* Wester Glentore
Glentre, *see* Clentre
Glenturk, NX4257; 597
Glenumquhair, *Crawfordmuir* [*Glenochar, NS9513], 102
Glenyla, Glenylay, Glenyly, *see* Glenisla
Glesfuird, *see* Glassford
Glesgow, *see* Glasgow
Gleuschart, *see* Clayshant
Glims Holm, *see* Holmes, isles of
Glook, *Methven collegiate kirk*, 296, 298
Goatmilk (Gaitmilk, Gaitmylk) [Fm, NT2499], 24, 29, 46, 51
Godscroft (Goddiscroft), NT7463; 166
Gogar, *Logie parish* [East, West Gogar, NT8395, NT8495], 290, 290n., 294, 295, 296, 348, 349
Gogar, *parish*, NT1772; parsonage of, 117; prebend (with Addiston) of Corstorphine collegiate kirk, 107, 107n., 112, 113, 115; vicarage of, 117

Gokstoun, *see* Cookston
Golford (Golfurd), NH9554; 469
Golspie, *see* Kilmalie
Gordon (Gordone, Gordoun), NT6442; 223, 227, 228, 229, 230, 232, 234, 235, 238, 240, 241; kirk of, 229, 234, 240; parish of, 234, 238, 241; parsonage of, 227; vicarage of, 227, 232, 235, 238, 241
Gordon East Mains (Eister Manes, Eister Mannis), NT6543; 234, 238, 241
Gordoun Manes [*Gordonmains Burn, NT6844], 228
Gorlabank, Gourlaband, Gourlawbank, Gurlawbankis, Gurlay Bankis, *with Traprain, St Andrews archbishopric*, 3, 6, 7; *see also* Gourlaw
Gorlaw, *see* Gourlaw
Gorpot, *see* Garpit
Gorsnewt, *Restalrig collegiate kirk*, 142
Gorton (Gortoun, Gourtoun) [Ho, NT2833], 142, 143
Gorvaldwode, *see* Garwelwod
Gosfuird, *Leith*, 128
Gospert [Easter, Wester Gospetry, NO1606], 69
Goukstoun, *see* Cookston
Gouria, *see* Gowrie
Gourlabank, Gourlawbank, *see* Gorlabank
Gourlaw (Cowirla), *with Traprain, Prestonkirk*, NT2761; Cowirla Bankis, 170; *see also* Gorlabank
Gourlaw (Gorlaw, Gowirlaw), NT2862; 142, 143
Gourtoun, *see* Gorton
Govan (Gevane, Govane), NS5664; 526, 533; chaplain of the Lady altar in, 526; chaplainry of the Lady altar in, 526; minister of, 533; parsonage of, 533; vicarage of, 533
Govis Hauch, *see* Lochtysyd
Gowirlaw, *see* Gourlaw
Gowishawcht, *see* Lochtysyd
Gowrie (Gouria, Gourye, Gowry) [Carse of, NO2020], 8, 16, 315, 322, 339, 538, 545; *see also* Carse of Gowrie
Goystintoun, *Netherfield, Cathkin*, 517
Graemsay (Gramsay), HY2505; 659
Grampians, 31n., 404n.
Gramsay, *see* Graemsay
Granden, Grandene, *near Persley,*

Aberdeen deanery, [*Mains of Grandhome, NJ9012], 422, 423
Grandtully (Grantillie), NN9152; 17
Grange of Aberbothrie (Grainge of Abirbothre, Grange of Abirbothre), NO2344; 354, 357, 369
Grange of Airlie (Grainge of Arlie, Grange of Erlie), NO3151; 354, 356, 357, 368, 369
Grange of Erlie, *see* Grange of Airlie
Grange, *Auchtertool*, 303
Grange, *Bass*, 211
Grange, *Calder*, 98
Grange, *Cambuskenneth abbey* [West Grange, NS8194], 544
Grange, *Carriden* [*Grangepans, NT0181], 157
Grange, *Dunbar*, 122
Grange, *Glasnick, Galloway bishopric*, 588, 590
Grange, *Lindores abbey*, NO2516; 29, 30, 33, 34, 35, 37
Grange, *North Berwick priory*, 146, 147, 167
Grange, *Penninghame*, 587, 588
Grange, *St Andrews priory, see* New Grange
Grange, *Ringwodfeild, Melrose abbey*, 259
Grange, *St Mary's Isle priory*, NX6847; 617, 619, 620; mill of, 617
Grange, *with Donibristle, Inchcolm abbey*, 62, 63, 64
Grangegreen (Grangegren, Grante Grein), NJ0058; 470
Granges (Grainges, Graingis), *Cambuskenneth abbey* [West Grange, NS8194], 537, 545, 547, 548
Grant, *Aberlour*, 489
Grante Grein, *see* Grangegreen
Grantillie, *see* Grandtully
Grantulye, *see* Gartly
Great Cross, NX6750; 617
Greenhead (Greenheid, Greinheid, Greneheid, Grenheid), NT4929; 223, 230, 230n., 233, 237, 239
Greenholm (Greinholme), NS5337; 564
Greenlaw (Greinelaw, Greinlaw, Grenelaw, Grenlaw, Grunlaw, Gryinlaw), *parish*, NT7146; 223, 226, 227, 227n., 228, 230, 232, 233, 234, 235, 237, 238, 239, 241; kirk

INDEX OF PLACES 815

of, 226, 228, 238, 241; parish of, 234, 235, 238, 241; vicarage of, 226, 227
Greenlaw (Grenelaw, Greynlaw), *Restalrig collegiate kirk* [Mains, NT2461], 142
Greenlawdean (Grenelawden, Grenelawdene, Grenladen), NT7046; 226, 226n.
Greenock (Greinok), NS2776; 533
Greenrig (Grenerig, Grenrig, Newrig, Nowrig), NS8542; 231, 231n., 232, 232n.
Greenwall, Greinwall, Greinweill, *Orkney*, 679, 680, 683
Grein Syd, *with Catmoss, Kelso abbey*, 226, 226n.
Greinand, *see* Grennan
Greinelaw, Greinlaw, *see* Greenlaw
Greinheid, *see* Greenhead
Greinholme, *see* Greenholm
Greinok, *see* Greenock
Greinsulle, Grindsulle, Grinsett, *with Northmavine, Orkney bishopric*, 658, 663, 670
Greistoun, *Peebles*, 254
Greithetlandis, *Stobo*, 249,
Grene End, Greyne End, *near Cot-town, Aberdeen deanery*, 422, 424
Grene Yaird, *Culross abbey*, 289
Greneheid, *see* Greenhead
Grenelaw, Grenlaw, *see* Greenlaw
Grenelawden, Grenelawdene, *see* Greenlawdean
Grenerig, *see* Greenrig
Grenheid, *Selkirk*, 230n.
Grenis, *Manuel priory*, 549
Grenladen, *see* Greenlawdean
Grenlaw, *see* Greenlaw
Grennan (Greinand), NX4145; 593
Grenok, *Dunblane bishopric* [*Braes of Greenock, NN6305], 349
Grenrig, *see* Greenrig
Greyne End, *see* Grene End
Greynlaw, *see* Greenlaw
Gribton (Gribtoun), NX9179; 269, 271, 276, 277, 623
Grindsulle, *see* Greinsulle
Grinsett, *see* Greinsulle
Growethe Hauch, *with Wideopen, Morebattle*, 214
Grubbit Hauch [*Grubbit Law, NT7923], 184

Grunlaw, *see* Greenlaw
Gryimschaw, *Eckford*, 220
Gryinlaw, *see* Greenlaw
Gueldres, Mary of, 404n.
Guillyhill (Guliehill, Gulliehill, Gulyhill), NX9678; 270, 276, 623
Guilschfeild, *see* Gelshfield
Guis croft, *St Ninians*, 546
Gulane, *see* Gullane
Gulatis, *see* Gellets
Gulen, *see* Gullane
Guliehill, *see* Guillyhill
Gullane (Gulane, Gulen), NT4882; 190, 197, 197n.; kirk of, 190, 197, 197n.; vicar of, 175; vicarage of, 175
Gullates, Gullatis, Gullattis, *see* Gellets
Gulliehill, *see* Guillyhill
Gulyhill, *see* Guillyhill
Gummeris Mylne, Gymmeris Mylne, Gynne Mylne, *Haddington priory*, 162, 163, 176, 178, 181
Gunisgrene [Gunsgreenhill, NT9463], 199
Gurdeis [Middle Gourdie, NO1142], 331
Gurlawbankis, Gurlay Bankis, *see* Gorlabank
Guthrie (Guthre), NO5650; 390, 394; reader at, 390; vicar of, 390; vicar pensionary of, 394
— collegiate kirk, 390, 391, 394; prebendary of, 390, 391; prebends of, 390; provostry of, 394; *see also* Hilton, Langlands
Gyraldweid, *see* Garwelwod
Gyrig, Gayrige, Zy Rig, *with Fogorig, Kelso abbey*, 223, 223n., 226, 226n.

Habaland, *see* Abbeyland
Hach, *see* Hauch
Hachtnis, *see* Hatchednize
Hacket, *Linlithgow*, 11
Hadden (Haldein, Halden, Yaldan), NT7836; 223, 230, 233, 237, 239
Hadderweik, Hadderwik, Haddirweik, *Montrose*, *see* Hedderwick
Haddington (Haddingtoun, Hadingtoun, Hadingtown, Hadyngtoun), NT5173; 8, 11, 16, 17, 19, 25, 42, 47, 53, 65 145, 165, 166, 174, 176, 178, 180, 188, 190, 297, 390, 450, 568; burgh of, 162, 176, 178, 180, 222, 259;

Dalzell's land, 180; Fish Market, 180; Gallowsyd, 180; High Street, 11; kirk in Nungate of, 163; kirk of, 16, 17, 163; Lawson's foreland, 180; mills of, 180; mains of the 'abbey' of, 176, 178; New Gait, 180; Nungate, 162, 162n., 163, 165, 177, 178, 181; Nunsyde, 180; Ogilbeis land, 180; parish kirk of, 163; Rankin's land, 180; Riglingtoun land, 180; St Martin's kirk of, 162, 162n.; waulk mill of, 181
—— priory, 157n., 161-165, 176-179, 180-182; bailie of, 163; chamberlain of, 164, 182; mains of, 165, 176, 178; nuns of, 163, 182; prioress of, 161, 164, 164n.
——, constabulary, 1n., 118n., 149n., 150n., 171, 172n., 186n., 249, 450n.
Haddirlik, *Fintray, see* Heatherwick
Haddo (Haddow, Haddowk), NO7473; 11, 18, 435
Haddowerk, *Fintray, see* Heatherwick
Haddowick, *Montrose, see* Hedderwick
Haddowk, *see* Haddo
Hadingtoun, Hadingtown, Hadyngtoun, *see* Haddington
Haggs (Haggis), *Monkland*, *NS7146; 102, 498
Hagmylne, *Monkland*, 102
Hailes (Hailles), *with Traprain* [Nether, Over, NT5775, NT5776], 6, 7, 170
Hailes, kirk of, 114
Haining Mains (Maynis of Hayning), NS4535; 531
Hairheid, *see* Harehead
Hairhoip, *see* Harehope
Hairwod, *see* Harewood
Halbaudrik, *see* Baldridge
Halbero, *see* Halcro
Halcro (Halbero), ND2360; 632, 657, 670
Haldain, Haldan, *see* Auldhame
Haldame, *see* Auldhame
Haldein, Halden, *see* Hadden
Halhill, NO2813; 33, 37
Halie Burne Home, Halybredhoill, Halybreidholme, *Mow*, 225, 225n.
Halieerik, *see* Halkirk
Halieweill, prebend of Dunglass collegiate kirk, 166
Halk Hill, *Glenbervie*, 405
Halkarton (Halkartoun), NO4448; 217

Halkburne (Balkburne, Halkburnam), NT4740; 207, 257, 259
Halkettil, Hoilkettill, *Lathrisk* [Kingskettle, NO3008], 15, 21
Halkirk (Halieerik), ND1358; 628, 631; kirk of, 631
Hall of Wistownis Hilend, Hall of Witstones Hill End, *Mearns*, 31, 35
Hall Teasses (Haltatis), NO4109; 55, 55n.; *see also* Hill Teasses
Hallhill, *Monkland*, 498
Hallhill, *Newburn*, 24, 26, 46, 48
Halliburton (Halyburtoun), NT6748; 187
Hallis Syd, *Morebattle*, 214
Hallrule (Harroull), NT5914; 220
Halltoun, *Kinglassie*, 29, 51
Halquhitsoun, *Ecclesgreig*, 16
Haltatis, *see* Hall Teasses
Haltonhill (Haltoun Hill, Haltounhill), NO2416; 29, 33, 35
Haltoun, *Ratho*, 100
Haltoun Linlour, *see* Linlour
Haltoun of Fintray, *see* Hatton of Fintray
Haltoun, *Carse of Gowrie, Perth Charterhouse*, 285
Haltoun, *Dunkeld chantory, see* Hiltoun
Haltoun, *Kinnell parish*, 366
Haltoun, prebend of Corstorphine collegiate kirk, *see* Hatton
Halwod, *see* Harewood
Halybredhoill, Halybreidholme, *see* Halie Burne Home
Halyburtoun Eistersyd [Halliburton, NT6748], 226
Halyburtoun Westersyd [Halliburton, NT6748], 226
Halyburtoun, *see* Halliburton
Halychtmoir, *see* Alichmore
Halyden, Halydene, *see* Holydean
Halyrudhous, Halyruidhous, *see* Holyroodhouse
Halywod, Halywood, *see* Holywood
Hame, *see* Holm
Hamilton (Hamiltoun), NS7255; 518, 519; vicar pensionary of, 519; vicarage pensionary of, 518
—— collegiate kirk, 518n.; provost of, 518, 518n.
Handerswod Eister and Wester, *Calder*, 98
Hangitschaw, *Traquair* [*Hangingshaw,

INDEX OF PLACES

NT3930], 496
Haning, *Selkirk*, 230n.
Harcarse (Hurkers), NT8148; 226
Hardenis, *Greenlaw*, 227
Hardhessillis, Hardhisselles, *St Bathans priory*, 192, 194
Hardingtoun [Ho, NS9630], 517
Hardis Mylne, *Hume*, 228
Hardmastoun, *see* Herdmanston
Hardwod Littill, *see* Little Harwood
Hardwod Mekill, *see* Harwood
Harehead (Airheid, Hairheid, Hareheid), NT6963; 223, 231, 233, 237, 240, 242
Harehope (Hairhoip), NT2044; 207, 258
Harelaw, Hordlaw, *Greenlaw*, 226, 235, 238, 241
Harewood (Hairwod, Halwod) [Burn, NY2798], 259, 259n.
Harlaw, *Merse*, 258
Harperdean (Harpardene, Harperden, Harperdene), NT5074; 162, 177, 178, 181
Harperrig, NT1061; 98
Harray, HY3020; 657, 659, 669; kirk of, 659; minister of, 659
Harris (Hereis, Heries), *NG9308; 674, 675
Harroull, *see* Hallrule
Harthvodmyris, *Selkirk*, 230n.
Hartisheid, *Leith*, 128
Hartside (Hartesyde, Hartsyd, Hartsyde), *parish*, 497, 497n., 513; parsonage of, 513
Hartside (Hartsyd, Hertsyd), *Melrose abbey*, *NT1555; 207, 258
Hartvodburne, *Selkirk*, 230n.
Harvieston (Harvestoun), NT3560; 176, 450
Harwod, *Hopkirk* [*Harwoodmill, NT5809], 220
Harwood (Hardwod Mekill), *Calder*, NT0162; 98
Hasildene, Hessilden, *St Andrews parish*, 13, 20, 21
Hassendean (Assindane, Hassinden), NT5420; 175, 207, 209, 211n., 257, 260, 261, 264, 264n.; kirk of, 209, 211n.; vicarage of, 175n., 264
Hassington Mains (Hassintoun Manes), NT7340; 258
Hassington (Hawssetoun), NT7341; 184
Hatchednize (Hachtnis, Hauchetnnes), NT8041; 186, 187
Hatherwik, *Dunbar*, *see* Hedderwick
Hathronesyd, *Hopkirk*, 220
Hatoun, *see* Hatton
Hatrik, *see* Hawick
Hatton (Haltoun), *Dallas*, NJ1253; 492
Hatton (Haltoun, Hatoun) [*Easter Hatton Mains, NT1469], prebend (with Dalmahoy) of Corstorphine collegiate kirk, 97, 107, 107n., 112, 121
Hatton (Hattoun), *Eassie*, NO3546; 412
Hatton of Fintray (Haltoun of Fintray), NJ8416; 31, 36
Hauch, parsonage of, 170, 170n., 174n. *See also* Linton, Prestonkirk
Hauch, Hach, *Fungarth, Dunkeld* [*Haughend, NO0342], 329, 365
Hauch, Hauche, *Lindores abbey*, 30, 32, 34, 35, 36
Hauch of Nether Monbennis [Upper Manbean, NJ1957], 453
Haucheid, Hewcheid, *with Caverton, Eckford*, 212, 215
Hauches *of Killiesmont, Kinloss abbey*, 464
Hauches, *Kinloss abbey* [*Haughs, NJ4151, NJ5049], 461
Hauchetnnes, *see* Hatchednize
Hauthornden (Hawthorneden, Hawthornedeyn, Hawthornedyne), NT2963; 142, 143
Hawestok, *Riccarton*, 531
Hawhill, *Lesmahagow*, 232
Hawick (Hatrik, Hawik, Hawyk), NT5014; 223, 223n., 263; parson of, 263; parsonage of, 263
Hawssetoun, *see* Hassington
Hawthorneden, Hawthornedeyn, Hawthornedyne, *see* Hauthornden
Hawyk, *see* Hawick
Hayperis croft, *Holywood abbey*, 623
Haytoun, *see* Heitoun
Hazeldean (Hessilden, Hiselden, Hisileden), NS7344; 515, 516; prebend of Bothwell collegiate kirk, 515, 516
Hazlehead (Heslie Hed), NJ8905; 423
Heatherwick (Haddirlik, Haddowerk), NJ8018; 31, 36
Heavin *of North Berwick, see* North Berwick, haven of
Hecfuird, *see* Eckford

Hectoures acre, *Haddington priory*, 181
Hedderwick (Hadderweik, Hadderwik, Haddirweik, Haddowick, Heddirweik), *Montrose*, NO7060; 383, 386, 388, 389
Hedderwick (Hatherwik), *Dunbar*, NT6377; 169
Hedelik, Hedelyk, *Lindores abbey*, *see* Edderlick
Hegatt, *Beith*, 574
Heid acres, *see* Hood acres
Heill, *see* Hill
Heislat, *Beith* [*Hessilhead, NS3852], 574; mains of, 574
Heitoun (Haytoun, Hevin, Hevine, Hewtoun, Heytoun, Hitowe, Hwtoun), NT7130; 223, 223n., 231, 234, 237, 240, 242
Heland, *with Godscroft, Dunglass collegiate kirk*, 166
Heleischowcht, *see* Elisheugh
Heling Hill, *St Andrews priory*, 18
Helmsdale (Helmisdaill), ND0215; 632; chaplain of St John's chaplainry of, 632; St John's chaplainry of, 632
Henwode, *see* Dodd
Hep Hill, *North Berwick priory*, 168
Herbert, *beside the Nith, Holywood abbey*, 277
Herdmanston (Firdmastoun, Hardmastoun) [Mains, NT4769], chaplainry of, 173
Hereis, Heries, *see* Harris
Hereot, *see* Heriot
Herheid, *Selkirk*, 230n.
Heriot (Hereot), NT3852; 102; kirk of, 102
Herringden, *see* Hirendean
Hertsyd, *see* Hartside
Heslie Hed, *see* Hazlehead
Hessilden, *Bothwell collegiate kirk*, *see* Hazeldean
Hessilden, *St Andrews parish*, *see* Hasildene
Hesyd, *see* Seaside
Hetspeth, *see* Hexpath
Heuch, *Polwarth*, 195
Heuch, Heuche, *North Berwick priory*, *see* Heugh
Heucheid, *see* Cleucheid
Heuchtbray, *Stobo*, 249
Heugh (Heuch, Heuche), *North Berwick priory*, NT5684; 146, 147, 169;

mains of, 167, 169, 169n.
Hevin, Hevine, *see* Heitoun
Hewcheid, *see* Haucheid
Hewlabutis, *Melrose abbey*, 207
Hewtoun, *see* Heitoun
Hexpath (Hetspeth, Hexpeth), NT6646; 227, 227n.
Heycorne, *Restalrig collegiate kirk*, 142
Heyndoun Towne, *see* Lindean
Heytoun, *see* Heitoun
Hielawis, *see* Highlaws
High Ersock (Over Ersslis), NX4437; 593
High Glasnick (Over Glasnycht), NX3562; 588
Highlaws (Hielawis), NT9363; 198
Hilend, *Ecclesgreig*, 16
Hilend, Hilheid, *Coldingham priory*, 198, 203
Hilend, *Holyhead*, 275
Hilheid, *Coldingham priory*, *see* Hilend
Hilheid, *Fungarth*, 329, 365
Hilhous, *Hartside*, 513
Hilhous, *Symington*, *see* Hillhouse
Hilhous, *with Balbuthie*, 146, 148, 167
Hill (Heill), *Fogo*, 226, 235, 235n., 238, 241
Hill Teasses (Hiltattis), NO4108; 55, 55n.; *see also* Hall Teasses
Hill, *Lindores abbey* [*Lindores Hill, NO2518], 32
Hill, *Musselburgh*, 25, 26, 43, 47, 48, 49, 53, 54
Hillhouse (Hilhous), *Symington, Failford*, NS3629; 564
Hillis of Towch, *St Ninians*, 546
Hillok Park, *Lindores abbey*, 30
Hills (Hillis), *Galloway bishopric*, 587, 603; *Holywood abbey*, 269, 270; *Whithorn priory*, 593; *Wigtown*, 597
Hilsyd, *see* Hillside
Hiltattis, *see* Hill Teasses
Hilton (Hiltone, Hiltoun), NO5551; prebend (with Langlands) of Guthrie collegiate kirk, 390
Hilton (Hiltoun), kirk of, *Berwickshire*, NT8850; 185
Hiltone, Hiltoun, Hyltoun, *of Craigie* [Craigie, NO4231], 30, 34, 35
Hiltoun, *Cadder*, 497, 498
Hiltoun, Haltoun, *Dunkeld chantory*, 288, 307
Hiltoun, Hiltoun of the Coilheuche,

INDEX OF PLACES 819

Inchcolm abbey, 62, 64
Hiltoun, Hirltoun, *Catterline, Brechin bishopric*, 383, 386, 389
Hirdinescheillis, *Calder*, 98
Hirendean (Herringden) [Castle, NT2951], 101
Hirltoun, *Catterline*, see Hiltoun
Hirsel (Hirsell, Hirsill), NT8240; 215; teinds of, 186
Hiselden, Hisileden, see Hazeldean
Hitowe, see Heitoun
Hobbetstoun, *Calder*, 98
Hobkirk (Hopkirk), NT5810; 219, 220; kirk of, 219, 220
Hobsburn (Hoppisburne), NT5712; 220
Hocht, Hoitt, Hote, *Kelso abbey*, 222, 230, 233, 237, 239
Hoddam, NY1572; parson of, 266; parsonage of, 266; vicarage of, 266
Hoilkettill, see Halkettil
Hoill Baudrik, Hoill Bawidrik, see Baldridge
Hoiltoun, Hoyltone, *Holywood abbey*, 275, 276
Hoipcartane, see Hopcarton
Hoisla, Hoislaw, see Horseley
Hoitt, see Hocht
Holden Nether, *Calder*, 98
Holden Over, *Calder*, 98
Holl Myln, *Inchcolm abbey*, 64
Hollynghirst, *Monkland, Newbattle abbey*, 102
Holm (Hame, Holme, Home), *Holywood abbey*, *NX9580; 271, 275, 623
Holm, *parish, Orkney*, HY4803; 657, 659, 662, 663, 669, 671; kirk of, 659; minister of, 659-660; parish of, 657, 662
Holmes, isles of [i.e. Glims Holm and Lamb Holm], *Orkney*, ND44799, HY4800; 659
Holme, *Croy and Moy*, 639
Holy Blood (and Nomine Jesu), chaplainry of, *St Giles' collegiate kirk*, 117-118
Holy Blood aisle, *St Giles' collegiate kirk*, 135
Holy Blood altar, *Cross kirk, Peebles*, 254, 254n.
Holy Blood altar, *Kirkcaldy*, 81, 84
Holy Blood altar, *Sanquhar parish kirk*, 273
Holy Blood altar, *St Giles' collegiate kirk*, 130, 131
Holy Blood chaplainry, *Glasgow cathedral*, 502
Holy Rood altar, *Brechin cathedral*, 400
Holy Rood altar, *Crail collegiate kirk*, 66
Holy Rood altar, *Glasgow cathedral*, 500
Holy Rood altar, *Kirkcaldy kirk*, 85
Holy Rood altar, *Manor parish kirk*, 255, 525
Holy Rood altar, *St Andrew's collegiate kirk, Peebles*, 254, 254n.
Holy Rood altar, *Stirling parish kirk*, 556
Holy Rood chaplainry, *Dumbarton parish kirk*, 540
Holy Rood chaplainry, *Glasgow cathedral*, 500, 502
Holydean (Halyden, Halydene) [Castle, NT5330], 223, 225, 231, 233, 237, 240
Holyroodhouse (Halyrudhous, Halyruidhous), 179, 183n., 202, 202n., 222, 222n., 261, 300, 459, 591, 591n., 602, 617n.; see also Canongate, Edinburgh
—— abbey, 91-93; abbot of, 591, 591n.; commendator of, 93, 93n., 222, 222n.
Holywood (Halywod, Halywood), NX9480; 269, 270, 271n., 272, 274, 274n., 275, 277, 622, 623, 624; minister of, 272, 277, 624
—— abbey, 269-272, 274-278, 622-624; abbot of, 271, 271n., 272; commendator of, 271n., 272, 277, 278, 278n., 624, 624n.; oeconomus of, 277; superior and minister of, 277
Home, *Holywood abbey*, see Holm
Home, *Kelso abbey*, see Hume
Homland *of Soulseat abbey*, 598
Hood acres (Heid, Hude, Hudis acres), *Cambuskenneth abbey*, 537, 544, 546, 547, 548
Hopcailzie, Hopcaltzie, Hopcalzie, Hopcalzo, Hopcalzocht, see Kailzie
Hopcarton (Hoipcartane, Hopcarten), NT1231; 203, 207, 258
Hopes (and Mount Lothian), kirk of, 91; see also Mount Lothian
Hopisland, *Carse of Gowrie, Perth*

Charterhouse, 285, 288
Hopkailzie, see Kailzie
Hopkirk, see Hobkirk
Hoppisburne, see Hobsburn
Hoprew Eister, see Easter Happrew
Hoprew Wester, see Wester Happrew
Hordlaw, see Harelaw
Horhart, *Dollar*, 24, 46
Horndean (Horndein, Horneden), NT8949; 224, 232, 234, 237, 240; kirk of, 224, 232; vicarage of, 234, 237
Horseley (Hoislaw, Horslaw, Hoslaw, Oislaw), NT8363; 223, 223n., 231, 231n., 234, 237, 240
Horsinpryg [*Horseupcleugh, NT6658], 184, 184n.
Horslaw, see Horseley
Hoslaw, see Horseley
Hospital of St Paul's work, see Leith
Hospitalitie of St Anthony, see Leith
Hospital, kirk of, see Spittal
Hote, see Hocht
Houdene, *Selkirk*, 230n.
Houndwood (Hunwod), NT8463; 199
Housbyre, Vousbyre, Wousbyre [*Easter, Wester Housebyres, NT5337, NT5336], 207, 257, 259
House o' Muir (Hous of the Muir), NT1264; 104
Houston (Houstoun), *Renfrewshire*, NS4066; 528; kirk of, 528; see also Killallan
Houston (Houstoun, Howesoun), *with Traprain* [Wood, NT0569], 3, 6, 7, 170
Houstoun, *Peebles*, 249
Hova [Howahill, NT7719], 220
Howburne Mylne [Howburn, NT8562], 204
Howden (Howdene), *Maxwell*, *NT6519; 225, 229
Howesoun, see Houston
Howlawheid [Easter, Wester Howlaws, NT7242], 227, 235, 238, 241
Howlawrig [Easter, Wester Howlaws, NT7242], 226
Howme, see Hume
Howmoir, *Uist*, 674; parsonage of, 674
Hownam (Hownem, Hownum, Hownumtowne), NT7719; 216, 219, 220; kirk of, 216, 219, 220
Hownam Grange (Hownoumgrange, Hunum Grange, Hunumgrange, Lounoumgrange), NT7822; 209, 258, 259; mill of, 207, 258
Hownoumgrange Mylne, see Hownam Grange
Howreik, see Auldbreck
Howtounhall, *Jedburgh abbey*, 216
Hoy, ND2598; 659; kirk of, 659; minister of, 659
Hoyltone, see Hoiltoun
Hude acres, see Hood acres
Hudis acres, see Hood acres
Humbie (Humby), *Kelso abbey*, see Keith-Humbie
Humbie, *Aberdour*, NO1986; 64
Humbie, *with Winchburgh, Kirknewton*, NT1167; 3, 104
Humby, *Eccles priory*, 184
Hume (Home, Howme), NT7041; 223, 228, 232, 234, 235, 238, 240, 241; kirk of, 232, 234, 238, 240; vicarage of, 228, 232, 235, 238, 241
Hundalee (Hundolie), NT6418; 219
Hunthall, *Riccarton*, 531
Hunthill, *Eccles priory*, NT7711; 184
Hunthill, *Jedburgh abbey*, NT6619; 219
Huntlawcoit, see Huntly Cot
Huntly (Huntlie), NJ5240; 492
Huntly Cot (Huntlawcoit), NT3052; 101
Huntlywood (Huntlie Wod), NT6143; 227
Huntrodland, *Dryburgh abbey*, 190
Hunum Grange, Hunumgrange, see Hownam Grange
Hunwod, see Houndwood
Hurkers, see Harcarse
Husbandtoun, *near Glenturk, Wigtown*, 597
Husland Tak Harlaw [*Harelaw, NS4133], 532
Hutron, see Hutton
Hutton (Hutron), NT9053; 266; parson of, 266; parsonage of, 266
Huttounhall [*Hutton Castle, NT8854], 200
Hwtoun, see Heitoun
Hyltoun *of Craigie*, see Hiltone
Hynhousfeild, Hynnoisfeild, *Jedburgh abbey*, 216, 219

Ibert (Ibertis), NN8825; 349
Iddirthane, see Edderton

Idvies (Idvie, Idvy, Idwye), NO5347; 376; mains of, 376; parson of, 375; parsonage of, 376
Idvy, see Idvies
Idwye, see Idvies
Iflie, see Evelaw
Ilay, see River Isla
Ile of Man, see Man, Isle of
Ile of Stromnys, see, Stroma, Isle of
Ile, Whithorn, see Whithorn, Isle of
Ilingtoun, mains of [*Helenton, NS3930], 564; see also Ellingtoun
Ilis, see Isles
Illa, see Islay
Illeray (Eillera), NF7863; 674
Illis, see Isles
Inch (Insch, Insche), parish, NX0962; 588, 588n., 590, 590n., 592, 607n.; kirk of, 588, 590, 592, 607; mill of, 588, 590; parsonage of, 607; vicarage of, 588, 607; vicarage pensionary of, 606
Inch (Insche), Whithorn priory, NX4447; 594
Inchaffray abbey (Inchaiffray, Inchearffray), NN9522; 347-348; commendator of, 348, 348n.
Inchaiden (Inchedene, Inche Kadene, Incheskadin, Incheskadyne) [Kenmore, NN7745], 301, 301n., 321, 321n.; benefice of, 321; parsonage of, 301, 301n.; vicarage of, 301, 301n.
Inchbrayock (Inchebryok), NO7156; 64; parsonage of, 64
Inchbrok [Mains of Inchbreck, NO7483], 405
Inchcailloch (Inchelyocht, Inschecalocht), NS4090; 540; parsonage of, 540; vicarage of, 540
Inchcolm (Sanct Colmis Inche, Sanctcolmisinche), NT1882; 62, 62n., 64n., 84, 85n., 88n., 300, 300n., 339n.
— abbey, 62-64, 88n., 339n.; abbot of, 62n., 63, 63n., 300, 300n.
Inchconans (Incheconane), NO2323; 285, 288
Inchdairnie (Inchdarnye, Inchedarnie, Inchedury, Inche Durne) [Inchdairniemuir Plantation, NT2497], 24, 29, 46, 51
Inche Durne, see Inchdairnie

Inche Kadene, see Inchaiden
Inche, Aberdeenshire, see Insch
Inchearffray, see Inchaffray
Inchebardy, Inchcolm abbey, 62
Inchebryok, see Inchbrayock
Incheconane, see Inchconans
Inchedarnie, see Inchdairnie
Inchedene, see Inchaiden
Inchedury, see Inchdairnie
Inchekery, Inchcolm abbey, 62
Inchelyocht, see Inchcailloch
Inchemagronocht, see Inchmagranachan
Inchemertine, Inchemertyne, see Inchmartine
Inchemichael, Inchemichaell, see Inchmichael
Inchemurtho Park, Inschemurtho park, Inchmurdo, St Andrews priory, 13, 13n., 20, 21
Inchery, see Inchrye
Inches, NO7682; 405, 492
Incheskadin, Incheskadyne, see Inchaiden
Inchestuir, see Inchture
Inchetheray, see Inchyra
Inchgotrick Easter (Inschgottrig Eister) [Inchgotrick, NS4133], 531
Inchimartene, see Inchmartine
Inchinnan (Inschinane), NS4769; 534; vicar of, 534; vicarage of, 534
Inchisture, Inchstoure, Inchstuir, see Inchture
Inchkeil (Ynchekyll), NJ1465; 482
Inchkeith, 95n.
Inchmagranachan (Inchemagronocht), NO0044; 310, 311; prebend of Dunkeld cathedral, 310, 311
Inchmahome (Inchemahome, Inchmahomo, Inschmahomo), NN5700; 190, 191, 290n., 343, 544, 548, 548n.
— priory, 190, 343, 544, 548-549; bailie of, 548; commendator of, 290n.; prior of, 548, 548n.
Inchmartine (Inchemertine, Inchemertyne, Inchimartene, Inchimertine), NO3835; 284, 285, 287, 288
Inchmichael (Inchemichael, Inchemichaell), NO2525; 237, 238, 240, 285, 286, 288
Inchmurdo, see Inchemurtho
Inchneuk (Inchnok) [Fm, NS7169], 102

Inchnok, *see* Inchneuk
Inchrory (Inscherory, Inschrory), NH5045; 644; chaplain of, 644; chaplainry of, 644
Inchrye (Inchery, Inchyray) *Lindores abbey*, [Abbey, NO2716], 32, 37
Inchture (Inchestuir, Inchisture, Inchstoure, Inchstuir, Insture), NO2828; 10, 16, 18, 317, 339; kirk of, 16; parson of, 317; vicar of, 339; vicarage of, 317, 339
Inchyra (Inchetheray, Inschyray), *with Inchture, St Andrews priory* [Ho NO1820], 10, 18, 18n.
Inchyray, *Lindores abbey*, *see* Inchrye
Inderraritie, *see* Inneraritie
Inertig, *see* Kirkcudbright-Innertig
Ingland, *see* England
Inglisberry Grange, *see* Englesberry Grange
Ingliston (Ynglistoun), *Eassie*, NO3345; 412
Inglistoun, *with Kilsture, Whithorn priory*, 594
Inner Bridge (Innerbrig), NO4519; 10, 19
Innerallane, *see* Inverallan
Inneraritie, Ennervarit, Inderraritie, *barony of Angus, Lindores abbey*, 30, 30n., 35
Inneraritie, *parish, see* Inverarity
Innerbundie, Innerbundy, *see* Inverboyndie
Innercarroun, *Fearn abbey*, 633
Innercochill (Innercochtkill), NN9138; 309
Innerdonet, Innerdovet, *see* Inverdovat
Inneresk, Innerresk, *see* Inveresk
Innergowrie, *see* Invergowrie
Innerkeddy, *see* Inverkeithing
Innerkeithing, *see* Inverkeithing
Innerkelour, *see* Inverkeilor
Innerkethine, *Fife, see* Inverkeithing
Innerkethnie, *Moray, see* Inverkeithny
Innerkip, *see* Inverkip
Innerleith, NO2811; 93, 127
Innerleithen (Innerlethaine, Innerlethame, Innerlethane, Innerletheim), NT3336; 229, 232, 234, 237, 240, 241, 243, 247; kirk of, 229, 232, 234, 237, 240, 241, 242; vicarage of, 248
Innerlochtie, *Dallas*, 492

Innermayth, Innermeth, *Auchterless*, 430
Innermessan (Innermessane), NX0863; 588, 588n., 590n., 607
Innernes, *see* Inverness
Innernisk, *see* Inveresk
Innernochtie, *see* Invernochty
Innerochty, *see* Invernochty
Innerork, *see* Denork
Innerowrie, Innerowry, *see* Inverurie
Innerrartie, *near Balcherry, Ross subdeanery*, 651
Innerritie, *see* Invereighty
Innerrychtny, *with Alvah, Coupar Angus abbey*, 355, 368, 369
Innertig, Innertige, *see* Kirkcudbright-Innertig
Innervak, Innerwak, *see* Invervack
Innerwick (Innerweck), NT7274; 529; kirk of, 529; vicarage of, 174
Insch (Inche, Insche), *Aberdeenshire*, NJ6328; 430, 431; glebe of, 431; kirk of, 32, 36; vicar of, 430-431; vicarage of, 430-431
Insch Clachan, *Tongland, Galloway bishopric* [High Clachan, NX6955], 590
Insch, Insche, *Wigtown*, *see* Inch
Inschecalocht, *see* Inchcailloch
Inschemurtho Park, *see* Inchemurtho Park
Inscherory, Inschrory, *see* Inchrory
Inschevin Wester, *see* Wester Incheschewin
Inschgottrig Eister, *see* Inchgotrick Easter
Inschinane, *see* Inchinnan
Inschmahomo, *see* Inchmahome
Inschurt, *Monkland*, 498
Inschyray, *with Inchture, see* Inchyra
Insture, *see* Inchture
Interaquas, *Lesmahagow*, 233
Interchallan, Interchallane, Interchallen, Interquhallan, *see* Inverchaolain
Inver (Inveir, Inweir), NO0142; 309, 336, 336n.; chaplain of, 336; chaplainry of, 336, 336n.; mill of, 336
Inverallan (Innerallane) [Ho, NJ0226], 466; kirk of, 466
Inverarity (Inneraritie), NO4343; 381, 382; parsonage of, 381-382
Inverboyndie (Innerbundie,

INDEX OF PLACES 823

Innerbundy), NJ6664; 359, 455
Inverboyndie (and Banff), vicarage of, 455; *see also* Banff
Inverchaolain (Interchallan, Interchallane, Interchallen, Interquhallan), NS0975; 558, 562, 564, 565; kirk of, 562, 564, 565; minister of, 558
Inverdovat (Innerdonet, Innerdovet), NO4327; 14, 20
Invereighty (Innerritie) [Ho, NO4345], 394
Inveresk (Inneresk, Innerresk, Innernisk), NT3572; 26, 38, 41, 43, 47, 48, 53; kirk of, 48; *see also* Pinkie
Invergowrie (Innergowrie), NO3430; 332, 334; kirk of, 332, 334
Inverkeilor (Innerkelour), NO6649; 83, 84n., 361, 412; kirk of, 361; vicar of, 83; vicarage of, 83
Inverkeithing (Innerkeddy, Innerkeithing), NT1381; 26, 38, 48, 68, 73, 76, 291; kirk of, 38; St John's altar in 73; vicar of, 76; vicarage of, 68, 76
Inverkeithny (Innerkethnie), NJ6247; 476; parson of, 476; parsonage of, 476
Inverkip (Innerkip), NS2072; 529, 533; kirk of, 529; vicar of, 533; vicarage of, 533
Inverness (Inneres), NH6645; 462, 463, 477, 483; chaplain of Our Lady in the Green, 477; chaplainry of Our Lady in the Green, 477; kirk of, 359; provost and bailies of, 463; vicarage of, 483
——, *sheriffdom*, 248n., 281n., 282, 456n., 636, 648
Invernochty (Innerochty) [Strathdon, NJ3512], 442; parson of, 442; parsonage of, 442; vicarage of, 442; *see also* Strathdon
Inverurie (Ennerawrie, Innerowrie, Innerowry), NJ7721; 32, 36, 427, 437; kirk of, 36; vicar of, 426, 437; vicarage of, 426, 437
Invervack (Innervak, Innerwak), NN8365; 354, 354n., 369
Inweir, *see* Inver
Inzievar (Ingzever, Inzefair), NT0288; 335

Iona (Ecolmkill, Ecomkill, Tempill-Ronaige), NM2724; 673, 674, 675; parsonage of, 675
—— abbey, 673-675
Irvine (Irvyn, Irvyne), NS3239; 576, 578, 579, 581, 582; burgess of, 581; kirk of, 576; parish of, 579; vicar of, 582; vicarage of, 582
Irvyn, Irvyne, *see* Irvine
Isaacstown (Eisauptoun, Eisauxtoun, Esauxtoun, Isaxtoun), *Brechin bishopric*, *NJ8021; 383, 384, 386, 387, 388, 389
Isaxtoun, *see* Isaacstown
Islay (Illa), NR3460; 674, 675
Isle of Man, *see*, Man, Isle of
Isle of Whithorn, *see* Whithorn, Isle of
Isles (Ilis, Illis, Ylis); 281n., 673, 675; bishopric of, 673-675
Iwey, *see* Evie
Iyle, *see* St Mary's Isle

Jargrais, Jargrayis, Gargrais, *with Gogar, Logie parish*, 290, 290n., 294
Jedburcht Syd, Jedburchtsyd [Jedburgh, NT6520], 216, 219
Jedburgh (Jedburcht), NT6520; 216, 217, 218, 219, 221, 259; kirk of, 216, 217, 218, 219; vicarage of, 217, 218
—— abbey, 216-217, 217n., 218-222, 218n., 221n., 222n., 282; bailie of, 216; canons of, 216; chamberlain of, 217; commendator of, 218, 218n., 356n.
Jefflie, *see* Yifflie
Jeindilkirk, *see* Channelkirk
Jervese-lands, *see* Bondhalff
Johnne Thornetounland, *see* Muncurris
Johnnestoun, *Balmerino abbey*, *see* Johnston
Johnnestounes Mylne, *Newburn*, 28, 50
Johnston (Johnnestoun), *Balmerino abbey*, NO2917; 58
Johnstone (Johnnestoun), *Eskdalemuir, Melrose abbey*, NT2400; 259
Jokkie Tailyouris satt, *Dallas*, 492
Jokstoun, *see* Cookstoun
Jonettis Scheil, *see* Newlandis *of Millegan*
Jonstoun, *see* Johnston

Jordanstone (Jurdistoun), NO2747; 306
Jurdistoun, see Jordanstone
Justice Port, see Aberdeen, Gallow Gait

Kailzie (Hopcailzie, Hopcaltzie, Hopcalzie, Hopcalzo, Hopcalzocht), NT2738; 150, 224, 232, 234, 237, 240, 241n., 243; kirk of, 224, 232, 234, 237, 240, 241, 241n., 243
Kaim of Duffus (Kayme, Keme), NJ1567; 468, 482
Kaithnes, Kaitnes, see Caithness
Kames (Kamyis), NT7845; 184
Kamyis, see Kames
Kansarne, see Kensary
Karinvath, see Carnwath
Karnbarrow, see Cairnborrow
Karrik, see Carrick
Katerling, see Catterline
Kathirlein, see Catterline
Kathlok, see Kedlock
Katkune, see Catcune
Kattely, Kettety, Kittuthy, *Leuchars*, 13, 20
Kayme, see Kaim of Duffus
Keallinne, *Mull*, 673
Kedlock (Kathlok), NO3819; 10, 18
Kedslie (Cadeslie, Cadislie, Cadlische, Caldslie), NT5540; 191, 197, 189, 192
Kega, see Keig
Keggieshoill, Keiggis Holl, *Abdie*, 30, 35
Keig (Kega), NJ6119; 2
Keiggis Holl, see Keggieshoill
Keilbakstar, *Trotternish*, 674
Keilchallumkill, see Kilcolmkill
Keilcheirran, see Kilkerran
Keilfeinchen, see Kilfinichen
Keillean, see Killean
Keilnoening, see Kilninian
Keilpedder, see Kilpheder
Keilvennie, *Islay*, 675
Keip, *Kirkcudbright*, 276n.
Keipny, see Kipney
Keir (Ker) [Ho, NS7798], 19, 343, 349, 507, 545, 546
Keir, *Holywood abbey* [Mill, NX8593], 270, 271, 276, 277, 623; Keirsyd, 275
Keir Mill (Mylne of the Keir), NX8593; 270

Keir, *Culross abbey*, 289
Keirsbeath (Baith Keir, Bayth Creir, Bayth Ker) [Fm, NT1389], 27, 49
Keirsyd, see Keir
Keith (Keyth), *Lothian* [Keith-Marischal, NT4464], 95, 95n.; parsonage of, 95; vicarage of, 95
Keith (Keyth), *Moray*, NJ4350; 465, 466, 467; kirk of, 466
Keith-Humbie, *Kelso abbey* [Humbie, NT4764], 224, 228n., 229, 230, 232, 233, 234, 237, 239, 240, 241, 243; vicarage of, 161
Keith-Marischal, see, Keith, *Lothian*
Keithick (Kethik), *Coupar Angus abbey*, NO2038; 303, 353, 356
Keithock (Kethik), *Brechin*, NO6063; 377, 384, 385, 389; waulk mill of, 377
Kelimore (Kilmoir), *parish, prebend of Brechin cathedral*, 378; parsonage of, 378
Kelleuch, see Coilheuche
Kelley, *with Balhalgardy, Lindores abbey*, 32, 36
Kelliesmonth, see Killiesmont
Kelline, Kellyne, see Killin
Kelloe (Kello), NT8453; 166, 200
— Castle, 200
Kellok, *with Windyedge, Dunfermline abbey*, 28, 50
Kells (Kellis), NX9457; 603; parson of, 603; parsonage of, 603; vicarage of, 603
Kellugrie, *Inverkeithny*, 476
Kelly Wod, see Kelty Wod
Kelour, *St Andrews Black friars* [*Keillor, NO2640], 89
Kelso (Calco), NT7234; 108, 108n., 222, 223, 223n., 224, 224n., 228, 229, 230, 230n., 232, 232n., 233, 234, 235, 235n., 236, 237, 238, 239, 240, 241, 242, 243, 243n.; kirk of, 224, 232, 234, 237, 240; vicarage of, 224; Toun croft, 222, 233, 237, 239
— abbey, 108n., 208n., 222-236, 224n., 236-243, 243n., 274n., 282; abbot of, 224, 224n., 225, 227; commendator of, 208n., 228, 243, 243n.
Kelton (Keltoun), NX7258; 91; kirk of, 91; parsonage of, 602
Kelty Wod (Kelly Wod, Kelte Wod)

INDEX OF PLACES 825

[Kelty, NO1494], 27, 49
Kemany, *see* Kemnay
Keme, *see* Kaim of Duffus
Kemnay (Kemany), NJ7316; 445; parish of, 445
Kemphill, NO2039; 352, 356
Kempland, *Haddington priory*, 180
Kencailyie, *see* Kennacoil
Kendalie, *Rothes*, 491
Kenmore, *see* Inchaiden
Kenmuir (Kenmure), NS6562; 498
Kennacoil (Kencailyie), NN9941; 309
Kenneddirschyre, *see* Kinneddar
Kennethmont (Kynnathmonth, Kynnathmound), NJ5328; 32, 36; kirk of, 32, 36
Kennetis, *Ross*, *see* Kinnettas
Kennettes, *Forfar*, *see* Kinnettles
Kennoway (Kennochie, Kennochquhy, Kennowy), NO3502; 15, 17, 83; kirk of, 15, 17; vicar of, 83; vicarage of, 83
Kensary (Kansame), ND2248; 632
Kepny, *see* Kipney
Keppelaw, *see* Kippilaw
Ker, *see* Keir
Kermuir, see Kirriemuir
Kernis, *see* Cairns
Kers of Gowrie, *see* Carse of Gowrie
Kers, *with Ogilface and Falkirktoun, Holyroodhouse abbey*, 91, 93, 93n.
Kers, *Whithorn priory*, 594
Kers, *Musselburgh*, *see* Cars
Kers, *see* Carse of Gowrie
Kerse (Kers), NS8141; 233
Kerse, *Perthshire*, 286n.
Kersie (Kerssie, Kersye) [Mains, NS8791], 538, 545
Kerskeith, *Tain and Edderton*, 652
Kersmyre, *Dryburgh abbey* [*Kersmains, NT7031], 190
Kethik, *Brechin*, *see* Keithock
Kethik, *Coupar Angus abbey*, *see* Keithick
Kethokis, *see* Catholris
Ketilstoun, *see* Kettlestoun
Kettelshiel (Kettill Cheyll), NT7051; 184
Kettemure, Kettymure, *see* Kittymuir
Kettestoun, *see* Kettlestoun
Kettety, *see* Kattely
Kettins (Keythins, Kithins, Kytins), NO2338; 249, 249n.; kirk of, 249,
249n.; parsonage of, 249; vicarage of, 249
Kettlestoun (Ketilstoun, Kettestoun, Kettilstoun, Kittilstoun) [Mains, NS9876], 538, 543, 545, 547, 548
Kevans (Crugleton Cawnis), NX4642; 593, 593n.
Kewbog, *see* Cowbog
Keyth, *see* Keith
Keythins, *see* Kettins
Kidcoldrum, *see* Kingoldrum
Kiddisdaill, *see* Kidsdale
Kidsdale (Kiddisdaill), NX4336; 593
Kilarstair [*Killaser Castle, NX0945], 594
Kilbane, Kilbein, *see* Kirkbean
Kilbarchan, NS4063; 529, 530, 534, 534n.; kirk of, 529, 530; vicar of, 534; vicarage, 534, 534n.
Kilbirnie (Kilbirny) [Ho, NS3054], 576; kirk of, 576
Kilbrackmont (Kilbrachmont, Kilbrauchmont), NO4706; 146, 167
Kilbrandane, *Mull*, 673
Kilbride (Kilbryd), *Argyll*, NR9638; 522
Kilbride (Kilbryd, Kylbryd) [West Kilbride, NS2048], 574, 576, 576n., 578, 579, 580; kirk of, 576, 576n., 578; parish of, 579; vicarage of, 580
Kilbride (Kilbryd) [East Kilbride, NS6454], parish of, 515; parsonage of, 499; vicarage of, 499
Kilbryde (Kilbryd), *Dunblane* [Castle, *NN7503], 343; parsonage of, 343; vicarage of, 343
Kilbucho (Kilbocho), NT0633; 247, 250; glebe of, 250; parsonage of, 247; vicarage of, 247; vicar pensionary of, 250; vicarage pensionary of, 250
Kilbuiack (Kilboyok), NJ0960; 462, 470
Kilburns (Kilburnis), NO3725; 58
Kilcalmonell (Calmanle), NR7557; 529; kirk of, 529
Kilchenzie (Skeirkenze), NR7145; parsonage of, 675
Kilchrist (Keilchrist) [Strath, NG6121], 675; parsonage of, 675
Kilcolmkill (Keilchallumkill), *Quinish, Mull*, parsonage of, 674
Kilconquhar (Kilconquhair, Kynnunquhar), NO4902; 77, 146, 167; kirk of, 146, 148, 166, 167;

mains of, 146; vicar of, 77; vicarage of, 77
Kilcormok, see Kirkcormack
Kildonan (Kildonane), ND9120; 631; parson of, 631, 631n.; parsonage of, 631; prebend of Dornoch cathedral, 631n.; vicarage of, 631
Kildrummy (Kildrummye, Kildrymmee, Kyldrymme), NJ4717; 425, 438, 449n.; kirk of, 425, 438; parish of, 449n.
Kildrymmee, see Kildrummy
Kilduncan (Kilduncane), NO5712; 177, 179n., 181
Kilelane, see Killallan
Kilfinichen (Keilfeinchen), NM4731; 674; parsonage of, 674
Kilgarth, NS7167; 102; see also Killyart
Kilgour, NO2208; 9, 15, 17, 17n., 21; kirk of, 9, 15, 17, 17n., 21
Kililane, see Killallan
Kilkerran (Keilcheirran, Kilkerrane), *Kintyre*, NR7219; 529; kirk of, 529; parsonage of, 674, 675
Killallan (Kilelane, Kililane, Killelan, Killelane), *Paisley abbey* [Houston, NS4066], 528, 529, 533; kirk of, 528, 529; vicar of, 533; vicarage of, 533
Killarne, see Killernie
Killean (Keillean), NM7129; parsonage of, 674
Killearn (Killern, Killerne), NS5285; 554; parson of, 554; parsonage of, 554; vicarage of, 554
Killearnan (Kilernane), NH5850; parish of, 626
Killernie (Killarne), NT0392; 24, 47
Killiemore (Kilyenar), NX3660; 588
Killiesmont (Kelliesmonth, Killeismonth), NJ4152; 461; haughs of, 464; over haughs of, 462, 464
Killin (Kelline, Kellyne), NN5732; 321; vicarage portionary of, 321
Killmallie, see Kilmalie
Killmane, Killmone, see Kilmun
Killnes (Kilnes), NX9479; 271, 275, 276, 623
Killochrig, *Dalry*, 605
Killore, *Inchcolm abbey* [*Kilrie, NT2489], 62
Killquhillie Drum, *Urray*, 635
Killwyning, see Kilwinning

Killyart, *Monkland* [*Kilgarth, NS7167], 498; see also Kilgarth
Killylung (Cullelung), NX9581; 271
Kilmacocharmik (Cill mo Charmaig, Kilmaharmull, McHarmik) [North Knapdale, NR7880], 573, 576
Kilmacolm (Kilmacalme, Kilmacholm, Kilmacolme), NS3669; 522, 529, 532; kirk of, 529; vicarage of, 532
Kilmadock (Kilmadok, Kilmodok), NN6705; 341, 544, 548, 549; kirk of, 544, 548, 549; vicarage of, 341
Kilmaharmull, see Kilmacocharmik
Kilmahog (Kilmahug), NN6108; 295, 348; kirk of, 295, 348
Kilmalemnock, St Andrew's kirk of, 466, 466n.; parish of, 468
Kilmalie (Killmallie, Kilmalie, Kilmaly) [Golspie, NC8300], 628, 629, 632, 632n.; parish of, 629
Kilmanitie, Over and Nether Sett of, *Kinloss abbey* [*Kinminintie, NJ4253], 461
Kilmardinny (Kilmerdeny), 541
Kilmarnock (Kilmarnak, Kilmarnok, Kilmernok), NS4237; 574, 576, 577, 578; kirk of, 574, 577; parish of, 576, 578; vicarage of, 578
Kilmaron (Kylmarone), NO3516; 14
Kilmaronock (Kilmarannok, Kilmaronoch, Kilmaronok) [Ho, NS4587], 538, 543, 545, 546; kirk of, 538, 545, 546
Kilmaurs (Kilmaris, Kilmaweris, Kilmawres, Kilmawris), NS4041; 224, 232, 234, 237, 240, 244, 580; vicar of, 580; vicarage of, 580
Kilmavernok, see Kilmoveonaig
Kilmawris, Myln of, *with Curling*, 129n.
Kilmerdeny, see Kilmardinny
Kilminning (Kylmynnane) [Castle, NO6208], 177, 178, 181n.
Kilmoir, see Kelimore
Kilmorack (Kilmorak), NH4944; 636, 641; vicar of, 636, 641; vicarage of, 636, 641
Kilmore, *Mull*, see Kilcolumkill *and* Kilmorie
Kilmorich (Kilmoreicht, Kilmorycht), NO0050; 303, 305, 342, 345
Kilmorie, *Mull* [*Kilmore, NM4249], 673
Kilmoveonaig (Kilmavernok), *Dunkeld*

INDEX OF PLACES 827

diocese, 319; curate of, 319;
  parsonage of, 319; vicarage of, 319
Kilmowir, Kilmowr, Kilmowre, *see*
  Kilmuir
Kilmuir-Easter (Kilmowir, Kilmowr,
  Kilmowre), NH7074; 626, 643;
  parson of, 643; parsonage of, 643
Kilmuir-Wester [Knockbain parish,
  NH6352], 639; vicar of, 639;
  vicarage of, 639
Kilmukles, *see* Kilmux
Kilmun (Killmane, Killmone), NS1781;
  529; kirk of, 529
Kilmuraig, *see* Kirkmering
Kilmux (Kilmukles, Kylmukles) [Fm,
  NO3605], 10, 18
Kilmyles, *Moray bishopric*, 465
Kilmyntutie, Hauches of, *Kinloss abbey*
  [*Kinminintie, NJ4253], 461
Kilnes, *see* Killnes
Kilninian (Kilneoning, Keilnoening),
  NM4046; 673, 674; parsonage of,
  674
Kilpatrick (Kilpatrik) [Old Kilpatrick,
  NS4672], 527, 528, 529, 530, 541;
  kirk of, 528, 530; parish of, 529;
  vicar of, 541; vicarage of, 541
Kilpheder (Keilpedder), NF7419; 674;
  parsonage of, 674
Kilphubbill, *Mull*, 673
Kilquhyis Wester, *see* Wester Kilwhiss
Kilrenny (Kilrennie, Kilrenry,
  Kilrynnie, Kylrynnie), NO5704; 75,
  190, 197, 197n.; kirk of, 190, 197,
  197n.; vicarage of, 75
Kilsalmound, *see* Kirkton of
  Culsalmond
Kilspindie (Kilspindy, Kinspindie,
  Kinspyndie), NO2125; 106, 110,
  307, 332, 333, 334, 334n., 346, 350;
  kirk of, 333, 334; vicar of, 110, 307;
  vicarage of, 307; vicarage portionary
  of, 110
Kilsture (Kilstuire), NX4348; 594
Kilsyth, NS7178; 552n.; *see also*
  Monyabroch
Kiltarlity (Kintallarlie, Kyntallartie)
  [Church, NH4943], 248, 248n., 456,
  456n., 460, 465; benefice of, 248,
  248n.; kirk of, 456, 456n.
Kiltearn (Kilterne, Kittarne), NH4867;
  635, 646; parson of, 635; parsonage
  of, 635; vicarage of, 646

Kilvickeon (Kilviceowin), NM4830;
  parsonage of, 674
Kilwinning (Kilwinging, Kilwining,
  Kilwyning, Kilwynning, Kylwining,
  Kylwinning, Kylwynning), NS3043;
  435, 435n., 510, 512, 513, 573, 574,
  575, 576, 578, 579; chaplain of, 510;
  kirk of, 576; parish of, 576, 578;
  vicarage of, 578
—— abbey, 282, 512, 513, 573-579;
  abbot of, 302, 302n.; chamberlain of,
  574, 575, 576, 577; commendator of,
  4, 4n., 575, 575n., 578, 579; monks
  of, 577
Kilyenar, *see* Killiemore
Kimmerghame (Kymmerghame)
  [Heugh, NT8252], 200
Kimmerghame Mains (Kymmerghame
  Maynes, Kymmermanis, Manes of
  Kimmgery), NT8050; 192, 200, 204,
  204n.
Kimmgery, mains of, *see* Kimmerghame
  Mains
Kinairney (Kynarny, Kynarnye),
  NJ1740; 426, 434; kirk of, 359; vicar
  of, 426; vicarage of, 426, 434
Kinaldy (Kynawdy, Kynnaldie),
  NO5110; 13, 55
Kinalty (Kynnalthie), NO3551; 356
Kinaroche, *see* Kirminnoch
Kincadin of Munteithit, *see* Kincardine
Kincairne, Kincarne, *see* Kincardine
Kincaldrowne, *see* Kingoldrum
Kincardine (Kincadin of Munteithit),
  NS7097; 538, 543, 545, 546; kirk of,
  538, 543, 546; mill of, 555; vicarage
  of, 343
Kincardine (Kincairdein, Kincairdin,
  Kincairne, Kincardin, Kincarne),
  Culross abbey, NN9387; 291, 291n.,
  292, 294, 294n.
Kincardine (Kincairdein, Kincardin,
  Kincairne), *Ross diocese*, NH6089;
  625, 626, 645; parson of, 645;
  parsonage of, 645
Kincardine (Kincardin, Kinkardin),
  sheriffdom, 1n., 281, 282, 283n.,
  371n., 403, 404, 408
Kincardine O'Neil (Kincardin Oneill,
  Kincardine Onyll. Kincardin,
  Kyncardene), NJ5899; 429, 430,
  430n., 439; parsonage of, 439;
  reader at, 430n.; vicar of, 430;

vicarage of, 430
Kincardine-on-Spey (Kincarn) [*Street of Kincardine, NH9417], 465, 465n.; vicar of, 465, 465n.; vicarage of, 465, 465n.
Kincarn, Kincarne, see Kincardine
Kincardine (Kincarne), *Crieff* [Fm, NN8721], 342, 345
Kincarne, Kincarneis, Kincarny, *Dunkeld bishopric* [*Kincairney, NO0844], 304, 308, 316
Kincavil, Kincavyll, see Kingscavil
Kinclaith, *Glasgow*, 516
Kinclaven (Kinclavin, Kynclavin), NO1537; 306, 307; parsonage of, 306; vicar pensionary of, 307; vicarage of, 306
Kincoldrum, see Kingoldrum
Kincragie, *Aberdeen Trinitarians*, 421
Kincraig, *Brechin parish*, NO6258; 389
Kincraig, *Kilconquhar*, NO4600; 146, 167
Kincraigie (Kincragie), *Dunkeld bishopric*, NN9849; 303, 304, 342, 345; mill of, 304
Kincraigie (Kyncragy), *Strathmiglo*, NO1910; 69
Kincreich (Kincreycht, Kyncreicht), NO4344; 368, 369; grange of, 354, 357
Kinderloch, see Lochindilloch
Kindrimy, see Chapelden of Kindrimy
Kindrochat, Kindroquhat, see Kintrockat
Kinfauns (Kinfawnis, Kynfawnis), NO1622; 327, 332, 333; kirk of, 332, 333; minister of, 333; reader at, 333; vicarage of, 332
King Edward (Kynedwart), *Deer abbey*, NJ7256; 458
Kingarro, *St Mary on the Rock collegiate kirk, St Andrews*, 55
Kingask (Kyngask), NO3816; 14, 65n.; prebend (with Kinglassie) of St Mary on the Rock collegiate kirk, St Andrews, 77
Kinghorn (Kingorne, Kynghorne, Kyngorne), NT2686; 25, 26, 28, 29, 39, 43, 48, 50, 51, 52, 53, 80; kirk of, 48, 52
Kinghorn Easter (Kingorne Eister, Kingorne Ester, Kyngorne Eister), 2, 80, 80n., 91; kirk of, 91; vicar of, 80; vicarage of, 80, 80n.
Kinghorn Wester (Kingorne Wester, Kyngorne Wester, Wester Kinghorne, Wester Kingorne), 24, 25, 26, 28, 38, 39, 40, 41, 42, 43, 44, 44n., 46, 47, 48, 50, 53, 54; mill of, 24, 46; kirk of, 26, 38; parish of, 26, 29, 38, 39, 48, 51
Kingildoris, Kingildouris, see Kingledores
Kingis Barnes, Kingis Bernis, see Kingsbarns
Kingis Carne, Kingiscarne, *Crail*, 177, 178, 181
Kingis Kettill, see Kingskettle
Kingis Medowis, see Kingsmeadow
Kingis Merksworthe, *Greenlaw*, 227, 258
Kingis Orchart, *Maxwell*, 225
Kingis Syd, *Lyne*, 252
Kingis Urray [Urray Beag, NH5052], 635
Kingis Wallis [*Kingswells, NJ8606], 423
Kingis Wark see Leith
Kingistoun, see Kingston
Kinglassie (Kinglassy, Kynglassie, Kynglassye), NT2398; 24, 25, 26, 29, 37, 38, 39, 40, 41, 43, 44, 46, 47, 48, 51, 52, 53, 54, 65n., 67, 77; parish of, 26, 29, 39, 41, 44, 51, 52, 53; prebend (with Kingask) of St Mary on the Rock collegiate kirk, St Andrews, 77; Brewland of, 24, 29, 46, 51; vicar of, 67; vicarage of, 67; waulk mill of, 24, 29, 46, 51; *see also* Kinglassieshire
Kinglassieshire (Kinglasschyre, Kynglassieschyre, Kynglassyschyre), 24, 46, 54; *see also* Kinglassie
Kingledores (Kingildoris, Kingildouris), NT1028; 207, 258
Kingoldrum (Kidcoldrum, Kincaldrowne, Kincoldrum, Kyncoldrum) [Kirkton of, NO3354], 351, 358, 379; vicarage of, 379
Kingorne, see Kinghorn
Kingorny, Kyngorny, *Catterline, Brechin bishopric*, 383, 384, 386, 389
Kingsbarns (Kingis Barnes), NO5912; 89, 177, 178, 181
Kingscavil (Kincavil, Kincavyll), NT0375; 184, 184n.

INDEX OF PLACES

Kingseat (Kingis Seitt, Kingis Sett) [Fm, NJ9019], 422, 423
Kingside (Kingsyd), NT2455; 258
Kingskettle (Kingis Kettill), NO3008; 15, 21
Kingsmeadow (Kingis Medowis), NT2639; 249
Kingston (Kingistoun, Kingstoun), NT5482; 147, 168n., 190
Kingussie (Kingusye, Kyngusy, Kynguyse), NH7500; 487, 491, 492; parson of, 487; parsonage of, 487; vicar of, 491, 492; vicarage of, 491-492
Kinhill, *Soulseat abbey*, 598
Kininmonthe, *see* Kinninmonth
Kinkardin, *see* Kincardine
Kinkas, *see* Knockhouse
Kinkell (Kynkell), *Aberdeenshire* [Church, NJ7819], 420, 420n., 445, 448; benefice of, 420nn.; parsonage of, 420, 445, 448
Kinkell, *Fife*, NO5414; prebend of St Mary on the Rock collegiate kirk, St Andrews, 80; prebendary of, 80
Kinketh Mylnes, Kynkaith Mylnes, *with Beanston, North Berwick parish*, 147, 167; crofts of, 147
Kinloch (Kinloche, Kinlocht, Kinloich, Kynloche, Kynlocht), *Lindores abbey*, NO2812; 30, 33, 35, 37
Kinloch, *Stirling parish*, 555
Kinloch, prebend of Dunkeld, *see* Lundeiff
Kinloss (Kinros, Kinlos, Kynlos), NJ0661; 456, 461, 461n., 462, 463, 463n., 464, 464n., 465, 650, 650n., 651
—— abbey, 456, 461-464, 465; abbot of, 464, 464n.; bailie of, 463; monks of, 463
Kinmuck (Kymmocht, Kymmok), NJ8119; 32, 36
Kinmundy (Kynmownd, Kynmund, Kynmundy), NJ8917; 422, 423
Kinnaird (Kinnard), *Dunkeld bishopric* [Ho, NO9849], 305
Kinnaird (Kinnard, Kynnard, Kynnarde) [Castle, NO6357], 324, 331, 382, 387, 388, 394, 403, 424, 425, 442
Kinnaird (Kynnaird, Kynnard, Kynnardie), *Abdie*, 32, 34, 37
Kinnaird (Kynnard), *pendicle of Inchture*, NO2428; 317, 339; *see also* Cookston
Kinnaird, *Monkland*, 102
Kinneddar (Kynnedoir, Kynnedour, Kynnedoure, Kynnedward), *Moray bishopric*, NJ2269; 465, 466, 467, 468, 484, 490; vicar of, 484, 490; vicarage of, 484, 490
Kinneddar (Kynnedder, Kynneddir, Kynnenndir), *Dunfermline abbey* [Mains, NT0291], 24, 38, 39, 46, 53; Brewland of, 24, 46; mill of, 24, 46; *see also* Kinneddarshire
Kinneddarshire (Kenneddirschyre, Kinneddderschyre, Kinneddirschyre, Kynnedderschyre), 24, 25, 47, 51; *see also* Kinneddar, *Dunfermline abbey*
Kinneff (Kynneff, Kynneiff), NO8574; 66, 67, 397; parish of, 67; kirk of, 66; vicarage of, 66 vicar pensionary of, 397
Kinneil (Kynneill, Kynnell), *West Lothian* [Church, NS9880], 91; kirk of, 91; vicar of, 154; vicarage of, 154
Kinnell (Kynnell), *Forfar*, NO6050; 366, 367, 377, 397; parish of, 367; parsonage of, 366; vicar of, 397; vicarage of, 397
Kinnellar (Kynnellar) [Ho, NJ8112], 445; parish of, 445
Kinneries (Kynneris), NO5245; 376
Kinnernie (Kynmarny), NJ7210; kirk of, 359
Kinnethmont, *see* Kennethmont
Kinnettes (Kennetis), *Ross*, 634; minister of, 635; parsonage of, 634; vicar of, 635; vicarage of, 634
Kinnettles (Kennettes, Kinnettelis, Kynnettes, Kynnettis), *Forfar* [Ho, NO4246], 390, 391, 394; parsonage of, 394; vicar of, 390; vicarage of, 390, 391
Kinninmonth (Kinnynmonth, Kynninmonthe, Kynnymonthe, Kynnynmont, Kynnynmonth, Kynnynmonthe, Kynnynmounth), NO4212; 11, 13, 19, 20, 21, 29, 51
Kinnoir (Kynnoir, Kynnour) [Corse of, NJ5543], 483; parsonage of, 483
Kinnoull (Kinnoule, Kynnoule) [Hill, NO1322], 323, 323n., 331;

parsonage of, 323, 331; vicar of,
  331; vicarage of, 323, 331
Kinpauch (Kinpache), NN8907; 349
Kinpunt, *Kirkliston*, 3
Kinross (Kinros, Kynros), NO1202; 26,
  38, 48, 88; kirk of, 26, 38, 48
—— and Orwell, vicar pensionary of, 88;
  vicarage pensionary of, 88; *see also*
  Orwell
Kinshaldy (Kynschawdy), NO4823; 14
Kinspindie, Kinspyndie, *see* Kilspindie
Kinsteary (Kinstarie, Kinstarry) [Lodge,
  NH9254], 469
Kintalarlie, *see* Kiltarlity
Kintore (Kintor), NJ7816; parish of, 445
Kintrockat (Kindrochat, Kindroquhat,
  Kyndrocat) [Ho, NO5659], 383, 387,
  388
Kintulloch (Kintuleoch, Kintuloch,
  Kintulocht), *Cambuskenneth abbey*,
  537, 545, 547, 548
Kintyre (Kintyr), *Argyll*, NR7023,
  NR7240; 529, 674, 675
Kintyre, *Whithorn priory*, 594
Kinvaid (Kynvaid), NO0630; 303, 323,
  342, 345
Kipney (Keipny, Kepny), NN9630; 360,
  361
Kippen (Cipane, Kippane, Kippin,
  Kippone), NS6594; 343, 538, 543,
  545, 546; kirk of, 538, 543, 545,
  546; vicarage of, 343
Kippen Ros, *Stirling parish*, 556
Kippilaw (Keppelaw, Kippelaw,
  Kippellaw, Kipperlawr, Kypellaw,
  Kyplaw, Tippilaw), NT5428; 223,
  225, 231, 231n., 233, 237, 240, 242
Kippo (Rippo), NO5710; 177, 178, 181
Kipps (Kyppis) [Fm, NS7366], 102
Kippsbyre (Kypbyre), NS7466; 102,
  498
Kir, 'Vissurdland callit the Kir', *see*
  Vissurdland
Kircaldie, *see* Kirkcaldy
Kircudbrycht, *see* Kirkcudbright
Kirk a Feild, Kirk Feild, Kirk of Feild,
  Kirkafeild, *see* Edinburgh, St Mary
  of the Fields' collegiate kirk
Kirk Hill, Kirkhill, *Linlithgow friary*,
  156
Kirk Marown (Kirkmarrow), *Isle of
  Man*, SC3278; parish kirk of, 594
Kirk O' Muir (Kirk of Mure),
  Stirlingshire, 553; chaplain of, 553;
  'vicarage of the chaplainry' of, 553;
  see also St Mary of Garvald,
  Kirkton, *St Ninians*
Kirk Urray [Urray Beag, NH5052], 635
Kirkaldie, *see* Kirkcaldy
Kirkandrews (Kirkcandreis), NX5948;
  595, 596; kirk of, 595, 596
Kirkapoll (Kirkapost), *Tiree*, NM0447;
  675; parsonage of, 675
Kirkapost, *Illeray* [*Kirkibost, NF7565],
  674
Kirkapost, *Tiree, see* Kirkapoll
Kirkbean (Kilbane, Kilbein, Kirkbeane,
  Kirkbene, Kirkbeyne, Kirkbyne),
  NX9859; 269, 277, 609, 610, 613,
  614; glebe of, 614; kirk of, 613;
  manse of, 614; parish of, 610; vicar
  of, 609-610, 614; vicarage of, 609-
  610, 614
Kirkbride (Kirkbryd), *Holywood abbey*,
  275
Kirkbride (Kirkbryd), *parish*, 278;
  parson of, 279; parsonage of, 278;
  vicarage of, 278
Kirkbuddo (Kirkbodo) [Ho, NO5043],
  390, 391; parson of, 390, 391;
  parsonage of, 391; vicar of, 390
Kirkburn (Kirkburne), *Cambuslang*,
  Lady Chapel of, 523; chaplain of
  Lady chapel of, 523
Kirkcaindelis, *see* Kirkandrews
Kirkcaldy (Kircaldie, Kirkaldie,
  Kirkcaldie), NT2791; 24, 25, 26, 28,
  29, 38, 39, 40, 41, 42, 43, 44, 46, 47,
  48, 50, 51, 52, 53, 54, 67, 81, 84;
  burgh of, 81, 84; chaplain of Holy
  Blood altar in, 81; chaplainry of
  Holy Blood altar in, 81, 84; Holy
  Rood altar in kirk of, 85; parish of,
  26, 29, 38, 39, 41, 48, 52; vicar of,
  67; vicarage of, 67
Kirkcandreis, *see* Kirkandrews
Kirkcarmok, *see* Kirkcormack
Kirkchrist (Kirkcryst), *Glasnick*,
  NX3659; 588
Kirkchrist (Kirkcryst),
  Kirkcudbrightshire, NX6751; 616,
  616n.; parson of, 616; parsonage of,
  616; vicarage of, 616
Kirkchrist (Kirkcryst), *with Nuntoun,
  Galloway bishopric*, NX6751; 587,
  589; fishing of, 587, 588, 590, 591,

591n., 592, 607
Kirkcolm (Kirkcum), NX0268; 611, 622; parsonage of, 611, 622; vicarage of, 611, 622
Kirkconnel (Kirkconell), *Holywood abbey*, NS7312; 271, 276, 277, 623; kirk of, 277
Kirkconnell, *Galloway bishopric*, NX6760; 585, 586, 587, 589, 607
Kirkcormack (Kilcormok, Kirkcarmok, Kirkcormo, Kirkcormok, Kirkcrumok), *parish*, 91, 604, 604n., 618; vicar of, 604, 618; vicarage of, 604, 618
Kirkcowan (Kirkcowane), NX3261; 602
Kirkcrumok, *see* Kirkcormack
Kirkcryst, *see* Kirkchrist
Kirkcudbright (Kircudbrycht, Kirkcudbrycht), NX6850; 91, 591, 602, 617, 617n.; vicar of, 617; vicarage of, 617
——, *stewartry*, 265, 265n., 276n., 279n., 610, 614, 616
Kirkcudbright-Innertig (Kircudbrycht, Kirkcudbrycht, Inertig, Innertige) [Ballantrae, NX0882], kirk of, 567, 567n.; parsonage of, 570; vicar of, 612; vicar pensionary of, 570; vicarage portionary of, 612-613; vicarage of, 570
Kirkcum, *see* Kirkcolm
Kirkdale (Kirkdaill), NX5153; 594; kirk of, 594
Kirkdauche, Kirkdawcht, *Brechin bishopric*, 384, 385, 389
Kirkdone, *Kirkpatrick-Juxta*, 615
Kirkdoun, *Crossraguel abbey*, 567
Kirkeich, *see* Kirkeoch
Kirkeoch (Kirkeich, Kirkoch, Kirkhauch, Kirkoch), NX6649; *Holywood abbey*, 269, 274; *Galloway bishopric*, 587, 589
Kirkettle (Carkettill), NT2661; 142; mill of, 142
Kirkfauld, *Eskdalemuir, Melrose abbey*, 259
Kirkforthar (Kirkforther, Kirkforthour), NO3004; 84, 86; curate of, 86; glebe of, 86; parsonage of, 86, vicar pensionary of, 84; vicarage of, 86; vicarage pensionary of 84; *see also* Forther Ramsay
Kirkgunzeon (Gargonzoun), NX8666; 604; benefice of, 604; vicar pensionary of, 604
Kirkhauch, *Dumfries-shire*, *see* Kirkeoch
Kirkhauch, *Ecclesgreig*, 16
Kirkheuch, *see* St Andrews, St Mary on the Rock collegiate kirk
Kirkhill, *Wranghame, Lindores abbey* [*Kirkhill of Kennethmont, NJ5328], 31, 36
Kirkhill, *Meigle*, 303, 342, 346
Kirkhill, *St Mary on the Rock collegiate kirk, St Andrews*, 55
Kirkhill, *Rothes*, NJ3051; 491
Kirkhill, *Scotlandwell priory*, 56
Kirkhillokkis, *Glenisla, Coupar Angus abbey*, 355
Kirkhope, NT3823; 258
Kirkinner (Kirkkineir, Kirkkynneir), NX4251; kirk of, 602
Kirkis croft, *see* Cuikis croft
Kirkkineir, Kirkkynneir, *see* Kirkinner
Kirkland [*East Kirkland, NX4356], 597
Kirklandbank (Kirkland Bank), *Alyth*, NO2349; 303, 342, 346
Kirklandhill, *Moorfoot*, 101
Kirklevingstoun [Livingston, NT0366, NT0668], 91; *see also* Livingston
Kirkliston (Kirklistoun), NT1274; 3; kirk of, 3; vicarage of, 3
Kirkmabreck (Kirkmabrek), NX4856; 606; minister of, 606; reader at, 606
Kirkmadrine (Kirkmadrin), *Whithorn priory*, NX0848; vicar of, 601; vicarage of, 601; *see also* Toskerton
Kirkmadrine (Kirkmadyne), *St Mary's Isle priory*, NX4748; kirk of, 617; *see also* Eggerness
Kirkmadyne, *see* Kirkmadrine, *St Mary's Isle priory*
Kirkmahoe (Kirkmaho), *Nithsdale*, 618; parsonage of, 618; vicar of, 618; vicarage of, 618
Kirkmaiden-in-Farines (Kirkmadin, Kirkmaidin), *Whithorn priory*, NX3639; 593, 596, 600; parish of, 593, 596; glebe of, 601; vicar of, 600-601; vicarage of, 600-601
Kirkmarkinche [Markinch, NO2901], 15
Kirkmarrow, *see* Kirk Marown
Kirkmering, Kilmuraig, *with Genoch, Soulseat abbey*, 598, 599

Kirkmichael (Kirkmichaell, Kirkmichell), *Whithorn priory, Carrick*, glebe of, 571; kirk of, 573, 594; manse of, 571; vicar pensionary of, 571

Kirkmichael (Kirkmichell), *Strathdon*, NJ1614; 485; curate of, 485; vicar of, 485; vicarage of, 485

Kirkmichael, *Dumfries*, parsonage of, 278, 278n.

Kirknes, *Portmoak priory*, 57

Kirknewton (Kirknewtoun), NT1066; 104, 105; parson of, 104-105; parsonage of, 104-105

Kirkoch, *see* Kirkeoch

Kirkoswald (Kirkoswell), NS2407; 567

Kirkpatrick, 586n.

Kirkpatrick Durham (Kirkpatrik Durand, Kirkpatrik Durane), NX7870; 611, 622; kirk of, 611; parsonage of, 622

Kirkpatrick-Juxta (Kirkpatrik Juxta), NT0000, NT0007; 615; parson of, 615; parsonage of, 615; vicarage of, 615

Kirkside (Kirksyd), NO7363; 12, 18

Kirkton (Kirktoun), *Beath*, 62, 64

Kirkton, *Cadder*, 497, 498

Kirkton, *Callander*, 349

Kirkton, *Dallas*, *NK1245; 492

Kirkton, *Duffus*, 482

Kirkton, *Eassie* [Eassie Church, NO3547], 412

Kirkton, *Errol, Perth Charterhouse*, 284, 285, 287, 289

Kirkton, *Ferne, Ruthven*, NO4380; 306

Kirkton, *Forgoun, see* Forgan

Kirkton, *Insch*, 31, 36

Kirkton, *Kinnell parish*, 366, 367

Kirkton, *Kinnettles*, NO4246; 394

Kirkton, *Lagganallachy*, 309

Kirkton, *Lundie*, NO2941; 376

Kirkton, *Moy*, *NK0152; 470

Kirkton, *Pitpointie or Dunkeld*, NO3437

Kirkton, *Premnay*, NJ6425; 31, 36

Kirkton, *St Andrews priory*, *NO4246, *NO4380; 10, 12

Kirkton, *St Ninians*, 553n.; *see also* Kirk O' Muir, St Mary of Garvald, St Ninians

Kirkton, *Strageath*, 349

Kirkton, *with Reard, Caithness*, 628

Kirkton of Clyne [Easter, Wester Clynekirkton, NH9006, NH8805], 630

Kirkton of Culsalmond (Colsamond, Culsalmonth, Culsalmound, Kilsalmound), NJ6432; 31, 36. *See also* Culsalmond

Kirkton of Glenisla (Kirktoun), NO2160; 354

Kirkton of Nevay (Kirktoun of Navay), NO3243; 367; *see also* Nevay

Kirkurd, NT1244; 250; parson of, 250; parsonage of, 250

Kirkwall, HY4510; 671

—— cathedral, 663, 664, 665, 666, 667, 668, 671; lector in, 671; prebend of St John in, 667; prebend of St Magnus in, 666; prebend of Woodwick in, 668; prebendary of St John in, 667; prebendary of St Magnus in, 666; prebendary of Woodwick in, 668n.; St Catherine's altar in, 663, 663n., 664, 665; *see also* Orkney cathedral

Kirkwood (Kirkwod), *NS7163; 102

Kirminnoch (Kinaroche, Kirmenoch, Kurmanocht), NX1258; 588, 590, 598

Kirriemuir (Kermuir, Kyrmuir), NO3853; 359, 396; kirk of, 359; vicar, 396; vicarage of, 396

Kirvennie (Crevane), NX4155; 597

Kissok, *Suddie and Kilmuir-Wester* [*Kessock Bridge, NH6647], 639

Kisterdaill, *Caithness bishopric*, 628

Kithins, *see* Kettins

Kittarne, *see* Kiltearn

Kittilstoun, *see* Kettleston

Kittuthy, *see* Kattely

Kittymuir (Kettemure, Kettymure), NS7548; prebend of Bothwell collegiate kirk, 121, 526

Kleland, *see* Cleland

Knapdale, *see* Kilmacocharmik

Knitis Pete, *Abernethy collegiate kirk* [*Wright's Pottie, NO1615], 322, 322n.

Knock (Knok), *Kirkmaiden*, NX3739; 593

Knock (Knok), *Kinloss abbey*, NJ5453; 461, 464

Knockando (Knokandoche), NJ1941; chaplainry of, Elgin cathedral, 478

Knockbain, *see* Kilmuir-Wester

INDEX OF PLACES 833

Knockdurn (Knonkburne), NJ5863; 437
Knockhill (Knokhill) [Ho, NO4225], 303
Knockhouse (Kinkas, Knokas), NT0786; 23, 27, 45, 49
Knok, *Dunfermline* [*Knock Hill, NT0593], 23n., 27, 45, 49
Knok, *see* Knock
Knokandoche, *see* Knockando
Knokas, Knokis, *see* Knockhouse
Knokdoron, *Dumbarton collegiate kirk*, 539
Knokhill, *see* Knockhill
Knoksudrattoun [Southerton, NT2691], 64
Knokydaff, *Fearn abbey*, 633
Knonkburne, *see* Knockdurn
Kow Inche, Kow Inchis, *Lindores abbey*, 30, 35
Krinslaw, Kyrnslaw, *Restalrig collegiate kirk*, 142, 143
Kudag, *see* Cudoig
Kurmanocht, *see* Kirminnoch
Kwyd, *see* Cawdor
Kyirpronte, *with Lundeis, Dunblane*, 343
Kylbryd, *see* Kilbride
Kyldrymme, *see* Kildrummy
Kyle, NS4821, NS5621, NS6519; 113, 531, 560, 564
Kylismure, Kylsmure, *Melrose abbey*, 210, 257,
Kylmarone, *see* Kilmaron
Kylmukles, *see* Kilmux
Kylmynnane, *see* Kilminning
Kylquhis Eister, *see* Easter Kilwhiss
Kylrynnie, *see* Kilrenny
Kylsmure, *see* Kylismure
Kyltyis, *see* Quilts
Kylwining, Kylwinning, Kylwynning, *see* Kilwinning
Kymmerghame Maynes, Kymmermanis, *see* Kimmerghame Mains
Kymmerghame, *see* Kimmerghame
Kymmermanis, *see* Kimmerghame Mains
Kymmocht, *see* Kinmuck
Kymmok, *see* Kinmuck
Kynarny, Kynarnye, *see* Kinairney
Kynawdy, *see* Kinaldy
Kyncardene, *kirk*, *see* Kincardine O'Neil
Kynclavin, *see* Kinclaven
Kynclydyn, *see* Lincluden

Kyncoldrum, *see* Kingoldrum
Kyncragy, *see* Kincraigie
Kyncreicht, *see* Kincreich
Kyndrocat, *see* Kintrockat
Kynebarch, *Cullicudden*, 646
Kynedwart, *Deer abbey*, *see* King Edward
Kynfawnis, *see* Kinfauns
Kynnedoir, Kynnedour, Kynnedoure, Kynnedward, *Elgin cathedral*, *see* Kinneddar
Kyngask, *see* Kingask
Kynglassie, Kynglassye, *see* Kinglassie
Kynglassieschyre, Kynglassyschyre, *see* Kinglassieshire
Kyngorne Eister, *see* Kinghorn
Kyngorne Wester, *see* Wester Kinghorn
Kyngorny, *see* Kingorny
Kyngusy, Kynguyse, *see* Kingussie
Kynkaith Mylnes, *see* Kinketh
Kynkell, *see* Kinkell
Kynloche, Kynlocht, *see* Kinloch
Kynlos, *see* Kinloss
Kynmarny, *see* Kinnernie
Kynmownd, Kynmund, Kynmundy, *see* Kinmundy
Kynnaird, Kynnard, Kynnarde, Kynnardie, *see* Kinnaird
Kynnaldie, *see* Kinaldy
Kynnalthie, *see* Kinalty
Kynnathmonth, Kynnathmound, *see* Kennethmont
Kynnedder, Kynneddir, *Dunfermline abbey*, *see* Kinneddar
Kynnedderschyre, *see* Kinneddarshire
Kynnedoir, Kynnedour, Kynnedoure, Kynnedward, *Elgin cathedral*, *see* Kinneddar
Kynneff, *see* Kinneff
Kynneid, *Brechin bishopric*, 383
Kynneiff, *see* Kinneff
Kynneill, *West Lothian*, *see* Kinneil
Kynneir [Wester, Easter Kinnear, NO4024, NO4023], 57, 58, 59, 60
Kynnell, *Forfar*, *see* Kinnell
Kynnell, *West Lothian*, *see* Kinneil
Kynnellar, *see* Kinnellar
Kynnenndir, *see* Kinneddar
Kynneris, *see* Kinneries
Kynnettes, Kynnettis, *see* Kinnettles
Kynninmonthe, Kynnymonthe, Kynnynmont, Kynnynmonth, Kynnynmonthe, Kynnynmounth, *see*

Kinninmonth
Kynnoir, Kynnour, *see* Kinnoir
Kynnoule, *see* Kinnoul
Kynnowdir, *Moray deanery*
    [*Kinnuddie, NH9055], 469
Kynnunquhar, *see* Kilconquhar
Kynpont, *Moorfoot, Newbattle abbey*, 101
Kynros, *see* Kinross
Kynschawdy, *see* Kinshaldy
Kynsleif Eister, *see* Easter Kinsleith
Kynsleif Wester, *see* Wester Kinsleith
Kynstar, *Leuchars*, 14
Kyntallartie, *see* Kiltarlity
Kynvaid, *see* Kinvaid
Kypbyre, *see* Kippsbyre
Kypchaplane, *Monkland*, 498
Kype [Muir, NS7139], 231
Kypellaw, *see* Kippilaw
Kyplaw, *see* Kippilaw
Kyppis, *see* Kipps
Kyppuithe, *Coupar Grange, Coupar Angus abbey*, 354
Kypsyd, *Lesmahagow* [Nether Kypeside, NS7541], 233
Kyrmuir, *see* Kirriemuir
Kyrnslaw, *see* Krinslaw
Kysidhill, *Moorfoot, Newbattle abbey*, 101
Kytins, *see* Kettins

Lachintille, Leichintullie, *Aberdeen Trinitarians*, 418, 421
Lachinvar, *see* Lochinvar
Lacorpe, *see* Lecropt
Ladeddie (Ladeddey, Ladeddy), NO4412; 13, 20, 21
Laderne, Laderny, *St Andrews parish*, 13, 20, 21
Ladhopemoor (Ladopemure, Laudopmuir), NT5039; 209, 257
Ladie Wigis [*Castlewigg, NX4343], 592
Ladopemure, *see* Ladhopemoor
Ladroun, *see* Latheron
Lady altar, *Govan*, 526
Lady Chapel of Kirkburn, *Cambuslang*, 523
Lady chaplainry, *Rossdhu*, 542
Lady Galloway prebend, *Lincluden collegiate kirk*, 273, 273n.
Lady Heuch, *see* St Andrews, St Mary on the Rock collegiate kirk
Lady kirk (Mary kirk, Our Lady), *parish, Sanday*, 660, 666n.; minister of, 660
Lady kirk (Mary kirk, Our Lady), *parish, Westray*, 655n., 660, 668
Lady kirk, *parish, South Ronaldsay*, 659
Lady Land, *Kinloss abbey*, 462
Ladyaird, *see* Ladyyard
Ladyland, *Riccarton*, 531
Ladyn Urquhart, *see* Leden Urquhart
Ladyton (Ladytoun), NS4079; 539
Ladyurd, NT1542; 211n.
Ladywall, *see* Ladywell
Ladywell (Ladywall), NO0241; 305, 342, 345
Ladyyard (Ladyaird), NS4729; 531
Lag, *Holywood abbey*, 272, 276
Lagand, *near Physgill, Whithorn priory*, 594
Laggan (Logan), NN6194; 486; vicar of, 486; vicarage of, 486
Lagganallachy (Logiallochie, Logialloquhy), NN9840; 309; benefice of, 308n.; kirk of, 309; mains of, 309; vicar pensionary of, 309; vicarage of, 309
Laichis, *Sweetheart abbey*, 611, 622
Laintymanniche, *Islay*, 674
Lairdmannoch (Larmanoch) [Lo, NX6559], 585, 589
Laklayok, *Carmichael*, 511
Lalathan (Lalathyne) [Fm, NO3404], 15
Lambden, NT7443; 227, 227n.
Lambieletham (Lamelethame) [North, South, NO5013, NO5012], prebend of St Mary on the Rock collegiate kirk, St Andrews, 86
Lamberton (Lambertoun), NT9657; 199, 202; kirk of, 199
Lambrid, Lambryd, *see* Lhanbryde
Lamb Holm, *see* Holmes, isles of
Lamelethame, *see* Lambieletham
Lamermuir, Lamermure, *see* Lammermuir
Lammermuir (Lamermuir, Lamermure, Lammermure, Lawmuirmoir), NT7458; 207, 257, 257n., 258
Lamont (Laumond, Laumund) [Port, NS0970], 558, 562, 565
Lamtrachen, *see* Lintrathen
Lanaile, *see* Lennel
Lanark (Lanrak, Lanrik), NS8843; 514;

burgh of, 231, 231n., 524; chaplain in, 521; chaplain of chaplainry in, 514; chaplainry of St Nicholas in, 514; deanery of, 513n.; glebe of, 508; St Nicholas' chapel in, 514; vicar of, 507, 508; vicarage of, 508
— parish kirk, 190, 191, 197, 197n., 506, 521; altarages in, 521; chaplains of altars in, 521
—, Franciscan friars of, 522
—, St Nicholas' kirk, 521, 524; altars in, 521; chaplain of Our Lady altar in, 524; chaplainry of Our Lady altar in, 524; chaplains of altars in, 521
—, *sheriffdom*, 71, 71n., 121, 121n., 213, 213n., 420n., 506, 509, 512, 520, 522, 522n., 523, 525n., 554n., 603n.
Landris, *Forgan*, 14
Lanell, *see* Lennel
Lang croft, *Glasgow*, 516
Lang Muirsyd, *Holywood abbey*, 276
Langcoit, *Kinloss abbey*, 462
Lange Gordoun, Langgardoun, Langgordoun, *Kelso abbey*, 233, 237, 239
Langforgound, Langforgung, *see* Longforgan
Langhope, NT4220; 258
Langhouswallis, *see* Banghouswallis
Langis Waird, *with Redy, Auchtermuchty, Lindores abbey*, 33
Langlandhill, *near Burnetland* [*Langlawhill, NT0938], 249
Langlands (Langlandis), NO5651; 390; prebend (with Hilton) of Guthrie collegiate kirk, 390
Langlee (Langlie) [Mains, NT5036], 207, 257, 259
Langmedo, Langmedow, *Melrose abbey*, 208, 209
Langnewtoun, *see* Longnewton
Langraw (Langrow), NO4913; 10, 13, 19, 20, 21
Langreynk, *see* Langruik
Langruik, Langreynk, *Lindean*, 224, 224n.
Langschaw, *Kilwinning abbey*, 573
Langshaw (Langschaw), *Melrose abbey*, NT5139; 207, 257, 259; mill of, 207
Langside (Langsyd), *Culross abbey*, NT9786; 290, 292
Langside (Langsyd), *Kelso abbey*, NT5528; 223
Langtoun, *Jedburgh abbey*, 219
Langton (Langtoun), *Kelso abbey*, NT7253; 192, 226, 232, 234, 238, 240, 242; kirk of, 226, 232, 234, 238, 240, 242; vicar of, 192; vicarage of, 192, 226
Langtounes lands, *with Lamberton*, 202
Langtownelaw, *Kelso abbey*, 223n.
Lanrak, Lanrik, *see* Lanark
Lany, *see* Lenzie
Lanzie, *see* Lenzie
Larbert (Larber, Larbeth, Lethbert), NS8582; 538, 543, 545, 546, 547; kirk of, 538, 543, 545, 546, 547
Larel (Larillis, Laroll) [Lower, Upper, ND1858, ND1857], 632, 633
Larg, *NX4365, *NX4858; 587, 600, 607
Largie (Largy), NJ6131; 31, 36
Largo (Largow), NO4203; 53, 53n., 54, 78, 89, 145, 146, 166, 167, 169; kirk of, 145, 146, 148, 166, 167; parish of, 145; vicar of, 78; vicarage perpetual of, 78-79
Largs (Largis), NS2059; 529; kirk of, 529
Larillis, *see* Larel
Larmanoch, *see* Lairdmannoch
Laroch, *see* Larroch
Laroll, *see* Larel
Larroch (Laroch), NX3740; 594
Lasoiddeis, Eister and Wester [Lassodie, NT1292], 27, 49, 50
Lassodie (Lossedy), NT1292; mill of, 62
Lassudden, *see* Lessudden
Lasswade (Laswaid, Leswaid), NT3065; 101, 115, 116, 118, 141, 143; vicar of, 115; vicarage of, 115
Lasualt, *see* Leswalt
Laswaid, *see* Lasswade
Laswalt, *see* Leswalt
Latcreiff, *see* Ledcrieff
Lathallan (Lathalland, Lathallane), *Kilconquhar*, 146, 167
Lathallan (Lathalland, Lathallane), *Newburn*, 24, 28, 29, 46, 50, 51
Lathalmond (Lathallmonth, Lathamonth, Lathamontht, Lathamounth, Lathawmounth, Lawthallmonth), NT0992; 23, 27, 45, 49, 50
Lathamen, *see* Letham

Lathane, *with Kipney, Arbroath abbey* [*Lethendy Cottage, NN9428], 360
Lathangy, *Dunfermline*, 23, 45
Latheron (Ladroun), ND1933; 629, 648; parish of, 629; vicar of, 648; vicarage of, 648
Lathindene, *see* Lethendy
Lathockar (Lathocker, Lathockker, Lathokker), NO4911; 13, 20, 21
Lathonis, *St Andrews parish*, 13, 20, 21
Lathrisk (Lauthrask, Lauthreisk, Lauthrisk, Lawthrisk) [Ho, NO2708], 9, 14, 17, 21, 75; kirk of, 9, 14, 17, 21; vicar of, 75; vicarage of, 75; Lauthrisk Wester, Lawthrisk Wester, 14, 21
Lauder, NT5247; 67; 190, 191, 194, 205; glebe of, 190, 191; kirk of, 191; parsonage of, 67, 190, 191, 194; vicarage of, 190, 191, 194
Laudopmuir, *see* Ladhopemoor
Laufeild, *see* Lawfield
Laumond, Laumund, *see* Lamont
Laurenstoun, *Ecclesgreig* [Mains of Lauriston, NO7766], 16
Lauthrask, Lauthreisk, *see* Lathrisk
Lauthrisk Wester, *see* Lathrisk
Laverokmure, *see* Lawerk Mure
Law Mylne, *see* Lawmyll
Law, *see* North Berwick Law
Lawch Clawethe, *Fearn abbey*, 634
Lawerk Mure, Laverokmure, *Linlithgow*, 155, 155n.
Lawers (Laweris) [Burn, NN6742], 58
Lawfield (Laufeild, Lawfeild), NO3310; 33, 37
Lawgrenis, *Newburn*, 24, 29, 46, 51
Lawmuirmoir, see Lammermuir
Lawmyll, Law Mylne, *St Andrews priory*, 10, 19
Lawreis land, Lowreis land, *Clarilaw, Kelso abbey*, 231, 233, 237, 240, 240n.
Lawrok, *see* St Ola
Lawsonestoun [Lacesston, NO1708], 69
Lawteis Hoill, *Moonzie*, 56
Lawthallmonth, *see* Lathalmond
Lawthrisk Eister, *see* Easter Lathrisk
Lawthrisk Wester, *see* Lathrisk
Lawthrisk, *see* Lathrisk
Lawtoun, *Dunkeld chancellory*, 308
Lay, Ley, *Restalrig collegiate kirk*, 142, 143

Leamicht, *see* Leanach
Leanach (Leamicht), NH7949; 639
Lecropt (Lacorpe, Lecrop, Lecrope, Licrept, Lucrope, Lycrope), NN8103; 303, 342, 346, 538, 543, 545, 547, 547n.; kirk of, 538, 543, 545, 547
Ledcasse, *Millhorn, Coupar Angus abbey*, 354
Ledcrieff (Latcreiff), NO2637; 376
Leddesdaill, *see* Liddesdale
Leddingame, Ledinghame [*Little Ledikin, NJ6529], 31, 36
Leden Urquhart (Ladyn Urquhart), NO1711; 69
Ledinghame, *see* Leddingame
Ledrassy, *with Pearsie, Coupar Angus abbey*, 369
Leffnoreis, *Hartside*, 514
Legaston (Ligistoun), NO5848; 376
Legattis Brig, *Dunfermline*, 23, 27, 45, 49
Legerwood (Lidgertwod, Lydgertwod, Lydgerwod), NT5943; 193, 529, 530; kirk of, 529, 530; vicar of, 193; vicarage of, 193
Leichintullie, *see* Lachintille
Leichstoun, *see* Bechestoun
Leidhill (i.e. Seidhill), *Paisley abbey*, 535, 535n.
Leidoun, Leidtoun, *see* Leyden
Leiffnoris (Luffnoris, Lufnoris), *Galston*, 564
Leightonhill (Lichtounhill, Lychtounhill), NO6361; 383, 384, 386, 388, 389
Leinzie, *see* Lenzie
Leis, *Carse of Gowrie, see* Leys
Leis, Leyis, *with Broadhaugh, Coldstream priory*, 186, 187
Leith (Leyth), NT2677; 10, 44, 45n., 86, 91, 92, 113, 114, 122, 128, 132, 133, 143, 150; Findguidis land, 128; Halkerstones land, 128; hospital of St Paul's Work, 95; hospital of St Anthony, 113, 114; King's Wark, 143; St Anthony's yard, 114; Todrikis land, 128
Leithen Hopes (Lethenehoippes, Lethinhopes), NT3344; 101, 103
Leitholm (Lethem), NT7844; 183, 183n.; Chapel Knowe of, 183n.
Lelileiff, *see* Lilliesleaf

INDEX OF PLACES

Lemlair (Lymmlar), NH5762; 643; parson of, 643; parsonage of, 643
Lempitlaw (Lemplaw), NT7832; 118
Lenbreyn, *see* Lymbarne
Lencluden, *see* Lincluden
Lenden, *see* Lindean
Leniestounlaw, Lennestounlaw, Lennestown Law, *Restalrig collegiate kirk*, 142, 143
Lennel (Lanaile, Lanell), NT8540; 186, 187, 188; kirk of, 186, 187; *see also* Coldstream
Lennestounlaw, Lennestown Law, *see* Leniestounlaw
Lennox [Lennoxtown, NS6277], 113
Lenzie (Lany, Lanzie, Leinzie), NS6571; 538, 543, 544, 545, 546, 548, 549; kirk of, 538, 543, 544, 545, 546, 548, 549
Leochel-Cushnie, *see* Cushnie
Leslie (Lesly), *Aberdeenshire*, NJ5924; 32, 36, 443; kirk of, 32, 36; mill of, 31, 36; vicarage of, 443-444
Leslie, *Fife*, NO2401; 62, 82, 300; kirk of, 300; vicarage of, 82
Lesmahagow (Lesmahago), NS8139; 231, 231n., 232, 232n., 235, 236, 236n., 241, 243; barony of, 229; kirk of, 229, 243, 244, 245; mains of, 231, 233, 245, 245n.; vicarage of, 229, 230, 244
— priory, 235, 235n., 236, 236n., 241, 243-245, 243n., 245n., 274n.; chamberlain of, 243
Lesmoir [Mains of, NJ4628], 475
Lessudden (Lassudden, Lessaden, Lessudden, Lossadden, Lossuddan) [St Boswells, NT5930], 189, 190, 191, 197, 197n., 207, 257; kirk of, 189, 197, 197n.
Lestalrig, see Restalrig
Leswaid, *Restalrig collegiate kirk*, *see* Lasswade
Leswaid, *Tongland*, *see* Leswalt
Leswalt (Lasualt, Laswalt, Leswaid, Leswane, Leswatt), NX0163; 588, 590, 592, 606, 607; kirk of, 590, 592; parsonage of, 588, 607; vicar of, 606; vicarage of, 588, 606, 607
Leswane, *see* Leswalt
Leswatt, *see* Leswalt
Letham (Lathamen), *Scoonie*, NO3704; 15, 21

Letham (Lethame), *Calder*, NT0668; 98
Letham (Lethame), *Fordell*, NT1483; 63
Lethbert, *see* Larbert
Lethem, *see* Leitholm
Lethen Bar (Leythynbar), NH9549; 469
Lethendy (Lathindene, Lethindene) [Kirkton of, NO1241], 308; parsonage of, 308
Lethenehoippes, Lethinhopes, *see* Leithen Hopes
Lethindene, *see* Lethendy
Lethington, 176n., 208n.
Lethinseid, *Kirknewton*, 104
Letho, *see* Lettoch
Lettirik, *Cambuslang* [Mid Letterickhills, NS6557], 508
Lettoch (Letho), NN9359; 284, 288
Leuchars (Leucheris, Leuchers, Leuchris, Lucheris, Luchris), NO4521; 9, 12, 13, 13n., 17, 20, 83; kirk of, 9, 13, 17, 20; parish of, 90n.; vicarage of, 83
Leuchatsbeath (Lowchat, Leuchquhattis Bayth), NO1592; 63, 64
Leuchie (Leuchin, Luquhen) [Ho, NT5783], 147, 168
Leuchland (Luchland), NO6259; 383, 384, 386, 388, 389
Leven (Levin, Levine, Leving, Levyn), NO3800; 15, 21, 56, 145, 167, 169
Levingstoun, *see* Livingston
Levyn, *see* Leven
Lewin aucht, *Dallas*, 492
Lewis (Loise), NB2022; 666, 667, 675
Ley, *see* Lay
Leyden (Leidoun, Leidtoun), NT0964; 104
Leyis, *see* Leis
Leyme Kill, *see* Limekilns
Leyne, *barony of*, 500
Leys (Leis), NN2624; 285, 288
Leyth, *see* Leith
Leythynbar, *see* Lethen Bar
Lhanbryde (Lambrid, Lambryd), NJ2761; 466n., 479, 480; kirk of, 466n.; parsonage of, 479; vicar pensionary, 480; vicarage of 479
Libberie, *see* Loweberry
Liberton (Libertoun), NT2788; 91, 93, 107; kirk of, 91, 93; minister of, 93; vicar of, 107; vicarage of, 107
Librie, *see* Loweberry
Lichtounhill, *see* Leightonhill

Licrept, *see* Lecropt
Liddesdale (Leddesdaill), NY4988, NX0461; 217
Lidgertwod, *see* Legerwood
Lidnocht, *see* Lynedoch
Lies, *with Clashendamer* [*Berryleys, NJ5461], 438
Liff (Lif, Lyf), NO3333; 332, 333, 334; kirk of, 334
Ligistoun, *see* Legaston
Lilisleif, *see* Lilliesleaf
Lilliesleaf (Lelileiff, Lilisleif, Lillisleiff), NT5325; 495, 499; kirk of, 499
Limekilns (Leyme Kill, Lymekillis, Lymkill), NT0783; 24, 26, 27, 46, 48, 49
Linbren, *see* Lymbarne
Lincluden (Kynclydyn, Lencluden, Linclouden, Linclowden, Lynclowdan, Lynclowden), NX9677; 21n., 265n., 274, 613; nunnery, 273n.
—— collegiate kirk, NX9678; 265, 265n., 273, 273n., 274n., 277, 277n., 538, 545, 613, 613n., 615, 620, 621; Lady Galloway prebend, 273, 273n.; mains of, 613; prebend of, 615, 620; prebendaries of, 621; prebendary of, 273, 615, 620; prebends of, 621; provost of, 10, 10n., 21n., 54, 97-98, 97n. 613, 613n., 621; provostry of, 613, 620
Lindean (Heyndoun Towne, Lenden, Lindene, Lyndane, Lyndein, Lynden, Lyndene, Lyndentoun, Lyndyne, Reddene), NT4931; 103, 103n., 223, 224, 224n., 231, 232, 233, 235, 237, 238, 239, 241, 242; mill of, 231, 234, 237, 240, 242, 264; kirk of, 232, 235, 238, 241, 242; parsonage, 224; reader at kirk of, 224n.; vicar of, 264; vicarage of, 224, 224n., 264
Linderties (Lundarteris), NO3351; 356
Lindores (Lindoris, Lundoris, Lundores), NO2617; 32, 34; *see also* Abdie, Auld Lindores, Newburgh
—— abbey, 29-37, 29n., 282, 378n.; abbot of, 29n.
Lingestoun, *Rothes*, 491
Linlithgow (Linlithqw, Linlythgow, Lynlithgw, Lynlythgow, Lynlythgu, Lynlythgw, Lythgow), NS9977; 8, 9n., 11, 16, 17, 105; kirk of, 8, 16, 17, 373, 549, 550; chaplainry of Allhallow altar in kirk of, 154; chaplain of St Bride's altar in kirk of, 154; chaplainry in, 157; chaplainry of St Bride's altar in kirk of, 154; East Port, 156; Kirk Hill, 156; St Michael's Wynd, 156; town of, 8; West Port, 156
——, Carmelite friary of, 155-156; prior of, 156
——, *sheriffdom*, 1n., 149n., 184n.
Linlour, Haltoun Linlour, *Dunkeld chantory* [*Meiklour, NO1539], 307, 307n.
Linross (Lunros), NO3549; 356
Lintalee (Lyntolie), NT6418; 219
Linton, *Haddington*, NT5977; kirk of, 170n., 174; chaplainry in kirk of, 174. *See also* Haugh, Prestonkirk
Linton, *Peebles*, NT1551; kirk of, 108, 108n.; parson of, 108, 253; parsonage of, 108, 253; vicar of, 251; vicarage of, 251
Linton, *Roxburgh*, NT7726; parson of, 212, 215; parsonage of, 215; vicar of, 215; vicarage of, 215
Lintrathen (Lamtrachen, Lunkathen) [Lodge, NO2853], 544, 548; kirk of, 544, 548
Lipot, *Aberdeen Trinitarians*, 418
Litilden, Littilden [*Littledeanlees, NT6823], 214, 215
Litilhope, *Stobo* [*Stobohope, NT1537], 249
Litillprestoun, *Cranston*, 3
Litiltoun, *see* Littleton
Litledon, *see* Littleton
Littilden, *see* Litilden
Little Abbey (Lytill Abay), *Kirkandrews*, 596
Little Airies (Lytill Areis, Lytill Aris, Lytill Arisch), NX4148; 594, 596; mill of, 593, 595
Little Ardres [*Airdrie, NX9658], 609
Little Areis, Aris, Arisch, *see* Little Airies
Little Arrow, *Glasserton* [High Arrow, NX4436], 593
Little Barnebowgall, *Leith*, 128
Little Brunquallis, *Dallas*, 492
Little Carcary (Litill Carrorie, Little Carrary), NO6355; 358, 365

Little Carorie, Carrary, *see* Little Carcary
Little Creiche, *Deer abbey* [Mains of Crichie, NJ9745], 457
Little Creicht, *Caithness bishopric*, 628
Little Culleloung, *see* Little Killylung
Little Dagathe, *see* Little Dalgaty
Little Dalgaty (Dagathe) [Dalgety, NO6159], 384, 385, 389
Little Dallas (Lytill Doles), NH6985; 652
Little Doles, *see* Little Dallas
Little Drostrie, *see* Little Troustrie
Little Drumquhendill [Drumwhindle, NJ9236], 435
Little Dunkeld (Letill Dunkeld, Lytildunkeldensis, Lytle Dunkeld, Lyttle Dunkeld, Lyttle Dunkelden), NO0342; 303, 342, 336n., 344, 346, 347; kirk of, 303, 342, 344, 346
Little Elrick (Lytill Elrik), NJ9244; 457
Little Fallsyd, *Tranent*, 91
Little Fordaill, Fordell, *Melrose abbey*, 207, 257, 259
Little Forter (Litill Forthir), NO1864; 355
Little Freirtoun, *Forgan* [Easter, South, Wester Friarton, NO4326, NO4325, NO4225], 14
Little Galtua [Galtway, NX7148], 617
Little Gledstanes, *Hobkirk*, 220
Little Goval (Litill Govill, Lytill Govyll), NJ8914; 422, 423
Little Guttoun, *see* Little Outoun
Little Harwood (Hardwod Littill), NT0161; 98
Little Hauche, *Coldstream priory*, 186
Little Killylung (Culleloung, Killelung) [Killylung, NX9581], 270, 275
Little Kinneir, *Balmerino abbey*, 58
Little Lour (Lytill Lowr), NO4744; 390
Little Lun (Littil Lone, Littil Lune, Littleron), NO3399; 10, 15, 18
Little Maisondieu (Massonden, Massondew, Messyndew) [Maisondieu, NO5861], 383, 387, 388
Little Methlick (Lytill Methlyk), NJ8537; 435
Little Monkton (Littill Monktoun, Littill Mounktoun), *Musselburgh* [Monktonhall, NO3701], 24, 26, 43, 43n., 47, 48, 49, 53

Little Myln, *Manuel priory*, 549
Little Mylne, *Brechin bishopric*, 384, 385, 389
Little Newtoun, *Kelso abbey*, 223, 224
Little Outoun, Outtoun, Guttoun, *with Gallows Outon, Whithorn priory*, 592, 594, 596
Little Park (Lytill Park), NX4565; 586, 589
Little Penick (Penik, Pennyk, Pynnik) [Penick, NH9356], 469
Little Perth, Pertht, *Coupar Angus abbey*, 354, 357, 368, 369, 409
Little Petforth [Mains of Pitforthie, NO6061], 400
Little Quhitsoun, *Ecclesgreig*, 16
Little Rany [Rhynie, NC8479], 634
Little Scamnocht [*Stennock, NX4737], 594
Little Sorn (Sornbeg, Soumertait), NS4933; 531, 531n.
Little Speddoch (Spadocht, Speddoche, Spoddoche) [Speddoch, NX8582], 269, 270, 274, 275, 622, 623
Little Trochrie (Letill Throchtkare), NN9840; 309
Little Troustrie (Drostrie, Trostrie) [Troustrie, NO5907], 177, 178, 179
Little Ullagrahame, *near Dorrery, Caithness bishopric*, 628
Little Waterston (Little Walterstoun, Littill Waterstoun) [Waterston, NO5159], 383, 387, 388
Little Wistones, Littill Wistownis, *Mearns, Lindores abbey*, 31, 35
Littlegill (Lytillgell), NS8144; 513
Littleron, *see* Little Lun
Littleton (Litiltoun), *Coupar Angus abbey*, NO3350; 356
Littleton (Litledon), *Dunkeld*, NO9944; 336, 336n.
Livingston (Levingstoun), NT0366, NT0668; 106, 107, 119; vicarage of, 106, 119-120; *see also* Kirklevingstoun
Lizard, *Monkland*, 102,
Loanhead (Loynheid), *Rathven*, NJ4566; 456
Loanhead (Loynheid), *Restalrig collegiate kirk*, NT2865; 141
Loanleven (Lonlavin, Lunlevin), NO0525; 296, 298, 298n.
Lobricht Hous alias Louchsyd, *see*

Lochside
Loch Leven (Lacum de Levin, Lochlavin, Lochleven, Lochlevin, Locht Levin), NO1401; 9, 9n., 17, 17n., 26, 26n., 48, 56, 89, 257n., 259, 259n., 546
Loch Ness (Lochnes), NH4314, NH4618, NH5023; 488, 489
Loch Tay (Locht Tay, Lochtay), NN6838; 284, 287, 288n.
Loche Heid, *Dunbar*, 122
Lochebochtsyd, *Glasgow cathedral*, 500
Lochindilloch (Lochindillocht, Lochkindeloch, Lokcindalocht, Lokindilloch, Kinderloch) [New Abbey, NX9666]; 612, 622; corn mill of, 622; parsonage of, 612, 622; vicarage of, 612, 622; waulk mill of, 622; *see also* New Abbey, Sweetheart
Lochinvar (Lachinvar), NX6585; 586, 586n., 587, 594, 600, 600n., 607
Lochirmakhous, *see* Longformacus
Lochiskyne, *see* Lochslin
Lochkindeloch, *see* Lochindilloch
Lochmair, *see* Lythmore
Lochmill (Lochmylne), *Balmerino abbey*, NO2216; 58
Lochrutton (Lochrutoun) [Loch, NX8973], 279, 613; kirk of, 613; vicarage of, 279
Lochside (Lobricht Hous alias Louchsyd), NT7928; 214
Lochslin (Lochiskyne), NH8480; 651
Lochtoun, *with Drumrae, Glasserton*, 593
Lochtower (Loycht Towir), NT8028; 214
Lochtullo, *Dunbar*, 249
Lochtysyd alias Gowishawcht, Buchthesyde, Govis Hauch [*Lochty, NO0625], 296, 296n., 298
Lochurd, NT1143; 211n.
Lochwinnoch (Lochvinock, Lochvinok), NS3559; 529; kirk of, 529; vicarage of, 529
Logan, *see* Laggan
Logaret, Logarrett, *see* Logierait
Loggy, Logy, *Dunfermline abbey*, 23, 46
Logiallochie, Logialloquhy, *see* Lagganallachy
Logie (Logy), *parish, Dunblane diocese,* *see* Logie-Atheron
Logie, *Balmerino*, *see* Logie-Murdoch
Logie, Logy, *Inchcolm abbey*, 62
Logie-Atheron (Logie, Logie-Wallach, Logy), NS8196; 290n., 295, 296, 313, 348; kirk of, 148, 148n., 166, 168, 348; parish of, 290n., 295; vicar of, 313; vicarage of, 313
Logie Dornoch, *see* Logie-Durno
Logie-Dundee (Logiedundie, Logydundy), 399; kirk of, 334; vicar pensionary of, 399; vicarage pensionary of, 399
Logie-Durno (Logie Dornoch, Logy Dornoche, Logydornoch, Logydornocht), NJ7026; kirk of, 32, 36, 448; vicar of, 448; vicarage of, 448
Logie-Easter (Logy), *Ross diocese*, NH7576; 636, 637; parson of, 637; parsonage of, 637; vicar of, 636; vicarage of, 636
Logie-Mar, kirk of, 425, 438
Logie-Montrose (Logymontros) [Logie-Pert, NO6664], 373, 374; minister of, 374; parson of, 374; parsonage of, 373; vicar of, 374; vicarage of, 374
Logie-Murdoch (Logymurtho) [Logie, NO4020], kirk of, 58; minister of, 59; vicar of, 79; vicarage of, 79
Logie-Pert, *see* Logie-Montrose
Logie-Wallach, *see* Logie-Atheron
Logie-Wester (Logy), *Ross diocese*, NH5957; 637; kirk of, 637
Logierait (Logaret, Logarrett, Logirate, Logiret, Logyrait, Logyrat), NN9751; 284, 287, 332, 333, 335; kirk of, 332; minister of, 333; vicarage of, 335; Port of, 303, 342, 345
Logy Buchan (Logybuchane) [Kirkton of, NJ9829], 425, 438; kirk of, 425, 438
Logy Dornoche, Logydornocht, *see* Logie-Durno
Logy Fyntray, Logyfintray, *Fintray, Lindores abbey*, 31, 36
Logy, *parish, Dunblane diocese*, *see* Logie-Atheron
Logy, *Inchcolm abbey*, *see* Logie
Logy, *Ross diocese*, *see* Logie-Easter, Logie-Wester

INDEX OF PLACES 841

Logy, *with Birnam, Dunkeld bishopric*, 303
Logydornoch, Logydornocht, *see* Logy-Durno
Logydundy, *see* Logie-Dundee
Logyhill, *with Easter Braikie, Kinnell*, 366
Logymar, Logymare, *see* Logie-Mar
Logymontros, *see* Logie-Montrose
Logymurtho, *see* Logie-Murdoch
Logyrait, Logyrat, *see* Logierait
Loise, *see* Lewis
Lokcindalocht, Lokindilloch, *see* Lochindilloch
Lonfannan, *see* Lumphanan
Longardy, *see* Luncarty
Longforgan (Langforgound, Langforgung), NO2732; 16, 18, 315; kirk of, 16, 18; vicar of, 315-316; vicarage of, 315-316
Longformacus (Lochirmakhous), NT6957; 184n., 185; kirk of, 185
Longfuird, *Calder*, 98
Longholm Myln, *Calder*, 98
Longley (Lunglie, Lungley), NJ8531; 359, 450; kirk of, 359
—— and Fetterangus, vicarage of, 450; *see also* Fetterangus
Longnewton (Langnewtoun), NT5827; 184, 216, 219; kirk of, 216, 219
Lonlavin, *see* Loanleven
Lonmay (Lunmey) [Sta, NJ0158], 427; parson of, 427; parsonage of, 427
Loppe Urquhart, *with Nether Urquhart, Strathmiglo*, 69
Loquhariot, NT3760; Wester Lochquhairat, 176
Lordis Manes, *Gordon*, 228
Lornie (Lorny), NO2221; 285, 286, 288
Lorrous, *Lesmahagow* [*Lauriesmuir, NS7839], 232
Lossadden, *see* Lessudden
Lossedy, *see* Lassodie
Lossuddan, *see* Lessudden
Loth (Lothe) [Sta, NC9510], 629; parish of, 629
Lothian (Lothiane, Loudiane, Louthian, Lowthian, Lowtheane, Lowthiane), 106, 113, 117, 161, 177, 178, 182, 276, 496; archdeacon of, 106; archdeaconry of, 117; Official of, 276
Lothrin, *see* Latheron

Louchsyd, *see* Lochside
Loudiane, *see* Lothian
Loudon (Lowdoun) [Kirk, NS4937], 576; kirk of, 576
Lounoumgrange, *see* Hownam Grange
Lour (Loure), NO4746; 217
Louthian, *see* Lothian
Low Ersock (Nether Ersik), NX4437; 593
Low Glasnick (Nethir Glasnycht), NX3461; 588
Lowchat, *see* Leuchatsbeath
Lowdoun, *see* Loudon
Loweberry (Libberie, Librie, Luberie, Luberrie, Lybrie), NX8781; 269, 271, 622, 623, 624
Lowes (Lowis), St Mary of the, *see* Yarrow
Lowfannan, *see* Lumphanan
Lowis, *see* Lowes
Lowreis land, *see* Lawreis land
Lowtheane, Lowthian, Lowthiane, *see* Lothian
Loycht Towir, *see* Lochtower
Loynheid, *see* Loanhead
Luberie, Luberrie, *see* Loweberry
Lucheris, Luchris, *see* Leuchars
Luchland, *see* Leuchland
Lucklaw (Luklaw), NO4120; 14
Lucrope, *see* Lecropt
Lude (Luid) [Kirkton of, NN9068], 319; curate of, 319; parson of, 319; parsonage of, 319; vicarage of, 319
Luffnoris, Lufnoris, *Galston*, 564
Lufnoris, *see* Luffnoris
Lughill, *Monkland*, 498
Lugton (Lugtoun), NT3367; 118
Lugtoun, *Airth*, 158
Luid, *see* Lude
Luklaw, *see* Lucklaw
Lumay, *see* Lunay
Lumbo, NO4814; 10, 12, 13, 19, 20
Lumloch (Lumocht, Lunlocht), NS6369; 497, 498
Lumocht, *see* Lumloch
Lumphanan (Lonfannan, Lowfannan), NJ5804; 439; kirk of, 439
Lumquhat, NO2413; 30, 33, 35
Lunan (Lunnane), NO6851; 361; kirk of, 361; minister of, 361; vicar of, 361
Lunay, Lumay, *Caithness, Orkney bishopric*, 657, 670

Lunay, *Shetland*, see Lunnasting
Luncarty (Longardy), NO0929; 324; curate of, 324; parson of, 324; parsonage of, 324; reader at, 324; vicarage of, 324
Lundarteris, *see* Linderties
Lundeiff (Lundeif) [Kinloch, NO1444], 323, 327, 337; parson of, 337; parsonage of, 327, 337; vicarage of, 327, 337
Lundeis [East, Mid, Wester Lundie, NN7304, NN7204, NN7104], 343
Lundie (Lundy), *St Andrews priory*, NO2936; 10, 17, 19, 376, 377; glebe of, 376; kirk of, 376, 377; parson of, 377; parsonage of, 376, 377; vicar of, 377; vicarage of, 376, 377
Lundie, Lundy, *Largo*, 145, 146, 148, 167, 168, 169
Lundores, Lundoris, *see* Lindores
Lungley, Lunglie, see Longley
Lunkathen, *see* Lintrathen
Lunlevin, *see* Loanleven
Lunlocht, *see* Lumloch
Lunmey, *see* Lonmay
Lunnane, see Lunan
Lunnasting (Lunay), HU4665; 658, 660
Lunros, *see* Linross
Luquhen, *see* Leuchie
Lurg, NT9586; 291, 292, 294, 294n.
Luscer Ewat, Luscerewert, Luscour Evert, Luscour Ewart, *see* East Luscar
Luscer Wester, *see* Wester Luscar
Luss (Lus), NS3592; 266, 541, 542, 577; chaplain of, 542; chaplainry of, 542; parson of, 266, 541, 577; parsonage of, 541; vicar pensionary of, 541, 542; vicarage of, 541; vicarage pensionary of, 542
Luthrie (Luthre), NO3319; 33
Lwne, *Bower*, 632
Lybrie, *see* Loweberry
Lycht, *see* Lyth
Lychtmoir, *see* Lythmore
Lychtounhill, *see* Leightonhill
Lycrope, *see* Lecropt
Lydenocht, *see* Lynedoch
Lydgertwod, Lydgerwod, *see* Legerwood
Lyf, *see* Liff
Lymbarne, Linbren, Lenbreyn, Lymbrane, *Paisley abbey*, 530, 530n.
Lymekilhill, Lymkilhill, *with Limekilns, Dunfermline abbey*, 26, 48
Lymekill, Lymkill, *Clachan, Galloway bishopric*, 586, 589
Lymekillis, *see* Limekilns
Lymkilhill, *see* Lymekilhill
Lymkill, *see* Limekilns
Lymmlar, *see* Lemlair
Lynclowdan, Lynclowden, Lyncluden, *see* Lincluden
Lyndane, Lyndein, Lynden, Lyndene, Lyndyne, *see* Lindean
Lyndentoun, *see* Lindean
Lyne, NT2041; 252; parson of, 252; parsonage of, 252; vicarage of, 252
Lynedoch (Lidnocht, Lydenocht) [Cottage, NO0328], 296, 298
Lynekirk, *with Cogle, Watten*, 632
Lynhous, *Calder*, 98
Lynlithgw, Lynlythgow, *see* Linlithgow
Lynnache, *Kinloss abbey*, 462
Lyntolie, see Lintalee
Lyth (Lycht), ND2863; 628, 632
Lythgow, *see* Linlithgow
Lythgw lands, *Lathrisk*, 14, 21
Lythmore (Lochmair, Lychtmoir), ND0566; mill of, 628
Lytildunkeldensis, *see* Little Dunkeld
Lytillgell, *see* Littlegill

McAlveris croft, McCaleis croft, McCalveris croft, *Holywood*, 270, 271, 276
Macawnis satt, *Dallas*, 492
McBurnistoun, *see* Makbrunstoun
McCairstoun, *see* Makerstoun
McCarsie, *see* Muckersie
McCarstoun, McCerstoun, *see* Makerstoun
McClychtan lands, *Glasgow Black friars*, 522
McClynnstoun, *see* Collieston
McCollestoun, *see* Collieston
McCullestoun, *see* Collieston
McGowin, *see* Merkgain
McHarmik, see Kilmacocharmik
McKirstoun, *see* Makerstoun
McQuhair, *near Genoch, Soulseat abbey*, 599
McQuhanrik, McQuhanrikis, *see* McWhanrick

McWhanrick (Makqualter, McQuhanrik, McQuharik, McQuhinrik), NX9381; 270, 275, 276, 623; McQuhanrikis, Makquhanrik Mekill and Littill, 271, 276
Magask Inferiour, *see* Nether Magask
Magask Superior, *see* Upper Magus
Magdalen chapel, *Edinburgh*, 127
Magdalene altar, *Trinity kirk, St Andrews*, 1n.
Magdalene chapel, *Dundee*, 400
Magdalene Chappeltoun, Magdalane Chapell, Chapeltoun, Chepeltoun, *Maryton*, 383, 384, 387, 388, 400
Magdalene hospital, *beside Musselburgh*, 39, 39n.
Magiscrist, *Monkland*, 498
Maidland, NX4354; 597
Maidlane parish, *see* St Magdalene
Maij, *see*, May, priory of
Mainport, *near Freugh, Soulseat abbey*, 599
Mains, *parish*, NO3836; 393; vicarage of, 393; *see also* Stradichtie
Mains, *Abernethy collegiate kirk* [Provost Mains, NO1816], 322
Mains, *Clunie* [West Mains, NO0943], 342, 345
Mains ('Brodland of Ogilbye alias Manis'), *Dunblane bishopric*, 349
Mains (Manys), *Kinnell*, 366
Mains of the Abbey (Abay, Abbay), *see* Abbey Mains
Mains of Abirlady, *see* Aberlady Mains
Mains of Airlie (Manis of Arlie), NO2951; 356
Mains of Arroll, *see* Mains of Errol
Mains of Dene, *Kilwinning abbey*, 573
Mains of Dillavaird (Manis), NO7482; 405
Mains of Errol (Mainis of Erroll, Manis of Arroll), NO2421; 285, 288
Mains of Farnell (Fernell, Fernewell, Fernvale, Fernvell), NO6255; 365, 385, 386, 389
Mains of Fernell, Fernewell, Fernvale, Fernvell, *see* Mains of Farnell
Mains of Hayning, *see* Haining Mains
Mains of Ilingtoun, *Symington, Kyle* [*Helenton, NS3930], 564; *see also* Ellingtoun Boig, Ilingtoun
Mains of Kimmgery, *see* Kimmerghame Mains

Mains of the Lang Coit, *Eddleston*, 496
Mains of Logy, *Monzie*, NO7063; 349
Mains of Methven (Manys of Methven, My lordis Manis), *with Loanleven, Methven collegiate kirk*, 296, 297, 298
Mains of Morphie (Manyis Morphy) [Morphie, NO7164], 16
Mairnes, *see* Mearns
Maislands, *see* Marislandis
Maison Dieu (Massindew), hospital, *Elgin Black friars*, 453, 453n.
Maison Dieu (Maisondew), *Stirling Black friars*, 554
Maison Dieu (Masondew), preceptory in Brechin, NO5861; 390
Maistertoun, *Dunfermline abbey, see* Mastertown
Maistertoun, *Newbattle abbey, see* Masterton
Makbrunstoun, McBurnistoun, *Holywood abbey*, 623
Makcarstoun, *see* Makerstoun
Makerstoun (Maccarstoun, Makcarstoun, McCairstoun, McCarstoun, McCerstoun, McKirstoun), NT6632; 213, 213n., 223, 223n., 224, 230, 232, 233, 234, 237, 238, 239, 240, 242; kirk of, 224, 226, 232, 234, 238, 240, 242; vicar of, 213; vicarage of, 213, 213n., 526
Makinrais set, *Dallas*, 492
Makqualter, *see* McWhanrick
Makquhanrik Mekill and Littill, *see* McWhanrick
Malcolestoun, *see* Collieston
Maldislie, *see* Mauldslie
Malestanis, *see* Mellerstain
Malingsyd, *see* Mellenside
Mamewlacht, Mamewlaycht, *near Cottown, Aberdeen deanery*, 422, 423
Man, Isle of (Ile of Man), 594
Manderston (Manderstoun) [Ho, NT8154], 166, 198, 198n.
Maner, *see* Manor
Manor (Maner, Mennar) [Manor Sware, NT2339], *parish kirk, Peebles*, 251, 253; beneficeof, 251; parsonage of, 251n., 153; prebend of the archdeaconry of Glasgow, 251n.; Holy Rood altar in, 255, 525; vicarage of, 251n., 253

Mantirope, Manturbe, Manturpe, *see* Monturpie
Manuel (Manuell, Manwell) [Ho, NS9676], 11, 549, 550n.; mains of, 549; mill of, 11
— priory, 549-550; prioress of, 550, 550n.
Manwell, *see* Manuel
Manygaff, *see* Minnigaff
Manyis, *see,* Mains
Manys, *see,* Mains
Mar (Mare), 354, 369, 435
Marboy, *Ecclesgreig,* 16
Marcaitnay [Over Marcartney, NX7672], 622
Marcarey, *see* Mertory
Maretoun, *see* Maryton
Margyr, *Glenbervie,* 405
Marie croft, Mary croft, *Lindores abbey,* 30, 33, 35, 37
Mariland, *see* Maryland
Marislandis (Virgin Mary's lands, Maislands), *Auchtermuchty,* 33, 33n.
Maristoun, *Lindores abbey,* 33, 37
Maritoun, *see* Maryton
Marjoriebanks (Marjoribankis), NY0883; 278
Markinch (Markinche, Markinsche), NO2901; 9, 10, 15, 17, 19; kirk of, 10, 15, 17
Markle (Markille) [Mains, NT5677], 161, 161n.; chaplain of, 161; chaplainry of, 161
Marnis, *see* Mearns
Marown, Old Church (Kirkmarrow), *see* Kirk Marown
Mars, *see* Merse
Martingtoun, Mertingtoun, Mertintoun, Mertyntoun, *Holywood abbey,* 270, 275, 276, 623
Martoun, *see* Morton
Marvik, *see* Marwick
Marway, *see* Marwick
Marwick (Marvik, Marway, Marwik), HY2324; 655, 659, 668; kirk of, 659; minister of, 659;
Mary croft, *see* Marie croft
Mary kirk, *Sanday, see* Lady kirk
Mary kirk, *Westray, see* Lady kirk
Maryland (Mariland), NS4177; 539
Maryton (Maretoun, Maritoun, Marytoun), NO6856; 383, 384, 385, 387, 388, 389; parish of, 388, 389

Masondew, *see* Maison Dieu
Massindew, *see* Maison Dieu
Masterton (Maistertoun), *Newbattle abbey* [Ho, NT3463], 101
Mastertown (Maistertoun), *Dunfermline abbey,* NT1284; 23, 27, 28, 45, 49, 50
Mathie, *see* Meathie
Mathill, *see* Methil
Mauchline (Mauchlein, Mauchling, Mauchlyne), NS4927; 210, 562, 564; kirk of, 210
Mauldslie (Maldislie), NT3053; 101
Maw, *with Little Lun, St Andrews priory,* 15
Mawr, *near Moracht,* 594
Maxpoffle (Maxpoppill, Moxpoffill), NT5530; 207, 257
Maxton (Maxtoun), NT6130; 189, 191, 197, 197n.; kirk of, 189, 197, 197n.
Maxwell (Maxwall) [Maxwellheugh, NT7333], kirk of, 225, 226, 234, 237, 240, 242, 269, 270, 276, 277; parsonage of, 225; St Thomas' chapel of, 225; vicar pensionary of, 226; vicarage of, 225, 226; *see also* Maxwellheugh
Maxwellheugh (Maxwellheuch), NT7333; 222, 222n., 225, 230, 233, 234, 237, 239, 240, 242; *see also* Maxwell
May, *Caithness, see* Mey
May, priory of, 332; *see also* Pittenweem
Maybole (Mayboill), NS2909; 523, 568, 570; kirk of, 148, 149, 166, 169; vicarage of, 523
— collegiate kirk, 568; prebendaries of, 568; prebends of, 568; provostry of, 568
Maynhill, *Monkland,* 498
Mearns (Maimes, Marnis, Mernes, Mernia, Mernis) [Howe of, NO6974], 10, 11, 16, 31, 34, 35, 66, 67, 363n., 372n., 375, 403, 405, 406, 408
Mearns (Mernis), *parish,* NS5455; 528, 529; kirk of, 528, 529
Meathie (Mathie) [Wester Meathie, NO4546], kirk of, 355, 369; parsonage of, 369; vicarage of, 370
Meddrawis [*North, South Medrox, NS7271, NS7370], 498; *see also*

INDEX OF PLACES 845

Midrois
Medo, *Coupar Angus abbey*, 370
Medowis, *Leuchars*, 14
Medowis, *Lindores abbey*, 30, 35
Medwig, *see* Midwig
Medy, *St Andrews Black friars*, 89
Meffen, *see* Methven
Megginch (Meginche, Melginche) [Castle, NO2424], 91, 285, 288; kirk of, 91
Megill, *see* Meigle
Meginche, *see* Megginch
Megle, *see* Meigle
Meidwig, *see* Midwig
Meigle (Megill, Megle), NO2844; 301, 337; kirk of, 301; vicarage of, 337
Meikillfeild, Meikilfeild, Mekilfeild, Mekill Feild, *Holywood abbey*, 270, 271, 275, 276
Meikillwig, Meikilwig [*Castlewigg, NX4343], 592, 594, 596
Meikle Airdrie [*Airdrie, NX9658], 609
Meikle Airies (Meikeill Areis, Meikill Areis), NX3948; 594, 596
Meikle Arow, Arrow [High Arrow, NX4436], 594, 596
Meikle Auchtchrydie [Auchreddie, NJ8846], 457
Meikle Carcary (Carrary, Carkorie) [Little Carcary, NO6355], 358, 365
Meikle Dalgety (Dagathe) [Dalgety, NO6159], 387, 388
Meikle Doles [Little Dallas, NH6985], 652
Meikle Dunrod, *Senwick*, 587
Meikle Feild, *see* Meikillfeild
Meikle Fethie [Fithie, NO6354], 358, 365
Meikle Forthir [Little Forter, NO1864], 355
Meikle Galtway [Galtway, NX7148], 619
Meikle Logie (Mekill Logye), NN9841; 309, 309n.
Meikle Maisondieu (Massonden, Massondew, Messindew) [Maisondieu, NO5861], 383, 386, 388
Meikle Moir, *Dunkeld* [*Meiklour, NO1539], 300
Meikle Moy [Little Moy, NH4855], 635
Meikle Penick (Penik, Pennyk) [Penick, NH9356], 469

Meikle Ranquallis, *Dallas*, 492
Meikle Rany, *Fearn abbey*, 633
Meikle Rig, *Haddington priory*, 180
Meikle Speddoch (Spadocht, Spedoche) [Speddoch, NX8582], 269, 270, 274, 275, 622, 623
Meikle Troustrie (Troistrie) [Troustrie, NO5907], 162
Meikle Waterston (Meikle Walterstoun, Meikle Waterstoun) [Waterston, NO5159], 383, 387, 388
Meirbank, *with Blainslie, Melrose abbey*, 257, 259
Meirden Vester, *Makerstoun*, 223n.
Mekilfeild, *see* Meikillfeild
Meldrum [Old Meldrum, NJ8027], 415
Meldrum's Mill (Meldrum Mylne, Meldrumes Mylne, Meldrumis Mylne), *Dunfermline*, 27, 49, 50
Melfort (Melphort), NM8314; 674
Melginche, *see* Megginch
Melgum, *Newburn*, 24, 46
Melgund [Castle, NO5456], 330, 393
Melingsyde, *see* Mellenside
Melleden, Mellenden, Meltondene, with *Easter Softlaw*, 234, 234n., 235, 238, 240, 241
Mellenside (Malingsyd, Melingsyde), NJ6530; 31, 36
Mellerstain (Malestanis, Mellerstanes, Melorstanis, Mullerstanes) [Ho, NT6439], 200, 204, 223, 231, 234, 237, 240
Melorstanis, *see* Mellerstain
Melphort, *see* Melfort
Melrose (Melros, Melrosland), NT5434; 193n., 207, 208, 209, 210, 211, 243, 243n., 257, 258, 260, 260n.
— abbey, 7, 207-211, 257-261, 264n.; abbot of, 257; commendator of, 207, 208n., 243, 243n., 257n., 260; monks of, 210, 260
Meltondene, *see* Melleden
Meltoun, *Fearn, Edderton, see* Milton
Melville (Melvile) [Mains, NT3067], 118; curate of, 118; minister of, 118; parson of, 118, 118n.; parsonage of, 118; vicarage of, 118
Meneabracht, *see* Monyabroch
Mengreowis, *see* Montgrew
Menisgrein, Menysgrene, *Collessie*, 33, 37
Menmuir [Kirkton of, NO5364], 303,

303n., 374, 375, 397; parson of, 303, 303n., 374-375; vicar of, 397; vicarage of, 397
Menmure, *Dunkeld bishopric*, 303, 342, 346
Mennar, *see* Manor
Menteith (Menteth, Munteithit), 343, 554, 555
Menteth, *see* Menteith
Menysgrene, *see* Menisgrein
Merbotill, Merbotle, Merbottill, *see* Morebattle
Merchiston, 94n.
Mereinethalis, *Caithness bishopric*, 628
Merkbradan [*Mark, NX1157], 597
Merkgain, McGowin [*Mark, NX1157], 598, 599
Mernes, Mernia, Mernis, *see* Mearns
Merse (Mars, Mers, Mires), 113, 128, 184, 207, 258, 302
Merseltoun, Mersiltoun, *see* Mersington, Mersington (Merseltoun, Mersiltoun), NT7744; 183n., 184, 214
Mertingtoun, Mertintoun, Mertyntoun, *see* Martingtoun
Mertory (Marcarey), *Mearns*, 31, 31n., 35
Mertoun (Netoun) [Ho, NT6131], 189, 191, 197, 197n.; kirk of, 189, 197, 197n.
Merwicht, *see* Moracht
Methil (Mathill), NT3699; 82; parsonage of, 82; vicarage of, 82
Methlick (Methlik, Methlyk), NJ8537; 435, 445; minister of, 435; parsonage of, 435; vicarage of, 435
Methven (Meffen, Methvene), NO0226; 283, 287, 296, 297, 298, 299, 379; mains of, 'My Lordis Manis', 296, 297, 298
— collegiate kirk, 283n., 299n.; chaplainries of, 299, 299n.; prebendaries of, 299; provost of, 283, 283n., 287, 296, 298, 298n., 299, 379; provostry of, 296-297
Mey (May), ND2872; 628
Mid Colingtoun, *see* Mid Killylung
Mid Crosswood (Corswod Midle), NT0556; 98
Mid Killylung (Mid Colingtoun, Myd Cuilelung, Mydekilliboung, Mydkillelong) [Killylung, NX9581], 270, 275, 623

Mid Port, *Stirling*, 546
Midchingill, Mydchingill, *fishing on the Dee, Aberdeen bishopric*, 416, 417
Middilgrang, *see* Middle Grange
Middle Baldridge (Middylbawdrik, Midle Bawdrick, Midle Bawdrik, Midle Bawidrik, Mydlbawidrik) [Easter, Wester Baldridge, NT0888], 23, 27, 28, 45, 49, 50;
Middle Bawdrick, Bawdrik, Bawidrik, *see* Middle Baldridge
Middle Callander (Myddell Callender) [Callander, NN6307], 349
Middle Dalguise (Middill Dulgus, Middle Dulgus), NN9947; 305, 342, 345
Middle Disblair [Disblair, NJ8619], 32, 36
Middle Drimmie (Middill Drymmy), NO1750; 354
Middle Dulgus, *see* Middle Dalguise
Middle Grange (Middilgrang), NT9889; 290, 293
Middle Growane, *Monzie*, 349
Middle Inchewan (Middill Inschevin), NO0341; 305
Middle Inschgotrig [Inchgotrick, NS4133], 531
Middle Newtoun, Midle Newtoun, *Forgan*, 14, 20
Middlefoodie (Midlfudie, Midlfyd, Mydlefudy, Mydil Fudie), NO4017; 2, 5, 14
Middlethird (Midethrid, Midlethrid, Mydlethrid), NT6743; 228, 234, 238, 241
Middleton (Midltoun), *Crichton*, NT3658; prebend of Crichton collegiate kirk, 108
Middleton (Myddiltoun), *Idvie*, NO5848; 376
Middylbawdrik, *see* Middle Baldridge
Midethrid, *see* Middlethird
Midgehope (Migiehoipe), NT2713; 258
Midilgany, *Fearn abbey*, 633
Midland, Myln land alias Multur croft, *Methven collegiate kirk*, 296, 298
Midlane, *see* Midlem
Midleburne, *Eskdalemuir, Melrose abbey*, 259
Midleholm, Mydleholm, *with Whiteside, Kelso abbey*, 231, 233
Midlem (Midlane, Medlanetoun,

INDEX OF PLACES 847

Midleme, Mydlame, Mydlane, Mydlem, Mydlen), NT5227; 223, 223n., 225, 231, 233, 237, 239, 242; mill of, 223, 225, 231, 233, 237, 239, 242
Midlemest Wallis, with *Easter and Wester Softlaw, Maxwell*, 225
Midlethrid, *see* Middlethird
Midlfudie, Midlfyd, *see* Middlefoodie
Midltoun, *see* Middleton
Midmar (Mydmar), NJ6807; 439; kirk of, 439
Midrois [*North, South Medrox, NS7271, NS7370], 102; *see also* Meddrawis
Midwig [*Castlewigg, NX4343], 592, 594, 596
Migiehoipe, *see* Midgehope
Migvie (Migvy, Mygvie), NJ4306; 8, 16, 17; kirk of, 8, 16, 17
Mikartschyre, Myckartschyre [*Muckhart Mill, NS9998], 2, 5
Mikillwig, *see* Meikillwig
Milcrist, *Monkland*, 498
Mildames, Mildamis, *see* Mildeans
Mildeans (Mildames, Mildamis), NT2401; 29, 51
Milhillis, Mylhillis, *Dunfermline*, 27, 49, 50
Mill Burn (Mylnburne), NO4773; 349
Mill of Furvy, Myln of Furvie, *Aberdeen Trinitarians*, 418, 421
Mill (Myln) of Glenbervie, NO7680; 405
Mill (Mylne) of Leslie, NJ5925; 31, 36
Millden (Mylden, Myldene) [Lodge, NO5478], 383, 385, 388
Millegan (Millegin), NJ5151; 461
Millhorn (Milnhorn, Milnhorne), NO2243; 354, 355
Milntoun, *Neutoun Freuchy, Coupar Angus abbey*, 355
Milrig, NS5034; 531
Milton (Meltoun, Myltoun), *Edderton*, NH7082; 633, 652
Milton (Miltoun, Myltoun), *Tullybeagles, Methven collegiate kirk*, NO0434; 297, 298
Milton (Mylntoun, Myltoun), *with Woodhouselee, Restalrig collegiate kirk* [*New Milton, NT2562], 142, 143
Milton (Myltoun), *Leuchars*, NO4420; 13, 20
Milton Fintray (Myll of Fyntray, Mylntoun of Fintray), NJ8316; 31, 36
Miltoun of Witstones, Myltoun of Wistounes, *Mearns*, 31, 35
Minabrocht, *see* Monyabroch
Minarbo Hill, Arbo Hill, *Crail*, 177, 178, 182
Mingaff, *see* Minnigaff
Miniabracht, *see* Monyabroch
Minigaff, *see* Minnigaff
Minnigaff (Manygaff, Mingaff, Minigaff, Monygaff), NX4166; 587, 590, 600, 607; kirk of, 590; parsonage of, 587, 607; vicar of, 600; vicarage of, 587, 600
Minto, 20n.
Mires, *see* Merse
Mirlabank, Morlabank, Morlay Bank, *Dunfermline*, 27, 28, 49, 50
Mochrum (Mowchtroum), NX3050; 593, 595, 597; 617, 619; glebe of, 597; kirk of, 593; vicarage of, 597
Mochrumtill, *Crossraguel abbey* [Mochrum, NS2609], 567
Moffat (Moffet), NT0805; 266; parson of, 266-267; parsonage of, 266
Moidart (Mwidort), NM7472; 675; parsonage of, 675
Moirtullich, *with Invervack, Coupar Angus abbey*, 354, 354n.
Moiye, *see* Moy
Molinis, *see* Mollins
Mollane, Molloun, *see* Moulin
Mollins (Molinis, Molnys), NS7171; 497, 498
Molnys, *see* Mollins
Moncreiffe (Moncreif), NO1121; 111
Moncur, NO2829; 19
Mondowrnocht, *see* Mundurno
Moneky, Moneyky, *see* Monikie
Monemeill, *see* Monimail
Moneydie (Monydie), NO0629; 87, 336, 336n., 350; parsonage of, 87, 336; prebend of Dunkeld cathedral, 336n.; vicarage of, 87, 336
Mongale Milln, Mongall Myln, *Manuel priory*, 549, 550
Mongarye, *see* Montgarrie
Moniabrok, *see* Monyabroch
Monifieth (Monifuith, Monifuth, Monyfuith, Monyfuth), NO4932;

352, 359, 360, 362, 363, 364, 393; kirk of, 352, 364; parsonage of, 363, 393; vicar pensionary of, 393; vicarage of, 360, 363, 393
Monikie (Moneky, Moneyky), NO4938; 351, 361, 395; kirk of, 352, 361; manse of, 395; vicar pensionary of, 395; vicarage pensionary of, 395
Monimail (Monemeill, Monymaill, Monymeill, Monymele), NO2914; 2, 3, 5, 74; kirk of 3; vicarage perpetual of, 74
Monk Perk, see Monkpark
Monkegy [Keithhall, NJ7821], 32, 36; kirk of, 32
Monkfauld, Monksfald, *Melrose abbey*, 208, 257
Monkis Medow, *Melrose abbey*, 209
Monkland (Munkland) [Old Monkland, NS7263], 496, 497, 497n., 518; parish, 102, 102n., 103, 497, 498n.; parish of, 102n.; vicar of, 496, 518; vicarage of, 518 ; see also Cadder
Monkland, *Carrick, Melrose abbey*, 210
Monkland, *Nithsdale*, see Dunscore
Monkpark, Monk Perk, *Melrose abbey*, 208, 209
Monksfald, see Monkfauld
Monkshill (Monkishill), NJ9125; 457, 458
Monkton (Munktoun), NS3527; 527, 529, 530, 562; kirk of, 529; vicar of, 562; vicarage of, 562
Monktonhall (Monktounhall, Mounktounhall), NT3471; 24, 25, 26, 38, 41, 42, 43, 47, 48, 53
Monsarne, see Munsary
Monteith, see Menteith
Montfleurie (Monthflowrie, Mounthflowry), NO3701; 15, 20
Montgarrie (Mongarye), NJ5717; 436
Montgrew (Mengreowis), NJ4551; 461
Month pairt, Monthis pairt, *of Strathkinness*, see Strathkinness
Month, Montht, see Mounth
Monthflowrie, see Montfleurie
Montlothiane, see Mount Lothian
Montquey (Motcay), NT2087; 63
Montquhanny, see Mountquhanie
Montrave (Monthris, Monthryis, Monthrys, Montreis, Mountreis, Munthreis), NO3706; 145, 148, 167, 168

Montrose (Montrois, Montros), NO7157; 351, 351n., 362, 364, 383, 384, 385, 386, 387, 388, 389; parish of, 388, 389
Monturpie (Mantirope, Manturbe, Manturpe), NO4303; 24, 26, 28, 29, 46, 47, 48, 50, 51
Monyabrik, see Monyabroch
Monyabroch (Meneabracht, Minabrocht, Miniabracht, Moniabrok, Monyabrik) [Kilsyth, NS7178], 506, 506n., 552; parsonage of, 552; vicar pensionary of, 506, 506nn.; vicarage of, 552; vicarage pensionary of, 506, 506nn.; *see also* Kilsyth
Monycabok, see Monykebbuck
Monydie, see Moneydie
Monyell, *Dunblane*, 343
Monyfuith, Monyfuth, see Monifieth
Monygaff, see Minnigaff
Monyhous, see Moy Hous
Monykebbuck (Monycabok), NJ8718; 422, 423
Monymaill, Monymeill, Monymele, see Monimail
Monymusk, NJ6815; 2, 118n., 418, 446, 446n.; minister at, 446; parson of, 19n.; parsonage of, 418, 446-447; vicar pensionary of, 446; vicarage of, 418
—— priory, 446-447
Monyvarde, Monyward, see Monzievaird
Monzie (Monze), NN8825; 295, 343, 348, 349; kirk of, 295, 348; parish of, 349
Monzievaird (Monyvarde, Monyward) [Brae of, NN8425], 300, 343; vicar of, 300; vicarage of, 300
Moonzie (Auchtermonsie, Auchtirmonsy, Auchtirmousall), NO3317; 56, 83; vicar pensionary of, 83; vicarage pensionary of, 83
Moonzie Mill (Munsymyll, Munsymylne), NO2419; 14, 20
Moorfoot (Morphet, Morphetland), NT2952; 101
Moracht (Merwicht, Murhycht), NX4635; 594
Moram, see Morham
Moray (Muray, Murray, Mwrray), 1nn., 114, 281n., 282, 282n., 283n., 302n., 306n., 318, 320n., 426, 449, 449n.,

456, 456n., 457, 461n., 465, 466, 467n., 468, 470, 471, 472, 473, 474, 475, 476, 477, 477n., 478, 479, 480, 481, 482, 483, 484, 485, 486, 487, 488, 489, 490, 492, 493, 494; archdeacon of, 487, 488, 494, 642, 683; archdeaconry of, 487-488, 494; bailie of bishopric of, 454n., 466; bishop of, 331n., 467, 467n., 474, 477, 477n.; bishopric of, 453n., 454n., 465-468, 490, 642; chamberlain of bishopric of, 469, 642; chancellor of, 486, 486n.; chancellory of, 486; chanter of, 479; chantory of, 479-480; chaplain of St Leonard's chaplainry in diocese of, 484; chaplainries of, 481, 482; chaplains of, 481, 482; dean of, 468; deanery of, 468-469; sheriff of, 469, 471, 472, 473; St Leonard's chaplainry in diocese of, 484; subchanter of, 470; subchantory of, 461n., 470; subdean of, 493
—— , cathedral of, see Elgin; cathedral dignitaries of, see Moray
Mordington (Mordingtoun) [Ho, NT9556], 185; kirk of, 185
Morebattle (Merbotill, Merbotle, Merbottill), NT7724; 214, 214n., 215; curate of, 214; parson of, 215; parsonage of, 214, 214n., 215; vicar pensionary of, 214; vicarage of, 214, 214n., 215; see also Mow
Moreslaw, see Muirhouselaw
Morestoun, *Melrose abbey*, see Morriston
Morestoun, Mornstoun, *Holywood abbey*, 274, 622, 623
Morestoun, *St Andrews priory*, 10, 18
Morham (Moram) [Mains, NT5571], minister of, 171; parsonage of, 170-171; reader of, 171
Morlabank, Morlay Bank, see Mirlabank
Mornichie [*Morange, NH7683], 651
Mornstoun, see Morestoun
Morphet, see Moorfoot
Morphetland, see Moorfoot
Morphy Fraser, *Ecclesgreig* [Morphie, NO7164], 16
Morriston (Morestoun) [Cott, NT5942], 207, 257
Morthlak, Morthlik, see Mortlach

Mortlach (Morthlak, Morthlik, Murthlaik, Murthlak, Murtlaik, Murtlake), NJ5045; 415, 444; parson of, 444; parsonage of, 444; vicarage of, 444
Morton (Martoun, Mortoun), *Crail*, NO5812; 177, 178, 181
Morton (Mortoun, Mortoune), *parish*, NS8898; 229, 232, 234, 237, 240, 241, 243, 244; kirk of, 229, 232, 234, 237, 240, 241, 243
Mortoun, *Lincluden collegiate kirk*, see Parton
Mortoun, Muirtoun, Murtoun, *Cambuskenneth abbey*, 537, 545
Mortoun, prebend of Corstorphine collegiate kirk, see Norton
Mortounhall, *with West Hope and East Hope, Haddington priory*, 176, 178
Moshoussis, see Mosshouses
Mosilie, see Mossilee
Mosmyning, *Lesmahagow*, 232
Mosshouses (Moshoussis), NT5339; 208, 209, 254, 258
Mosside (Mosseyde, Mossyd), *Holywood abbey*, 270, 271, 274, 276, 623
Mossilee (Mosilie), NT4735; 224
Motcay, see Montquey
Moulin (Mollane, Molloun), NN9459; 26, 38, 48; kirk of, 26, 38, 48
Mounkland, Mureland, Murland, Nunland, *Haddington priory*, 162, 165, 177, 178, 181
Mounktoun, see Monktonhall
Mounktounhall, see Monktonhall
Mount Lothian (Montlothiane), NT2756; 91
—— and the Hopes, kirk of, 91; see also Hopes
Mounth (Month, Montht), 18, 31, 31n., 32, 404, 404n., 408
Mounthflowry, see Montfleurie
Mounthis pairt *of Strathkinness*, see Strathkinness
Mountquhanie (Ballqhany, Balquhany, Montquhanny) [Ho, NO3421], 15, 15n., 60
Mountreis, see Montrave
Mouswald, NY0672; 265; parson of, 265; parsonage of, 265-266
Mow Haucht, see Mowhaugh
Mow [Morebattle, NT7724], 225, 225n.,

232, 232n., 234, 235, 238, 240, 241; kirk of, 232, 234, 238, 240; parsonage of, 225; vicarage of, 225, 232, 235, 238, 241; *see also* Morebattle
Mowchtroum, *see* Mochrum
Mowhaugh (Mow Haucht), NT8120; 225
Mowmanis, *Eckford*, 220
Mowtoun Mains [*Mowhaugh, NT8120], 225
Moxpoffill, *see* Maxpoffle
Moy (Moiye, Moye) [*Moy, NJ0059; *Moy and Dalarossie parish, NH7630], 465, 470, 471, 639; kirk of, 639; minister of, 471; parson of, 470, 471; parsonage of, 470; vicarage of, 471
—— and Croy, kirks of, 639; parson of, 639; *see also* Croy
Moy Hous, Moyhous, Monyhous, *Kinghorn*, 24, 28, 46, 50
Moyness (Mynes) [Mains of, NH9554], 469
Mucarsy, *see* Muckersie
Muckairn (Muckarn), NN0038; 674
Muckersie (McCarsie, Mucarsy, Muckersye, Mukersie) [Wester Muckersie, NO0615], 303, 305, 326, 340; parson of, 340; parsonage of, 326, 340; vicar pensionary of, 326; vicarage of, 326, 340
Muckhart (Mukart, Mukert, Mukertie) [Mill, NS9998], 338; parson of, 338; parsonage of, 338; vicarage of, 338
Mudy, *Peebles*, 249
Muirake (Muraik), NJ5657; 438
Muiralehouse, *see* Muirton
Muircambus (Muircambes, Murecambuies), NO4602; 146, 167
Muircammes Myln [Muircambus, NO4602], 291
Muiredge (Mureege), NO2322; 285, 288
Muirhall, NS9952; 520; chaplain of, 520; chaplainry of, 520
Muirhous, Muirhoussis, *parish*, *see* Murroes
Muirhouselaw (Moreslaw, Muirhouslaw, Murehouslaw), NT6228; 189, 207, 257
Muirsyd, *Holywood abbey*, 271
Muirsyd, *Methven collegiate kirk*, *see* Myreside

Muirton (Muirtoun, Murtoun), *Brechin bishopric*, NO6462; 384, 385, 389
Muirton (Muirtoun), *Cambuskenneth abbey*, 547
Muirton (Murtownes), *with Tullyfergus, Coupar Angus abbey*, NO2247; 369
Muirton, *Leuchars*, 14, 20
Muirton, Muntros, *Kelton*, 602
Muirton, Murtoun, *Kinglassie*, 29, 51
Muirton, Murtoun, *Kirkcaldy*, 28, 50
Muirtoun, *Restenneth*, *see* Murton
Muirton, *St Ninians*, NS8189; chapel of, 553n.; vicarage of, 553, 553n.
Muiryfold (Murifald), NJ4852; 461
Mukart, *see* Muckhart
Mukersie, *see* Muckersie
Mukert, Mukertie, *see* Muckersie
Mukray Easter [*Muckcroft, NS6871], 497
Mukray Wester [*Muckcroft, NS6871], 497
Mulderg (Muldarg), NH8378; 633
Mull (Mulle), NM5435, NM6531; 673, 674
Mullen, *Dundurcas*, 456
Mullerstanes, *see* Mellerstain
Mulroun, *Calder*, 98
Multur croft, 'Myln land alias Multur croft', *see* Midland
Mundurno (Mondowrnocht), NJ9412; 423
Munkland, *see* Monkland
Munktoun, *see* Monkton
Munnraw, *see* Nunraw
Munsary (Monsarne), ND2145; 632
Munsymyll, Munsymylne, *see* Moonzie Mill
Munteithit, *see* Menteith
Munthreis, *see* Montrave
Muntoun, *see* Nunton,
Muntros, *Kelton*, *see* Muirtoun
Muraik, *see* Muirake
Murcowr [Newton of Murcar, NJ9412], 423
Mure, *see* Muir
Murecambuies, *see* Muircambus
Mureege, *see* Muiredge
Murehall, *Monimail*, 2, 5
Murehous, *parish*, *see* Murroes
Murehouslaw, *see* Muirhouselaw
Mureland, *see* Mounkland
Murhoussis, *see* Murroes
Murhycht, *see* Moracht

Murie (Murei) [Mains of, NN2322], 285, 288
Murifald, see Muiryfold
Murischot, *Greenlaw*, 227
Murisdykis, *Calder*, 98
Muristonis [*Muirieston Ho, NT0664], 98
Murland, see Mounkland
Murntoun, *Inchcolm abbey*, 62
Murnyll, *Kinnell* [*Muirmills, NO6154], 367
Murray, see Moray
Murrestoun, *Moray bishopric*, 466
Murroes (Muirhous, Muirhoussis, Murehous, Murhoussis), NO4437; 354, 357, 361, 398; kirk of, 361; vicar of, 398; vicarage of, 398
Murryell, '*Chapelden of Kindrimy*', 449
Murtes Hill, *Glenbervy*, 405
Murthill, *Aberdeen bishopric*, 414, 417
Murthlaik, Murthlak, Murtlaik, Murtlake, see Mortlach
Murthlak, see Murthlie
Murthly (Murthlak, Murthlie), NO0938; 341, 354, 369
Murton (Muirtoun), *Restenneth priory*, NO4951; 217
Murtoun, *Brechin bishopric*, see Muirtoun
Murtoun, see Muirtoun
Murtowneis, *with Tullyfergus*, see Muirton
Musselburgh (Mussilbrucht, Mussilburche, Mussilburcht, Mussilburgh, Mussilburghe), NT3573; 24, 25, 25n., 26, 29, 37, 38, 39, 40, 40n., 41, 41n., 47, 47n., 48, 49, 51, 52, 53, 102, 103, 104n., 114, 120; Magdalene hospital beside, 39, 39n.; New Mylne of, 41, 42, 53; parish of, 25, 26, 29, 38, 39, 47, 48, 51, 52; Sands of, 26, 43n., 49; vicarage of, 120; *see also* Musselburghshire
Musselburghshire (Mussilburchtschyre, Mussilburgheschyre, Mussilburghschyr, Mussilburghschyre), 26, 40, 41, 42, 43, 44, 47, 48, 49, 53, 54; *see also* Musselburgh
Mutehill (Muthill), *St Mary's Isle priory*, NX6848; 620
Muthill (Muthil), *Dunblane bishopric*, NN8617; 294, 295, 296, 326, 340, 348, 349; kirk of, 296, 348; parish of, 349; vicarage of, 295, 326, 340, 349
Muthill, *St Mary's Isle priory*, see Mutehill
Mwicheleische, *Islay*, 674
Mwidort, see Moidart
Mwrray, see Moray
Myckartschyre, see Mikartschyre
Myd Cuilelung, see Mid Killylung
Mydblak Craigis, see Blackcraig
Mydchingill, see Midchingill
Myddilsteid, *Selkirk*, 230n.
Myddiltoun, see Middleton
Mydekilliboung, see Mid Killylung
Mydern, *Dairsie, St Andrews priory*, 14
Mydil Fudie, see Middlefoodie
Mydkillelong, see Mid Killylung,
Mydlame, Mydlane, see Midlem
Mydlawheid [*Midlaw Burn, NT1615], 259
Mydlbawidrik, see Middle Baldridge
Mydlefudy, see Middlefoodie
Mydleholm, see Midleholm
Mydlem, Mydlen, see Midlem
Mydlethrid, see Middlethird
Mydmar, see Midmar
Mygvie, see Migvie
Myladewy, Myll Audevye, Mylne of Auchdowy [*Ardovie, NO5956], 384, 385, 388
Mylburn, *Newtown, Bothwell collegiate kirk*, 510
Mylcroft, *Fearn abbey*, 633
Mylden, Myldene, see Millden
Mylerlandis, *Kirkbean*, 609
Mylhill, *Collessie*, 33, 37
Mylhillis, *Dunfermline*, see Milhillis
Mylis Tarvet [Tarvit Mill, NO3612], 55
Mylkinwood, *Cambuslang*, 509
Myll of Kincragie [Milton of Kincraigie, NN9948], 304
Myll Audevye, see Myladewy
Myll of Fyntray, see Milton Fintray
Myln land alias Multur croft, see Midland
Myln of Boupie, *near Fechil, Deer abbey*, 458
Myln of Furvie, see Mill of Furvy
Myln of Morocht, *Ross bishopric*, 625
Myln of Petmowy [Pitmuies Mill Fm, NO5849], 376

Myln, *Glenbervie, see* Mill of Glenbervie
Mylnburne, *see* Mill Burn
Mylne of Auchdowy, *see* Myladewy
Mylne of Crawmond, *Inchcolm abbey*, 62
Mylne of Leslie, *see* Mill of Leslie
Mylne of Lyntoun, *North Berwick priory*, 147, 168, 168n.
Mylne of Newgrange, *Melrose abbey*, *see* Newgrange Mylne
Mylne of Newtoun, *see* Newtoun Mylne
Mylne of the Abay, of the Abbay, *see* Abbey Mill
Mylne of the Keir, *see* Keir Mill
Mylneburne, *Ross treasurership*, 637
Mylnerig, *Belses, Jedburgh abbey*, 220
Mylntoun *of Craigie, see* Myltoun
Mylntoun of Fintray, *see* Milton Fintray
Mylntoun, *Lesmahagow*, 231
Mylntoun, Myltoun, *Kirkcaldy*, 24, 28, 46, 50
Mylntoun, Myltoun, *with Woodhouselee, Restalrig collegiate kirk, see* Milton
Myltoun (Mylntoun) *of Craigie* [Craigie, NO4231], 30, 34, 35
Myltoun of Wistounes, *see* Miltoun of Witstones
Myltoun, *Fearn, Edderton, see* Milton
Myltoun, *Kinnell*, 366
Myltoun, *Leuchars, see* Milton
Myltoun, *Markinch*, 15
Myltoun, *Tullybeagles, Methven collegiate kirk*, see Milton
Myltree of Clus, *Lundie*, 376
Mynes, *see* Moyness
Myre, *Belses, Jedburgh abbey*, 220
Myrehead (Myre Heidis), NS9677; 549
Myrelandhorn (Myrelandnorne), ND2758; 628n.
Myreside (Muirsyde, Myresyd), NO0425; 296, 298
Myris Over and Nether [Nethermyers, NO2410], 33
Myvat, *Monkland barony* [*North Myvot, NS7372], 102
Mywetis, *see* North Myot

Naikit Feild, Nakit Feild, *Crail*, 177, 178, 181
Nairn (Narne), NH8856; 462, 469, 485

——, *sheriffdom*, 485n.
Naitoun, Natoun, Nattoun, *Musselburgh*, 24, 25, 26, 38, 41, 43, 47, 48, 53
Name of Jesus altar, *Glasgow cathedral*, 522
Nanewair, Nanewarie, *see* Ninewar
Nantharne, Nantherne, *see* Nenthorn
Naperis Cros, *Holywood abbey*, 623
Narne, *see* Nairn
Narrine, *Rothes*, 491
Natherholden, *see* Netherhowden
Nathermure, *Leuchars*, 14
Nathertoun of Bathedloskis (Bothedillach, Bothedlach), *Inchcolm abbey*, 64, 64n.; *see also* Easter Bothylokis, Nethertoun
Natoun, *see* Naitoun
Nattoun, *see* Naitoun
Nauchtane [Naughton Ho, NO3724], 14, 58
Nava, Navay, *see* Nevay
Nedryis Wynd, *see* Niddry's Wynd
Nefang, *see* Nesting
Neilston (Neilstoun), NS4757; 529; kirk of, 529
Neisbit, *see* Nisbet
Nenthorn (Nanetharne, Nantharne, Nantherne, Nenthorne, Newtharne), NT6737; 223, 224, 232, 233, 234, 235, 237, 238, 239, 240, 241, 242; kirk of, 224, 232, 234, 237, 240, 242; vicarage of, 224, 232, 235, 238, 241
Nentoun, *Holywood abbey*, 623
Nes, *see* River Ness
Nesbites, *see* Nisbet
Nesting (Nefang, Nestinge), *parish*, HU4460; 658, 658n., 660, 670; kirk of, 660; minister of, 660
Nether Ancrum (Nather Ancrum) [Ancrum, NT6224], mains of, 236, 366
Nether Auchendinny, Auchindony, Auchindonye, Auchindunnye, Auchindunye [Auchendinny, NT2562], 142, 143
Nether Auchinleische [Auchinleish, NO1960], 354, 355
Nether Aytoun [Ayton, NT9261], 199
Nether Baith, Bayth, *Dunfermline*, 23, 45
Nether Balmerosche, *see* Nether

Bancrosche
Nether Bancrosche, Nether Balmerosche [Barncrosh, NX7059], 586, 589
Nether Barr (Nether Bar), NX4263; 587, 590
Nether Bartcapill, Bartcappill [Barcaple, NX6757], 585, 589, 592
Nether Calquha [Culquha, NX6958], 585, 589
Nether Campsy [Campsie, NO1223], 352
Nether Cantlie [Meikle Cantly, NJ4750], 461, 464
Nether Chatto [Chatto, NT7717], 220
Nether Coittes, *Monkland*, 102
Nether Craigo [Craigo, NO6764], 374
Nether Craling [Crailing, NT6824], 219
Nether Crammond [Cramond, NT1876], 150
Nether Dalmeyn, *with Garscadden, Paisley abbey*, 529
Nether Elrik [Alrick, NO1961], 354
Nether Ersik, *see* Low Ersock
Nether Flemyngtoun [Flemington, NT9460], 199
Nether Glasnycht, *see* Low Glasnick
Nether Glene, *see* Netherglen
Nether Glengouer, *Crawfordmuir, Newbattle abbey*, 102
Nether Grange, *Kinghorn* [*Grange, NT2688], 26, 40, 44, 48
Nether Halkarston [Halkerston, NT3458], 111
Nether Holden, *see* Netherhowden
Nether Ingzever [Inzievar, NT0288], 290
Nether Killelung [Killylung, NX9581], 276
Nether Kincraig (Kincragy, Kyncrage, Kyncraig) [Kincraig, NO6258], 383, 384, 387, 388
Nether Kingorne [Kinghorn, NT2686], 43
Nether Kinneddar (Kinnedder, Kinneddir), NT0291; 24, 46
Nether Kylmanedy, *Kinloss abbey* [*Kinminintie, NJ4253], 461
Nether Kynmonth [Upper Kinmonth, NO7782], 405
Nether Lassodie (Lasoiday, Lasoiddy) [Lassodie, NT1292], 23, 45
Nether Magask (Magask Inferiour, Nather Magus), NO4415; 12, 20, 21

Nether Magus, *see* Nether Magask
Nether Mains, *Chirnside*, 166
Nether Markinche [Markinch, NO2901], 15
Nether Monbennis, Hauch of [Upper Manbean, NJ1957], 453
Nether Murtoun, *with Tullyfergus, Coupar Angus abbey* [Muirton, NO2247], 354, 357, 369
Nether Myln *of Auchindanery, Kinloss abbey* [Nethermills, NJ5050], 461
Nether Nisbit (Nether Nisbitis) [Nisbet, NT6725], 219; *see also* Upper Nisbet
Nether Petforthie, Nether Pitforthie, Nether Pitforthye [Mains of Pitforthie, NO6161], 383, 385, 388; *see also* Petforky
Nether Pratis (Pratas, Prateris, Prateris), NO3905; 145, 148, 167, 168
Nether Quhitsoun, *Ecclesgreig*, 16
Nether Quhytsyd, *see* Nether Whiteside
Nether Rankeilour [Rankeilour, NO3211], 71n., 93n., 154n., 175n., 196n., 229n., 248n., 264n., 286n., 372n., 411n., 420n., 476n., 513n., 534n., 552n., 575n., 600n., 611n., 649n.
Nether Reres [Rires, NO4604], 146, 167
Nether Sannik [Senwick Church, NX6546], 587
Nether Scheilfeild [Sheilfield Wood, NT6038], 190, 191
Nether Smychtstoun, *Lundie*, 376
Nether Staddaill, *with Cogle, Caithness archdeaconry*, 632
Nether Stroquhan (Straquhan, Strauchane) [Stroquhan, NX8483], 269, 623
Nether Urquhart (Urquhard), NO1808; 69
Nether Whiteside (Quhytsyd) [Moor, NX8183], 270, 275, 622, 623
Nether Wistownis, Nedder Witstones [*Nether Woodston, NO7565], 31, 35
Netherbyres (Nether Byre), NT9463; 199
Netherfield (Netherfeild), prebend of Bothwell collegiate kirk, 517
Netherglen (Nether Glene), NJ2454; 491
Netherhowden (Natherholden, Nethirholden, Nethir Howden),

NT5053; 231, 233, 237, 240, 242
Netherland Norne [*Myrelandhorn, ND2758], 628
Netherlands (Netherland), NS4733; 531
Netherton (Nethertown), *Carmichael*, NS9241; 511
Nethertoun, *Inchcolm abbey*, 62; *see also* Nathertoun of Bathedloskis
Netherurd, *Melrose abbey*, NT1144; 211n.
Netherurd, *Cathkin*, 511
Nethir Sett of Kilmanitie, *Kinloss abbey* [*Kinminintie, NJ4253], 461
Netoun, *see* Mertoun
Neubiggyne, *Coupar Angus abbey*, 357
Neucalsay, *see* Newcalsy
Neutoun Freuchy, *see* Newton
Nevay (Nava, Navay) [Kirkton of, NO3243], 367; parson of, 368; parsonage of, 367; *see also* Kirkton of Nevay
New Abbey (New Abay), NX9666; 612, 622; abbot of, 612; mill of the barony of, 612; *see also* Lochindilloch, Sweetheart
New Aberdeen (New Abirdene, Newtoun of Abirdene) [Aberdeen, NJ9206], 413, 446; minister of, 446, 447; *see also* Aberdeen
New Burcht, *see* Newburgh
New College [of St Mary and St Anne], *Glasgow*, *see* Glasgow, New kirk
New College, *St Andrews*, *see* St Andrews, St Mary's college
New Comonis, *Riccarton*, 531
New Fuird Haucht, Norfurdhaucht, *Melrose abbey*, 257, 259
New Grange (Grange, Newgrange), *St Andrews priory* [*Grange Ho, NO5115], 10, 13, 18, 20, 21
New Hevin, *see* Newhaven
New kirk [collegiate kirk of St Mary and St Anne], *see* Glasgow, New kirk
New Landis, *see* Newlands
New Myll, *Dunkeld bishopric*, 305
New Myln crofts, *Aberdeen deanery*, 423
New Myln, *Fordyce*, 437
New Myln, Newmyln, *Kirkchrist*, 587, 589
New Mylne, *Dairsie*, *see* Newmill
New Mylne, *Musselburgh*, 41, 42, 52, 53

New Mylne, Newmylne, *St Andrews*, 10
New Orcheard, *Dryburgh abbey*, 189
Newbattle (Newbotill, Newbotle, Newbottell, Newbottill, Newbottle), NT3365; 25, 25n., 40n., 42, 47, 53, 68, 92, 100, 101, 102, 103, 103n., 116, 116n.; benefice of, 116n.; chaplainry of St Catherine, 68, 144; parsonage of, 102; vicarage of, 102
—— abbey, 92, 100-103, 282, 496n., 497n.; commendator of, 103, 103n.
Newbigging (Newbiging), *Kinnell*, NO6450; 366
Newbigging, *Oxnam*, NT7015; 219
Newbigging, *Restalrig collegiate kirk*, NT2760; 142, 143
Newbiging, Newbiggin, Newbiggyne, *Brechin bishopric*, 383, 384, 387, 388
Newbirne, Newbirnetoun, Newbirntoun, *see* Newburn
Newbirneschyre, Newbirnschyre, *see* Newburnshire
Newbotill, Newbotle, Newbottell, Newbottle, *see* Newbattle
Newburgh (New Burcht, Newburcht), NO2318; 30, 33, 34, 35, 37, 90, 378; chaplainry of St Catherine in, 378; manse of chaplainry in, 378; St Catherine's kirk in, 378; *see also* Abdie, Auld Lindores, Lindores
Newburn (Newbirne, Newbirnetoun, Newbirntoun, Newburne, Newburnetoun, Newbyrn, Newbyrne), NO4304, 24, 25, 26, 28, 29, 37, 38, 39, 41, 42, 43, 46, 47, 48, 50, 51, 51n., 52, 53, 54, 69; kirk of, 53; mill of, 24, 46; parish of, 25, 26, 29, 38, 39, 41, 47, 48, 51, 52; parsonage and vicarage of, 53; reader at 69; vicar of, 69; vicarage of, 69; *see also* Newburnshire
Newburnshire (Newbirneschyre, Newbirnschyre, Newburneschyre, Newbyrneschyre), 24, 25, 28, 42, 43, 44, 46, 47, 50, 53, 54; *see also* Newburn
Newbyggyne, *Glenbervie*, 405
Newbyre, *Eskdalemuir, Melrose abbey*, 259
Newbyre, *Newbattle abbey* [*Newbyres Castle, NT3461], 101
Newbyrn, Newbyrne, *see* Newburn

INDEX OF PLACES 855

Newbyrneschyre, see Newburnshire
Newcalsy, Neucalsy, *Coupar Angus abbey*, 353
Newdosk, *parish*, 382, 382n.; parson of, 382; parsonage of, 382
Newgrange Mylne, Mylne of Newgrange, *Lammermuir, Melrose abbey*, 207, 257
Newgrange, *Melrose abbey*, 208
Newgrange, *St Andrews priory, see* New Grange
Newhall, *Crail*, 177, 178, 181
Newhall, *Kelso abbey*, 223
Newhaven (New Hevin, Newhevin), NT2577; 108, 128; chapel of Our Lady and St James, 108, 108n.; chaplainry of, 108n.
Newland *of Easter Cranokis* [*Crannach, NJ4954], 461
Newland *of Fortry*, 461
Newlandis *of Millegan*, 'callit Jonettis Scheill', 462
Newlandis, *with Humbie*, 104
Newlands (New Landis, Newlandis), *Dunfermline*, 26, 28, 48, 50
Newlands (Newland, Newlandis), *Arbroath abbey*, *NO4447; 351, 358
Newlands (Newlandis), *Haddington priory*, NT5666; 162, 165, 176, 178, 181
Newlands (Newlandis), *parish*, NT1749; 251; parsonage of, 251
Newliston (Newlistoun), NT1073; 3
Newmill (Newmyll, New Mylne, Newmylne), *of Dairsie*, NO3915; 10, 14, 18
Newmontros, *Ecclesjohn* [Montrose, NO7175], 372
Newmore (Newmoir), NH6872; 647; chaplainry in Tain collegiate kirk, 647
Newmylne, *St Bathans priory*, 192
Newraw, *Methven collegiate kirk*, NN9521; 296, 297, 298
Newrig, *see* Greenrig
Newstead (Newsteid), NT5634; 209, 257
Newtharne, *see* Nenthorn
Newton (Little Newtoun, Newtone, Newtoun), *Crail*, 162, 165, 176, 177, 178, 179, 181, 182
Newton (Neutoun Freuchy), NO2360; 355

Newton (Newtoun), *Collessie*, NO2912; 33, 37
Newton (Newtoun), *Inchcolm priory*, NT1792; 62, 63
Newton (Newtoun), *with Smiddyhill, Brechin bishopric*, NO5964; 383, 385, 388
Newton (Newtoun), *with Wrangham, Lindores abbey*, NJ6324; 31, 36
Newton (Newtown), *with Lochslin, Ross subdeanery*, NH8481; 651
Newton of Airlie (Newtoun), NO3250; 356
Newton of Falkland (Newtoun Falkland, Newtoun of Falkland), NO2607; 15, 21
Newton (Newtoun), *Kirkchrist*, NX6550; 587, 598, 607
Newtoun East, *Dunbar*, 122
Newtoun Easter, *Forgan*, 14
Newtoun Markinche, Newtoun Markinsche [Markinch, NO2901], 15
Newtoun Mylne, Mylne of Newtoun, *Melrose abbey*, 207, 257
Newtoun of Abirdene, *see* New Aberdeen
Newtoun Wester, *Forgan*, 14
Newtoun, *Holywood abbey*, 270, 271
Newtoun, *Lessudden, Dryburgh abbey*, 189, 191
Newtoun, *Melrose abbey*, 207, 230, 233, 237, 239, 257
Newtoun, *Newbattle abbey* [*Newtongrange, NT3464], 101, 103
Newtoun, *Randerston, Haddington priory*, 162
Newtoun, *Rires, North Berwick priory*, 146, 167
Newtoun, *Aberdeen, see* New Aberdeen
Newtounes Easter and Wester, *Kennoway*, 15
Newtown (Newtoun, New-town), prebend of Bothwell collegiate kirk, *NS6660; 510
Newtown, *Cambuslang*, 508
Newtyle (Newtyld), NO2941; 30, 30n., 34, 35, 359, 396; kirk of, 359; vicar of, 396; vicarage of, 396
Neyth, *see* River Nith
Niddisdaill, Niddsdaill, *see* Nithsdale
Niddry (Nudry) [Castle, NS0974], 91, 93

Nidisdaill, *see* Nithsdale
Nigg (Nig, Nyg), *Ross bishopric*, NH8071; 625, 626, 627; curate of, 626; parish of, 625, 626; preacher at, 627
Nigg (Nyg), *Kincardineshire*, NJ9402; 420, 420n.; benefice of, 420nn.; vicar of, 420; vicarage of, 420
Ninewar (Nanewair, Nanewarie), NT8055; 173, 192
Nisbet (Neisbit, Nesbit), NT6725; 217, 219; kirk of, 217, 219
Nithe, *see* River Nith
Nithsdale (Niddisdaill, Niddsdaill, Nidisdaill, Nithisdaill, Nyddisdaill, Nythdisdaill), NX8990; 113, 210, 257, 272, 274n., 496, 558, 562; commissioner of, 272
Nomine Jesu altar, *St Giles' collegiate kirk, Edinburgh*, *see* St Blaise
Nomine Jesu and Holy Blood chaplainry, *St Giles' collegiate kirk, Edinburgh*, 117-118, 117n.; *see also* St Blaise
Norfurdhaucht, New Fuird Haucht, *Melrose abbey*, 257, 259
Norlethenis, *see* North Lethans
Normaven, *see* Northmavine
North Bank (Northbank), NO4810; 10, 12, 18, 20, 21
North Berwick (North Beruik, North Berwik, Northberuick, Northberuik, Northberwik), NT5585; 77, 147, 167, 168; crofts of, 147, 168; haven of, 147; kirk of, 147, 148, 166, 167; mains of, 147; parish of, 147; vicarage of, 147
— priory, 77, 145-149, 166, 168; chamberlain of, 148; nuns of, 148; prioress of, 77, 77n., 148, 148n., 166, 166n., 168
North Berwick Law (Law), NT5584; 147, 167, 169
North Bowhill, *Selkirk*, 230n.
North Ferrie, North Ferry, Northe Ferrie, Northferry, *see* North Queensferry,
North Iles, 658
North Killimster (North Kilmster), *Caithness bishopric*, ND3255; 628
North Knapdale, *see* Kilmacocharmik
North Lethans (Norlethenis, Northlethemes, Northlethinis, Northlethunes, Northtlethenes, Northtlethenis), NT0595; 23, 27, 28, 45, 49, 50
North Medow, Northmedow, *with Heugh, North Berwick priory*, 147, 167, 169
North Muire (Northmure), *Strathmiglo*, 70
North Myot (Mywetis), *Glasgow subdeanery*, NS7372; 498
North pairt quheit feild, *Dunfermline*, 28, 50
North Queensferry (North Ferrie, North Ferry, Northe Ferrie, Northferry), NT1380; 27, 38, 49, 50
North Ronaldsay, *Orkney*, HY7553; 660
Northbrig, *Aberfoyle*, 344
Northfad, Northfaid, *see* Northfod
Northferry, *Arbroath abbey* [*Broughty Ferry, NO4630], 352, 356, 362
Northferry, *Dunfermline*, *see* North Queensferry
Northfield (Northfeild, Worthfeild), NT9167; 198, 203, 204
Northfod (Northfad, Northfaid, Foirrdis South and North, Fouddeis South and North, Fouddis, Foudeis), NT1288; 23, 27, 45, 49, 50
Northhouse (Northous), *Melrose abbey*, NT4307; 259
Northlethemes, Northlethinis, Northlethunes, Northtlethenes, Northtlethenis, *see* North Lethans
Northmavine (Normaven, Northmawing, Northtmavin, Northmoving), HU3082; 658, 659, 660, 663, 670; kirk of, 660; minister of, 660
Norton (Mortoun, Nortoun) [Mains, NT1472], prebend (with Rathobyres) of Corstorphine collegiate kirk, 107, 107n., 109, 113, 130
Norwallis land, *Stirling*, 556
Noryis land, *see* Pyncartonis
Nountlaw, *see* Nunraw
Nowrig, *see* Greenrig
Nudreis Wynd, *see* Niddry's Wynd
Nudry Foster, *Eddleston*, 496
Nudry, *see* Niddry
Nueke, *see* Nuke
Nuke, Nueke, *with Catmoss, Kelso abbey*, 227, 227n.
Nunbank, *Eccles*, 184

Nunbothis, *Glasgow*, 523
Nunland, *see* Mounkland
Nunraw (Munnraw, Nountlaw), NT5970; 163, 163n., 177, 178; Quhytcastell alias Munnraw, 162
Nunton (Muntoun), *Kirkchrist* [High, NX6449; Lower, NX6548], 587, 589, 616
Nyddisdaill, *see* Nithsdale
Nyddreis Wynd, *see* Niddry's Wynd
Nyg, *see* Nigg
Nygdaill, *Caithness bishopric*, 628
Nyth, *see* River Nith
Nythdisdaill, *see* Nithsdale

Obstule (Obstull), chaplainry, *Fortrose cathedral*, 637, 637n.
Occlositter, *see* Collessetter
Ochiltre, Ochiltrie, *St Andrews priory*, 11, 19
Ochiltree (Ouchiltrie, Uchiltrie), *parish*, NS5021; 210; kirk of, 210
Ochterderay, Ochtirdery, *see* Auchterderran
Ochtermerny, *see* Auchtermairnie
Ochtermonthie, *see* Auchtermuchty
Ochtertyre (Auchtirtyir), NN8423; 300
Ochtirles, *see* Auchterless
Ochtrelure (Auchirlair, Auchtirluir), NX0559; 598, 599
Ogilface, *beside Linlithgow, St Andrews priory*, 11
Ogilface, *with Kers and Falkirktoun, Holyroodhouse abbey*, 91, 93
Ogilvie (Ogilbye, Ogilvy), Brodland of, alias Manis [*Mill of Ogilvie, NN8908], 349; barony of, 349; mains of, 367
Ogston (Ogstoun, Ugstoun), *Moray* [Drainie, NJ2168], 466; kirk of, 466; vicar of, 466; *see also* Drainie
Ogstounis Over and Nether, *Kelso abbey*, 223
Oislaw, *see* Horseley
Olbrek, *see* Auldbreck
Old Aberdeen (Auld Abirdene), NJ9408; 413
Old Cambus (Aldcambuies, Auldcambies, Auldcambuies), NT8069; 166, 199; kirk of, 199
Old Goval (Ald Govill, Auld Govyll), NJ8815; 422, 423

Old Greenlaw (Auld Greinlaw), NT7144; 226
Old Jeddart (Auld Jedburcht), NT6614; 219
Old Liston (Aldlistoun), NT1172; 3, 97
Old Melrose (Auld Melros), NT5834; 208, 257, 259
Old Montrose (Auld Montrois), *Maryton*, NO6756; 383, 384, 386, 388, 389
Oldhamstocks (Aldhamestok, Auldhamstoks), NT7470; 166; parson of, 171; parsonage of, 171
Olrig (Alrig) [Ho, ND1866], 648; parson of, 648; parsonage of, 648; vicarage of, 648
Ones Clois Croft, *see* Ovenscloss
Ones croft, *see* Ovenscloss
Onst, *see* Unst
Orchard lands, *Sciennes priory, Edinburgh*, 94
Orchardton (Orcharttoun), *St Mary's Isle priory*, NX4549; 619
Orchartoun, *Sweetheart abbey*, 622
Orcheart, *Carsgrange, Coupar Angus abbey*, 354
Ordefull, *see* Ordiquhill
Ordehuissis, Ordenhuissis, Ordinhuissis, *Aberdeen chapter* [*Ordens, NJ5160], 437, 438
Ordies (Ordeyis), *Kinloss abbey*, NJ1061; 462
Ordinhuissis, *see* Ordehuissis
Ordiquhill (Ordefull, Tordiquhill), *Aberdeen chapter*, NJ5243; 437, 438; curate of, 438
Ore Mills (Orres Myllis), NT3097; 70
Orkie (Orky), NO2907; 14
Orkney (Orknay), HY3615; 282, 491, 628n., 655, 656n., 657, 658, 659, 660, 661, 662, 663, 664, 666, 667, 668, 670, 675, 677, 678, 679, 680, 681, 682, 683; bailie of bishopric of, 659, 671; bishop of, 491, 491n., 574n., 659, 659n., 660, 661, 677-683; bishopric, 495n., 655-663, 655n., 668-671; diocese of, 114, 659n. *See also* Shetland, archdeaconry of
—— cathedral (St Magnus), chancellor of, 666n.; provost of, 659, 659n.; subchanter of, 664, 664n., 677, 677n.; subchantory of, 664, treasurer

of, 660, 660n.; *see also* Kirkwall
    cathedral, *and* Shetland,
    archdeaconry of
Orky, *see* Orkie
Ormestoun Hill, *Kirknewton*,
    [*Ormiston, NT0966], 104
Ormiston (Ormestoun), *Eckford* [Mill,
    NT7027], 220
Ormiston (Ormestoun), *Fife*, NO2416;
    29, 33, 35
Ormiston (Ormestoun), *Trinity
    collegiate kirk*, 175
Orphir (Orpher, Orphar), HY3406; 657,
    659, 669; kirk of, 659; minister of,
    659
Orquhat, *see* Urquhart
Orres Myllis, *see* Ore Mills
Orrock (Orrok), *Kinghorn Wester*
    NT2288; 24, 28, 46, 50
Orwell (Overquhill, Uruell, Uruquhill,
    Urwale), NO1504; 26, 38, 48, 88;
    kirk of, 26, 38, 48
Orwell (and Kinross), vicar pensionary
    of, 88; vicarage pensionary of, 88;
    see also Kinross
Otterston (Otterstoun), *Inchcolm abbey*
    [Loch, NT1685], 63
Ouchiltrie, *see* Ochiltree
Oucht, *see* Outh
Ouchtergavin, Ouchtergawin, *see*
    Auchtergaven
Ouchterhous, *see* Auchterhouse
Ouchtermuchtie, *see* Auchtermuchty
Ouchtertule, Ouchtertulle, Ouchtertullie,
    see Auchtertool
Our Lady altar, *Ayr parish kirk*, 571
Our Lady altar, *Cambuskenneth abbey*,
    551
Our Lady altar, *Crail collegiate kirk*, 65
Our Lady altar, *Dalmeny kirk*, 154, 155
Our Lady altar, *Dunkeld cathedral*, 323
Our Lady altar, *Sanquhar parish kirk*,
    273
Our Lady altar, *St Nicholas' kirk,
    Lanark*, 524
Our Lady and St James, chapel of, *see*
    Newhaven
Our Lady chapel, *Eccles*, 183n.
Our Lady chapel, *Peebles*, 253, 254
Our Lady chaplainry, *Dunblane
    cathedral*, 314
Our Lady chaplainry, *Elgin cathedral*,
    476, 476n.

Our Lady in the Green chaplainry,
    Inverness, 477
Our Lady kirk, *Sanday, see* Lady kirk
Our Lady kirk, *South Ronaldsay, see*
    Lady kirk
Our Lady kirk, *Westray, see* Lady kirk
Our Lady of Pity altar, *Glasgow
    cathedral*, 522
Our Lady servant, prebend of Crail
    collegiate kirk, 66
Our Lady, chaplainry of, *St Catherine's
    kirk, Cupar*, 81
Our Lady, prebend at the High altar,
    Crail collegiate kirk, 66
Our Lady, prebend of St Mary of the
    Fields' collegiate kirk, Edinburgh,
    125
Outh (Oucht), NT0694; 23, 27, 28, 45,
    49, 50
Outoun Burges, *see* Burgess Outon
Outoun Chapell, Outoun Chappell, *see*
    Chapel Outon
Outoun Gallous, *see* Gallows Outon
Outtoun Burges, *see* Burgess Outon
Outtoun Chappell, *see* Chapel Outon
Ovenscloss (Ones Clois Croft, Ones
    croft, Unes Croft, Uniscroft,
    Unysclos, Vinsclos), NT4730; 223,
    230, 231, 231n., 233, 237, 239
Over Ancrum (Ovir Ancrame) [Ancrum,
    NT6224], 236, 366
Over and Nether Cantlie, *Kinloss abbey*
    [Meikle Cantly, NJ4750], 461, 464
Over and Nether Dalmeynis, *with
    Garscadden, Paisley abbey*, 529
Over and Nether Halkarstonis,
    Borthwick [Halkerston, NT3458],
    111
Over and Nether Holdenis, *see*
    Overhowden, Netherhowden
Over and Nether Nisbitis, *see* Upper
    Nisbet, Nether Nisbet
Over Auchendinny, Auchindonye,
    Auchyndunye, *Restalrig collegiate
    kirk* [Auchendinny, NT2562], 142
Over Auchinleische, *Coupar Angus
    abbey* [Auchinleish, NO1960], 354
Over Baddinheycht, *Dunblane
    bishopric*, 349
Over Balbrogy, *Coupar Angus abbey*
    [Balbrogie, NO2442], 356
Over Balmerosche, Over Balmorsch, *see*
    Over Bancrosche

INDEX OF PLACES 859

Over Bancrosche, Over Balmerosche, Over Balmorsch, *Galloway bishopric* [Barncrosh, NX7059], 586, 589, 592
Over Bar, *see* Upper Barr
Over Barnton, 114n.
Over Bartcapill, Bartcappill, *Galloway bishopric* [Barcaple, NX6757], 585, 589
Over Broomrig (Over Brumrig), *Holywood abbey*, NX9679; 276
Over Calquha, *Galloway bishopric* [Culquha, NX6958], 585, 589
Over Campsie (Over Camspsy alias the Wolf Hill), *Coupar Angus abbey* [Campsie, NO1223] 352, 352n., 356
Over Chatto, *Jedburgh abbey* [Chatto, NT7717], 220
Over Craling [Crailing, NT6824], 219
Over Cromlikis, *Dunblane bishopric* [Cromlix Ho, NN7806], 349
Over Ersslis, *see* High Ersock
Over Glen, *Rothes* [Netherglen, NJ2454], 491
Over Glengouer, *Crawfordmuir, Newbattle abbey*, 102
Over Grange, *Culross abbey* [Middle Grange, NT9889], 293
Over Grange (Overgrange, Uver Grang), *Kinghorn* [*Grange, NT2688], 24, 24n., 28, 46, 50
Over Hauches *of Killiesmont, Kinloss abbey*, 462, 464
Over Howden, *see* Overhowden
Over Ilrik, *Coupar Angus abbey* [Alrik, NO1961], 354
Over Kilbacho, '*Chapelden of Kindrimy*', Aberdeenshire, 449
Over Killelung, *Holywood abbey* [Killylung, NX9581], 276
Over Kilmenedy, *Kinloss abbey* [*Kinminintie, NJ4253], 464
Over Kincraig (Over Kincraigye, Uver Kincraig, Wver Kyngraig), *Brechin bishopric* [Kincraig, NO6258], 383, 386, 389
Over Kinnedder, Kinneddir, *see* Upper Kinneddar
Over Kynmonth, *see* Upper Kinmonth
Over Lasoidey, Uver Lasoiddy, *Dunfermline abbey* [Lassodie, NT1292], 23, 45
Over Lassuaid, Laswaid, Lessuaid,

Restalrig collegiate kirk [Lasswade, NT3065], 141, 143
Over Magus, *St Andrews*, *see* Upper Magus
Over Mains, *Monkton, Paisley abbey*, 529
Over Mains, *Ulston, Jedburgh abbey*, 219
Over Monbennis, *see* Upper Manbean
Over Muretoun, Murtoun, *with Tullyfergus, Coupar Angus abbey*, [Muirton, NO2247], 354, 357
Over Myln *of Auchindanery, Kinloss abbey* [*Nethermills, NJ5050] 461
Over Neisbit, Nisbit, *see* Upper Nisbet
Over Newlistoun, *St Andrews archbishopric* [Newliston, NT1073], 3
Over Petcarne, Pitcarne, *Dunkeld bishopric* [Pitcairn, NN8850], 342, 345
Over Petteddedrie, *Rothes*, 491
Over Pitforthie (Over Pitforthe, Wver Petforthie, Wver Pitforthye), *Brechin bishopric* [Mains of Pitforthie, NO6161], 384, 385, 389; *see also* Petforky
Over Pratas, Over Prateris, *North Berwick priory* [Pratis, NO3806], 145, 148, 167, 168
Over Quhittoun, *see* Over Whitton
Over Quhytsyd, Overquhytsyd, *see* Upper Whiteside
Over Reres [Rires, NO4604],146n., 167, 169
Over Sannik [Senwick Church, NX6546], 587
Over Sculdaill, *with Cogle, Caithness archdeaconry*, 632
Over Sett and Nethir Sett of Kilmanitie, *Kelso abbey* [*Kinminintie, NJ4253], 461
Over Smychtstoun, *Ballumbie, Lundie*, 376
Over Speddoch [Speddoch, NX8582], 275
Over Urquhard, *see* Upper Urquhart
Over Whitton (Over Quhittoun), NT7519; 220
Over Williamston (Williamstoun Over), NT0561; 98
Overhous, *Newbattle abbey* [Overhouses, NS6336], 102

Overhowden (Overholden, Over Howden), NT4852; 231, 233, 237, 240, 242
Overland (Overlond), NS4833; 531
Overquhill, see Orwell
Overton (Overtoun), *Culross abbey*, NT9989; 290
Overton (Overtoun), *Kirknewton*, NT1066; 104, 105
Overtoun, *Oxnam*, 219
Ovirtown, *Carmichael* [Netherton, NS9241], 511
Owyn, Owyne, see Oyne
Oxemon, see Oxnam
Oxinsyd, *Morebattle* [*Oxnam, NT6918], 214
Oxmansyd, *Morebattle* [*Oxnam, NT6918], 214
Oxmuir (Oxmure), *Hume, Kelso abbey*, NT7141; 228
Oxnam (Exnem, Oxemon, Oxnen, Oxnum), NT6918; 216, 219; kirk of, 216, 219; mains of, 219
Oxnen, see Oxnam
Oxnum, see Oxnam
Oylandis, *Netherfield, Bothwell collegiate kirk*, 517
Oyne (Owyn, Owyne), NJ6725; 428, 429; kirk of, 433n.; vicar pensionary of, 429

Pacakis, *Glasgow*, 516
Paisley (Paislay, Paslay, Pasley), NS4863; 282, 527, 528, 529, 530, 530n., 531, 532, 535; town of, 529; vicarage of, 529
—— abbey, 282, 527-531, 532, 534n.; abbot of, 6, 527; commendator of, 6n., 527n., 534, 534n.
—— parish kirk, 528, 535; chaplain of altar of St Mirin and St Colm in, 535
Paithnick (Pethnik), NJ4753; 461, 464
Pakyis Land, 'Smyddyhill alias Pakyis Land', see Smyddehill
Palhous, *Cambuskenneth abbey*, see Powis
Pallbethe, *Calder*, see Polbeth
Palmallet (Polmowert), NX4742; 595
Palwhilly (Balquhilly), NX4460; 587
Pambryde, see Panbride
Panbride (Pambryde, Panbrid), NO5735; 361, 372, 373; kirk of, 361; vicarage of, 372-373
Pancland, see Pincland
Pandell, Pandellis Craigis, see Randell Craigis
Panes, Pannes, *Musselburgh* [Prestonpans, NT3874], 25, 47
Panscheillis, *Lammermuir, Melrose abbey*, see Penshiel
Pansterstoun, *Calder* [*Pumpherston, NT0669]; 98
Papa Stronsay (Papy Sonsay), HY6629; 660; kirk of, 660; minister of, 660
Papa Westray (Papa Wastray, Papea Westray, Papy), HY4952; 655, 660, 668; parsonage of, 655, 668
Papea Westray, see Papa Westray
Paplay, Papplay, *Holm parish*, 659, 659n., 680, 683
Papy Sonsay, see Papa Stronsay
Papy, see Papa Westray
Parbroath (Perbroithe), NO3217; 21n., 33
Pareis, see Paris
Paremounth, Paremounthe, *of Strathkinness*, see Strathkinness
Paris (Pareis), *France*, 169, 435n.
Park Auld, *Moray deanery* [*Park, NH9053], 469
Park, *with Bow Inche, Lindores abbey*, 35
Parkhill, *Abdie*, 32, 37
Parkhill, *Belses, Jedburgh abbey*, 220
Parkland [? Parkhead], *Penpont*, 273
Parkley (Parklye, Perkly), *St Andrews priory* [Place, NT0176], 10, 18
Partarrak, see Portrack
Partick (Partik), NS5467; mill of, 496
Parton (Mortoun, Partoun, Partown), NX6970; 274, 274n., 616; parson of, 616; parsonage of, 616; reader of, 274; vicarage of, 616
Pascar Mylne, *Inchcolm abbey*, 62
Paslay, Pasley, see Paisley
Patefeild, *Crail*, 162
Patequhen, *Monkland*, 498
Patortheis, see Pitcorthies
Pattoun Towne, *Lagganallachy, Dunkeld archdeaconry*, 309
Paxton (Paxtoun), NT9353; 166, 173, 199, 202
Pearsie (Perseis), NO3659; 369
Peblis, see Peebles
Pedagogie, *St Andrews*, see St Andrews,

INDEX OF PLACES 861

St Mary's college
Pedderburne, *Newbattle abbey*
  [*Petersburn, NS7764]; 102
Peddrethy, *see* Pitdrichie
Pedwethy, *see* Pitwathie
Peebles (Peblis, Peiblis, Piblis),
  NT2540; 128, 149, 249, 249n.,
  250n., 251, 251n., 252, 253, 253n.,
  254, 254n., 259, 525; benefice of,
  251; Brig End, 254; chaplain of Our
  Lady chaplainry of, 254; chaplainry
  of Our Lady in, 253-254; Cross Gait,
  525; dean of the Christianity of,
  250n.; Gleidstanes land, 254;
  Lydgait, 254; Mathesounes land,
  254; Melwingis land, 254; minister
  of, 251; North Port, 254; Our Lady
  chapel in, 253, 254; parsonage of,
  251n., 253, 253n.; Rowcastle's land,
  254; Rud Mylne, 254; vicar of, 251;
  vicarage of, 251-252, 253, 253n.;
  Watersyd, 254
———, Cross Kirk, 249, 249n., 254,
  254n.; Holy Blood altar in, 254,
  254n.
———, St Andrew's collegiate kirk, 254,
  254n.; Holy Rood altar in, 254,
  254n.
———, Trinitarians, 'minister' of, 249,
  249n.; 'ministry' of, 249
———, *sheriffdom*, 1n., 108n., 223, 247,
  248n., 251, 500, 525n.
Peelwalls (Peilwallis), NT9160; 199
Peiblis, *see* Peebles
Peilwallis, *see* Peelwalls
Peirstoun, *see* Perceton
Peitcleocht, *Lagganallachy, Dunkeld
  archdeaconry*, 309
Peitinweyme, see Pittenweem
Peitirugy, *see* Peterugie
Pencaitland (Penciatland) [Easter,
  Wester], NT4469; 190, 197, 197n.;
  kirk of, 190, 197, 197n.; vicar of,
  165; vicarage of, 165
Pendicle Hill, *Maxwell, Kelso abbey*
  [*Pinnacle, NT5825]; 225
Pendrech, *Newbattle abbey*, 92
Pendreche, *Brechin bishopric*, *see*
  Pittendreich
Penfillan (Penfillane), NX8592; 275
Penick (Penik, Penneik, Pennyk,
  Pynnik), NH9356; 493
Penicuik (Penny Cuik), NT2360; 119;

parsonage of, 119
Peninghame, *see* Penninghame
Penmurtie, *Holywood abbey*, 275
Pennakle, *see* Pinnacle
Pennanguschoip, *Ringwodfeild, Melrose
  abbey*, 259
Penneik, Pennyk, *see* Penick
Penninghame (Peninghame) [Mains,
  NX4060], 601, 607; benefice of,
  601, 601n., 602; mill of, 587, 588,
  590, 592, 601, 601n., 602n.; reader
  of, 602n.; vicar of, 602; vicar
  pensioner of, 602n.
Penninghame Mains, NX4060; 587
Penny Cuik, *see* Penicuik
Penpont (Penponait), NX8494; 271,
  273, 277, 623; kirk of, 271, 277,
  623; vicar of, 273; vicarage of, 273
Penshiel (Panscheillis), *Lammermuir,
  Melrose abbey*, NT6463; 258
Pentland [Mains, NT2665], 116;
  parsonage and vicarage of, 116-117
Pentoscall, Pentoskyll, Pettoskell,
  Brechin bishopric, 383, 387, 388
Pepper croft, *Haddington priory*, 180
Perbroithe, *see* Parbroath
Perkly, *see* Parkley
Perceton (Peirstoun, Perstoun), NS3440;
  kirk of, 576, 578
Perseis, *see* Pearsie
Persley (Perslie), NJ9009; 422, 423
Perstoun, *see* Perceton
Perth (Perthe, Pertht, Sanct Johnnesoun,
  Sanct Johnnestoun, Sanctjhon),
  NO1122; 11, 19, 26, 31, 35, 35n.,
  38, 42, 48, 53, 58, 74, 74n., 259,
  283, 284, 285, 286, 287, 291, 304,
  306, 306n., 315, 317, 320, 323, 326,
  326n., 327, 331, 342, 346, 355, 362,
  369, 409, 538, 545; altarage in, 320;
  bailies of, 306; burgess of, 326;
  burgh of, 355; Camerone land, 327;
  Castle Gait Brig, 306; Foirgait, 31,
  35; Gilkannies land, 283, 286;
  Kirkgait, 283, 287; Magdalannis,
  Magdalenis, 284, 287; Myllikynnis
  land, 12; Newraw, 283, 284, 286,
  287; Northgait, 283, 286, 287;
  Pincland (Pancland), 12; provost of
  burgh of, 306; St Mary Magdalene's
  hospital in, 284, 287; St Paul's
  chapel in, 283, 286; town of, 331;
  Tung yard, 284, 287; Turat Brig,

283, 286; vicar of, 315; vicar pensionary of, 74, 74n., 315, 326n.; vicarage pensionary of, 74, 74n.; Water Gait, 31, 35, 283, 287
——, Charterhouse of (Charterhous, Chartous), 241, 283-289, 284n., 289n., 306; prior of, 283, 285, 286, 289, 289n.
——, St John's kirk (kirk of St John the Baptist), 38, 48, 52, 53, 74, 74n., 306, 315; chaplainry of St Stephen in, 306
——, *sheriffdom*, 1n., 35n., 36, 86n., 87, 87n., 88, 88n., 110, 110n., 282, 283n., 310, 317, 320n., 321, 323, 331, 337, 338, 339n., 341, 365, 365n., 392n., 522, 522n.
Pertmork, *see* Portmoak
Petblad, *see* Pitbladdo
Petcastell, *see* Pitcastle
Petchadlie, *Fordyce*, 437
Petcokis, *see* Pitcox
Petcorthy, *see* Pitcorthie
Petcraigy, *see* Pitcraigie
Petcrowye, *see* Pitcruvie
Petcullane, *see* Pitcullen
Petcullo, *see* Pitcullo
Petcuntlie, Petcuntly, *lordship of Monimail, St Andrews archbishopric*, 2, 5
Petcur, *see* Pitcur
Peterculter (Petirculter, Petircultir, Pettirculter, Petturculter), NJ8410; 224, 232, 234, 237, 240, 242; kirk of, 224, 232, 234, 237, 240, 242, 427
Peterhead (Peterheid), NK1346; 457; *see also* Peterugie
Peterugie (Peiterugy, Peterugy) [Peterhead, NK1346], 431, 432, 457, 458; curate of, 432; parish of, 457, 458; vicar of, 432; vicarage of, 431-432; *see also* Peterhead
Petfoddellis, 416n.
Petforky, Pitforkeis, Pytforkeis, *Brechin bishopric* [*Mains of Pitforthie, NO6161], 383, 387, 388; *see also* Over Pitforthie, Nether Petforthie
Petgerello, *Tain and Edderton*, 651
Pethnik, *see* Paithnick
Petincreifis, *see* Pettincreif
Petirculter, Petircultir, *see* Peterculter
Petirugy, *see* Peterugie
Petkuillen, *see* Pitcullen

Petlethie, *see* Pitlethie
Petlyell, *Lundie*, 376
Petmaddane, *see* Pitmedden
Petmeddy, *see* Pitmedden
Petmillie, *see* Pitmilly
Petmolye, *St Andrews Black friars*, 89
Petmukkistoun, *Aberdeen deanery* [*Pettymuck, NJ9024]; 423
Petpoint, *see* Pitpointie
Petpullox, Pitpollox, Pytpollox, *Brechin bishopric*, 383, 387, 388
Petpunty, Pitpuntie, *St Andrews priory*, 11, 19
Petquhiwin, *near Penick, Moray deanery*, 469
Petrochie, *see* Pitreuchie
Petruifnye, *see* Botriphnie
Petslego, *see* Pitsligo
Pettegar, *Atholl, Perth Charterhouse*, 284, 288
Pettendrech, *Lundie*, 376
Pettermo, *see* Pitermo
Petticommone, Petticommonis, Petticommoun, *Culross abbey*, 291, 291n.
Pettillo, *see* Pittillock
Pettinain (Pettinane, Pettynane, Pittinane), NS9542; 190, 191, 197, 197n.; kirk of, 197, 197n.
Pettinbrog, Pettinbroge, prebend of Abernethy collegiate kirk, 340
Pettincreif (Petincreifis, Pettincreiff, Pettincreiffis), *Atholl, Perth Charterhouse*, 284, 288; mill of 284, 287, 288
Pettincreif, *Fife, see* Pittencrieff
Pettincrevy, Pittavy, *Methven collegiate kirk*, 296, 298
Pettindreich, *Restalrig collegiate kirk*, 141, 142, 143
Pettindynie, *see* Pittendynie
Pettinragorie, *Atholl, Perth Charterhouse*, 284, 288n.
Pettinweme, Pettinweyme, *see* Pittenweem
Pettirculter, Petturculter, *see* Peterculter
Pettlochery, Pettlochry, *see* Pitlochry
Pettogarty [Mid Pithogarty, NH8082], 651
Pettoskell, *see* Pentoscall
Peyll, *Eccles*, 184
Pharinhirst, Pherniherst, *see* Ferniehirst
Pherne, *abbey, see* Fearn

Pherne, *parish, see* Fern
Philgour, *Hownam*, 220
Phillophauch, *Selkirk*, 230n.
Philorth (Phillorcht), NK0064; 424, 459; parsonage of, 459
Physgill (Fisgill, Fysgill) [Ho, NX4236], 594, 596
Piblis, *see* Peebles
Pikin, Pinkin, Pynkin, *Musselburgh*, 24, 24n., 38, 47, 48; *see also* Cars, Pinkie
Pilmure, *St Andrews priory*, 10
Pincartoun, Pinkartoun, *Crail*, 177, 179, 182
Pincland (Pancland), *Perth*, 12
Pinkerton (Pinkertoun), NT6974; prebend of Dunbar collegiate kirk, 171
Pinkie (Pynkie) [Mains, NT3672], 24, 25, 26, 26n., 38, 42, 53; *see also* Inveresk
Pinkie et Cars, *see* Cars and Pikin
Pinkie feild [Pinkie Ho, NT3472], 80, 586n.
Pinkin, *see* Pikin
Pinnacle (Pennakle), NT5825; 220
Pirn (Pyrin, Pyrm), NT4447; 253, 254
Pitarro, *see* Pittarow
Pitbauchlie (Pitbaulie, Pitbawillie, Pitbawkie, Pitbawlie), NT1186; 23, 27, 28, 45, 49, 50
Pitbladdo (Petblad), *St Andrews priory*, NO3617; 14
Pitcairn (Pitcarne), NN8850; mill of, 342, 345
Pitcarthie, *see* Pitcrathie
Pitcastle (Petcastell), NN9755; 284, 288
Pitconmark (Pitcormokt, Pittinmark, Pittinmork, Pittonmark, Pittormok), *with Torbain, Dunfermline abbey*, 28, 43, 44, 50, 54
Pitconochie (Pitconchy, Pitconnochy, Pitconochy, Pitconquhy), NT0586; 27, 28, 49, 50
Pitcormokt, *see* Pitconmark
Pitcorthie (Petcorthy), *Crail* [*East, West Pitcorthie, NO5707, NO5706], 162, 177, 179
Pitcorthies (Patortheis, Pitcortheis Easter and Wester, Pitcorthis, Pitcorthris), *Dunfermline abbey* [*Easter, Wester Pitcorthie, NO1186, NO1085], 23, 27, 28, 45, 49, 50
Pitcowie, Pitcowy, Pitlowy, *Crail*, 177, 179, 182
Pitcox (Petcokis), NT6475; prebend of Dunbar collegiate kirk, 175
Pitcraigie (Petcraigy), NJ2551; 491
Pitcrathie, Pitcrathie, *Cambuskenneth abbey*, 544, 547
Pitcruvie (Petcrowye, Pitcurvie), NO4104; 145, 145n, 167, 167n.
Pitcullen (Petcullane, Petkuillen), *Perth Charterhouse*, NJ6402; 283, 286
Pitcullo (Petcullo, Pitcullocht), *St Andrews priory*, NO4119; 13, 20
Pitcur (Petcur), NT2537; 249, 249n.
Pitcurvie, *see* Pitcruvie
Pitdrichie (Peddrethy), *Aberdeen deanery*, NO7982; 405
Pitduneis, Pitdunes Easter and Wester, Pitdunes, Pitdunis Easter and Wester, *Dunfermline abbey* [*Pitdinnie, NT0487], 27, 28, 49, 50
Pitermo (Pettermo), NO3036; 376
Pitfaris, Pitferris, *Culross abbey*, 290, 294
Pitfirrane (Pitferren), *Dunfermline abbey*, NO0686; 26, 27, 49, 50
Pitforkeis, *see* Petforky
Pitgorno (Pitgornow), NO1910; 58, 69
Pithowcher, *see* Pitteuchar
Pitkerrie (Pitkory), NC8679; 633
Pitkory, *see* Pitkerrie
Pitlair (Pitlare), NO3112; 33, 37
Pitlethie (Petlethie, Potlathy), NO4522; 10, 19
Pitliver (Pitlever, Pitlover) [Ho, NT0685], 27, 28, 49, 50
Pitlochie (Pitlochy), *Strathmiglo*, NO1709; 69
Pitlochie (Pitlowchy), *Kinglassie*, NT2397; 24, 29, 46, 51
Pitlochrie, *Glenisla, Coupar Angus abbey*, 354
Pitlochry (Pettlochery, Pettlochry), *Atholl*, NN9458; 284, 288
Pitlochy, *Kilconquhar*, 146
Pitlover, *see* Pitliver
Pitlowchy, *see* Pitlochie
Pitlowy, *see* Pitcowie
Pitmedden (Petmaddane, Petmeddy, Pittmedden), NO2214; 338; prebend of Abernethy collegiate kirk, 338; prebendary of, 338

Pitmilly (Petmillie, Pitmillie, Pitmyllie, Potmulye), NO5813; 11, 19, 66, 177, 178, 181
Pitpointie (Petpoint), NO3537; 309
Pitpollox, see Petpullox
Pitpuntie, see Petpunty
Pitreavie (Pitrovey, Pittoravy, Pittrave, Pittravie, Pittravy, Pittrayvie), NT1184; 27, 28, 49, 50
Pitreuchie (Petrochie), NO4749; 217
Pitrovey, see Pitreavie
Pitsligo (Petslego), *parish*, NJ9365; 421
Pittamous, Pittenhous, Pittennus, *Mearns, Lindores abbey*, 31, 31n., 35
Pittangus, Pittargus, *Mearns, Lindores abbey*, 31, 31n., 35
Pittareis, *Mearns, Lindores abbey*, 31n.
Pittarow (Pitarro, Pittarro), NO7275; 5n., 61, 61n., 73n., 94n., 152n. 264n., 269n., 286n., 351n., 391n., 406, 407, 411n., 423n., 464, 464n., 465n., 543n., 564n., 575n., 579, 579n., 651n.
Pittavy, see Pettincrevy
Pittencrieff (Pettincreiff), *Cupar*, NO3715; 14
Pittencrieff (Pettincreiff, Pittincreif, Pittincreiff), *Dunfermline abbey* [Park, NT0887], 27, 28, 49, 50
Pittendreich (Pendreche, Pittindreich, Pittindreicht), *Brechin bishopric*, NO5761; 383, 386, 388
Pittendreich, 248n.
Pittendynie (Pettindynie, Pittindinie, Pittindyny), NO0529; 303, 342, 345
Pittennus, Pittenhous, see Pittamous
Pittenweem (Peitinweyme, Pettinweyme, Peitinweyme, Pettinweme, Pittenweim Pittinweme), NO5402; 22, 23, 46
— priory, 22; see also May
Pitteuchar (Pithowcher, Pityoquhair, Pityowquhair), NT2899; 24, 29, 46, 51
Pittillock (Pettillo), NO2705; 15
Pittinane, see Pettinain
Pittinmark, Pittinmork, Pittonmark, see Pitconmark
Pittlesheugh (Putellseucht), NT7543; 258
Pittmedden, see Pitmedden
Pittoravy, see Pitreavie

Pittormie (Pottormie), NO4118; 14
Pittormok, see Pitconmark
Pittrave, Pittravie, Pittravy, Pittrayvie, see Pitreavie
Pittrefeyne, see Botriphnie
Pitwathie (Pedwethy), NO7784; 405
Pityoquhair, Pityowquhair, see Pitteuchar
Plaiddes, *Tain parish*, 651
Plandergaist, *Bathgate*, 102
Platt (Plat), prebend (with Bonnington) of Corstorphine collegiate kirk, 107, 107n., 116, 121
Plenderleith (Prenderleth, Prenderleyth), NT7311; 219; kirk of, 217
Plewland, with *Camieston, Melrose abbey*, 208, 257
Pluscarden priory (Pluscardin), NJ1457; 471, 472, 473, 471-474, 474n., 491n.; chamberlain of, 471, 472, 473; kirk of, 472, 473; monks of, 472; oeconomus of, 473
Poffilis, Poffillis, *Culross abbey*, 290, 292, 294
Poffill *of Strathkinness, St Andrews priory*, 18
Pokihaucht, Polkehaugh, Pokye Houich, with *Greenlawdean, Kelso abbey*, 226, 226n.
Pokye Houich, see Pokihaucht
Polbeth (Pallbethe), NT0364; 98
Polcalk (Polcak), *Millhorn, Coupar Angus abbey*, NO2346; 354, 357, 369
Poldrait, *Balmerino abbey*, 58
Poldroch, *Haddington*, 11
Polduff, with *Boarhills, St Andrews archbishopric*, 2, 5, 13, 20, 21
Poldune, *Arbroath abbey*, 362
Polkak, *Carse of Gowrie, Perth Charterhouse*, 285, 289
Polkehaugh, see Pokihaucht
Polmowert, see Palmallet
Polton (Poltoun), *Lothian*, NT2964; 97, 142; St Leonard's chapel, 97
Poltoun, *Whithorn priory*, see Powton
Poltoun Myln, *Whithorn priory*, see Powton
Polwarth (Polwart), NT7450; 146, 147, 148, 149, 166, 168, 195, 200; parson of, 195; parsonage of, 195
Pople, *St Bathans priory*, 192
Port Lamont, see Lamont

Port, *of Logierait*, 303, 304, 342, 345
Port of Menteith (Porte), NN5801; kirk of, 544, 548; vicarage of, 548
Porta Latronum, *see* Aberdeen, Gallow Gait
Portcarrik, *see* Portyerrock
Portmoak priory (Pertmork, Portmook, Portmork), NO1500; 16, 17, 56, 57, 57n.; kirk of, 16, 17; vicarage of, 56, 57
Portrack (Partarrak, Porterak, Potterak) [Ho, NX9382], 270, 271, 275, 276, 277, 623
Portyerrock (Portcarrik), NX4738; 593; mill of, 593
Pot, Pott, *fishing on the Dee, Aberdeen bishopric*, 416, 417, 420
Potdothwan, *Monkland*, 102,
Pothowan, *Crawfordmuir, Newbattle abbey*, 102,
Potlathy, *see* Pitlethie
Potmulye, *see* Pitmilly
Potterak, *see* Portrack
Pottie (Potye), *with Dunbarney and Moncreiffe*, [*Pottiehill, NO1513]; teinds of, 111
Pottinfar, *Strathspey, Elgin Black friars*, [*Peterfair, NJ1937]; 453
Pottormie, *see* Pittormie
Potye, *see* Pottie
Powcleif, Powcleis, *Eskdalemuir, Melrose abbey*, 259, 259n.
Powdono, *see* Rodono
Powis (Palhous), *Cambuskenneth abbey* [Mains, NS8295], 537, 544, 545
Powis (Powys), *Culross abbey*, *NS8295; 291
Powmonk, *Eskdalemuir, Melrose abbey*, 259
Pownell, Pownill, *Lesmahagow*, 231, 231n., 233
Powquhyit, Powquhyt, *see* Pulwhite
Powton (Poltoun), NX4645; 594, 595, 596; mill of, 593; mill croft of, 593; waulk mill of, 593, 595, 596
Powys, *see* Powis
Prandergaist, *see* Prenderguest
Preisthauch, *see* Priesthaugh
Preistis Croun, *Eckford*, 212, 215
Preistis Hauches, *Eckford*, 215
Preistisfeild, *Jedburgh abbey*, 216
Preistlaw, *Restalrig collegiate kirk*, *see* Eistraw
Preistlaw, *Lammermuir*, *see* Priestlaw
Preistoun, *Kelso abbey*, *see* Prieston
Preistoun, *with Bunkle*, *see* Preston
Premeth, *see* Premnay
Premnay (Premeth), *parish*, NJ6124; 32, 36; kirk of, 32, 36
Prenderguest (Prandergaist), NT9159; 198
Prenderleth, Prenderleyth, *see* Plenderleith
Press [Mains, NT8765], 198
Pressock (Pressoik), NO5649; 376
Preston (Preistoun, Prestoun), [Bunkle and Preston parish, NT8057]; 302, 304, 309, 342, 346; kirk of, 302, 342, 346; vicarage of, 185; *see also* Bunkle
Preston (Prestoun), salt pans of [*Prestonpans, NT3874], 209, 260
Preston, *Kelso abbey*, *see* Prieston
Prestones land, *Musselburgh*, 114
Prestonkirk, NT5977; parish of, 170n., 174; chaplainry of Our Lady in kirk of, 174. *See also* Hauch, Linton
Prestrie, NX4637; 593, 596
Prestwick (Prestuik), NS3425; 529; kirk of, 529
Priesthaugh (Preisthauch), NT4604; 259
Priestlaw (Preistlaw) *Lammermuir, Melrose abbey*, *NT6463, 258
Prieston (Preistoun, Prestoun), *Kelso abbey*, NT5228; 223, 225, 230, 231, 234, 237, 240, 242
Primrose (Primrois, Primros, Prymrois, Prymros), NT1084; 23, 27, 28, 45, 49, 50
Primside (Primsyd), NT8026; 214
Primsidemill (Primsyd Myln, Prymsyd Mylne), NT8126; 214
Prinlaws (Prinles), NO2401; 62
Prinles, *see* Prinlaws
Prior Lathyne, Prior Lathynen, *see* Priorletham
Prior Wall, *see* Priorwell
Prior's Croft, *Crail*, 11, 18
Priorletham (Prior Lathyne, Prior Lathynen, Priorlethame), NO4012; 10, 13, 18
Priorwell (Prior Wall), NO3523; 10, 19
Priour Wod, Pryourwod, *Melrose abbey*, 208, 258
Priourhill, Forowhill, *Lesmahagow* [North, South Priorhill, NS7540,

NS7539], 231, 231n.
Prok, *see* Purroch
Prymrois, Prymros, *see* Primrose
Prymsyd Mylne, *see* Primsidemill
Pryourwod, *see* Priour Wod
Puddingrow, *with Lambden, Greenlaw*, 227, 227n.
Puderneis Easter and Wester, *Kirkcaldy*, 81
Pulmior, *with Inchture, St Andrews priory* [*Pilmore, NO3229], 18
Pulwhite (Powquhyit, Powquhyt), NJ6532; 31, 36
Pumpherston (Pansterstoun), *Calder*, NT0669; 98
Pure Heuch, *Culross abbey*, 291
Purin, NO2606; 15
Purroch (Prok), NS4635; 531
Pursk, *see* Pusk
Pusk (Pursk), NO4320; 13, 20
Putellseucht, *see* Pittlesheugh
Pynkie, *see* Pinkie
Pynkin and Cars, Pynkin and Kers, *see* Cars and Pikin
Pynkin, *see* Pikin
Pynnik, *see* Penick
Pyrin, see Pirn
Pyrm, *see* Pirn
Pytforkeis, *see* Petforky
Pytpollox, *see* Petpullox

Quaigis de Strathalloin, *Dunblane cathedral*, 343n.
Qualsay, *see* Whalsay
Quarrypits (Correllpittis, Quarrell Pites, Quarrell Pittis), NT5374; 162, 176, 178, 181
Quarter Waterston (Quarter Waterstoun) [Waterston, NO5159], 386
Queen's college, *see* Edinburgh, Trinity collegiate kirk
Queltoun, *see* Coylton
Quhalsay, *see* Whalsay
Quheilland, *Liddesdale, Jedburgh abbey*, 217
Quheit Hill, Quheithill, *Musselburgh*, *see* Whitehill
Quheitbank, see Whitebank
Quheitsyde, *see* Whiteside
Quheityaird, *Melrose abbey*, 208
Quhiggis, *Dunblane cathedral*, 343, 343n.

Quhiltoun, *see* Whitton
Quhincarstanes, *see* Whinkerstones
Quhirlfuird, *Riccarton*, 531
Quhitarne, Quhiterne, Quhithorne, Quhitorne, Quhittarne, Quhittern, Quhitterne, Quhittorn, Quhittorne, *see* Whithorn
Quhite Croft, *Leuchars, see* White Croft
Quhithill, *Kinnell parish*, 366
Quhithill, *Musselburgh, see* Whitehill
Quhithorne, *see* Whithorn
Quhitkirk, *see* Whitekirk
Quhitlawhous, Quhytlawhous [Over, Nether Whitelaw, NT5129], 223, 224n., 225, 231, 233, 237, 239, 242
Quhitlie, *see* Whitelee
Quhitmuirtoun, *see* Whitmuir
Quhitrig, *Mertoun parish*, 189
Quhitrigmerschell, *Riccarton* [*Wheatriggs, NS4435], 531
Quhitstoun, Quhitsome, Quhitsum, *see* Whitsome
Quhitsyd, *Musselburgh, see* Whiteside
Quhittane [*Whyntie Wood, NJ6264], 359
Quhitterne, *Dryburgh abbey*, 190
Quhittinghame, *see* Whittingehame
Quhothorny, *Tullynessle*, 436
Quhyt Chester, *see* Whitchester
Quhyt Croft, *Leuchars, see* White Croft
Quhyt Mure Hall, *Selkirk*, 230n.
Quhytbank, *see* Whitebank
Quhytcastell alias Munnraw, *see* Nunraw
Quhytcroft, *St Mary's Isle priory*, 619
Quhytehill, Quhythill, *Musselburgh, see* Whitehill
Quhytfald, *St Mary's Isle priory*, 619
Quhytfeild, *see* Whitfeild
Quhytfuird, *Paisley abbey*, 529
Quhythill, *Inchcolm abbey, see* Whitehill
Quhythillis, *see* Whitehills
Quhytkirk, *see* Whitekirk
Quhytkirk, 655
Quhytlaw Hous, *see* Quhitlawhous
Quhytlaw, *Edrom, see* Whitelaw
Quhytlaw, *Ecclesmachan*, 151
Quhytlawhous, *see* Quhitlawhous
Quhytmuirhall, *see* Whitmuir Hall
Quhytmuirtoun, *see* Whitmuir
Quhytquarrellhoip, *fishing, Balmerino abbey*, 58

Quhytrig, *Ayton parish*, 199
Quhytsyd, *Musselburgh, see* Whiteside
Quhytsyd, *see* Whiteside
Quickiswod, *see* Quixwood
Quihtorne, see Whithorn
Quilts (Kyltyis, Quyltis), NO0211; 290, 292, 294, 343
Quinish (Quyneise), NM4155; 674
Quithel (Cuthill), NO7884; 405
Quithorn, *see* Whithorn
Quixwood (Quickiswod, Quixwod), NT7863; 192, 194
Quylt Myln, [Quilts, NO0211]; 290
Quyltis, *see* Quilts
Quyneise, *see* Quinish

Raasay (Rasay), NG5644; 675
Raburne, *see* Raeburn
Racht, *see* Raith
Radshaw, *see* Fasschaw
Raeburn (Raburn), NY2971; 259
Raeshaw (Raschaw), NT3651; 102
Rafflat, *see* Rawflat
Rafford (Rafferd, Rafforte, Raforte, Raforth), NJ0656; 465, 467, 470; kirk of, 470
Raforte, Raforth, *see* Rafford
Ragortoun, *see* Redgorton
Rahill, Grange of, *Deer abbey*, 457, 458
Raik, Rak, *fishing on the Dee, Aberdeen bishopric*, 416, 417
Railstoun, *Paisley abbey* [*Ralstonhill, NS4438], 529
Railstoun, prebend of St Giles' collegiate kirk, *see* Ravelston
Raine, *see* Rayne
Rait, NO2226; kirk of, 334, 334n.; *see also* Kilspindie
Raith (Racht, Raithe, Rathe, Ratht, Raucht) [Ho, NT2591], 24, 28, 41, 46, 50
Rak, *see* Raik
Ralaldsay [South Ronaldsay, ND4489], 659
Ralnabreche, *see* Balnabriech
Ralsay, *see* Rousay
Ramboy, *see* Rameldry
Rameldry (Ramboy, Ramelry), NO3206; 15, 21
Ramornie (Ramorgny), NO3209; 14, 21
Ramseycleuch (Ramsecleucht, Ravynniscleuch), NT2714; 258, 258n.
Ramsky, Ramskye, *see* Rosemarkie
Ranalstoun, *Beith*, 573
Randall, Randell, *Orkney, see* Rendall
Randell Craigis, Randellis Craigis (Pandell, Pandellis Craigis), *Dunfermline*, 27, 28, 49, 50
Randelstoun, *Crail, see* Randerston
Randelstoun, Rannaldsoun, *Carse of Gowrie, Perth Charterhouse*, 285, 288
Randerston (Randerstoun, Randelstoun), *Crail*, NO6010; 177, 178, 181; Newtoun of, 162
Randfurd, *Stirling*, 545
Randintracht, *Fordyce*, 437
Rane Patrik, *see* Redkirk
Ranfrew, *see* Renfrew
Rankeilour (Rankelour) [Mains, NO3212], 78
Rannaldsay Besouth, *see* South Ronaldsay
Rannaldsoun, *see* Randelstoun
Rannestoun, *see* Rennieston
Raperlaw (Repperlaw), NT5523; 216
Rapoche, [*Rippachie, NJ4111], 449
Raquhausyd, *St Bathans priory* [*Roadside Woods, NT7865] 192
Rasay, *see* Raasay
Raschaw, *Melrose abbey, see* Fasschaw
Raschaw, *Heriot, see* Raeshaw
Rascobe, Rascobie, *see* Roscobie
Raslok, *see* Roscobie
Rasneth, *see* Rosneath
Rassallis croft, *Riccarton*, 531
Rassyth, *see* Rosyth
Rathabyeris, *see* Ratho Byres
Rathe, *see* Raith
Rathen (Rathin), NK0060; 425, 438; kirk of, 425
Rathernn-Striviling alias Quaigis de Strathalloin, 343n.
Rathibeath, *St Andrews Black friars*, 89
Rathmuriel, *see* Christskirk
Ratho (Rotho), NT1370; 97, 100, 123, 130, 131; minister of, 131; vicar of, 123, 131; vicar pensionary of, 123; vicarage of, 123, 131
Ratho Byres (Byres, Rathabyeris, Ratho Byeris, Rathobyares, Rathobyers, Rothobiares), NT1471; prebend (with Norton) of Corstorphine collegiate kirk, 107, 107n., 109, 113,

128, 130
Ratht, *see* Raith
Rathven (Rothven, Rothwen), NJ4465; 455, 456, 456n., 460; bedesmen of the hospital of, 456; benefice of, prebend of Aberdeen cathedral, 248, 248n.; hospital of, 456n.; kirk of, 359, 438; parson of, 456n.; parsonage of, 456, 456n.; vicarage of, 455
Ratownray, Rattownraw, *see* Rottenrow
Rattray, NO1845; 328; parsonage of, 328
Raucht, *see* Raith
Ravelston (Railstoun), NT2274; prebend of St Giles' collegiate kirk, Edinburgh, 111-112
Ravenstone (Ralveinstoun), *Sorbie* [Mains, NX4144], mill croft of, 593
Ravynniscleuch, *see* Ramseycleuch
Raw, *see* Rawes
Rawes (Raw), NN3028; 285, 286, 288
Rawflat (Rafflat), NT5824; 220
Ray, *see* Reay
Rayne (Raine, Rayn) [Old Rayne, NJ6728], 414, 417, 418, 432, 432n.; parsonage of, 432; vicar pensionary of, 432; vicarage of, 432
Raynpatrick, *see* Redkirk
Reard, *Caithness bishopric*, 628
Reay (Ray), ND9564; 628, 648; parish of, 628; vicarage of, 648
Redbrais East Mains, *Polwarth*, 195
Redden (Reddane, Reddein, Reiddayne), NT7737; 223, 230, 233, 237, 239
Reddene, *see* Lindean
Rede Croce, Reid Croce, Reidcroce, Croce, 352, 353, 354, 356, 357, 368, 369
Redgorton (Ragortoun), NO0628; 332, 334; kirk of, 332, 334
Redhall (Reidhall), *Ayton*, NT9462; 199
Redheugh (Reidheuch), NY4990; 199
Redie, *Airlie parish*, 356
Redkirk (Rane Patrik, Raynpatrick), NY3065; 274, 274n.
Redpath (Reidpethe), *Melrose abbey*, NT5835; 200, 204, 207, 257
Redrowll, *see* Bedrule
Redwells (Reidwellis, Rudwellis), NO5508; 177, 179, 182
Redy, *Auchtermuchty* [*Reedieleys, NO2310], 33

Reedleyis, *see* Braidlevis
Regeboll, *with Skiall, Caithness bishopric*, 627
Reid Croce, Reidcroce, *see* Rede Croce
Reid Inche, Reid Inchis, Reidinche, *Lindores abbey*, 30, 35
Reidbraes, *St Bathans priory*, 192
Reiddayne, *see* Redden
Reidhall (Reidhaw), *Lothian*, 117n., 118, 118n., 179, 183n.
Reidhall, *Ayton, see* Redhall
Reidheuch, *see* Redheugh
Reidpeth, Rydpeth, *Fogo, Kelso abbey*, 226, 226n.
Reidpethe, *Melrose abbey, see* Redpath
Reidspittal, *see* Spittall
Reidwellis, *see* Redwells
Rendall (Randall, Randell), HY4850; 657, 659, 670; kirk of, 659; minister of, 659
Renfrew (Ranfrew), NS4966; 527, 532, 535; parsonage of, 532; vicarage of, 532
Renmure (Rynmure), NO6452; 366
Rennieston (Rannestoun), NT7120; 219
Renton (Rentoun), *Coldingham priory* [Barns, NT8265], 203; mains of, 198; mill of, 203
Repperlaw, *see* Raperlaw
Reres, *see* Rires
Rescobe, Rescoby, *Dunfermline abbey, see* Roscobie
Rescobie (Rescoby), NO5052; kirk of, 66; parish of, 67; vicarage of, 66
Rescoby, lordship of, *St Andrews priory* [*Rescobie, NO5052; *Roscobie, NT0892], 4
Reskyne, *see* Rosskeen
Reslein, *see* Restenneth
Respine, Respyn, Respyne, *see* Rispain
Restalrig (Lestalrig, Restalrige), NT2874; 133, 133n., 443, 443n.; glebe of, 119; manse of, 119; parson of, 119; parsonage of, 119;
—— collegiate kirk, 103n., 116, 141, 143, 144, 195, 443; chaplainries of, 143-144; chaplains of, 143-144; dean of, 116, 116n., 141, 443, 443n.; deanery of, 103, 116, 141-143; prebendaries of, 143-144; prebends of, 143-144; prebend of St Triduana's aisle, 143, 143n.
Restenneth (Reslein, Restant, Restenet,

INDEX OF PLACES

Restenit, Restennet, Restennethe, Restennett, Restennot, Resteyneth, Restrant, Restrint), NO4751; 217, 217n., 218, 218n., 220, 221, 356, 356n., 370, 409, 412; kirk of, 217; mains of, 217; vicarage of, 217
—— priory, 217-218, 218n, 220, 221, 370; 'abbot' of, 356; chamberlain of, 218; commendator of, 356n.
Restis land, *Stirling*, 556
Reuchtwen, *Glasserton* [*High Rauchan, NX4138], 593
Rhins of Galloway, NX0860, 605
Rhynd (Rynd), *Leuchars*, NO4623; 14, 20
Rhynd (Ryndis Easter and Wester), *parish*, NO1520; 22, 321; vicarage of, 321
Rhynie (Ryne), NJ4927; 486; parson of, 486; parsonage of, 486
Ribbislaw, *Aberdeen deanery*, 423
Riccarton (Richertoun, Rilkertoun, Ritchartoun), NS4236; 528, 531; benefice of, 531-532; kirk of, 528
Richartoun Holms, *Riccarton*, 531
Richertoun, *see* Riccarton
Riffan, Riffanis, *see* Ruthven
Riggis (Riggs), NO2807; 14, 21
Rilkertoun, see Riccarton
Ringwodfeild, *Melrose abbey*, 208, 211, 259
Ringwodhaucht, *Ringwodfeild*, 259
Rippo, *see* Kippo
Rires (Reres), NO4604; 146, 148; Newtoun of, 146, 167
Rispain (Respine, Respyn, Respyne), NX4339; 588, 590, 592, 593
Ritchartoun, *see* Riccarton
River Clyde (Clyd), NS5964; 231
River Cree (Crie), NX3081, NX3575, NX4263; 585n., 587, 588, 589, 590, 591, 619, 619n.
River Dee (Walter of Die, Watter of Dee, Wattir of Dee), *Aberdeenshire*, NJ8600, NN9894, NO1290, NO4097, NO7396; 416, 417, 418
River Dee (Water of Dee), *Wigtownshire*, NX7157; 586, 589
River Don (Waltir of Done, Watter of Done, Wattir of Done), NJ2309, NJ4513, NJ8115; 415, 416, 417, 432
River Ericht (Waltirris, Wateris of Areicht), NN5061, NO1551; 354, 357, 369
River Findhorn (Waltir of Findorne, Watter of Findorne, Wattir of Findorne), NH7016, NH9444; 462
River Forth, NS6396, NS9088; 341, 342, 344, 345, 346, 347
River Isla (Waltirris of Ylay, Wateris of Ylay, Water of Ilay, Water of Ylay, Watter of Ylay, Wattersyd of Yle, Ylay), NJ3945, NJ5049, NO1873, NO2160, NO2443; 301, 305, 331, 333, 354, 357, 369
River Ness (Nes), NH6441; 626
River Nith (Neyth, Nithe, Nyth), NS7113, NX8890, NY0062; 272, 277, 609, 610, 621
River Spey, NJ3151, NH7801, NH9620, NJ0830, NN4595; 466
River Tay (Walter of Tay, Wattir of Tay), NN7947, NO0441, NO1521; 30, 35, 283, 287, 289
River Thurso (Watter of Thurseth), ND1154; 628
River Tweed (Tueid, Tuyde, Twid), NT1435, NT4136, NT7738, NT9351, NT9351, NT9651; 222, 230, 233, 236, 239, 258
Rivole de Paislay et Blakstoun, *Paisley abbey*, 530n.
Riyardis, *Monkland* [*Rawyards, NS7766], 498
Robbers' Gate, *see* Aberdeen, Gallow Gait
Robertland, NS4447; 573, 580
Roberton (Robertoun, Roberttoun) [Wiston and Roberton parish, NS9130], 213, 213n., 224, 232, 234, 237, 240, 242, 244; kirk of, 224, 232, 234, 237, 240, 242, 243; vicarage of, 213, 213n., 526
Robroyston (Robrestoun) [Mains, NS6369], 497
Rochesterik, Rochesterrig, *Greenlaw* [*Rowchester Ho, NT7343], 235, 238, 241
Rochsoles (Rouchsollis, Ruscholis), NS7567; 102, 498
Rochtnakenzeis, *see* Rothmackenzie
Rodel (Roidill), NG0483; parsonage of, 675
Rodono (Powdono, Rowdono), *Melrose abbey* [Hotel, NT2321], 208, 259
Rogerhill, NS7843; 233

Roidill, see Rodel
Romanograrige [Romanno House Fm, NT1648], 101
Rome (Roome, Rowme), *Italy*, 63, 159n., 370, 374, 409, 435n., 507, 650
Rona na nav, *bishopric of the Isles*, 675
Ronaldsay, see North Ronaldsay, South Ronaldsay
Rood altar, *Manor parish kirk*, 255, 525
Roome, see Rome
Roresounis Croft (Waltir-McRoreiscroftis), *Fearn abbey*, 634, 634n.
Ros, see Ross
Ros, *with Gallowflat, Perth Charterhouse*, 285, 288
Roscobie (Rascobie, Rescobe, Rescoby), *Dunfermline*, NT0892; 23, 27, 28, 46, 49, 50
Rosdew, see Rossdhu
Rosehill, NS8540; 232n.
Roseisle (Rossill, Rossyill), NJ1467; 468, 482
Rosemarkie (Bosmerkin, Ramsky, Ramskye, Rosmaky, Rosmerkie, Rosmerky), NH7357; 626, 634, 637, 642, 643; kirk of, 634, 637, 642; minister of, 635; vicar of, 635, 643
Roskyne, see Rosskeen
Roslin (Roslein, Rosling, Roslyng), NT2763; 141; waulk mill of, 142
—— collegiate kirk, 116; provost of, 117
Roslyng Wodheid [Roslin, NT2763], 142
Rosmaky, Rosmerkie, Rosmerkin, Rosmerky, see Rosemarkie
Rosneath (Rasneth, Rosneth), NS2583; 529, 530; kirk of, 529, 530
Ross (Ros, Rossansis, Rossensis), 461n, 491n, 481, 482, 491, 625, 626, 627, 634, 635, 636, 637, 638, 638n., 639, 640, 642, 643, 644, 645, 646, 651, 652, 653; archdeacon of, 638; archdeaconry of, 638; bishop of, 7, 7n., 8, 8n., 114, 143, 143n., 491, 491n., 512, 512n., 574, 574n., 577, 577n., 579, 579n., 627; bishopric of, 282, 302n., 625-627; chamberlain of bishopric of, 625, 626; chancellor of, 634-635; chancellory of, 634-635; chanter of, 642; chantory of, 642; chaplains of, 635; dean of, 642; deanery of, 642; subdean of, 651, 652; subdeanery of, 159, 159n., 651-652; treasurership of, 637
—— cathedral, see Fortrose
Rossdhu (Rosdew) [Ho, NS3689], 542; chaplain of Lady chaplainry of, 542
Rosse Easter, *Lindores abbey* [Wester Rossie, NO2512], 33, 37
Rosse Wester, see Wester Rossie
Rosse, *bishopric of the Isles*, 673, 674, 675
Rossie (Rossy), *parish*, 16, 18, 77, 78; kirk of, 16, 18; vicar of, 77; vicarage of, 77-78, 412
Rossill, Rossyill, see Roseisle
Rosskeen (Reskyne, Roskyne), NH6174; 644; Rosskeen, parson of, 644; parsonage of, 644
Rossyclero, Rossythletay, *St Andrews priory*, 11, 19
Rossyth, see Rosyth
Rossythletay, see Rossyclero
Roswod, *Lesmahagow priory* [*Rosehill, *Woodfoot, NS8540], 232, 232n.
Rosyth (Rassyth, Rossyth), NT1183; 62, 62n.
Rothart Holme, *with Skellyhill, Kelso abbey*, 231n.
Rothes (Rothos), NJ2749; 449, 474n., 475, 475n., 491, 491n., 494; mains of, 449, 491; parish of, 491; parson of, 475, 491; parsonage of, 449, 475, 491; vicarage of, 475, 491
Rothesay (Rothsay), NS0865; 143
Rothiemay (Rothemay, Rothomey) [Castle, NJ5548], 428, 466; kirk of, 466
Rothiemurchus (Rothemurcus, Rothymurcus), NH8908, NH9408; 465, 465n., 466; kirk of, 466; vicar of, 465, 465n.; vicarage, 465, 465n.
Rothmackenzie (Rochtnakenzeis), NJ5758; 438
Rothmell, see Rotmell
Rotho, see Ratho
Rothobiares, see Ratho Byres
Rothos, see Rothes
Rothven, see Rathven
Rotmell (Rothmell, Bot Mell) [Fm, NO0047], 303, 305, 342, 345
Rotmethe, *Dallas*, 492
Rouchsollis, see Rochsoles
Rouchsollo, Rouchsolloch, see

INDEX OF PLACES 871

Ruchsellocht
Rousay (Ralsay, Rowsay, Ralsay), HY4030; 655, 656, 660, 662, 669, 670; kirk of, 660; minister of, 660; parsonage of, 655, 656, 669
Row of Menteith, *Stirling*, 554
Rowallan (Rowalane, Rowallone), NS4342; mains of, 573, 579
Rowanstoun, Rowenstones, Rowenstounes, Rowistoun, *Gordon*, 227, 227n., 235, 238
Rowdono, *see* Rodono
Rowens (Rowenes), ND2347; 632
Rowenstones, *see* Rowanstoun
Rowistoun, *see* Rowanstoun
Rowny, *see* Dounie, *Edderton parish*
Rowsay, *see* Rousay
Roxburgh (Roxburcht), NT6930; 222, 230, 233, 236, 239; St James' kirk in, 224, 232, 234, 237, 240
——, *sheriffdom*, 1nn., 159n., 175n., 213, 213n., 215, 217n., 223, 245n., 264n., 365n.
Ruchinving, *Monkland parish*, 498
Ruchsellocht (Rouchsollo, Rouchsolloch), *Monkland*, 102, 498
Rud Croft, *beside Stirling*, 556
Rudwellis, *see* Redwells
Ruffanis, *see* Ruthven
Ruffell (Ruffill), NO0941; prebend of Dunkeld cathedral, 316; prebendary of, 316
Ruglen, *see* Rutherglen
Ruid acres, *see* Hood acres
Ruid Kirk, *Stirling*, 39
Rumbleton (Rymmiltoun), NT6845; 227
Rumbletonlaw (Rymiltounlaw, Rymmiltounlaw, Rymyltounlaw), NT6745; 228, 234, 238, 241
Rumelzeiris, *see* Drumelzier
Rumgally (Rumgallie) [Ho, NO4014], prebend (with Dura) of St Mary on the Rock collegiate kirk, St Andrews, 71
Runynnour, *Dallas*, 492
Ruschaw Easter, *Calder*, 98
Ruschaw Wester, *Calder*, 98
Ruscholis, *see* Rochsoles
Russellis Mylne, *Cupar* [*Russell Mains, NO3512], 14
Rutherford (Rutherfuird), *Lanark*, 517, 517n.
Rutherford (Rutherfuird), *Maxton*, NT6430; 189, 191
Rutherglen (Ruglen), NS5862; 519, 528; kirk of, 528; vicar of, 519; vicarage of, 519
Ruthrieston (Rutherstoun), NJ9204; 423
Ruthven (Riffan, Riffanis, Ruffanis), *Brechin diocese*, NO2848; 305, 353, 359; mains of, 305; vicar of, 305; vicarage of, 305-306
Ruthven, *Badenoch*, NN7699; 428, 474; kirk of, 474; parsonage of, 474; vicar of, 428, 493; vicarage of, 428, 493
Rydane, *see* Ryden
Ryden (Rydane) [Mains, NS7468], 102
Rydpeth, *see* Reidpeth
Rymiltounlaw, *see* Rumbletonlaw
Rymmiltoun, *see* Rumbleton
Rymmiltounlaw, *see* Rumbletonlaw
Rymyltounlaw, *see* Rumbletonlaw
Rynd, *see* Rhynd
Ryndis Easter and Wester, *see* Rhynd
Ryndislaw, Ryndslaw, *with Crafrae, Haddington priory*, 162, 165, 165n., 176, 178, 181
Ryne, *see* Rhynie
Rynmure, *see* Renmure
Rynnis of Galloway, *see* Rhins
Ryschill, with *West Nisbet, Edrom*, 200
Ryslawrig [Ryslaw, NT7948], 226
Ryslawtoun [Ryslaw, NT7948], 226

Sacrista Major, prebend of Glasgow, New kirk, 523
Sadilhillok, *Ruthven*, 305
Sagy, *see* Seggie
St Agnes chaplainry, *Brechin cathedral*, 372
St Andrews (Sanctandrois, Sanctiandree), NO5016; 2, 5, 6n., 10, 11, 13, 18, 19, 30, 35, 55, 58, 65n., 71, 77, 80, 86, 87, 89, 96, 104n., 105, 110, 156n., 170, 170n., 282, 302, 307, 317, 321, 323, 331, 339, 360, 367, 373, 381, 382, 395, 396, 398, 403, 404, 407, 408, 512n.; archbishop of, 6, 6n., 7, 24, 104, 104n., 105, 170, 170n., 218, 218n., 302, 302nn., 360, 367, 373, 373n., 381, 381n., 527; archbishop of, 218; archbishopric of, 1-7, 90, 104n., 282, 495n.; archdeacon of, 65, 380n.; archdeaconry, 3, 7n., 65n., 66, 66n.,

67, 67n.; city of, 2; dean of the Christianity of, 5, 81; diocese, 67n., 68n., 72n., 78, 79n., 80, 81, 83n., 104n., 194, 195, 217; great customs of city of, 2; kirk of, 9, 12, 17, 20, 21; Official of, 323, 331, 403; Trinity chaplainry in, 512n.; vicarage of, 86; Newmylne of, 18
—— 'abbey kirk', 65; altar of St John the Baptist in, 65; altar of St John the Evangelist in, 65
——, Black friars (Dominicans) of, 89
—— castle, 3
—— priory, 8, 9, 10, 11, 12, 16, 17, 19, 20, 20n., 21, 77n., 86n., 282; prior of, 9, 8-21; see also 'Abbey kirk' above
——, St Leonard's college, 395
——, St Mary on the Rock collegiate kirk (Kirkheuch, Lady Heuch), 55, 55n., 65n., 71, 77, 80, 86, 87; provostry of, 55, 55n.; see also Cairns, Dura, Kingask, Kinglassie, Kinkell, Lambeletham, Rumgally
——, St Mary's (New) college, 64, 65, 364n.; altar of St John the Evangelist in, 65
——, St Nicholas' hospital in, 13, 20, 21
——, St Salvator's college, 96; see also Cranstoun Riddel
——, Trinity kirk (Ecclesia Trinitatis), 1n., 9, 9n., 12, 17, 20, 21, 71, 71n., 86; altar of St Anthony in, 65; Magdalene altar in, 1n.; Trinity chaplainry in, 71, 71n., 512n.
—— university, 64, 364, 364n., 395; rector of, 307
St Andrews Lhanbryde, see Lhanbryde
St Andrew's altar, *Elgin cathedral*, 478, see also Clyne
St Andrew's altar, *Stirling parish kirk*, 551
St Andrew's chapel, *Garvock*, 82
St Andrew's chaplainry, *Caithness*, 632; chaplain of, 632
St Andrew's collegiate kirk, *Peebles*, 254, 254n.
St Andrew's kirk, at Kilmalemnock, 466, 466n.; parish of, 468
St Andrew's parish (Sandrowis, Sanctandros, Sanct Androis), *Orkney bishopric*, HY5106; parish of (Tankerness), 657, 660, 662, 669;

minister of, 660
St Anne (and St Mary), collegiate kirk, see Glasgow, New kirk
St Anthony's altar, *Trinity kirk, St Andrews*, 65
St Anthony's hospital, *Leith*, 113, 114
St Anthony's yard, *Leith*, 114
St Barbara's altar, *Dundee kirk*, 399
St Bathans (St Bothanes, St Bothanis, St Bothenis, St Bothans) [Abbey St Bathans, NT7662], 170, 170n., 192, 193, 193n., 194; corn mill of, 192; waulk mill of, 192
St Bathans priory, 192-193, 194, 207n.; prioress of, 170, 170n., 192
St Blaise or Nomine Jesu altar, *St Giles' collegiate kirk, Edinburgh*, 117n.; see also Nomine Jesu and Holy Blood chaplainry
St Blaise's chaplainry, *Dunblane cathedral*, 314, 328
St Bothanes, St Bothanis, St Bothans, St Bothenis, see St Bathans
St Bride's altar, *Linlithgow kirk*, 154
St Catherine's (Katherinis, Katreinis) chapel, *Aberdeen*, 419, 421
St Catherine's altar, *Crail collegiate kirk*, 66
St Catherine's altar, *Edinburgh*, 134
St Catherine's altar, *Forgandenny*, 329
St Catherine's altar, *Kirkwall cathedral*, 663, 663n., 664, 665
St Catherine's chapel, *Dirleton*, 190-191
St Catherine's chaplainry, *Elgin cathedral*, 490
St Catherine's chaplainry, *Newbattle*, 68, 144
St Catherine's chaplainry, *Newburgh*, 378
St Catherine's Hill, *Aberdeen*, 419, 421
St Catherine's kirk, *Newburgh*, 378
St Catherine's kirk, *Cupar*, 81; chaplainry of Our Lady in, 81
St Colm (and St Mirin), altar of, *Paisley parish kirk*, 535
St Colme (Sanct Cowme), *Sanday*, 660n., 663n. 664, 664n. 666n.; kirk of, 664, 664n.; parson of, 664; parsonage of, 664, 664n.
St Colme's altar, *St Giles' collegiate kirk, Edinburgh*, 123
St Cosmo and St Damian chaplainry, *Bothans collegiate kirk*, 176

INDEX OF PLACES 873

St Cosmo chaplainry, *Bothans collegiate kirk*, 176
St Cuthbert's altar, *Dalmeny kirk*, 149, 150, 159
St Cuthbert's kirk, beside Edinburgh, *see* Edinburgh, St Cuthbert's kirk
St Cuthbert's *parish, see* Eccles
St Damian (and St Cosmo), chaplainry, *Bothans collegiate kirk*, 176
St Denis' altar, *St Giles' collegiate kirk, Edinburgh*, 132
St Duthac's altar, *St Giles' collegiate kirk, Edinburgh*, 124
St Duthac's chaplainry, *Arbroath*, 380
St Duthac's chaplainry, *Elgin cathedral*, 476, 476n.
St Edmund's prebend, *Bothans collegiate kirk*, 175
St Eloy's altar, *St Giles' collegiate kirk, Edinburgh*, 134, 135
St Fillan's chaplainry, *Doune* [Doune, NN7301], 345
St Francis' altar, *St Giles' collegiate kirk, Edinburgh*, 132
St Geill, Gelis, *see* Edinburgh, St Giles' collegiate kirk
St George's hospital, *Dunkeld*, 304n.
St James' altar, *Crail collegiate kirk*, 66, 85
St James' (and Our Lady's) chapel, *see* Newhaven
St James' chaplainry, *Glasgow cathedral*, 503
St James' kirk, *Kelso*, 224, 225n., 232, 234, 237, 240
St Jeill, Jeillis, Jellis, *see* Edinburgh, St Giles' collegiate kirk
St John the Baptist kirk, *see* Perth
St John the Baptist's altar, *Crail collegiate kirk*, 66
St John the Baptist's altar, *St Andrews abbey kirk*, 65
St John the Baptist's altar, *Stirling parish kirk*, 555
St John the Evangelist's altar, *Crail collegiate kirk*, 65
St John the Evangelist's altar, *St Andrews abbey kirk*, 65
St John the Evangelist's altar, *St Mary's college, St Andrews*, 65
St John the Evangelist's chaplainry, *St Giles' collegiate kirk, Edinburgh*, 138

St John's aisle, *St Giles' collegiate kirk, Edinburgh*, 140
St John's altar, *Inverkeithing*, 73
St John's chapel, *Barras*, 408
St John's chapel, *Dryburgh abbey*, 189, 191, 197
St John's chapel, *Eccles priory*, 183, 183n.
St John's chapel, with Kedslie, *Dryburgh abbey*, 189, 191, 197
St John's chaplainry, *Helmsdale*, 632
St John's croft, *Sorbie*, 593; *Soulseat abbey*, 598
St John's kirk, *see* Perth
St John's prebend, *Kirkwall cathedral*, 667
St Johnstone, *see* Perth
St Laurence (and St Stephen), altar of, *Glasgow cathedral*, 500
St Laurence croft, *Stirling*, 545
St Laurence House, *St Andrews priory*, 17
St Laurence House, *Sciennes priory, Edinburgh*, 94
St Laurence's altar, *St Giles' collegiate kirk*, 125-126, 132
St Laurence's altar, *Stirling parish kirk*, 545
St Laurence's chaplainry, *Dingwall*, 636
St Leonard's chapel, *Polton*, 97
St Leonard's chaplainry, *Moray*, 484
St Leonard's hospital, *Edinburgh*, 92
St Leonard's, hospital, *Peebles*, 253
St Machar's cathedral, Aberdeen, vicarage of, 423, 424
St Madoes (Sanct Modice, Sanctmodoce), *parish*, NO1920; 345; parson of, 345; parsonage of, 345; vicarage of, 345
St Magdalene's chapel (Maidlane), *Eccles priory*, 183, 183n.
St Magnus' prebend, *Kirkwall cathedral*, 666
St Margaret's Stone (Sanct Margaretis, Margarettis Stane), NT1084; 23, 27, 28, 45, 49, 50
St Martin in the Nungate, chapel of (Sanct Martines kirk), *Haddington priory*, 162, 162n.
St Martins (Sanct Martenis), NH6463; 646
St Mary and St Anne collegiate kirk, *see* Glasgow, New Kirk

St Mary in the Forest, *see* Yarrow
St Mary kirk, *Stronsay*, 660n.
St Mary of Farmainishop, *see* Yarrow
St Mary of Garvald, chapel of, 553n.; *see also* Kirk O' Muir, Kirkton, St Ninians
St Mary of the Fields' collegiate kirk, *see* Edinburgh
St Mary of the Lowes, *see* Yarrow
St Mary on the Rock collegiate kirk, *see* St Andrews
St Mary's chapel, Edinburgh, *see* Barras
St Mary's chaplainry, *Cocklarachy, Aberdeen*, 447, 447n.
St Mary's college, *see* St Andrews
St Mary's Isle (Sanct Marie Iyle, Sancte Marie de Traill, Sanct Maries Ile); NX6749; 601, 617, 618, 620
— priory, 617, 618-620; prior of, 617
St Matthew, prebend of St Mary of the Fields' collegiate kirk, Edinburgh, 139
St Michael's altar, *Crail collegiate kirk*, 65
St Michael's altar, *Crieff*, 337
St Michael's altar, *Stirling parish kirk*, 554
St Michael's altarage, *St Giles' collegiate kirk, Edinburgh*, 140
St Michael's chaplainry, *Dunblane cathedral*, 335
St Michael's chaplainry, *Glasgow cathedral*, 501, 515, 615
St Michael's Cross, *see* Crossmichael
St Mirin and St Colm, altar of, *Paisley parish kirk*, 535
St Monance (Sanct Monanis, Sanctmonanis), *Fife*, NO5201; 17, 89, 92, 146, 156, 167; friary of, 89; Crowe Sanct Monanis, 167
St Monance, *Ross*, 635; chaplainry of, 639
St Mungo's altar, *Berwick*, 553
St Mungo's altar, *Culross abbey kirk*, 522
St Mungo's altar, *Currie*, 139; chaplain of, 139
St Mungo's chaplainry, *beside Glasgow*, 502n., 503
St Mungo's chaplainry, *Glasgow cathedral*, 501
St Nicholas, NO5115; hospital of, St Andrews, 13, 20, 21, 89

St Nicholas' altar, *Crail collegiate kirk*, 66
St Nicholas' altar, *Glasgow cathedral*, 525
St Nicholas' altar, *St Giles' collegiate kirk, Edinburgh*, 137
St Nicholas' chapel, *Lanark*, 514
St Nicholas' chaplainry, *Arbroath*, 400
St Nicholas' chaplainry, *Dunblane cathedral*, 327
St Nicholas' kirk, *Aberdeen*, 447
St Nicholas' kirk, *Lanark*, 521, 524
St Nicholas' kirk, *Stronsay*, 660, 660n., 664, 666; minister of, 660; parson of, 666; parsonage of, 666
St Ninian's acre, *Coupar Angus abbey*, 353
St Ninian's altar, *Dunkeld cathedral*, 320
St Ninian's chapel, *Darnley*, 501, 501n.
St Ninian's chapel, *Scone abbey*, 332
St Ninian's croft, *Stirling*, 546
St Ninians (Kirkton), 553n.; *see also* Kirk O' Muir, St Mary of Garvald
St Ola (Lawrok, Sanct Awlaw, Sanct Awlay, Sowlowis, Sanct Awlauis, St Olais), *parish*, HY4510; 657, 662, 663, 669; parish of, 657, 662, 663, 669, 678, 679
St Patrick's altar, *St Giles' collegiate kirk, Edinburgh*, 134-135
St Paul's chapel, *Perth*, 283, 286
St Paul's work, hospital of, *Leith*, 95
St Peter's altar, *Aberdeen*, 419, 422
St Peter's altar, *Ayr parish kirk*, 557
St Peter's chaplainry, *Glasgow cathedral*, 501
St Peter's kirk, *South Ronaldsay*, 656, 659, 660, 669
St Peter's kirk, *Stronsay*, 656, 660, 660n., 669; minister of, 660; parsonage of, 656, 669
St Quivox, NS3724; 529; kirk of, 529
St Roche's altar, *St Giles' collegiate kirk, Edinburgh*, 138-139
St Roche's chaplainry, *beside Glasgow*, 514
St Roche's kirk, *beside Glasgow*, 514
St Rule (Sanct Regule), *Ross*, chaplain of, 644; chaplainry of, 644
St Salvator's altar, *St Giles' collegiate kirk, Edinburgh*, 96, 129, 130
St Salvator's altarage, *St Giles'*

collegiate kirk, Edinburgh, 141
St Salvator's chaplainry in Holy Blood aisle, *St Giles' collegiate kirk, Edinburgh*, 135-137
St Salvator's college, *see* St Andrews
St Serf's Island, NO1500; 56
St Stephen and St Laurence altar, *Glasgow cathedral*, 500
St Stephen's chaplainry, *Dundee kirk*, 400
St Stephen's chaplainry, *Glasgow cathedral*, 502
St Stephen's chaplainry, *Perth parish kirk*, 306
St Thomas the Evangelist's altar, *Aberdeen*, 419, 421
St Thomas' altar, *Cramond kirk*, 144, 149-150
St Thomas' altar, *Elgin cathedral*, 477
St Thomas' chapel, *Maxwell*, 225
St Thomas' chaplainry, *Stirling*, 551n.
St Thomas' land, *Stirling*, 556
St Triduana's Aisle (Sentredwallis Yle), *Restalrig collegiate kirk*, 143, 143n.
St Vigeans, NO6342; 360; vicar of, 360
St Vincent, or songschool, prebend of St Mary of the Fields' collegiate kirk, Edinburgh, 124
Salbarry, *Lesmahagow*, 231
Salfet, *Melrose abbey*, 258
Saline (Sawling), NT0292; 300, 335; kirk of, 300; parsonage of, 300; 335; vicarage of, 335
Salisbery [*Salisbury Craigs, NT2673], 92
Sallway, *see* Swona
Salpinschaw, *see* Shapinsay
Salseitt, *see* Soulseat
Salt Pans, *Kirkcaldy*, 24, 24n., 46
Saltcottis, *Duffus*, 482
Saltoun [Hall, NT4668], 190, 191, 197, 197n.; kirk of, 190, 191, 197, 197n.; vicarage of, 174
Samieston (Samuelstoun), NT7221; 219
Samuelston (Samuelstoun), *NT4870; 77, 169
Samuelstoun, *Jedburgh abbey*, *see* Samieston
Sancour, *see* Sanquhar
Sanct Allauis, *see* St Ola
Sanct Awlaw, Sanct Awlay, *see* St Ola
Sanct Colmis Inche, *see* Inchcolm
Sanct Cowme, *see* St Colm, Sanday

Sanct Geiligrainge, *of the Sciennes*, 500
Sanct Johnnesoun, Sanct Johnnestoun, *see* Perth
Sanct John's croft, *Soulseat abbey*, 598
Sanct Marie Iyle, *see* St Mary's Isle
Sanct Modice, Sanctmodoce, *see* St Madoes
Sanct Phillane, *see* St Fillans
Sanct Regule, *see* St Rule
Sanctandrie, Sanctandrois, *see* St Andrews
Sanctandrois kirk, *see* St Andrews, Trinity kirk
Sanctandrois, Sanctandros, *Moray*, *see* Lhanbryde
Sanctcolmisinche, *see* Inchcolm
Sancte Crucis, *see* Holyroodhouse
Sancte Marie de Traill, *see* St Mary's Isle
Sancti Kevoci, *see* St Quivox
Sanctiandree, *see* St Andrews
Sanctjhon, *see* Perth
Sanctmonanis, *see* St Monance
Sand, *Uist*, parsonage of, 674
Sanday (Sanda), HY6640; 656, 660, 660n., 662, 663n., 664, 664n., 666, 666n., 667, 669, 671; vicar of, 666, 667; vicarage of, 666, 666n., 667; *see also* Cross kirk, Lady kirk, St Colme
Sande Riggis, Sanderigis, *with Dalkeith, Newbattle abbey*, 103, 103n., 116, 116n., 141, 142
Sande, Sandis, *Musselburgh*, 26, 43n., 49
Sandefuird, Sandfuird, Balbuthie, 146, 148, 167
Sandfurd Hay, *Forgan* [*Sandford Hill, NO4024], 14, 20
Sandfurd Nachtan, Sandfurd Narne, Sandfurd Nauchlame, *Forgan* [*Sandford Hill, NO4024], 14, 20
Sandis, *see* Sandside
Sandlandis, *Kinglassie, see* Smiddeland
Sandrowis, *see* St Andrews parish
Sands (Sandis) [House, NT9586], 289, 291
Sandside, *Deirness*, HU5906; 655, 655n., 662, 671
Sandsting (Standsting), *parish*, HU3153; 658, 658n., 660, 670
—— and Aithsting, parish of, 660; *see also* Aithsting

Sandwick (Sandwik, Sandwike, Sandwill), *parish*, HY2519; 657, 659, 668, 670, 670n.; kirk of, 659; minister of, 659
— and Stromness, vicarage of, 668; *see also* Stromness
Sandwill, *see* Sandwick
Sandyruidis, *Kirkcaldy*, 24, 46
Sanehill, *see* Stoneyhill
Sannik, Sannyk, *see* Senwick
Sanquhar (Sancour), NS7809; 271, 271n., 273, 276, 276n., 623; parson of, 276
— parish kirk, altar of Holy Blood in, 273; altar of Our Lady in, 273; chaplain of Holy Blood altar in, 273; chaplain of Our Lady altar in, 273
Sansyd, *see* Sandside
Sanwik, *see* Senwick
Sastoun, *see* Cookston
Sauchop (Savope), *Methven collegiate kirk*, 296, 298
Sauchope (Sauchop), *NO6208; 162, 177, 178, 181n., 182
Sauchquhytoun, *see* Windehillis
Sauchtonehall, *Edinburgh*, 132
Sauchy town, *see* Windehillis
Sauldseitt, Saullseitt, Saulseitt, *see* Soulseat
Savok, Sawik, *Coldstream priory*, 186, 187
Savope, *see* Sauchop
Sawik, *Coldstream priory*, *see* Savok
Sawilk, vicarage of, *see* Senwick
Sawling, *see* Saline
Scabcleuch (Scabecleucht), NT2414; 258
Scahyntie, *with Crosshouse, Restalrig collegiate kirk*, 142
Scalpinsay, Scalpinschaw, *see* Shapinsay
Scamnoch, Scamnocht [*Stennock, NX4737], 594, 596
Scampstar [*Stemster, ND1761], 632
Scarmclate (Scarmalat, Scarmlat, Scarneclati), ND2059; 632; mill of, 633
Scarpinsay, *see* Shapinsay
Scatirrun [*Strathtyrum Ho, NO4817], 66
Scatsta (Skatsay, Skatstay, Skattay), HU3872; 658, 660, 670; kirk of, 660
Scaythmure, *see* Skaithmuir

Schabray, Shabray [*Skelbrae, HY6743], 664
Schailpinsay, *see* Shapinsay
Schaipnysay, *see* Shapinsay
Schalowchtbroun, *Glasserton*, 593
Schanwell, see Shanwell
Schavitinchame, *Dunblane*, 343
Schaw, *Selkirk*, 230n.
Schawhill, *see* Shawhill
Scheariscroft, *Monkland parish*, 498
Scheil Hopdykis, *Coldingham priory*, 198,
Scheilbank, *Stobo*, 249
Scheilhill, *see* Shielhill
Scheillis, *Collessie*, 33, 37
Scheitland, *see* Shetland
Schelhill, *Calder*, 98
Schelis, Cheles, *Inchcolm abbey*, 62, 64
Schenan, *Tongland, Galloway bishopric*, 607
Schetaucht Ley, *see* Sheddocksley
Schethin, NJ8832; 435
Schettoun, *Markinch*, 15
Schewsay, *see* Switha
Schichel, *see* Stichill
Schilstounbreist [*Shielstockbraes, NT7523], 214
Schirefhall, *see* Sheriffhall
Schirefstell, *Kinloss abbey*, 462
Schiris, *Forgan*, 14
Schort Cleuch [North Shortcleuch, NS9217], 102
Schortchope, *see* Shorthope
Schortles, *see* Shortlees
Schotsoun Croft [*Scotstoun, NT1377], 155
Schottiscraig, *see* Scotscraig
Schoulderflatt, *Beith*, 573
Schyirdaill, *see* Sheardale
Schyris Mylne, *see* Shires Mill
Sciennes, *see* Edinburgh, priory of Sciennes
Sclintheuchis, *Calder*, 98
Scockness (Stoknes), HY4532; 655
Scone (Scounis, Scune) [Old Scone, NO1226], 23, 46, 282, 283, 287, 306, 306n., 331, 332, 333, 334n., 355; kirk of, 333
— abbey, 282, 331-334, 631n.; commendator of, 306, 306n., 307, 331n.; prior of, 631n.
Scony, *see* Scoonie
Scoonie (Scony, Scowny, Scuny,

Scunye, Skoniycht, Skony, Skuny), NO3801; 9, 15, 17, 17n., 21, 71, 71n., 504; kirk of, 9, 15, 17, 17n., 21; vicarage of, 71, 71n.
Scotland, 60, 111, 302, 330, 401, 495n., 558, 678
Scotlandwell, NO1801; 56; priory of, 56
Scotscraig (Schottiscraig, Scotiscraig, Scottiscraig), NO4428; 2, 5, 13, 20, 90, 90n.
Scottistoun, *Aberdeen deanery* [*Mains of Scotstown, NJ9210], 422, 423
Scottistoun, *Ecclesgreig*, 16
Scottistoun, *Mearns* [*Scotston of Kirkside, NO7463], 31, 35
Scounis, *see* Scone
Scowny, *see* Scoonie
Scrabster (Strybustar), ND1070; 657, 670
Scraesburgh (Scraisburcht), NT6718; 219
Scroggs, NY1681; 500
Scune, *see* Scone
Scuny, Scunye, *see* Scoonie
Seaside, Hesyd, Sesyd, *Inchcolm abbey*, 63, 63n.
Seaton (Setoun) [Ho, NJ9409], 423
Sedacht, *see* Sedoch
Sedhill, Seidhill, *see* Leidhill
Sedoch, Sedocht, Sedacht [*Shaddock, NX4739], 594, 596
Seggie (Sagy, Segy), NO4419; 13, 20
Segis, *Lindores abbey*, 32
Segy, *see* Seggie
Seidhill, Sedhill, *see* Leidhill
Seillebalbie, Seillybalbie, *see* Sellebair
Seinis, *see* Edinburgh, priory of Sciennes
Seipseis, Seipsis, *see* Sypsies
Selkirk (Selkrik, Selrig), NT4728; 1n., 175, 223, 224n., 230, 230n., 232, 232n., 234, 237, 240, 242, 259, 264; benefice of, 224n.; kirk of, 230, 230n.; parsonage of, 230, 234, 237, 240, 242; vicarage of, 230, 232, 234, 237, 240, 242
——, *sheriffdom*, 1nn., 175n., 264n.
Selkrik, *see* Selkirk
Sellebair, Seillebalbie, Seillybalbie, Sellebalbe, Sellebalbie, Sellebare [*Silverbarton, NT2286], 24, 28, 46, 50
Sellebalbe, Sellebalbie, *see* Sellebair

Selletoun Halkheid, Selletoun Hallheid, Stelletoun Hakheid, *Dunfermline*, 27, 28n., 49, 50
Selletoun Sanctandrois, *see* Selletoun Sandis
Selletoun Sandis, Selletoun Sanctandrois, Sillietoun-Sanderis, *Dunfermline*, 27, 27n., 28, 28n., 49, 50
Selletoun, *Dunfermline*, 23, 27, 45, 49
Selrig, *see* Selkirk
Semaisters lands, *see* Femaisteris lands
Semple (Sempill) [Castle Semple, NS3760], 531
—— collegiate kirk, 531; provost of, 531
Semprim, *see* Simprim
Sennikin, *see* Senwick
Sennis, *see* Edinburgh, priory of Sciennes
Sentredwallis Yle, *see* St Triduana's Aisle
Senwick (Sannik, Sannyk, Sanwik, Sawilk, Sennikin) [Church, NX6546], 587, 590, 592, 598, 604, 607; kirk of, 587, 589, 592; parish of, 587; parsonage of, 607; mill croft of, 587; reader at, 604; vicarage of, 604
Seraffeat, *St Andrews priory*, 11
Seres, *see* Ceres
Sesfuird, *see* Cessford
Sessnak, *see* Cessnock
Sesyd, *see* Seaside
Seton (Seytoun), NT4174; 99
—— collegiate kirk, 99; provost of, 99-100; provostry of, 99-100
Setoun, *Aberdeen*, *see* Seaton
Sey Ile, *see* Fair Yle
Seytoun, *see* Seton
Shabray, *see* Schabray
Shannabank (Channonbank), NT7562; 166
Shanwell (Schanwell), NO4726; 13, 20
Shapinsay (Salpinschaw, Scalpinsay, Scalpinschaw, Scarpinsay, Schailpinsay, Schaipnysay, Schalpinsay, Schalpinschaw), HY5017; 656, 660, 662, 669; kirk of, 660; minister of, 660
Shawhill (Schawhill), NS4537; 531
Sheardale (Schyirdaill), NS9596; 24, 46
Sheddocksley (Schetaucht Ley), NJ8907; 423

Sheriffhall (Schirefhall), NT3167; 26, 43, 48, 49, 53, 54
Shetland (Scheitland, Zeitland, Zetland), HU4167; 114, 282, 655, 658, 659, 660, 662, 663, 664, 665, 670, 671; archdeacon of, 660, 665; archdeaconry of, 665
Shielhill (Scheilhill), NO1133; 309
Shirmers (Chirmis), NX6574; 603
Shires Mill (Schyris Mylne), NT9987; 290
Shorthope (Schortchope), NT2212; 258
Shortlees (Schortles), NS4335; 531
Sibbaldbie (Sibbalbie, Sibbilbe), NY1487; 216, 221; kirk of, 216
Sillietoun-Sanderis, *see* Selletoun Sandis
Simprim (Semprim, Simpreay, Sumpryne, Symphryn, Symprein, Sympriane, Symprine, Sympryin, Symprym, Sympryn), NT8545; 186, 223, 226, 232, 233, 234, 235, 237, 238, 239, 240, 241, 242; curate of, 226; kirk of, 226, 232, 234, 238, 240, 242; vicarage of, 226, 232, 235, 238, 241
Sipittall, *see* Spital
Sipseis, *see* Sypsies
Sisterpath (Systerpeth, Fyscer Peth), NT7548; 226, 226n.
Skaill, *see* Skiall
Skaithmuir (Scaythmure), NT8443; 186
Skalmorlie, *see* Skelmorlie
Skardy, *Tain parish*, 651
Skarris, *see* Skerries
Skatsay, Skatstay, Skattay, *see* Scatsta
Skaw, *Ancrum*, 236, 366
Skebo, *see* Skelbo
Skeirachnaheie, *Lewis*, 675
Skeirdrostane, *see* Skirdustan
Skeirkenze, Skeirkenzie, *Kintyre*, 674, 675
Skeirreis, *see* Skerries
Skelbo (Skebo), *Caithness bishopric*, NH7995; 627
Skellarie (Stelary) [Rock, NX4551], 619
Skellocht, *see* Stellock
Skellyhill (Skaillihill, Skellehill), NS7937; 231, 231n., 233
Skelmorlie (Skalmorlie), NS1967; 500
Skemesay, *see* Stronsay
Skenand, *see* Skinnet
Skene (Skeyn) [Ho, NJ7609], 445; parish of, 445
Skeog (Skeoch, Skyocht), NX4539; 598, 599
Skerries (Skarris, Skeirreis), *with Nesting, Shetland*, HU6771; 658, 670
Skeyn, *see* Skene
Skiall (Skaill), ND0267; 627
Skinford (Skynfuird, Skyufuird), NX8483; 622, 623
Skinneris landis, *Dalmeny*, 150
Skinnet (Skenand), ND1261; 631; kirk of, 631
Skirdustan (Skeirdrostane) [Aberlour, NJ2743], 460
—— and Botriphnie, parsonage of, 320, 320n.; *see also* Aberlour, Botriphnie
Skirling (Skraling), NT0739; 7n., 76n., 150, 180, 248n., 250, 253, 537n., 609n.; parson of, 253; parsonage of, 253; vicarage of, 253
Skoniycht, Skony, *see* Scoonie
Skoretholm, with *Skellyhil, Kelso abbey*, 233
Skraling, *see* Skirling
Skubster, *see* Scrabster
Skuny, *see* Scoonie
Skuyfuird, *see* Skinford
Skynfuird, *see* Skinford
Skynthy, *see* Switha
Skyocht, *see* Skeog
Skyrcht [*Broughton Skeog, NX4544], 592
Slaid, Slayid, *Haddington priory*, 162, 165, 177, 178, 181
Slains (Slainis, Slanis) [Castle, NK0529], 425, 441; kirk of, 425, 441
Slamannan (Slamananmure, Sleimanane, Slenannamure), NS8573; 552; glebe of, 552; manse of, 552; parsonage of, 552; vicar pensionary of, 552; vicarage of, 552; vicarage pensionary of, 552
Slanis, *see* Slains
Slatyr, *near Cot-town, Aberdeen deanery*, 424
Sleat (Sleatt), NG6010; 674, 675; parsonage of, 675
Slegden (Clegdenie, Sligdene), NT7347; 227
Slegna, *see* Slognaw
Sleimanane, *see* Slamannan

Slenannamure, *see* Slamannan
Sligdene, *see* Slegden
Slogary, Slograry, *with Beoch*, 585, 585n.
Slognaw (Slegna), NX7458; 602
Smailholm (Smalahame), NT6536; 189, 197, 197n.; kirk of, 189, 197, 197n.; Smalahame Craiges, 189; Smalahame lands, 190, 191; Smalahame Spittell, 190, 191
Smalahame, *see* Smailholm
Smeaton (Smetoun), NT3569; 24, 25, 26, 38, 47, 48
Smeaton (Smetoun), NT5978; 147
Smedan Hous, *Riccarton*, 531
Smeithstoun [*Smithston, NS4112], 571
Smetoun, *see* Smeaton
Smiddeland, Smyddie Land, Smyddy Land, Sandlandis, *Kinglassie*, 24, 29, 46, 51
Smiddie acres, *near Quarrypits, Haddington priory*, 181
Smiddie Landis, Syndry landis, *Kirkcaldy*, 24, 46
Smiddyhill (Smiddehill, Smyddehill), NO6165; 383, 385, 388
Smithis Landis, *Fearn abbey*, 634
Smithwood (Smythwode), NS9510; 102
Smychttie Grene, *Rothes*, 491
Smyddehill, Smyddyhill alias Pakyis land, *with Loanleven, Methven collegiate kirk*, 296, 298
Smyddehill, *Brechin bishopric, see* Smiddyhill
Smythtoun, *Holywood abbey*, 275
Smythwode, *see* Smithwood
Snadoun, *see* Snawdon
Snaw, *see* Snow
Snawdon (Snadoun, Snawdoun), NT5867; 162, 165, 176, 178, 181
Snisport, *see* Snizort
Snizort (Snisport), NG4155; 675; parsonage of, 675
Snow (Snaw), *parish*, 425, 442; curate of, 442; kirk of, 425; parsonage of, 442
Soccach (Sochowche), NJ1248; 492
Sochowche, *see* Soccach
Softlaw [Easter, Wester], NT7532, NT7330; 223, 223n., 231, 234, 237, 240, 242; *see also* Easter, Wester Softlaw
Soiribie, *see* Sorobie

Sokkoth, Sokocht, Sokoth, *with Kincraigie, Dunkeld bishopric*, 303, 304, 342, 345
Sorbie, NX4346; 593, 594, 603; parish of, 593; vicar of, 603; vicarage of, 603
Sornbeg, *see* Little Sorn
Sornebege, Sornibegis, *Galston*, 562, 564
Sornhill, NS5134; 531
Sorobie [Sorobaidh Bay, NL9942], 674; parsonage of, 674
Sorrowlessfield (Soroleisfeild, Sorrowlesfeill) [Mains, NT5367], 208, 257
Soudoun, *see* Southdean
Soulseat (Salseitt, Sauldseitt, Saullseitt, Saulseitt) [Loch, NX1058], 598, 599, 599n.; mains of, 598
—— abbey, 598-600; commendator of, 599, 599n.; chamberlain of, 599, 599n.
Soumertait, *see* Little Sorn
Sounay, *see* Swona
Sourhope (Sowrope), NT8420; 258
Soutarhousis, Soutarhoussis, *Coupar Angus abbey*, 352, 356
Soutercroft, Sowtercroft, *Melrose abbey*, 257, 259
Souterhous, *Monkland*, 102
South Bowhill, *Selkirk*, 230n.
South Cote (Southait, Southcoit, Southgait, Sowthcoit), NT7921; 209, 209n., 258, 259
South Lethans (South Lethemes, South Lethenes, Southlethinis, Southtlothenis), NT0694; 23, 28, 46, 50
South Queensferry (Southferrie, Southferry, Southfery, Southe and Northe the Quene Ferreis, South Queinis Ferry), NT1278; 25, 25n., 38, 47, 51, 155
South Ronaldsay (Rannaldsay Besouth), ND4489; 657, 659n., 669; *see also* Lady kirk, St Peter's kirk
South Sydie, *see* Soutside
South Wod, Southwod, *with Woodhead, Lindores abbey*, 30, 32, 34, 35, 37
Southdean (Soudoun), NT6309; 213; parson of, 213; parsonage of, 213; vicarage of, 213
Southdean Rig (Sowdenrig), NT4305;

259
Southe and Northe the Quene Ferreis, see North Queensferry, South Queensferry
Southeild, see Southfield
Southfad, Southfaid, see Southfod
Southferrie, Southferry, Southfery, see South Queensferry
Southfield (Southeild, Southfeild), with Lucklaw, St Andrews priory, NO4321; 14
Southfield (Southfeild), Lesmahagow, NS7944; 233
Southfod (Foirrdis South and North, Fouddeis South and North, Fouddis, Foudeis, Southfad, Southfaid, South Ford), NT1287; 23, 23n., 27, 45, 49, 50
Southgait, see South Cote
Southlethinis, Southtlothenis, see South Lethans
Southside (South Sydie), Newbattle abbey, NT3663; 101
Southwod, see South Wod
Sowdenrig, see Southdean Rig
Sowlowis, see St Ola
Sowna, see Swona
Sowrope, see Sourhope
Sowtercroft, see Soutercroft
Sowterland, Kinnell parish, 367
Sowtha, see Switha
Sowthcoit, see South Cote
Spadocht, see Speddoch
Speddoch (Spadocht, Speddoche, Spedoche, Spoddoche), Holywood abbey, NX8582; 269, 274
Spedoche, see Speddoch
Spettall, Spettell, Jedburgh abbey, see Spittal
Spey yarde, 'besyd Perth', 306
Spey, see River Spey
Spirelands, Moorfoot, 101
Spital (Sipittall, Spittal, Spittall), ND1654; 631; parson of, 631; parsonage of, 631
Spittal (Spettall, Spettell), Jedburgh abbey, *NT5819; 216; mains, 216, 218
Spittal (Spittaill), parish, Aberdeen, 426, 426n., 434; kirk of, 'called hospital', 434; parsonage of, 426, 434
Spittal Mains, Jedburgh abbey, 216, 218
Spittalhill (Spittal), NS4033; 562, 564

Spittall, Spittell, Reidspittall, prebend of Dunglass collegiate kirk, 166, 166n., 173
Spittalrig (Spittelrig), NT4773; 94
Spittelfeild, with Kipney, Arbroath abbey, 360
Spittell, Cambuslang, NS6758; 508
Spittil, Markinch, 15
Spoddoche, see Speddoch
Spoddoche, see Speddoch
Spott (Spote), NT6775; 171; parson of, 171; parsonage of, 171; prebend of Dunbar collegiate kirk, 171; vicarage of, 171
Spottiswoode (Spottiswod), NT6049; 196, 227
Sprostoun, see Sproustoun
Sprotland, Haddington priory, 180
Sproustoun (Sprostoun, Sproustoun), Kelso abbey, NT7535; 196, 222, 223, 224, 225, 228, 229, 232, 233, 234, 235, 236, 237, 238, 239, 240, 241; kirk of, 224, 225, 232; mains of, 233, 235, 238, 239, 241, 242; parish of, 234, 235, 237, 240, 241; vicar of, 196; vicarage of, 196-197, 228;
Sproustoun, Perth Charterhouse, 284, 287
Spydhill, with Canderside, Kelso abbey, 233
Spynie (Spyne, Spynnie, Spynye), NJ2365; 465, 466, 467, 468, 488; parson of, 488; parsonage of, 488
Stabilgordoun, Stabilgortoun, Stablegordoun, see Staplegordon
Stainhous, Lanark, see Stonehouse
Stainhous, Kilwinning abbey, 573
Staintoun, see Stenton
Stainunsteir [Stemster, ND0365], 628
Stakithort, Dumbarton collegiate kirk, 539
Stalling croftis, near Beoch, Galloway bishopric, 585n., 589
Stalyis, Lesmahagow, 232
Standhous, Stanehous, Stanhous, Kinghorn, see Stenhouse
Standsting, see Sandsting
Stane Morphy, see Stone of Morphie
Stanebyres, see Stonebyre
Stanehall, see Stoneyhill
Stanehill, Staniehill, Stanihill, Stanyhill, see Stoneyhill

Stanehous, Stanhous, *Orkney*, see Stenness
Stanehous, Staynhous, *Edinburgh*, see Stenhouse
Stanmaquhy, *near Mains of Farnell, Brechin deanery*, 365
Stannuk Carbot, Stennok-Corbak, Stennok-Corbat, *near Burgess Outon, Whithorn priory*, 592, 592n.
Stanstil (Stanestall, Stansall), ND2760; 628, 632
Stanyiris, *Kinloss abbey*, 462
Stanywod, see Stoneywood
Staplegordon (Stabilgordoun, Stabilgortoun, Stablegordoun, Stowgordoun), NY3588; 224, 232, 234, 237, 240; kirk of, 224, 232, 234, 237, 240
Starmonthe, Stermonth, Stermontht, Stormoth, *with South Queensferry*, 25, 25n., 38, 47, 51
Starwod, *Duffus*, 482
Stathvey, see Stravithie
Staucherlat, *Riccarton*, 531
Stayhill, see Stoneyhill
Staynhous, see Stenhouse
Stayntoun, see Stenton
Stealstoun, see Steilston
Stedingis, *Garvald*, 162
Stedinurland, *Strathmiglo*, 69
Steidrig, see Todrig
Steilston (Stealstoun, Steilstoun, Stelistoun, Steylistoun), NX9081; 269, 269n., 271, 275, 622, 622n.
Steinlethe, *Hobkirk*, 220
Steinsoun, see Stevenston
Steintoun, see Stenton
Steipfuird, see Stepford
Stelary, see Skellarie
Stelistoun, see Steilston
Stell of Kissok [*Kessock Bridge, NH6647], 639
Stelletoun Hakheid, see Selletoun Halkheid
Stelletoun Sanctandrois, see Selletoun Sandis
Stelling Trie, Stellintrie, Stellyntrie, *Holywood abbey*, 270, 271, 275, 276, 623
Stellock (Skellocht), NX3741; 594
Stenehous, *'in the west land'*, see Stanehous
Stenhous, *Lanark*, see Stonehouse

Stenhous, *Orkney*, see Stenness
Stenhouse (Standhous, Stanehous, Stanhous), *Kinghorn* [Fm, NT2188], 28, 44, 50
Stenhouse (Stanehous, Staynhous, Stenehous), NT2171; 137, 143, 290
Stenness (Stanehous, Stanhous, Stenhous), *Orkney*, HY3111; 657, 659, 668, 670, 670n.; kirk of, 659
Stenstanes, Stenstanis, *Dunfermline abbey*, 23, 45
Stenton (Staintoun, Stayntoun, Steintoun), NT6374; 162, 177, 178
Stentoun, *St Bathans priory*, 192
Stenton (Stentoun, Stentounes), *Kinglassie* [Over, Nether, NT2799, NT2699], 24, 29, 46, 51
Stepford (Steipfuird, Stepfuird), NX8681; 269, 270, 275, 622, 623
Stermonth, Stermontht, see Starmonthe
Stevenston (Steinsoun, Stevinsoun, Stevinstoun), NS2642; 540, 574, 577; kirk of, 577; parish of, 574; reader at, 540; vicarage of, 540
Stewartfeild, *Jedburgh abbey*, 219
Stewarton (Stewartoun), *Kilwinning abbey*, NS4246; 573, 576, 578; kirk of, 576, 578; parish of, 573
Stewarton (Stewartoun), *St Mary's Isle priory*, NX4349; 619
Steylistoun, see Steilston
Stichill (Sthichel, Stitchell, Stritchell), NT7138; 195, 200, 204; kirk of, 200, 204; vicar of, 195; vicarage of, 195
Stirkfield (Stirkfeild), NT1040; 249
Stirling (Stirveling, Stirviling, Stirvilling, Stirvling, Striveling, Striviling, Strivilling, Styrviling), NS7994; 23, 26, 38, 39, 46, 48, 229n., 291, 323, 323n., 538, 543, 544, 545, 546, 547, 551, 555, 560, 602, 603; Akeyn's lands, 555; almshouse in, 551; Brig of, 545; Burrow Myln of, 555; chaplain of St Thomas' chaplainry in, 551; chaplainry of St Thomas and almshouse in, 551; glebe of, 23, 46; kirk of, 26, 38, 39, 48; reader of, 550; schoolmaster of, 545; Haldan's land, 291; vicarage of, 550
—— parish kirk, 545, 551, 554, 555, 555n., 556; altarage of St John the Baptist in, 555; altarage of St

Michael in, 554; chaplain of Holy
Rood altar in, 556; chaplain of St
Laurence's altar in, 545; chaplain of
St Michael's altar in, 554; chaplainry
of St Laurence's altar in, 545; Holy
Rood altar in, 556; St Andrew's altar
in, 551
——, Black friars (Dominicans) of, 554-
555
——, Chapel Royal, 263n., 323, 323n.,
560, 602, 603; prebendaries of, 560
—— (Strivelingschyre), *sheriffdom*, 88n.,
113, 148n., 183, 282, 538n., 554
Stitchell, *see* Stichill
Stobicote (Stobecoit), NT4607; 259
Stobis, *Leith*, 114
Stobo, NT1838; 249, 250, 250n., 252,
495; parsonage of, 249-250; parson
of, 250; prebend of Glasgow
cathedral, 250n.; vicarage portionary
of, 252
Stockbriggs (Stokbrig), NS7936; 233
Stodrik, *see* Todrig
Stoirdaill, *Caithness bishopric*, 628
Stok and Garthe, *fishing beside Perth,
Balmerino abbey*, 58
Stokbrig, *see* Stockbriggs
Stoknes, *see* Scockness
Stone of Morphie (Stane Morphy),
NO7162; 16
Stonebyres (Stanebyres) [Holdings,
NS8343], 229
Stonehaven, NO8786; 31n., 404n.
Stonehouse (Stainhous, Stenhous),
NS7546; 511, 512, 513; glebe of,
513; manse of, 513; prebend of
Bothwell collegiate kirk, 511-512;
vicarage of, 513
Stoneyhill (Sanehill, Stanehall,
Stanehill, Staniehill, Stanihill,
Stanyhill, Stayhill), NT3272; 24, 26,
41, 43, 47, 48, 49, 53, 54
Stoneywood (Stanywod), NJ8911; 423
Stormnes, *see* Stromness
Stormoth, *see* Starmonthe
Stow, NT4644; kirk of, 3; parish of, 3;
reader of, 120; vicar pensionary of,
120; vicarage pensionary of, 120
Stowgordoun, *see* Staplegordon
Straarlie, Straayrlie, *see* Strathairly
Strablane, *see* Strathblane
Strabrok, *see* Strathbrock
Straburne, *see* Strathburn

Strabusteir, Strabuster [Scrabster,
ND0970], 628
Stracathro (Strachathro, Strathcathro,
Strethcathro) [Ho, NO6265], 380,
384, 386, 388, 389; waulk mill of,
384, 386, 389; vicar of, 380;
vicarage of, 380
Strachan (Straquhain), NJ6792; kirk of,
380; minister of, 380; vicarage of,
380
Stradichin, *see* Stradichtie
Stradichtie (Stradichin) [Mains,
NO3837], 361; *see also* Mains
Stradoun, *see* Strathdon
Strafontanes, *Dunglass collegiate kirk*,
166
Strafuthy, *see* Stravithie
Strageath (Strogeicht, Strogeith,
Strogeycht, Strogeyth), NN8818;
295, 296, 348, 349; Brodland of,
349; kirk of, 295, 348; parish of, 349
Straib Knowe (Straib Know), *Kinloss
abbey*, NJ5150; 462
Straithmeglo, *see* Strathmiglo
Straithspey of Cardell, *Elgin Black
friars*, 453
Straithspey, *see* Strathspey
Straithylay, *see* Strath Isla
Straiton (Straitoun), NS3804; 567, 569,
582; kirk of, 567; minister of, 569;
vicar of, 569, 582; vicarage of, 569,
582
Strakinnes, Strakynnes, Strakynnis, *see*
Strathkinness
Straloch (Stralocht, Straloycht), NJ8621;
422, 423
Stralrynnes, Stralrynnis, *see*
Strathkinness
Strameglo, Stramiglo, *see* Strathmiglo
Straquhain, *Forfar*, *see* Strachan
Straquhan, Straquhen, Straquhon,
Dumfries-shire, *see* Stroquhan
Strath Isla (Straithylay, Straythylay),
NJ4149, NJ4850; 461, 463; mains
of, 461
Strathairly (Straarlie, Straayrlie),
NO4303; 145, 166
Strathallan (Strathalloin), 'Quaigis de',
343n.
Strathardle (Straithardil, Strathardill),
NO0955; 26, 38, 48; kirk of, 26, 38,
48
Strathaven (Avendaill, Avondale,

Evindale, Strathabane), NS7044; 229, 233, 364n., 603, 603n.; vicarage pensionary of, 603
Strathawradall, *Skye*, 675
Strathblane (Strablane), NS5679; 539; kirk of, 539
Strathbogie (Strathboggie), NJ6737; 478; deanery of, 478
Strathbrock (Strabrok) [Uphall, NT0571], minister of, 153; parsonage of, 153; parson of, 153; vicar of, 152; vicarage of, 152
Strathburn (Straburne), NO4223; 13, 20
Strathcathro, *see* Stracathro
Strathdon (Stradoun, Stradowyne), NJ3512; 432, 485; *see also* Invernochty
Strathearn (Stratherne), NN9717; 31, 35, 35n., 45n., 113, 295, 349
Strathkinness (Strakinnes, Strakynnes, Stralrynnes, Strakynnis, Stralrynnes, Stralrynnis, Strathkynnes), NO4616; 10, 12, 18, 20, 21; Month pairt, Monthis pairt, Mounthis pairt, Paremounth, Paremounthe, 10, 12, 18, 20, 21; Poffill of, 18
Strathmartine (Strathmerting, Straythmertyne, Stricmartine) [Kirkton of, NO3735], 19, 79, 412; vicarage of, 79
Strathmiglo (Straithmeglo, Strameglo, Stramiglo), NO2110; 69, 70, 303, 342, 346; benefice of, 69; kirk of, 303, 346
Strathnawer, *Caithness archdeaconry*, 632
Strathore (Strathoire), NT2697; 70
Strathspey (Straithspey), NH7800, NJ0629, NJ0931, NJ1437; 465
Strauchane, *see* Stroquhan
Stravathy, Stravethy, *see* Stravithie
Stravithie (Stathvey, Strafuthy, Stravathy, Stravethy), NO5311; 10, 13, 18, 70
Straythylay, *see* Strath Isla
Strethcathro, *see* Stracathro
Stricmartine, *see* Strathmartine
Stritchell, *see* Stichill
Striveling, Striviling, Strivilling, *see* Stirling
Strivelingshyre, *see* Stirling
Strogeicht, Strogeith, Strogeycht, Strogeyth, see Strageath

Stroma (Stromnys), Island of, *Pentland Firth*, ND3577; 418
Stromness (Stormnes, Stromnes), *Orkney bishopric*, HY2509; 657, 659, 664, 668, 669, 677; kirk of, 659; minister of, 659; parish of, 677; parson of, 659
—— (and Sandwick), vicarage of, 668; *see also* Sandwick
Stromnys, Isle of, *see* Stroma
Stronsay (Skemesay, Stromsay, Stronsay), HY6021, HY6624; 656, 660, 660n., 666, 669; *see also* St Mary, St Nicholas, St Peter
Stroquhan (Straquhan, Straquhen, Straquhon, Strauchane, Strowquhon), NX8483; 270, 275, 277
Strowan (Strowane), *Dunblane diocese* [Ho, NN8121], 313; vicar of, 313; vicarage of, 313
Strowquhones, *see* Stroquhan
Struan (Strowane), *Dunkeld diocese*, NN8065; 319; curate of, 319; parsonage of, 319; vicarage of, 319
Strybustar, *see* Scrabster
Stynkand Styll, *see* Stinkand Style
Styrviling, *see* Stirling
Suaay, *see* Swona
Suansfeild, *see* Swansfield
Subcelleris land, Subsellarisland, Subtelleris land, *Melrose abbey*, 208, 208n.
Subdean's croft and mains, *Ross*, 652
Subtelleris land, *see* Subcelleris land
Subunster, *Caithness bishopric*, 628
Suddie (Suddy, Suddye) [Church, NH6654], 634, 639; minister of, 635; parsonage of, 634; vicar of, 639; vicarage of, 639
Sueithart, Sueitharte, Sueithartis, *see* Sweetheart
Summerhill (Symmerhill), NO5646; 367
Sumpryne, *see* Simprim
Sunay, *see* Swona
Sunderland Hall and Towne, *Selkirk*, 230n.
Sutherland (Suthirland), 629
Suynesyd, *see* Swinside
Suynewod, *see* Swinwood
Suynie, *see* Swinnie
Suyntoun, *see* Swinton
Swailend (Swaillend, Swayll End),

NJ8816; 422, 423
Swansfield (Suansfeild), NT8462; 187, 198, 203
Swayll End, see Swailend
Sweetheart abbey (Sueithart, Sueitharte, Sueirhartis, Sweithart) [New Abbey, NX9666], 611, 612, 612n., 622, 622n.; abbot of, 612, 612n., 622, 622n.; benefice of, 622; see also Lochindilloch, New Abbey
Swinnie (Suynie), NT6316; 219
Swinside (Suynesyd) [Hall, NT7216], 219
Swinton (Suyntoun, Swyntoun), NT8347; 187, 200, 204; kirk of, 200, 204
Swinwood (Suynewod, Swounwod, Swynwode) [Mill, NT8962], 198, 201, 203
Switha (Schewsay, Skynthy, Sowtha), ND3690; 656, 659, 662, 663
Swona (Sallway, Sounay, Sowna, Suaay, Sunay), ND3884; 656, 659, 662, 663, 668
Swounwod, see Swinwood
Swyinshauche, Swynishauch, Scoonie, 15, 21
Swyntoun, see Swinton
Swynwode, see Swinwood
Syde (Syid), NO6164; 383, 385, 388
Sydserf (Sydserff, Syndsorff), NT5481; 147, 168
Syid, see Syde
Sykis, Oxnam, 219
Symington (Symontoun), Fail, NS3831; 558, 562, 563, 564, 569; minister of, 558, 564; parish of, 558, 562, 563, 564; reader at, 563; vicar of, 563, 569; vicarage of, 563, 564, 569
Symington (Symintown, Symontoun, Symontoune, Symontown, Symountoun), Clydesdale, NS9935; 224, 232, 234, 237, 240, 242, 244, 517, 517n.; kirk of, 224, 232, 234, 237, 240, 242, 244; vicar of, 517, 517n.; vicarage of, 517
Symmerhill, see Summerhill
Symontoun, Symontoune, Symountoun, see Symington
Symphryn, Symprein, Symprene, Sympriane, Symprim, Symprine, Sympryin, Symprym, Sympryme, Sympryn, see Simprim

Syndry landis, Kirkcaldy, see Smiddie Landis
Syndsorff, see Sydserf
Synton (Syntoun), NT4822; 212
Sypsies (Seipseis, Seipsis, Sipseis), NO6008; 177, 178, 181
Systerpeth, see Sisterpath

Tain (Tayne, Than, Thane, Thayne, Thyne), NH7882; 636, 640, 651, 652; parson of, 651; parsonage of, 651; town of, 652; vicarage of, 640
— collegiate kirk, 463, 636, 640, 647; chaplain in, 647; chaplainry of Newmore in, 647; chaplainry of Tarlogie in, 647; prebend of, 636-637; prebendary of, 636; provost of, 640; provostry of, 640-641
Talaquhy, see Tarlogie
Talgaty, see Dalgety
Tallaracie, see Dalarossie
Tallo (Tallow), 122, 122n.
Talraquhy, see Tarlogie
Talrogy, see Tarlogie
Tambutoun, see Thankerton
Tankertoun, see Thankerton
Tannachie (Tannaquhy), Glenbervie, NO7883; 405
Tannadice (Tannades, Tennedais), NO4758; 64, 364, 364n.; curate of, 65; kirk of, 64, 364; reader at, 65
Tannoquhy, Kinloss abbey, 462
Tantallon (Tantalloun) [Castle, NT5985], 147, 148, 168, 169; mains of, 169
Tarbat (Terbat), Ross bishopric, NH7773; 625, 626, 627; kirk of, 626; curate of, 626; preacher at, 627
Tarbolton (Carboltoun, Tarboltoun), NS4327; 580; parson of, 580-581; parsonage of, 580; vicarage of, 581
Tarland (Tarlane, Terland), NJ4804; 8, 16, 17; kirk of, 8, 16, 17
Tarlaquhy, see Tarlogie
Tarlogie (Carlaquhy, Talaquhy, Tarlaquhy), NH7583; 647, 651, 652; chaplain of, 647; chaplainry of, in Tain collegiate kirk, 647
Tarmore (Toirmoir), NJ4152; 461
Tarthorall, Tarthorrall, see Torthorwald
Tartus McKer, [Torhousekie, NX3756]; 597, 597n.

Tartyis McCulloch [Torhousekie, NX3756], 597, 597n.
Tartyus Mur [Torhousemuir, NX3957], 597, 597n.
Tarves (Terves), NJ8631; 351, 358, 359, 448, 448n.; kirk of, 359; vicar of, 448; vicarage of, 448
Tarvit (Tervat), *parish* [Tarvit Mill, NO3612], 73, 88; parson of, 88; parsonage of, 73, 88; vicarage of, 73, 88
Tathis, *see* Teaths
Tattis Mylne, *see* Teassesmill
Tay, *see* River Tay
Tayne, *see* Tain
Tayok, *Brechin bishopric*, 383, 384, 387, 388, 389
Tealing (Teling), NO4138; 308, 308n., 364; benefice of, 308; vicarage of, 364
Teassesmill (Tattis Mylne), NO4010; 55
Teaths (Tathis, Thauthtis), NS8542; 231, 232
Teinzie, *see* Lenzie
Teirie, *see* Tiree
Teling, *see* Tealing
Tempilhill, *near Keithock, Brechin cathedral* [Templewood, NO6162], 400
Tempill land *of Keithock, Brechin cathedral* [Templewood, NO6162], 400
Tempill-Ronaige, *see* Iona
Tempillall, *see* Templehall
Tempilland, Templand, *Maxton parish*, 189, 191,
Temple (Tempill), NT3258; 110; vicarage of, 110
Templehall (Tempillall), NT2088; 63
Templeton (Tempiltoun), *Nevay*, NO3142; 367
Templum, *Logie-Montrose*, 373
Tenenbray, *see* Denbrae
Tenenheid, *see* Denhead
Tennedais, *see* Tannadice
Terbacks, *Ross bishopric*, 625
Terbat, *see* Tarbat
Tereglis, *see* Terregles
Terland, *see* Tarland
Terregles (Tereglis), NX9377; 613, 624n.; kirk of, 613
Terreris croft, Terrouris croft, *Musselburgh*, 25, 47

Terrouris acres, *Kinghorn*, 28, 50
Terrouris acres, *Newburn*, 28, 29, 50
Tertelell, *see* Torthorwald
Tervat, *see* Tarvit
Terves, *see* Tarves
Teviotdale (Teviotdaill, Teviot Daill, Toviotdaill), NT4815, NT5822, NT6123; 113; archdeacon of, 214, 214n., 215; archdeaconry of, 215
Than, Thane, *see* Tain
Thankerton (Tambutoun, Tanbutoun, Tankertoun, Thankertoun), NS9738; 499, 518; parsonage of, 499, 518; vicarage of, 499, 518
Thauthtis, *see* Teaths
Thayne, *see* Tain
Thesaureris feild, *Caithness treasureship*, 630
Thessie, *Melrose abbey*, 208
Thirlestane (Thirlistane, Thirlstane), NT5647; 21n., 54, 258
Thobane, *Lagganallachy* [Meikle Tombane, NN9440], 309
Thomastoun, *Cupar*, 14
Thomgarne [*Tomnagairn, NN9439], 309
Thomgrow, *Lagganallachy* [*Tomgarrow, NO0040; *Tomnagrew, NN9439], 309
Thomrandy, *Dunblane bishopric*, 349
Thornedykis, *Gordon*, 227
Thornisdaill, *Caithness*, 628
Thornton (Thornitoun), *Strath Isla*, NJ4851; 461
Thornton (Thornntoun, Throntoun), *St Andrews priory*, 17, 18
Thoumegrow, *Dunblane bishopric* [*Tomgarrow, NO0040; *Tomnagrew, NN9439], 349
Thraiplandis, Treplandis, *Newburgh*, 32, 34, 36
Thraipwod, *see* Threepwood
Three Kings' of Cologne altar, *Dundee kirk*, 381, 381n.
Three Merkland (Thrie Merk Land, Thrie Merkland), *Holywood abbey*, NX8067; 623
Threepwood (Thraipwod, Threipwod, Threipwode), *Melrose abbey*, *NT5142; 207, 209, 257
Threepwood (Threipwod), *Kilwinning abbey*, *NS5234; 579; mains of, 573
Threepwood (Trypwod) *Lesmahagow*,

NS8147; 233
Threthrevis [High, Middle, Low Threave, NX3758, NX3759], 588, 588n.
Thriddis Croft, *Aberdeen Trinitarians*, 421
Thrie Merkland, see Three Merkland
Thripwod [West Threepwood, NS5134], 531
Throntoun, see Thornton
Throsk (Tros), NS8591; 92, 92n., 538, 545, 547, 548
Thurseth, see Thurso
Thurso (Thurseth, Thursoch, Thursoche, Thursocht), ND1168; 628, 631, 648; parish of, 628; vicar of, 648; vicarage of, 631-632, 648
Thursoch, Thursoche, Thursocht, see Thurso
Thyne, see Tain
Tibbermallocht, Tibbermellocht, Tibbirmello, see Tippermallo
Tibbermore (Tibbermuir, Tibbermure, Tibbirmuir, Tibbirmure), NO0523; 302, 303, 342, 345, 346; kirk of, 302, 342, 346
Tibermallocht, Tibermellocht, see Tippermallo
Tilleculter, see Tillicoultry
Tillicoultry (Dillycoutry, Tilleculter, Tulliculie, Tulycultie), NS9197; 538, 543, 545, 546; kirk of, 538, 543, 545, 546
Tillykerrie (Tullie Carie, Tyllikere), NJ8321; 32, 36
Tillymorgan (Tullimorgoun, Tullymorgoun) [Mains of, NJ6634], 31, 36
Tillynaught (Tulmacht), NJ6061; 437
Tingwall, HU4048; kirk of, 660; minister of, 660
Tinnyghame, see Tyninghame
Tippermallo (Tibbermallocht, Tibbermellocht, Tibbirmello, Tibermallocht, Tibermellocht) [Mains of, NO0224], 296, 297, 298
Tippilaw, see Kippilaw
Tiree (Teirie), NL9944; 673, 674, 675
Todhills (Todhillis), NT3068; 101
Todis Clois, see Toddis Clois
Todrig (Cutrig, Steidrig, Stodrik, Todrige, Todrik, Totrige), *Kelso abbey*, NT7043; 223, 223n., 227, 227n., 228, 230n., 233, 233n., 237, 239
Todrig, *Coldstream priory*, NT7942; 186, 187
Todshawhill (Todschawhill, Treshawhill), *Eskdalemuir, Melrose abbey*, *NY2293; 259, 259n.
Toftingall (Toftingaill), *Caithness archdeaconry*, ND1754; 632
Toirmoir, see Tarmore
Toirrasa, see Torosay
Tokles, *St Mary on the Rock collegiate kirk, St Andrews*, 55
Tongland (Tungland), NX6954; 91, 93, 93n., 279n., 585, 586, 587, 588, 589, 590, 591, 592, 607, 616n.; abbacy of, 586-592, 606-607; abbey of, 279n.; mains of, 586, 589; parsonage of, 607; vicarage of, 587, 607
Tonigask, see Dunnygask
Tor, *Cupar* [Tor of Moonzie, NO3517], 14
Torbain (Turbane) [Fm, NT2493], 28, 44, 50, 54
Torbanes alias Treis, see Trees
Torcraik, NT3659; 111
Tordiquhill, see Ordiquhill
Torosay (Toirrasa), NM6134; 674
Torphichen (Torphechin, Torphichane), NS9672; 11, 110n., 176n., 450n.; preceptor of, 110; vicar of, 152; vicarage of, 152
Torphynis [East, West Torphin, NT0361], 98
Torquhoik [*Torchuaig Hill, NO0039], 309
Torrance (Torrence, Torrens), *Kilbride*, NS6173; 515, 516; parson of, 515; parsonage of, 515, 516
Torreff, see Turriff
Torrie, see, Torry
Torrinichtrache, *Brolass, bishopric of the Isles*, 673
Torrs (Torris) [Moor], NX6846; 617, 619
Torry [*Torry Ho, NT0186], 291; minister of, 73; parson of, 73n.; parsonage of, 73, 73n.; vicarage of, 73
Torry, Torrye, *Arbroath abbey*, 351, 358
Torrywald, Torvald, *Lagganallachy*, 305, 309
Torschaw, *Kilwinning abbey*, 574

Torthorwald (Tarthorall, Tarthorrall, Tertelell, Torthoraill), NY0378; 558, 562, 563; kirk of, 558, 562; minister of, 558; reader at, 563
Torvald, see Torrywald
Torwodheid, Cambuskenneth abbey, 546
Toscartoun, see Toskerton
Toskerton (Toscartoun, Toskartoun), kirk of, 594; vicar of, 605; vicarage of, 605; see also Kirkmadrine, Whithorn
Totrige, see Todrig
Touch [Ho], NS7592; 544, 547, 548, 549
Touch (Towch), Hills of, St Ninians, 546
Touch Myln, Dunfermline abbey, 23, 28, 45, 50
Touch, Dysart, 70
Touch, parish, see Tough
Touch, Toucht, Dunfermline abbey, 23, 27, 28, 45, 49, 50
Tough (Touch, Towch), parish [*Kirkton, NJ6113], 425; parson of, 424; parsonage of, 424, 425; vicar pensionary of, 424; vicarage of, 424, 425
Toun croft, Towne croft, Kelso abbey, 222, 233, 237, 239
Toun Heid, Holywood abbey, 276
Toun Mains, Hume, 228
Toun of Roull, see Town-o'-rule
Tounfeild, see Downfield
Toux (Towkis), NJ9850; 457
Towch, parish, see Tough
Towch, Hills of, see Touch Hills
Towchyis, Aberchirder, 481
Towie (Towyis), NJ4312; 449
Towir, see Fenton Tower
Towkis, see Toux
Town-o'-rule (Toun of Roull), NT5813; 220
Towne croft, see Toun croft
Towrmonth, Restalrig collegiate kirk, 142, 143
Towyis, see Towie
Trabroun, NT4674; 153
Trailflat (Trainflate), NY0485; 224, 232, 234, 237, 240, 244; kirk of, 232, 234, 237, 240, 244
Trailtrow, NY1471; 'preceptory' of, 274
Trainflate, see Trailflat

Tranent, NT4072; 91, 144; kirk of, 91; vicarage of, 144
Traprain (Trapern, Traperne, Trapren), NT5975; 3, 6, 7, 170
Traquair, (Traquhair), Glasgow bishopric, NT3334; 207; kirk of, 496
Traqueir, Tongland, see Troqueer
Treaton (Trottoun), NO3202; 15
Trees (Torbanes alias Treis) [House, NS9465], 92, 92n.
Treis, see Trees
Trem, see Drem
Treplandis, see Thraiplandis
Treshawhill, see Todshawhill
Trinity altar, St Cuthbert's kirk, Edinburgh, 128, 129
Trinity chaplainry, Dunblane cathedral, 324, 324n.
Trinity chaplainry, St Andrews, 512n.
Trinity chaplainry, Trinity kirk, St Andrews, 71, 71n.
Trinity collegiate kirk (Trinitie College), Edinburgh, 94, 111, 118, 118n., 119n., 175, 404, 404n.
Trinity kirk (Ecclesia Trinitatis, Sanctandrois kirk), see St Andrews
Trinity parish, see St Andrews, Trinity kirk
Trondra (Troundray), HU3937; 660; kirk of, 660; minister of, 660
Trone, see Trows
Tronehill, NT5623; 236, 366
Troqueer (Traqueir, Troquair, Troqueir, Troweir), Tongland, NX9775; 279, 587, 589, 607, 616, 624; kirk of, 589; parsonage of, 587, 607; vicar of, 279, 616; vicarage of, 279, 587, 616
Troquhain (Troquhoun), NX6879; 585
Tros, see Throsk
Trostrie, see Troustrie
Trotternish (Trouterneiss, Troutirneise), NG4153; 674, 675
Trottoun, see Treaton
Troundray, see Trondra
Troustrie (Trostrie), NO5907; 182
Trouterneiss, Troutirneise, see Trotternish
Troweir, see Troqueer
Trows (Trone, Trow), Ugginnis, Melrose abbey, *NT6832; 258
Trumcarro, see Drumcarro
Trypwod, see Threepwood

Tueddaill, *see* Tweeddale
Tueid, *see* River Tweed
Tueidmure, *see* Tweedsmuir
Tuinrige, *see* Crumrig
Tulemulie, Tullimulie, *Dunkeld bishopric*, 303, 342, 345
Tulenessene, *see* Tullynessle
Tulibarand, Tulibardin, *see* Tullibardine
Tulibardin, *see* Tullibardin
Tulibodie, *see* Tullibody
Tuliboll, *see* Tullibole
Tulilum, *see* Tullilum
Tulipourie, Tulipowrie, *see* Tullypowrie
Tullach, *Lagganallachy*, 309
Tullacht, *Aberdeen deanery, see* Tulloch
Tullaracie, *see* Dalarossie
Tullegowneis, *Methlick*, 435
Tullemoran, *Monzie parish*, 349
Tulliallan (Tulleallon, Tulleallone, Tulliallon, Tulliallone, Tullialloun, Tullyallone) [Castle, NS9388], 295, 296, 297, 348, 349; kirk of, 295, 296, 348, 349
Tullibaglis, *see* Tullybeagles
Tullibardine (Tulibarand, Tulibardin, Tullibairdin, Tullibardin, Tullybardin), NN9214; 7n., 76, 76n., 99n., 248, 248n., 281n., 402, 402n., 495n., 537n., 573, 573n., 574, 574n., 579, 606, 609n., 610, 659n., 662, 662n., 663, 664
Tullibody (Tulibodie, Tullebody, Tullybody), NS8695; 538, 543, 545, 546, 548; kirk of, 538, 543, 545, 546, 548
Tullibole (Tuliboll, Tullibuill), *parish*, 290, 315; *see also* Fossoway
Tullibrek, *see* Tullybreck
Tullich (Tulloch), NC8576; 634
Tullichane, *Atholl, Coupar Angus abbey*, 354, 354n.
Tulliculie, *see* Tillicoultry
Tullie Carie, *see* Tillykerrie
Tullifergus [East, West Tullyfergus, NO2149, NO2148], 354, 357, 369
Tullilum, *Perth*, Carmelite friary at, 303
Tullimorgoun, Tullymorgoun, *see* Tillymorgan
Tullimulie, *see* Tulemulie
Tulliocht, *with Toux, Deer abbey*, 457
Tullipourie, *see* Tullypowrie
Tulloch (Tullacht), *Aberdeen deanery*, *NJ8509; 423

Tulloch (Tullocht), *Coupar Angus abbey* [Tulloch Hill, NO8664], 369
Tulloch, *see* Tullich
Tulloquhair, *Eskdalemuir, Melrose abbey*, 259
Tullyallone, *see* Tulliallan
Tullybeagles (Tullibagles, Tulybaiglis) [Lodge, NO0136], 296, 298
Tullybreck (Tullibrek), NO3198; 15
Tullymoran (Tullemoran), NN9729; 349
Tullynessle (Tulenessene, Tullenessin, Tullenessill, Tullinessill, Tullynessill, Tulynessill), NJ5519; 414, 417, 418, 427, 436, 437; parson of, 426, 437; parsonage of, 426, 436
Tullypowrie (Tulipourie, Tulipowrie, Tullipourie), NN9154; 324; mill of, 324
—— and Fyndynate, chaplain of, 324; Fyndynate, chaplainry of, 324
Tulmacht, *see* Tillynaught
Tulybaiglis, *see* Tullybeagles
Tulycultie, *see* Tillicoultry
Tulynessill, *see* Tullynessle
Tunegask, *see* Dunnygask
Tungland, *see* Tongland
Tunmeth, *see* Dalmeath
Tunnynghame, *see* Tyninghame
Tunrik, *with Lambden, Greenlaw*, 227n.
Tunruchan, *see* Dunruchan
Turbane, *see* Torbain
Turnlaw (Turnlay), NS6458; 508
Turnpike (Turnpyk), *Stirling*, 546
Turriff (Torreff, Tureff), NJ7250; 431, 433, 440, 445, 451; parson of, 431, 433, 440, 445, 451; parsonage of, 433, 445; vicar of, 433, 445; vicarage of, 433, 445
Turtestoun, *see* Curtestoun
Tusbust, *Bower*, 632
Tuyde, *see* River Tweed
Twa Balyeachanis, *Atholl, Perth Charterhouse* [Ballechin, NN9353], 284, 288
Tweddell, *see* Tweeddale
Tweeddale (Tueddaill, Tweddell, Tweiddaill), 258
Tweeddale (Tweiddaill) Traquair [Traquair, NT3334], 207; *see also* Traquair
Tweedsmuir (Tueidmure), NT0924; 252
Tweiddaill, *see* Tweeddale
Twid, *see* River Tweed

Twnmeth, *see* Dalmeath
Twynem, *see* Twynholm
Twynholm (Twynem), NX6654; 91; kirk of, 91
Tyllikere, *see* Tillykerrie
Tylmuth Hauch, *Coldstream priory* [*Tillmouth Fm, NT8844], 186
Tymmerhill, Tynunerhill, *Eskdalemuir, Melrose abbey*, 259, 259n.
Tyninghame (Tinnyghame, Tunnynghame, Tynninghame), NT6079; 2, 65; kirk of, 65; parsonage and vicarage of, 65
Tynninghame, *see* Tyninghame
Tynram, *see* Tynron
Tynron (Tynram, Tynrun), NX8093; 272, 276, 277, 623; kirk of, 277, 623
Tynynerhill, *see* Tymmerhill
Tyrie (Tyrye), NJ9363; 429; parson of, 429; parsonage of, 429; vicarage of, 429
Tyrye, *see* Tyrie

Uchiltrie, *see* Ochiltree
Udny (Udnie), NJ9024; 463
Ugginnis, Ugingis, Wyingis, *Melrose abbey*, 208, 208n., 258
Ugingis, *see* Ugginnis
Ugstoun, *Dryburgh abbey* [*Oxton, NT4953], 190, 191, 197
Ugstoun, *see* Ogston
Uist (Veist, Weist) [North, NF8369; South, NF7618], 673, 674
Ullischewin, *Capo, Brechin bishopric*, 384
Ulston (Ulstoun, Ustoun, Wstoun), NT6621; 216, 217, 218, 219
Unes croft, *see* Ovenscloss
Unganab, *Uist*, 674
Ungiltoun, *near Beanstoun, Haddington priory*, 163
Uniscroft, *see* Ovenscloss
Unnertown, *Cambuslang*, 508
Unst (Onst, Wnster), HP6109; 658, 660, 670; kirk of, 660; minister of, 660
Unthank, *Brechin bishopric*, NO6061; 383, 384, 385, 386, 387n., 388
Unthank, *St Andrews priory*, *see* Winthank
Unthank, *parish*, NJ1667, 475; parson of, 475; parsonage of, 475; *see also* Duffus

Unysclos, *see* Ovenscloss
Upper Barr (Over Bar), NX4262; 587, 590
Upper Kinmonth (Ovir Kynmonth), NO7782; 405
Upper Kinneddar (Over Kinnedder, Over Kinneddir), NT0392; 24, 46
Upper Magus (Magask Superiour, Over Magus), NO4414; 12, 21
Upper Manbean (Over Monbennis), NJ1957; 453
Upper Nisbet (Over Neisbit, Over Nisbit), NT6727; 216, 219
Upper Urquhart (Ovir Urquhard), NO1908; 69
Upper Whiteside (Over Quhytsyd, Overquhytsyd) [Moor], NX8082; 274, 275, 622, 623
Upsatlingtoun Scheillis, [Ladykirk Shiels, NT8645], 166
Upsetlington (Upsettingtoun), NT8846; 185; kirk of, 185; parson of, 186n.
Urdis [Ladyurd, Lochurd, Netherurd, NT1542, NT1143, NT1144], 211
Urquhart (Orquhat), *Inchcolm abbey*, NT0786; 62
Urquhart (Urquhard, Urquharde, Urquherde), *Pluscarden priory*, NJ2862; 471, 472, 473; kirk of, 472
Urquhart, *Ross*, NH5553; 637, kirk of, 637
Urr (Ur), *Holyroodhouse abbey*, 91; kirk of, 91
Urray (Vurray) [Beag, NH5052], 635, 640; parsonage of, 635; vicar of, 640; vicarage of, 640
Urriewall, *with Birnam, Dunkeld bishopric*, 303
Urschous Mylne, *see* Westoussis Myln
Uruell, *see* Orwell
Uruquhill, *see* Orwell
Urwale, *see* Orwell
Ustoun, *see* Ulston
Uttershill (Wowterischill, Wowterschill), *Restalrig deanery*, NT2459; 142, 143
Uver Grang, *see* Over Grange
Uver Kincraig, *see* Over Kincraig
Uver Lasoiddy, *see* Lassodie
Uverquhill, *see* Orwell

Vactin, *see* Watten

Vairdis, Wairdis, *Melrose abbey*, 257, 259
Valafeild, Valafeld, *see* Valleyfield
Valeyfeild, *see* Valleyfield
Valleyfield (Valafeild, Valafeld, Valeyfeild), *Culross abbey* [High, Low, NO0086, NO0085], 290, 293, 294
Vanglaw, *see* Bangly
Vanles Aikaris, *Culross abbey*, 291
Vannis, *see* Walls
Vatstirker, *see* Westerkirk
Veist, *see* Uist
Vesterraw, *see* Westerhaw
Vesthous Mylne, *see* Westoussis Myln
Vicarland, *Kinnell parish*, 367
Viliamlaw, *see* William Law
Vindleis, *see* Winless
Vinsclos, *see* Ovenscloss
Virgin Mary's lands, *see* Marislandis
Vissurdland, 'callit the Kir', *Dunblane deanery* [*Keir Mains, NS7799], 315
Vitfute, *see* Weetfoot
Vithank, *Netherfield, Cathkin*, 517
Vogarie, *see* Vogrie
Vogrie (Vogarie) [Grange, NT3762], prebend of Crichton collegiate kirk, 109
Voulschill, *see* Wolfhill
Vouplaw, *see* Wooplaw
Vousbyre, *see* Housbyre
Voustoun, *see* Wiston
Vrmistoun, *see* Wiston
Vurray, *see* Urray

Wachopdaill [Wauchope, NT5808], 113, 221
Wachtoun, *see* Waughton
Waichtoun [*Waughton, NT5680], 170
Waiffuird, Wyffand, *Fail*, 558, 562
Wailstoun, see Walston
Wais, *see* Walls
Waistbust, *with Dunn, Caithness archdeaconry*, 632
Waisthaill, *see* Westhall
Wakkeris Burn, *see* Walkerburn
Walderstoun, *Ecclesmachan* [*Waterstone, NT0574], toun of, 151
Walk Mylne, Walkmylne, *Kinglassie*, 24, 29, 46, 51
Walkarland, Walkerland, *Dunfermline abbey*, 23, 27, 28, 45, 49, 50
Walker croft, *Melrose abbey*, 258
Walkerburn (Wakkeris Burn), *Morebattle*, NT3637; 214
Walkmyll, Walkmylne, *Dairsie*, 10, 18
Walkmyln croft, *Holywood abbey*, 270, 275, 623
Walkmyln toun, Walkmylntoun, *Manuel priory* [Waukmilton, NS9877], 549
Walkmylne, *near Quarrypits, Haddington priory*, 162, 176, 178
Walls (Vannis, Wais, Wanyis), *Shetland*, HU2449; 658, 660, 670; kirk of, 660; minister of, 660
Walls (Wais, Wawis), *Orkney*, ND2793; 657, 659, 669; kirk of, 659; minister of, 659
Wallyford (Wallyfurd, Wellifuird), *Dunfermline abbey*, NT3772; 40, 53
Walston (Wailstoun, Welstoun, Wolstoun), *parish*, 499, 513; kirk of, 499; vicar of, 513; vicarage of, 513
Walter of Die, *see* River Dee
Walter of Tay, *see* River Tay
Waltersame, *see* Baltersan
Walterstoun, *see* Mekill Walterstoun
Waltir of Done, *see* River Don
Waltir of Findorne, *see* River Findhorn
Waltir-McRoreis-croftis, *see* Roresounis Croft
Waltirris of Areicht, *see* River Ericht
Waltirris of Ylay, *see* River Isla
Wamfray [*Wamphraygate, NY1296], 599
Wandel Burn (Wandelburne, Wandellburne), NS9626; 513
Wanyis, *see* Walls
Warakstoun, *see* Warrackston
Ward of Kynmonde, *near Seaton, Aberdeen deanery*, 423
Wardlaw [Kirkhill, NH5545], 466; kirk of, 466, 650
Wardraptoun, *Ecclesgreig* [*Warburton, NO7363], 16
Warkwik Hill, *see* Warwickhill
Warlaw, *see* Wardlaw
Warrackston (Warakstoun), NJ5520; 436
Warwickhill (Warkwik Hill, Warwikhill), NS3739; 582, 583
Wastray, Wastrey, *see* Westray
Water *of Dee*, *see* River Dee
Water *of Don*, *see* River Don

INDEX OF PLACES 891

Water *of Findhorn, see* River Findhorn
Water *of Tay, see* River Tay
Water of Thurseth, *see* River Thurso
Water *of Tweed* (Tueid), *see* River Tweed
Water, Watter *of Isla, see* River Isla
Wateris of Areicht, *see* River Ericht
Wateris of Ylay and Areicht, *see* River Isla and River Ericht
Waters of Ylay, *see* River Isla
Waterston (Waterstoun), NO5159; *see* Little Waterston; Meikle Waterston; Quarter Waterston
Watistoun, *see* Watston
Watston (Watistoun), *Kelso abbey*, NS7646; 231
Watten (Vactin), ND2454; 632, 633; parish of, 632; vicarage of, 633
Watterrocat, Watterrocat Grange, *Eskdalemuir, Melrose abbey* [*Watcarrick, NY2496]; 259
Wattersyd of Yle, *see* River Isla
Watterybuttis, *Carsgrange, Coupar Angus abbey*, 357
Wauchope (Wauchop) [Fm, NT5808], 216, 220; kirk of, 216
Wauchtoun [*Waughton, NT5680], 249
Wauchtoun, *see* Waughton
Waughton (Wachtoun, Wauchtoun), NT5680; 192, 193, 194
Waustoun, *Maybole*, 568
Wawis, *see* Walls
Wedderburn (Wedderburne) [Castle, NT8052], 195, 200
Weddersbie, NO2612; 33, 37
Weddirhill, Woddirhill, *Culross abbey*, 290, 291
Wedwik, *see* Woodwick
Weem (Weyme), NN8449; 318; parsonage of, 318; vicar pensionary of, 318; vicarage of, 318
Weens (Weindis), NT5812; 220
Weetfoot (Weitfit, Vitfute), *Greenlaw* [Bog, NT6752], 226, 226n.
Weik, *see* Wick
Weindis, *see* Weens
Weinsyde, *Eccles*, 184
Weist, *see* Uist
Weitfit, *see* Weetfoot
Weitland, *Fearn abbey*, 633
Wellifuird, *see* Wallyford
Welmedo, *Paisley*, 535
Welmerstoun, *see* Wormistone

Welschis Croft, *Holywood abbey*, 623
Welstoun, *see* Walston
Welstoun, *see* Wiston
Weltoun, *Dunfermline abbey* [Walton, *NT2186, *NT2090], 24, 28, 46, 50
Weltoun, *see* Wilton
Wemes, *Kinloss abbey*, 463
Wendeage, *see* Windyedge
Wendmethill, *see* Windmillhill
West Barns (West Barnes), *Dunbar*, NT6578; 122
West Barns (West Barnes, West Barnis), *Crail*, NT6578; 177, 178, 179
West Binny (Bynnie, Wester Binning, Wester Bynnyng), NT0372; 8, 16, 17; *see also* Binning
West Boundis, *see* Bowndis
West Carse (Wast Kers, West Kers), NS7394; 549, 550
West Church (Westkirk), NT9786; 290
West croft, Westeroff, *Kelso abbey*, 222, 230, 233, 234, 237, 239, 240
West Feild, *near Scottistoun, Aberdeen deanery*, 423
West Fudy, *see* Wester Craigfoodie
West Gordoun, *Gordon parish* [Gordon, NT6442], 227
West Gordoun, *Merse*, 128
West Gormack (Wester Gormok), NO1447; 308
West Grange (Wast Grainge), *Cambuskenneth abbey*, NS8194; 547
West Grange (Westergrang, Westir Grange), *Culross abbey*, NT9889; 290, 293
West Greenless (Wastir Greinleis), NS6359; 508
West Hope, West Hopes, Westhopes, Westschoipes, *Haddington priory* [Hope, NT4062], 162, 165, 176, 178, 181
West Hous, West Houssis, *Newbattle abbey* [Easthouses, NT3465], 101, 103
West Kers, *see* West Carse
West Kilbride, *see* Kilbride
West Lothian, 159n.
West Lumsdaine (West Lummisdan, Wester Lummisden) [Lumsdaine, NT8769], 198, 203
West Mains (Wester Manes, Wester Mannis), *Gordon*, NT6442; 234, 238, 241

West Marche, *Restalrig collegiate kirk*, 142
West Mathers (Wester Mathouris), NO7665; 16
West Medow, *Kelso abbey*, 222
West Medow, *Lindores abbey*, 35
West Mephen [*Wester Meathie, NO4546], 89
West Myln, West Mylne, *of Kirkcaldy*, 24, 28, 46, 50
West Mylne, *Inchcolm abbey*, 64
West Nisbet (West Nisbit), NT6725; 200
West Restoun [East Reston, NT9061], 198, 203, 204; mill of, 203
West Wod, alias the Bowmanis land, *see* Bowmannis land
Wester Baith, *Dunfermline abbey*, 23, 45
Wester Baldridge (Bawdrick, Bawdrik), NT0888; 23, 45
Wester Balgray [Midton of Balgray, NS4434], 531
Wester Balrymonth (Balrymonthe, Balrymounth, Balrymounthe Wester), NO5014; 10, 13, 20, 21
Wester Balyalnach (Balliealyenoch, Balyealyenoch) [Ballyalnach, NN9453], 284, 288
Wester Bawdrick, Bawdrik, *see* Wester Baldridge
Wester Bengour [East Bangour, NT0471], 151
Wester Binning, Wester Bynnyng, *see* West Binny
Wester Blak Craigis, *see* Blackcraig
Wester Bochlaweis [Easter Bucklyvie, NT1788], 62
Wester Bogsyde, *Coupar Angus abbey*, 355
Wester Bracklinn (Westir Brokland), NN6508, 349
Wester Braikie (Westerbroky), NO6251; 366
Wester Cairnie (Wester Cairny), NO0318; 349
Wester Callander [Callander, NN6307], 349
Wester Cantna, *Croy and Moy*, 639
Wester Cash (Wester Casche), NO2209; 69
Wester Cotyardis, *Millhorn, Coupar Angus abbey*, 354, 357, 369

Wester Craigduky, *see* Craigduckie Wester
Wester Craigfoodie (West Fudy), NO4017; 14
Wester Cranokis, *Kinloss abbey* [Crannach, NJ4954], 461
Wester Crombie (Wester Cumrie), *Culross abbey* [Crombie, NT 0485], 290
Wester Croylettis, *Kinloss abbey* [Croylet, NJ5056], 461
Wester Culbo, *Cullicudden* (Wastir Culboll) [Wood, NH6459], 646
Wester Cumrie, *Culross abbey*, *see* Wester Crombie
Wester Daldolicht, *Croy and Moy* [Mains of Daltulich, NH7341], 639
Wester Daldowie (Daldowe Waster), NS6662; 497
Wester Deanshouses (Wester Denshoussis), NT2151; 101
Wester Denfallanceis, Wester Dunfallances, *Atholl, Perth Charterhouse* [Dunfallandy Ho, NO9556], 284, 287
Wester Denhead (Wester Denhede, Westere Denhede), NO2340; 353, 356
Wester Denshoussis, *see* Wester Deanshouses
Wester Disblair (Wester Disblaire, Wester Displair), NJ8420; 32, 36
Wester Drymmy [East Drimmie, NO1749], 354
Wester Dulguis, Wester Dulgus [Easter Dalguise, NN9947], 303, 305, 342, 345
Wester Farnie, *Edderton*, 652
Wester Feddal (Wester Feddellis), NN8208; 31, 35
Wester Fetteris, Wester Fotteris, *Leuchars* [*Fetterdale, NO4725], 14, 20
Wester Fintray (Wester Fyntray), NJ8116; 31, 36
Wester Fotteris, *see* Wester Fetteris
Wester Fowmerden [Fourmartdean, NT7927], 214
Wester Fyntray, *see* Wester Fintray
Wester Gany, *Fearn abbey*, 633
Wester Garquhone, *Dunblane bishopric* [Gartchonzie, NN6007], 349
Wester Geddis [Meikle Geddes,

NH8752], 469
Wester Gellet (Waster Gullet, Wester Gullet), NT0984; 26, 27, 49; *see also* Gellets
Wester Glentore (Glenter Waster), NS8171; 498
Wester Gormok, *see* West Gormack
Wester Grange, *see* West Grange
Wester Greinleis, *see* West Greenlees
Wester Growan, *Monzie*, 349
Wester Guideltoun, *Morebattle*, 214
Wester Gullet, *see* Wester Gellet
Wester Happrew (Hoprew Wester), NT1741; 249
Wester Hartsyd, *Hartside parish*, 513
Wester Howlaws (Wester Howlaw), NT7242; 227, 235, 238, 241
Wester Incheschewin, Inschevin Wester [Middle Inchewan, NO0341], 305, 342, 345
Wester Ingzever [Inzievar, NT0288], 290
Wester Inneraritie, *Kinloss abbey* [Inverharity, NO1963], 354, 355
Wester Inschgottrig [Inchgotrick, NS4133], 532
Wester Keith (Wester Kethe), NO2837; 376
Wester Ker, *see* Westerkirk
Wester Killis, *Dallas*, 492
Wester Kilwhiss (Kilquhyis Wester), NO2510; 33
Wester Kinghorne, Wester Kingorne, *see* Kinghorn Wester
Wester Kinsleith (Kynsleif Wester), NO3219; 34
Wester Lochtquharrut, *Crichton collegiate kirk* [Loquhariot, NT3760], 450
Wester Lowstonis, *with Kinvaid, Dunkeld cathedral*, 323
Wester Luscar (Luscer Wester), *Dunfermline abbey* [East Luscar, NT0589], 23, 27, 45, 49, 50
Wester Manes, Wester Mannis, *Gordon*, *see* West Mains
Wester Markinch [Markinch, NO2901], 15, 56
Wester Mathouris, *see* West Mathers
Wester Nunraw [Nunraw, NT5970], 181
Wester Pearsie (Wester Persey), NO3458; 354
Wester Petmowie [Pitmuies Mill Fm,

NO5849], 376
Wester Raik, *see* Westerfuird
Wester Rossie (Rosse Wester), NO2512; 33, 37
Wester Softlaw, NT7330; 225; *see also* Softlaw
Wester Tyre, *Atholl, Perth Charterhouse* [Eastertyre, NN9552], 284, 288n.
Wester Wooden (Wester Woddein), NT7025; 225
Wester, kirk of, *see* Westerkirk
Westerbroky, *see* Wester Braikie
Westerfuird, Wester Raik, *with Housbyre, Melrose abbey*, 207, 257
Westergrang, *see* West Grange
Westerhaw, Vesterraw, *Greenlaw*, 226, 226n.
Westerkirk (Vatstirker, Wester Ker), NT3097; 209, 209n., 259, 259n.; kirk of, 209, 209n.
Westeroff, *see* West croft
Westertoun, *Glenbervie*, 405
Westerwod, *see* Bowmannis land
Westhall (Waisthaill), *Aberdeen diocese* [Old Westhall, NJ6725], 433; chaplainry of, 433; chaplains of, 433
Westhopes, Westschoipes, *see* West Hope
Westhorn, Westhorne, *Carsgrange, Coupar Angus abbey*, 354, 357
Westis Ferme, *Fearn abbey*, 633
Westkirk, *see* West Church
Westleis, *Hobkirk*, 220
Westoussis Myln, Urschous Mylne, Vesthous Mylne, *Melrose abbey*, 208, 258, 258n.
Westownes, *see* Witstones
Westray (Wastray, Wastrey), HY4446; 655, 655n., 660, 660n., 663, 666, 667, 668; parsonage of, 655, 668; *see also* Cross kirk, Lady kirk
Westray, Westrey, *Fearn abbey*, 633, 652
Weyme, *see* Weem
Whalsay (Qualsay, Quhalsay), HU5663; 658, 660, 670; kirk of, 660; minister of, 660
Whinkerstones (Quhincarstanes), NT7647; 226
Whitchester (Quhyt Chester), NT7158; 186
White Croft (Quhite Croft, Quhyt

Croft), *Leuchars*, 14, 20
Whitebank (Quheitbank, Quhytbank), NO0025; 296, 298
Whitehill (Quheit Hill, Quheithill, Quhithill, Quhytehill, Quhythill), *Musselburgh*, NT3566; 26, 39, 39n., 41, 43, 48, 49, 53, 54
Whitehill (Quhythill), *Inchcolm abbey*, NT1885; 62, 63
Whitehills (Quhythillis), *Sorbie*, NX4546; 593, 594; mill of, 593
Whitekirk (Quhitkirk, Quhytkirk), NT5981; 91; vicar of, 169; vicarage of, 169
Whitelaw (Quhytlaw), *Edrom*, NT8352; 199
Whitelee (Quhitlie), *Melrose abbey*, NT4639; 207, 257, 259
Whiteside (Quheitsyde, Quhitsyd, Quhytsyd), *Greenlaw*, NS7937; 227, 227n., 231, 233
Whiteside (Quhitsyd, Quhytsyd), *Musselburgh*, 25, 26, 43, 47, 48, 49, 53, 54
Whitfeild (Quhytfeild), *Coldingham priory*, NT9063; 203
Whithorn (Quhitarne, Quhiterne, Quhithorne, Quhitorne, Quhittarne, Quhittern, Quhitterne, Quhittorn, Quhittorne, Quihtorne), NX4440; 505, 573, 576, 588, 590, 592, 593, 594, 595, 595n., 596, 598, 599, 601n.; parish of, 592, 596
—— priory, 282, 573, 576, 592-596; commendator of, 595 595n.; manse of prior of, 594; prior of, 573, 573n., 596
Whithorn, Isle of, (Ile), NX4836; 593, 594, 596
Whitmuir (Quhitmuirtoun, Quhytmuirtoun, Quhytmure Towne), *Kelso abbey*, NT4926; 223, 230, 230n., 233, 239, 242
Whitmuir Hall (Quhytmuirhall), *Kelso abbey*, NT5027; 223, 230, 233, 237, 239
Whitsome (Quhitstoun, Quhitsome, Quhitsum), NT8650; 171, 171n., 195, 336; parson of, 195-196; parsonage of, 195-196
Whittingehame (Quhittinghame) [Ho, NT6073], 97; croft of, 174; manse of, 174; vicarage pensionary of, 174

Whitton (Quhiltoun), *Morebattle*, NT7622; 214
Wick (Weik), ND3650; 628, 629, 648; parish of, 629; vicar of, 648-649; vicarage of, 648-649
Wideopen (Wyd Hoppin), *Morebattle*, NT8130; 214
Wiggingairne, *Whithorn priory* [*Castlewigg, NX4343], 592, 594, 596
Wigtown (Wigton, Wigtoun, Wigttoun), NX4355; 591, 597, 600, 601, 603, 606; minister of, 597; parsonage of, 597; reader at, 597; vicar pensionary of, 597; vicarage of, 597
——, Black friars (Dominicans) of, 606
—— *sheriffdom*, 590n., 598n.
Wigtown Bay, NX5408; 619n.
Wilkieston (Wilkestoun, Wilkiestoun), NO4412; 12, 20, 21, 66
Willdunhop, *Selkirk*, 230n.
William Law (Viliamlaw, Williamlaw, Williame Law), *Melrose abbey*, NT4739; 207, 257, 257n., 259.
Williamcraigs (Williame Craigis), *Manuel priory*, NS9875; 549
Williamhope, Williame Bischope, *Newbattle abbey*, 103, 103n.
Williamlaw, *see* William Law
Williamston (Williamstoun, Williamestoun), *Lindores abbey* [Ho, NJ6431], 31, 36
Williamstoun Over, *see* Over Williamston
Wilmerstoun, *see* Wormistone
Wilstoun, *see* Wiston
Wilton (Weltoun), NT4915; 212; parsonage of, 212
Winchburgh (Wincheburcht), NT0874; 3
Windehillis called Sauchquhytoun, Wynde Hillis called the Sauchy Town, *Kinloss abbey* [*Wester Windyhills, NJ4857], 462, 464
Windeleis, *see* Winless
Windihall, *Riccarton*, 531
Windingtoun, *see* Wyndingtoun
Windmillhill (Wendmethill, Windmilhill, Wondmilhill, Wondmylhill, Wyndmilhill), *Dunfermline*, 23, 27, 28, 45, 49, 50
Windyaige, *Coupar Angus abbey*, 355
Windyedge (Wendeage, Windeage,

INDEX OF PLACES 895

Wyndeage), *Dunfermline abbey*, NT1392; 23, 27, 28, 45, 49, 50
Winless (Vindleis, Windeleis), ND3054; 628, 628n.; mill of, 628
Winterscheildykip, *Lammermuir, Melrose abbey*, 258
Winthank (Unthank), *St Andrews priory*, NO4713; 10, 18
Wiston (Voustoun, Vrmistoun, Welstoun, Wilstoun, Wistoun, Wolstoun, Wostoun, Wystoun), NS9532; 224, 232, 234, 237, 240, 242, 244, 244n., 507, 508; kirk of, 224, 232, 234, 237, 240, 242, 244; vicar of, 507, 508; vicarage of, 508
Wistownis Hilend, Witstones Hill End, Hall of, *Lindores abbey*, 31, 35
Witstones, *Lindores abbey*, 31, 35
Wnster, *see* Unst, 660
Wnsterdaill, *Caithness bishopric*, 628
Wobstar croft, Wobstaris croft, *Tullybeagles, Methven collegiate kirk*, 296, 298
Wod End, Wodend, *Ruthven*, 305
Wod Nuk, *Holywood abbey*, 623
Wod, with Glenturk, *Wigtown*, 597
Wodaker, Woidaker, *Dunfermline abbey*, 28, 50
Wodderlie, *Gordon*, 227
Woddirhill, *see* Weddirhill
Wodend, *see* Woodend
Wodhavin, *see* Woodhaven
Wodheid, *Ancrum, see* Woodhead
Wodheid, *Greenlaw*, 227
Wodheid, *see* Woodhead
Wodhouslie, *see* Woodhouselee
Wodmylne, *see* Woodmill
Wodrufhill, Wodruif, Wodrwff Hill, *Lindores abbey*, 32, 34, 36
Wodsyd, *Morebattle*, 214
Woidaker, *see* Wodaker
Woisterrie, *Ringwodfeild, Melrose abbey*, 259
Wolfclid, Wolfclyd, Wolfurde, *Melrose abbey*, 207, 207n., 258
Wolfhill (Voulschill, Woulfhill alias Over Campsie, Wolf Hill, Ower Camspsy), *Coupar Angus abbey*, NO1433; 352, 352n., 356, 370
Wollis Over and Nether, *Jedburgh abbey* [*Over, Nether Wells, NT6820, NT6920], 219
Wollis, *Hobkirk*, 220

Wollscheit, *Paisley abbey*, 530, 530n.
Wolmet, *see* Woolmet
Wolstoun, *see* Walston
Wolstoun, *see* Wiston
Womet, *see* Woolmet
Wondmilhill, Wondmylhill, *see* Windmillhill
Wonet, *Dunfermline abbey* [*Woolmet, NT3169], 26, 48
Wonet Bank, Wowmetbank, *Dunfermline abbey* [*Woolmet, NT3169], 24, 24n., 47
Woodend (Wodend), *Haddington priory*, 176, 178, 181
Woodfoot, *Lesmahagow*, NS8540; 232n.
Woodhaven (Wodhavin), *St Andrews priory*, NO4126; 14
Woodhead (Wodheid), *Cullicudden*, NH6460; 646
Woodhead (Wodheid), *Ancrum*, NT6125; 236, 366
Woodhead (Wodheid), *Lindores abbey*, NO2615; 30, 35
Woodhouselee (Wodhouslie), *Restalrig collegiate kirk*, NT2364; 142, 143
Woodmill (Wodmylne), *Lindores abbey*, NO2714; 32, 37
Woodside (Wodsyd), *Kilwinning abbey*, NS3455; 573
Woodwick (Wedwik), HY3832; 668, 668n.; prebend of Kirkwall cathedral, 668, 668n.
Woolmet (Wolmet, Womet, Wowmet), NT3169; 24, 24n., 42, 47
Wooplaw (Vouplaw, Wouplaw), NT4942; 207, 257, 259
Worestoun, *Cambuslang*, 508
Wormistone (Welmerstoun, Wilmerstoun), NO6109; 177, 178, 181
Worthfeild, *see* Northfield
Wostoun, *see* Wiston
Woulfhill, *see* Wolfhill
Wouplaw, *see* Wooplaw
Wousbyre, *see* Housbyre
Wowmet, *see* Woolmet
Wowmetbank, *see* Wonet Bank
Wowterischill, Wowterschill, *see* Uttershill
Wraes (Wraichis), *Riccarton*, 531
Wraichis, *see* Wraes
Wrangham (Wranghame), *Lindores*

abbey, NJ6331; 31, 34, 36
Wrangholme, *Smailholm*, 189
Wrynnes croft, *Holywood abbey*, 623
Wstoun, *see* Ulston
Wver Kyngraig, *see* Over Kincraig
Wver Petforthie, Wver Pitforthy, *see* Over Pitforthie
Wyd Hoppin, see Wideopen
Wyffand, *see* Waiffuird
Wyingis, *see* Ugginnis
Wyle, *Tiree*, 673
Wynde Hillis called the Sauchy Town, *see* Windehillis
Wyndeage, *Dunfermline abbey*, *see* Windyedge
Wyndehege, *Monkland parish*, 498
Wyndingtoun, Windingtoun, *Jedburgh abbey*, 216, 216n.
Wyndmilhill, *see* Windmillhill
Wyndmilne Croce, *Culross abbey*, 292
Wynstoun, *Seton collegiate kirk*, 100
Wystoun, *see* Wiston

Yairdmyll, Yaird Mylne, *St Andrews priory*, 10, 19
Yaldan, *see* Hadden
Yallowstruther, *Calder*, 98
Yare, *Selkirk*, 230n.
Yarrow (St Mary of the Lowes, St Mary in the Forest, St Mary of Farmainishop), NT3527; kirk of, 263, 263n.
Yell, HU4890; 658, 660, 670; kirk of, 660; minister of, 660
Yester, *see* Bothans
Yetbyre, *Eskdalemuir, Melrose abbey*, 259
Yetholm (Yettem) [Kirk Yetholm, NT8228], 243; parson of, 243; parsonage of, 243
Yettem, *see* Yetholm
Yettoun Orchart, *St Mary's Isle priory*, 619
Yettoun, *Kirkmadrine*, 601
Yifflie, Jefflie, *Gordon* [*Evelaw, NT6653], 227, 227n.
Ylay, *see* River Isla
Yle of Crawmond, *see* Cramond Island
Yle, *see* River Isla
Yles, Ylis, *see* Isles
Ylis, *Rothes*, 491
Ylyscheuch, *see* Elisheugh

Ynchekyll, *see* Inchkeil
Ynglistoun, *see* Ingliston
Yourek, *see* Auldbreck

Zetland, Zeitland, *see* Shetland
Zy Rig, *see* Gyrig

RECORDS OF SOCIAL AND ECONOMIC HISTORY
(New Series)

I. *Charters of the Honour of Mowbray 1107–1191*. D. E. Greenway. 1972
II. *The Lay Subsidy of 1334*. R. E. Glasscock. 1975
III. *The Diary of Ralph Josselin, 1616–1683*. A. Macfarlane. 1976
IV. *Warwickshire Grazier and London Skinner, 1532–1555. The Account Book of Peter Temple and Thomas Heritage*. N. W. Alcock. 1981
V. *Charters and Custumals of the Abbey of Holy Trinity, Caen*. M. Chibnall. 1982
VI. *The Cartulary of the Knights of St. John of Jerusalem in England, Secunda Camera, Essex*. M. Gervers. 1982
VII. *The Correspondence of Sir John Lowther of Whitehaven, 1693–1698*. D. R. Hainsworth. 1983
VIII. *The Farming and Memorandum Books of Henry Best of Elmswell, 1642*. D. Woodward. 1984
IX. *The Making of King's Lynn*. D. M. Owen. 1984
X. *The Compton Census of 1676. A Critical Edition*. A. Whiteman. 1986
XI. *The Early Records of Medieval Coventry*. P. R. Coss. 1986
XII. *Markets and Merchants of the Late Seventeenth Century. The Marescoe-David Letters, 1668–1680*. H. Roseveare. 1987
XIII. *The Diary of Bulstrode Whitelocke, 1605–1675*. R. Spalding. 1990
XIV. *Contemporaries of Bulstrode Whitelocke, 1605–1675. Biographies, Illustrated by Letters and other Documents*. R. Spalding. 1990
XV. *The Account Book of Richard Latham, 1724–1767*. L. Weatherill. 1990
XVI. *An American Quaker in the British Isles. The Travel Journals of Jabez Maud Fisher, 1775–1779*. K. Morgan. 1992
XVII. *Household Accounts from Medieval England, Part 1*. C. M. Woolgar. 1992
XVIII. *Household Accounts from Medieval England, Part 2*. C. M. Woolgar. 1993
XIX. *The Warwickshire Hundred Rolls of 1279–80: Stoneleigh and Kineton Hundreds*. T. John. 1992
XX. *Lordship and Landscape in Norfolk, 1250–1350. The Early Records of Holkam*. W. Hassall & J. Beauroy. 1993
21. *The Books of Assumption of the Thirds of Benefices. Scottish Ecclesiastical Rentals at the Reformation*. J Kirk. 1995
22. *Charters and Custumals of the Abbey of Holy Trinity, Caen, Part 2: The French Estates*. J. Walmsley. 1994

RECORDS OF SOCIAL AND ECONOMIC HISTORY
(New Series)

I. *Charters of the Honour of Mowbray 1107–1191.* D. E. Greenway. 1972
II. *The Lay Subsidy of 1334.* R. E. Glasscock. 1975
III. *The Diary of Ralph Josselin, 1616–1683.* A. Macfarlane. 1976
IV. *Warwickshire Grazier and London Skinner, 1532–1555. The Account Book of Peter Temple and Thomas Heritage.* N. W. Alcock. 1981
V. *Charters and Custumals of the Abbey of Holy Trinity, Caen.* M. Chibnall. 1982
VI. *The Cartulary of the Knights of St. John of Jerusalem in England, Secunda Camera, Essex.* M. Gervers. 1982
VII. *The Correspondence of Sir John Lowther of Whitehaven, 1693–1698.* D. R. Hainsworth. 1983
VIII. *The Farming and Memorandum Books of Henry Best of Elmswell, 1642.* D. Woodward. 1984
IX. *The Making of King's Lynn.* D. M. Owen. 1984
X. *The Compton Census of 1676. A Critical Edition.* A. Whiteman. 1986
XI. *The Early Records of Medieval Coventry.* P. R. Coss. 1986
XII. *Markets and Merchants of the Late Seventeenth Century. The Marescoe-David Letters, 1668–1680.* H. Roseveare. 1987
XIII. *The Diary of Bulstrode Whitelocke, 1605–1675.* R. Spalding. 1990
XIV. *Contemporaries of Bulstrode Whitelocke, 1605–1675. Biographies, Illustrated by Letters and other Documents.* R. Spalding. 1990
XV. *The Account Book of Richard Latham, 1724–1767.* L. Weatherill. 1990
XVI. *An American Quaker in the British Isles. The Travel Journals of Jabez Maud Fisher, 1775–1779.* K. Morgan. 1992
XVII. *Household Accounts from Medieval England, Part 1.* C. M. Woolgar. 1992
XVIII. *Household Accounts from Medieval England, Part 2.* C. M. Woolgar. 1993
XIX. *The Warwickshire Hundred Rolls of 1279–80: Stoneleigh and Kineton Hundreds.* T. John. 1992
XX. *Lordship and Landscape in Norfolk, 1250–1350. The Early Records of Holkam.* W. Hassall & J. Beauroy. 1993
21. *The Books of Assumption of the Thirds of Benefices. Scottish Ecclesiastical Rentals at the Reformation.* J Kirk. 1995
22. *Charters and Custumals of the Abbey of Holy Trinity, Caen, Part 2: The French Estates.* J. Walmsley. 1994